Analytic Number Theory, Approximation Theory, and Special Functions

Gradimir V. Milovanović • Michael Th. Rassias
Editors

Analytic Number Theory, Approximation Theory, and Special Functions

In Honor of Hari M. Srivastava

Springer

Editors
Gradimir V. Milovanović
Mathematical Institute
Serbian Academy of Sciences and Arts
Belgrade, Serbia

Michael Th. Rassias
ETH-Zürich
Department of Mathematics
Zürich, Switzerland

ISBN 978-1-4939-0257-6 ISBN 978-1-4939-0258-3 (eBook)
DOI 10.1007/978-1-4939-0258-3
Springer New York Heidelberg Dordrecht London

Library of Congress Control Number: 2014931380

Mathematics Subject Classification (2010): 05Axx, 11xx, 26xx, 30xx, 31Axx, 33xx, 34xx, 39Bxx, 41xx, 44Axx, 65Dxx

Printed on acid-free paper

Springer is part of Springer Science+Business Media (www.springer.com)

Preface

The volume "Analytic Number Theory, Approximation Theory, and Special Functions" consists of 35 articles written by eminent scientists from the international mathematical community, who present both research and survey works. The volume is dedicated to professor Hari M. Srivastava in honor of his outstanding work in mathematics.

Professor Hari Mohan Srivastava was born on July 5, 1940, at Karon in District Ballia of the province of Uttar Pradesh in India. He studied at the University of Allahabad, India, where he obtained his B.Sc. in 1957 and M.Sc. in 1959. He received his Ph.D. from the Jodhpur University (now Jai Narain Vyas University), India, in 1965. He begun his university-level teaching career at the age of 19.

H. M. Srivastava joined the Faculty of Mathematics and Statistics of the University of Victoria in Canada as associate professor in 1969 and then as a full professor in 1974. He is an Emeritus Professor at the University of Victoria since 2006.

Professor Srivastava has held several visiting positions in universities of the USA, Canada, the UK, and many other countries.

He has published 21 books and monographs and edited volumes by well-known international publishers, as well as over 1,000 scientific research journal articles in pure and applied mathematical analysis He has served or currently is an active member of the editorial board of several international journals in mathematics. It is worth mentioning that he has published jointly with more than 385 scientists, including mathematicians, statisticians, physicists, and astrophysicists, from several parts of the world. Professor Srivastava has supervised several graduate students in numerous universities towards their master's and Ph.D. degrees.

Professor Srivastava has been bestowed several awards including most recently the NSERC 25-Year Award by the University of Victoria, Canada (2004), the Nishiwaki Prize, Japan (2004), Doctor Honoris Causa from the Chung Yuan Christian University, Chung-Li, Taiwan, Republic of China (2006), and Doctor Honoris Causa from the University of Alba Iulia, Romania (2007).

The name of professor Srivastava has been associated with several mathematical terms, including Carlitz–Srivastava polynomials, Srivastava–Panda multivariable

H-function, Srivastava–Agarwal basic (or q-) generating function, Srivastava–Buschman polynomials, Chan–Chyan–Srivastava polynomials, Srivastava–Wright operators, Choi–Srivastava methods in Analytic Number Theory, and Wu–Srivastava inequality for higher transcendental functions.

In this volume dedicated to professor Srivastava an attempt has been made to discuss essential developments in mathematical research in a variety of problems, most of which have occupied the interest of researchers for long stretches of time. Some of the characteristic features of this volume can be summarized as follows:

- Presents mathematical results and open problems in a simple and self-contained manner.
- Contains new results in rapidly progressing areas of research.
- Provides an overview of old and new results, methods, and theories towards the solution of long-standing problems in a wide scientific field.

The book consists of the following five parts:

1. Analytic Number Theory, Combinatorics, and Special Sequences of Numbers and Polynomials
2. Analytic Inequalities and Applications
3. Approximation of Functions and Quadratures
4. Orthogonality, Transformations, and Applications
5. Special and Complex Functions and Applications

Part I consists of nine contributions. A. Ivić presents a survey with a detailed discussion of power moments of the Riemann zeta function $\zeta(s)$, when s lies on the "critical line" $\Re s = 1/2$. It includes early results, the mean square and mean fourth power, higher moments, conditional results, and some open problems. M. Hassani gives some explicit upper and lower bounds for γ_n, where $0 < \gamma_1 < \gamma_2 < \gamma_3 < \cdots$ are consecutive ordinates of nontrivial zeros $\rho = \beta + \mathrm{i}\gamma$ of the Riemann zeta function, including the asymptotic relation $\gamma_n \log^2 n - 2\pi n \log n \sim 2\pi n \log\log n$ as $n \to \infty$. Y. Ihara and K. Matsumoto prove an unconditional basic result related to the value-distributions of $\{(L'/L)(s,\chi)\}_\chi$ and of $\{(\zeta'/\zeta)(s + i\tau)\}_\tau$, where χ runs over Dirichlet characters with prime conductors and τ runs over $\mathbb{R}$. J. Choi presents a survey on recent developments and applications of the simple and multiple gamma functions Γ_n, including results on multiple Hurwitz zeta functions and generalized Goldbach–Euler series. A. A. Bytsenko and E. Elizalde consider a partition function of hyperbolic three-geometry and associated Hilbert schemes. In particular, the role of (Selberg-type) Ruelle spectral functions of hyperbolic geometry for the calculation of partition functions and associated q-series is discussed. Y. Simsek considers families of twisted Bernoulli numbers and polynomials and their applications. Several relationships between Bernoulli functions, Euler functions, some arithmetic sums, Dedekind sums, Hardy Berndt sums, DC-sums, trigonometric sums, and Hurwitz zeta function are given. A. K. Agarwal and M. Rana discuss a combinatorial interpretation of a generalized basic series. Namely, using a bijection between the Bender–Knuth matrices and the n-color partitions established by Agarwal (ARS Combinatoria **61**, 97–117, 2001), they extend a recent result to a 3-way infinite family of combinatorial identities. A. Sofo proves some identities for reciprocal binomial coefficients, and M. Merca shows that the q-Stirling numbers can be expressed in terms of the q-binomial coefficients and vice versa.

Part II is dedicated to analytic inequalities and several applications. Ibrahim and Dragomir give a survey of some recent results for the celebrated Cauchy–Bunyakovsky–Schwarz inequality for functions defined by power series with nonnegative coefficients. G. D. Anderson, M. Vuorinen, and X. Zhang provide a survey of recent results in special functions of classical analysis and geometric function theory, in particular the circular and hyperbolic functions, the gamma function, the elliptic integrals, the Gaussian hypergeometric function, power series, and mean values. M. Merkle gives a collection of some selected facts about the completely monotone (*CM*) functions that can be found in books and papers devoted to different areas of mathematics. In particular, he emphasizes the role of representation of a *CM* function as the Laplace transform of a measure and also presents and discusses a little known connection with log-convexity. S. Abramovich considers results on superquadracity, especially those related to Jensen, Jensen–Steffensen, and Hardy's inequalities. S. Ding and Y. Xing present an up-to-date account of the advances made in the study of L^p theory of Green's operator applied to differential forms, including L^p-estimates, Lipschitz and *BMO* norm inequalities, as well as inequalities with $L^p(\log L)^\alpha$ norms. B. Yang uses methods of weight coefficients and techniques of real analysis to derive a multidimensional discrete Hilbert-type inequality with a best possible constant factor. F. Qi, Q.-M. Luo, and B.-N. Guo establish sufficient and necessary conditions such that the function $(e^{\alpha t} - e^{\beta t})/(e^{\lambda t} - e^{\mu t})$ is monotonic, logarithmic convex, logarithmic concave, 3-log-convex, and 3-log-concave on $\mathbb{R}$. P. Cerone obtains approximation and bounds of the Gini mean difference and provides a review of recent developments in the area. Finally, M. A. Noor considers the parametric nonconvex variational inequalities and parametric nonconvex Wiener–Hopf equations. Using the projection technique, he establishes the equivalence between them.

Approximation of functions and quadratures is treated in Part III. N. K. Govil and V. Gupta discuss Stancu-type generalization of operators introduced by Srivastava and Gupta (Math. Comput. Modelling **37**, 1307–1315, 2003). M. Mursaleen and S. A. Mohiuddine prove the Korovkin-type approximation theorem for functions of two variables, using the notion of statistical summability $(C, 1, 1)$, recently introduced by Moricz (J. Math. Anal. Appl. **286**, 340–350, 2003). In 1961, Baker, Gammel, and Wills formulated their famous conjecture that if a function f is meromorphic in the unit ball and analytic at 0, then a subsequence of its diagonal Padé approximants converges uniformly in compact subsets to f. This conjecture was disproved in 2001, but it generated a number of related unresolved conjectures. D. S. Lubinsky reviews their status. A. R. Hayotov, G. V. Milovanović, and K. M. Shadimetov construct the optimal quadrature formulas in the sense of Sard, as well as interpolation splines minimizing the semi-norm in the space $K_2(P_2)$, where $K_2(P_2)$ is a space of functions φ which φ' is absolutely continuous and φ'' belongs to $L_2(0, 1)$ and $\int_0^1 (\varphi''(x) + \omega^2 \varphi(x))^2 dx < \infty$. Finally, a survey on some specific nonstandard methods for numerical integration of highly oscillating functions, mainly based on some contour integration methods and applications of some kinds of Gaussian quadratures, including complex oscillatory weights, is presented by G. V. Milovanović and M. P. Stanić.

In Part IV, C. Ferreira, J. L. López, and E. Pérez Sinusía study asymptotic reductions between the Wilson polynomials and the lower-level polynomials of the Askey scheme, and K. Castillo, L. Garza, and F. Marcellán analyze a perturbation of a nontrivial probability measure $d\mu$ supported on an infinite subset on the real line, which consists of the addition of a time-dependent mass point. A. F. Loureiro and S. Yakubovich consider special cases of Boas–Buck-type polynomial sequences and analyze some examples of generalized hypergeometric-type polynomials. P. W. Karlsson gives an analysis of Goursat's hypergeometric transformations, and A. Kılıçman considers partial differential equations (PDE) with convolution term and proposes a new method for solving PDE. Finally, using the fixed point method, C. Park proves the Hyers–Ulam stability of the orthogonally additive–additive functional equation.

Special functions and complex functions with several applications are presented in Part V. Á. Baricz, P. L. Butzer, and T. K. Pogány have provided a generalization of the complete Butzer–Flocke–Hauss (BFH) Ω-function in a natural way by using two approaches and obtain several interesting properties. Á. Baricz and T. K. Pogány consider properties of the product of modified Bessel functions and establish discrete Chebyshev-type inequalities for sequences of modified Bessel functions of the first and second kind. S. Porwal and D. Breaz investigate the mapping properties of an integral operator involving Bessel functions of the first kind on a subclass of analytic univalent functions. V. V. Mityushev introduces and uses the Poincaré $\boldsymbol{\alpha}$-series ($\boldsymbol{\alpha} \in \mathbb{R}^n$) for classical Schottky groups in order to solve Riemann–Hilbert problems for n-connected circular domains. He also gives a fast algorithm for the computation of Poincaré series for disks that are close to each other. N. E. Cho considers inclusion properties for certain classes of meromorphic multivalent functions. He introduces several new subclasses of meromorphic multivalent functions and investigates various inclusion properties of these subclasses. Finally, I. Lahiri and A. Banerjee discuss the influence of Gross-problem on the set sharing of entire and meromorphic functions. It is hope that the book will be particularly useful to researchers and graduate students in mathematics, physics, and other computational and applied sciences.

Finally, we wish to express our deepest appreciation to all the mathematicians from the international mathematical community, who contributed their papers for publication in this volume dedicated to Hari M. Srivastava, as well as to the referees for their careful reading of the manuscripts. A thank also goes to professor Marija Stanić (University of Kragujevac, Serbia), for her help during the technical preparation of the manuscript. Last but not least, we are very thankful to Springer for its generous support for the publication of this volume.

Belgrade, Serbia Gradimir V. Milovanović
Zürich, Switzerland Michael Th. Rassias

Contents

Part IV Orthogonality, Transformations, and Applications

Part V Special and Complex Functions and Applications

Part I
Analytic Number Theory, Combinatorics, and Special Sequences of Numbers and Polynomials

The Mean Values of the Riemann Zeta-Function on the Critical Line

Aleksandar Ivić

Dedicated to Professor Hari M. Srivastava

Abstract In this overview we give a detailed discussion of power moments of $\zeta(s)$, when s lies on the "critical line" $\operatorname{Re} s = \frac{1}{2}$. The survey includes early results, the mean square and mean fourth power, higher moments, conditional results and some open problems.

1 Introduction

The classical Riemann zeta-function

$$\zeta(s) = \sum_{n=1}^{\infty} n^{-s} = \prod_{p} (1 - p^{-s})^{-1} \qquad (s = \sigma + it,\ \sigma, t \in \mathbb{R},\ \sigma > 1) \tag{1}$$

admits analytic continuation to $\mathbb{C}$. It is regular on $\mathbb{C}$ except for a simple pole at $s = 1$. The product representation in (1) shows that $\zeta(s)$ does not vanish for $\sigma > 1$. The Laurent expansion of $\zeta(s)$ at $s = 1$ reads

$$\zeta(s) = \frac{1}{s-1} + \gamma_0 + \gamma_1(s-1) + \gamma_2(s-1)^2 + \cdots,$$

where the so-called Stieltjes constants γ_k are given by

$$\gamma_k = \frac{(-1)^k}{k!} \lim_{N\to\infty} \left(\sum_{m\leqslant N} \frac{\log^k m}{m} - \frac{\log^{k+1} N}{k+1} \right) \qquad (k = 0, 1, 2, \ldots), \tag{2}$$

A. Ivić (✉)
Katedra Matematike, Rudarsko-geološki Fakultet, Universitet u Beogradu,
Djušina 7, 11000 Beograd, Serbia
e-mail: ivic@rgf.bg.ac.rs; aivic_2000@yahoo.com

G.V. Milovanović and M.Th. Rassias (eds.), *Analytic Number Theory, Approximation Theory, and Special Functions*, DOI 10.1007/978-1-4939-0258-3_1,

and $\gamma = \gamma_0 = -\Gamma'(1) = 0.5772157\ldots$ is the Euler constant. It was Euler [19] who first introduced $\zeta(s)$, albeit only for real values of the variable s. Riemann, in his epoch making memoir [91] of 1859, was the first to consider $\zeta(s)$ as a function of the complex variable s. Thus $\zeta(s)$ justly bears the name the *Riemann zeta-function.*

The product in (1) is called the *Euler product.* As usual, p denotes prime numbers, so that by its very essence $\zeta(s)$ represents an important tool for the investigation of prime numbers. This is even more evident from the relation

$$-\frac{\zeta'(s)}{\zeta(s)} = \sum_{n=1}^{\infty} \Lambda(n)n^{-s} \qquad (\sigma > 1),$$

which follows by logarithmic differentiation of (1), where *the von Mangoldt function* $\Lambda(n)$ is defined as

$$\Lambda(n) = \begin{cases} \log p & \text{if } n = p^\alpha, \\ 0 & \text{if } n \neq p^\alpha. \end{cases} \qquad (\alpha \in \mathbb{N})$$

The zeta-function can be also used to generate many other important arithmetic functions (see, e.g.. Chap. 1 of the author's book [52]). For example, one has, for a given $k \in \mathbb{N}$,

$$\zeta^k(s) = \sum_{n=1}^{\infty} d_k(n)n^{-s} \qquad (\sigma > 1), \tag{3}$$

where the (general) divisor function $d_k(n)$ represents the number of ways n can be written as a product of k factors, so that in particular $d_1(n) \equiv 1$ and

$$d(n) \equiv d_2(n) = \sum_{\delta|n} 1$$

is the number of positive divisors of n. The function $d_k(n)$ is a multiplicative function of n ($d_k(mn) = d_k(m)d_k(n)$ if m and n are coprime), and

$$d_k(p^\alpha) = (-1)^\alpha \binom{-k}{\alpha} = \frac{k(k+1)\cdots(k+\alpha-1)}{\alpha!}$$

for primes p and $\alpha \in \mathbb{N}$.

Another significant aspect of $\zeta(s)$ is that it can be generalized to many other similar Dirichlet series (or L-functions); see, e.g., the paper of Bombieri [9]. In fact, $\zeta(s)$ can be considered as a prototype of such functions. There exist many generalizations of the zeta-function in the literature, notably to the so-called Selberg class of L-functions. This class was introduced in 1989 by Selberg (see [93]), and for a good overview the reader is referred to Kaczorowski [70].

As remarked at the beginning, the Riemann zeta-function admits analytic continuation to $\mathbb{C}$, where the complex variable has traditionally the notation $s = \sigma + it$ ($\sigma, t \in \mathbb{R}$). There are many ways to see this. For example, for $x > 1$, one has

$$\sum_{n \leqslant x} n^{-s} = \int_{1-0}^{x} u^{-s} \mathrm{d}[u] = [x]x^{-s} + s \int_{1}^{x} [u]u^{-s-1}\mathrm{d}u$$

$$= O(x^{1-\sigma}) + s \int_{1}^{x} ([u] - u)u^{-s-1}\mathrm{d}u + \frac{s}{s-1} - \frac{sx^{1-s}}{s-1}.$$

If $\sigma > 1$ and $x \to \infty$, it follows that

$$\zeta(s) = \frac{s}{s-1} + s \int_{1}^{\infty} ([u] - u)u^{-s-1}\mathrm{d}u.$$

By using the customary notation $\psi(x) = x - [x] - 1/2$, this relation can be written as

$$\zeta(s) = \frac{1}{s-1} + \frac{1}{2} - s \int_{1}^{\infty} \psi(u)u^{-s-1}\mathrm{d}u. \tag{4}$$

Since $\int_y^{y+1} \psi(u)\mathrm{d}u = 0$ for any real y, integration by parts shows that (4) provides the analytic continuation of $\zeta(s)$ to the half-plane $\sigma > -1$, and in particular it shows that $\zeta(0) = -1/2$. On successive integrations by parts of the integral in (4) one can obtain the analytic continuation of $\zeta(s)$ to $\mathbb{C}$.

A notable feature of $\zeta(s)$, whose analogues are true for many Dirichlet series, is the functional equation, proved first by Riemann [91]. In a symmetric form it says that for $s \in \mathbb{C}$,

$$\pi^{-s/2}\zeta(s)\Gamma(\tfrac{1}{2}s) = \pi^{-(1-s)/2}\zeta(1-s)\Gamma(\tfrac{1}{2}(1-s)). \tag{5}$$

For a proof of this fundamental result, see, e.g., the monographs of Karatsuba–Voronin [71], the author [52], and in particular the classical work of Titchmarsh [98], which contains seven different proofs of the functional equation. Alternatively we can write (5) as

$$\zeta(s) = \chi(s)\zeta(1-s), \tag{6}$$

where

$$\chi(s) := \frac{\Gamma(\frac{1}{2}(1-s))}{\Gamma(\frac{1}{2}s)}\pi^{s-1/2},$$

and $\Gamma(s)$ is the familiar gamma-function. This expression can be put into other equivalent forms. For example, we have

$$\chi(s) = 2^s\pi^{s-1}\sin(\tfrac{1}{2}\pi s)\Gamma(1-s) = \frac{(2\pi)^s}{2\Gamma(s)\cos(\pi s/2)}, \tag{7}$$

where we used the well-known identities

$$\Gamma(s)\Gamma(1-s) = \frac{\pi}{\sin(\pi s)}, \quad \Gamma(s)\Gamma(s+\tfrac{1}{2}) = 2^{1-2s}\sqrt{\pi}\,\Gamma(2s). \tag{8}$$

We note that (6) gives the identity

$$\chi(s)\chi(1-s) = 1. \tag{9}$$

All identities (5)–(9) hold for $s \in \mathbb{C}$.

A fundamental problem in the theory of $\zeta(s)$ is the study of its zeros. From (5) it follows that $\zeta(-2n) = 0$ for $n \in \mathbb{N}$. These zeros are the only real zeros of $\zeta(s)$ and are called the *trivial zeros* of $\zeta(s)$. Riemann [91] in 1859 calculated a few complex zeros of $\zeta(s)$ and found that they lie on the line $\operatorname{Re} s = \frac{1}{2}$, which is called the *critical line* in the theory of $\zeta(s)$. The first four pairs of complex zeros (arranged in size according to their absolute value) are (see, e.g., Haselgrove [28])

$$\frac{1}{2} \pm i14.134725\ldots, \ \frac{1}{2} \pm i21.022039\ldots, \ \frac{1}{2} \pm i25.010857\ldots, \ \frac{1}{2} \pm i30.424876\ldots.$$

The number of complex zeros $\rho = \beta + i\gamma$ of $\zeta(s)$ with $0 < \gamma \leqslant T$ (multiplicities included) is denoted by $N(T)$. The asymptotic formula for $N(T)$ is the famous *Riemann–von Mangoldt formula*. It was enunciated by Riemann [91] in 1859, but proved by von Mangoldt [101] in 1895. It says the following (see [52, 71] or [98] for a proof). Let

$$S(T) := \frac{1}{\pi} \arg \zeta(\tfrac{1}{2} + iT). \tag{10}$$

Then

$$N(T) = \frac{T}{2\pi} \log\Big(\frac{T}{2\pi}\Big) - \frac{T}{2\pi} + \frac{7}{8} + S(T) + O\Big(\frac{1}{T}\Big), \tag{11}$$

where the O-term is a continuous function of T and

$$S(T) = O(\log T). \tag{12}$$

Here $\arg \zeta(\frac{1}{2} + iT)$ is evaluated by continuous variation starting from $\arg \zeta(2) = 0$ and proceeding along straight lines, first up to $2+iT$ and then to $1/2+iT$, assuming that T is not an ordinate of a zeta zero. If T is an ordinate of a zero, then we set $S(T) = S(T+0)$.

The Riemann hypothesis (henceforth RH for short) is the conjecture, stated by Riemann in [91], that *very likely all complex zeros of* $\zeta(s)$ *have real parts equal to* 1/2. Mainly for this reason the line $\sigma = 1/2$ is called the "critical line" in the theory of $\zeta(s)$. Notice that Riemann was rather cautious in formulating the RH and that he used the wording "very likely" ("sehr wahrscheinlich" in the German original) in connection with it. Riemann goes on to say in his paper: "One would of course like to have a rigorous proof of this, but I have put aside the search for such a proof after some fleeting vain attempts because it is not necessary for the immediate objective of my investigation". The RH is undoubtedly one of the most celebrated and difficult open problems in whole Mathematics. Its proof (or disproof) would have very important consequences in multiplicative number theory, especially in problems involving the distribution of primes. It would also very likely lead to generalizations to many other zeta-functions (Dirichlet series) having similar properties as $\zeta(s)$. For a comprehensive account on the RH the reader should consult the paper [8] of Bombieri and the monograph [12] of Borwein et al. Despite the impressive numerical evidence in favour of the RH, there are certain arguments against its truth; see, e.g., the author's paper in [12].

The RH implies (see, e.g., [52, 98] for a proof) that

$$\zeta(\tfrac{1}{2}+it) \ll \exp\left(\frac{C\log t}{\log\log t}\right) \qquad (C>0). \tag{13}$$

A slightly weaker bound than (13), which in practice can often replace the RH, is the bound

$$\zeta(\tfrac{1}{2}+it) \ll_\varepsilon (|t|+1)^\varepsilon, \tag{14}$$

which is known as the *Lindelöf hypothesis* (LH for short). It is also unproved, and it is not known whether (14) implies the RH, although this is not very likely.

The aim of this paper is to give an account on the mean values of $|\zeta(\frac{1}{2}+it)|$. This is one of the central themes in the theory of $\zeta(s)$. There are two monographs dedicated solely to it: the author's [42] and that of Ramachandra [90]. However, since the time of writing of these works, there have been new developments, and they will be discussed in this paper. Of course, the moments on any σ-line are also of interest. If $\sigma > 1$, this is easy in view of the absolute convergence of the series in (1). When $\sigma = 1$, see the author's recent paper [58] for a sharp asymptotic formula for the second moment in question and [7] for general moments. When σ lies in the *critical strip* $\frac{1}{2} < \sigma < 1$, there are many results in the literature; see, e.g., the survey paper of Matsumoto [79] concerning mean square results. For general moments of $|\zeta(\sigma+it)|$ see Chap. 8 of [52] and the author's paper [53]. Due to the restrictions on the length of this paper, complete proofs of all the lemmas and theorems will not be given, but relevant references are given where the interested reader can find all the details.

And finally when $\sigma < \frac{1}{2}$ this case is essentially reduced to the case $\sigma > \frac{1}{2}$ in view of the functional equation (6) and the asymptotic formula

$$\chi(s) = \left(\frac{2\pi}{t}\right)^{\sigma+it-1/2} \mathrm{e}^{i(t+\pi/4)}\left(1+O\left(\frac{1}{t}\right)\right), \tag{15}$$

which easily follows from the classical Stirling formula for the gamma-function $\Gamma(s)$. Note that the term $O(1/t)$ admits an asymptotic expansion in terms of the descending powers of t.

Notation. Owing to the nature of this text, absolute consistency in notation could not be attained, although whenever possible standard notation is used. By $\mathbb{N}, \mathbb{Z}, \mathbb{R}, \mathbb{C}$ we denote the set of natural numbers, integers, real numbers and complex numbers, respectively. The symbol ε will denote arbitrarily small positive numbers, not necessarily the same ones at each occurrence. The symbols $f(x) = O(g(x))$ and $f(x) \ll g(x)$ both mean that $|f(x)| \leqslant Cg(x)$ for some constants $C > 0$ and $x \geqslant x_0 > 0$. By $f(x) \ll_{a,b,\ldots} g(x)$ we mean that the constant implied by the $\ll$-symbol depends on $a, b, \ldots$.

2 Early Results

When one has a (complex-valued) function $F(t)$ integrable on $[0, T]$, the mean value integral $\int_0^T |F(t)|^2 \mathrm{d}t$ provides information about the mean value of $F(t)$ and the distribution of its values. In zeta-function theory, of particular interest are the moments

$$I_k(T) := \int_0^T |\zeta(\tfrac{1}{2} + it)|^{2k} \mathrm{d}t. \tag{16}$$

Although the right-hand side of (16) makes sense if $\operatorname{Re} k > 0$, usually one takes $k \in \mathbb{N}$. Namely in this case one can use the obvious identity $|z|^2 = z \cdot \bar{z}$, and if one has a "good" expression for $z = \zeta(\frac{1}{2} + it)$, then by multiplying the expressions for z and $\bar{z}$ one can tackle $I_k(T)$, at least in principle.

Historically, the first significant results concern the asymptotic evaluation of $I_1(T)$ and $I_2(T)$ at the beginning of the twentieth century. This was achieved by Hardy–Littlewood [25, 26] and Ingham [39], respectively. The results are contained in

Theorem 2.1. *We have*

$$\begin{aligned} I_1(T) &= \int_0^T |\zeta(\tfrac{1}{2} + it)|^2 \mathrm{d}t = T \log T + O(T), \\ I_2(T) &= \int_0^T |\zeta(\tfrac{1}{2} + it)|^4 \mathrm{d}t = \frac{1}{2\pi^2} T \log^4 T + O(T \log^3 T). \end{aligned} \tag{17}$$

We shall, in essence, indicate here the proofs of these classical results. Before this is done, however, some remarks are in order. First, note that both asymptotic formulas displayed in (17) are "weak" in the sense that the error term is only by a log-factor smaller than the main term. Secondly, observe that the first formula shows that the average value of $|\zeta(\frac{1}{2} + it)|$ in $[0, T]$ is about $\log T$, while the second one shows that it is larger. This phenomenon, which is ubiquitous in the theory of $\zeta(s)$, shows the complexity/irregularity of $|\zeta(\frac{1}{2} + it)|$. Indeed, at the time of the writing of this text, we do not know (unconditionally) the true order of magnitude of this function nor the exact distribution of its (real) zeros. It is clear that no asymptotic formula for $I_k(T)$ is known when $k > 2$. On the other hand, nowadays there is plenty of information about the fundamental functions $I_1(T)$ and $I_2(T)$. This will be discussed in detail in Sects. 3 and 4, respectively.

To find an appropriate expression for $z = \zeta(\frac{1}{2}+it)$ above, one uses the so-called approximate functional equations (henceforth AFE for short) for suitable power of $\zeta(s)$ (see, e.g., Chap. 4 of [42, 52] and [59]). In general an AFE for an L-function $F(s) = \sum_{n=1}^{\infty} f(n)n^{-s}$ is an expression for $F(s)$, when s lies outside of the region of absolute convergence of $F(s)$, involving finite sums of $f(n)n^{-s}$ and $f(n)n^{s-1}$. Some of the most common AFEs for $\zeta(s)$ are given below.

Theorem 2.2. *For $0 < \sigma_0 \leqslant \sigma \leqslant 2, x \geqslant |t|/\pi, s = \sigma + it$,*

$$\zeta(s) = \sum_{n \leqslant x} n^{-s} + \frac{x^{1-s}}{s-1} + O(x^{-\sigma}). \tag{18}$$

Theorem 2.3. *Let $0 \leqslant \sigma \leqslant 1$; $x, y, t > C > 0$; $2\pi xy = t$. Then uniformly in σ we have*

$$\zeta(s) = \sum_{n \leqslant x} n^{-s} + \chi(s) \sum_{n \leqslant y} n^{s-1} + O(x^{-\sigma}) + O(t^{1/2-\sigma} y^{\sigma-1}). \tag{19}$$

Theorem 2.4. *Let $0 < \sigma < 1$; $x, y, t > C > 0$; $4\pi^2 xy = t^2$. Then uniformly in σ we have*

$$\zeta^2(s) = \sum_{n \leqslant x} d(n)n^{-s} + \chi^2(s) \sum_{n \leqslant y} d(n)n^{s-1} + O(x^{1/2-\sigma} \log t). \tag{20}$$

Theorem 2.2 is elementary, and its proof will be given shortly. It does not require the functional equation (6)–(7), while (19) and (20) do. These formulas were proved in a classic paper by Hardy–Littlewood [27], and their proof is more involved. In fact, Theorem 2.3 is a weakened form of the result known in the literature as the "Riemann–Siegel" formula (see the work of Siegel [94]). This is one of the deepest results on zeta-function theory, obtained by Siegel (op. cit.) after looking at Riemann's notes, which are kept in the library of the Göttingen University. Both error terms, in the general case, are best possible; see [52] or [42] for this.

Proof of Theorem 2.2. We have, for $\operatorname{Re} s > 1$ and $N \geqslant 2$,

$$\begin{aligned}\sum_{n>N} n^{-s} &= \int_N^{\infty} \tau^{-s} \mathrm{d}[\tau] = -N^{1-s} + s \int_N^{\infty} [\tau]\tau^{-s-1} \mathrm{d}\tau \\ &= \frac{N^{1-s}}{s-1} - \tfrac{1}{2}N^{-s} - s \int_N^{\infty} \psi(\tau)\tau^{-s-1} \mathrm{d}\tau,\end{aligned}$$

where $\psi(x) = x - [x] - \frac{1}{2}$. Therefore

$$\begin{aligned}\zeta(s) &= \sum_{n \leqslant N} n^{-s} + \sum_{n>N} n^{-s} \\ &= \sum_{n \leqslant N} n^{-s} + \frac{N^{1-s}}{s-1} - \tfrac{1}{2}N^{-s} - s\int_N^\infty \psi(\tau)\tau^{-s-1}\mathrm{d}\tau,\end{aligned} \tag{21}$$

and by analytic continuation (21) is valid for $\sigma > 0$, the last summand being $\ll (1+|t|)N^{-\sigma}$. If $u \geqslant x\ (\geqslant 1)$, we set

$$A(u) := \sum_{x<n\leqslant u} n^{-it},$$

and apply the following elementary, standard lemma (see [52, Chap. 1]) from the theory of exponential sums ($\mathrm{e}(x) := \exp(2\pi i x)$):

Lemma 2.1. *Let $f(x)$ be a real-valued function on the interval $[a,b]$ and let $f'(x)$ be continuous and monotonic on $[a,b]$ and $|f'(x)| \leqslant \delta < 1$. Then*

$$\sum_{a<n\leqslant b} \mathrm{e}(f(n)) = \int_a^b \mathrm{e}(f(x))\mathrm{d}x + O\Big((1-\delta)^{-1}\Big).$$

We apply Lemma 2.1 with

$$f(x) = \frac{1}{2\pi}|t|\log x, \quad \delta = \frac{1}{2},$$

provided that $x \geqslant |t|/\pi$. This gives

$$A(u) = \int_x^u y^{-it}\mathrm{d}y + O(1) = \frac{u^{1-it} - x^{1-it}}{1-it} + O(1).$$

For $x \leqslant N$ partial summation gives

$$\begin{aligned}\sum_{x<n\leqslant N} n^{-s} &= \sigma\int_x^N u^{-\sigma-1}A(u)\mathrm{d}u + A(N)N^{-\sigma} \\ &= \sigma\int_x^N \frac{u^{-s} - u^{-\sigma-1}x^{1-it}}{1-it}\mathrm{d}u + O(x^{-\sigma} + xN^{-\sigma}) + \frac{N^{1-\sigma-it}}{1-it} \\ &= \frac{N^{1-s}}{1-s} - \frac{x^{1-s}}{1-s} + O(x^{-\sigma} + xN^{-\sigma}).\end{aligned}$$

Substituting this expression in (21) we finally have, for $\sigma_0 \leqslant \sigma \leqslant 2$,

$$\zeta(s) = \sum_{n \leqslant x} n^{-s} + \frac{x^{1-s}}{1-s} + O(x^{-\sigma}) + O(xN^{-\sigma} + |t|N^{-\sigma}).$$

If we let $N \to \infty$, we obtain (18). Note that the basic idea of the preceding proof was to estimate the tails in the series for $\zeta(s)$ by Lemma 2.1. This, however, is particular to the sums of n^{-s}, when the corresponding integral of x^{-s} can be easily evaluated. Thus Theorem 2.2 cannot be easily generalized to other L-functions. Again, note that the functional equation for $\zeta(s)$ was not used in the proof, which is one way to show the elementary nature of Theorem 2.2.

Proof of Theorem 2.1. We begin the proof of the first formula in (17). In (18) we take $s = \frac{1}{2} + it, T/2 \leqslant t \leqslant T, x = T$ to obtain

$$\zeta(\tfrac{1}{2} + it) = S + O(T^{-1/2}), \quad S := \sum_{n \leqslant T} n^{-1/2-it}.$$

This gives

$$\begin{aligned} \int_{T/2}^{T} |\zeta(\tfrac{1}{2} + it)|^2 \mathrm{d}t &= \int_{T/2}^{T} |S|^2 \mathrm{d}t + \int_{T/2}^{T} S \cdot O(T^{-1/2}) \mathrm{d}t + O(1) \\ &= \int_{T/2}^{T} |S|^2 \mathrm{d}t + O(T). \end{aligned} \tag{22}$$

since trivially $S \ll T^{1/2}$.

The first formula in (17) follows easily from (22) and the general following result, well known as the *mean value theorem for Dirichlet polynomials*. This is formulated here as (see, e.g., Chap. 5 of [52] for a proof)

Lemma 2.2. *Let $a_1, \ldots, a_N$ be arbitrary complex numbers. Then*

$$\int_0^T \Big|\sum_{n \leqslant N} a_n n^{it}\Big|^2 \mathrm{d}t = T \sum_{n \leqslant N} |a_n|^2 + O\Big(\sum_{n \leqslant N} n|a_n|^2\Big), \tag{23}$$

and (23) remains valid if $N = \infty$, provided that the series on the right-hand side converge.

Applying (23), once with T and once with $T/2$ and subtracting the resulting expressions, we obtain

$$\int_{T/2}^{T} |S|^2 \mathrm{d}t = \int_{T/2}^{T} \sum_{n \leqslant T} \frac{1}{n} + O\Big(\sum_{n \leqslant T} 1\Big) = \tfrac{1}{2} T \log T + O(T). \tag{24}$$

Replacing in (24) T by $T2^{-j}$ and summing over $j = 0, 1, 2, \ldots$ we obtain the first formula in (17).

The proof of the second formula in (17) is more involved, but certainly less difficult than Ingham's original proof in [39]. It is the one given in Chap. 5 of [52] and is based on the work of Ramachandra [89]. The essential arithmetic ingredient that is used is

Lemma 2.3. *For a suitable constant* $c(a)$ (> 0) *we have*

$$\sum_{n\leqslant x} d^2(n)n^a = \begin{cases} c(a)x^{1+a}\log^3 x + O(x^{1+a}\log^2 x) & (a > -1), \\ (4\pi^2)^{-1}\log^4 x + O(\log^3 x) & (a = -1). \end{cases} \tag{25}$$

Proof of Lemma 2.3. One obtains (25) by partial summation from

$$\sum_{n\leqslant x} d^2(n) = \pi^{-2}x\log^3 x + O(x\log^2 x), \tag{26}$$

hence the proof of (25) reduces to establishing (26). To this end, note that we have the Dirichlet series representations

$$\sum_{n=1}^{\infty} d^2(n)n^{-s} = \frac{\zeta^4(s)}{\zeta(2s)}, \quad \sum_{n=1}^{\infty}\mu(n)n^{-2s} = \frac{1}{\zeta(2s)}, \tag{27}$$

which are valid for $\operatorname{Re} s > 1$ and $\operatorname{Re} s > 1/2$, respectively, where $\mu(n)$ is the familiar Möbius function, generated by $1/\zeta(s)$. The first identity in (27) follows from $d(p^j) = j + 1$ and

$$\begin{aligned} \frac{\zeta^4(s)}{\zeta(2s)} &= \prod_p \frac{1-p^{-2s}}{(1-p^{-s})^4} = \prod_p \frac{1+p^{-s}}{(1-p^{-s})^3} \\ &= \prod_p (1+p^{-s})\Big(1+\sum_{j=1}^{\infty}\tfrac12(j+1)(j+2)p^{-js}\Big) \\ &= \prod_p \Big(1+\sum_{j=1}^{\infty}(j+1)^2p^{-js}\Big) = \sum_{n=1}^{\infty} d^2(n)n^{-s}. \end{aligned}$$

From the first identity in (27) we have, on equating the coefficients of the Dirichlet series appearing in the identity,

$$d^2(n) = \sum_{k\ell^2=n} d_4(k)\mu(\ell).$$

This gives, setting $n = k\ell^2$,

$$\begin{aligned}\sum_{n\leqslant x} d^2(n) &= \sum_{k\ell^2\leqslant x} d_4(k)\mu(\ell) = \sum_{\ell\leqslant x^{1/2}} \mu(\ell) \sum_{k\leqslant x\ell^{-2}} d_4(k)\\ &= \sum_{\ell\leqslant x^{1/2}} \mu(\ell)\left(\tfrac{1}{6}x\ell^{-2}\log^3(x\ell^{-2}) + O(x\ell^{-2}\log^2 x)\right)\\ &= \pi^{-2}x\log^3 x + O(x\log^2 x).\end{aligned}$$

Here we used the weak asymptotic formula (since the generating function $\zeta^4(s)$ of $d_4(n)$ has a pole of degree four at $s = 1$)

$$\sum_{n\leqslant x} d_4(n) = \tfrac{1}{6}x\log^3 x + O(x\log^2 x)$$

and the identity

$$\sum_{n=1}^{\infty} \mu(n)n^{-2} = \frac{1}{\zeta(2)} = \frac{6}{\pi^2}.$$

Recall the inversion formula (this is (A.7) of [52])

$$\mathrm{e}^{-x} = \frac{1}{2\pi i}\int_{c-i\infty}^{c+i\infty} \Gamma(s)x^{-s}\mathrm{d}s \qquad (c, x > 0). \tag{28}$$

Setting $s = \frac{1}{2} + it, T/2 \leqslant t \leqslant T$ and using (28) we infer that

$$\begin{aligned}\sum_{n=1}^{\infty} d(n)\mathrm{e}^{-n/T}n^{-s} &= \frac{1}{2\pi i}\int_{\mathrm{Re}\, w=2} \zeta^2(s+w)\Gamma(w)T^w\mathrm{d}w\\ &= \zeta^2(s) + O(T^{-c}) + \frac{1}{2\pi i}\int_{\mathrm{Re}\, w=-3/4} \chi^2(s+w)\zeta^2(1-s-w)\Gamma(w)T^w\mathrm{d}w\\ &= \zeta^2(s) + O(T^{-c}) + \frac{1}{2\pi i}\int_{\mathrm{Re}\, w=-3/4} \chi^2(s+w)\sum_{n=1}^{\infty} d(n)n^{w+s-1}\Gamma(w)T^w\mathrm{d}w\\ &= \zeta^2(s) + O(T^{-c}) + \frac{1}{2\pi i}\int_{\mathrm{Re}\, w=-3/4} \chi^2(s+w)\sum_{n>T} d(n)n^{w+s-1}\Gamma(w)T^w\mathrm{d}w\\ &\quad - \chi^2(s)\sum_{n\leqslant T} d(n)n^{s-1} + O(T^{-c})\\ &\quad + \frac{1}{2\pi i}\int_{\mathrm{Re}\, w=1/4} \chi^2(s+w)\sum_{n\leqslant T} d(n)n^{w+s-1}\Gamma(w)T^w\mathrm{d}w.\end{aligned} \tag{29}$$

Here $c > 0$ can be arbitrarily large, and we used the residue theorem, the functional equation for $\zeta(s)$ and finally Stirling's formula for $\Gamma(s)$, which makes the contribution of the residue of $\zeta^2(s+w)\Gamma(w)T^w$ at the double pole $w = 1 - s$ small. From (6) and (7) we have

$$\zeta^2(\tfrac{1}{2}+it)\chi^{-1}(\tfrac{1}{2}+it) = |\zeta(\tfrac{1}{2}+it)|^2, \tag{30}$$

so we can deduce from (29) and (30) that

$$|\zeta(\tfrac{1}{2}+it)|^2 = \sum_{k=1}^{6} J_k(t) + O(T^{-c}) \qquad (T/2 \leqslant t \leqslant T) \tag{31}$$

for any fixed $c > 0$, where with $J_k \equiv J_k(t)$ we have

$$\begin{aligned}
J_2 = \bar{J}_1 &= \chi(\tfrac{1}{2}+it)\sum_{n\leqslant T} d(n)n^{-1/2+it},\\
J_3 &= \chi^{-1}(\tfrac{1}{2}+it)\sum_{n>T} d(n)\mathrm{e}^{-n/T}n^{-1/2-it},\\
J_4 &= \chi^{-1}(\tfrac{1}{2}+it)\sum_{n\leqslant T} d(n)(\mathrm{e}^{-n/T}-1)n^{-1/2-it},\\
J_5 &= -\frac{1}{2\pi i}\chi^{-1}(\tfrac{1}{2}+it)\int_{\mathrm{Re}\,w=-3/4, |\,\mathrm{Im}\,w|\leqslant \log^2 T}\chi^2(\tfrac{1}{2}+it+w)\\
&\quad\times\sum_{n>T} d(n)n^{w-1/2+it}\Gamma(w)T^w\mathrm{d}w,\\
J_6 &= -\frac{1}{2\pi i}\chi^{-1}(\tfrac{1}{2}+it)\int_{\mathrm{Re}\,w=1/4, |\,\mathrm{Im}\,w|\leqslant \log^2 T}\chi^2(\tfrac{1}{2}+it+w)\\
&\quad\times\sum_{n\leqslant T} d(n)n^{w-1/2+it}\Gamma(w)T^w\mathrm{d}w.
\end{aligned} \tag{32}$$

This seemingly complicated procedure allows us, when we integrate (31)–(32), to use (25) for $a = -1$ in the mean square integral of J_1 and for $a > -1$ the other formula for the ensuing integrals. In this way we obtain the second formula in (17). More precisely, we have

$$\begin{aligned}
\int_{T/2}^{T}|\zeta(\tfrac{1}{2}+it)|^4\mathrm{d}t &= 2\int_{T/2}^{T}|J_1|^2\mathrm{d}t + \int_{T/2}^{T}(J_1^2+J_2^2)\mathrm{d}t\\
&+ O\left(\sum_{k=3}^{6}\int_{T/2}^{T}|J_k|^2\mathrm{d}t\right) + O\left(\sum_{k=3}^{6}\left|\int_{T/2}^{T}(J_1+J_2)J_k\mathrm{d}t\right|\right) + O(1).
\end{aligned} \tag{33}$$

Now we apply Lemmas 2.2 and 2.3, obtaining first

$$2\int_{T/2}^{T} |J_1|^2 \mathrm{d}t = T\sum_{n\leqslant T} d^2(n)n^{-1}+O\Big(\sum_{n\leqslant T} d^2(n)\Big)$$

$$= (4\pi^2)^{-1} T\log^4 T+O(T\log^3 T), \tag{34}$$

since $|\chi(\frac{1}{2}\pm it)| = 1$. Therefore (34) indeed contributes the main term to our formula, since at the end we replace T by $T2^{-j}$ and sum the resulting expressions. Using (18) with $a > -1$ we have, again by Lemma 2.2,

$$\int_{T/2}^{T} |J_3|^2 \mathrm{d}t \ll T\sum_{n>T} d^2(n)\mathrm{e}^{-2n/T}n^{-1} + \sum_{n>T} d^2(n)\mathrm{e}^{-2n/T} \ll 1,$$

$$\int_{T/2}^{T} |J_4|^2 \mathrm{d}t \ll T\sum_{n\leqslant T} d^2(n)(\mathrm{e}^{-2n/T}-1)^2 n^{-1} + + \sum_{n\leqslant T} d^2(n)(\mathrm{e}^{-2n/T}-1)^2$$

$$\ll T^{-1}\sum_{n\leqslant T} d^2(n)n + T^{-2}\sum_{n\leqslant T} d^2(n)n^2 \ll T\log^3 T,$$

$$\int_{T/2}^{T} |J_5|^2 \mathrm{d}t \ll T^{5/2}\sum_{n>T} d^2(n)n^{-5/2} + T^{3/2}\sum_{n>T} d^2(n)n^{-3/2} \ll T\log^3 T,$$

$$\int_{T/2}^{T} |J_6|^2 \mathrm{d}t \ll T^{1/2}\sum_{n\leqslant T} d^2(n)n^{-1/2} + T^{-1/2}\sum_{n\leqslant T} d^2(n)n^{1/2} \ll T\log^3 T.$$

Next, we write

$$i\int_{T/2}^{T} J_1^2 \mathrm{d}t = \int_{1/2+iT/2}^{1/2+iT} J_1^2(s)\mathrm{d}s, \quad J_1(s) := \chi^{-1}(s)\sum_{n\leqslant T} d(n)n^{-s}, \tag{35}$$

and consider the last integral as an integral of the complex variable s. To avoid Lemma 2.3 with $a = -1$ we replace, by Cauchy's theorem, the segment of integration in (35) by segments joining the points

$$\tfrac{1}{2}+\tfrac{1}{2}iT,\ \ \tfrac{1}{4}+\tfrac{1}{2}iT,\ \ \tfrac{1}{4}+iT,\ \ \tfrac{1}{2}+iT.$$

On using (15) it is seen that the integrals over the horizontal segments are $\ll T\log^3 T$, as desired. On the other hand, again by the mean value theorem for Dirichlet polynomials,

$$\int_{1/4+iT/2}^{1/4+iT} J_1^2(s)\mathrm{d}s \ll T^{-1/2}\int_{T/2}^{T}\Big|\sum_{n\leqslant T} d(n)n^{-1/4+it}\Big|^2 \mathrm{d}t \ll T\log^3 T.$$

The same procedure may be applied to the integral of J_2^2 to yield

$$\int_{T/2}^{T} (J_1^2 + J_2^2)\mathrm{d}t \ll T\log^3 T.$$

The remaining integrals in (33) are written as

$$i\int_{1/2+iT/2}^{1/2+iT} J_1(s)J_k(s)\mathrm{d}s + i\int_{1/2+iT/2}^{1/2+iT} J_2(s)J_k(s)\mathrm{d}s \qquad (k=3,4,5,6),$$

and are treated similarly. In the integrals with $J_1(s)$, the segment of integration $[\frac{1}{2}+\frac{1}{2}iT,\ \frac{1}{2}+iT]$ is being replaced by the segment of integration $[\frac{3}{8}+\frac{1}{2}iT,\ \frac{3}{8}iT]$ with an error $\ll T\log^3 T$, while in the integrals containing $J_2(s)$, this segment is replaced by the segment $[\frac{5}{8}+\frac{1}{2}iT,\ \frac{5}{8}+iT]$, again with an error $\ll T\log^3 T$. Applying the Cauchy–Schwarz inequality, Lemma 2.2, and collecting all the estimates we finally obtain

$$\int_{T/2}^{T} |\zeta(\tfrac{1}{2}+it)|^4\mathrm{d}t = (4\pi^2)^{-1}T\log^4 T + O(T\log^3 T),$$

which yields the second asymptotic formula in (17) on replacing T by $T2^{-j}$ and summing the resulting expressions over $j=1,2\ldots$.

Later research revealed that there is another main term in the first formula in (17). Namely, if one defines

$$\int_0^T |\zeta(\tfrac{1}{2}+it)|^2\mathrm{d}t \;=\; T\Big(\log\frac{T}{2\pi}+2\gamma-1\Big)+E(T), \tag{36}$$

where γ is Euler's constant, then $E(T)$ is a true error term in (36) in the sense that

$$E(T) \;=\; o(T) \qquad (T\to\infty). \tag{37}$$

In fact, Titchmarsh [98] gives in Chap. 7, by the method of Atkinson [1], the proof that

$$E(T) \;=\; O_\varepsilon(T^{1/2+\varepsilon}), \tag{38}$$

and states that Ingham [39] obtained $O(T^{1/2}\log T)$. Later Titchmarsh [97] improved further the bound in (38) to $O(T^{5/12}\log^2 T)$, by using van der Corput's theory of exponential sums (see the monograph [20] of Graham and Kolesnik for a detailed account). However the pioneering work in this field was done by F.V. Atkinson (see [1–3]). His most important achievement was made in [4] in 1949, when he produced an explicit formula for $E(T)$, which serves as the basis of modern research of this function. This will be discussed in detail in the next section.

3 The Second Moment

Atkinson's explicit formula for the function $E(T)$, defined by (36), is one of the most important results of zeta-function theory. It is given here as

Theorem 3.1. *Let $0 < A < A'$ be any two fixed constants such that $AT < N < A'T$ and let $N' = N'(T) = T/(2\pi) + N/2 - (N^2/4 + NT/(2\pi))^{1/2}$. Then*

$$E(T) = \Sigma_1(T) + \Sigma_2(T) + O(\log^2 T), \tag{39}$$

where

$$\sum\nolimits_1(T) = 2^{1/2}(T/(2\pi))^{1/4} \sum_{n\leqslant N} (-1)^n d(n) n^{-3/4} e(T,n) \cos(f(T,n)), \tag{40}$$

$$\sum\nolimits_2(T) = -2 \sum_{n\leqslant N'} d(n) n^{-1/2} (\log T/(2\pi n))^{-1} \cos(g(T,n)), \tag{41}$$

with

$$\begin{aligned} f(T,n) &= 2T\operatorname{arsinh}\left(\sqrt{\pi n/(2T)}\right) + \sqrt{2\pi n T + \pi^2 n^2} - \pi/4 \\ &= -\frac{1}{4}\pi + 2\sqrt{2\pi n T} + \frac{1}{6}\sqrt{2\pi^3} n^{3/2} T^{-1/2} \\ &\quad + a_5 n^{5/2} T^{-3/2} + a_7 n^{7/2} T^{-5/2} + \cdots, \\ g(T,n) &= T \log T/(2\pi n) - T + \pi/4, \end{aligned} \tag{42}$$

where

$$\begin{aligned} e(T,n) &= (1 + \pi n/(2T))^{-1/4} \left\{(2T/\pi n)^{1/2} \operatorname{arsinh}\left(\sqrt{\pi n/(2T)}\right)\right\}^{-1} \\ &= 1 + O(n/T) \qquad (1 \leqslant n < T), \end{aligned} \tag{43}$$

and $\operatorname{arsinh} x = \log(x + \sqrt{1 + x^2})$.

Remark 3.1. If we estimate trivially the sums in (40) and (41) (using the inequality $|\sum F(n)| \leqslant \sum |F(n)|$), we obtain the bound $E(T) \ll T^{1/2} \log T$. Of course, using exponential sum techniques is possible to obtain better results. It was indicated in Corollary 15.4 of [52] how one obtains

$$E(T) \ll_\varepsilon T^{35/108+\varepsilon} \qquad \left(\frac{35}{108} = 0.32\overline{407}\right).$$

The latest record is due to Watt [105], who proved that

$$E(T) \ll_\varepsilon T^{131/416+\varepsilon} \qquad \left(\frac{131}{416} = 0.314903\ldots\right). \tag{44}$$

His proof is based on a refined version of the *Bombieri–Iwaniec method* for the estimation of exponential sums. For this, see [10,11] and the survey article of Huxley and the author [38].

Remark 3.2. There is another explicit formula for $E(T)$, also with the error term $O(\log^2 T)$. Obtained almost 30 years after Atkinson's work, this result is due to Balasubramanian [5] and says that

$$E(T) = 2\sum_{n\leqslant K}\sum_{m\leqslant K, m\neq n}\frac{\sin(T\log n/m)}{\sqrt{mn}\log n/m}$$

$$+2\sum_{n\leqslant K}\sum_{m\leqslant K, m\neq n}\frac{\sin(2\theta_1 - T\log mn)}{\sqrt{mn}\,(2\theta_1' - \log mn)} + O(\log^2 T), \tag{45}$$

where $\theta_1 = \theta_1(T) = \frac{1}{2}T\log(T/(2\pi)) - \frac{1}{2}T - \frac{1}{8}\pi, K = \sqrt{T/(2\pi)}$. Both Atkinson's formula and Balasubramanian's (45) contain two exponential sums, but (45) contains *double sums*, whereas Atkinson's formula contain one-dimensional sums with the divisor function $d(n)$. The proof of (45) consists of the integration of the sharp version of the Riemann–Siegel formula [see (19)].

Remark 3.3. A proof of Atkinson's formula, different from [4], was obtained in 1987 by Motohashi [82]. His error term $O(\log T)$ is better than Atkinson's $O(\log^2 T)$, and the proof is based on his approximate functional equation for $\zeta^2(\frac{1}{2} + it)$. There are two other different proofs of Atkinson's formula. They are due to Jutila [68,69] and Lukkarinen [78] and are both based on the use of Laplace transforms of $|\zeta(\frac{1}{2} + ix)|^2$.

We are not going to give a full proof of Atkinson's formula (for this see [4,42, 52]), but we shall only indicate the salient points of the argument. Atkinson starts from the decomposition

$$\zeta(u)\zeta(v) = \sum_{m=1}^{\infty}\sum_{n=1}^{\infty} m^{-u}n^{-v} = \zeta(u+v) + f(u,v) + f(v,u), \tag{46}$$

where

$$f(u,v) := \sum_{r=1}^{\infty}\sum_{s=1}^{\infty} r^{-u}(r+s)^{-v} \quad (\operatorname{Re} u > 1, \operatorname{Re} v > 1).$$

The "diagonal" terms $m = n$ give rise to $\zeta(u+v)$, while the "non-diagonal" terms $m \neq n$ contribute to $f(u,v) + f(v,u)$. The main problem, appearing several times in the course of the proof, is to obtain analytic continuation of a certain expression. In the case at hand, we seek the representation in (46) to hold outside the region of absolute convergence of the complex variables u and v. Applying partial summation to the sum $\sum_{s=1}^{\infty}(r+s)^{-v}$ it is seen that

$$f(u,v) - (v-1)^{-1}\zeta(u+v-1) + \tfrac{1}{2}\zeta(u+v)$$

is regular for $\operatorname{Re}(u+v) > 0$. Thus (46) holds by analytic continuation when u, v both lie in the critical strip, apart from the poles at $v = 1, u+v = 1$ and $u+v = 2$. In the case when $\operatorname{Re} u < 0, \operatorname{Re}(u+v) > 2$, one uses the familiar *Poisson summation formula* (see (A.25) of [52]), namely,

$$\sum_{a\leqslant n\leqslant b}{}' f(n) = \int_a^b f(x)\mathrm{d}x + 2\sum_{n=1}^{\infty}\int_a^b f(x)\cos(2\pi nx)\mathrm{d}x,$$

which holds if $f'(x)$ is bounded on $[a,b]$, and the dash $'$ in summation means that $\frac{1}{2}f(a)$ and $\frac{1}{2}f(b)$ are to be taken instead of $f(a)$ and $f(b)$, respectively, if a and b are integers. This gives

$$\begin{aligned}\sum_{r=1}^{\infty} r^{-u}(r+s)^{-v} = s^{1-u-v}\Big\{&\int_0^{\infty} y^{-u}(1+y)^{-v}\mathrm{d}y\\ &+2\sum_{m=1}^{\infty}\int_0^{\infty} y^{-u}(1+y)^{-v}\cos(2\pi msy)\mathrm{d}y\Big\}.\end{aligned}$$

Summation over s shows that

$$\begin{aligned}g(u,v) &:= f(u,v) - \Gamma(u+v-1)\Gamma(1-u)\Gamma^{-1}(v)\zeta(u+v-1)\\ &= 2\sum_{s=1}^{\infty} s^{1-u-v}\sum_{m=1}^{\infty}\int_0^{\infty} y^{-u}(1+y)^{-v}\cos(2\pi msy)\mathrm{d}y. \qquad (47)\end{aligned}$$

To investigate the convergence of the last expression, we note that for

$$\operatorname{Re} u < 1,\ \operatorname{Re}(u+v) > 0,\ n \geqslant 1,\ \mathrm{e}(u) = \exp(2\pi i u)$$

we have

$$\begin{aligned}2\int_0^{\infty} y^{-u}(1+y)^{-v}\cos(2\pi msy)\mathrm{d}y &= n^{u-1}\int_0^{\infty} y^{-u}(1+y/n)^{-v}\big(\mathrm{e}(y)+\mathrm{e}(-y)\big)\mathrm{d}y\\ &= n^{u-1}\int_0^{i\infty} y^{-u}(1+y/n)^{-v}\mathrm{e}(y)\mathrm{d}y\\ &\quad + n^{u-1}\int_0^{-i\infty} y^{-u}(1+y/n)^{-v}\mathrm{e}(-y)\mathrm{d}y\\ &\ll \frac{n^{\operatorname{Re} u-1}}{|u-1|}.\end{aligned}$$

This bound holds uniformly for bounded u and v, which follows after integration by parts. Thus the double series in (47) is absolutely convergent for $\operatorname{Re} u < 0, \operatorname{Re} v > 1$, $\operatorname{Re}(u+v) > 0$, by comparison with $\sum_{s=1}^{\infty}|s^{-v}|\sum_{m=1}^{\infty}|m^{u-1}|$, and represents an analytic function of both variables in this region. Hence (47) holds throughout this region, and grouping together terms with $ms = n$ together we have

$$g(u,v) = 2\sum_{n=1}^{\infty}\sigma_{1-u-v}(n)\int_0^{\infty} y^{-u}(1+y)^{-v}\cos(2\pi ny)\mathrm{d}y, \tag{48}$$

where $\sigma_k(n) = \sum_{d|n} d^k$ is the sum of the kth powers of divisors of n, so that $\sigma_0(n) \equiv d(n)$. Therefore, if $g(u,v)$ is the analytic continuation of the function given by (47), then for $0 < \mathrm{Re}\, u < 1, 0 < \mathrm{Re}\, v < 1, u+v \neq 1$ we have

$$\begin{aligned}\zeta(u)\zeta(v) &= \zeta(u+v) + \zeta(u+v-1)\Gamma(u+v-1)\left(\frac{\Gamma(1-u)}{\Gamma(v)} + \frac{\Gamma(1-v)}{\Gamma(u)}\right)\\ &\quad + g(u,v) + g(v,u).\end{aligned} \tag{49}$$

It is, however, the exceptional case $u+v=1$, in which we are interested. Here we may use the fact that $g(u,v)$ is continuous. We write $u+v = 1+\delta, 0 < |\delta| < \frac{1}{2}$ and let $\delta \to 0$. It follows that, for $0 < \mathrm{Re}\, u < 1$,

$$\begin{aligned}\zeta(u)\zeta(1-u) = \frac{1}{2}&\left(\frac{\Gamma'(1-u)}{\Gamma(1-u)} + \frac{\Gamma'(u)}{\Gamma(u)}\right)\\ &+ 2\gamma - \log 2\pi + g(u,1-u) + g(1-u,u),\end{aligned} \tag{50}$$

with a view to the eventual application $u = \frac{1}{2} + it$ in mind. Reasoning as in (48) we have, for $\mathrm{Re}\, u < 0$,

$$g(u,1-u) = 2\sum_{n=1}^{\infty} d(n)\int_0^{\infty} y^{-u}(1+y)^{u-1}\cos(2\pi ny)\mathrm{d}y, \tag{51}$$

and so what we need is the analytic continuation of (51) valid when $\mathrm{Re}\, u = \frac{1}{2}$. At this point the *Voronoï formula* for $\Delta(x)$ comes into play, where

$$\Delta(x) := {\sum_{n\leqslant x}}' d(n) - x(\log x + 2\gamma - 1) - 1/4, \tag{52}$$

and the dash $'$, similarly as in Poisson's summation formula, means that the last term in the sum in (52) is to be halved if $x \in \mathbb{N}$. At the beginning of the twentieth century Voronoï [102] proved the exact, explicit formula

$$\Delta(x) = -2\pi^{-1}x^{1/2}\sum_{n=1}^{\infty} d(n)n^{-1/2}\Big\{K_1(4\pi\sqrt{nx}) + \tfrac{1}{2}\pi Y_1(4\pi\sqrt{nx})\Big\}, \tag{53}$$

where K_1, Y_1 are *Bessel functions* in standard notation (see, e.g., the treatise [103] of Watson) and the series above is boundedly, but not absolutely, convergent. In the present context it is more expedient to use a truncated version of (53), namely

$$\Delta(x) = (\pi\sqrt{2})^{-1}x^{1/4}\sum_{n=1}^{\infty} d(n)n^{-3/4}$$
$$\times\left\{\cos(4\pi\sqrt{nx}) - 3(32\pi\sqrt{nx})^{-1}\sin(4\pi\sqrt{nx})\right\} + O(x^{-3/4}). \quad (54)$$

It is not difficult to see that the definitions of $E(T)$ and $\Delta(x)$ are similar and that there are also similarities between Atkinson's formula for $E(T)$ and Voronoï's formula for $\Delta(x)$. Indeed, there is a deep connection between these two problems, and for a study of this phenomenon, the reader is referred to [54, 55, 67, 68].

Returning to the discussion of the proof of Theorem 3.1, let $N \in \mathbb{N}$ and

$$h(u,x) := 2\int_0^{\infty} y^{-u}(1+y)^{u-1}\cos(2\pi xy)\mathrm{d}y.$$

Then we have, with $D(x) := \sum_{n\leqslant x} d(n)$,

$$\begin{aligned}\sum_{n>N} d(n)h(u,n) &= \int_{N+1/2}^{\infty} h(u,x)\mathrm{d}D(x)\\ &= \int_{N+1/2}^{\infty} (\log X + 2\gamma)h(u,x)\mathrm{d}x + \int_{N+1/2}^{\infty} h(u,x)\mathrm{d}\Delta(x)\\ &= -\Delta(N+\tfrac{1}{2})h(u,N+\tfrac{1}{2}) + \int_{N+1/2}^{\infty}(\log x + 2\gamma)h(u,x)\mathrm{d}x\\ &\quad - \int_{N+1/2}^{\infty}\frac{\partial h(u,x)}{\partial x}\mathrm{d}x.\end{aligned}$$

Hence (51) becomes

$$\begin{aligned}g(u,1-u) &= \sum_{n\leqslant N} h(u,n)d(n) - \Delta(N+\tfrac{1}{2})h(u,N+\tfrac{1}{2})\\ &\quad + \int_{N+1/2}^{\infty}(\log x + 2\gamma)h(u,x)\mathrm{d}x - \int_{N+1/2}^{\infty}\frac{\partial h(u,x)}{\partial x}\mathrm{d}x\\ &= g_1(u) - g_2(u) + g_3(u) - g_4(u), \qquad (55)\end{aligned}$$

say. Here $g_1(u)$ and $g_2(u)$ are analytic functions in the region $\mathrm{Re}\, u < 1$. Consider next $g_4(u)$. We have

$$h(u,x) = \int_0^{i\infty} y^{-u}(1+y)^{u-1}\mathrm{e}(xy)\mathrm{d}y + \int_0^{-i\infty} y^{-u}(1+y)^{u-1}\mathrm{e}(-xy)\mathrm{d}y,$$

which gives after differentiation over x that

$$\frac{\partial h(u,x)}{\partial x} \ll x^{\mathrm{Re}\, u - 2}$$

for $\operatorname{Re} u \leqslant 1$ and bounded u. Using only the estimate $\Delta(x) \ll_\varepsilon x^{1/3+\varepsilon}$ (see [52] or [38]) it is seen that the integral defining $g_4(u)$ is an analytic function of u, at any rate when $\operatorname{Re} u < 2/3$.

It remains to consider $g_3(u)$. Let, for brevity, $X = N + \frac{1}{2}$. Then

$$g_3(u) = \int_X^\infty (\log x + 2\gamma)\Big\{\int_0^{i\infty} y^{-u}(1+y)^{u-1}\mathrm{e}(xy)\mathrm{d}y + \int_0^{-i\infty} y^{-u}(1+y)^{u-1}\mathrm{e}(-xy)\mathrm{d}y\Big\}\mathrm{d}x. \tag{56}$$

For $\operatorname{Re} u < 0$ an integration by parts shows that the first two integrals in (56) are equal to

$$\begin{aligned}
&-(2\pi i)^{-1}(\log X + 2\gamma)\int_0^{i\infty} y^{-u-1}(1+y)^{u-1}\mathrm{e}(Xy)\mathrm{d}y\\
&-(2\pi i)^{-1}\int_X^\infty \mathrm{d}x\int_0^{i\infty} y^{-u-1}(x+y)^{u-1}\mathrm{e}(y)\mathrm{d}y\\
&= -(2\pi i)^{-1}(\log X + 2\gamma)\int_0^{i\infty} y^{-u-1}(1+y)^{u-1}\mathrm{e}(Xy)\mathrm{d}y\\
&+(2\pi i)^{-1}\int_0^{i\infty} y^{-u-1}(X+y)^{u}\mathrm{e}(y)\mathrm{d}y.
\end{aligned}$$

In the last integral above, the line of integration may be taken as $[0, \infty)$ and the variable y replaced by $y = Xz$. The other two integrals in (56) are treated similarly, and the results may be combined to produce

$$g_3(u) = -\pi^{-1}(\log X + 2\gamma)\int_0^\infty y^{-u-1}(1+y)^{u-1}\sin(2\pi Xy)\mathrm{d}y + (\pi u)^{-1}\int_0^\infty y^{-u-1}(1+y)^{u-1}\sin(2\pi Xy)\mathrm{d}y. \tag{57}$$

Noting that the integrals in (57) are uniformly convergent when $\operatorname{Re} u \leqslant 1 - \varepsilon$, it follows that (57) provides us with an analytic continuation which is valid when $\operatorname{Re} u = \frac{1}{2}$, and thus we may proceed to integrate (50). When $u = \frac{1}{2} + it$ we have

$$\zeta(u)\zeta(1-u) = |\zeta(\tfrac{1}{2} + it)|^2,$$

so that the integration of (50) gives

$$\begin{aligned}
2iI(T) &= \int_{1/2-iT}^{1/2+iT} \zeta(u)\zeta(1-u)\mathrm{d}u\\
&= \tfrac{1}{2}\Big(-\log\Gamma(1-u) + \log\Gamma(u)\Big)\Big|_{1/2-iT}^{1/2+iT} + 2iT(\gamma - \log 2\pi)
\end{aligned}$$

$$+\int_{1/2-iT}^{1/2+iT}\Big(g(u,1-u)+g(1-u,u)\Big)\mathrm{d}u$$

$$=\log\frac{\Gamma(\frac{1}{2}+iT)}{\Gamma(\frac{1}{2}-iT)}+2iT(2\gamma-\log 2\pi)+2\int_{1/2-iT}^{1/2+iT}g(u,1-u)\mathrm{d}u.$$

If one uses Stirling's formula for the gamma-function (see, e.g., [18]), namely

$$\Gamma(s)=\sqrt{2\pi}|t|^{\sigma-1/2}\exp\left\{-\tfrac{1}{2}\pi|t|+i\left(t\log|t|-t+\frac{\pi t}{2|t|}\left(\sigma-\frac{1}{2}\right)\right)\right\}$$

$$\times\left(1+\frac{i}{2t}(\sigma-\sigma^2-\tfrac{1}{6})+O(|t|^{-2})\right), \tag{58}$$

this becomes

$$I(T)=T\Big(\log\frac{T}{2\pi}+2\gamma\Big)-i\int_{1/2-iT}^{1/2+iT}g(u,1-u)\mathrm{d}u$$

$$=T\Big(\log\frac{T}{2\pi}+2\gamma\Big)+I_1-I_2+I_3-I_4+O(1), \tag{59}$$

where for $n=1,2,3,4$

$$I_n:=-i\int_{1/2-iT}^{1/2+iT}g_n(u)\mathrm{d}u,$$

so that

$$I_1=4\sum_{n\leqslant N}d(n)\int_0^\infty\frac{\sin(T\log(1+y)/y)\cos(2\pi ny)}{y^{1/2}(1+y)^{1/2}\log(1+y)/y}\mathrm{d}y,$$

$$I_2=4\Delta(x)\int_0^\infty\frac{\sin(T\log(1+y)/y)\cos(2\pi Xy)}{y^{1/2}(1+y)^{1/2}\log(1+y)/y}\mathrm{d}y,$$

$$I_3=\frac{2}{\pi}(\log X+2\gamma)\int_0^\infty\frac{\sin(T\log(1+y)/y)\sin(2\pi Xy)}{y^{3/2}(1+y)^{1/2}\log(1+y)/y}\mathrm{d}y,$$

$$+\frac{1}{i\pi}\int_0^\infty\frac{\sin(2\pi Xy)}{y}\mathrm{d}y\int_{1/2-iT}^{1/2+iT}(1+y^{-1})^u u^{-1}\mathrm{d}u,$$

$$I_4=\int_X^\infty\Delta(x)\mathrm{d}x\int_{1/2-iT}^{1/2+iT}\frac{\partial h(u,x)}{\partial x}\mathrm{d}u, \tag{60}$$

where $N \in \mathbb{N}$, $X = N + \frac{1}{2}$, and as in the formulation of Theorem 3.1 we restrict N to the range $AT < N < A'T$, $0 < A < A'$. One can transform the last integral in (60) to obtain

$$I_4 = 4\Delta(x)\int_X^\infty \mathrm{d}x \int_0^\infty \frac{\cos(2\pi xy)}{y^{1/2}(1+y)^{3/2}\log(1+y)/y} \times \left\{T\cos\Big(T\log\frac{1+y}{y}\Big) - \sin\Big(T\log\frac{1+y}{y}\Big)\Big(\frac{1}{2}+\log^{-1}\frac{1+y}{y}\Big)\right\}\mathrm{d}y. \tag{61}$$

The main difficulty lies now in the evaluation of the integrals in (60) and (61) which represent I_n. Since the integrals in question are exponential integrals, one needs two lemmas (see Lemmas 15.1 and 15.2 of [52]) for the evaluation of such integrals. This is rather technical and involved, and the details are therefore omitted. The transformations of I_n lead ultimately to the expressions in (39)–(42).

Atkinson's formula lay forgotten for almost 30 years until Heath-Brown [30, 31] used it to obtain important results. In [30] he proved a mean square result which is stated here as

Theorem 3.2. *We have*

$$\int_2^T E^2(t)\mathrm{d}t = CT^{3/2} + O(T^{5/4}\log^2 T), \quad C = \frac{2}{3}(2\pi)^{-1/2}\sum_{n=1}^\infty d^2(n)n^{-3/2}. \tag{62}$$

Remark 3.4. By using (27) one can rewrite the constant C as

$$C = \frac{2}{3}(2\pi)^{-1/2}\frac{\zeta^4(3/2)}{\zeta(3)} = 10.3047\ldots.$$

Remark 3.5. If $f(x) = \Omega(g(x))$ as usual means that $\lim_{x\to\infty} f(x)/g(x) \neq 0$, then from (62) it clearly follows that

$$E(T) = \Omega(T^{1/4}). \tag{63}$$

The omega result (63) is far from the upper bound (44).

Problem 1. What is the true order of $E(T)$?

The mean square formula (62) suggests that $E(T)$ is "small" on the average; in particular, it is commonly conjectured that one has

$$E(T) = O_\varepsilon(T^{1/4+\varepsilon}), \tag{64}$$

but this is certainly beyond reach by present methods.

Proof of Theorem 3.2. Writing $R(T)$ for the error term in Atkinson's formula, we obtain

$$\int_T^{2T} E^2(t)\mathrm{d}t = \int_T^{2T} \sum\nolimits_1^2(t)\mathrm{d}t + 2\int_T^{2T} \sum\nolimits_1(t)\Big(\sum\nolimits_2(t) + R(t)\Big)\mathrm{d}t + \int_T^{2T} \Big(\sum\nolimits_2(t) + R(t)\Big)^2 \mathrm{d}t. \tag{65}$$

We choose $N = T$ in Atkinson's formula (40) to obtain that

$$\int_T^{2T} \sum\nolimits_1^2(t)\mathrm{d}t = 2\int_T^{2T} \sqrt{t/(2\pi)} \sum_{m\leqslant N}\sum_{n\leqslant N} \cdots \mathrm{d}t.$$

Here the diagonal terms $m = n$ clearly contribute

$$\frac{2}{3}(2\pi)^{-1/2}\sum_{n=1}^{\infty} d^2(n)n^{-3/2}\Big((2T)^{3/2} - T^{3/2}\Big) + O_\varepsilon(T^{1+\varepsilon}).$$

The non-diagonal terms $m \neq n$ give rise to an expression of the type

$$\sum_{m\neq n\leqslant T} \frac{1}{4}(-1)^{m+n}d(m)d(n)\int_T^{2T} g(t)\cos(f(t))\mathrm{d}t,$$

where, with $f(T,n)$ given by (42), we have

$$f(t) = f(t,m) \mp f(t,n), \quad g(t) = g_1(t)g_2(t)g_3(t)g_4(t),$$

$$g_1(t) = \left(\frac{t}{2\pi m} + \frac{1}{4}\right)^{-1/4}, \quad g_2(t) = \left(\frac{t}{2\pi n} + \frac{1}{4}\right)^{-1/4},$$

$$g_3(t) = \left(\operatorname{ar\,sinh}\sqrt{\pi m/2t}\right)^{-1}, \quad g_4(t) = \left(\operatorname{ar\,sinh}\sqrt{\pi n/2t}\right)^{-1}.$$

The functions $g_i(t)$ are monotonic, and we may apply the standard *first derivative test*, namely (see, e.g., Chap. 2 of [52]).

Lemma 3.1. *Let $F(x)$ be a real differentiable function such that $F'(x)$ is monotonic and $|F'(x)| > m > 0$ in $[a, b]$, and $G(x)$ is a positive, monotonic function such that $|G(x)| \leqslant G$ in $[a,b]$. Then*

$$\left|\int_a^b G(x)\mathrm{e}^{iF(x)}\mathrm{d}x\right| \ll G/m. \tag{66}$$

Since $f'(t,n) = 2\mathrm{ar\,sinh}\sqrt{\pi n/2t}$, the contribution of the terms $m \neq n$ is then, on using (66),

$$\ll T \sum_{m \neq n \leqslant T} d(m)d(n)(mn)^{-3/4}|n^{1/2} - m^{1/2}|^{-1} \ll_\varepsilon T^{1+\varepsilon}.$$

The contribution of the remaining terms in (65) is estimated similarly. In evaluating the mean square of $\sum_2(T)$ we shall obtain

$$\int_T^{2T} \sum\nolimits_2^2(t)\mathrm{d}t \ll T\log^4 T,$$

keeping in mind that

$$N' = N'(t) = t/(2\pi) + N/2 - (N^2/4 + nt/(2\pi))^{1/2} \asymp T,$$

where $N = T$ depends on t. Further, on using the Cauchy–Schwarz inequality for integrals, we obtain

$$\int_T^{2T} \sum\nolimits_1(t)\sum\nolimits_2(t)\mathrm{d}t \ll \left\{\int_T^{2T} \sum\nolimits_1^2(t)\mathrm{d}t \int_T^{2T} \sum\nolimits_2^2(t)\mathrm{d}t\right\}^{1/2} \ll T^{5/4}\log T,$$

and the integrals with $R(t)$ are estimated on using Atkinson's bound $R(T) \ll \log^2 T$. Collecting the preceding estimates we have

$$\int_T^{2T} E^2(t)\mathrm{d}t = C\left((2T)^{3/2} - T^{3/2}\right) + O(T^{5/4}\log T),$$

and the assertion of Theorem 3.2 follows if we replace T by $T2^{-j}$, $j = 1, 2, \ldots$ and sum the resulting expressions.
One can improve on Theorem 3.2 and obtain

$$\int_2^T E^2(t)\mathrm{d}t = CT^{3/2} + O(T\log^4 T). \tag{67}$$

The bound in (67) was obtained independently by Preissman [86] and the author [42] in Chap. 2. The best current result is due to Lau and Tsang [76] who proved that

$$F(T) = O(T\log^3 T\log\log T), \quad F(T) := \int_2^T E^2(t)\mathrm{d}t - CT^{3/2}. \tag{68}$$

It should be also noted that Lee and Tsang [77] proved that

$$\int_0^T F(t)\mathrm{d}t = -3\pi^{-2}T^2\log^2 T\log\log T + O(T^2\log^2 T), \tag{69}$$

which in particular implies that

$$F(T) = \Omega_-(T\log^2 T\log\log T).$$

Problem 2. What is the true order of $F(T)$?

In view of the result of (69) of [77], $F(T)$ is probably of the order

$$T\log^2 T\log\log T.$$

Their formula (69) is significant from another aspect. Namely it shows a difference in behaviour between $E(T)$ and $\Delta(x)$, the error term in the classical Dirichlet divisor problem. If $L(x)$ is the analogue of $F(T)$ in (68) for $\Delta(x)$, then we have the result of Lau and Tsang [74] that

$$\int_0^X L(x)\mathrm{d}x = -(8\pi^2)^{-1}X^2\log^2 X + cX^2\log X + O(X^2),$$

which is different from (69). One can compare this with the author's result [45], who proved the Laplace transform formula

$$\int_0^\infty \Delta^2(x)\mathrm{e}^{-x/T}\mathrm{d}x = C_1T^{3/2} + (A_1\log^2 T + A_2\log T + A_3)T + O_\varepsilon(T^{2/3+\varepsilon}),$$

where $C_1 > 0, A_1 > 0$. On the other hand, using integration by parts and (69), it is not difficult to see that

$$\int_0^\infty E^2(x)\mathrm{e}^{-x/T}\mathrm{d}x = D_1T^{3/2} + B_1T\log^2 T\log\log T + O(T\log^2 T),$$

which has a different structure than the above formula. For a detailed analysis on results and problems involving $\Delta(x)$ and $E(T)$, see the paper of Tsang [100].

Problem 3. Find an exact asymptotic formula for $\int_0^\infty E^2(x)\mathrm{e}^{-x/T}\mathrm{d}x$.

Continuing our discussion on $E(T)$, note that the defining relation (36) yields

$$E(T+x) - E(T) \geqslant -2Cx\log T \qquad (0 \leqslant x \leqslant T,\ T \geqslant T_0).$$

Hence integration gives

$$\begin{aligned}\int_T^{T+x} E(t)\mathrm{d}t &= xE(T) + \int_0^x (E(T+u) - E(T))\mathrm{d}u\\ &\geqslant xE(T) - 2C\log T\int_0^x u\mathrm{d}u\\ &= xE(T) - Cx^2\log T. \qquad (70)\end{aligned}$$

Therefore we obtain

$$E(T) \leqslant x^{-1}\int_T^{T+x} E(t)\mathrm{d}t + Cx\log T \qquad (0 < x \leqslant T,\ T \geqslant T_0), \tag{71}$$

and in a similar fashion we also have

$$E(T) \geqslant x^{-1}\int_{T-x}^{T} E(t)\mathrm{d}t - Cx\log T \qquad (0 < x \leqslant T,\ T \geqslant T_0). \tag{72}$$

Combining (71) and (72) we have

$$|E(T)| \leqslant x^{-1}\int_{T-x}^{T+x} |E(t)|\mathrm{d}t + 2Cx\log T \qquad (0 < x \leqslant T,\ T \geqslant T_0). \tag{73}$$

From this inequality one can relate the order of $E(T)$ to the mean square formula for $E(T)$. Namely, if $F(T)$ denotes the error term in (62), then from (73) we obtain

$$\begin{aligned}
|E(T)| &\ll x^{-1/2}\Big(\int_{T-x}^{T+x} E^2(t)\mathrm{d}t\Big)^{1/2} + x\log T \\
&\ll x^{-1/2}\Big(C(T+x)^{3/2} - C(T-x)^{3/2} + F(T+x) - F(T-x)\Big)^{1/2} + x\log T \\
&\ll T^{1/4} + x^{-1/2}\max_{T-x\leqslant t\leqslant T+x}|F(t)|^{1/2} + x\log T.
\end{aligned}$$

Choosing $x = (\max\cdots)^{1/3}(\log T)^{-2/3}$ this leads to

Theorem 3.3. *If $F(T)$ denotes the error term in* (62)*, then*

$$E(T) \ll T^{1/4} + \Big(\max_{T-\sqrt{T}\leqslant t\leqslant T+\sqrt{T}}|F(t)|\Big)^{1/3}\log^{4/3} T. \tag{74}$$

From (67) and (74) it follows that

$$E(T) \ll T^{1/3}\log^{8/3} T,$$

which is a non-trivial result, but the sharpest exponent of Watt [105] requires much more effort.

One can improve the omega result (63). Namely, Hafner and the author [21, 22] showed that there exist positive constants C and D such that

$$E(T) = \Omega_+\Big\{(T\log T)^{1/4}(\log\log T)^{(3+\log 4)/4}\exp(-C\sqrt{\log\log\log T})\Big\} \tag{75}$$

and

$$E(T) = \Omega_-\left\{T^{1/4}\exp\Big(D(\log\log T)^{1/4}(\log\log\log T)^{-3/4}\Big)\right\}. \tag{76}$$

Lau and Tsang [75] used the method of Soundararajan [95], who like Hafner–Ivić proved the analogous result for $\Delta(x)$ (the error term in the divisor problem), namely they proved

$$E(T) = \Omega\left\{(T\log T)^{\frac{1}{4}}(\log\log T)^{\frac{3}{4}(2^{\frac{4}{3}}-1)}(\log\log\log T)^{-\frac{5}{8}}\right\}.$$

Soundararajan's method does not yield an Ω_+ or Ω_- result, but just an ω result. However, it improves either (75) or (76), but one cannot tell which one. A quantitative omega result was obtained by the author [41]. It was shown there that there exist constants $B, C > 0$ such that every interval $[T, T + C\sqrt{T}]$, for $T \geqslant T_0$, contains numbers τ_1, τ_2 such that

$$E(\tau_1) > B\tau_1^{1/4}, \qquad E(\tau_2) < -B\tau_2^{1/4}. \tag{77}$$

The conjecture that $E(T) = O_\varepsilon(T^{1/4+\varepsilon})$ is supported not only by omega results and the mean square formula for $E(T)$ but also by the results on higher moments and the distribution of values of $E(t)$ (see the review paper of Tsang [100]). The author [40] proved (see also Chap. 15 of [52])

$$\int_0^T |E(t)|^A \mathrm{d}t \ll_\varepsilon \begin{cases} T^{(A+4+\varepsilon)/4} & \text{if } \ 0 \leqslant A \leqslant 35/4, \\ T^{(35A+38+\varepsilon)/108} & \text{if } \ A \geqslant 35/4. \end{cases} \tag{78}$$

The proof depended on the use of the best-known exponent for the order of $|\zeta(\frac{1}{2} + it)|$, which has been since improved to $32/205+\varepsilon = 0.15609\ldots+\varepsilon$, due to Huxley [37]. This enabled Heath-Brown [32] to extend the first bound in (78) to $A \leqslant 28/3$, which one expects to be the true order of magnitude of the integral in question for all $k \geqslant 0$. He also showed, by using properties of *almost periodic functions*, that there exists a well-behaved distribution function $f(t)$ such that ($\mu(\cdot)$ denotes measure)

$$X^{-1}\mu\left\{x \in [1, X] \,:\, x^{-1/4}E(x) \in \mathcal{J}\right\} \to \int_{\mathcal{J}} f(t)\mathrm{d}t$$

as $X \to \infty$ for any interval $\mathcal{J}$. He then deduced that the moments

$$c_k := \lim_{T\to\infty} T^{-1-k/4}\int_0^T |E(t)|^k \mathrm{d}t$$

exist for all real $k \in [0,\ 9]$, as do the odd moments

$$d_k := \lim_{T\to\infty} T^{-1-k/4}\int_0^T E^k(t)\mathrm{d}t \qquad (k = 1, 3, 5, 7, 9).$$

Several results of the type

$$\int_0^T E^k(t)\mathrm{d}t = d_k T^{1+k/4} + O_\varepsilon(T^{1+k/4-\delta_k+\varepsilon}) \tag{79}$$

have been obtain for various integral k and suitable $\delta_k > 0$. We mention the works of Tsang [99], the author [47], Sargos and the author [64], and the works of Zhai [107, 108] who obtained asymptotic formulas of the form (79) for $k \leqslant 9\,(k \in \mathbb{N})$. In all known cases it turned out that d_k for odd k is positive (this is trivial for even k), and so one may ask:

Problem 4. Does d_k exist for all $k \in \mathbb{N}$? If yes, is it true that $d_k > 0$ for all odd k?

In [63] te Riele and the author investigated the distribution of the zeros of $E(T)$, which is a continuous function for $T > 0$. For example, they showed that there exists a constant $c > 0$ such that $E(t) - \pi$ has a zero of odd order in $[T, T + c\sqrt{T}]$ for $T \geqslant T_0$. If t_n denotes the nth zero of $E(t) - \pi$, then the first ten t_n's are

$$t_1 = 1.199593,\ t_2 = 4.757482,\ t_3 = 9.117570,\ t_4 = 13.545429,\ t_5 = 17.685444,$$
$$t_6 = 22.098708, t_7 = 27.706900, t_8 = 31.884578, t_9 = 35.337567,$$
$$t_{10} = 40.500321.$$

One has then

$$t_{n+1} - t_n \;=\; O(t_n^{1/2}), \tag{80}$$

and in the opposite direction Heath-Brown and Tsang [33] proved the following result. Let $\delta > 0$ be any given small quantity. Then for any $T \geqslant T_0(\delta)$, there are at least $c_4\delta\sqrt{T}\log^5 T$ disjoint subintervals of length $c_5\delta\sqrt{T}\log^{-5} T$ in $[T, 2T]$, such that $|E(t)| > (\frac{1}{2}c_3 - \delta)t^{1/4}$ whenever t lies in any of these intervals. In particular, $E(t)$ does not change sign in any of these intervals, so that

$$t_{n+1} - t_n \;=\; \Omega(t_n^{1/2}\log^{-5} t_n). \tag{81}$$

Thus, up to a logarithmic factor, (80) and (81) settle the question of the order of the gap between the consecutive zeros of $E(T)$.

The reason why the zeros of $E(t) - \pi$ are considered, and not simply the zeros of $E(t)$, is that π is the mean value of $E(t)$. Namely Hafner and the author [21, 22] proved that

$$G(T) = O(T^{3/4}),\ G(T) = \Omega_\pm(T^{3/4}),\ \int_2^T G^2(t)\mathrm{d}t = BT^{5/2} + O(T^2),$$

where $G(T) := E(T) - \pi$ and $B > 0$ is an explicit constant. As usual, $F(x) = \Omega_\pm(H(x))$ means that $F(x) = \Omega_+(H(x))$ and $F(x) = \Omega_-(H(x))$ both hold.

These results determine, up to the value of the numerical constants that are involved, the true order of the function $G(T)$. In fact one has the explicit formula

$$G(T) = \frac{1}{2}\left(\frac{2T}{\pi}\right)^{3/4} \sum_{n \leqslant T} d(n) n^{-5/4} e_2(T,n) \sin(f(T,n))$$
$$-2 \sum_{n \leqslant c_0 T} d(n) n^{-1/2} (\log T/(2\pi n))^{-2} \sin(g(T,n)) + O(T^{1/4}),$$

which is clearly an integrated version of Atkinson's formula. Here

$$c_0 = \frac{1}{2\pi} + \frac{1}{2} - \sqrt{\frac{1}{4} + \frac{1}{2\pi}},$$
$$e_2(T,n) = \left(1 + \frac{\pi n}{2T}\right)^{-1/4} \left\{\left(\frac{2T}{\pi n}\right)^{1/2} \operatorname{ar\,sinh}\left(\frac{\pi n}{2T}\right)^{1/2}\right\}^{-2}.$$

4 The Fourth Moment of $|\zeta(\frac{1}{2} + it)|$

For more than fifty years Ingham's formula (17) for the fourth moment withstood improvements. Then in 1979 Heath-Brown [31] obtained a substantial sharpening of (17). He proved that (γ is Euler's constant)

$$\int_0^T |\zeta(\tfrac{1}{2} + it)|^4 \mathrm{d}t = a_4 T \log^4 T + a_3 T \log^3 T$$
$$+ a_2 T \log^2 T + a_1 T \log T + a_0 T + O_\varepsilon(T^{7/8+\varepsilon}) \qquad (82)$$

with

$$a_4 = (2\pi^2)^{-1}, \quad a_3 = 2(4\gamma - 1 - \log 2\pi - 12\zeta'(2)\pi^{-2})\pi^{-2}.$$

His proof uses a new approximate functional equation for $|\zeta(\frac{1}{2} + it)|^{2k}$, as well as results involving the asymptotic formula for $\sum_{n \leqslant x} d(n) d(n + r)$ (the so-called additive divisor problem, where r ($\in \mathbb{N}$) is not necessarily fixed, but may vary with x). This in turn depended on estimates for the Kloosterman sums

$$S(m,n;c) := \sum_{1 \leqslant d \leqslant c} e\left(\frac{md + nd'}{c}\right) \qquad (e(z) = \exp(2\pi i z)),$$

where $dd' \equiv 1 \bmod c$. Sums of Kloosterman sums have become very important in many problems from analytic number theory in the last 35 years, due to the efforts of R. Bruggeman, H. Iwaniec, N. Kuznetsov, Y. Motohashi and others.

Heath-Brown also indicated how one can evaluate in closed form the remaining coefficients a_j in (82), but his expressions are cumbersome. If the main term for the fourth moment in (82) is written as $Tp_4(L), L = \log(T/(2\pi))$, where p_4 is a polynomial of degree four, then Conrey [13] and the author [43] independently evaluated the coefficients of $p_4(x)$. Conrey, whose analysis is based on [31], has shown that $p_4(x) = g_0(x) + g_1(x)$, where

$$g_0(x) = \operatorname*{Res}_{s=0} \frac{2x^s\zeta^4(s+1)}{s(s+1)\zeta(2s+2)}$$

and

$$g_1(x) = \left(\frac{\mathrm{d}}{\mathrm{d}s}\right)^2 \frac{(x\mathrm{e}^{2\gamma})^s\left\{\frac{1}{2}\zeta^2(s+1) - s^{-1}\zeta(2s+1) - \zeta(2s+2)\right\}}{(s+1)\zeta(s+2)}\Bigg|_{s=0}.$$

Numerical calculation shows that

$$Tp_4(L) = T\Big(0.050660L^4 + 0.496227L^3 + 0.937279L^2 + 1.35334L - 0.040924\Big),$$

where the coefficients are accurate within six decimal places. The forthcoming papers of Hiary and Odlyzko [34], and Rubinstein and Yamagishi [92] contain many numerical results concerning various moments of $|\zeta(\frac{1}{2}+it)|$.

An important moment in the theory of the fourth moment of $|\zeta(\frac{1}{2}+it)|$ is Motohashi's work [83], later expounded in his monograph [85]. It yields an explicit formula for the weighted integral

$$I(T,\Delta) := (\Delta\sqrt{\pi})^{-1}\int_{-\infty}^{\infty} |\zeta(\tfrac{1}{2}+i(T+t))|^4 \mathrm{e}^{-(t/\Delta)^2}\mathrm{d}t \qquad (0 < \Delta < T/\log T). \tag{83}$$

Remark 4.1. The range $0 < \Delta < T/\log T$ is very wide and is sufficient for applications.

Remark 4.2. The *Gaussian exponential factor* $\mathrm{e}^{-(t/\Delta)^2}$ inserted in the integrand facilitates convergence problems and analytic continuation essential to Motohashi's method of proof. It is, of course, not the only weight function which may be used in this context, as clearly shown in [85] (see also Theorem 4.1 below). On the other hand, it is clear that the resulting expression(s) for the fourth moment will not be the analogue(s) of Atkinson's formula for the mean square of $|\zeta(\frac{1}{2}+it)|$.

Problem 5. Does there exist an analogue of Atkinson's formula for the mean square of $|\zeta(\frac{1}{2}+it)|$, also for the fourth moment of $|\zeta(\frac{1}{2}+it)|$?

No one has ever found such a formula, so the answer is probably not, but this has not been proved.

To formulate Motohashi's result on (83), we need some notation from the spectral theory of the non-Euclidean Laplacian. As usual, $H_j(s)$ is the Hecke series

$$H_j(s) = \sum_{n=1}^{\infty} t_j(n)n^{-s} = \prod_p (1 - t_j(p)p^{-s} + p^{-2s})^{-1} \qquad (\sigma > 1),$$

associated with the *Maass wave form* $\psi_j(z)$, where $\rho_j(1)t_j(n) = \rho_j(n)$ and $\rho_j(n)$ is the nth Fourier coefficient of $\psi_j(z)$. The function $H_j(s)$ can be continued analytically to an entire function on $\mathbb{C}$. It satisfies the functional equation, similar to the functional equation for $\zeta(s)$, namely

$$H_j(s) = 2^{2s-1}\pi^{2s-2}\Gamma(1-s+i\kappa_j)\Gamma(1-s-i\kappa_j)(\varepsilon_j \cosh(\pi\kappa_j) - \cos(\pi s))H_j(1-s),$$

where ε_j $(= \pm 1)$ is the so-called parity sign of $\psi_j(z)$ $(z = x + iy)$. This means that $\varepsilon_j = 1$ if $\psi_j(z)$ is an even function of x and $\varepsilon_j = -1$ if $\psi_j(z)$ is an odd function of x. By

$$\left\{\lambda_j = \kappa_j^2 + \frac{1}{4}\right\} \cup \{0\}$$

we denote the eigenvalues (discrete spectrum) of the *hyperbolic Laplacian*

$$\Delta = -y^2\left(\left(\frac{\partial}{\partial x}\right)^2 + \left(\frac{\partial}{\partial y}\right)^2\right)$$

acting over the Hilbert space composed of all Γ-automorphic functions which are square integrable with respect to the hyperbolic measure ($\Gamma = \mathrm{PSL}(2, \mathbb{Z})$). Further $\alpha_j = |\rho_j(1)|^2(\cosh \pi\kappa_j)^{-1}$, where $\rho_j(1)$ is the first Fourier coefficient of the Maass wave form corresponding to the eigenvalue λ_j to which the Hecke L-function $H_j(s)$ is attached. We have $H_j(\frac{1}{2}) \geqslant 0$ and

$$\sum_{\kappa_j \leqslant T} \alpha_j H_j^3(\tfrac{1}{2}) \ll T^2 \log^8 T. \tag{84}$$

Motohashi's general formula for weighted integral of the fourth moment of $|\zeta(\frac{1}{2} + it)|$ will involve not only Maass wave forms (non-holomorphic cusp forms) but holomorphic cusp forms as well. Thus let

$$\{f_{j,2k}(z)\},\ 1 \leqslant j \leqslant d_{2k},\ k \geqslant 6$$

be the orthonormal basis, which consists of eigenfunctions of Hecke operators $T_{2k}(n)$, of the *Petersson unitary space* of holomorphic cusp forms of weight $2k$ for the full modular group. Hence for every $n \in \mathbb{N}$ there is a $t_{j,2k}(n)$ such that

$$T_{2k}(n)(\varphi_{j,2k}(z)) = n^{-1/2} \sum_{ad=n, d>0} \left(\frac{a}{d}\right)^k \sum_{b\,(\mathrm{mod}\, d)} \varphi_{j,2k}\left(\frac{az+b}{d}\right) = t_{j,2k}(n)\varphi_{j,2k}(z).$$

The corresponding Hecke series

$$H_{j,2k}(s) = \sum_{n=1}^{\infty} t_{j,2k}(n)n^{-s}$$

as in the case of $H_j(s)$ converges absolutely at least for $\sigma > 2$. The function $H_{j,2k}(s)$ is entire and is $\ll_j k^c$ for some $c > 0$. If $\rho_{j,2k}(1)$ is the first Fourier coefficient of $H_{j,2k}(s)$, then one defines

$$\alpha_{j,2k} := (2k-1)!2^{2-4k}\pi^{-2k-1}|\rho_{j,2k}(1)|^2.$$

With this notation we can formulate Motohashi's fundamental result. Since it is of a fairly general nature, we need to assume certain conditions on the weight function $g(r)$: the function $g(r)$ takes real values on the real axis; and there exists a large positive constant A such that $g(r)$ is regular and $O((|r|+1)^{-A})$ in the horizontal strip $|\operatorname{Im} r| \leqslant A$. With

$$\mathcal{I}_k(g) := \int_{-\infty}^{\infty} |\zeta(\tfrac{1}{2}+it)|^{2k} g(t)\mathrm{d}t \qquad (k \in \mathbb{N}) \tag{85}$$

we have the exact formula, which we state here as

Theorem 4.1.

$$\mathcal{I}_2(g) = \{\mathcal{I}_{2,r} + \mathcal{I}_{2,d} + \mathcal{I}_{2,c} + \mathcal{I}_{2,h}\}(g), \tag{86}$$

where

$$\mathcal{I}_{2,r}(g) := \int_{-\infty}^{\infty} \sum_{a,b,k,l\geqslant 0; ak+bl\leqslant 4} c(a,k;b,l)\operatorname{Re}\left[\left(\frac{\Gamma^{(a)}}{\Gamma}\right)^k \left(\frac{\Gamma^{(b)}}{\Gamma}\right)^l (\tfrac{1}{2}+it)\right] g(t)\mathrm{d}t$$
$$-2\pi\Big\{(\gamma - \log(2\pi))g(\tfrac{1}{2}i) + \tfrac{1}{2}ig'(\tfrac{1}{2}i)\Big\} \tag{87}$$

with effectively computable real, absolute constants $c(a,k;b,l)$ *and*

$$\mathcal{I}_{2,d}(g) := \sum_{j=1}^{\infty} \alpha_j H_j^3(\tfrac{1}{2})\Lambda(\kappa_j;g),$$

$$\mathcal{I}_{2,c}(g) := \frac{1}{\pi}\int_{-\infty}^{\infty} \frac{|\zeta(\frac{1}{2}+ir)|^6}{|\zeta(1+2ir)|^2}\Lambda(r;g)\mathrm{d}r$$

$$\mathcal{I}_{2,h}(g) := \sum_{k=6}^{\infty}\sum_{j=1}^{d_{2k}} \alpha_{j,2k} H_{j,2k}^3(\tfrac{1}{2})\Lambda((\tfrac{1}{2}-2k)i;g). \tag{88}$$

Here γ *is Euler's constant and*

$$\Lambda(r;g) = \int_0^{\infty} (y(1+y))^{-1/2} g_c(\log(1+1/y))$$

$$\times \operatorname{Re}\Big[y^{-1/2-ir}\Big(1+\frac{i}{\sinh(\pi r)}\Big)\frac{\Gamma^2(\frac{1}{2}+ir)}{\Gamma(1+2ir)}F(\tfrac{1}{2}+ir,\tfrac{1}{2}+ir;1+2ir;-1/y)\Big]\mathrm{d}y$$

with the hypergeometric function F and

$$g_c(x) := \int_{-\infty}^{\infty} g(t)\cos(xt)\mathrm{d}t.$$

Remark 4.3. The function F appearing above is the *Gauss hypergeometric function* ${}_2F_1(a;b;c;z)$. In standard notation it is

$${}_2F_1(a;b;c;z) = \frac{\Gamma(c)}{\Gamma(a)\Gamma(b)}\sum_{n=0}^{\infty}\frac{\Gamma(a+n)\Gamma(b+n)}{\Gamma(c+n)}\frac{z^n}{n!} \qquad (|z|<1).$$

It is clear that the Gaussian exponential weight is a good candidate for the function g appearing in Theorem 4.1. With this function, after several simplifications, one is led to Motohashi's explicit formula with a logarithmic error term. This is

Theorem 4.2. *Let D be an arbitrary positive constant and let us assume that*

$$T^{1/2}(\log T)^{-D} \leqslant \Delta \leqslant T(\log T)^{-1}. \tag{89}$$

Then we have, in the notation of (83),

$$I(T,\Delta) = \frac{\pi}{\sqrt{2T}}\sum_{j=1}^{\infty}\alpha_j H_j^3(\tfrac{1}{2})\kappa_j^{-1/2}\sin\left(\kappa_j\log\frac{\kappa_j}{4\mathrm{e}T}\right)\mathrm{e}^{-\frac{1}{4}(\Delta\kappa_j/T)^2} + O(\log^{3D+9}T), \tag{90}$$

where the O-constant depends only on D.

The proofs of Theorems 4.1 and 4.2 are more difficult than the proof of Atkinson's formula and will not be given here. Sums of Kloosterman sums naturally arise in the course of the proof, and their transformation via the *Bruggeman–Kuznetsov trace formula* (see, e.g., [85]) plays an important rôle. However, we shall indicate how one can derive some consequences from (90) involving the function $E_2(T)$, the error term in the asymptotic formula (82). Hence,

$$\begin{aligned}E_2(T) = &\int_0^T |\zeta(\tfrac{1}{2}+it)|^4\mathrm{d}t\\ &-\left(a_4T\log^4T + a_3T\log^3T + a_2T\log^2T + a_1T\log T + a_0T\right),\end{aligned} \tag{91}$$

and $E_2(T) = o(T)$ $(T\to\infty)$. At a first glance the range for Δ in (89) looks rather restrictive, but it will turn out that this is not the case. Another problem is that (90) is not an explicit formula for $E_2(T)$, but for a weighted integral.

The smooth structure and fast decay of the Gaussian weight function $\mathrm{e}^{-\frac{1}{4}(\Delta\kappa_j/T)^2}$ in (90) will enable us to pass from $I(T,\Delta)$ to $E_2(T)$. First we note that, similarly to

Theorem 4.2, integrating the expression in Theorem 4.1, one obtains

$$\int_0^T I(t,\Delta)\mathrm{d}t = TP_4(\log T) + S(T,\Delta) + R(T,\Delta), \quad P_4(x) = \sum_{j=0}^4 a_j x^j, \quad (92)$$

where

$$S(T,\Delta) := \pi\sqrt{T/2}\sum_{j=1}^\infty \alpha_j\kappa_j^{-3/2}H_j^3(\tfrac12)\cos\left(\kappa_j\log\frac{\kappa_j}{4\mathrm{e}T}\right)\mathrm{e}^{-\frac14(\Delta\kappa_j/T)^2},$$
$$R(T,\Delta) := T^{1/2}\log^{C(D)}T, \quad (93)$$

assuming that (89) holds. Suppose henceforth that $T^\varepsilon \leqslant \Delta \leqslant T\exp(-\sqrt{\log T})$ and put first $T_1 = T - \Delta\log T$, $T_2 = 2T + \Delta\log T$. Then

$$\int_{T_1}^{T_2} I_4(t,\Delta)\mathrm{d}t = \int_{-\infty}^{\infty}|\zeta(\tfrac12+ix)|^4\left(\frac{1}{\Delta\sqrt{\pi}}\int_{T_1}^{T_2}\mathrm{e}^{-(t-x)^2/\Delta^2}\mathrm{d}t\right)\mathrm{d}x$$
$$\geqslant \int_T^{2T}|\zeta(\tfrac12+ix)|^4\left(\frac{1}{\Delta\sqrt{\pi}}\int_{T-\Delta\log T}^{2T+\Delta\log T}\mathrm{e}^{-(t-x)^2/\Delta^2}\mathrm{d}t\right)\mathrm{d}x.$$

But for $T \leqslant u \leqslant 2T$ we have, by the change of variable $t - x = \Delta v$,

$$\frac{1}{\Delta\sqrt{\pi}}\int_{T-\Delta\log T}^{2T+\Delta\log T}\mathrm{e}^{-(t-x)^2/\Delta^2}\mathrm{d}t = \frac{1}{\sqrt{\pi}}\int_{(T-x)/\Delta-\log T}^{(2T-x)/\Delta+\log T}\mathrm{e}^{-v^2}\mathrm{d}v$$
$$= \frac{1}{\sqrt{\pi}}\int_{-\infty}^{\infty}\mathrm{e}^{-v^2}\mathrm{d}v + O\Big(\int_{\log T}^{\infty}\mathrm{e}^{-v^2}\Big)\mathrm{d}v$$
$$= 1 + O\Big(\mathrm{e}^{-\log^2 T}\Big).$$

By the same technique we can bound from above $\int_{T_1}^{T_2} I_4(t,\Delta)\mathrm{d}t$ with $T_1 = T + \Delta\log T$, $T_2 = 2T - \Delta\log T$. The results are contained in

Lemma 4.1. *For $0 < \varepsilon < 1$ fixed and $T^\varepsilon \leqslant \Delta \leqslant T\exp(-\sqrt{\log T})$, we have with the above notation*

$$\int_T^{2T}|\zeta(\tfrac12+it)|^4\mathrm{d}t \leqslant 2TP_4(\log 2T) - TP_4(\log T) + O(\Delta\log^5 T)$$
$$+ S(2T+\Delta\log T,\Delta) - S(T-\Delta\log T,\Delta)$$
$$+ R(2T+\Delta\log T,\Delta) - R(T-\Delta\log T,\Delta) \quad (94)$$

and

$$\int_T^{2T} |\zeta(\tfrac{1}{2}+it)|^4 \mathrm{d}t \geqslant 2TP_4(\log 2T) - TP_4(\log T) + O(\Delta \log^5 T)$$
$$+ S(2T - \Delta \log T, \Delta) - S(T + \Delta \log T, \Delta)$$
$$+ R(2T - \Delta \log T, \Delta) - R(T + \Delta \log T, \Delta). \quad (95)$$

To obtain an upper bound for $E_2(T)$ from Lemma 4.1 note that, for $\tau \asymp T$, we have uniformly for $A > 0$ sufficiently large

$$S(\tau, \Delta) = \pi\sqrt{\tau/2} \sum_{\kappa_j \leqslant AT\Delta^{-1}\sqrt{\log T}} \alpha_j \kappa_j^{-3/2} H_j^3(\tfrac{1}{2}) + O(1).$$

But using (84), (94)–(95) and partial summation it follows that

$$\int_T^{2T} |\zeta(\tfrac{1}{2}+it)|^4 \mathrm{d}t = 2TP_4(\log 2T) - TP_4(\log T) + O(\Delta \log^5 T)$$
$$+ O(T^{1/2}\log^C T) + O(T\Delta^{-1/2}\log^C T).$$

Choosing $\Delta = T^{2/3}\log^C T$ it follows that we have obtained $E_2(2T) - E_2(T) \ll T^{2/3}\log^C T$, which implies

Theorem 4.3. *There is a constant $C > 0$ such that*

$$E_2(T) \ll T^{2/3}\log^C T. \quad (96)$$

Remark 4.4. The bound in (96) was obtained by Motohashi and the author [62]. Motohashi in his monograph [85] obtained the value $C = 8$. In [62] it was also proved that

$$E_2(T) = \Omega(\sqrt{T}). \quad (97)$$

The proof uses the fact that not all G_h vanish, where for a fixed μ_h

$$G_h := \sum_{\kappa_j = \mu_h} \alpha_j H_j^3(\tfrac{1}{2}).$$

It was also indicated (see also [42]) that, if one could prove a certain linear independence of $\alpha_j H_j^3(\frac{1}{2})$ over the integers, one would obtain

$$\limsup_{T\to\infty} \frac{|E_2(T)|}{\sqrt{T}} = +\infty,$$

which is stronger than (97). In [84] (see also [85]), Motohashi proved that

$$E_2(T) = \Omega_\pm(\sqrt{T}), \tag{98}$$

but there is still a large gap between (98) and the upper bound in (96).

Problem 6. What is the true order of magnitude of $E_2(T)$? It is reasonable to conjecture that

$$E_2(T) \ll_\varepsilon T^{1/2+\varepsilon}. \tag{99}$$

The conjectural bound in (99) is supported by two mean value results, proved by Motohashi and the author [60, 61]. In [60] it was shown that

$$\int_0^T E_2(t)\mathrm{d}t \ll T^{3/2},$$

while the result of [61] on the mean square of $E_2(T)$ is stated as

Theorem 4.4. *There is a constant $C > 0$ such that*

$$\int_0^T E_2^2(t)\mathrm{d}t \ll T^2 \log^C T. \tag{100}$$

The value $C = 22$ in (100) is worked out by Motohashi in [85]. On the other hand, the author [48] proved that

$$\int_0^T E_2^2(t)\mathrm{d}t \gg T^2, \tag{101}$$

so that (100) and (101) determine, up to a logarithmic factor, the true order of the integral in question.

Problem 7. What is the true order of magnitude of the integral in (100)? Is it perhaps true that there exists a constant $A > 0$ such that

$$\int_0^T E_2^2(t)\mathrm{d}t = AT^2 + F(T), \quad F(T) = o(T^2) \ (T \to \infty)? \tag{102}$$

One may further conjecture (see [46]) that $F(T) = O_\varepsilon(T^{3/2+\varepsilon})$ and $F(T) = \Omega(T^{3/2-\delta})$ hold (for any given $\delta, \varepsilon > 0$) if (102) holds. The upper bound for $F(T)$ is very strong. Namely, similarly to (73), one obtains

$$|E_2(T)| \leqslant x^{-1} \int_{T-x}^{T+x} |E_2(t)|\mathrm{d}t + 2Cx \log^4 T \quad (0 < x \leqslant T, T \geqslant T_0),$$

whence by the Cauchy–Schwarz inequality for integrals it follows that

$$E_2(T) \ll x^{-1/2}\left(\int_{T-x}^{T+x} E_2^2(t)\mathrm{d}t\right)^{1/2} + x\log^4 T. \tag{103}$$

Suppose now that $F(T) = O_\varepsilon(T^{3/2+\varepsilon})$ holds. Then (102) implies

$$E_2(T) \ll x^{-1/2}\Big(xT + F(T+x) - F(T-x)\Big)^{1/2} + x\log^4 T \ll_\varepsilon T^{1/2+\varepsilon}, \tag{104}$$

and even the weak $F(T) \ll T^2\log^C T$ (implied by Theorem 4.4) gives the bound in Theorem 4.3. One also obtains from $F(T) = O_\varepsilon(T^{3/2+\varepsilon})$ the (hitherto unproved) bound

$$\zeta(\tfrac{1}{2} + it) \ll_\varepsilon |t|^{1/8+\varepsilon}. \tag{105}$$

This follows from (see Lemma 7.1 of [42])

Lemma 4.2. *Let $k \in \mathbb{N}$ be fixed and $T/2 \leqslant t \leqslant 2T$. Then*

$$|\zeta(\tfrac{1}{2} + it)|^k \ll \log T\left(1 + \int_{-\log^2 T}^{\log^2 T} |\zeta(\tfrac{1}{2} + it + iT)|^k \mathrm{e}^{-|v|}\mathrm{d}v\right). \tag{106}$$

Proof. Let $\mathfrak{D}$ be the rectangle with vertices $\pm c \pm i\log^2 T$, where $c = 1/\log T$. With $s = \frac{1}{2} + it$ we have, by the residue theorem,

$$\zeta^k(s) = \frac{1}{2\pi i}\int_{\mathfrak{D}} \zeta^k(s+z)\Gamma(z)\mathrm{d}z. \tag{107}$$

As $T \to \infty$ the integrals over the horizontal sides of $\mathfrak{D}$ are $o(1)$. By the functional equation (6) and (15)

$$|\zeta(\tfrac{1}{2} - c + it)| \ll |\zeta(\tfrac{1}{2} + c + it)|T^c \ll |\zeta(\tfrac{1}{2} + c + it)|.$$

Since $s = 0$ is a simple pole of $\Gamma(s)$, then for any real v we have

$$\Gamma(\pm c \pm iv) \ll \mathrm{e}^{-|v|}(c + |v|)^{-1}. \tag{108}$$

Hence (107) and (108) yield

$$\zeta^k(s) \ll 1 + \int_{-\log^2 T}^{\log^2 T} |\zeta(\tfrac{1}{2} + c + it + iv)|^k \mathrm{e}^{-|v|}(c + |v|)^{-1}\mathrm{d}v. \tag{109}$$

Now set $s' = s + c + iv = \frac{1}{2} + c + it + iv$. By (28) we have

$$\frac{1}{2\pi i}\int_{1-i\infty}^{1+i\infty} \zeta^k(s'+w)\Gamma(w)\mathrm{d}w = \sum_{n=1}^{\infty} d_k(n)\mathrm{e}^{-n}n^{-s'} \ll 1.$$

In the last integral we shift the line of integration to $\operatorname{Re} w = -c$ and use again the residue theorem and Stirling's formula. There are poles at $w = 0$ and $w = 1 - s'$ with residues $\zeta^k(s')$ and $O(1)$. We obtain

$$\zeta^k(s') \ll 1 + \int_{-\infty}^{\infty} |\zeta(\tfrac{1}{2} + it + iv)|^k \mathrm{e}^{-|v|}(c + |v|)^{-1}\mathrm{d}v. \tag{110}$$

Hence from (109) and (110) we have

$$\zeta^k(\tfrac{1}{2} + it) \ll 1 + \int_{-\log^2 T}^{\log^2 T} \mathrm{e}^{-|u|}\Big(1 + \int_{-\infty}^{\infty} |\zeta(\tfrac{1}{2} + it + iu + iv)|^k \times \mathrm{e}^{-|v|}(c + |v|)^{-1}\mathrm{d}v\Big)(c + |u|)^{-1}\mathrm{d}u.$$

To estimate the right-hand side of the above expression first note that trivially

$$\int_{-\log^2 T}^{\log^2 T} \mathrm{e}^{-|u|}(c + |u|)^{-1}\mathrm{d}u \ll c^{-1} = \log T.$$

In the remaining integral we make the substitution $v = x - u$ and invert the order of integration. This gives

$$\zeta^k(\tfrac{1}{2} + it) \ll \log T + \int_{-\infty}^{\infty} |\zeta(\tfrac{1}{2} + it + ix)|^k \times \left(\int_{-\infty}^{\infty} \mathrm{e}^{-|u|-|x-u|}(c + |u|)^{-1}(c + |x-u|)^{-1}\mathrm{d}u\right)\mathrm{d}x,$$

and the proof of the lemma will be finished if we can show that

$$\int_{-\infty}^{\infty} \mathrm{e}^{-|u|-|x-u|}(c + |u|)^{-1}(c + |x-u|)^{-1}\mathrm{d}u \ll c^{-1}\mathrm{e}^{-|u|}.$$

This is obvious when $x = 0$, and since the cases $x > 0$ and $x < 0$ are treated analogously, only the case $x > 0$ will be considered. Write

$$\int_{-\infty}^{\infty} \mathrm{e}^{-|u|-|x-u|}(c + |u|)^{-1}(c + |x-u|)^{-1}\mathrm{d}u = \int_{-\infty}^{0} + \int_{0}^{x} + \int_{x}^{\infty} = I_1 + I_2 + I_3,$$

say. Then

$$I_1 = \int_0^\infty \mathrm{e}^{-x}(c+v)^{-1}(c+x+v)^{-1}$$
$$\ll \mathrm{e}^{-x}\left(\int_0^c c^{-2}\mathrm{d}v + \int_c^\infty v^{-2}\mathrm{d}v\right) \ll \mathrm{e}^{-x}c^{-1}.$$

In a similar vein it is proved that

$$I_2 \ll \mathrm{e}^{-x}c^{-1}, \qquad I_3 \ll \mathrm{e}^{-x}c^{-1},$$

and Lemma 4.2 follows. □

Having at our disposal (106) with $k = 4$, we easily obtain

$$\begin{aligned} |\zeta(\tfrac{1}{2}+it)|^4 &\ll \log t\Big(1 + I_2(t+\log^2 t) - I_2(t-\log^2 t)\Big) \\ &\ll \log t\Big(\log^6 t + \max_{t-\log^2 t\leqslant x\leqslant t+\log^2 t} |E_2(x)|\Big). \end{aligned} \tag{111}$$

Therefore $F(T) = O_\varepsilon(T^{3/2+\varepsilon})$ implies $E_2(T) = O_\varepsilon(T^{1/2+\varepsilon})$, which in turn [by (111)] implies the bound (105).

The omega result (96) was sharpened by the author [44] to the following: there exist constants $A > 0$, $B > 1$ such that, for $T \geqslant T_0$, every interval $[T,\ BT]$ contains points T_1, T_2 such that

$$E_2(T_1) > AT_1^{1/2}, \qquad E_2(T_2) < -AT_2^{1/2}.$$

This is the analogue of the omega result contained in (77) for $E(T)$. Further, in [46] it was proved that the same interval contains also points T_3, T_4 for which

$$\int_0^{T_3} E_2(t)\mathrm{d}t > AT_3^{3/2}, \qquad \int_0^{T_3} E_2(t)\mathrm{d}t < -AT_3^{3/2}.$$

For the integral of $E_2(t)$ one can derive an explicit formula, as was done in [49,50]. We have

Theorem 4.5. *Let*

$$\eta(T) := (\log T)^{3/5}(\log\log T)^{-1/5},$$

$$R_1(\kappa_h) := \sqrt{\frac{\pi}{2}}\left(2^{-i\kappa_h}\frac{\Gamma(\frac14 - \frac12 i\kappa_h)}{\Gamma(\frac14 + \frac12 i\kappa_h)}\right)^3 \Gamma(2i\kappa_h)\cosh(\pi\kappa_h).$$

Then there exists a constant $C > 0$ *such that*

$$\int_0^T E_2(t)\mathrm{d}t = 2T^{\frac{3}{2}}\,\mathrm{Re}\left\{\sum_{j=1}^{\infty}\alpha_j H_j^3(\tfrac{1}{2})\frac{T^{i\kappa_j}}{(\frac{1}{2}+i\kappa_j)(\frac{3}{2}+i\kappa_j)}R_1(\kappa_j)\right\} + O(T^{\frac{3}{2}}e^{-C\eta(T)}). \tag{112}$$

From Stirling's formula for the gamma-function it follows that $R_1(\kappa_j) \ll \kappa_j^{-1/2}$, hence by partial summation it follows that the series on the right-hand side of (112) is absolutely convergent, and it can be also shown (see [42, 49, 51]) that $\mathrm{Re}\{\ldots\}$ is also $\Omega_\pm(1)$. Thus from Theorem 4.5 we can easily deduce all previously known Ω-results for $E_2(T)$. The error term in (112) is similar to the error term in the strongest known form of the prime number theorem (see, e.g., [42, Chap. 12]). This is by no means a coincidence.

In concluding this discussion on the fourth moment of $|\zeta(\frac{1}{2}+it)|$, let us mention that many bounds, including the pointwise for $E_2(T)$, ultimately depend on the exponential sum

$$\sum_{K<\kappa_j\leqslant K'\leqslant 2K}\alpha_j H_j^3(\tfrac{1}{2})\exp\left(i\kappa_j\log\left(\frac{T}{\kappa_j}\right)\right) \qquad (1 \ll K \leqslant T^{1/2}).$$

However, at present, all that appears possible seems to be trivial estimation, coming from the bound (84).

5 Higher Moments

We have seen in the previous two sections that we have plenty of information about $I_1(T)$ and $I_2(T)$ [see (16)]. Unfortunately, the situation changes with $I_k(T)$ when $k > 2$. The most important result on higher moments is due to Heath-Brown [29]. He proved (with $C = 17$)

Theorem 5.1.

$$I_6(T) = \int_0^T |\zeta(\tfrac{1}{2}+it)|^{12}\mathrm{d}t \ll T^2(\log T)^C. \tag{113}$$

From Theorem 5.1, the bound $I_2(T) \ll T\log^4 T$ and Hölder's inequality we obtain

Corollary 5.1. *With some* $C = C(A)$ *we have*

$$\int_0^T |\zeta(\tfrac{1}{2}+it)|^A\mathrm{d}t \ll T^{(A+4)/8}\log^C T \qquad (4 \leqslant A \leqslant 12). \tag{114}$$

In (114) A is a fixed constant, but does not have to be an integer. For $A > 12$ there are some results (see Chap. 8 of [52]), but essentially the best one can do is

$$\int_0^T |\zeta(\tfrac{1}{2}+it)|^A \mathrm{d}t \ll_\varepsilon T^{1+(A-4)\mu(1/2)+\varepsilon} \qquad (A > 12), \tag{115}$$

where

$$\mu(\sigma) := \limsup_{t\to\infty} \frac{\log|\zeta(\sigma+it)|}{\log t} \qquad (\sigma \in \mathbb{R}),$$

so that the best-known result is $\mu(1/2) \leqslant 32/205$ $(= 0.15609\ldots)$, due to Huxley [37]. The function $\mu(\sigma)$ is continuous, convex downwards and satisfies $\mu(\sigma) = 0$ for $\sigma \geqslant 1$ and $\mu(\sigma) = 1/2-\sigma$ for $\sigma \leqslant 0$, the second assertion being a consequence of the first assertion and the functional equation for $\zeta(s)$. The Lindelöf hypothesis is that $\mu(1/2) = 0$ or, equivalently, that $\mu(\sigma) = 0$ for $\sigma \geqslant 1/2$. It is easy to show, by using (106), that the Lindelöf hypothesis is equivalent to the statement that $I_k(T) \ll_{k,\varepsilon} T^{1+\varepsilon}$ for each k. For several other equivalent statements to the Lindelöf hypothesis, see Chap. 13 of Titchmarsh's book [98]. The Lindelöf hypothesis is a consequence of the Riemann hypothesis (op. cit.), since the latter implies the bound

$$\zeta(\tfrac{1}{2}+it) \ll \exp\left(C\frac{\log t}{\log\log t}\right) \qquad (C > 0). \tag{116}$$

It is not known whether the Lindelöf hypothesis implies the Riemann hypothesis, but this does not seem to be likely.

We shall proceed now to prove Theorem 5.1. The original proof of (113) in [29] is based on the use of Atkinson's formula and the use of the Gaussian exponential weight as a truncating device. In this sense Heath-Brown's proof represents an important application of Atkinson's formula. The proof given below is from [61]. It uses Theorem 4.4, showing incidentally its strength, and gives first

Theorem 5.2. *Let $T \leqslant t_1 < t_2 < \ldots < t_R \leqslant 2T$ be points such that $t_{r+1} - t_r \geqslant \Delta$ $(r = 1, \ldots, R-1)$ with $\log T \ll \Delta \ll T/\log T$. Then*

$$\sum_{r=1}^{R} \int_{t_r}^{t_r+\Delta} |\zeta(\tfrac{1}{2}+it)|^4 \mathrm{d}t \ll R\Delta\log^4 T + R^{-1/2}\Delta^{-1/2} T\log^C T. \tag{117}$$

Proof. Theorem 5.2 improves (replacing "ε" by a log-factor) on a result of Iwaniec [65]. The expected term for the left-hand side of (117) is $R\Delta\log^4 T$, and the sum in question is clearly $\gg R\Delta\log^4 T$. To obtain (117), let $f(t)$ be a smooth function with support in $[-2\Delta, 2\Delta]$ such that $f(t) = 1$ for $-\Delta \leqslant t \leqslant \Delta$. Using the definition of $E_2(T)$ we obtain [see (129)]

$$\int_{t_r}^{t_r+\Delta} |\zeta(\tfrac{1}{2}+it)|^4 \mathrm{d}t \leqslant \int_{-\infty}^{\infty} f(t)|\zeta(\tfrac{1}{2}+it_r+it)|^4 \mathrm{d}t$$

$$= \int_{-\infty}^{\infty} f(t)\left\{Q_k + Q_k'\right\}(\log(t+t_r))\mathrm{d}t - \int_{t_r-2\Delta}^{t_r+2\Delta} f'(t-t_r)E_2(t)\mathrm{d}t$$

$$\ll \Delta \log^4 T + \Delta^{-1}\int_{t_r-2\Delta}^{t_r+2\Delta} |E_2(t)|\mathrm{d}t.$$

Therefore it follows that

$$\sum_{r=1}^{R}\int_{t_r}^{t_r+\Delta} |\zeta(\tfrac{1}{2}+it)|^4 \mathrm{d}t \ll R\Delta\log^4 T + \Delta^{-1}\sum_{r=1}^{R}\int_{t_r-2\Delta}^{t_r+2\Delta} |E_2(t)|\mathrm{d}t. \quad (118)$$

Note that the intervals $[t_r - 2\Delta, t_r + 2\Delta]$ $(r = 1, 2, \ldots, R-1)$ are not necessarily disjoint. However, if we split the sum on the right-hand side of (118) into five sums $\sum_j$ $(0 \leqslant j \leqslant 4)$, where in $\sum_j$ summation is over the points $\tau_r := t_{5r+j}$, then the intervals $[\tau_r - 2\Delta, \tau_r + 2\Delta]$ are disjoint. This is because $t_{r+1} - t_r \geqslant \Delta$ implies that

$$\tau_{r+1} - 2\Delta = t_{5r+5+j} - 2\Delta \geqslant t_{5r+j} + 5\Delta - 2\Delta = \tau_r + 3\Delta > \tau_r + 2\Delta.$$

Thus, we obtain, by the Cauchy–Schwarz inequality for integrals, with some suitable R_j such that $R_j \leqslant R$,

$$\sum_{r=1}^{R}\int_{t_r-2\Delta}^{t_r+2\Delta} |E_2(t)|\mathrm{d}t \ll \max_j \sum_{r=1}^{R}\left(\int_{\tau_r-2\Delta}^{\tau_r+2\Delta} |E_2(t)|^2\mathrm{d}t\right)^{1/2}(4\Delta)^{1/2}$$

$$\ll \Delta^{1/2}\max_j\left(\sum_{r=1}^{R}\int_{\tau_r-2\Delta}^{\tau_r+2\Delta} |E_2(t)|^2\mathrm{d}t\right)^{1/2} R_j^{1/2}$$

$$\ll (\Delta R)^{1/2}\left(\int_{T/2}^{5T/2} E_2^2(t)\mathrm{d}t\right)^{1/2}$$

$$\ll (\Delta R)^{1/2} T\log^C T,$$

where we used (99). This completes the proof of Theorem 5.2. □

Proof of Theorem 5.1. By (106) of Lemma 4.2, we have

$$|\zeta(\tfrac{1}{2}+it)|^4 \ll \log t\left(1 + \int_{t-\log^2 t}^{t+\log^2 t} |\zeta(\tfrac{1}{2}+iu)|^4 \mathrm{e}^{-|u|}\mathrm{d}u\right). \quad (119)$$

Suppose that $|\zeta(\frac{1}{2}+it_r)| \geqslant V \geqslant \log^4 T$ for a system of points $\{t_r\}$ such that

$$T+1/3 \leqslant t_1 < t_2 < \ldots < t_R \leqslant 2T-1/3, t_{r+1}-t_r \geqslant \log^3 T \ (r = 1,2,\ldots,R-1).$$

Then (119) gives

$$V^4 \ll \log T \int_{t_r-\log^2 T}^{t_r+\log^2 T} |\zeta(\tfrac{1}{2}+iu)|^4 \mathrm{d}u \qquad (r = 1,2,\ldots,R).$$

If we consider separately points $\{t_r\}$ with even and odd indices and denote their number by R_0 and R_1, respectively, then with a slight abuse of notation summation gives

$$R_j V^4 \ll \log T \sum_{r=1}^{R_j} |\zeta(\tfrac{1}{2}+iu)|^4 \mathrm{d}u \qquad (j = 0,1).$$

If we apply Theorem 5.2 to the last sum, it follows that

$$RV^4 \ll R\Delta \log^5 T + R^{1/2}\Delta^{-1/2} T \log^C T, \tag{120}$$

and then, with $\Delta = \delta V^4 \log^{-5} T$ and $\delta > 0$ sufficiently small, (120) yields

$$R \ll T^2 V^{-12} \log^C T \qquad (t_{r+1}-t_r \geqslant \log^3 T, r = 1,2,\ldots,R_1). \tag{121}$$

Finally we note that

$$\int_T^{2T} |\zeta(\tfrac{1}{2}+it)|^{12}\mathrm{d}t \leqslant \sum_{r=1}^{R} \int_{t_r}^{t_r+\Delta} |\zeta(\tfrac{1}{2}+it)|^{12}\mathrm{d}t,$$

with $t_r := T+(r-1)\log^3 T$, $R \ll T$. Those t_r where $|\zeta(\frac{1}{2}+it)| \leqslant \log^C T$ make a negligible contribution, and the remaining points (again relabelling them by picking even and odd indices) are chosen to satisfy $V \leqslant |\zeta(\frac{1}{2}+it)| \leqslant 2V, \log^C T \leqslant V \ll T^{1/6}$ (since $\zeta(\frac{1}{2}+it) \ll t^{1/6}$). By using (121) for each such system of points we infer that

$$\int_T^{2T} |\zeta(\tfrac{1}{2}+it)|^{12}\mathrm{d}t \ll \log T \max_V V^{12} T^2 V^{-12} \log^C T = T^2 \log^{C+1} T.$$

Replacing T by $T2^{-j}$ and summing the resulting expressions over j we obtain Theorem 5.1. More careful analysis leads to an explicit value of C in (113).

6 The Conjectural Formula for $I_k(T)$

The conjectural formula for $I_k(T)$ in question is due to Conrey et al. [14, 15]. It is of the form

$$I_k(T) = \int_0^T |\zeta(\tfrac{1}{2}+it)|^{2k}\,\mathrm{d}t = (1+o(1))\int_0^T P_k\left(\log\Big(\frac{t}{2\pi}\Big)\right)\mathrm{d}t \quad (T\to\infty). \tag{122}$$

In (122) $P_k(x)$ is the polynomial, all of whose coefficients depend on the (fixed) integer $k \geqslant 1$, given explicitly by the $2k$-fold residue

$$\begin{aligned} P_k(x) &= \frac{(-1)^k}{k!^2}\frac{1}{(2\pi i)^{2k}}\int_{|z_1|=\varepsilon_1}\cdots\int_{|z_{2k}|=\varepsilon_{2k}} \\ &\times \frac{G(z_1,\dots,z_{2k})\Delta^2(z_1,\dots,z_{2k})}{\prod_{i=1}^{2k} z_i^{2k}}\exp\Big\{\tfrac{1}{2}x\sum_{i=1}^{k}(z_i - z_{i+k})\Big\}\mathrm{d}z_1\dots\mathrm{d}z_{2k}, \end{aligned} \tag{123}$$

where the ε_i's are small positive numbers. We have

$$\Delta(z_1,\dots,z_m) = \prod_{1\leqslant i<j\leqslant m}(z_j - z_i) = |z_i^{j-1}|_{m\times m},$$

which is the Vandermonde determinant,

$$G(z_1,\dots,z_{2k}) = A_k(z_1,\dots,z_{2k})\prod_{i=1}^{k}\prod_{j=1}^{k}\zeta(1+z_i-z_{j+k}),$$

and finally A_k is the Euler product ($\mathrm{e}(\theta) := \exp(2\pi i\theta)$)

$$\begin{aligned} A_k(z_1,\dots,z_{2k}) &= \prod_p\prod_{i,j=1}^{k}(1-p^{-1-z_i+z_{k+j}}) \\ &\times \int_0^1\prod_{j=1}^{k}\Big(1-\mathrm{e}(\theta)p^{-1/2-z_j}\Big)^{-1}\Big(1-\mathrm{e}(-\theta)p^{-1/2+z_{k+j}}\Big)^{-1}\mathrm{d}\theta. \end{aligned}$$

The authors actually conjecture, in all cases, an error term in (122) of the order $O_{k,\varepsilon}(T^{1/2+\varepsilon})$, which in the general case this author finds too optimistic. In fact the paper brings forth a conjecture (via characteristic polynomials from random matrix theory (see, e.g., the work [80] of Mehta on random matrices) for the complete main term in the asymptotic formulas for a wide class of L-functions. Coefficients of $P_k(x)$ are given when $2 \leqslant k \leqslant 7$, and numerically computed moments are compared to the values obtained for the main term by the moments conjecture.

As already mentioned, further numerical calculations involving $P_k(x)$ were carried out by Hiary and Odlyzko [34], and Rubinstein and Yamagishi [92]. In all cases when an asymptotic formula for the moment was rigorously proved, the main term in question coincided with the expression predicted by the authors, which renders the conjectural formulas quite important. Heretofore there have been several conjectures for particular L-functions, to mention here just the works of Keating and Snaith [72], and Conrey and Gonek [16] on the conjectural formula for $I_k(T)$.

The "recipe" for obtaining explicitly the conjectural formula for $I_k(T)$ is best explained by using the function

$$Z(t) := \zeta(\tfrac{1}{2}+it)\big(\chi(\tfrac{1}{2}+it)\big)^{-1/2}, \quad \zeta(s) = \chi(s)\zeta(1-s), \tag{124}$$

commonly called *Hardy's function*. Used originally by Hardy (see [23, 24]) to prove that there are infinitely many zeros of $\zeta(s)$ on the "critical line" $\operatorname{Re} s = \frac{1}{2}$, this function has several remarkable properties, easily derived from the functional equation and the definition (7) of $\chi(s)$. It turns out conveniently that $Z(t) \in \mathbb{R}$ when $t \in \mathbb{R}$ and that $|\chi(\frac{1}{2}+it)| = 1$, hence $|\zeta(\frac{1}{2}+it)| = |Z(t)|$ when $t \in \mathbb{R}$. In [14] one looks first at the "shifted moment"

$$M(\alpha_1,\dots,\alpha_{2k}) := \int_0^T Z(\tfrac{1}{2}+t+\alpha_1)\dots Z(\tfrac{1}{2}+t+\alpha_{2k})\mathrm{d}t,$$

where the α_j are distinct complex numbers with $\operatorname{Re}\alpha_j > -1/4$, so that the integrand becomes $|\zeta(\frac{1}{2}+it)|^{2k}$ when $\alpha_1 = \cdots = \alpha_{2k} = 0$. In $M(\alpha_1,\dots,\alpha_{2k})$ we substitute each factor by the approximate expression

$$Z(s) = \chi^{-1/2}(s) \sum_{n\leqslant\sqrt{t/(2\pi)}} n^{-s} + \chi^{-1/2}(1-s) \sum_{n\leqslant\sqrt{t/(2\pi)}} n^{s-1} + O(t^{-\sigma/2}), \tag{125}$$

where $s = \sigma + it, 0 < \sigma < 1$, and in (18) of Theorem 2.3 we take $x = y = \sqrt{t/(2\pi)}$. If $s = z + it, t > 1$ with z bounded (but not necessarily real), then [see (7) and (15)]

$$\chi(s) = \left(\frac{t}{2\pi}\right)^{1/2-s} \mathrm{e}^{it+\pi i/4}\left(1+O\left(\frac{1}{t}\right)\right),$$

$$\chi(1-s) = \left(\frac{t}{2\pi}\right)^{s-1/2} \mathrm{e}^{-it-\pi i/4}\left(1+O\left(\frac{1}{t}\right)\right).$$

These formulas are used to determine which products containing $\chi(s)$ and $\chi(1-s)$ are oscillating, when (125) is inserted in $M(\alpha_1,\dots,\alpha_{2k})$. Then one proceeds heuristically as follows:

1. The error term $O(t^{-\sigma/2})$ is ignored everywhere, and the product of Z-values is expanded, producing 2^{2k} terms.

2. Of these 2^{2k} terms, only the terms with the same number of s's and $1-s$'s are considered. The reasoning is that $\chi(s)$ is highly oscillating and there should be a lot of cancellation unless each s gets paired with a $1-s$.
3. For any such terms, the main contribution comes from the "diagonal" term (when $m_1 m_2 \ldots m_k = n_1 n_2 \ldots n_k$) when the sums are multiplied out.
4. The truncated diagonal sums are extended to infinity, and the sums which diverge are replaced by the corresponding analytic continuation (the assumption $\operatorname{Re}\alpha_j > -1/4$ is used in this process).
5. Setting finally $\alpha_j \to 0$ $(\forall j)$ one eventually obtains the expression (123) for $E_k(x)$.

It is not easy to justify the heuristic steps (1)–(4). In fact, the terms which are omitted in this process cannot be neglected individually. It appears that some sort of cancellation takes place among these terms (as they are taken without absolute values signs) so that at the end the above steps lead to the correct asymptotic formula for $I_k(T)$. Perhaps the strongest reason that gives credence to the conjecture is that, as already stated, in all moments involving L-functions when an asymptotic formula exists, it coincides with the prediction given by [14]. This paper contains explicit examples of unitary, symplectic and orthogonal families of L-functions with relevant conjectures involving the moments.

If one writes $P_k(x)$ as

$$P_k(x) = \sum_{r=0}^{k^2} c_r(k) x^{k^2-r}, \tag{126}$$

then all the coefficients $c_r(k)$ can be evaluated explicitly. In particular, the leading coefficient $c_0(k)$ of $P_k(x)$ is given as

$$c_0(k) = a_k \prod_{j=0}^{k-1} \frac{j!}{(j+k)!},$$

where the "arithmetic" part a_k is given by

$$a_k = \prod_p (1-p^{-1})^{k^2} {}_2F_1(k,k;1;1/p),$$

and the "geometric" part is the product over j. The conjectural formula in fact provides no information about the error term

$$E_k(T) := I_k(T) - \int_0^T P_k\left(\log\left(\frac{t}{2\pi}\right)\right) \mathrm{d}t, \tag{127}$$

where it is assumed that $P_k(x)$ is given by (126) and $E_k(T)$ is the error term in the sense that, for fixed $k \in \mathbb{N}$, one has

$$E_k(T) = o(T) \qquad (T \to \infty). \tag{128}$$

Note that integration by parts reveals that

$$\int_0^T P_k\left(\log\left(\frac{t}{2\pi}\right)\right) \mathrm{d}t = TQ_k(\log T), \tag{129}$$

where $Q_k(x)$ is another polynomial of degree k^2, all of whose coefficients depend on k. We have seen in Sects. 3 and 4 that (127) and (128) hold true when $k = 1$ and $k = 2$, in which cases much more than (128) is known. But as discussed in Sect. 5, no asymptotic formula for $I_k(T)$ when $k > 2$ is known, so that one can only speculate about the size of $E_k(T)$ in the general case.

It seems plausible to the author that

$$\int_0^T |\zeta(\tfrac{1}{2} + it)|^6 \mathrm{d}t = TQ_3(\log T) + E_3(T),$$

$$E_3(T) = O_\varepsilon(T^{3/4+\varepsilon}), \quad E_3(T) = \Omega(T^{3/4})$$

holds, where the main term $Q_3(x)$ is an explicit polynomial in x of degree nine, as given by (126) and (129).

In what concerns the true order of higher moments of $|\zeta(\frac{1}{2} + it)|$, the situation is even more unclear. Already for the eighth moment it is hard to ascertain what goes on, much less for the higher moments. The main term for the general $2k$th moment should involve a main term of the type suggested by Conrey et al. [14], but it could turn out that the error term $E_k(T)$ in the general case (when $k \geqslant 4$) contains expressions which make it *larger* than the term $TQ_k(\log T)$. For this see the discussion in [57] (also [85, pp. 218–219]). Essentially the argument is as follows. In general, from the knowledge about the order of $E_k(T)$, one can deduce a bound for $\zeta(\frac{1}{2} + iT)$ via the elementary estimate

$$\zeta(\tfrac{1}{2} + iT) \ll_k (\log T)^{(k^2+1)/(2k)} + \left(\log T \max_{t\in[T-1,T+1]} |E_k(t)|\right)^{1/(2k)}, \tag{130}$$

which is the proved analogously to (111) (see Lemma 4.2 of [42]). The conjectured bounds

$$E_k(T) \ll_{\varepsilon,k} T^{k/4+\varepsilon} \qquad (k \leqslant 4) \tag{131}$$

by (130) all imply $\zeta(\frac{1}{2} + it) \ll_\varepsilon |t|^{1/8+\varepsilon}$, which is out of reach at present, but is still much weaker than the Lindelöf hypothesis that $\zeta(\frac{1}{2} + it) \ll_\varepsilon |t|^\varepsilon$. On the other hand, we know that the omega result

$$E_k(T) = \Omega(T^{k/4}) \tag{132}$$

holds for $k = 1, 2$, and as already explained (see [57]), there are reasons to believe that (132) holds for $k = 3$. Perhaps it holds for $k = 4$ also, but the truth of (132) for any $k > 4$ would imply that the Lindelöf hypothesis is false and ipso facto the falsity of the Riemann hypothesis (RH). Namely, it is well known that the RH implies (116), which is stronger than the Lindelöf hypothesis, namely, that $\log|\zeta(\frac12+it)| \ll_\varepsilon \varepsilon \log|t|$. The reason why, in general, (131) makes sense is that a bound $E_k(T) \ll T^{c_k}$ for some fixed k (> 4) with $c_k < k/4$ would imply [by (130)] the bound $\zeta(\frac12+it) \ll_\varepsilon |t|^{c_k/(2k)+\varepsilon}$ with $c_k/(2k) < 1/8$. But the most one can get [by using (130)] from the error term in the mean square and the fourth moment of $|\zeta(\frac12+it)|$ is the bound

$$\zeta(\tfrac12+it) \ll_\varepsilon |t|^{1/8+\varepsilon}.$$

It does not appear likely to me that, say from the twelfth moment ($k = 6$), one will get a better pointwise estimate from (130) for $\zeta(\frac12+it)$ than what one can get from the mean square formula ($k = 1$). Nothing, of course, precludes yet that this does not happen, just that it appears not to be likely. As in all such dilemmas, only rigorous proofs will reveal in due time the real truth.

7 Lower Bound and the Upper Bound Under the RH

There are many mean value results for the lower bound of powers of $\zeta(s)$, some of which are easily generalized to more general L-functions. We present now one such result, which holds in the wide range $\sigma \geqslant \frac12$. This is

Theorem 7.1. *If $k \geqslant 1$ is a fixed integer, $\sigma \geqslant \frac12$ is fixed,*

$$12\log\log T \leqslant H \leqslant T, \quad T \geqslant T_0 > 0,$$

then uniformly in σ

$$\int_{T-H}^{T+H} |\zeta(\sigma+it)|^k \mathrm{d}t \gg H. \tag{133}$$

Proof. Let $\sigma_1 = \sigma + 2, s_1 = \sigma_1 + it, T - \frac12 H \leqslant t \leqslant T + \frac12 H$. Then $\zeta(s_1) \gg 1$ and therefore

$$\int_{T-\frac12 H}^{T+\frac12 H} |\zeta(\sigma_1+it)|^k \mathrm{d}t \gg H. \tag{134}$$

Let now $\mathcal{E}$ be the rectangle with vertices $\sigma + iT \pm iH, \sigma_2 + iT \pm iH$ ($\sigma_2 = \sigma + 3$) and let X be a parameter which satisfies

$$T^{-c} \leqslant X \leqslant T^{c} \tag{135}$$

for some constant $c > 0$. The residue theorem gives then

$$\frac{1}{\mathrm{e}}\zeta^k(s_1) = \frac{1}{2\pi i}\int_{\mathcal{E}} \frac{\zeta^k(w)}{w-s_1}\exp\left(-\cos\left(\frac{w-s_1}{3}\right)\right)X^{s_1-w}\mathrm{d}w.$$

On $\mathcal{E}$ we have $|\operatorname{Re}\frac{1}{3}(w-s_1)| \leqslant 1$, and on its horizontal sides

$$\left|\operatorname{Im}\left(\frac{w-s_1}{3}\right)\right| \geqslant \frac{1}{3}\cdot\frac{H}{2} \geqslant 2\log\log T.$$

Note that for $w = u + iv$ $(u, v \in \mathbb{R})$ we have

$$\begin{aligned}|\exp(-\cos w)| &= \left|\exp\left(-\tfrac{1}{2}(\mathrm{e}^{iw} + \mathrm{e}^{-iw})\right)\right| \\ &= \left|\exp\left(-\tfrac{1}{2}(\mathrm{e}^{iu}\mathrm{e}^{-v} + \mathrm{e}^{-iu}\mathrm{e}^{v})\right)\right| = \exp(-\cos u \cdot \cosh v).\end{aligned}$$

The above function $\exp(-\cos w)$ sets the limit to the lower bound for H (a multiple of $\log\log T$) in Theorem 7.1. Observe now that, if w lies on the horizontal sides of $\mathcal{E}$, we have

$$\begin{aligned}&\left|\exp\left(-\cos\left(\frac{w-s_1}{3}\right)\right)\right| \\ &\leqslant \exp\left(-\tfrac{1}{2}\cos 1\exp(2\log\log T)\right) = \exp\left(-\tfrac{1}{2}\cos 1(\log T)^2\right).\end{aligned}$$

Therefore the condition (135) ensures that, for suitable $C, c_1 > 0$,

$$\begin{aligned}\zeta^k(\sigma_1 + it) \ll X^2 &\int_{T-H}^{T+H} |\zeta(\sigma + iv)|^k \exp\left(-c_1\mathrm{e}^{|v-t|/3}\right)\mathrm{d}v \\ &+ X^{-1}\int_{T-H}^{T+H} \exp\left(-c_1\mathrm{e}^{|v-t|/3}\right)\mathrm{d}v + \mathrm{e}^{-C\log^2 T}.\end{aligned}$$

Integrating this estimate over $t \in [T - \frac{1}{2}H,\ T + \frac{1}{2}H]$ and using (134) we obtain

$$\begin{aligned}H \ll X^2 &\int_{T-H}^{T+H} |\zeta(\sigma + iv)|^k \mathrm{d}v\left(\int_{T-\frac{1}{2}H}^{T+\frac{1}{2}H} \exp\left(-c_1\mathrm{e}^{|v-t|/3}\right)\mathrm{d}t\right) \\ &+X^{-1}\int_{T-H}^{T+H} \mathrm{d}v\left(\int_{T-\frac{1}{2}H}^{T+\frac{1}{2}H} \exp\left(-c_1\mathrm{e}^{|v-t|/3}\right)\mathrm{d}t\right) \\ \ll X^2 &\int_{T-H}^{T+H} |\zeta(\sigma + iv)|^k \mathrm{d}v + X^{-1}H.\end{aligned} \tag{136}$$

Let now

$$I := \int_{T-H}^{T+H} |\zeta(\sigma + iv)|^k \mathrm{d}v,$$

and choose $X = H^{\varepsilon}$. Then (136) gives $I \gg H^{1-2\varepsilon}$, showing that I cannot be too small. Then we choose $X = H^{1/3}I^{-1/3}$, so that (since $\zeta(\frac{1}{2} + it) \ll |t|^{1/6}$) trivially

$$T^{-k/18} \ll X \ll H \ll T,$$

and (135) is satisfied. With this choice of X, (136) reduces to $H \ll H^{2/3}I^{1/3}$, and (133) follows. Slightly sharper results than (133), involving powers of $\log\log T$, are known. They are extensively discussed, e.g., by Ramachandra in [89, 90]. In what concerns power moments on $\sigma = \frac{1}{2}$, namely $I_k(T)$, it was proved (op. cit.) that unconditionally one has

$$I_k(T) \gg T(\log T)^{k^2} \tag{137}$$

for any fixed integer $k \geqslant 1$. The lower bound furnished by (137) is of the same order of magnitude as the conjectural formula (122). Recently Radziwill and Soundararajan [88] showed that (unconditionally)

$$\int_0^T |\zeta(\tfrac{1}{2} + it)|^{2k} \mathrm{d}t \geqslant \mathrm{e}^{-30k^4} T(\log T)^{k^2}$$

holds for any real $k > 1$ and $T \geqslant T_0$. This bound not only holds for any $k > 1$, but also it is explicit and at the same time continuous in k, although the constant e^{-30k^4} is certainly not the best one possible. □

Unfortunately, it is the upper bound for $I_k(T)$ that is much more difficult to attain. Even under the RH one cannot, at present, obtain an upper bound of the form $I_k(T) \ll T(\log T)^{k^2}$ for all $k > 2$. Radziwill [87], however, has succeeded recently in establishing this bound (under the RH) for $k \leqslant 2.18$, and this is where the matter stands at present. This shows the great difficulty of the evaluation of $I_k(T)$.

However, recently Soundararajan [96] complemented (137) by obtaining, under the RH, the non-trivial upper bound

$$\int_0^T |\zeta(\tfrac{1}{2} + it)|^{2k} \mathrm{d}t \ll_{\varepsilon} T(\log T)^{k^2+\varepsilon}, \tag{138}$$

which is valid for any fixed $k > 0$ and any given $\varepsilon > 0$. In view of (137) this result, apart from "ε", is therefore best possible. His method of proof is based on a large values estimate for $\log|\zeta(\frac{1}{2} + it)|$, and the author [56] recently generalized it to replace "ε" by an explicit function and to include the bound for "short" intervals of the type $[T - H, T + H]$. This is, in the notation of (122),

Theorem 7.2. *Let $H = T^\theta$ where $0 < \theta \leqslant 1$ is a fixed number, and let k be a fixed positive number. Then, under the RH, we have*

$$I_k(T+H) - I_k(T-H) = \int_{T-H}^{T+H} |\zeta(\tfrac{1}{2}+it)|^{2k}\,dt \ll H(\log T)^{k^2\left(1+O(1/\log_3 T)\right)}. \tag{139}$$

Theorem 7.2 will be deduced from a large values estimate for $\log|\zeta(\frac{1}{2}+it)|$. Setting $\log_2 T := \log(\log T), \log_3 T := \log(\log_2 T)$, we have

Theorem 7.3. *Let $H = T^\theta$ where $0 < \theta \leqslant 1$ is a fixed number, and let $\mu(T, H, V)$ denote the measure of points t from $[T-H,\ T+H]$ such that*

$$\log|\zeta(\tfrac{1}{2}+it)| \geqslant V, \quad 10\sqrt{\log_2 T} \leqslant V \leqslant \frac{3\log 2T}{8\log_2(2T)}. \tag{140}$$

Then, under the RH, for $10\sqrt{\log_2 T} \leqslant V \leqslant \log_2 T$, we have

$$\mu(T,H,V) \ll H\frac{V}{\sqrt{\log_2 T}}\exp\left(-\frac{V^2}{\log_2 T}\left(1-\frac{7}{2\theta\log_3 T}\right)\right), \tag{141}$$

for $\log_2 T \leqslant V \leqslant \frac{1}{2}\theta\log_2 T\log_3 T$, we have

$$\mu(T,H,V) \ll H\exp\left(-\frac{V^2}{\log_2 T}\left(1-\frac{7V}{4\theta\log_2 T\log_3 T}\right)^2\right), \tag{142}$$

and for $\frac{1}{2}\theta\log_2 T\log_3 T \leqslant V \leqslant \frac{3\log 2T}{8\log_2(2T)}$, we have

$$\mu(T,H,V) \ll H\exp(-\tfrac{1}{20}\theta V\log V). \tag{143}$$

To see how Theorem 7.3 implies Theorem 7.2, first note that the contribution of t satisfying $\log|\zeta(\frac{1}{2}+it)| \leqslant \frac{1}{2}k\log_2 T$ to the left-hand side of (139) is

$$\leqslant H\left\{(\log T)^{k/2}\right\}^{2k} = H(\log T)^{k^2}. \tag{144}$$

Likewise the bound (139) holds, by (141) and (142), for the contribution of t satisfying $\log|\zeta(\frac{1}{2}+it)| \geqslant 10k\log_2 T$. Thus we can consider only the range

$$V + \frac{j-1}{\log_3 T} \leqslant \log|\zeta(\tfrac{1}{2}+it)| \leqslant V + \frac{j}{\log_3 T}, \tag{145}$$

where $1 \leqslant j \ll \log_3 T, V = 2^{\ell-\frac{1}{2}}k\log_3 T,\ 1 \leqslant \ell \leqslant \frac{3}{2} + \left[\frac{\log 10}{\log 2}\right]$. If we set

$$U = U(V, j; T) := V + \frac{j-1}{\log_3 T}, \tag{146}$$

then we have

$$I_k(T+H) - I_k(T-H) \ll H(\log T)^{k^2} + \\ + \log_3 T \max_U \mu(T, H, U) \exp\Big(2k\big(U + \frac{1}{\log_3 T}\big)\Big), \tag{147}$$

where $\mu(T, H, U)$ is the measure of $t \in [T-H,\ T+H]$ for which $\log|\zeta(\frac{1}{2}+it)| \geqslant U$ and the maximum is over U satisfying (146)–(147). If we use (141) and (142) of Theorem 7.3, then in the relevant range for U, we obtain

$$\mu(T, H, U) \exp\big(2k(U + 1/\log_3 T)\big) \ll H \log_2 T \exp\Big(2kU - U^2 G(T)\Big),$$

$$G(T) := \frac{1}{\log_2 T}\Big(1 + O\big(\frac{1}{\log_3 T}\big)\Big).$$

Since $\varphi(U) = 2kU - U^2 G(T)$ attains its maximal value at $U = k/G(T)$, we have

$$\mu(T, H, U) \exp\Big(2k\big(U + \frac{1}{\log_3 T}\big)\Big) \ll H \log_2 T\ \exp\Big(k^2\big(1 + O\big(\frac{1}{\log_3 T}\big)\big) \log_2 T\Big)$$
$$= H(\log T)^{k^2(1+O(1/\log_3 T))},$$

so that (147) yields then (139) of Theorem 7.2.

The proof of Theorem 7.3 is based on the following lemmas, whose proofs may be found in Soundararajan [96] (see also [57]).

Lemma 7.1. *Assume the RH. Let $T \leqslant t \leqslant 2T, T \geqslant T_0, 2 \leqslant x \leqslant T^2$. If $\lambda_0 = 0.4912\ldots$ denotes the unique positive real number satisfying $\mathrm{e}^{-\lambda_0} = \lambda_0 + \frac{1}{2}\lambda_0^2$, then for $\lambda \geqslant \lambda_0$ we have*

$$\log|\zeta(\tfrac{1}{2}+it)| \leqslant \mathrm{Re}\left\{\sum_{2\leqslant n\leqslant x} \frac{\Lambda(n)}{n^{\frac{1}{2}+\frac{\lambda}{\log x}+it}\log n}\frac{\log(x/n)}{\log x}\right\} + \frac{(1+\lambda)}{2}\frac{\log T}{\log x} + O\Big(\frac{1}{\log x}\Big).$$

Lemma 7.2. *Assume the RH. If $T \leqslant t \leqslant 2T, 2 \leqslant x \leqslant T^2, \sigma \geqslant \frac{1}{2}$, then*

$$\sum_{2\leqslant n\leqslant x, n\neq p} \frac{\Lambda(n)}{n^{\sigma+it}\log n}\frac{\log(x/n)}{\log x} \ll \log_3 T.$$

Lemma 7.3. *Let $2 \leqslant x \leqslant T$, $T \geqslant T_0$. Let $1 \ll H \leqslant T$ and $r \in \mathbb{N}$ satisfy $x^r \leqslant H$. For any complex numbers $a(p)$ (p denotes primes) we have*

$$\int_T^{T+H}\Big|\sum_{p\leqslant x}\frac{a(p)}{p^{\frac12+it}}\Big|^{2r}\mathrm{d}t \ \ll\ Hr!\Big(\sum_{p\leqslant x}\frac{|a(p)|^2}{p}\Big)^r.$$

Now we can give a sketch of the proof of Theorem 7.3. We assume the RH and let

$$x \ =\ H^{A/V}, \quad z\ =\ x^{1/\log_2 T}, \quad A = A(T,V)\ (\geqslant 1),$$

where A will be suitably chosen below. We follow the method of proof of [57, 96] and accordingly consider three cases.

Case 1. When $10\sqrt{\log_2 T} \leqslant V \leqslant \log_2 T$, we take $A = \frac12 \log_3 T$.
Case 2. When $\log_2 T \leqslant V \leqslant \frac12\theta \log_2 T \log_3 T$, we take $A = \frac{\log_2 T\log_3 T}{2V}$.
Case 3. When $\frac12\theta\log_2 T\log_3 T \leqslant V \leqslant (3\log 2T)/(8\log_2 2T)$ we take $A = 2/\theta$.

Note that the last bound for V comes from the bound (135) with $C = 3/8$ (under the RH). Suppose that $\log|\zeta(\frac12+it)| \geqslant V \geqslant 10\sqrt{\log_2 T}$ holds. Then Lemmas 7.1 and 7.2 yield

$$V \leqslant S_1(t) + S_2(t) + \frac{1+\lambda_0}{2A\theta}V + O(\log_3 T), \tag{148}$$

where we set

$$S_1(t) := \Big|\sum_{p\leqslant z}\frac{\log(x/p)}{\log x}p^{-\frac12-\frac{\lambda_0}{\log x}-it}\Big|,\ S_2(t) := \Big|\sum_{z<p\leqslant x}\frac{\log(x/p)}{\log x}p^{-\frac12-\frac{\lambda_0}{\log x}-it}\Big|. \tag{149}$$

This means that either

$$S_1(t)\ \geqslant\ V_1 = V\left(1-\frac{7}{8A\theta}\right) \tag{150}$$

or

$$S_2(t)\ \geqslant\ \frac{V}{8A\theta}, \tag{151}$$

since we easily get a contradiction if neither (150) nor (151) holds. Let now $\mu_i(T,H,V)$ $(i=1,2)$ denote the measure of the set of points $t\in[T-H,\,T+H]$ for which (150) and (151) hold, respectively. Supposing that (150) holds then, by using Lemma 7.3 with $a(p) = \frac{\log(x/p)}{\log x}p^{-\lambda_0/\log x}$, we obtain

$$\mu_1(T,H,V)V_1^{2r} \leqslant \int_T^{T+H}|S_1(t)|^{2r}\mathrm{d}t \ll Hr!\Big(\sum_{p\leqslant z}\frac1p\Big)^r. \tag{152}$$

The condition in Lemma 7.3 ($x^r \leqslant H$ with $x = z$) is equivalent to

$$\frac{Ar}{V\log_2 T} \leqslant 1. \tag{153}$$

Recalling that

$$\sum_{p\leqslant X}\frac{1}{p} = \log_2 X + O(1),$$

it follows that

$$\log z = \frac{\log x}{\log_2 T} = \frac{A\theta}{V\log_2 T}\log T \leqslant \frac{\log T}{\log_2 T},$$

since $A \leqslant V$ in all cases. Therefore we have

$$\sum_{p\leqslant z}\frac{1}{p} \leqslant \log_2 T \qquad (T \geqslant T_0). \tag{154}$$

Noting that Stirling's formula yields $r! \ll r^r\sqrt{r}\mathrm{e}^{-r}$, we infer from (152) and (154) that

$$\mu_1(T,H,V) \ll H\sqrt{r}\Big(\frac{r\log_2 T}{\mathrm{e}V_1^2}\Big)^r. \tag{155}$$

In Cases 1 and 2 and also in Case 3 when $V \leqslant \frac{2}{\theta}\log_2^2 T$, one chooses

$$r = \left[\frac{V_1^2}{\log_2 T}\right]\ \Big(\geqslant 1\Big).$$

With this choice of r, it is readily seen that (153) is satisfied and (155) gives

$$\mu_1(T,H,V) \ll H\frac{\sqrt{V}}{\log_2 T}\exp\Big(-\frac{V_1^2}{\log_2 T}\Big). \tag{156}$$

Finally in Case 3 when $\frac{2}{\theta}\log_2^2 T \leqslant V \leqslant (3\log 2T)/(8\log_2 2T)$ and $A = 2/\theta$, we have

$$V_1 = V\left(1 - \frac{7}{8A\theta}\right) = V\left(1 - \frac{7}{16}\right) > \frac{V}{2}.$$

Thus with the choice $r = [V/2]$ we see that (153) is again satisfied and

$$\sqrt{r}\Big(\frac{r\log_2 T}{\mathrm{e}V_1^2}\Big)^r \leqslant \sqrt{V}\left(\frac{2\log_2 T}{\mathrm{e}V}\right)^r \leqslant V^{\frac{1}{2}-\frac{r}{4}} \ll \exp(-\tfrac{1}{10}V\log V),$$

giving in this case

$$\mu_1(T,H,V) \ll H\exp(-\tfrac{1}{10}V\log V). \tag{157}$$

We bound $\mu_2(T,H,V)$ in a similar way by using (151). It follows, again by Lemma 7.3, that

$$\begin{aligned}\left(\frac{V}{8A\theta}\right)^{2r}\mu_2(T,H,V) &\leqslant \int_T^{T+H}|S_2(t)|^{2r}\,\mathrm{d}t\\ &\ll Hr!\Big(\sum_{z<p\leqslant x}\frac{1}{p}\Big)^r = Hr!\big(\log_2 x-\log_2 z+O(1)\big)^r\\ &\ll H\Big\{r\Big(\log_3 T+O(1)\Big)\Big\}^r.\end{aligned}$$

We obtain

$$\mu_2(T,H,V) \ll H\Big(\frac{8A}{V}\Big)^{2r}(2r\log_3 T)^r \ll H\exp\left(-\frac{V}{2A}\log V\right). \tag{158}$$

Namely the second inequality in (158) is equivalent to

$$\left(\frac{A}{V}\right)^2 r\log_3 T \ll \exp\left(-\frac{V}{2rA}\log V\right). \tag{159}$$

In all Cases 1–3 we take

$$r = \left[\frac{V}{A}-1\right]\ \big(\geqslant 1\big).$$

The condition $x^r \leqslant H$ in Lemma 7.3 is equivalent to $rA \leqslant V$, which is trivial with the above choice of r. To establish (159) note first that

$$\left(\frac{A}{V}\right)^2 r\log_3 T \leqslant \frac{A}{V}\log_3 T. \tag{160}$$

In Case 1 the second expression in (160) equals $\log_3^2 T/(2V)$, while

$$\exp\left(-\frac{V}{2rA}\log V\right) = \exp\big(-(\tfrac12+o(1))\log V\big) = V^{-1/2+o(1)}.$$

Therefore it suffices to have

$$\frac{\log_3^2 T}{V} \ll V^{-1/2+o(1)},$$

which is true since $10\sqrt{\log_2 T} \leqslant V$. In Case 2 the analysis is similar. In Case 3 we have $A = 2/\theta$, hence $(A \log_3 T)/V \ll (\log_3 T)/V$ and

$$\exp\left(-\frac{V}{2rA}\log V\right) = \exp\Big(-(\tfrac{1}{2}\theta + o(1))\log V\Big) = V^{-(\frac{1}{2}\theta + o(1))},$$

so that (159) follows again. Thus we have shown that in all cases

$$\mu_2(T, H, V) \ll H \exp\left(-\frac{V}{2A}\log V\right). \tag{161}$$

Theorem 7.3 follows now from (155), (156) and (161). Namely, in Case 1, we have

$$\frac{V_1^2}{\log_2 T} = \frac{V^2\Big(1 - \frac{7}{4\theta \log_3 T}\Big)^2}{\log_2 T} \leqslant V\frac{\log V}{\log_3 T} = \frac{V \log V}{2A},$$

which gives (141). If Case 2 holds, we have again

$$\frac{V_1^2}{\log_2 T} = \frac{V^2\Big(1 - \frac{7V}{4\theta \log_2 T \log_3 T}\Big)^2}{\log_2 T} \leqslant \frac{V^2 \log V}{\log_2 T \log_3 T} = \frac{V \log V}{2A},$$

and (142) follows. In Case 3 when $\frac{1}{2}\theta \log_2 T \log_3 T \leqslant V \leqslant \frac{2}{\theta}\log_2^2 T$ we have

$$\begin{aligned}\mu(T, H, V) &\ll H \exp\Big(-\frac{V_1^2}{\log_2 T}\Big) + H \exp(-\theta V \log V)\\ &\ll H \exp(-\frac{\theta}{20} V \log V),\end{aligned}$$

since

$$\frac{V_1^2}{\log_2 T} \geqslant \frac{V^2}{4\log_2 T} \geqslant \frac{\theta V \log_2 T \log_3 T}{8 \log_2 T} \geqslant \frac{\theta}{20} V \log V.$$

In the remaining range of Case 3 we have

$$\begin{aligned}\mu(T, H, V) &\ll H \exp(-\tfrac{1}{10} V \log V) + H \exp\Big(-\frac{V}{2A}\log V\Big)\\ &\ll H \exp(-\tfrac{1}{10} V \log V) + \exp(-\tfrac{\theta}{4} V \log V),\end{aligned}$$

and (143) follows. The proof of Theorem 7.3 is complete.

8 Miscellaneous Results

A natural way to bound high moments of $|\zeta(\frac{1}{2}+it)|$ is to try to bound

$$\int_0^T |\zeta(\tfrac{1}{2}+it)|^{2k} \Big| \sum_{N<n\leqslant 2N} a_n n^{-1/2-it} \Big|^2 \mathrm{d}t \qquad (a_n \ll_\varepsilon N^\varepsilon), \tag{162}$$

in conjunction with approximate functional equations for $\zeta(\frac{1}{2}+it)$. This is especially useful when $k = 1$ or $k = 2$, with the aim of obtaining bounds for the sixth or eighth moment.

The first result in this direction is due to Iwaniec [66] in 1980. He proved that, if $N \leqslant T^{1/10}$,

$$\int_0^T |\zeta(\tfrac{1}{2}+it)|^2 |N(it)|^2 \mathrm{d}t \ll_\varepsilon T^{1+\varepsilon} \sum_{n\leqslant N} |a_n|^2, \quad N(s) := \sum_{n\leqslant N} a_n n^{-s}. \tag{163}$$

The proof depended on the *Laplace transform* of $|\zeta(\frac{1}{2}+ix)|^2$ and the estimate of Weil [106] (in a somewhat modified form) for the *Kloosterman sums*. This is the bound

$$|S(m,n;c)| \leqslant (m,n,c)^{1/2} c^{1/2} d(c),$$

$$S(m,n;c) = \sum_{ad\equiv 1 \,(\mathrm{mod}\, c)} \exp\Big(\frac{2\pi i(dm+an)}{c}\Big),$$

obtained by Weil as a consequence of the RH for curves over finite fields.

A little later Deshouillers and Iwaniec [17] studied, in the above notation,

$$I(T,N) := \frac{1}{T}\int_0^T |\zeta(\tfrac{1}{2}+it)|^4 |N(it)|^2 \mathrm{d}t. \tag{164}$$

For $I(T,N)$ in (164) they proved that

$$I(T,N) \ll_\varepsilon T^\varepsilon (1 + N^2 T^{-1/2} + N^{5/4} T^{-1/4}) \sum_{n\leqslant N} |a_n|^2, \tag{165}$$

which may be compared to the conjectured bound

$$I(T,N) \ll_\varepsilon T^\varepsilon (1 + NT^{-1}) \sum_{n\leqslant N} |a_n|^2,$$

which is incidentally equivalent to the Lindelöf hypothesis (if $\log N \ll \log T$). An improvement of (165) was obtained by Watt [104] who proved the bound

$$I(T,N) \ll_\varepsilon T^\varepsilon(1 + N^2T^{-1/2}) \max_{n\leqslant N} |a_n|^2.$$

The proof is based on technical refinements within the circle of ideas to be found in [17].

The asymptotic formula for

$$J(T,N) := \int_0^T |\zeta(\tfrac{1}{2}+it)|^2|N(it)|^2\mathrm{d}t$$

was obtained by Balasubramanian et al. [6].

Namely, they proved that, for $\log N \ll \log T, a(n) \ll_\varepsilon n^\varepsilon$,

$$J(T,N) = T\sum_{h,k\leqslant N} \frac{a(h)}{h}\frac{\overline{a(k)}}{k}(h,k)\left(\log\frac{T(h,k)^2}{2\pi hk} + 2\gamma - 1\right)$$
$$+ O(T\log^{-B} T) + O_\varepsilon(N^2T^\varepsilon)$$

for any given constant $B > 0$. Thus one has an asymptotic formula if $N \ll T^{1/2-\delta}$, where δ is any positive constant, and the formula is conjectured to hold even for $N \ll T^{1-\delta}$. In any case, on Hooley's hypothesis R^* on incomplete Kloosterman sums (see Hooley [35]), the bound for N can be extended to $T^{4/7-\delta}$. Moreover, if $a(n) = \mu(n)F(n)$, where F is a function satisfying $F(x) \ll 1$ and $F'(x) \ll x^{-1}$ for $1 \leqslant x \leqslant N$, then the bound is $\ll T^{9/17-\delta}$, unconditionally.

An interesting special case, related to classical work of Atle Selberg, arises when $a(n) = \mu(n)(1 - \log n/\log N)$; then $J(T,N) \sim T(1 + \log T/\log N)$ for $N \ll_\varepsilon T^{9/17-\varepsilon}$ as $T \to \infty$. This implies that $\sum(\beta - \frac{1}{2}) \leqslant (0.0845 + o(1))T$, where the sum is over all zeros $\beta + i\gamma$ of $\zeta(s)$ such that $\beta > \frac{1}{2}$ and $0 < \gamma < T$. The deepest result needed in [6] is Weil's estimate for Kloosterman sums [106]. Their result is complemented by the result of Motohashi [81], whose method is based on the method of Atkinson, used in the proof of his famous formula (Theorem 3.1). Motohashi replaces the error terms in the formula for $J(T,N)$ by $O_\varepsilon(T^{1/3+\varepsilon}N^{4/3})$. None of the above results, unfortunately, are strong enough to bring improvements on the best-known bounds for the sixth or eighth moment of $|\zeta(\frac{1}{2}+it)|$, namely [see (114)],

$$I_3(T) \ll T^{5/4}\log^C T, \qquad I_4(T) \ll T^{3/2}\log^C T.$$

The more difficult case of the asymptotic evaluation of the integral

$$K(T,N) := \int_0^T |\zeta(\tfrac{1}{2}+it)|^4|N(it)|^2\mathrm{d}t \tag{166}$$

was done by Hughes and Young [36]. They established that an asymptotic formula indeed holds for $K(T,N)$ if the length N of the Dirichlet polynomial satisfies $N \leqslant T^{1/11-\varepsilon}$. To deal with $K(T,N)$ they considered the more general "twisted fourth moment integral", namely

$$I(h,k) := \int_{-\infty}^{\infty}\left(\frac{h}{k}\right)^{-it}\zeta(\tfrac12+\alpha+it)\zeta(\tfrac12+\beta+it)\zeta(\tfrac12+\gamma-it)\zeta(\tfrac12+\delta-it)w(t)\mathrm{d}t,$$

where $w(t)$ is a suitable smooth function, $(h,k) = 1$, and $\alpha,\beta,\gamma,\delta$ are complex numbers $\ll 1/\log T$, with the idea of letting eventually $\alpha,\beta,\gamma,\delta$ all tend to zero. Then, summing the resulting expression over suitable h,k, one obtains the asymptotic formula for $K(T,N)$ in (166). The relatively short range for N, namely $N \leqslant T^{1/11-\varepsilon}$, is compensated by the fact that one indeed obtains an asymptotic formula and not just an upper bound as was done in previous works. For example, Watt [104] obtains the desired upper bound for $N \ll T^{1/4}$, but his method does not produce an asymptotic formula for the integral $K(T,N)$.

It is also of interest to evaluate the *Laplace transforms* of powers of $|\zeta(\frac12+it)|$. To this end let

$$L_k(s) := \int_0^{\infty} |\zeta(\tfrac12+ix)|^{2k}\mathrm{e}^{-sx}\mathrm{d}x \qquad (k\in\mathbb{N},\ \sigma = \operatorname{Re} s > 0). \tag{167}$$

A classical result of Kober from 1936 [73] says that

$$L_1(2\sigma) = \frac{\gamma - \log(4\pi\sigma)}{2\sin\sigma} + \sum_{n=0}^{N} c_n\sigma^n + O(\sigma^{N+1}) \quad (\sigma\to 0+)$$

for any given integer $N \geqslant 1$, where the c_n's are effectively computable constants and γ is Euler's constant.

For complex values of s the function $L_1(s)$ was studied by Atkinson [1] and more recently by Jutila [69]. Jutila noted that Atkinson's argument actually gives

$$L_1(s) = -i\mathrm{e}^{\frac12 is}\left[\log(2\pi) - \gamma + \left(\frac{\pi}{2} - s\right)i\right] + 2\pi\mathrm{e}^{-\frac12 is}\sum_{n=1}^{\infty} d(n)\exp\left(-2\pi i n\mathrm{e}^{-is}\right) + \lambda_1(s)$$

in the strip $0 < \operatorname{Re} s < \pi$, where the function $\lambda_1(s)$ is holomorphic in the strip $|\operatorname{Re} s| < \pi$. Moreover, in any strip $|\operatorname{Re} s| \leqslant \theta$ with $0 < \theta < \pi$, we have

$$\lambda_1(s) \ll_\theta (|s|+1)^{-1}.$$

Atkinson [2] obtained the asymptotic formula, as $\sigma \to 0+$,

$$L_2(\sigma) = \frac{1}{\sigma}\left(A\log^4\frac{1}{\sigma} + B\log^3\frac{1}{\sigma} + C\log^2\frac{1}{\sigma} + D\log\frac{1}{\sigma} + E\right) + \lambda_2(\sigma), \quad (168)$$

where

$$A = \frac{1}{2\pi^2},\; B = \frac{1}{\pi^2}\Big(2\log(2\pi) - 6\gamma + \frac{24\zeta'(2)}{\pi^2}\Big),$$

and

$$\lambda_2(\sigma) \ll_\varepsilon \left(\frac{1}{\sigma}\right)^{\frac{13}{14}+\varepsilon},$$

and indicated how the exponent 13/14 can be replaced by 8/9. This is of historical interest, since it is one of the first applications of Kloosterman sums to zeta-function theory. Atkinson in fact showed that ($\sigma = \operatorname{Re} s > 0$)

$$L_2(s) = 4\pi e^{-\frac{1}{2}s}\sum_{n=1}^{\infty} d_4(n)K_0(4\pi i\sqrt{n}e^{-\frac{1}{2}s}) + \phi(s), \quad (169)$$

where $d_4(n)$ is the divisor function generated by $\zeta^4(s)$, K_0 is the Bessel function, and the series in (169) as well as $\phi(s)$ are both analytic in the region $|s| < \pi$. When $s = \sigma \to 0+$ one can use the asymptotic formula

$$K_0(z) = \tfrac{1}{2}\sqrt{\pi}z^{-1/2}e^{-z}\left(1 - 8z^{-1} + O(|z|^{-2})\right) \quad \left(|\arg z| < \theta < \frac{3\pi}{2},\ |z| \geqslant 1\right)$$

to make a simplification of (169).

The author [43] gave explicit, albeit complicated expressions for the remaining coefficients C, D and E in (168). More importantly, he established that

$$\lambda_2(\sigma) \ll \sigma^{-1/2} \qquad (\sigma \to 0+), \quad (170)$$

which is actually best possible. In [43] it was also proved that

$$\lambda_k\Big(\frac{1}{T}\Big) = O(T^{c_k - 1})$$

where in general one defines, for a fixed $k \in \mathbb{N}$,

$$\lambda_k\Big(\frac{1}{T}\Big) = \int_0^\infty |\zeta(\tfrac{1}{2} + it)|^{2k} e^{-t/T}\,dt - TQ_{k^2}(\log T)$$

for a suitable polynomial $Q_{k^2}(x)$ in x of degree k, provided that

$$\int_0^T E_k(t)\mathrm{d}t = O(T^{c_k}) \qquad (c_k > 0).$$

Here $E_k(T)$ is defined by (127). Since, by the bound (see [60–62])

$$\int_0^T E_2(t)\mathrm{d}t = O(T^{3/2})$$

we have $c_2 = 3/2$, it follows that

$$\lambda_2\Big(\frac{1}{T}\Big) = O(T^{1/2}),$$

which is equivalent to $\lambda_2(\sigma) \ll \sigma^{-1/2}$ $(\sigma \to 0+)$. Moreover the coefficients of $Q_{k^2}(y)$ can be expressed as linear combinations of the coefficients of $P_{k^2}(y)$.

The author [49] obtained in fact the following result on $L_2(s)$.

Theorem 8.1. *Let* $0 \leqslant \phi < \frac{\pi}{2}$ *be given. Then for* $0 < |s| \leqslant 1$ *and* $|\arg s| \leqslant \phi$ *we have*

$$L_2(s)=\frac{1}{s}(A\log^4\frac{1}{s} + B\log^3\frac{1}{s} + C\log^2\frac{1}{s} + D\log\frac{1}{s} + E) + G_2(s)$$
$$+s^{-\frac{1}{2}}\left\{\sum_{j=1}^{\infty}\alpha_j H_j^3(\tfrac{1}{2})\left(s^{-i\kappa_j}R(\kappa_j)\Gamma(\tfrac{1}{2}+i\kappa_j)+s^{i\kappa_j}R(-\kappa_j)\Gamma(\tfrac{1}{2}-i\kappa_j)\right)\right\}, \tag{171}$$

where

$$R(y) := \sqrt{\frac{\pi}{2}}\left(2^{-iy}\frac{\Gamma(\frac{1}{4}-\frac{i}{2}y)}{\Gamma(\frac{1}{4}+\frac{i}{2}y)}\right)^3\Gamma(2iy)\cosh(\pi y) \tag{172}$$

and in the above region $G_2(s)$ *is a regular function satisfying* ($C > 0$ *is a suitable constant)*

$$G_2(s) \ll |s|^{-1/2}\exp\left\{-\frac{C\log(|s|^{-1}+20)}{(\log\log(|s|^{-1}+20))^{2/3}(\log\log\log(|s|^{-1}+20))^{1/3}}\right\}. \tag{173}$$

Remark 8.1. The constants A, B, C, D, E in (171) are the same ones as in Atkinson's (165).

Remark 8.2. From Stirling's formula for the gamma-function it follows that $R(\kappa_j) \ll \kappa_j^{-1/2}$. In view of (84) this means that the series in (171) is absolutely convergent and uniformly bounded in s when $s = \sigma$ is real. Therefore, when $s = \sigma \to 0+$, (171) gives a refinement of (170).

Remark 8.3. From (4) and (7) it transpires that $\lambda(\sigma)$ is an error term when $0 < \sigma < 1$. For this reason we considered the values $0 < |s| \leqslant 1$ in (171), although one could treat the case $|s| > 1$ as well.

Remark 8.4. From (171) and elementary properties of the Laplace transform one can easily obtain the Laplace transform of

$$E_2(T) := \int_0^T |\zeta(\tfrac{1}{2} + it)|^4 \mathrm{d}t - TP_4(\log T), \quad P_4(x) = \sum_{j=0}^{4} a_j x^j,$$

where $a_4 = 1/(2\pi^2)$.

References

1. Atkinson, F.V.: The mean value of the zeta-function on the critical line. Quart. J. Math. Oxford **10**, 122–128 (1939)
2. Atkinson, F.V.: The mean value of the zeta-function on the critical line. Proc. London Math. Soc. **47**, 174–200 (1941)
3. Atkinson, F.V.: A mean value property of the Riemann zeta-function. J. London Math. Soc. **23**, 128–135 (1948)
4. Atkinson, F.V.: The mean value of the Riemann zeta-function. Acta Math. **81**, 353–376 (1949)
5. Balasubramanian, R.: An improvement on a theorem of Titchmarsh on the mean square of $|\zeta(\frac{1}{2} + it)|$. Proc. London Math. Soc. **36**(3), 540–576 (1978)
6. Balasubramanian, R., Conrey, J.B., Heath-Brown, D.R.: Asymptotic mean square of the product of the Riemann zeta-function and a Dirichlet polynomial. J. Reine Angew. Math. **357**, 161–181 (1985)
7. Balasubramanian, R., Ivić, A., Ramachandra, K.: An application of the Hooley-Huxley contour. Acta Arith. **65**, 45–51 (1993)
8. Bombieri, E.: Riemann Hypothesis: The Millennium Prize Problems, pp. 107–124. Clay Mathematics Institute, Cambridge (2006)
9. Bombieri, E.: The classical theory of zeta and L-functions. Milan J. Math. **78**, 11–59 (2010)
10. Bombieri, E., Iwaniec, H.: On the order of $\zeta(\frac{1}{2} + it)$. Ann. Scuola Norm. Sup. Pisa Cl. Sci. **13**(4), 449–472 (1986)
11. Bombieri, E., Iwaniec, H.: Some mean value theorems for exponential sums. Ann. Scuola Norm. Sup. Pisa Cl. Sci. **13**(4), 473–486 (1986)
12. Borwein, P., Choi, S., Rooney B., Weirathmueller, A.: The Riemann Hypothesis: A Resource for the Afficionado and the Virtuoso Alike. CMS Books in Mathematics. Springer, Berlin (2008)
13. Conrey, J.B.: A note on the fourth power moment of the Riemann zeta-function. In: Berndt, B.C., et al. (eds.) Analytic Number Theory, vol. 1. Proceedings of a Conference in Honor of Heini Halberstam, Urbana. Birkhäuser, Boston (1995); Prog. Math. **138**, 225–230 (1996)
14. Conrey, J.B., Farmer, D.W., Keating, J.P., Rubinstein, M.O., Snaith, N.C.: Integral moments of L-functions. Proc. London Math. Soc. **91**(3), 33–104 (2005)

15. Conrey, J.B., Farmer, D.W., Keating, J.P., Rubinstein, M.O., Snaith, N.C.: Lower order terms in the full moment conjecture for the Riemann zeta function. J. Number Theory **128**, 1516–1554 (2008)
16. Conrey, J.B., Gonek, S.M.: High moments of the Riemann zeta-function. Duke Math. J. **107**, 577–604 (2001)
17. Deshouillers, J.-M., Iwaniec, H.: Power mean values of the Riemann zeta-function. Mathematika **29**, 202–212 (1982); Acta Arith. II **48**, 305–312 (1984)
18. Erdélyi, A., Magnus, W., Oberhettinger, F., Tricomi, F.G.: Higher Transcendental Functions, vol. I. McGraw-Hill, New York (1953)
19. Euler, L.: Remarques sur un beau rapport entre les séries des puissances tout directes que réciproques. Mém. Acad. Roy. Sci. Belles Lettres **17**, 83–106 (1768)
20. Graham, S.W., Kolesnik, G.: Van der Corput's Method for Exponential Sums. London Mathematical Society Lecture Note Series, vol. **126**. Cambridge University Press, Cambridge (1991)
21. Hafner, J.L., Ivić, A.: On some mean value results for the Riemann zeta-function. In: Proceedings International Number Theory Conference Québec 1987, pp. 348–358. Walter de Gruyter and Co., Berlin (1989)
22. Hafner, J.L., Ivić, A.: On the mean square of the Riemann zeta-function on the critical line. J. Number Theory **32**, 151–191 (1989)
23. Hardy, G.H.: On the zeros of Riemann's zeta-function. Proc. London Math. Soc. Ser. 2 **13**(records of proceedings at meetings) (1914)
24. Hardy, G.H.: Sur les zéros de la fonction $\zeta(s)$ de Riemann. Comptes Rendus Acad. Sci. (Paris) **158**, 1012–1014 (1914)
25. Hardy, G.H., Littlewood, J.E.: Contributions to the theory of the Riemann zeta-function and the distribution of primes. Acta Math. **41**, 119–196 (1917)
26. Hardy, G.H., Littlewood, J.E.: The approximate functional equation in the thepry of the zeta-function, with applications to the divisor problems of Dirichlet and Piltz. Proc. London Math. Soc. **21**(2), 39–74 (1922)
27. Hardy, G.H., Littlewood, J.E.: The approximate functional equation for $\zeta(s)$ and $\zeta^2(s)$. Proc. London Math. Soc. **29**(2), 81–97 (1929)
28. Haselgrove, C.B.: Tables of the Riemann Zeta Function. Cambridge University Press, Cambridge (1960)
29. Heath-Brown, D.R.: The twelfth power moment of the Riemann zeta-function. Quart. J. Math. (Oxford) **29**, 443–462 (1978)
30. Heath-Brown, D.R.: The mean value theorem for the Riemann zeta-function. Mathematika **25**, 177–184 (1978)
31. Heath-Brown, D.R.: The fourth power moment of the Riemann zeta function. J. London Math. Soc. **38**(3), 385–422 (1979)
32. Heath-Brown, D.R.: The distribution and moments of the error term in the Dirichlet divisor problems. Acta Arith. **60**(4), 389–415 (1992)
33. Heath-Brown, D.R., Tsang, K.-M.: Sign changes of $E(T)$, $\Delta(x)$ and $P(x)$. J. Number Theory **49**, 73–83 (1994)
34. Hiary, G.A., Odlyzko, A.M.: The zeta function on the critical line: numerical evidence for moments and random matrix theory models. Math. Comp. **81**, 1723–1752 (2012)
35. Hooley, C.: On the Brun-Titchmarsh theorem. J. Reine Angew. Math. **225**, 60–79 (1972)
36. Hughes, C.P., Young, M.P.: The twisted fourth moment of the Riemann zeta function. J. Reine Angew. Math. **641**, 203–236 (2010)
37. Huxley, M.N.: Exponential sums and the Riemann zeta function V. Proc. London Math. Soc. **90**(3), 1–41 (2005)
38. Huxley, M.N., Ivić, A.: Subconvexity for the Riemann zeta-function and the divisor problem. Bulletin CXXXIV de l'Académie Serbe des Sciences et des Arts – Classe des Sciences mathématiques et naturelles, Sciences mathématiques **32**, 13–32 (2007)
39. Ingham, A.E.: Mean-value theorems in the theory of the Riemann zeta-function. Proc. London Math. Soc. **27**(2), 273–300 (1926)

40. Ivić, A.: Large values of the error term in the divisor problem. Invent. Math. **71**, 513–520 (1983)
41. Ivić, A.: Large values of certain number-theoretic error terms. Acta Arith. **56**, 135–159 (1990)
42. Ivić, A.: Mean Values of the Riemann Zeta-Function LN's, vol. 82. Tata Institute of Fundamental Research, Bombay (1991) (distr. by Springer, Berlin)
43. Ivić, A.: On the fourth moment of the Riemann zeta-function. Publications Inst. Math. (Belgrade) **57**(71), 101–110 (1995)
44. Ivić, A.: Some problems on mean values of the Riemann zeta-function. J. Théorie des Nombres Bordeaux **8**, 101–122 (1996)
45. Ivić, A.: The Laplace transform of the square in the circle and divisor problems. Studia Scient. Math. Hungarica (Budapest) **32**, 181–205 (1996)
46. Ivić, A.: The Mellin transform and the Riemann zeta-function. In: Nowak, W.G., Schoißengeier, J. (eds.) Proceedings of the Conference on Elementary and Analytic Number Theory (Vienna, 18–20 July 1996), pp. 112–127. Universität Wien, Universität für Bodenkultur, Vienna (1996)
47. Ivić, A.: On some problems involving the mean square of $|\zeta(\frac{1}{2}+it)|$. Bulletin CXVI de l'Académie Serbe des Sciences et des Arts - Classe des Sciences mathématiques et naturelles, Sciences mathématiques **23**, 71–76 (1998)
48. Ivić, A.: On the error term for the fourth moment of the Riemann zeta-function. J. London Math. Soc. **60**(2), 21–32 (1999)
49. Ivić, A.: The Laplace transform of the fourth moment of the zeta-function. Univ. Beograd. Publ. Elektrotehn. Fak. Ser. Mat. **11**, 41–48 (2000)
50. Ivić, A.: Some mean value results for the Riemann zeta-function. In: Jutila, M., Metsänkylä, T. (eds.) Number Theory and Diophantine Analysis. Proceedings of the Turku Symposium on Number Theory in Memory of K. Inkeri (1999), pp. 145–161. Walter de Gruyter, Berlin (2000)
51. Ivić, A.: On the integral of the error term in the fourth moment of the Riemann zeta-function. Functiones et Approximatio **28**, 105–116 (2000)
52. Ivić, A.: The Riemann Zeta-Function. Wiley, New York (1985) (reissue, Dover, Mineola, New York, 2003)
53. Ivić, A.: On mean values of some zeta-functions in the critical strip. J. Théorie des Nombres de Bordeaux **15**, 163–178 (2003)
54. Ivić, A.: On the Riemann zeta function and the divisor problem. Central Euro. J. Math. **4**(2), 1–15 (2004); Central Euro. J. Math. II **2**(3), 203–214 (2005)
55. Ivić, A.: On the Riemann zeta-function and the divisor problem IV. Uniform Distrib. Theory **1**, 125–135 (2006)
56. Ivić, A.: On the mean square of the zeta-function and the divisor problem. Ann. Acad. Scien. Fennicae Math. **23**, 1–9 (2007)
57. Ivić, A.: On the moments of the Riemann zeta-function in short intervals. Hardy-Ramanujan J. **32**, 4–23 (2009)
58. Ivić, A.: The mean value of the zeta-function on $\sigma = 1$. The Ramanujan J. **26**, 209–227 (2011)
59. Ivić, A.: Lectures on Hardy's Z-function. Cambridge University Press, Cambridge (2012)
60. Ivić, A., Motohashi, Y.: A note on the mean-value of the zeta and L-functions VII. Proc. Japan Acad. Ser. A **66**, 150–152 (1990)
61. Ivić, A., Motohashi, Y.: The mean square of the error term for the fourth moment of the zeta-function. Proc. London Math. Soc. **69**(3), 309–329 (1994)
62. Ivić, A., Motohashi, Y.: On the fourth power moment of the Riemann zeta-function. J. Number Theory **51**, 16–45 (1995)
63. Ivić, A., te Riele, H.: On the zeros of the error term for the mean square of $|\zeta(\frac{1}{2}+it)|$. Math. Computation **56**(193), 303–328 (1991)
64. Ivić, A., Sargos, P.: On the higher power moments of the error term in the divisor problem. Illinois J.Math. **81**, 353–377 (2007)

65. Iwaniec, H.: Fourier coefficients of cusp forms and the Riemann zeta function. Exposé No. 18, Séminaire de Théorie des Nombres, Université Bordeaux (1979/1980)
66. Iwaniec, H.: On mean values for Dirichlet's polynomials and the Riemann zeta-function. J. London Math. Soc. **22**(2), 39–45 (1980)
67. Jutila, M.: Riemann's zeta-function and the divisor problem. Arkiv Mat. **21**, 75–96 (1983); ibid. II **31**, 61–70 (1993)
68. Jutila, M.: Mean values of Dirichlet series via Laplace transforms. In: Motohashi, Y. (ed.) Analytic Number Theory. London Mathematical Society LNS, vol. 247, pp. 169–207. Cambridge University Press, Cambridge (1997)
69. Jutila, M.: Atkinson's formula revisited. In: Voronoï's Impact on Modern Science, Book 1, pp. 137–154. Institute of Mathematics, National Academy of Sciences of Ukraine, Kyiv (1998)
70. Kaczorowski, J.: Axiomatic theory of L-functions: the Selberg class. In: Perelli, A., Viola, C. (eds.) Analytic Number Theory, pp. 133–209. Springer, Berlin (2006)
71. Karatsuba, A.A., Voronin, S.M.: The Riemann Zeta-Function. Walter de Gruyter, Berlin (1992)
72. Keating, J.P., Snaith, N.C.: Random matrix theory and $\zeta(\frac{1}{2}+it)$. Comm. Math. Phys. **214**, 57–89 (2000)
73. Kober, H.: Eine Mittelwertformel der Riemannschen Zetafunktion. Compositio Math. **3**, 174–189 (1936)
74. Lau, Y.-K., Tsang, K.-M.: Mean square of the remainder term in the Dirichlet divisor problem. J. Théorie Nombres Bordeaux **7**, 75–92 (1995)
75. Lau, Y.-K., Tsang, K.-M.: Omega result for the mean square of the Riemann zeta function. Manuscr. Math. **117**, 373–381 (2005)
76. Lau, Y.-K., Tsang, K.-M.: On the mean square formula of the error term in the Dirichlet divisor problem. Math. Proc. Camb. Phil. Soc. **146**(2), 227–287 (2009)
77. Lee, K.-Y., Tsang, K.-M.: On a mean value theorem for the second moment of the Riemann zeta-function. Comment. Math. Univ. Sancti Pauli **60**, 188–209 (2011)
78. Lukkarinen, M.: The Mellin transform of the square of Riemann's zeta-function and Atkinson's formula. Doctoral Dissertation, Annales Academiae Scientiarum Fennicae, vol. 140, 74 pp., Helsinki (2005)
79. Matsumoto, K.: Recent developments in the mean square theory of the Riemann zeta and other zeta-functions. In: Number Theory. Trends Math., pp. 241–286. Birkhäuser, Basel (2000)
80. Mehta, M.L.: Random matrices. In: Pure and Applied Mathematics, vol. 142, 3rd edn. Elsevier/Academic, Amsterdam (2004)
81. Motohashi, Y.: A note on the mean value of the zeta and L-functions V. Proc. Japan Acad. Ser. A **62**, 399–401 (1986)
82. Motohashi, Y.: Riemann–Siegel formula. Lecture Notes, University of Colorado, Boulder (1987)
83. Motohashi, Y.: An explicit formula for the fourth power mean of the Riemann zeta-function. Acta Math. **170**, 181–220 (1993)
84. Motohashi, Y.: A relation between the Riemann zeta-function and the hyperbolic Laplacian. Annali Scuola Norm. Sup. Pisa, Cl. Sci. Ser. IV **22**, 299–313 (1995)
85. Motohashi, Y.: Spectral Theory of the Riemann Zeta-Function. Cambridge University Press, Cambridge (1997)
86. Preissmann, E.: Sur la moyenne de la fonction zêta. In: Nagasaka, K. (ed.) Analytic Number Theory and Related Topics, pp. 119–125. Proceedings of the Symposium, Tokyo, Japan, 11–13 November 1991. World Scientific, Singapore (1993)
87. Radziwill, M.: The 4.36th moment of the Riemann zeta-function. Int. Math. Res. Not. IMRN **18**, 4245–4252 (2012)
88. Radziwill, M., Soundararajan, K.: Continuous lower bounds for moments of zeta and L-functions. Mathematika **59**, 119–128 (2013)
89. Ramachandra, K.: Applications of a theorem of Montgomery and Vaughan to the zeta-function. J. London Math. Soc. **10**(2), 482–486 (1975)

90. Ramachandra, K.: On the Mean-Value and Omega-Theorems for the Riemann Zeta-Function. LN's vol. 85. Tata Institute of Fundamental Research, Bombay (1995) (distributed by Springer, Berlin)
91. Riemann, B.: Über die Anzahl der Primzahlen unter einer gegebener Grösse. Monatshefte Preuss. Akad. Wiss., pp. 671–680 (1859–1860)
92. Rubinstein, M.O., Yamagishi, S.: Computing the moment polynomials of the zeta function (to appear). Preprint available at arXiv:1112.2201
93. Selberg, A.: Selected Papers, vol. I. Springer, Berlin (1989); vol. II, Springer, Berlin (1991)
94. Siegel, C.L.: Über Riemanns Nachlaß zur analytischen Zahlentheorie. Quell. Stud. Gesch. Mat. Astr. Physik **2**, 45–80 (1932) [also in Gesammelte Abhandlungen, Band I, pp. 275–310. Springer, Berlin (1966)]
95. Soundarajan, K.: Omega results for the divisor and circle problems, Int. Math. Res. Not. **36**, 1987–1998 (2003)
96. Soundarajan, K.: Moments of the Riemann zeta function, Ann. Math. **170**, 981–993 (2010)
97. Titchmarsh, E.C.: On van der Corput's method and the zeta-function of Riemann. Quart. J. Math. (Oxford) **5**, 195–210 (1934)
98. Titchmarsh, E.C.: The Theory of the Riemann Zeta-Function, 2nd edn. Oxford University Press, Oxford (1986)
99. Tsang, K.-M.: Higher-power moments of $\Delta(x)$, $E(t)$ and $P(x)$. Proc. London Math. Soc. **65**, 65–84 (1992)
100. Tsang, K.-M.: Recent progress on the Dirichlet divisor problem and the mean square of the Riemann zeta-function. Sci. China Math. **53**, 2561–2572 (2010)
101. von Mangoldt, H.: Zu Riemann's Abhandlung "Über die Anzahl ...". Crelle's J. **114**, 255–305 (1895)
102. Voronoï, G.F.: Sur une fonction transcendante et ses applications à la sommation de quelques séries. Ann. École Normale **21**(3), 207–268 (1904); ibid. **21**(3), 459–534 (1904)
103. Watson, G.N.: A Treatise on the Theory of Bessel Functions, 2nd edn. Cambridge University Press, Cambridge (1944)
104. Watt, N.: Kloosterman sums and a mean value theorem for Dirichlet polynomials. J. Number Theory **53**, 179–210 (1995)
105. Watt, N.: A note on the mean square of $|\zeta(\frac{1}{2}+it)|$. J. London Math. Soc. **82**(2), 279–294 (2010)
106. Weil, A.: On some exponential sums. Proc. Nat. Acad. Sci. USA **34**, 258–284 (1948)
107. Zhai, W.: On higher-power moments of $E(t)$. Acta Arith. **115**(4), 329–348 (2004)
108. Zhai, W.: On higher-power moments of $\Delta(x)$. Acta Arith. **112**(4), 367–395 (2004); Acta Arith. II **114**(1), 35–54 (2004); Acta Arith. III **118**(3), 263–281 (2005)

Explicit Bounds Concerning Non-trivial Zeros of the Riemann Zeta Function

Mehdi Hassani

Dedicated to Professor Hari M. Srivastava

Abstract In this paper, we get explicit upper and lower bounds for γ_n, where $0 < \gamma_1 < \gamma_2 < \gamma_3 < \cdots$ are consecutive ordinates of non-trivial zeros $\rho = \beta + \mathrm{i}\gamma$ of the Riemann zeta function. Meanwhile, we obtain the asymptotic relation $\gamma_n \log^2 n - 2\pi n \log n \sim 2\pi n \log\log n$ as $n \to \infty$.

1 Introduction

The Riemann zeta function is defined for $\mathrm{Re}(s) > 1$ by $\zeta(s) = \sum_{n=1}^{\infty} n^{-s}$ and extended by analytic continuation to the complex plan with a simple pole at $s = 1$ with residues 1. It is known [3, 7] that

$$N(T) := \sum_{\substack{0<\gamma\leqslant T \\ \zeta(\beta+\mathrm{i}\gamma)=0}} 1 = \frac{T}{2\pi}\log\frac{T}{2\pi \mathrm{e}} + O(\log T). \tag{1}$$

As a consequence of (1) we get

$$\sum_{\substack{0<\gamma\leqslant T \\ \zeta(\beta+\mathrm{i}\gamma)=0}} \frac{1}{\gamma} = \frac{1}{4\pi}\log^2 T - \frac{\log(2\pi)}{2\pi}\log T + E(T),$$

with $E(T) = O(1)$. Recently, we obtained an explicit form of this approximate formula by proving that $\frac{3}{50} < E(T) < \frac{109}{250}$ for any $T \geqslant \gamma_1$ (see [1, Theorem 1]), where

M. Hassani (✉)
Department of Mathematics, University of Zanjan, University Blvd., 45371-38791, Zanjan, Iran
e-mail: mehdi.hassani@znu.ac.ir

G.V. Milovanović and M.Th. Rassias (eds.), *Analytic Number Theory, Approximation Theory, and Special Functions*, DOI 10.1007/978-1-4939-0258-3_2,

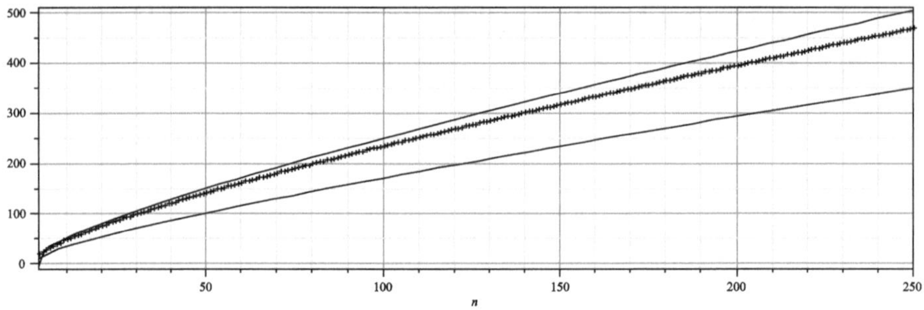

Fig. 1 Graph of the point set (n, γ_n) for $2 \leqslant n \leqslant 250$ and functions $\frac{2\pi n}{\log n}(1 + a\frac{\log\log n}{\log n})$ with $a = 3/4$ and $a = 5/2$

$$\gamma_1 = \min\{\gamma > 0 : \zeta(\beta + \mathrm{i}\gamma) = 0\} \approxeq 14.134725141734693790457251983562.$$

More generally, we set $0 < \gamma_1 < \gamma_2 < \gamma_3 < \cdots$ to be consecutive ordinates of the imaginary parts of non-trivial zeros $\rho = \beta + \mathrm{i}\gamma$ of $\zeta(s)$. Another consequence of (1) is

$$\gamma_n \sim \frac{2\pi n}{\log n}, \quad \text{as } n \to \infty.$$

Our intention in writing this note is to obtain explicit forms of this approximate formula. More precisely, we show the following.

Theorem 1.1. *For any integer $n \geqslant 5$ we have*

$$\frac{2\pi n}{\log n}\left(1 + \frac{3}{4}\frac{\log\log n}{\log n}\right) \leqslant \gamma_n \leqslant \frac{2\pi n}{\log n}\left(1 + \frac{5}{2}\frac{\log\log n}{\log n}\right). \tag{2}$$

Figure 1 shows graph of the point set (n, γ_n) for $2 \leqslant n \leqslant 250$, and lower and upper bounds appeared in (2). We note that the left-hand side of (2) is valid for $2 \leqslant n \leqslant 4$, too.

One may obtain better bounds for γ_n by using numerical information, which we obtain during proof of Theorem 1.1. More precisely, by considering Tables 1 and 2, we have the following.

Theorem 1.2. *Assume that we choose pairs λ and n_λ from Table 1, and also we choose pairs η and n_η from Table 2. Then, we have*

$$\frac{2\pi n}{\log n}\left(1 + \frac{\lambda}{2\pi}\frac{\log\log n}{\log n}\right) \leqslant \gamma_n \quad \textit{and} \quad \gamma_n \leqslant \frac{2\pi n}{\log n}\left(1 + \frac{\eta}{2\pi}\frac{\log\log n}{\log n}\right),$$

respectively, for $n \geqslant n_\lambda$ and for $n \geqslant n_\eta$.

Table 1 Some values of λ and n_λ for which the inequality (11) is valid for $n \geqslant n_\lambda$

λ	$n_\lambda \approxeq$	λ	$n_\lambda \approxeq$
-2π	3.9	$3\pi/2$	4984.5
$-\pi$	5.1	$5\pi/3$	392062.1
-1	7.8	$7\pi/4$	138610176.5
0	10.7	$9\pi/5$	2499273431483.9
1	16.7	$11\pi/6$	1095110510643676001902 50.3
π	97.1	5.795	876581819433015771165641491644046075.5

Table 2 Some values of η and n_η for which the inequality (12) is valid for $n \geqslant n_\eta$

η	$n_\eta \approxeq$	η	$n_\eta \approxeq$
20π	8.8	6π	1197.1
10π	11.7	5π	26245.8
8π	64.3	4π	80727920.5
7π	217.7	3π	742199235320620698359223515347 87.7

On the other hand, we mention that the constants $\frac{3}{4}$ and $\frac{5}{2}$ in Theorem 1.1, as more as, the constants $\frac{\lambda}{2\pi}$ and $\frac{\eta}{2\pi}$ in Theorem 1.2, are not optimal. More precisely, if we let

$$R_n := \frac{\frac{\gamma_n}{\frac{2\pi n}{\log n}} - 1}{\frac{\log\log n}{\log n}}, \tag{3}$$

then Theorem 1.1 yields that $\frac{3}{4} \leqslant R_n \leqslant \frac{5}{2}$ for any integer $n \geqslant 5$. But, the proof of above theorems includes an argument in its heart, which implies that $\lim_{n\to\infty} R_n = 1$. Indeed, we show the following.

Theorem 1.3. *Let*

$$\Lambda_n = \frac{\gamma_n \log^2 n - 2\pi n \log n}{n \log\log n}. \tag{4}$$

Then, we have $\lim_{n\to\infty} \Lambda_n = 2\pi$.

Corollary 1.1. *For any real* $\varepsilon \in (0, 1)$, *there exists positive integer* n_ε *such that for* $n \geqslant n_\varepsilon$ *we have*

$$\frac{2\pi n}{\log n}\left(1 + (1-\varepsilon)\frac{\log\log n}{\log n}\right) \leqslant \gamma_n \leqslant \frac{2\pi n}{\log n}\left(1 + (1+\varepsilon)\frac{\log\log n}{\log n}\right).$$

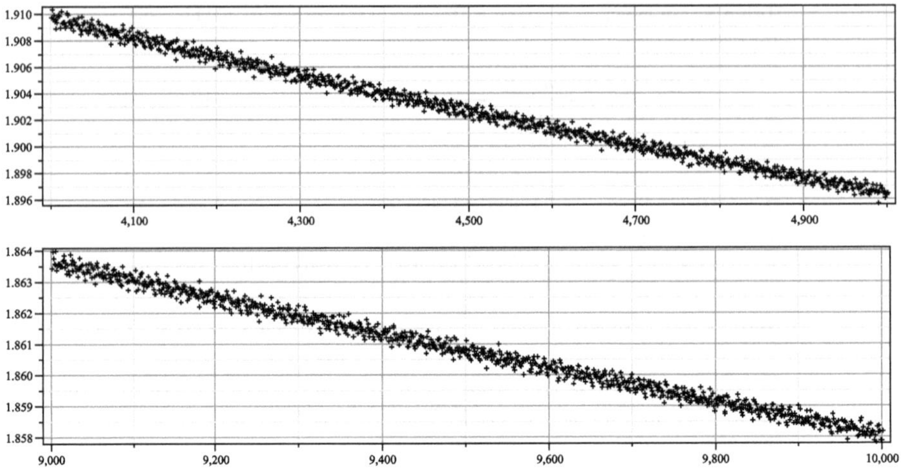

Fig. 2 Graph of the pointset (n, R_n) for $4000 \leqslant n \leqslant 5000$ and $9000 \leqslant n \leqslant 10000$, where R_n is defined by (3)

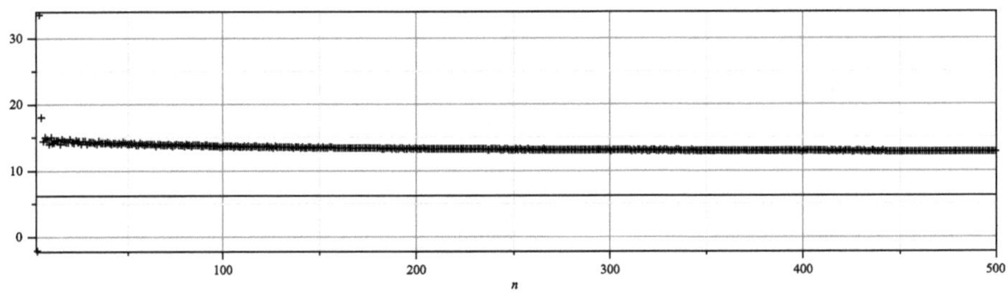

Fig. 3 Graph of the point set (n, Λ_n) for $2 \leqslant n \leqslant 500$, where Λ_n is defined by (4), and horizontal line at height 2π

Remark 1.1. Figure 2 pictures some values of R_n for several values of n. As our computations show, one may have the inequality $R_n > 1$ for $n \geqslant 3$. This means that one may have the validity of the left-hand side of (2) with 1 instead of $\frac{3}{4}$, for any integer $n \geqslant 3$. This conjecture is pictured in Fig. 3 in another point of view, where we plot values of Λ_n for $2 \leqslant n \leqslant 500$ and horizontal line at height 2π. Also, it seems that there exists a positive integer $m \approxeq 250$ such that $R_{n+m} \geqslant R_n$ for any integer $n \geqslant 3$.

Remark 1.2. The truth of Corollary 1.1 asserts that as $n \to \infty$ we have

$$\gamma_n = \frac{2\pi n}{\log n}\left(1 + (1 + o(1))\frac{\log\log n}{\log n}\right).$$

One may ask for such asymptotic expansions with more precise terms.

In the next two sections, we prove our results. To generate figures which appeared on present paper, as more as, during proofs, we will do several computations running over the numbers γ_n, all of which have been done by using Maple software and are based on the tables of zeros of the Riemann zeta function due to Odlyzko [4].

2 Lambert W Function, the Key of Proof

The main idea to get explicit results similar to (2) is applying an explicit version of the Riemann–von Mangoldt formula (1). This can be found in the following result due to Rosser, which is Theorem 19 of [6].

Proposition 2.1. *For any $T \geqslant 2$ we have $|N(T) - F(T)| \leqslant R(T)$, with*

$$F(T) = \frac{T}{2\pi}\log\frac{T}{2\pi e} + \frac{7}{8} \quad \textit{and} \quad R(T) = \frac{137}{1000}\log T + \frac{443}{1000}\log\log T + \frac{397}{250}.$$

For our purpose, we need to modify the truth of above proposition as follows. For the whole text, we set

$$\ell = \frac{14}{25} \quad \text{and} \quad u = \frac{11}{50}.$$

Lemma 2.1. *Let*

$$L(T) = \frac{1}{2\pi}T\log T - \ell T \quad \textit{and} \quad U(T) = \frac{1}{2\pi}T\log T - uT. \tag{5}$$

Then, for $T \geqslant \gamma_1 - 10^{-5}$ we have

$$L(T) \leqslant N(T) \leqslant U(T). \tag{6}$$

Moreover, $U(T)$ and $L(T)$ are strictly increasing for $T \geqslant e^{2\pi u-1} \approxeq 1.465653$ and $T \geqslant e^{2\pi\ell-1} \approxeq 12.411008$, respectively.

Proof. We consider Proposition 2.1 to write $F(T)-R(T) \leqslant N(T) \leqslant F(T)+R(T)$ for $T \geqslant 2$. On the other hand, for $T \geqslant \gamma_1 - 10^{-5}$ we have $F(T)+R(T) \leqslant U(T)$ and $L(T) \leqslant F(T) - R(T)$. This proves both sides of (6). Monotonicity of the functions $U(T)$ and $L(T)$ is straightforward. □

The following lemma brings lower and upper bounds for $N(T)$ to bounds for γ_n in terms of inverses of mentioned bounds for $N(T)$.

Lemma 2.2. *Assume that $L(T)$ and $U(T)$ are defined as in (5), and denote by $L^{-1}(T)$ and $U^{-1}(T)$ their inverses, respectively. Then, for any integer $n \geqslant 1$ we have*

$$U^{-1}(n) \leqslant \gamma_n \leqslant L^{-1}(n). \tag{7}$$

Proof. Assume that $n \geqslant 1$ is any arbitrary integer and $\delta \in (0, 1)$ is any arbitrary real. We have $N(\gamma_n) = n$. Thus, we get $N(\gamma_n + \delta) \geqslant n$ and $N(\gamma_n - \delta) \leqslant n - 1$. Therefore, we obtain

$$1 + N(\gamma_n - \delta) \leqslant n \leqslant N(\gamma_n + \delta). \tag{8}$$

Right-hand sides of (6) and (8) give $\gamma_n + \delta \geqslant U^{-1}(n)$. Thus, we get $\gamma_n \geqslant U^{-1}(n)$. Similarly, left-hand sides of (6) and (8) give $L(\gamma_n - \delta) \leqslant N(\gamma_n - \delta) \leqslant n - 1 < n$. So, we get $\gamma_n - \delta \leqslant L^{-1}(n)$, and this implies validity of $\gamma_n \leqslant L^{-1}(n)$. □

In order to use inequalities (7), we need formulas for the inverses of the functions $L(T)$ and $U(T)$. This may be done in terms of the Lambert W function $W(x)$, which is defined by the relation $W(x)\mathrm{e}^{W(x)} = x$ for $x \in [-\mathrm{e}^{-1}, +\infty)$. The Lambert W function has the asymptotic expansion $W(x) = \log x + O(\log\log x)$ as $x \to \infty$, (see [5, p. 111]). The following lemma summarizes what we need about the inverses of the functions $L(T)$ and $U(T)$.

Lemma 2.3. *Assume that a and b are some positive real numbers, and let*

$$f(T) = \frac{1}{a} T \log T - bT.$$

We denote the inverse function of f by f^{-1}. Then, for $T \geqslant \mathrm{e}^{ab-1}$ the function f is strictly increasing and we have

$$f^{-1}(T) = \frac{aT}{W(a\mathrm{e}^{-ab}T)}. \tag{9}$$

In particular, as $T \to +\infty$, we obtain $f^{-1}(T) \sim aT/\log T$.

Proof. Assume that $T > 0$. Then, by definition of the Lambert W function, we imply that $f(\mathrm{e}^{W(a\mathrm{e}^{-ab}T)+ab}) = T$ or equivalently $f^{-1}(T) = \mathrm{e}^{W(a\mathrm{e}^{-ab}T)+ab}$. Definition of the Lambert W function also gives that $aT = W(a\mathrm{e}^{-ab}T)\mathrm{e}^{W(a\mathrm{e}^{-ab}T)+ab}$. Thus, we obtain (9). The asymptotic relation comes from $W(a\mathrm{e}^{-ab}T) \sim \log T$, which is valid as $T \to +\infty$. □

Finally, to get our desired explicit results, we need some explicit bounds for the Lambert W function. The following proposition, which is Theorem 2.8 of [2], offers such sharp bounds.

Proposition 2.2. *Assume that $\alpha > 0$ is real, and let*

$$\omega_\alpha(x) := \log x - \log\log x + \alpha \frac{\log\log x}{\log x}.$$

Then, for every $x \geqslant \mathrm{e}$ we have

$$\omega_{\frac{1}{2}}(x) \leqslant W(x) \leqslant \omega_{\frac{\mathrm{e}}{\mathrm{e}-1}}(x), \tag{10}$$

with equality only for $x = \mathrm{e}$.

3 Proof of Results

3.1 *Proof of the Left-Hand Side of* (2)

We let $c_u = 2\pi \mathrm{e}^{-2\pi u}$. By applying the validity of Lemma 2.3, considering the left-hand side of (7), and considering the right-hand side of (10), we obtain

$$\gamma_n \geqslant U^{-1}(n) = \frac{2\pi n}{W(c_u n)} \geqslant \frac{2\pi n}{\omega_{\frac{\mathrm{e}}{\mathrm{e}-1}}(c_u n)},$$

for $c_u n \geqslant \mathrm{e}$ or equivalently for $n \geqslant \frac{\mathrm{e}}{c_u} \approxeq 1.7$. Moreover, by computation, for any integer $n \geqslant 1$ we get

$$\gamma_n \geqslant \frac{2\pi n}{\omega_{\frac{\mathrm{e}}{\mathrm{e}-1}}(c_u n)} := g(n),$$

say. We let

$$h(n) := \frac{g(n) - \frac{2\pi n}{\log n}}{\frac{n \log\log n}{\log^2 n}}.$$

Now, we note that the function $h : (\mathrm{e}, +\infty) \longrightarrow (-\infty, 2\pi)$ defined by $h(n)$ is continuous and strictly increasing. Moreover, we have

$$\lim_{n \to \mathrm{e}^+} h(n) = -\infty \quad \text{and} \quad \lim_{n \to +\infty} h(n) = 2\pi.$$

Therefore, for any real $\lambda \in (-\infty, 2\pi)$, there exists unique $n_\lambda \in (\mathrm{e}, +\infty)$ such that $h(n) \geqslant \lambda$ for $n \geqslant n_\lambda$ with equality only for $n = n_\lambda$. Hence, for $n \geqslant n_\lambda$ we obtain

$$\gamma_n \geqslant \frac{2\pi n}{\log n} + \lambda \frac{n \log\log n}{\log^2 n}. \tag{11}$$

In Table 1 we list some values of λ and related values of n_λ. We use information of this table choosing $\lambda = \frac{3\pi}{2}$, from which we obtain the inequality

$$\gamma_n \geqslant \frac{2\pi n}{\log n} + \frac{3\pi n \log\log n}{2\log^2 n},$$

for $n \geqslant 4985$. By computation, we confirm validity of it for $2 \leqslant n \leqslant 4984$, too. This completes the proof of left-hand side of (2).

3.2 *Proof of the Right-Hand Side of* (2)

Let $c_\ell = 2\pi \mathrm{e}^{-2\pi\ell}$. We use the validity of Lemma 2.3, the right-hand side of (7), and the left-hand side of (10) to get

$$\gamma_n \leqslant L^{-1}(n) = \frac{2\pi n}{W(c_\ell n)} \leqslant \frac{2\pi n}{\omega_{1/2}(c_\ell n)},$$

for $c_\ell n \geqslant \mathrm{e}$ or equivalently for $n \geqslant \frac{\mathrm{e}}{c_\ell} \approxeq 14.6$. As more as, by computation, for any integer $n \geqslant 8$, we obtain

$$\gamma_n \leqslant \frac{2\pi n}{\omega_{1/2}(c_\ell n)} := v(n),$$

say. We set

$$z(n) := \frac{v(n) - \frac{2\pi n}{\log n}}{\frac{n \log\log n}{\log^2 n}}.$$

Also, we let

$$y_1 := \lim_{n \to 1/c_\ell +} \frac{1}{z(n)} = \frac{\log(-\log c_\ell)}{2\pi \log c_\ell} \approxeq -0.049167.$$

We note that the function $y : (1/c_\ell, +\infty) \to (y_1, 1/(2\pi))$ defined by $y(n) = 1/z(n)$ is continuous and strictly increasing. Thus, there exists unique $n_0 > 1/c_\ell$ such that $y(n_0) = 0$. By computation, we observe that $n_0 \approxeq 7.745051$. Now, we note that the function $z : (n_0, +\infty) \to (2\pi, +\infty)$ defined by $z(n)$ is continuous and strictly decreasing. Moreover, we have

$$\lim_{n \to n_0^+} z(n) = +\infty \quad \text{and} \quad \lim_{n \to +\infty} z(n) = 2\pi.$$

Therefore, for any $\eta \in (2\pi, +\infty)$, there exists unique $n_\eta \in (n_0, +\infty)$ such that $z(n) \leqslant \eta$ for $n \geqslant n_\eta$ with equality only for $n = n_\eta$, and consequently, for $n \geqslant n_\eta$ we get

$$\gamma_n \leqslant \frac{2\pi n}{\log n} + \eta \frac{n \log\log n}{\log^2 n}. \tag{12}$$

Table 2 includes some values of η and related values of n_η. Considering our computational tools, we choose $\eta = 5\pi$ from this table, from which for $n \geqslant 26246$ we obtain the inequality

$$\gamma_n \leqslant \frac{2\pi n}{\log n} + \frac{5\pi n \log\log n}{\log^2 n}.$$

By computation, we confirm validity of it for $5 \leqslant n \leqslant 26245$, too. This completes the proof of right-hand side of (2).

3.3 *Proof of Theorem 1.3*

We note that inequalities (11) and (12) imply

$$\liminf_{n\to\infty} \Lambda_n \geqslant 2\pi \quad \text{and} \quad \limsup_{n\to\infty} \Lambda_n \leqslant 2\pi,$$

respectively. This gives assertion of Theorem 1.3.

References

1. Hassani, M.: On a sum related by non-trivial zeros of the Riemann zeta function. Appl. Math. E-Notes **12**, 1–4 (2012)
2. Hoorfar, A., Hassani, M.: Inequalities on the lambert W function and hyperpower function. J. Inequal. Pure Appl. Math. **9**(2), 5 pp. (2008) (Article 51)
3. Ivić, A.: The Riemann Zeta Function. Wiley, New York (1985)
4. Odlyzko, A.M.: http://www.dtc.umn.edu/~odlyzko/zeta-tables/index.html
5. Olver, F.W.J., Lozier, D.W., Boisvert, R.F., Clark, C.W.: NIST Handbook of Mathematical Functions. Cambridge University Press, Cambridge (2010)
6. Rosser, J.B.: Explicit bounds for some functions of prime numbers. Amer. J. Math. **63**, 211–232 (1941)
7. Titchmarsh, E.C.: The Theory of the Riemann Zeta Function, 2nd edn. Oxford University Press, Oxford (1986) (Revised by Heath-Brown, D.R.)

On the Value-Distribution of Logarithmic Derivatives of Dirichlet L-Functions

Yasutaka Ihara and Kohji Matsumoto

Dedicated to Professor Hari M. Srivastava

Abstract We shall prove an unconditional basic result related to the value- distributions of $\{(L'/L)(s,\chi)\}_\chi$ and of $\{(\zeta'/\zeta)(s+i\tau)\}_\tau$, where χ runs over Dirichlet characters with prime conductors and τ runs over $\mathbf{R}$. The result asserts that the expected density function common for these distributions are in fact the density function in an appropriate sense. Under the generalized Riemann hypothesis, stronger results have been proved in our previous articles, but our present result is unconditional.

1 Introduction and Statement of the Result

The present paper is a part of authors' research on the value-distribution of L-functions over global fields and is regarded as a supplement of our former papers [3, 7]. In [3], we defined and studied the "would-be density function" $M_\sigma(w)$ ($\sigma > 1/2$) for the value-distribution of $L'/L(s,\chi)$ on the complex plane $\mathbf{C}$ for certain family of L-functions over any global field (s: fixed with $\mathrm{Re}(s) = \sigma$) and established the expected connection under some restrictive hypothesis. This was generalized and strengthened in [8] under GRH, the generalized Riemann hypothesis. In [7] we treated the analogous "would-be" density function $\mathcal{M}_\sigma(w)$ for the log L case, and in this case, when the base field is the rational number field

Y. Ihara
RIMS, Kyoto University, Kitashirakawa-Oiwakecho, Sakyo-ku, Kyoto, 606-8502, Japan
e-mail: ihara@kurims.kyoto-u.ac.jp

K. Matsumoto (✉)
Graduate School of Mathematics, Nagoya University, Furocho, Chikusa-ku, Nagoya, 464-8602, Japan
e-mail: kohjimat@math.nagoya-u.ac.jp

G.V. Milovanović and M.Th. Rassias (eds.), *Analytic Number Theory, Approximation Theory, and Special Functions*, DOI 10.1007/978-1-4939-0258-3_3,

$\mathbf{Q}$, we were able to obtain an unconditional result on the expected connection. The purpose of the present paper is to show that a parallel unconditional result for the L'/L case over $\mathbf{Q}$ can be obtained with but small modifications of the methods used in [7].

Let $s = \sigma + i\tau$ be a complex variable, $\zeta(s)$ be the Riemann zeta function, χ a Dirichlet character with prime conductor, and $L(s,\chi)$ the associated Dirichlet L-function. We study the value-distribution of $(L'/L)(s,\chi)$ when χ varies or $(\zeta'/\zeta)(s+i\tau')$ when τ' varies. In the latter case, defining $\chi_{\tau'}(n) = n^{-i\tau'}$ ($\tau' \in \mathbf{R}$, $n = 1, 2, \ldots$), we may regard that $\zeta(s+i\tau') = L(s,\chi_{\tau'})$ and the "character" $\chi_{\tau'}$ varies. Therefore our object consists of two types of infinite families of characters, (FI) all Dirichlet characters χ of prime conductors, or (FII) characters of the form $\chi_{\tau'}$, $\tau' \in \mathbf{R}$.

Let $M_\sigma(w)$ for $\sigma > 1/2$ be the function of $w \in \mathbf{C}$ defined in [3]. The construction of $M_\sigma(w)$ will be reviewed at the beginning of Sect. 2. Here we take $K = \mathbf{Q}$ (in terms of [8], this corresponds to the function $M_\sigma(w)$ for "Case 1", $K = \mathbf{Q}$, $P_\infty = (\infty)$).

We shall prove the following theorem.

Theorem 1.1. *Let $s = \sigma + i\tau \in \mathbf{C}$ be fixed, with $\sigma = \mathrm{Re}\, s > 1/2$. Then the equality*

$$\mathrm{Avg}_\chi \Phi\left(\frac{L'}{L}(s,\chi)\right) = \int_{\mathbf{C}} M_\sigma(w)\Phi(w)|dw| \tag{1}$$

holds simultaneously for both families (FI) *and* (FII), *where $|dw| = du\,dv/2\pi$ for $w = u + iv$, the meaning of* Avg_χ *is defined below, and the test function Φ is one of the following:*

(i) *Φ is any continuous bounded function.*
(ii) *Φ is the characteristic function of either a compact subset of $\mathbf{C}$ or the complement of such a subset.*

Finally, when $s = 1$, (at least) in case of the family (FI), *the test function Φ can be any continuous function of at most polynomial growth.*

The above statement for $\sigma > 1$ and stronger but conditional results for $\sigma > 1/2$ under GRH [over more general base fields for the family (FI)] were already shown in [3, 6, 8] (cf. also a survey article [5]). The purpose of the present paper is to prove this theorem unconditionally for any $\sigma > 1/2$.

The definition of Avg_χ is as follows.

Case (FI). For any prime $f (> 2)$, let $X(f)$ denote the set of all primitive Dirichlet characters whose conductor is precisely f, and $X'(f) = X'(f,s)$ be the subset of $X(f)$ consisting of all χ such that $L(s,\chi) \neq 0$ for our fixed s. By a theorem of Montgomery [13] it satisfies

$$\lim_{f\to\infty} \frac{|X'(f)|}{|X(f)|} = 1. \tag{2}$$

(For any finite set A we denote by $|A|$ its cardinality.) For any complex-valued function $\phi(\chi)$ on $X'(f)$, we define the averages

$$\mathrm{Avg}_{X'(f)}\phi(\chi) = \frac{1}{|X(f)|}\sum_{\chi\in X'(f)}\phi(\chi), \tag{3}$$

$$\mathrm{Avg}_{f\leq m}\phi(\chi) = \frac{1}{\pi(m)}\sum_{f\leq m}\mathrm{Avg}_{X'(f)}\phi(\chi), \tag{4}$$

where m is any positive integer, f runs over all odd prime numbers up to m, and $\pi(m)$ denotes the number of prime numbers up to m. Now define

$$\mathrm{Avg}_{\chi}\phi(\chi) = \lim_{m\to\infty}\left(\mathrm{Avg}_{f\leq m}\phi(\chi)\right). \tag{5}$$

When we state a formula for Avg_χ, it will always include the claim that the limit exists. We remark here that the main statement of the theorem deals only with the averages of those $\phi(\chi)$ which are *bounded* on the union of $X'(f)$ over all f (because the test function Φ is bounded). Therefore, if we replace $X'(f)$ by a smaller subset preserving the condition (2), the average (3) [resp. (4)] changes only by a quantity which tends to 0 as $f\to\infty$ (resp. $m\to\infty$), hence the limit average (5) remains the same (e.g., the subset "$X'(f)$" in [7] or the subset denoted by $X''(f)$ defined below in Sect. 2 used for the proof). As regards the additional statement for $s=1$, note that $X'(f,1)=X(f)$.

Case (FII). The definition of Avg_χ in this case is simply

$$\mathrm{Avg}_{\chi}\phi(\chi_{\tau'}) = \lim_{T\to\infty}\frac{1}{2T}\int_{-T}^{T}\phi(\chi_{\tau'})d\tau', \tag{6}$$

for any integrable function $\phi(\chi_{\tau'})$ of τ'.

A closely related problem is the study on the value-distribution of $\log L(s,\chi)$. In [7], we have constructed a continuous nonnegative density function $\mathcal{M}_\sigma(w)$ parametrized by $\sigma>1/2$ and established the following theorem.

Theorem 1.2 ([7]). *For any $s\in\mathbf{C}$ with $\sigma=\mathrm{Re}(s)>1/2$,*

$$\mathrm{Avg}_{\chi}\Phi(\log L(s,\chi)) = \int_{\mathbf{C}}\mathcal{M}_\sigma(w)\Phi(w)|dw| \tag{7}$$

holds simultaneously for both families (FI) *and* (FII) *for a suitable choice of the branch of the logarithm, a suitable definition of the average* Avg_χ, *where Φ is as in Theorem 1.1.*

Our Theorem 1.1 implies that the exact analogue of Theorem 1.2 holds in the L'/L case.

To prove these unconditional results, our method is to apply several mean value results on L-functions. As for the $\log L$ case, such mean value theorems were obtained in [7] to prove Theorem 1.2. It is possible to use the same mean value theorems in our present situation because L'/L can be written as an integral involving $\log L$ in the integrand, by using the Cauchy integral formula. Note that the idea of applying the Cauchy integral formula in such a situation already appeared in Kershner and Wintner [11] in the (FII) case (see Remark 3.1).

In the following sections we will prove Theorem 1.1. Since the basic structure of the proof is similar to those developed in [3,7], we will only point out the differences from those and omit the details.

2 Proof in the Case (FI)

First of all, we review how to construct the density function $M_\sigma(w)$ (in the case $K = \mathbf{Q}$) in [3]. Let p be a prime number and

$$c_{\sigma,p} = \frac{-\log p}{p^{2\sigma}-1}, \qquad r_{\sigma,p} = \frac{p^\sigma \log p}{p^{2\sigma}-1}.$$

Write $w \in \mathbf{C}$ as $w = c_{\sigma,p} + re^{i\theta}$, where $r \geq 0$ and $\theta \in \mathbf{R}$, and define $M_{\sigma,p}$ by

$$M_{\sigma,p}(w) = \frac{p^{2\sigma}-1}{|p^\sigma - e^{i\theta}|^2} \cdot \frac{\delta(r - r_{\sigma,p})}{r},$$

where $\delta(\cdot)$ stands for the usual one-dimensional Dirac delta function. Let $P = P_y$ be the set of all prime numbers not greater than y, and define $M_{\sigma,P}$ as the convolution product of $M_{\sigma,p}$ $(p \in P)$ with respect to $|dw|$. Then, for $\sigma > 1/2$, $M_{\sigma,P}(w)$ converges uniformly to a nonnegative real-valued C^∞-function as $y \to \infty$ [3, Theorem 2], which we denote by $M_\sigma(w)$.

Now we start the proof of Theorem 1.1. As mentioned in Sect. 1, the assertion of Theorem 1.1 was already shown in [3] when $\sigma > 1$. Therefore it is sufficient to consider the case $1/2 < \sigma \leq 1$. The final statement for $s = 1$ then follows directly by combining [9, Sect. 5] (Theorem 5) with [8, Sect. 5] (Lemma A).

As in [7, Sect. 7], take a number σ_0 satisfying $1/2 < \sigma_0 < 1$ and $\sigma_0 \leq \sigma$, and let $0 < 3\varepsilon_1 < \sigma_0 - 1/2$, $\alpha_0 = \sigma_0 - \varepsilon_1$, $\alpha_1 = \sigma_0 - 2\varepsilon_1$, $\alpha_2 = 1/2 + \varepsilon_1$. Then $1/2 < \alpha_2 < \alpha_1 < \alpha_0 < \sigma_0 < 1$. These constants are regarded to be fixed, and the implied constants of Landau's O-symbol or Vinogradov's symbol below may depend on them.

Let $T = |\tau|+2$, and let $X''(f)$ be the set of all $\chi \in X(f)$ for which $L(s',\chi) \neq 0$ for any $s' = \sigma' + i\tau'$ in the region $\sigma' \geq \sigma_0$, $|\tau'| \leq T$. Then obviously, $X''(f) \subset X'(f)$ and Proposition 2.1 of [7] (which is based on a theorem of Montgomery [13]) asserts that

$$\lim_{f\to\infty}\frac{|X''(f)|}{|X(f)|}=1. \tag{8}$$

So it suffices to prove the theorem where the average is defined with respect to $X''(f)$.

We study the case $\Phi=\psi_z$ first, where $z\in\mathbf{C}$ and ψ_z is the additive character of $\mathbf{C}$ defined by $\psi_z(w)=\exp(i\,\mathrm{Re}(\bar{z}w))$. When once this case is established, we can deduce the assertion of the case (FI) of Theorem 1.1 for general Φ satisfying (i) and (ii), quite similarly to the argument in [7, Sect. 9] (see also Remark 3.2).

In the case $\Phi=\psi_z$, the right-hand side of (1) is equal to

$$\int_{\mathbf{C}} M_\sigma(w)\psi_z(w)|dw|=\tilde{M}_\sigma(z),$$

the Fourier dual of $M_\sigma(z)$ (see Theorem 3 of [3]). Since ψ_z is bounded, the average (3) [and so (4), (5)] does not change if we replace $X'(f)$ by $X''(f)$. Therefore, noting $|X(f)|=f-2$ for any odd prime f, we find that what we have to prove in this case is

$$\lim_{m\to\infty}\frac{1}{\pi(m)}\sum_{f\le m}\frac{1}{f-2}\sum_{\chi\in X''(f)}\psi_z\left(\frac{L'}{L}(s,\chi)\right)=\tilde{M}_\sigma(z). \tag{9}$$

First we introduce the "finite truncation" of L-functions. Let $1<y<m$, $P=P_y$ as above, and write $P=\{p_1,\dots,p_r\}$, $r=\pi(y)\sim y/\log y$. Define

$$L_P(s,\chi)=\prod_{p\in P}(1-\chi(p)p^{-s})^{-1}$$

and

$$\log L_P(s,\chi)=-\sum_{p\in P}\mathrm{Log}(1-\chi(p)p^{-s}),$$

where "Log" means the principal branch. By [3], $M_{\sigma,P}(w)$ is the density function for the value-distribution of $(L'_P/L_P)(s,\chi)$ and let $\tilde{M}_{\sigma,P}(z)$ be its Fourier dual.

The starting point of the proof of (9) is the following inequality:

$$\left|\frac{1}{\pi(m)}\sum_{f\le m}\frac{1}{f-2}\sum_{\chi\in X''(f)}\psi_z\left(\frac{L'}{L}(s,\chi)\right)-\tilde{M}_\sigma(z)\right|$$

$$\le\left|\frac{1}{\pi(m)}\sum_{f\le m}\frac{1}{f-2}\sum_{\chi\in X''(f)}\left\{\psi_z\left(\frac{L'}{L}(s,\chi)\right)-\psi_z\left(\frac{L'_P}{L_P}(s,\chi)\right)\right\}\right|$$

$$+\left|\frac{1}{\pi(m)}\sum_{f\le m}\frac{1}{f-2}\sum_{\chi\in X''(f)}\psi_z\left(\frac{L'_P}{L_P}(s,\chi)\right)-\tilde{M}_{\sigma,P}(z)\right|$$
$$+\left|\tilde{M}_{\sigma,P}(z)-\tilde{M}_\sigma(z)\right|$$
$$=X_P^{ld}(z)+Y_P^{ld}(z)+Z_P^{ld}(z), \tag{10}$$

say. This is an analogue of [7, (125)], and "ld"s (which stand for the "logarithmic derivative") are attached only for the purpose of distinguishing our notation from that in [7].

In order to estimate $X_P^{ld}(z)$, we first introduce some more notation. For each Dirichlet character χ, from the half plane $\{s' \mid \sigma' > 1/2\}$, we exclude all the segments of the form $\{\sigma' + i\,\mathrm{Im}\,\rho \mid 1/2 < \sigma' \le \mathrm{Re}\,\rho\}$ (for all possible zeros ρ of $L(s',\chi)$ with $\mathrm{Re}\,\rho > 1/2$) and denote the remaining region by G_χ. In the region G_χ, we can define the value of $\log L(s',\chi)$ by the analytic continuation along the horizontal path $\{\sigma'' + i\tau' \mid \sigma'' \ge \sigma'\}$. Define

$$R_P(s',\chi)=\log L(s',\chi)-\log L_P(s',\chi)$$

for $s' \in G_\chi(\alpha_1) = G_\chi \cap \{\sigma' > \alpha_1\}$. Let c and δ be fixed small positive numbers, and let $\beta_0 = \beta_0(\delta) > 1$, $\beta_1 = \beta_1(\delta) = 2\beta_0$, $H(\tau)$, $Q_0(\tau)$, $Q_1(\tau)$, $f_P(s',\chi)$, $F_P(\tau,\chi)$ be as in [7, Sect. 7]. The distance between the boundaries of the two sets $Q_0(\tau)$ and $Q_1(\tau)$ is $\varepsilon_2 = \min\{\varepsilon_1, c\}$. Let $X_1(f)$ be the set of all $\chi \in X''(f)$ such that

$$F_P(\tau,\chi)\ge\pi\left(\frac{\varepsilon_2}{2}\right)^2\left(\frac{\delta}{2}\right)^2, \tag{11}$$

and $X_2(f)$ its complement in $X''(f)$, that is, all those $\chi \in X''(f)$ satisfying

$$F_P(\tau,\chi)<\pi\left(\frac{\varepsilon_2}{2}\right)^2\left(\frac{\delta}{2}\right)^2. \tag{12}$$

We divide

$$\sum_{\chi\in X''(f)}\left\{\psi_z\left(\frac{L'}{L}(s,\chi)\right)-\psi_z\left(\frac{L'_P}{L_P}(s,\chi)\right)\right\}$$
$$=\sum_{\chi\in X_1(f)}+\sum_{\chi\in X_2(f)}=S_1^{ld}(f)+S_2^{ld}(f), \tag{13}$$

say.

Consider $S_2^{ld}(f)$. First, using the fact $|\psi_z(w)-\psi_z(w')| \le |z|\cdot|w-w'|$ [3, (6.5.19)], we obtain

$$|S_2^{ld}(f)|\le|z|\sum_{\chi\in X_2(f)}\left|\frac{L'}{L}(s,\chi)-\frac{L'_P}{L_P}(s,\chi)\right|. \tag{14}$$

Since (12) holds for $\chi \in X_2(f)$, by Lemma 7.2 of [7] we obtain

$$|f_P(s', \chi)| < \delta/2 \qquad (s' \in Q_0(\tau)). \tag{15}$$

Therefore by Lemma 7.1 of [7] we find that $H(\tau) \subset G_\chi(\alpha_1)$ (especially $L(s', \chi) \neq 0$ for $s' \in H(\tau)$) and $|R_P(s', \chi)| < \delta$ for $s' \in H(\tau)$.

Let $U = U(s)$ be the circle of radius $\varepsilon_2/2$ whose center is s. Then $U \subset H(\tau)$ (because $\sigma - \varepsilon_2/2 \geq \sigma_0 - \varepsilon_2/2 > \sigma_0 - \varepsilon_1 = \alpha_0$), and so $(L'/L)(s', \chi)$ is holomorphic on and inside U. Therefore by the Cauchy integral formula we have

$$\begin{aligned} \frac{L'}{L}(s, \chi) &= (\log L(s, \chi))' = \frac{1}{2\pi i} \int_{U(s)} \frac{\log L(s', \chi)}{(s' - s)^2} ds' \\ &= \frac{1}{\pi \varepsilon_2} \int_0^{2\pi} \log L\left(s + \frac{\varepsilon_2}{2} e^{i\theta}, \chi\right) e^{-i\theta} d\theta, \end{aligned} \tag{16}$$

and similarly

$$\frac{L_P'}{L_P}(s, \chi) = \frac{1}{\pi \varepsilon_2} \int_0^{2\pi} \log L_P\left(s + \frac{\varepsilon_2}{2} e^{i\theta}, \chi\right) e^{-i\theta} d\theta. \tag{17}$$

Substituting (16) and (17) into (14), we obtain

$$|S_2^{ld}(f)| \leq \frac{|z|}{\pi \varepsilon_2} \int_0^{2\pi} \sum_{\chi \in X_2(f)} \left| R_P\left(s + \frac{\varepsilon_2}{2} e^{i\theta}, \chi\right) \right| d\theta. \tag{18}$$

Here we note that $U \subset Q_0(\tau)$. In fact, we have already seen that $U \subset H(\tau)$, and also we see $U \subset \{\sigma' < \beta_0\}$ because β_0 is large. Therefore (15) holds for $s' \in U$. This implies, as is shown in the proof of Lemma 7.1 of [7],

$$|R_P(s', \chi)| \leq 2|f_P(s', \chi)| \qquad (s' \in U). \tag{19}$$

Combining (18) and (19) and using Schwarz's inequality, we have

$$\begin{aligned} |S_2^{ld}(f)| &\leq \frac{2|z|}{\pi \varepsilon_2} \int_0^{2\pi} \sum_{\chi \in X_2(f)} \left| f_P\left(s + \frac{\varepsilon_2}{2} e^{i\theta}, \chi\right) \right| d\theta \\ &\ll |z| f^{1/2} \int_0^{2\pi} \left(\sum_{\chi \in X_2(f)} \left| f_P\left(s + \frac{\varepsilon_2}{2} e^{i\theta}, \chi\right) \right|^2 \right)^{1/2} d\theta. \end{aligned} \tag{20}$$

Since $\sigma' = \mathrm{Re}(s + (\varepsilon_2/2)e^{i\theta}) > \alpha_0 > \alpha_1$ for $s' = \sigma' + i\tau' \in U$, using [7, (133)] (this is the point where a mean-value result on L-functions is necessary) we obtain

$$|S_2^{ld}(f)| \ll |z| f^{1/2} A(\tau', f, y)^{1/2} \ll |z| f^{1/2} A(\tau, f, y)^{1/2}, \tag{21}$$

where

$$A(\tau, f, y) = fy^{1-2\alpha_1} + f^{(1-\alpha_1)/(1-\alpha_2)} \exp\left(B_0 \frac{y^{1-\alpha_2}}{\log y}\right)\left(1 + \frac{|\tau|+1}{f^{2\alpha_2}}\right) \quad (22)$$

with a certain absolute positive constant B_0.

The treatment of $S_1^{ld}(f)$ can be done exactly in the same manner as in the argument around [7, (135), (136)]. We have $|S_1^{ld}(f)| \ll A(\tau, f, y)$, and, combining this with (21), we obtain

$$X_P^{ld}(z) \ll \frac{1}{\pi(m)} \sum_{f \leq m} \frac{1}{f}(|z| f^{1/2} A(\tau, f, y)^{1/2} + A(\tau, f, y)). \quad (23)$$

This is the L'/L-analogue (exactly the same form!) of Proposition 7.4 of [7].

Now we consider $Y_P^{ld}(z)$. Divide

$$\frac{1}{\pi(m)} \sum_{f \leq m} \frac{1}{f-2} \sum_{\chi \in X''(f)} \psi_z\left(\frac{L'_P}{L_P}(s, \chi)\right)$$

into $J_0^{ld(m)} + J_1^{ld(m)} + J_2^{ld(m)}$, analogously to the decomposition of [7, (137)]. The treatment of $J_0^{ld(m)}$ and $J_2^{ld(m)}$ is exactly the same as that of $J_0^{(m)}$ and $J_2^{(m)}$ in [7]. As for $J_1^{ld(m)}$, we first note that, when the conductor f of χ is larger than y, it holds that

$$\psi_z\left(\frac{L'_P}{L_P}(s, \chi)\right) = \sum_{\mathbf{n}_P \in \mathbf{Z}_P} A_{\sigma,P}^{ld}(\mathbf{n}_P; z, \bar{z}) \chi_P^{\mathbf{n}_P} P^{-i\tau\mathbf{n}_P}, \quad (24)$$

where $\mathbf{Z}_P = \prod_{p \in P} \mathbf{Z}$, and for $\mathbf{n}_P = (n_p)_{p \in P} \in \mathbf{Z}_P$,

$$\chi_P^{\mathbf{n}_P} = \prod_{p \in P} \chi(p)^{n_p}, \quad P^{-i\tau\mathbf{n}_P} = \prod_{p \in P} p^{-i\tau n_p}$$

and $A_{\sigma,P}^{ld}(\mathbf{n}_P; z, \bar{z})$ is given by [3, (5.1.7)] (without "ld"). This follows from [3, (1.5.4) and (5.1.6)] and is the L'/L-analogue of [7, (138)]. Starting from (24), we proceed similarly to the argument around [7, (139)–(147)]. (On this occasion we note that $\sum_{n_p \in \mathbf{Z}}$ is missing after the product symbol $\prod_{p \in P}$ in the first line of [7, (147)].) We use [3, (5.1.14)] instead of [7, (89)] and [3, (3.1.10)] instead of [7, (32)]. Proposition 5.3 of [7] includes the present L'/L case, and so we can apply it. Then, instead of $\eta(y)$ in [7] (see [7, (116)]),

$$\eta^{ld}(y) = \eta^{ld}(\sigma, y) = \begin{cases} y^{1-\sigma} \text{ if } 1/2 < \sigma < 1, \\ \log y \text{ if } \sigma = 1. \end{cases} \quad (25)$$

appears. The conclusion is that $Y_P^{ld}(z)$ satisfies the same inequality as that in Proposition 7.5 of [7] (with replacing $\eta(y)$ by $\eta^{ld}(y)$).

Finally we choose $y = (\log m)^{\omega_2}$ with $0 < \omega_2 < 2$. Then we find that $X_P^{ld}(z)$, $Y_P^{ld}(z)$ tend to 0 as $m \to \infty$. Also Theorem 3 of [3] implies that $Z_P^{ld}(z) \to 0$ as $m \to \infty$. Therefore we now complete the proof of (9). Moreover this convergence is uniform in $|z| \le R$ for any $R > 0$.

3 Proof in the Case (FII)

As in the case (FI), it is enough to consider the case $\Phi = \psi_z$, i.e., to prove

$$\lim_{T\to\infty} \frac{1}{2T}\int_{-T}^{T} \psi_z\left(\frac{\zeta'}{\zeta}(\sigma + i\tau')\right) d\tau' = \tilde{M}_\sigma(z) \tag{26}$$

(cf. [7, (92)]). Similarly to [7, (95)], we begin with the inequality

$$\begin{aligned}
&\left|\frac{1}{2T}\int_{-T}^{T} \psi_z\left(\frac{\zeta'}{\zeta}(\sigma + i\tau')\right) d\tau' - \tilde{M}_\sigma(z)\right| \\
&\le \left|\frac{1}{2T}\int_{-T}^{T} \left\{\psi_z\left(\frac{\zeta'}{\zeta}(\sigma + i\tau')\right) - \psi_z\left(\frac{\zeta_P'}{\zeta_P}(\sigma + i\tau')\right)\right\} d\tau'\right| \\
&\quad + \left|\frac{1}{2T}\int_{-T}^{T} \psi_z\left(\frac{\zeta_P'}{\zeta_P}(\sigma + i\tau')\right) d\tau' - \tilde{M}_{\sigma,P}(z)\right| \\
&\quad + \left|\tilde{M}_{\sigma,P}(z) - \tilde{M}_\sigma(z)\right| \\
&= X_P^{ld}(z) + Y_P^{ld}(z) + Z_P^{ld}(z),
\end{aligned} \tag{27}$$

say. Note that the meaning of these $X_P^{ld}(z)$, $Y_P^{ld}(z)$, $Z_P^{ld}(z)$ is different from that in Sect. 2.

The method of evaluating $X_P^{ld}(z)$ is a little different from the argument in [7]; rather, we follow the idea in Sect. 2. Noting $|\psi_z| = 1$ we have

$$\begin{aligned}
X_P^{ld}(z) &\le \frac{1}{2T}\int_{-2}^{2} 2d\tau' \\
&\quad + \frac{1}{2T}\int_{I(T)} \left|\psi_z\left(\frac{\zeta'}{\zeta}(\sigma + i\tau')\right) - \psi_z\left(\frac{\zeta_P'}{\zeta_P}(\sigma + i\tau')\right)\right| d\tau',
\end{aligned} \tag{28}$$

where $I(T) = [-T, -2] \cup [2, T]$. Let $I_1(T)$ (resp. $I_2(T)$) be the set of all $\tau' \in I(T)$ for which (11) [resp. (12)], with replacing τ by τ' and putting $\chi = \mathbf{1}$ (the trivial character), holds. Decompose the second integral on the right-hand side of (28) as $X_1^{ld} + X_2^{ld}$, where X_j^{ld} denotes the integral on $I_j(T)$ $(j = 1, 2)$. Then

$$X_P^{ld}(z) \le \frac{4}{T} + \frac{1}{2T}(X_1^{ld} + X_2^{ld}). \tag{29}$$

Consider X_2^{ld}. When $\tau' \in I_2(T)$, as in Sect. 2, we see that $\zeta(s'') \neq 0$ and $|R_P(s'',\mathbf{1})| \leq 2|f_P(s'',\mathbf{1})| < \delta$ for any $s'' \in H(\tau')$. Therefore $(\zeta'/\zeta)(s'')$ is holomorphic on and inside the circle U' of radius $\varepsilon_2/2$ whose center is $\sigma + i\tau'$, so

$$\frac{\zeta'}{\zeta}(\sigma + i\tau') = \frac{1}{2\pi i}\int_{U'} \frac{\log\zeta(s'')}{(s''-\sigma-i\tau')^2}ds''. \tag{30}$$

Similarly to (20), we obtain

$$X_2^{ld} \ll |z|T^{1/2}\int_0^{2\pi}\left(\int_{I_2(T)}\left|f_P\left(\sigma+i\tau'+\frac{\varepsilon_2}{2}e^{i\theta},\mathbf{1}\right)\right|^2 d\tau'\right)^{1/2} d\theta. \tag{31}$$

A mean square estimate of $|f_P|$ was obtained in Lemma 5 of [12] (see also [7, (102), (106)]). Applying this lemma, we have

$$\frac{1}{2T}X_2^{ld} \ll |z|\left\{y^{1-2\alpha_1+\varepsilon} + T^{1-2\alpha_1+\varepsilon}\exp\left(C_1\left(\frac{y}{\log y}\right)^{1/2}\right)\right\}, \tag{32}$$

for any small $\varepsilon > 0$, where C_1 is an absolute positive constant.

As for X_1^{ld}, we first use $|\psi_z| = 1$ to obtain

$$X_1^{ld} \leq 2\mathrm{meas}(I_1(T)), \tag{33}$$

where meas(A) means the one-dimensional Lebesgue measure of the set A. Using (11) for $\tau' \in I_1(T)$, we have

$$\begin{aligned}\mathrm{meas}(I_1(T)) &\ll \int_{I_1(T)} F_P(\tau',\mathbf{1})d\tau' \\ &= \int_{\alpha_1}^{\beta_1} d\sigma'' \int_{-T-2c}^{T+2c} |f_P(\sigma''+i\tau'',\mathbf{1})|^2 d\tau'' \int_{J_1(\tau'')} d\tau',\end{aligned} \tag{34}$$

where $J_1(\tau'') = I_1(T) \cap [\tau''-2c, \tau''+2c]$. The innermost integral is $\leq 4c$ and is equal to 0 if $\tau'' \in (-2+2c, 2-2c)$. Therefore we can apply Lemma 5 of [12] [7, (102), (106)] to the right-hand side of (34). Combining with (33), we obtain

$$\begin{aligned}\frac{1}{2T}X_1^{ld} &\ll \int_{\alpha_1}^{2}\left\{y^{1-2\alpha_1+\varepsilon} + T^{1-2\alpha_1+\varepsilon}\exp\left(C_1\left(\frac{y}{\log y}\right)^{1/2}\right)\right\} d\sigma'' \\ &\quad + \int_2^{\beta_1}\left\{\frac{1}{\sigma''}y^{1-2\sigma''+\varepsilon} + \frac{1}{\sigma'' T}y^{2-2\sigma''+\varepsilon}\right\} d\sigma'' \\ &\ll y^{1-2\alpha_1+\varepsilon} + T^{1-2\alpha_1+\varepsilon}\exp\left(C_1\left(\frac{y}{\log y}\right)^{1/2}\right) \\ &\quad + y^{-3+\varepsilon}\log\beta_1 + \frac{1}{T}y^{-2+\varepsilon}\log\beta_1.\end{aligned} \tag{35}$$

Since the factor $\log \beta_1$ can be absorbed into the implied constant, from (29), (32), and (35), we obtain

$$X_P^{ld}(z) \ll (|z|+1)\left\{y^{1-2\alpha_1+\varepsilon} + T^{1-2\alpha_1+\varepsilon}\exp\left(C_1\left(\frac{y}{\log y}\right)^{1/2}\right)\right\} + \frac{1}{T} + y^{-3+\varepsilon}. \tag{36}$$

The way of evaluating $Y_P^{ld}(z)$ is almost the same as that around [7, (109)–(122)]; only replace $\eta(y)$ by $\eta^{ld}(y)$. As an analogue of Proposition 6.2 of [7], we obtain

$$Y_P^{ld}(z) \ll \frac{1}{T}\exp\left(C_3\left(|z|y^{3/2-\sigma} + \frac{y}{\log y}\right)\right) \tag{37}$$

with an absolute constant $C_3 > 0$.

Choosing $y = (\log T)^{\omega_1}$ $(0 < \omega_1 < 1)$, from (36), (37), and Theorem 3 of [3], we find, as in [7], that $X_P^{ld}(z)$, $Y_P^{ld}(z)$, and $Z_P^{ld}(z)$ tend to 0 as $T \to \infty$, uniformly in $|z| \leq R$ for any $R > 0$. This proves (26).

Remark 3.1. Bohr and Jessen [2] proved the case (FII) of Theorem 1.2 for Φ with (ii), and Jessen and Wintner [10] reformulated the result in terms of asymptotic distribution functions. Kershner and Wintner [11] then proved that the analogue of the Jessen-Wintner theory is valid in the $\zeta'/\zeta(s)$ case (see also [1, 14]). Therefore the case (FII) of our Theorem 1.1, for Φ with (ii), is essentially included in Kershner and Wintner [11], though the density function is not explicitly given in their paper. The general (FII) case can be deduced from their result by the argument suggested in Remark 9.1 of [7]. Our method in the present paper is rather different from theirs and has advantages such as the unified treatment of both the cases (FI) and (FII) and the explicit construction of the density function $M_\sigma(w)$. In fact, the function $M_\sigma(w)$ and its Fourier dual themselves are interesting objects of research (see [4, 8]).

Remark 3.2. To show the general conclusion of our theorem from the special case $\Phi = \psi_z$, we can apply the method given in [7, Sect. 9], as indicated at the beginning of Sect. 2. This step can be explained as a consequence of a general theorem on weak convergence of probability measures.

Here we show how to deduce the case (i) of Theorem 1.1 from the case $\Phi = \psi_z$. In case (FI), the left-hand side of (1) is

$$\lim_{m\to\infty}\frac{1}{\pi(m)}\sum_{f\leq m}\frac{1}{X(f)}\sum_{\chi\in X'(f)}\Phi\left(\frac{L'}{L}(s,\chi)\right) = \lim_{m\to\infty}\frac{1}{\pi(m)-1}\sum_{f\leq m}\frac{1}{X'(f)}\sum_{\chi\in X'(f)}\Phi\left(\frac{L'}{L}(s,\chi)\right). \tag{38}$$

Let δ_w be the complex Dirac measure which is nonzero only at w, and define

$$\mu_m = \frac{1}{\pi(m)-1}\sum_{f\le m}\frac{1}{X'(f)}\sum_{\chi\in X'(f)}\delta_{L'/L(s,\chi)}.$$

Then this is a probability measure, and the right-hand side of (38) can be written as

$$\lim_{m\to\infty}\int_{\mathbf{C}}\Phi(w)d\mu_m(w).$$

Therefore (1) for any continuous bounded Φ is nothing but the weak convergence of probability measures μ_m to $M_\sigma(w)|dw|$. It is a well-known fact that the weak convergence of probability measures can be verified if we can check the special case $\Phi = \psi_z$.

In case (FII), we define the probability measure

$$\mu_T(A) = \frac{1}{2T}\mathrm{meas}\{\tau' \in [-T,T] \mid (L'/L)(s+i\tau') \in A\}$$

(where A is any Borel subset of $\mathbf{C}$) and proceed similarly. The above argument was pointed out by Professor Philippe Biane and Professor Katusi Fukuyama, to whom the authors express their sincere gratitude.

References

1. Bohr, H.: Über die Funktion $\zeta'/\zeta(s)$. J. Reine Angew. Math. **141**, 217–234 (1912)
2. Bohr, H., Jessen, B.: Über die Werteverteilung der Riemannschen Zetafunktion I, II. Acta Math. **54**, 1–35 (1930); **58**, 1–55 (1932)
3. Ihara, Y.: On "M-functions" closely related to the distribution of L'/L-values. Publ. Res. Inst. Math. Sci. **44**, 893–954 (2008)
4. Ihara, Y.: On certain arithmetic functions $\widetilde{M}(s; z_1, z_2)$ associated with global fields: Analytic properties. Publ. Res. Inst. Math. Sci. **47**, 257–305 (2011)
5. Ihara, Y., Matsumoto, K.: On the value-distribution of $\log L$ and L'/L. In: Steuding, R., Steuding, J. (eds.) New Directions in Value-Distribution Theory of Zeta and L-Functions. Würzburg Conference (Oct 2008), pp. 85–97. Shaker, Maastricht (2009)
6. Ihara, Y., Matsumoto, K.: On L-functions over function fields: Power-means of error-terms and distribution of L'/L-values. In: Nakamura, H., et al. (eds.) Algebraic Number Theory and Related Topics 2008, vol. B19, pp. 221–247. RIMS Kokyuroku Bessatsu, Kyoto (2010)
7. Ihara, Y., Matsumoto, K.: On certain mean values and the value-distribution of logarithms of Dirichlet L-functions. Quart. J. Math. (Oxford) **62**, 637–677 (2011)
8. Ihara, Y., Matsumoto, K.: On $\log L$ and L'/L for L-functions and the associated "M-functions": Connections in optimal cases. Moscow Math. J. **11**, 73–111 (2011)
9. Ihara, Y., Murty, V.K., Shimura, M.: On the logarithmic derivatives of Dirichlet L-functions at $s = 1$. Acta Arith. **137**, 253–276 (2009)
10. Jessen, B., Wintner, A.: Distribution functions and the Riemann zeta function. Trans. Amer. Math. Soc. **38**, 48–88 (1935)

11. Kershner, R., Wintner, A.: On the asymptotic distribution of $\zeta'/\zeta(s)$ in the critical strip. Amer. J. Math. **59**, 673–678 (1937)
12. Matsumoto, K.: Asymptotic probability measures of zeta-functions of algebraic number fields. J. Number Theory **40**, 187–210 (1992)
13. Montgomery, H.L.: Topics in Multiplicative Number Theory. Lecture Notes in Mathematics, vol. 227. Springer, Berlin (1971)
14. van Kampen, E.R., Wintner, A.: Convolutions of distributions on convex curves and the Riemann zeta function. Amer. J. Math. **59**, 175–204 (1937)

Multiple Gamma Functions and Their Applications

Junesang Choi

Dedicated to Professor Hari M. Srivastava

Abstract The double Gamma function Γ_2 and the multiple Gamma functions Γ_n were defined and studied systematically by Barnes in about 1900. Before their investigation by Barnes, these functions had been introduced in a different form by, for example, Hölder, Alexeiewsky, and Kinkelin. Although these functions did not appear in the tables of the most well-known special functions, yet the double Gamma function was cited in the exercises by Whittaker and Watson's book and recorded also by Gradshteyn and Ryzhik's book. In about the middle of the 1980s, these functions were revived in the study of the determinants of the Laplacians on the n-dimensional unit sphere $\mathbf{S}^n$. Here, in this expository paper, from the middle of the 1980s until today, we aim at giving an eclectic review for recent developments and applications of the simple and multiple Gamma functions.

1 Introduction and Preliminaries

The double Gamma function Γ_2 and the multiple Gamma functions Γ_n were defined and studied systematically by Barnes [11–14] in around 1900. Before their investigation by Barnes, these functions had been introduced in a different form by, for example, Hölder [80], Alexeiewsky [5], and Kinkelin [83]. Although these functions did not appear in the tables of the most well-known special functions, yet the double Gamma function was cited in the exercises by Whittaker and Watson [124, p. 264] and recorded also by Gradshteyn and Ryzhik [77, p. 661, Entry 6.441(4); p. 937, Entry 8.333]. In about the middle of the 1980s, these functions were revived in the study of the determinants of the Laplacians on the

J. Choi (✉)
Department of Mathematics, Dongguk University, Gyeongju 780-714, Republic of Korea
e-mail: junesang@mail.dongguk.ac.kr

G.V. Milovanović and M.Th. Rassias (eds.), *Analytic Number Theory, Approximation Theory, and Special Functions*, DOI 10.1007/978-1-4939-0258-3_4,

n-dimensional unit sphere $\mathbf{S}^n$ (see [25,30,89,100,103,118,121]). Shintani [110] also used the double Gamma function to prove the classical Kronecker limit formula. Friedman and Ruijsenaars [73] showed that Shintani's work on multiple Zeta and Gamma functions can be simplified and extended by making use of difference equations. Its p-adic analytic extension appeared in a formula of Cassou–Noguès [21] for the p-adic L-functions at the point 0. Choi et al. (see [31,41,42]) used these functions in order to evaluate some families of series involving the Riemann Zeta function as well as to compute the determinants of the Laplacians. Choi et al. [31] addressed the converse problem and applied various formulas for series associated with the Zeta and related functions with a view to developing the corresponding theory of multiple Gamma functions. Adamchik [4] discussed some theoretical aspects of the multiple Gamma functions and their applications to summation of series and infinite products. Matsumoto [92] proved several asymptotic expansions of the Barnes double Zeta function and the double Gamma function and presented an application to the Hecke L-functions of real quadratic fields. Ruijsenaars [107] showed how various known results concerning the Barnes multiple Zeta and Gamma functions can be obtained as specializations of the simple features shared by a quite remarkably extensive class of functions.

The main object of this expository paper is to give an eclectic review of certain recent developments and applications of the classical Gamma function, the multiple Gamma functions, and their related functions.

1.1 Gamma Function

The origin of the *Gamma function* can be traced back to two letters from Leonhard Euler (1707–1783) to Christian Goldbach (1690–1764), just as a simple desire to extend factorials to values between the integers. The first letter (dated October 13, 1729) dealt with the interpolation problem, while the second letter (dated January 8, 1730) dealt with integration and tied the two together.

The Gamma function $\Gamma(z)$ developed by Euler is usually defined by

$$\Gamma(z) := \int_0^\infty \mathrm{e}^{-t}\, t^{z-1}\, \mathrm{d}t \quad (\mathrm{Re}(z) > 0). \tag{1}$$

We also present here several equivalent forms of the Gamma function $\Gamma(z)$, one by Weierstrass:

$$\Gamma(z) = \frac{\mathrm{e}^{-\gamma z}}{z} \prod_{k=1}^{\infty} \left\{ \left(1 + \frac{z}{k}\right)^{-1} \mathrm{e}^{z/k} \right\} \quad (z \in \mathbb{C}\setminus\mathbb{Z}_0^-;\ \mathbb{Z}_0^- := \{0, -1, -2, \ldots\}), \tag{2}$$

where $\mathbb{C}$ is the set of complex numbers and γ denotes the Euler–Mascheroni constant defined by (see, e.g., [27,37,50,74,79,85,105])

$$\gamma := \lim_{n\to\infty}\left(\sum_{k=1}^{n}\frac{1}{k} - \log n\right) \cong 0.57721\,56649\,01532\,86060\,6512\ldots, \quad (3)$$

and the other by Gauss:

$$\begin{aligned}\Gamma(z) &= \lim_{n\to\infty}\left\{\frac{(n-1)!\,n^z}{z(z+1)\cdots(z+n-1)}\right\}\\ &= \lim_{n\to\infty}\left\{\frac{n!\,n^z}{z(z+1)\cdots(z+n)}\right\} \quad \left(z\in\mathbb{C}\setminus\mathbb{Z}_0^-\right),\end{aligned} \quad (4)$$

since

$$\lim_{n\to\infty}\frac{n}{z+n} = 1 = \lim_{n\to\infty}\frac{n^z}{(n+1)^z}.$$

In terms of the Pochhammer symbol $(\lambda)_\nu$ or the *shifted factorial*, since

$$(1)_n = n! \quad (n\in\mathbb{N}_0 := \mathbb{N}\cup\{0\};\ \mathbb{N} := \{1, 2, 3, \cdots\}),$$

which is defined (for $\lambda,\ \nu\in\mathbb{C}$), in terms of the familiar Gamma function Γ, by

$$\begin{aligned}(\lambda)_\nu &:= \begin{cases}1, & \nu=0;\ \lambda\in\mathbb{C}\setminus\{0\},\\ \lambda(\lambda+1)\cdots(\lambda+n-1), & \nu=n\in\mathbb{N};\ \lambda\in\mathbb{C},\end{cases}\\ &= \frac{\Gamma(\lambda+\nu)}{\Gamma(\lambda)} \quad \left(\nu\in\mathbb{N}_0;\ \lambda\in\mathbb{C}\setminus\mathbb{Z}_0^-\right),\end{aligned}$$

it being *understood conventionally* that $(0)_0 := 1$, the definition (4) can easily be written in an equivalent form:

$$\Gamma(z) = \lim_{n\to\infty}\frac{(n-1)!\,n^z}{(z)_n} \quad \left(z\in\mathbb{C}\setminus\mathbb{Z}_0^-\right).$$

For a complex number z, we have the following asymptotic expansion:

$$\log\Gamma(z) = \left(z-\frac{1}{2}\right)\log z - z + \frac{1}{2}\log(2\pi) + \sum_{k=1}^{n}\frac{B_{2k}}{2k(2k-1)z^{2k-1}} + O\left(z^{-2n-1}\right), \quad (5)$$

for $|z|\to\infty$; $|\arg(z)| \leqq \pi-\epsilon$ $(0<\epsilon<\pi)$; $n\in\mathbb{N}_0$, where $B_k(x)$ are the Bernoulli polynomials defined by the generating function

$$\frac{ze^{xz}}{e^z-1} = \sum_{k=0}^{\infty}B_k(x)\frac{z^k}{k!} \quad (|z|<2\pi)$$

and the Bernoulli numbers are defined by $B_k := B_k(0)$. Taking exponentials on each side of (5) yields an asymptotic formula for the Gamma function:

$$\begin{aligned}\Gamma(z) =& z^z e^{-z} \sqrt{\frac{2\pi}{z}} \left[1 + \frac{1}{12\,z} + \frac{1}{288\,z^2} - \frac{139}{51840\,z^3} - \frac{571}{2488320\,z^4}\right.\\ &\left.+ \frac{163879}{209018880\,z^5} + \frac{50043869}{75246796800\,z^6} + O\left(z^{-7}\right)\right]\\ &(|z| \to \infty;\ |\arg(z)| \leqq \pi - \epsilon\ (0 < \epsilon < \pi)).\end{aligned} \tag{6}$$

The asymptotic formula (6), in conjunction with the recurrence relation

$$\Gamma(z+1) = z\,\Gamma(z)$$

is useful in computing the numerical values of $\Gamma(z)$ for large real values of z.

Some useful consequences of (5) or (6) include the asymptotic expansions

$$\begin{aligned}&\log \Gamma(z+\alpha) = \left(z + \alpha - \frac{1}{2}\right) \log z - z + \frac{1}{2} \log(2\pi) + O\left(z^{-1}\right)\\ &(|z| \to \infty;\ |\arg(z)| \leqq \pi - \epsilon;\ |\arg(z+\alpha)| \leqq \pi - \epsilon;\ 0 < \epsilon < \pi),\end{aligned}$$

and

$$\begin{aligned}&\frac{\Gamma(z+\alpha)}{\Gamma(z+\beta)} = z^{\alpha-\beta} \left[1 + \frac{(\alpha-\beta)(\alpha+\beta-1)}{2z} + O\left(z^{-2}\right)\right]\\ &(|z| \to \infty;\ |\arg(z)| \leqq \pi - \epsilon;\ |\arg(z+\alpha)| \leqq \pi - \epsilon;\ 0 < \epsilon < \pi),\end{aligned}$$

where α and β are bounded complex numbers.

The *Psi* (or *Digamma*) *function* $\psi(z)$ defined by

$$\psi(z) := \frac{d}{dz}\{\log\,\Gamma(z)\} = \frac{\Gamma'(z)}{\Gamma(z)} \quad \text{or} \quad \log\,\Gamma(z) = \int_1^z \psi(t)\,dt \tag{7}$$

possesses the following properties:

$$\psi(z) = \lim_{n\to\infty} \left(\log n - \sum_{k=0}^{n} \frac{1}{z+k}\right); \tag{8}$$

$$\psi(z) = -\gamma - \frac{1}{z} + \sum_{n=1}^{\infty} \frac{z}{n(z+n)} = -\gamma + (z-1) \sum_{n=0}^{\infty} \frac{1}{(n+1)(z+n)}, \tag{9}$$

where γ is the Euler–Mascheroni constant defined by (3).

These results clearly imply that $\psi(z)$ is meromorphic (i.e., analytic everywhere in the bounded complex z-plane except for poles) with simple poles at $z = -n$ ($n \in \mathbb{N}_0$) with its residue -1. Also we have

$$\psi(1) = -\gamma,$$

which follows at once from (9).

The *Polygamma functions* $\psi^{(n)}(z)$ ($n \in \mathbb{N}$) are defined by

$$\psi^{(n)}(z) := \frac{\mathrm{d}^{n+1}}{\mathrm{d}z^{n+1}} \log \Gamma(z) = \frac{\mathrm{d}^n}{\mathrm{d}z^n} \psi(z) \quad (n \in \mathbb{N}_0;\ z \in \mathbb{C} \setminus \mathbb{Z}_0^-).$$

In terms of the generalized (or Hurwitz) Zeta function $\zeta(s,a)$ (see Sect. 2.2), we can write

$$\psi^{(n)}(z) = (-1)^{n+1}\, n! \sum_{k=0}^{\infty} \frac{1}{(k+z)^{n+1}} = (-1)^{n+1}\, n!\, \zeta(n+1, z) \quad (n \in \mathbb{N};\ z \in \mathbb{C} \setminus \mathbb{Z}_0^-),$$

which may be used to deduce the properties of $\psi^{(n)}(z)$ ($n \in \mathbb{N}$) from those of $\zeta(s,z)$ ($s = n+1;\ n \in \mathbb{N}$).

For various other properties of Gamma function and its related functions, refer to [1, 6, 7, 9, 19, 20, 48, 49, 70, 81, 84, 88, 90, 94, 98, 99, 101, 102, 122, 123].

1.2 Double and Multiple Gamma Functions

Barnes [11] defined the double Gamma function $\Gamma_2 = 1/G$ satisfying each of the following properties:

(a) $G(z+1) = \Gamma(z)G(z)$ ($z \in \mathbb{C}$);
(b) $G(1) = 1$;
(c) Asymptotically,

$$\begin{aligned} \log G(z+n+2) =& \frac{n+1+z}{2} \log(2\pi) + \left[\frac{n^2}{2} + n + \frac{5}{12} + \frac{z^2}{2} + (n+1)z\right] \log n \\ & - \frac{3n^2}{4} - n - nz - \log A + \frac{1}{12} + O\left(n^{-1}\right) \quad (n \to \infty), \end{aligned} \tag{10}$$

where Γ is the Gamma function given in (1) and A is called the Glaisher–Kinkelin constant defined by (see [76])

$$\log A = \lim_{n\to\infty} \left\{ \sum_{k=1}^{n} k \log k - \left(\frac{n^2}{2} + \frac{n}{2} + \frac{1}{12}\right) \log n + \frac{n^2}{4} \right\}, \tag{11}$$

the numerical value of A being given by

$$A \cong 1.282427130\cdots.$$

From this definition, Barnes [11] deduced several explicit Weierstrass canonical product forms of the double Gamma function Γ_2, one of which is recalled here in the form

$$\begin{aligned}\{\Gamma_2(z+1)\}^{-1} &= G(z+1)\\ &= (2\pi)^{\frac{1}{2}z} \exp\left(-\frac{1}{2}z-\frac{1}{2}(\gamma+1)z^2\right)\prod_{k=1}^{\infty}\left\{\left(1+\frac{z}{k}\right)^k \exp\left(-z+\frac{z^2}{2k}\right)\right\},\end{aligned}$$

where γ denotes the Euler–Mascheroni constant given by (3).

Barnes [11] also gave the following two more equivalent forms of the double Gamma function Γ_2:

$$\begin{aligned}\{\Gamma_2(z+1)\}^{-1} = G(z+1) =&(2\pi)^{\frac{1}{2}z} \exp\left(-\frac{1}{2}z(z+1)-\frac{1}{2}\gamma z^2\right)\\ &\times\prod_{k=1}^{\infty}\frac{\Gamma(k)}{\Gamma(z+k)}\exp\left[z\psi(k)+\frac{1}{2}z^2\psi'(k)\right];\end{aligned}$$

$$\begin{aligned}\{\Gamma_2(z+1)\}^{-1} = G(z+1) = &(2\pi)^{\frac{1}{2}z} \exp\left[\left(\gamma-\frac{1}{2}\right)z-\left(\frac{\pi^2}{6}+1+\gamma\right)\frac{z^2}{2}\right]\Gamma(z+1)\\ &\times\prod_{m=0}^{\infty}\prod_{n=0}^{\infty}{}'\left(1+\frac{z}{m+n}\right)\exp\left(-\frac{z}{m+n}+\frac{z^2}{2(m+n)^2}\right),\end{aligned}$$

where the prime denotes the exclusion of the case $n = m = 0$ and the Psi (or Digamma) function ψ is given by (7). Each form of these products is convergent for all finite values of $|z|$, by the Weierstrass factorization theorem (see Conway [60, p. 170]).

The double Gamma function satisfies the following relations:

$$G(1) = 1 \quad \text{and} \quad G(z+1) = \Gamma(z)G(z) \quad (z \in \mathbb{C}).$$

For sufficiently large real x and $a \in \mathbb{C}$, we have the Stirling formula for the G-function:

$$\begin{aligned}\log G(x+a+1) = &\frac{x+a}{2}\log(2\pi) - \log A + \frac{1}{12} - \frac{3x^2}{4} - ax\\ &+\left(\frac{x^2}{2}-\frac{1}{12}+\frac{a^2}{2}+ax\right)\log x + O\left(x^{-1}\right) \quad (x \to \infty).\end{aligned} \tag{12}$$

The following special values of G (see Barnes [11]) may be recalled here:

$$G\left(\frac{1}{2}\right) = 2^{\frac{1}{24}} \cdot \pi^{-\frac{1}{4}} \cdot e^{\frac{1}{8}} \cdot A^{-\frac{3}{2}}; \tag{13}$$

$$G(n+2) = 1!\,2! \cdots n! \quad \text{and} \quad G(n+1) = \frac{(n!)^n}{1 \cdot 2 \cdot 3^2 \cdot 4^3 \cdots n^{n-1}} \quad (n \in \mathbb{N}).$$

There are two known ways to define the n-ple Gamma functions Γ_n. First of all, Barnes [14] (see also Vardi [118]) defined Γ_n by using the n-ple Hurwitz Zeta functions given in Sect. 2 (see, e.g., [35,53], [114, Chap. 2]). Secondly, a recurrence relation of the Weierstrass canonical product forms of the n-ple Gamma functions Γ_n was given by Vignéras [119] who used the theorem of Dufresnoy and Pisot [64] which provides the existence, uniqueness, and expansion of the series of Weierstrass satisfying a certain functional equation.

By making use of the aforementioned Dufresnoy–Pisot theorem and starting with

$$f_1(x) = -\gamma\, x + \sum_{n=1}^{\infty} \left[\frac{x}{n} - \log\left(1 + \frac{x}{n}\right)\right],$$

Vignéras [119] obtained a recurrence relation of Γ_n $(n \in \mathbb{N})$ which is stated here as Theorem 1.1 below (see, e.g., [4, 35, 48, 52, 53]).

Theorem 1.1. *The n-ple Gamma functions Γ_n are defined by*

$$\Gamma_n(z) = [G_n(z)]^{(-1)^{n-1}} \quad (n \in \mathbb{N}),$$

where $G_n(z+1) = \exp[f_n(z)]$ and the functions $f_n(z)$ are given by

$$f_n(z) = -zA_n(1) + \sum_{k=1}^{n-1} \frac{p_k(z)}{k!} \left[f_{n-1}^{(k)}(0) - A_n^{(k)}(1)\right] + A_n(z),$$

with

$$A_n(z) = \sum_{m \in \mathbb{N}_0{}^{n-1} \times \mathbb{N}} \left[\frac{1}{n}\left(\frac{z}{L(m)}\right)^n - \frac{1}{n-1}\left(\frac{z}{L(m)}\right)^{n-1} + \cdots \right.$$
$$\left. + (-1)^{n-1} \frac{z}{L(m)} + (-1)^n \log\left(1 + \frac{z}{L(m)}\right)\right],$$

where $L(m) = m_1 + m_2 + \cdots + m_n$ if $m = (m_1, m_2, \ldots, m_n) \in \mathbb{N}_0{}^{n-1} \times \mathbb{N}$ and the polynomials $p_n(z)$ given by

$$p_n(z) := \begin{cases} 1^n + 2^n + 3^n + \cdots + (N-1)^n & (z = N;\ N \in \mathbb{N} \setminus \{1\}), \\ \dfrac{B_{n+1}(z) - B_{n+1}}{n+1} & (z \in \mathbb{C}), \end{cases}$$

satisfy the following relations:

$$p_n'(z) = \frac{B_{n+1}'(z)}{n+1} = B_n(z) \quad and \quad p_n(0) = 0,$$

$B_n(z)$ being the Bernoulli polynomial of degree n in z.

By analogy with the Bohr–Mollerup theorem (see [10, p. 14]; see also [114, p. 13]), which guarantees the uniqueness of the Gamma function Γ, one can give, for the double Gamma function and (more generally) for the multiple Gamma functions of order n ($n \in \mathbb{N}$), a definition of Artin [10] by means of the following theorem (see Vignéras [119, p. 239]).

Theorem 1.2. *For all $n \in \mathbb{N}$, there exists a unique meromorphic function $G_n(z)$ satisfying each of the following properties:*

(a) $G_n(z+1) = G_{n-1}(z)G_n(z)$ $(z \in \mathbb{C})$;
(b) $G_n(1) = 1$;
(c) *For $x \geqq 1$, $G_n(x)$ are infinitely differentiable and*

$$\frac{\mathrm{d}^{n+1}}{\mathrm{d}x^{n+1}} \{\log G_n(x)\} \geqq 0;$$

(d) $G_0(x) = x$.

It is not difficult to verify (see, e.g., [114, pp. 40–41]) that $\{\Gamma_n(z)\}^{-1}$ is an entire function with zeros at $z = -k$ $(k \in \mathbb{N}_0)$ with multiplicities

$$\binom{n+k-1}{n-1} \quad (n \in \mathbb{N};\ k \in \mathbb{N}_0). \tag{14}$$

In our earlier investigations, we gave explicit forms of the multiple Gamma functions Γ_n ($n = 3, 4, 5$) (see, e.g., [31, 57]). Now, by observing (14), we can present the following explicit form of the multiple Gamma functions Γ_n ($n \in \mathbb{N}$) for a potential and easier future use.

Theorem 1.3. *The n-ple Gamma functions Γ_n in Theorem 1.1 can be written in a more explicit form as follows:*

$$\Gamma_n(1+z) = \exp\left[Q_n(z)\right] \prod_{k=1}^{\infty} \left\{ \left(1 + \frac{z}{k}\right)^{-\binom{n+k-2}{n-1}} \exp\left[\binom{n+k-2}{n-1} \left(\sum_{j=1}^{n} \frac{(-1)^{j-1}}{j} \frac{z^j}{k^j} \right) \right] \right\},$$

where $Q_n(z)$ is a polynomial in z of degree n given by

$$Q_n(z) := (-1)^{n-1}\left[-zA_n(1) + \sum_{k=1}^{n-1} \frac{p_k(z)}{k!}\left(f_{n-1}^{(k)}(0) - A_n^{(k)}(1)\right)\right],$$

$$f_n(z) := -z\,A_n(1) + \sum_{k=1}^{n-1} \frac{p_k(z)}{k!}\left[f_{n-1}^{(k)}(0) - A_n^{(k)}(1)\right] + A_n(z),$$

$$A_n(z) := \sum_{k=1}^{\infty}(-1)^{n-1}\binom{n+k-2}{n-1}\left[-\log\left(1+\frac{z}{k}\right) + \sum_{j=1}^{n}\frac{(-1)^{j-1}}{j}\frac{z^j}{k^j}\right], \quad (15)$$

and

$$p_n(z) = \frac{1}{n+1}\sum_{k=1}^{n+1}\binom{n+1}{k}B_{n+1-k}z^k \quad (n \in \mathbb{N}).$$

Remark 1.1. In order to get explicit forms of the multiple Gamma functions $\Gamma_n(1+z)$ in Theorem 1.3, it is indispensable to compute $A_n(1)$ explicitly. In fact, by using the Taylor–Maclaurin expansion of $\log(1+t)$ in (15) and certain series involving Zeta functions, Choi et al. [31] found that

$$A_n(z) = \frac{1}{(n-1)!}\sum_{j=0}^{n-1} s(n-1,j)\left[\sum_{k=0}^{j}(-1)^k\binom{j}{k}\zeta'(-k,1+z)z^{j-k} + (-1)^{j+1}\zeta'(-j)\right.$$
$$\left. -\sum_{\ell=0}^{j-1}(-1)^{\ell}\frac{\zeta(-\ell)}{j-\ell}z^{j-\ell} + \frac{z^{j+1}}{j+1}(H_j+\gamma) - \sum_{k=2}^{n-j}(-1)^k\frac{\zeta(k)}{k+j}z^{k+j}\right], \quad (16)$$

where $\zeta(s,a)$ and $\zeta(s)$ are the generalized (or Hurwitz) Zeta function and the Riemann Zeta function, respectively, and $s(n,k)$ denotes the Stirling numbers of the first kind (see [114, pp. 56–57]) and H_n denotes the harmonic numbers given by

$$H_n := \sum_{k=1}^{n}\frac{1}{k} \quad (n \in \mathbb{N}).$$

Now, by applying (16) in Theorem 1.3, we can give explicit forms of the multiple Gamma functions Γ_n $(n \in \mathbb{N})$ whose cases $(n = 3, 4, 5)$ are recalled here as the following corollary (see [31]).

Corollary 1.1. *Each of the following expressions holds true:*

$$\Gamma_3(1+z) = \exp\left(c_1 z + c_2 z^2 + c_3 z^3\right) \times \prod_{k=1}^{\infty} \left\{ \left(1+\frac{z}{k}\right)^{-\binom{k+1}{2}} \exp\left[\binom{k+1}{2}\left(\frac{z}{k} - \frac{z^2}{2k^2} + \frac{z^3}{3k^3}\right)\right]\right\},$$

where

$$c_1 = \frac{3}{8} - \frac{1}{4}\log(2\pi) - \log A, \quad c_2 = \frac{1}{8} + \frac{1}{4}\log(2\pi) + \frac{\gamma}{4}, \quad c_3 = -\frac{1}{4} - \frac{\pi^2}{36} - \frac{\gamma}{6};$$

$$\Gamma_4(1+z) = \exp\left(d_1 z + d_2 z^2 + d_3 z^3 + d_4 z^4\right) \times \prod_{k=1}^{\infty} \left\{ \left(1+\frac{z}{k}\right)^{-\binom{k+2}{3}} \exp\left[\binom{k+2}{3}\left(\frac{z}{k} - \frac{z^2}{2k^2} + \frac{z^3}{3k^3} - \frac{z^4}{4k^4}\right)\right]\right\},$$

where

$$d_1 = \frac{7}{24} - \log A - \frac{1}{2}\log B - \frac{1}{6}\log(2\pi), \quad d_2 = -\frac{1}{144} + \frac{\gamma}{6} + \frac{1}{4}\log(2\pi) + \frac{1}{2}\log A,$$

$$d_3 = -\frac{2}{9} - \frac{\gamma}{6} - \frac{1}{12}\log(2\pi) - \frac{\pi^2}{54}, \quad d_4 = \frac{11}{144} + \frac{\gamma}{24} + \frac{\pi^2}{48} + \frac{\zeta(3)}{12};$$

$$\Gamma_5(1+z) = \exp\left(e_1 z + e_2 z^2 + e_3 z^3 + e_4 z^4 + e_5 z^5\right) \times \prod_{k=1}^{\infty} \left\{ \left(1+\frac{z}{k}\right)^{-\binom{k+3}{4}} \exp\left[\binom{k+3}{4}\left(\frac{z}{k} - \frac{z^2}{2k^2} + \frac{z^3}{3k^3} - \frac{z^4}{4k^4} + \frac{z^5}{5k^5}\right)\right]\right\},$$

where

$$e_1 = \frac{409}{1728} - \frac{1}{8}\log(2\pi) - \frac{11}{12}\log A - \frac{1}{6}\log C - \frac{3\,\zeta(3)}{16\pi^2} + \frac{1}{20}\zeta(4) - \frac{1}{20}\zeta(5),$$

$$e_2 = -\frac{1}{16} + \frac{\gamma}{8} + \frac{11}{48}\log(2\pi) + \frac{3}{4}\log A + \frac{\zeta(3)}{16\pi^2},$$

$$e_3 = -\frac{149}{864} - \frac{11}{72}\gamma - \frac{1}{8}\log(2\pi) - \frac{1}{6}\log A - \frac{1}{12}\zeta(2),$$

$$e_4 = \frac{7}{64} + \frac{1}{16}\gamma + \frac{1}{48}\log(2\pi) + \frac{11}{96}\zeta(2) + \frac{1}{16}\zeta(3),$$

$$e_5 = -\frac{5}{288} - \frac{1}{120}\gamma - \frac{1}{20}\zeta(2) - \frac{11}{120}\zeta(3) - \frac{1}{20}\zeta(4).$$

For various other properties and applications of the double and multiple Gamma functions, refer to [15, 16, 18, 22–24, 29, 34, 36, 47, 59, 71, 72, 75, 76, 86, 87, 91, 96, 97, 106].

2 Multiple Hurwitz Zeta Functions

In this section, we first introduce (and investigate the various properties and relationships satisfied by) the multiple Hurwitz Zeta function $\zeta_n(s,a)$ $(n \in \mathbb{N})$ and consider its relatively more familiar special case when $n = 1$, that is, the Hurwitz (or generalized) Zeta function $\zeta(s,a)$. We then deal with the Riemann Zeta function.

2.1 *Multiple Hurwitz Zeta Functions*

Barnes [14] introduced and studied the generalized multiple Hurwitz Zeta function $\zeta_n(s, a \mid w_1, \cdots, w_n)$ defined, for $\mathrm{Re}(s) > n$, by the n-ple series

$$\zeta_n(s, a \mid w_1, \ldots, w_n) := \sum_{m_1,\ldots,m_n=0}^{\infty} \frac{1}{(a+\Omega)^s} \quad (\mathrm{Re}(s) > n;\ n \in \mathbb{N}), \tag{17}$$

where $\Omega = m_1 w_1 + \cdots + m_n w_n$ and the general conditions for a and the parameters $w_1, \ldots, w_n$ are given in Barnes [14] who used it in the study of the multiple Gamma functions (see Sect. 1). We consider only the simple case of (17) when $w_j = 1$ $(j = 1, \ldots, , n;\ j, n \in \mathbb{N})$ and

$$\zeta_n(s,a) := \sum_{k_1,\ldots,k_n=0}^{\infty} (a + k_1 + \cdots + k_n)^{-s} \quad (\mathrm{Re}(s) > n;\ a \in \mathbb{C} \setminus \mathbb{Z}_0^-), \tag{18}$$

which is also referred to as *n-ple* (or, simply, *multiple*) *Hurwitz Zeta function*. We shall give some known properties and characteristics of the function $\zeta_n(s,a)$ in (18), including its analytic continuation.

Let $s = \sigma + it$ $(\sigma, t \in \mathbb{R})$ where $\mathbb{R}$ denotes the set of real numbers. First, for convergence, we consider $\zeta_n(s,a)$ in (18) for the case when $a > 0$:

$$\zeta_n(s,a) = \sum_{k_1,\ldots,k_n=0}^{\infty} (a + k_1 + \cdots + k_n)^{-s} \quad (\mathrm{Re}(s) = \sigma > n;\ a > 0). \tag{19}$$

Theorem 2.1. *The series for $\zeta_n(s,a)$ in (19) converges absolutely for $\sigma > n$. The convergence is uniform in every half-plane $\sigma \geqslant n + \delta$ $(\delta > 0)$, so $\zeta_n(s,a)$ is an analytic function of s in the half-plane $\sigma > n$.*

Next we present an integral representation of $\zeta_n(s,a)$, which is given by

Theorem 2.2. *If* $\mathrm{Re}(s) = \sigma > n$, *then*

$$\Gamma(s)\zeta_n(s,a) = \int_0^\infty \frac{x^{s-1}\mathrm{e}^{-ax}}{(1-\mathrm{e}^{-x})^n}\,\mathrm{d}x \quad (\mathrm{Re}(s) > n;\ n \in \mathbb{N}). \tag{20}$$

In order to extend $\zeta_n(s,a)$ to the half-plane on the left of the line $\sigma = n$, we derive another representation in terms of a contour integral. The contour C is essentially a Hankel's loop (cf., e.g., Whittaker and Watson [124, p. 245]), which starts from ∞ along the upper side of the positive real axis, encircles the origin once in the positive (counterclockwise) direction, and then returns to ∞ along the lower side of the positive real axis.

Theorem 2.3. *If* $a > 0$, *then the function defined by the following contour integral:*

$$I_n(s,a) = -\frac{1}{2\pi\mathrm{i}}\int_\infty^{(0+)} \frac{(-z)^{s-1}\mathrm{e}^{-az}}{(1-\mathrm{e}^{-z})^n}\,\mathrm{d}z \tag{21}$$

is an entire function of s. *Moreover,*

$$\zeta_n(s,a) = \Gamma(1-s)\,I_n(s,a) \quad (\mathrm{Re}(s) = \sigma > n). \tag{22}$$

In (22), valid for $\sigma > n$, the function $I_n(s,a)$ is an entire function of s, and $\Gamma(1-s)$ is analytic for every complex s for $s \in \mathbb{C} \setminus \mathbb{N}$. We, therefore, can use this equation to define $\zeta_n(s,a)$ for $\sigma \leqq n$, that is, outside $\sigma > n$ as desired.

Definition 2.1. If $\mathrm{Re}(s) = \sigma \leqq n$, we define $\zeta_n(s,a)$ by

$$\zeta_n(s,a) := \Gamma(1-s)I_n(s,a), \tag{23}$$

where $I_n(s,a)$ is given in (21).

This equation (23) provides the analytic continuation of $\zeta_n(s,a)$ to the whole complex s-plane.

Theorem 2.4. *The function* $\zeta_n(s,a)$ *defined by* (23) *is analytic for all* s *except for simple poles at* $s = k$ $(1 \leqq k \leqq n)$, *with their respective residues given by*

$$\operatorname*{Res}_{s=k} \zeta_n(s,a) = \frac{1}{(n-k)!(k-1)!}\lim_{z\to 0}\frac{\mathrm{d}^{n-k}}{\mathrm{d}z^{n-k}}\frac{z^n\mathrm{e}^{-az}}{(1-\mathrm{e}^{-z})^n} \quad (k = 1, \cdots, n;\ n \in \mathbb{N}).$$

In particular, when $s = n$, *its residue is* $1/(n-1)!$.

2.2 Relationship Between $\zeta_n(s,x)$ and $B_n^{(\alpha)}(x)$

The *generalized Bernoulli polynomials* $B_n^{(\alpha)}(x)$ of degree n in x are defined by the generating function

$$\left(\frac{z}{e^z-1}\right)^{\alpha} e^{xz} = \sum_{n=0}^{\infty} B_n^{(\alpha)}(x)\frac{z^n}{n!} \quad (|z|<2\pi;\ 1^{\alpha}:=1) \tag{24}$$

for arbitrary (real or complex) parameter α.

The value of $\zeta_n(-\ell,x)$ can be calculated explicitly for $\ell \in \mathbb{N}_0$. Taking $s=-\ell$ in the relation (22) with a replaced by x, we find that

$$\zeta_n(-\ell,x) = \Gamma(1+\ell)I_n(-\ell,x) = \ell!\,I_n(-\ell,x). \tag{25}$$

Now, from (24) and (25), we have the desired relationship:

$$\zeta_n(-\ell,x) = (-1)^n\frac{\ell!}{(n+\ell)!}B_{n+\ell}^{(n)}(x) \quad (\ell \in \mathbb{N}_0). \tag{26}$$

Setting $n=1$ in (26), we have the well-known result:

$$\zeta(-\ell,x) = -\frac{B_{\ell+1}(x)}{\ell+1} \quad (\ell \in \mathbb{N}_0), \tag{27}$$

where $\zeta(s,x) := \zeta_1(s,x)$ is the Hurwitz (or generalized) Zeta function [see (34)].

It is known (see [26]) that $\zeta_n(s,x)$ is expressible as a finite combination of the generalized Zeta function $\zeta(s,x)$ with polynomial coefficients in x:

$$\zeta_n(s,x) = \sum_{j=0}^{n-1} p_{n,j}(x)\zeta(s-j,x), \tag{28}$$

where

$$p_{n,j}(x) = \frac{1}{(n-1)!}\sum_{\ell=j}^{n-1}(-1)^{n+1-j}\binom{\ell}{j}s(n,\ell+1)x^{\ell-j}$$

and $s(n,k)$ are the Stirling numbers of the first kind.

Since $\zeta(s,x)$ can be continued analytically to a meromorphic function having a simple pole at $s=1$ with its residue 1, the representation (28) shows that $\zeta_n(s,x)$ is analytic for all s except for simple poles only at $s=k$ $(k=1,\ldots,n;\ n\in\mathbb{N})$ with their respective residues given by

$$\operatorname*{Res}_{s=k}\zeta_n(s,x) = p_{n,k-1}(x) \quad (k=1,\ldots,n;\ n\in\mathbb{N}).$$

From (28), $\zeta_n(s,x)$ can be expressed explicitly for the first few values of n:

$$\begin{aligned}
\zeta_2(s,x) =&(1-x)\zeta(s,x)+\zeta(s-1,x),\\
\zeta_3(s,x) =&\frac{1}{2}\left(x^2-3x+2\right)\zeta(s,x)+\left(\frac{3}{2}-x\right)\zeta(s-1,x)+\frac{1}{2}\zeta(s-2,x),\\
\zeta_4(s,x) =&\frac{1}{6}\left\{\left(-x^3+6x^2-11x+6\right)\zeta(s,x)+\left(3x^2-12x+11\right)\zeta(s-1,x)\right.\\
&\left.-(3x-6)\zeta(s-2,x)+\zeta(s-3,x)\right\}.
\end{aligned} \tag{29}$$

2.3 The Vardi–Barnes Multiple Gamma Functions

Vardi [118, p. 498] gave another expression for the multiple Gamma functions $\Gamma_n(a)$ whose general form was also studied by Barnes [14]:

$$\Gamma_n(a)=\left[\prod_{m=1}^{n} R_{n-m+1}^{(-1)^m\binom{a}{m-1}}\right] G_n(a) \quad (n\in\mathbb{N}), \tag{30}$$

where

$$G_n(a):=\exp\left[\zeta_n'(0,a)\right] \quad \text{with} \quad \zeta_n'(s,a)=\frac{\partial}{\partial s}\zeta_n(s,a)$$

and

$$R_m:=\exp\left(\sum_{k=1}^{m}\zeta_k'(0,1)\right) \quad \text{with} \quad R_0=1.$$

In particular, the special cases of (30) when $n=1$ and $n=2$ give other forms of the simple and double Gamma functions $\Gamma_1=\Gamma$ and Γ_2:

$$\Gamma(a)=\exp\left[-\zeta'(0)+\zeta'(0,a)\right]=\sqrt{2\pi}\,\exp\left[\zeta'(0,a)\right], \tag{31}$$

where $\zeta(s):=\zeta(s,1)$ is the Riemann Zeta function [see (43)];

$$\Gamma_2(a)=A(2\pi)^{\frac{1}{2}-\frac{1}{2}a}\exp\left(-\frac{1}{12}+\zeta_2'(0,a)\right), \tag{32}$$

where we have used (29) and the known identity (see Voros [121, p. 462, Eq. (A.11)]):

$$\log A=\frac{1}{12}-\zeta'(-1).$$

Here we can give another proof of the multiplication formula for Γ_2 different from that of Barnes [11] by using (32) (see Choi and Quine [35]). We consider

$$\begin{aligned}\sum_{\ell=0}^{n-1}\sum_{j=0}^{n-1}\zeta_2\left(s, a+\frac{\ell+j}{n}\right) &= \sum_{\ell=0}^{n-1}\sum_{j=0}^{n-1}\sum_{k_1,k_2=0}^{\infty}\left(a+\frac{\ell+j}{n}+k_1+k_2\right)^{-s}\\ &= n^s\sum_{k_1,k_2=0}^{\infty}\sum_{\ell=0}^{n-1}\sum_{j=0}^{n-1}(na+\ell+j+nk_1+nk_2)^{-s}\\ &= n^s\sum_{k_1,k_2=0}^{\infty}(na+k_1+k_2)^{-s} = n^s\zeta_2(s,na),\end{aligned}$$

which, upon differentiating with respect to s, yields

$$\sum_{\ell=0}^{n-1}\sum_{j=0}^{n-1}\zeta_2'\left(s, a+\frac{\ell+j}{n}\right) = (\log n)n^s\zeta_2(s,na) + n^s\zeta_2'(s,na).$$

By virtue of (32), we readily obtain the following multiplication formula for Γ_2:

$$\prod_{\ell=0}^{n-1}\prod_{j=0}^{n-1}\Gamma_2\left(a+\frac{\ell+j}{n}\right) = \mathcal{C}(n)(2\pi)^{-\frac{1}{2}n(n-1)a}n^{-\frac{n^2a^2}{2}+na}\Gamma_2(na), \qquad (33)$$

where

$$\mathcal{C}(n) := A^{n^2-1}\cdot e^{\frac{1}{12}(1-n^2)}\cdot(2\pi)^{\frac{1}{2}(n-1)}\cdot n^{\frac{5}{12}}.$$

An interesting identity is also obtained from (33):

$$\prod_{\ell=0}^{n-1}\prod_{j=0}^{n-1}{}'\,\Gamma_2\left(\frac{\ell+j}{n}\right) = \frac{\mathcal{C}(n)}{n},$$

where the prime denotes the exclusion of the case when $\ell = 0 = j$.

2.4 *The Hurwitz (or Generalized) Zeta Function*

The *Hurwitz* (or *generalized*) *Zeta function* $\zeta(s,a)$ is defined by

$$\zeta(s,a) := \sum_{k=0}^{\infty}(k+a)^{-s} \quad (\mathrm{Re}(s) > 1;\ a \in \mathbb{C}\setminus\mathbb{Z}_0^-). \qquad (34)$$

It is easy to see that $\zeta(s,a) = \zeta_1(s,a)$ for the case when $n = 1$ in (18). Thus we can deduce many properties of $\zeta(s,a)$ from those of $\zeta_n(s,a)$. Indeed, the series for $\zeta(s,a)$ in (34) converges absolutely for $\mathrm{Re}(s) = \sigma > 1$. The convergence is uniform in every half-plane $\sigma \geqq 1+\delta$ ($\delta > 0$), so $\zeta(s,a)$ is an analytic function of s in the half-plane $\mathrm{Re}(s) = \sigma > 1$. Setting $n = 1$ in (20), we have the integral representation

$$\begin{aligned}\Gamma(s)\zeta(s,a) &= \int_0^\infty \frac{x^{s-1}\mathrm{e}^{-ax}}{1-\mathrm{e}^{-x}}\,\mathrm{d}x = \int_0^\infty \frac{x^{s-1}\mathrm{e}^{-(a-1)x}}{\mathrm{e}^x-1}\,\mathrm{d}x \\ &= \int_0^1 \frac{x^{a-1}}{1-x}\left(\log\frac{1}{x}\right)^{s-1}\mathrm{d}x \quad (\mathrm{Re}(s)>1;\ \mathrm{Re}(a)>0). \qquad (35)\end{aligned}$$

Moreover, $\zeta(s,a)$ can be continued meromorphically to the whole complex s-plane (except for a simple pole at $s = 1$ with its residue 1) by means of the contour integral representation (see Theorem 2.3):

$$\zeta(s,a) = -\frac{\Gamma(1-s)}{2\pi\mathrm{i}}\int_C \frac{(-z)^{s-1}\mathrm{e}^{-az}}{1-\mathrm{e}^{-z}}\,\mathrm{d}z,$$

where the contour C is the Hankel loop of Theorem 2.3. The connection between $\zeta(s,a)$ and the Bernoulli polynomials $B_n(x)$ is also given in (27).

From the definition (34) of $\zeta(s,a)$, it easily follows that

$$\zeta(s,a) = \zeta(s,n+a) + \sum_{k=0}^{n-1}(k+a)^{-s} \quad (n \in \mathbb{N}); \qquad (36)$$

$$\zeta\left(s,\frac{1}{2}a\right) - \zeta\left(s,\frac{1}{2}a+\frac{1}{2}\right) = 2^s\sum_{n=0}^{\infty}(-1)^n(a+n)^{-s}.$$

2.5 Hurwitz's Formula for $\zeta(s,a)$

The series expression $\zeta(s,a)$ was originally meaningful for $\sigma > 1$ ($s = \sigma + \mathrm{i}t$). Hurwitz obtained another series representation for $\zeta(s,a)$ valid in the half-plane $\sigma < 0$:

$$\zeta(1-s,a) = \frac{\Gamma(s)}{(2\pi)^s}\left\{\mathrm{e}^{-\frac{1}{2}\pi\mathrm{i}s}L(a,s) + \mathrm{e}^{\frac{1}{2}\pi\mathrm{i}s}L(-a,s)\right\} \qquad (37)$$

$$(0 < a \leqq 1,\ \sigma = \mathrm{Re}(s) > 1;\ 0 < a < 1,\ \sigma > 0),$$

where the function $L(x, s)$ is defined by

$$L(x,s) := \sum_{n=1}^{\infty} \frac{e^{2\pi i n x}}{n^s} \quad (x \in \mathbb{R};\ \sigma = \mathrm{Re}(s) > 1), \tag{38}$$

which is often referred to as the *periodic* (or *Lerch*) *Zeta function*.

We note that the Dirichlet series in (38) is a periodic function of x with period 1 and that $L(1, s) = \zeta(s)$, the Riemann Zeta function (see Sect. 2.3). The series in (38) converges absolutely for $\sigma > 1$. Yet, if $x \notin \mathbb{Z}$, $\mathbb{Z}$ being the set of integers, the series can also be seen to converge conditionally for $\sigma > 0$. So the formula (37) is also valid for $\sigma > 0$ if $a \neq 1$.

If we take $a = p/q$ in the Hurwitz formula (37), we obtain

$$\zeta\left(1-s, \frac{p}{q}\right) = \frac{2\Gamma(s)}{(2\pi q)^s} \sum_{r=1}^{q} \cos\left(\frac{\pi s}{2} - \frac{2\pi r p}{q}\right) \zeta\left(s, \frac{r}{q}\right) \quad (1 \leqq p \leqq q;\ p, q \in \mathbb{N}),$$

which holds true, by the principle of analytic continuation, for all admissible values of $s \in \mathbb{C}$.

For other interesting properties of $\zeta(s, a)$, see [66–69].

2.6 Hermite's Formula for $\zeta(s, a)$

We recall *Hermite's formula* for $\zeta(s, a)$:

$$\zeta(s,a) = \frac{1}{2} a^{-s} + \frac{a^{1-s}}{s-1} + 2 \int_0^{\infty} \left(a^2 + y^2\right)^{-\frac{1}{2}s} \left\{\sin\left(s \arctan \frac{y}{a}\right)\right\} \frac{dy}{e^{2\pi y} - 1}. \tag{39}$$

We note that the integral involved in (39) converges for all admissible values of $s \in \mathbb{C}$. Moreover, the integral is an entire function of s. A special case of the formula (39) when $a = 1$ is attributed to Jensen.

Setting $s = 0$ in (39), we have

$$\zeta(0, a) = \frac{1}{2} - a.$$

It is also known that

$$\psi(s) = \log s - \frac{1}{2s} - 2 \int_0^{\infty} \frac{t\, dt}{(t^2 + s^2)(e^{2\pi t} - 1)} \quad (\mathrm{Re}(s) > 0). \tag{40}$$

Taking the limit in (39) as $s \to 1$, by virtue of the uniform convergence of the integral in (39), we get

$$\lim_{s\to 1}\left\{\zeta(s,a)-\frac{1}{s-1}\right\}=\lim_{s\to 1}\frac{a^{1-s}-1}{s-1}+\frac{1}{2a}+2\int_0^\infty \frac{y\,\mathrm{d}y}{(a^2+y^2)\,(e^{2\pi y}-1)},$$

which, in view of (40), yields

$$\lim_{s\to 1}\left\{\zeta(s,a)-\frac{1}{s-1}\right\}=-\frac{\Gamma'(a)}{\Gamma(a)}=-\psi(a). \tag{41}$$

Differentiating (39) with respect to s and setting $s=0$ in the resulting equation, we have

$$\left\{\frac{\mathrm{d}}{\mathrm{d}s}\zeta(s,a)\right\}_{s=0}=\left(a-\frac{1}{2}\right)\log a-a+2\int_0^\infty \frac{\arctan\left(\frac{y}{a}\right)}{\mathrm{e}^{2\pi y}-1}\,\mathrm{d}y,$$

which yields

$$\left.\frac{\mathrm{d}}{\mathrm{d}s}\zeta(s,a)\right|_{s=0}=\log\Gamma(a)-\frac{1}{2}\log(2\pi), \tag{42}$$

which is equivalent to the identity (31). In addition to (42), it is easy to find from the definition (34) of $\zeta(s,a)$ that

$$\frac{\partial}{\partial a}\zeta(s,a)=-s\zeta(s+1,a).$$

The respective special cases of (41) and (42) when $a=1$ become

$$\lim_{s\to 1}\left\{\zeta(s,a)-\frac{1}{s-1}\right\}=\lim_{\epsilon\to 0}\left\{\zeta(1+\epsilon,a)-\frac{1}{\epsilon}\right\}=\gamma$$

and

$$\zeta'(0)=-\frac{1}{2}\log(2\pi),$$

where $\zeta(s)$ is the Riemann Zeta function [see definition (43)].

2.7 The Riemann Zeta Function

The *Riemann Zeta function* $\zeta(s)$ is defined by

$$\zeta(s):=\begin{cases}\displaystyle\sum_{n=1}^\infty\frac{1}{n^s}=\frac{1}{1-2^{-s}}\sum_{n=1}^\infty\frac{1}{(2n-1)^s} & (\mathrm{Re}(s)>1),\\ \displaystyle\frac{1}{1-2^{1-s}}\sum_{n=1}^\infty\frac{(-1)^{n-1}}{n^s} & (\mathrm{Re}(s)>0;\ s\neq 1).\end{cases} \tag{43}$$

It is easy to see from the definitions (43) and (34) that

$$\zeta(s) = \zeta(s,1) = (2^s - 1)^{-1}\,\zeta\left(s, \frac{1}{2}\right) = 1 + \zeta(s,2) \tag{44}$$

and

$$\zeta(s) = \frac{1}{m^s - 1}\sum_{j=1}^{m-1}\zeta\left(s, \frac{j}{m}\right) \quad (m \in \mathbb{N}\setminus\{1\}).$$

In view of (44), we can deduce many properties of $\zeta(s)$ from those of $\zeta(s,a)$ given in the previous section. In fact, the series $\zeta(s) = \sum_{n=1}^{\infty} n^{-s}$ in (43) represents an analytic function of s in the half-plane $\mathrm{Re}(s) = \sigma > 1$. Setting $a = 1$ in (35), we have an integral representation of $\zeta(s)$ in the form

$$\begin{aligned}\Gamma(s)\zeta(s) &= \int_0^\infty \frac{x^{s-1}\mathrm{e}^{-x}}{1-\mathrm{e}^{-x}}\,\mathrm{d}x = \int_0^\infty \frac{x^{s-1}}{\mathrm{e}^x - 1}\,\mathrm{d}x \\ &= \int_0^1 \frac{1}{1-x}\left(\log\frac{1}{x}\right)^{s-1}\mathrm{d}x \quad (\mathrm{Re}(s) > 1).\end{aligned}$$

Furthermore, just as $\zeta(s,a)$, $\zeta(s)$ can be continued meromorphically to the whole complex s-plane (except for a simple pole at $s = 1$ with its residue 1) by means of the contour integral representation:

$$\zeta(s) = -\frac{\Gamma(1-s)}{2\pi\mathrm{i}}\int_C \frac{(-z)^{s-1}\mathrm{e}^{-z}}{1-\mathrm{e}^{-z}}\,\mathrm{d}z,$$

where the contour C is the Hankel loop of Theorem 2.3.

Here, for later use, we choose to recall some properties and relationships of $\zeta(s)$:

$$\zeta(s) = \zeta(s, n+1) + \sum_{k=1}^{n} k^{-s} \quad (n \in \mathbb{N}_0).$$

The connection between $\zeta(s)$ and the Bernoulli numbers is given as follows:

$$\zeta(-n) = \begin{cases} -\dfrac{1}{2} & (n = 0) \\[2ex] -\dfrac{B_{n+1}}{n+1} & (n \in \mathbb{N}). \end{cases}$$

The *Riemann's functional equation* for $\zeta(s)$ is

$$\zeta(1-s) = 2(2\pi)^{-s}\Gamma(s)\cos\left(\frac{1}{2}\pi s\right)\zeta(s) \tag{45}$$

or, equivalently,

$$\zeta(s) = 2(2\pi)^{s-1}\,\Gamma(1-s)\,\sin\left(\frac{1}{2}\pi s\right)\,\zeta(1-s). \tag{46}$$

Taking $s = 2n+1$ ($n \in \mathbb{N}$) in (45), the factor $\cos\left(\frac{1}{2}\pi s\right)$ vanishes and we find that

$$\zeta(-2n) = 0 \quad (n \in \mathbb{N}), \tag{47}$$

which are often referred to as the *trivial zeros* of $\zeta(s)$.

We have the well-known identity

$$\zeta(2n) = (-1)^{n+1}\frac{(2\pi)^{2n}}{2(2n)!}B_{2n} \quad (n \in \mathbb{N}_0),$$

which enables us to list the following special values:

$$\zeta(2) = \frac{\pi^2}{6}, \quad \zeta(4) = \frac{\pi^4}{90}, \quad \zeta(6) = \frac{\pi^6}{945},$$
$$\zeta(8) = \frac{\pi^8}{9450}, \quad \zeta(10) = \frac{\pi^{10}}{93555}, \ldots.$$

It is easy to derive from (46) and (47) that (cf., e.g., Srivastava [113, p. 387, Eq. (1.15)])

$$\zeta'(-2n) = \lim_{\epsilon\to 0}\frac{\zeta(-2n+\epsilon)}{\epsilon} = (-1)^n\frac{(2n)!}{2(2\pi)^{2n}}\zeta(2n+1) \quad (n \in \mathbb{N}).$$

3 A Set of Mathematical Constants

There are some classes of mathematical constants involved naturally in the Gamma and multiple Gamma functions. Here we introduce those well-known mathematical constants associated with the Gamma and multiple Gamma functions and show how they are involved, if possible (see [28]; see also [115, Chap. 7]).

We begin by noting that γ *is a constant so chosen that* $\Gamma(1) = 1$ in the Weierstrass product form of the Gamma function $\Gamma(z)$ [see (2)] and the constant γ is the very Euler constant in (2).

In fact, the function

$$f(z) = \prod_{n=1}^{\infty}\left(1+\frac{z}{n}\right)e^{-\frac{z}{n}}$$

can be seen to be an entire function having zeros at the negative integers. The function $f(z-1)$ is an entire function having zeros at the origin as well as at the negative integers.

It is known that

$$f(z-1) = e^{g(z)} z f(z), \tag{48}$$

where $g(z)$ is some entire function. The logarithmic derivative of (48) gives

$$\sum_{n=1}^{\infty} \left(\frac{1}{n+z-1} - \frac{1}{n} \right) = g'(z) + \frac{1}{z} + \sum_{n=1}^{\infty} \left(\frac{1}{n+z} - \frac{1}{n} \right). \tag{49}$$

The sum on the left side of (49) can be expressed as

$$\begin{aligned} \left(\frac{1}{z}-1\right)+\sum_{n=2}^{\infty}\left(\frac{1}{n+z-1}-\frac{1}{n}\right) &= \frac{1}{z}-1+\sum_{n=1}^{\infty}\left\{\left(\frac{1}{n+z}-\frac{1}{n}\right)+\left(\frac{1}{n}-\frac{1}{n+1}\right)\right\} \\ &= \frac{1}{z}-1+\sum_{n=1}^{\infty}\left(\frac{1}{n+z}-\frac{1}{n}\right)+1 \\ &= \frac{1}{z}+\sum_{n=1}^{\infty}\left(\frac{1}{n+z}-\frac{1}{n}\right). \end{aligned}$$

Setting this in (49) yields $g'(z) \equiv 0$. Thus $g(z)$ is a constant and let $g(z) = \gamma$. To determine γ, putting $z = 1$ in (48) gives

$$1 = f(0) = e^{\gamma} \prod_{n=1}^{\infty} \left(1 + \frac{1}{n}\right) e^{-\frac{1}{n}}.$$

Therefore, taking the natural logarithm yields the *Euler constant* (also called *Euler–Mascheroni constant*) given in (3) (see, e.g., [33, 125]).

The Glaisher–Kinkelin constant A given in (11) is a constant which involves naturally in the theory of the double Gamma function $\Gamma_2 = 1/G$ (see, e.g., (10), (12) and (13)).

We introduce two interesting mathematical constants, in addition to the Glaisher–Kinkelin constant A, by recalling the Euler–Maclaurin summation formula (cf. Hardy [78, p. 318]):

$$\sum_{k=1}^{n} f(k) \sim C_0 + \int_a^n f(x)\, dx + \frac{1}{2} f(n) + \sum_{r=1}^{\infty} \frac{B_{2r}}{(2r)!} f^{(2r-1)}(n),$$

where C_0 is an arbitrary constant to be determined in each special case and

$$B_0 = 1,\ B_1 = -\frac{1}{2},\ B_2 = \frac{1}{6},\ B_4 = -\frac{1}{30},\ B_6 = \frac{1}{42},\ B_8 = -\frac{1}{30},\ B_{10} = \frac{5}{66}, \ldots,$$

and $B_{2n+1} = 0$ ($n \in \mathbb{N}$) are the Bernoulli numbers. For another useful summation formula, see Edwards [65, p. 117].

Letting $f(x) = x^2 \log x$ and $f(x) = x^3 \log x$ in (68) with $a = 1$, we obtain

$$\log B = \lim_{n\to\infty}\left[\sum_{k=1}^{n} k^2 \log k - \left(\frac{n^3}{3} + \frac{n^2}{2} + \frac{n}{6}\right)\log n + \frac{n^3}{9} - \frac{n}{12}\right] \tag{50}$$

and

$$\log C = \lim_{n\to\infty}\left[\sum_{k=1}^{n} k^3 \log k - \left(\frac{n^4}{4} + \frac{n^3}{2} + \frac{n^2}{4} - \frac{1}{120}\right)\log n + \frac{n^4}{16} - \frac{n^2}{12}\right], \tag{51}$$

respectively; here B and C are constants whose approximate numerical values are given by

$$B \cong 1.03091\,675\ldots \quad \text{and} \quad C \cong 0.97955\,746\ldots.$$

The constants B and C were considered recently by Choi and Srivastava [40, 42].

Bendersky [17] (see also [2, p. 199]) presented a set of constants including B and C defined, respectively, by (50) and (51): There exist constants D_k defined by

$$\log D_k := \lim_{n\to\infty}\left(\sum_{m=1}^{n} m^k \log m - p(n,k)\right) \quad (k \in \mathbb{N}_0), \tag{52}$$

where the definition of $p(n,k)$ in Adamchik [2, p. 198, Eq. (20)] is *corrected* here as follows:

$$\begin{aligned} p(n,k) :=\, & \frac{n^k}{2}\log n + \frac{n^{k+1}}{k+1}\left(\log n - \frac{1}{k+1}\right) \\ & + k!\sum_{j=1}^{k} \frac{n^{k-j}\,B_{j+1}}{(j+1)!(k-j)!}\left[\log n + (1-\delta_{kj})\sum_{\ell=1}^{j}\frac{1}{k-\ell+1}\right] \end{aligned}$$

and δ_{kj} is the Kronecker symbol defined by $\delta_{kj} = 0\ (k \neq j)$ and $\delta_{kj} = 1\ (k = j)$.

For the constants D_k $(k \in \mathbb{N}_0)$ defined in (52), we can show that

$$D_0 = (2\pi)^{\frac{1}{2}}, \quad D_1 = A, \quad D_2 = B, \quad D_3 = C$$

and

$$\log D_k = \frac{B_{k+1}H_k}{k+1} - \zeta'(-k) \quad (k \in \mathbb{N}_0),$$

where B_n are the Bernoulli numbers and H_n are the harmonic numbers.

The constants introduced in this section can be seen to be involved in the theory of multiple Gamma functions. For example, the following log-multiple Gamma integral (see [31, p. 523, Eq. (2.50)])

$$\int_0^{\frac{3}{2}} \log \Gamma_3(t+2)\,\mathrm{d}t = \int_0^1 \log G(t+1)\,\mathrm{d}t + \int_0^1 \log \Gamma_3(t+1)\,\mathrm{d}t + \int_0^{\frac{1}{2}} \log \Gamma(t+1)\,\mathrm{d}t$$
$$+ 2\int_0^{\frac{1}{2}} \log G(t+1)\,\mathrm{d}t + \int_0^{\frac{1}{2}} \log \Gamma_3(t+1)\,\mathrm{d}t$$
$$= -\frac{259}{768} - \frac{29}{1920}\log 2 + \frac{9}{16}\log \pi - \frac{15}{16}\log A - \frac{5}{4}\log B + \frac{15}{16}\log C.$$

For other analogous or generalized classes of mathematical constants and their applications, see [32, 40, 45, 58], [46, Theorem 2, p. 403].

4 Series Associated with the Zeta Functions

A rather classical (over two centuries old) theorem of Christian Goldbach (1690–1764), which was stated in a letter dated 1729 from Goldbach to Daniel Bernoulli (1700–1782), was revived in 1986 by Shallit and Zikan [109] as the following problem:

$$\sum_{\omega \in \mathcal{S}} (\omega - 1)^{-1} = 1, \tag{53}$$

where $\mathcal{S}$ denotes the set of all nontrivial integer kth powers, that is,

$$\mathcal{S} := \{n^k \mid n, k \in \mathbb{N} \setminus \{1\}\}.$$

Goldbach's theorem (53) assumes the elegant form (cf. Shallit and Zikan [109, p. 403])

$$\sum_{\omega \in \mathcal{S}} (\omega - 1)^{-1} = \sum_{k=2}^{\infty} \{\zeta(k) - 1\} = 1 \tag{54}$$

or, equivalently,

$$\sum_{k=2}^{\infty} \mathcal{F}(\zeta(k)) = 1,$$

where $\mathcal{F}(x) := x - [x]$ denotes the *fractional* part of $x \in \mathbb{R}$. As a matter of fact, it is fairly straightforward to observe also that

$$\sum_{k=2}^{\infty} (-1)^k \mathcal{F}(\zeta(k)) = \frac{1}{2}, \quad \sum_{k=1}^{\infty} \mathcal{F}(\zeta(2k)) = \frac{3}{4} \quad \text{and} \quad \sum_{k=1}^{\infty} \mathcal{F}(\zeta(2k+1)) = \frac{1}{4}.$$

The subject of closed-form summation of series involving the Zeta functions has been remarkably widely investigated (see [8, 31, 41, 42, 44, 46, 111, 112, 114, 115]). Among the various methods and techniques used in the vast literature on the subject, Srivastava and Choi [114, 115] gave reasonably detailed accounts of those using the binomial theorem, generating functions, multiple Gamma functions (see [11–14, 41, 42, 89, 118]), and hypergeometric identities, presented a rather extensive collection of closed-form sums of series involving the Zeta functions, and showed that many of those summation formulas find their applications in the evaluations of the determinants of the Laplacians for the n-dimensional sphere $\mathbf{S}^n$ with the standard metric (see [25, 31, 41, 61, 89, 100, 103, 108, 114, 115, 118, 121]).

Here we choose to recall some closed-form summations of series involving the Zeta functions expressed (or evaluated) by means of the Gamma and multiple Gamma functions and their related functions (see [114, 115]; see also references [38, 39, 43, 51, 56, 82]):

$$\sum_{k=2}^{\infty} \zeta(k,a)\frac{t^k}{k} = \log\Gamma(a-t) - \log\Gamma(a) + t\psi(a) \quad (|t|<|a|);$$

$$\sum_{k=2}^{\infty} \zeta(k,a)t^{k-1} = -\psi(a-t) + \psi(a) \quad (|t|<|a|); \tag{55}$$

$$\sum_{k=2}^{\infty} \{\zeta(k)-1\}\frac{t^k}{k} = \log\Gamma(2-t) + (1-\gamma)t \quad (|t|<2);$$

$$\begin{aligned}\sum_{k=2}^{\infty} \zeta(k,a)\frac{z^{k+1}}{k+1} &= \frac{\psi(a)}{2}z^2 + z\log\Gamma(a-z) + \int_0^{-z}\log\Gamma(a+t)\,\mathrm{d}t \\ &= [2a-1-\log(2\pi)]\frac{z}{2} + [\psi(a)-1]\frac{z^2}{2} + (a-1)\log\Gamma(a-z) \\ &\quad - \log G(a-z) + (1-a)\log\Gamma(a) + \log G(a) \quad (|z|<|a|);\end{aligned}$$

$$\sum_{k=1}^{\infty} \{\zeta(2k)-1\}\frac{z^{2k+1}}{2k+1} = \frac{1}{2}\left\{[3-\log(2\pi)]z + \log\frac{G(2+z)\Gamma(2-z)}{G(2-z)\Gamma(2+z)}\right\} \quad (|z|<2);$$

$$\begin{aligned}\sum_{k=2}^{\infty} \zeta(k,a)\frac{z^{k+2}}{k+2} &= \frac{1-a}{2}[1-2a+\log(2\pi)]z + [1-\log(2\pi)]\frac{z^2}{4} + [\psi(a)-1]\frac{z^3}{3} \\ &\quad + (a-1)^2\log\Gamma(a-z) - (z+a-1)\log G(a-z) - (a-1)^2\log\Gamma(a) \\ &\quad + (a-1)\log G(a) - \int_0^{-z}\log G(t+a)\,\mathrm{d}t \quad (|z|<|a|).\end{aligned}$$

We also recall a known general formula for the series associated the Zeta functions (see [114, p. 149, Theorem 3.1]; see also [44]) asserted by the following theorem.

Theorem 4.1. *For every nonnegative integer n,*

$$\sum_{k=2}^{\infty} \frac{\zeta(k,a)}{(k)_{n+1}} t^{n+k} = \frac{(-1)^n}{n!}[\zeta'(-n,a-t)-\zeta'(-n,a)]$$
$$+\sum_{k=1}^{n} \frac{(-1)^{n+k}}{n!}\binom{n}{k}\left[(H_n-H_{n-k})\,\zeta(k-n,a)-\zeta'(k-n,a)\right]t^k$$
$$+[H_n+\psi(a)]\frac{t^{n+1}}{(n+1)!} \quad (|t|<|a|;\ n\in\mathbb{N}_0),$$

where $\zeta'(s,a) = \frac{\partial}{\partial s}\zeta(s,a)$.

5 Generalized Goldbach–Euler Series

We first present the corrected expression for a certain widely recorded generalized Goldbach–Euler series. The corrected forms are then shown to be connected with the problem of closed-form evaluation of series involving the Zeta functions, which happens to be an extensively investigated subject since the time of Euler as, for example, in the Goldbach theorem (see (53) and (54)). It should be remarked in passing that the arguments in this section is based on the paper [55].

The generalized Goldbach–Euler series has been widely investigated and recorded in the following form (see [95, p. 59, Eq. (9)]; see also [77, p. 894, Entry 8.363 (7)] and [93, p. 88, Eq. (5)]):

$$\sum_{k=2}^{\infty}\sum_{n=0}^{\infty} \frac{1}{(pn+r)^k-1} = \frac{1}{p}\left[\psi\left(\frac{r}{p}\right)-\psi\left(\frac{r-1}{p}\right)\right] \tag{56}$$

$$(p\in\mathbb{N};\ r=p\neq 1 \text{ or } r=p+1).$$

For convenience, we rewrite the cases $r = p \neq 1$ and $r = p+1$ $(p \in \mathbb{N})$ of the generalized Goldbach–Euler series (56) in the following separate forms:

$$\sum_{k=2}^{\infty}\sum_{n=1}^{\infty} \frac{1}{(pn)^k-1} = \frac{1}{p}\left[\psi(1)-\psi\left(1-\frac{1}{p}\right)\right] \quad (p\in\mathbb{N}\setminus\{1\}) \tag{57}$$

and

$$\sum_{k=2}^{\infty}\sum_{n=1}^{\infty} \frac{1}{(pn+1)^k-1} = 1+\frac{1}{p}\left[\psi\left(\frac{1}{p}\right)-\psi(1)\right] \quad (p\in\mathbb{N}). \tag{58}$$

The special case of the generalized Goldbach–Euler series (58) when $p = 1$ is recorded in [93, p. 88] (see also [95, p. 59, Eq. (10)]):

$$\sum_{k=2}^{\infty}\sum_{n=2}^{\infty}\frac{1}{n^k-1}=1. \tag{59}$$

By using *Mathematica* (Version 6) with a view to finding an approximate numerical value of the series in (59), we observed that

$$\sum_{k=2}^{10}\sum_{n=2}^{10}\frac{1}{n^k-1}\simeq 1.02912 \quad \text{and} \quad \sum_{k=2}^{100}\sum_{n=2}^{100}\frac{1}{n^k-1}\simeq 1.1204.$$

Obviously, we can also show the error in the expression in (59) [and so in (57) and (58) as well] by observing that

$$\sum_{k=2}^{\infty}\sum_{n=2}^{\infty}\frac{1}{n^k-1}>\sum_{k=2}^{\infty}\sum_{n=2}^{\infty}\frac{1}{n^k}=\sum_{n=2}^{\infty}\frac{1}{n(n-1)}=1.$$

5.1 *Corrected Expressions for the Generalized Goldbach–Euler Series*

The duly corrected forms of the generalized Goldbach–Euler series (57) and (58) are asserted by Theorem 5.1 below (see [55, Theorem 1]).

Theorem 5.1. *Each of the following results holds true:*

$$\sum_{\omega\in\mathcal{S}_{p,0}}(\omega-1)^{-1}=\frac{1}{p}\left[\psi(1)-\psi\left(1-\frac{1}{p}\right)\right] \quad (p\in\mathbb{N}\setminus\{1\}) \tag{60}$$

and

$$\sum_{\omega\in\mathcal{S}_{p,1}}(\omega-1)^{-1}=1+\frac{1}{p}\left[\psi\left(\frac{1}{p}\right)-\psi(1)\right] \quad (p\in\mathbb{N}), \tag{61}$$

where the set $\mathcal{S}_{p,0}$ is defined (for fixed $p\in\mathbb{N}\setminus\{1\}$) by

$$\mathcal{S}_{p,0}:=\left\{(pn)^k : n\in\mathbb{N} \text{ and } k\in\mathbb{N}\setminus\{1\}\right\} \tag{62}$$

and the set $\mathcal{S}_{p,1}$ is defined (for fixed $p\in\mathbb{N}$) by

$$\mathcal{S}_{p,1}:=\left\{(pn+1)^k : n\in\mathbb{N} \text{ and } k\in\mathbb{N}\setminus\{1\}\right\}. \tag{63}$$

5.2 Closed-Form Evaluation of Series Involving Zeta Functions

Just as the Zeta-function series in (54), the series (60) and (61) can be expressed as series involving the Riemann and Hurwitz (or generalized) Zeta functions (see [55, Theorem 2]).

Theorem 5.2. *Each of the following results holds true:*

$$\sum_{\omega \in \mathcal{S}_{p,0}} (\omega - 1)^{-1} = \sum_{k=2}^{\infty} \frac{\zeta(k)}{p^k} = \frac{1}{p}\left[\psi(1) - \psi\left(1 - \frac{1}{p}\right)\right] \quad (p \in \mathbb{N} \setminus \{1\}) \quad (64)$$

and

$$\sum_{\omega \in \mathcal{S}_{p,1}} (\omega - 1)^{-1} = \sum_{k=2}^{\infty} \frac{1}{p^k} \zeta\left(k, 1 + \frac{1}{p}\right) = 1 + \frac{1}{p}\left[\psi\left(\frac{1}{p}\right) - \psi(1)\right] \quad (p \in \mathbb{N}), \quad (65)$$

where the sets $\mathcal{S}_{p,0}$ *and* $\mathcal{S}_{p,1}$ *are defined by* (62) *and* (63)*, respectively.*

Remark 5.1. In view of the identity (36), the special case of (64) when $p = 1$ yields the Goldbach theorem (54). Each of the series involving the Zeta function in (64) and (65) is an obvious special case of (55).

6 Determinants of Laplacians

During the last two decades, the problem of evaluation of the determinants of the Laplacians on Riemann manifolds has received considerable attention by many authors including (among others) D'Hoker and Phong [61, 62], Sarnak [108], and Voros [121], who computed the determinants of the Laplacians on compact Riemann surfaces of constant curvature in terms of special values of the Selberg Zeta function. Although the first interest in the determinants of the Laplacians arose mainly for Riemann surfaces, it is also interesting and potentially useful to compute these determinants for classical Riemannian manifolds of higher dimensions, such as spheres. In this chapter, we are particularly concerned with the evaluation of the functional determinant for the n-dimensional sphere $\mathbf{S}^n$ with the standard metric.

In computations of the determinants of the Laplacians on manifolds of constant curvature, an important rôle is played by the closed-form evaluations of the series involving the Zeta function given in Chap. 3 (cf., e.g., Choi and Srivastava [41, 42], and Choi et al. [31]) as well as the theory of the multiple Gamma functions presented in Sect. 1 (cf., e.g., Voros [121], Vardi [118], Choi [25], and Quine and Choi [103]).

6.1 The n-Dimensional Problem

Let $\{\lambda_n\}$ be a sequence such that

$$0 = \lambda_0 < \lambda_1 \leqq \lambda_2 \leqq \cdots \leqq \lambda_n \leqq \cdots; \quad \lambda_n \uparrow \infty \quad (n \to \infty); \tag{66}$$

henceforth we consider only such nonnegative increasing sequences. Then we can show that

$$Z(s) := \sum_{n=1}^{\infty} \frac{1}{\lambda_n^s}, \tag{67}$$

which is known to converge absolutely in a half-plane $\mathrm{Re}(s) > \sigma$ for some $\sigma \in \mathbb{R}$.

Definition 6.1 (cf. Osgood et al. [100]). The determinant of the Laplacian Δ on the compact manifold M is defined to be

$$\det{}' \Delta := \prod_{\lambda_k \neq 0} \lambda_k, \tag{68}$$

where $\{\lambda_k\}$ is the sequence of eigenvalues of the Laplacian Δ on M.

The sequence $\{\lambda_k\}$ is known to satisfy the condition as in (66), but the product in (68) is always divergent; so, in order for the expression (68) to make sense, some sort of regularization procedure must be used (see e.g., [63, 104, 116, 120]). It is easily seen that, formally, $\mathrm{e}^{-Z'(0)}$ is the product of nonzero eigenvalues of Δ. This product does not converge, but $Z(s)$ can be continued analytically to a neighborhood of $s = 0$. Therefore, we can give a *meaningful* definition:

$$\det{}' \Delta := \mathrm{e}^{-Z'(0)},$$

which is called the *Functional Determinant of the Laplacian* Δ on M.

Definition 6.2. The *order* μ of the sequence $\{\lambda_k\}$ is defined by

$$\mu := \inf\left\{\alpha > 0 \;\middle|\; \sum_{k=1}^{\infty} \frac{1}{\lambda_k^\alpha} < \infty\right\}.$$

The analogous and shifted analogous Weierstrass canonical products $E(\lambda)$ and $E(\lambda, a)$ of the sequence $\{\lambda_k\}$ are defined, respectively, by

$$E(\lambda) := \prod_{k=1}^{\infty} \left\{ \left(1 - \frac{\lambda}{\lambda_k}\right) \exp\left(\frac{\lambda}{\lambda_k} + \frac{\lambda^2}{2\lambda_k^2} + \cdots + \frac{\lambda^{[\mu]}}{[\mu]\lambda_k^{[\mu]}} \right) \right\}$$

and

$$E(\lambda,a) := \prod_{k=1}^{\infty}\left\{\left(1-\frac{\lambda}{\lambda_k+a}\right)\exp\left(\frac{\lambda}{\lambda_k+a}+\cdots+\frac{\lambda^{[\mu]}}{[\mu]\,(\lambda_k+a)^{[\mu]}}\right)\right\},$$

where $[\mu]$ denotes the greatest integer part in the order μ of the sequence $\{\lambda_k\}$.

There exists the following relationship between $E(\lambda)$ and $E(\lambda,a)$ (see Voros [121]):

$$E(\lambda,a) = \exp\left(\sum_{m=1}^{[\mu]}\mathcal{R}_{m-1}(-a)\frac{\lambda^m}{m!}\right)\frac{E(\lambda-a)}{E(-a)},$$

where, for convenience,

$$\mathcal{R}_{[\mu]}(\lambda-a) := \frac{\mathrm{d}^{[\mu]+1}}{\mathrm{d}\lambda^{[\mu]+1}}\{-\log E(\lambda,a)\}.$$

The shifted series $Z(s,a)$ of $Z(s)$ in (67) by a is given by

$$Z(s,a) := \sum_{k=1}^{\infty}\frac{1}{(\lambda_k+a)^s}.$$

Formally, indeed, we have

$$Z'(0,-\lambda) = -\sum_{k=1}^{\infty}\log(\lambda_k-\lambda),$$

which, if we define

$$D(\lambda) := \exp\left[-Z'(0,-\lambda)\right], \tag{69}$$

immediately implies that

$$D(\lambda) = \prod_{k=1}^{\infty}(\lambda_k-\lambda).$$

In fact, Voros [121] gave the relationship between $D(\lambda)$ and $E(\lambda)$ as follows:

$$D(\lambda) = \exp[-Z'(0)]\,\exp\left[-\sum_{m=1}^{[\mu]}\mathrm{FP}\,Z(m)\frac{\lambda^m}{m}\right]$$

$$\times\exp\left[-\sum_{m=2}^{[\mu]}C_{-m}\left(\sum_{k=1}^{m-1}\frac{1}{k}\right)\frac{\lambda^m}{m!}\right]E(\lambda),$$

where an empty sum is interpreted to be nil and the *finite part* prescription is applied (as usual) as follows (cf. Voros [121, p. 446]):

$$\mathrm{FP}\, f(s) := \begin{cases} f(s), & \text{if } s \text{ is not a pole,} \\ \lim\limits_{\epsilon \to 0} \left(f(s+\epsilon) - \dfrac{\text{Residue}}{\epsilon} \right), & \text{if } s \text{ is a simple pole,} \end{cases}$$

and

$$Z(-m) = (-1)^m m! C_{-m}.$$

Now consider the sequence of eigenvalues on the standard Laplacian Δ_n on $\mathbf{S}^n$. It is known from the work of Vardi [118] (see also Terras [117]) that the standard Laplacian Δ_n ($n \in \mathbb{N}$) has eigenvalues

$$\mu_k := k(k+n-1) \tag{70}$$

with multiplicity

$$\begin{aligned} \beta_k^n := \binom{k+n}{n} - \binom{k+n-2}{n} &= \frac{(2k+n-1)\,(k+n-2)!}{k!\,(n-1)!} \\ &= \frac{2k+n-1}{(n-1)!} \prod_{j=1}^{n-2} (k+j) \quad (k \in \mathbb{N}_0). \end{aligned} \tag{71}$$

From now on we consider the shifted sequence $\{\lambda_k\}$ of $\{\mu_k\}$ in (70) by $\left(\frac{n-1}{2}\right)^2$ as a fundamental sequence. Then the sequence $\{\lambda_k\}$ is written in the following simple and tractable form:

$$\lambda_k = \mu_k + \left(\frac{n-1}{2}\right)^2 = \left(k + \frac{n-1}{2}\right)^2 \tag{72}$$

with the same multiplicity as in (71).

We will exclude the zero mode, that is, start the sequence at $k = 1$ for later use. Furthermore, with a view to emphasizing n on $\mathbf{S}^n$, we choose the notations $Z_n(s)$, $Z_n(s,a)$, $E_n(\lambda)$, $E_n(\lambda,a)$, and $D_n(\lambda)$ instead of $Z(s)$, $Z(s,a)$, $E(\lambda)$, $E(\lambda,a)$, and $D(\lambda)$, respectively.

We readily observe from (69) that

$$D_n\left(\left(\frac{n-1}{2}\right)^2\right) = \det{}' \Delta_n,$$

where $\det' \Delta_n$ denote the *determinants of the Laplacians* on $\mathbf{S}^n$ ($n \in \mathbb{N}$).

Choi [25] computed the determinants of the Laplacians on the n-dimensional unit sphere $\mathbf{S}^n$ $(n = 1, 2, 3)$ by factorizing the analogous Weierstrass canonical product of a shifted sequence $\{\lambda_k\}$ in (72) of eigenvalues of the Laplacians on $\mathbf{S}^n$ into multiple Gamma functions, while Choi and Srivastava [41, 42] and Choi et al. [31] made use of some closed-form evaluations of the series involving Zeta function given in [114, Chap. 3] for the computation of the determinants of the Laplacians on $\mathbf{S}^n$ $(n = 2, 3, 4, 5, 6, 7)$. Quine and Choi [103] made use of Zeta-regularized products to compute $\det' \Delta_n$ and the determinant of the conformal Laplacian, $\det(\Delta_{\mathbf{S}^n} + n(n-2)/4)$.

In the following sections we compute the determinants of the Laplacians on $\mathbf{S}^n$ $(n = 1, 2 \text{ and } 3)$.

6.2 *Factorizations into Simple and Multiple Gamma Functions*

We begin by expressing $E_n(\lambda)$ $(n = 1, 2, 3)$ as the simple and multiple Gamma functions. Our results are summarized in the following proposition (see Choi [25]; see also Voros [121]).

Theorem 6.1. *The analogous Weierstrass canonical products* $E_n(\lambda)$ $(n = 1, 2, 3)$ *of the shifted sequence* $\{\lambda_k\}$ *in* (72) *can be expressed in terms of the simple and multiple Gamma functions as follows:*

$$E_1(\lambda) = \frac{1}{\left\{\Gamma\left(1 - \sqrt{\lambda}\right)\Gamma\left(1 + \sqrt{\lambda}\right)\right\}^2},$$

$$E_2(\lambda) = \frac{\left\{\Gamma_2\left(\frac{1}{2}\right)\right\}^4 e^{c_1\lambda}\,\Gamma\left(\frac{1}{2} - \sqrt{\lambda}\right)\Gamma\left(\frac{1}{2} + \sqrt{\lambda}\right)}{\pi\left(1 - 2\sqrt{\lambda}\right)\left(1 + 2\sqrt{\lambda}\right)\left\{\Gamma_2\left(\frac{1}{2} - \sqrt{\lambda}\right)\Gamma_2\left(\frac{1}{2} + \sqrt{\lambda}\right)\right\}^2},$$

and

$$E_3(\lambda) = \frac{e^{c_2\lambda}}{1 - \lambda}\,\frac{\Gamma_2\left(1 - \sqrt{\lambda}\right)\Gamma_2\left(1 + \sqrt{\lambda}\right)}{\left\{\Gamma_3\left(1 - \sqrt{\lambda}\right)\Gamma_3\left(1 + \sqrt{\lambda}\right)\right\}^2},$$

where, for convenience,

$$c_1 := 2(\gamma - 1 + 2\log 2) \quad \text{and} \quad c_2 := \log(2\pi) - \tfrac{3}{2},$$

and γ *is the Euler–Mascheroni constant defined by* (3).

6.3 Evaluations of $\det' \Delta_n$ $(n = 1, 2, 3)$

By making use of Theorem 6.1, we can compute the determinants of the Laplacians on $\mathbf{S}^n$ $(n = 1, 2, 3)$ explicitly.

Theorem 6.2. *The following evaluations hold true:*

$$\begin{aligned} \det{}'\Delta_1 &= 4\pi^2, \\ \det{}'\Delta_2 &= \exp\left[\tfrac{1}{2} - 4\,\zeta'(-1)\right] = e^{\frac{1}{6}}\, A^4 = 3.195311496\cdots, \\ \det{}'\Delta_3 &= \pi \exp\left[\frac{\zeta(3)}{2\pi^2}\right] = 3.338851215\ldots, \end{aligned}$$

where A is the Glaisher–Kinkelin's constant defined by (11).

Acknowledgements This research was supported by the *Basic Science Research Program* through the *National Research Foundation of Korea* funded by the Ministry of Education, Science and Technology of the Republic of Korea (2010-0011005).

References

1. Abramowitz, M., Stegun, I.A. (eds.): Handbook of Mathematical Functions with Formulas, Graphs, and Mathematical Tables. Tenth Printing, National Bureau of Standards, Applied Mathematics Series, vol. 55. National Bureau of Standards, Washington (1972); Reprinted by Dover Publications, New York (1965) (see also [99])
2. Adamchik, V.S.: Polygamma functions of negative order. J. Comput. Appl. Math. **100**, 191–199 (1998)
3. Adamchik, V.S.: On the Barnes function. In: Mourrain, B. (ed.) Proceedings of the 2001 International Symposium on Symbolic and Algebraic Computation, pp. 15–20. ACM, New York (2001) (London, Ontario, 22–25 July 2001)
4. Adamchik, V.S.: The multiple gamma function and its application to computation of series. Ramanujan J. **9**, 271–288 (2005)
5. Alexeiewsky, W.P.: Über eine Classe von Funktionen, die der Gammafunktion analog sind. Leipzig Weidmannsche Buchhandlung **46**, 268–275 (1894)
6. Alzer, H.: Inequalities involving $\Gamma(x)$ and $\Gamma(1/x)$. J. Comput. Appl. Math. **192**, 460–480 (2006)
7. Alzer, H.: Sub-and superadditive property of Euler's Gamma function. Proc. Amer. Math. Soc. **135**, 3641–3648 (2007)
8. Apostol, T.M.: Some series involving the Riemann Zeta function. Proc. Amer. Math. Soc. **5**, 239–243 (1954)
9. Apostol, T.M.: Introduction to Analytic Number Theory. Springer, New York (1976)
10. Artin, E.: The Gamma Function. Holt, Rinehart and Winston, New York (1964)
11. Barnes, E.W.: The theory of the G-function. Quart. J. Math. **31**, 264–314 (1899)
12. Barnes, E.W.: The genesis of the double Gamma function. Proc. London Math. Soc. (Ser. 1) **31**, 358–381 (1900)
13. Barnes, E.W.: The theory of the double Gamma function. Philos. Trans. Roy. Soc. London Ser. A **196**, 265–388 (1901)

14. Barnes, E.W.: On the theory of the multiple Gamma functions. Trans. Cambridge Philos. Soc. **19**, 374–439 (1904)
15. Batir, N.: Inequalities for the double Gamma function. J. Math. Anal. Appl. **351**, 182–185 (2009)
16. Batir, N., Cancan, M.: A double inequality for the double Gamma function. Internat. J. Math. Anal. **2**, 329–335 (2008)
17. Bendersky, L.: Sur la fonction Gamma généralisée, Acta Math. **61**, 263–322 (1933)
18. Billingham, J., King, A.C.: Uniform asymptotic expansions for the Barnes double Gamma function. Proc. Roy. Soc. London Ser. A **453**, 1817–1829 (1997)
19. Campbell, R.: Les Intégrals Eulériennes Et Leurs Applications. Dunod, Paris (1966)
20. Carrier, G.F., Krook, M., Pearson, C.E.: Functions of a Complex Variable. Hod Books, Ithaca (1983)
21. Cassou-Noguès, P.: Analogues p-adiques des fonctions Γ-multiples. In: Journées Arithmétiques de Luminy (International Colloquium of the CNRS, Centre University of Luminy, Luminy, 1978), pp. 43–55. Astérisque, vol. 61. Société mathématique de France, Paris (1979)
22. Chen, C.-P.: Inequalities associated with Barnes G-function. Exposition. Math. **29**, 119–125 (2011)
23. Chen, C.-P.: Glaisher-Kinkelin constant. Integr. Transf. Spec. Funct. **23**, 785–792 (2012) (iFirst)
24. Chen, C.-P., Srivastava, H.M.: Some inequalities and monotonicity properties associated with the Gamma and Psi functions and the Barnes G-function. Integr. Transf. Spec. Funct. **22**, 1–15 (2011)
25. Choi, J.: Determinant of Laplacian on S^3. Math. Japon. **40**, 155–166 (1994)
26. Choi, J.: Explicit formulas for the Bernolli polynomials of order n. Indian J. Pure Appl. Math. **27**, 667–674 (1996)
27. Choi, J.: Integral and series representations for the Euler's constant. In: Proceedings of the Seventh Conference on Real and Complex Analysis, pp. 43–55. Hiroshima University, Japan (2003)
28. Choi, J.: Some mathematical constants. Appl. Math. Comput. **187**, 122–140 (2007)
29. Choi, J.: A set of mathematical constants arising naturally in the theory of the multiple Gamma functions. Abstract and Applied Analysis, vol. **2012**, Article ID 121795, p. 11 (2012)
30. Choi, J.: Determinants of the Laplacians on the n-dimensional unit sphere $\mathbf{S}^n$ $(n = 8, 9)$.
31. Choi, J., Cho, Y.J., Srivastava, H.M.: Series involving the Zeta function and multiple Gamma functions. Appl. Math. Comput. **159**, 509–537 (2004)
32. Choi, J., Lee, J.: Closed-form evaluation of a class of series associated with the Riemann zeta function. In: Proceedings of the 11th International Conference on Finite or Infinite Dimensional Complex Analysis and Applications, pp. 36–53 (2003)
33. Choi, J., Lee, J., Srivastava, H.M.: A generalization of Wilf's formula. Kodai Math. J. **26**, 44–48 (2003)
34. Choi, J., Nash, C.: Integral representations of the Kikelin's constant A. Math. Japon. **45**, 223–230 (1997)
35. Choi, J., Quine, J.R.: E. W. Barnes' approach of the multiple Gamma functions. J. Korean Math. Soc. **29**, 127–140 (1992)
36. Choi, J., Seo, T.Y.: The double Gamma function. East Asian Math. J. **13**, 159–174 (1997)
37. Choi, J., Seo, T.Y.: Integral formulas for Euler's constant. Commun. Korean Math. Soc. **13**, 683–689 (1998)
38. Choi, J., Seo, T.Y.: Identities involving series of the Riemann zeta function. Indian J. Pure Appl. Math. **30**, 649–652 (1999)
39. Choi, J., Srivastava, H.M.: Sums associated with the Zeta function. J. Math. Anal. Appl. **206**, 103–120 (1997)
40. Choi, J., Srivastava, H.M.: Certain classes of series involving the Zeta function. J. Math. Anal. Appl. **231**, 91–117 (1999)
41. Choi, J., Srivastava, H.M.: An application of the theory of the double Gamma function. Kyushu J. Math. **53**, 209–222 (1999)

42. Choi, J., Srivastava, H.M.: Certain classes of series associated with the Zeta function and multiple Gamma functions. J. Comput. Appl. Math. **118**, 87–109 (2000)
43. Choi, J., Srivastava, H.M.: A certain class of series associated with the Zeta function. Integr. Transf. Spec. Funct. **12**, 237–250 (2001)
44. Choi, J., Srivastava, H.M.: A certain family of series associated with the Zeta and related functions. Hiroshima Math. J. **32**, 417–429 (2002)
45. Choi, J., Srivastava, H.M.: A family of log-Gamma integrals and associated results. J. Math. Anal. Appl. **303**, 436–449 (2005)
46. Choi, J., Srivastava, H.M.: Certain families of series associated with the Hurwitz-Lerch Zeta function. Appl. Math. Comput. **170**, 399–409 (2005)
47. Choi, J., Srivastava, H.M.: A note on a multiplication formula for the multiple Gamma function Γ_n. Italian J. Pure Appl. Math. **23**, 179–188 (2008)
48. Choi, J., Srivastava, H.M.: Some applications of the Gamma and Polygamma functions involving convolutions of the Rayleigh functions, multiple Euler sums and log-sine integrals. Math. Nachr. **282**, 1709–1723 (2009)
49. Choi, J., Srivastava, H.M.: Integral representations for the Gamma function, the Beta function, and the double Gamma function. Integr. Transf. Spec. Funct. **20**, 859–869 (2009)
50. Choi, J., Srivastava, H.M.: Integral representations for the Euler-Mascheroni constant γ. Integr. Transf. Spec. Funct. **21**, 675–690 (2010)
51. Choi, J., Srivastava, H.M.: Mathieu series and associated sums involving the Zeta functions. Comput. Math. Appl. **59**, 861–867 (2010)
52. Choi, J., Srivastava, H.M.: Asymptotic formulas for the triple Gamma function Γ_3 by means of its integral representation. Appl. Math. Comput. **218**, 2631–2640 (2011)
53. Choi, J., Srivastava, H.M.: The multiple Hurwitz Zeta function and the multiple Hurwitz-Euler Eta function. Taiwanese J. Math. **15**, 501–522 (2011)
54. Choi, J., Srivastava, H.M.: Some two-sided inequalities for multiple Gamma functions and related results. Appl. Math. Comput. **219**, 10343–10354 (2013)
55. Choi, J., Srivastava, H.M.: Series involving the Zeta functions and a family of generalized Goldbach-Euler series. Accepted for publication at Amer. Math. Monthly (2013)
56. Choi, J., Srivastava, H.M., Quine, J.R.: Some series involving the Zeta function. Bull. Austral. Math. Soc. **51**, 383–393 (1995)
57. Choi, J., Srivastava, H.M., Quine, J.R.: Some series involving the Zeta function. Bull. Austral. Math. Soc. **51**, 383–393 (1995)
58. Choi, J., Srivastava, H.M., Zhang, N.-Y.: Integrals involving a function associated with the Euler-Maclaurin summation formula. Appl. Math. Comput. **93**, 101–116 (1998)
59. Choi, J., Srivastava, H.M., Adamchik, V.S.: Multiple Gamma and related functions. Appl. Math. Comput. **134**, 515–533 (2003)
60. Conway, J.B.: Functions of One Complex Variable, 2nd edn. Springer, New York (1978)
61. D'Hoker, E., Phong, D.H.: On determinant of Laplacians on Riemann surface. Comm. Math. Phys. **104**, 537–545 (1986)
62. D'Hoker, E., Phong, D.H.: Multiloop amplitudes for the bosonic polyakov string. Nucl. Phys. B **269**, 204–234 (1986)
63. Dittrich, W., Reuter, M.: Effective QCD-Lagrangian with ξ-function regularization. Phys. Lett. B **128**, 321–326 (1983)
64. Dufresnoy, J., Pisot, C.: Sur la relation fonctionnelle $f(x+1) - f(x) = \varphi(x)$. Bull. Soc. Math. Belg. **15**, 259–270 (1963)
65. Edwards, J.: A treatise on the integral calculus with applications: Examples and problems, Vol. 1, 2. Chelsea Publishing Company, New York (1954)
66. Elizalde, E.: Derivative of the generalized Riemann Zeta function $\zeta(z,q)$ at $z = -1$. J. Phys. A: Math. Gen. **18**, 1637–1640 (1985)
67. Elizalde, E.: An asymptotic expansion for the first derivative of the generalized Riemann Zeta function. Math. Comput. **47**, 347–350 (1986)
68. Elizalde, E., Romeo, A.: An integral involving the generalized Zeta function. Internat. J. Math. Math. Sci. **13**, 453–460 (1990)

69. Elizalde, E., Odintsov, S.D., Romeo, A., Bytsenko, A.A., Zerbini, S.: Zeta Regularization Techniques with Applications. World Scientific Publishing Company, Singapore (1994)
70. Euler, L.: Comm. Acad. Petropol. **7**, 156 (1734–1735)
71. Ferreira, C.: An asymptotic expansion of the double Gamma function. J. Approx. Theory **111**, 298–314 (2001)
72. Finch, S.R.: Mathematical Constants. Cambridge University Press, Cambridge (2003)
73. Friedman, E., Ruijsenaars, S.: Shintani-Barnes Zeta and Gamma functions. Adv. Math. **187**, 362–395 (2004)
74. Glaisher, J.W.L.: On the history of Euler's constant. Messenger Math. **1**, 25–30 (1872)
75. Glaisher, J.W.L.: On the product $1^1 \cdot 2^2 \cdot 3^3 \cdots n^n$. Messenger Math. **7**, 43–47 (1877)
76. Gosper, Jr., R.W.: $\int_{n/4}^{m/6} \ln \Gamma(s)\, ds$. Fields Inst. Comm. **14**, 71–76 (1997)
77. Gradshteyn, I.S., Ryzhik, I.M. (eds.): Tables of Integrals, Series, and Products (Corrected and Enlarged edition prepared by A. Jeffrey), 6th edn. Academic, New York (2000)
78. Hardy, G.H.: Divergent Series. Clarendon (Oxford University) Press, Oxford (1949); 2nd (Textually Unaltered) edn, Chelsea Publishing Company, New York (1991)
79. Havil, J.: Gamma (Exploring Euler's Constant). Princeton University Press, Princeton (2003)
80. Hölder, O.: Über eine Transcendente Funktion, vol. 1886, pp. 514–522. Dieterichsche Verlags-Buchhandlung, Göttingen (1886)
81. Hölder, O.: Über eine von Abel untersuchte Transzendente und eine merkwürdige Funktionalbeziehung. In: Berichte über die Verhandlungen der Saechsischen Akademie der Wissenschaften zu Leipzig, Mathematisch-Physische Klasse, vol. 80, pp. 312–325 (1928)
82. Kanemitsu, S., Kumagai, H., Yoshimoto, M.: Sums involving the Hurwitz zeta function. Ramanujan J. **5**, 5–19 (2001)
83. Kinkelin, V.H.: Über eine mit der Gamma Funktion verwandte Transcendente und deren Anwendung auf die Integralrechnung. J. Reine Angew. Math. **57**, 122–158 (1860)
84. Knopp, K.: Theory and Application of Infinite Series. Hafner Publishing Company, New York (1951) (2nd English edn., translated from the 2nd German edn. revised in accordance with the 4th German edn. by R.C.H. Young)
85. Knuth, D.E.: Euler's constant to 1271 places. Math. Comput. **16**, 275–281 (1962)
86. Koumandos, S.: On Ruijsenaars' asymptotic expansion of the logarithm of the double gamma function. J. Math. Anal. Appl. **341**, 1125–1132 (2008)
87. Koumandos, S., Pedersen, H.L.: Completely monotonic functions of positive order and asymptotic expansions of the logarithm of Barnes double gamma function and Euler's gamma function. J. Math. Anal. Appl. **355**, 33–40 (2009)
88. Koumandos, S., Pedersen, H.L.: Absolutely monotonic functions related to Euler's gamma function and Barnes' double and triple gamma function. Monatsh. Math. **163**, 51–69 (2011)
89. Kumagai, H.: The determinant of the Laplacian on the n-sphere. Acta Arith. **91**, 199–208 (1999)
90. Lewin, L.: Polylogarithms and Associated Functions. Elsevier (North-Holland), New York (1981)
91. Matsumoto, K.: Asymptotic series for double Zeta and double Gamma functions of Barnes. RIMS Kōkyūroku **958**, 162–165 (1996)
92. Matsumoto, K.: Asymptotic series for double Zeta, double Gamma, and Hecke L-functions. Math. Proc. Cambridge Philos. Soc. **123**, 385–405 (1998); Corrigendum and Addendum: Math. Proc. Cambridge Philos. Soc. **132**, 377–384 (2002)
93. Melzak, Z.A.: Companion to Concrete Mathematics: Mathematical Techniques and Various Applications, vol. I. Wiley, New York (1973)
94. Mortici, C.: New improvements of the Stirling formula. Appl. Math. Comput. **217**, 699–704 (2010)
95. Nielsen, N.: Handbuch der Theorie der Gammafunktion. Druck und Verlag von B.G. Teubner, Leipzig (1906); Reprinted by Chelsea Publishing Company, New York (1965)
96. Nishizawa, M.: On a q-analogue of the multiple Gamma functions. Lett. Math. Phys. **37**, 201–209 (1996)

97. Nishizawa, M.: Multiple Gamma function, its q- and elliptic analogue. Rocky Mountain J. Math. **32**, 793–811 (2002)
98. Olver, F.W.J.: Asymptotics and Special Functions. A.K. Peters, Wellesley, Massachusetts (1997)
99. Olver, F.W.J., Lozier, D.W., Boisvert, R.F., Clark, C.W. (eds.): NIST Handbook of Mathematical Functions [With 1 CD-ROM (Windows, Macintosh and UNIX)]. U.S. Department of Commerce, National Institute of Standards and Technology, Washington (2010); Cambridge University Press, Cambridge (2010) (see also [1])
100. Osgood, B., Phillips, R., Sarnak, P.: Extremals of determinants of Laplacians. J. Funct. Anal. **80**, 148–211 (1988)
101. Qi, F., Cui, R.-Q., Chen, C.-P., Guo, B.-N.: Some completely monotonic functions involving Polygamma functions and an application. J. Math. Anal. Appl. **310**, 303–308 (2005)
102. Qi, F., Chen, S.-X., Cheung, W.-S.: Logarithmically completely monotonic functions concerning gamma and digamma functions. Integr. Transf. Spec. Funct. **18**, 435–443 (2007)
103. Quine, J.R., Choi, J.: Zeta regularized products and functional determinants on spheres. Rocky Mountain J. Math. **26**, 719–729 (1996)
104. Quine, J.R., Heydari, S.H., Song, R.Y.: Zeta regularized products. Trans. Amer. Math. Soc. **338**, 213–231 (1993)
105. Ramanujan, S.: A series for Euler's constant γ. Messenger Math. **46**, 73–80 (1916/1917)
106. Ruijsenaars, S.: First order analytic difference equations and integrable quantum systems. J. Math. Phys. **38**, 1069–1146 (1997)
107. Ruijsenaars, S.: On Barnes' multiple zeta and gamma functions. Adv. Math. **156**, 107–132 (2000)
108. Sarnak, P.: Determinants of Laplacians. Comm. Math. Phys. **110**, 113–120 (1987)
109. Shallit, J.D., Zikan, K.: A theorem of Goldbach. Amer. Math. Monthly **93**, 402–403 (1986)
110. Shintani, T.: A proof of the classical Kronecker limit formula. Tokyo J. Math. **3**, 191–199 (1980)
111. Srivastava, H.M.: A unified presentation of certain classes of series of the Riemann Zeta function. Riv. Mat. Univ. Parma Ser. 4 **14**, 1–23 (1988)
112. Srivastava, H.M.: Sums of certain series of the Riemann Zeta function. J. Math. Anal. Appl. **134**, 129–140 (1988)
113. Srivastava, H.M.: Some rapidly converging series for $\zeta(2n+1)$. Proc. Amer. Math. Soc. **127**, 385–396 (1999)
114. Srivastava, H.M., Choi, J.: Series Associated with the Zeta and Related Functions. Kluwer Academic Publishers, Dordrecht (2001)
115. Srivastava, H.M., Choi, J.: Zeta and q-Zeta Functions and Associated Series and Integrals. Elsevier Science Publishers, Amsterdam (2012)
116. Steiner, F.: On Selberg's Zeta function for compact Riemann surfaces. Phys. Lett. B **188**, 447–454 (1987)
117. Terras, A.: Harmonic Analysis on Symmetric Spaces and Applications. vol. I, Springer, Berlin (1985)
118. Vardi, I.: Determinants of Laplacians and multiple Gamma functions. SIAM J. Math. Anal. **19**, 493–507 (1988)
119. Vignéras, M.-F.: L'équation fonctionnelle de la fonction zêta de Selberg du groupe moudulaire $PSL(2, Z)$. In: Journées Arithmétiques de Luminy (International Colloquium of the CNRS, Centre University of Luminy Luminy, 1978), pp. 235–249. Astérisque, vol. 61. Société mathématique de France, Paris (1979)
120. Voros, A.: The Hadamard factorization of the Selberg Zeta function on a compact Riemann surface. Phys. Lett. B **180**, 245–246 (1986)
121. Voros, A.: Special functions, spectral functions and the Selberg Zeta function. Comm. Math. Phys. **110**, 439–465 (1987)
122. Wade, W.R.: An Introduction to Analysis, 3rd edn. Pearson Prentice Hall, Upper Saddle River (2004)

123. Walfisz, A.: Weylsche Exponentialsummen in der Neueren Zahlentheorie, pp. 114–115. B.G. Teubner, Leipzig (1963)
124. Whittaker, E.T., Watson, G.N.: A Course of Modern Analysis: An Introduction to the General Theory of Infinite Processes and of Analytic Functions; with an Account of the Principal Transcendental Functions, 4th edn. Cambridge University Press, Cambridge (1963)
125. Wilf, H.S.: Problem 10588. Amer. Math. Monthly **104**, 456 (1997)

On Partition Functions of Hyperbolic Three-Geometry and Associated Hilbert Schemes

A.A. Bytsenko and E. Elizalde

Dedicated to Professor Hari M. Srivastava

Abstract Highest-weight representations of infinite-dimensional Lie algebras and Hilbert schemes of points are considered, together with the applications of these concepts to partition functions, which are most useful in physics. Partition functions (elliptic genera) are conveniently transformed into product expressions, which may inherit the homology properties of appropriate (poly)graded Lie algebras. Specifically, the role of (Selberg-type) Ruelle spectral functions of hyperbolic geometry in the calculation of partition functions and associated q-series is discussed. Examples of these connections in quantum field theory are considered (in particular, within the AdS/CFT correspondence), as the AdS_3 case where one has Ruelle/Selberg spectral functions, whereas on the CFT side, partition functions and modular forms arise. These objects are here shown to have a common background, expressible in terms of Euler-Poincaré and Macdonald identities, which, in turn, describe homological aspects of (finite or infinite) Lie algebra representations. Finally, some other applications of modular forms and spectral functions (mainly related with the congruence subgroup of $SL(2, \mathbb{Z})$) to partition functions, Hilbert schemes of points, and symmetric products are investigated by means of homological and K-theory methods.

A.A. Bytsenko
Departamento de Física, Universidade Estadual de Londrina, Caixa Postal 6001, Londrina-Paraná, Brazil
e-mail: abyts@uel.br

E. Elizalde (✉)
Consejo Superior de Investigaciones Científicas (ICE/CSIC) and Institut d'Estudis Espacials de Catalunya (IEEC), Campus UAB, Facultat de Ciències, Torre C5-Par-2a, 08193 Bellaterra, Barcelona, Spain
e-mail: elizalde@ieec.uab.es

G.V. Milovanović and M.Th. Rassias (eds.), *Analytic Number Theory, Approximation Theory, and Special Functions*, DOI 10.1007/978-1-4939-0258-3_5,

1 Introduction

This paper deals with highest-weight representations of infinite-dimensional Lie algebras and Hilbert schemes of points and considers applications of these concepts to partition functions, which are ubiquitous and very useful in physics. The role of (Selberg-type) Ruelle spectral functions of hyperbolic geometry in the calculation of the partition functions and associated q-series will be discussed. Among mathematicians, partition functions are commonly associated with new mathematical invariants for spaces, while for physicists they are one-loop partition functions of quantum field theories. In fact, partition functions (elliptic genera) can be conveniently transformed into product expressions, which may inherit the homological properties of appropriate (poly)graded Lie algebras. In quantum field theory the connection referred to above is particularly striking in the case of the so-called AdS/CFT correspondence. In the anti-de Sitter space AdS_3, for instance, one has Ruelle/Selberg spectral functions, whereas on the conformal field theory (CFT) side, on the other hand, one encounters partition functions and modular forms. What we are ready to show here is that these objects do have a common background, expressible in terms of Euler-Poincaré and Macdonald identities which, in turn, describe homological aspects of (finite or infinite) Lie algebra representations. This is both quite remarkable and useful.

Being more specific, we will be dealing in what follows with applications of modular forms and spectral functions (mainly related to the congruence subgroup of $SL(2,\mathbb{Z})$) to partition functions, Hilbert schemes of points, and symmetric products. Here are the contents of the paper. In Sect. 2 we shortly discuss the case of two-geometries and then present Thurston's list of three-geometries. This list has been organized in terms of the corresponding compact stabilizers being isomorphic to $SO(3)$, $SO(2)$, or the trivial group $\{1\}$, respectively. The analogue list of four-geometries and the corresponding stabilizer subgroups are also considered in this section. Special attention is paid to the important case of hyperbolic three-geometry.

In Sect. 3 we introduce the Petterson-Selberg and the Ruelle spectral functions of hyperbolic three-geometry. Then, in Sect. 3.1, we consider examples for which we explicitly show that the respective partition functions can be written in terms of Ruelle's spectral functions associated with q-series, although the hyperbolic side remains still to be explored. In Sect. 4 we briefly explain the relationship existing between the Heisenberg algebra and its representation and with the Hilbert scheme of points in Sect. 4.1. This allows to construct a representation of products of Heisenberg and Clifford algebras on the direct sum of homology groups of all components associated with schemes. Hilbert schemes of points of surfaces are discussed in Sect. 4.2; we rewrite there the character formulas and Göttsche's formula in terms of Ruelle's spectral functions.

In Sect. 4.3 we pay attention to the special case of algebraic structures of the K-groups $K_{\tilde{H}\Gamma_N}(X^N)$ of Γ_N-equivariant Clifford supermodules on X^N, following the lines of [1]. This case is important since the direct sum $\mathcal{F}_{\Gamma}^{-}(X) = \oplus_N^{\infty} K_{\tilde{H}\Gamma_N}(X^N)$ naturally carries a Hopf algebra structure, and it is isomorphic to the Fock space of a twisted Heisenberg superalgebra with $K_{\tilde{H}\Gamma_N}(X) \cong K_\Gamma(X)$. In terms of the Ruelle spectral function we represent the dimension of a direct sum of the equivariant K-groups (related to a suitable supersymmetric algebra). We analyze elliptic genera for generalized wreath and symmetric products on N-folds; these cases are examples of rather straightforward applications of the machinery of modular forms and spectral functions discussed above. Finally, in the conclusions, Sect. 5, we briefly outline some issues and further perspectives for the analysis of partition functions in connection with deformation quantization.

2 Classification of Low-Dimensional Geometries

The problem of the classification of geometries is most important in complex analysis and in mathematics as a whole and also plays a fundamental role in physical theories. Indeed, in quantum field theory, functional integration over spaces of metrics can be separated into an integration over all metrics for some volume of a definite topology, followed by a sum over all topologies. But even for low-dimensional spaces (say, e.g., the three-dimensional case) of fixed topology, the moduli space of all metric diffeomorphisms is infinite dimensional. And this leads back to the deep mathematical task associated with the classification problem. In this section we present a brief discussion of the classification (uniformization) issue and of the sum over the topology for low-dimensional cases, along the lines of [2–4].

All curves of genus zero can be uniformized by rational functions, those of genus one by elliptic functions, and those of genus higher than one by meromorphic functions, defined on proper open subsets of $\mathbb{C}$. These results, due to Klein, Poincaré, and Koebe, are among the deepest achievements in mathematics. A complete solution of the uniformization problem has not yet been obtained (with the exception of the one-dimensional complex case). However, there have been essential advances in this problem, which have brought to the foundation of topological methods, covering spaces, existence theorems for partial differential equations, existence and distortion theorems for conformal mappings, etc.

Three-Geometries. According to Thurston's conjecture [5], there are eight model spaces in three dimensions:

$$X = G/\Gamma = \left\{ \begin{array}{l} \mathbb{R}^3 \text{ (Euclidean space)},\ S^3 \text{ (spherical space)},\ H^3 \text{ (hyperbolic space)} \\ H^2\times\mathbb{R},\quad S^2\times\mathbb{R},\quad \widetilde{SL(2,\mathbb{R})},\quad \text{Nil}^3,\quad \text{Sol}^3 \end{array} \right\}.$$

An important remark is in order. This conjecture follows from considering the identity component of the isotropy group, $\Gamma \equiv \Gamma_x$ of X, through a point, x. Γ is a compact, connected Lie group, and one must distinguish the three different cases: $\Gamma = SO(3)$, $SO(2)$ and $\{1\}$.

Table 1 List of the four-geometries

Stabilizer-subgroup Γ	Space X
$SO(4)$	$\mathbb{R}^4,\ S^4,\ H^4$
$U(2)$	$\mathbb{C}P^2, \mathbb{C}H^2$
$SO(2) \times SO(2)$	$S^2 \times \mathbb{R}^2,\ S^2 \times S^2,\ S^2 \times H^2,\ H^2 \times \mathbb{R}^2,\ H^2 \times H^2$
$SO(3)$	$S^3 \times \mathbb{R},\ H^3 \times \mathbb{R}$
$SO(2)$	$\mathrm{Nil}^3 \times \mathbb{R},\ \widetilde{PSL}(2,\mathbb{R}) \times \mathbb{R},\ \mathrm{Sol}^4$
S^1	F^4
Trivial	$\mathrm{Nil}^4,\ \mathrm{Sol}^4_{m,n}$ (including $\mathrm{Sol}^3 \times \mathbb{R}$), Sol^4_1

1. $\Gamma = SO(3)$. In this case the space X has constant curvature: $\mathbb{R}^3$, S^3 (modeled on $\mathbb{R}^3$), or H^3 (which can be modeled on the half-space $\mathbb{R}^2 \times \mathbb{R}^+$).
2. $\Gamma = SO(2)$. In this case there is a one-dimensional subspace of TX that is invariant under Γ, which has a complementary plane field P_x. If the plane field P_x is integrable, then X is a product $\mathbb{R}\times S^2$ or $\mathbb{R}\times H^2$. If the plane field P_x is non-integrable, then X is a nontrivial fiber bundle with fiber S^1: $S^1 \hookrightarrow X \twoheadrightarrow \Sigma_{g\geq 2}$ ($\widetilde{SL(2,\mathbb{R})}$-geometry), Σ_g stands for a surface of genus g, $S^1 \hookrightarrow X \twoheadrightarrow T^2$ (Nil^3-geometry) or $S^1 \hookrightarrow X \twoheadrightarrow S^2$ (S^3-geometry).
3. $\Gamma = \{1\}$. In this case we have the three-dimensional Lie groups: $\widetilde{SL(2,\mathbb{R})}$, Nil^3, and Sol^3.

The first five geometries are familiar objects, so we briefly discuss the last three of them. The group $\widetilde{SL(2,\mathbb{R})}$ is the universal covering of $SL(2,\mathbb{R})$, the three-dimensional Lie group of all 2×2 real matrices with determinant equal to 1. The geometry of Nil is the three-dimensional Lie group of all 3×3 real upper triangular matrices endowed with ordinary matrix multiplication. It is also known as the nilpotent Heisenberg group. The geometry of Sol is the three-dimensional (solvable) group.

Four-Geometries. The list of Thurston three-geometries has been organized in terms of the compact stabilizers Γ. The analogue list of four-geometries can also be organized using connected groups of isometries only (Table 1). Here we have the four irreducible four-dimensional Riemannian symmetric spaces: sphere S^4, hyperbolic space H^4, complex projective space $\mathbb{C}P^2$, and complex hyperbolic space $\mathbb{C}H^2$ (which we may identify with the open unit ball in $\mathbb{C}^2$, with an appropriate metric). The other cases are more specific and we shall illustrate them for the sake of completeness only.

The nilpotent Lie group Nil^4 can be presented as the split extension $\mathbb{R}^3 \rtimes_U \mathbb{R}$ of $\mathbb{R}^3$ by $\mathbb{R}$, where the real three-dimensional representation U of $\mathbb{R}$ has the form $U(t) = \exp(tB)$ with

$$B = \begin{pmatrix} 0 & 1 & 0 \\ 0 & 0 & 1 \\ 0 & 0 & 0 \end{pmatrix}.$$

In the same way, the soluble Lie groups $\mathrm{Sol}^4_{m,n} = \mathbb{R}^3 \rtimes_{T_{m,n}} \mathbb{R}$, on real three-dimensional representations $T_{m,n}$ of $\mathbb{R}$, $T_{m,n}(t) = \exp(tC_{m,n})$, where $C_{m,n} = \mathrm{diag}(\alpha, \beta, \gamma)$ and $\alpha + \beta + \gamma = 0$ for $\alpha > \beta > \gamma$. Furthermore e^α, e^β, and e^γ are the roots of $\lambda^3 - m\lambda^2 + n\lambda - 1 = 0$, with m, n positive integers. If $m = n$, then $\beta = 0$ and $\mathrm{Sol}^4_{m,n} = \mathrm{Sol}^3 \times \mathbb{R}$. In general, if $C_{m,n} \propto C_{m',n'}$, then $\mathrm{Sol}^4_{m,n} \cong \mathrm{Sol}^4_{m',n'}$. It gives infinitely many classes of equivalence. When $m^2n^2 + 18 = 4(m^3 + n^3) + 27$, one has a new geometry, Sol^4_0, associated with the group $SO(2)$ of isometries rotating the first two coordinates. The soluble group Sol^4_1 is most conveniently represented as the matrix group

$$\left\{ \begin{pmatrix} 1 & b & c \\ 0 & \alpha & a \\ 0 & 0 & 1 \end{pmatrix} \; : \; \alpha, a, b, c \in \mathbb{R}, \alpha > 0 \right\}.$$

Finally, the geometry F^4 is associated with the isometry group $\mathbb{R}^2 \rtimes PSL(2,\mathbb{R})$ and stabilizer $SO(2)$. Here the semidirect product is taken with respect to the action of the group $PSL(2,\mathbb{R})$ on $\mathbb{R}^2$. The space F^4 is diffeomorphic to $\mathbb{R}^4$ and has alternating signs in the metric. A connection of these geometries with complex and Kählerian structures (preserved by the stabilizer Γ) can be found in [3].

We conclude this section with some comments. In two-dimensional quantum theory it is customary to perform the sum over all topologies. Then, any functional integral of fixed genus g can be written in the form

$$\int [Dg] = \sum_{g=0}^{\infty} \int_{\text{(fixed genus)}} [Dg].$$

A necessary first step to implement this in the three dimensional case is the classification of all possible three-topologies (by Kleinian groups). Provided Thurston's conjecture is true, every compact closed three-dimensional manifold can be represented as $\bigcup_{\ell=1}^{\infty} G_{n_\ell}/\Gamma_{n_\ell}$, where $n_\ell \in (1,\ldots,8)$ represents one of the eight geometries, and Γ is the (discrete) isometry group of the corresponding geometry. It has to be noted that gluing the above geometries, characterizing different coupling constants by a complicated set of moduli, is a very difficult task. Perhaps this could be done, however, with a bit of luck, but the more important contribution to the vacuum persistence amplitude should be given by the compact hyperbolic geometry, the other geometries appearing only for a small number of exceptions [6]. Indeed, many three-manifolds are hyperbolic (according to a famous theorem by Thurston [5]). For example, the complement of a knot in S^3 admits a hyperbolic structure unless it is a torus or satellite knot. Moreover, according to the Mostow Rigidity Theorem [7], any geometric invariant of a hyperbolic three-manifold is a topological invariant. Our special interest here is directed towards hyperbolic spaces. Some examples of partition functions and elliptic genera written in terms of spectral functions of H^3 spaces and their quotients by a subgroup of the isometry group $PSL(2,\mathbb{C}) \equiv SL(2,\mathbb{C})/\{\pm 1\}$ can be found in [8, 9].

3 The Ray-Singer Norm: Hyperbolic Three-Geometry

If $\mathfrak{L}_p$ is a self-adjoint Laplacian on p-forms, then the following results hold. There exists $\varepsilon, \delta > 0$ such that for $0 < t < \delta$ the heat kernel expansion for Laplace operators on a compact manifold X is given by $\mathrm{Tr}\left(e^{-t\mathfrak{L}_p}\right) = \sum_{0\leq \ell \leq \ell_0} a_\ell(\mathfrak{L}_p)t^{-\ell} + O(t^\varepsilon)$. The zeta function of $\mathfrak{L}_p$ is the Mellin transform

$$\zeta(s|\mathfrak{L}_p) = \mathfrak{M}\left[\mathrm{Tr}\, e^{-t\mathfrak{L}_p}\right] = [\Gamma(s)]^{-1}\int_{\mathbb{R}_+} \mathrm{Tr}\, e^{-t\mathfrak{L}_p}t^{s-1}dt. \tag{1}$$

This function equals $\mathrm{Tr}\left(\mathfrak{L}_p^{-s}\right)$ for $s > (1/2)\dim X$. Let χ be an orthogonal representation of $\pi_1(X)$. Using the Hodge decomposition, the vector space $H(X;\chi)$ of twisted cohomology classes can be embedded into $\Omega(X;\chi)$ as the space of harmonic forms. This embedding induces a norm $|\cdot|^{RS}$ on the determinant line $\det H(X;\chi)$. The Ray-Singer norm $||\cdot||^{RS}$ on $\det H(X;\chi)$ is defined by [10]

$$||\cdot||^{RS} \stackrel{\mathrm{def}}{=} |\cdot|^{RS}\prod_{p=0}^{\dim X}\left[\exp\left(-\frac{d}{ds}\zeta(s|\mathfrak{L}_p)|_{s=0}\right)\right]^{(-1)^p p/2}, \tag{2}$$

where the zeta function $\zeta(s|\mathfrak{L}_p)$ of the Laplacian acting on the space of p-forms orthogonal to the harmonic forms has been used. For a closed connected orientable smooth manifold of odd dimension, and for the Euler structure $\eta \in \mathrm{Eul}(X)$, the Ray-Singer norm of its cohomological torsion $\tau_{\mathrm{an}}(X;\eta) = \tau_{\mathrm{an}}(X) \in \det H(X;\chi)$ is equal to the positive square root of the absolute value of the monodromy of χ along the characteristic class $c(\eta) \in H^1(X)$ [11]: $||\tau_{\mathrm{an}}(X)||^{RS} = |\det{}_\chi c(\eta)|^{1/2}$. And in the special case where the flat bundle χ is acyclic (namely the vector space $H^q(X;\chi)$ of twisted cohomology is zero), we have

$$[\tau_{\mathrm{an}}(X)]^2 = |\det{}_\chi c(\eta)|\prod_{p=0}^{\dim X}\left[\exp\left(-\frac{d}{ds}\zeta(s|\mathfrak{L}_p)|_{s=0}\right)\right]^{(-1)^{p+1}p}. \tag{3}$$

3.1 Spectral Functions of Hyperbolic Three-Geometry

For a closed oriented hyperbolic three-manifold of the form $X_\Gamma = H^3/\Gamma$ and for acyclic χ, the analytic torsion reads [12–14]: $[\tau_{\mathrm{an}}(X_\Gamma)]^2 = \mathcal{R}(0)$, where $\mathcal{R}(s)$ is the Ruelle function.[1] A Ruelle-type zeta function, for $\mathrm{Re}\, s$ large, can be defined as the

[1]Vanishing theorems for the type $(0,q)$ cohomology of locally symmetric spaces can be found in [15]. Again, if χ is acyclic ($H(X;\chi) = 0$), the Ray-Singer norm (3) is a topological invariant: it

product over prime closed geodesics γ of factors det $(I - \xi(\gamma)e^{-s\ell(\gamma)})$, where $\ell(\gamma)$ is the length of γ, and can be continued meromorphically to the entire complex plane $\mathbb{C}$ [16]. The function $\mathcal{R}(s)$ is an alternating product of more complicated factors, each of which is a Selberg zeta function $Z_\Gamma(s)$. The relation between the Ruelle and Selberg functions is

$$\mathcal{R}(s) = \prod_{p=0}^{\dim X - 1} Z_\Gamma(p+s)^{(-1)^j}. \tag{4}$$

The Ruelle function associated with closed oriented hyperbolic three-manifold X_Γ has the form $\mathcal{R}(s) = Z_\Gamma(s)Z_\Gamma(2+s)/Z_\Gamma(1+s)$.

We would like here to shed light on some aspects of the so-called AdS_3/CFT_2 correspondence, which plays a very important role in quantum field theory. Indeed, it is known that the geometric structure of three-dimensional gravity allows for exact computations, since its Euclidean counterpart is locally isomorphic to constant curvature hyperbolic space. Because of the AdS_3/CFT_2 correspondence, one expects a correspondence between spectral functions related to Euclidean AdS_3 and modular-like functions (Poincaré series).[2] One assumes this correspondence to occur when the arguments of the spectral functions take values on a Riemann surface, viewed as the conformal boundary of AdS_3. According to the holographic principle, strong ties exist between certain field theory quantities on the bulk of an AdS_3 manifold and related quantities on its boundary at infinity. To be more precise, the classes of Euclidean AdS_3 spaces are quotients of the real hyperbolic space by a discrete group (a Schottky group). The boundary of these spaces can be compact oriented surfaces with conformal structure (compact complex algebraic curves). A general formulation of the "Holography Principle" states that there is a correspondence between a certain class of fields, their properties and their correlators in the bulk space, where gravity propagates, and a class of primary fields, with their properties and correlators of the conformal theory on the boundary. More precisely, the set of scattering poles in $3D$ coincides with the zeros of a Selberg-type spectral function [14, 17]. Thus, encoded on a Selberg function is the spectrum of a three-dimensional model. In the framework of this general principle, we would like to illustrate the correspondence between spectral functions of hyperbolic three-geometry (its spectrum being encoded in the Petterson-Selberg spectral functions) and Poincaré series associated with the conformal structure in two dimensions.

Let us consider a three-geometry with an orbifold description H^3/Γ. The complex unimodular group $G = SL(2, \mathbb{C})$ acts on the real hyperbolic three-space

---

does not depend on the choice of the metric on X and χ used in the construction. If X is a complex manifold (smooth C^∞-manifold or topological space), then $\mathbb{E} \to X$ is the induced complex (or smooth, or continuous) vector bundles. We write $H^{p,q}(X; \mathbb{E}) \simeq H^{0,q}(X; \Lambda^{p,0}X \otimes \mathbb{E})$ holonomic vector bundles $\Lambda^{p,0}X \to X$ (see [15] for details).

[2] The modular forms in question are the forms for the congruence subgroup of $SL(2, \mathbb{Z})$, which is viewed as the group that leaves fixed one of the three nontrivial spin structures on an elliptic curve.

H^3 in a standard way, namely for $(x, y, z) \in H^3$ and $g \in G$, one has $g \cdot (x, y, z) = (u, v, w) \in H^3$. Thus, for $r = x + iy$, $g = \begin{bmatrix} a & b \\ c & d \end{bmatrix}$,

$$u + iv = [(ar + b)\overline{(cr + d)} + a\bar{c}z^2] \cdot [|cr + d|^2 + |c|^2 z^2]^{-1},$$

$$w = z \cdot [|cr + d|^2 + |c|^2 z^2]^{-1}.$$

Here the bar denotes complex conjugation. Let $\Gamma \in G$ be the discrete group of G be defined as

$$\Gamma = \{\mathrm{diag}(e^{2n\pi(\mathrm{Im}\,\tau + i\mathrm{Re}\,\tau)}, e^{-2n\pi(\mathrm{Im}\,\tau + i\mathrm{Re}\,\tau)}) : n \in \mathbb{Z}\} = \{\mathfrak{g}^n : n \in \mathbb{Z}\},$$
$$\mathfrak{g} = \mathrm{diag}(e^{2\pi(\mathrm{Im}\,\tau + i\mathrm{Re}\,\tau)}, e^{-2\pi(\mathrm{Im}\,\tau + i\mathrm{Re}\,\tau)}). \tag{5}$$

One can define a Selberg-type zeta function for the group $\Gamma = \{\mathfrak{g}^n : n \in \mathbb{Z}\}$ generated by a single hyperbolic element of the form $\mathfrak{g} = \mathrm{diag}(e^z, e^{-z})$, where $z = \alpha + i\beta$ for $\alpha, \beta > 0$. In fact, we will take $\alpha = 2\pi\mathrm{Im}\,\tau$, $\beta = 2\pi\mathrm{Re}\,\tau$. For the standard action of $SL(2, \mathbb{C})$ on H^3, one has

$$\mathfrak{g}\begin{bmatrix} x \\ y \\ z \end{bmatrix} = \begin{bmatrix} e^\alpha & 0 & 0 \\ 0 & e^\alpha & 0 \\ 0 & 0 & e^\alpha \end{bmatrix}\begin{bmatrix} \cos(\beta) & -\sin(\beta) & 0 \\ \sin(\beta) & \cos(\beta) & 0 \\ 0 & 0 & 1 \end{bmatrix}\begin{bmatrix} x \\ y \\ z \end{bmatrix}. \tag{6}$$

Therefore, $\mathfrak{g}$ is the composition of a rotation in $\mathbb{R}^2$, with complex eigenvalues $\exp(\pm i\beta)$ and a dilatation $\exp(\alpha)$. The Patterson-Selberg spectral function $Z_\Gamma(s)$ is meromorphic on $\mathbb{C}$ and can be attached to H^3/Γ. It is given, for $\mathrm{Re}\, s > 0$, by the formulas [17–19]

$$Z_\Gamma(s) := \prod_{k_1, k_2 \geq 0} [1 - (e^{i\beta})^{k_1}(e^{-i\beta})^{k_2} e^{-(k_1 + k_2 + s)\alpha}], \tag{7}$$

$$\log Z_\Gamma(s) = -\frac{1}{4}\sum_{n=1}^{\infty} \frac{e^{-n\alpha(s-1)}}{n[\sinh^2\left(\frac{\alpha n}{2}\right) + \sin^2\left(\frac{\beta n}{2}\right)]}. \tag{8}$$

The zeros of $Z_\Gamma(s)$ are the complex numbers $\zeta_{n,k_1,k_2} = -(k_1 + k_2) + i\,(k_1 - k_2)\,\beta/\alpha + 2\pi i n/\alpha$, $n \in \mathbb{Z}$ (for details, see [18]). It can be also shown that the zeta function $Z_\Gamma(s)$ is an entire function of order three and finite type. It is bounded in absolute value, for $\mathrm{Re}\, s \geq 0$ as well as for $\mathrm{Re}\, s \leq 0$, and can be estimated as follows:

$$|Z_\Gamma(s)| \leq \prod_{k_1 + k_2 \leq |s|}^{\infty} e^{|s|\ell} \prod_{k_1 + k_2 \geq |s|}^{\infty} (1 - e^{(|s| - k_1 - k_2)\ell}) \leq C_1 e^{C_2 |s|^3}, \tag{9}$$

where C_1 and C_2 are suitable constants. The first factor on the right hand side of (9) yields exponential growth, while the second factor is bounded, what proves the required growth estimate. The spectral function $Z_\Gamma(s)$ is an entire function of order three and finite type and can be written as a Hadamard product [17]

$$Z_\Gamma(s) = e^{Q(s)} \prod_{\zeta\in\Sigma} (1 - \frac{s}{\zeta}) \exp\left(\frac{s}{\zeta} + \frac{s^2}{2\zeta^2} + \frac{s^3}{3\zeta^3}\right), \tag{10}$$

where $\zeta \equiv \zeta_{n,k_1,k_2}$, and we denote the set of such numbers by Σ, $Q(s)$ being a polynomial of degree at most three. It follows from the Hadamard product representation of $Z_\Gamma(s)$ (10) that

$$\frac{d}{ds} \log Z_\Gamma(s) = \frac{d}{ds} Q(s) + \sum_{\zeta\in\Sigma} \frac{(s/\zeta)^3}{s-\zeta}. \tag{11}$$

Let us define $\Xi(y \pm i\xi) := (d/ds) \log Z_\Gamma(s)$ for $s = y \pm i\xi$. Then

$$\Xi(y \pm i\xi) = \frac{d}{ds} Q(s = y \pm i\xi) + i^{-1} \sum_{y\pm i\varepsilon\in\Sigma} \frac{(y \pm i\xi)^3}{(y \pm i\varepsilon)^3(\pm\xi - \varepsilon)}. \tag{12}$$

Generating Functions. Using the equality

$$\sinh^2\left(\frac{\alpha n}{2}\right) + \sin^2\left(\frac{\beta n}{2}\right) = |\sin(n\pi\tau)|^2 = \frac{|1-q^n|^2}{4|q|^n}$$

and (8), we get

$$\log \prod_{m=\ell}^{\infty} (1 - q^{m+\varepsilon}) = \sum_{m=\ell}^{\infty} \log(1 - q^{m+\varepsilon}) = -\sum_{n=1}^{\infty} \frac{q^{(\ell+\varepsilon)n}(1-\overline{q}^n)|q|^{-n}}{4n|\sin(n\pi\tau)|^2}$$

$$= \log\left[\frac{Z_\Gamma(\xi(1-it))}{Z_\Gamma(\xi(1-it)+1+it)}\right], \tag{13}$$

$$\log \prod_{m=\ell}^{\infty} (1 - \overline{q}^{m+\varepsilon}) = \sum_{m=\ell}^{\infty} \log(1 - \overline{q}^{m+\varepsilon}) = -\sum_{n=1}^{\infty} \frac{\overline{q}^{(\ell+\varepsilon)n}(1-q^n)|q|^{-n}}{4n|\sin(n\pi\tau)|^2}$$

$$= \log\left[\frac{Z_\Gamma(\xi(1+it))}{Z_\Gamma(\xi(1+it)+1-it)}\right], \tag{14}$$

$$\log \prod_{m=\ell}^{\infty} (1 + q^{m+\varepsilon}) = \sum_{m=\ell}^{\infty} \log(1 + q^{m+\varepsilon}) = -\sum_{n=1}^{\infty} \frac{(-1)^n q^{(\ell+\varepsilon)n}(1-\overline{q}^n)|q|^{-n}}{4n|\sin(n\pi\tau)|^2}$$

$$= \log\left[\frac{Z_\Gamma(\xi(1-it)+i\eta(\tau))}{Z_\Gamma(\xi(1+it)+1-it+i\eta(\tau))}\right], \tag{15}$$

$$\log \prod_{m=\ell}^{\infty}(1+\overline{q}^{m+\varepsilon}) = \sum_{m=\ell}^{\infty}\log(1+\overline{q}^{m+\varepsilon}) = -\sum_{n=1}^{\infty}\frac{(-1)^n q^{(\ell+\varepsilon)n}(1-q^n)|q|^{-n}}{4n|\sin(n\pi\tau)|^2}$$

$$= \log\left[\frac{Z_\Gamma(\xi(1+it)+i\eta(\tau))}{Z_\Gamma(\xi(1+it)+1-it+i\eta(\tau))}\right], \tag{16}$$

where $\ell \in \mathbb{Z}_+$, $\varepsilon \in \mathbb{C}$, $t = \mathrm{Re}\,\tau/\mathrm{Im}\,\tau$, $\xi = \ell+\varepsilon$, and $\eta(\tau) = \pm(2\tau)^{-1}$. Let us next introduce some well-known functions and their modular properties under the action of $SL(2,\mathbb{Z})$. The special cases associated with (13) and (14) are (see [20])

$$f_1(q) = q^{-\frac{1}{48}}\prod_{m>0}(1-q^{m+\frac{1}{2}}) = \frac{\eta_D(q^{\frac{1}{2}})}{\eta_D(q)}, \tag{17}$$

$$f_2(q) = q^{-\frac{1}{48}}\prod_{m>0}(1+q^{m+\frac{1}{2}}) = \frac{\eta_D(q)^2}{\eta_D(q^{\frac{1}{2}})\eta_D(q^2)}, \tag{18}$$

$$f_3(q) = q^{\frac{1}{24}}\prod_{m>0}(1+q^{m+1}) = \frac{\eta_D(q^2)}{\eta_D(q)}, \tag{19}$$

where $\eta_D(q) \equiv q^{1/24}\prod_{n>0}(1-q^n)$ is the Dedekind η-function. The linear span of $f_1(q)$, $f_2(q)$, and $f_3(q)$ is $SL(2,\mathbb{Z})$-invariant [20] $\left(g \in \begin{bmatrix} a & b \\ c & d \end{bmatrix}, g\cdot f(\tau) = f\left(\frac{a\tau+b}{c\tau+d}\right)\right)$. As $f_1(q)\cdot f_2(q)\cdot f_3(q) = 1$, we get

$$\mathcal{R}(s = 3/2-(3/2)it)\cdot\mathcal{R}(\sigma = 3/2-(3/2)it+i\eta(\tau))\cdot\mathcal{R}(\sigma = 2-2it+i\eta(\tau)) = 1\,.$$

For a closed oriented hyperbolic three-manifold of the form $X = H^3/\Gamma$ (and any acyclic orthogonal representation of $\pi_1(X)$) a set of useful generating functions is collected in Table 2.

4 Generalized Symmetric Products of N-Folds

4.1 Hilbert Schemes and Heisenberg Algebras

Before entering the discussion of the main topic in this section, a short comment about Heisenberg algebras and Hilbert schemes will be in order. Preliminary to the subject of symmetric products and their connection with spectral functions, we briefly explain the relation between the Heisenberg algebra and its representations, and the Hilbert scheme of points, mostly following the lines of [21].

Table 2 List of generating functions

$$\prod_{n=\ell}^{\infty}(1-q^{n+\varepsilon}) = \left[\frac{Z_\Gamma(\xi(1-it))}{Z_\Gamma(\xi(1-it)+1+it)}\right] = \mathcal{R}(s=\xi(1-it))$$

$$\prod_{n=\ell}^{\infty}(1-\overline{q}^{n+\varepsilon}) = \left[\frac{Z_\Gamma(\xi(1+it))}{Z_\Gamma(\xi(1+it)+1-it)}\right] = \mathcal{R}(\overline{s}=\xi(1+it))$$

$$\prod_{n=\ell}^{\infty}(1+q^{n+\varepsilon}) = \left[\frac{Z_\Gamma(\xi(1-it)+i\eta(\tau))}{Z_\Gamma(\xi(1-it)+i\eta(\tau)+1+it)}\right] = \mathcal{R}(\sigma=\xi(1-it)+i\eta(\tau))$$

$$\prod_{n=\ell}^{\infty}(1+\overline{q}^{n+\varepsilon}) = \left[\frac{Z_\Gamma(\xi(1+it)+i\eta(\tau))}{Z_\Gamma(\xi(1+it)+i\eta(\tau)+1-it)}\right] = \mathcal{R}(\overline{\sigma}=\xi(1+it)+i\eta(\tau))$$

$$\prod_{n=\ell}^{\infty}(1-q^{n+\epsilon})^n = \mathcal{R}(s=\xi(1-it))^\ell \prod_{n=\ell}^{\infty}\mathcal{R}(s=(n+\varepsilon+1)(1-it))$$

$$\prod_{n=\ell}^{\infty}(1-\overline{q}^{n+\epsilon})^n = \mathcal{R}(\overline{s}=\xi(1+it))^\ell \prod_{n=\ell}^{\infty}\mathcal{R}(s=(n+\varepsilon+1)(1+it))$$

$$\prod_{n=\ell}^{\infty}(1+q^{n+\epsilon})^n = \mathcal{R}(\sigma=\xi(1-it)+i\eta(\tau))^\ell \prod_{n=\ell}^{\infty}\mathcal{R}(\sigma=(n+\varepsilon+1)(1-it)+i\eta(\tau))$$

$$\prod_{n=\ell}^{\infty}(1+\overline{q}^{n+\epsilon})^n = \mathcal{R}(\overline{\sigma}=\xi(1+it)+i\eta(\tau))^\ell \prod_{n=\ell}^{\infty}\mathcal{R}(\overline{\sigma}=(n+\varepsilon+1)(1+it)+i\eta(\tau))$$

To be more specific, note that the infinite-dimensional Heisenberg algebra (or, simply, the Heisenberg algebra) plays a fundamental role in the representation theory of affine Lie algebras. An important representation of the Heisenberg algebra is the Fock space representation on the polynomial ring of infinitely many variables. The degrees of polynomials (with different degree variables) give a direct sum decomposition of the representation, which is called weight space decomposition.

The Hilbert scheme of points on a complex surface appears in algebraic geometry. The Hilbert scheme of points decomposes into infinitely many connected components according to the number of points. Betti numbers of the Hilbert scheme have been computed in [22]. The sum of the Betti numbers of the Hilbert scheme of N-points is equal to the dimension of the subspaces of the Fock space representation of degree N.

Algebraic Preliminaries. Let $R=\mathbb{Q}[p_1,p_2,\ldots]$ be the polynomial ring of infinite many variables $\{p_j\}_{j=1}^{\infty}$. Define $P[j]$ as $j\partial/\partial p_j$ and $P[-j]$ as a multiplication of p_j for each positive j. Then, the commutation relation holds: $[P[i],\,P[j]]=i\delta_{i+j,0}\,\mathrm{Id}_R$, $i,j\in\mathbb{Z}/\{0\}$. We define the infinite-dimensional Heisenberg algebra as the Lie algebra generated by $P[j]$ and K with defining relation

$$[P[i],\,P[j]]=i\delta_{i+j,0}K_R,\quad [P[i],\,K]=0,\quad i,j\in\mathbb{Z}/\{0\}. \tag{20}$$

The above R labels the representation. If $1 \in R$ is the constant polynomial, then $P[i]1 = 0$, $i \in \mathbb{Z}_+$ and

$$R = \mathrm{Span}\{P[-j_1]\cdots P[-j_k]1 \mid k \in \mathbb{Z}_+ \cup \{0\},\ j_1,\ldots,j_k \in \mathbb{Z}_+\}. \tag{21}$$

1 is a highest-weight vector. This is known in physics as the *bosonic Fock space*. The operators $P[j]\,(j < 0)$ $(P[j]\,(j > 0))$ are the *creation (annihilation) operators*, while 1 is the *vacuum vector*. Define the degree operator $\mathfrak{D} : R \to R$ by $\mathfrak{D}(p_1^{m_1}p_2^{m_2}\cdots) \stackrel{\mathrm{def}}{=} (\sum_i im_i)p_1^{m_1}p_2^{m_2}\ldots$. The representation R has $\mathfrak{D}$ eigenspace decomposition; the eigenspace with eigenvalue N has a basis $p_1^{m_1}p_2^{m_2}\cdots(\sum_i im_i) = N$. Recall that a partition of N is defined by a nonincreasing sequence of nonnegative integers $\nu_1 \geq \nu_2 \geq \cdots$ such that $\sum_\ell \nu_\ell = N$. One can represent ν as $(1^{m_1}, 2^{m_2}, \ldots)$ (where 1 appears m_1-times, 2 appears m_2-times, and so on, in the sequence). Therefore, elements of the basis correspond bijectively to a partition ν. The generating function of eigenspace dimensions, or the *character* in the terminology of representation theory, is known to have the form

$$\mathrm{Tr}_R\, q^{\mathfrak{D}} \stackrel{\mathrm{def}}{=} \sum_{N\geq 0} q^N \dim\{r \in R \mid \mathfrak{D}r = Nr\} = \prod_{n=1}^{\infty}(1-q^n)^{-1}. \tag{22}$$

Let us define now the Heisenberg algebra associated with a finite-dimensional $\mathbb{Q}$-vector space V with nondegenerate symmetric bilinear form $(\,,\,)$. Let $W = (V \otimes t\,\mathbb{Q}[t]) \oplus (V \otimes t^{-1}\,\mathbb{Q}[t^{-1}])$, then define a skew-symmetric bilinear form on W by $(r \otimes t^i, s \otimes t^j) = i\delta_{i+j,0}(r,s)$. The Heisenberg algebra associated with V can be defined as follows: take the quotient of the free algebra $A(W)$ divided by the ideal $\mathfrak{I}$ generated by $[r,s] - (r,s)1$ $(r,s \in W)$. It is clear that when $V = \mathbb{Q}$ one has the above Heisenberg algebra. For an orthogonal basis $\{r_j\}_{j=1}^p$ the Heisenberg algebra associated with V is isomorphic to the tensor product of p-copies of the above Heisenberg algebra.

Let us consider next the *super*-version of the Heisenberg algebra, known as the super-Heisenberg algebra. The initial data are constituted by a vector space, V, with decomposition $V = V_{\mathrm{even}} \oplus V_{\mathrm{odd}}$, and a nondegenerate bilinear form satisfying $(r,s) = (-1)^{|r||s|}(r,s)$. In this formula, r,s are either elements of V_{even} or V_{odd}, while $|r| = 0$ if $r \in V_{\mathrm{even}}$ and $|r| = 1$ if $r \in V_{\mathrm{odd}}$. As above, we can define W, the bilinear form on W, and $A(W)/\mathfrak{I}$, where now we replace the Lie bracket $[\,,\,]$ by the super-Lie bracket. In addition, to construct the free-super-Lie algebra in the tensor algebra, we set

$$(r \otimes t^i, s \otimes t^j) = (r \otimes t^i)(s \otimes t^j) + (s \otimes t^j)(r \otimes t^i)$$

for $r,s \in V_{\mathrm{odd}}$. By generalizing the representation on the space of polynomials of infinitely many variables one can get a representation of the super-Heisenberg algebra on the symmetric algebra $R = S^*(V \otimes t\,\mathbb{Q}[t])$ of the positive degree part $V \otimes t\,\mathbb{Q}\,[t]$. As above, we can define the degree operator $\mathfrak{D}$. The following character formulas hold:

$$\mathrm{Tr}_R\, q^{\mathfrak{D}} = \prod_{n=1}^{\infty} \frac{(1+q^n)^{\dim V_{\rm odd}}}{(1-q^n)^{\dim V_{\rm even}}} = \frac{\mathcal{R}(\sigma = 1 - it + i\eta(\tau))^{\dim V_{\rm odd}}}{\mathcal{R}(s = 1 - it)^{\dim V_{\rm even}}}\,, \tag{23}$$

$$\mathrm{STr}_R\, q^{\mathfrak{D}} = \prod_{n=1}^{\infty} (1-q^n)^{\dim V_{\rm odd} - \dim V_{\rm even}} = \mathcal{R}(s = 1 - it)^{\dim V_{\rm odd} - \dim V_{\rm even}}\,, \tag{24}$$

where we have counted the odd degree part by -1 and replaced the usual trace by the super-trace.[3]

If we consider the generating function of the Poincaré polynomials associated with sets of points, we get the character of the Fock space representation of the Heisenberg algebra. This is, in general, the integrable highest-weight representation of the corresponding affine Lie algebra and is known to have modular invariance, as was proven in [23]. This occurrence is naturally explained through the relation to partition functions of conformal field theory on a torus. In this connection, the affine Lie algebra has a close relationship to conformal field theory.

4.2 One-Dimensional Higher Variety

Let us consider the N-fold symmetric product $\mathfrak{S}^N X$ of a Kähler manifold X, that is, the $\mathfrak{S}^N X = [X^N/\mathfrak{S}_N] := \underbrace{X \times \cdots \times X}_{N}/\mathfrak{S}_N$ orbifold space, $\mathfrak{S}_N$ being the symmetric group of N elements. Objects of the category of the orbispace $[X^N/\mathfrak{S}_N]$ are the N-tuples $(x_1, \ldots, x_N)$ of points in X; arrows are elements of the form $(x_1, \ldots, x_N; \sigma)$, where $\sigma \in \mathfrak{S}_N$. In addition, the arrow $(x_1, \ldots, x_N; \sigma)$ has as its source $(x_1, \ldots, x_N)$ and as its target $(x_{\sigma(1)}, \ldots, x_{\sigma(N)})$. This category is a groupoid for the inverse of $(x_1, \ldots, x_N; \sigma)$ is $(x_{\sigma(1)}, \ldots, x_{\sigma(N)}; \sigma^{-1})$. (The orbispace as a groupoid has been described in [24, 25].) For a one-dimensional higher variety (i.e., for a surface) the following results hold:

[3] In the case when V has the one-dimensional odd degree part only (the bilinear form is $(r,r) = 1$ for a nonzero vector $r \in V$) and the above condition is not satisfied, we can modify the definition of the corresponding super-Heisenberg algebra by changing the bilinear form on W as $(r \otimes t^i, r \otimes t^j) = \delta_{i+j,0}$. The resulting algebra is termed an *infinite-dimensional Clifford algebra*. The above representation R is the *fermionic Fock space* in physics and it can be modified as follows: the representation of the even degree part was realized as the space of polynomials of infinitely many variables; the Clifford algebra is realized on the exterior algebra $R = \Lambda^*(\bigoplus_j \mathbb{Q}dp_j)$ of a vector space with a basis of infinitely many vectors. For $j > 0$ we define $r \otimes t^{-j}$ as an exterior product of dp_j and $r \otimes t^j$ as an interior product of $\partial/\partial p_j$.

- For a Riemann surface ($\dim X = 1$) $\mathfrak{S}^N X$ and X^N are isomorphic under the Hilbert-Chow morphism.
- If X is a nonsingular quasi-projective surface, the Hilbert-Chow morphism $\pi : X^{[N]} \to \mathfrak{S}^N X$ gives a resolution of the singularities of the symmetric product $\mathfrak{S}^N X$ [26]. In particular, $X^{[N]}$ is a nonsingular quasi-projective variety of dimension $2N$.
- If X has a symplectic form, $X^{[N]}$ has also a symplectic form. For $N = 2$ this has been proven in [27], and for N general in [28].
- The generating function of the Poincaré polynomials $P_r(X^{[N]})$ of $X^{[N]}$ is given by

$$\sum_{N=0}^{\infty} q^N P_r(X^{[N]}) = \prod_{n=1}^{\infty} \frac{(1+r^{2n-1}q^n)^{b_1(X)}(1+r^{2n+1}q^n)^{b_3(X)}}{(1-r^{2n-2}q^n)^{b_0(X)}(1-r^{2n}q^n)^{b_2(X)}(1-r^{2n+2}q^n)^{b_4(X)}}$$

$$= \frac{\prod_{j=1,2} \mathcal{R}(\sigma = \xi_{2j-1}(1-it) + i\eta(\tau))^{b_{2j-1}(X)}}{\prod_{j=1,2,3} \mathcal{R}(s = \xi_{2j-2}(1-it))^{b_{2j-2}(X)}}, \qquad (25)$$

where $\xi_{2j-1} = j - 1/2$, $\xi_{2j-2} = j - 1$ and $r = \exp(\pi i \tau)$.

4.3 Equivariant K-Theory, Wreath Products

We study here a direct sum of the equivariant K-groups $\mathcal{F}_\Gamma(X) := \oplus_{N\geq 0}\underline{K}_{\Gamma_N}(X^N)$ associated with a topological Γ-space [1]. Γ is a finite group and the wreath (semidirect) product $\Gamma_N \rtimes \mathfrak{S}_N$ acts naturally on the Nth Cartesian product X^N. One can calculate the torsion free part of $K^\bullet_\Gamma(Y)$ (where Γ acts on Y and Γ is a finite group) by localizing on the prime ideals of $R(\Gamma)$, the representation ring of Γ (for details, see [29]) $\underline{K}^\bullet_\Gamma(Y) \cong \bigoplus_{\{\gamma\}} \underline{K}^\bullet(Y^\gamma)^{\Gamma_\gamma}$, where $\underline{K}_{\Gamma_N}(X^N) \equiv K_{\Gamma_N}(X^N) \otimes \mathbb{C}$. Here $\{\gamma\}$ runs over the conjugacy classes of elements in Γ, Y^γ are the fixed point loci of γ, and Γ_γ is the centralizer of γ in Γ. The fixed point set $\{X^N\}^\gamma$ is isomorphic to $X^N = X^{\sum_n N_n}$, $\gamma \in \mathfrak{S}_N$, and $\Gamma_\gamma \cong \prod_n \mathfrak{S}_{N_n} \ltimes (\mathbb{Z}/n)^{N_n}$. The cyclic groups $\mathbb{Z}/n$ act trivially in $K^\bullet(X^N)$, and therefore, the following decomposition for the $\mathcal{R}_N$-equivariant K-theory holds [30]:

$$\underline{K}^\bullet(\mathfrak{S}^N X) \cong \bigoplus_{\{\gamma\}} \underline{K}^\bullet((\mathfrak{S}^N)^\gamma))^{\Gamma_\gamma} \cong \bigoplus_{\sum nN_n = N} \bigotimes_n \underline{K}^\bullet(\mathfrak{S}^{N_n})^{\mathfrak{S}_{N_n}}. \qquad (26)$$

As an example of such K-group we here analyze the group $K^-_{\tilde{H}\Gamma_N}(X^N)$ which has been introduced in [1]. The semidirect product Γ_N can be extended to the action of a larger finite supergroup $\tilde{H}\Gamma_N$, which is a double cover of the semidirect product $(\Gamma \times \mathbb{Z}_2)^N \rtimes \mathfrak{S}_N$. The category of $\tilde{H}\Gamma_N$-equivariant spin vector superbundles over

X^N is the category of Γ_N-equivariant vector bundles E over X^N such that E carries a supermodule structure with respect to the complex Clifford algebra of rank N.[4]

It has been shown [1] that the following statements hold:

1. The direct sum $\mathcal{F}_{\Gamma}^{-}(X) := \oplus_{N=0}^{\infty} \underline{K}_{\tilde{H}\Gamma_N}(X^N)$ carries naturally a Hopf algebra structure.
2. It is isomorphic to the Fock space of a twisted Heisenberg superalgebra (in this section *super* means $\mathbb{Z}_2$-graded) associated with $K_{\tilde{H}\Gamma_N}^{-}(X) \cong K_{\Gamma}(X)$.
3. If X is a point, the K-group $\underline{K}_{\tilde{H}\Gamma_N}(X^N)$ becomes the Grothendieck group of spin supermodules of $\tilde{H}\Gamma_N$.

Such a twisted Heisenberg algebra has played an important role in the theory of affine Kac-Moody algebras [31]. The structure of the space $\mathcal{F}_{\Gamma}^{-}(X)$ under consideration can be modeled on the ring $\Omega_{\mathbb{C}}$ of symmetric functions with a linear basis given by the so-called Schur Q-functions (or equivalently on the direct sum of the spin representation ring of $\tilde{H}\Gamma_N$ for all N). The graded dimension of the ring $\Omega_{\mathbb{C}}$ is given by the denominator $\prod_{n=0}(1-q^{2n-1})^{-1}$. On the basis of Göttsche's formula [22] it has been conjectured [32] that the direct sum $\mathcal{H}(S)$ of the homology groups for Hilbert scheme $S^{[N]}$ of N-points on a (quasi-)projective surface S should carry the structure of the Fock space of a Heisenberg algebra, which was realized subsequently in a geometric way [21, 33]. Parallel algebraic structures such as Hopf algebra, vertex operators, and Heisenberg algebra as part of vertex algebra structures [31, 34] have naturally showed up in $\mathcal{H}(S)$ as well as in $\mathcal{F}_{\Gamma}(X)$. If S is a suitable resolution of singularities of an orbifold X/Γ, there appears close connections between $\mathcal{H}(S)$ and $\mathcal{F}_{\Gamma}(X)$ [1]. In fact the special case of Γ trivial is closely related to the analysis considered in [35]. It would be interesting to find some applications of results discussed above in string theory.

The Generating Function. The orbifold Euler number $\mathbf{e}(X,\Gamma)$ was introduced in [36] in the study of orbifold string theory and it has been interpreted as the Euler number of the equivariant K-group $K_{\Gamma}(X)$ [37]. Define the Euler number of the generalized symmetric product to be the difference

$$\mathbf{e}(X^N, \tilde{H}\Gamma_N) := \dim K_{\tilde{H}\Gamma_n}^{-,0}(X^N) - \dim K_{\tilde{H}\Gamma_N}^{-,1}(X^N),$$

the series $\sum_{N=0}^{\infty} q^N \mathbf{e}(X^N, \tilde{H}\Gamma_N)$ can be written in terms of spectral functions:

$$\sum_{N=0}^{\infty} q^N \mathbf{e}(X^N, \tilde{H}\Gamma_N) = \prod_{n=1}^{\infty}(1-q^{2n-1})^{-\mathbf{e}(X,\Gamma)} = \left[q^{-\frac{25}{24}}(q-1)f_3(q)\right]^{\mathbf{e}(X,\Gamma)}$$

$$= \mathcal{R}(s = 1/2 - (1/2)it)^{-\mathbf{e}(X,\Gamma)}. \tag{27}$$

[4] A fundamental example of $\tilde{H}\Gamma_N$-vector superbundles over X^N (X compact) is the following: given a Γ-vector bundle V over X, consider the vector superbundle $V \oplus V$ over X with the natural $\mathbb{Z}_2$-grading. One can endow the Nth outer tensor product $(V \oplus V)^{\boxtimes N}$ with a natural $\tilde{H}\Gamma_N$-equivariant vector superbundle structure over X^N.

One can give an explicit description of $\mathfrak{F}_\Gamma^-(X)$ as a graded algebra. Indeed, the following statement holds [1]: as a $(\mathbb{Z}_+ \times \mathbb{Z}_2)$-graded algebra, $\mathfrak{F}_\Gamma^-(X,q)$ is isomorphic to the supersymmetric algebra $\mathfrak{S}\left(\bigoplus_{N=1}^\infty q^{2N-1}\underline{K}_\Gamma(X)\right)$. In particular,

$$\dim_q \mathfrak{F}_\Gamma^-(X) = \prod_{n=1}^\infty \frac{(1+q^{2n-1})^{\dim K_\Gamma^1(X)}}{(1-q^{2n-1})^{\dim K_\Gamma^0(X)}}$$

$$= \frac{[\mathcal{R}(\sigma = 1/2 - (1/2)it + (1/2)i\eta(\tau))]^{\dim K_\Gamma^1(X)}}{[\mathcal{R}(s = 1/2 - (1/2)it)]^{\dim K_\Gamma^0(X)}}, \tag{28}$$

where the supersymmetric algebra is equal to the tensor product of the symmetric algebra $\mathfrak{S}\left(\bigoplus_{N=1}^\infty q^{2N-1}\underline{K}_\Gamma^0(X)\right)$ and the exterior algebra $\Lambda\left(\bigoplus_{N=1}^\infty q^{2N-1}\underline{K}_\Gamma^1(X)\right)$. In the case when X_{pt} is a point we have

$$\sum_{N\geq 0} q^N \dim \mathfrak{F}_\Gamma^-(X_{pt}) = \prod_{n=1}^\infty (1-q^{2n-1})^{-|\Gamma_*|} = [\mathcal{R}(s = 1/2 - (1/2)it)]^{-|\Gamma_*|}\,. \tag{29}$$

In (29) Γ is a finite group with $r+1$ conjugacy classes; $\Gamma^* := \{\gamma_j\}_{j=0}^r$ is the set of complex irreducible characters, where γ_0 denotes the trivial character. By Γ_* we denote the set of conjugacy classes.

5 Conclusions

Having advocated in this paper the basic role of modular forms and spectral functions with their connection to Lie algebra cohomologies and K-theory methods, we are now, in concluding, naturally led to other crucial problems, related to the deformation procedure. For instance, one might ask whether the form of the partition functions could be algebraically interpreted by means of infinitesimal deformations of the corresponding Lie algebras. No doubt this analysis will require a new degree of mathematical sophistication. Perhaps all the concepts of what could eventually be called the "deformation theory of everything" might be possibly tested in the case of associative algebras, which are algebras over operads [38]. In many examples dealing with algebras over operads, arguments of the universality of associative algebras are called forth. This strongly suggests that a connection between the deformation theory (deformed partition functions) and algebras over operads might exist. We expect to be able to discuss this key problem in forthcoming work.

Acknowledgements AAB would like to acknowledge the Conselho Nacional de Desenvolvimento Científico e Tecnológico (CNPq, Brazil) and Fundaçao Araucaria (Parana, Brazil) for the financial support. EE's research has been partly supported by MICINN (Spain), contract

PR2011-0128 and projects FIS2006-02842 and FIS2010-15640, and by the CPAN Consolider Ingenio Project and by AGAUR (Generalitat de Catalunya), contract 2009SGR-994. EE's research was done in part while visiting the Department of Physics and Astronomy, Dartmouth College, NH, USA.

References

1. Wang, W.: Equivariant K-theory, generalized symmetric products, and twisted Heisenberg algebra. Commun. Math. Phys. **234**, 101–127 (2003). arXiv:math.QA/0104168v2
2. Filipkiewicz, R.P.: Four-dimensional geometries. Ph.D. thesis, University of Warwick (1984)
3. Wall, C.T.C.: Geometries and geometric structures in real dimension 4 and complex dimension 2. In: Geometry and Topology. Lectures Notes in Mathematics, vol. 1167, pp. 268–292. Springer, Berlin (1986)
4. Bytsenko, A.A., Cognola, G., Elizalde, E., Moretti, V., Zerbini, S.: Analytic Aspects of Quantum Fields. World Scientific, Singapore (2003)
5. Thurston, W.: Three-dimensional manifolds, Kleinian groups and hyperbolic geometry. Bull. Amer. Math. Soc. (N.S.) **6**, 357–381 (1982)
6. Besse, A.: Geométrie Riemannianne Eu Dimension 4. Séminaire Arthur Besse, 1978/1979. Cedic/Fernand Nathan, Paris (1981)
7. Mostow, G.D.: Quasi-conformal mapping in n-space and the rigidity of hyperbolic space forms. Inst. Hautes Etudes Sci. Publ. Math. **34**, 53–104 (1968)
8. Bonora, L., Bytsenko, A.A.: Fluxes, brane charges and chern morphisms of hyperbolic geometry. Class. Quant. Grav. **23**, 3895–3916 (2006). arXiv:hep-th/0602162
9. Bonora, L., Bytsenko, A.A.: Partition functions for quantum gravity, black holes, elliptic genera and Lie algebra homologies. Nucl. Phys. B **852**, 508–537 (2011). arXiv:hep-th/1105.4571
10. Ray, D., Singer, I.: R-torsion and the Laplacian on Riemannian manifolds. Adv. Math. **7**, 145–210 (1971)
11. Farber, M., Turaev, V.: Poincaré-reidemaister metric, Euler structures, and torsion. J. reine angew. Math. (Crelles Journal) **2000**, 195–225 (2006)
12. Fried, D.: Analytic torsion and closed geodesics on hyperbolic manifolds. Invent. Math. **84**, 523–540 (1986)
13. Bytsenko, A.A., Vanzo, L., Zerbini, S.: Ray-Singer torsion for a hyperbolic 3-manifold and asymptotics of Chern-Simons-Witten invariant. Nucl. Phys. **B505**, 641–659 (1997). arXiv:hep-th/9704035v2
14. Bytsenko, A.A., Vanzo, L., Zerbini, S.: Semiclassical approximation for Chern-Simons theory and 3-hyperbolic invariants. Phys. Lett. **B459**, 535–539 (1999). arXiv:hep-th/9906092v1
15. Williams, F.L.: Vanishing theorems for type $(0, q)$ cohomology of locally symmetric spaces. Osaka J. Math. **18**, 147–160 (1981)
16. Deitmar, A.: The Selberg trace formula and the Ruelle zeta function for compact hyperbolics. Abh. Math. Se. Univ. Hamburg **59**, 101–106 (1989)
17. Perry, P., Williams, F.: Selberg zeta function and trace formula for the BTZ black hole. Int. J. Pure Appl. Math. **9**, 1–21 (2003)
18. Bytsenko, A.A., Guimãraes, M.E.X., Williams, F.L.: Remarks on the spectrum and truncated heat Kernel of the BTZ black hole. Lett. Math. Phys. **79**, 203–211 (2007). arXiv:hep-th/0609102
19. Bytsenko, A.A., Guimãraes, M.E.X.: Expository remarks on three-dimensional gravity and hyperbolic invariants. Class. Quantum Grav. **25**, 228001 (2008). arXiv:hep-th/0809.5179
20. Kac, V.G.: Infinite Dimensional Lie Algebras, 3rd edn. Cambridge University Press, Cambridge (1990)

21. Nakajima, H.: Lectures on Hilbert Schemes of Points on Surface. AMS University Lectures Series, vol. 18. American Mathematical Society, Providence (1999) [MR 1711344 (2001b:14007)]
22. Göttsche, L.: The Betti numbers of the Hilbert scheme of points on a smooth projective surface. Math. Ann. **286**, 193–207 (1990)
23. Kac, V., Peterson, D.: Infinite dimensional Lie algebras, theta functions and modular forms. Adv. Math. **53**, 125–264 (1984)
24. Kontsevich, M.: Enumeration of rational curves via torus actions. In: The Moduli Space of Curves (Texel Island, Netherlands, 1994). Progress in Mathematics, vol. 129, pp. 335–368. Birkhäuser Boston (1995). arXiv:hep-th/9405035v2 [MR: MR1363062 (97d:14077)]
25. Moerdijk, I.: Orbifolds as groupoids: an introduction. In: Orbifolds in Mathematics and Physics (Madison, 2001). Contemporary Mathematics, vol. 310. American Mathematical Society, Providence (2002). arXiv:math.DG/0203100v1 (205, MR 1 950 948)
26. Fogarty, J.: Algebraic families on an algebraic surface. Amer. J. Math. **90**, 511–521 (1968)
27. Fujiki, A.: On primitive symplectic compact Kähler v-manifolds of dimension four. In: Ueno, K. (ed.) Classification of Algebraic and Analytic Manifolds. Progress in Mathematics, vol. 39, pp. 71–125. Birkhäuser, Boston (1983)
28. Beauville, A.: Complex Algebraic Surfaces. LMS Student Texts, vol. 34. Cambridge University Press, Cambridge (1996)
29. Segal, G.: Equivariant K-theory. Inst. Hautes Études Sci. Publ. Math. **34**, 129–151 (1968) [MR MR0234452 (38 # 2769)]
30. Lupercio, E., Uribe, B., Xicotencatl, M.A.: The loop orbifold of the symmetric product. J. Pure Appl. Algebra (ISSN 0022–4049) **211** 293–306 (2007). arXiv:math.AT/0606573
31. Frenkel, I., Lepowsky, J., Meurman, A.: Vertex Operator Algebras and the Monster. Academic, New York (1988)
32. Vafa, C., Witten, E.: A strong coupling test of S-duality. Nucl. Phys. B **431**, 3–77 (1994). arXiv:hep-th/9408074v2
33. Grojnowski, I.: Instantons and affine algebras I: the Hilbert scheme and vertex operators. Math. Res. Lett. **3**, 275–291 (1996). arXiv:alg-geom/9506020v1
34. Borcherds, R.E.: Vertex algebras, Kac-Moody algebras, and the Monster. Proc. Natl. Acad. Sci. USA **83**, 3068–3071 (1986)
35. de Boer, J., Cheng, M.C.N., Dijkgraaf, R., Manschot, J., Verlinde, E.: A farey tail for attractor black holes. JHEP **0611** 024 (2006). arXiv:hep-th/0608059
36. Dixon, L., Harvey, J.A., Vafa, C., Witten, E.: Strings on orbifolds. Nucl. Phys. B **261** 678–686 (1985)
37. Atiyah, M., Singer, I.M.: The index of elliptic operators: III. Ann. Math. **87** 546–604 (1968)
38. Kontsevich, M., Soibelman, Y.: Deformations of algebras over operads and Deligne's conjecture. arXiv:math/0001151v2

Families of Twisted Bernoulli Numbers, Twisted Bernoulli Polynomials, and Their Applications

Yilmaz Simsek

Dedicated to Professor Hari M. Srivastava

Abstract This chapter is motivated by the fact that the theories and applications of the many methods and techniques used in dealing with some different families of the twisted Bernoulli numbers, the twisted Bernoulli polynomials, and their families of interpolation functions, which are the family of twisted zeta functions, the family of twisted L-functions. By using the p-adic Volkenborn integral, twisted (h,q)-Bernoulli numbers and twisted (h,q)-Bernoulli polynomials are introduced. The p-adic meromorphic functions, which interpolation twisted (h,q)-Bernoulli numbers and twisted (h,q)-Bernoulli polynomials, associated with the p-adic Volkenborn integral, are presented. Furthermore relationships between Bernoulli functions, Euler functions, some arithmetic sums, Dedekind sums, Hardy Berndt sums, DC-sums, trigonometric sums, and Hurwitz zeta function are given.

1 Introduction

The Bernoulli numbers are named after the great Swiss mathematician Jacob Bernoulli (1654–1705), who used these numbers in the power sums, which is defined by

$$S(p,n) = \sum_{k=0}^{n} k^p.$$

Y. Simsek (✉)
Department of Mathematics, University of Akdeniz Faculty Science, 07058 Antalya, Turkey
e-mail: ysimsek@akdeniz.edu.tr

G.V. Milovanović and M.Th. Rassias (eds.), *Analytic Number Theory, Approximation Theory, and Special Functions*, DOI 10.1007/978-1-4939-0258-3_6,

The sum $S(p,n)$ is the so-called Faulhaber's formula, named after Johann Faulhaber, a German mathematician (5 May 1580–10 September 1635). The sum $S(p,n)$ is given explicitly by the Bernoulli polynomials as follows:

$$S(p,n) = \frac{B_{p+1}(n+1) - B_{p+1}(0)}{p+1}.$$

Therefore, the history of the Bernoulli numbers and the Bernoulli polynomials goes back to Bernoulli in the sixteenth century. From Bernoulli to this time, the Bernoulli numbers can be defined in many different ways and areas. Thus many applications of these numbers and their generating functions have been looked for by many authors in the literature. Many different special functions are used to construct generating functions for the Bernoulli numbers and the Bernoulli polynomials. The Bernoulli polynomials are associated with many of special functions, for example, in particular the Riemann zeta function, the Hurwitz zeta function, the family of L-functions, and trigonometric functions. The Bernoulli polynomials are an Appell sequence, i.e., a Sheffer sequence for the ordinary derivative operator. The Bernoulli numbers and the Bernoulli polynomials have many applications in Analytic Number Theory. We summarize this chapter as follows: In Sect. 1, we give definitions and some properties of the Bernoulli numbers and the Bernoulli polynomials. We present definitions of the twisted Bernoulli numbers and the twisted Bernoulli polynomials. We introduce various properties of these numbers and polynomials. We also consider the twisted Bernoulli polynomials and numbers in terms of a Dirichlet character. In Sect. 2, we present q-analogue of the generating functions for Bernoulli-type numbers and polynomials. We define the modified twisted q-Bernoulli numbers and the modified twisted q-Bernoulli polynomials. We also introduce generalized modified twisted q-Bernoulli numbers and polynomials. We give many series representations and relations related to these numbers and polynomials. In Sect. 3, we introduce three families of interpolation functions of the modified twisted q-Bernoulli numbers and modified twisted q-Bernoulli polynomials. These interpolation functions are related to three families of modified twisted q-extension Riemann zeta function, modified twisted q-extension Hurwitz zeta function, and modified twisted q-extension Dirichlet L-functions. We also present a p-adic interpolation function of the modified twisted generalized q-Bernoulli numbers. In Sect. 4, we define twisted (h,q)-Bernoulli numbers and polynomials by using p-adic Volkenborn integral. We also give two variable p-adic meromorphic functions associated with the p-adic Volkenborn integral. These functions interpolate twisted (h,q)-Bernoulli numbers and polynomials at negative integers. In Sect. 5, we give functional equation of Dedekind eta functions which include Dedekind sums. We also give some arithmetic sums related to the Hardy–Berndt sums and trigonometric functions and Bernoulli functions. Finally we present DC-sums which are related to not only the Hardy–Berndt sums but also trigonometric functions and Euler functions and the $Y(h,k)$ sums.

Throughout of this chapter, we use the following notations:

$$\mathbb{N} := \{1, 2, 3, \ldots\}, \quad \mathbb{Z}^- = \{-1, -2, -3, \ldots\}, \quad \text{and} \quad \mathbb{N}_0 := \mathbb{N} \cup \{0\}.$$

Here $\mathbb{Z}$ denotes the set of integers, $\mathbb{R}$ denotes the set of real numbers, $\mathbb{R}^+$ denotes the set of positive real numbers, and $\mathbb{C}$ denotes the set of complex numbers. We also assume that $\ln(z)$ denotes the *principal branch of the many-valued function* $\ln(z)$ with the imaginary part $\mathrm{Im}(\ln(z))$ constrained by $-\pi < \mathrm{Im}(\ln(z)) \leqq \pi$. Finally,

$$\binom{n}{k} = \frac{m!}{n!(m-n)!}$$

and $[x]_G$ denotes the largest integer $\leq x$.

2 The Family of the Bernoulli Polynomials and Numbers

The *Bernoulli polynomials* $B_n(x)$ are defined by means of the following generating function:

$$F_B(t, x) = \frac{te^{tx}}{e^t - 1} = \sum_{n=0}^{\infty} B_n(x) \frac{t^n}{n!} \quad (|t| < 2\pi). \tag{1}$$

The numbers $B_n = B_n(0)$ are called the *Bernoulli numbers*, which are defined by means of the following generating function:

$$F_B(t) = \frac{t}{e^t - 1} = \sum_{n=0}^{\infty} B_n \frac{t^n}{n!} \quad (|t| < 2\pi). \tag{2}$$

By using (1) and (2), we get the following functional equation:

$$F_B(t, x) = e^{tx} F_B(t).$$

It easily follows from the above functional equation that

$$\sum_{n=0}^{\infty} B_n(x) \frac{t^n}{n!} = \left(\sum_{n=0}^{\infty} x^n \frac{t^n}{n!} \right) \left(\sum_{n=0}^{\infty} B_n \frac{t^n}{n!} \right).$$

Therefore

$$B_n(x) = \sum_{k=0}^{n} \binom{n}{k} x^{n-k} B_k. \tag{3}$$

By using (1), we obtain the following functional equation:

$$F_B(t, x+1) - F_B(t, x) = te^{tx}.$$

From the above equation, one can easily obtain

$$B_n(x+1) - B_n(x) = nx^{n-1}, \tag{4}$$

where $n \in \mathbb{N}_0$. It easily follows from (4) that

$$B_n(0) = B_n(1), \tag{5}$$

where $n \in \mathbb{N} \setminus \{1\}$.

We are now ready to give recursion formula for computing Bernoulli numbers. Substituting $x = 1$ into (3), in view of (5), we have

$$B_n = \sum_{k=0}^{n} \binom{n}{k} B_k, \tag{6}$$

where $B_0 = 1$ and $n > 1$

By using the above recursion formula, we have the following list for the first few of the Bernoulli numbers:

$$B_0 = 1,\ B_1 = -\frac{1}{2},\ B_2 = \frac{1}{6},\ B_4 = -\frac{1}{30},\ B_6 = \frac{1}{42},\ B_8 = -\frac{1}{30},$$

$$B_{10} = \frac{5}{66},\ B_{12} = -\frac{691}{2730},\ B_{14} = \frac{7}{6},\ B_{16} = -\frac{3617}{510},\ B_{18} = \frac{43867}{798},\ \ldots$$

and $B_{2n+1} = 0$ $(n \in \mathbb{N})$.

By using (3) and (6), we have the following list for the first few of the Bernoulli polynomials:

$$B_0(x) = 1,\ B_1(x) = x - \frac{1}{2},\ B_2(x) = x^2 - x + \frac{1}{6},\ B_3(x) = x^3 - \frac{3}{2}x^2 + \frac{1}{2}x,$$

$$B_4(x) = x^4 - 2x^3 + x^2 - \frac{1}{30},\ B_5(x) = x^5 - \frac{5}{2}x^4 + \frac{5}{3}x^3 - \frac{1}{6}x,\ \ldots.$$

By using (1), one can easily derive the following well-known basic identities for the Bernoulli polynomials:

By (1), we obtain the following partial differential equation:

$$\frac{\partial}{\partial x} F_B(x,t) = tF_B(x,t).$$

By using the above partial differential equation, we have

$$\frac{d}{dx} B_n(x) = nB_{n-1}(x).$$

We set $F_B(t, 1-x) = F_B(-t, x)$. By using the above functional equation, one can easily get

$$B_n(1-x) = (-1)^n B_n(x).$$

One also has

$$(-1)^n B_n(-x) = B_n(x) + nx^{n-1}.$$

We remark that the Bernoulli polynomials are the Appell polynomials (cf. [20, 76, 77]). **The von Staudt–Clausen Theorem:**

This theorem was given by Karl Georg Christian von Staudt and Thomas Clausen independently in 1840. This theorem is given as follows (cf. [87, p. 55, Theorem 5.10]):

Theorem 2.1. *Let n be even and positive integer. Then*

$$B_n + \sum_{(p-1)|n} \frac{1}{p}$$

is an integer. The sum extends over all primes p for which $p-1$ divides $2n$.

A consequence of this is that the denominator of B_{2n} is given by the product of all primes p for which $p-1$ divides $2n$. Thus, these denominators are square-free and divisible by 6 (cf. [21, 22, 76, 77, 87]).

Asymptotic Approximation

By the Stirling formula, as n goes to infinity,

$$|B_{2n}| \sim 4\sqrt{n\pi}\left(\frac{n}{\pi e}\right)^{2n}$$

(cf. [76, 77]).

2.1 The Twisted Bernoulli Polynomials and Numbers

The twisted Bernoulli numbers are defined by means of the following generating function:

$$F_\xi(t) = \frac{t}{\xi e^t - 1} = \sum_{n=0}^{\infty} B_{n,\xi} \frac{t^n}{n!}, \tag{7}$$

where $\xi^r = 1$, $\xi \neq 1$, and $r \in \mathbb{Z}^+$ are the set of positive integers (cf. [15, 33, 43, 45, 61, 63, 66, 68]).

The twisted Bernoulli numbers are related to the Frobenius–Euler numbers, which are defined as follows:

Let u be an algebraic number. For $u \in \mathbb{C}$ with $|u| > 1$, the Frobenius–Euler numbers $H_n(u)$ belonging to u are defined by means of the following generating function:

$$\frac{1-u}{e^t - u} = \sum_{n=0}^{\infty} H_n(u) \frac{t^n}{n!}. \tag{8}$$

By using the above equation and following the usual convention of symbolically replacing $H^n(u)$ by $H_n(u)$, we have

$$H_0 = 1$$

and for $n \geq 1$,

$$(H(u) + 1)^n = uH_n(u).$$

We also note that

$$H_n(-1) = E_n,$$

where E_n denotes the classical Euler numbers, which are defined by (94) (cf. [33, 42, 76, 77, 79]).

By using (7) and (8), a relation between the twisted Bernoulli numbers and the Frobenius–Euler numbers is given by

$$B_{n,\xi} = \frac{n}{\xi - 1} H_{n-1}(\xi^{-1}), \quad (\xi \neq 1). \tag{9}$$

The twisted Bernoulli polynomials $B_{n,\xi}(x)$ are defined by means of the following generating function:

$$F_{\xi}(t,x) = F_{\xi}(t)e^{tx}, \tag{10}$$

that is,

$$F_{\xi}(t,x) = \frac{te^{tx}}{\xi e^t - 1} = \sum_{n=0}^{\infty} B_{n,\xi}(x) \frac{t^n}{n!} \tag{11}$$

(cf. [33, 41, 42, 57, 62, 63, 68, 79]).

We note that the twisted Bernoulli numbers are special case of the Apostol–Bernoulli numbers (cf. [77]).

By using (7) and (10), we easily have

$$B_{n,\xi}(x) = \sum_{k=0}^{n} \binom{n}{k} x^{n-k} B_{k,\xi}. \tag{12}$$

Setting the following functional equation, which is used to get the difference equation for the twisted Bernoulli polynomials, we have

$$\xi F_{\xi}(t, x+1) - F_{\xi}(t, x) = te^{tx}. \tag{13}$$

It is easily from (11) and (13) that

$$\xi B_{n,\xi}(x+1) - B_{n,\xi}(x) = nx^{n-1},$$

which yields

$$\xi B_{n,\xi}(1) = B_{n,\xi}(0), \tag{14}$$

where $n > 1$ and

$$\xi B_{n,\xi}(0) - B_{n,\xi}(-1) = n\,(-1)^{n-1}.$$

2.2 Computing the Twisted Bernoulli Numbers

Let $\xi^r - 1$ ($\xi \in \mathbb{Z}^+$); $\xi \neq 1$. By using umbral calculus convention in (7), we get $t = \xi e^{(B_\xi+1)t} - e^{B_\xi t}$. From the above equation, we have

$$t = \sum_{n=0}^{\infty} \left(\xi(B_\xi+1)^n - B_{n,\xi}\right) \frac{t^n}{n!}.$$

Therefore, we get the following recursion formula for the twisted Bernoulli numbers:

$$B_{0,\xi} = 0 \tag{15}$$

and

$$\xi(B_\xi+1)^n - B_{n,\xi} = \begin{cases} 1/(\xi-1), & \text{if } n = 1, \\ 0, & \text{if } n > 1, \end{cases}$$

where $(B_\xi)^n$ is replaced by $B_{n,\xi}$.

Substituting $x = 1$ into (12), in view of (14), we also get the another recursion formula, which is given as follows:

$$B_{n,\xi} = \sum_{k=0}^{n} \binom{n}{k} \xi B_{k,\xi},$$

where $n \geq 2$ and $B_{1,\xi} = 1/(\xi - 1)$.

By using the above recursion formula, we have the following list for the first few of the twisted Bernoulli numbers:

$$B_{0,\xi} = 0, \quad B_{1,\xi} = \frac{1}{\xi - 1}, \quad B_{2,\xi} = -\frac{2\xi}{(\xi - 1)^2}, \quad B_{3,\xi} = \frac{9\xi - 3\xi^2}{(\xi - 1)^3},$$

$$B_{4,\xi} = \frac{4\xi - 48\xi^2 + 20\xi^3}{(\xi - 1)^4}, \quad \ldots .$$

The first few of the twisted Bernoulli polynomials are given below:

$$B_{0,\xi}(x) = 0, \quad B_{1,\xi}(x) = \frac{1}{\xi - 1}, \quad B_{2,\xi}(x) = \frac{2}{\xi - 1}x - \frac{2\xi}{(\xi - 1)^2},$$

$$B_{3,\xi} = \frac{3}{\xi - 1}x^2 - \frac{6\xi}{(\xi - 1)^2}x + \frac{9\xi - 3\xi^2}{(\xi - 1)^3}, \quad \ldots .$$

Because of (15), it is not difficult to derive the following important results for the twisted Bernoulli polynomials.

If $\xi \neq 1$, then *degree of the twisted Bernoulli polynomials* $B_{n,\xi}(x)$ *is* $n - 1$.

Observe that if $x = 0$, then (11) reduces to (7). Also, if $\xi \to 1$, then (7) reduces to (2).

Integral Formula

Integrating equation (10) with respect to x from y to z, we have

$$\int_{y}^{z} F_{\xi}(t, x)dx = F_{\xi}(t) \int_{y}^{z} e^{tx}dx = \frac{1}{t} F_{\xi}(t) \left(e^{tz} - e^{ty}\right).$$

From the above integral equation, we get the following integral formula:

$$\int_{y}^{z} B_{n,\xi}(x)dx = \frac{B_{n+1,\xi}(z) - B_{n+1,\xi}(y)}{n+1}.$$

Multiplication Formula

We set the following functional equation: Let d be a positive integer.

$$F_{\xi}(t, x) = \frac{1}{d} \sum_{j=0}^{d-1} \xi^{j} F_{\xi^d}\left(dt, \frac{x+j}{d}\right).$$

By using this equation, we get the following multiplication formula:

$$B_{n,\xi}(dx) = d^{n-1}\sum_{j=0}^{d-1}\xi^j B_{n,\xi^d}\left(x+\frac{j}{d}\right).$$

Addition Formula

By using (11), we arrive at the following addition formula:

$$B_{n,\xi}(x+y) = \sum_{k=0}^{n}\binom{n}{k} y^{n-k} B_{k,\xi}(x).$$

2.3 *Convolution of the Twisted Bernoulli Polynomials*

Here, we firstly give a generating function for the generalized twisted Bernoulli polynomials $B_{n,\xi}^{(m)}(x)$ in x as follows:

$$F_{\xi}^{(m)}(t,x) = \left(\frac{t}{\xi e^t - 1}\right)^m e^{tx} = \sum_{n=0}^{\infty} B_{n,\xi}^{(m)}(x)\frac{t^n}{n!}, \tag{16}$$

$m \in \mathbb{N}$. It is easily observed that

$$B_{n,\xi}^{(1)}(x) = B_{n,\xi}(x) \quad \text{and} \quad B_{n,\xi}^{(1)} = B_{n,\xi}.$$

The generalized twisted Bernoulli numbers $B_{n,\xi}^{(m)}$ are defined by means of the following generating function:

$$F_{\xi}^{(m)}(t) = \left(\frac{t}{\xi e^t - 1}\right)^m = \sum_{n=0}^{\infty} B_{n,\xi}^{(m)}\frac{t^n}{n!}$$

From the above function, we derive

$$F_{\xi}^{(m+1)}(t) = F_{\xi}^{(m)}(t)F_{\xi}(t).$$

By using this equation, we give the following formula, which is computed the numbers $B_{n,\xi}^{(m)}$:

$$B_{n,\xi}^{(m+1)} = \sum_{k=0}^{n}\binom{n}{k} B_{k,\xi}^{(m)} B_{n-k,\xi}.$$

The first few of the generalized twisted Bernoulli numbers $B_{n,\xi}^{(m)}$ are given below:

$$B_{0,\xi}^{(2)} = 0, \ B_{1,\xi}^{(2)} = 0, \ B_{2,\xi}^{(2)} = \frac{2}{(\xi-1)^2}, \ B_{3,\xi}^{(2)} = -\frac{12\xi}{(\xi-1)^3}, \ \ldots ;$$

$$B_{0,\xi}^{(3)} = 0, \ B_{1,\xi}^{(3)} = 0, \ B_{2,\xi}^{(3)} = 0, \ B_{3,\xi}^{(3)} = \frac{6}{(\xi-1)^3}, \ \ldots .$$

By using (16), we have

$$B_{n,\xi}^{(m)}(x) = \sum_{k=0}^{n} \binom{n}{k} x^{n-k} B_{k,\xi}^{(m)}.$$

By using the above formula, the first few of the generalized twisted Bernoulli polynomials $B_{n,\xi}^{(m)}(x)$ are given below:

$$B_{0,\xi}^{(2)}(x) = 0, \ B_{1,\xi}^{(2)}(x) = 0, \ B_{2,\xi}^{(2)}(x) = \frac{2}{(\xi-1)^2}, \ B_{3,\xi}^{(2)} = \frac{6}{(\xi-1)^2}x - \frac{12\xi}{(\xi-1)^3}, \ \ldots ;$$

$$B_{0,\xi}^{(3)}(x) = 0, \ B_{1,\xi}^{(3)}(x) = 0, \ B_{2,\xi}^{(3)}(x) = 0, \ B_{3,\xi}^{(3)}(x) = \frac{6}{(\xi-1)^3}, \ \ldots .$$

Substituting $m = 2$ into (16), we have

$$\sum_{n=0}^{\infty} B_{n,\xi}^{(2)}(x+y)\frac{t^n}{n!} = \left(\frac{te^{tx}}{\xi e^t - 1}\right)\left(\frac{te^{ty}}{\xi e^t - 1}\right).$$

By using (11) in the above equation, we obtain

$$B_{n,\xi}^{(2)}(x+y) = \sum_{k=0}^{n} \binom{n}{k} B_{k,\xi}(x) B_{n-k,\xi}(y). \tag{17}$$

We derive the following well-known differential equation:

$$\xi \left(F_\xi(t, x+y)\right)^2 = (x+y-1)tF_\xi(t, x+y-1) - t^2\frac{d}{dt}\left(\frac{e^{t(x+y-1)}}{\xi e^t - 1}\right).$$

By using the above equation, we obtain the following identity:

$$\xi B_{n,\xi}^{(2)}(x+y) = n(x+y-1)B_{n-1,\xi}(x+y-1) + (n-1)B_{n,\xi}(x+y-1). \tag{18}$$

Combining (17) and (18), we get convolution formula of the twisted Bernoulli polynomials by the following theorem:

Theorem 2.2.

$$\sum_{k=0}^{n}\binom{n}{k} B_{k,\xi}(x)B_{n-k,\xi}(y) = n(x+y-1)B_{n-1,\xi}(x+y-1)+(n-1)B_{n,\xi}(x+y-1).$$

Secondly, we give a generating function for the twisted Euler polynomials $E_{n,\xi}(x)$ as follows:

$$g_{\xi}(t,x) = \frac{2e^{tx}}{\xi e^{t}+1} = \sum_{n=0}^{\infty} E_{n,\xi}(x)\frac{t^n}{n!}. \tag{19}$$

We derive the following functional equation:

$$F_{\xi^2}\left(2t,\frac{x+y}{2}\right) = F_{\xi}(t,x+y) - \frac{t}{2}g_{\xi}(t,x+y).$$

By using this equation with (11) and (19), we give a relation between the twisted Bernoulli polynomials and the twisted Euler polynomials as follows:

Theorem 2.3.

$$B_{n,\xi^2}\left(\frac{x+y}{2}\right) = 2^{-n}B_{n,\xi}\left(x+y\right) - n2^{-n-1}E_{n-1,\xi}\left(x+y\right),$$

or

$$\sum_{k=0}^{n}\binom{n}{k} B_{k,\xi}(x)E_{n-k,\xi}(y) = B_{n,\xi}\left(x+y\right) - \frac{n}{2}E_{n-1,\xi}\left(x+y\right).$$

We note that the special case $x = y = 0$ of Theorems 2.2, we get twisted version of the Euler identity, which is given by

$$\sum_{k=0}^{n}\binom{n}{k} B_{k,\xi}B_{n-k,\xi} = -nB_{n-1,\xi}(-1) + (n-1)B_{n,\xi}(-1).$$

If $\xi \to 1$, we have Euler identity:

$$\sum_{k=2}^{n-2}\binom{n}{k} B_{k}B_{n-k} = -(n+1)B_{n}$$

(cf. [1, 16, 18, 76, 77]).

2.4 The Twisted Bernoulli Polynomials and Numbers in Terms of a Dirichlet Character

Here, in terms of a Dirichlet character χ of conductor $f \in \mathbb{Z}^+$, the sets of positive integer and the twisted generalized Bernoulli numbers and polynomials are defined, respectively, by means of the following generating functions:

$$G_{\chi,\xi}(t) = \sum_{a=1}^{f} \frac{\chi(a)\xi^a te^{at}}{\xi^f e^{ft} - 1} = \sum_{n=0}^{\infty} B_{n,\chi,\xi} \frac{t^n}{n!} \tag{20}$$

and

$$G_{\chi,\xi}(t,x) = \sum_{a=1}^{f} \frac{\chi(a)\xi^a te^{(x+a)t}}{\xi^f e^{ft} - 1} = \sum_{n=0}^{\infty} B_{n,\chi,\xi}(x) \frac{t^n}{n!}. \tag{21}$$

Clearly, we have

$$B_{n,\chi,\xi}(0) = B_{n,\chi,\xi}.$$

Observe that if $\xi = 1$, then $G_{\chi,1}(t,x) = F_\chi(t,x)$ and $B_{n,\chi,1} = B_{n,\chi}$, which are certain algebraic numbers (cf. [21–91]).

From (21), we have $G_{\chi,\xi}(t,x) = G_{\chi,\xi}(t)e^{tx}$. From this functional equation, we get

$$B_{n,\chi,\xi}(x) = \sum_{k=0}^{n} \binom{n}{k} x^{n-k} B_{k,\chi,\xi}.$$

By using (11) and (21), we get the following functional equation:

$$G_{\chi,\xi}(t,x) = \frac{1}{f} \sum_{a=1}^{f} \chi(a)\xi^a F_{\xi^f}\left(ft, \frac{x+a}{f}\right).$$

By using this equation, we have

$$\sum_{n=0}^{\infty} B_{n,\chi,\xi}(x) \frac{t^n}{n!} = \sum_{n=0}^{\infty} \left(f^{n-1} \sum_{a=1}^{f} \chi(a)\xi^a B_{n,\xi^f}\left(\frac{x+a}{f}\right)\right) \frac{t^n}{n!}.$$

Comparing the coefficients of $t^n/n!$ on both sides of the above equation, we arrive at the following result:

Theorem 2.4.

$$B_{n,\chi,\xi}(x) = f^{n-1} \sum_{a=1}^{f} \chi(a)\xi^a B_{n,\xi^f}\left(\frac{x+a}{f}\right)$$

and

$$B_{n,\chi,\xi} = f^{n-1} \sum_{a=1}^{f} \chi(a)\xi^a B_{n,\xi^f}\left(\frac{a}{f}\right). \tag{22}$$

By using (22), we give the first few of the numbers $B_{n,\chi,\xi}$:

$$B_{0,\xi} = 0,$$

$$B_{1,\chi,\xi} = \frac{1}{\xi^f - 1} \sum_{a=1}^{f} \chi(a)\xi^a,$$

$$B_{2,\xi} = \frac{1}{f^2\left(\xi^f - 1\right)} \sum_{a=1}^{f} a\chi(a)\xi^a - \frac{2\xi^f}{f\left(\xi^f - 1\right)^2} \sum_{a=1}^{f} \chi(a)\xi^a, \quad \ldots .$$

3 q-Analogue Generating Functions

We assume that q is a real number with $0 < q < 1$. We denote

$$[x : q] = \lfloor x \rfloor = \frac{1-q^x}{1-q}.$$

Note that

$$\lim_{q \to 1} [x : q] = x.$$

To obtain an integral expression of the q-zeta function, Wakayama and Yamasaki [85] defined the following q-functions:

$$\mathcal{L}_{q,+}^{(v)}(t,z) = t^v \sum_{n=0}^{\infty} q^{-v(n+z)} \exp\left([-(n+z)]t\right) \tag{23}$$

and

$$\mathcal{L}_{q,-}^{(v)}(t,z) = t^v \sum_{n=1}^{\infty} q^{v(n-z)} \exp\left([n-z]t\right) \tag{24}$$

when

$$-\pi \leq \arg(t) < \pi$$

$$z \in D_q = \left\{ z \in \mathbb{C} : |\mathrm{Im}(z)| < \frac{\pi}{2\,|\log q|} \right\}$$

and

$$R_q(z) = \left\{ t \in \mathbb{C} : |\arg(t) - (\mathrm{Im}(z)) \log q| < \frac{\pi}{2} \right\}.$$

For any $z \in D_q$, $\mathbb{R}^+ \subset R_q(z)$, $\mathbb{R}^+$ denotes the positive real axis.

Lemma 3.1 ([85]). *We assume that* $-\pi \leq \arg(t) < \pi$ *and* $z \in D_q$. *The function* $\mathcal{L}_{q,+}^{(v)}(t,z)$ *is holomorphic in* $R_q(z)$. *The function* $\mathcal{L}_{q,-}^{(v)}(t,z)$ *is entire. The functions* $\mathcal{L}_{q,\pm}^{(v)}(t,z)$ *satisfy, respectively,*

$$\mathcal{L}_{q,\pm}^{(v)}(qt,z) = e^{-t} \left(\mathcal{L}_{q,\pm}^{(v)}(t,z) \pm \left(q^{1-z}t\right)^v e^{t[1-z]} \right).$$

Lemma 3.2 ([85]). *Put* $z = x + \sqrt{-1}y \in D_q$ $(x, y \in \mathbb{R})$ *and* $\beta_y = \cos(y \log q)$. *For* $t > 0$, *we have*

$$\left| \mathcal{L}_{q,+}^{(v)}(t,z) \right| \leq \exp\left(-t \frac{q^{-x}\beta_y - 1}{1-q} \right) \left((q^{-x}t)^v + \frac{\left(\frac{ve^{-1}}{\beta_y} \right)^v}{1 - e^{-tq^{-x}\beta_y}} \right).$$

Further, suppose that $z \in J_q = \left\{ z = x + \sqrt{-1}y \in D_q : x > 0, q^{-x} \cos(y \log q) > 1 \right\}$. *The function* $t^\alpha \mathcal{L}_{q,+}^{(v)}(t,z)$ *is an integrable function on* $[0,\infty)$ *provided* $\mathrm{Re}(\alpha) > 0$.

Wakayama and Yamasaki [85] also defined the following q-function, which is linear combination of the functions $\mathcal{L}_{q,+}^{(v)}(t,z)$ and $\mathcal{L}_{q,-}^{(v)}(t,z)$:

$$\mathcal{L}_{q,0}^{(v)}(t,z) = \mathcal{L}_{q,+}^{(v)}(t,z) + \mathcal{L}_{q,-}^{(v)}(t,z) = t^v \sum_{n=-\infty}^{\infty} q^{v(n+z)} \exp\left(-tq^{-(n+z)}[n+z]t \right),$$

where $t \in R_q(z)$. One can easily see that the function $\mathcal{L}_{q,0}^{(v)}(t,z)$ is periodic, that is,

$$\mathcal{L}_{q,0}^{(v)}(t,z+1) = \mathcal{L}_{q,0}^{(v)}(t,z).$$

Thus, this function has the following Fourier expansion:

Proposition 3.1 ([85]). *Let $z \in D_q$. For $t \in R_q(z)$, we have*

$$\mathcal{L}_{q,0}^{(v)}(t,z) = -\frac{(1-q)^v}{\log q} e^{\frac{t}{1-q}} \sum_{n=-\infty}^{\infty} \left(\frac{1-q}{t}\right)^{m\delta_q} \Gamma(v+m\delta_q) e^{2\pi\sqrt{-1}mz},$$

where

$$\delta_q = \frac{2\pi\sqrt{-1}}{\log q}.$$

Proof. By applying Fourier transform to

$$f_q^{(v)}(\eta,z) = q^{-v(\eta+z)} \exp\left(-tq^{-(\eta+z)}[\eta+z]\right),$$

we have

$$\widetilde{f_q^{(v)}}(w,z) = \int_{-\infty}^{\infty} f_q^{(v)}(\eta,z) e^{-2\pi\sqrt{-1}\eta w} d\eta.$$

After some elementary calculations, we have

$$\widetilde{f_q^{(v)}}(w,z) = -\frac{(1-q)^v}{\log q} e^{\frac{t}{1-q}+2\pi\sqrt{-1}wz} \left(\frac{1-q}{t}\right)^{v+w\delta_q} \Gamma(v+w\delta_q).$$

Applying the Poisson summation formula to the above function $f_q^{(v)}(\eta,z)$, we have

$$\mathcal{L}_{q,0}^{(v)}(t,z) = t^v \sum_{n=-\infty}^{\infty} f_q^{(v)}(n,z) = t^v \sum_{m=-\infty}^{\infty} \widetilde{f_q^{(v)}}(m,z).$$

See for detail [85, Proof of Proposition 2.4]. Thus we arrive at the desired result.

3.1 q-Analogues of Bernoulli Polynomials

Wakayama and Yamasaki [85] defined q-analogues of Bernoulli polynomials by means of the following generating function:

$$G_q^{(v)}(t,z) = -\frac{(1-q)^v(v-1)!}{\log q} e^{\frac{t}{1-q}} - \mathcal{L}_{q,-}^{(v)}(t,z) = \sum_{n=0}^{\infty} \tilde{B}_n^{(v)}(z,q)\frac{t^n}{n!} \quad (0<q<1,\ t\in\mathbb{C}).$$

The function $G_q^{(v)}(t,z)$ is holomorphic at $t=0$. This function satisfies the following q-difference equation:

$$G_q^{(v)}(qt,z) = e^{-t}\left(G_q^{(v)}(t,z) + \left(q^{1-z}t\right)^v e^{t[1-z]}\right).$$

Substituting $t = 0$ into the above equation, we have

$$G_q^{(v)}(0,z) = -\frac{(1-q)^v(v-1)!}{\log q}.$$

By using the above q-difference equation, recursion formula for the polynomials $\tilde{B}_n^{(v)}(z,q)$ is given by [85]

$$\tilde{B}_0^{(v)}(z,q) = -\frac{(1-q)^v(v-1)!}{\log q},$$

$$\sum_{k=0}^{n}(-1)^k\binom{n}{k}q^k\tilde{B}_k^{(v)}(z,q) = (-1)^n\tilde{B}_n^{(v)}(z,q) + v!\binom{n}{v}q^{(1-z)v}[1-z]^{n-v}.$$

Remark 3.1. For $t \in (0, 2\pi)$, we have

$$\lim_{q\uparrow 1} G_q^{(v)}(t,z) = \frac{t^v}{e^t-1}e^{(1-z)t}.$$

This gives us generating function for the classical Bernoulli polynomials for $v = 1$. That is,

$$\lim_{q\uparrow 1}\tilde{B}_n^{(1)}(z,q) = B_n(z)$$

which is defined in (1). The polynomials $\tilde{B}_n^{(1)}(z,q)$ have been studied by Tsumura [82] and also the author [68] which are given in the next section.

3.2 The Modified Twisted q-Bernoulli Numbers

In [68], the author studied the modified twisted Bernoulli numbers and polynomials and their interpolation functions. In complex s-plane, the generating function of the modified twisted q-Bernoulli numbers is given by [68]

$$\mathcal{H}_{q,\xi}(t) = \sum_{n=0}^{\infty} B_{n,\xi}^*(q)\frac{t^n}{n!} = B_{0,\xi}^*(q) + B_{1,\xi}^*(q)\frac{t}{1!} + \cdots + B_{n,\xi}^*(q)\frac{t^n}{n!} + \cdots,$$

where $\xi^r = 1$ ($r \in \mathbb{Z}^+$); $\xi \neq 1$ and $q \in \mathbb{R}$ with $0 < q < 1$.

This function is the unique solution of the following q-difference equation:

$$\mathcal{H}_{q,\xi}(t) = \xi e^t\mathcal{H}_{q,\xi}(qt) - t. \tag{25}$$

By using the above q-difference equation, the recurrence formula of the modified twisted q-Bernoulli numbers is given by

$$B_{0,\xi}^{*}(q) = \frac{q-1}{\log q^{\xi}}$$

and

$$\xi \sum_{k=0}^{n} \binom{n}{k} q^{k} B_{k,\xi}^{*}(q) - B_{k,\xi}^{*}(q) = \begin{cases} 1, & \text{if } n = 1, \\ 0, & \text{if } n > 1. \end{cases}$$

Lemma 3.3 ([68]). *Suppose that $q \in \mathbb{R}$ with $0 < q < 1$, for $r \in \mathbb{Z}^{+}$ and $\xi \neq 1$, $\xi^{r} = 1$. Then*

$$\mathcal{H}_{q,\xi}(t) = t \sum_{n=1}^{\infty} \xi^{-n} q^{-n} \exp\left(-q^{-n}[n]t\right). \tag{26}$$

Proof. The right-hand side of (26) is uniformly convergent in the wider sense and satisfies q-difference equation in (25).

Observe that for $q \in \mathbb{R}$ with $0 < q < 1$, we have $\xi^{r} = 1$:

$$\mathcal{H}_{q,1}(t) = \mathcal{H}_{q}(t) = t \sum_{n=1}^{\infty} q^{-n} \exp\left(-q^{-n}[n]t\right)$$

(cf. [63, 64, 68, 71, 82]). By using Lemma 3.3, we have

$$\sum_{m=0}^{\infty} B_{m,\xi}^{*}(q) \frac{t^{m}}{m!} = \sum_{m=1}^{\infty} \left(\frac{m}{(q-1)^{m-1}} \sum_{k=0}^{m-1} \binom{m-1}{k} \frac{(-1)^{k}}{\xi q^{m-k} - 1} \right) \frac{t^{m}}{m!}.$$

By comparing the coefficients $\frac{t^m}{m!}$ on both sides of the above equation, we obtain

$$B_{m,\xi}^{*}(q) = \frac{m}{(q-1)^{m-1}} \sum_{k=0}^{m-1} \binom{m-1}{k} \frac{(-1)^{k}}{\xi q^{m-k} - 1}.$$

To define interpolation functions of the twisted q-Bernoulli numbers and polynomials, we need Lemma 3.3; for $q \in \mathbb{R}$ with $0 < q < 1$, $r, k \in \mathbb{Z}^{+}$ and $\xi \neq 1$, $\xi^{r} = 1$, we have

$$\frac{d^{k}}{dt^{k}} \mathcal{H}_{\xi}(t, q) \mid_{t=0} = B_{k,\xi}^{*}(q) = -(-1)^{k} k \sum_{n=1}^{\infty} \xi^{-n} q^{-kn} [n]^{k-1}. \tag{27}$$

3.3 The Modified Generalized Twisted q-Bernoulli Numbers

Now we define the following generating function which is a generalization of (26). Let χ be a Dirichlet character of conductor $f \in \mathbb{Z}^{+}$; the modified twisted generalized q-Bernoulli numbers are defined by means of the following generating function:

For $q \in \mathbb{R}$ with $0 < q < 1$, for $r \in \mathbb{Z}^{+}$ and $\xi \neq 1$,

$$\xi^{r} = 1,$$

we define

$$\mathcal{K}_{\xi,\chi}(t,q) = \sum_{n=1}^{\infty} \xi^{-n} \chi(n) q^{-n} \exp\left(-q^{-n}[n]t\right) = \sum_{n=0}^{\infty} B^{*}_{n,\xi,\chi}(q) \frac{t^{m}}{m!}. \tag{28}$$

The right-hand side of (28) is uniformly convergent in the wider sense. Hence, by using derivative operator $\frac{d^k}{dt^k}$ to Eq. (28), we obtain

$$\frac{d^k}{dt^k} \mathcal{H}_{\xi,\chi}(t,q) \mid_{t=0} = B^{*}_{k,\xi,\chi}(q) = (-1)^{k+1} k \sum_{n=1}^{\infty} \xi^{-n} \chi(n) q^{-kn} [n]^{k-1}, \tag{29}$$

where $q \in \mathbb{R}$ with $0 < q < 1$, $r \in \mathbb{Z}^{+}$ and $\xi \neq 1$, $\xi^{r} = 1$.

Observe that if $\chi \equiv 1$, then (28) reduces to (26).

3.4 The Modified Twisted q-Bernoulli Polynomials

The modified twisted q-Bernoulli polynomials $B^{*}_{m,\xi}(x,q)$ are defined by means of the following generating function [68]:

Let $q \in \mathbb{R}$ with $0 < q < 1$, $r \in \mathbb{Z}^{+}$ and $\xi \neq 1$,

$$\xi^{r} = 1.$$

We set

$$\mathfrak{H}_{q,\xi}(t,x) = t \sum_{n=1}^{\infty} \xi^{-n} q^{-n} \exp\left(-q^{-n}[n+x]t\right) = \sum_{m=0}^{\infty} B^{*}_{m,\xi}(x,q) \frac{t^{m}}{m!} \tag{30}$$

or

$$\mathfrak{H}_{q,\xi}(t,x) = \mathcal{H}_{q,\xi}(t,q) e^{-t[x]}.$$

From (30), we have

$$\sum_{m=0}^{\infty} B_{m,\xi}^{*}(x,q)\frac{t^m}{m!} = e^{-t[x]}\sum_{m=0}^{\infty} B_{m,\xi}^{*}(q)\frac{t^m}{m!}.$$

By using Taylor series of the e^x function in the above, we have

$$\sum_{m=0}^{\infty} B_{m,\xi}^{*}(x,q)\frac{t^m}{m!} = \left(\sum_{m=0}^{\infty} B_{m,\xi}^{*}(q)\frac{t^m}{m!}\right)\left(\sum_{m=0}^{\infty}(-1)^m[x]^m\frac{t^m}{m!}\right).$$

By using Cauchy product in the right side of the above equation, we get

$$\sum_{m=0}^{\infty} B_{m,\xi}^{*}(x,q)\frac{t^m}{m!} = \sum_{m=0}^{\infty}\left(\sum_{k=0}^{m}(-1)^{m-k}\binom{m}{k}[x]^{m-k}B_{k,\xi}^{*}(q)\right)\frac{t^m}{m!}.$$

By comparing the coefficients of $t^n/n!$ in both sides of the above equation, we arrive at the following theorem:

Theorem 3.1 ([68]). *Let* $\xi^r = 1$, $\left(r \in \mathbb{Z}^+\right)$; $\xi \neq 1$. *We have*

$$B_{m,\xi}^{*}(x,q) = \sum_{k=0}^{m}(-1)^{m-k}\binom{m}{k}[x]^{m-k}B_{k,\xi}^{*}(q). \tag{31}$$

By using (31), we have

$$B_{m,\xi}^{*}(x,q) = \frac{m}{(q-1)^{m-1}}\sum_{k=0}^{m-1}(-1)^k\binom{m-1}{k}\frac{q^{xj}}{\xi q^{m-k}-1}.$$

From (30), we have

$$\begin{aligned}\mathfrak{H}_{q,\xi}(t,x) &= t\sum_{n=1}^{\infty}\xi^{-n}q^{-n}\exp\left(-q^{-n}([n]+q^n[x])t\right) \qquad (32)\\ &= t\sum_{n=1}^{\infty}\xi^{-n}q^{-n}\exp\left(-q^{-n}([n+x])t\right) = \sum_{m=0}^{\infty} B_{m,\xi}^{*}(x,q)\frac{t^m}{m!}.\end{aligned}$$

We now give multiplication (Raabe) relation of the twisted q-Bernoulli polynomials as follows.

Theorem 3.2. *Let $n, r \in \mathbb{Z}^+$, $(\xi^r = 1)$, $\xi \neq 1$. If $r \nmid k$, then we have*

$$B^*_{n,\xi}(x,q) = [k]^{n-1} \sum_{j=0}^{k-1} \xi^{-j} q^{-nj} B^*_{n,\xi^k}\left(\frac{x+j}{k}, q^k\right). \tag{33}$$

Proof. By (32), we have

$$\begin{aligned}
&\sum_{n=0}^{\infty} B^*_{n,\xi}(x,q)\frac{t^n}{n!} \\
&\quad = \sum_{m=1}^{\infty} t\xi^{-m} q^{-m} \exp\left(-q^{-m}([m+x])t\right) \\
&\quad = \sum_{j=0}^{k-1} \xi^{-j} q^{-j} \sum_{y=1}^{\infty} t\xi^{-yk} q^{-yk} \exp\left(-q^{-yk}\left(\left[y + \frac{x+j}{k}, q^k\right]\right)(q^{-j}[k]t)\right) \\
&\quad = \frac{1}{[k]} \sum_{j=0}^{k-1} \xi^{-j} \sum_{n=0}^{\infty} B^*_{n,\xi^k}\left(\frac{x+j}{k}, q^k\right) \frac{(q^{-j}[k]t)^n}{n!}.
\end{aligned}$$

By comparing coefficients $t^n/n!$ on both sides of the above equation, we arrive at the desired result.

Corollary 3.1. *Let $n \in \mathbb{Z}^+$. $\xi^r = 1$ $(r \in \mathbb{Z}^+)$; $\xi \neq 1$. If $r \mid k$, then we have*

$$B^*_{n,\xi}(x,q) = [k]^{n-1} \sum_{j=0}^{k-1} \xi^{-j} q^{-nj} B^*_n\left(\frac{x+j}{k}, q^k\right), \tag{34}$$

where $B^*_n(x,q)$ is defined by

$$B^*_n(x,q) = \sum_{k=0}^{n} (-1)^{n-k} \binom{n}{k} [x]^{n-k} B^*_k(q). \tag{35}$$

Replacing x by kx into (33), we have

$$B^*_{n,\xi}(kx,q) = [k]^{n-1} \sum_{j=0}^{k-1} \xi^{-j} q^{-nj} B^*_{n,\xi^k}\left(x + \frac{j}{k}, q^k\right).$$

Thus, we have multiplication formula as follows:

Corollary 3.2. *Let $n, r \in \mathbb{Z}^+$, $\xi^r = 1$, $\xi \neq 1$. If $r \nmid k$, then we have*

$$\frac{1}{[k]}\sum_{j=0}^{k-1} \xi^{-j} q^{-nj} B^*_{n,\xi^k}\left(x + \frac{j}{k}, q^k\right) = [k]^{-n} B^*_{n,\xi}(kx, q), \tag{36}$$

and if $r \mid k$, then we have

$$\frac{1}{[k]}\sum_{j=0}^{k-1} \xi^{-j} q^{-nj} B^*_{n}\left(x + \frac{j}{k}, q^k\right) = [k]^{-n} B^*_{n,\xi}(kx, q).$$

We now find derivative of the twisted q-Bernoulli polynomials:

Theorem 3.3.

$$\frac{d}{dx} B^*_{n,\xi}(x, q) = \log q^{\frac{n}{1-q}} B^*_{n-1,\xi}(x, q).$$

Proof. By applying derivative operator to (31), we have

$$\frac{d}{dx} B^*_{m,\xi}(x, q) = \frac{\log q}{1-q}\sum_{k=0}^{m}(-1)^{m-1-k}\binom{m-1}{k} m[x]^{m-1-k} B^*_{k,\xi}(q).$$

After some elementary calculations, we arrive at the desired result.

3.5 *The Modified Generalized Twisted q-Bernoulli Polynomials*

The modified generalized twisted q-Bernoulli polynomials are defined by means of the following generating function:

Let $q \in \mathbb{R}$ with $0 < q < 1$, for $r \in \mathbb{Z}^+$ and $\xi \neq 1$, $\xi^r = 1$. We define

$$\mathfrak{K}_{\xi,\chi}(x, t, q) = \sum_{n=1}^{\infty} \xi^{-n} \chi(n) q^{-n} \exp\left(-\left(q^{-n}[n] + [x]\right) t\right) = \sum_{m=0}^{\infty} B^*_{m,\xi,\chi}(x, q)\frac{t^m}{m!}, \tag{37}$$

or

$$\mathfrak{K}_{\xi,\chi}(x, t, q) = \mathcal{K}_{\xi,\chi}(t, q) e^{-t[x]},$$

where χ is a Dirichlet character of conductor f (cf. [68]).

By using (37), we have

$$B^*_{n,\xi,\chi}(x, q) = \sum_{k=0}^{n}\binom{n}{k}[x]^{n-k} B^*_{n,\xi,\chi}(q).$$

We derive the following functional equation:

$$\mathfrak{K}_{\xi,\chi}(x,t,q) = \frac{1}{[f]}\sum_{k=1}^{f} \xi^{-k}\chi(k)\mathfrak{H}_{q^f,\xi^f}\left(\frac{[f]}{q^k}t, \frac{x+k}{[f]}\right). \tag{38}$$

By using the above equation and (30), we have the following theorem:

Theorem 3.4.

$$B^*_{n,\xi,\chi}(x,q) = [f]^{n-1}\sum_{k=1}^{f} \xi^{-k}q^{-nk}\chi(k)B^*_{n,\xi^f}\left(\frac{x+k}{[f]}, q^f\right). \tag{39}$$

By substituting $x = 0$ into (39), we have the following result:

$$B^*_{n,\xi,\chi}(q) = [f]^{n-1}\sum_{k=1}^{f} \xi^{-k}q^{-kn}\chi(k)B^*_{n,\xi^f}\left(\frac{k}{[f]}, q^f\right). \tag{40}$$

4 Modified q-Zeta Functions

By the integral expression of the gamma function $\Gamma(s)$, Hurwitz's zeta function

$$\zeta(s,x) = \sum_{n=0}^{\infty} \frac{1}{(x+n)^s}$$

is obtained by the Mellin transform of the generating function for the Bernoulli polynomials $F_B(t, 1-x)$ in (1) (cf. [76,77,85,88]):

$$\zeta(s,x) = \frac{1}{\Gamma(s)}\int_0^{\infty} t^{s-1}F_B(-t, 1-x)\frac{dt}{t} \quad (\mathrm{Re}(s) > 1).$$

For any $0 < a \leq +\infty$ and $0 < \varepsilon < \min\{a, 2\pi\}$, $\zeta(s,x)$ is represented by the following the integral expression, which is called meromorphic continuation of $\zeta(s,x)$ (cf. [85]):

$$\zeta(s,x) = \frac{\Gamma(1-s)}{2\pi\sqrt{-1}}\int_{C(\varepsilon,a)} \frac{(-t)^s e^{(1-x)t}}{e^t - 1}\frac{dt}{t} + \frac{1}{\Gamma(s)}\int_0^{\infty} t^{s-1}F_B(-t, 1-x)\frac{dt}{t},$$

where $C(\varepsilon, a)$ is a contour along the real axis from a to ε, counterclockwise around the circle of radius ε with center at the origin, and then along the real axis from ε to a. This integral expression is related to Riemann around 1859. By using Cauchy residue theorem, we can easily see that $\zeta(s,x)$ has a simple pole at $s = 1$:

$$\operatorname*{Res}_{s=1}\{\zeta(s,x)\} = B_0(x) = 1,$$

one can also show that for $m \in \mathbb{N}$,

$$\zeta(1-m,x) = -\frac{B_m(x)}{m}.$$

Let $0 < q < 1$. Dirichlet-type q-series has been defined by

$$\zeta_q(s,t,z) = \sum_{n=1}^{\infty} \frac{q^{(n+z)t}}{[n+z]^s}$$

where $\mathrm{Re}(t) > 0$ (cf. [43, 57, 68, 82, 85]).

In [85], Wakayama and Yamasaki defined the following Dirichlet-type q-series:

$$\zeta_q^{(v)}(s,z) = \zeta_q(s,s-v,z) = \sum_{n=1}^{\infty} \frac{q^{(n+z)(s-v)}}{[n+z]^s} \quad (v \in \mathbb{N}).$$

The meromorphic continuation of the above function was obtained not only by the binomial expansion but also by the Euler–MacLaurin summation formula.

Since $z \in J_q$ and $\mathrm{Re}(s) > v+1$, $t^{s-v-1}\mathcal{L}_{q,+}^{(v)}(t,z)$ is integrable on $[0,\infty)$, by Lemma 3.2. We are now ready to apply Mellin transformation to (23). Thus we have

$$\zeta_q^{(v)}(s,x) = \frac{1}{\Gamma(s)} \int_0^{\infty} t^{s-v} \mathcal{L}_{q,+}^{(v)}(t,z) \frac{dt}{t},$$

where $\mathrm{Re}(s) > v+1$ (cf. [85]).

By using the Mellin transformation and Lemma 3.3, we define modified twisted q-zeta function. Applying the Mellin transformations to the equation (26), we find the q-analogue of the twisted Riemann zeta functions as follows:

Definition 4.1 ([68]). Let $s \in \mathbb{C}$ and $q \in \mathbb{R}$ with $0 < q < 1$. Let $r \in \mathbb{Z}^+$, $\xi^r = 1$, $\xi \neq 1$. We define

$$\zeta_{\xi,q}(s) = \sum_{n=1}^{\infty} \frac{\xi^{-n} q^{-n}}{(q^{-n}[n])^s}. \tag{41}$$

The right-hand side of this series converges when $Re(s) > 1$. Analytic continuation of the q-analogue of the twisted Riemann zeta function was given by (cf. [68, 72]); see also cf. [86, 88, 92].

We now give analytic continuation of the q-analogue of the twisted Riemann zeta function. Firstly we need the Euler–MacLaurin summation formula cf. [86, 88, 92]:

Let $f(x)$ be any (complex-valued) $\mathbb{C}^{\infty}$ function on $[1, \infty)$ and let m and N be two positive integers. Then the Euler–MacLaurin summation formula is defined as follows:

$$\sum_{n=1}^{m} f(n) = \int_{1}^{m} f(x)dx + \sum_{k=0}^{N}(-1)^k \frac{B_{k+1}}{(k+1)!}\left(f^{(k-1)}(m) - f^{(k-1)}(1)\right)$$
$$-\frac{(-1)^{N+1}}{(N+1)!}\int_{1}^{m} \overline{B}_{N+1}(x) f^{(N+1)}(x)dx, \tag{42}$$

where $\overline{B}_k(x)$ is the kth Bernoulli functions, which are defined by

$$\overline{B}_n(x) = B_n(x - [x]_G) = \begin{cases} 0, & \text{if } n = 1,\ x \in \mathbb{Z}, \\ \overline{B}_n(\{x\}), & \text{otherwise,} \end{cases} \tag{43}$$

where $\{x\}$ is the fractional part of x and $B_n(x)$ denotes Bernoulli polynomials. Since $\overline{B}_n(x+1) = \overline{B}_n(x)$, $\overline{B}_n(x)$ is the periodic function. Hence $\overline{B}_n(x)$ remains bounded over the whole interval $[1, \infty)$. Fourier expansion of this function is given by

$$\overline{B}_n(x) = -n! \sum_{j \in \mathbb{Z}\setminus\{0\}} \frac{e^{2\pi i j x}}{(2\pi i j)^n} \tag{44}$$

(cf. [86, 88]). By substituting $n = 2$ into (44), we obtain

$$\overline{B}_2(x) = \frac{1}{\pi^2}\sum_{y=1}^{\infty}\frac{\cos(2\pi y x)}{y^2}, \tag{45}$$

where $\overline{B}_2(x) = (x - [x]_G)^2 - (x - [x]_G) + \frac{1}{6}$. For every positive integer $N \geq 2$, Zhao [92] obtained upper bounded of the function $\overline{B}_M(x)$ as follows:

Lemma 4.1. *For every positive integer $N \geq 2$ and $x > 1$, we have*

$$|\overline{B}_N(x)| \leq \frac{4N!}{(2\pi)^N}.$$

Set

$$f(x) = \frac{\xi^{-x} q^{-x}}{(q^{-x}[x])^s}.$$

From the above equation, we find the following derivatives:

$$\frac{d}{dx} f(x) = \frac{(1-q)^s \left(\xi^{-1} q^{s-1}\right)^x (s - 1 + q^x)\log q}{(1-q^x)^{s+1}},$$

and

$$\frac{d^2}{dx^2}f(x) = \frac{(1-q)^s\left(\xi^{-1}q^{s-1}\right)^x(\log q)^2((1-q^x)^2+s(s+1)-3s(1-q^x))}{(1-q^x)^{s+2}}.$$

By substituting $m \to \infty$, $N = 1$, $\frac{d}{dx}f(x)$, and $\frac{d^2}{dx^2}f(x)$ into (42), we obtain the q-analogue of the twisted Riemann zeta function by the following theorem:

Theorem 4.1. *Let $s \in \mathbb{C}$ with $Re(s) > 1$ and let $r \in \mathbb{Z}^+$, $\xi^r = 1$, $\xi \neq 1$. Thus we have*

$$\zeta_{\xi,q}(s) = \sum_{j=0}^{\infty}\binom{s+j-1}{j}\frac{(1-q)^s q^{s+j-1}}{\xi(1-s-j)\log q} + \frac{q^{s-1}}{2\xi} + \frac{q^{s-1}(s-1+q)\log q}{12\xi(q-1)} + Y_{\xi,q}(s), \tag{46}$$

where

$$Y_{\xi,q}(s) = -\frac{(1-q)^s}{2}\sum_{j=0}^{\infty}\binom{s+j+1}{j}\int_1^{\infty}((x-[x]_G)^2-\left(x-[x]_G)+\frac{1}{6}\right)\left(\xi^{-1}q^{s-1+j}\right)^x$$
$$\times(\log q)^2((1-q^x)^2+s(s+1)-3s(1-q^x))dx. \tag{47}$$

From the above theorem, the right-hand side of the above formula (46) defines a meromorphic function on $\mathbb{C}$ whose only singularity is a simple pole of order 1 at $s = 1$. We now calculate residue of the function $\zeta_{w,q}(s)$ at $s = 1$. By using (46), we have

$$\mathrm{res}_{s=1}(\zeta_{\xi,q}(s)) = \frac{q-1}{\xi\log q}.$$

Thus we arrive at the following corollary:

Corollary 4.1. *The function $\zeta_{\xi,q}(s)$ is analytically continued to the whole complex plane, except for a simple pole at $s = 1$ with residue $(q-1)/(\xi\log q)$.*

Remark 4.1. In [72], Simsek and Srivastava studied some properties of the family of zeta functions. As already observed by (among others) the author [68], the q-Riemann zeta function is defined by

$$\zeta_q(s) := \sum_{n=1}^{\infty}\frac{q^{-n}}{(q^{-n}[n])^s} =: \zeta_{\xi,q}(s)_{\xi=1},$$

where $q \in \mathbb{R}$ with $0 < q < 1$, $r \in \mathbb{Z}^+$, $\xi^r = 1$, $\xi \neq 1$. The function $\zeta_q(s)$ is a meromorphic function on $\mathbb{C}$ with simple pole at $s = 1$. Moreover, in its limit case when $q \to 1$, (41) reduces to the twisted Riemann zeta function $\zeta_{\xi}^{*}(s)$ defined by $(\mathrm{Re}(s) > 1; \xi^r = 1\ (r \in \mathbb{Z}^+), \xi \neq 1)$,

$$\zeta_{\xi}^{*}(s) := \sum_{n=1}^{\infty} \frac{\xi^{-n}}{n^s} =: Li_s(\xi^{-1}) \tag{48}$$

in terms of the polylogarithm function $Li_s(z)$ defined below. For a set of complex numbers $\{c_n\}$, Tsumura [82] defined a q-extension of Dirichlet series as follows:

$$\mathfrak{f}(s) = \sum_{n=1}^{\infty} \frac{c_n q^{-n}}{(q^{-n}[n])^s} \quad (\mathrm{Re}(s) > 1),$$

which, in the special case when $n \in \mathbb{Z}$,

$$c_n = \xi^{-n},$$

yields the modified twisted Riemann zeta function $\zeta_{\xi,q}(s)$ defined by (41). Furthermore, as already observed by Tsumura [82], the q-Riemann zeta function $\zeta_q(s)$, that is,

$$\zeta_q(s) = \zeta_{\xi,q}(s)\,\big|_{\xi=1}\,,$$

can also be continued analytically to the whole complex s-plane, except for a simple pole at $s = 1$ with residue:

$$\frac{q-1}{\log q}.$$

Remark 4.2. In [64], the author defined generating functions similar to (26). By using the Mellin transformation to these functions, we constructed not only q-zeta function and q-L-functions but also q-Dedekind-type sums.

4.1 Modified Twisted *q*-Extension Hurwitz Zeta Function

Here, we define a modified twisted q-extension of the Hurwitz zeta function.

In terms of the generating function $\mathfrak{H}_{q,\xi}(t,x)$ occurring in (32), we have the following integral representation for the modified twisted q-extension of the Hurwitz zeta q-function $\zeta_{\xi,q}(s,x)$ defined by (50), which involves the Mellin transformation:

For $q \in \mathbb{R}$ with $0 < q < 1$, $\xi^r = 1\ (r \in \mathbb{Z}^+)$, $\xi \neq 1$,

$$\frac{1}{\Gamma(s)} \int_0^{\infty} t^{s-1} \mathfrak{H}_{q,\xi}(t,x) \frac{dt}{t} = \zeta_{\xi,q}(s,x), \tag{49}$$

and $\min\{\mathrm{Re}(s), \mathrm{Re}(x)\} > 0$, where the additional constraint $\mathrm{Re}(x) > 0$ is required for the convergence of the infinite integral occurring in (49). By using the above integral expression, we are to define a modified twisted q-extension of the Hurwitz zeta function as follows:

Definition 4.2 ([68]). For a given positive integer r, let $\xi^r = 1$ $(r \in \mathbb{Z}^+)$ and $\xi \neq 1$. Suppose also that $q \in \mathbb{R}$ with $0 < q < 1$, $0 < x \leq 1$, and $s \in \mathbb{C}$. Then we define a modified twisted q-extension of the Hurwitz zeta function by

$$\zeta_{w,q}(s,x) = \sum_{n=0}^{\infty} \frac{\xi^{-n} q^{-n}}{\left(q^{-n}[n+x]\right)^s}. \tag{50}$$

Remark 4.3. In its limit case when $q \to 1$, then $\zeta_{\xi,q}(s,x)$ yields the twisted Hurwitz zeta function $\zeta_\xi(s,z)$, which is defined by [68]: $(\mathrm{Re}(s) > 1; \xi^r = 1$ $(r \in \mathbb{Z}^+), \xi \neq 1)$,

$$\zeta_\xi(s,z) = \sum_{n=0}^{\infty} \frac{\xi^{-n}}{(n+z)^s} =: \Phi(\xi^{-1}, s, z), \tag{51}$$

where $\Phi(z,s,x)$ denotes the familiar Hurwitz–Lerch zeta function defined by (cf. e.g., [76, p. 121 et seq.], [27, 77])

$$\Phi(z,s,a) = \sum_{n=0}^{\infty} \frac{z^n}{(n+a)^s},$$

which converges for ($a \in \mathbb{C} \setminus \mathbb{Z}_0^-$, $s \in \mathbb{C}$ when $|z| < 1$; $\mathrm{Re}(s) > 1$ when $|z| = 1$) where, as usual,

$$\mathbb{Z}_0^- := \mathbb{Z}^- \cup \{0\}.$$

The above-defined general Hurwitz–Lerch zeta function $\Phi(z,s,a)$ contains, as its special cases, not only the Riemann and Hurwitz (or generalized) zeta functions

$$\zeta(s) = \Phi(1,s,1) \quad \text{and} \quad \zeta(s,a) = \Phi(1,s,a)$$

and the Lerch zeta function

$$\ell_s(\tau) = \sum_{n=1}^{\infty} \frac{e^{2\pi i n \tau}}{n^s} = e^{2\pi i \tau} \Phi(e^{2\pi i \tau}, s, 1),$$

where $\tau \in \mathbb{R}$; $\mathrm{Re}(s) > 1$, but also such other important functions of Analytic Function Theory as the polylogarithm [occurring already in (48)]:

$$Li_s(z) = \sum_{n=1}^{\infty} \frac{z^n}{n^s} = z\Phi(z,s,1),$$

where for $s \in \mathbb{C}$ when $|z| < 1$; $\mathrm{Re}(s) > 1$ when $|z| = 1$ and the Lipschitz–Lerch zeta function (cf. [76, p. 122, Eq. (2.5) (11)], [72, p. 2980, Remark 3])

$$\phi(\tau, a, s) = \sum_{n=1}^{\infty} \frac{e^{2\pi i n \tau}}{(n+a)^s} = \Phi(e^{2\pi i \tau}, s, a) =: L(\tau, a, s)$$

($a \in \mathbb{C} \diagdown \mathbb{Z}_0^-$; $\mathrm{Re}(s) > 0$ when $\tau \in \mathbb{R} \backslash \mathbb{Z}$; $\mathrm{Re}(s) > 1$ when $\tau \in \mathbb{Z}$), which was first studied by Rudolf Lipschitz (1832–1903) and Matyas Lerch (1860–1922) in connection with Dirichlet's famous theorem on primes in arithmetic progressions (cf. [72, p. 2980, Remark 3]), and the Dirichlet's eta function $\eta(s)$

$$\zeta_{-1,1}(s, 1) = \Phi(-1, s, 1) = \sum_{n=1}^{\infty} \frac{(-1)^{n-1}}{n^s},$$

where $\mathrm{Re}(s) > 1$ (cf. [15, 27], [72, p. 2980, Remark 3], [76, 77]).

In [72, p. 2980, Remark 4], due to Simsek and Srivastava, some interesting *multiparameter* generalizations of the Hurwitz–Lerch zeta function $\Phi(z, s, a)$ were investigated by Garf et al. [23], Lin et al. [48], and Choi et al. [14].

This function interpolates the modified twisted Bernoulli polynomials $B^*_{k,\xi}(x, q)$ at negative integers, which is given by the next theorem [68].

By using Definition 4.2 and (32), we arrive at the following theorem:

Theorem 4.2. *If $k \in \mathbb{Z}^+$, then we have*

$$\zeta_{\xi,q}(1-k, x) = \frac{(-1)^{k+1}}{k} B^*_{k,\xi}(x, q).$$

4.2 Modified Twisted q-L-Functions

We recall work of Iwasawa [30] and Koblitz [45] that Dirichlet L-series is defined as follows:

Let $f : \mathbb{Z} \to \mathbb{C}$ be a periodic function with period d, $f(x+d) = f(x)$. Then the Dirichlet L-series is defined by

$$L(s, f) = \sum_{n=1}^{\infty} \frac{f(n)}{n^s}$$

for $\mathrm{Re}(s) > 1$ and extended by analytic continuation to other $s \in \mathbb{C}$.

The generalized Bernoulli numbers are

$$B_{n,f} = n!.\text{coefficient of } t^n \text{ in } \sum_{a=1}^{d} \frac{f(a)te^{at}}{e^{dt}-1}.$$

One can see that for n a positive integer $L(-n, f) = -B_{n,f}/n$ (cf. [30,45]).

When $f = \chi$ is a character, i.e., a homomorphism $\chi : (\mathbb{Z}/d\mathbb{Z}) \to \mathbb{C}^*$ from the multiplicative group of integers mod$d$ (where χ is extended $\chi(n) = 0$ for all n having a common factor with d), the L-function equals the following **Euler product** if $\mathrm{Re}(s) > 1$:

$$L(s,\chi) = \prod \left(1 - \frac{\chi(p)}{p^n}\right)^{-1},$$

where the product is taken over all prime p.

It is well known that L-functions occur in many situations in Analytic Number Theory. For example, the class number h of an imaginary quadratic field $\mathbb{Q}(\sqrt{-d})$ of discriminant $-d$ is given by

$$h = \frac{w\sqrt{d}}{2\pi} L(1,\chi) = -\frac{w}{2d} \sum_{a=1}^{d-1} a\chi(a),$$

where $w = 2, 4$, or 6 is the number of roots of unity in $\mathbb{Q}(\sqrt{-d})$ and $\chi : (\mathbb{Z}/d\mathbb{Z}) \to \{-1, 1\}$ is the Legendre symbol or quadratic residue symbol (cf. [45, p. 25]).

We shall also want to consider modified two variable **twisted** q-analogue L-functions. We assume that $q \in \mathbb{R}$ with $0 < q < 1$, $\xi^r = 1$ $(r \in \mathbb{Z}^+)$, $\xi \neq 1$.

Upon substituting from (49) into the right-hand side of (38), we obtain the following formula involving the Mellin transformation:

$$\begin{aligned} L_{\xi,q}(s,x,\chi) &= \frac{1}{[f]\Gamma(s)} \sum_{k=1}^{f} \xi^{-k}\chi(k) \int_0^\infty t^{s-1} \mathfrak{H}_{q^f,\xi^f}\left(\frac{[f]}{q^k}t, \frac{x+k}{[f]}\right)\frac{dt}{t} \\ &= [f]^{s-1} \sum_{k=1}^{f} (\xi q)^{-k} \chi(k) \zeta_{q^f,\xi^f}\left(s, \frac{x+k}{[f]}\right), \end{aligned}$$

with

$$\min\{\mathrm{Re}(s), \mathrm{Re}(x)\} > 0.$$

From the above equation, modified two variable **twisted** q-analogue L-function related to the modified twisted q-extension of the Hurwitz zeta function is provided by the next theorem:

Theorem 4.3. *Let $s \in \mathbb{C}$. Let χ be a Dirichlet character of conductor f and let $q \in \mathbb{R}$ with $0 < q < 1$, $\xi^r = 1$ $(r \in \mathbb{Z}^+)$, $\xi \neq 1$. Then*

$$L_{\xi,q}(s,x,\chi) = \sum_{k=1}^{f} (\xi q)^{-k} \chi(k) \zeta_{q^f,\xi^f}\left(s, \frac{x+k}{[f]}\right). \tag{52}$$

By substituting (50) into (52), for $q \in \mathbb{R}$ with $0 < q < 1$, $\xi^r = 1$ $(r \in \mathbb{Z}^+)$, $\xi \neq 1$, modified two variable twisted q-analogue L-function is explicitly given by

$$L_{\xi,q}(s,x,\chi) = \sum_{n=0}^{\infty} \frac{\xi^{-n}\chi(n)q^{-n}}{(q^{-n}[n+x])^s}. \tag{53}$$

If we substitute $x = 1$ into (53), we have modified twisted q-analogue Dirichlet L-function as follows:

Definition 4.3 ([68]). Let $s \in \mathbb{C}$. Let χ be a Dirichlet character of conductor f and let $q \in \mathbb{R}$ with $0 < q < 1$, $\xi^r = 1$ $(r \in \mathbb{Z}^+)$, $\xi \neq 1$. We define

$$L_{\xi,q}(s,\chi) = \sum_{n=1}^{\infty} \frac{\xi^{-n}\chi(n)q^{-n}}{(q^{-n}[n])^s}, \quad \mathrm{Re}(s) > 1. \tag{54}$$

Remark 4.4. When $\chi \equiv 1$, (54) reduces to (41) and

$$\lim_{q \to 1} L_{\xi,q}(s,\chi) = L_\xi(s,\chi),$$

which is a Dirichlet L-function [41, 43, 66, 67]. In [43], Koblitz defined twisted L-functions as follows: Let $r \in \mathbb{Z}^+$, set of positive integers, let χ be a Dirichlet character of conductor $f \in \mathbb{Z}^+$, and let $\xi^r = 1$, $\xi \neq 1$. Twisted L-functions are defined by

$$L(s,\chi,w) = \sum_{n=1}^{\infty} \frac{\chi(n)\xi^n}{n^s}.$$

Since the function $n \to \chi(n)\xi^n$ has period fr, this is a special case of the Dirichlet L-functions. This function interpolates Carlitz's q-Bernoulli numbers at nonpositive integers.

Relation between $\zeta_{\xi,q}(s,x)$ and $L_{\xi,q}(s,\chi)$ is given as follows:

Theorem 4.4 ([68]). *Let $s \in \mathbb{C}$. Let χ be a Dirichlet character of conductor f and let $q \in \mathbb{R}$ with $0 < q < 1$, $\xi^r = 1$ $(r \in \mathbb{Z}^+)$, $\xi \neq 1$. Then*

$$L_{q,\xi}(s,\chi) = \frac{1}{[f]^s} \sum_{a=1}^{f} \xi^{-a} q^{(s-1)a} \chi(a) \zeta_{\xi^f,q^f}\left(s, \frac{a}{[f]}\right). \tag{55}$$

Proof. Substituting $n = a + mf$, where $m = 0, 1, \ldots, \infty$ and $a = 1, 2, \ldots, f$ into Definition 4.3, we obtain

$$L_{q,w}(s,\chi) = \sum_{a=1}^{f} w^a q^{-a} \chi(a) \sum_{m=0}^{\infty} \frac{w^{-mf} q^{-mf}}{(q^{-a-mf}[a+mf])^s}.$$

By using

$$[a+mf] = [a] + q^a [f][m : q^f]$$

in the above equation, after elementary calculations, we obtain the desired result. □

By substituting $s = 1-n$, n is a positive integer, into Theorem 4.4 and using Theorem 4.2, we have

$$L_{q,\xi}(1-n,\chi) = \frac{(-1)^{n+1}[f]^{n-1}}{n} \sum_{k=1}^{f} \xi^{-a} q^{-an} \chi(a) B^*_{n,\xi^f}\left(\frac{a}{f}, q^f\right).$$

By substituting (40) into the above equation, we arrive at the following theorem:

Theorem 4.5 ([68]). *If $n \geq 1$, where n is a positive integer, then we have*

$$L_{q,\xi}(1-n,\chi) = \frac{(-1)^{n+1}}{n} B^*_{n,\xi,\chi}(q).$$

In the next sections, we give modified twisted partial q-zeta function and p-adic interpolation function of the modified twisted q-Bernoulli polynomials, which are constructed by Simsek and Srivastava [72].

4.3 A Class of Modified Twisted Partial q-Zeta Function

Simsek and Srivastava [72] defined twisted partial q-zeta functions, which are given here in detail. We assume that $q \in \mathbb{C}$ with $|q| < 1$. Then a class of modified twisted partial q-zeta function is defined as follows:

Definition 4.4 ([72]). Let $s \in \mathbb{C}$ and $r, n \in \mathbb{Z}^+$. Let χ be a Dirichlet character of conductor f and let $q \in \mathbb{R}$ with $0 < q < 1$, $\xi^r = 1$, and $\xi \neq 1$. Also let a and F be integers with $0 < a < F$. We define

$$H_{\xi,q}(s,a:F) = \sum_{n \equiv a (\mathrm{mod}\, F)}^{\infty} \frac{q^{-n(s-1)} \xi^{-n}}{[n]^s}. \tag{56}$$

From (56), we have

$$H_{\xi,q}(s,a:F)=\sum_{m=0}^{\infty}\frac{q^{-(a+mF)(s-1)}\xi^{-(a+mF)}}{[a+mF]^s}$$

so that by writing $[a+mF]=[F][\frac{a}{F}+m:q^F]$ and after some elementary calculations, we find that

$$H_{\xi,q}(s,a:F)=\frac{\xi^{-a}q^{a(1-s)}}{[F]^s}\sum_{m=0}^{\infty}\frac{q^{mF(1-s)}\xi^{-mF}}{[\frac{a}{F}+m:q^F]^s}.$$

By substituting (50) into the above equation, then we obtain the following relationship between $H_{\xi,q}(s,a:F)$ and $\zeta_{\xi,q}(s,x)$:

$$H_{\xi,q}(s,a:F)=\frac{q^{a(1-s)}}{\xi^a[F]^s}\zeta_{\xi^F,q^F}\left(s,\frac{a}{F}\right). \tag{57}$$

Substituting $s=1-n$ $(n\in\mathbb{Z}^+)$ into (57), if make use of Theorem 4.2 and (31), we arrive at the following relation:

$$H_{\xi,q}(1-n,a:F)=\frac{q^{an}[a]^n}{n\xi^a}\sum_{k=0}^{n}(-1)^{1+k}\binom{n}{k}\frac{[F]^{k-1}}{[a]^k}B^*_{k,\xi^F}(q^F). \tag{58}$$

We now modify the twisted partial q-zeta function explicitly by the following theorem:

Theorem 4.6. *Let $s\in\mathbb{C}$ and $r,n\in\mathbb{Z}^+$. Let χ be a Dirichlet character of conductor f and let $q\in\mathbb{R}$ with $0<q<1$, $\xi^r=1$ and $\xi\neq 1$. Then*

$$H_{\xi,q}(s,a:F)=\frac{q^{a(1-s)}[a]^{1-s}}{(1-s)\xi^a}\sum_{k=0}^{\infty}(-1)^{1+k}\binom{1-s}{k}\frac{[F]^{k-1}}{[a]^k}B^*_{k,\xi^F}(q^F).$$

Proof. By substituting $s=1-n$ into (58) and using (31), then we arrive at the desired result.

Corollary 4.2. *Let $\xi^r=1$ $(r\in\mathbb{Z}^+)$ and $\xi\neq 1$. We have*

$$H_{\xi,q}(0,a:F)=\frac{q^a}{\xi^a}\left(B^*_{1,\xi^F}(q^F)-\frac{[a]}{[F]}B^*_{0,\xi^F}(q^F)\right).$$

The function $H_{\xi,q}(s,a:F)$ is analytically continued to the whole complex s-plane except for a simple pole at $s=1$ with residue

$$\frac{B^*_{0,\xi^F}(q^F)}{[F]\xi^a}.$$

By using (55) in the above, we find relation between $H_{\xi,q}(s,a:F)$ and $L_{\xi,q}(s,\chi)$ as follows [72]:

Theorem 4.7. *Let χ be a Dirichlet character with conductor $f=f_\chi$ and $f_\chi|F$ and let $\xi^r=1$ $(r\in\mathbb{Z}^+)$ and $\xi\neq 1$. We have*

$$L_{\xi,q}(s,\chi)=\sum_{a=1}^{F}\chi(a)H_{\xi,q}(s,a:F). \tag{59}$$

Theorem 4.8 ([72]). *Let χ is a Dirichlet character with conductor $f=f_\chi$ and $f_\chi|F$ and $s\in\mathbb{C}$ and let $\xi^r=1$ $(r\in\mathbb{Z}^+)$ and $\xi\neq 1$. Then we have*

$$L_{\xi,q}(s,\chi)=\frac{1}{1-s}\sum_{a=1}^{F}\chi(a)q^{a(1-s)}\xi^{-a}[a]^{1-s}$$
$$\times\sum_{k=0}^{\infty}(-1)^{1+k}\binom{1-s}{k}\frac{[F]^{k-1}}{[a]^k}B^*_{k,\xi^F}(q^F).$$

We now give some applications of Theorem 4.8.

If we substitute $s=0$ into Theorem 4.8, then we easily arrive at the following corollary [72]:

Corollary 4.3. *Let χ is a Dirichlet character with conductor $f=f_\chi$ and $f_\chi|F$ and let $\xi^r=1$ $(r\in\mathbb{Z}^+)$ and $\xi\neq 1$. Then we have*

$$L_{\xi,q}(0,\chi)=\sum_{a=1}^{F}\frac{\chi(a)q^a}{\xi^a}\left(B^*_{1,\xi^F}(q^F)-\frac{[a]}{[F]}B^*_{0,\xi^F}(q^F)\right).$$

If we substitute $s=2$ into Theorem 4.8, then we easily arrive at the following corollary [72]:

Corollary 4.4. *Let χ be a Dirichlet character with conductor $f=f_\chi$ and $f_\chi|F$ and let $\xi^r=1$ $(r\in\mathbb{Z}^+)$ and $\xi\neq 1$. Then we have*

$$L_{\xi,q}(2,\chi)=\sum_{k=0}^{\infty}\sum_{a=1}^{F}\frac{\chi(a)[F]^{k-1}B^*_{k,\xi^F}(q^F)}{(\xi q)^a\,[a]^{k+1}}. \tag{60}$$

4.4 A p-Adic Interpolation Function of the Modified Twisted Generalized q-Bernoulli Numbers

Here we give p-adic interpolation function for the modified twisted generalized q-Bernoulli numbers. This function was constructed by Simsek and Srivastava [72].

Throughout this section, $q \in \mathbb{C}_p$, with $\mid q-1 \mid_p < p^{-\frac{1}{p-1}}$ and $\mid \xi - 1 \mid_p \leq 1$. Consequently, we note that $q^x = \exp(x \log q)$ for $q \in \mathbb{C}_p$, with $\mid x \mid_p < 1$.

We give some notations which are related to (among others) Washington [87], Kim [37, 68] and the author [70] and Simsek and Srivastava [72].

Let the integer p^* be defined by $p^* = p$ if $p > 2$ and $p^* = 4$ if $p = 2$ (cf. [17, 21, 22, 29, 30, 35, 37, 38, 43–45, 70, 79, 87]). Let w denote the Teichmüller character, having conductor $f_w = p^*$. For an arbitrary character χ, we define $\chi_n = \chi w^{-n}$, where $n \in \mathbb{Z}$, in the sense of the product of characters. Let F be a positive integral multiple of $f_w = p^*$ and $f = f_\chi$. If $q \in \mathbb{C}_p$, then we assume $\mid 1-q \mid_p < p^{-\frac{1}{p-1}}$. Let

$$< a > = < a, q > = w^{-1}(a)[a] = \frac{[a]}{w(a)}.$$

We note that $< a > \equiv 1 (\text{mod } p^* p^{-\frac{1}{p-1}})$ cf. [29, 37, 38].

Thus, we have $< a + p^* t > = w^{-1}(a + p^* t)[a + p^* t] = w^{-1}(a) \left([a] + q^a [p^* t]\right) \equiv 1 (\text{mod } p^* p^{-\frac{1}{p-1}})$, here $t \in \mathbb{C}_p$, with $\mid 1-q \mid_p \leq 1$:

$$D = \left\{ s \in \mathbb{C}_p : \mid s \mid_p \leq p^* p^{-\frac{1}{p-1}} \right\},$$

where $D \subset \mathbb{C}_p$ (cf. [87]). For $\mid q-1 \mid_p < p^{-\frac{1}{p-1}}$, we note that $< a >^{p^N} \equiv 1 (\text{mod} p^*)$. Then $\log_p a = \log_p < a >$ (cf. [87]).

By using (58) and Theorem 4.6, we now define a p-adic meromorphic function $H_{\xi,p,q}(s, a : F)$ on D as follows (cf. [72]):

Definition 4.5. Let $p^* \mid F$ and $p^* \nmid a$. Let $s \in D$. Then we define

$$H_{\xi,p,q}(s, a : F) = \frac{(q^a < a >)^{1-s}}{(1-s)\xi^a} \sum_{k=0}^{\infty} (-1)^{1+k} \binom{1-s}{k} \frac{[F]^{k-1}}{[a]^k} B^*_{k,\xi^F}(q^F). \quad (61)$$

According to Washington [87], $< a, q^{\frac{1}{F}} >$ and

$$\sum_{k=0}^{\infty} (-1)^{1+k} \binom{1-s}{k} \frac{[F]^{k-1}}{[a]^k} B^*_{k,\xi^F}(q^F)$$

are analytic in D.

Theorem 4.9 ([72]). *Suppose that* $p^* \mid F$ *and* $p \nmid a$. *Then the* p*-adic meromorphic function* $H_{\xi,p,q}(s,a:F)$ *satisfies the following relation:*

$$H_{\xi,p,q}(1-n,a:F) = w^{-n}(a)H_{\xi,q}(1-n,a:F),\ n \geq 1. \tag{62}$$

Proof. Substituting $s = 1-n$ into (61), then, we have

$$\begin{aligned} &H_{\xi,p,q}(1-n,a:F) \\ &= \frac{(q^a < a >)^n}{n\xi^a} \sum_{k=0}^{n} (-1)^{n-k} \binom{n}{k} \frac{[F]^{k-1}}{[a]^k} B^*_{k,\xi^F}(q^F) \\ &= \frac{q^{an} w^{-n}(a)[a]^n}{n\xi^a} \sum_{k=0}^{n} (-1)^{n-k} \binom{n}{k} \frac{[F]^{k-1}}{[a]^k} B^*_{k,\xi^F}(q^F) \\ &= w^{-n}(a)H_{\xi,q}(1-n,a:F). \end{aligned}$$

Thus the proof of theorem is completed. □

The p-adic meromorphic function $H_{\xi,p,q}(s,a:F)$ interpolates modified twisted Bernoulli numbers.

By substituting (58) into (62), we arrive at the following corollary:

Corollary 4.5 ([72]). *Suppose that* $p^* \mid F$ *and* $p^* \nmid a$. *Let* $n \in \mathbb{Z}^+$; *we have*

$$H_{\xi,p,q}(1-n,a:F) = \frac{(-1)^{n+1} w^{-n}(a) B^*_{k,\xi^F}(\frac{a}{F}, q^F)}{n}. \tag{63}$$

Now, we are ready to define p-adic interpolation function of the modified twisted generalized q-Bernoulli numbers at negative integer. This function is denoted by $L_{\xi,p,q}(s,\chi)$, which is defined as follows [72]:

Definition 4.6. Let χ be a Dirichlet character of conductor f and let F be any multiple of p^* and f. A p-adic meromorphic (analytic if $\chi \neq 1$) function $L_{\xi,p,q}(s,\chi)$ on D is defined by

$$L_{\xi,p,q}(s,\chi) = \sum_{\substack{a=1 \\ p \nmid a}}^{F} \chi(a) H_{\xi,p,q}(s,a:F). \tag{64}$$

Thus we give the main theorem in this section as follows [72]:

Theorem 4.10. *Let* χ *be a Dirichlet character of conductor* f *and let* F *be any multiple of* p^* *and* f. *There exists a* p*-adic meromorphic (analytic if* $\chi \neq 1$*) function* $L_{\xi,p,q}(s,\chi)$ *on* D *such that*

$$L_{\xi,p,q}(1-n,\chi) = \frac{(-1)^{n+1}}{n}(B^*_{n,\xi,\chi w^{-n}}(q) - [p]^{n-1}\chi w^{-n}(p) B^*_{n,\xi^p,\chi w^{-n}}(q^p)).$$

If $\chi = 1$, then $L_{\xi,p,q}(s,\chi)$ is analytic except for a pole at $s = 1$ with residue:

$$\frac{1}{[F]}\sum_{\substack{a=1\\ p\nmid a}}^{F}\chi(a)\xi^{-a}B^{*}_{0,\xi^F}(q^F).$$

Furthermore, we have the formula

$$L_{\xi,p,q}(s,\chi)=\frac{1}{(1-s)[F]^s}\sum_{\substack{a=1\\ p\nmid a}}^{F}\chi(a)\xi^{-a}\left(q^a<a>\right)^{1-s}\sum_{k=0}^{\infty}(-1)^{1+k}\binom{1-s}{k}[a,q^{\frac{1}{F}}]^{-k}B^{*}_{k,\xi^F}(q^F)$$

Proof. By using (64), we give analytic property of this function as follows. At $s = 1$, $L_{p,q,\xi}(s,\chi)$ has residue:

$$\lim_{s\to 1}(1-s)L_{\xi,p,q}(s,\chi)=\sum_{\substack{a=1\\ p\nmid a}}^{F}\chi(a)\lim_{s\to 1}(1-s)H_{\xi,p,q}(s,a:F)=\frac{1}{[F]}\sum_{\substack{a=1\\ p\nmid a}}^{F}\chi(a)\xi^{-a}B^{*}_{0,\xi^F}(q^F),$$

where we use (61) in the above. If $\chi \neq 1$ and $\xi \neq 1$, then $L_{\xi,p,q}(s,\chi)$ has no pole at $s = 1$. If $n \geq 1$, then we have

$$L_{\xi,p,q}(1-n,\chi)=\sum_{\substack{a=1\\ p\nmid a}}^{F}\chi(a)H_{\xi,p,q}(1-n,a:F).$$

By substituting (63) into the above, we have

$$L_{\xi,p,q}(1-n,\chi)=\frac{(-1)^{n+1}[F]^{n-1}}{n}\sum_{\substack{a=1\\ p\nmid a}}^{F}\chi w^{-n}(a)q^{na}\xi^{-a}B^{*}_{n,\xi^F}\left(\frac{a}{F},q^F\right).$$

From the above, we obtain

$$\begin{aligned}L_{p,q,\xi}(1-n,\chi)&=\frac{(-1)^{n+1}[F]^{n-1}}{n}\sum_{a=1}^{F}\chi w^{-n}(a)q^{na}\xi^{-a}B^{*}_{n,\xi^F}\left(\frac{a}{F},q^F\right)\\&\quad-\frac{(-1)^{n+1}[\frac{pF}{p}]^{n-1}}{n}\sum_{b=1}^{\frac{F}{p}}\chi w^{-n}(bp)q^{nbp}\xi^{-bp}B^{*}_{n,\xi^F}\left(\frac{bp}{F},q^F\right)\\&=-\frac{(-1)^{n+1}}{n}\left(B^{*}_{n,\chi w^{-n},\xi}(q)-\chi w^{-n}(p)[p]^{n-1}B^{*}_{n,\chi w^{-n},\xi^p}(q^p)\right).\end{aligned}$$

This completes the proof of theorem. □

5 Twisted (h, q) Bernoulli Numbers and Polynomials

Here, we define new type Bernoulli numbers and polynomials by using p-adic Volkenborn integral. We firstly give some notations and definitions.

Let v_p be the normalized exponential valuation of $\mathbb{C}_p$ with $|p|_p = p^{-v_p(p)} = p^{-1}$. When one talks of q-extension, q is variously considered as an indeterminate, a complex number $q \in \mathbb{C}$, or p-adic number $q \in \mathbb{C}_p$. If $q \in \mathbb{C}_p$, then we assume that $|1-q|_p < p^{-\frac{1}{p-1}}$, so that $q^x = \exp(x \log q)$ for $|x|_p \leq 1$. If $q \in \mathbb{C}$, then we assume $|q| < 1$ (cf. [33, 35, 37]).

For $f \in \mathrm{UD}(\mathbb{Z}_p, \mathbb{C}_p) = \{f \mid f : \mathbb{Z}_p \to \mathbb{C}_p \text{ is uniformly differentiable function}\}$, the p-adic q-integral (q-Volkenborn integration) is defined by Kim (cf. [36]):

$$I_q(f) = \int_{\mathbb{Z}_p} f(x) d\mu_q(x) = \lim_{N\to\infty} \frac{1}{[p^N]_q} \sum_{x=0}^{p^N-1} q^x f(x), \tag{65}$$

where μ_q denotes p-adic q-Haar distribution which is originally introduced by Kim [36],

$$\mu_q(a + p^N \mathbb{Z}_p) = \frac{q^a}{[p^N]_q}, \quad N \in \mathbb{Z}^+.$$

If $q \to 1$ in (65), then we have

$$I_1(f) = \int_{\mathbb{Z}_p} f(x) d\mu_1(x) = \lim_{N\to\infty} \frac{1}{p^N} \sum_{x=0}^{p^N-1} f(x) \tag{66}$$

(cf. [2, 58, 84]).

Observe that in (66), we easily see that $I_1(f) = \lim_{q\to 1} I_q(f)$,

$$I_1(f_1) = I_1(f) + f'(0), \tag{67}$$

where $f_1(x) = f(x+1)$ and $f'(0) = \frac{d}{dx} f(x) \mid_{x=0}$ (cf. [36, 58]).

Let p be a fixed prime. For a fixed positive integer f with $(p, f) = 1$, we set

$$\mathbb{X} = \mathbb{X}_f = \lim_{\overleftarrow{N}} \mathbb{Z}/\mathbb{Z} f p^N, \quad \mathbb{X}_1 = \mathbb{Z}_p, \quad \mathbb{X}^* = \bigcup_{\substack{0 < a < fp \\ (a,p)=1}} a + fp\mathbb{Z}_p$$

and

$$a + fp^N \mathbb{Z}_p = \{x \in X \mid x \equiv a (\text{mod } fp^N)\},$$

where $a \in \mathbb{Z}$ satisfies the condition $0 \leq a < fp^N$. For $f \in \mathrm{UD}(\mathbb{Z}_p, \mathbb{C}_p)$,

$$\int_{\mathbb{Z}_p} f(x) d\mu_1(x) = \int_{\mathbb{X}} f(x) d\mu_1(x), \tag{68}$$

(cf. [33, 35–37]).

By using (67), one can easily see that

$$I_1(f_b) = I_1(f) + \sum_{j=0}^{b-1} f'(j), \tag{69}$$

where $f_b(x) = f(x+b)$, $(b \in \mathbb{Z}^+)$ (cf. [37]).

According to [60, 89], for each integer $N \geq 0$, C_{p^N} denotes the multiplicative group of the primitive p^Nth roots of unity in $\mathbb{C}_p^* = \mathbb{C}_p \setminus \{0\}$. Let

$$\mathbb{T}_p = \left\{\xi \in \mathbb{C}_p : \xi^{p^N} = 1, \text{ for } N \geq 0\right\} = \underset{N \geq 0}{\cup} C_{p^N}.$$

The dual of $\mathbb{Z}_p$, in the sense of p-adic Pontrjagin duality, is $\mathbb{T}_p = C_{p^\infty}$, the direct limit (under inclusion) of cyclic groups C_{p^N} of order p^N with $N \geq 0$, with the discrete topology. $\mathbb{T}_p$ admits a natural $\mathbb{Z}_p$-module structure which we shall write exponentially, viz., ξ^x for $\xi \in \mathbb{T}_p$ and $x \in \mathbb{Z}_p$. $\mathbb{T}_p$ can be embedded discretely in $\mathbb{C}_p$ as the multiplicative p-torsion subgroup, and we choose, for once and all, one such embedding. If $\xi \in \mathbb{T}_p$, then $\phi_\xi : (\mathbb{Z}_p, +) \to (\mathbb{C}_p, \cdot)$ is the locally constant character, $x \to \xi^x$, which is the locally analytic character if $\xi \in \left\{\xi \in \mathbb{C}_p : v_p(\xi - 1) > 0\right\}$. Then ϕ_ξ has continuation to a continuous group homomorphism from $(\mathbb{Z}_p, +)$ to $(\mathbb{C}_p, \cdot)$ (cf. [33, 40, 60, 66, 70, 89]); see also the references cited in each of these earlier works.

Substituting $f(x) = \phi_\xi(x) q^{hx} e^{tx}$ into (67), then we have

$$\xi q^h e^t I_1(\phi_\xi(x) q^{hx} e^{tx}) = I_1(\phi_v(x) q^{hx} e^{tx}) + h \log q + t.$$

Therefore

$$F_{w,q}^{(h)}(t) = I_1(\phi_w(x) q^{hx} e^{tx}) = \frac{\log q^h + t}{w q^h e^t - 1} = \sum_{n=0}^{\infty} B_{n,w}^{(h)}(q) \frac{t^n}{n!} \tag{70}$$

for $|t| < p^{-1/(p-1)}$ and h is an integer.

The twisted (h, q)-extension of Bernoulli numbers $B_{n,w}^{(h)}(q)$ are defined by means of the generating function:

$$F_{\xi,q}^{(h)}(t) = \frac{\log q^h + t}{\xi q^h e^t - 1} = \sum_{n=0}^{\infty} B_{n,\xi}^{(h)}(q) \frac{t^n}{n!}, \quad | t + \log\left(\xi q^h\right) | < 2\pi, \tag{71}$$

where $\xi^r = 1$ ($r \in \mathbb{Z}^+$); $\xi \neq 1$. By applying the *umbral calculus* convention in the above equation, and the usual convention of symbolically replacing $(B_{\xi}^{(h)}(q))^n$ by $B_{n,\xi}^{(h)}(q)$, then we have

$$B_{0,\xi}^{(h)}(q) = \frac{\log q^h}{\xi q^h - 1} \tag{72}$$

$$\xi q^h (B_{\xi}^{(h)}(q) + 1)^n - B_{n,\xi}^{(h)}(q) = \delta_{1,n},\ n \geq 1,$$

where $\delta_{1,n}$ is denoted by Kronecker symbol.

Remark 5.1. If $\xi = 1$, then (71) reduces to the following generating function:

$$\frac{\log q^h + t}{q^h e^t - 1} = \sum_{n=0}^{\infty} B_{n,1}^{(h)}(q) \frac{t^n}{n!},$$

(cf. [37]).

Remark 5.2. In recent years, many authors have studied on various interesting unification of the classical Bernoulli numbers B_n and the Apostol–Bernoulli numbers $\mathcal{B}_n(\lambda)$, which are defined by means of the following generating function:

$$\frac{t}{\lambda e^t - 1} = \sum_{n=0}^{\infty} \mathcal{B}_n(\lambda) \frac{t^n}{n!}$$

(cf. [4, 37, 49, 50, 52, 75, 77, 79]). The twisted (h, q)-Bernoulli numbers are related to the the Apostol–Bernoulli numbers, that is,

$$\sum_{n=0}^{\infty} B_{n,\xi}^{(h)}(q) \frac{t^n}{n!} = \frac{\log q^h}{\xi q^h e^t - 1} + \frac{t}{\xi q^h e^t - 1}$$
$$= \sum_{n=0}^{\infty} \left(\log\left(\xi^n q^{nh}\right) \mathcal{B}_{n-1}(q^h) + \mathcal{B}_n(\xi q^h)\right) \frac{t^n}{n!}.$$

Therefore, we obtain relation between the Apostol–Bernoulli numbers and the twisted (h, q)-Bernoulli numbers:

$$B_{n,\xi}^{(h)}(q) = \log\left(\xi^n q^{nh}\right) \mathcal{B}_{n-1}(q^h) + \mathcal{B}_n(\xi q^h).$$

Remark 5.3. If $q \to 1$ in the above, then we have (2). If $q \to 1$, then (71) reduces to (7).

We now give relation between the twisted (h, q)-extension of Bernoulli numbers and the Frobenius–Euler numbers as follows:

By (71), we have

$$\sum_{n=0}^{\infty} B_{n,\xi}^{(h)}(q)\frac{t^n}{n!} = \left(\frac{\log q^h}{\xi q^h - 1}\right)\left(\frac{1-\xi^{-1}q^{-h}}{e^t - \xi^{-1}q^{-h}}\right) + \left(\frac{t}{\xi q^h - 1}\right)\left(\frac{1-\xi^{-1}q^{-h}}{e^t - \xi^{-1}q^{-h}}\right)$$
$$= \left(\frac{1}{\xi q^h - 1}\right)\sum_{n=0}^{\infty}\left(\left(\log q^h\right) H_n(\xi^{-1}q^{-h}) + nH_{n-1}(\xi^{-1}q^{-h})\right)\frac{t^n}{n!}.$$

By comparing the coefficients of $t^n/n!$ on both sides of the above, we easily obtain

$$B_{n,\xi}^{(h)}(q) = \frac{\left(\log q^h\right) H_n(\xi^{-1}q^{-h}) + nH_{n-1}(\xi^{-1}q^{-h})}{\xi q^h - 1}. \tag{73}$$

If $q \to 1$ in (73), then we have

$$B_{n,\xi} = \frac{nH_{n-1}(\xi^{-1})}{\xi - 1}, \quad n \geq 1$$

(cf. [33, 70]).

The Witt's formula for $B_{n,\xi}^{(h)}(q)$ is given by the following theorem:

Theorem 5.1. *For $h \in \mathbb{Z}$ and $q \in \mathbb{C}_p$ with $| q - 1 |_p < p^{-\frac{1}{p-1}}$, we have*

$$B_{n,\xi}^{(h)}(q) = \int_{\mathbb{Z}_p} \phi_\xi(x) q^{hx} x^n d\mu_1(x). \tag{74}$$

Proof. By using Taylor series of e^{tx} in (70), we have

$$I_1\left(\phi_\xi(x)q^{hx}\sum_{n=1}^{\infty}\frac{x^n t^n}{n!}\right) = \sum_{n=1}^{\infty}\left(I_1(\phi_\xi(x)q^{hx}x^n)\right)\frac{t^n}{n!} = \sum_{n=0}^{\infty} B_{n,\xi}^{(h)}(q)\frac{t^n}{n!}.$$

By comparing coefficients $t^n/n!$ in the above equation, we arrive at the desired result.

Twisted (h,q)-extension of Bernoulli polynomials $B_{n,\xi}^{(h)}(z,q)$ is defined by means of the following generating function:

$$F_{\xi,q}^{(h)}(t,z) = \frac{(t + \log q^h)e^{tz}}{\xi q^h e^t - 1} = I_1(\phi_\xi(x)q^{hx}e^{t(z+x)}) = \sum_{n=0}^{\infty} B_{n,\xi}^{(h)}(z,q)\frac{t^n}{n!}. \tag{75}$$

We note that

$$B_{n,\xi}^{(h)}(0,q) = B_{n,\xi}^{(h)}(q), \quad B_{n,1}^{(h)}(z,q) = B_n^{(h)}(z,q)$$

(cf. [37, 66]).

If $q \to 1$ in (75), then we arrive at (1).

Twisted version of Witt's formula for $B_{n,\xi}^{(h)}(z,q)$ is given by the following theorem:

Theorem 5.2. *For $h \in \mathbb{Z}$ and $q \in \mathbb{C}_p$ with $\mid q-1\mid_p < p^{-\frac{1}{p-1}}$, we obtain*

$$B_{n,\xi}^{(h)}(z,q) = \int_{\mathbb{Z}_p} \phi_\xi(x) q^{hx} (x+z)^n d\mu_1(x). \tag{76}$$

Theorem 5.3. *For $n \geq 0$ any positive integer k, we have*

$$B_{n,\xi}^{(h)}(z,q) = k^{n-1} \sum_{a=0}^{k-1} \phi_\xi(a) q^{ha} B_{n,\xi^k}^{(h)}\left(\frac{a+z}{k}, q^k\right).$$

Proof. By using (68) and (76), it is easy to see that

$$\begin{aligned} B_{n,\xi}^{(h)}(z,q) &= \int_{\mathbb{X}} \phi_\xi(x) q^{hx} (x+z)^n d\mu_1(x) \\ &= \lim_{N\to\infty} \frac{1}{kp^N} \sum_{x=0}^{kp^N-1} \xi^x q^{hx} (x+z)^n \\ &= \frac{1}{k} \lim_{N\to\infty} \frac{1}{p^N} \sum_{a=0}^{k-1} \sum_{x=0}^{p^N-1} \xi^{a+kx} q^{h(a+kx)} (a+kx+z)^n. \end{aligned}$$

By using (76) in the above equation, we obtain the desired result.

The polynomials $B_{n,\xi}^{(h)}(z,q)$ are given explicitly by the following theorem:

Theorem 5.4. *For $n \geq 0$, we have*

$$B_{n,\xi}^{(h)}(z,q) = \sum_{k=0}^{n} \binom{n}{k} z^{n-k} B_{k,\xi}^{(h)}(q).$$

Proof. By using Taylor series of e^{tz} in (75), we have

$$\sum_{n=0}^{\infty} \left(B_{n,\xi}^{(h)}(z,q) \frac{1}{n!} \right) t^n = \sum_{n=0}^{\infty} \left(\sum_{k=0}^{n} B_{k,\xi}^{(h)}(q) \frac{z^{n-k}}{k!(n-k)!} \right) t^n.$$

By comparing coefficients t^n in the above equation, we obtain the desired result.

Remark 5.4. By using (74) and the binomial theorem in (76), and after some elementary calculations, we easily arrive at the another proof of Theorem 5.4.

Let χ be a Dirichlet character with conductor f. The generalized twisted (h,q)-extension of Bernoulli numbers is defined by means of the generating function:

$$F_{\chi,\xi,q}^{(h)}(t) = \sum_{a=1}^{f} \frac{\chi(a)\phi_\xi(a)q^{ha}e^{at}(t+\log q^h)}{\xi^f q^{hf} e^{ft} - 1} = \sum_{n=0}^{\infty} B_{n,\chi,\xi}^{(h)}(q)\frac{t^n}{n!}.$$

Note that

$$B_{n,\chi,1}^{(h)}(q) = B_{n,\chi}^{(h)}(q)$$

(cf. [37]).

Theorem 5.5. *Let χ be a Dirichlet character of conductor $f \in \mathbb{Z}^+$. We obtain*

$$B_{n,\chi,\xi}^{(h)}(q) = \int_{\mathbb{Z}_p} \chi(x)\phi_\xi(x)q^{hx}x^n d\mu_1(x). \tag{77}$$

Proof. If we take $f(x) = \chi(x)\phi_\xi(x)q^{hx}e^{tx}$ in (69), we get

$$\begin{aligned}\xi^f q^{fh} e^{ft} I_1(\phi_\xi(x)\chi(x)q^{hx}e^{tx}) &= I_1(\phi_\xi(x)\chi(x)q^{hx}e^{tx}) \\ &\quad + \sum_{a=0}^{f-1} q^{ha}\chi(a)\phi_\xi(a)e^{ta}\left(\log q^h + t\right).\end{aligned}$$

After some elementary calculations in the above equation, we have

$$\begin{aligned} F_{\chi,\xi,q}^{(h)}(t) &= \int_{\mathbb{Z}_p} \chi(x)\phi_\xi(x)q^{hx}e^{tx}d\mu_1(x) \\ &= \sum_{a=0}^{f-1} \frac{\chi(a)\phi_\xi(a)q^{ha}e^{at}(t+\log q^h)}{\xi^f q^{hf} e^{ft} - 1} \\ &= \sum_{n=0}^{\infty} B_{n,\chi,\xi}^{(h)}(q)\frac{t^n}{n!}. \end{aligned} \tag{78}$$

By using the Taylor series of e^{tz} in (78), we obtain

$$\begin{aligned} \sum_{n=0}^{\infty} B_{n,\chi,\xi}^{(h)}(q)\frac{t^n}{n!} &= \int_{\mathbb{Z}_p} \chi(x)\phi_\xi(x)q^{hx}e^{tx}d\mu_1(x) \\ &= \sum_{n=0}^{\infty}\left(\int_{\mathbb{Z}_p} \chi(x)\phi_\xi(x)q^{hx}x^n d\mu_1(x)\right)\frac{t^n}{n!}. \end{aligned}$$

By comparing coefficients $t^n/n!$ in the above equation, we easily arrive at the desired result. □

We define

$$f(x) = \chi(x)\phi_\xi(x)q^{hx}e^{t(x+z)}.$$

By substituting the above function into (69), then we obtain the generalized twisted (h,q)-extension of Bernoulli polynomials $B_{n,\chi,\xi}^{(h)}(z,q)$, which are given by means of the following generating function:

$$\begin{aligned} F_{\chi,\xi,q}^{(h)}(t,z) &= \int_{\mathbb{Z}_p} \chi(x)\phi_\xi(x)q^{h}e^{t(x+z)}d\mu_1(x) \\ &= \sum_{a=1}^{f} \frac{\chi(a)\phi_\xi(a)q^{ha}e^{(z+a)t}(t+\log q^h)}{\xi^f q^{hf}e^{ft}-1} \\ &= \sum_{n=0}^{\infty} B_{n,\chi,\xi}^{(h)}(z,q)\frac{t^n}{n!}. \end{aligned} \tag{79}$$

Note that substituting $z = 0$ into (79), we have $B_{n,\chi,\xi}^{(h)}(0,q) = B_{n,\chi,\xi}^{(h)}(q)$. If $q \to 1$ in (79), we arrive at (10).

Relation between $B_{n,\chi,\xi}^{(h)}(z,q)$ and $B_{k,\chi,\xi}^{(h)}(q)$ is given by the following theorem:

Theorem 5.6. *Let χ be a Dirichlet character of conductor $f \in \mathbb{Z}^+$. We have*

$$B_{n,\chi,\xi}^{(h)}(z,q) = \sum_{k=0}^{n}\binom{n}{k} z^{n-k} B_{k,\chi,\xi}^{(h)}(q).$$

Proof. By using the Taylor series of $e^{t(x+z)}$ in (79), we get

$$\sum_{n=0}^{\infty} B_{n,\chi,\xi}^{(h)}(z,q)\frac{t^n}{n!} = \sum_{n=0}^{\infty}\left(\sum_{k=0}^{n}\binom{n}{k} z^{n-k}\int_{\mathbb{Z}_p}\chi(x)\phi_\xi(x)q^{hx}x^k d\mu_1(x)\right)\frac{t^n}{n!}.$$

By using (77) in the above equation and comparing coefficients $t^n/n!$, we easily arrive at the desired result.

Remark 5.5. Integral representation of the generalized twisted (h,q)-Bernoulli polynomials is easily given as follows:

$$B_{n,\chi,\xi}^{(h)}(z,q) = \int_{\mathbb{Z}_p} \chi(x)\phi_\xi(x)q^{hx}(x+z)^n d\mu_1(x), \tag{80}$$

where $q \in \mathbb{C}_p, \mid q-1 \mid_p < p^{-\frac{1}{p-1}}$.

Theorem 5.7. *For any positive integer n, we have*

$$B_{n,\chi,\xi}^{(h)}(z,q) = f^{n-1}\sum_{a=0}^{f-1}\chi(a)\phi_{\xi}(a)q^{ha}B_{n,\xi^f}^{(h)}\left(\frac{a+z}{f},q^f\right). \tag{81}$$

Proof. By using (68) and (80), it is easy to see that

$$\begin{aligned} B_{n,\chi,\xi}^{(h)}(z,q) &= \int_{\mathbb{X}}\chi(x)\phi_{\xi}(x)q^{hx}(x+z)^n d\mu_1(x) \\ &= \lim_{N\to\infty}\frac{1}{fp^N}\sum_{x=0}^{fp^N-1}\chi(x)\xi^x q^{hx}(x+z)^n \\ &= \frac{1}{f}\lim_{N\to\infty}\frac{1}{p^N}\sum_{a=0}^{f-1}\sum_{x=0}^{p^N-1}\chi(a+fx)\xi^{a+fx}q^{h(a+fx)}(a+fx+z)^n \\ &= f^{n-1}\sum_{a=0}^{f-1}\chi(a)\xi^a q^{ha}\int_{\mathbb{X}}\phi_{\xi^f}(x)q^{fhx}\left(x+\frac{a+z}{f}\right)^n d\mu_1(x). \end{aligned}$$

By using (76) in the above equation, we obtain the desired result. □

5.1 *The Family of the Twisted (h, q)-Zeta Functions and (h, q)-L-Function*

Here, we assume that $q\in\mathbb{C}$ with $\mid q\mid<1$ and $s\in\mathbb{C}$. Let $\xi^r=1$ $(r\in\mathbb{Z}^+)$; $\xi\neq 1$.

By applying the Mellin transformation to (70) and (75), we have the following integral representations:

$$\int_0^\infty t^{s-1}e^{-t}F_{\xi,q}^{(h)}(-t)\frac{dt}{t} = \Gamma(s)\zeta_{\xi,q}^{(h)}(s) \tag{82}$$

and by using similar method in the above, we have

$$\int_0^\infty t^{s-2}F_{\xi,q}^{(h)}(-t,x)dt = \Gamma(s)\zeta_{\xi,q}^{(h)}(s,x). \tag{83}$$

By using (82) and (83), we define new twisted (h,q)-zeta functions as follows (cf. [66]):

Definition 5.1. Let $s\in\mathbb{C}$, $x\in\mathbb{R}^+$. We define

$$\zeta_{\xi,q}^{(h)}(s) = \sum_{n=1}^{\infty}\frac{\xi^{n-1}q^{(n-1)h}}{n^s} - \frac{h\log q}{s-1}\sum_{n=1}^{\infty}\frac{\xi^{n-1}q^{(n-1)h}}{n^{s-1}},$$

and

$$\zeta_{\xi,q}^{(h)}(s,x) = \sum_{n=0}^{\infty} \frac{\xi^{n-1} q^{(n-1)h}}{(n+x)^s} - \frac{h \log q}{s-1} \sum_{n=0}^{\infty} \frac{\xi^{n-1} q^{nh}}{(n+x)^{s-1}}.$$

Remark 5.6. Observe that when $q \to 1$ and $\zeta_q^{(h)}(s)$ reduces to

$$\zeta(s) = \sum_{n=1}^{\infty} \frac{1}{n^s},$$

Riemann zeta function and $\zeta_q^{(h)}(s,x)$ reduces to

$$\zeta(s,x) = \sum_{n=1}^{\infty} \frac{1}{(n+x)^s},$$

Hurwitz zeta function (cf. [1, 4–91]). We also note that $\zeta_{\xi,q}^{(h)}(s)$ are analytically continued for $\mathrm{Re}(s) > 1$.

The value of twisted (h,q)-zeta function at negative integers is given explicitly by the following theorem:

Theorem 5.8. *Let $n \in \mathbb{Z}^+$. We obtain [66]*

$$\zeta_{\xi,q}^{(h)}(1-n) = -\frac{B_{n,\xi}^{(h)}(q)}{n}.$$

Proof. Proof of this theorem is similar to that of Theorem 8 in [79]. In view of (82), we define $y(s)$ by the following contour integral:

$$y(s) = \int_C z^{s-2} e^{-z} F_{\xi,q}^{(h)}(-z) dz, \tag{84}$$

where C is Hankel's contour along the cut joining the points $z = 0$ and $z = \infty$ on the real axis, which starts from the point at ∞, encircles the origin ($z = 0$) once in the positive (counterclockwise) direction, and returns to the point at ∞ . Here, as usual, we interpret z^s to mean $\exp(s \log z)$, where we assume log to be defined by $\log t$ on the top part of the real axis and by $\log t + 2\pi i$ on the bottom part of the real axis. We thus find from definition (84) that

$$y(s) = (e^{2\pi i s} - 1) \int_{\varepsilon}^{\infty} t^{s-2} e^{-t} F_{\xi,q}^{(h)}(-t) dt + \int_{C_\varepsilon} z^{s-2} e^{-z} F_{\xi,q}^{(h)}(-z) dz,$$

where C_ε denotes a circle of radius $\varepsilon > 0$ (and centered at the origin), which is described in the positive (counterclockwise) direction. Assume first that $\mathrm{Re}(s) > 1$. Then

$$\int_{C_\varepsilon} \to 0 \text{ as } \varepsilon \to 0,$$

so we have

$$y(s) = (e^{2\pi is} - 1)\int_0^\infty t^{s-2}e^{-t}F_{\xi,q}^{(h)}(-t)dt,$$

which, upon substituting from (70) into it, yields

$$y(s) = (e^{2\pi is} - 1)\Gamma(s)\zeta_{\xi,q}^{(h)}(s).$$

Consequently,

$$\zeta_{\xi,q}^{(h)}(s) = \frac{y(s)}{(e^{2\pi is} - 1)\Gamma(s)}, \tag{85}$$

which, by analytic continuation, holds true for all $s \neq 1$. This evidently provides us with an analytic continuation of $\zeta_{w,q}^{(h)}(s)$.

Let $s \to 1 - n$ in (85), where n is a positive integer. Since

$$e^{2\pi is} = e^{2\pi i(1-n)} = 1 \quad (n \in \mathbb{Z}^+),$$

we have

$$\begin{aligned}\lim_{s\to 1-n}\{(e^{2\pi is} - 1)\Gamma(s)\} &= \lim_{s\to 1-n}\left\{\frac{(e^{2\pi is} - 1)}{\sin(\pi s)}\frac{\pi}{\Gamma(1-s)}\right\} \\ &= \frac{2\pi i(-1)^{n-1}}{(n-1)!} \quad (n \in \mathbb{Z}^+)\end{aligned} \tag{86}$$

by means of the familiar reflection formula for $\Gamma(s)$. Furthermore, since the integrand in (84) has simple pole order $n + 1$ at $z = 0$, it can also be found from definition (84) with $s = 1 - n$ that

$$\begin{aligned}y(1-n) &= \int_C z^{-n-1}e^{-z}F_{\xi,q}^{(h)}(-z)dz \\ &= 2\pi i \operatorname*{Res}_{z=0}\left\{z^{-n-1}e^{-z}F_{\xi,q}^{(h)}(-z)\right\} = (2\pi i)\frac{(-1)^n}{n!}B_{n,\xi}^{(h)}(q),\end{aligned} \tag{87}$$

where we have made of the power-series representation in (70). Thus, by Cauchy residue theorem, we easily arrive at the desired result upon suitably combining (86) and (87) with (85).

Remark 5.7. The value of $\zeta_{\xi,q}^{(h)}(s,x)$ function at negative integers is given explicitly as follows: for $n \in \mathbb{Z}^+$,

$$\zeta_{\xi,q}^{(h)}(1-n,x) = -\frac{B_{n,\xi}^{(h)}(x,q)}{n}. \tag{88}$$

Proof of (88) runs parallel to that of Theorem 5.8, so we choose to omit the details involved.

The twisted (h,q)-L-function is defined as follows:

Definition 5.2 ([66]). Let $s \in \mathbb{C}$. Let χ be a Dirichlet character of conductor $f \in \mathbb{Z}^+$. We define

$$L_{\xi,q}^{(h)}(s,\chi) = \sum_{n=1}^{\infty} \frac{q^{nh}\xi^n \chi(n)}{n^s} - \frac{\log q^h}{s-1} \sum_{n=1}^{\infty} \frac{q^{nh}\xi^n \chi(n)}{n^{s-1}}.$$

Remark 5.8. Observe that if $\xi \to 1$ in the above equation, we have

$$L_{q}^{(h)}(s,\chi) = \sum_{n=1}^{\infty} \frac{q^{nh} \chi(n)}{n^s} - \frac{\log q^h}{s-1} \sum_{n=1}^{\infty} \frac{q^{nh} \chi(n)}{n^{s-1}}.$$

If $q \to 1$ in the above equation, then we have

$$L(s,\chi) = \sum_{n=1}^{\infty} \frac{\chi(n)}{n^s},$$

where $L(s,\chi)$ is the Dirichlet L-function (cf. [37, 51, 68, 79, 90]).

Relation between $\zeta_{\xi,q}^{(h)}(s,z)$ and $L_{\xi,q}^{(h)}(s,\chi)$ is given by the following theorem:

Theorem 5.9. *Let $s \in \mathbb{C}$. Let χ be a Dirichlet character of conductor $f \in \mathbb{Z}^+$. We have [66]*

$$L_{\xi,q}^{(h)}(s,\chi) = \frac{1}{f^s} \sum_{a=1}^{f} q^{ha}\xi^a \chi(a) \zeta_{\xi^f,q^f}^{(h)}\left(s, \frac{a}{f}\right). \tag{89}$$

Proof. Substituting $n = a + mf$, where $m = 0, 1, \ldots, \infty$ and $a = 1, 2, \ldots, f - 1$ into Definition 4, we obtain

$$L_{w,q}^{(h)}(s,\chi) = \sum_{a=1}^{f} q^{ha}\xi^{a}\chi(a)\sum_{m=0}^{\infty}\frac{q^{mfh}\xi^{fm}}{(a+mf)^{s}}$$

$$-\frac{\log q^{h}}{s-1}\sum_{a=1}^{f} q^{ha}\xi^{a}\chi(a)\sum_{m=0}^{\infty}\frac{q^{mfh}\xi^{fm}}{(a+mf)^{s-1}}$$

$$=\frac{1}{f^{s}}\sum_{a=1}^{f} q^{ha}\xi^{a}\chi(a)\left(\sum_{m=0}^{\infty}\frac{q^{mfh}\xi^{fm}}{\left(m+\frac{a}{f}\right)^{s}}-\frac{\log q^{fh}}{s-1}\sum_{m=1}^{\infty}\frac{q^{mfh}\xi^{fm}}{\left(m+\frac{a}{f}\right)^{s-1}}\right).$$

By using Definition 3 in the above equation, we obtain the desired result. □

The value of twisted (h,q)-L-function at negative integers is given explicitly by the following theorem:

Theorem 5.10. *Let χ be a Dirichlet character of conductor $f \in \mathbb{Z}^{+}$. Let $n \in \mathbb{Z}^{+}$. We have [66]*

$$L_{\xi,q}^{(h)}(-n,\chi) = -\frac{B_{n+1,\chi,\xi}^{(h)}(q)}{n+1}. \tag{90}$$

Proof. Substituting $s = 1 - n$, $n \in \mathbb{Z}^{+}$ into (89), we have

$$L_{\xi,q}^{(h)}(1-n,\chi) = f^{n-1}\sum_{a=1}^{f} q^{ha}\xi^{a}\chi(a)\zeta_{\xi^{f},q^{f}}^{(h)}\left(1-n,\frac{a}{f}\right).$$

By using (88) in the above equation, we obtain

$$L_{\xi,q}^{(h)}(1-n,\chi) = -\frac{f^{n-1}}{n}\sum_{a=1}^{f} q^{ha}\xi^{a}\chi(a)B_{n,\xi^{f}}^{(h)}\left(\frac{a}{f},q^{f}\right) = -\frac{1}{n}B_{n,\chi,\xi}^{(h)}(q).$$

By substituting (81) into the above equation, we arrive at the desired result. □

5.2 *p-Adic (h, q)-Interpolation Function*

Let p be an odd prime. Let $v_p : \mathbb{C}_p \to \mathbb{Q} \cup \{\infty\}$ denote the p-adic valuation of $\mathbb{C}_p$ normalized so that $v_p(p) = 1$. The absolute value on $\mathbb{C}_p$ is denoted by $| \,.\, |_p$, and $| x |_p = p^{-v_p(x)}$ for $x \in \mathbb{C}_p$. The integer p^* is defined by

$$p^* = \begin{cases} p, & \text{if } p > 2, \\ 4, & \text{if } p = 2. \end{cases}$$

Let w denote the Teichmüller character, having conductor $f_w = p^*$. For an arbitrary character χ, we define $\chi_n = \chi w^{-n}$, where $n \in \mathbb{Z}$, in the sense of the product of characters. Let

$$\mathfrak{D} = \left\{ s \in \mathbb{C}_p : \mid s \mid_p \leq \mid p^* \mid^{-1} p^{-\frac{1}{p-1}} \right\}$$

and

$$\mid p^* \mid^{-1} p^{-\frac{1}{p-1}} > 1$$

(cf. [30, 37, 44, 70, 87, 91]); see also the references cited in each of these earlier works.

We recall [30,46,91] that p-adic analogue of (90) is the Kubota–Leopoldt p-adic L-function $L_p(s,\chi)$, which is unique analytic function on $\mathfrak{D}$ (except for a simple pole at $s = 1$ when $\chi \equiv 1$) for which

$$L_p(1-n,\chi) = -\frac{(1-\chi_n(p)p^{n-1})B_{n,\chi_n}}{n},$$

where $n \in \mathbb{Z}^+$ and χ_n denotes the Dirichlet character χw^{-n}.

Here, we can use some notations which are due to Washington [87], Koblitz [43], and Kim [37]. Let w denote the Teichmüller character, having conductor $f_w = p^*$. For an arbitrary character χ, we define $\chi_n = \chi w^{-n}$, where $n \in \mathbb{Z}$, in the sense of the product of characters. If $q \in \mathbb{C}_p$, then we assume that $\mid 1 - q \mid_p < p^{-\frac{1}{p-1}}$. Let $< a >= w^{-1}(a)a = a/w(a)$. We note that $< a >\equiv 1 (\bmod\, p^*\mathbb{Z}_p)$. Thus, we see that

$$< a + p^*t > = w^{-1}(a + p^*t)(a + p^*t)$$

$$= w^{-1}(a)a + w^{-1}(a)(p^*t) \equiv 1 (\bmod\, p^*\mathbb{Z}_p[t]),$$

where $t \in \mathbb{C}_p$ with$\mid t \mid_p \leq 1$, $(a,p) = 1$.

We are ready to give p-adic analogues of the twisted **two variable** q-L-function. Let F be a positive integral multiple of p^* and $f = f_\chi$.

We define

$$L_{\xi,p,q}^{(h)}(s,t,\chi) = \frac{1}{(s-1)F} \sum_{\substack{a=1 \\ (a,p)=1}}^{F} \chi(a) < a + p^*t >^{1-s} q^{ha} \xi^a$$

$$\times \sum_{k=0}^{\infty} \binom{1-s}{k} \left(\frac{F}{a+p^*t} \right)^k B_{k,\xi^F}^{(h)}(q^F), \tag{91}$$

where $\xi \in \mathbb{T}_p$ and χ is a Dirichlet character of conductor f and F be any multiple of p^* and f. Then $L^{(h)}_{\xi,p,q}(s,t,\chi)$ is analytic for $t \in \mathbb{C}_p$ with $\mid t \mid_p \leq 1$, provided $s \in \mathfrak{D}$, except $s = 1$ when $\chi \neq 1$.

In [70], we proved the following theorem:

Theorem 5.11. *Let $\xi \in \mathbb{T}_p$. Let χ be a Dirichlet character of conductor f and F be any multiple of p^* and f. Let $s \in \mathfrak{D}$. Then we have*

$$L^{(h)}_{\xi,p,q}(s,t,\chi) = \frac{1}{(s-1)F} \sum_{\substack{a=1 \\ (a,p)=1}}^{F} \chi(a) < a + p^* t >^{1-s} q^{ha} \xi^a$$

$$\times \sum_{k=0}^{\infty} \binom{1-s}{k} \left(\frac{F}{a+p^*t} \right)^k B^{(h)}_{k,\xi^F}(q^F).$$

Then $L^{(h)}_{\xi,p,q}(s,t,\chi)$ is analytic for $h \in \mathbb{Z}^+$ and $t \in \mathbb{C}_p$ with $\mid t \mid_p \leq 1$, provided $s \in \mathfrak{D}$, except $s = 1$. Also, if $t \in \mathbb{C}_p$ with $\mid t \mid_p \leq 1$, this function is analytic for $s \in \mathfrak{D}$ when $\chi \neq 1$ and meromorphic for $s \in \mathfrak{D}$, with simple pole at $s = 1$ having residue

$$\frac{\log q^h}{q^h \xi - 1} \left(\frac{1 - q^{hF} \xi^F}{1 - q^h \xi} - \frac{1 - q^{hpF}}{1 - q^{ph}} \right)$$

when $\chi = 1$. In addition, for each $n \in \mathbb{Z}^+$, we have

$$L^{(h)}_{\xi,p,q}(1-n,t,\chi) = -\frac{B^{(h)}_{n,\chi_n,\xi}(p^*t,q) - \chi_n(p) p^{n-1} B^{(h)}_{n,\chi_n,1}(p^{-1}p^*t,q^p)}{n}.$$

Proof of this theorem is the same as that of Theorem 4.10.

Remark 5.9. Observe that if $\xi = 1$, then

$$L^{(h)}_{1,p,q}(s,t,\chi) = L^{(h)}_{p,q}(s,t,\chi)$$

(cf. [37]).

$$\lim_{q \to 1} L_{p,q}(s,\chi) = L_p(s,\chi)$$

(cf. [17,21,22,30,37,40,44,70,87,87,91]).

We now give some applications related to the twisted p-adic interpolation function for the (h,q)-extension of the generalized twisted Bernoulli polynomials.

Let

$$Y = \{ q \in \mathbb{C}_p : \mid q - 1 \mid < 1 \},$$

and let $\overline{Y} = \mathbb{C}_p \setminus Y$ be the complement of the open unit disc around 1. According to Kim [35], if $q \in \overline{Y}$ and $\operatorname{ord}_p(1-q) \neq -\infty$, then

$$\mu_q(a + dp^N\mathbb{Z}_p) = \frac{q^a}{[dp^N]_q}$$

is the measure. We assume that $q \in \overline{Y}$ and $\operatorname{ord}_p(1-q) \neq -\infty$ (cf. see also [40]).

By using (80), we modify twisted p-adic interpolation function as follows:

$$L^{(h)}_{\xi,p,q}(s,\chi) = \frac{1}{s-1}\int_{\mathbb{X}^*} \chi(x)\phi_\xi(x) < x >^{-s} q^{hx} d\mu_q(x),$$

where $\xi \in \mathbb{T}_p$ and $q \in \overline{Y}$, with $\operatorname{ord}_p(1-q) \neq -\infty$, and χ is a Dirichlet character of conductor f and F is any multiple of p^* and f and $s \in \mathfrak{D}$.

Substituting $s = 1-n$, $n \in \mathbb{Z}^+$ into the above, after some calculations, we obtain

$$\begin{aligned}
&L^{(h)}_{\xi,p,q}(1-n,\chi)\\
&= -\frac{1}{n}\int_{\mathbb{X}^*} \phi_\xi(x)\chi(x) < x >^{n-1} q^{hx} d\mu_q(x)\\
&= -\frac{1}{n}\left(\int_{\mathbb{X}} \chi_n(x)\phi_\xi(x) q^{hx} x^n d\mu_q(x) - \int_{p\mathbb{X}} \chi_n(px)\phi_\xi(px) q^{phx} x^n d\mu_{q^p}(x)\right)\\
&= -\frac{1}{n}\left(B^{(h)}_{n,\chi,\xi}(q) - \chi_n(p)p^{n-1}B^{(h)}_{n,\chi,1}(q^p)\right).
\end{aligned}$$

Consequently, we arrive at the following theorem:

Theorem 5.12. *Let $\xi \in \mathbb{T}_p$ and $q \in \overline{Y}$, with $\operatorname{ord}_p(1-q) \neq -\infty$. Let χ be a Dirichlet character of conductor f and F be any multiple of p^* and f. Let $s \in \mathfrak{D}$; then we have [70]*

$$L^{(h)}_{\xi,p,q}(s,\chi) = \frac{1}{s-1}\int_{\mathbb{X}^*} \chi(x)\phi_\xi(x) < x >^{-s} q^{hx} d\mu_q(x).$$

For $n \in \mathbb{Z}^+$, we have

$$\begin{aligned}
L^{(h)}_{\xi,p,q}(1-n,\chi) &= -\frac{1}{n}\int_{\mathbb{X}^*} \phi_\xi(x)\chi(x) < x >^{n-1} q^{hx} d\mu_q(x)\\
&= -\frac{1}{n}\left(B^{(h)}_{n,\chi,\xi}(q) - \chi_n(p)p^{n-1}B^{(h)}_{n,\chi,1}(q^p)\right).
\end{aligned}$$

Remark 5.10. By using the function $L^{(h)}_{\xi,p,q}(s,t,\chi)$, Kummer congruence of the generalized (h,q)-twisted Bernoulli numbers can be obtained.

Most of the congruence relations for the Bernoulli numbers and the generalized Bernoulli numbers follow from p-adic L-function. We recall from work of Washington [87, p. 60, Corollary 5.13] that for $m, n \in \mathbb{Z}$ and $\chi \neq 1$ and $pq \nmid f$, then

$$L_p(m, \chi) \equiv L_p(n, \chi) (\bmod\, p),$$

and both members are p-integral. **Kummer's Congruences**

Suppose $m \equiv n \not\equiv 0\ (\bmod\, p - 1)$ are positive even integers. Then

$$\frac{B_m}{m} \equiv \frac{B_n}{n} (\bmod\, p)$$

(cf. [87, p. 61, Corollary 5.14]).

6 Bernoulli Functions and Arithmetic Sums

The history of the Dedekind sums can be traced back, respectively, to Julius Wilhelm Richard Dedekind (1831–1916), who did important work in abstract algebra (particularly ring theory), algebraic number theory, and the foundations of the real numbers, and Hans Adolph Rademacher (1892–1969), who also did important work in mathematical analysis and number theory. It is well known that Dedekind sums, named after Richard Dedekind, are certain sums of products of a sawtooth function. Dedekind introduced them to express the functional equation of the Dedekind eta function. They have subsequently been much studied in number theory and have occurred in some problems of topology and other branches of mathematics. Although two-dimensional Dedekind sums have been around since the nineteenth century and higher-dimensional Dedekind sums have been explored since the 1950s, it is only recently that such sums have figured prominently in so many different areas. The Dedekind sums have also many applications in some realms such as number theory, modular forms, random numbers, the Riemann–Roch theorem, and the Atiyah–Singer index theorem.

In many applications of elliptic modular functions to number theory, the eta function plays a central role. It was introduced by Dedekind in 1877 and is defined by the upper half plane

$$\mathbb{H} = \{\tau \in \mathbb{C} : \mathrm{Im}(z) > 0\}$$

by the following equation:

$$\eta(\tau) = e^{\pi i \tau/12} \prod_{m=1}^{\infty} \left(1 - e^{2\pi i m \tau}\right).$$

The infinite product has the form $\prod_{n=1}^{\infty}(1-x^n)$ where $x = e^{2\pi i\tau}$. If $\tau \in \mathbb{H}$, then $|x| < 1$, so the product converges absolutely and is nonzero. Furthermore, since the convergence is uniform on compact subsets of $\mathbb{H}$, $\eta(\tau)$ is analytic on $\mathbb{H}$. The function $\eta(\tau)$ is related to analysis, number theory, combinatorics, q-series, Weierstrass elliptic functions, modular forms, and Kronecker limit formula. The behavior of this function under the modular group $\Gamma(1)$, defined by

$$\Gamma(1) = \left\{A = \begin{bmatrix} a & b \\ c & d \end{bmatrix} : ad - bc = 1, a, b, c, d \in \mathbb{Z}\right\},$$

is given by the following functional equation:

Theorem 6.1. *Let* $A = \begin{bmatrix} a & b \\ c & d \end{bmatrix} \in \Gamma(1)$. *Then*

$$\log\eta(Az) = \log\eta(z) + \frac{\pi i(a+d)}{12c} - \pi i\left(s(d,c) - \frac{1}{4}\right) + \frac{1}{2}\log(cz+d),$$

where $s(d,c)$ *is called the Dedekind sums, which are defined by* (92).

Let h and k be coprime integers with $k > 0$; the classical Dedekind sum $s(h,k)$, which firstly arose in the transformation formula of the Dedekind eta function, is defined as follows:

$$s(h,k) = \sum_{a=1}^{k-1}\left(\left(\frac{a}{k}\right)\right)\left(\left(\frac{ha}{k}\right)\right), \tag{92}$$

where $((x))$ denotes the sawtooth function $((x)) : \mathbb{R} \to \mathbb{R}$,

$$((x)) = \begin{cases} x - [x]_G - \frac{1}{2}, & \text{if } x \in \mathbb{R}\backslash\mathbb{Z}, \\ 0, & \text{if } x \in \mathbb{Z}, \end{cases}$$

(cf. [6, 26]).

Using the theory of elliptic functions, Dedekind showed that a certain reciprocity formula holds for these sums, that is,

$$s(h,k) + s(k,h) = -\frac{1}{4} + \frac{1}{12}\left(\frac{h}{k} + \frac{k}{h} + \frac{1}{hk}\right),$$

where $(h,k) = 1$ and $h, k \in \mathbb{N}$ (cf. [6, 26]).

In 1950, Apostol [3] gave relation between Dedekind sums and Bernoulli polynomials and functions. He generalized the Dedekind sums as follows:

$$S_p(h,k) = \sum_{a \bmod k} \frac{a}{k}\overline{B}_p\left(\frac{ah}{k}\right),$$

where $(h,k) = 1$ and $h,k \in \mathbb{N}$ and $\overline{B}_p(x)$ is the pth Bernoulli function, which is defined by (43) and (44).

We consider now some arithmetic sums. We need some properties of the Euler functions, which are given below.

The *first kind* mth Euler function $\overline{E}_m(x)$ is defined as follows:

$$\overline{E}_m(x) = \frac{2(m!)}{(\pi i)^{m+1}} \sum_{n=-\infty}^{\infty} \frac{e^{(2n+1)\pi i x}}{(2n+1)^{m+1}}, \tag{93}$$

where $m \in \mathbb{N}$, $0 \leq x < 1$ (cf. [39, 69, 74, 76, 77]).

Observe that if $0 \leq x < 1$, then (1) reduces to the *first kind* nth Euler polynomials $E_n(x)$ which are defined by means of the following generating function:

$$\frac{2e^{tx}}{e^t + 1} = \sum_{n=0}^{\infty} E_n(x) \frac{t^n}{n!} \quad (|t| < \pi).$$

Observe that $E_n(0) = E_n$ denotes the first kind Euler number which is given by the following recurrence formula:

$$E_0 = 1 \quad \text{and} \quad E_n = -\sum_{k=0}^{n} \binom{n}{k} E_k. \tag{94}$$

Some of them are given by 1, $-1/2$, 0, $1/4$, ..., $E_n = 2^n E_n(1/2)$ and $E_{2n} = 0$ ($n \in \mathbb{N}$) (cf. [1, 6–92]) and see also the references cited in each of these earlier works.

From (93) it is easy to see that

$$\sum_{n=0}^{\infty} \frac{1}{(2n+1)^{2m+2}} = \frac{(-1)^{m+1} \pi^{2m+2} E_{2m+1}}{4(2m+1)!}, \tag{95}$$

(cf. [1, 6–92]) and see also the references cited in each of these earlier works.

The *second kind* Euler numbers, E_m^* are defined by means of the following generating functions:

$$\text{sech} x = \frac{1}{\cosh x} = \frac{2e^x}{e^{2x}+1} = \sum_{n=0}^{\infty} E_n^* \frac{x^n}{n!} \quad \left(|x| < \frac{\pi}{2}\right) \tag{96a}$$

(cf. [1, 6–92]) and see also the references cited in each of these earlier works. By (96a), it is easy to see that

$$E_m^* = \sum_{n=0}^{m} \binom{m}{n} 2^n E_n,$$

From the above $E_0^* = 1$, $E_1^* = 0$, $E_2^* = -1$, $E_3^* = 0$, $E_4^* = 5, \ldots,$ and $E_{2m+1}^* = 0$ (cf. [39, 69, 77]).

The first and the second kind Euler numbers are also related to $\tan z$ and $\sec z$.

$$\tan z = \frac{e^{2iz}}{2i}\left(\frac{2}{e^{2iz}+1}\right) - \frac{e^{-2iz}}{2i}\left(\frac{2}{e^{-2iz}+1}\right).$$

From the above equation, we have

$$\tan z = \sum_{j=0}^{\infty}(-1)^n 2^{2j+1}\left(\sum_{k=0}^{2j+1}\binom{2j+1}{k}E_k\right)\frac{z^{2j+1}}{(2j+1)!}.$$

By using (94), one can find that

$$\tan z = \sum_{n=0}^{\infty}(-1)^{n+1}\frac{2^{2n+1}E_{2n+1}}{(2n+1)!}z^{2n+1} \quad \left(|z| < \frac{\pi}{2}\right) \tag{97}$$

(cf. [39, 69, 77]).

$$i \tan z = \frac{e^{iz}-e^{-iz}}{e^{iz}+e^{-iz}} = 1 - \frac{2}{e^{2iz}-1} + \frac{4}{e^{4iz}-1}.$$

From the above equation, we have

$$z \tan z = \sum_{n=0}^{\infty}(-1)^n\frac{4^n(1-4^n)B_{2n}}{(2n)!}z^{2n};$$

see also (cf. [39, 69, 77]) and the references cited in each of these earlier works. By using the above, we arrive at (97):

$$\sec z = \sum_{n=0}^{\infty}(-1)^n\frac{E_{2n}^*}{(2n)!}z^{2n} \quad \left(|z| < \frac{\pi}{2}\right)$$

(cf. [39, 69, 77]) and see also the references cited in each of these earlier works.

Kim [39] and the author [69] have studied on *the Dedekind-type DC* (*Daehee-Changhee*) *sums*, which are defined as follows:

Definition 6.1 ([39]). Let h and k be coprime integers with $k > 0$. Then

$$T_m(h,k) = 2\sum_{j=1}^{k-1}(-1)^{j-1}\frac{j}{k}\overline{E}_m\left(\frac{hj}{k}\right), \tag{98}$$

where $\overline{E}_m(x)$ denotes the mth (first kind) Euler function.

We now modify (93) as follows:

$$\overline{E}_m(x) = \begin{cases} \dfrac{2(m!)}{(\pi i)^{m+1}} \displaystyle\sum_{n=0}^{\infty} \dfrac{\sin((2n+1)\pi x)}{(2n+1)^{m+1}}, & \text{if } m+1 \text{ is odd,} \\ \dfrac{2(m!)}{(\pi i)^{m+1}} \displaystyle\sum_{n=0}^{\infty} \dfrac{\cos((2n+1)\pi x)}{(2n+1)^{m+1}}, & \text{if } m+1 \text{ is even.} \end{cases} \tag{99}$$

If $m+1$ is even, then m is odd; consequently, (99) reduces to the following relation: For $y \in \mathbb{N}$ and $m = 2y-1$, we have

$$\overline{E}_{2y-1}(x) = 4(-1)^y \frac{(2y-1)!}{\pi^{2y}} \sum_{n=0}^{\infty} \frac{\cos((2n+1)\pi x)}{(2n+1)^{2y}}.$$

If $m+1$ is odd, then m is odd; hence (99) reduces to the following relation: For $m = 2y$, $y \in \mathbb{N}$, we have

$$\overline{E}_{2y}(x) = 4(-1)^y \frac{(2y)!}{\pi^{2y+1}} \sum_{n=0}^{\infty} \frac{\sin((2n+1)\pi x)}{(2n+1)^{2y+1}}.$$

Hence, from the above equation, we arrive at the following Lemma.

Lemma 6.1 ([69]). *Let $y \in \mathbb{N}$ and $0 \le x \le 1$. Then we have*

$$\overline{E}_{2y-1}(x) = \frac{(-1)^y 4(2y-1)!}{\pi^{2y}} \sum_{n=0}^{\infty} \frac{\cos((2n+1)\pi x)}{(2n+1)^{2y}} \tag{100}$$

and

$$\overline{E}_{2y}(x) = \frac{(-1)^y 4(2y)!}{\pi^{2y+1}} \sum_{n=1}^{\infty} \frac{\sin((2n+1)\pi x)}{(2n+1)^{2y+1}}. \tag{101}$$

We now modify the sum $T_m(h,k)$ for odd and even integer m:

Definition 6.2 ([69]). Let h and k be coprime integers with $k > 0$. Then

$$T_{2y-1}(h,k) = 2\sum_{j=0}^{k-1} (-1)^{j-1} \frac{j}{k} \overline{E}_{2y-1}\left(\frac{hj}{k}\right) \tag{102}$$

and

$$T_{2y}(h,k) = 2\sum_{j=0}^{k-1} (-1)^{j-1} \frac{j}{k} \overline{E}_{2y}\left(\frac{hj}{k}\right), \tag{103}$$

where $\overline{E}_{2y-1}(x)$ and $\overline{E}_{2y}(x)$ denote the Euler functions.

By substituting equation (100) into (102), we have

$$T_{2y-1}(h,k) = -\frac{8(-1)^y(2y-1)!}{k\pi^{2y}} \sum_{j=1}^{k-1}(-1)^j j \sum_{n=0}^{\infty} \frac{\cos\left(\frac{\pi hj(2n+1)}{k}\right)}{(2n+1)^{2y}}. \tag{104}$$

From the above we have

$$T_{2y-1}(h,k) = -\frac{8(-1)^y(2y-1)!}{k\pi^{2y}} \sum_{n=0}^{\infty} \frac{1}{(2n+1)^{2y}} \sum_{j=1}^{k-1}(-1)^j j \cos\left(\frac{\pi hj(2n+1)}{k}\right). \tag{105}$$

We next recall from [12, 24] that

$$\sum_{j=1}^{k-1} je^{\frac{(2n+1)\pi hij}{k}} = \begin{cases} \dfrac{k}{e^{\frac{(2n+1)\pi ih}{k}} - 1}, & \text{if } 2n+1 \not\equiv 0\,(k), \\ \dfrac{k(k-1)}{2}, & \text{if } 2n+1 \equiv 0\,(k). \end{cases}$$

From the above, it is easy to get

$$\sum_{j=1}^{k-1}(-1)^j je^{\frac{(2n+1)\pi hij}{k}} = \frac{k}{e^{\frac{(k+(2n+1)h)\pi i}{k}} - 1}.$$

By using an elementary calculations, we have

$$\sum_{j=1}^{k-1}(-1)^j j \cos\left(\frac{(2n+1)\pi hj}{k}\right) = -\frac{k}{2} \tag{106}$$

and

$$\sum_{j=1}^{k-1}(-1)^j j \sin\left(\frac{(2n+1)\pi hj}{k}\right) = \frac{k\tan\left(\frac{\pi h(2n+1)}{2k}\right)}{2}, \tag{107}$$

where $2n+1 \not\equiv 0\,(k)$. By substituting (106) into (105) and after some elementary calculations, we obtain

$$T_{2y-1}(h,k) = \frac{8(-1)^y(2y-1)!}{k\pi^{2y}} \sum_{n=0}^{\infty} \frac{1}{(2n+1)^{2y}}.$$

By substituting (95) into the above, we easily arrive at the following theorem.

Theorem 6.2. *Let $y \in \mathbb{N}$, then we have [69]*

$$T_{2y-1}(h,k) = 4E_{2y-1}.$$

By substituting equation (101) into (103), we have

$$T_{2y}(h,k) = \frac{8(-1)^y(2y)!}{k\pi^{2y+1}} \sum_{j=1}^{k-1}(-1)^j j \sum_{n=0}^{\infty} \frac{\sin\left(\frac{(2n+1)hj\pi}{k}\right)}{(2n+1)^{2y+1}}. \tag{108}$$

By substituting (107) into the above, we arrive at the following theorem.

Theorem 6.3. *Let h and k be coprime positive integers. Let $y \in \mathbb{N}$, then we have [69]*

$$T_{2y}(h,k) = \frac{4(-1)^y(2y)!}{\pi^{2y+1}} \sum_{\substack{n=0 \\ 2n+1 \not\equiv 0 (\mathrm{mod}\, k)}}^{\infty} \frac{\tan\left(\frac{h\pi(2n+1)}{2k}\right)}{(2n+1)^{2y+1}}. \tag{109}$$

The DC-sums related to many special functions (cf. [69]). In [74], Srivastava proved the following formulae which are related to Hurwitz zeta function, trigonometric functions, and Euler polynomials:

$$E_{2y-1}\left(\frac{p}{q}\right) = (-1)^y \frac{4(2y-1)!}{(2q\pi)^{2y}} \sum_{j=1}^{q} \zeta\left(2y, \frac{2j-1}{q}\right) \cos\left(\frac{\pi p(2j-1)}{q}\right),$$

where $y, q \in \mathbb{N}$, $p \in \mathbb{N}_0$; $0 \le p \le q$, and

$$E_{2y}\left(\frac{p}{q}\right) = (-1)^y \frac{4(2y)!}{(2q\pi)^{2y+1}} \sum_{j=1}^{q} \zeta\left(2y+1, \frac{2j-1}{2q}\right) \sin\left(\frac{\pi p(2j-1)}{q}\right),$$

where $y, q \in \mathbb{N}$, $p \in \mathbb{N}_0$; $0 \le p \le q$ and $\zeta(s,x)$ denotes the Hurwitz zeta function. By substituting $p = 0$ in the above, then we have

$$E_{2y-1}(0) = (-1)^y \frac{4(2y-1)!}{(2q\pi)^{2y}} \sum_{j=1}^{q} \zeta\left(2y, \frac{2j-1}{q}\right).$$

By using the above equation, we modify the sum $T_{2y-1}(h,k)$ as follows:

Corollary 6.1. *Let $y, q \in \mathbb{N}$. Then we have [69]*

$$T_{2y-1}(h,k) = (-1)^y \frac{4(2y-1)!}{(2q\pi)^{2y}} \sum_{j=1}^{q} \zeta\left(2y, \frac{2j-1}{q}\right).$$

The famous property of the all arithmetic sums is the **reciprocity law**. By using contour integration, we prove reciprocity law of (109).

The initial different proof of the following reciprocity theorem is due to Kim [39], for the sum $T_y(h,k)$.

Theorem 6.4. *Let h, k, $y \in \mathbb{N}$ with $h \equiv 1 \pmod 2$ and $k \equiv 1 \pmod 2$ and $(h,k) = 1$. Then we have [69]*

$$kh^{2y+1}T_{2y}(h,k) + hk^{2y+1}T_{2y}(k,h)$$
$$= \frac{(-1)^y \pi^{2y-1}\Gamma(2y+1)}{2\Gamma(4y+2)}E_{4y+1} + 4\pi^2(2y)!\sum_{a=0}^{y-1}\frac{E_{2a+1}E_{2y-2a-1}h^{2a+2}k^{2y-2a}}{(2a+1)!(2y-2a+1)!},$$

where $\Gamma(n+1) = n!$ and E_n denote Euler gamma function and first kind Euler numbers, respectively.

Proof. We shall give just a brief sketch as the details are similar to those in [9, Theorem 4.2], [12, Theorem 3], [25] or [26]. For the proof we use contour integration method. So we define

$$F_y(z) = \frac{\tan \pi hz \tan \pi kz}{z^{2y+1}}.$$

Let C_N be a positive oriented circle of radius R_N, with $1 \leq N < \infty$, centered at the origin. Assume that the sequence of radii R_N is increasing to ∞. R_N is chosen so that the circles always at a distance greater than some fixed positive integer number from the points $\frac{m}{2h}$ and $\frac{n}{2k}$, where m and n are integers.

Let

$$I_N = \frac{1}{2\pi i}\int_{C_N}\frac{\tan \pi hz \tan \pi kz}{z^{2y+1}}dz.$$

From the above, we get

$$I_N = \frac{1}{2\pi}\int_0^{2\pi}\frac{\tan\left(\pi hR_Ne^{i\theta}\right)\tan\left(\pi kR_Ne^{i\theta}\right)}{\left(R_Ne^{i\theta}\right)^{2y}}d\theta.$$

By C_N, if $R_N \to \infty$, then $\tan\left(R_Ne^{i\theta}\right)$ is bounded. Consequently, we easily see that

$$\lim_{N\to\infty} I_N = 0 \quad \text{as} \quad R_N \to \infty.$$

Thus, on the interior C_N, the integrand of I_N that is $F_y(z)$ has simple poles at $z_1 = \frac{2m+1}{2h}$, $-\infty < m < \infty$, and $z_2 = \frac{2n+1}{2k}$, $-\infty < n < \infty$. If we calculate the residues at the z_1 and z_2, we easily obtain, respectively, as follows:

$$-\frac{2^{2y+1}k^{2y}}{\pi(2m+1)^{2y+1}}\tan\left(\frac{(2m+1)\pi h}{2k}\right), \quad -\infty < m < \infty$$

and

$$-\frac{2^{2y+1}h^{2y}}{\pi(2n+1)^{2y+1}}\tan\left(\frac{(2n+1)\pi k}{2h}\right), \quad -\infty < n < \infty.$$

If h and k are odd integers, then $F_y(z)$ has double poles at $z_3 = \frac{2j+1}{2}$, $-\infty < j < \infty$. Thus the residue is easily found to be

$$-\frac{(2y+1)2^{2y+1}}{2(2j+1)^{4y+2}\pi^2 hk}, \quad -\infty < j < \infty.$$

The integrand of I_N has pole of order $2y+1$ at $z_4 = 0$, $y \in \mathbb{N}$. Recall the familiar Taylor expansion of $\tan z$ in (97). By straightforward calculation, we find the residues at the z_4 as follows:

$$(-1)^y (2\pi)^{2y+2} \sum_{a=1}^{y-1} \frac{E_{2a+1}E_{2y-2a-1}h^{2a+1}k^{2y-2a-1}}{(2a+1)!(2j-2a-1)!}.$$

Now we are ready to use residue theorem; hence we find that

$$I_N = -\frac{2^{2y+1}h^{2y}}{\pi} \sum_{|\frac{2m+1}{2h}|<R_N} \frac{\tan\left(\frac{(2m+1)\pi k}{2h}\right)}{(2m+1)^{2y+1}} - \frac{2^{2y+1}k^{2y}}{\pi} \sum_{|\frac{2n+1}{2k}|<R_N} \frac{\tan\left(\frac{(2n+1)\pi k}{2h}\right)}{(2n+1)^{2y+1}}$$

$$-\frac{(2y+1)2^{2y}}{\pi^2 hk} \sum_{j=-\infty}^{\infty} \frac{1}{(2j+1)^{4y+2}} + (-1)^y (2\pi)^{2y+2}$$

$$\sum_{a=0}^{y-1} \frac{E_{2a+1}E_{2y-2a-1}h^{2a+1}k^{2y-2a-1}}{(2a+1)!(2y-2a-1)!}.$$

By using (95) and letting $N \to \infty$ into the above, after straightforward calculations, we arrive at the desired result.

Remark 6.1. We also recall from [55, p. 20, Eqs. (11.2) and (11-3)] that

$$\tan z = \sum_{k=1}^{\infty} \mathfrak{T}_k \frac{z^{2k-1}}{(2k-1)!}, \tag{110}$$

where

$$\mathfrak{T}_k = (-1)^{k-1}\frac{B_{2k}}{(2k)}(2^{2k}-1)2^{2k}.$$

The integrand of I_N has pole of order $2y+1$ at $z_4 = 0$, $y \in \mathbb{N}$. Recall the familiar Taylor expansion of $\tan z$ in (110). By straightforward calculation, we find the residues at the z_4 as follows:

$$\pi^{2y}\sum_{a=0}^{y+1}\frac{\mathfrak{T}_a\mathfrak{T}_{y-a+1}}{(2a-1)!(2y-2a-1)!}h^{2a-1}k^{2y-2a+1}.$$

Thus we modify Theorem 6.4 as follows:

$$kh^{2y+1}T_{2y}(h,k)+hk^{2y+1}T_{2y}(k,h)=\frac{(-1)^y\pi^{2y-1}\Gamma(2y+1)}{2\Gamma(4y+2)}E_{4y+1}$$

$$+\frac{(-1)^y\,(2y)!}{4^y}\sum_{a=0}^{y+1}\frac{\mathfrak{T}_a\mathfrak{T}_{y-a+1}}{(2a-1)!(2y-2a-1)!}h^{2a-1}k^{2y-2a+1}.$$

We now give relation between Hurwitz zeta function, $\tan z$, and the sum $T_{2y}(h,k)$.

Hence, substituting $n = rk+j, 0 \le r \le \infty, 1 \le j \le k$ into (109), and recalling that $\tan(\pi+\alpha) = \tan\alpha$, then we have

$$T_{2y}(h,k) = \frac{4(-1)^y(2y)!}{\pi^{2y+1}}\sum_{j=1}^{k}\sum_{r=0}^{\infty}\frac{\tan\left(\frac{\pi h 2(rk+j)+1}{2k}\right)}{(2(rk+j)+1)^{2y+1}}$$

$$= \frac{4(-1)^y(2y)!}{\pi^{2y+1}(2k)^{2y+1}}\sum_{j=1}^{k}\tan\left(\frac{\pi h(2j+1)}{2k}\right)\sum_{r=0}^{\infty}\frac{1}{\left(r+\frac{2j+1}{2k}\right)^{2y+1}}.$$

Thus we arrive at the following theorem:

Theorem 6.5. *Let h and k be coprime positive integers. Let $y \in \mathbb{N}$. Then we have [69]*

$$T_{2y}(h,k) = \frac{4(-1)^y(2y)!}{(2k\pi)^{2y+1}}\sum_{j=1}^{k}\tan\left(\frac{\pi h(2j+1)}{2k}\right)\zeta\left(2y+1,\frac{2j+1}{2k}\right), \quad (111)$$

where $\zeta(s,x)$ denotes the Hurwitz zeta function.

We set

$$\frac{e^{2iz}}{1+e^{2iz}} = i\tan z + \frac{e^{-2iz}}{1+e^{-2iz}}, \quad (112)$$

where $i^2 = -1$.

Hence setting $2iz = h\pi i(2n+1)/k$, with $(h,k) = 1$, $n \in \mathbb{N}$ in (111) with (109), we obtain the following corollary:

Corollary 6.2. *Let h and k be coprime positive integers. Let $y \in \mathbb{N}$; then we have [69]*

$$T_{2y}(h,k)\frac{4i(-1)^{y+1}(2y)!}{(2k\pi)^{2y+1}} \sum_{\substack{n=1\\ 2n+1\not\equiv 0\,(\mathrm{mod}\,k)}}^{\infty} \frac{1}{(2n+1)^{2y+1}}\left(\frac{e^{\frac{h\pi i(2n+1)}{k}}}{1+e^{\frac{h\pi i(2n+1)}{k}}} - \frac{e^{-\frac{h\pi i(2n+1)}{k}}}{1+e^{-\frac{h\pi i(2n+1)}{2k}}}\right). \tag{113}$$

In (109) if h and k are odd and $y = 0$, then $T_{2y}(h,k)$ reduces to the Hardy–Berndt sum $s_5(h,k)$ which is defined by:

Let h and k be integers with $(h,k) = 1$. Then

$$s_5(h,k) = \sum_{j=1}^{k}(-1)^{j+[hj/k]_G}\left(\left(\frac{j}{k}\right)\right). \tag{114}$$

From the above, recall from [12] that we have

$$s_5(h,k) = \sum_{j=1}^{k}(-1)^{j}\frac{j}{k}(-1)^{[hj/k]_G}. \tag{115}$$

By substituting the well-known Fourier expansion (cf. [12, 24])

$$(-1)^{[x]_G} = \frac{4}{\pi}\sum_{n=0}^{\infty}\frac{\sin((2n+1)\pi x)}{2n+1}$$

into (115), we get

$$s_5(h,k) = \frac{4}{k\pi}\sum_{n=0}^{\infty}\frac{1}{2n+1}\sum_{j=1}^{k}(-1)^{j} j \sin\left(\frac{(2n+1)\pi hj}{k}\right).$$

By substituting (108) into the above, we immediately find the following result:

Lemma 6.2. *Let h and k be odd with $(h,k) = 1$. Then we have [69]*

$$T_0(h,k) = 2s_5(h,k).$$

By using Lemma 6.2 and Theorem 6.3, we arrive at the following theorem:

Theorem 6.6. *Let h and k be odd with $(h,k) = 1$. Then we have [69]*

$$\mathcal{S}_5(h,k;y) = \frac{T_{2y}(h,k)}{2}.$$

Remark 6.2. Substituting $y = 0$ into Theorem 6.6, we get $s_5(h,k) = 2\mathcal{S}_5(h,k;0)$. Consequently, the sum $T_{2y}(h,k)$ gives us generalized Hardy–Berndt sum $s_5(h,k)$.

In [71], we define the following $Y(h,k)$ sum:

$$Y(h,k) = 4ks_5(h,k),$$

where h and k are odd with $(h,k) = 1$. Thus from Lemma 6.2, we have the following corollary:

Corollary 6.3. *Let h and k be odd with $(h,k) = 1$. Then we have [66]*

$$T_0(h,k) = \frac{Y(h,k)}{2k}.$$

Observe that the sum $T_{2y}(h,k)$ also gives us generalization of the sum $Y(h,k)$.

Acknowledgements This paper was supported by the Scientific Research Project Administration Akdeniz University.

References

1. Agoh, T., Dilcher, K.: Convolution identities and lacunary recurrences for Bernoulli numbers. J. Number Theory **124**, 105–122 (2007)
2. Amice, Y.: Integration p-adique, selon A. Volkenborn, Seminaire Delange-Pisot-Poitou. Theorie des Nombres **13**(2), G4, G1–G9 (1971–1972)
3. Apostol, T.M.: Generalized Dedekind sums and transformation formulae of certain Lambert series. Duke Math. J. **17**, 147–157 (1950)
4. Apostol, T.M.: On the Lerch zeta function. Pacific J. Math. **1**, 161–167 (1951)
5. Apostol, T.M.: Theorems on generalized Dedekind sums. Pacific J. Math. **2**, 1–9 (1952)
6. Apostol, T.M.: Introduction to Analytic Number Theory. Springer, New York (1976)
7. Beck, M.: Dedekind cotangent sums. Acta Arith. **109**(2), 109–130 (2003)
8. Berndt, B.C.: On the Hurwitz zeta-function. Rocky Mountain J. Math. **2**(1), 151–157 (1972)
9. Berndt, B.C.: Dedekind sums and a paper of G. H. Hardy. J. London Math. Soc. **13**(1), 129–137 (1976)
10. Berndt, B.C.: Reciprocity theorems for Dedekind sums and generalizations. Adv. Math. **23**(3), 285–316 (1977)
11. Berndt, B.C.: Analytic Eisenstein series, theta-functions and series relations in the spirit of Ramanujan. J. Reine Angew. Math. **303/304**, 332–365 (1978)
12. Berndt, B.C., Goldberg, L.A.: Analytic properties of arithmetic sums arising in the theory of the classical theta functions. SIAM J. Math. Anal. **15**(1), 143–150 (1984)
13. Choi, J.: Some applications of the Gamma and polygamma functions involving convolutions of the Rayleigh functions, multiple Euler sums and log-sine integrals. Math. Nachr. **282**, 1709–1723 (2009)
14. Choi, J., Jang, D.S., Srivastava, H.M.: A generalization of the Hurwitz-Lerch zeta function. Integral Transforms Spec. Funct. **19**, 65–79 (2008)
15. Choi, J., Anderson, P.J., Srivastava, H.M.: Some q-extensions of the Apostol-Bernoulli and the Apostol-Euler polynomials of order n, and the multiple Hurwitz zeta function. Appl. Math. Comput. **199**(2), 723–737 (2008)

16. Chu, W., Zhou, R.R.: Convolutions of Bernoulli and Euler Polynomials. Sarajevo J. Math. **6**(18), 147–163 (2010)
17. Diamond, J.: The p-adic log gamma function and p-adic Euler constant. Trans. Am. Math. Soc. **233**, 321–337 (1977)
18. Dilcher, K.: Sums of products of Bernoulli numbers. J. Number Theory **60**, 23–41 (1996)
19. Dedekind, R.: Erlauterungen zu zwei Fragmenten von Riemann. Bernhard Riemann's Gesammelte Mathematische werke, pp. 466–472, 2nd edn. B. G. Teubner, Leipzig (1892)
20. Djordjevic, L.N., Milosevic, D.M., Milovanovic, G.V., Srivastava, H.M.: Some finite summation formulas involving multivariable hypergeometric polynomials. Integral Transform. Spec. Funct. **14**, 349–361 (2003)
21. Ferrero, B., Greenberg, R.: On the behavior of p-adic L-functions at $s = 0$. Invent. Math. **50**, 91–102 (1978)
22. Fox, G.J.: A p-adic L-function of two variables. Enseign. Math. II. Sér. **46**(3–4), 225–278 (2000)
23. Garg, M., Jain, K., Srivastava, H. M.: Some relationships between the generalized Apostol-Bernoulli polynomials and Hurwitz-Lerch Zeta functions. Integral Transforms Spec. Funct. **17**, 803–815 (2006)
24. Goldberg, L.A.: Transformations of theta-functions and analogues of Dedekind sums . Ph.D. Thesis, Vassar College, Urbana, Illinois (1975)
25. Grosswald, E.; Dedekind-Rademacher sums. Am. Math. Monthly **78**, 639–644 (1971)
26. Grosswald, E., Rademacher, H.: Dedekind Sums. Carus Monograph, vol. 16. Mathematical Association of America, Washington (1972)
27. Guillera, J., Sondow, J.: Double integrals and infinite products for some classical constants via analytic continuations of Lerch's transcendent. Ramanujan J. **16**, 247–270 (2008). arXiv:math/0506319v3 [math.NT]
28. Gupta, P.L., Gupta, R.C., Ong, S.-H., Srivastava, H.M.: A class of Hurwitz-Lerch Zeta distributions and their applications in reliability. Appl. Math. Comput. **196**, 521–531 (2008)
29. Jang, L.-C.: On a q-analogue of the p-adic generalized twisted L-functions and p-adic q-integrals. J. Korean Math. Soc. **44**(1), 1–10 (2007)
30. Iwasawa, K.: Lectures on p-adic L-functions. Princeton University Press, Princeton (1972)
31. Kac, V., Cheung, P.: Quantum Calculus. Springer, New York (2001)
32. Kawagoe, K., Wakayama, M., Yamasaki, Y.: q-Analogues of the Riemann zeta, the Dirichlet L-functions, and a crystal zeta function. Forum Math. **20**(1), 1–26 (2008)
33. Kim, T.: An analogue of Bernoulli numbers and their applications. Rep. Fac. Sci. Engrg. Saga Univ. Math. **22**, 21–26 (1994)
34. Kim, T.: On explicit formulas of p-adic q-L-functions. Kyushu J. Math. **48**(1), 73–86 (1994)
35. Kim, T.: On a q-analogue of the p-adic log gamma functions and related integrals. J. Number Theory **76**(2), 320–329 (1999)
36. Kim, T.: q-Volkenborn integration. Russ. J. Math Phys. **19**, 288–299 (2002)
37. Kim, T.: A new approach to p-adic q-L-functions. Adv. Stud. Contemp. Math. **12**(1), 61–72 (2006)
38. Kim, T.: On a p-adic interpolation function for the q-extension of the generalized Bernoulli polynomials and its derivative. Discrete Math. **309**, 1593–1602 (2008)
39. Kim, T.: Note on Dedekind type DC sums. Adv. Stud. Contemp. Math. (Kyungshang) **18**(2), 249–260 (2009)
40. Kim, M.-S., Son, J.-W.: Analytic properties of the q-Volkenborn integral on the ring of p-adic integers. Bull. Korean Math. Soc. **44**(1), 1–12 (2007)
41. Kim, T., Jang, L.-C., Rim, S.-H., Pak, H. K.: On the twisted q-zeta functions and q-Bernoulli polynomials. Far East J. Appl. Math. **13**(1), 13–21 (2003)
42. Kim, T., Rim, S.-H., Simsek, Y., Kim, D.: On the analogs of Bernoulli and Euler numbers, related identities and zeta and L-functions. J. Korean Math. Soc. **45**(2), 435–453 (2008)
43. Koblitz, N.: p-adic Numbers p-adic Analysis and Zeta Functions. Springer, New York (1977)
44. Koblitz, N.: A new proof of certain formulas for p-adic L-functions. Duke Math. J. **40**, 455–468 (1979)

45. Koblitz, N.: p-Adic Analysis: A Short Course on Recent Work. London Mathematical Society Lecture Note Series, vol. 46. Cambridge University Press, Cambridge (1980)
46. Kubota, T., Leopoldt, H.W.: Eine p-adische theorie der zetawerte I. J. Reine Angew. Math. **214/215**, 328–339 (1964)
47. Lin, S.-D., Srivastava, H. M.: Some families of the Hurwitz-Lerch Zeta functions and associated fractional derivative and other integral representations. Appl. Math. Comput. **154**, 725–733 (2004)
48. Lin, S.-D., Srivastava, H.M., Wang, P.-Y.: Some expansion formulas for a class of generalized Hurwitz-Lerch Zeta functions. Integral Transforms Spec. Funct. **17**, 817–822 (2006)
49. Luo, Q.-M., Srivastava, H.M.: Some generalizations of the Apostol-Genocchi polynomials and the Stirling numbers of the second kind. Appl. Math. Comput. **217**, 5702–5728 (2011)
50. Ozden, H.: Unification of generating function of the Bernoulli, Euler and Genocchi numbers and polynomials. In: Proceedings of the International Conference on Numerical Analysis and Applied Mathematics, American Institute of Physics Conference Proceedings, vol. 1281, pp. 1125–1128 (2010)
51. Ozden, H., Simsek, Y.: A new extension of q-Euler numbers and polynomials related to their interpolation functions. Appl. Math. Lett. **21**, 934–939 (2008)
52. Ozden, H., Simsek, Y., Srivastava, H.M.: A unified presentation of the generating functions of the generalized Bernoulli, Euler and Genocchi polynomials. Comput. Math. Appl. **60**, 2779–2787 (2010)
53. Rademacher, H.: Reciprocitatsformel a Modulfüggevenyek Elmeleteböl. Mat. Fiz. Lapok **40**, 24–34 (1933)
54. Rademacher, H.: Die reziprozitatsformel für Dedekindsche Summen. Acta Sci. Math. (Szeged) **12(B)**, 57–60 (1950)
55. Rademacher, H.: Topics in Analytic Number Theory. Die Grundlehren der mathematischen Wissenschaften, Band 169. Springer, Berlin (1973)
56. Răducanu, D., Srivastava, H.M.: A new class of analytic functions de ned by means of a convolution operator involving the Hurwitz-Lerch Zeta function. Integral Transforms Spec. Funct. **18**, 933–943 (2007)
57. Satho, J.: q-analogue of Riemann's ς-function and q-Euler numbers. J. Number Theory **31**, 346–362 (1989)
58. Schikhof, W. H.: Ultrametric Calculus: An Introduction to p-adic Analysis. Cambridge University Press, Cambridge (1984)
59. Shiratani, K., Yamamoto, S.: On a p-adic interpolation function for the Euler numbers and its derivative. Mem. Fac. Kyushu Uni. **39**, 113–125 (1985)
60. Shratani, K., Yokoyama, S.: An Application of p-adic convolutions. Mem. Fac. Sci. Kyushu Univ. Ser. A, Math. **36**(1), 73–83 (1982)
61. Simsek, Y.: On q-analogue of the twisted L-functions and q-twisted Bernoulli numbers. J. Korean Math. Soc. **40**(6), 963–975 (2003)
62. Simsek, Y.: q-analogue of the twisted l-series and q-twisted Euler numbers. J. Number Theory **110**(2), 267–278 (2005)
63. Simsek, Y.: Theorems on twisted L-functions and twisted Bernoulli numbers. Adv. Stud. Contep. Math. **11**(2), 205–218 (2005)
64. Simsek, Y.: q-Dedekind type sums related to q-zeta function and basic L-series. J. Math. Anal. Appl. **318**(1), 333–351 (2006)
65. Simsek, Y.: On p-adic twisted q-L-functions related to generalized twisted Bernoulli numbers. Russ. J. Math Phys. **13**(3), 340–348 (2006)
66. Simsek, Y.: Twisted (h,q)-Bernoulli numbers and polynomials related to twisted (h,q)-zeta function and L-function. J. Math. Anal. Appl. **324**, 790–804 (2006)
67. Simsek, Y.: On twisted q-Hurwitz zeta function and q-two-variable L-function. Appl. Math. Comput. **187**, 466–473 (2007)
68. Simsek, Y.: Generating functions of the twisted Bernoulli numbers and polynomials associated with their interpolation functions. Adv. Stud. Contep. Math. **16**(2), 251–278 (2008)

69. Simsek, Y.: Special functions related to Dedekind-type DC-sums and their applications. Russ. J. Math. Phys. **17**(4), 495–508 (2010)
70. Simsek, Y.: Twisted p-adic (h, q)-L-functions. Comput. Math. Appl. **59**, 2097–2110 (2010)
71. Simsek, Y.: On Analytic properties and character analogs of Hardy Sums. Taiwanese J. Math. **13**(1), 253–268 (2009)
72. Simsek, Y., Srivastava, H.M.: A family of p-adic twisted interpolation functions associated with the modifed Bernoulli numbers. Appl. Math. Comput. **216**, 2976–2987 (2010)
73. Srivastava, H.M.: A note on the closed-form summation of some trigonometric series. Kobe J. Math. **16**(2), 177–182 (1999)
74. Srivastava, H.M.: Some formulas for the Bernoulli and Euler polynomials at rational arguments. Math. Proc. Cambridge Philos. Soc. **129**, 77–84 (2000)
75. Srivastava, H.M.: Some generalizations and basic (or q-) extensions of the Bernoulli, Euler and Genocchi polynomials. Appl. Math. Inform. Sci. **5**, 390–444 (2011)
76. Srivastava, H.M., Choi, J.: Series Associated with the Zeta and Related Functions. Kluwer Acedemic Publishers, Dordrecht (2001)
77. Srivastava, H.M., Choi, J.: Zeta and q-Zeta Functions and Associated Series and Integrals. Elsevier Science Publishers, Amsterdam (2012)
78. Srivastava, H.M., Pinter, A.: Remarks on some relationships between the Bernoulli and Euler polynomials. Appl. Math. Lett. **17**(4), 375–380 (2004)
79. Srivastava, H.M., Kim, T., Simsek, Y.: q-Bernoulli numbers and polynomials associated with multiple q-zeta functions and basic L-series. Russ. J. Math Phys. **12**(2), 241–268 (2005)
80. Srivastava, H.M., Garg, M., Choudhary, S.: A new generalization of the Bernoulli and related polynomials. Russian J. Math. Phys. **17**, 251–261 (2010)
81. Srivastava, H.M., Saxena, R.K., Pogany, T.K., Saxena, R.: Integral and computational representations of the extended Hurwitz-Lerch Zeta function. Integral Transforms Spec. Funct. **22**, 487–506 (2011)
82. Tsumura, H.: A note on q-analogue of the Dirichlet Series and q-Bernoulli numbers. J. Number Theory **39**, 251–256 (1991)
83. Titchmarsh, E.C.: The Theory of the Riemann Zeta-Function, 2nd edn. Clarendon (Oxford University) Press, Oxford (1951) (Revised by D. R. Heath-Brown, 1986)
84. Volkenborn, A.: On generalized p-adic integration. Memoires de la S. M. F. **39–40**, 375–384 (1974)
85. Wakayama, M., Yamasaki, Y.: Integral representation of q-analogues of the Hurwitz zeta function. Monatsh. Math. **149**, 141–154 (2006)
86. Waldschmidt, M., Moussa, P., Luck, J.M., Itzykson, C.: From Number. Theory to Physics. Springer, New York (1995)
87. Washington, L.C.: Introduction to Cyclotomic Fields. Springer, New York (1982)
88. Whittaker, E.T., Watson, G.N.: A Course of Modern Analysis, 4th edn. Cambridge University Press, Cambridge (1962)
89. Woodcock, C.F.: Convolutions on the ring of p-adic integers. J. London Math. Soc. **20**(2), 101–108 (1979)
90. Yıldırım, C.Y.: Zeros of derivatives of Dirichlet L-function. Turkish J. Math. **20**(4), 521–534 (1996)
91. Young, P.T.: On the behavior of some two-variable p-adic L-function. J. Number Theory **98**, 67–86 (2003)
92. Zhao, J.: Multiple q-zeta functions and multiple q-polylogarithms. Ramanujan J. **14**, 189–221 (2007)

Combinatorial Interpretation of a Generalized Basic Series

A.K. Agarwal and M. Rana

Dedicated to Professor Hari M. Srivastava

Abstract Recently Goyal and Agarwal (ARS Combinatoria, to appear) have interpreted a generalized basic series as a generating function for a colour partition function and a weighted lattice path function. This resulted in an infinite family of combinatorial identities. Using a bijection between the Bender–Knuth matrices and the n-colour partitions established by the first author in Agarwal (ARS Combinatoria, **61**, 97–117, 2001), in this paper we extend the main result of Goyal and Agarwal to a 3-way infinite family of combinatorial identities. We illustrate by two examples that our main result has the potential of yielding many Rogers–Ramanujan–MacMahon type combinatorial identities.

1 Introduction, Definitions and the Main Result

A series involving factors like rising q-factorial $(a;q)_n$ defined by

$$(a;q)_n = \prod_{i=0}^{\infty} \frac{(1-aq^i)}{(1-aq^{n+i})}$$

is called basic series (or q-series or Eulerian series). The following two "sum-product" basic series identities are known as Rogers–Ramanujan identities:

A.K. Agarwal (✉)
Center for Advanced Study in Mathematics, Panjab University, Chandigarh 160014, India
e-mail: aka@pu.ac.in

M. Rana
School of Mathematics and Computer Applications, Thapar University, Patiala, Punjab, India
e-mail: mrana@thapar.edu

G.V. Milovanović and M.Th. Rassias (eds.), *Analytic Number Theory, Approximation Theory, and Special Functions*, DOI 10.1007/978-1-4939-0258-3_7,

$$\sum_{n=0}^{\infty} \frac{q^{n^2}}{(q;q)_n} = \prod_{n=1}^{\infty} (1-q^{5n-1})^{-1}(1-q^{5n-4})^{-1}, \tag{*}$$

$$\sum_{n=0}^{\infty} \frac{q^{n^2+n}}{(q;q)_n} = \prod_{n=1}^{\infty} (1-q^{5n-2})^{-1}(1-q^{5n-3})^{-1}. \tag{**}$$

They were first discovered by Rogers [20] and rediscovered by Ramanujan in 1913. MacMahon [19] gave the following partition theoretic interpretations of $(*)$ and $(**)$, respectively:

Theorem A. *The number of partitions of n into parts with minimal difference* 2 *equals the number of partitions of n into parts which are congruent to* $\pm 1 \pmod 5$.

Theorem B. *The number of partitions of n into parts with minimal part* 2 *and minimal difference* 2 *equals the number of partitions of n into parts which are congruent to* $\pm 2 \pmod 5$.

Partition theoretic interpretations of many more q-series identities like $(*)$ and $(**)$ have been given by several mathematicians. See, for instance, Göllnitz [13,14], Gordon [15], Connor [12], Hirschhorn [18], Agarwal and Andrews [6], Subbarao [22] and Subbarao and Agarwal [23].

In all these results ordinary partitions were used. In [7] n-colour partitions were defined. Using these partitions several more basic series identities were interpreted combinatorially (see, for instance, [1–4, 16]). Recently in [17] the basic series

$$\sum_{n=0}^{\infty} \frac{q^{n(n+k-1)}(-q;q^2)_n}{(q^4;q^4)_n},$$

where k is a positive integer, was interpreted as generating function of two different combinatorial objects, viz., an n-colour partition function and a weighted lattice path function. This led to an infinite family of combinatorial identities. Our objective here is to extend the main result of [17] by using Bender and Knuth matrices. This gives us an infinite family of 3-way identities which have the potential of yielding many Rogers–Ramanujan–MacMahon type combinatorial identities like Theorems A and B. First we recall the following definitions from [7]:

Definition 1.1. A partition with "$n+t$ copies of n" (also called an $(n+t)$-colour partition), $t \geq 0$, is a partition in which a part of size n, $n \geq 0$, can occur in $(n+t)$ different colours denoted by subscripts: $n_1, n_2, \ldots, n_{n+t}$. For example, the partitions of 2 with "$n+1$ copies of n" are

$$2_1,\ 2_1+0_1,\ 1_1+1_1,\ 1_1+1_1+0_1,$$

$$2_2,\ 2_2+0_1,\ 1_2+1_1,\ 1_2+1_1+0_1,$$

$$2_3,\ 2_3+0_1,\ 1_2+1_2,\ 1_2+1_2+0_1.$$

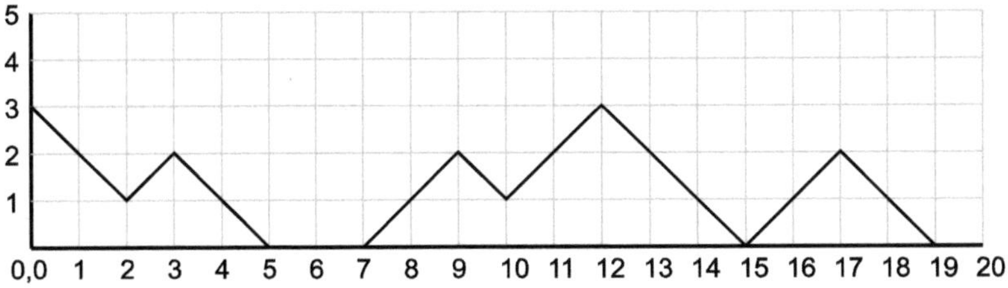

Fig. 1 Lattice paths

Note that zeros are permitted if and only if t is greater than or equal to one. Also, in no partition are zeros permitted to repeat.

Definition 1.2. The weighted difference of two parts $m_i, n_j, m \geq n$ is defined by $m - n - i - j$ and denoted by $((m_i - n_j))$.

Definition 1.3. We reproduce the following definitions of lattice paths from [8]: All paths will be of finite length lying in the first quadrant. They will begin on the y-axis and terminate on the x-axis. Only three moves are allowed at each step:

Northeast: from (i, j) to $(i + 1, j + 1)$
Southeast: from (i, j) to $(i + 1, j - 1)$, only allowed if $j > 0$
Horizontal: from $(i, 0)$ to $(i + 1, 0)$, only allowed along x-axis

All our lattice paths are either empty or terminate with a southeast step: from $(i, 1)$ to $(i + 1, 0)$.

The following terminology will be used in describing lattice paths (Fig. 1):

PEAK: Either a vertex on the y-axis which is followed by a southeast step or a vertex preceded by a northeast step and followed by a southeast step.

VALLEY: A vertex preceded by a southeast step and followed by a northeast step. Note that a southeast step followed by a horizontal step followed by a northeast step does not constitute a valley.

MOUNTAIN: A section of the path which starts on either the x-axis or y-axis, which ends on the x-axis, and which does not touch the x-axis anywhere in between the end points. Every mountain has at least one peak and may have more than one.

PLAIN: A section of the path consisting of only horizontal steps which starts either on the y-axis or at a vertex preceded by a southeast step and ends at a vertex followed by a northeast step.

The HEIGHT of a vertex is its y-coordinate. The WEIGHT of a vertex is its x-coordinate. The WEIGHT of a path is the sum of the weights of its peaks.

Example 1.1. The following path has five peaks, three valleys, three mountains and one plain.

In this example, there are two peaks of height three and three of height two, two valley of height one and one of height zero.

The weight of this path is $0 + 3 + 9 + 12 + 17 = 41$.

Definition 1.4. A plane partition π of a positive integer ν is an array of non-negative integers for which $\sum_{i,j} n_{i,j} = \nu$ and rows and columns are arranged in non-increasing order. The non-zero entries $n_{i,j}$ are called the parts of π.

$$\begin{array}{|cccc} \hline n_{1,1} & n_{1,2} & n_{1,3} & \cdots \\ n_{2,1} & n_{2,2} & n_{2,3} & \cdots \\ \vdots & \vdots & \vdots & \end{array}$$

Bender and Knuth [11] proved the following:

Theorem (Bender and Knuth). *There is a one-to-one correspondence between plane partitions of* ν*, on the other hand, and infinite matrices* $a_{i,j}$ $(i, j \geq 1)$ *of non-negative integer entries which satisfy*

$$\sum_{r\geq 1} r \left\{ \sum_{i+j=r+1} a_{i,j} \right\} = \nu,$$

on the other.

Note. For the definition and other details of the one-to-one correspondence of this theorem which is denoted by ϕ the reader is referred to [9].

Corresponding to every non-negative integer ν we shall call the matrices of the above theorem BK_ν-matrices (BK for Bender and Knuth). These are infinite matrices but will be represented in the sequel by the largest possible square matrices whose last row (column) is non-zero. Thus, for example, we will represent six BK_3-matrices by

$$(3), \begin{pmatrix} 1 & 0 \\ 1 & 0 \end{pmatrix}, \begin{pmatrix} 1 & 1 \\ 0 & 0 \end{pmatrix}, \begin{pmatrix} 0 & 0 \\ 0 & 1 \end{pmatrix}, \begin{pmatrix} 0 & 0 & 1 \\ 0 & 0 & 0 \\ 0 & 0 & 0 \end{pmatrix}, \begin{pmatrix} 0 & 0 & 0 \\ 0 & 0 & 0 \\ 1 & 0 & 0 \end{pmatrix}.$$

We give here three more definitions:

Definition 1.5. We define a matrix $E_{i,j}$ as an infinite matrix whose (i, j)th entry is 1 and the other entries are all zeros. We call $E_{i,j}$ distinct units of BK_ν-matrix.

Definition 1.6. In the set of all units the order is defined as follows: If $r+s < p+q$ then $E_{r,s} < E_{p,q}$ and if $r + s = p + q$, then $E_{r,s} < E_{p,q}$, where $r < p$. Thus, the units satisfy the order:

$$E_{1,1} < E_{1,2} < E_{2,1} < E_{1,3} < E_{2,2} < E_{3,1} < E_{1,4} < E_{2,3} < E_{3,2} < \cdots.$$

Definition 1.7. The order difference of two units $E_{p,q}, E_{r,s}$ $(p+q \geq r+s)$ is defined by $q-s-2r$ and is denoted by $\{\{E_{p,q} - E_{r,s}\}\}$.

The following result was proved in [17].

Theorem 1.1. *For a positive integer k, let $A_k(\nu)$ denote the number of n-colour partitions of ν such that*

(1.1.a) *the parts are greater than or equal to k,*
(1.1.b) *the parts are of the form $(2l-1)_1$ or $(2l)_2$, if k is an odd and of the form $(2l-1)_2$ or $(2l)_1$, if k is an even,*
(1.1.c) *if m_i is the smallest or the only part in the partition, then $m \equiv i+k-1 \pmod 4$ and*
(1.1.d) *the weighted difference between any two consecutive parts is non-negative and is $\equiv 0 \pmod 4$.*

Let $B_k(\nu)$ denote the number of lattice paths of weight ν which start at $(0,0)$, such that

(1.1.e) *they have no valley above height 0,*
(1.1.f) *there is a plain of length $\equiv k-1 \pmod 4$ in the beginning of the path; other plains, if any, are of length which are multiples of 4 and*
(1.1.g) *the height of each peak of odd (resp., even) weight is 1 (resp. 2) if k is odd and 2 (resp., 1) if k is even. Then*

$$A_k(\nu) = B_k(\nu), \quad \textit{for all } \nu, \tag{1}$$

and

$$\sum_{\nu=0}^{\infty} A_k(\nu) q^{\nu} = \sum_{\nu=0}^{\infty} B_k(\nu) q^{\nu} = \sum_{n=0}^{\infty} \frac{q^{n(n+k-1)}(-q;q^2)_n}{(q^4;q^4)_n}. \tag{2}$$

In this paper we shall prove the following result which provides a 3-way extension of Theorem 1.1:

Theorem 1.2. *For $k, \nu \geq 1$, let $C_k(\nu)$ denote the number of Bk_ν-matrices Δ such that*

(1.2.a) *if k is odd (resp. even), then even (resp. odd) columns are zero,*
(1.2.b) *all rows after the second row are zero,.*
(1.2.c) *if $E_{i,j}$ is the (i,j)th entry in Δ such that either it is the only non-zero entry or $i+j$ is minimum, then $j \equiv k \pmod 4$,*
(1.2.d) *the order difference of any two units of Δ is non-negative and is $\equiv 0 \pmod 4$,*
(1.2.e) *for odd $k > 1$, the first $(k-1)/2$ odd columns are zeroand*
(1.2.f) *for even $k > 2$, the first $(k-2)/2$ even columns are zero, then*

$$A_k(\nu) = C_k(\nu), \quad \textit{for all } k \textit{ and } \nu.$$

Note. In view of (1.2.d) the entries in $\triangle$ cannot exceed 1.

Example 1.2. $A_1(5) = 2$, since the relevant n-colour partitions are 5_1, $4_2 + 1_1$; $C_1(5) = 2$, since the relevant BK_5-matrices are

$$\begin{pmatrix} 0&0&0&0&1 \\ 0&0&0&0&0 \\ 0&0&0&0&0 \\ 0&0&0&0&0 \\ 0&0&0&0&0 \end{pmatrix}, \quad \begin{pmatrix} 1&0&0 \\ 0&0&1 \\ 0&0&0 \end{pmatrix}.$$

Example 1.3. $A_3(8) = 2$, in this case the relevant n-colour partitions are 8_2, 5_1+3_1; $C_3(8) = 2$, since the relevant BK_8-matrices in this case are

$$\begin{pmatrix} 0&0&0&0&0&0&0 \\ 0&0&0&0&0&0&1 \\ 0&0&0&0&0&0&0 \\ 0&0&0&0&0&0&0 \\ 0&0&0&0&0&0&0 \end{pmatrix}, \quad \begin{pmatrix} 0&0&1&0&1 \\ 0&0&0&0&0 \\ 0&0&0&0&0 \\ 0&0&0&0&0 \\ 0&0&0&0&0 \end{pmatrix}.$$

Remark 1.1. Theorem 1.2 extends the Identity (1) to a 3-way identity

$$A_k(\nu) = B_k(\nu) = C_k(\nu). \tag{3}$$

Using Agarwal's bijection [5] between BK_ν-matrices and n-colour partitions of ν, we shall prove Theorem 1.2 in the next section. In Sect. 3 we discuss two particular cases and obtain new combinatorial interpretations of two well-known basic series identities, viz.,

$$\sum_{n=0}^{\infty} \frac{q^{n^2}(-q;q^2)_n}{(q^4;q^4)_n} = \frac{(-q;q^2)_\infty (q^3;q^3)_\infty (q^3;q^6)_\infty}{(q^2;q^2)_\infty}, \tag{4}$$

$$\sum_{n=0}^{\infty} \frac{q^{n^2+2n}(-q;q^2)_n}{(q^4;q^4)_n} = \frac{(-q;q^2)_\infty (q^6;q^6)_\infty (q;q^6)_\infty (q^5;q^6)_\infty}{(q^2;q^2)_\infty}. \tag{5}$$

Identity (4) is due to Slater [21, p. 154, Eq. (25)] and Identity (5) was given by Andrews [10, p. 105].

Here we recall Agarwal's bijection from [5] for clarity. Let

$$\triangle = a_{1,1}E_{1,1} + a_{1,2}E_{1,2} + \cdots + a_{2,1}E_{2,1} + a_{2,2}E_{2,2} + \cdots \tag{6}$$

be BK_ν-matrices, where $a_{i,j}$ are non-negative integers which denote the multiplicities of $E_{i,j}$.

We map each unit $E_{p,q}$ of $\triangle$ to a single part m_i of an n-colour partition of ν. The mapping denoted by f is defined as

$$f : E_{p,q} \to (p+q-1)_p, \tag{7}$$

and the inverse mapping f^{-1} is easily seen to be

$$f^{-1} : m_i \to E_{i,m-i+1}. \tag{8}$$

For $\nu = 3$, this bijection is illustrated in the following table:

BK_3-matrices Δ	$f(\Delta)$
$(3) = 3E_{1,1}$	$3 1_1 = 1_1 + 1_1 + 1_1$
$\begin{pmatrix} 1 & 1 \\ 0 & 0 \end{pmatrix} = E_{1,1} + E_{1,2}$	$1_1 + 2_1$
$\begin{pmatrix} 1 & 0 \\ 1 & 0 \end{pmatrix} = E_{1,1} + E_{2,1}$	$1_1 + 2_2$
$\begin{pmatrix} 0 & 0 & 1 \\ 0 & 0 & 0 \\ 0 & 0 & 0 \end{pmatrix} = E_{1,3}$	3_1
$\begin{pmatrix} 0 & 0 \\ 0 & 1 \end{pmatrix} = E_{2,2}$	3_2
$\begin{pmatrix} 0 & 0 & 0 \\ 0 & 0 & 0 \\ 1 & 0 & 0 \end{pmatrix} = E_{3,1}$	3_3

2 Proof of Theorem 1.2

We shall prove that if $\triangle$ is a matrix enumerated by $C_k(\nu)$, then the n-colour partition $f(\triangle)$ is enumerated by $A_k(\nu)$, and conversely, if π is an n-colour partition enumerated by $A_k(\nu)$, then the BK_ν-matrix $f^{-1}(\pi)$ is enumerated by $C_k(\nu)$.

Let the matrix enumerated by $E_k(\nu)$ has representation (6). Clearly, in view of the note given after Theorem 1.2, each $a_{i,j} = 1$ or 0. Let $E_{p,q}$, $E_{r,s}(p+q \geq r+s)$ be two units of $\triangle$ which correspond to two n-colour parts m_i, n_j of $f(\triangle)$. Then $m_i = (p+q-1)_p$ and $n_j = (r+s-1)_r$ by (7). Since $(p+q \geq r+s)$, therefore $m \geq n$ and

$$((m_i - n_j)) = (p+q-1) - p - (r+s-1) - r = q - s - 2r = \{\{E_{p,q} - E_{r,s}\}\},$$

which is non-negative and $\equiv 0 \pmod 4$. This shows that (1.2.d) implies (1.1.d). Since $f(E_{i,j}) = (i+j-1)_i = m_i$ (say), by (7), so if $E_{i,j}$ is the only non-zero entry in $\triangle$ or $i + j$ is minimum, it means that in $f(\triangle)$ either m_i is the only part or the least part. Thus (1.2.c) implies (1.1.c).

Next, we see that if k is odd, then by (1.2.a) even columns in $\triangle$ are zero which means that in $E_{p,q}$, q is odd. Further since $p \leq 2$ by (1.2.b), we conclude that

$$f(E_{p,q}) = \begin{cases} q_1, & \text{if } p = 1, \\ (q+1)_2, & \text{if } p = 2. \end{cases}$$

This shows that in $f(\triangle)$ the parts are of the form $(2l-1)_1$ or $(2l)_2$. Similarly, we can show that if k is even, then in $f(\triangle)$ the parts are of the form $(2l-1)_2$ or $(2l)_1$. Thus (1.2.a) and (1.2.b) imply (1.1.b). Finally, when k is odd, say, $(2l-1)$, the first $(l-1)$ odd columns, that is, 1st, 3rd, ..., $(2l-3)$th are zero by $(1.5d)$ and since $E_{1,2l-3} = (2l-3)_1$ and $E_{2,2l-3} = (2l-2)_2$, we see that in $f(\triangle)$ the parts are $\geq k$. Thus (1.2.e) implies (1.1.a) when k is odd. Similarly, we can show that (1.2.f) implies (1.1.a) when k is even. Thus $f(\triangle)$ is enumerated by $A_k(\nu)$.

To see the reverse implication, let π be an n-colour partition of ν enumerated by $A_k(\nu)$. We shall prove that the BK_ν-matrix $f^{-1}(\pi)$ is enumerated by $C_k(\nu)$.

Let $m_i, n_j (m \geq n)$ be two parts of π such that $f^{-1}(m_i) = E_{p,q}$ and $f^{-1}(n_j) = E_{r,s}$. Then $E_{p,q} = E_{i,m-i+1}$ and $E_{r,s} = E_{j,n-j+1}$ by (8). Since $(m \geq n)$, we have $p + q = m + 1 \geq n + 1 = r + s$, and

$$\begin{aligned} \{\{E_{p,q} - E_{r,s}\}\} &= \{\{E_{i,m-i+1} - E j, n - j + 1\}\} \\ &= (m - i + 1) - (n - j + 1) - 2j \\ &= m - n - i - j \\ &= ((m_i - nj)). \end{aligned}$$

Thus (1.1.d) implies (1.2.d) since $f^{-1}(m_i) = E_{i,m-i+1} = E_{i,j}$ (say) [by (8)], so if m_i is the only part or the least part of π, it means that in $f^{-1}(\pi)$ either $E_{i,j}$ is the only non-zero entry or $i + j$ is minimum. Thus (1.1.c) implies (1.2.c).

To prove (1.2.a), (1.2.b), (1.2.e) and (1.2.f), we first consider the case when k is odd. Since $f^{-1}((2l-1)_i) = E_{1,2l-1}$ and $f^{-1}((2l)_2) = E_{2,2l-1}$, we see that in $f^{-1}(\pi)$ even columns are zero and all rows after the second row are zero. This proves (1.2.a) and (1.2.b). Furthermore, by (1.1.a) we see that in $f^{-1}((2l-1)_1) = E_{1,2l-1}$, $(2l-1) \geq k$ and in $f^{-1}((2l)_2) = E_{2,2l-1}$, $(2l) \geq k$, that is, (1.2.e) and (1.2.f) are satisfied. Similarly, we can prove the case when k is even. This completes the proof of Theorem 1.2. □

3 Identities (4) and (5) and Their Combinatorial Meanings

By a little series manipulations, Identities (4) and (5) can be written in the following forms, respectively:

$$\prod_{\substack{n=1 \\ n\equiv \pm 1, \pm 2 \pmod 6}}^{\infty} \frac{1}{1-q^n} = \left(\sum_{n=0}^{\infty} \frac{q^{n^2}(-q;q^2)}{(q^4;q^4)_n}\right)\left(\prod_{\substack{n=1 \\ n\equiv \pm 2, \pm 3, 6 \pmod{12}}}^{\infty} \frac{1}{1-q^n}\right) \tag{9}$$

and

$$\prod_{\substack{n=1 \\ n\equiv \pm 2, 3 \pmod 6}}^{\infty} \frac{1}{1-q^n} = \left(\sum_{n=0}^{\infty} \frac{q^{n^2+2n}(-q;q^2)}{(q^4;q^4)_n}\right)\left(\prod_{n=1}^{\infty} \frac{1}{1-q^{4n-2}}\right). \tag{10}$$

Now an appeal to Theorem 1.2 with $k = 1$ and $k = 3$ gives the following 4-way combinatorial interpretations of the identities (9) and (10), respectively:

Theorem 3.1. *Let $D_1(\nu)$ and $E_1(\nu)$ denote the number of partitions of ν into parts $\equiv \pm 2, \pm 3, 6 \pmod{12}$ and the number of partitions of ν into parts $\equiv \pm 1, \pm 2 \pmod 6$, respectively. Then*

$$E_1(\nu) = \sum_{k=0}^{\nu} A_1(k) D_1(\nu - k) = \sum_{k=0}^{\nu} B_1(k) D_1(\nu - k) = \sum_{k=0}^{\nu} C_1(k) D_1(\nu - k).$$

Theorem 3.2. *Let $D_3(\nu)$ denote the number of partitions of ν into parts $\equiv 2 \pmod 4$, and let $E_3(\nu)$ denote the number of partitions of ν into parts $\equiv \pm 2, 3 \pmod 6$. Then*

$$E_3(\nu) = \sum_{k=0}^{\nu} A_3(k) D_3(\nu - k) = \sum_{k=0}^{\nu} B_3(k) D_3(\nu - k) = \sum_{k=0}^{\nu} C_3(k) D_3(\nu - k).$$

Remark 3.1. Each of Theorems 3.1 and 3.2 yields six combinatorial identities in the usual sense.

Remark 3.2. A different combinatorial interpretation of Identity (10) was given by Alladi and Berkovich in [9].

4 Conclusion

The work done in this paper shows a nice interaction between the theory of basic series and combinatorics. Theorem 1.2 in conjunction with Theorem 1.1 gives a 3-way identity for each value of k. Thus we get infinitely many combinatorial identities. In particular cases, viz., $k = 1$ and $k = 3$, we get 4-way combinatorial interpretations of two well-known basic series identities of L. J. Slater and G. E. Andrews. It would be of interest if more applications of Theorems 1.1 and 1.2 are found.

Acknowledgements The first author is an emeritus scientist of the Council of Scientific and Industrial Research (CSIR), Government of India. He was supported by CSIR Research Scheme No. 21(0879)/11/EMR-II.

References

1. Agarwal, A.K.: Rogers-Ramanujan identities for n-color partitions. J. Number Theory **28**, 299–305 (1988)
2. Agarwal, A.K.: New combinatorial interpretations of two analytic identities. Proc. Am. Math. Soc. **107**(2), 561–567 (1989)
3. Agarwal, A.K.: q-functional equations and some partition identities, combinatorics and theoretical computer science (Washington, 1989). Discrete Appl. Math. **34**(1–3), 17–26 (1991)
4. Agarwal, A.K.: New classes of infinite 3-way partition identities. ARS Combinatoria **44**, 33–54 (1996)
5. Agarwal, A.K.: n- color analogues of Gaussian polynomials. ARS Combinatoria **61**, 97–117 (2001)
6. Agarwal, A.K., Andrews, G.E.: Hook differences and lattice paths. J. Statist. Plann. Inference **14**(1), 5–14 (1986)
7. Agarwal, A.K., Andrews, G.E.: Rogers-Ramanujan identities for partitions with "N copies of N". J. Combin. Theory Ser. A. **45**(1), 40–49 (1987)
8. Agarwal, A.K., Bressoud, D.M.: Lattice paths and multiple basic hypergeometric series. Pacific J. Math. **136**(2), 209–228 (1989)
9. Alladi, K., Berkovich, A.: Gollnitz-Gordon partitions with weights and parity conditions. In: Aoki, T., Kanemitsu, S., Nakahara, M., Ohno, Y. (eds.) Zeta Functions, Topology and Quantum Physics. Developments in Mathematics, vol. 14, pp. 1–17. Springer, New York (2005)
10. Andrews, G.E.: An introduction to Ramanujan's "Lost" notebook. Am. Math. Monthly **86**, 89–108 (1979)
11. Bender, E.A., Knuth, D.E.: Enumeration of plane partitions. J. Combin. Theory (A) **13**, 40–54 (1972)
12. Conner, W.G.: Partition theorems related to some identities of Rogers and Watson. Trans. Am. Math. Soc. **214**, 95–111 (1975)
13. Göllnitz, H.: Einfache partitionen (unpublished). Diplomarbeit W.S., 65 pp., Gotttingen (1960)
14. Göllnitz, H.: Partitionen mit Differenzenbedingungen. J. Reine Angew. Math. **225**, 154–190 (1967)
15. Gordon, B.: Some continued fractions of the Rogers-Ramanujan type. Duke J. Math. **32**, 741–748 (1965)
16. Goyal, M., Agarwal, A.K.: Further Rogers-Ramanujan identities for n-color partitions. Utilitas Mathematica (to appear)
17. Goyal, M., Agarwal, A.K.: On a new class of combinatorial identities. ARS Combinatoria (to appear)
18. Hirschhorm, M.D.: Some partition theorems of the Rogers-Ramanujan type. J. Combin. Theory, Ser. A **27**(1), 33–37 (1979)
19. MacMahan, P.A.: Combinatory Analysis, vol. 2. Cambridge University Press, London and Newyork (1916)
20. Rogers, L.J.: Second memoir on the expansion of certain infinite products. Proc. Lond. Math. Soc. **25**, 318–343 (1894)

21. Slater, L.J.: Further identities of the Rogers-Ramanujan type. Proc. London Math. Soc. **54**, 147–167 (1952)
22. Subbarao, M.V.: Some Rogers-Ramanujan type partition theorems. Pacific J. Math. **120**, 431–435 (1985)
23. Subbarao, M.V., Agarwal, A.K.: Further theorems of the Rogers-Ramanujan type. Canad. Math. Bull. **31**(2), 210–214 (1988)

Identities for Reciprocal Binomials

Anthony Sofo

Dedicated to Professor Hari M. Srivastava

Abstract Euler's results related to the sum of the ratios of harmonic numbers and binomial coefficients are investigated in this paper. We give a particular example involving quartic binomial coefficients.

1 Introduction and Preliminaries

In the paper [24], Sofo and Srivastava studied the expression of infinite sums of harmonic numbers in closed form. In this paper the author extends the results given in [24]; we continue with the study of representations in closed form of the sum

$$\sum_{n\geq 1} \frac{H_n}{\binom{n+k}{k}^p}; \qquad k,\ p \in \mathbb{N} := \{1, 2, 3, \ldots\}$$

in terms of zeta functions. For the harmonic numbers H_n and the generalized harmonic numbers $H_n^{(s)}$ defined by

$$H_n := H_n^{(1)} \quad \text{and} \quad H_n^{(s)} := \sum_{k=1}^{n} \frac{1}{k^s} \qquad (s \in \mathbb{C};\ n \in \mathbb{N}),$$

the following elegant formulas:

$$\sum_{n=1}^{\infty} \frac{H_n^{(1)}}{(n+1)^2} = \zeta(3) \quad \text{and} \quad \sum_{n=1}^{\infty} \frac{H_n^{(1)}}{n^3} = \frac{5}{4}\zeta(4)$$

A. Sofo (✉)
Victoria University, PO Box 14428, Melbourne City 8001, VIC, Australia
e-mail: Anthony.Sofo@vu.edu.au

G.V. Milovanović and M.Th. Rassias (eds.), *Analytic Number Theory, Approximation Theory, and Special Functions*, DOI 10.1007/978-1-4939-0258-3_8,

were discovered by Euler in relation to *Euler sums*. In terms of the Riemann zeta function, $\zeta(s)$ is defined by (see [26])

$$\zeta(s) = \begin{cases} \displaystyle\sum_{k=1}^{\infty} \frac{1}{k^s} = \frac{1}{1-2^{-s}} \sum_{k=1}^{\infty} \frac{1}{(2k-1)^s}, & \mathrm{Re}(s) > 1, \\ \displaystyle\frac{1}{1-2^{1-s}} \sum_{k=1}^{\infty} \frac{(-1)^{k-1}}{k^s}, & \mathrm{Re}(s) > 0;\ s \neq 1, \end{cases}$$

and there is also a recurrence formula

$$(2n+1)\,\zeta(2n) = 2\sum_{r=1}^{n-1} \zeta(2r)\,\zeta(2n-2r),$$

which shows that in particular, for $n = 2$, $5\zeta(4) = 2(\zeta(2))^2$ and that more generally $\zeta(2n)$ is a rational multiple of $(\zeta(n))^2$. Another elegant recursion known to Euler was

$$2\sum_{n=1}^{\infty} \frac{H_n^{(1)}}{n^q} = (q+2)\,\zeta(q+1) - \sum_{r=1}^{q-2} \zeta(r+1)\,\zeta(q-r).$$

Also in terms of the psi function,

$$H_n = \int_0^1 \frac{1-t^n}{1-t}\,dt = \sum_{r=1}^{n} \frac{1}{r} = \gamma + \psi(n+1), \quad H_n := 0, \tag{1}$$

where γ denotes the Euler–Mascheroni constant defined by

$$\gamma = \lim_{n\to\infty}\left(\sum_{r=1}^{n} \frac{1}{r} - \log(n)\right) = -\psi(1) \approx 0.5772156649\ldots,$$

and where $\psi(z)$ denotes the psi, or digamma function defined by

$$\psi(z) = \frac{d}{dz}\log\Gamma(z) = \frac{\Gamma'(z)}{\Gamma(z)} = \sum_{n=0}^{\infty}\left(\frac{1}{n+1} - \frac{1}{n+z}\right) - \gamma,$$

and the gamma function

$$\Gamma(z) = \int_0^{\infty} u^{z-1} e^{-u}\,du,$$

for $\text{Re}(z) > 0$. For a complex number a and a nonnegative integer n let $(a)_n$ denote the rising factorial defined by $(a)_0 = 1$ and

$$(a)_n = a(a+1)(a+2)\cdots(a+n-1) = \frac{\Gamma(a+n)}{\Gamma(a)}$$

for $n > 0$.

The polygamma function $\psi^{(\alpha)}(z)$ is defined as

$$\psi^{(\alpha)}(z) = \frac{d^{\alpha+1}}{dz^{\alpha+1}}[\log\Gamma(z)] = \frac{d^{\alpha}}{dz^{\alpha}}[\psi(z)], \quad z \neq \{0,-1,-2,-3,\ldots\}.$$

To evaluate $H_{z-1}^{(\alpha)}$ we have available a relation in terms of the polygamma function $\psi^{(\alpha)}(z)$, for real arguments z,

$$H_{z-1}^{(\alpha+1)} = \zeta(\alpha+1) + \frac{(-1)^{\alpha}}{\alpha!}\psi^{(\alpha)}(z). \tag{2}$$

Euler sums have been recently considered by Liu and Wang [13], and recently Dil and Kurt [10] evaluated various binomial identities involving power sums with harmonic numbers. Further work in the summation of harmonic numbers and binomial coefficients has also been done by Basu [2], Choi [5], Choi and Srivastava [8], Chu [9], and Munarini [14]. The identity

$$\sum_{n=1}^{\infty}\frac{H_n^{(1)}}{\binom{n+k}{k}} = \frac{k}{(k-1)^2}$$

for $k > 1$ is alluded to by Cloitre as reported in [25] and later proved in [19]. In [19], the author also gave the identity

$$\sum_{n=1}^{\infty}\frac{\left(\frac{1}{2}\right)^n H_n^{(1)}}{\binom{n+k}{k}}$$
$$= (-1)^{k+1}k\left[\ln^2 2 + 2\ln 2\sum_{r=1}^{k-1}\binom{k-1}{r}\frac{(-1)^r}{r} + 2\sum_{r=1}^{k-1}\binom{k-1}{r}\frac{(-1)^r(1-2^r)}{r^2}\right].$$

Similarly in [21] the author proved the identities

$$\sum_{n=1}^{\infty}\frac{H_n^{(2)}}{\binom{n+k}{k}} = \frac{k}{k-1}\left(\zeta(2) - H_{k-1}^{(2)}\right)$$

and

$$\sum_{n=1}^{\infty}\frac{(H_n)^2}{\binom{n+k}{k}} = \frac{k}{k-1}\left(\zeta(2) + \frac{2}{(k-1)^2} - H_{k-1}^{(2)}\right).$$

Specifically, we investigate closed form representations for sums of harmonic numbers and binomial coefficients; we then give a particular example involving quartic binomial coefficients. The works [1, 17–20, 22, 23] and [27], and references therein, also investigate various representations of binomial sums and zeta functions in simpler form by the use of the beta function and other techniques. Srivastava has also contributed many works on series and harmonic numbers (see, e.g., [4, 6, 7, 11, 12, 15, 16, 28, 29]).

2 Some Lemmas

The following lemmas are required for the proof of the main theorem:

Lemma 2.1. *Let k, n, and $p \in \mathbb{N}$. Then,*

$$\frac{1}{\binom{n+k}{k}^p} = \frac{(k!)^p}{(n+1)^p} \sum_{r=2}^{k} \sum_{m=1}^{p} \frac{A_{p-m}(k,r)}{(n+r)^m}, \tag{3}$$

where

$$A_{p-m}(k,r) = \frac{1}{(p-m)!} \lim_{n \to -r} \frac{d^{p-m}}{dn^{p-m}} \left\{ \frac{(n+r)^p}{\prod\limits_{r=2}^{k} (n+r)^p} \right\}. \tag{4}$$

Proof. By partial fraction expansion

$$\begin{aligned} \frac{1}{\binom{n+k}{k}^p} &= \left(\frac{k!}{n+1} \right)^p \frac{1}{(n+2)_{k+1}^p} \\ &= \left(\frac{k!}{n+1} \right)^p \frac{1}{\prod\limits_{r=2}^{k} (n+r)^p} \\ &= \frac{(k!)^p}{(n+1)^p} \sum_{r=2}^{k} \sum_{m=1}^{p} \frac{A_{p-m}(k,r)}{(n+r)^m} \end{aligned}$$

and $A_{p-m}(k,r)$ is given by (4). □

Corollary 2.1. *For $p = 4$ we have*

$$\frac{1}{\binom{n+k}{k}^4} = \frac{(k!)^4}{(n+1)^4} \sum_{r=2}^{k} \left\{ \frac{A_3(k,r)}{n+r} + \frac{A_2(k,r)}{(n+r)^2} + \frac{A_1(k,r)}{(n+r)^3} + \frac{A_0(k,r)}{(n+r)^4} \right\},$$

where

$$A_0(k,r) = \lim_{n\to -r}\left\{\frac{(n+r)^4}{\prod_{r=2}^{k}(n+r)^4}\right\} = \left[\frac{2}{k!}\binom{k}{r}\binom{r}{2}\right]^4;$$

$$A_1(k,r) = \lim_{n\to -r}\frac{d}{dn}\left\{\frac{(n+r)^4}{\prod_{r=2}^{k}(n+r)^4}\right\} = -4A_0(k,r)\,Y_-^{(1)}(k,r),$$

and $Y_\pm^{(q)}(k,r) = H_{k-r}^{(q)} \pm H_{r-2}^{(q)}$ *for* $q \in \mathbb{N},\ r \geq 2;$

$$A_2(k,r) = \frac{1}{2!}\lim_{n\to -r}\frac{d^2}{dn^2}\left\{\frac{(n+r)^4}{\prod_{r=2}^{k}(n+r)^4}\right\}$$

$$= 2A_0(k,r)\left[4\left(Y_-^{(1)}(k,r)\right)^2 + Y_+^{(2)}(k,r)\right];$$

$$A_3(k,r) = \frac{1}{3!}\lim_{n\to -r}\frac{d^3}{dn^3}\left\{\frac{(n+r)^4}{\prod_{r=2}^{k}(n+r)^4}\right\}$$

$$= -\frac{2}{3}A_0(k,r)\left[16\left(Y_-^{(1)}(k,r)\right)^3 + 12Y_-^{(1)}(k,r)\,Y_+^{(2)}(k,r) + 2Y_-^{(3)}(k,r)\right].$$

Proof. The proof follows by the evaluation of $A_{p-m}(k,r)$ for $p = 4$ and $m = 1,2,3,4$. □

Lemma 2.2. *Let a and b be positive real numbers,* $p,q \in \mathbb{N}$. *Then for* $a \neq b$

$$R(a,b) = \sum_{n\geq 1}\frac{H_n}{(n+a)(n+b)} = \frac{1}{2(a-b)}\left[H_{a-1} - H_{b-1} + H_{a-1}^{(2)} - H_{b-1}^{(2)}\right], \tag{5}$$

and for $b = a$

$$R(a,a) = \sum_{n\geq 1}\frac{H_n}{(n+a)^2} = \zeta(3) + \zeta(2)\,H_{a-1} - H_{a-1}H_{a-1}^{(2)} - H_{a-1}^{(3)}. \tag{6}$$

Further, for $p \geqslant 2$,

$$R_{p-1}(a,b) = \sum_{n\geq 1} \frac{H_n}{(n+a)^p (n+b)} = \frac{(-1)^{p-1}}{(p-1)!} \frac{\partial^{p-1}}{\partial a^{p-1}} R(a,b), \tag{7}$$

and for $p = 1$, $R_0(a,b) = R(a,b)$. Finally, for $p, q \geq 2$,

$$R_{p-1,q-1}(a,b) = \sum_{n\geq 1} \frac{H_n}{(n+a)^p (n+b)^q}$$

$$= \frac{(-1)^{p+q}}{(p-1)!\,(q-1)!} \frac{\partial^{p+q-2}}{\partial a^{p-1} \partial b^{q-1}} R(a,b), \tag{8}$$

and for $p = q = 1$, $R_{0,0}(a,b) = R(a,b)$.

Proof. From [3], for $a \neq b$

$$\sum_{n\geq 1} \frac{\psi(n)}{(n+a)(n+b)} = \frac{1}{2(a-b)} \left[\psi'(b+1) - \psi'(a+1) - \psi^2(b+1) + \psi^2(a+1)\right]$$

$$= \sum_{n\geq 1} \frac{H_n - \frac{1}{n} - \gamma}{(n+a)(n+b)}.$$

We can also evaluate

$$\sum_{n\geq 1} \frac{1}{(n+a)(n+b)} = \frac{1}{a-b}(H_a - H_b),$$

$$\sum_{n\geq 1} \frac{1}{n(n+a)(n+b)} = \frac{1}{ab(a-b)}(aH_b - bH_a).$$

We also have the property

$$H_n^{(\alpha)} = \frac{1}{n^\alpha} + H_{n-1}^{(\alpha)}.$$

Hence, using the identity of the polygamma functions (2), we attain (5). To prove (6) we begin with the identity given in [3]:

$$\sum_{n\geq 1} \frac{\psi(n)}{(n+a)^{v+2}} = \sum_{n\geq 1} \frac{H_n - \frac{1}{n} - \gamma}{(n+a)^{v+2}} \tag{9}$$

$$= \frac{(-1)^{v+1}}{(v+1)!} \left(\frac{1}{2}\psi^{(v+2)}(a+1) - \sum_{r=0}^{v} \psi^{(v-r)}(a+1)\, \psi^{(r+1)}(a+1)\right)$$

and choose $v = 0$ so that

$$\sum_{n\geq 1} \frac{H_n - \frac{1}{n} - \gamma}{(n+a)^2} = \psi(a+1)\,\psi'(a+1) - \frac{1}{2}\psi''(a+1). \tag{10}$$

Now we notice that

$$\sum_{n\geq 1} \frac{\gamma}{(n+a)^2} = \gamma\psi'(a+1),$$

$$\sum_{n\geq 1} \frac{1}{n(n+a)^2} = \frac{\gamma + \psi(a+1)}{a^2} - \frac{\psi'(a+1)}{a},$$

substituting in (10) and simplifying and using the polygamma identity (2) we obtain (6). To prove (7) we utilize (5) and differentiate with respect to a, $(p-1)$ times. Similarly to prove (8) we utilize (5) and differentiate $(p-1)$ times with respect to a and $(q-1)$ times with respect to b. □

Corollary 2.2. *From Lemma 2.2 for* $p = 4$, $a = 1$, $b = r$, $q = 1, 2, 3$, *and* m *a positive integer, let*

$$V_{\pm}^{(m)}(r-1) = H_{r-1}^{(m)} \pm H_{r-1}^{m}.$$

Then

$$R_{3,3}(1,r) = \frac{1}{3!3!}\frac{\partial^6}{\partial a^3 \partial b^3} R(a,b)\bigg|_{(a=1,b=r)} = \sum_{n\geq 1} \frac{H_n}{(n+1)^4 (n+r)^4} \tag{11}$$

$$= \frac{4\zeta(5)}{(r-1)^4} - \frac{2\zeta(2)\zeta(3)}{(r-1)^4} + \frac{\zeta(4) H_{r-1}}{(r-1)^4} + \zeta(3)\left(\frac{20}{(r-1)^6} + \frac{4H_{r-1}}{(r-1)^5} + \frac{H_{r-1}^{(2)}}{(r-1)^4}\right)$$

$$+\zeta(2)\left(\frac{10H_{r-1}}{(r-1)^6} + \frac{4H_{r-1}^{(2)}}{(r-1)^5} + \frac{H_{r-1}^{(3)}}{(r-1)^4}\right) - \frac{10\left(H_{r-1}H_{r-1}^{(2)} + H_{r-1}^{(3)} + V_{+}^{(2)}(r-1)\right)}{(r-1)^6}$$

$$-\frac{2\left(\left(H_{r-1}^{(2)}\right)^2 + 2H_{r-1}H_{r-1}^{(3)} + 3H_{r-1}^{(4)}\right)}{(r-1)^5} - \frac{H_{r-1}^{(2)}H_{r-1}^{(3)} + H_{r-1}H_{r-1}^{(4)} + 2H_{r-1}^{(5)}}{(r-1)^4},$$

$$R_{3,2}(1,r) = -\frac{1}{3!2!}\frac{\partial^5}{\partial a^3 \partial b^2} R(a,b)\bigg|_{(a=1,b=r)} \tag{12}$$

$$= \sum_{n\geq 1} \frac{H_n}{(n+1)^4 (n+r)^3} = \frac{2\zeta(5)}{(r-1)^3} - \frac{\zeta(2)\zeta(3)}{(r-1)^3}$$

$$-\frac{\zeta(4)}{2(r-1)^4}+\zeta(3)\left(\frac{10}{(r-1)^5}+\frac{H_{r-1}}{(r-1)^4}\right)+\zeta(2)\left(\frac{H_{r-1}^{(2)}}{(r-1)^4}+\frac{4H_{r-1}}{(r-1)^5}\right)$$

$$+\frac{5V_{+}^{(2)}(r-1)}{(r-1)^6}-\frac{\left(H_{r-1}^{(2)}\right)^2+2H_{r-1}H_{r-1}^{(3)}+3H_{r-1}^{(4)}}{(r-1)^4}-\frac{4\left(H_{r-1}H_{r-1}^{(2)}+H_{r-1}^{(3)}\right)}{(r-1)^4},$$

$$R_{3,1}(1,r)=\frac{1}{3!1!}\frac{\partial^4}{\partial a^3\partial b}R(a,b)\Big|_{(a=1,b=r)}=\sum_{n\geq1}\frac{H_n}{(n+1)^4(n+r)^2} \tag{13}$$

$$=\frac{2\zeta(5)}{(r-1)^2}-\frac{\zeta(2)\zeta(3)}{(r-1)^2}-\frac{\zeta(4)}{2(r-1)^3}+\frac{4\zeta(3)}{(r-1)^4}$$

$$+\frac{\zeta(2)H_{r-1}}{(r-1)^4}-\frac{H_{r-1}H_{r-1}^{(2)}+H_{r-1}^{(3)}}{(r-1)^4}+\frac{2V_{-}^{(2)}(r-1)}{(r-1)^5},$$

and

$$R_{3,0}(1,r)=-\frac{1}{3!0!}\frac{\partial^3}{\partial a^3}R(a,b)\Big|_{(a=1,b=r)}=\sum_{n\geq1}\frac{H_n}{(n+1)^4(n+r)} \tag{14}$$

$$=\frac{2\zeta(5)}{r-1}-\frac{\zeta(2)\zeta(3)}{36(r-1)}-\frac{2\zeta(4)}{4(r-1)^2} \tag{15}$$

$$+\frac{\zeta(3)}{(r-1)^3}+\frac{35\zeta(2)}{72(r-1)^4}+\frac{V_{-}^{(2)}(r-1)}{2(r-1)^4}.$$

Proof. Consider (11),

$$\sum_{n\geq1}\frac{H_n}{(n+1)^4(n+r)^4} \tag{16}$$

$$=\sum_{n\geq1}\frac{H_n}{(r-1)^4}\left[\frac{1}{(n+1)^4}-\frac{4}{(r-1)(n+1)^3}+\frac{10}{(r-1)^2(n+1)^2}\right.$$

$$\left.+\frac{1}{(n+r)^4}+\frac{4}{(r-1)(n+r)^3}+\frac{10}{(r-1)^2(n+r)^2}-\frac{20}{(r-1)^2(n+1)(n+r)}\right].$$

Now each of the sums can be listed as follows. From (9) and after some simplifications,

$$\sum_{n\geq 1}\frac{H_n}{(n+1)^4}=2\zeta(5)-\zeta(2)\zeta(3),$$

$$\sum_{n\geq 1}\frac{H_n}{(n+1)^3}=2\frac{\zeta(4)}{4},$$

$$\sum_{n\geq 1}\frac{H_n}{(n+1)^2}=2\zeta(3),$$

$$\sum_{n\geq 1}\frac{H_n}{(n+1)(n+r)}=\frac{V_+^{(2)}(r-1)}{2(r-1)},\quad r\neq 1,$$

$$\sum_{n\geq 1}\frac{H_n}{(n+r)^2}=2\zeta(3)+\zeta(2)H_{r-1}-H_{r-1}H_{r-1}^{(2)}-H_{r-1}^{(3)},$$

$$\sum_{n\geq 1}\frac{H_n}{(n+r)^3}$$

$$=\frac{\zeta(4)}{4}+\zeta(3)H_{r-1}+\zeta(2)H_{r-1}^{(2)}-\frac{\left(H_{r-1}^{(2)}\right)^2}{2}-H_{r-1}H_{r-1}^{(3)}-\frac{3H_{r-1}^{(4)}}{2}$$

and

$$\sum_{n\geq 1}\frac{H_n}{(n+r)^4}=2\zeta(5)-\zeta(2)\zeta(3)+\zeta(4)H_{r-1}+\zeta(3)H_{r-1}^{(2)}+\zeta(2)H_{r-1}^{(3)}$$

$$-H_{r-1}^{(2)}H_{r-1}^{(3)}-H_{r-1}H_{r-1}^{(4)}-2H_{r-1}^{(5)}.$$

Putting each of these sums into (16) and collecting like terms we attain the identity (11).

Similar analysis leads us to the sum identities (12), (13), and (14). □

Theorem 2.1. *Let $k,p\in\mathbb{N}$. Then for $k\geq 2$*

$$\sum_{n\geq 1}\frac{H_n}{\binom{n+k}{k}^p}=\sum_{r=2}^{k}\sum_{m=1}^{p}(k!)^p A_{p-m}(k,r)\,R_{p-1,m-1}(1,r),\tag{17}$$

and for $k = 1$

$$2\sum_{n\geq 1}\frac{H_n}{(n+1)^p} = p\zeta(p+1) - \sum_{m=1}^{p-2}\zeta(m+1)\zeta(p-m), \tag{18}$$

where $A_{p-m}(k,r)$ *is defined in* (4) *and* $R_{p-1,m-1}(1,r)$ *is defined in* (8).

Proof. Expand

$$\sum_{n\geq 1}\frac{H_n}{p} = \sum_{n\geq 1}\frac{(k!)^p\ H_n}{(n+1)^p\,(n+2)_{k+1}^p} = \sum_{n\geq 1}\frac{(k!)^p\ H_n}{(n+1)^p\prod_{r=2}^{k}(n+r)^p}$$

$$= \sum_{n\geq 1}\frac{(k!)^p\ H_n}{(n+1)^p}\sum_{r=2}^{k}\sum_{m=1}^{p}\frac{(k!)^p A_{p-m}(k,r)}{(n+r)^m},$$

where $A_{p-m}(k,r)$ is given by (4). Now interchanging the order of summation

$$\sum_{n\geq 1}\frac{H_n}{\binom{n+k}{k}^p} = \sum_{r=2}^{k}\sum_{m=1}^{p}(k!)^p A_{p-m}(k,r)\sum_{n\geq 1}\frac{H_n}{(n+1)^p(n+r)^m}$$

$$= \sum_{r=2}^{k}\sum_{m=1}^{p}(k!)^p A_{p-m}(k,r)\ R_{p-1,m-1}(1,r)$$

by Lemma 2.2. For $k = 1$ the identity (18) comes from [23]; hence, the proof is complete. □

Corollary 2.3. *Let* $p = 4$. *Then*

$$\sum_{n\geq 1}\frac{H_n}{\binom{n+k}{k}^4} = \sum_{r=2}^{k}\sum_{m=1}^{4}(k!)^4 A_{4-m}(k,r)\,R_{3,m-1}(1,r)$$

$$= \sum_{r=2}^{k}(k!)^4\left(A_3(k,r)\,R_{3,0}(1,r) + A_2(k,r)\,R_{3,1}(1,r)\right.$$

$$\left. + A_1(k,r)\,R_{3,2}(1,r) + A_0(k,r)\,R_{3,3}(1,r)\right).$$

Proof. From Theorem 2.1, we choose $p = 4$ and we have the evaluations of $A_0(k,r)$, $A_1(k,r)$, $A_2(k,r)$ and $A_3(k,r)$ from Corollary 2.1. Similarly $R_{3,0}(1,r)$, $R_{3,1}(1,r)$, $R_{3,2}(1,r)$, and $R_{3,3}(1,r)$ are evaluated in Corollary 2.2; hence, the proof is complete. □

Example 2.1. We have

$$\sum_{n\geq 1} \frac{H_n}{\binom{n+3}{3}^4} = 2916\zeta(5) - 1458\zeta(3)\zeta(2) + \frac{2835}{2}\zeta(4) + \frac{42525}{4}\zeta(3) + \frac{77355}{8}\zeta(2) - \frac{485757}{16}.$$

References

1. Alzer, H., Karaannakis D., Srivastava, H.M.: Series representations of some mathematical constants. J. Math. Anal. Appl. **320**, 145–162 (2006)
2. Basu, A.: A new method in the study of Euler sums. Ramanujan J. **16**, 7–24 (2008)
3. Brychkov, Yu.A.: Handbook of Special Functions: Derivatives, Integrals, Series and Other Formulas. Chapman and Hall/CRC Press, Boca Raton (2008)
4. Cho, Y., Jung, M., Choi, J., Srivastava, H.M.: Closed-form evaluations of definite integrals and associated infinite series involving the Riemann zeta function. Int. J. Comput. Math. **83**, 461–472 (2006)
5. Choi, J.: Certain summation formulas involving harmonic numbers and generalized harmonic numbers. Appl. Math. Comput. **218**, 734–740 (2011)
6. Choi, J., Srivastava, H.M.: Sums associated with the zeta function. J. Math. Anal. Appl. **206**, 103–120 (1997)
7. Choi, J., Srivastava, H.M.: Explicit evaluation of Euler and related sums. Ramanujan J. **10**, 51–70 (2005)
8. Choi, J., Srivastava, H.M.: Some summation formulas involving harmonic numbers and generalized harmonic numbers. Math. Comp. Modelling. **54**, 2220–2234 (2011)
9. Chu, W.: Summation formulae involving harmonic numbers. Filomat **26**, 143–152 (2012)
10. Dil, A., Kurt, V.: Polynomials related to harmonic numbers and evaluation of harmonic number series I. Integers **12**, A38 (2012)
11. Lin, S., Tu, S., Hsieh, T., Srivastava, H.M.: Some finite and infinite sums associated with the digamma and related functions. J. Fract. Calc. **22**, 103–114 (2002)
12. Lin, S., Hsieh, T., Srivastava, H.M.: Some families of multiple infinite sums associated with the digamma and related functions. J. Fract. Calc. **24**, 77–85 (2003)
13. Liu, H., Wang, W.: Harmonic number identities via hypergeometric series and Bell polynomials. Integral Transforms Spec. Funct. **23**(1), 49–68 (2012)
14. Munarini, E.: Riordan matrices and sums of harmonic numbers. Appl. Anal. Discrete Math. **5**, 176–200 (2011)
15. Petojević, A., Srivastava, H.M.: Computation of Euler's type sums of the products of Bernoulli numbers. Appl. Math. Lett. **22**, 796–801 (2009)
16. Rassias, T.M., Srivastava, H.M.: Some classes of infinite series associated with the Riemann zeta and polygamma functions and generalized harmonic numbers. Appl. Math. Comput. **131**, 593–605 (2002)
17. Sofo, A.: Computational Techniques for the Summation of Series. Kluwer Academic/Plenum Publishers, New York (2003)
18. Sofo, A.: Integral forms of sums associated with Harmonic numbers. Appl. Math. Comput. **207**, 365–372 (2009)
19. Sofo, A.: Harmonic numbers and double binomial coefficients. Integral Transforms Spec. Funct. **20**(11), 847–857 (2009)
20. Sofo, A.: Sums of derivatives of binomial coefficients. Adv. Appl. Math. **42**, 123–134 (2009)
21. Sofo, A.: Harmonic sums and integral representations. J. Appl. Anal. **16**, 265–277 (2010)

22. Sofo, A.: Summation formula involving harmonic numbers. Anal. Math. **37**(1), 51–64 (2011)
23. Sofo, A., Cvijovic, D.: Extensions of Euler harmonic sums. Appl. Anal. Discrete Math. **6**, 317–328 (2012)
24. Sofo, A., Srivastava, H.M.: Identities for the harmonic numbers and binomial coefficients. Ramanujan J. **25**, 93–113 (2011)
25. Sondow, J., Weisstein, E.W.: Harmonic number. From MathWorld-A Wolfram Web Resources. http://mathworld.wolfram.com/HarmonicNumber.html
26. Srivastava, H.M., Choi, J.: Series Associated with the Zeta and Related Functions. Kluwer Academic Publishers, London (2001)
27. Wei, C., Gong, D., Wang, Q.: Chu-Vandermonde convolution and harmonic number identities. Integral Transforms and Spec. Funct. (2012). doi: 10.1080/10652469.2012.689762
28. Wu, T., Leu, S., Tu, S., Srivastava, H.M.: A certain class of infinite sums associated with digamma functions. Appl. Math. Comput. **105**, 1–9 (1999)
29. Wu, T., Tu, S., Srivastava, H.M.: Some combinatorial series identities associated with the digamma function and harmonic numbers. Appl. Math. Lett. **13**, 101–106 (2000)

A Note on q-Stirling Numbers

Mircea Merca

Dedicated to Professor Hari M. Srivastava

Abstract The q-Stirling numbers of both kinds are specializations of the complete or elementary symmetric functions. In this note, we use this fact to prove that the q-Stirling numbers can be expressed in terms of the q-binomial coefficients and vice versa.

1 Introduction

We start with the q-analogue of the classical binomial coefficients which are called the q-binomial coefficients and are defined by

$$\binom{n}{k}_q = \begin{cases} \dfrac{[n]_q!}{[k]_q![n-k]_q!}, & \text{for } k \in \{0,\ldots,n\}, \\ 0, & \text{otherwise,} \end{cases}$$

where

$$[n]_q! = [n]_q[n-1]_q \cdots [1]_q,$$

is q-factorial, with $[0]_q! = 1$ and

$$[n]_q = 1 + q + \cdots + q^{n-1} = \frac{1-q^n}{1-q}, \quad q \neq 1$$

is q-number.

M. Merca (✉)
Department of Mathematics, University of Craiova, A. I. Cuza 13, Craiova, 200585 Romania
e-mail: mircea.merca@profinfo.edu.ro

G.V. Milovanović and M.Th. Rassias (eds.), *Analytic Number Theory, Approximation Theory, and Special Functions*, DOI 10.1007/978-1-4939-0258-3_9,

The q-Stirling numbers of the first kind $s(n,k)_q$ and the second kind $S(n,k)_q$ are a natural extensions of the classical Stirling numbers [3]. These are the coefficients in the expansions

$$(x)_{n,q} = \sum_{k=0}^{n} s(n,k)_q x^k$$

and

$$x^n = \sum_{k=0}^{n} S(n,k)_q (x)_{k,q},$$

where

$$(x)_{n,q} = \prod_{k=0}^{n-1} (x - [k]_q),$$

with $(x)_{0,q} = 1$.

In this paper, we prove:

Theorem 1.1. *Let k and n be two positive integers. Then*

1. $s(n+1, n+1-k)_q = (1-q)^{-k} \sum_{i=0}^{k} (-1)^{k-i} q^{\binom{i+1}{2}} \binom{n-i}{k-i} \binom{n}{i}_q;$

2. $\binom{n}{k}_q = q^{-\binom{k+1}{2}} \sum_{i=0}^{k} (1-q)^i \binom{n-i}{k-i} s(n+1, n+1-i)_q;$

3. $S(n+1+k, n+1)_q = (1-q)^{-k} \sum_{i=0}^{k} (-q)^i \binom{n+k}{k-i} \binom{n+i}{i}_q;$

4. $\binom{n+k}{k}_q = q^{-k} \sum_{i=0}^{k} (q-1)^i \binom{n+k}{k-i} S(n+1+i, n+1)_q.$

2 Proof of Theorem 1.1

In order to indicate that

$$\lambda = [\lambda_1, \lambda_2, \ldots, \lambda_r] \qquad \text{or} \qquad \lambda = [1^{t_1} 2^{t_2} \ldots n^{t_n}]$$

is an integer partition [1] of n, i.e.,

$$n = \lambda_1 + \lambda_2 + \cdots + \lambda_r \qquad \text{and} \qquad \lambda_1 \geqslant \lambda_2 \geqslant \cdots \geqslant \lambda_r > 0$$

or

$$n = t_1 + 2t_2 + \cdots + nt_n,$$

we use the notation $\lambda \vdash n$. We denote by $l(\lambda)$ the number of parts of λ, i.e.,

$$l(\lambda) = r \qquad \text{or} \qquad l(\lambda) = t_1 + t_2 + \cdots + t_n.$$

For each partition $\lambda \vdash k$, the Schur function s_λ in n variables can be defined as the ratio of two $n \times n$ determinants as follows [4, I.3]:

$$s_\lambda(x_1, x_2, \ldots, x_n) = \frac{\det\left(x_i^{\lambda_j+n-j}\right)_{1\leqslant i,j\leqslant n}}{\det\left(x_i^{n-j}\right)_{1\leqslant i,j\leqslant n}},$$

where we consider that $\lambda_j = 0$ for $j > l(\lambda)$.

To prove the theorem we use the following Schur function formula [4, p. 47]:

$$s_\lambda(x_1+1, \ldots, x_n+1) = \sum_{\mu\subseteq\lambda} d_{\lambda\mu} s_\mu(x_1, \ldots, x_n), \tag{1}$$

where

$$d_{\lambda\mu} = \det\binom{\lambda_i + n - i}{\mu_j + n - j}_{1\leqslant i,j\leqslant n}.$$

If $\lambda = [1^k]$, it is well known that s_λ is the kth elementary symmetric function e_k,

$$e_k(x_1, x_2, \ldots, x_n) = \sum_{1\leqslant i_1<i_2<\cdots<i_k\leqslant n} x_{i_1} x_{i_2} \ldots x_{i_k}.$$

This implies

$$e_k(x_1+1, \ldots, x_n+1) = \sum_{m=0}^{k} d_{[1^k][1^m]} e_m(x_1, \ldots, x_n). \tag{2}$$

According to [2, Theorem 1], it is an easy exercise to show that

$$d_{[1^k][1^m]} = \binom{n-m}{k-m}.$$

Taking into account that the q-Stirling numbers of the first kind and the q-binomial coefficients are specializations of the elementary symmetric functions, i.e.,

$$s(n,n-k)_q = (-1)^k e_k\left([1]_q,[2]_q,\ldots,[n-1]_q\right)$$

and

$$\binom{n}{k}_q = q^{-\binom{k}{2}} e_k(1,q,\ldots,q^{n-1}),$$

we can write

$$\begin{aligned} s(n+1,n+1-k)_q &= (-1)^k e_k\left(\frac{1-q}{1-q},\frac{1-q^2}{1-q},\ldots,\frac{1-q^n}{1-q}\right) \\ &= \left(-\frac{1}{1-q}\right)^k \sum_{i=0}^{k}\binom{n-i}{k-i} e_i(-q,-q^2,\ldots,-q^n) \\ &= \left(-\frac{1}{1-q}\right)^k \sum_{i=0}^{k}(-q)^i\binom{n-i}{k-i} e_i(1,q,\ldots,q^{n-1}) \\ &= (1-q)^{-k}\sum_{i=0}^{k}(-1)^{k-i} q^{\binom{i+1}{2}}\binom{n-i}{k-i}\binom{n}{i}_q \end{aligned}$$

and the first identity is proved.

It is immediate from the relation (2) that

$$e_k(x_1,\ldots,x_n) = \sum_{m=0}^{k}(-1)^{k-m}\binom{n-m}{k-m} e_m(1+x_1,\ldots,1+x_n).$$

Now

$$\begin{aligned} \binom{n}{k}_q &= q^{-\binom{k}{2}}(-q)^{-k} e_k(-q,-q^2,\ldots,-q^n) \\ &= q^{-\binom{k}{2}}(-q)^{-k}\sum_{i=0}^{k}(-1)^{k-i}\binom{n-i}{k-i} e_i(1-q,\ldots,1-q^n) \\ &= q^{-\binom{k+1}{2}}\sum_{i=0}^{k}(-1)^i(1-q)^i\binom{n-i}{k-i} e_i\left(\frac{1-q}{1-q},\ldots,\frac{1-q^n}{1-q}\right) \\ &= q^{-\binom{k+1}{2}}\sum_{i=0}^{k}(1-q)^i\binom{n-i}{k-i} s(n+1,n+1-i)_q \end{aligned}$$

and the second identity is proved.

On the other hand, for $\lambda = [k]$ it is well known that s_λ is the kth complete symmetric function h_k,

$$h_k(x_1, x_2, \ldots, x_n) = \sum_{1 \leqslant i_1 \leqslant i_2 \leqslant \ldots \leqslant i_k \leqslant n} x_{i_1} x_{i_2} \ldots x_{i_k}.$$

By (1), we obtain

$$h_k(x_1 + 1, \ldots, x_n + 1) = \sum_{m=0}^{k} d_{[k][m]} h_m(x_1, \ldots, x_n), \tag{3}$$

where

$$d_{[k][m]} = \binom{n-1+k}{k-m}.$$

Taking into account that

$$\binom{n+k}{k}_q = h_k(1, q, \ldots, q^n)$$

and

$$S(n+k, n)_q = h_k[1]_q, [2]_q, \ldots, [n]_q,$$

we have

$$\begin{aligned} S(n+1+k, n+1)_q &= h_k\left(\frac{1-q}{1-q}, \frac{1-q^2}{1-q}, \ldots, \frac{1-q^{n+1}}{1-q}\right) \\ &= (1-q)^{-k} \sum_{i=0}^{k} \binom{n-k}{k-i} h_i(-q, -q^2, \ldots, -q^{n+1}) \\ &= (1-q)^{-k} \sum_{i=0}^{k} (-q)^i \binom{n-k}{k-i} \binom{n+i}{n}_q. \end{aligned}$$

The third identity is proved.

By (3), we deduce that

$$h_k(x_1, \ldots, x_n) = \sum_{m=0}^{k} (-1)^k \binom{n-1+k}{k-m} h_m(1 + x_1, \ldots, 1 + x_n).$$

We have

$$\begin{aligned}
\binom{n+k}{n}_q &= (-q)^{-k} h_k(-q, -q^2, \ldots, -q^{n+1}) \\
&= (-q)^{-k} \sum_{i=0}^{k} (-1)^{k-i} \binom{n+k}{k-i} h_i(1-q, \ldots, 1-q^{n+1}) \\
&= q^{-k} \sum_{i=0}^{k} (-1)^i (1-q)^i \binom{n+k}{k-i} h_i\left(\frac{1-q}{1-q}, \ldots, \frac{1-q^{n+1}}{1-q}\right) \\
&= q^{-k} \sum_{i=0}^{k} (q-1)^i \binom{n+k}{k-i} S(n+1+i, n+1)_q
\end{aligned}$$

and the last identity is proved.

Acknowledgements The author expresses his gratitude to Oana Merca for the careful reading of the manuscript and helpful remarks.

References

1. Andrews, G.E.: The Theory of Partitions. Addison-Wesley Publishing, Reading (1976)
2. Call, G.S., Velleman, D.J.: Pascal's matrices. Amer. Math. Monthly **100**(4), 372–376 (1993)
3. Ernst, T.: q-Stirling numbers, an umbral approach. Adv. Dyn. Syst. Appl. **3**(2), 251–282 (2008)
4. Macdonald, I.G.: Symmetric Functions and Hall Polynomials, 2nd edn. Clarendon, Oxford (1995)

Part II
Analytic Inequalities and Applications

A Survey on Cauchy–Bunyakovsky–Schwarz Inequality for Power Series

Alawiah Ibrahim and Silvestru Sever Dragomir

Dedicated to Professor Hari M. Srivastava

Abstract In this paper, we present a survey of some recent results for the celebrated Cauchy–Bunyakovsky–Schwarz inequality for functions defined by power series with nonnegative coefficients. Particular examples for fundamental functions of interest are presented. Applications for some special functions are given as well.

1 Introduction

The *Cauchy–Bunyakovsky–Schwarz inequality*, or for short the CBS inequality, is also known in the literature as the *Cauchy's*, the *Schwarz's*, or the *Cauchy–Schwarz's inequality*. It plays an important role in different branches of modern mathematics such as Hilbert space theory, probability and statistics, classical real and complex analysis, numerical analysis, qualitative theory of differential equations and their applications.

It is well known that the classical CBS inequality has been generalized, refined and applied by a remarkably large number of researchers for different and various motivations. For the detail, see particularly the survey paper [9], the relevant

A. Ibrahim (✉)
School of Engineering and Science, Victoria University, PO Box 14428,
Melbourne City, MC 8001, Australia
e-mail: alawiah.ibrahim@live.vu.edu.au

S.S. Dragomir
School of Engineering and Science, Victoria University, PO Box 14428,
Melbourne City, MC 8001, Australia

School of Computational and Applied Mathematics, University of The Witwatersrand,
Private Bag 3, Johannesburg 2050, South Africa
e-mail: sever.dragomir@vu.edu.au

G.V. Milovanović and M.Th. Rassias (eds.), *Analytic Number Theory, Approximation Theory, and Special Functions*, DOI 10.1007/978-1-4939-0258-3_10,

chapters in the books [10], [12, Chap. 2], [11, Chap. 1] and the numerous references which are cited therein.

The main aim of this paper is to survey some recent results obtained by the authors, that is, to identify the inequalities for power series that are related to the CBS inequality. Utilizing the classical results that have been available in the literature such as the de Bruijn inequality and the Buzano and Schwarz results, we provide some refinements and improvement of the CBS inequality for functions defined by power series with nonnegative coefficients. Particular examples that are related to some fundamental complex functions such as exponential, logarithm, trigonometric and hyperbolic functions are presented. Some applications for special functions such as polylogarithm, hypergeometric, Bessel and modified Bessel functions of the first kind are presented as well.

This paper contains seven sections including Sect. 1 as above. Section 2 provides the basic inequalities of the CBS type in real and complex numbers and in inner product spaces. The corresponding version of the CBS inequality for functions defined by power series is also given. Some results related to the celebrated CBS type such as the de Bruijn inequality and the Buzano and Schwarz results are mentioned as a foundation for the next sections. These are followed by the recent results on the CBS inequality for functions defined by power series with nonnegative coefficients on utilizing the de Bruijn, the Buzano, and the Schwarz results and by making use of the different techniques based on the continuity of the modulus (see Sects. 3–6, respectively). The inequalities in Sects. 3–6 can be applied for some fundamental complex functions and for the special functions such as polylogarithm, hypergeometric, Bessel and modified Bessel functions as well, which are given in Sect. 7.

2 The CBS Type Inequalities

2.1 CBS Inequality for Real and Complex Numbers

In the following, we state the *Cauchy–Bunyakovsky–Schwarz's* (*CBS*) inequality for real numbers which is also known in the literature as the Cauchy's inequality [16, p. 83] (see also [10, p. 1]).

Theorem 2.1. *If* $a = (a_1, a_2, \ldots, a_n)$ *and* $b = (b_1, b_2, \ldots, b_n)$ *are sequences of real numbers, then*

$$\left(\sum_{k=1}^{n} a_k b_k\right)^2 \leq \sum_{k=1}^{n} a_k^2 \sum_{k=1}^{n} b_k^2 \tag{1}$$

with equality if and only if the sequences a *and* b *are proportional, i.e., there is a real number* $r \in \mathbb{R}$ *such that* $a_k = r b_k$ *for each* $k \in \{1, 2, \ldots, n\}$.

The following version of the CBS inequality for complex numbers also holds.

Theorem 2.2. *If $a = (a_1, a_2, \ldots, a_n)$ and $b = (b_1, b_2, \ldots, b_n)$ are sequences of complex numbers, then*

$$\left|\sum_{k=1}^{n} a_k b_k\right|^2 \leq \sum_{k=1}^{n} |a_k|^2 \sum_{k=1}^{n} |b_k|^2 \tag{2}$$

with equality if and only if the sequences a and $\overline{b}$ are proportional, i.e., there is a complex number $c \in \mathbb{C}$ such that $a_k = c\overline{b}_k$ for any $k \in \{1, 2, \ldots, n\}$.

Remark 2.1. The inequality

$$\left|\sum_{k=1}^{n} p_k a_k b_k\right|^2 \leq \sum_{k=1}^{n} p_k |a_k|^2 \sum_{k=1}^{n} p_k |b_k|^2, \tag{3}$$

where $p_k \geq 0$, while $a_k, b_k \in \mathbb{C}, k \in \{1, 2, \ldots, n\}$ is called the weighted version of the CBS inequality (2).

Remark 2.2. By the CBS inequality for real numbers (1) and the generalized triangle inequality for complex numbers, i.e.,

$$\left|\sum_{k=1}^{n} z_k\right| \leq \sum_{k=1}^{n} |z_k|, \quad z_k \in \mathbb{C},\ k \in \{1, 2, \ldots, n\}, \tag{4}$$

we also have

$$\left|\sum_{k=1}^{n} a_k b_k\right|^2 \leq \left(\sum_{k=1}^{n} |a_k b_k|\right)^2 \leq \sum_{k=1}^{n} |a_k|^2 \sum_{k=1}^{n} |b_k|^2. \tag{5}$$

2.2 De Bruijn Inequality

There are a large number of refinements of the CBS inequalities (1) and (2) in the literature (see [2, 8, 17] and the reference cited therein). For instance, in 1960, de Bruijn ([5], [16, p. 89], [10, p. 48]) established the following refinement of the classical CBS inequality.

Theorem 2.3 ([5]). *If $b = (b_1, b_2, \ldots, b_n)$ is a sequence of real numbers and $z = (z_1, z_2, \ldots, z_n)$ is a sequence of complex numbers, then*

$$\left|\sum_{k=1}^{n} b_k z_k\right|^2 \leq \frac{1}{2}\sum_{k=1}^{n} b_k^2 \left(\sum_{k=1}^{n} |z_k|^2 + \left|\sum_{k=1}^{n} z_k^2\right|\right)$$

$$\left(\leq \sum_{k=1}^{n} b_k^2 \sum_{k=1}^{n} |z_k|^2\right). \tag{6}$$

Equality holds in (6) if and only if for $k \in \{1, 2, \ldots, n\}$, $b_k = \operatorname{Re}(\lambda z_k)$, *where* λ *is a complex number such that the quantity* $\lambda^2 \sum_{k=1}^{n} z_k^2$ *is nonnegative real number.*

The proof of Theorem 2.3 can also be found in the book of Mitrinovic et al. [16, p. 89] and Dragomir [10, p. 48].

Remark 2.3. The weighted version of the de Bruijn inequality also holds, namely

$$\left|\sum_{k=1}^{n} p_k b_k z_k\right|^2 \leq \frac{1}{2} \sum_{k=1}^{n} p_k b_k^2 \left(\sum_{k=1}^{n} p_k |z_k|^2 + \left|\sum_{k=1}^{n} p_k z_k^2\right|\right), \tag{7}$$

where $p_k \geq 0$, $b_k \in \mathbb{R}$, $z_k \in \mathbb{C}$, $k \in \{1, 2, \ldots, n\}$.

2.3 A Generalization for Power Series

The following result holds [7].

Theorem 2.4. *Let* $F : (-r, r) \to \mathbb{R}$, $F(x) = \sum_{k=0}^{\infty} \alpha_k x^k$ *with* $\alpha_k \geq 0$, $k \in \mathbb{N}$. *If* $a = (a_1, a_2, \ldots, a_n)$ *and* $b = (b_1, b_2, \ldots, b_n)$ *are sequences of real numbers such that*

$$a_j b_j, a_j^2, b_j^2 \in (-r, r) \text{ for any } j \in \{1, 2, \ldots, n\},$$

then one has the inequality

$$\left(\sum_{j=1}^{n} F(a_j b_j)\right)^2 \leq \sum_{j=1}^{n} F\left(a_j^2\right) \sum_{j=1}^{n} F\left(b_j^2\right). \tag{8}$$

Particular inequalities of (8) for some fundamental functions hold [7] and are given as follows:

1. If a and b are sequences of real numbers, then one has the inequality

$$\left(\sum_{k=1}^{n} \exp(a_k b_k)\right)^2 \leq \sum_{k=1}^{n} \exp\left(a_k^2\right) \sum_{k=1}^{n} \exp\left(b_k^2\right),$$

$$\left(\sum_{k=1}^{n} \sinh(a_k b_k)\right)^2 \leq \sum_{k=1}^{n} \sinh\left(a_k^2\right) \sum_{k=1}^{n} \sinh\left(b_k^2\right),$$

$$\left(\sum_{k=1}^{n}\cosh\left(a_k b_k\right)\right)^2 \le \sum_{k=1}^{n}\cosh\left(a_k^2\right)\sum_{k=1}^{n}\cosh\left(b_k^2\right).$$

2. If a and b are such that $a_k b_k \in (-1,1)\,, k \in \{1,2,\ldots,n\}$, then one has the inequality

$$\left(\sum_{k=1}^{n}\tan\left(a_k b_k\right)\right)^2 \le \sum_{k=1}^{n}\tan\left(a_k^2\right)\sum_{k=1}^{n}\tan\left(b_k^2\right),$$

$$\left(\sum_{k=1}^{n}\arcsin\left(a_k b_k\right)\right)^2 \le \sum_{k=1}^{n}\arcsin\left(a_k^2\right)\sum_{k=1}^{n}\arcsin\left(b_k^2\right),$$

$$\left\{\sum_{k=1}^{n}\frac{1}{\left(1-a_k b_k\right)^m}\right\}^2 \le \sum_{k=1}^{n}\frac{1}{\left(1-a_k^2\right)^m}\sum_{k=1}^{n}\frac{1}{\left(1-b_k^2\right)^m}, \qquad m>0.$$

2.4 CBS Inequality in Inner Product Spaces and the Related Results

Let H be a linear space over the real or complex number field $\mathbb{K}$. The functional $\langle\cdot,\cdot\rangle : H \times H \to \mathbb{K}$ is called an *inner product* on H if it satisfies the following conditions:

(i) $\langle x,x\rangle \ge 0$ for any $x \in H$ and $\langle x,x\rangle = 0$ if and only if $x=0$.
(ii) $\langle \alpha x+\beta y,z\rangle = \alpha\langle x,z\rangle + \beta\langle y,z\rangle$ for any $\alpha,\beta \in \mathbb{K}$ and $x,y,z \in H$.
(iii) $\langle y,x\rangle = \overline{\langle x,y\rangle}$ for any $x,y \in H$.

If we denote $\|x\| := \sqrt{\langle x,x\rangle}$, $x \in H$, then one may state the following properties:

(a) $\|x\| \ge 0$ for any $x \in H$ and $\|x\| = 0$ if and only if $x=0$.
(b) $\|\alpha x\| = |\alpha|\,\|x\|$ for any $\alpha \in \mathbb{K}$ and $x \in H$.
(c) $\|x+y\| \le \|x\| + \|y\|$ for any $x,y \in H$ (the triangle inequality).

That is, $\|\cdot\|$ is a *norm* on H.

A fundamental consequence of the above properties (a)–(c) is called the *Schwarz inequality*, that is,

$$|\langle x,y\rangle|^2 \le \langle x,x\rangle\langle y,y\rangle \tag{9}$$

for any $x,y \in H$. The equality holds in (9) if and only if the vectors x and y are *linearly dependent*, i.e., there exists a nonzero constant $\alpha \in \mathbb{K}$ so that $x=\alpha y$.

In [3], Buzano obtained the following extension of the celebrated Schwarz's inequality (9) for a real or complex inner product space $(H;\langle\cdot,\cdot\rangle)$.

Theorem 2.5. *Let $a, b, x \in H$. Then*

$$|\langle a, x\rangle \langle x, b\rangle| \leq \frac{1}{2}\left[\|a\| \cdot \|b\| + |\langle a, b\rangle|\right] \|x\|^2, \tag{10}$$

with equality holds if and only if there exists a scalar $\lambda \in \mathbb{K}$ ($\mathbb{R}$ or $\mathbb{C}$) such that $x = \lambda a$.

It is clear that for $a = b$, the above inequality (10) becomes the standard Schwarz inequality (9).

Further, in 1985, Dragomir [6] has obtained the following refinement of the Schwarz inequality in inner product space $(H; \langle\cdot,\cdot\rangle)$ over the real and complex number field $\mathbb{K}$.

Theorem 2.6. *For any $x, y \in H$ and $e \in H$ with $\|e\| = 1$, the following refinement of the Schwarz inequality holds:*

$$\|x\| \, \|y\| \geq |\langle x, y\rangle - \langle x, e\rangle \langle e, y\rangle| + |\langle x, e\rangle \langle e, y\rangle| \geq |\langle x, y\rangle|. \tag{11}$$

Remark 2.4. If in the first inequality of (11) we choose $e = \frac{z}{\|z\|}$, $z \in H \setminus \{0\}$, then we get

$$\|x\| \, \|y\| \, \|z\|^2 - |\langle x, z\rangle \langle z, y\rangle| \geq \left|\langle x, y\rangle \|z\|^2 - \langle x, z\rangle \langle z, y\rangle\right| \tag{12}$$

for any $x, y, z \in H$.

2.5 CBS Inequality for Power Series

Let

$$f(z) = \sum_{n=0}^{\infty} a_n z^n \tag{13}$$

be a power series convergent on the disk $D(0, R)$, $R > 0$. If the coefficients a_n in (13) are complex numbers and applying the well-known CBS inequality (2), then we can deduce that

$$|f(z)|^2 = \left|\sum_{n=0}^{\infty} a_n z^n\right|^2 \leq \sum_{n=0}^{\infty} |a_n|^2 \sum_{n=0}^{\infty} |z|^{2n} = \frac{1}{1 - |z|^2} \sum_{n=0}^{\infty} |a_n|^2 \tag{14}$$

for any $z \in D(0, R) \cap D(0, 1)$.

If we assume that the coefficients in the representation function (13) are nonnegative and utilizing the weighted version of the *CBS* inequality (3), then we can state that

$$|f(zw)|^2 = \left|\sum_{n=0}^{\infty} a_n z^n w^n\right|^2 \leq \sum_{n=0}^{\infty} a_n |z|^{2n} \sum_{n=0}^{\infty} a_n |w|^{2n} = f\left(|z|^2\right) f\left(|w|^2\right) \quad (15)$$

for any $z, w \in \mathbb{C}$ with $zw, |z|^2, |w|^2 \in D(0, R)$.

In the power series (13) with real coefficients a_n, we can naturally construct another power series which will have as coefficients the absolute values of the coefficient of the original series, namely

$$f_A(z) = \sum_{n=0}^{\infty} |a_n| z^n, \quad z \in D(0,1), \quad (16)$$

where $a_n = |a_n| \operatorname{sgn}(a_n)$, $n \in \{0, 1, 2, \ldots\}$ with $\operatorname{sgn}(x)$ is the real signum function defined to be 1 if $x > 0$, -1 if $x < 0$ and 0 if $x = 0$. It is obvious that the new power series $f_A(z)$ in (16) have the same radius of convergence as the original power series $f(z)$ in (13).

3 Applications of the De Bruijn Inequality for Power Series

On utilizing the de Bruijn inequality (6), Cerone and Dragomir [4] established some inequalities for power series (13) with nonnegative coefficients as follows:

Theorem 3.1 ([4]). *Let $f(z) = \sum_{n=0}^{\infty} a_n z^n$ be an analytic function defined by a power series with nonnegative coefficients a_n, $n \geq 0$, and convergent on the open disk $D(0, R) \subset \mathbb{C}$, $R > 0$. If a is a real number and z a complex number such that $az, a^2, z^2, |z|^2 \in D(0, R)$, then*

$$|f(az)|^2 \leq \frac{1}{2} f\left(a^2\right)\left[f\left(|z|^2\right) + \left|f(z^2)\right|\right]. \quad (17)$$

Proof. First of all, notice that by the de Bruijn inequality (6) for the choice $b_k = \sqrt{a_k} c_k$, $z_k = \sqrt{a_k} w_k$ with $a_k \geq 0$, $c_k \in \mathbb{R}$, and $w_k \in \mathbb{C}$, $k \in \{0, 1, \ldots, n\}$, we can state the weighted inequality:

$$\left|\sum_{k=1}^{n} a_k c_k w_k\right|^2 \leq \frac{1}{2} \sum_{k=1}^{n} a_k c_k^2 \left(\sum_{k=1}^{n} a_k |w_k|^2 + \left|\sum_{k=1}^{n} a_k w_k^2\right|\right). \quad (18)$$

Now, on making use of (18) for the partial sums of the function $f(z) = \sum_{k=0}^{n} a_k z^k$ with nonnegative coefficients a_k, we are able to state that

$$\left|\sum_{k=1}^{n} a_k a^k z^k\right|^2 \leq \frac{1}{2} \sum_{k=1}^{n} a_k a^{2k} \left(\sum_{k=1}^{n} a_k |z|^{2k} + \left|\sum_{k=1}^{n} a_k z^{2k}\right|\right) \quad (19)$$

for any $n \geq 0$, $a \in \mathbb{R}$, $z \in \mathbb{C}$ with $az, a^2, z^2, |z|^2 \in D(0, R)$.

Taking the limit as $n \to \infty$ in (19) and noticing that all the involved series in (19) are convergent on $D(0,R)$, we deduce the desired inequality (17). □

The inequality (17) is a valuable source of particular inequalities for real numbers a and complex numbers z as will be outlined in the following.

1. If in (17) we choose the fundamental power series $f(z) = \frac{1}{1-z} = \sum_{n=0}^{\infty} z^n$, $z \in D(0,1)$, then we can state that

$$\left(1 - |z|^2 + \left|1 - z^2\right|\right) |1 - az|^2 \geq 2\left(1 - a^2\right)\left(1 - |z|^2\right)\left|1 - z^2\right| \tag{20}$$

for any $a \in (-1,1)$ and $z \in D(0,1)$.
2. If in (17) we choose the function $f(z) = \exp(z) = \sum_{n=0}^{\infty} \frac{1}{n!} z^n$, $z \in \mathbb{C}$, then we have

$$|\exp(az)|^2 \leq \frac{1}{2} \exp\left(a^2\right) \left[\exp\left(|z|^2\right) + \left|\exp(z^2)\right|\right] \tag{21}$$

for any $a \in \mathbb{R}$ and $z \in \mathbb{C}$. In particular, the choice $z = i$ in (21) generates the following simple and interesting result:

$$|\exp(ia)|^2 \leq \frac{e^2 + 1}{2e} \exp\left(a^2\right) \tag{22}$$

for any $a \in \mathbb{R}$.
3. Now, if we choose the power series $f(z) = -\ln(1-z) = \sum_{n=0}^{\infty} \frac{1}{n} z^n$, $z \in D(0,1)$ and apply the inequality (17), then we get the following inequality for logarithms:

$$|\ln(1 - az)|^2 \leq \frac{1}{2} \ln\left(\frac{1}{1 - a^2}\right) \left[\ln\left(\frac{1}{1 - |z|^2}\right) + \left|\ln\left(1 - z^2\right)\right|\right] \tag{23}$$

for any $a \in (-1,1)$ and $z \in D(0,1)$. Moreover, if in (23) we choose $z = \pm ib$ with $b \in (-1,1)$, then we obtain the simpler result:

$$|\ln(1 \pm iab)|^2 \leq \frac{1}{2} \ln\left(\frac{1}{1 - a^2}\right) \ln\left(\frac{1 + b^2}{1 - b^2}\right) \tag{24}$$

for any $a, b \in (-1,1)$.
4. Further, if we utilize the following function as power series representations with nonnegative coefficients:

$$\cosh(z) = \sum_{n=0}^{\infty} \frac{1}{(2n)!} z^{2n}, \qquad z \in \mathbb{C},$$
$$\sinh(z) = \sum_{n=0}^{\infty} \frac{1}{(2n+1)!} z^{2n+1}, \qquad z \in \mathbb{C},$$
$$\frac{1}{2}\ln\left(\frac{1+z}{1-z}\right) = \sum_{n=1}^{\infty} \frac{1}{2n-1} z^{2n-1}, \qquad z \in D(0,1),$$
$$\sin^{-1}(z) = \sum_{n=0}^{\infty} \frac{\Gamma\left(n+\frac{1}{2}\right)}{\sqrt{\pi}\,(2n+1)\,n!} z^{2n+1}, \; z \in D(0,1),$$
$$\tanh^{-1}(z) = \sum_{n=1}^{\infty} \frac{1}{2n-1} z^{2n-1}, \qquad z \in D(0,1),$$

where Γ is the *gamma function*, then we can state the following inequalities:

$$|\cosh(az)|^2 \le \frac{1}{2}\cosh\left(a^2\right)\left[\cosh\left(|z|^2\right) + \left|\cosh(z^2)\right|\right],$$

$$|\sinh(az)|^2 \le \frac{1}{2}\sinh\left(a^2\right)\left[\sinh\left(|z|^2\right) + \left|\sinh(z^2)\right|\right]$$

for all $a \in \mathbb{R}$, $z \in \mathbb{C}$ and

$$\left|\ln\left(\frac{1+az}{1-az}\right)\right|^2 \le \frac{1}{2}\ln\left(\frac{1+a^2}{1-a^2}\right)\left[\ln\left(\frac{1+|z|^2}{1-|z|^2}\right) + \left|\ln\left(\frac{1+z^2}{1-z^2}\right)\right|\right],$$

$$\left|\sin^{-1}(az)\right|^2 \le \frac{1}{2}\sin^{-1}\left(a^2\right)\left[\sin^{-1}\left(|z|^2\right) + \left|\sin^{-1}(z^2)\right|\right],$$

$$\left|\tanh^{-1}(az)\right|^2 \le \frac{1}{2}\tanh^{-1}\left(a^2\right)\left[\tanh^{-1}\left(|z|^2\right) + \left|\tanh^{-1}(z^2)\right|\right],$$

for $a \in (-1,1)$, $z \in D(0,1)$.

Cerone and Dragomir [4] have also proved an analogous inequality of (17) for functions defined by the power series with real coefficients.

Theorem 3.2. *Let $f(z) = \sum_{n=0}^{\infty} a_n z^n$ be a function defined by power series with real coefficients and convergent on the open disk $D(0,R) \subset \mathbb{C}$, $R > 0$. If $a \in \mathbb{R}$ and $z \in \mathbb{C}$ are such that $az, a^2, z^2, |z|^2 \in D(0,R)$, then*

$$|f(az)|^2 \le \frac{1}{2} f_A\left(a^2\right)\left[f_A\left(|z|^2\right) + \left|f_A(z^2)\right|\right]. \tag{25}$$

Proof. Again, utilizing the de Bruijn inequality with positive weights (18), we have

$$\begin{aligned} |f(az)|^2 &= \left|\sum_{n=0}^{\infty} |a_n| \operatorname{sign}(a_n) a^n z^n\right| \\ &\le \frac{1}{2}\sum_{n=0}^{\infty} |a_n| [\operatorname{sign}(a_n)]^2 a^{2n} \left[\sum_{n=0}^{\infty} |a_n| |z|^{2n} + \left|\sum_{n=0}^{\infty} |a_n| z^{2n}\right|\right] \\ &= \frac{1}{2} f_A(a^2) \left[f_A\left(|z|^2\right) + |f_A(z^2)|\right] \end{aligned} \tag{26}$$

for any $a \in \mathbb{R}$, $z \in \mathbb{C}$ with $az, a^2, z^2, |z|^2 \in D(0, R)$. □

In the following examples, we exemplify how the above inequality (25) may be used to establish some inequalities for real and complex numbers:

1. If we take the function

$$f(z) = \sin(z) = \sum_{n=0}^{\infty} \frac{(-1)^n}{(2n+1)!} z^{2n+1}, \quad z \in \mathbb{C},$$

then

$$f_A(z) = \sum_{n=0}^{\infty} \frac{1}{(2n+1)!} z^{2n+1} = \sinh(z) = \frac{1}{2}(e^z - e^{-z})$$

for $z \in \mathbb{C}$. Applying the inequality (25) for this function will produce the result

$$|\sin(az)|^2 \le \frac{1}{2}\sinh(a^2)\left[\sinh\left(|z|^2\right) + |\sinh(z^2)|\right] \tag{27}$$

for any $a \in \mathbb{R}$ and $z \in \mathbb{C}$. Now, if in (27) we choose $z = ib$ with $b \in \mathbb{R}$, then we obtain the inequality

$$|\sin(iab)|^2 \le \sinh\left(a^2\right)\sinh\left(b^2\right) \tag{28}$$

for any $a, b \in \mathbb{R}$.

2. The function

$$f(z) = \cos(z) = \sum_{n=0}^{\infty} \frac{(-1)^n}{(2n)!} z^{2n}, \quad z \in \mathbb{C},$$

has the transform

$$f_A(z) = \cosh(z) = \frac{1}{2}(e^z + e^{-z})$$

for $z \in \mathbb{C}$. Utilizing the inequality (25) for $f(z)$ as above gives

$$|\cos(az)|^2 \leq \frac{1}{2}\cosh\left(a^2\right)\left[\cosh\left(|z|^2\right) + \left|\cosh(z^2)\right|\right] \tag{29}$$

for any $a \in \mathbb{R}$ and $z \in \mathbb{C}$. In particular, we have from (29) with $z = ib$, $b \in \mathbb{R}$,

$$|\cos(iab)|^2 \leq \cosh\left(a^2\right)\cosh\left(b^2\right)$$

for each $a, b \in \mathbb{R}$.

The following result is also obtained in [4], which shows a connection between two power series, one having positive coefficients.

Theorem 3.3. *Let* $f(z) = \sum_{n=0}^{\infty} a_n z^n$ *and* $g(z) = \sum_{n=0}^{\infty} b_n z^n$ *be two power series with* $a_n > 0$ *and* $b_n \in \mathbb{R}$, $n \geq 0$. *If* f *is convergent on* $D(0, R_1)$, g *is convergent on* $D(0, R_2)$ *and the numerical series* $\sum_{n=0}^{\infty} b_n^2/a_n$ *is convergent, then we have the inequality*

$$|g(z)|^2 \leq \frac{1}{2}\sum_{n=0}^{\infty} b_n^2/a_n \left[f\left(|z|^2\right) + |f(z^2)|\right] \tag{30}$$

for any $z \in \mathbb{C}$ *with* $z \in D(0, R_2)$ *and* $a, |z|^2 \in D(0, R_1)$.

Proof. Utilizing the de Bruijn weighted inequality (18) we can state that

$$\begin{aligned} |g(z)|^2 &= \left|\sum_{n=0}^{\infty} \frac{b_n}{a_n} a_n z^n\right|^2 \\ &\leq \frac{1}{2}\sum_{n=0}^{\infty} a_n \left(\frac{b_n}{a_n}\right)^2 \left[\sum_{n=0}^{\infty} a_n |z|^{2n} + \left|\sum_{n=0}^{\infty} a_n z^{2n}\right|\right] \\ &= \frac{1}{2}\sum_{n=0}^{\infty} \frac{b_n^2}{a_n}\left[f\left(|z|^2\right) + |f(z^2)|\right] \end{aligned} \tag{31}$$

for any $z \in \mathbb{C}$ with $z, z^2, |z|^2 \in D(0, R_1) \cap D(0, R_2)$. □

Remark 3.1. The above inequality (30) is useful in comparing different functions for which a bound for the numerical series $\sum_{n=0}^{\infty} b_n^2/a_n$ is known.

The following corollaries hold [4].

Corollary 3.1. *Let* $g(z) = \sum_{n=0}^{\infty} b_n z^n$ *be a power series with real coefficients and convergent on* $D(0, R)$. *If the numerical series* $\sum_{n=0}^{\infty} b_n^2$ *is convergent, then*

$$|g(z)|^2 \le \frac{1}{2}\sum_{n=0}^{\infty} b_n^2 \cdot \frac{1-|z|^2+\left|1-z^2\right|}{\left(1-|z|^2\right)|1-z^2|}, \tag{32}$$

for any $z \in \mathbb{C}$ *with* $z \in D(0,R)$ *and* $z, |z|^2 \in (0,1)$.

Corollary 3.2. *Let* $g(z)$ *be as in Corollary* 3.1. *If the numerical series* $\sum_{n=0}^{\infty}\left(n!b_n^2\right)$ *is convergent, then*

$$|g(z)|^2 \le \frac{1}{2}\sum_{n=0}^{\infty}\left(n!b_n^2\right)\left[\exp\left(|z|^2\right)+\left|\exp(z^2)\right|\right], \tag{33}$$

for any $z \in D(0,R)$.

If we consider the series expansion

$$\frac{1}{z}\ln\left(\frac{1}{1-z}\right)=\sum_{n=0}^{\infty}\frac{z^n}{n+1}, \quad z \in D(0,1)\setminus\{0\},$$

then utilizing the inequality (32) for the choice $b_n = 1/(n+1)$ and taking into account that

$$\sum_{n=0}^{\infty}\frac{1}{(1+n)^2}=\sum_{n=1}^{\infty}\frac{1}{n^2}=\zeta(2)=\frac{\pi^2}{6}, \tag{34}$$

where ζ is the Riemann zeta function, we can state the following inequality:

$$|\ln(1-z)|^2 \le \frac{\pi^2}{12}|z|^2\left(\frac{1-|z|^2+\left|1-z^2\right|}{\left(1-|z|^2\right)|1-z^2|}\right) \tag{35}$$

for any $z \in D(0,1)$.

4 Power Series Inequality via the Buzano Result

S. S. Dragomir has observed that from [5], on utilizing the Buzano inequality (10) in the complex inner product space $(H;\langle\cdot,\cdot\rangle)$, one can obtain the discrete inequality

$$\left|\sum_{j=1}^{n} p_j c_j \overline{x}_j \sum_{j=1}^{n} p_j x_j \overline{b}_j\right|$$
$$\leq \frac{1}{2}\left[\left(\sum_{j=1}^{n} p_j \left|c_j\right|^2 \sum_{j=1}^{n} p_j \left|b_j\right|^2\right)^{1/2} + \left|\sum_{j=1}^{n} p_j c_j \overline{b}_j\right|\right] \sum_{j=1}^{n} p_j \left|x_j\right|^2, \quad (36)$$

where $p_j \geq 0$, $x_j, b_j, c_j \in \mathbb{C}$, $j \in \{1, \ldots, n\}$. If we take in (36) $b_j = \overline{c}_j$, for $j \in \{1, 2, \ldots, n\}$, then we get

$$\left|\sum_{j=1}^{n} p_j c_j \overline{x}_j \sum_{j=1}^{n} p_j c_j x_j\right| \leq \frac{1}{2}\left[\sum_{j=1}^{n} p_j \left|c_j\right|^2 + \left|\sum_{j=1}^{n} p_j c_j^2\right|\right] \sum_{j=1}^{n} p_j \left|x_j\right|^2, \quad (37)$$

for any $p_j \geq 0, x_j, c_j \in \mathbb{C}, j \in \{1, 2, \ldots, n\}$.

As pointed out in [4], if $x_j, j \in \{1, 2, \ldots, n\}$ are real numbers, then (36) generates the de Bruijn refinement of the celebrated weighted CBS inequality

$$\left|\sum_{j=1}^{n} p_j x_j z_j\right|^2 \leq \frac{1}{2} \sum_{j=1}^{n} p_j x_j^2 \left[\sum_{j=1}^{n} p_j \left|z_j\right|^2 + \left|\sum_{j=1}^{n} p_j z_j^2\right|\right], \quad (38)$$

where $p_j \geq 0, x_j \in \mathbb{R}, z_j \in \mathbb{C}, j \in \{1, 2, \ldots, n\}$.

The following result has been obtained in [13] by Ibrahim and Dragomir.

Theorem 4.1. *Let $f(z) = \sum_{n=0}^{\infty} a_n z^n$ be a power series with nonnegative coefficients a_n and convergent in the open disk $D(0, R)$. If $x, \alpha, \beta \in \mathbb{C}$ so that $\alpha\overline{x}$, $\overline{\beta}x$, $|\alpha|^2$, β^2, $a\overline{\beta}$, $|x|^2 \in D(0, R)$, then*

$$\left|f(\alpha\overline{x}) f\left(\overline{\beta}x\right)\right| \leq \frac{1}{2}\left[\left[f\left(|\alpha|^2\right) f\left(|\beta|^2\right)\right]^{1/2} + \left|f\left(\alpha\overline{\beta}\right)\right|\right] f\left(|x|^2\right). \quad (39)$$

Proof. On utilizing the inequality (36), for the choices $p_n = a_n, c_n = \alpha^n, x_n = x^n$, $b_n = \beta^n, n \geq 0$, we have

$$\left|\sum_{n=0}^{m} a_n \alpha^n (\overline{x})^n \sum_{n=0}^{m} a_n \left(\overline{\beta}\right)^n x^n\right|$$
$$\leq \frac{1}{2}\left[\left(\sum_{n=0}^{m} a_n |\alpha|^{2n} \sum_{n=0}^{m} a_n |\beta|^{2n}\right)^{1/2} + \left|\sum_{n=0}^{m} a_n \alpha^n (\overline{\beta})^n\right|\right] \sum_{n=0}^{m} a_n |x|^{2n}, \quad (40)$$

for any $m \geq 0$.

Since $\alpha\overline{x}$, $\overline{\beta}x$, $|\alpha|^2$, $|\beta|^2$, $\alpha\overline{\beta}$, $|x|^2$ belong to the convergence disk $D(0, R)$, hence the series in (40) are convergent and letting $m \to \infty$, we deduce the desired inequality (39). □

A particular case of interest is as follows:

Corollary 4.1. *Let $f(z)$ be as in Theorem 4.1 and $z, x \in \mathbb{C}$ with $z\overline{x}$, zx, $|z|^2$, z^2, $|x|^2 \in D(0, R)$. Then*

$$|f(z\overline{x}) f(zx)| \leq \frac{1}{2}\left[f\left(|z|^2\right) + \left|f\left(z^2\right)\right|\right] f\left(|x|^2\right). \tag{41}$$

This follows from (39) by choosing $\alpha = z$, $\beta = \overline{z}$.

Remark 4.1. In particular, if $x = a \in \mathbb{R}$, then from (41) we deduce the inequality (17) [4].

The above result (39) has some natural applications for particular complex functions of interest as follows:

1. If we apply the inequality (39) for $f(z) = \frac{1}{1-z}$, $z \in D(0, 1)$, then we get

$$\left|\frac{1}{1-\alpha\overline{x}} \cdot \frac{1}{1-\overline{\beta}x}\right| \leq \frac{1}{2}\left[\left(\frac{1}{1-|\alpha|^2} \cdot \frac{1}{1-|\beta|^2}\right)^{1/2} + \left|\frac{1}{1-\alpha\overline{\beta}}\right|\right] \frac{1}{1-|x|^2}, \tag{42}$$

for any $x, \alpha, \beta \in D(0, 1)$. This is equivalent with

$$\begin{aligned} &2\left(1-|x|^2\right)\left|1-\alpha\overline{\beta}\right|\sqrt{\left(1-|\alpha|^2\right)(1-|\beta|)^2} \\ &\leq |1-\alpha\overline{x}|\left|1-\overline{\beta}x\right|\left[\left|1-\alpha\overline{\beta}\right| + \sqrt{\left(1-|\alpha|^2\right)(1-|\beta|)^2}\right], \end{aligned} \tag{43}$$

for $x, \alpha, \beta \in D(0, 1)$. In particular, if $\beta = \overline{\alpha}$, then we get from (43) that

$$\begin{aligned} &2(1-|x|)\left(1-|\alpha|^2\right)\left|1-\alpha^2\right| \\ &\leq |1-\alpha\overline{x}|\,|1-\alpha x|\left[\left|1-\alpha^2\right| + 1 - |\alpha|^2\right], \end{aligned} \tag{44}$$

for any $x, \alpha \in D(0, 1)$.

2. If we apply (39) for $f(z) = \exp(z)$, $z \in \mathbb{C}$, then we get the inequality

$$\begin{aligned} &\left|\exp\left(\alpha\overline{x} + \overline{\beta}x\right)\right| \\ &\leq \frac{1}{2}\left[\left(\exp\left(|\alpha|^2 + |\beta|^2\right)\right)^{1/2} + \left|\exp\left(\alpha\overline{\beta}\right)\right|\right] \exp\left(|x|^2\right), \end{aligned} \tag{45}$$

for any $\alpha, \beta, x \in \mathbb{C}$. In particular, if $\alpha = \overline{\beta}$, then we get from (45) that

$$|\exp(2\alpha \operatorname{Re}(x))| \leq \frac{1}{2}\left[\left(\exp(2|\alpha|^2)\right)^{1/2} + \left|\exp\left(\alpha^2\right)\right|\right] \exp\left(|x|^2\right), \tag{46}$$

for any $\alpha, x \in \mathbb{C}$.

3. If we apply (39) for the Koebe function $f(z) = z/(1-z)^2$, $z \in D(0,1)$, then we get

$$\begin{aligned}&\left|\frac{\alpha\overline{\beta}\,|x|^2}{(1-\alpha\overline{x})^2\left(1-\overline{\beta}x\right)^2}\right| \\ &\leq \frac{1}{2}\left(\frac{|\alpha\beta|}{\left(1-|\alpha|^2\right)\left(1-|\beta|^2\right)} + \left|\frac{\alpha\overline{\beta}}{\left(1-\alpha\overline{\beta}\right)^2}\right|\right)\frac{|x|^2}{\left(1-|x|^2\right)^2},\end{aligned} \tag{47}$$

for any $x, \alpha, \beta \in D(0,1)$. If we simplify (47), then we get

$$\frac{1-|x|^2}{\left|(1-\alpha\overline{x})\left(1-\overline{\beta}x\right)\right|} \leq \left(\frac{1}{2\left(1-|\alpha|^2\right)\left(1-|\beta|^2\right)} + \frac{1}{2\left|1-\alpha\overline{\beta}\right|^2}\right)^{1/2}, \tag{48}$$

for any $\alpha, \beta, x \in D(0,1)$. In particular, if $\beta = \overline{\alpha}$, then we get from (48) that

$$\frac{1-|x|^2}{|(1-\alpha\overline{x})(1-\alpha x)|} \leq \left(\frac{1}{2\left(1-|\alpha|^2\right)^2} + \frac{1}{2\left|1-\alpha^2\right|^2}\right)^{1/2}, \tag{49}$$

for any $\alpha, x \in D(0,1)$.

4. If we apply the same inequality (39) for the function

$$f(z) = \cosh(z) = \sum_{n=0}^{\infty} \frac{z^{2n}}{(2n)!}, \; z \in \mathbb{C},$$

then we obtain

$$\begin{aligned}&\left|\cosh\left(\alpha\overline{x} + \overline{\beta}x\right) + \cosh\left(\alpha\overline{x} - \overline{\beta}x\right)\right| \\ &\leq \left(\left[\frac{1}{2}\left(\cosh\left(|\alpha|^2 + |\beta|^2\right) + \cosh\left(|\alpha|^2 - |\beta|^2\right)\right)\right]^{1/2} + \left|\cosh\left(\alpha\overline{\beta}\right)\right|\right) \\ &\quad \times \cosh\left(|x|^2\right),\end{aligned} \tag{50}$$

for any $x, \alpha, \beta \in \mathbb{C}$. In particular, for $\beta = \overline{\alpha}$, we get that from (50)

$$\begin{aligned} &|\cosh(2\alpha \operatorname{Re}(x)) + \cosh(2i\alpha \operatorname{Im}(x))| \\ &\quad \leq \left[\cosh\left(|\alpha|^2\right) + \left|\cosh\left(\alpha^2\right)\right|\right] \cosh\left(|x|^2\right) \end{aligned} \tag{51}$$

that holds for any $\alpha, x \in \mathbb{C}$.

The following result contains an inequality which connects the power series function $f(z)$ with its transform $f_A(z)$ proved by Ibrahim and Dragomir in [13].

Theorem 4.2. *Let $f(z) = \sum_{n=0}^{\infty} a_n z^n$ be a function defined by a power series with real coefficients and convergent on the open disk $D(0,R) \subset \mathbb{C}$, $R > 0$. If α, β, x are complex numbers such that $\alpha\overline{x}, \overline{\beta}x, \alpha\overline{\beta}, |\alpha|^2, |\beta|^2, |x^2| \in D(0,R)$, then*

$$\left|f(\alpha\overline{x}) f\left(\overline{\beta}x\right)\right| \leq \frac{1}{2}\left(\left[f_A\left(|\alpha|^2\right) f_A\left(|\beta|^2\right)\right]^{1/2} + \left|f_A\left(\alpha\overline{\beta}\right)\right|\right) f_A\left(|x|^2\right). \tag{52}$$

Proof. By choosing $p_n = |a_n|$, $c_n = \alpha^n$, $b_n = \beta^n$ and $x_n = \operatorname{sgn}(a_n) x^n$, $n \geq 0$ in (36) we have

$$\begin{aligned} &\left|\sum_{n=0}^{m} a_n (\alpha\overline{x})^n \sum_{n=0}^{m} a_n \left(\overline{\beta}x\right)^n\right| \\ &= \left|\sum_{n=0}^{m} |a_n| \operatorname{sgn}(a_n) \alpha^n (\overline{x})^n \sum_{n=0}^{m} |a_n| \operatorname{sgn}(a_n) x^n \left(\overline{\beta}\right)^n\right| \\ &\leq \frac{1}{2}\left(\left[\sum_{n=0}^{m} |a_n| |\alpha|^{2n} \sum_{n=0}^{m} |a_n| |\beta|^{2n}\right]^{1/2} + \left|\sum_{n=0}^{m} |a_n| \left(\alpha\overline{\beta}\right)^n\right|\right) \sum_{n=0}^{m} |a_n| |x|^{2n} \end{aligned} \tag{53}$$

for any $\alpha, \beta, x \in \mathbb{C}$ with $\alpha\overline{x}, \overline{\beta}x, \alpha\overline{\beta}, |\alpha|^2, |\beta|^2, |x|^2 \in D(0,R)$. Taking the limit as $m \to \infty$ in (53) and noticing that all the involved series are convergent, then we deduce the desired inequality (52). □

In what follows we provide some applications of the inequality (52) for particular functions of interest:

1. If we take the function

$$f(z) = \frac{1}{1+z} = \sum_{n=0}^{\infty} (-1)^n z^n, \quad z \in D(0,1),$$

then

$$f_A(z) = \sum_{n=0}^{\infty} z^n = \frac{1}{1-z}, \quad z \in D(0,1).$$

Applying Theorem 4.2, we get the following inequality:

$$2\left|1-\alpha\overline{\beta}\right|\left(1-|x|^2\right)\left[\left(1-|\alpha|^2\right)\left(1-|\beta|^2\right)\right]^{1/2}$$
$$\leq |1+\alpha\overline{x}|\left|1+\overline{\beta}x\right|\left(\left|1-\alpha\overline{\beta}\right|+\left[\left(1-|\alpha|^2\right)\left(1-|\beta|^2\right)\right]^{1/2}\right), \quad (54)$$

for any $\alpha, \beta, x \in D(0,1)$. In particular, if $\alpha = \overline{\beta}$, then from (54) we obtain

$$2\left|1-\alpha^2\right|(1-|\alpha|^2)(1-|x|^2) \leq |1+\alpha x|\,|1+\alpha\overline{x}|\left(1-|\alpha|^2+\left|1-\alpha^2\right|\right), \quad (55)$$

for any $\alpha, x \in D(0,1)$.

2. For the function

$$f(z) = e^{-z} = \sum_{n=0}^{\infty} \frac{(-1)^n}{n!} z^n, \quad z \in \mathbb{C},$$

we have the transform

$$f_A(z) = \sum_{n=0}^{\infty} \frac{z^n}{n!} = e^z, \quad z \in \mathbb{C}.$$

Utilizing the inequality (52) we obtain

$$\frac{1}{|\exp(\alpha\overline{x}+\overline{\beta}x)|} \leq \frac{1}{2}\left\{\exp\left[\frac{1}{2}(|\alpha|^2+|\beta|^2)+|x|^2\right]+\exp(|x|^2)\left|\exp(\alpha\overline{\beta})\right|\right\}, \quad (56)$$

for any $\alpha, \beta, x \in \mathbb{C}$. In particular, if $\alpha = \overline{\beta}$ in (56), then we get

$$\frac{1}{|\exp(2\alpha\,\mathrm{Re}(x))|} \leq \frac{1}{2}\left[\exp\left(|\alpha|^2+|\beta|^2\right)+\exp\left(|x|^2\right)\left|\exp\left(\alpha^2\right)\right|\right], \quad (57)$$

for any $\alpha, x \in \mathbb{C}$.

3. If in (52) we choose the function

$$f(z) = \cos z = \sum_{n=0}^{\infty} \frac{(-1)^n}{(2n)!} z^{2n}, \quad z \in \mathbb{C},$$

then

$$f_A(z) = \sum_{n=0}^{\infty} \frac{1}{(2n)!} z^{2n} = \cosh(z) = \frac{1}{2}\left(e^z + e^{-z}\right).$$

Applying the inequality (52) will produce the result

$$\begin{aligned} &\left|\cos(\alpha\overline{x})\cos\left(\overline{\beta}x\right)\right| &(58)\\ &\leq \frac{1}{2}\left(\left[\cosh\left(|\alpha|^2\right)\cosh\left(|\beta|^2\right)\right]^{1/2} + \left|\cosh\left(\alpha\overline{\beta}\right)\right|\right)\cosh\left(|x|^2\right), \end{aligned}$$

for any $\alpha, \beta, x \in \mathbb{C}$. In particular, if we choose $\alpha = \overline{\beta}$ in (58), then we obtain the inequality

$$|\cos(\alpha\overline{x})\cos(\alpha x)| \leq \frac{1}{2}\left[\cosh\left(|\alpha|^2\right) + \left|\cosh\left(\alpha^2\right)\right|\right]\cosh\left(|x|^2\right), \tag{59}$$

for any $\alpha, x \in \mathbb{C}$.

Ibrahim and Dragomir [13] have proved the following result, which connects two power series, one having positive coefficients.

Theorem 4.3. *Let $g(z) = \sum_{n=0}^{\infty} g_n z^n$ and $f(z) = \sum_{n=0}^{\infty} a_n z^n$ be two power series with $g_n \in \mathbb{C}$ and $a_n > 0$ for $n \geq 0$. If f and g are convergent on $D(0, R_1)$ and $D(0, R_2)$, respectively, and the numerical series $\sum_{n=0}^{\infty} |g_n|^2 / a_n$ is convergent, then we have the inequality:*

$$|g(z)\, g(\overline{z})| \leq \frac{1}{2}\sum_{n=0}^{\infty} \frac{|g_n|^2}{a_n}\left[f\left(|z|^2\right) + \left|f\left(z^2\right)\right|\right] \tag{60}$$

for any $z \in \mathbb{C}$ with $z, z^2, |z|^2 \in D(0, R_1) \cap D(0, R_2)$.

Proof. On utilizing the inequality (37) for the choices $p_n = a_n$, $c_n = z^n$, $x_n = g_n / a_n$, $n \geq 0$, we have

$$\left|\sum_{n=0}^{m} g_n z^n \sum_{n=0}^{m} \overline{g}_n z^n\right| \leq \frac{1}{2}\left[\sum_{n=0}^{m} a_n\left(|z|^2\right)^n + \left|\sum_{n=0}^{m} a_n\left(z^2\right)^n\right|\right]\sum_{n=0}^{m} \frac{|g_n|^2}{a_n}, \tag{61}$$

for any $m \geq 0$.

Observe that $\sum_{n=0}^{m} \overline{g}_n z^n = \overline{\sum_{n=0}^{m} g_n (\overline{z})^n}$ and then $\left|\sum_{n=0}^{m} \overline{g}_n z^n\right| = \left|\sum_{n=0}^{m} g_n (\overline{z})^n\right|$. Replacing this in (61) we get

$$\left|\sum_{n=0}^{m} g_n z^n \sum_{n=0n}^{m} g_n (\overline{z})^n\right| \leq \frac{1}{2}\sum_{n=0}^{m} \frac{|g_n|^2}{a_n}\left[\sum_{n=0}^{m} a_n\left(|z|^2\right)^n + \left|\sum_{n=0}^{m} a_n\left(z^2\right)^n\right|\right]. \tag{62}$$

Since $z, z^2, |z|^2 \in D(0, R_1) \cap D(0, R_2)$, hence the series in (62) are convergent and letting $m \to \infty$, we deduce the desired inequality (60). □

Remark 4.2. If the coefficients g_n, $n \geq 0$ are real, then we recapture the inequality (30) (or the inequality (27) from the paper [4]).

Corollary 4.2. *Let $g(z) = \sum_{n=0}^{\infty} g_n z^n$ be a power series with complex coefficients and convergent on the open disk $D(0, R)$. If the numerical series $\sum_{n=0}^{\infty} |g_n|^2$ is convergent, then*

$$|g(z)g(\bar{z})| \leq \frac{1}{2}\sum_{n=0}^{\infty} |g_n|^2 \left[\frac{1 - |z|^2 + |1 - z^2|}{\left(1 - |z|^2\right)|1 - z^2|}\right], \tag{63}$$

for any $z \in D(0, 1) \cap D(0, R)$.

This follows from (60) for $f(z) = 1/(1 - z)$, $z \in D(0, 1)$.

If we consider the series expansion

$$\frac{1}{iz}\ln\left(\frac{1}{1 - iz}\right) = \sum_{n=0}^{\infty} \frac{i^n}{n + 1} z^n; \quad z \in D(0, 1) \setminus \{0\},$$

then, on utilizing the inequality (63) for the choice $g_n = i^n/(n + 1)$ and taking into account the identity (34) we can state the following inequality:

$$\left|\ln\left(\frac{1}{1 - iz}\right)\ln\left(\frac{1}{1 - i\bar{z}}\right)\right| \leq \frac{\pi^2}{12}\left(\frac{|z|^2}{1 - |z|^2}\right)\left(\frac{1 - |z|^2 + |1 - z^2|}{|1 - z^2|}\right), \tag{64}$$

for $z \in D(0, 1)$.

Corollary 4.3. *Let $g(z) = \sum_{n=0}^{\infty} g_n z^n$ be a power series with complex coefficients and convergent on the open disk $D(0, R)$. If the numerical series $\sum_{n=0}^{\infty} n! |g_n|^2$ is convergent, then*

$$|g(z)g(\bar{z})| \leq \frac{1}{2}\sum_{n=0}^{\infty} n! |g_n|^2 \left[\exp\left(|z|^2\right) + \left|\exp\left(z^2\right)\right|\right], \tag{65}$$

for any $z \in D(0, R)$.

This follows from Theorem 4.3 by choosing $f(z) = \exp(z)$.

Some applications of the inequality (65) are as follows:

1. If we apply the inequality (65) for the function $\sin(iz) = \sum_{n=0}^{\infty} \frac{i}{(2n+1)!} z^{2n+1}$, then we obtain the inequality

$$|\sin(iz)\sin(i\bar{z})| \le \frac{1}{2}\sum_{n=0}^{\infty}\frac{n!}{[(2n+1)!]^2}\left[\exp\left(|z|^2\right)+\left|\exp\left(z^2\right)\right|\right], \tag{66}$$

for any $z \in \mathbb{C}$.

2. If we apply the inequality (65) for the function $\sinh(iz) = \sum_{n=0}^{\infty}\frac{(-1)^n i}{(2n+1)!}z^{2n+1}$, then we obtain the inequality

$$|\sin(z)|^2 \le \frac{1}{2}\sum_{n=0}^{\infty}\frac{n!}{[(2n+1)!]^2}\left[\exp\left(|z|^2\right)+\left|\exp\left(z^2\right)\right|\right], \tag{67}$$

for any $z \in \mathbb{C}$. Indeed, observing that

$$|\sinh(iz)\sinh(i\bar{z})| = |i\sin(z)\cdot i\sin(\bar{z})| = |\sin(z)\sin(\bar{z})| = |\sin z|^2 \tag{68}$$

and by (65) we have

$$|\sinh(iz)\sinh(i\bar{z})| \le \frac{1}{2}\sum_{n=0}^{\infty}\frac{n!}{[(2n+1)!]^2}\left[\exp\left(|z|^2\right)+\left|\exp\left(z^2\right)\right|\right], \tag{69}$$

then we deduce desired inequality (67).

The next result is also given in [13].

Theorem 4.4. *Let* $g(z) = \sum_{n=0}^{\infty} g_n z^n$, $h(z) = \sum_{n=0}^{\infty} h_n z^n$ *and* $f(z) = \sum_{n=0}^{\infty} a_n z^n$ *be three power series with* $g_n, h_n \in \mathbb{C}$ *and* $a_n > 0$ *for* $n \ge 0$. *If* f, g *and* h *are convergent on* $D(0,R_1)$, $D(0,R_2)$ *and* $D(0,R_3)$, *respectively, and the numerical series* $\sum_{n=0}^{\infty} |g_n|^2/a_n$, $\sum_{n=0}^{\infty} |h_n|^2/a_n$ *and* $\sum_{n=0}^{\infty} g_n\overline{h_n}/a_n$ *are convergent, then we have the inequality:*

$$|g(z)h(z)| \le \frac{1}{2}\left(\left[\sum_{n=0}^{\infty}\frac{|g_n|^2}{a_n}\sum_{n=0}^{\infty}\frac{|h_n|^2}{a_n}\right]^{1/2} + \left|\sum_{n=0}^{\infty}\frac{g_n\overline{h_n}}{a_n}\right|\right) f\left(|z|^2\right), \tag{70}$$

for any $z \in \mathbb{C}$ *with* $z, |z|^2 \in D(0,R_1)\cap D(0,R_2)\cap D(0,R_3)$.

Proof. On utilizing the Buzano inequality (36) for the choices $p_n = a_n$, $c_n = g_n/a_n$, $b_n = h_n/a_n$, $x_n = z^n$, $n \ge 0$, we can state that

$$|g(z)h(z)| = \left|\sum_{n=0}^{\infty} a_n\left(\frac{g_n}{a_n}\right)z^n \sum_{n=0}^{\infty} a_n\left(\frac{h_n}{a_n}\right)z^n\right|, \tag{71}$$

$$\leq \frac{1}{2}\left(\left[\sum_{n=0}^{\infty} a_n \left|\frac{g_n}{a_n}\right|^2 \sum_{n=0}^{\infty} a_n \left|\frac{h_n}{a_n}\right|^2\right]^{1/2} + \left|\sum_{n=0}^{\infty} a_n \left(\frac{g_n}{a_n}\right)\overline{\left(\frac{h_n}{a_n}\right)}\right|\right) \times \sum_{n=0}^{\infty} a_n \left(|z|^2\right)^n,$$

$$= \frac{1}{2}\left(\left[\sum_{n=0}^{\infty} \frac{|g_n|^2}{a_n} \sum_{n=0}^{\infty} \frac{|h_n|^2}{a_n}\right]^{1/2} + \left|\sum_{n=0}^{\infty} \frac{g_n \overline{h}_n}{a_n}\right|\right) f\left(|z|^2\right),$$

for any $z \in \mathbb{C}$ with $z, |z|^2 \in D(0, R_1) \cap D(0, R_2) \cap D(0, R_3)$. □

Remark 4.3. In particular, if $g_n = h_n$, then from (70) we have

$$|g(z)|^2 \leq f\left(|z|^2\right) \sum_{n=0}^{\infty} \frac{|g_n|^2}{a_n}, \tag{72}$$

for any $z, |z|^2 \in D(0, R_1) \cap D(0, R_2)$.

Remark 4.4. Also if $h_n = \overline{g}_n$, then we get the following inequality:

$$|g(z)g(\overline{z})| \leq \frac{1}{2}\left(\sum_{n=0}^{\infty} \frac{|g_n|^2}{a_n} + \left|\sum_{n=0}^{\infty} g_n^2\right|\right) f\left(|z|^2\right), \tag{73}$$

for any $z, |z|^2 \in D(0, R_1) \cap D(0, R_2)$.

Corollary 4.4. *Let $g(z)$ and $h(z)$ be power series as in Theorem 4.4. If the numerical series $\sum_{n=0}^{\infty} |g_n|^2$, $\sum_{n=0}^{\infty} |h_n|^2$ and $\sum_{n=0}^{\infty} \left|g_n \overline{h}_n\right|$ are convergent, then*

$$|g(z)h(z)| \leq \frac{1}{2\left(1 - |z|^2\right)}\left(\left[\sum_{n=0}^{\infty} |g_n|^2 \sum_{n=0}^{\infty} |h_n|^2\right]^{1/2} + \left|\sum_{n=0}^{\infty} g_n \overline{h}_n\right|\right), \tag{74}$$

for any $z \in D(0, 1) \cap D(0, R_2) \cap D(0, R_3)$.

If we consider the series

$$\frac{1}{iz} \ln\left(\frac{1}{1 - iz}\right) = \sum_{n=0}^{\infty} \frac{i^n}{(n+1)} z^n, \quad z \in D(0, 1) \setminus \{0\}$$

and

$$\ln\left(\frac{1}{1 + iz}\right) = \sum_{n=1}^{\infty} \frac{(-i)^n}{n} z^n, \quad z \in D(0, 1),$$

then on utilizing the inequality (74) for the choices $g_0 = h_0 = 0$, $g_n = i^n/(n+1)$, $h_n = (-i)^n/n$, $n \geq 1$ and taking into account that

$$\sum_{n=0}^{\infty} g_n \overline{h}_n = \sum_{n=1}^{\infty} \frac{1}{n(n+1)} = 1 \tag{75}$$

and the equality (34), we obtain the following inequality:

$$\left| \ln\left(\frac{1}{1-iz}\right) \ln\left(\frac{1}{1+iz}\right) \right| \leq \frac{\pi^2+6}{12} \left(\frac{|z|}{1-|z|^2}\right), \tag{76}$$

for any $z \in D(0,1)$.

5 Power Series Inequality via a Refinement of the Schwarz Inequality

If we write the inequality (12) for the particular inner product space $(\mathbb{K}^n; \langle \cdot, \cdot \rangle)$, where

$$\langle x, y \rangle_p = \sum_{j=1}^{n} p_j x_j \overline{y}_j$$

for $x = (x_1, x_2, \ldots, x_n)$, $y = (y_1, y_2, \ldots, y_n) \in \mathbb{K}^n$ and $p = (p_1, p_2, \ldots, p_n)$ with $p_j \geq 0$, $j \in \{1, 2, \ldots, n\}$, then we get the discrete inequality

$$\begin{aligned}
&\left(\sum_{j=1}^{n} p_j |x_j|^2\right)^{1/2} \left(\sum_{j=1}^{n} p_j |y_j|^2\right)^{1/2} \sum_{j=1}^{n} p_j |z_j|^2 - \left|\sum_{j=1}^{n} p_j x_j \overline{z}_j \sum_{j=1}^{n} p_j z_j \overline{y}_j\right| \\
&\geq \left|\sum_{j=1}^{n} p_j x_j \overline{y}_j \sum_{j=1}^{n} p_j |z_j|^2 - \sum_{j=1}^{n} p_j x_j \overline{z}_j \sum_{j=1}^{n} p_j z_j \overline{y}_j\right|,
\end{aligned} \tag{77}$$

where $p_j \geq 0$, $x_j, y_j, z_j \in \mathbb{K}$, $j \in \{1, 2, \ldots, n\}$. In particular, if we take in (77) $y_j = \overline{x}_j$ for $j \in \{1, 2, \ldots, n\}$, then we obtain

$$\sum_{j=1}^{n} p_j |x_j|^2 \sum_{j=1}^{n} p_j |z_j|^2 - \left|\sum_{j=1}^{n} p_j x_j \overline{z}_j \sum_{j=1}^{n} p_j x_j z_j\right| \tag{78}$$

$$\geq \left| \sum_{j=1}^{n} p_j x_j^2 \sum_{j=1}^{n} p_j |z_j|^2 - \sum_{j=1}^{n} p_j x_j \bar{z}_j \sum_{j=1}^{n} p_j x_j z_j \right|,$$

for $p_j \geq 0, x_j, z_j \in \mathbb{K}, j \in \{1, 2, \ldots, n\}$.

On applying the inequality (77) for power series, Ibrahim and Dragomir [14] established the following result.

Theorem 5.1. *Let $f(z) = \sum_{n=0}^{\infty} a_n z^n$ be a power series with nonnegative coefficients a_n and convergent on the open disk $D(0, R)$. If $x, y, z \in \mathbb{C}$, so that $|x|^2, |y|^2, |z|^2, x\bar{z}, z\bar{y}, x\bar{y} \in D(0, R)$, then*

$$\left[f\left(|x|^2\right) f\left(|y|^2\right) \right]^{1/2} f\left(|z|^2\right) - |f(x\bar{z}) f(z\bar{y})| \tag{79}$$
$$\geq \left| f(x\bar{y}) f\left(|z|^2\right) - f(x\bar{z}) f(z\bar{y}) \right|.$$

Proof. If we choose $p_n = a_n$, $x_n = x^n$, $y_n = y^n$, $z_n = z^n$, $n \in \{0, 1, 2, \ldots, m\}$ in (77), then we have

$$\left[\sum_{n=0}^{m} a_n \left(|x|^2\right)^n \right]^{1/2} \left[\sum_{n=0}^{m} a_n \left(|y|^2\right)^n \right]^{1/2} \sum_{n=0}^{m} a_n \left(|z|^2\right)^n$$
$$- \left| \sum_{n=0}^{m} a_n (x\bar{z})^n \sum_{n=0}^{m} a_n (z\bar{y})^n \right|$$
$$\geq \left| \sum_{n=0}^{m} a_n (x\bar{y})^n \sum_{n=0}^{m} a_n \left(|z|^2\right)^n - \sum_{n=0}^{m} a_n (x\bar{z})^n \sum_{n=0}^{m} a_n (x\bar{y})^n \right|. \tag{80}$$

Since $|x|^2, |y|^2, |z|^2, x\bar{z}, z\bar{y}, x\bar{y}$ belong to the convergence disk $D(0, R)$ and taking the limit as $m \to \infty$ in (80), we deduce the desired result (79). □

Some examples for particular functions that are generated by power series with nonnegative coefficients are as follows:

1. If we choose in the above inequality (79) for $f(z) = 1/(1-z)$, $z \in D(0, 1)$, then we have

$$\frac{|(1 - x\bar{z})(1 - z\bar{y})|}{\left[\left(1 - |x|^2\right)\left(1 - |y|^2\right)\right]^{1/2}\left(1 - |z|^2\right)} - 1 \geq \left| \frac{(1 - x\bar{z})(1 - z\bar{y})}{(1 - x\bar{y})\left(1 - |z|^2\right)} - 1 \right|, \tag{81}$$

for any $x, y, z \in D(0, 1)$. In particular for $z = \bar{x}$ in (81) we get

$$\frac{\left|\left(1-x^{2}\right)\left(1-\overline{xy}\right)\right|}{\left(1-|x|^{2}\right)^{3/2}\left(1-|y|^{2}\right)^{1/2}}-1\geq\left|\frac{\left(1-x^{2}\right)\left(1-\overline{xy}\right)}{\left(1-x\overline{y}\right)\left(1-|x|^{2}\right)}-1\right|, \tag{82}$$

for any $x, y \in D(0,1)$. Also, if $z = a \in \mathbb{R}$ and $x, y \in \mathbb{C}$, then from (81), we obtain

$$\frac{\left|\left(1-ax\right)\left(1-a\overline{y}\right)\right|}{\left[\left(1-|x|^{2}\right)\left(1-|y|^{2}\right)\right]^{1/2}\left(1-a^{2}\right)}-1\geq\left|\frac{\left(1-ax\right)\left(1-a\overline{y}\right)}{\left(1-x\overline{y}\right)\left(1-a^{2}\right)}-1\right|, \tag{83}$$

for any $x, y \in D(0,1)$ and $a \in (-1,1)$.

2. If we apply (79) for $f(z) = \exp(z)$, $z \in \mathbb{C}$, then we get

$$\exp\left(\frac{|x|^{2}+|y|^{2}}{2}+|z|^{2}\right)-\left|\exp\left(x\overline{z}+z\overline{y}\right)\right|$$
$$\geq\left|\exp\left(x\overline{y}+|z|^{2}\right)-\exp\left(x\overline{z}+z\overline{y}\right)\right|, \tag{84}$$

for any $x, y, z \in \mathbb{C}$. In particular, for $z = \overline{x}$ in (84), we get

$$\exp\left(\frac{3|x|^{2}+|y|^{2}}{2}\right)-\left|\exp\left(x^{2}+\overline{xy}\right)\right|$$
$$\geq\left|\exp\left(x\overline{y}+|x|^{2}\right)-\exp\left(x^{2}+\overline{xy}\right)\right|, \tag{85}$$

for any $x, y \in \mathbb{C}$. Also, if $z = a \in \mathbb{R}$ and $x, y \in \mathbb{C}$, then from (84),

$$\exp\left(\frac{|x|^{2}+|y|^{2}}{2}+a^{2}\right)-\left|\exp\left[a\left(x+\overline{y}\right)\right]\right|$$
$$\geq\left|\exp\left(x\overline{y}+a^{2}\right)-\exp\left[a\left(x+\overline{y}\right)\right]\right|, \tag{86}$$

for any $x, y \in \mathbb{C}$ and $a \in \mathbb{R}$.

3. For the *Koebe function* $f(z) = z/(1-z)^2$, $z \in D(0,1)$, we get from (79) the following inequality:

$$\frac{1}{\left(1-|x|^{2}\right)\left(1-|y|^{2}\right)\left(1-|z|^{2}\right)^{2}}-\frac{1}{\left|\left[\left(1-x\overline{z}\right)\left(1-z\overline{y}\right)\right]^{2}\right|}$$
$$\geq\left|\frac{1}{\left[\left(1-x\overline{y}\right)\left(1-|z|^{2}\right)\right]^{2}}-\frac{1}{\left[\left(1-x\overline{z}\right)\left(1-z\overline{y}\right)\right]^{2}}\right|, \tag{87}$$

for any $x, y, z \in D(0,1)$. In particular, for $y = \overline{x}$ in (87), we get

$$\frac{1}{\left[\left(1-|x|^{2}\right)\left(1-|z|^{2}\right)\right]^{2}}-\frac{1}{\left|[(1-x\overline{z})(1-xz)]^{2}\right|}$$

$$\geq\left|\frac{1}{\left[(1-x^{2})\left(1-|z|^{2}\right)\right]^{2}}-\frac{1}{[(1-x\overline{z})(1-xz)]^{2}}\right|, \tag{88}$$

for any $x, z \in D(0,1)$. Also for $z = \overline{x}$, we have from (87)

$$\frac{1}{\left(1-|x|^{2}\right)^{3}\left(1-|y|^{2}\right)}-\frac{1}{\left|[(1-x^{2})(1-\overline{xy})]^{2}\right|}$$

$$\geq\left|\frac{1}{\left[(1-x\overline{y})\left(1-|x|^{2}\right)\right]^{2}}-\frac{1}{[(1-x^{2})(1-\overline{xy})]^{2}}\right|, \tag{89}$$

for any $x, y \in D(0,1)$. If $z = a \in \mathbb{R}$ and $x, y \in \mathbb{C}$, then from (87)

$$\frac{1}{\left(1-|x|^{2}\right)\left(1-|y|^{2}\right)(1-a^{2})^{2}}-\frac{1}{\left|[(1-ax)(1-a\overline{y})]^{2}\right|}$$

$$\geq\left|\frac{1}{[(1-x\overline{y})(1-a^{2})]^{2}}-\frac{1}{[(1-ax)(1-a\overline{y})]^{2}}\right|, \tag{90}$$

for any $x, y \in D(0,1)$ and $a \in (-1,1)$.

Remark 5.1. If $z = 0$, then from (79) we obtain

$$\left[f\left(|x|^{2}\right)f\left(|y|^{2}\right)\right]^{1/2}-|f(0)|\geq|f(x\overline{y})-f(0)|, \tag{91}$$

where $f(0) = a_0 > 0$, $|x|^2, |y|^2, x\overline{y} \in D(0,R)$.

Some applications of the inequality (91) for particular functions of interest are as follows:

1. If we apply the inequality (91) for the function $f(z) = \exp(z)$, $z \in \mathbb{C}$, then we obtain the inequality

$$\exp\left(\frac{|x|^{2}+|y|^{2}}{2}\right)-1\geq|\exp(x\overline{y})-1|, \tag{92}$$

for any $x, y \in \mathbb{C}$. Moreover, if $y = \overline{x}$, then from (92) we get

$$\exp\left(|x|^2\right) - 1 \geq \left|\exp\left(x^2\right) - 1\right|,$$

for any $x \in \mathbb{C}$.

2. If we apply the same inequality (91) for the function $f(z) = \cos(z)$, $z \in \mathbb{C}$, then we get the following inequality:

$$\left[\cos\left(|x|^2\right)\cos\left(|y|^2\right)\right]^{1/2} - 1 \geq \left|\cos\left(x\overline{y}\right) - 1\right|, \tag{93}$$

for any $x, y \in \mathbb{C}$. Also, if $y = \overline{x}$, then from (93) we get

$$\cos\left(|x|^2\right) - 1 \geq \left|\cos\left(x^2\right) - 1\right|,$$

for any $x \in \mathbb{C}$.

3. For the function $f(z) = 1/(1-z)$, $z \in D(0,1)$ and applying the inequality (91) we obtain

$$\frac{1}{\left[\left(1 - |x|^2\right)\left(1 - |y|^2\right)\right]^{1/2}} - 1 \geq \left|\frac{x\overline{y}}{1 - x\overline{y}}\right|,$$

for any $x, y \in \mathbb{C}$ with $|x|^2, |y|^2, x\overline{y} \in D(0,1)$.

Remark 5.2. If $y = \overline{x}$ in (79), then we get

$$f\left(|x|^2\right) f\left(|z|^2\right) - |f(xz) f(x\overline{z})| \geq \left|f\left(x^2\right) f\left(|z|^2\right) - f(xz) f(x\overline{z})\right|, \tag{94}$$

for $x, z \in \mathbb{C}$ with $|x|^2, |z|^2, x\overline{z}, zx \in D(0,R)$. Moreover, for $z = a \in \mathbb{R}$, from (94), we deduce

$$f\left(|x|^2\right) f\left(a^2\right) - |f(ax)|^2 \geq \left|f\left(x^2\right) f\left(a^2\right) - f^2(ax)\right|, \tag{95}$$

for any $x \in \mathbb{C}$, $a \in \mathbb{R}$. If we choose in (95) $a = 1$, then we have the inequality

$$f\left(|x|^2\right) f(1) - |f(x)|^2 \geq \left|f\left(x^2\right) f(1) - f^2(x)\right|, \tag{96}$$

for any $x \in \mathbb{C}$.

For some applications, we apply the inequality (96) for the function $f(z) = \exp(z)$; then we have

$$\exp\left(|x|^2 + 1\right) - |\exp(2x)| \geq \left|\exp\left(x^2 + 1\right) - \exp(2x)\right|, \tag{97}$$

for any $x \in \mathbb{C}$. Since $|\exp(2x)| \neq 0$, then (97) is equivalent with

$$\frac{\exp\left(|x|^2+1\right)}{|\exp(2x)|} - 1 \geq \left|\exp(x-1)^2 - 1\right|,$$

for any $x \in \mathbb{C}$.

Remark 5.3. If $z = \overline{x}$ in (79), then we get

$$\begin{aligned}&\left[f\left(|x|^2\right) f\left(|y|^2\right)\right]^{1/2} f\left(|x|^2\right) - \left|f\left(x^2\right) f(xy)\right| \\ &\geq \left|f(x\overline{y}) f\left(|x|^2\right) - f\left(x^2\right) f\left(\overline{xy}\right)\right|,\end{aligned} \tag{98}$$

for $x, y \in \mathbb{C}$ with $x^2, xy, |x|^2, |y|^2 \in D(0, R)$.

If we apply the inequality (98) for the function $f(z) = \exp(z)$, $z \in \mathbb{C}$, then we get

$$\exp\left(\frac{3|x|^2+|y|^2}{2}\right) - \left|\exp\left[x^2+\overline{xy}\right]\right| \geq \left|\exp\left(x\overline{y}+|x|^2\right) - \exp\left(x^2+\overline{xy}\right)\right|, \tag{99}$$

for any $x, y \in \mathbb{C}$. Moreover, if $x = a \in \mathbb{R}$, then from (99) we obtain

$$\exp\left(\frac{3a^2+|y|^2}{2}\right) \geq \left|\exp\left[a(a+y)\right]\right|, \tag{100}$$

for any $y \in \mathbb{C}, a \in \mathbb{R}$.

Theorem 5.2 ([14]). *Let $f(z) = \sum_{n=0}^{\infty} a_n z^n$ be a power series with real coefficients a_n and convergent on $D(0,R) \subset \mathbb{C}$, $R > 0$. If $x, y, z \in \mathbb{C}$, so that $|x|^2, |y|^2, |z|^2$, $x\overline{z}$, $z\overline{y}$, $x\overline{y} \in D(0,R)$, then*

$$\begin{aligned}&\left[f_A\left(|x|^2\right) f_A\left(|y|^2\right)\right]^{1/2} f_A\left(|z|^2\right) - \left|f(x\overline{z}) f(z\overline{y})\right| \\ &\geq \left|f_A(x\overline{y}) f_A\left(|z|^2\right) - f(x\overline{z}) f(z\overline{y})\right|.\end{aligned} \tag{101}$$

Proof. By choosing $p_n = |a_n| \geq 0$, $x_n = x^n$, $y_n = y^n$, $z_n = \operatorname{sgn}(a_n) z^n$, $n \geq 0$ in (77), we have

$$\begin{aligned}&\left|\sum_{n=0}^{m} a_n (x\overline{z})^n \sum_{n=0}^{m} a_n (z\overline{y})^n\right| \\ &= \left|\sum_{n=0}^{m} |a_n| \operatorname{sgn}(a_n)\, x^n (\overline{z})^n \sum_{n=0}^{m} |a_n| \operatorname{sgn}(a_n)\, z^n (\overline{y})^n\right|\end{aligned}$$

$$\leq \left(\sum_{n=0}^{m} |a_n| |x|^{2n}\right)^{1/2} \left(\sum_{n=0}^{m} |a_n| |y|^{2n}\right)^{1/2} \sum_{n=0}^{m} |a_n| |\operatorname{sgn}(a_n) z^n|^2$$

$$- \left| \sum_{n=0}^{m} |a_n| x^n (\overline{y})^n \sum_{n=0}^{m} |a_n| |\operatorname{sgn}(a_n) z^n|^2 \right.$$

$$\left. - \sum_{n=0}^{m} |a_n| x^n [\operatorname{sgn}(a_n) (\overline{z})^n] \times \sum_{n=0}^{m} |a_n| [\operatorname{sgn}(a_n) z^n] (\overline{y})^n \right|$$

$$= \left(\sum_{n=0}^{m} |a_n| \left(|x|^2\right)^n\right)^{1/2} \left(\sum_{n=0}^{m} |a_n| \left(|y|^2\right)^n\right)^{1/2} \sum_{n=0}^{m} |a_n| \left(|z|^2\right)^n$$

$$- \left| \sum_{n=0}^{m} |a_n| (x\overline{y})^n \sum_{n=0}^{m} |a_n| \left(|z|^2\right)^n - \sum_{n=0}^{m} a_n (x\overline{z})^n \sum_{n=0}^{m} a_n (z\overline{y})^n \right| \tag{102}$$

for any $x, y, z \in \mathbb{C}$ with $x\overline{y}, x\overline{z}, z\overline{y}, |x|^2, |y|^2, |z|^2 \in D(0, R)$. Taking the limit as $m \to \infty$ in (102), then we deduce the desired inequality (101). □

In what follows we provide some applications of the inequality (101) for particular functions of interest:

1. If we take the function

$$f(z) = \frac{1}{1+z} = \sum_{n=0}^{\infty} (-1)^n z^n, \quad z \in D(0,1),$$

then

$$f_A(z) = \sum_{n=0}^{\infty} z^n = \frac{1}{1-z}, \quad z \in D(0,1).$$

Applying the inequality (101), we can state that

$$\left[\frac{1}{1-|x|^2} \cdot \frac{1}{1-|y|^2}\right]^{1/2} \left(\frac{1}{1-|z|^2}\right) - \left|\frac{1}{1+x\overline{z}} \cdot \frac{1}{1+z\overline{y}}\right|$$

$$= \frac{1}{\left[\left(1-|x|^2\right)\left(1-|y|^2\right)\right]^{1/2} \left(1-|z|^2\right)} - \frac{1}{|(1+x\overline{z})(1+z\overline{y})|},$$

$$= \frac{|(1+x\bar{z})(1+z\bar{y})| - \left[\left(1-|x|^2\right)\left(1-|y|^2\right)\right]^{1/2}\left(1-|z|^2\right)}{\left[\left(1-|x|^2\right)\left(1-|y|^2\right)\right]^{1/2}\left(1-|z|^2\right)|(1+x\bar{z})(1+z\bar{y})|}$$

$$\geq \left|\frac{1}{1-x\overline{y}} \cdot \frac{1}{1-|z|^2} - \frac{1}{1+x\overline{z}} \cdot \frac{1}{1+z\overline{y}}\right|$$

$$= \frac{\left|(1+x\bar{z})(1+z\bar{y}) - (1-x\bar{y})\left(1-|z|^2\right)\right|}{\left|(1-x\bar{y})\left(1-|z|^2\right)\right||(1+x\bar{z})(1+z\bar{y})|}.$$

Hence we have

$$\frac{|(1+x\bar{z})(1+z\bar{y})|}{\left[\left(1-|x|^2\right)\left(1-|y|^2\right)\right]^{1/2}\left(1-|z|^2\right)} - 1 \geq \left|\frac{(1+x\bar{z})(1+z\bar{y})}{(1-x\bar{y})\left(1-|z|^2\right)} - 1\right|, \quad (103)$$

for any $x, y, z \in D(0,1)$. In particular, if $y = \overline{x}$, $z = a \in \mathbb{R}$, then from (103) we get

$$\frac{|1+ax|^2}{\left(1-|x|^2\right)(1-a^2)} - 1 \geq \left|\frac{(1+ax)^2}{(1-x^2)(1-a^2)} - 1\right|,$$

for any $x \in D(0,1)$, $a \in \mathbb{R}$.

2. For the function $f(z) = \exp(-z) = \sum_{n=0}^{\infty} \frac{(-1)^n}{n!} z^n$, $z \in \mathbb{C}$, we have the transform

$$f_A(z) = \sum_{n=0}^{\infty} \frac{1}{n!} z^n = \exp(z), \quad z \in \mathbb{C}.$$

Utilizing the inequality (101) we obtain

$$\exp\left(\frac{|x|^2+|y|^2}{2} + |z|^2\right)|\exp(x\bar{z}+z\bar{y})| - 1 \geq \left|\exp\left(x\bar{y}+|z|^2+x\bar{z}+z\bar{y}\right) - 1\right|, \quad (104)$$

for any $x, y, z \in \mathbb{C}$. In particular, if $y = \overline{x}$, $z = a \in \mathbb{R}$, then from (104) we get

$$\exp\left(|x|^2+a^2\right)|\exp(2ax)| - 1 \geq \left|\exp\left(|x|^2+a+2ax\right) - 1\right|,$$

for any $x \in \mathbb{C}$, $a \in \mathbb{R}$.

In the following result, we state a connection between two power series, one having positive coefficients while the other having complex coefficients.

Theorem 5.3 ([14]). *Let $g(z) = \sum_{n=0}^{\infty} g_n z^n$ and $f(z) = \sum_{n=0}^{\infty} a_n z^n$ be two power series with $g_n \in \mathbb{C}$ and $a_n > 0$, $n \geq 0$. If f and g are convergent on $D(0, R_1)$ and $D(0, R_2)$, respectively, and the numerical series $\sum_{n=0}^{\infty} \frac{|g_n|^2}{a_n}$ is convergent, then we have the inequality*

$$\sum_{n=0}^{\infty} \frac{|g_n|^2}{a_n} f\left(|z|^2\right) - |g(z)g(\overline{z})| \geq \left| \sum_{n=0}^{\infty} \frac{|g_n|^2}{a_n} f\left(z^2\right) - g(z)\overline{g(\overline{z})} \right|, \tag{105}$$

for any $z \in \mathbb{C}$ with $z, z^2|z|^2 \in D(0, R_1) \cap D(0, R_2)$.

Proof. On utilizing the inequality (78) for the choices $p_n = a_n$, $x_n = z^n$, $z_n = \frac{g_n}{a_n}$, $n \in \{0, 1, 2, \ldots, m\}$, we have

$$\begin{aligned}
&\sum_{n=0}^{m} a_n |z|^{2n} \sum_{n=0}^{m} a_n \left| \frac{g_n}{a_n} \right|^2 - \left| \sum_{n=0}^{m} a_n z^n \overline{\left(\frac{g_n}{a_n} \right)} \sum_{n=0}^{n} a_n z^n \left(\frac{g_n}{a_n} \right) \right| \\
&= \sum_{n=0}^{m} a_n \left(|z|^2\right)^n \sum_{n=0}^{m} \frac{|g_n|^2}{a_n} - \left| \sum_{n=0}^{m} \overline{g}_n z^n \sum_{n=0}^{n} g_n z^n \right|, \\
&\geq \left| \sum_{n=0}^{m} a_n z^{2n} \sum_{n=0}^{m} a_n \left| \frac{g_n}{a_n} \right|^2 - \sum_{n=0}^{m} a_n z^n \overline{\left(\frac{g_n}{a_n} \right)} \sum_{n=0}^{n} a_n z^n \left(\frac{g_n}{a_n} \right) \right| \\
&= \left| \sum_{n=0}^{m} a_n \left(z^2\right)^n \sum_{n=0}^{m} \frac{|g_n|^2}{a_n} - \sum_{n=0}^{m} \overline{g}_n z^n \sum_{n=0}^{n} g_n z^n \right|,
\end{aligned} \tag{106}$$

for any $m \geq 0$. Observe that $\sum_{n=0}^{m} \overline{g}_n z^n = \overline{\sum_{n=0}^{m} g_n (\overline{z})^n}$ and then

$$\left| \sum_{n=0}^{m} \overline{g}_n z^n \right| = \left| \sum_{n=0}^{m} g_n (\overline{z})^n \right|. \tag{107}$$

Replacing (107) in (106) we get

$$\begin{aligned}
&\sum_{n=0}^{m} \frac{|g_n|^2}{a_n} \sum_{n=0}^{m} a_n \left(|z|^2\right)^n - \left| \sum_{n=0}^{m} g_n (\overline{z})^n \sum_{n=0}^{n} g_n z^n \right| \\
&\qquad \geq \left| \sum_{n=0}^{m} a_n \left(z^2\right)^n \sum_{n=0}^{m} \frac{|g_n|^2}{a_n} - \sum_{n=0}^{m} \overline{g}_n z^n \sum_{n=0}^{n} g_n z^n \right|.
\end{aligned} \tag{108}$$

Since $z, z^2|z|^2 \in D(0, R_1) \cap D(0, R_2)$, hence the series in (108) are convergent and letting $m \to \infty$, we deduce the desired inequality (105). □

Remark 5.4. If the coefficients $g_n, n \geq 0$ are real, then we have the inequality

$$\sum_{n=0}^{\infty} \frac{g_n^2}{a_n} f\left(|z|^2\right) - |g(z)g(\overline{z})| \geq \left|\sum_{n=0}^{\infty} \frac{g_n^2}{a_n} f\left(z^2\right) - g^2(z)\right|, \tag{109}$$

for any $z \in \mathbb{C}$ with $z, z^2, |z|^2 \in D(0, R_1) \cap D(0, R_2)$.

Corollary 5.1. *Let $g(z) = \sum_{n=0}^{\infty} g_n z^n$ be a power series with complex coefficients and convergent on the open disk $D(0, R)$. If the numerical series $\sum_{n=0}^{\infty} |g_n|^2$ is convergent, then*

$$\left(\frac{1}{1-|z|^2}\right) \sum_{n=0}^{\infty} |g_n|^2 - |g(z)g(\overline{z})| \geq \left|\left(\frac{1}{1-z^2}\right) \sum_{n=0}^{\infty} |g_n|^2 - g(z)\overline{g(\overline{z})}\right|, \tag{110}$$

for any $z \in D(0,1) \cap D(0, R)$.

This follows from (105) for $f(z) = 1/(1-z)$, $z \in D(0,1)$.

If we consider the series expansion

$$\frac{1}{iz} \ln\left(\frac{1}{1-iz}\right) = \sum_{n=0}^{\infty} \frac{i^n}{n+1} z^n, \quad z \in D(0,1) \setminus \{0\},$$

then on utilizing the inequality (110) for the choice $g_n = i^n/(n+1)$ and taking into account the equality (34), we can state the following inequality:

$$\frac{\pi^2}{6}\left(\frac{|z|^2}{1-|z|^2}\right) - \left|\ln\left(\frac{1}{1-iz}\right) \ln\left(\frac{1}{1-i\overline{z}}\right)\right|$$
$$\geq \left|\frac{\pi^2}{6}\left(\frac{z^2}{1-z^2}\right) + \ln\left(\frac{1}{1-iz}\right) \ln\left(\frac{1}{1+iz}\right)\right|, \tag{111}$$

for any $z \in D(0,1)$.

Corollary 5.2. *Let $g(z) = \sum_{n=0}^{\infty} g_n z^n$ be a power series with complex coefficients and convergent on the open disk $D(0, R)$. If the numerical series $\sum_{n=0}^{\infty} n! |g_n|^2$ is convergent, then*

$$\sum_{n=0}^{\infty} n! |g_n|^2 \exp\left(|z|^2\right) - |g(z)g(\overline{z})| \geq \left|\sum_{n=0}^{\infty} n! |g_n|^2 \exp\left(z^2\right) - g(z)\overline{g(\overline{z})}\right|, \tag{112}$$

for any $z \in D(0, R)$.

This follows from Theorem 5.3 by choosing $f(z) = \exp(z)$.

If we apply the inequality (112) for the function

$$\sin(iz) = \sum_{n=0}^{\infty} \frac{i}{(2n+1)!} z^{2n+1},$$

then we obtain the inequality

$$\sum_{n=0}^{\infty} \frac{n!}{(2n+1)!} \exp\left(|z|^2\right) - |\sin(iz)\sin(i\bar{z})| \qquad (113)$$

$$\geq \left| \sum_{n=0}^{\infty} \frac{n!}{(2n+1)!} \exp\left(z^2\right) - \sin^2(iz) \right|,$$

for any $z \in \mathbb{C}$.

6 More Power Series Inequalities on CBS Type

Utilizing a different technique based on the continuity properties of modulus, in this section, we provide more inequalities for power series related to the CBS inequality. These results contain in [15]. We begin by proving the following result.

Theorem 6.1. *Assume that the power series* $f(z) = \sum_{n=0}^{\infty} p_n z^n$ *with real coefficients is convergent on the disk* $D(0,R)$, $R > 0$. *If* $x, z \in \mathbb{C}$ *are such that* $x, xz, |x||z|^2 \in D(0,R)$, *then we have the inequality*

$$f_A\left(|x||z|^2\right) f_A(|x|) - |f_A(|x|z)|^2 \geq |f(x) f(x|z|z) - f(xz) f(x|z|)| \geq 0. \qquad (114)$$

Proof. If $z \in D(0,R)$, then

$$\left|z^n - z^j\right|^2 = \left|z^n - z^j\right| \left|z^n - z^j\right| \geq \left|z^n - z^j\right| \left||z|^n - |z|^j\right| \qquad (115)$$

for any $n, j \in \mathbb{N}$. We also have

$$\left|z^n - z^j\right|^2 = |z^n|^2 - 2\operatorname{Re}\left(z^n \bar{z}^j\right) + \left|z^j\right|^2 = |z|^{2n} - 2\operatorname{Re}\left(z^n \bar{z}^j\right) + |z|^{2j}$$

and

$$\left|z^n - z^j\right| \cdot \left||z|^n - |z|^j\right| = \left|z^n |z|^n + z^j |z|^j - z^n |z|^j - z^j |z|^n\right|$$

for any $n, j \in \mathbb{N}$.

Utilizing (115) we get the inequality

$$|z|^{2n} - 2\operatorname{Re}\left(z^n \bar{z}^j\right) + |z|^{2j} \geq \left|z^n |z|^n + z^j |z|^j - |z|^j z^n - |z|^n z^j\right| \qquad (116)$$

for any $n, j \in \mathbb{N}$. If we multiply (116) by $|p_n| |x|^n |p_j| |x|^j \geq 0$ where $x \in D(0, R)$ and $n, j \in \mathbb{N}$, then we have

$$\begin{aligned} &|p_n| |x|^n |z|^{2n} p_j |x|^j + |p_n| |x|^n p_j |x|^j |z|^{2j} - 2\operatorname{Re}\left(|p_n| |x|^n z^n |p_j| |x|^j \bar{z}^j\right) \\ &\geq \left| p_n x^n |z|^n z^n p_j x^j + p_n x^n p_j x^j |z|^j z^j - p_n x^n z^n p_j x^j |z|^j - p_n x^n |z|^n p_j x^j z^j \right| \end{aligned} \tag{117}$$

for any $n, j \in \mathbb{N}$.

Summing over n and j from 0 to k and utilizing the triangle inequality for the modulus, we have from (117)

$$\begin{aligned} &\sum_{n=0}^{k} |p_n| |x|^n |z|^{2n} \sum_{j=0}^{k} |p_j| |x|^j + \sum_{n=0}^{k} |p_n| |x|^n \sum_{j=0}^{k} |p_j| |x|^j |z|^{2j} \\ &\quad - 2\operatorname{Re}\left(\sum_{n=0}^{k} |p_n| |x|^n z^n \sum_{j=0}^{k} |p_j| |x|^j (\bar{z})^j\right) \\ &\geq \left| \sum_{n=0}^{k} p_n x^n |z|^n z^n \sum_{j=0}^{k} p_j x^j + \sum_{n=0}^{k} p_n x^n \sum_{j=0}^{k} p_j x^j |z|^j z^j \right. \\ &\quad \left. - \sum_{n=0}^{k} p_n x^n z^n \sum_{j=0}^{k} p_j x^j |z|^j - \sum_{n=0}^{k} p_n x^n |z|^n \sum_{j=0}^{k} p_j x^j z^j \right|. \end{aligned} \tag{118}$$

Since $\sum_{j=0}^{k} |p_j| |x|^j (\bar{z})^j = \overline{\sum_{n=0}^{k} |p_n| |x|^n z^n}$, then

$$\operatorname{Re}\left(\sum_{n=0}^{k} |p_n| |x|^n z^n \sum_{j=0}^{k} |p_j| |x|^j (\bar{z})^j\right) = \left|\sum_{n=0}^{k} |p_n| |x|^n z^n\right|^2. \tag{119}$$

Hence, from the inequality (118), we have

$$\begin{aligned} &\sum_{n=0}^{k} |p_n| |x|^n |z|^{2n} \sum_{n=0}^{k} |p_n| |x|^n - \left|\sum_{n=0}^{k} |p_n| |x|^n z^n\right|^2 \\ &\quad \geq \left| \sum_{n=0}^{k} p_n x^n \sum_{n=0}^{k} p_n x^n |z|^n z^n - \sum_{n=0}^{k} p_n x^n z^n \sum_{n=0}^{k} p_n x^n |z|^n \right|. \end{aligned} \tag{120}$$

Since all the series whose partial sums are involved in (120) are convergent, then by taking the limit over $k \to \infty$ in (120), we deduce the desired inequality (114). □

Corollary 6.1. *If* $\sum_{n=0}^{\infty} |p_n| < \infty$, *i.e.,* $f_A(1) < \infty$, *then for any* $\zeta \in \mathbb{C}$ *with* $|\zeta| = 1$, *we have*

$$f_A\left(|z|^2\right) f_A(1) - |f_A(z)|^2 \geq |f(\zeta) f(\zeta |z| z) - f(\zeta z) f(\zeta |z|)| \geq 0. \quad (121)$$

In particular, for $\zeta = 1$, *we have*

$$f_A\left(|z|^2\right) f_A(1) - |f_A(z)|^2 \geq |f(1) f(|z| z) - f(z) f(|z|)| \geq 0 \quad (122)$$

for any z, $|z|^2 \in D(0, R)$.

Some applications of the inequalities (114) and (122) are as follows:

1. If we apply the inequality (114) for the function $f(z) = \frac{1}{1-z}$, $z \in D(0, 1)$, then we get

$$\frac{|1-z|}{(1-|x|)\left(1-|x||z|^2\right)|1-|x|z|^2} \geq \left|\frac{1-|z|}{(1-x)(1-xz)(1-x|z|)(1-x|z|z)}\right| \quad (123)$$

for any $x, z \in \mathbb{C}$ with $x, |x||z|^2 \in D(0, 1)$.

2. If we apply the inequality (114) for the function $f(z) = \frac{1}{1+z} = \sum_{n=0}^{\infty} (-1)^n z^n$, $z \in D(0, 1)$, then we get the inequality

$$\frac{|1-z|}{(1-|x|)\left(1-|x||z|^2\right)|1-|x|z|^2} \geq \left|\frac{1-|z|}{(1+x)(1+xz)(1+x|z|)(1+x|z|z)}\right| \quad (124)$$

for any $x, z \in \mathbb{C}$ with $x, xz, |x||z|^2 \in D(0, 1)$.

3. If we apply the inequality (122) for the function $f(z) = \exp(z)$, $z \in \mathbb{C}$, then we get the inequality

$$\exp\left(|z|^2 + 1\right) - |\exp(z)|^2 \geq |\exp(z|z| + 1) - \exp(z + |z|)| \quad (125)$$

for any $z \in \mathbb{C}$.

Remark 6.1. The inequality (114) can also be written in the form

$$\det\begin{bmatrix} f_A\left(|x||z|^2\right) & f_A(|x|z) \\ f_A(|x|\bar{z}) & f_A(|x|) \end{bmatrix} \geq \left|\det\begin{bmatrix} f(x) & f(xz) \\ f(x|z|) & f(x|z|z) \end{bmatrix}\right| \quad (126)$$

for any $x, z \in \mathbb{C}$ with $x, xz, |x||z|^2 \in D(0, R)$.

Theorem 6.2 ([15]). *Assume that the power series $f(z) = \sum_{n=0}^{\infty} p_n z^n$ with real coefficients is convergent on the disk $D(0,R)$, $R > 0$. If $x, z \in \mathbb{C}$ are such that $x, xz, |x|\,|z|^2 \in D(0,R)$, then we have the inequality*

$$\begin{aligned} & f_A(|x|)\, f_A\left(|x|\,|z|^2\right) - \operatorname{Re}\left[f_A^2(|x|\,z)\right] \\ & \geq \frac{1}{2}\left|f(x) f(x\,|z|\,z) + f(x) f(x\,|z|\,\overline{z}) - f(xz) f(x\,|z|) - f(x\,|z|)\, f(x\overline{z})\right|. \end{aligned} \tag{127}$$

Proof. If $z \in D(0,R)$, then

$$\begin{aligned} \left|z^n - (\overline{z})^j\right|^2 &= \left|z^n - (\overline{z})^j\right| \left|z^n - (\overline{z})^j\right| \\ &\geq \left|z^n - (\overline{z})^j\right| \left|\,|z|^n - |z|^j\right| \\ &= \left|\,|z|^n z^n + (\overline{z})^j |z|^j - |z|^j z^n - |z|^n (\overline{z})^j\right| \end{aligned} \tag{128}$$

for any $n, j \in \mathbb{N}$. We also have

$$\left|z^n - (\overline{z})^j\right|^2 = |z^n|^2 - 2\operatorname{Re}\left(z^n z^j\right) + \left|z^j\right|^2 = |z|^{2n} - 2\operatorname{Re}\left(z^n z^j\right) + |z|^{2j} \tag{129}$$

for any $n, j \in \mathbb{N}$. Utilizing (128) we have the inequality

$$|z|^{2n} - 2\operatorname{Re}\left(z^n z^j\right) + |z|^{2j} \geq \left|\,|z|^n z^n + (\overline{z})^j z^n - |z|^j z^n - |z|^n (\overline{z})^j\right| \tag{130}$$

for any $n, j \in \mathbb{N}$. Now, on utilizing a similar argument to the one in the proof of Theorem 6.1, we deduce the desired result (127). The details are omitted. □

Corollary 6.2. *If $z = \overline{x}$ in (127), then we have*

$$\begin{aligned} & f_A(|x|)\, f_A\left(|x|^3\right) - \operatorname{Re}\left[f_A^2(|x|\,\overline{x})\right] \\ & \geq \frac{1}{2}\left|f(x) f\left(|x|^3\right) + f(x) f\left(|x|\,x^2\right) - f(|x|^2) f(|x|\,x) - f(|x|\,x)\, f\left(x^2\right)\right| \end{aligned} \tag{131}$$

for any $x \in \mathbb{C}$ such that $x, |x|\,x, |x|\,x^2 \in D(0,R)$.

In the following, we give some applications of above inequality (131) for particular complex functions of interest:

1. If we take the function $f(z) = \dfrac{1}{1+z}$, $z \in D(0,1)$, then we have $f_A(z) = \dfrac{1}{1-z}$, $z \in D(0,1)$. Applying (131), we get the following inequality:

$$\frac{1}{(1-|x|)\left(1-|x|^3\right)} - \operatorname{Re}\left(\frac{1}{1-|x|\,\overline{x}}\right)^2$$
$$\geq \frac{1}{2}\left|\frac{2+|x|\,x^2+|x|^3}{(1+x)\left(1+|x|^3\right)(1+|x|\,x^2)} - \frac{2+x^2+|x|^2}{(1+x^2)(1+|x|\,x)\left(1+|x|^2\right)}\right| \tag{132}$$

for any $x, |x|\,x, |x|\,x^2 \in D(0,1)$.

2. If we apply the inequality (131) for the function $f(z) = \exp(z)$, $z \in \mathbb{C}$, then we get

$$\exp\left(|x| + |x|^3\right) - \operatorname{Re}\left[\exp\left(2\,|x|\,\overline{x}\right)\right]$$
$$\geq \frac{1}{2}\left|\exp(x+|x|^3)+\exp(x+|x|\,x^2)-\exp(|x|^2+|x|\,x)-\exp(|x|\,x+x^2)\right| \tag{133}$$

for any $x \in \mathbb{C}$.

Theorem 6.3 ([15]). *Assume that the power series $f(z) = \sum_{n=0}^{\infty} p_n z^n$ with real coefficients is convergent on the disk $D(0,R)$, $R > 0$. If $x, y \in \mathbb{C}$ are such that $|x|^2, |y|^2 < R$, then we have the inequality*

$$f_A\left(|x|^2\right) f_A\left(|y|^2\right) - |f_A(xy)|^2\, |f(|x|\,x) f(|y|\,\overline{y}) - f(|y|\,x) f(|x|\,\overline{y})|. \tag{134}$$

Proof. If $x, y \in \mathbb{C}$, then we have

$$\left|x^n(\overline{y})^j - x^j(\overline{y})^n\right|^2 = \left|x^n(\overline{y})^j - x^j(\overline{y})^n\right|\left|x^n(\overline{y})^j - x^j(\overline{y})^n\right|$$
$$\geq \left|x^n(\overline{y})^j - x^j(\overline{y})^n\right|\left||x|^n|y|^j - |x|^j|y|^n\right| \tag{135}$$

for any $n, j \in \mathbb{N}$. We have upon simple calculations that

$$|x|^{2n}|y|^{2j} - 2\operatorname{Re}\left(x^n y^n(\overline{x})^j(\overline{y})^j\right) + |y|^{2n}|x|^{2j}$$
$$\geq \left||x|^n x^n |y|^j(\overline{y})^j + |y|^n(\overline{y})^n|x|^j x^j - |y|^n x^n|x|^j(\overline{y})^j - |x|^n(\overline{y})^n|y|^j x^j\right| \tag{136}$$

for any $n, j \in \mathbb{N}$.

If we multiply the inequality (136) with $|p_n|\left|p_j\right| \geq 0$ and summing over n and j from 0 to k, then we have

$$\sum_{n=0}^{k}|p_n|\,|x|^{2n}\sum_{j=0}^{\infty}|p_j|\,|y|^{2j} + \sum_{n=0}^{k}|p_n|\,|y|^{2n}\sum_{j=0}^{k}|p_j|\,|x|^{2j}$$

$$
\begin{aligned}
&-2\,\mathrm{Re}\left(\sum_{n=0}^{k}|p_n|\,x^n y^n \sum_{j=0}^{k}|p_j|\,(\overline{x})^j\,(\overline{y})^j\right)\\
&\geq \left|\sum_{n=0}^{k} p_n\,|x|^n\,x^n \sum_{j=0}^{k} p_j\,|y|^j\,(\overline{y})^j + \sum_{n=0}^{k} p_n\,|y|^n\,(\overline{y})^n \sum_{j=0}^{k} p_j\,|x|^j\,x^j\right.\\
&\quad \left. - \sum_{n=0}^{k} p_n\,|y|^n\,x^n \sum_{j=0}^{k} p_j\,|x|^j\,(\overline{y})^j - \sum_{n=0}^{k} p_n\,|x|^n\,(\overline{y})^n \sum_{j=0}^{k} p_j\,|y|^j\,x^j\right|. \qquad (137)
\end{aligned}
$$

Due to the fact that $\sum_{n=0}^{k}|p_n|\,x^n y^n \sum_{j=0}^{\infty}|p_j|\,(\overline{x})^j\,(\overline{y})^j = \left|\sum_{n=0}^{k}|p_n|\,x^n y^n\right|^2$, the inequality (137) is equivalent with

$$
\begin{aligned}
&\sum_{n=0}^{k}|p_n|\,|x|^{2n}\sum_{n=0}^{k}|p_n|\,|y|^{2n} - \left|\sum_{n=0}^{k}|p_n|\,x^n y^n\right|^2\\
&\quad\geq \left|\sum_{n=0}^{k} p_n\,|x|^n\,x^n \sum_{n=0}^{k} p_n\,|y|^n\,(\overline{y})^n - \sum_{n=0}^{k} p_n\,|y|^n\,x^n \sum_{n=0}^{k} p_n\,|x|^n\,(\overline{y})^n\right|. \qquad (138)
\end{aligned}
$$

Since all the series with the partial sums involved in (138) are convergent, then by taking the limit over $k \to \infty$ in (138), we deduce the desired result from (134). □

Remark 6.2. The inequality (134) is also equivalent to

$$
\det\begin{bmatrix} f_A\left(|x|^2\right) & f_A(xy) \\ f_A(\overline{xy}) & f_A\left(|y|^2\right)\end{bmatrix} \geq \left|\det\begin{bmatrix} f(|x|\,x) & f(|y|\,x) \\ f(|x|\,\overline{y}) & f(|y|\,\overline{y})\end{bmatrix}\right| \qquad (139)
$$

for any $x, y \in \mathbb{C}$ with $|x|^2, |y|^2 < R$.

The inequality (134) has some applications for particular complex functions of interest which will be pointed out as follows:

1. If we apply the inequality (134) for the function $f(z) = \dfrac{1}{1-z}$, $z \in D(0,1)$, then we get

$$
\begin{aligned}
&\frac{1}{\left(1-|x|^2\right)\left(1-|y|^2\right)} - \frac{1}{|1-xy|^2}\\
&\quad\geq \left|\frac{1}{(1-x\,|x|)\,(1-|y|\,\overline{y})} - \frac{1}{(1-|y|\,x)\,(1-|x|\,\overline{y})}\right| \qquad (140)
\end{aligned}
$$

for any $x, y \in \mathbb{C}$. In particular, if in (140) we choose $y = 0$, then we obtain the simpler inequality

$$\frac{1}{1-|x|^2} - 1 \geq \left| \frac{1}{1-x\,|x|} - 1 \right| \tag{141}$$

for any $x \in \mathbb{C}$.

2. If we apply the inequality (134) for the function $f(z) = \exp(z)$, $z \in \mathbb{C}$, then we have

$$\exp\left(|x|^2 + |y|^2\right) - |\exp(xy)|^2 \geq |\exp(x\,|x| + |y|\,\overline{y}) - \exp(|y|\,x + |x|\,\overline{y})| \tag{142}$$

for any $x, y \in \mathbb{C}$. In particular, if in (142) we choose $y = 0$, then we get

$$\exp\left(|x|^2\right) - 1 \geq |\exp(x\,|x|) - 1| \tag{143}$$

for any $x \in \mathbb{C}$.

3. If we take the function $f(z) = \cos(z)$, $z \in \mathbb{C}$, then we have $f_A(z) = \cosh(z)$, $z \in \mathbb{C}$. Utilizing the inequality (134) for $f(z)$ as above gives

$$\begin{aligned} &\cosh\left(|x|^2\right)\cosh\left(|y|^2\right) - |\cosh(xy)|^2 \\ &\geq |\cos(x\,|x|)\cos(|y|\,\overline{y}) - \cos(|y|\,x)\cos(|x|\,\overline{y})| \end{aligned} \tag{144}$$

for any $x, y \in \mathbb{C}$. In particular, we have, with $y = 0$ in (144),

$$\cosh\left(|x|^2\right) - 1 \geq |\cos(x\,|x|) - 1| \tag{145}$$

for any $x \in \mathbb{C}$.

Theorem 6.4 ([15]). *Assume that the power series $f(z) = \sum_{n=0}^{\infty} p_n z^n$ with real coefficients is convergent on the disk $D(0, R)$, $R > 0$. If $x, y \in \mathbb{C}$ are such that $|x|^2, |y|^2 < R$, then we have the inequality*

$$\begin{aligned} &f_A\left(|x|^2\right) f_A\left(|y|^2\right) - \operatorname{Re}\left[f_A^2(x\overline{y})\right] \\ &\geq \frac{1}{2}\,|f(|x|\,x) f(|y|\,\overline{y}) + f(|x|\,\overline{x}) f(|y|\,y) - f(|x|\,y) f(|y|\,\overline{x}) - f(|y|\,x) f(|x|\,\overline{y})|\,. \end{aligned} \tag{146}$$

Proof. If $x, y \in D(0, R)$, then we have

$$\begin{aligned} \left|x^n(\overline{y})^j - (\overline{x})^j y^n\right|^2 &= \left|x^n(\overline{y})^j - (\overline{x})^j y^n\right|\left|x^n(\overline{y})^j - (\overline{x})^j y^n\right| \\ &\geq \left|x^n(\overline{y})^j - (\overline{x})^j y^n\right|\left||x|^n|y|^j - |x|^j|y|^n\right| \end{aligned} \tag{147}$$

for any $n, j \in \mathbb{N}$. Doing simple calculations we get that

$$|x|^{2n}|y|^{2j} - 2\operatorname{Re}\left[x^n(\overline{y})^j x^j (\overline{y})^n\right] + |x|^{2j}|y|^{2n}$$
$$\geq \left||x|^n x^n |y|^j (\overline{y})^j + |x|^j (\overline{x})^j |y|^n y^n - |x|^n y^n |y|^j (\overline{x})^j - |y|^n x^n |x|^j (\overline{y})^j\right| \tag{148}$$

for any $n, j \in \mathbb{N}$.

If we multiply (148) with $|p_n|\,|p_j| \geq 0$ and summing over n and j from 0 to k, then we get

$$\sum_{n=0}^{k}|p_n|\,|x|^{2n}\sum_{j=0}^{k}|p_j|\,|y|^{2j} - 2\operatorname{Re}\left[\sum_{n=0}^{k}|p_n|\,x^n(\overline{y})^n\sum_{j=0}^{k}|p_j|\,x^j(\overline{y})^j\right]$$
$$+\sum_{n=0}^{k}|p_n|\,|y|^{2n}\sum_{j=0}^{k}|p_j|\,|x|^{2j}$$
$$\geq \left|\sum_{n=0}^{k}p_n|x|^n x^n\sum_{j=0}^{k}p_j|y|^j(\overline{y})^j + \sum_{n=0}^{k}p_n y^n|y|^n\sum_{j=0}^{k}p_j|x|^j(\overline{x})^j\right.$$
$$\left.-\sum_{n=0}^{k}p_n|x|^n y^n\sum_{j=0}^{k}p_j|y|^j(\overline{x})^j - \sum_{n=0}^{k}p_n|y|^n x^n\sum_{j=0}^{k}p_j|x|^j(\overline{y})^j\right| \tag{149}$$

for any $n, j \in \mathbb{N}$.

Since all the series whose partial sums are involved in (149) are convergent, then by taking the limit over $k \to \infty$ in (149), we deduce the desired result (146). □

The inequality (146) is also a valuable source of particular inequalities for complex functions of interest that will be outlined in the following:

1. In (146), we take the function $f(z) = \exp(z)$, $z \in \mathbb{C}$, then we can state that

$$\exp\left(|x|^2 + |y|^2\right) - \operatorname{Re}\left[\exp\left(2x\overline{y}\right)\right]$$
$$\geq \frac{1}{2}\left|\exp\left(|x|\,x + |y|\,\overline{y}\right) + \exp\left(|x|\,\overline{x}\right) + |y|\,y\right)$$
$$-\exp\left(|x|\,y\right) + |y|\,\overline{x}) - \exp\left(|y|\,x + |x|\,\overline{y}\right)| \tag{150}$$

for any $x, y \in \mathbb{C}$. If in (150) we choose $y = 0$, then we obtain the simpler result:

$$\exp\left(|x|^2\right) - 1 \geq \frac{1}{2}\left|\exp\left(|x|\,x\right) + \exp\left(|x|\,\overline{x}\right) - 2\right| \tag{151}$$

for any $x \in \mathbb{C}$.

2. If we apply the inequality (146) for the function $f(z) = \cos(z)$, $z \in \mathbb{C}$, with its transform $f_A(z) = \cosh(z)$, $z \in \mathbb{C}$, then we get

$$\begin{aligned}
&\cosh\left(|x|^2\right)\cosh\left(|y|^2\right) - \operatorname{Re}\left[\cosh^2(x\overline{y})\right] \\
&\geq \frac{1}{2}\left|\cos(|x|\,x)\cos(|y|\,\overline{y}) + \cos(|x|\,\overline{x})\cos(|y|\,y)\right. \\
&\quad \left. - \cos(|x|\,y)\cos(|y|\,\overline{x}) - \cos(|y|\,x)\cos(|x|\,\overline{y})\right|, \qquad (152)
\end{aligned}$$

for any $x, y \in \mathbb{C}$. In particular, if in (152) we choose $y = 0$, then we obtain that

$$\cosh\left(|x|^2\right) - 1 \geq \frac{1}{2}\left|\cos(|x|\,x) + \cos(|x|\,\overline{x}) - 2\right| \qquad (153)$$

for any $x \in \mathbb{C}$.

7 Applications to Special Functions

7.1 Definition and Basic Concepts

In this section, we give some inequalities for special functions such as polylogarithm, hypergeometric, Bessel and modified Bessel functions of the first kind. Before we state our results for these special functions that can be obtained on utilizing the de Bruijn, the Buzano and the Schwarz inequality, we recall here the definitions and some basic concepts.

The *polylogarithm* $Li_n(z)$ also known as *Jonquiére's function* is a function defined by the power series:

$$Li_n(z) = \sum_{k=1}^{\infty} \frac{z^k}{k^n}. \qquad (154)$$

This series (154) converges absolutely for all complex values of the order n and the argument z where $z \in D(0,1)$.

The polylogarithm of nonnegative order n arises in the sums of the form

$$Li_{-n}(r) = \sum_{k=1}^{\infty} k^n r^k = \frac{1}{(1-r)^{n+1}} \sum_{i=0}^{s} E_{n,i} r^{n-i},$$

where $E_{n,i}$ is an *Eulerian number*, namely, we recall that

$$E_{n,k} := \sum_{j=0}^{k+1} (-1)^j \binom{n+1}{j} (k-j+1)^n .$$

Polylogarithms also arise in sum of *generalized harmonic numbers* $H_{n,r}$ as

$$\sum_{n=1}^{\infty} H_{n,r} z^n = \frac{Li_r(z)}{1-z},$$

for $z \in D(0,1)$, where $H_{n,r} := \sum_{k=1}^{\infty} \frac{1}{k^r}$ and $H_{n,1} := H_n = \sum_{k=1}^{\infty} \frac{1}{k}$.

The polylogarithm function involves the ordinary logarithm for $n = 1$, i.e., $Li_1(z) = \ln\left(\frac{1}{1-z}\right)$, while for $n = 2$, we have

$$Li_2(z) = \sum_{k=1}^{\infty} \frac{z^k}{k^2}, \quad z \in D(0,1), \tag{155}$$

which is called the *dilogarithm* or *Spence's function.* Other special forms of low-order polylogarithms include

$$Li_{-2}(z) = \frac{z(z+1)}{(1-z)^3}, \quad Li_{-1}(z) = \frac{z}{(1-z)^2}, \quad Li_0(z) = \frac{z}{1-z}$$

for all $z \in D(0,1)$. The polylogarithm also has relationship to other functions for the special cases of argument z. For instance

$$Li_n(1) = \zeta(n), \quad Li_n(-1) = -\eta(n), \quad Li_n(\pm i) = \frac{1}{2^s}\eta(n) \pm i\beta(n),$$

where $\zeta(n)$, $\eta(n)$ and $\beta(n)$ are the *Riemann zeta, Dirichlet eta* and *Dirichlet beta function*, respectively.

The *hypergeometric function* ${}_2F_1(a,b;c;z)$ is defined for all $|z| < 1$ by the series

$${}_2F_1(a,b;c;z) = \sum_{n=0}^{\infty} \frac{(a)_n (b)_n}{(c)_n} \frac{z^n}{n!} \tag{156}$$

for arbitrary $a, b, c \in \mathbb{R}$ with $c \neq 0, -1, -2, \ldots$, and the $(t)_n$, $n \in \{0, 1, 2, \ldots\}$ is a *Pochhammer* symbol which is defined by

$$(t)_n = \begin{cases} 1, & \text{if } n = 0, \\ t(t+1)\cdots(t+n-1), & \text{if } n > 0. \end{cases} \tag{157}$$

Hypergeometric function (156) with particular arguments of a, b, and c reduces to elementary functions, for example,

$$
\begin{aligned}
{}_2F_1\left(1,1;1;z\right) &= \frac{1}{1-z} & &= {}_2F_1\left(1,2;2;z\right),\\
{}_2F_1\left(1,2;1;z\right) &= \frac{1}{\left(1-z\right)^2}, & {}_2F_1\left(a,b;b;z\right) &= \frac{1}{\left(1-z\right)^a},\\
{}_2F_1\left(1,1;2;z\right) &= \frac{1}{z}\ln\left(\frac{1}{1-z}\right), & {}_2F_1\left(1,1;2;-z\right) &= \frac{1}{z}\ln\left(1+z\right).
\end{aligned}
$$

The *Bessel functions* of the first kind $J_\alpha(z)$ are defined as the solutions to the *Bessel differential equation*, i.e.,

$$z^2\frac{d^2y}{dz^2}+z\frac{dy}{dz}+\left(z^2-\alpha^2\right)y=0 \tag{158}$$

for an arbitrary real or complex order α. This solution of (158) is an analytic function of z in $\mathbb{C}$, except for a point $z=0$ when α is not an integer. These solutions, denoted by $J_\alpha(z)$, are defined by Taylor series expansion around the origin [1, p. 360], i.e.,

$$J_\alpha(z)=\sum_{n=0}^{\infty}\frac{(-1)^n}{\Gamma\left(n+\alpha+1\right)n!}\left(\frac{z}{2}\right)^{2n+\alpha}, \tag{159}$$

where $\Gamma(x)$ is the *gamma function*.

For non-integer order α, $J_\alpha(z)$ and $J_{-\alpha}(z)$ are linearly independent and therefore the two solutions of the differential equation (158). The $J_\alpha(z)$ and $J_{-\alpha}(z)$ are linearly dependent for α integer; hence, the following relationship is valid [1, p. 358]:

$$J_{-\alpha}(z)=(-1)^\alpha J_\alpha(z). \tag{160}$$

If z in (158) is replaced by the arguments $\pm iz$, then the solutions of the second-order differential equation, $I_\alpha(z)$, are called the *modified Bessel functions* of the first kind. It is easy to verify from (159) that the modified Bessel function is defined by the following power series [1, p. 375]:

$$I_\alpha(z)=\sum_{n=0}^{\infty}\frac{1}{\Gamma\left(n+\alpha+1\right)n!}\left(\frac{z}{2}\right)^{2n+\alpha} \tag{161}$$

for $\alpha, z\in\mathbb{C}$. We observe that the function $I_\alpha(z)$ has all the nonnegative coefficients. Similar to Bessel functions, the modified Bessel function (161) also satisfies the following relations:

$$I_\alpha(-z)=(-1)^\alpha I_\alpha(z) \text{ and } I_{-\alpha}(z)=I_\alpha(z)$$

for $\alpha \in \mathbb{Z}, z \in \mathbb{C}$. The modified Bessel functions of the first kind of order α, $I_\alpha(z)$, can be expressed by the Bessel function of the first kind, that is,

$$J_\alpha(iz) = i^\alpha I_\alpha(z). \tag{162}$$

7.2 Some Inequalities for Polylogarithm Functions

It is clearly seen that the polylogarithm function (154) is the power series with nonnegative coefficients and convergent on the open disk $D(0,1)$, so that all the results in Sects. 3–6 hold true. Therefore, for instance from (17), (39), (79), (114) and (134), we have the following corollaries.

Corollary 7.1. *If $Li_n(z)$ is the polylogarithm function, then we have*

$$|Li_n(az)|^2 \leq \frac{1}{2} Li_n(a^2)\left[Li_n\left(|z|^2\right) + \left|Li_n(z^2)\right|\right] \tag{163}$$

for $a \in (-1,1)$, $z \in \mathbb{C}$ with $az, a^2, z^2, |z|^2 \in D(0,1)$ and n is a negative or a positive integer.

Corollary 7.2. *If $Li_n(z)$ is the polylogarithm function, then we have*

$$\left|Li_n(\alpha\overline{x})\, Li_n(\overline{\beta}x)\right| \leq \frac{1}{2}\left[\left[Li_n(|\alpha|^2) Li_n(|\beta|^2)\right]^{1/2} + \left|Li_n(\alpha\overline{\beta})\right|\right] Li_n(|x|^2) \tag{164}$$

for any $\alpha, \beta, x \in C$ with $|\alpha|^2, |\beta|^2, |x|^2, \alpha\overline{\beta}, \alpha\overline{x}, \overline{\beta}x \in D(0,1)$ and n is a negative or a positive integer.

Corollary 7.3. *If $Li_n(z)$ is the polylogarithm function, then we have*

$$\begin{aligned} &\left[Li_n\left(|x|^2\right) Li_n\left(|y|^2\right)\right]^{1/2} Li_n\left(|z|^2\right) - |Li_n(x\overline{z})\, Li_n(z\overline{y})| \\ &\geq \left|Li_n(x\overline{y})\, Li_n\left(|z|^2\right) - Li_n(x\overline{z})\, Li_n(z\overline{y})\right| \end{aligned} \tag{165}$$

for any $x, y, z \in \mathbb{C}$ such that $|x|^2, |y|^2, |z|^2, x\overline{z}, z\overline{y}, x\overline{y} \in D(0,1)$ and n is a negative or a positive integer.

Corollary 7.4. *If $Li_n(z)$ is the polylogarithm function, then we have*

$$Li_n\left(|x|\,|z|^2\right) f_A(|x|) - |Li_n(|x|\,z)|^2 \geq |Li_n(x) Li_n(x\,|z|\,z) - Li_n(xz) Li_n(x\,|z|)| \tag{166}$$

and

$$Li_n\left(|x|^2\right)Li_n\left(|y|^2\right) - |Li_n(xy)|^2 \geq |Li_n(|x|\,x)Li_n(|y|\,\overline{y}) - Li_n(|y|\,x)Li_n(|x|\,\overline{y})|, \tag{167}$$

for any $x, y, z \in C$ *with* $x, xz, |x|\,|z|^2, |x|^2, |z|^2 \in D(0,1)$ *and* n *is a negative or a positive integer.*

In the following, we present some results that connect different order polylogarithms.

Theorem 7.1 ([4]). *Let* $Li_n(z)$ *be the polylogarithm function,* $a \in (-1,1)$, $z \in D(0,1)$ *and* p, q, r *integers such that the following series exist. Then*

$$\left|Li_{r+p+q}(az)\right|^2 \leq \frac{1}{2} Li_{r+2p}\left(a^2\right)\left[Li_{r+2q}\left(|z|^2\right) + \left|Li_{r+2q}(z^2)\right|\right]. \tag{168}$$

Proof. Utilizing the de Bruijn inequality (18) with positive weights we have

$$\begin{aligned}
\left|Li_{r+p+q}(az)\right|^2 &= \left|\sum_{k=1}^{\infty} \frac{1}{k^r} \cdot \frac{a^k}{k^p} \cdot \frac{z^k}{k^q}\right|^2 \\
&\leq \frac{1}{2}\left(\sum_{k=1}^{\infty} \frac{1}{k^r} \cdot \frac{a^k}{k^p}\right)\left[\sum_{k=1}^{\infty} \frac{1}{k^r} \cdot \frac{|z|^{2k}}{k^{2q}} + \left|\sum_{k=1}^{\infty} \frac{1}{k^r} \cdot \frac{z^{2k}}{k^{2q}}\right|\right] \\
&= \frac{1}{2} Li_{r+2p}\left(a^2\right)\left[Li_{r+2q}\left(|z|^2\right) + \left|Li_{r+2q}(z^2)\right|\right]
\end{aligned} \tag{169}$$

and the inequality (168) is proved. □

Theorem 7.2 ([13]). *Let* $\alpha, \beta, x \in \mathbb{C}$, $\alpha\overline{x}, \overline{\beta}x, |\alpha|^2, |\beta|^2, \alpha\overline{\beta}, |x|^2 \in D(0,1)$ *and* p, q, r *integers such that the following series exist. Then*

$$\begin{aligned}
&\left|Li_{r+p+q}(\alpha\overline{x})\, Li_{r+p+q}\left(\overline{\beta}x\right)\right| \\
&\quad \leq \frac{1}{2}\left(\left[Li_{r+2q}\left(|\alpha|^2\right) Li_{r+2q}\left(|\beta|^2\right)\right]^{1/2} + \left|Li_{r+2q}\left(\alpha\overline{\beta}\right)\right|\right) Li_{r+2p}\left(|x|^2\right).
\end{aligned} \tag{170}$$

Proof. Utilizing the Buzano inequality (36) for $p_k = 1/k^r$, $c_k = \alpha^k/k^q$, $b_k = \beta^k/k^q$ and $x_k = x^k/k^p$, we have

$$\left|Li_{r+p+q}(\alpha\overline{x})\, Li_{r+p+q}\left(\overline{\beta}x\right)\right| = \left|\sum_{k=1}^{\infty} \frac{1}{k^r} \frac{\alpha^k}{k^q} \frac{(\overline{x})^k}{k^p} \sum_{k=1}^{\infty} \frac{1}{k^r} \frac{\left(\overline{\beta}\right)^k}{k^q} \frac{(x)^k}{k^p}\right|,$$

$$\leq \frac{1}{2}\left(\left[\sum_{k=1}^{\infty} \frac{1}{k^r} \frac{\left(|\alpha|^2\right)^k}{k^{2q}} \sum_{k=1}^{\infty} \frac{1}{k^r} \frac{\left(|\beta|^2\right)^k}{k^{2q}}\right]^{1/2} + \left|\sum_{k=1}^{\infty} \frac{1}{k^r} \frac{\left(\alpha\overline{\beta}\right)^k}{k^{2q}}\right|\right) \sum_{k=1}^{\infty} \frac{1}{k^r} \frac{\left(|x|^2\right)^k}{k^{2p}},$$

$$= \frac{1}{2}\left(\left[Li_{r+2q}\left(|\alpha|^2\right) Li_{r+2q}\left(|\beta|^2\right)\right]^{1/2} + \left|Li_{r+2q}\left(\alpha\overline{\beta}\right)\right|\right) Li_{r+2p}\left(|x|^2\right), \quad (171)$$

and the inequality is proved. □

On making use of the above result (170), we can get some simpler inequalities. For instance, if $\alpha = z$, $\beta = \overline{z}$, then from (170) we can state that

$$\left|Li_{r+p+q}\left(z\overline{x}\right) Li_{r+p+q}\left(zx\right)\right| \le \frac{1}{2}\left[Li_{r+2q}\left(|z|^2\right) + \left|Li_{r+2q}\left(z^2\right)\right|\right] Li_{r+2p}\left(|x|^2\right). \quad (172)$$

Moreover, if $x = a \in R$, then from (172) we deduce the inequality (163) or the inequality (33) in paper [4].

Theorem 7.3. *Let $x, y, z \in \mathbb{C}$ with $|x|^2, |y|^2, |z|^2, x\overline{z}, z\overline{y}, x\overline{y} \in D(0,1)$ and p, q, r integers such that the following series exist. Then*

$$\left[Li_{r+2q}\left(|x|^2\right) Li_{r+2q}\left(|y|^2\right)\right]^{1/2} Li_{r+2p}\left(|z|^2\right) - \left|Li_{r+p+q}\left(x\overline{z}\right) Li_{r+p+q}\left(z\overline{y}\right)\right| \ge \left|Li_{r+2q}\left(x\overline{y}\right) Li_{r+2p}\left(|z|^2\right) - Li_{r+p+q}\left(x\overline{z}\right) Li_{r+p+q}\left(z\overline{y}\right)\right|. \quad (173)$$

Proof. Utilizing the discrete inequality (77) for $p_k = \frac{1}{k^r}$, $x_k = \frac{x^k}{k^q}$, $y_k = \frac{y^k}{k^q}$, $z_k = \frac{z^k}{k^p}$, $k \in \{1, 2, \ldots, m\}$, we have

$$\left(\sum_{k=1}^{m} \frac{1}{k^r}\left|\frac{x^k}{k^q}\right|^2\right)^{1/2} \left(\sum_{k=1}^{m} \frac{1}{k^r}\left|\frac{y^k}{k^q}\right|^2\right)^{1/2} \sum_{k=1}^{m} \frac{1}{k^r}\left|\frac{z^k}{k^p}\right|^2 - \left|\sum_{k=1}^{n} \frac{1}{k^r}\frac{x^k}{k^q}\frac{(\overline{z})^k}{k^p} \sum_{k=1}^{n} \frac{1}{k^r}\frac{z^k}{k^p}\frac{(\overline{y})^k}{k^q}\right|$$
$$\ge \left|\sum_{k=1}^{m} \frac{1}{k^r}\frac{x^k}{k^q}\frac{(\overline{y})^k}{k^q} \sum_{k=1}^{n} \frac{1}{k^r}\left|\frac{z^k}{k^p}\right|^2 - \sum_{k=1}^{m} \frac{1}{k^r}\frac{x^k}{k^q}\frac{(\overline{z})^k}{k^p} \sum_{k=1}^{m} \frac{1}{k^r}\frac{z^k}{k^p}\frac{(\overline{y})^k}{k^q}\right|;$$

hence

$$\left(\sum_{k=1}^{m} \frac{1}{k^{r+2q}}\left(|x|^2\right)^k\right)^{1/2} \left(\sum_{k=1}^{m} \frac{1}{k^{r+2q}}\left(|y|^2\right)^k\right)^{1/2} \sum_{k=1}^{m} \frac{1}{k^{r+2p}}\left(|z|^2\right)^k - \left|\sum_{k=1}^{m} \frac{1}{k^{r+p+q}}\left(x\overline{z}\right)^k \sum_{k=1}^{m} \frac{1}{k^{r+p+q}}\left(z\overline{y}\right)^k\right|$$
$$\ge \sum_{k=1}^{m} \frac{(x\overline{y})^k}{k^{r+2q}} \sum_{k=1}^{m} \frac{\left(|z|^2\right)^k}{k^{r+2p}} - \sum_{k=1}^{m} \frac{(x\overline{z})^k}{k^{r+p+q}} \sum_{k=1}^{m} \frac{(z\overline{y})^k}{k^{r+p+q}}, \quad (174)$$

for $m \ge 0$.

Taking the limit as $m \to \infty$ in (174), then we deduce the desired inequality (173). □

On making use of the above result (173), we can get some simpler inequalities; for instance, if $y = \overline{x}$, then from (173) we can state that

$$\begin{aligned} & Li_{r+2q}\left(|x|^2\right) Li_{r+2p}\left(|z|^2\right) - \left|Li_{r+p+q}(xz)\, Li_{r+p+q}(x\overline{z})\right| \\ & \geq \left|Li_{r+2q}\left(x^2\right) L_{r+2p}\left(|z|^2\right) - Li_{r+p+q}(xz)\, Li_{r+p+q}(x\overline{z})\right|, \end{aligned} \tag{175}$$

for $x, z \in \mathbb{C}$. Moreover, if $z = a \in \mathbb{R}$, then from (175) we deduce the inequality

$$\begin{aligned} & Li_{r+2q}\left(|x|^2\right) Li_{r+2p}\left(a^2\right) - \left|Li_{r+p+q}(ax)\right|^2 \\ & \geq \left|Li_{r+2q}\left(x^2\right) Li_{r+2p}\left(a^2\right) - Li^2_{r+p+q}(ax)\right|, \end{aligned} \tag{176}$$

for any $x \in \mathbb{C}$, $a \in \mathbb{R}$.

From a different perspective, we can state the following inequality which incorporates the Riemann zeta function, ζ:

Corollary 7.5. *Let $z \in D(0,1)$ and p, q, r integers such that $r + 2p > 1$. Then*

$$\left|Li_{r+p+q}(az)\right|^2 \leq \frac{1}{2}\zeta(r+2p)\left[Li_{r+2q}\left(|z|^2\right) + \left|Li_{r+2q}(z^2)\right|\right]. \tag{177}$$

Proof. The proof follows by Theorem 7.1 for $a = 1$. □

Corollary 7.6. *Let $\alpha, \beta \in D(0,1)$ and p, q, r integers such that $r + 2p > 1$. Then*

$$\begin{aligned} & \left|Li_{r+p+q}(-\alpha i)\, Li_{r+p+q}\left(\overline{\beta} i\right)\right| \\ & \leq \frac{1}{2}\zeta(r+2p)\left(\left[Li_{r+2q}\left(|\alpha|^2\right) Li_{r+2q}\left(|\beta|^2\right)\right]^{1/2} + \left|Li_{r+2q}\left(\alpha\overline{\beta}\right)\right|\right). \end{aligned} \tag{178}$$

The proof follows by Theorem 7.2 for $x = i$.

Corollary 7.7. *Let $x \in D(0,1)$ and p, q, r integers. Then*

$$\begin{aligned} & \zeta(r+2p)\, Li_{r+2q}\left(|x|^2\right) - \left|Li_{r+p+q}(x)\right|^2 \\ & \geq \left|\zeta(r+2p)\, Li_{r+2q}\left(x^2\right) - Li^2_{r+p+q}(x)\right|. \end{aligned} \tag{179}$$

The proof follows by (176) for $a = 1$.

On utilizing the inequalities (177)–(179) and taking into account that some particular values of $\zeta(n)$ are known, such as (34) and $\zeta(4) = \pi^4/90$, then we can state the following inequalities:

1. $\left|Li_{q+1}(z)\right|^2 \le \frac{\pi^2}{12}\left[Li_{2q}\left(|z|^2\right) + \left|Li_{2q}(z^2)\right|\right]$

2. $\left|Li_{q+2}(z)\right|^2 \le \frac{\pi^4}{180}\left[Li_{2q}\left(|z|^2\right) + \left|Li_{2q}(z^2)\right|\right]$

3. $\left|Li_{q+1}(-\alpha i)\, Li_{q+1}\left(\overline{\beta} i\right)\right|$
$\le \frac{\pi^2}{12}\left(\left[Li_{2q}\left(|\alpha|^2\right) Li_{2q}\left(|\beta|^2\right)\right]^{1/2} + \left|Li_{2q}\left(\alpha\overline{\beta}\right)\right|\right)$

4. $\left|Li_{q+2}(-\alpha i)\, Li_{q+2}\left(\overline{\beta} i\right)\right|$
$\le \frac{\pi^2}{180}\left(\left[Li_{2q}\left(|\alpha|^2\right) Li_{2q}\left(|\beta|^2\right)\right]^{1/2} + \left|Li_{2q}\left(\alpha\overline{\beta}\right)\right|\right)$

5. $\left|Li_{q+3}(-\alpha i)\, Li_{q+3}\left(\overline{\beta} i\right)\right|$
$\le \frac{\pi^4}{180}\left(\left[Li_{2(q+1)}\left(|\alpha|^2\right) Li_{2(q+1)}\left(|\beta|^2\right)\right]^{1/2} + \left|Li_{2(q+1)}\left(\alpha\overline{\beta}\right)\right|\right)$

6. $\left|\frac{\pi^2}{6} Li_{2q}\left(x^2\right) - Li^2_{q+1}(x)\right| \le \frac{\pi^2}{6} Li_{2q}\left(|x|^2\right) - \left|Li_{q+1}(x)\right|^2$

7. $\left|\frac{\pi^4}{90} Li_{2q}\left(x^2\right) - Li^2_{q+2}(x)\right| \le \frac{\pi^4}{90} Li_{2q}\left(|x|^2\right) - \left|Li_{q+2}(x)\right|$

8. $\left|\frac{\pi^4}{90} Li_{2(q+1)}\left(x^2\right) - Li^2_{q+3}(x)\right| \le \frac{\pi^4}{90} Li_{2(q+1)}\left(|x|^2\right) - \left|Li_{q+3}(x)\right|^2$

for any $\alpha, \beta, x, z \in D(0,1)$ and q an integer.

7.3 Some Inequalities for Hypergeometric Functions

It is clearly seen that the hypergeometric function (156) is the power series with nonnegative coefficients and convergent on the open disk $D(0,1)$, so that all the results in Sects. 3–6 hold true. Therefore, for instance from (134), we have the following corollaries (see [15]).

Corollary 7.8. *Let ${}_2F_1\,(a,b;c;z)$ be the hypergeometric function. Then*

$$
\begin{aligned}
&{}_2F_1\left(a,b;c;|x|^2\right){}_2F_1\left(a,b;c;|y|^2\right)-\left|{}_2F_1\,(a,b;c;xy)\right|^2\\
&\quad\geq \left|{}_2F_1\,(a,b;c;|x|\,x)\,{}_2F_1\,(a,b;c;|y|\,\overline{y})-{}_2F_1\,(a,b;c;|y|\,x)\,{}_2F_1\,(a,b;c;|x|\,\overline{y})\right|
\end{aligned}
\tag{180}
$$

for any $a,b,c\in\mathbb{R}$, with $c\neq 0,-1,-2,\ldots$ and $x,y\in\mathbb{C}$ such that $|x|,|y|<1$.

Corollary 7.9. *If in* (180) *we choose $c=b$, then we have*

$$
\begin{aligned}
&\frac{1}{\left[\left(1-|x|^2\right)\left(1-|y|^2\right)\right]^a}-\frac{1}{|1-xy|^{2a}}\\
&\quad\geq\left|\frac{1}{\left[(1-|x|\,x)\,(1-|y|\,\overline{y})\right]^a}-\frac{1}{\left[(1-|y|\,x)\,(1-|x|\,\overline{y})\right]^a}\right|
\end{aligned}
\tag{181}
$$

for any $a\in\mathbb{R}$, $x,y\in\mathbb{C}$ such that $|x|,|y|<1$.

Remark 7.1. For $a=1$, the inequality (181) reduces to (140).

Corollary 7.10. *If in* (180) *we choose $a=b=1$, $c=2$, then we have*

$$
\begin{aligned}
&\ln\left(\frac{1}{1-|x|^2}\right)\ln\left(\frac{1}{1-|y|^2}\right)-\left|\ln\left(\frac{1}{1-xy}\right)\right|^2\\
&\quad\geq\left|\ln\left(\frac{1}{1-|x|\,x}\right)\ln\left(\frac{1}{1-|y|\,\overline{y}}\right)-\ln\left(\frac{1}{1-|y|\,x}\right)\ln\left(\frac{1}{1-|x|\,\overline{y}}\right)\right|
\end{aligned}
\tag{182}
$$

for $x,y\in\mathbb{C}$ with $|x|,|y|<1$.

7.4 Some Inequalities for Bessel Functions

Similar to the polylogarithm and hypergeometric function, the modified Bessel function (161) is also the power series with nonnegative coefficients and convergent on the open disk $D(0,1)$, so that all the results in Sects. 3–6 hold true. Therefore, for instance from (134), we have the following corollaries (see [15]).

Corollary 7.11. *If $J_\alpha\,(x)$ and $I_\alpha\,(x)$ are the Bessel function and modified Bessel function for the first kind, respectively, then we have*

$$
I_\alpha\left(|x|^2\right)I_\alpha\left(|y|^2\right)-\left|I_\alpha\,(xy)\right|^2\geq\left|J_\alpha(|x|\,x)J_\alpha\,(|y|\,\overline{y})-J_\alpha(|y|\,x)J_\alpha(|x|\,\overline{y})\right|
\tag{183}
$$

for $\alpha\in\mathbb{R}$, $x,y\in\mathbb{C}$ with $|x|,|y|<1$.

In particular, if $y = \alpha = 0$ *in* (183)*, then for any* $|x| < 1$*, we obtain that*

$$I_0\left(|x|^2\right) - 1 \geq |J_0(|x|\,x) - 1)| \tag{184}$$

where

$$J_0(z) = \sum_{k=0}^{\infty} \frac{(-1)^n}{(k!)^2}\left(\frac{z}{2}\right)^{2k} \quad \text{and} \quad I_0(z) = \sum_{k=0}^{\infty} \frac{1}{(k!)^2}\left(\frac{z}{2}\right)^{2k}, \quad |z| < 1.$$

References

1. Abramowitz, M., Stegun, I. (eds.): Handbook of Mathematical Functions with Formulas, Graphs and Mathematical Tables. Dover, New York (1972)
2. Alzer, H.: A refinement of the Cauchy-Schwarz inequality. J. Math. Anal. Appl. **168**, 596–604 (1992)
3. Buzano, M.L.: Generalizzazione della diseguaglianza di Cauchy-Schwarz (Italian). Rend. Sem. Mat. Univ. e Politech. Torino. **31**(1971/73), 405–409 (1974)
4. Cerone, P., Dragomir, S.S.: Some applications of de Bruijn's inequality for power series. Integral Transform. Spec. Funct. **18**(6), 387–396 (2007)
5. de Bruijn, N.G.: Problem 12. Wisk. Opgaven **21**, 12–14 (1960)
6. Dragomir, S.S.: Some refinements of Schwarz inequality. In: Simpozionul de Mathematişi Aplicaţii, Timişoara, pp. 13–16, Romania, 1–2 Noiembrie 1985
7. Dragomir, S.S.: Inequalities of Cauchy–Bunyakovsky–Schwarz's type for positive linear functionals (Romanian). Gaz. Mat. Metod. (Bucharest) **9**, 162–164 (1988)
8. Dragomir, S.S.: Some refinements of Cauchy-Bunyakovsky-Schwarz inequality for sequences. In: Proceedings of the Third Symposium of Mathematics and Its Applications, pp. 78–82 (1989)
9. Dragomir, S.S.: A survey on Cauchy–Bunyakovsky–Schwarz Type discrete inequalities. J. Inequal. Pure Appl. Math. **4**(3, 63), 1–140 (2003)
10. Dragomir, S.S.: Discrete Inequalities of the Cauchy Bunyakovsky Schwarz Type. Nova Science Publishers Inc., New York (2004)
11. Dragomir, S.S.: Advances in Inequality of the Schwarz, Grüss and Bessel Type in Inner Product Spaces. Nova Science Publishers Inc., New York (2005)
12. Dragomir, S.S.: Advances in Inequality of the Schwarz, Triangle and Heisenberg Type in Inner Product Space. Nova Science Publishers Inc., New York (2007)
13. Ibrahim, A., Dragomir, S.S.: Power series inequalities via Buzano's result and applications. Integral Transform. Spec. Funct. **22**(12), 867–878 (2011)
14. Ibrahim, A., Dragomir, S.S.: Power series inequalities via a refinement of Schwarz inequality. Integral Transform. Spec. Funct. **23**(10), 769–781 (2012)
15. Ibrahim, A., Dragomir, S.S., Darus, M.: Some inequalities for power series with applications. Integral Transform. Spec. Funct. **24**(5), 364–376 (2013)
16. Mitrinović, D.S., Pečarić, J.E., Fink, A.M.: Classical and New Inequalities in Analysis. Kluwer Academic Publishers, Dordrecht (1993)
17. Zheng, L.: Remark on a refinement of the Cauchy–Schwarz inequality. J. Math. Anal. Appl. **218**, 13–21 (1998)

Topics in Special Functions III

Glen D. Anderson, Matti Vuorinen, and Xiaohui Zhang

Dedicated to Professor Hari M. Srivastava

Abstract The authors provide a survey of recent results in special functions of classical analysis and geometric function theory, in particular, the circular and hyperbolic functions, the gamma function, the elliptic integrals, the Gaussian hypergeometric function, power series, and mean values.

1 Introduction

The study of quasiconformal maps led the first two authors in their joint work with Vamanamurthy to formulate open problems or questions involving special functions [14, 16]. During the past two decades, many authors have contributed to the solution of these problems. However, most of the problems posed in [14] are still open.

The present paper is the third in a series of surveys by the first two authors, the previous papers [20, 23] being written jointly with the late Vamanamurthy. The aim of this series of surveys is to review the results motivated by the problems in [14, 16] and related developments during the past two decades. In the first of these we studied classical special functions, and in the next we focused on special functions occurring in the distortion theory of quasiconformal maps. Regretfully, Vamanamurthy passed

G.D. Anderson
Department of Mathematics, Michigan State University, East Lansing, MI 48824, USA
e-mail: anderson@math.msu.edu

M. Vuorinen • X. Zhang (✉)
Department of Mathematics and Statistics, University of Turku, Assistentinkatu 7, FIN-20014, Finland
e-mail: vuorinen@utu.fi; xiazha@utu.fi

G.V. Milovanović and M.Th. Rassias (eds.), *Analytic Number Theory, Approximation Theory, and Special Functions*, DOI 10.1007/978-1-4939-0258-3_11,

away in 2009, and the remaining authors acknowledge his crucial role in our joint work. For an update to the bibliographies of [20, 23] the reader is referred to [12].

In 1993 the following monotone rule was derived [17, Lemma 2.2]. Though simple to state and easy to prove by means of the Cauchy Mean Value Theorem, this l'Hôpital Monotone Rule (LMR) has had wide application to special functions by many authors. Vamanamurthy was especially skillful in the application of this rule. We here quote the rule as it was restated in [21, Theorem 2].

Theorem 1.1 (l'Hôpital Monotone Rule). *Let $-\infty < a < b < \infty$, and let $f, g : [a,b] \to \mathbb{R}$ be continuous functions that are differentiable on (a,b), with $f(a) = g(a) = 0$ or $f(b) = g(b) = 0$. Assume that $g'(x) \neq 0$ for each $x \in (a,b)$. If f'/g' is increasing (decreasing) on (a,b), then so is f/g.*

Theorem 1.1 assumes that a and b are finite, but the rule can be extended easily by similar methods to the case where a or b is infinite. The LMR has been used effectively in the study of the monotonicity of a quotient of two functions. For instance, Pinelis' note [146] shows the potential of the LMR. As a complement to Pinelis' note, the paper [21] contains many applications of LMR in calculus. Also the history of LMR is reviewed there.

In this survey we give an account of the work in the special functions of classical analysis and geometric function theory since our second survey. In many of these results the LMR was an essential tool. Because of practical constraints, we have had to exclude many fine papers and have limited our bibliography to those papers most closely connected to our work.

The aim of our work on special functions has been to solve open problems in quasiconformal mapping theory. In particular, we tried to settle Mori's conjecture for quasiconformal mappings [127] (see also [118, p. 68]). For the formulation of this problem, let $K > 1$ be fixed and let $M(K)$ be the least constant such that

$$|f(z_1) - f(z_2)| \leqslant M(K)|z_1 - z_2|^{1/K}, \quad \text{for all } z_1, z_2 \in B,$$

for every K-quasiconformal mapping $f : B \to B$ of the unit disk B onto itself with $f(0) = 0$. A. Mori conjectured in 1956 that $M(K) \leqslant 16^{1-1/K}$. This conjecture is still open in 2013. Some of the open problems that we found will be discussed in the last section.

2 Generalizations of Jordan's Inequality

The LMR application list, begun in [21], led to the Master's thesis of Visuri, on which [109] is based. Furthermore, applications of LMR to trigonometric inequalities were given in [109]. Numerous further applications to trigonometric functions were found by many authors, and some of these papers are reviewed in this section and the next.

By elementary geometric methods one can prove that

$$\frac{2}{\pi} \leqslant \frac{\sin x}{x} < 1, \;\; 0 < x \leqslant \frac{\pi}{2},$$

a result known as *Jordan's inequality*. In a recent work, Klén et al. [109] have obtained the inequalities

$$\cos^2\frac{x}{2} \leqslant \frac{\sin x}{x} \leqslant \cos^3\frac{x}{3}, \;\; x \in (-\sqrt{27/5}, \sqrt{27/5})$$

and

$$\cosh^{1/4} x < \frac{\sinh x}{x} < \cosh^{1/2} x, \;\; x \in (0, 1).$$

Inspired by these results, Lv, Wang, and Chu [121] proved that, for $a = (\log(\pi/2))/\log\sqrt{2} \approx 1.30299$,

$$\cos^{4/3}\frac{x}{2} < \frac{\sin x}{x} < \cos^a\frac{x}{2}, \;\; x \in (0, \pi/2),$$

where $4/3$ and a are best constants and that for $b = (\log\sinh 1)/(\log\cosh 1) \approx 0.372168$,

$$\cosh^{1/3} x < \frac{\sinh x}{x} < \cosh^b x, \;\; x \in (0, 1),$$

where $1/3$ and b are best constants.

Many authors have generalized or sharpened Jordan's inequality, either by replacing the bounds by finite series or hyperbolic functions or by obtaining analogous results for other functions such as hyperbolic or Bessel functions. The comprehensive survey paper by Qi et al. [150] gives a clear picture of these developments as of 2009. For example, in 2008 Niu et al. [143] obtained the sharp inequality

$$\frac{2}{\pi} + \sum_{k=1}^{n} \alpha_k(\pi^2 - 4x^2)^k \leqslant \frac{\sin x}{x} \leqslant \frac{2}{\pi} + \sum_{k=1}^{n} \beta_k(\pi^2 - 4x^2)^k, \;\; 0 < x \leq \pi/2,$$

for each natural number n, with best possible constants α_k and β_k. That same year Wu and Srivastava [198] obtained upper and lower estimates on $(0, \pi/2]$ for $(\sin x)/x$ that are finite series in powers of $(x - \theta)$, where $\theta \in [x, \pi/2]$, while Zhu [211] obtained bounds as finite series in powers of $(\pi^2 - 4x^2)$. Zhu [210] obtained bounds for $(\sin x)/x$ as finite series in powers of $(r^2 - x^2)$ for $0 < x \leqslant r \leqslant \pi/2$, yielding a new infinite series

$$\frac{\sin x}{x} = \sum_{n=0}^{\infty} a_n (r^2 - x^2)^n, \quad \text{for } 0 < |x| \leqslant r \leqslant \pi/2.$$

Yang [199] showed that a function f admits an infinite series expansion of the above type if and only if f is analytic and even.

In 2011 Huo et al. [97] obtained the following generalization of Jordan's inequalities:

$$\sum_{k=1}^{n} \mu_k (\theta^t - x^t)^k \leqslant \frac{\sin x}{x} - \frac{\sin \theta}{\theta} \leqslant \sum_{k=1}^{n} \omega_k (\theta^t - x^t)^k$$

for $t \geqslant 2$, $n \in \mathbb{N}$, and $0 < x \leq \theta < \pi$, where the coefficients μ_k and ω_k are defined recursively and are best possible.

More recently, in 2012, Chen and Debnath [74] have proved that, for $0 < x \leqslant \pi/2$,

$$f_1(x) \leqslant \frac{\sin x}{x} \leqslant f_2(x),$$

where

$$f_1(x) = \frac{2}{\pi} + \frac{2\pi^{-\theta-1}}{\theta}(\pi^\theta - (2x)^\theta) + \frac{(-\pi^2 + 4 + 4\theta)\pi^{-2\theta-1}}{4\theta^2}(\pi^\theta - (2x)^\theta)^2$$

and

$$f_2(x) = \frac{2}{\pi} + \frac{2\pi^{-\theta-1}}{\theta}(\pi^\theta - (2x)^\theta) + \frac{((\pi-2)\theta - 2)\pi^{-2\theta-1}}{\theta}(\pi^\theta - (2x)^\theta)^2,$$

for any $\theta \geqslant 2$, with equality when $x = \pi/2$.

In a recent work Sándor [164] (see also [165, p. 9]) proved that $h(x) \equiv [\log(x/\sin x)]/\log((\sinh x)/x)$ is strictly increasing on $(0, \pi/2)$. He used this result to prove that the best positive constants p and q for which

$$\left(\frac{\sinh x}{x}\right)^p < \frac{x}{\sin x} < \left(\frac{\sinh x}{x}\right)^q$$

is true are $p = 1$ and $q = [\log(\pi/2)]/\log((\sinh(\pi/2))/(\pi/2)) \approx 1.18$.

In an unpublished manuscript, Barbu and Pişcoran [28] have proved, in particular, that

$$(1 - x^2/3)^{-1/4} < \frac{\sinh x}{x} < 1 + \frac{x^2}{5}, \quad x \in (0, 1).$$

Kuo [116] has developed a method of obtaining an increasing sequence of lower bounds and a decreasing sequence of upper bounds for $(\sin x)/x$, and he has conjectured that the two sequences converge uniformly to $(\sin x)/x$.

Since there is a close connection between the function $(\sin x)/x$ and the Bessel function $J_{1/2}(x)$ (cf. [219]), it is natural for authors to seek analogs of the Jordan inequality for Bessel and closely related functions. Baricz and Wu [35, 40], Zhu [219, 220], and Niu et al. [144] have produced inequalities of this type. Zhu [221] has also obtained Jordan-type inequalities for $((\sin x)/x)^p$ for any $p > 0$. Wu and Debnath [195] have generalized Jordan's inequality to functions $f(x)/x$ on $[0, \theta]$ such that f is $(n+1)$-times differentiable, $f(0) = 0$, and either n is a positive even integer with $f^{(n+1)}$ increasing on $[0, \theta]$ or n is a positive odd integer with $f^{(n+1)}$ decreasing on $[0, \theta]$.

3 Other Inequalities Involving Circular and Hyperbolic Functions

3.1 Redheffer

In 1968 Redheffer [157] proposed the problem of showing that

$$\frac{\sin \pi x}{\pi x} \geqslant \frac{1-x^2}{1+x^2}, \quad \text{for all real } x \tag{1}$$

or, equivalently, that

$$\frac{\sin x}{x} \geqslant \frac{\pi^2 - x^2}{\pi^2 + x^2}, \quad \text{for all real } x. \tag{2}$$

A solution of this problem was provided by Williams [192], using infinite products, who also proved the stronger inequality

$$\frac{\sin \pi x}{\pi x} \geqslant \frac{1-x^2}{1+x^2} + \frac{(1-x)^2}{x(1+x^2)}, \quad \text{for } x \geqslant 1.$$

Later, using Erdös-Turán series and harmonic analysis, Li and Li [120] proved the double inequality

$$\frac{(1-x^2)(4-x^2)(9-x^2)}{x^6 - 2x^4 + 13x^2 + 36} \leqslant \frac{\sin \pi x}{\pi x} \leqslant \frac{1-x^2}{\sqrt{1+3x^4}}, \quad \text{for } 0 < x < 1.$$

They also found a method for obtaining new bounds from old for $(\sin x)/x$, but Kuo [116] gave an example to show that the new bounds are not necessarily stronger.

In 2003 Chen et al. [76], using mathematical induction and infinite products, found analogs of the Redheffer inequality for $\cos x$:

$$\cos x \geqslant \frac{\pi^2 - 4x^2}{\pi^2 + 4x^2}, \quad \text{for } |x| \leqslant \frac{\pi}{2},$$

and for hyperbolic functions

$$\frac{\sinh x}{x} \leqslant \frac{\pi^2 + x^2}{\pi^2 - x^2}, \quad \text{for } 0 < |x| \leqslant \pi; \quad \cosh x \leqslant \frac{\pi^2 + 4x^2}{\pi^2 - 4x^2}, \quad \text{for } |x| \leqslant \frac{\pi}{2}.$$

In 2008, inspired by the inequalities above, Zhu and Sun [224] proved that

$$\left(\frac{\pi^2 - 4x^2}{\pi^2 + 4x^2}\right)^{\alpha} \leqslant \cos x \leqslant \left(\frac{\pi^2 - 4x^2}{\pi^2 + 4x^2}\right)^{\beta}, \quad \text{for } 0 \leqslant x \leqslant \frac{\pi}{2},$$

with best possible constants $\alpha = 1$ and $\beta = \pi^2/16$, and

$$\left(\frac{\pi^2 - x^2}{\pi^2 + x^2}\right)^{\gamma} \leqslant \frac{\sin x}{x} \leqslant \left(\frac{\pi^2 - x^2}{\pi^2 + x^2}\right)^{\delta}, \quad \text{for } 0 < x < \pi,$$

with best possible constants $\gamma = 1$ and $\delta = \pi^2/12$. They obtained similar results for the hyperbolic sine and cosine functions. In 2009 Zhu [216] showed that

$$\left(\frac{\pi^2 - x^2}{\sqrt{\pi^4 + 3x^4}}\right)^{\alpha} \leqslant \frac{\sin x}{x} \leqslant \left(\frac{\pi^2 - x^2}{\sqrt{\pi^4 + 3x^4}}\right)^{\beta}, \quad 0 < x \leqslant \pi,$$

holds if and only if $\alpha \geqslant \pi^2/6$ and $\beta \leqslant 1$, with analogous results for $\cos x$ and $(\tan x)/x$. In 2009 Baricz and Wu [41] and in 2011 Zhu [222] proved Redheffer-type inequalities for Bessel functions.

3.2 *Cusa-Huygens*

The inequality

$$\frac{\sin x}{x} < \frac{2 + \cos x}{3}, \quad 0 < x < \pi/2$$

was discovered by N. de Cusa in the fifteenth century (cf. [71]) and proved rigorously by Huygens [98] in the seventeenth century. In 2009 Zhu [218] obtained the following inequalities of Cusa-Huygens type:

$$\left(\frac{\sin x}{x}\right)^{\alpha} < \frac{2}{3} + \frac{1}{3}(\cos x)^{\alpha}, \quad 0 < x < \frac{\pi}{2}, \quad \alpha \geqslant 1,$$

and

$$\left(\frac{\sinh x}{x}\right)^{\alpha} < \frac{2}{3} + \frac{1}{3}(\cosh x)^{\alpha}, \quad x > 0, \quad \alpha \geqslant 1.$$

That same year Zhu [214] discovered a more general set of inequalities of Cusa type, from which many other types of inequalities for circular functions can be derived. He proved the following: Let $0 < x < \pi/2$. If $p \geqslant 1$, then

$$(1-\alpha) + \alpha(\cos x)^p < \left(\frac{\sin x}{x}\right)^p < (1-\beta) + \beta(\cos x)^p \tag{3}$$

if and only if $\beta \leqslant 1/3$ and $\alpha \geqslant 1 - (2/\pi)^p$. If $0 \leqslant p \leqslant 4/5$, then (3) holds if and only if $\alpha \geqslant 1/3$ and $\beta \leqslant 1 - (2/\pi)^p$. If $p < 0$, then the second inequality in (3) holds if and only if $\beta \geqslant 1/3$. In a later paper [219] Zhu obtained estimates for $(\sin x)/x$ and $(\sinh x)/x$ that led to new infinite series for these functions. For some similar results see also [194].

In 2011 Chen and Cheung [71] obtained the sharp Cusa-Huygens-type inequality

$$\left(\frac{2+\cos x}{3}\right)^{\alpha} < \frac{\sin x}{x} < \left(\frac{2+\cos x}{3}\right)^{\beta},$$

for $0 < x < \pi/2$, with best possible constants $\alpha = (\log(\pi/2))/\log(3/2) \approx 1.11$ and $\beta = 1$.

In 2011 Neuman and Sándor [142] discovered a pair of optimal inequalities for hyperbolic and trigonometric functions, proving that, for $0 < x < \pi/2$, the best positive constants p and q in the inequality

$$\frac{1}{(\cosh x)^p} < \frac{\sin x}{x} < \frac{1}{(\cosh x)^q}$$

are $p = (\log(\pi/2))/\log\cosh(\pi/2) \approx 0.49$ and $q = 1/3$ and that for $x \neq 0$ the best positive constants p and q in the inequality

$$\left(\frac{\sinh x}{x}\right)^p < \frac{2}{\cos x + 1} < \left(\frac{\sinh x}{x}\right)^q$$

are $p = 3/2$ and $q = (\log 2)/\log[(\sinh(\pi/2))/(\pi/2)] \approx 1.82$.

3.3 Becker-Stark

In 1978 Becker and Stark [49] obtained the double inequality

$$\frac{8}{\pi^2 - 4x^2} < \frac{\tan x}{x} < \frac{\pi^2}{\pi^2 - 4x^2}, \quad 0 < x < \frac{\pi}{2},$$

where the numerator constants 8 and π^2 are best possible.

In 2008 Zhu and Sun [224] showed that

$$\left(\frac{\pi^2+4x^2}{\pi^2-4x^2}\right)^{\alpha} \leqslant \frac{\tan x}{x} \leqslant \left(\frac{\pi^2+4x^2}{\pi^2-4x^2}\right)^{\beta}, \quad 0<x<\frac{\pi}{2},$$

holds if and only if $\alpha \leqslant \pi^2/24$ and $\beta \geqslant 1$.

In 2010 Zhu and Hua [223] sharpened the Becker-Stark inequality by proving that

$$\frac{\pi^2+\alpha x^2}{\pi^2-4x^2} < \frac{\tan x}{x} < \frac{\pi^2+\beta x^2}{\pi^2-4x^2}, \quad 0<x<\frac{\pi}{2},$$

where $\alpha = 4(8-\pi^2)/\pi^2 \approx -0.76$ and $\beta = \pi^2/3-4 \approx -0.71$ are the best possible constants. They also developed a systematic method for obtaining a sequence of sharp inequalities of this sort.

In 2011 Ge [88] obtained

$$\frac{8}{\pi^2-4x^2} + \left(1-\frac{8}{\pi^2}\right) < \frac{\tan x}{x} < \frac{\pi^4}{12}\frac{1}{\pi^2-4x^2} + \left(1-\frac{\pi^2}{12}\right),$$

for $0<x<\pi/2$. That same year Chen and Cheung [71] proved the sharp Becker-Stark-type inequality

$$\left(\frac{\pi^2}{\pi^2-4x^2}\right)^{\alpha} < \frac{\tan x}{x} < \left(\frac{\pi^2}{\pi^2-4x^2}\right)^{\beta},$$

with best possible constants $\alpha = \pi^2/12 \approx 0.82$ and $\beta = 1$.

3.4 Wilker

In 1989 Wilker [190] posed the problem of proving that

$$\left(\frac{\sin x}{x}\right)^2 + \frac{\tan x}{x} > 2, \quad \text{for } 0<x<\frac{\pi}{2} \tag{4}$$

and of finding

$$c \equiv \inf_{0<x<\pi/2} \frac{\left(\frac{\sin x}{x}\right)^2 + \frac{\tan x}{x} - 2}{x^3 \tan x}. \tag{5}$$

Anglesio et al. [191] showed that the function in (5) is decreasing on $(0,\pi/2)$, that the value of c in (5) is $16/\pi^4$, and that, moreover, the supremum of the expression in (5) on $(0,\pi/2)$ is $8/45$. Hence

$$2+\frac{16}{\pi^4}x^3\tan x \leqslant \left(\frac{\sin x}{x}\right)^2+\frac{\tan x}{x} \leqslant 2+\frac{8}{45}x^3\tan x, \tag{6}$$

for $0 < x < \pi/2$, where $16/\pi^4 \approx 0.164$ and $8/45 \approx 0.178$ are best possible constants. (Note: [21] erroneously quoted [191] as saying that the function in (5) is increasing.) In 2007 Wu and Srivastava [197] proved the Wilker-type inequality

$$\left(\frac{x}{\sin x}\right)^2+\frac{x}{\tan x} > 2, \quad \text{for } 0 < x < \frac{\pi}{2}. \tag{7}$$

However, Baricz and Sándor [39] discovered that (7) is implied by (4).

In 2009 Zhu [218] generalized (4) and obtained analogs for hyperbolic functions, showing that, for $0 < x < \pi/2, \alpha \geqslant 1$,

$$\left(\frac{\sin x}{x}\right)^{2\alpha}+\left(\frac{\tan x}{x}\right)^{\alpha} > \left(\frac{x}{\sin x}\right)^{2\alpha}+\left(\frac{x}{\tan x}\right)^{\alpha} > 2$$

and that, for $x > 0, \alpha \geqslant 1$,

$$\left(\frac{\sinh x}{x}\right)^{2\alpha}+\left(\frac{\tanh x}{x}\right)^{\alpha} > \left(\frac{x}{\sinh x}\right)^{2\alpha}+\left(\frac{x}{\tanh x}\right)^{\alpha} > 2.$$

These two results of Zhu are special cases of a recent lemma due to Neuman [138, Lemma 2].

In 2012 Sándor [162] has proved that, for $0 < x \leq \pi/2, \alpha > 0$,

$$\left(\frac{x}{\sin x}\right)^{2\alpha}+\left(\frac{x}{\sinh x}\right)^{\alpha} > \left(\frac{\sinh x}{x}\right)^{2\alpha}+\left(\frac{\sin x}{x}\right)^{\alpha} > 2.$$

Using power series, Chen and Cheung [72] obtained the following sharper versions of (6):

$$\frac{16}{315}x^5\tan x < \left(\frac{\sin x}{x}\right)^2+\frac{\tan x}{x}-\left[2+\frac{8}{45}x^4\right] < \left(\frac{2}{\pi}\right)^6 x^5\tan x, \tag{8}$$

and

$$\frac{104}{4725}x^7\tan x < \left(\frac{\sin x}{x}\right)^2+\frac{\tan x}{x}-\left[2+\frac{8}{45}x^4+\frac{16}{315}x^6\right] < \left(\frac{2}{\pi}\right)^8 x^7\tan x. \tag{9}$$

The constants $16/315 \approx 0.051$ and $(2/\pi)^6 \approx 0.067$ in (8) and $104/4725 \approx 0.022$ and $(2/\pi)^8 \approx 0.027$ in (9) are best possible. For $0 < x < \pi/2$, Chen and Cheung also obtained upper estimates complementary to (7):

$$\left(\frac{x}{\sin x}\right)^2 + \frac{x}{\tan x} < 2 + \frac{2}{45}x^3 \tan x$$

and

$$\left(\frac{x}{\sin x}\right)^2 + \frac{x}{\tan x} < 2 + \frac{2}{45}x^4 + \frac{8}{945}x^5 \tan x,$$

where the constants $2/45$ and $8/945$ are best possible.

In 2012, Sándor [164] has shown that

$$\frac{\sin x}{x} + q\frac{\sinh x}{x} > q + 1,\ x \neq 0$$

and

$$\left(\frac{\sinh x}{x}\right)^q + \frac{\sin x}{x} > 2,\ 0 < x < \frac{\pi}{2},$$

where $q = [\log(\pi/2)]/\log[(\sinh(\pi/2))/(\pi/2)] \approx 1.18$.

Extensions of the generalized Wilker inequality for Bessel functions were obtained by Baricz and Sándor [39] in 2008.

3.5 *Huygens*

An older inequality due to Huygens [98] is similar in form to (4):

$$2\left(\frac{\sin x}{x}\right) + \frac{\tan x}{x} > 3, \quad \text{for } 0 < |x| < \frac{\pi}{2} \tag{10}$$

and actually implies (4) (see [141]). In 2009, Zhu [217] obtained the following inequalities of Huygens type:

$$(1-p)\frac{\sin x}{x} + p\frac{\tan x}{x} > 1 > (1-q)\frac{\sin x}{x} + q\frac{\tan x}{x}$$

for all $x \in (0, \pi/2)$ if and only if $p \geqslant 1/3$ and $q \leqslant 0$;

$$(1-p)\frac{\sinh x}{x} + p\frac{\tanh x}{x} > 1 > (1-q)\frac{\sinh x}{x} + q\frac{\tanh x}{x}$$

for all $x \in (0, \infty)$ if and only if $p \leqslant 1/3$ and $q \geqslant 1$;

$$(1-p)\frac{x}{\sin x} + p\frac{x}{\tan x} > 1 > (1-q)\frac{x}{\sin x} + q\frac{x}{\tan x}$$

for all $x \in (0, \pi/2)$ if and only if $p \leqslant 1/3$ and $q \geqslant 1 - 2/\pi$; and

$$(1-p)\frac{x}{\sinh x} + p\frac{x}{\tanh x} > 1 > (1-q)\frac{x}{\sinh x} + q\frac{x}{\tanh x}$$

for all $x \in (0, \infty)$ if and only if $p \geqslant 1/3$ and $q \leqslant 0$.

In 2012 Sándor [162] has showed that, for $0 < x \leq \pi/2, \alpha > 0$,

$$2\left(\frac{\sinh x}{x}\right)^{\alpha} + \left(\frac{\sin x}{x}\right)^{\alpha} > 2\left(\frac{x}{\sin x}\right)^{\alpha} + \left(\frac{x}{\sinh x}\right)^{\alpha} > 3.$$

Chen and Cheung [72] also found sharper versions of (10) as follows: For $0 < x < \pi/2$,

$$3 + \frac{3}{20}x^3 \tan x < 2\left(\frac{\sin x}{x}\right) + \frac{\tan x}{x} < 3 + \left(\frac{2}{\pi}\right)^4 x^3 \tan x \tag{11}$$

and

$$\frac{3}{56}x^5 \tan x < 2\left(\frac{\sin x}{x}\right) + \frac{\tan x}{x} - \left[3 + \frac{3}{20}x^4\right] < \left(\frac{2}{\pi}\right)^6 x^5 \tan x, \tag{12}$$

where the constants $3/20 = 0.15$ and $(2/\pi)^4 \approx 0.16$ in (11) and $3/56 \approx 0.054$ and $(2/\pi)^6 \approx 0.067$ in (12) are best possible.

Recently Hua [96] have proved the following sharp inequalities: For $0 < |x| < \pi/2$,

$$3 + \frac{1}{40}x^3 \sin x < \frac{\sin x}{x} + 2\frac{\tan(x/2)}{x/2} < 3 + \frac{80 - 24\pi}{\pi^4}x^3 \sin x,$$

where the constants $1/40$ and $(80 - 24\pi)/\pi^4$ are best possible, and, for $x \neq 0$,

$$3 + \frac{3}{20}x^3 \tanh x < 2\frac{\sinh x}{x} + \frac{\tanh x}{x} < 3 + \frac{3}{20}x^3 \sinh x,$$

where the constant $3/20$ is best possible.

3.6 Shafer

The problem of proving

$$\arctan x > \frac{3x}{1 + 2\sqrt{1 + x^2}}, \quad x > 0,$$

was proposed by Shafer [166] in 1966. Solutions were obtained by Grinstein, Marsh, and Konhauser [169] in 1967. In 2011 Chen, Cheung, and Wang [73] found, for each $a > 0$, the largest number b and the smallest number c such that the inequalities

$$\frac{bx}{1 + a\sqrt{1 + x^2}} \leqslant \arctan x \leqslant \frac{cx}{1 + a\sqrt{1 + x^2}}$$

are valid for all $x \geqslant 0$. Their answer to this question is indicated in the following table:

a	Largest b	Smallest c
$0 < a \leqslant \pi/2$	$b = \pi a/2$	$c = 1 + a$
$\pi/2 < a \leqslant 2/(\pi - 2)$	$b = 4(a^2 - 1)/a^2$	$c = 1 + a$
$2/(\pi - 2) < a < 2$	$b = 4(a^2 - 1)/a^2$	$c = \pi a/2$
$2 \leqslant a < \infty$	$b = 1 + a$	$c = \pi a/2$

In 1974, in a numerical analytical context [167], Shafer presented the inequality

$$\arctan x \geqslant \frac{8x}{3 + \sqrt{25 + (80/3)x^2}}, \quad x > 0,$$

which he later proved analytically [168]. In [213] Zhu proved that the constant $80/3$ in Shafer's inequality is best possible and also obtained the complementary inequality

$$\arctan x < \frac{8x}{3 + \sqrt{25 + (256/\pi^2)x^2}}, \quad x > 0,$$

where $256/\pi^2$ is the best possible constant.

3.7 *Fink*

In [132, p. 247], there is a lower bound for $\arcsin x$ on $[0, 1]$ that is similar to Shafer's for $\arctan x$. In 1995 Fink [87] supplied a complementary upper bound. The resulting double inequality is

$$\frac{3x}{2 + \sqrt{1 - x^2}} \leqslant \arcsin x \leqslant \frac{\pi x}{2 + \sqrt{1 - x^2}}, \quad 0 \leqslant x \leqslant 1, \tag{13}$$

and both numerator constants are best possible. Further refinements of these inequalities, along with analogous ones for $\operatorname{arcsinh} x$, were obtained by Zhu [212] and by Pan with Zhu [145]. We note that, for $0 < x < 1$, this double inequality is equivalent to

$$\frac{2+\cos x}{\pi} < \frac{\sin x}{x} < \frac{2+\cos x}{3}, \quad 0 < x < \frac{\pi}{2},$$

in which the second relation is the Cusa inequality.

3.8 *Carlson*

In 1970 Carlson [67, (1.14)] proved the inequality

$$\frac{6\sqrt{1-x}}{2\sqrt{2}+\sqrt{1+x}} < \arccos x < \frac{\sqrt[3]{4}\cdot\sqrt{1-x}}{(1+x)^{1/6}}, \quad 0 \leqslant x < 1. \tag{14}$$

In 2012, seeking to sharpen and generalize (14), Chen and Mortici [75] determined, for each fixed $c > 0$, the largest number a and smallest number b such that the double inequality

$$\frac{a\sqrt{1-x}}{c+\sqrt{1+x}} \leqslant \arccos x \leqslant \frac{b\sqrt{1-x}}{c+\sqrt{1+x}}$$

is valid for all $x \in [0, 1]$. Their answer to this question is indicated in the following table:

c	Largest a	Smallest b
$0 < x < (2\pi-4)/(4-\pi)$	$(1+a)\pi/2$	$2+\sqrt{2}a$
$(2\pi-4)/(4-\pi) \le x \le (4-\pi)/(\pi-2\sqrt{2})$	$8(a^2-2)/a^2$	$2+\sqrt{2}a$
$(4-\pi)/(\pi-2\sqrt{2}) < x < 2\sqrt{2}$	$4(a^2-1)/a^2$	$(1+a)\pi/2$
$2\sqrt{2} \le x < \infty$	$2+\sqrt{2}a$	$(1+a)\pi/2$

These authors also proved that, for all $x \in [0, 1]$, the inequalities

$$\frac{\sqrt[3]{4}\cdot\sqrt{1-x}}{a+(1+x)^{1/6}} \leqslant \arccos x \leqslant \frac{\sqrt[3]{4}\cdot\sqrt{1-x}}{b+(1+x)^{1/6}}$$

hold on $[0, 1]$, with best constants $a = (2\sqrt[3]{4}-\pi)/\pi \approx 0.01$ and $b = 0$.

Moreover, in view of the right side of (14), in 2011 Chen, Cheung, and Wang [73] considered functions of the form

$$f(x) \equiv \frac{r(1-x)^p}{(1+x)^q}$$

on [0, 1] and determined the values of p, q, r such that $f(x)$ is the best third-order approximation of $\arccos x$ in a neighborhood of the origin. The answer is that, for $p = (\pi + 2)/\pi^2$, $q = (\pi - 2)/\pi^2$, $r = \pi/2$, one has

$$\lim_{x \to 0} \frac{\arccos x - f(x)}{x^3} = \frac{\pi^2 - 8}{6\pi^2}.$$

With the values of p, q, r stated above, the authors were led to a new lower bound for arccos:

$$\arccos x \geqslant \frac{(\pi/2)(1-x)^{(\pi+2)/\pi^2}}{(1+x)^{(\pi-2)/\pi^2}}, \quad 0 < x \leqslant 1.$$

3.9 *Lazarević*

In [117] Lazarević proved that, for $x \neq 0$,

$$\left(\frac{\sinh x}{x}\right)^q > \cosh x$$

if and only if $q \geqslant 3$. Zhu improved upon this inequality in [215] by showing that if $p > 1$ or $p \leqslant 8/15$, then

$$\left(\frac{\sinh x}{x}\right)^q > p + (1-p)\cosh x$$

for all $x > 0$ if and only if $q \geqslant 3(1-p)$. For some similar results see also [194].

In 2008 Baricz [34] extended the Lazarević inequality to modified Bessel functions and also deduced some Turán- and Lazarević-type inequalities for the confluent hypergeometric functions.

3.10 *Neuman*

Neuman [137] has recently established several inequalities involving new combinations of circular and hyperbolic functions. In particular, he has proved that if $x \neq 0$, then

$$(\cosh x)^{2/3} < \frac{\sinh x}{\arcsin(\tanh x)} < \frac{1 + 2\cosh x}{3},$$

$$[(\cosh 2x)^{1/2} \cosh^2 x]^{1/3} < \frac{\sinh x}{\operatorname{arcsinh}(\tanh x)} < \frac{(\cosh 2x)^{1/2} + 2\cosh x}{3},$$

and

$$[(\cosh 2x)\cosh x]^{1/3} < \frac{\sinh x}{\arctan(\tanh x)} < \frac{2(\cosh 2x)^{1/2} + \cosh x}{3}.$$

4 Euler's Gamma Function

For $\operatorname{Re} z > 0$ the *gamma function* is defined by

$$\Gamma(z) \equiv \int_0^\infty t^{z-1} e^{-t} dt,$$

and the definition is extended by analytic continuation to the entire complex plane minus the set of nonpositive integers. This function, discovered by Leonhard Euler in 1729, is a natural generalization of the factorial, because of the functional identity

$$\Gamma(z + 1) = z\Gamma(z).$$

The gamma function is one of the best-known and most important special functions in mathematics and has been studied intensively.

We begin our treatment of this subject by considering an important special constant discovered by Euler and related to the gamma function.

4.1 The Euler-Mascheroni Constant and Harmonic Numbers

The *Euler-Mascheroni constant* $\gamma = 0.5772156649\ldots$ is defined as

$$\gamma \equiv \lim_{n\to\infty} \gamma_n, \tag{15}$$

where $\gamma_n \equiv H_n - \log n$, $n \in \mathbb{N}$ and where H_n are the *harmonic numbers*

$$H_n \equiv \sum_{k=1}^{n} \frac{1}{k} = \int_0^1 \frac{1 - x^n}{1 - x} dx. \tag{16}$$

The number γ is one of the most important constants in mathematics and is useful in analysis, probability theory, number theory, and other branches of pure and applied mathematics. The numerical value of γ is known to $29,844,489,545$ decimal places, thanks to computation by Yee and Chan in 2009 [201] (see [77, p. 273]).

The sequence γ_n converges very slowly to γ, namely with order $1/n$. By replacing $\log n$ in this sequence by $\log(n+1/2)$, DeTemple [84] obtained quadratic convergence (see also [69]). In [130] Mortici made a careful study of how convergence is affected by changes in the logarithm term. He introduced new sequences

$$M_n \equiv H_n - \log\frac{P(n)}{Q(n)},$$

where P and Q are polynomials with leading coefficient 1 and $\deg P - \deg Q = 1$. By judicious choice of the degrees and coefficients of P and Q he was able to produce sequences M_n tending to γ with convergence of order $1/n^4$ and $1/n^6$. He also gave a recipe for obtaining sequences converging to γ with order $1/n^{2k+2}$, where k is any positive integer. This study is based on the author's lemma, proved in [129], that connects the rate of convergence of a convergent sequence $\{x_n\}$ to that of the sequence $\{x_n - x_{n+1}\}$.

In 1997 Negoi [134] showed that if $T_n \equiv H_n - \log(n + 1/2 + 1/(24n))$, then $T_n + [4n^3]^{-1}$ is strictly decreasing to γ and $T_n + [48(n+1)]^{-3}$ is strictly increasing to γ, so that $[48(n+1)]^{-3} < \gamma - T_n < [48n^3]^{-1}$. In 2011 Chen [70] established sharper bounds for $\gamma - T_n$ by using a lemma of Mortici [129].

Using another approach, in 2011 Chlebus [77] developed a recursive scheme for modifying the sequence $H_n - \log n$ to accelerate the convergence to γ to any desired order. The first step in Chlebus' scheme is equivalent to the DeTemple [84] approximation, while the next step yields a sequence that closely resembles the one due to Negoi [134].

In [8] Alzer studied the harmonic numbers (16), obtaining several new inequalities for them. In particular, for $n \geqslant 2$, he proved that

$$\alpha\frac{\log(\log n + \gamma)}{n^2} \leqslant H_n^{1/n} - H_{n+1}^{1/(n+1)} < \beta\frac{\log(\log n + \gamma)}{n^2},$$

where $\alpha = (6\sqrt{6} - 2\sqrt[3]{396})/(3\log(\log 2 + \gamma)) \approx 0.014$ and $\beta = 1$ are the best possible constants and γ is the Euler-Mascheroni constant.

4.2 Estimates for the Gamma Function

In [14, Lemma 2.39] Anderson, Vamanamurthy, and Vuorinen proved that

$$\lim_{x\to\infty}\frac{\log\Gamma\left(\frac{x}{2}+1\right)}{x\log x} = \frac{1}{2} \tag{17}$$

and that the function $(\log \Gamma(1 + x/2))/x$ is strictly increasing from $[2, \infty)$ onto $[0, \infty)$. In [13] Anderson and Qiu showed that $(\log \Gamma(x + 1))/(x \log x)$ is strictly increasing from $(1, \infty)$ onto $(1 - \gamma, 1)$, where γ is the Euler-Mascheroni constant defined by (15), thereby obtaining the strict inequalities

$$x^{(1-\gamma)x-1} < \Gamma(x) < x^{x-1}, \quad x > 1. \tag{18}$$

They also conjectured that the function $(\log \Gamma(x + 1))/(x \log x)$ is concave on $(1, \infty)$, and this conjecture was proved by Elbert and Laforgia in [85, Sect. 3]. One should note that in 1989 Sándor [159] proved that the function $(\Gamma(x + 1))^{1/x}$ is strictly concave for $x \geq 7$.

Later Alzer [4] was able to extend (18) by proving that, for $x \in (0, 1)$,

$$x^{\alpha(x-1)-\gamma} < \Gamma(x) < x^{\beta(x-1)-\gamma}, \tag{19}$$

with best possible constants $\alpha = 1 - \gamma = 0.42278\ldots$ and $\beta = (\pi^2/6 - \gamma)/2 = 0.53385\ldots$. For $x \in (1, \infty)$ Alzer was able to sharpen (18) by showing that (19) holds with best possible constants $\alpha = (\pi^2/6 - \gamma)/2 \approx 0.534$ and $\beta = 1$. His principal new tool was the convolution theorem for Laplace transforms.

Another type of approximation for $\Gamma(x)$ was derived by Ivády [102] in 2009:

$$\frac{x^2 + 1}{x + 1} < \Gamma(x + 1) < \frac{x^2 + 2}{x + 2}, \quad 0 < x < 1. \tag{20}$$

In 2011 Zhao, Guo, and Qi [207] simplified and sharpened (20) by proving that the function

$$Q(x) = \frac{\log \Gamma(x + 1)}{\log(x^2 + 1) - \log(x + 1)}$$

is strictly increasing from $(0, 1)$ onto $(\gamma, 2(1-\gamma))$, where γ is the Euler-Mascheroni constant. As a consequence, they proved that

$$\left(\frac{x^2 + 1}{x + 1}\right)^{\alpha} < \Gamma(x + 1) < \left(\frac{x^2 + 1}{x + 1}\right)^{\beta}, \quad 0 < x < 1,$$

with best possible constants $\alpha = 2(1 - \gamma)$ and $\beta = \gamma$.

Very recently Mortici [133] has determined by numerical experiments that the upper estimate in (18) is a better approximation for $\Gamma(x)$ than the lower one when x is very large. Hence, he has sought estimates of the form $\Gamma(x) \approx x^{a(x)}$, where $a(x)$ is close to $x - 1$ as x approaches infinity. For example, he proves that

$$x^{(x-1)a(x)} < \Gamma(x) < x^{(x-1)b(x)}, \quad x \geqslant 20,$$

where $a(x) = 1 - 1/\log x + 1/(2x) - (1 - (\log 2\pi)/2)/(x \log x)$ and where $b(x) = 1 - 1/\log x + 1/(2x)$. The left inequality is valid for $x \geqslant 2$. Mortici has also obtained a pair of sharper inequalities of this type, valid for $x \geqslant 2$, and has showed how lower and upper estimates of any desired accuracy may be obtained. His proofs are based on an approximation for $\log \Gamma(x)$ in terms of series involving Bernoulli numbers [25, p. 29] and on truncations of an asymptotic series for the function $(\log \Gamma(x))/((x-1)\log x)$. These results provide improvements of (18).

4.3 Factorials and Stirling's Formula

The well-known *Stirling's formula* for $n!$,

$$\alpha_n \equiv \left(\frac{n}{e}\right)^n \sqrt{2\pi n}, \tag{21}$$

discovered by the precocious homeschooled and largely self-taught eighteenth-century Scottish mathematician James Stirling, approximates $n!$ asymptotically in the sense that

$$\lim_{n\to\infty} \frac{n!}{\alpha_n} = 1.$$

Because of the importance of this formula in probability and statistics, number theory, and scientific computations, several authors have sought to replace (21) by a simple sequence that approximates $n!$ more closely (see the discussions in [47,48]). For example, Burnside [63] proved in 1917 that

$$n! \sim \beta_n \equiv \sqrt{2\pi}\left(\frac{n+1/2}{e}\right)^{n+1/2}, \tag{22}$$

that is, $\lim_{n\to\infty}(n!/\beta_n) = 1$. In 2008, Batir [47] determined that the best constants a and b such that

$$\frac{n^{n+1}e^{-n}\sqrt{2\pi}}{\sqrt{n-a}} \leqslant n! < \frac{n^{n+1}e^{-n}\sqrt{2\pi}}{\sqrt{n-b}} \tag{23}$$

are $a = 1 - 2\pi e^{-2} \approx 0.1497$ and $b = 1/6 \approx 0.1667$. Batir offers a numerical table illustrating that his upper bound formula $n^{n+1}e^{-n}\sqrt{2\pi}/\sqrt{n-1/6}$ gives much better approximations to $n!$ than does either (21) or (22).

In a later paper [48] Batir observed that many of the improvements of Stirling's formula take the form

$$n! \sim e^{-a}\left(\frac{n+a}{e}\right)^n \sqrt{2\pi(n+b)} \tag{24}$$

for some real numbers a and b. Batir sought the pair of constants a and b that would make (24) optimal. He proved that the best pairs (a,b) are (a_1,b_1) and (a_2,b_2), where

$$a_1 = \frac{1}{3} + \frac{\lambda}{6} - \frac{1}{6}\sqrt{6-\lambda^2+4/\lambda} \approx 0.54032,\ \ b_1 = a_1^2 + 1/6 \approx 0.45861$$

and

$$a_2 = \frac{1}{3} + \frac{\lambda}{6} + \frac{1}{6}\sqrt{6-\lambda^2+4/\lambda} \approx 0.95011,\ \ b_2 = a_2^2 + 1/6 \approx 1.06937,$$

where $\lambda = \sqrt{2+2^{2/3}+2^{4/3}} \approx 2.47128$ and a_1 and a_2 are the real roots of the quartic equation $3x^4 - 4x^3 + x^2 + 1/12 = 0$.

Ramanujan [156] sought to improve Stirling's formula (21) by replacing $\sqrt{2n}$ in the formula by the sixth root of a cubic polynomial in n:

$$\Gamma(n+1) \approx \sqrt{\pi}\left(\frac{n}{e}\right)^n \sqrt[6]{8n^3+4n^2+n+\frac{1}{30}}. \tag{25}$$

In this connection there appears in the record also his double inequality, for $x \geqslant 1$,

$$\sqrt[6]{8x^3+4x^2+x+\frac{1}{100}} < \frac{\Gamma(x+1)}{\sqrt{\pi}\left(\frac{x}{e}\right)^x} < \sqrt[6]{8x^3+4x^2+x+\frac{1}{30}}. \tag{26}$$

Motivated by this inequality of Ramanujan, the authors of [18] defined the function $h(x) \equiv u(x)^6 - (8x^3+4x^2+x)$, where $u(x) = (e/x)^x\Gamma(x+1)/\sqrt{\pi}$, and conjectured that $h(x)$ is increasing from $(1,\infty)$ into $(1/100, 1/30)$. In 2001 Karatsuba [106] settled this conjecture by showing that $h(x)$ is increasing from $[1,\infty)$ onto $[h(1), 1/30)$, where $h(1) = e^6/\pi^3 - 13 \approx 0.011$.

In an unpublished document, E. A. Karatsuba suggested modifying Ramanujan's approximation formula (25) by replacing the radical with the $2k$th root of a polynomial of degree k and determining the best such asymptotic approximation. Such a program was partially realized by Mortici [132] in 2011, who proposed formula (27) below for $k = 4$, but the more general problem suggested by Karatsuba remains an open problem. Mortici's proposed Ramanujan-type asymptotic approximation is as follows:

$$\Gamma(n+1) \approx \sqrt{\pi}\left(\frac{n}{e}\right)^n \sqrt[8]{16n^4+\frac{32}{3}n^3+\frac{32}{9}n^2+\frac{176}{405}n-\frac{128}{1215}}. \tag{27}$$

In connection with (27), he defined the function

$$g(x) \equiv u(x)^8 - \left(16x^4+\frac{32}{3}x^3+\frac{32}{9}x^2+\frac{176}{405}x\right),$$

where $u(x) = (e/x)^x \Gamma(x+1)/\sqrt{\pi}$, and proved that $g(x)$ is strictly decreasing from $[3,\infty)$ onto $(g(\infty), g(3)]$, where $g(\infty) = -128/1215 \approx -0.105$ and $g(3) = 256e^{24}/(43046721\pi^4) - 218336/135 \approx -0.088$. Mortici's method for proving monotonicity was simpler than Karatsuba's, because he employed an excellent result of Alzer [3] concerning complete monotonicity (see Sect. 4.6 below for definitions). Mortici claimed that his method would also simplify Karatsuba's proof in [106]. Finally, he proved that, for $x \geqslant 3$,

$$R(x,\alpha) < \frac{\Gamma(x+1)}{\sqrt{\pi}\left(\frac{x}{e}\right)^x} \leqslant R(x,\beta),$$

where $R(x,t) \equiv \sqrt[8]{16x^4 + \frac{32}{3}x^3 + \frac{32}{9}x^2 + \frac{176}{405}x - t}$, and $\alpha = 128/1215$, $\beta = g(3)$ are the best possible constants.

In 2012 Mahmoud, Alghamdi, and Agarwal [124] deduced a new family of upper bounds for $\Gamma(n+1)$ of the form

$$\Gamma(n+1) < \sqrt{2\pi n}\left(\frac{n}{e}\right)^n e^{M_n^{[m]}}, \quad n \in \mathbb{N},$$

$$M_n^{[m]} \equiv \frac{1}{2m+3}\left[\frac{1}{4n} + \sum_{k=1}^{m} \frac{2m-2k+2}{2k+1} 2^{-2k}\zeta(2k, n+1/2)\right], \quad n \in \mathbb{N},$$

where ζ is the *Hurwitz zeta function*

$$\zeta(s,q) \equiv \sum_{k=0}^{\infty} \frac{1}{(k+q)^s}.$$

These upper bounds improve Mortici's inequality (27).

4.4 Volume of the Unit Ball

The volume Ω_n of the unit ball in $\mathbb{R}^n$ is given in terms of the gamma function by the formula

$$\Omega_n = \frac{\pi^{n/2}}{\Gamma(n/2+1)}, \quad n \in \mathbb{N}.$$

Whereas the volume of the unit cube is 1 in all dimensions, the numbers Ω_n strictly increase to the maximum $\Omega_5 = 8\pi^2/15$ and then strictly decrease to 0 as $n \to \infty$ (cf. [60, p.264]). Anderson, Vamanamurthy, and Vuorinen [14] proved that $\Omega_n^{1/n}$ is strictly decreasing and that the series $\sum_{n=2}^{\infty} \Omega_n^{1/\log n}$ is convergent. In [13] Anderson and Qiu proved that $\Omega_n^{1/(n \log n)}$ is strictly decreasing with limit $e^{-1/2}$ as $n \to \infty$.

In 2008 Alzer published a collection of new inequalities for combinations of different dimensions and powers of Ω_n [7, Sect. 3]. We quote several of them below:

$$a\frac{(2\pi e)^{n/2}}{n^{(n-1)/2}} \leqslant (n+1)\Omega_n - n\Omega_{n+1} < b\frac{(2\pi e)^{n/2}}{n^{(n-1)/2}}, \quad n \geqslant 1, \tag{28}$$

where the best possible constants are $a = (4 - 9\pi/8)(2/(\pi e))^{1/2}/e = 0.0829\ldots$ and $b = \pi^{-1/2} = 0.5641\ldots$;

$$a\frac{(2\pi e)^{n}}{n^{n+2}} \leqslant \Omega_n^2 - \Omega_{n-1}\Omega_{n+1} < b\frac{(2\pi e)^{n}}{n^{n+2}}, \quad n \geqslant 2, \tag{29}$$

with best possible constant factors $a = (4/e^2)(1 - 8/(3\pi)) = 0.0818\ldots$ and $b = 1/(2\pi) = 0.1591\ldots$;

$$\frac{a}{\sqrt{n}} \leqslant \frac{\Omega_n}{\Omega_{n-1} + \Omega_{n+1}} < \frac{b}{\sqrt{n}}, \quad n \geqslant 2, \tag{30}$$

with best possible constants $a = 3\sqrt{2}\pi/(6 + 4\pi) = 0.7178\ldots$ and $b = \sqrt{2\pi} = 2.5066\ldots$; and

$$\frac{a}{\sqrt{n}} \leqslant (n+1)\frac{\Omega_{n+1}}{\Omega_n} - n\frac{\Omega_n}{\Omega_{n-1}} < \frac{b}{\sqrt{n}}, \quad n \geqslant 2, \tag{31}$$

with best possible constants $a = (4 - \pi)\sqrt{2} = 1.2139\ldots$ and $b = \sqrt{2\pi}/2 = 1.2533\ldots$.

Alzer's work in [7] includes a number of new results about the gamma function and its derivatives.

In 2010 Mortici [128], improving on some earlier work of Alzer [5, Theorem 1], obtained, for $n \geqslant 1$ on the left and for $n \geqslant 4$ on the right,

$$\frac{a}{\sqrt[2n]{2\pi}} \leqslant \frac{\Omega_n}{\Omega_{n+1}^{n/(n+1)}} < \frac{\sqrt{e}}{\sqrt[2n]{2\pi}},$$

where $a = 64\cdot 720^{11/12}\cdot 2^{1/22}/(10395\cdot\pi^{5/11}) = 1.5714\ldots$. He sharpened the work of Alzer [5, Theorem 2] and Qiu and Vuorinen [154] in the following result, valid for $n \geqslant 1$:

$$\sqrt{\frac{2n+1}{4\pi}} < \frac{\Omega_{n-1}}{\Omega_n} < \sqrt{\frac{2n+1}{4\pi} + \frac{1}{16\pi n}}.$$

Mortici also proved, in [128, Theorem 4], that, for $n \geqslant 4$,

$$\left(1 + \frac{1}{n}\right)^{\frac{1}{2} - \frac{1}{4n}} < \frac{\Omega_n^2}{\Omega_{n-1}\Omega_{n+1}} < \left(1 + \frac{1}{n}\right)^{\frac{1}{2}}.$$

This result improves a similar one by Alzer [5, Theorem 3, valid for $n \geqslant 1$], where the exponent on the left is the constant $2-\log_2 \pi$. Very recently, Yin [202] improved Mortici's result as follows: For $n \geq 1$,

$$\frac{(n+1)(n+1/6)}{(n+\beta)^2} < \frac{\Omega_n^2}{\Omega_{n-1}\Omega_{n+1}} < \frac{(n+1)(n+\beta/2)}{(n+1/3)^2},$$

where $\beta = (391/30)^{1/3}$.

4.5 Digamma and Polygamma Functions

The logarithmic derivative of the gamma function, $\psi(x) \equiv \frac{d}{dx}\log \Gamma(x) = \Gamma'(x)/\Gamma(x)$, is known as the *digamma function*. Its derivatives $\psi^{(n)}, n \geqslant 1$, are known as the *polygamma functions* ψ_n. These functions have the following representations [1, pp. 258–260] for $x > 0$ and each natural number n:

$$\psi(x) = -\gamma + \int_0^\infty \frac{e^{-t} - e^{-xt}}{1-e^{-t}}dt = -\gamma - \frac{1}{x} + \sum_{n=1}^\infty \frac{x}{n(x+n)}$$

and

$$\psi_n(x) = (-1)^{n+1}\int_0^\infty \frac{t^n e^{-xt}}{1-e^{-t}}dt = (-1)^{n+1}n!\sum_{n=0}^\infty (x+k)^{-n-1}.$$

Several researchers have studied the properties of these functions. In 2007, refining the left inequality in [6, Theorem 4.8], Batir [45] obtained estimates for ψ_n in terms of ψ or ψ_k, with $k < n$. In particular, he proved, for $x > 0$ and $n \in \mathbb{N}$:

$$(n-1)!\exp(-n\psi(x+1/2)) < |\psi_n(x)| < (n-1)!\exp(-n\psi(x)),$$

and, for $1 \leqslant k \leqslant n-1$, $x > 0$,

$$(n-1)!\left(\frac{\psi_k(x+1/2)}{(-1)^{k-1}(k-1)!}\right)^{n/k} < |\psi_n(x)| < (n-1)!\left(\frac{\psi_k(x)}{(-1)^{k-1}(k-1)!}\right)^{n/k}.$$

He also proved, for example, the difference formula

$$\alpha < \left((-1)^{n-1}\psi_n(x+1)\right)^{-1/n} - \left((-1)^{n-1}\psi_n(x)\right)^{-1/n} < \beta,$$

where $\alpha = (n!\zeta(n+1))^{-1/n}$ and $\beta = ((n-1)!)^{-1/n}$ are best possible, and the sharp estimates

$$-\gamma < \psi(x) + \log(e^{1/x} - 1) < 0,$$

where γ is the Euler-Mascheroni constant.

In 2010 Mortici [131] proved the following estimates, for $x > 0$ and $n \geqslant 1$, refining the work of Guo, Chen, and Qi [89]:

$$-\frac{1}{720}\frac{(n+3)!}{x^{n+4}} < |\psi_n(x)| - \left[\frac{(n-1)!}{x^n} + \frac{1}{2}\frac{n!}{x^{n+1}} + \frac{1}{12}\frac{(n+1)!}{x^{n+2}}\right] < 0.$$

4.6 Completely Monotonic Functions

A function f is said to be *completely monotonic* on an interval I if $(-1)^n f^{(n)}(x) \geqslant 0$ for all $x \in I$ and all nonnegative integers n. If this inequality is strict, then f is called *strictly completely monotonic*. Such functions occur in probability theory, numerical analysis, and other areas. Some of the most important completely monotonic functions are the gamma function and the digamma and polygamma functions. The Hausdorff-Bernstein-Widder theorem [189, Theorem 12b, p. 161] states that f is completely monotonic on $[0, \infty)$ if and only if there is a nonnegative measure μ on $[0, \infty)$ such that

$$f(x) = \int_0^\infty e^{-xt}\, d\mu(t)$$

for all $x > 0$. There is a well-written introduction to completely monotonic functions in [125].

In 2008 Batir [46] proved that the following function $F_a(x)$ related to the gamma function is completely monotonic on $(0, \infty)$ if and only if $a \geqslant 1/4$ and that $-F_a(x)$ is completely monotonic if and only if $a \leqslant 0$:

$$F_a(x) \equiv \log \Gamma(x) - x \log x + x - \frac{1}{2}\log(2\pi) + \frac{1}{2}\psi(x) + \frac{1}{6(x-a)}.$$

As a corollary he was able to prove, for $x > 0$, the inequality

$$\exp\left(-\frac{1}{2}\psi(x) - \frac{1}{6(x-\alpha)}\right) < \frac{\Gamma(x)}{x^x e^{-x}\sqrt{2\pi}} < \exp\left(-\frac{1}{2}\psi(x) - \frac{1}{6(x-\beta)}\right),$$

with best constants $\alpha = 1/4$ and $\beta = 0$, improving his earlier work with Alzer [9].

In 2010 Mortici [131] showed that for every $n \geqslant 1$, the functions $f, g : (0, \infty) \to \mathbb{R}$ given by

$$f(x) \equiv |\psi_n(x)| - \frac{(n-1)!}{x^n} - \frac{1}{2}\frac{n!}{x^{n+1}} - \frac{1}{12}\frac{(n+1)!}{x^{n+2}} + \frac{1}{720}\frac{(n+3)!}{x^{n+4}}$$

and

$$g(x) \equiv \frac{(n-1)!}{x^n} + \frac{1}{2}\frac{n!}{x^{n+1}} + \frac{1}{12}\frac{(n+1)!}{x^{n+2}} - |\psi_n(x)|$$

are completely monotonic on $(0,\infty)$. As a corollary, since $f(x)$ and $g(x)$ are positive, he obtained estimates for $|\psi_n(x)|$ as finite series in negative powers of x.

Anderson and Qiu [13], as well as some other authors (see [2]), have studied the monotonicity properties of the function $f(x) \equiv (\log \Gamma(x+1))/x$. In 2011 Adell and Alzer [2] proved that f' is completely monotonic on $(-1,\infty)$.

In the course of pursuing research inspired by [13, 14] (see [53]), in 2012 Alzer [7] discussed properties of the function

$$f(x) \equiv \left(1 - \frac{\log x}{\log(1+x)}\right) x \log x,$$

which Qi and Guo [149] later conjectured to be completely monotonic on $(0,\infty)$. In [53] Berg and Pedersen proved this conjecture.

In 2001 Berg and Pedersen [50] proved that the derivative of the function

$$f(x) \equiv \frac{\log \Gamma(x+1)}{x \log x}, \quad x > 0$$

is completely monotonic (see also [51]). This result extends the work of [13, 85]. Very recently, Berg and Pedersen [52] showed that the function

$$F_a(x) \equiv \frac{\log \Gamma(x+1)}{x \log(ax)}$$

is a Pick function when $a \geq 1$, that is, it extends to a holomorphic function mapping the upper half plane into itself. The authors also considered the function

$$f(x) \equiv \left(\frac{\pi^{x/2}}{\Gamma(1+x/2)}\right)^{1/(x \log x)}$$

and proved that $\log f(x+1)$ is a Stieltjes function and hence that $f(x+1)$ is completely monotonic on $(0,\infty)$.

5 The Hypergeometric Function and Elliptic Integrals

The classical *hypergeometric function* is defined by

$$F(a,b;c;x) \equiv {}_2F_1(a,b;c;x) = \sum_{n=0}^{\infty} \frac{(a,n)(b,n)}{(c,n)} \frac{x^n}{n!}, \quad |x| < 1,$$

where $(a,n) \equiv a(a+1)(a+2)\cdots(a+n-1)$ for $n \in \mathbb{N}$ and $(a,0) = 1$ for $a \neq 0$. This function is so general that for proper choice of the parameters a, b, c, one obtains logarithms, trigonometric functions, inverse trigonometric functions, elliptic integrals, or polynomials of Chebyshev, Legendre, Gegenbauer, Jacobi, and so on (see [1, Chap. 15]).

5.1 Hypergeometric Functions

The Bernoulli inequality [126, p. 34] may be written as

$$\log(1+ct) \leqslant c\log(1+t), \tag{32}$$

where $c > 1, t > 0$. In [111] some Bernoulli-type inequalities have been obtained.

It is well known that in the zero-balanced case $c = a + b$ the hypergeometric function $F(a,b;c;x)$ has a logarithmic singularity at $x = 1$ (cf. [18, Theorem 1.19(6)]). Moreover, as a special case [1, 15.1.3],

$$xF(1,1,2;x) = \log\frac{1}{1-x}. \tag{33}$$

Because of this connection, Vuorinen and his collaborators [110] have generalized versions of (32) to a wide class of hypergeometric functions. In the course of their investigation they have studied monotonicity and convexity/concavity properties of such functions. For example, for positive a, b let $g(x) \equiv xF(a,b;a+b;x)$, $x \in (0,1)$. These authors have proved that $G(x) \equiv \log g(e^x/(1+e^x))$ is concave on $(-\infty,\infty)$ if and only if $1/a + 1/b \geqslant 1$. And they have shown that, for fixed $a, b \in (0,1]$ and for $x \in (0,1)$, $p > 0$, the function

$$\left(\frac{x^p}{1+x^p}F\left(a,b;a+b;\frac{x^p}{1+x^p}\right)\right)^{1/p}$$

is increasing in p. In particular,

$$\frac{\sqrt{r}}{1+\sqrt{r}}F\left(a,b;a+b;\frac{\sqrt{r}}{1+\sqrt{r}}\right) \leq \left(\frac{r}{1+r}F\left(a,b;a+b;\frac{r}{1+r}\right)\right)^{1/2}.$$

Motivated by the asymptotic behavior of $F(x) = F(a,b;c;x)$ as $x \to 1^-$, Simić and Vuorinen have carried the above work further in [170], finding best possible bounds, when $a,b,c > 0$ and $0 < x, y < 1$, for the quotient and difference

$$\frac{F(a,b;c;x) + F(a,b;c;y)}{F(a,b;c;x+y-xy)}, \quad F(x) + F(y) - F(x+y-xy).$$

In 2009 Karp and Sitnik [108] obtained some inequalities and monotonicity of ratios for the generalized hypergeometric function. The proofs hinge on a generalized Stieltjes representation of the generalized hypergeometric function.

5.2 Complete Elliptic Integrals

For $0 < r < 1$, the *complete elliptic integrals of the first and second kind* are defined as

$$\mathcal{K}(r) \equiv \int_0^{\pi/2} \frac{dt}{\sqrt{1 - r^2 \sin^2 t}} = \int_0^1 \frac{dt}{\sqrt{(1-t^2)(1-r^2t^2)}} \tag{34}$$

and

$$\mathcal{E}(r) \equiv \int_0^{\pi/2} \sqrt{1 - r^2 \sin^2 t}\, dt = \int_0^1 \sqrt{\frac{1-r^2t^2}{1-t^2}}\, dt, \tag{35}$$

respectively. Letting $r' \equiv \sqrt{1-r^2}$, we often denote

$$\mathcal{K}'(r) = \mathcal{K}(r'), \quad \mathcal{E}'(r) = \mathcal{E}(r').$$

These elliptic integrals have the hypergeometric series representations

$$\mathcal{K}(r) = \frac{\pi}{2} F\left(\tfrac{1}{2}, \tfrac{1}{2}; 1; r^2\right), \quad \mathcal{E} = \frac{\pi}{2} F\left(\tfrac{1}{2}, -\tfrac{1}{2}; 1; r^2\right). \tag{36}$$

5.3 The Landen Identities

The functions $\mathcal{K}$ and $\mathcal{E}$ satisfy the following identities due to Landen [64, 163.01, 164.02]:

$$\mathcal{K}\left(\frac{2\sqrt{r}}{1+r}\right) = (1+r)\mathcal{K}(r), \quad \mathcal{K}\left(\frac{1-r}{1+r}\right) = \frac{1}{2}(1+r)\mathcal{K}'(r),$$

$$\mathcal{E}\left(\frac{2\sqrt{r}}{1+r}\right) = \frac{2\mathcal{E}(r) - r'^2\mathcal{K}(r)}{1+r}, \quad \mathcal{E}\left(\frac{1-r}{1+r}\right) = \frac{\mathcal{E}'(r) + r\mathcal{K}'(r)}{1+r}.$$

Using Landen's transformation formulas, we have the following identities [177, Lemma 2.8]: For $r \in (0,1)$, let $t = (1-r)/(1+r)$. Then

$$
\begin{aligned}
\mathcal{K}(t^2) =& \frac{(1+r)^2}{4}\mathcal{K}'(r^2), \quad \mathcal{K}'(t^2) = (1+r)^2\mathcal{K}(r^2),\\
\mathcal{E}(t^2) =& \frac{\mathcal{E}'(r^2) + (r + r^2 + r^3)\mathcal{K}'(r^2)}{(1+r)^2},\\
\mathcal{E}'(t^2) =& \frac{4\mathcal{E}(r^2) - (3 - 2r^2 - r^4)\mathcal{K}(r^2)}{(1+r)^2}.
\end{aligned}
$$

Generalizing a Landen identity, Simić and Vuorinen [171] have determined the precise regions in the ab-plane for which a Landen inequality holds for zero-balanced hypergeometric functions. They proved that for all $a, b > 0$ with $ab \leqslant 1/4$ the inequality

$$F\left(a, b; a+b; \frac{4r}{(1+r)^2}\right) \leqslant (1+r)F\left(a, b; a+b; r^2\right)$$

holds for $r \in (0,1)$, while for $a, b > 0$ with $1/a + 1/b \leqslant 4$, the following reversed inequality is true for each $r \in (0,1)$:

$$F\left(a, b; a+b; \frac{4r}{(1+r)^2}\right) \geqslant (1+r)F\left(a, b; a+b; r^2\right).$$

In the rest of the ab-plane neither of these inequalities holds for all $r \in (0,1)$. These authors have also obtained sharp bounds for the quotient

$$\frac{(1+r)F(a, b; a+b; r^2)}{F(a, b; a+b; 4r/(1+r)^2)}$$

in certain regions of the ab-plane.

Some earlier results on Landen inequalities for hypergeometric functions can be found in [152]. Recently, Baricz obtained Landen-type inequalities for generalized Bessel functions [29, 37].

Inspired by an idea of Simić and Vuorinen [171], Wang, Chu, and Jiang [188] obtained some inequalities for zero-balanced hypergeometric functions which generalize Ramanujan's cubic transformation formulas.

5.4 Legendre's Relation and Generalizations

It is well known that the complete elliptic integrals satisfy the *Legendre relation* [64, 110.10]:

$$\mathcal{E}\mathcal{K}' + \mathcal{E}'\mathcal{K} - \mathcal{K}\mathcal{K}' = \frac{\pi}{2}.$$

This relation has been generalized in various ways. Elliott [86] proved the identity

$$F_1F_2 + F_3F_4 - F_3F_2 = \frac{\Gamma(1+\lambda+\mu)\Gamma(1+\mu+\nu)}{\Gamma(\lambda+\mu+\nu+\frac{3}{2})\Gamma(\frac{1}{2}+\mu)},$$

where

$$F_1 = F(\tfrac{1}{2}+\lambda, -\tfrac{1}{2}-\nu; 1+\lambda+\mu; x),\quad F_2 = F(\tfrac{1}{2}-\lambda, \tfrac{1}{2}+\nu; 1+\mu+\nu; 1-x),$$
$$F_3 = F(\tfrac{1}{2}+\lambda, \tfrac{1}{2}-\nu; 1+\lambda+\mu; x),\quad F_4 = F(-\tfrac{1}{2}-\lambda, \tfrac{1}{2}+\nu; 1+\mu+\nu; 1-x).$$

Elliott proved this formula by a clever change of variables in multiple integrals. Another proof, based on properties of the hypergeometric differential equation, was suggested without details in [25, p. 138], and the missing details were provided in [20]. It is easy to see that Elliott's formula reduces to the Legendre relation when $\lambda = \mu = \nu = 0$ and $x = r^2$.

Another generalization of the Legendre relation was given in [19]. With the notation

$$u = u(r) = F(a-1, b; c; r),\ v = v(r) = F(a, b; c; r),$$
$$u_1 = u(1-r),\ v_1 = v(1-r),$$

the authors considered the function

$$\mathcal{L}(a, b, c, r) = uv_1 + u_1v - vv_1,$$

proving, in particular, that

$$\mathcal{L}(a, 1-a, c, r) = \frac{\Gamma^2(c)}{\Gamma(c+a-1)\Gamma(c-a+1)}.$$

This reduces to Elliott's formula in case $\lambda = \nu = 1/2 - a$ and $\mu = c + a - 3/2$. In [19] it was conjectured that for $a, b \in (0,1)$, $a+b \le 1 (\ge 1)$, $\mathcal{L}(a, b, c, r)$ is concave (convex) as a function of r on $(0,1)$. In [107] Karatsuba and Vuorinen determined, in particular, the exact regions of abc-space in which the function $\mathcal{L}(a, b, c, r)$ is concave, convex, constant, positive, negative, zero, and where it attains its unique extremum.

In [27] Balasubramanian, Naik, Ponnusamy, and Vuorinen obtained a differentiation formula for an expression involving hypergeometric series that implies Elliott's identity. This paper contains a number of other significant results, including a proof that Elliott's identity is equivalent to a formula of Ramanujan [54, p. 87, Entry 30] on the differentiation of quotients of hypergeometric functions.

5.5 *Some Approximations for $\mathcal{K}(r)$ by* arth *r*

Anderson, Vamanamurthy, and Vuorinen [15] approximated $\mathcal{K}(r)$ by the inverse hyperbolic tangent function arth, obtaining the inequalities

$$\frac{\pi}{2}\left(\frac{\operatorname{arth} r}{r}\right)^{1/2} < \mathcal{K}(r) < \frac{\pi}{2}\frac{\operatorname{arth} r}{r}, \tag{37}$$

for $0 < r < 1$. Alzer and Qiu [11] refined (37) as

$$\frac{\pi}{2}\left(\frac{\operatorname{arth} r}{r}\right)^{3/4} < \mathcal{K}(r) < \frac{\pi}{2}\frac{\operatorname{arth} r}{r}, \tag{38}$$

with the best exponents $3/4$ and 1 for $(\operatorname{arth} r)/r$ on the left and right, respectively. Seeking to improve the exponents in (38), they conjectured that the double inequality

$$\frac{\pi}{2}\left(\frac{\operatorname{arth} r}{r}\right)^{3/4+\alpha r} < \mathcal{K}(r) < \frac{\pi}{2}\left(\frac{\operatorname{arth} r}{r}\right)^{3/4+\beta r} \tag{39}$$

holds for all $0 < r < 1$, with best constants $\alpha = 0$ and $\beta = 1/4$. Very recently Chu et al. [81] gave a proof for this conjecture.

András and Baricz [24] presented some improved lower and upper bounds for $\mathcal{K}(r)$ involving the Gaussian hypergeometric series.

5.6 *Approximations for $\mathcal{E}(r)$*

In [90] Guo and Qi have obtained new approximations for $\mathcal{E}(r)$ as well as for $\mathcal{K}(r)$. For example, they showed that, for $0 < r < 1$,

$$\frac{\pi}{2} - \frac{1}{2}\log\frac{(1+r)^{1-r}}{(1-r)^{1+r}} < \mathcal{E}(r) < \frac{\pi-1}{2} + \frac{1-r^2}{4r}\log\frac{1+r}{1-r}.$$

In recent work [82, 178, 185] Chu et al. have obtained estimates for $\mathcal{E}(r)$ in terms of rational functions of the arithmetic, geometric, and root-square mean, implying new inequalities for the perimeter of an ellipse.

5.7 *Generalized Complete Elliptic Integrals*

For $0 < a < \min\{c, 1\}$ and $0 < b < c \leqslant a + b$, define the *generalized complete elliptic integrals of the first and second kind* on $[0, 1]$ by [95]

$$\mathcal{K}_{a,b,c} = \mathcal{K}_{a,b,c}(r) \equiv \frac{B(a,b)}{2} F(a,b;c;r^2), \tag{40}$$

$$\mathcal{E}_{a,b,c} = \mathcal{E}_{a,b,c}(r) \equiv \frac{B(a,b)}{2} F(a-1,b;c;r^2), \tag{41}$$

$$\mathcal{K}'_{a,b,c} = \mathcal{K}_{a,b,c}(r') \quad \text{and} \quad \mathcal{E}'_{a,b,c} = \mathcal{E}_{a,b,c}(r'), \tag{42}$$

for $r \in (0,1)$, $r' = \sqrt{1-r^2}$. The end values are defined by limits as r tends to $0+$ and $1-$, respectively. Thus,

$$\mathcal{K}_{a,b,c}(0) = \mathcal{E}_{a,b,c}(0) = \frac{B(a,b)}{2}$$

and

$$\mathcal{E}_{a,b,c}(1) = \frac{1}{2} \frac{B(a,b)B(c,c+1-a-b)}{B(c+1-a,c-b)}, \quad \mathcal{K}_{a,b,c}(1) = \infty.$$

Note that the restrictions on the parameters a, b, and c ensure that the function $\mathcal{K}_{a,b,c}$ is increasing and unbounded, whereas $\mathcal{E}_{a,b,c}$ is decreasing and bounded, as in the classical case $a = b = 1/2$, $c = 1$.

Heikkala, Lindén, Vamanamurthy, and Vuorinen [94, 95] derived several differentiation formulas and obtained sharp monotonicity and convexity properties for certain combinations of the generalized elliptic integrals. They also constructed a conformal mapping $\mathrm{sn}_{a,b,c}$ from a quadrilateral with internal angles $b\pi$, $(c-b)\pi$, $(1-a)\pi$, and $(1-c+a)\pi$ onto the upper half plane. These results generalize the work of [19]. For some particular parameter triples (a,b,c), there are very recent results by many authors [37, 181, 206, 209].

With suitable restrictions on the parameters a, b, c, Neuman [135] has obtained bounds for $\mathcal{K}_{a,b,c}$ and $\mathcal{E}_{a,b,c}$ and for certain combinations and products of them. He has also proved that these generalized elliptic integrals are logarithmically convex as functions of the first parameter.

In 2007 Baricz [33, 36, 38] established some Turán-type inequalities for Gaussian hypergeometric functions and generalized complete elliptic integrals. He also studied the generalized convexity of the zero-balanced hypergeometric functions and generalized complete elliptic integrals [31] (see also [30, 32, 37]). Very recently, Kalmykov and Karp [103, 104] have studied log-convexity and log-concavity for series involving gamma functions and derived many known and new inequalities for the modified Bessel and Kummer and generalized hypergeometric functions and ratios of the Gauss hypergeometric functions. In particular, they improved and generalized Baricz's Turán-type inequalities.

5.8 *The Generalized Modular Function and Generalized Linear Distortion Function*

Let $a, b, c > 0$ with $a+b \geq c$. A *generalized modular equation of order* (or *degree*) $p > 0$ is

$$\frac{F(a,b;c;1-s^2)}{F(a,b;c;s^2)} = p\frac{F(a,b;c;1-r^2)}{F(a,b;c;r^2)}, \quad 0 < r < 1. \tag{43}$$

The *generalized modulus* is the decreasing homeomorphism $\mu_{a,b,c} : (0,1) \to (0,\infty)$, defined by

$$\mu_{a,b,c}(r) \equiv \frac{B(a,b)}{2}\frac{F(a,b;c;1-r^2)}{F(a,b;c;r^2)}. \tag{44}$$

The generalized modular equation (43) can be written as

$$\mu_{a,b,c}(s) = p\mu_{a,b,c}(r).$$

With $p = 1/K$, $K > 0$, the solution of (43) is then given by

$$s = \varphi_K^{a,b,c}(r) \equiv \mu_{a,b,c}^{-1}(\mu_{a,b,c}(r)/K).$$

Here $\varphi_K^{a,b,c}$ is called the (a,b,c)-*modular function* with *degree* $p = 1/K$ [19, 94, 95]. Clearly the following identities hold:

$$\mu_{a,b,c}(r)\mu_{a,b,c}(r') = \left(\frac{B(a,b)}{2}\right)^2,$$

$$\varphi_K^{a,b,c}(r)^2 + \varphi_{1/K}^{a,b,c}(r')^2 = 1.$$

In [94], the authors generalized the functional inequalities for the modular functions and Grötzsch function μ proved in [19] to hold also for the generalized modular functions and generalized modulus in the case $b = c - a$. For instance, for $0 < a < c \leq 1$ and $K > 1$, the inequalities

$$\mu_{a,c-a,c}(1-\sqrt{(1-u)(1-t)}) \leq \frac{\mu_{a,c-a,c}(u) + \mu_{a,c-a,c}(t)}{2} \leq \mu_{a,c-a,c}(\sqrt{ut}) \tag{45}$$

hold for all $u, t \in (0,1)$, with equality if and only if $u = t$, and

$$r^{1/K} < \varphi_K^{a,c-a,c}(r) < e^{(1-1/K)R(a,c-a)/2}r^{1/K}, \tag{46}$$

$$r^K > \varphi_{1/K}^{a,c-a,c}(r) > e^{(1-K)R(a,c-a)/2}r^K. \tag{47}$$

For the special case of $a = 1/2$ and $c = 1$ the readers are referred to [18]. Wang et al. [181] presented several sharp inequalities for the generalized modular functions with specific choice of parameters $c = 1$ and $b = 1 - a$.

A linearization for the generalized modular function is also presented in [94] as follows: Let $p : (0,1) \to (-\infty,\infty)$ and $q : (-\infty,\infty) \to (0,1)$ be given by $p(x) = 2\log(x/x')$ and $q(x) = p^{-1}(x) = \sqrt{e^x/(e^x+1)}$, respectively, and for $a \in (0,1)$, $c \in (a,1]$, $K \in (1,\infty)$, let $g, h : (-\infty,\infty) \to (-\infty,\infty)$ be defined by $g(x) = p(\varphi_K^{a,c-a,c}(q(x)))$ and $h(x) = p(\varphi_{1/K}^{a,c-a,c}(q(x)))$. Then

$$g(x) \geq \begin{cases} Kx, & \text{if } \; x \geq 0, \\ x/K, & \text{if } \; x < 0, \end{cases} \quad \text{and} \quad h(x) \leq \begin{cases} x/K, & \text{if } \; x \geq 0, \\ Kx, & \text{if } \; x < 0. \end{cases}$$

In the same paper the authors also studied how these generalized functions depend on the parameter c. Corresponding results for the case $c = 1$ can be found in the articles [19, 153, 204].

Recently Bhayo and Vuorinen [55] have studied the Hölder continuity and submultiplicative properties of $\varphi_K^{a,b,c}(r)$ in the case where $c = 1$ and $b = 1 - a$ and have obtained several sharp inequalities for $\varphi_K^{a,1-a,1}(r)$.

For $x, K \in (0,\infty)$, define

$$\eta_K^a(x) \equiv \left(\frac{s}{s'}\right)^2, \; s = \varphi_K^{a,1-a,1}(r), \; r = \sqrt{\frac{x}{1+x}},$$

and the *generalized linear distortion function*

$$\lambda(a,K) \equiv \left(\frac{\varphi_K^{a,1-a,1}(1/\sqrt{2})}{\varphi_{1/K}^{a,1-a,1}(1/\sqrt{2})}\right)^2, \quad \lambda(a,1) = 1.$$

For $a = 1/2$, these two functions reduce to the well-known special case denoted by $\eta_K(x)$ and $\lambda(K)$, respectively, which play a crucial role in quasiconformal theory. Several inequalities for these functions have been obtained as an application of the monotonicity and convexity of certain combinations of these functions and some elementary functions; see [55, 80, 122, 123, 180, 203]. For instance, the following chain of inequalities is proved in [80]: for $a \in (0,1/2]$, $K \in (1,\infty)$ and $x, y \in (0,\infty)$,

$$\begin{aligned} &\max\left\{\frac{2\eta_K^a(x)\eta_K^a(y)}{\eta_K^a(x)+\eta_K^a(y)}, \eta_K^a\left(\frac{2xy}{x+y}\right)\right\} \leq \eta_K^a(\sqrt{xy}) \\ &\quad \leq \sqrt{\eta_K^a(x)\eta_K^a(y)} \leq \min\left\{\frac{\eta_K^a(x)+\eta_K^a(y)}{2}, \eta_K^a\left(\frac{x+y}{2}\right)\right\}, \end{aligned}$$

with equality if and only if $x = y$.

6 Inequalities for Power Series

The following theorem [95, Theorem 4.3] is an interesting tool in simplified proofs for monotonicity of the quotient of two power series:

Theorem 6.1. *Let $\sum_{n=0}^{\infty} a_n x^n$ and $\sum_{n=0}^{\infty} b_n x^n$ be two real power series converging on the interval $(-R, R)$. If the sequence $\{a_n/b_n\}$ is increasing (decreasing) and $b_n > 0$ for all n, then the function*

$$f(x) = \frac{\sum_{n=0}^{\infty} a_n x^n}{\sum_{n=0}^{\infty} b_n x^n}$$

is also increasing (decreasing) on $(0, R)$. In fact, the function

$$f'(x)\left(\sum_{n=0}^{\infty} b_n x^n\right)^2$$

has positive Maclaurin coefficients.

A more general version of this theorem appears in [58] and [147, Lemma 2.1]. This kind of rule also holds for the quotient of two polynomials instead of two power series (cf. [95, Theorem 4.4]):

Theorem 6.2. *Let $f_n(x) = \sum_{k=0}^{n} a_k x^k$ and $g_n(x) = \sum_{k=0}^{n} b_k x^k$ be two real polynomials, with $b_k > 0$ for all k. If the sequence $\{a_k/b_k\}$ is increasing (decreasing), then so is the function $f_n(x)/g_n(x)$ for all $x > 0$. In fact, $g_n f_n' - f_n g_n'$ has positive (negative) coefficients.*

In 1997 Ponnusamy and Vuorinen [147] refined Ramanujan's work on asymptotic behavior of the hypergeometric function and also obtained many inequalities for the hypergeometric function by making use of Theorem 6.1. Many well-known results of monotonicity and inequalities for complete elliptic integrals have been extended to the generalized elliptic integrals in [94, 95].

Motivated by an open problem of Anderson et al. [16], in 2006 Baricz [30] considered ratios of general power series and obtained the following theorem. Note the similarity of the last inequality in Theorem 6.3 with the left-hand side of the inequality (45).

Theorem 6.3. *Suppose that the power series $f(x) = \sum_{n=0}^{\infty} a_n x^n$ with $a_n > 0$ for all $n \geq 0$ is convergent for all $x \in (0, 1)$ and also that the sequence $\{(n + 1)a_{n+1}/a_n - n\}_{n\geq 0}$ is strictly decreasing. Let the function $m_f : (0, 1) \to (0, \infty)$ be defined as $m_f(r) = f(1 - r^2)/f(r^2)$. Then*

$$\sqrt[k]{\prod_{i=1}^{k} m_f(r_i)} \leq m_f\left(\sqrt[k]{\prod_{i=1}^{k} r_i}\right),$$

for all $r_1, r_2, \ldots, r_k \in (0,1)$*, where equality holds if and only if* $r_1 = r_2 = \cdots = r_k$*. In particular, for* $k = 2$ *the inequalities*

$$\sqrt{m_f(r_1)m_f(r_2)} \leq m_f(\sqrt{r_1 r_2}),$$

$$\frac{1}{m_f(r_1)} + \frac{1}{m_f(r_2)} \geq \frac{2}{m_f(\sqrt{r_1 r_2})},$$

$$m_f(r_1) + m_f(r_2) \geq 2m_f\left(\sqrt{1 - \sqrt{(1-r_1^2)(1-r_2^2)}}\right)$$

hold for all $r_1, r_2 \in (0,1)$*, and in all these inequalities equality holds if and only if* $r_1 = r_2$*.*

The following Landen-type inequality for power series is also due to Baricz [29].

Theorem 6.4. *Suppose that the power series* $f(x) = \sum_{n=0}^{\infty} a_n x^n$ *with* $a_n > 0$ *for all* $n \geq 0$ *is convergent for all* $x \in (0,1)$ *and that for a given* $\delta > 1$ *the sequence* $\{n!a_n/(\log \delta)^n\}_{n\geq 0}$ *is decreasing. If* $\lambda_f(x) = f(x^2)$*, then*

$$\lambda_f\left(\frac{2\sqrt{r}}{1+r}\right) < \rho \lambda_f(r)$$

holds for all $r \in (0,1)$ *and* $\rho \geq \delta^{4\sqrt{2}-5}$*.*

Anderson, Vamanamurthy, and Vuorinen [22] studied generalized convexity and gave sufficient conditions for generalized convexity of functions defined by Maclaurin series. These results yield a class of new inequalities for power series which improve some earlier results obtained by Baricz. More inequalities for power series can be found in [37, 80].

In 1928 T. Kaluza gave a criterion for the signs of the power series of a function that is the reciprocal of another power series.

Theorem 6.5 ([105]). *Let* $f(x) = \sum_{n\geq 0} a_n x^n$ *be a convergent Maclaurin series with radius of convergence* $r > 0$*. If* $a_n > 0$ *for all* $n \geq 0$ *and the sequence* $\{a_n\}_{n\geq 0}$ *is log-convex, that is, for all* $n \geq 0$

$$a_n^2 \leq a_{n-1}a_{n+1}, \tag{48}$$

then the coefficients b_n *of the reciprocal power series* $1/f(x) = \sum_{n\geq 0} b_n x^n$ *have the following properties:* $b_0 = 1/a_0 > 0$ *and* $b_n \leq 0$ *for all* $n \geq 1$*.*

In 2011 Baricz, Vesti, and Vuorinen [42] showed that the condition (48) cannot be replaced by the condition

$$a_n \leq \left(\frac{a_{n-1}^t + a_{n+1}^t}{2}\right)^{1/t},$$

for any $t > 0$. However, it is not known whether the condition (48) is necessary.

In 2009 Koumandos and Pedersen [115, Lemma 2.2] proved the following interesting result, which deals with the monotonicity properties of the quotient of two series of functions.

Theorem 6.6. *Suppose that $a_k > 0$, $b_k > 0$ and that $\{u_k(x)\}$ is a sequence of positive C^1-functions such that the series*

$$\sum_{k=0}^{\infty} a_k u_k^{(l)}(x) \quad \text{and} \quad \sum_{k=0}^{\infty} b_k u_k^{(l)}(x), \quad l = 0, 1,$$

converge absolutely and uniformly over compact subsets of $[0, \infty)$. Define

$$f(x) \equiv \frac{\sum_{k=0}^{\infty} a_k u_k(x)}{\sum_{k=0}^{\infty} b_k u_k(x)}.$$

1. *If the logarithmic derivatives $u_k'(x)/u_k(x)$ form an increasing sequence of functions and if a_k/b_k decreases (resp. increases), then $f(x)$ decreases (resp. increases) for $x \geq 0$.*
2. *If the logarithmic derivatives $u_k'(x)/u_k(x)$ form a decreasing sequence of functions and if a_k/b_k decreases (resp. increases), then $f(x)$ increases (resp. decreases) for $x \geq 0$.*

For inequalities of power series as complex functions, see [99–101] and the references therein.

7 Means

A *homogeneous bivariate mean* is defined as a continuous function $\mathcal{M} : \mathbb{R}^+ \times \mathbb{R}^+ \to \mathbb{R}$ satisfying $\min\{x, y\} \leq \mathcal{M}(x, y) \leq \max\{x, y\}$ and $\mathcal{M}(\lambda x, \lambda y) = \lambda \mathcal{M}(x, y)$ for all $x, y, \lambda > 0$. Important examples are the *arithmetic mean* $A(a, b)$, the *geometric mean* $G(a, b)$, the *logarithmic mean* $L(a, b)$, the *identric mean* $I(a, b)$, the *root-square mean* $Q(a, b)$, and the *power mean* $M_r(a, b)$ *of order r* defined, respectively, by

$$A(a,b)=\frac{a+b}{2},\quad G(a,b)=\sqrt{ab},$$

$$L(a,b)=\frac{a-b}{\log a-\log b},\quad I(a,b)=\frac{1}{e}\left(\frac{a^a}{b^b}\right)^{1/(a-b)},$$

$$Q(a,b)=\sqrt{\frac{a^2+b^2}{2}},\quad M_r(a,b)=\sqrt[r]{\frac{a^r+b^r}{2}}.$$

7.1 Power Means

The *weighted power means* are defined by

$$M_\lambda(\omega;a,b)\equiv\left(\omega a^\lambda+(1-\omega)b^\lambda\right)^{1/\lambda}\qquad(\lambda\neq 0),$$

$M_0(\omega;a,b)\equiv a^\omega b^{1-\omega}$, with *weights* $\omega, 1-\omega>0$. The *power means* are the equally weighted means $M_\lambda(a,b)=M_\lambda(1/2;a,b)$. As a special case, we have $M_0(1/2;a,b)=G(a,b)$.

In [114] Kouba studied the ratio of differences of power means

$$\rho(s,t,p;a,b)\equiv\frac{M_s^p(a,b)-G^p(a,b)}{M_t^p(a,b)-G^p(a,b)},$$

finding sharp bounds for $\rho(s,t,p;a,b)$ in various regions of stp-space with a,b positive and $a\neq b$. This work extends the results of Alzer and Qiu [10], Trif [175], Kouba [113], Wu [193], and Wu and Debnath [196]. Kouba also extended the range of validity of the following inequality, due to Wu and Debnath [196]:

$$\frac{2^{-p/r}-2^{-p/s}}{2^{-p/t}-2^{-p/s}}<\frac{M_r^p(a,b)-M_s^p(a,b)}{M_t^p(a,b)-M_s^p(a,b)}<\frac{r-s}{t-s}$$

to the set of real numbers r,t,s,p satisfying the conditions $0<s<t<r$ and $0<p\leqslant(4t+2s)/3$.

7.2 Toader Means

If $p:\mathbb{R}^+\to\mathbb{R}^+$ is a strictly monotonic function, then define

$$f(a,b;p,n)\equiv\begin{cases}\dfrac{1}{2\pi}\int_0^{2\pi}p((a^n\cos^2\theta+b^n\sin^2\theta)^{1/n})d\theta & \text{if}\quad n\neq 0,\\[2ex] \dfrac{1}{2\pi}\int_0^{2\pi}p(a^{\cos^2\theta}b^{\sin^2\theta})d\theta & \text{if}\quad n=0,\end{cases}$$

where a, b are positive real numbers. The *Toader mean* [174] of a and b is defined as $T(a,b;p,n) \equiv p^{-1}(f(a,b;p,n))$. It is easy to see that the Toader mean is symmetric. For special choices of p, let $T_{q,n}(a,b) = T(a,b;p,n)$ if $p(x) = x^q$ with $q \neq 0$ and $T_{0,n}(a,b) = T(a,b;p,n)$ if $p(x) = \log x$. The means $T_{q,n}$ belong to a large family of means called the *hypergeometric means*, which have been studied by Carlson and others [62,65,68]. It is easy to see that $T_{q,n}$ is homogeneous. In particular, we have

$$T_{0,2}(a,b) = A(a,b), \quad T_{-2,2}(a,b) = G(a,b), \quad T_{2,2}(a,b) = Q(a,b).$$

Furthermore, the Toader means are related to the complete elliptic integrals: for $a \geq b > 0$,

$$T_{-1,2}(a,b) = \frac{\pi a}{2\mathcal{K}(\sqrt{1-(b/a)^2})} \quad \text{and} \quad T_{1,2}(a,b) = \frac{2a}{\pi}\mathcal{E}(\sqrt{1-(b/a)^2}).$$

In 1997 Qiu and Shen [151] proved that, for all $a, b > 0$ with $a \neq b$,

$$M_{3/2}(a,b) < T_{1,2}(a,b).$$

This inequality had been conjectured by Vuorinen [176]. Alzer and Qiu [10] proved the following best possible power mean upper bound:

$$T_{1,2}(a,b) < M_{\log 2/\log(\pi/2)}(a,b).$$

Very recently, Chu and his collaborators [78,79,83] obtained several bounds for $T_{1,2}$ with respect to some combinations of various means.

7.3 *Seiffert Means*

The *Seiffert means* S_1 and S_2 are defined by

$$S_1(a,b) \equiv \frac{a-b}{2\arcsin\frac{a-b}{a+b}}, \; a \neq b, \; S_1(a,a) = a,$$

and

$$S_2(a,b) \equiv \frac{a-b}{2\arctan\frac{a-b}{a+b}}, \; a \neq b, \; S_2(a,a) = a.$$

It is well known that

$$\sqrt[3]{G^2 A} < L < \frac{2G+A}{3}.$$

Sándor proved similar results for Seiffert means [160, 161]:

$$\sqrt[3]{A^2G} < S_1 < \frac{G+2A}{3} < I \tag{49}$$

and

$$\sqrt[3]{Q^2A} < S_2 < \frac{A+2Q}{3}. \tag{50}$$

The inequalities (49) and (50) are special cases of more general results obtained by Neuman and Sándor [139, 140].

7.4 Extended Means

Let $a, b \in (0, \infty)$ be distinct and $s, t \in \mathbb{R} \setminus \{0\}$, $s \neq t$. We define the *extended mean* [172] with parameters s and t by

$$E_{s,t}(a,b) \equiv \left(\frac{t}{s}\frac{a^s - b^s}{a^t - b^t}\right)^{1/(s-t)}$$

and also

$$E_{s,s}(a,b) \equiv \exp\left(\frac{1}{s} + \frac{a^s \log a - b^s \log b}{a^s - b^s}\right),$$

$$E_{s,0}(a,b) \equiv \left(\frac{a^s - b^s}{s \log(x/y)}\right)^{1/s} \quad \text{and} \quad E_{0,0}(a,b) \equiv \sqrt{ab}.$$

We see that all the classical means belong to the family of extended means. For example, $E_{2,1} = A$, $E_{0,0} = G$, $E_{-1,-2} = H$, and $E_{1,0} = L$ and, more generally, $M_\lambda = E_{2\lambda,\lambda}$ for $\lambda \in \mathbb{R}$. The reader is referred to the survey [148] for many interesting results on the extended mean.

In 2002 Hästö [91] studied a certain monotonicity property of ratios of extended means and Seiffert means, which he called a *strong inequality*. These strong inequalities were shown to be related to the so-called relative metric [92, 93].

7.5 Means and the Circular and Hyperbolic Functions

It is easy to check the following identities:

$$A(1+\sin x, 1-\sin x) = 1, \quad G(1+\sin x, 1-\sin x) = \cos x, \tag{51}$$

$$Q(1+\sin x, 1-\sin x) = \sqrt{1+\sin^2 x},\;\; S_1(1+\sin x, 1-\sin x) = \frac{\sin x}{x}, \tag{52}$$

$$A(e^x, e^{-x}) = \cosh x,\;\; G(e^x, e^{-x}) = 1,\;\; Q(e^x, e^{-x}) = \sqrt{\cosh 2x}, \tag{53}$$

$$L(e^x, e^{-x}) = \frac{\sinh x}{x},\;\; I(e^x, e^{-x}) = e^{x\coth x - 1}, \tag{54}$$

$$S_1(e^x, e^{-x}) = \frac{\sinh x}{\arcsin(\tanh x)},\;\; S_2(e^x, e^{-x}) = \frac{\sinh x}{\arctan(\tanh x)}. \tag{55}$$

One can get many inequalities by combining the above identities and inequalities between means. For example, combining (49) and (52), we have

$$\sqrt[3]{\cos x} < \frac{\sin x}{x} < \frac{\cos x + 2}{3},$$

where the second inequality is the well-known Cusa-Huygens inequality, and combining (50), (53), and (55), we have

$$\sqrt[3]{(\cosh 2x)(\cosh x)} < \frac{\sinh x}{\arctan(\tanh x)} < \frac{\cosh x + 2\sqrt{\cosh 2x}}{3}.$$

More inequalities on mean values and trigonometric and hyperbolic functions can be found in [136, 163, 165, 200, 208] and references therein.

7.6 *Means and Hypergeometric Functions*

In 2005 Richards [158] obtained sharp power mean bounds for the hypergeometric function: Let $0 < a, b \le 1$ and $c > \max\{-a, b\}$. If $c \ge \max\{1-2a, 2b\}$, then

$$M_\lambda(1-b/c; 1, 1-r) \le F(-a, b; c; r)^{1/a}$$

if and only if $\lambda \le \frac{a+c}{1+c}$. If $c \le \min\{1-2a, 2b\}$, then

$$M_\mu(1-b/c; 1, 1-r) \le F(-a, b; c; r)^{1/a}$$

if and only if $\mu \ge \frac{a+c}{1+c}$. These inequalities generalize earlier results proved by Carlson [66].

For hypergeometric functions of form $F(1/2-s, 1/2+s; 1; 1-r^p)^q$, Borwein et al. [61] exhibited explicitly iterations similar to the arithmetic-geometric mean. Barnard et al. [43] presented sharp bounds for hypergeometric analogs of the arithmetic-geometric mean as follows: For $0 < \alpha \le 1/2$ and $p > 0$,

$$M_\lambda(\alpha; 1, r) \le F(\alpha, 1-\alpha; 1; 1-r^p)^{-1/(\alpha p)} \le M_\mu(\alpha; 1, r)$$

if and only if $\lambda \le 0$ and $\mu \ge p(1-\alpha)/2$.

Some other inequalities involving hypergeometric functions and bivariate means can be found in the very recent survey [44].

For any two power means M_λ and M_μ, a function f is called $M_{\lambda,\mu}$-*convex* if it satisfies

$$f(M_\lambda(x, y)) \le M_\mu(f(x), f(y)).$$

Recently many authors have proved that the zero-balanced Gaussian hypergeometric function is $M_{\lambda,\lambda}$-convex when $\lambda \in \{-1, 0, 1\}$. For details see [22,26,37,80]. Baricz [31] generalized these results to the $M_{\lambda,\lambda}$-convexity of zero-balanced Gaussian hypergeometric functions with respect to a power mean for $\lambda \in [0, 1]$. Zhang et al. [205] extended these results to the case of $M_{\lambda,\mu}$-convexity with respect to two power means: For all $a, b > 0$, $\lambda \in (-\infty, 1]$, and $\mu \in [0, \infty)$ the hypergeometric function $F(a, b; a+b; r)$ is $M_{\lambda,\mu}$-convex on $(0, 1)$.

The following interesting open problem is presented by Baricz [36]:

Open Problem. *If m_1 and m_2 are bivariate means, then find conditions on $a_1, a_2 > 0$ and $c > 0$ for which the inequality*

$$m_1(F_{a_1}(r), F_{a_2}(r)) \le (\ge) F_{m_2(a_1,a_2)}(r)$$

holds true for all $r \in (0, 1)$, where $F_a(r) = F(a, c-a; c; r)$.

7.7 Means and Quasiconformal Analysis

Special functions have always played an important role in the distortion theory of quasiconformal mappings. Anderson, Vamanamurthy, and Vuorinen [18] have systematically investigated classical special functions and their extensive applications in the theory of conformal invariants and quasiconformal mappings. Some functional inequalities for special functions in quasiconformal mapping theory involve the arithmetic mean, geometric mean, or harmonic mean. For example, for the well-known Grötzsch ring function μ and the Hersch-Pfluger distortion function φ_K, the following inequalities hold for all $s, t \in (0, 1)$ with $s \ne t$:

$$\sqrt{\mu(s)\mu(t)} < \mu(\sqrt{st}),$$

and

$$\sqrt{\varphi_K(s)\varphi_K(t)} < \varphi_K(\sqrt{st}) \text{ for } K > 1.$$

Recently, Wang, Zhang, and Chu [182, 183] have extended these inequalities as follows:

$$M_\lambda(\mu(s),\mu(t)) < \mu(M_\lambda(s,t)) \text{ if and only if } \lambda \le 0,$$
$$M_\lambda(\varphi_K(s),\varphi_K(t)) < \varphi_K(M_\lambda(s,t)) \text{ if and only if } \lambda \ge 0 \text{ and } K > 1,$$

and

$$M_\lambda(\varphi_K(s),\varphi_K(t)) > \varphi_K(M_\lambda(s,t)) \text{ if and only if } \lambda \ge 0 \text{ and } 0 < K < 1.$$

Some similar results for the generalized Grötzsch function, generalized modular function, and other special functions related to quasiconformal analysis can be found in [155, 179, 184, 186, 187].

8 Epilogue and a View Toward the Future

In earlier work we have listed many open problems. See especially [14, pp. 128–131] and [18, p. 478]. Many of these problems are still open. In Sects. 4, 6, and 7 above, we have also mentioned some open problems.

Finally, we wish to suggest some ideas for further research. In a frequently cited paper [119] Lindqvist introduced in 1995 the notion of *generalized trigonometric functions* such as $\sin_p$, and presently there is a large body of literature about this topic. For the case $p = 2$ the classical functions are obtained. In 2010, Biezuner et al. [59] developed a practical numerical method for computing values of $\sin_p$. Recently, Takeuchi [173] has gone a step further, introducing functions depending on two parameters p and q that reduce to the p-functions of Lindqvist when $p = q$. In [56, 57, 112] the authors have continued the study of this family of generalized functions and have suggested that many properties of classical functions have a counterpart in this more general setting. It would be natural to generalize the properties of trigonometric functions cited in this survey to the (p, q)-trigonometric functions of Takeuchi.

Acknowledgements The authors wish to thank Á. Baricz, C. Berg, E. A. Karatsuba, C. Mortici, E. Neuman, H. L. Pedersen, S. Ponnusamy, and G. Tee for careful reading of this paper and for many corrections and suggestions. The research of Matti Vuorinen was supported by the Academy of Finland, Project 2600066611. Xiaohui Zhang is indebted to the Finnish National Graduate School of Mathematics and its Applications for financial support.

References

1. Abramowitz, M., Stegun, I.A.: Handbook of Mathematical Functions with Formulas, Graphs, and Mathematical Tables. Dover Publications, New York (1970)
2. Adell, J.A., Alzer, H.: A monotonicity property of Euler's gamma function. Publ. Math. Debrecen **78**, 443–448 (2011)
3. Alzer, H.: On some inequalities for the gamma and psi functions. Math. Comput. **66**, 373–389 (1997)
4. Alzer, H.: Inequalities for the gamma function. Proc. Am. Math. Soc. **128**, 141–147 (1999)
5. Alzer, H.: Inequalities for the volume of the unit ball in $\mathbb{R}^n$. J. Math. Anal. Appl. **252**, 353–363 (2000)
6. Alzer, H.: Sharp inequalities for digamma and polygamma functions. Forum Math. **16**, 181–221 (2004)
7. Alzer, H.: Inequalities for the volume of the unit ball in $\mathbb{R}^n$ II. Mediterr. J. Math. **5**, 395–413 (2008)
8. Alzer, H.: Inequalities for the harmonic numbers. Math. Z. **267**, 367–384 (2011)
9. Alzer, H., Batir, N.: Monotonicity properties of the gamma function. Appl. Math. Lett. **20**, 778–781 (2007)
10. Alzer, H., Qiu, S.-L.: Inequalities for means in two variables. Arch. Math. (Basel) **80**, 201–215 (2003)
11. Alzer, H., Qiu, S.-L.: Monotonicity theorems and inequalities for the complete elliptic integrals. J. Comput. Appl. Math. **172**, 289–312 (2004)
12. Anderson, G.D., Vuorinen, M.: Reflections on Ramanujan's mathematical gems. Math. Newsl. **19**, 87–108 (2010). Available via arXiv:1006.5092v1 [math.CV]
13. Anderson, G.D., Qiu, S.-L.: A monotoneity property of the gamma function. Proc. Am. Math. Soc. **125**, 3355–3362 (1997)
14. Anderson, G.D., Vamanamurthy, M.K., Vuorinen, M.: Special functions of quasiconformal theory. Expo. Math. **7**, 97–136 (1989)
15. Anderson, G.D., Vamanamurthy, M.K., Vuorinen, M.: Functional inequalities for hypergeometric functions and complete elliptic integrals. SIAM J. Math. Anal. **23**, 512–524 (1992)
16. Anderson, G.D., Vamanamurthy, M.K., Vuorinen, M.: Hypergeometric functions and elliptic integrals. In: Srivastava, H.M., Owa, S. (eds.) Current Topics in Analytic Function Theory, pp. 48–85. World Scientific Publishing Co., Singapore (1992)
17. Anderson, G.D., Vamanamurthy, M.K., Vuorinen, M.: Inequalities for quasiconformal mappings in space. Pacific J. Math. **160**, 1–18 (1993)
18. Anderson, G.D., Vamanamurthy, M.K., Vuorinen, M.: Conformal Invariants, Inequalities, and Quasiconformal Maps. Wiley, New York (1997)
19. Anderson, G.D., Qiu, S.-L., Vamanamurthy, M.K., Vuorinen, M.: Generalized elliptic integrals and modular equations. Pacific J. Math. **192**, 1–37 (2000)
20. Anderson, G.D., Vamanamurthy, M.K., Vuorinen, M.: Topics in special functions. In: Papers on Analysis: A Volume Dedicated to Olli Martio on the Occasion of his 60th Birthday, vol. 83, pp. 5–26. Report University of Jyväskylä (2001)
21. Anderson, G.D., Vamanamurthy, M.K., Vuorinen, M.: Monotonicity rules in calculus. Am. Math. Monthly **133**, 805–816 (2006)
22. Anderson, G.D., Vamanamurthy, M.K., Vuorinen, M.: Generalized convexity and inequalities. J. Math. Anal. Appl. **335**, 1294–1308 (2007)
23. Anderson, G.D., Vamanamurthy, M.K., Vuorinen, M.: Topics in special functions II. Conform. Geom. Dyn. **11**, 250–271 (2007)
24. András, S., Baricz, Á.: Bounds for complete elliptic integrals of the first kind. Expo. Math. **28**, 357–364 (2010)
25. Andrews, G., Askey, R., Roy, R.: Special Functions. Encyclopedia of Mathematics and its Applications, vol. 71. Cambridge University Press, Cambridge (1999)

26. Balasubramanian, R., Ponnusamy, S., Vuorinen, M.: Functional inequalities for the quotients of hypergeometric functions. J. Math. Anal. Appl. **218**, 256–268 (1998)
27. Balasubramanian, R., Naik, S., Ponnusamy, S., Vuorinen, M.: Elliott's identity and hypergeometric functions. J. Math. Anal. Appl. **271**, 232–256 (2002)
28. Barbu, C., Pişcoran, L.-I.: On Panaitopol and Jordan type inequalities (unpublished manuscript)
29. Baricz, Á.: Landen-type inequalities for Bessel functions. Comput. Methods Funct. Theory **5**, 373–379 (2005)
30. Baricz, Á.: Functional inequalities involving special functions. J. Math. Anal. Appl. **319**, 450–459 (2006)
31. Baricz, Á.: Convexity of the zero-balanced Gaussian hypergeometric functions with respect to Hölder means. JIPAM. J. Inequal. Pure Appl. Math. **8**, 9 (2007) (Article 40)
32. Baricz, Á.: Functional inequalities involving special functions II. J. Math. Anal. Appl. **327**, 1202–1213 (2007)
33. Baricz, Á.: Turán type inequalities for generalized complete elliptic integrals. Math. Z. **256**, 895–911 (2007)
34. Baricz, Á.: Functional inequalities involving Bessel and modified Bessel functions of the first kind. Expo. Math. **26**, 279–293 (2008)
35. Baricz, Á.: Jordan-type inequalities for generalized Bessel functions. JIPAM. J. Inequal. Pure Appl. Math. **9**, 6 (2008) (Article 39)
36. Baricz, Á.: Turán type inequalities for hypergeometric functions. Proc. Am. Math. Soc. **136**, 3223–3229 (2008)
37. Baricz, Á.: Generalized Bessel functions of the first kind. Lecture Notes in Mathematics 1994. Springer, Berlin (2010)
38. Baricz, Á.: Landen inequalities for special functions. Proc. Am. Math. Soc. Available via arXiv:1301.5255 [math.CA] (to appear)
39. Baricz, Á., Sándor, J.: Extensions of the generalized Wilker inequality to Bessel functions. J. Math. Inequal. **2**, 397–406 (2008)
40. Baricz, Á., Wu, S.-H.: Sharp Jordan type inequalities for Bessel functions. Publ. Math. Debrecen **74**, 107–126 (2009)
41. Baricz, Á., Wu, S.-H.: Sharp exponential Redheffer-type inequalities for Bessel functions. Publ. Math. Debrecen **74**, 257–278 (2009)
42. Baricz, Á., Vesti, J., Vuorinen, M.: On Kaluza's sign criterion for reciprocal power series. Ann. Univ. Mariae Curie-Skłodowska Sect A **65**, 1–16 (2011)
43. Barnard, R.W., Richards, K.C.: On inequalities for hypergeometric analogues of the arithmetic-geometric mean. JIPAM. J. Inequal. Pure Appl. Math. **8**, 5 (2007) (Article 65)
44. Barnard, R.W., Richards, K.C., Tiedeman, H.C.: A survey of some bounds for Gauss' hypergeometric function and related bivariate means. J. Math. Inequal. **4**, 45–52 (2010)
45. Batir, N.: On some properties of digamma and polygamma functions. J. Math. Anal. Appl. **328**, 452–465 (2007)
46. Batir, N.: On some properties of the gamma function. Expo. Math. **26**, 187–196 (2008)
47. Batir, N.: Sharp inequalities for factorial n. Proyecciones **27**, 97–102 (2008)
48. Batir, N.: Improving Stirling's formula. Math. Commun. **16**, 105–114 (2011)
49. Becker, M., Stark, E.L.: On a hierarchy of quolynomial inequalities for $\tan x$. Univ. Beograd. Publ. Elektrotehn. Fak. Ser. Mat. Fiz. **602–633**, 133–138 (1978)
50. Berg, C., Pedersen, H.L.: A completely monotone function related to the gamma function. J. Comput. Appl. Math. **133**, 219–230 (2001)
51. Berg, C., Pedersen, H.L.: Pick functions related to the gamma function. Rocky Mountain J. Math. **32**, 507–525 (2002)
52. Berg, C., Pedersen, H.L.: A one-parameter family of Pick functions defined by the gamma function and related to the volume of the unit ball in n-space. Proc. Am. Math. Soc. **139**, 2121–2132 (2011)
53. Berg, C., Pedersen, H.L.: A completely monotonic function used in an inequality of Alzer. Comput. Methods Funct. Theory **12**, 329–341 (2012)

54. Berndt, B.C.: Ramanujan's Notebooks, Part II. Springer, New York (1987)
55. Bhayo, B.A., Vuorinen, M.: On generalized complete elliptic integrals and modular functions. Proc. Edinburgh Math. Soc. **55**, 591–611 (2012)
56. Bhayo, B.A., Vuorinen, M.: On generalized trigonometric functions with two parameters. J. Approx. Theory **164**, 1415–1426 (2012)
57. Bhayo, B.A., Vuorinen, M.: Inequalities for eigenfunctions of the p-Laplacian. Issues Anal. **2**(20), 14–37 (2013)
58. Biernacki, M., Krzyż, J.: On the monotonicity of certain functionals in the theory of analytic functions. Ann. Univ. M. Curie-Skłodowska **2**, 134–145 (1995)
59. Biezuner, R.J., Ercole, G., Martins, E.M.: Computing the $\sin_p$ function via the inverse power method. Comput. Methods Appl. Math. **2**, 129–140 (2012)
60. Böhm, J., Hertel, E.: Polyedergeometrie in n-Dimensionalen Räumen Konstanter Krümmung. Birkhäuser, Basel (1981)
61. Borwein, J.M., Borwein, P.B., Garvan, F.: Hypergeometric analogues of the arithmetic-geometric mean iteration. Constr. Approx. **9**, 509–523 (1993)
62. Brenner, J.L., Carlson, B.C.: Homogeneous mean values: weights and asymptotics. J. Math. Anal. Appl. **123**, 265–280 (1987)
63. Burnside, W.: A rapidly convergent series for log $N!$. Messenger Math. **46**, 157–159 (1917)
64. Byrd, P.F., Friedman, M.D.: Handbook of Elliptic Integrals for Engineers and Scientists, 2nd edn. Die Grundlehren der mathematischen Wissenschaften, vol. 67. Springer, Berlin (1971)
65. Carlson, B.C.: A hypergeometric mean value. Proc. Am. Math. Soc. **16**, 759–766 (1965)
66. Carlson, B.C.: Some Inequalities for hypergeometric functions. Proc. Am. Math. Soc. **16**, 32–39 (1966)
67. Carlson, B.C.: Inequalities for a symmetric elliptic integral. Proc. Am. Math. Soc. **25**, 698–703 (1970)
68. Carlson, B.C., Tobey, M.D.: A property of the hypergeometric mean value. Proc. Am. Math. Soc. **19**, 255–262 (1968)
69. Chen, C.-P.: Inequalities for the Euler-Mascheroni constant. Appl. Math. Lett. **23**, 161–164 (2010)
70. Chen, C.-P.: Sharpness of Negoi's inequality for the Euler-Mascheroni constant. Bull. Math. Anal. Appl. **3**, 134–141 (2011)
71. Chen, C.-P., Cheung, W.-S.: Sharp Cusa and Becker-Stark inequalities. J. Inequal. Appl. **2011**, 6 (2011) (Article 136)
72. Chen, C.-P., Cheung, W.-S.: Sharpness of Wilker and Huygens type inequalities. J. Inequal. Appl. **2012**, 11 (2012) (Article 72)
73. Chen, C.-P., Cheung, W.-S., Wang, W.-S.: On Shafer and Carlson inequalities. J. Inequal. Appl. **2011**, 10 (2011) (Article ID 840206)
74. Chen, C.-P., Debnath, L.: Sharpness and generalization of Jordan's inequality and its application. Appl. Math. Lett. **25**, 594–599 (2012)
75. Chen, C.-P., Mortici, C.: Generalization and sharpness of Carlson's inequality for the inverse cosine function (unpublished manuscript)
76. Chen, C.-P., Zhao, J.-W., Qi, F.: Three inequalities involving hyperbolically trigonometric functions. RGMIA Res. Rep. Coll. **6**(3), 437–443 (2003) (Article 4)
77. Chlebus, E.: A recursive scheme for improving the original rate of convergence to the Euler-Mascheroni constant. Am. Math. Monthly **118**, 268–274 (2011)
78. Chu, Y.-M., Wang, M.-K.: Inequalities between arithmetic-geometric, Gini, and Toader Means. Abstr. Appl. Anal. **2012**, 11 (2012) (Article ID 830585)
79. Chu, Y.-M., Wang, M.-K.: Optimal Lehmer mean bounds for the Toader mean. Results Math. **61**, 223–229 (2012)
80. Chu, Y.-M., Wang, G.-D., Zhang, X.-H., Qiu, S.-L.: Generalized convexity and inequalities involving special functions. J. Math. Anal. Appl. **336**, 768–776 (2007)
81. Chu, Y.-M., Wang, M.-K., Qiu, Y.-F.: On Alzer and Qiu's conjecture for complete elliptic integral and inverse hyperbolic tangent function. Abstr. Appl. Anal. **2011**, 7 (2011) (Article ID 697547)

82. Chu, Y.-M., Wang, M.-K., Qiu, S.-L., Jiang, Y.-P.: Bounds for complete elliptic integrals of the second kind with applications. Comput. Math. Appl. **63**, 1177–1184 (2012)
83. Chu, Y.-M., Wang, M.-K., Qiu, S.-L.: Optimal combinations bounds of root-square and arithmetic means for Toader mean. Proc. Indian Acad. Sci. Math. Sci. **122**, 41–51 (2012)
84. DeTemple, D.W.: Convergence to Euler's constant. Am. Math. Monthly **100**, 468–470 (1993)
85. Elbert, Á., Laforgia, A.: On some properties of the gamma function. Proc. Am. Math. Soc. **128**, 2667–2673 (2000)
86. Elliott, E.B.: A formula including Legendre's $\mathcal{E}\mathcal{K}' + \mathcal{K}\mathcal{E}' - \mathcal{K}\mathcal{K}' = \frac{1}{2}\pi$. Messenger Math. **33**, 31–40 (1904)
87. Fink, A.M.: Two inequalities. Univ. Beograd. Publ. Elektrotehn. Fak. Ser. Mat. **6**, 49–50 (1995)
88. Ge, H.-F.: New sharp bounds for the Bernoulli numbers and refinement of Becker-Stark inequalities. J. Appl. Math. **2012**, 7 (2012) (Article ID 137507)
89. Guo, B.-N., Chen, R.-J., Qi, F.: A class of completely monotonic functions involving the polygamma functions. J. Math. Anal. Approx. Theory **1**, 124–134 (2006)
90. Guo, B.-N., Qi, F.: Some bounds for the complete elliptic integrals of the first and second kinds. Math. Inequal. Appl. **14**, 323–334 (2011)
91. Hästö, P.A.: A monotonicity property of ratios of symmetric homogeneous means. JIPAM. J. Inequal. Pure Appl. Math. **3**, 23 (2002) (Article 71)
92. Hästö, P.A.: A new weighted metric: the relative metric I. J. Math. Anal. Appl. **274**, 38–58 (2002)
93. Hästö, P.A.: A new weighted metric: the relative metric II. J. Math. Anal. Appl. **301**, 336–353 (2005)
94. Heikkala, V., Lindén, H., Vamanamurthy, M. K., Vuorinen, M.: Generalized elliptic integrals and the Legendre M -function. J. Math. Anal. Appl. **338**, 223–243 (2008)
95. Heikkala, V., Vamanamurthy, M. K., Vuorinen, M.: Generalized elliptic integrals. Comput. Methods Funct. Theory **9**, 75–109 (2009)
96. Hua, Y.: Refinements and sharpness of some new Huygens type inequalities. J. Math. Inequal. **6**, 493–500 (2012)
97. Huo, Z.-H., Niu, D.-W., Cao, J., Qi, F.: A generalization of Jordan's inequality and an application. Hacet. J. Math. Stat. **40**, 53–61 (2011)
98. Huygens, C.: Oeuvres Completes. Société Hollondaise des Science, Haga (1888–1940)
99. Ibrahim, A., Dragomir, S.S.: Power series inequalities via Buzano's result and applications. Integral Transforms Spec. Funct. **22**, 867–878 (2011)
100. Ibrahim, A., Dragomir, S.S., Cerone, P., Darus, M.: Inequalities for power series with positive coefficients. J. Inequal. Spec. Funct. **3**, 1–15 (2012)
101. Ibrahim, A., Dragomir, Darus, M.: Some inequalities for power series with applications. Integral Transforms Spec. Funct. iFirst, 1–13 (2012)
102. Ivády, P.: A note on a gamma function inequality. J. Math. Inequal. **3**, 227–236 (2009)
103. Kalmykov, S.I., Karp, D.B.: Log-concavity for series in reciprocal gamma functions and applications. Integral Transforms Spec. Funct. Available via arXiv:1206.4814v1 [math.CA] (2013)
104. Kalmykov, S.I., Karp, D.B.: Log-convexity and log-concavity for series in gamma ratios and applications. J. Math. Anal. Appl. **406**, 400–418 (2013)
105. Kaluza, T.: Über die Koeffizienten reziproker Potenzreihen. Math. Z. **28**, 161–170 (1928)
106. Karatsuba, E.A.: On the asymptotic representation of the Euler gamma function by Ramanujan. J. Comput. Appl. Math. **135**, 225–240 (2001)
107. Karatsuba, E.A., Vuorinen, M.: On hypergeometric functions and generalizations of Legendre's relation. J. Math. Anal. Appl. **260**, 623–640 (2001)
108. Karp, D., Sitnik, S.M.: Inequalities and monotonicity of ratios for generalized hypergeometric function. J. Approx. Theory **161**, 337–352 (2009)
109. Klén, R., Visuri, M., Vuorinen, M.: On Jordan type inequalities for hyperbolic functions. J. Inequal. Appl. **2010**, 14 (2010) (Article ID 362548)

110. Klén, R., Manojlovic, V., Simić, S., Vuorinen, M.: Bernoulli inequality and hypergeometric functions. Proc. Am. Math. Soc. **142**, 559–573 (2014)
111. Klén, R., Manojlović, V., Vuorinen, M.: Distortion of normalized quasiconformal mappings. Available via arXiv:0808.1219 [math.CV]
112. Klén, R., Vuorinen, M., Zhang, X.-H.: Inequalities for the generalized trigonometric and hyperbolic functions. J. Math. Anal. Appl. **409**, 521–529 (2014)
113. Kouba, O.: New bounds for the identric mean of two arguments. JIPAM. J. Inequal. Pure Appl. Math. **9** (2008) (Article 71)
114. Kouba, O.: Bounds for the ratios of differences of power means in two arguments. Math. Inequal. Appl. Available via arXiv:1006.1460v1 [math.CA] (to appear)
115. Koumandos, S., Pedersen, H.L.: On the asymptotic expansion of the logarithm of Barnes triple gamma function. Math. Scand. **105**, 287–306 (2009)
116. Kuo, M.-K.: Refinements of Jordan's inequality. J. Inequal. Appl. **2011**(130), 6 (2011)
117. Lazarević, I.: Neke nejednakosti sa hiperbolickim funkčijama. Univ. Beograd. Publ. Elektrotehn. Fak. Ser. Mat. **170**, 41–48 (1966)
118. Lehto, O., Virtanen, K.I.: Quasiconformal Mappings in the Plane, 2nd edn. Die Grundlehren der mathematischen Wissenschaften, Band 126. Springer, New York (1973)
119. Lindqvist, P.: Some remarkable sine and cosine functions. Ric. Mat. **44**, 269–290 (1995)
120. Li, J.-L., Li, Y.-L.: On the strengthened Jordan's inequality. J. Inequal. Appl. **2007**, 8 (2007) (Article ID 74328)
121. Lv, Y.-P., Wang, G.-D., Chu, Y.-M.: A note on Jordan type inequalities for hyperbolic functions. Appl. Math. Lett. **25**, 505–508 (2012)
122. Ma, X.-Y., Qiu, S.-L., Zhong, G.-H., Chu, Y.-M.: Some inequalities for the generalized linear distortion function. Appl. Math. J. Chinese Univ. Ser. B **27**, 87–93 (2012)
123. Ma, X.-Y., Qiu, S.-L., Zhong, G.-H., Chu, Y.-M.: The Hölder continuity and submultiplicative properties of the modular function. Appl. Math. J. Chinese Univ. Ser. A **27**, 481–487 (2012)
124. Mahmoud, M., Alghamdi, M.A., Agarwal, R.P.: New upper bounds of $n!$. J. Inequal. Appl. **2012**, 9 (2012) (Article 27)
125. Miller, K.S., Samko, S.G.: Completely monotonic functions. Integral Transforms Spec. Funct. **12**, 389–402 (2001)
126. Mitrinović, D.S.: Analytic Inequalities. Springer, Berlin (1970)
127. Mori, A.: On an absolute constant in the theory of quasiconformal mappings. J. Math. Soc. Jpn. **8**, 156–166 (1956)
128. Mortici, C.: Monotonicity properties of the volume of the unit ball in $\mathbb{R}^n$. Optim. Lett. **4**, 457–464 (2010)
129. Mortici, C.: New approximations of the gamma function in terms of the digamma function. Appl. Math. Lett. **23**, 97–100 (2010)
130. Mortici, C.: On new sequences converging towards the Euler-Mascheroni constant. Comput. Math. Appl. **59**, 2610–2614 (2010)
131. Mortici, C.: Very accurate estimates of the polygamma functions. Asymptot. Anal. **68**, 125–134 (2010)
132. Mortici, C.: Ramanujan's estimate for the gamma function via monotonicity arguments. Ramanujan J. **25**, 149–154 (2011)
133. Mortici, C.: Gamma function by x^{x-1}. Carpathian J. Math. (to appear)
134. Negoi, T.: A faster convergence to Euler's constant. Math. Gaz. **83**, 487–489 (1999)
135. Neuman, E.: Inequalities and bounds for generalized complete elliptic integrals. J. Math. Anal. Appl. **373**, 203–213 (2011)
136. Neuman, E.: A note on a certain bivariate mean. J. Math. Inequal. **6**, 637–643 (2012)
137. Neuman, E.: Inequalities involving hyperbolic functions and trigonometric functions. Bull. Int. Math. Virt. Instit. **2**, 87–92 (2012)
138. Neuman, E.: On Wilker and Huygens type inequalities. Math. Inequal. Appl. **15**, 271–279 (2012)
139. Neuman, E., Sándor, J.: On the Schwab-Borchardt mean. Math. Pannon. **14**, 253–266 (2003)
140. Neuman, E., Sándor, J.: On the Schwab-Borchardt mean II. Math. Pannon. **17**, 49–59 (2006)

141. Neuman, E., Sándor, J.: On some inequalities involving trigonometric and hyperbolic functions with emphasis on the Cusa-Huygens, Wilker, and Huygens inequalities. Math. Inequal. Appl. **13**, 715–723 (2010)
142. Neuman, E., Sándor, J.: Optimal inequalities for hyperbolic and trigonometric functions. Bull. Math. Anal. Appl. **3**, 177–181 (2011)
143. Niu, D.-W., Huo, Z.-H., Cao, J., Qi, F.: A general refinement of Jordan's inequality and a refinement of L. Yang's inequality. Integral Transforms Spec. Funct. **19**, 157–164 (2008)
144. Niu, D.-W., Cao, J., Qi, F.: Generalizations of Jordan's inequality and concerned relations. Politehn. Univ. Bucharest Sci. Bull. Ser. A Appl. Math. Phys. **72**, 85–98 (2010)
145. Pan, W.-H., Zhu, L.: Generalizations of Shafer-Fink-type inequalities for the arc sine function. J. Inequal. Appl. **2009**, 6 (2009) (Article ID 705317)
146. Pinelis, I.: L'Hospital rules for monotonicity and the Wilker-Anglesio inequality. Am. Math. Monthly **111**, 905–909 (2004)
147. Ponnusamy, S., Vuorinen, M.: Asymptotic expansions and inequalities for hypergeometric functions. Mathematika **44**, 278–301 (1997)
148. Qi, F.: The extended mean values: definition, properties, monotonicities, comparison, convexities, generalizations, and applications. RGMIA Res. Rep. Coll. **5**, 19 (2001)
149. Qi, F., Guo, B.-N.: Monotonicity and logarithmic convexity relating to the volume of the unit ball. Optim. Lett. **7**, 1139–1153 (2013)
150. Qi, F., Niu, D.-W., Guo, B.-N.: Refinements, generalizations, and applications of Jordan's inequality and related problems. J. Inequal. Appl. **2009**, 52 (2009) (Article ID 271923)
151. Qiu, S.-L., Shen, J.-M.: On two problems concerning means. J. Hangzhou Inst. Electronic Engg. **17**, 1–7 (1997)
152. Qiu, S.-L., Vuorinen, M.: Landen inequalities for hypergeometric functions. Nagoya Math. J. **154**, 31–56 (1999)
153. Qiu, S.-L., Vuorinen, M.: Duplication inequalities for the ratios of hypergeometric functions. Forum Math. **12**, 109–133 (2000)
154. Qiu, S.-L., Vuorinen, M.: Some properties of the gamma and psi functions, with applications. Math Comput. **74**, 723–742 (2004)
155. Qiu, S.-L., Qiu, Y.-F., Wang, M.-K., Chu, Y.-M.: Hölder mean inequalities for the generalized Grötzsch ring and Hersch-Pfluger functions. Math. Inequal. Appl. **15**, 237–245 (2012)
156. Ramanujan, S.: The Lost Notebook and Other Unpublished Papers, with an Introduction by George E. Andrews. Narosa Publishing House, New Delhi (1988)
157. Redheffer, R.: Problem 5642. Am. Math. Monthly **76**, 422 (1969)
158. Richards, K.C.: Sharp power mean bounds for the Gaussian hypergeometric function. J. Math. Anal. Appl. **308**, 303–313 (2005)
159. Sándor, J.: Sur la fonction gamma. Publ. Centre Rech. Math. Pures (I) **21**, 4–7 (1989)
160. Sándor, J.: On certain inequalities for means III. Arch. Math. (Basel) **76**, 34–40 (2001)
161. Sándor, J.: Über zwei Mittel von Seiffert. Wurzel **36**, 104–107 (2002)
162. Sándor, J.: On some new Wilker and Huygens type trigonometric-hyperbolic inequalities. Proc. Jangjeon Math. Soc. **15**, 145–153 (2012)
163. Sándor, J.: On Huygens' inequalities and the theory of means. Int. J. Math. Math. Sci. **2012**, 9 (2012) (Article ID 597490)
164. Sándor, J.: Two sharp inequalities for trigonometric and hyperbolic functions. Math. Inequal. Appl. **15**, 409–413 (2012)
165. Sándor, J.: Trigonometric and hyperbolic inequalities. Available via arXiv:1105.0859v1 [math.CA]
166. Shafer, R.E.: Problem E 1867. Am. Math. Monthly **73**, 309–310 (1966)
167. Shafer, R.E.: On quadratic approximation. SIAM J. Numer. Anal. **11**, 447–460 (1974)
168. Shafer, R.E.: Analytic inequalities obtained by quadratic approximation. Univ. Beograd. Publ. Elektrotehn. Fak. Ser. Mat. Fiz. **577–598**, 96–97 (1977)
169. Shafer, R.E., Grinstein, L.S., Marsh, D.C.B., Konhauser, J.D.E.: Problems and solutions: an inequality for the inverse tangent: E 1867. Am. Math. Monthly **74**, 726–727 (1967)
170. Simić, S., Vuorinen, M.: On quotients and differences of hypergeometric functions. J. Inequal. Appl. **2011**, 10 (2011) (Article 141)

171. Simić, S., Vuorinen, M.: Landen inequalities for zero-balanced hypergeometric functions. Abstr. Appl. Anal. **2012**, 11 (2012) (Article ID 932061)
172. Stolarsky, K.B.: Generalizations of the logarithmic mean. Math. Mag. **48**, 87–92 (1975)
173. Takeuchi, S.: Generalized Jacobian elliptic functions and their application to bifurcation problems associated with p-Laplacian. J. Math. Anal. Appl. **385**, 24–35 (2012)
174. Toader, Gh.: Some mean values related to the arithmetic-geometric mean. J. Math. Anal. Appl. **218**, 358–368 (1998)
175. Trif, T.: Note on certain inequalities for means in two variables. JIPAM. J. Ineq. Pure Appl. Math. **6** (2005) (Article 43)
176. Vuorinen, M.: Hypergeometric functions in geometric function theory. In: Srinivasa Rao, K., Jagannathan, R., Vanden Berghe, G., Van der Jeugt, J. (eds.) Special Functions and Differential Equations, pp. 119–126. Proceedings of a workshop held at The Institute of Mathematical Sciences, Madras, India, Jan 13–24, 1997. Allied Publishers (1998)
177. Vuorinen, M., Zhang, X.-H.: On exterior moduli of quadrilaterals and special functions. J. Fixed Point Theory Appl. **13**, 215–230 (2013)
178. Wang, M.-K., Chu, Y.-M.: Asymptotical bounds for complete elliptic integrals of the second kind. J. Math. Anal. Appl. **402**, 119–126 (2013)
179. Wang, G.-D., Qiu, S.-L., Zhang, X.-H., Chu, Y.-M.: Approximate convexity and concavity of generalized Grötzsch ring function. Appl. Math. J. Chinese Univ. Ser. B **21**, 203–206 (2006)
180. Wang, G.-D., Zhang, X.-H., Qiu, S.-L., Chu, Y.-M.: The bounds of the solutions to generalized modular equations. J. Math. Anal. Appl. **321**, 589–594 (2006)
181. Wang, G.-D., Zhang, X.-H., Chu, Y.-M.: Inequalities for the generalized elliptic integrals and modular functions. J. Math. Anal. Appl. **331**, 1275–1283 (2007)
182. Wang, G.-D., Zhang, X.-H., Chu, Y.-M.: A Hölder mean inequality for the Hersch-Pfluger distortion function. Sci. Sin. Math. **40**, 783–786 (2010)
183. Wang, G.-D., Zhang, X.-H., Chu, Y.-M.: A power mean inequality for the Grötzsch ring function. Math. Inequal. Appl. **14**, 833–837 (2011)
184. Wang, G.-D., Zhang, X.-H., Jiang, Y.-P.: Concavity with respect to Hölder means involving the generalized Grötzsch function. J. Math. Anal. Appl. **379**, 200–204 (2011)
185. Wang, M.-K., Chu, Y.-M., Qiu, S.-L., Jiang, Y.-P.: Bounds for the perimeter of an ellipse. J. Approx. Theory **164**, 928–937 (2012)
186. Wang, G.-D., Zhang, X.-H., Jiang, Y.-P.: Hölder concavity and inequalities for Jacobian elliptic functions. Integral Transforms Spec. Funct. **23**, 337–345 (2012)
187. Wang, G.-D., Zhang, X.-H., Chu, Y.-M.: A power mean inequality involving the complete elliptic integrals. Rocky Mountain J. Math. (to appear)
188. Wang, M.-K., Chu, Y.-M., Jiang, Y.-P.: Ramanujan's cubic transformation inequalities for zero-balanced hypergeometric functions. Available via arXiv:1210.6126v1 [math.CA]
189. Widder, D.V.: The Laplace Transform. Princeton University Press, Princeton (1941)
190. Wilker, J.B.: Problem E 3306. Am. Math. Monthly **96**, 55 (1989)
191. Wilker, J.B., Sumner, J.S., Jagers, A.A., Vowe, M., Anglesio, J.: Problems and solutions: solutions of elementary problems: E 3306. Am. Math. Monthly **98**, 264–267 (1991)
192. Williams, J.P.: Solutions of advanced problems: a delightful inequality 5642. Am. Math. Monthly **76**, 1153–1154 (1969)
193. Wu, S.-H.: Generalization and sharpness of the power means inequality and their applications. J. Math. Anal. Appl. **312**, 637–652 (2005)
194. Wu, S.-H., Baricz, Á.: Generalizations of Mitrinović, Adamović and Lazarević's inequalities and their applications. Publ. Math. Debrecen **75**, 447–458 (2009)
195. Wu, S.-H., Debnath, L.: Jordan-type inequalities for differentiable functions and their applications. Appl. Math. Lett. **21**, 803–809 (2008)
196. Wu, S.-H., Debnath, L.: Inequalities for differences of power means in two variables. Anal. Math. **37**, 151–159 (2011)
197. Wu, S.-H., Srivastava, H.M.: A weighted and exponential generalization of Wilker's inequality and its applications. Integral Transforms Spec. Funct. **18**, 529–535 (2007)

198. Wu, S.-H., Srivastava, H.M.: A further refinement of a Jordan type inequality and its application. Appl. Math. Comput. **197**, 914–923 (2008)
199. Yang, S.-J.: Absolutely (completely) monotonic functions and Jordan-type inequalities. Appl. Math. Lett. **25**, 571–574 (2012)
200. Yang, Z.-H.: New sharp bounds for identric mean in terms of logarithmic mean and arithmetic mean. J. Math. Inequal. **6**, 533–543 (2012)
201. Yee, A.J.: Large computations. Available at http://www.numberworld.org/nagisa_runs/computations.html (2010)
202. Yin, L.: Several inequalities for the volume of the unit ball in $\mathbb{R}^n$. Bull. Malays. Math. Sci. Soc. (2) (to appear)
203. Zhang, X.-H., Wang, G.-D., Chu, Y.-M., Qiu, S.-L.: Monotonicity and inequalities for the generalized η-distortion function. (Chinese) Chinese Ann. Math. Ser. A **28**, 183–190 (2007) (translation in Chinese J. Contemp. Math. **28**, 141–148, 2007)
204. Zhang, X.-H., Wang, G.-D., Chu, Y.-M.: Some inequalities for the generalized Grötzsch function. Proc. Edinb. Math. Soc. (2) **51**, 265–272 (2008)
205. Zhang, X.-H., Wang, G.-D., Chu, Y.-M.: Convexity with respect to Hölder mean involving zero-balanced hypergeometric functions. J. Math. Anal. Appl. **353**, 256–259 (2009)
206. Zhang, X.-H., Wang, G.-D., Chu, Y.-M.: Remarks on generalized elliptic integrals. Proc. Roy. Soc. Edinburgh Sect. A **139**, 417–426 (2009)
207. Zhao, J.-L., Guo, B.-N., Qi, F.: A refinement of a double inequality for the gamma function. Publ. Math. Debrecen **50**, 1–10 (2011)
208. Zhao, T.-H., Chu, Y.-M., Liu, B.-Y.: Some best possible inequalities concerning certain bivariate means. Available via arXiv:1210.4219v1 [math.CA]
209. Zhou, L.-M., Qiu, S.-L., Wang, F.: Inequalities for the generalized elliptic integrals with respect to Hölder means. J. Math. Anal. Appl. **386**, 641–646 (2012)
210. Zhu, L.: A general form of Jordan's inequalities and its applications. Math. Inequal. Appl. **11**, 655–665 (2008)
211. Zhu, L.: A general refinement of Jordan-type inequality. Comput. Math. Appl. **55**, 2498–2505 (2008)
212. Zhu, L.: New inequalities of Shafer-Fink type for arc hyperbolic sine. J. Inequal. Appl. **2008**, 5 (2008) (Article ID 368275)
213. Zhu, L.: On a quadratic estimate of Shafer. J. Math. Inequal. **2**, 571–574 (2008)
214. Zhu, L.: A source of inequalities for circular functions. Comput. Math. Appl. **58**, 1998–2004 (2009)
215. Zhu, L.: Generalized Lazarević's inequality and its applications: Part II. J. Inequal. Appl. **2009**, 4 (2009) (Article ID 379142)
216. Zhu, L.: Sharpening Redheffer-type inequalities for circular functions. Appl. Math. Lett. **22**, 743–748 (2009)
217. Zhu, L.: Some new inequalities of the Huygens type. Comput. Math. Appl. **58**, 1180–1182 (2009)
218. Zhu, L.: Some new Wilker-type inequalities for circular and hyperbolic functions. Abstr. Appl. Anal. **2009**, 9 (2009) (Article ID 485842)
219. Zhu, L.: A general form of Jordan-type double inequality for the generalized and normalized Bessel functions. Appl. Math. Comput. **215**, 3802–3810 (2010)
220. Zhu, L.: Jordan type inequalities involving the Bessel and modified Bessel functions. Comput. Math. Appl. **59**, 724–736 (2010)
221. Zhu, L.: An extended Jordan's inequality in exponential type. Appl. Math. Lett. **24**, 1870–1873 (2011)
222. Zhu, L.: Extension of Redheffer type inequalities to modified Bessel functions. Appl. Math. Comput. **217**, 8504–8506 (2011)
223. Zhu, L., Hua, J.-K.: Sharpening the Becker-Stark inequalities. J. Inequal. Appl. **2010**, 4 (2010) (Article ID 931275)
224. Zhu, L., Sun, J.-J.: Six new Redheffer-type inequalities for circular and hyperbolic functions. Comput. Math. Appl. **56**, 522–529 (2008)

Completely Monotone Functions: A Digest

Milan Merkle

Dedicated to Professor Hari M. Srivastava

Abstract This work has a purpose to collect selected facts about the completely monotone (*CM*) functions that can be found in books and papers devoted to different areas of mathematics. We opted for lesser known ones and for those which may help in determining whether or not a given function is completely monotone. In particular, we emphasize the role of representation of a *CM* function as the Laplace transform of a measure, and we present and discuss a little-known connection with log-convexity. Some of presented methods are illustrated by several examples involving Gamma and related functions.

1 Introduction

A positive function defined on $(0, +\infty)$ of the class C^{∞}, such that the sequence of its derivatives alternates signs at every point, is called completely monotone (*CM*) function. A brief search in MathSciNet reveals a total of 286 items that mention this class of functions in the title from 1932 till the end of the year 2011; 98 of them have been published since the beginning of 2006.

This vintage topic was developed in 1920s/1930s by S. Bernstein, F. Hausdorff, and V. Widder, originally with relation to the so-called moment problem, cf. [3, 13, 14, 26, 27]. The much-cited (but perhaps not that much read) Widder's book [28] contains a detailed account on properties of *CM* functions and their characterizations. The second volume of Feller's probability book [8] discusses *CM* functions through their relationship with infinitely divisible measures, which are

M. Merkle (✉)
University of Belgrade, P.O. Box 35-54, 11000 Belgrade, Serbia

Union University, Računarski fakultet, Kneza Mihaila 6, 11000 Belgrade, Serbia
e-mail: emerkle@etf.rs; mmerkle@raf.edu.rs

G.V. Milovanović and M.Th. Rassias (eds.), *Analytic Number Theory, Approximation Theory, and Special Functions*, DOI 10.1007/978-1-4939-0258-3_12,

fundamental in defining Lévy processes. In the past several decades, Lévy processes have gained popularity in financial models, as well as in biology and physics; this is probably a reason for increased interest in *CM* functions, too. There are also other interesting topics in probability and statistics where *CM* functions play a role; see [16] for one such topic. Aside from probability and measure theory, *CM* and related functions appear in the field of approximations of functions, as documented in [6]. Finally, they are naturally linked to various inequalities; several general inequalities for *CM* functions can be found in [17]; for a quite recent contribution in this area, see [2].

This text has a purpose to collect well-known facts about *CM* functions, together with some less-known ones, which may help in determining whether or not a given function is completely monotone. In that sense, this work can be thought of as being an extension and supplement to another paper in the same spirit—[24] by Miller and Samko. In particular, we emphasize the role of representation of a *CM* function as a Laplace transform of a measure, and we present and discuss a little-known (and even less being used) connection between *CM* function and log-convexity. Some of the methods discussed in Sects. 2–5 are illustrated by several examples involving Gamma and related functions in Sect. 6. References and examples reflect author's preferences and are by no means complete; the same can be said for the selection of topics that are discussed in this work.

2 Representations of Completely Monotone Functions

We start with a classical definition of *CM* functions, and we present two possible representations in terms of integral transforms of measures and alternative representations for Stieltjes transforms and *CM* probability densities.

2.1 Integral Representations

Definition 2.1. A function f defined on $(0, +\infty)$ is completely monotone if it has derivatives of all orders and

$$(-1)^k f^{(k)}(t) > 0, \quad t \in (0, +\infty), \quad k = 0, 1, 2, \dots . \tag{1}$$

In particular, this implies that each *CM* function on $(0, +\infty)$ is positive, decreasing, and convex, with concave first derivative.

By (1), there exist limits of $f^{(k)}(x)$ as $x \to 0$ for any $k \geq 0$; if those limits are finite, then f can be extended to $[0, +\infty)$ and (1) will also hold for $x = 0$ (with strict inequality for all k). Limits at zero need not be finite, as in $f(x) = 1/x$, for example.

Clearly, $\lim_{x\to+\infty} f^{(k)}(x) = 0$ for all $k \geq 1$. The limit of $f(x)$ at $+\infty$ must be finite, and if it is non-zero, then it has to be positive (e.g., $f(x) = 1 + \mathrm{e}^{-x}$).

Lemma 2.1. *The function f is CM if and only if* [28]

$$f(x) = \int_{[0,+\infty)} \mathrm{e}^{-xt}\, \mathrm{d}\mu(t), \tag{2}$$

where $\mu(t)$ is a positive measure on Borel sets of $[0, +\infty)$ (i.e., $\mu(B) \geq 0$ for every Borel set $B \in \mathbb{R}_+$) and the integral converges for $0 < x < +\infty$.

In other words, completely monotone functions are real one-side Laplace transforms of a positive measure on $[0, +\infty)$. If the measure μ has an atom at $t = 0$, then $\lim_{x\to+\infty} f(x) > 0$. The measure μ is a probability measure if and only if $\lim_{x\to 0_+} f(0) = 1$ (by monotone convergence theorem).

The Lebesgue integral in (2) can be expressed as a Lebesgue-Stieltjes integral

$$f(x) = \int_{[0,+\infty)} \mathrm{e}^{-xt}\, \mathrm{d}g(t), \tag{3}$$

where $g(t) = \mu([0,t])$ is the distribution function of μ, with $g(0_-) = 0$. For a positive measure μ, the function g is non-decreasing, and by change of variables $t = -\log s$ we get the following result:

Lemma 2.2. *The function f is completely monotone on $(0, +\infty)$ if and only if*

$$f(x) = \int_{[0,1]} s^x\, \mathrm{d}h(s), \tag{4}$$

where $h(s) = -g(-\log s)$ is a non-decreasing function.

If f is a *CM* function which is the Laplace transform of a measure μ, as in (2), we write $f = \mathcal{L}(\mathrm{d}\mu)$ or $f(x) = \mathcal{L}(\mathrm{d}\mu(t))$. Similarly, the relation (3) between f and a distribution function g can be denoted as $f = \mathcal{L}(\mathrm{d}g)$. If μ has a density h with respect to Lebesgue measure, we write $f(x) = \mathcal{L}(h(t)\,\mathrm{d}t)$ or only $f = \mathcal{L}(h)$. It follows from inversion formulas that each *CM* f determines one positive measure μ via relation $f = \mathcal{L}(\mathrm{d}\mu)$ and it is of interest in many applications to find that measure.

Remark 2.1. Since measures are determined by their Laplace transforms, if $f = \mathcal{L}(\mathrm{d}\mu)$, then f is *CM* if and only if μ is a positive measure. If there exists a continuous density h of μ, then f is *CM* if and only if $h(t) \geq 0$ for all $t \geq 0$.

Let us now observe a subclass of *CM* functions which contains all functions f that can be represented as Stieltjes transform of some positive measure μ, that is,

$$f(x) = \int_{[0,+\infty)} \frac{\mathrm{d}\mu(s)}{x+s}. \tag{5}$$

It is easy to verify that each function of the form (5) with a positive measure μ is *CM*; hence $f = \mathcal{L}(\nu)$, where ν is a positive measure. To find ν, we start with

$$\frac{1}{x+s} = \int_{[0,+\infty)} \mathrm{e}^{-(x+s)u}\,\mathrm{d}u,$$

and, after a change of order of integration, we arrive at the following result.

Lemma 2.3. *The Stieltjes transform of a positive measure μ as defined by (5) can be represented as a Laplace transform*

$$f(x) = \int_{[0,+\infty)} \mathrm{e}^{-xu}\left(\int_{[0,+\infty)} \mathrm{e}^{-su}\,\mathrm{d}\mu(s)\right)\mathrm{d}u.$$

That is, $f = \mathcal{L}(\nu)$, where the measure ν is absolutely continuous with respect to Lebesgue measure, with a density $\mathcal{L}(\mathrm{d}\mu)$.

Stieltjes transforms f have the property that $-f$ is reciprocally convex (in terminology introduced in [21], a function $g(x)$ is reciprocally convex if it is defined for $x > 0$ and concave there, whereas $g(1/x)$ is convex). As proved in [21], each reciprocally convex function generates an increasing sequence of quasi-arithmetic means, and hence *CM* functions that are also Stieltjes transforms are interesting as a tool for generating means.

2.2 Completely Monotone Probability Densities

Let f be a probability density with respect to Lebesgue measure on $[0, +\infty)$, that is,

$$\int_0^{+\infty} f(x)\,\mathrm{d}x = 1 \quad \text{and} \quad f(x) \geq 0 \quad \text{for all } x \geq 0.$$

Then f is a *CM* function if and only if (2) holds, which, after integration with respect to $x \in (0, +\infty)$, gives (via Fubini theorem for $f \geq 0$)

$$1 = \int_0^{+\infty} \frac{1}{t}\,\mathrm{d}\mu(t).$$

Defining a new probability measure ν by $\nu(B) = \int_B \frac{1}{t}\,\mathrm{d}\mu(t)$, we have that

$$f(x) = \int_{[0,+\infty)} t\mathrm{e}^{-xt}\,\mathrm{d}\nu(t) = \int_{[0,+\infty)} t\mathrm{e}^{-xt}\,\mathrm{d}G(t), \tag{6}$$

where G is the distribution function for ν. The function $x \mapsto t\mathrm{e}^{-xt}$ is the density of exponential distribution Exp(t). Therefore, a density f of a probability measure on $(0, +\infty)$ is a *CM* function if and only if it is a mixture of exponential densities. Note that (6) can be written as $f(x) = \mathrm{E}(T\mathrm{e}^{xT})$, where T is a random variable with distribution function G; by letting $S = 1/T$, we find that

$$f(x) = \mathrm{E}\left(\frac{1}{S}\mathrm{e}^{x/S}\right) = \int_{[0,+\infty)} \frac{1}{s}\mathrm{e}^{-x/s}\,\mathrm{d}H(s), \tag{7}$$

where H is the distribution function of S. The latter form is taken as a definition of what is meant by a *CM* density in [17, 18.B.5]; this is more natural than (6) because the mixing measure H is defined on values of expectations (s) of exponential distributions in the mixture, rather than on their reciprocal values as in (6).

3 Further Properties and Connection with Infinitely Divisible Measures

Starting from the mentioned representations of *CM* functions, an interesting criterion for equality of two *CM* functions is derived in [7]:

Lemma 3.1. *If f and g are CM functions and if $f(x_n) = g(x_n)$ for a positive sequence $\{x_n\}$ such that the series $\sum_n 1/x_n$ diverges, then $f(x) = g(x)$ for all $x \geq 0$.*

As a corollary to Lemma 3.1, we can see that if *CM* functions f and g agree in any proper subinterval of $(0, +\infty)$, then $f(x) = g(x)$ for all $x \geq 0$. A converse result, which is also proved in [7], is more surprising: if f is *CM* and if the series $\sum_n 1/x_n$ converges, then there exists another *CM* function $g \neq f$, such that $f(x_n) = g(x_n)$ for all n.

3.1 Convolution and Infinitely Divisible Measures

Given measures μ and ν on $[0, +\infty)$ and their distribution functions g_μ and g_ν, we define the convolution $\mu * \nu$ as a measure with the distribution function defined by

$$g_{\mu*\nu}(t) = \int_{[0,t]} g_\mu(t-u)\,\mathrm{d}g_\nu(u) = \int_{[0,t]} g_\nu(t-v)\,\mathrm{d}g_\mu(v). \tag{8}$$

To show equality of integrals above, we use the formula for integration by parts in Lebesgue-Stieltjes integral (see [15] or [4]) and note that the function $u \mapsto g_\mu(t-u)$ is continuous from the left, while $u \mapsto g_\nu(u)$ is continuous from the right; hence the additional term due to discontinuities in the integration by parts' formula equals zero, that is,

$$\int_{[0,t]} g_\mu(t-u)\,\mathrm{d}g_\nu(u) = -\int_{[0,t]} g_\nu(u)\,\mathrm{d}g_\mu(t-u)$$

and then we apply change of variables in the last integral, $u = t - v$.

Repeated convolution is defined by induction, using associativity. In particular, the nth convolution power of a measure μ, denoted by μ^{n*}, is defined by $n-1$ repeated convolutions $\mu * \mu * \cdots * \mu$.

A measure μ is called infinitely divisible (*ID*) if for every natural number n there exists a measure μ_n such that $\mu = \mu_n^{n*}$.

In the next two lemmas we collect some basic properties of *CM* functions. For a collection of other properties we refer to [24].

Lemma 3.2. *If f and g are CM functions with $f = \mathcal{L}(\mathrm{d}\mu)$ and $g = \mathcal{L}(\mathrm{d}\nu)$, then for $a > 0$,*

$$af = \mathcal{L}(\mathrm{d}(a\mu)), \quad f + g = \mathcal{L}(\mathrm{d}(\mu+\nu)), \quad fg = \mathcal{L}(\mathrm{d}(\mu * \nu)).$$

Therefore, if f, g are CM, then $af + bg$ $(a, b > 0)$ and fg are also CM.

Proof. The first two properties follow from the definition of Laplace transform. The third property for arbitrary positive measures is proved in [8, p. 434]. □

Lemma 3.3. (i) *If g' is CM, then the function $x \mapsto f(x) = \mathrm{e}^{-g(x)}$ is CM.*
(ii) *If $\log f$ is CM, then f is CM (the converse is not true).*
(iii) *If f is CM and g is a positive function with a CM derivative, then $x \mapsto f(g(x))$ is CM.*

Proof. To prove (i), let $h(x) = \mathrm{e}^{-g(x)}$ and note that $h > 0$ and $h' = -g'h < 0$. Then by induction, using Leibniz chain rule, it follows that $(-1)^n h^{(n)} > 0$. In particular, if $\log f$ is *CM*, then $(-\log f)'$ is also *CM*, and (ii) follows from (i) with $g = -\log f$. The function $x \mapsto \mathrm{e}^{-x}$ is a *CM* function but its logarithm is not the one, so the converse does not hold. For (iii), we note that $f = \mathcal{L}(\mathrm{d}\mu)$ for some positive measure μ; hence

$$\frac{\mathrm{d}}{\mathrm{d}x} f(g(x)) = -g'(x) \int_0^{+\infty} \mathrm{e}^{-g(x)t}\, t \,\mathrm{d}\mu(t). \tag{9}$$

By part (i), the function $x \mapsto \mathrm{e}^{-g(x)t}$ is *CM* for every $t > 0$, and so the function $x \mapsto g'(x)\mathrm{e}^{-g(x)t}$ is also *CM* as a product of two *CM* functions. Then from representation (9) it follows that the first derivative of $-f(g(x))$ is *CM*, which together with positivity of f yields the desired assertion. □

Note that if we can find measures ν and ν_t in representations $g'(x) = \mathcal{L}(\mathrm{d}\nu)$ and $\mathrm{e}^{-g(x)t} = \mathcal{L}(\mathrm{d}\nu_t)$, then from (9) we find that

$$\frac{\mathrm{d}}{\mathrm{d}x} f(g(x)) = -\int_0^{+\infty} t \int_0^{+\infty} \mathrm{e}^{-ux}\, \mathrm{d}(\nu * \nu_t)(u)\, \mathrm{d}\mu(t). \tag{10}$$

It turns out that *CM* functions f of the form as in (i) of Lemma 3.3 are Laplace transforms of *ID* measures. If $f(0) = 1$, the associated measure is a probability measure, which is the case that is of interest in applications. Proofs of statements of the next lemma can be found in [8].

Lemma 3.4. (i) *A function f is the Laplace transform of an ID probability measure if and only if*

$$f(x) = \mathrm{e}^{-g(x)}, \tag{11}$$

where g is a positive function with a CM derivative and $g(0) = 0$. Equivalently, f is the Laplace transform of an ID positive measure if and only if $f(x) > 0$ for all $x > 0$, and the function $x \mapsto -\log f(x)$ has a CM derivative. This measure is a probability measure if and only if $f(0_+) = 1$.

(ii) *A function f is the Laplace transform of an ID probability measure if and only if*

$$-\log f(x) = \int_0^{+\infty} \frac{1 - \mathrm{e}^{-xt}}{t}\, \mathrm{d}\mu(t), \tag{12}$$

where μ is a positive measure such that

$$\int_1^{+\infty} \frac{1}{t}\, \mathrm{d}\mu(t) < +\infty. \tag{13}$$

Remark 3.1. 1° If $\log f$ is *CM*, then $\log f$ has a *CM* derivative and by Lemma 3.4(i), $f = \mathcal{L}(\mathrm{d}\mu)$, where μ is an *ID* positive measure. By *CM* property of $\log f$, we have that $\log f = \mathcal{L}(\mathrm{d}\nu)$, where ν is some other positive measure. Note that positivity of ν implies that $\log f(0) > 0$, that is, $\mu([0, +\infty)) = f(0) > 1$, and so, μ cannot be a probability measure.

2° Non-negative functions with a *CM* first derivative have a special name—*Bernstein functions*; Lemmas 3.3 and 3.4 explain their role in probability theory; more about this class of functions can be found in [25].

4 Majorization, Convexity and Logarithmic Convexity

A good source for studying all three topics that are very much interlaced is the book [17]. In this short digest we include only necessary definitions and results that one can need for understanding a connection with *CM* functions.

For a vector $\mathbf{x} \in \mathbb{R}^n$ define $x_{[i]}$ to be the ith largest coordinate of $\mathbf{x}$, so that

$$x_{[1]} \geq x_{[2]} \geq \cdots \geq x_{[n]}.$$

We say that $\mathbf{x}$ is majorized by $\mathbf{y}$ in notation $\mathbf{x} \prec \mathbf{y}$ if

$$\sum_{i=1}^{k} x_{[i]} \leq \sum_{i=1}^{k} y_{[i]} \quad \text{for } k = 1, 2, \ldots, n-1 \quad \text{and} \quad \sum_{i=1}^{n} x_{[i]} = \sum_{i=1}^{n} y_{[i]}.$$

For example, $(1, 1, 1) \prec (2, 1, 0)$. Clearly, majorization is invariant to permutations of coordinates of vectors.

A function f which is defined on a symmetric set $S \subset \mathbb{R}^n$ (S is symmetric if $\mathbf{x} \in S$ implies that $\mathbf{y} \in S$ where $\mathbf{y}$ is any vector obtained by permuting the coordinates of $\mathbf{x}$) is called Schur-convex if for any $\mathbf{x}, \mathbf{y} \in S$,

$$\mathbf{x} \prec \mathbf{y} \Rightarrow f(\mathbf{x}) \leq f(\mathbf{y}). \tag{14}$$

The following result, due to Fink [9], reveals an interesting relationship between concepts of Schur-convexity and complete monotonicity.

Lemma 4.1. *For a CM function f and a non-negative integer vector of a dimension $d > 1$, $\mathbf{m} = (m_1, m_2, \ldots, m_d)$, let*

$$u_x(\mathbf{m}) = (-1)^{m_1} f^{(m_1)}(x)(-1)^{m_2} f^{(m_2)}(x) \cdots (-1)^{m_d} f^{(m_d)}(x).$$

Then $u_x(\mathbf{m})$ is a Schur-convex function on $\mathbf{m}$ for every $x > 0$ and $d > 1$.

An important corollary of 4.1 is with $d = 2$, taking $\mathbf{m} = (1, 1)$ and $\mathbf{n} = (2, 0)$. Clearly, $\mathbf{m} \prec \mathbf{n}$ and from the above definition of Schur-convexity we get that $u_x(1, 1) \leq u_x(0, 2)$, that is, $(f'(x))^2 \leq f(x)f''(x)$, which is, knowing that $f(x) > 0$, equivalent to $(\log f(x))'' \geq 0$. We formulate this result as a separate lemma.

Lemma 4.2. *Any CM function f is log-convex, i.e., the function* $\log f(x)$ *is convex.*

A converse does not hold, for example, the Gamma function restricted to $(0, +\infty)$ is log-convex, but it is not *CM*. However, the fact that each *CM* function is also log-convex helps us to search for possible candidates for complete monotonicity only among functions that are log-convex. In addition, there is a very rich theory that produces inequalities using convexity or Schur-convexity, and we can use it for *CM* functions.

Log-convexity of *CM* functions is equivalent to decreasing of the ratio $f'(x)/f(x)$, and (arguing that $f^{(2k)}$ and $-f^{(2k+1)}$ are *CM*) this implies

Corollary 4.1. *If f is a CM function, then the ratio*

$$x \mapsto \left| \frac{f^{(k+j)}(x)}{f^{(k)}(x)} \right|$$

is decreasing for every integers k, j.

In the next lemma we give two consequences of convexity and log-convexity of *CM* functions. Similar inequalities for *CM* functions can be found in [16], but with more involved proofs.

Lemma 4.3. *If f is completely monotone, then*

$$f(x) + f(y) \leq f(x-\varepsilon) + f(y+\varepsilon) \leq f(0) + f(x+y), \tag{15}$$

$$f(x)f(y) \leq \; f(x-\varepsilon)f(y+\varepsilon) \; \leq f(0)f(x+y) \tag{16}$$

where $0 \leq \varepsilon < x < y$, assuming that $f(0)$ is defined as $f(0_+)$ (as in Sect. 2.1, finite or not).

Proof. If φ is a convex function, then the divided difference

$$\Delta_{\varphi,\varepsilon}(x) = \frac{\varphi(x) - \varphi(x-\varepsilon)}{\varepsilon}$$

is increasing with x; hence in the present setup,

$$\Delta_{f,\varepsilon}(x) \leq \Delta_{f,\varepsilon}(y+\varepsilon) \quad \text{and} \quad \Delta_{f,x-\varepsilon}(x-\varepsilon) \leq \Delta_{f,x-\varepsilon}(x+y),$$

which proves (15). The same proof holds for (16), but with $\log f$ in place of f. □

Let us note that under assumptions of Lemma 4.3,

$$(x,y) \prec (x-\varepsilon, y+\varepsilon) \prec (0, x+y),$$

and so we have just proved that the functions $(x,y) \mapsto f(x) + f(y)$ and $(x,y) \mapsto f(x)f(y)$ are Schur-convex on $\mathbb{R}_+ \times \mathbb{R}_+$. More generally, for any f being *CM*, the functions of n variables

$$\sum_{i=1}^{n} f(x_i) \quad \text{and} \quad \prod_{i=1}^{n} f(x_i) \tag{17}$$

are Schur-convex on $\mathbb{R}_+^n$. For a proof of this statement see [17].

Finally, the fact that f' is concave (i.e., $f''' < 0$) is equivalent to each of three inequalities in the next lemma [19,20].

Lemma 4.4. *For a CM function f, it holds*

$$\frac{f'(x) + f'(y)}{2} < \frac{f(y) - f(x)}{y - x} < f'\left(\frac{x+y}{2}\right), \quad \textit{for all } x, y > 0, \tag{18}$$

$$\frac{f(y) - f(x)}{y - x} < \frac{f(y-\varepsilon) - f(x+\varepsilon)}{y - x - 2\varepsilon}, \quad \textit{for } 0 < x < y \textit{ and } 0 < \varepsilon < \frac{y-x}{2}. \tag{19}$$

5 Inversion Formulas

It is sometimes easier to find a measure μ that corresponds to function f via Laplace transform in (3) than to show that f is *CM* by verifying the definition; in view of applications, it is definitely useful and desirable to know the associated measure. In many cases we can use properties of Laplace transform and the tables that can be found in textbooks. In many applications the Laplace transform is not limited to real argument, and it is more common to define $f(z)$ by (3), where complex argument z belongs to some half space $\operatorname{Re} z \geq a$, for some positive a. We may use the power of complex Laplace transform calculus applied to real function of real argument, due to well-known properties of regular functions.

Due to similarity between Fourier transform, complex Laplace transform, and real Laplace transform, we may use inversion formulas for all three mentioned classes, whenever it is appropriate. In probability theory, for a random variable Z, the function $x \mapsto \mathrm{E}\,\mathrm{e}^{ixZ}$ (which corresponds to Fourier transform, except the sign in the exponent) is called the characteristic function, whereas the real Laplace transform (mind the sign!) $x \mapsto \mathrm{E}\,\mathrm{e}^{xZ}$ is called the moment generating function. There are several formulas that can be found in textbooks, but we will mention here only a not widely known inversion theorem that enables finding a finite measure μ defined on Borel sets of $\mathbb{R}$, provided that we know its characteristic function

$$\varphi(x) = \int_{-\infty}^{+\infty} \mathrm{e}^{itx}\,\mathrm{d}F(t), \tag{20}$$

where $F(t) = \mu\{(-\infty, t]\}$. The following result (given here in a slightly generalized version) is due to Gil-Pelaez [10].

Lemma 5.1. *For φ and F as in (20), with $\varphi(0)$ being finite, we have that, for all $t \in \mathbb{R}$,*

$$\frac{F(t) + F(t_-)}{2} = \frac{\varphi(0)}{2} - \frac{1}{\pi}\int_0^{+\infty} \operatorname{Re}\left(\frac{\mathrm{e}^{-itx}\varphi(x)}{ix}\right)\mathrm{d}x. \tag{21}$$

Note that the underlying measure here need not necessarily be restricted to the positive part of the real axis. As an example of how (21) can be used to determine a measure μ such that $f = \mathcal{L}(\mu)$, consider a simple case $f(x) = \mathrm{e}^{-x}$, where we already know that the measure is Dirac at $t = 1$. Supposing that we wish to use (21) to derive this, note that if f is the Laplace transform of μ, then its characteristic function is $\varphi(x) = f(-ix) = \mathrm{e}^{ix}$, and (21) yields (assuming that t is a point of continuity of F)

$$F(t) = \frac{1}{2} - \frac{1}{\pi}\int_0^{+\infty} \frac{\sin x(1-t)}{x}\,\mathrm{d}x. \tag{22}$$

Knowing that

$$\int_0^{+\infty} \frac{\sin(ax)}{x}\,\mathrm{d}x = \frac{\pi}{2}\operatorname{sgn} a,$$

we find that $F(t) = 0$ for $t < 1$ and $F(t) = 1$ for $t > 1$; hence (by right-continuity and non-decreasing of F), the corresponding measure μ is indeed a Dirac measure at $t = 1$.

For other formulas and methods, including numerical evaluation of inverse, see [5]. In the next lemma we complement some examples from [24] by effectively finding the corresponding measure.

Lemma 5.2. *We have the following representations:*

$$\mathrm{e}^{-ax} = \mathcal{L}(\,\mathrm{d}\delta_a(t)), \tag{23}$$

where δ_a is the probability measure with unit mass (Dirac measure) at $a \geq 0$;

$$\frac{1}{(ax+b)^c} = \mathcal{L}\left(\mathrm{e}^{-bt/a}\frac{t^{c-1}}{a^c\Gamma(c)}\right), \quad a,b,c \geq 0,\ a^2+b^2 > 0; \tag{24}$$

$$\log\left(a+\frac{b}{x}\right) = \mathcal{L}(\,\mathrm{d}\mu(t)) \qquad a \geq 1,\ b > 0, \tag{25}$$

where the measure μ is determined by its distribution function

$$\mu([0,t]) = \log a + \int_0^x \frac{1-\mathrm{e}^{-bs/a}}{s}\,\mathrm{d}s;$$

$$\frac{\log(1+x)}{x} = \mathcal{L}(E_1(t)), \tag{26}$$

where (see [1, p. 56]*) E_1 is exponential integral*

$$E_1(t) = \int_1^{+\infty} \mathrm{e}^{-tu}\frac{\mathrm{d}u}{u};$$

$$\mathrm{e}^{a/x} = \mathcal{L}\left(\mathrm{d}\delta_0(t) + \frac{aI_1(2\sqrt{at})}{\sqrt{at}}\,\mathrm{d}t\right), \tag{27}$$

where I_1 is a modified Bessel function as defined in [1].

Proof. The relation (23) is obvious, and (24) is a consequence of standard rules for (complex) Laplace transform:

$$\mathcal{L}\left(\mathrm{e}^{-bt/a}\frac{t^{c-1}}{a^c\Gamma(c)}\right)=\frac{1}{a^c\Gamma(c)}\mathcal{L}(t^{c-1})(x-b/a)=\frac{1}{a^c\Gamma(c)}\cdot\frac{\Gamma(c)}{(x-b/a)^c}=\frac{1}{(ax+b)^c}$$

To prove (25), denote its left side by f, and observe that, by (24),

$$f'(x)=\frac{a}{ax+b}-\frac{1}{x}=\mathcal{L}\left(\mathrm{e}^{-bt/a}-1\right).$$

Now we use the rule

$$\mathcal{L}\left(\frac{g(t)}{t}\right)=\int_x^{+\infty}\mathcal{L}(g(t))[y]\,\mathrm{d}y$$

to conclude that

$$f(x)=\log a-\int_x^{+\infty}f'(y)\,\mathrm{d}y=\mathcal{L}(\log a\,\mathrm{d}\delta_0)-\mathcal{L}\left(\frac{\mathrm{e}^{-bt/a}-1}{t}\right),$$

which yields (25). To prove (27), we note that

$$\frac{a^k}{k!x^k}=\mathcal{L}\left(\frac{a^kt^{k-1}}{k!(k-1)!}\right),$$

which tells us that

$$\mathrm{e}^{\frac{a}{x}}=1+\mathcal{L}\left(\sum_{k=1}^{+\infty}\frac{a^kt^{k-1}}{(k-1)!k!}\right).$$

Now we observe that

$$\sum_{k=1}^{+\infty}\frac{a^kt^{k-1}}{(k-1)!k!}=\frac{aI_1(2\sqrt{at})}{\sqrt{at}},$$

and (27) follows.

The simplest way to prove (26) would be to perform an integration on the right-hand side and show that it yields the left side. However, in order to show the derivation, we start with the observation that

$$f(x):=\frac{\log(1+x)}{x}=F(1,1,2;-x),$$

where $F(a,b,c;\cdot)={}_2F_1(a,b,c;\cdot)$ is a Gauss' hypergeometric function; hence there is the following integral representation [1]:

$$f(x) = \int_0^1 \frac{\mathrm{d}s}{1+sx}.$$

Now we use (25) to find that

$$\frac{1}{1+sx} = \frac{1}{s}\int_0^{+\infty} \mathrm{e}^{-xt}\mathrm{e}^{-t/s}\,\mathrm{d}t,$$

and, exchanging the order of integration, we find that

$$f(x) = \int_0^{+\infty} \mathrm{e}^{-xt}\left(\int_0^1 \mathrm{e}^{-t/s}\frac{\mathrm{d}s}{s}\right)\mathrm{d}t.$$

Finally, a change of variables $1/s = u$ in the inner integral shows that it is equal to $E_1(t)$, and the formula is proved. □

6 Some Examples Related to the Gamma Function

Functions related to the Gamma function are good candidates to be *CM*, and there is a plenty of such results in literature. The function $g(x) = \log\Gamma(x)$ is a unique convex solution of Krull's functional equation

$$g(x+1) - g(x) = f(x), \qquad x > 0, \tag{28}$$

with $f(x) = \log x$ and with $g(1) = 0$. The same equation, but with $f(x) = (\log x)^{(n+1)}$, $n = 0, 1, 2, \ldots$, has for its solutions functions $\Psi^{(n)}(x) = (\log\Gamma(x))^{(n+1)}$. Although $\log x$ is not *CM*, all its derivatives are monotone functions, which automatically implies the same property for $\Psi^{(n)}(x), n \geq 2$, and alike functions via the following result (see [22]).

Lemma 6.1. *Suppose that $x \mapsto f(x)$ is a function of the class $C^\infty(0,+\infty)$ with all derivatives being monotone functions, with $f'(x) \to 0$ as $x \to +\infty$. Then there is a unique (up to an additive constant) solution g of (28) in the class C^∞, with*

$$g'(x) = \lim_{n\to+\infty}\left(f(x+n) - \sum_{k=0}^{n} f'(x+k)\right) \tag{29}$$

and

$$g^{(j)}(x) = -\sum_{k=0}^{+\infty} f^{(j)}(x+k) \qquad (j \geq 2). \tag{30}$$

From (29) and (30) it follows that if $\pm f$ is *CM* (or if only $\pm f''$ is such), then $\mp g''$ is a *CM*, while $\pm g$ and $\pm g'$ need not be *CM* . Our first example is formulated in the form of a lemma, and its proof provides a pattern that can be used in many similar cases.

Lemma 6.2. *The function*

$$W(x) = -\big(\log \Gamma(x) - (x-1)\log x\big)'' = \frac{1}{x} + \frac{1}{x^2} - \Psi'(x)$$

has the following integral representation:

$$W(x) = \int_0^{+\infty} \left(1 + t - \frac{t}{1 - \mathrm{e}^{-t}}\right) \mathrm{e}^{-xt}\, \mathrm{d}t \tag{31}$$

and it is a CM function.

Proof. The integral representation follows from

$$\mathcal{L}\left((-1)^{n+1}\frac{t^n}{1-\mathrm{e}^{-t}}\right) = \Psi^{(n)}(x), \qquad n = 1, 2, \ldots, \tag{32}$$

and

$$\mathcal{L}(t^a) = \frac{\Gamma(a+1)}{x^{a+1}}, \qquad a > -1. \tag{33}$$

The *CM* property follows from positivity of the function under integral sign, which is equivalent to the inequality $\mathrm{e}^t > 1 + t$ for $t > 0$. □

Remark 6.1. The function $g(x) = \log \Gamma(x) - (x-1)\log x$ satisfies the functional equation (28) with $f(x) = \log\left(x/(x+1)\right)^x$; it can be easily checked that f'' is *CM*; hence from Lemma 6.1 we can conclude without any additional work that W as defined above is *CM*.

Lemma 6.3. *For $a \geq 0$ and $x > 0$, let*

$$G_a(x) = \log \Gamma(x) - \left(x - \frac{1}{2}\right)\log x - \frac{1}{12}\Psi'(x+a) + x - \frac{1}{2}\log(2\pi).$$

The following representation holds:

$$G_a(x) = \int_0^{+\infty} \frac{t - 2 + (2+t)\mathrm{e}^{-t} - (t^3/6)\mathrm{e}^{-at}}{2t^2(1-\mathrm{e}^{-t})}\mathrm{e}^{-xt}\, \mathrm{d}t. \tag{34}$$

The function $x \mapsto G_a(x)$ is CM if and only if $a \geq 1/2$ and the function $x \mapsto -G_a(x)$ is CM if and only if $a = 0$.

Proof. Starting with

$$G_a''(x) = \Psi'(x) - \frac{1}{12}\Psi'''(x+a) - \frac{2}{x} + \frac{x-1/2}{x^2},$$

it is easy to show (in a similar way as in Lemma 6.2) that

$$G_a''(x) = \int_0^{+\infty} \frac{t-2+(2+t)\mathrm{e}^{-t} - (t^3/6)\mathrm{e}^{-at}}{2(1-\mathrm{e}^{-t})} \mathrm{e}^{-xt}\,\mathrm{d}t. \tag{35}$$

Further, we have that

$$\lim_{x\to+\infty} G_a(x) = \lim_{x\to+\infty} G_a'(x) = 0,$$

and

$$G_a(x) = \int_x^{+\infty}\int_v^{+\infty} G_a''(u)\,\mathrm{d}u\,\mathrm{d}v,$$

hence (34) holds. The complete monotonicity is related to the sign of the function

$$h_a(t) = t-2+(2+t)\mathrm{e}^{-t} - \frac{t^3}{6}\mathrm{e}^{-at}. \tag{36}$$

The function G_a is *CM* if and only if $h_a(t) \geq 0$ for all $t \geq 0$. From (36) we see that this is equivalent to

$$a \geq \frac{\log 6 + \log((2+t)\mathrm{e}^{-t} + t - 2) - 3\log t}{-t} := u(t). \tag{37}$$

Using standard methods, we can find that u is a decreasing function; hence

$$u(t) \leq \lim_{t\to 0_+} u(t) = \frac{1}{2},$$

and so, (37) holds if and only if $a \geq 1/2$.

Further, $-G_a$ is *CM* if and only if $h_a(t) \leq 0$ for all $t \geq 0$, which is equivalent to

$$a \leq u(t), \tag{38}$$

where $u(t)$ is defined in (37). Since u is decreasing, we have that $u(t) \geq \lim_{t\to+\infty} u(t) = 0$, and so, (38) holds if and only if $a \leq 0$, that is, $a = 0$. □

Remark 6.2. Let

$$F_a(x) = \log \Gamma(x) - \left(x - \frac{1}{2}\right)\log x - \frac{1}{12}\Psi'(x+a), \qquad a \geq 0, x > 0.$$

This function is studied in [18, Theorem 1], where it is shown that $x \mapsto F_0(x)$ is concave on $x > 0$ and that $x \mapsto F_a(x)$ is convex on $x > 0$ for $a \geq \frac{1}{2}$. Since $F_a(x)'' = G_a''(x)$ where G_a is defined as above, this example gives much stronger statement.

Lemma 6.4. *For $b \geq 0$ and $c \geq 0$, let*

$$f_{b,c}(x) = \frac{e^x \Gamma(x+b)}{x^{x+c}}, \qquad x > 0. \tag{39}$$

The function

$$\varphi_{b,c}(x) = \log f_{b,c}(x) = x + \log \Gamma(x+b) - (x+c)\log x \tag{40}$$

is CM if and only if $b \geq \frac{1}{2} + \frac{1}{\sqrt{12}}$ and $c = b - \frac{1}{2}$ and then it has the representation

$$\varphi_{b,b-\frac{1}{2}}(x) = \int_{[0,+\infty)} \frac{1}{t^2}\left(\frac{te^{-bt}}{1-e^{-t}} + t\left(b - \frac{1}{2}\right) - 1\right) dt. \qquad x > 0. \tag{41}$$

Proof. By expanding $\log \Gamma(x+b)$ in (40) by means of Stirling's formula [1, p. 258], it follows that for $\delta = b - c \neq \frac{1}{2}$,

$$\lim_{x \to +\infty} \varphi_{b,c}(x) = \left(\delta - \frac{1}{2}\right) \cdot (+\infty),$$

so $\varphi_{b,c}$ is not a *CM* function (see Sect. 2.1). Let $\delta = 1/2$ and let

$$G_b(x) := \varphi_{b,b-\frac{1}{2}}(x) = x + \log \Gamma(x+b) - \left(x + b - \frac{1}{2}\right) \log x.$$

Further, we find without difficulties that

$$\lim_{x \to +\infty} G_b(x) = \lim_{x \to +\infty} G_b'(x) = 0 \tag{42}$$

and that

$$G_b''(x) = \Psi'(x+b) - \frac{1}{x} + \frac{b - \frac{1}{2}}{x^2}.$$

In the same way as shown in Lemma 6.2, we find that $G_b''(x) = \mathcal{L}(h_b(t)\,dt)$, where

$$h_b(t) = \frac{te^{-bt}}{1-e^{-t}} + t\left(b - \frac{1}{2}\right) - 1. \tag{43}$$

By standard methods we find that

$$h(t) = \left(\frac{b^2}{2} - \frac{b}{2} + \frac{1}{12}\right)t^2 + o(t^2) \qquad (t \to 0), \tag{44}$$

so the Laplace transform $G_b(x)$ of the function $t \mapsto g(t)/t^2$ exists for all $x > 0$ and applying Fubini theorem as in Lemma 6.3 and using (42) we find that

$$G_b(x) = \int_x^{+\infty} \int_v^{+\infty} G_a''(u)\,\mathrm{d}u\,\mathrm{d}v = \int_{[0,+\infty)} \frac{h(t)}{t^2} \mathrm{e}^{-tx}\,\mathrm{d}t,$$

which is the representation (41). Then G_b will be *CM* if and only if $h(t) \geq 0$ for each $t \geq 0$ (see Remark 2.1). By (44) we have that $h(0) < 0$ for $b \in (b_1, b_2)$, where $b_{1,2} = \frac{1}{2} \pm \frac{1}{\sqrt{12}}$; further, $c = b - 1/2 > 0$ gives $b > 1/2$, so only $b \geq b_2$ remains as a possibility. It is straightforward to check that $\frac{\partial h_b(t)}{\partial b} > 0$ for all $t \geq 0$, so it suffices to show that $h_{b_2}(t) \geq 0$ for $t \geq 0$, which can be done along the lines of [11]. □

Remark 6.3. Complete monotonicity of functions $f_{b,c}$ and $\varphi_{b,c}$ for various values of parameters was discussed in [11,12]. Let us remark that, by Lemma 3.3, the function $f_{b,c}$ is *CM* whenever $\varphi_{b,c}$ is the one.

Let us mention that the Barnes function $G(x)$ satisfies the relation

$$\log G(x+1) - \log G(x) = \log \Gamma(x), \quad x > 0,$$

which is (28) with $g = \log \Gamma$. Here also the function $x \mapsto (\log G(x))'' = 2\Psi'(x) + (x-1)\Psi''(x)$ is *CM*. More details about the properties of the G-function as a solution of Krull's equation can be found in [23].

Acknowledgement This work is supported by Ministry of Education and Science of Serbia under projects 174024 and III 44006.

References

1. Abramowitz, M., Stegun, I.A.: A Handbook of Mathematical Functions. National Bureau of Standards, New York. http://people.math.sfu.ca/~cbm/aands/ (1964)
2. Audenaert, K.M.R.: Trace inequalities for completely monotone functions and Bernstein functions. Linear Algebra Appl. **437**, 601–611 (2012)
3. Bernstein, S.: Sur les fonctions absolument monotones. Acta Math. **52**, 1–66 (1929)
4. Carter, M., van Brunt, B.: The Lebesgue-Stieltjes Integral: A Practical Introduction. Springer, New York (2000)
5. Cohen, A.M.: Numerical Methods for Laplace Transform Inversion. Springer, New York (2007)
6. Fasshauer, G.: Meshfree Approximation Methods with MATLAB. World Scientific Publishing, Hackensack (2007)

7. Feller, W.: On Müntz' theorem and completely monotone functions. Am. Math. Monthly **75**, 342–350 (1968)
8. Feller, W.: An Introduction to Probability Theory and Its Applications, vol. II. Wiley, New York (1970)
9. Fink, A.M.: Kolmogorov-Landau inequalities for monotone functions. J. Math. Anal. Appl. **90**, 251–258 (1982)
10. Gil-Pelaez, J.: Note on the inversion theorem. Biometrika **38**, 481–482 (1951)
11. Guo, S., Srivastava, H.M.: A class of logarithmically completely monotonic functions. Appl. Math. Lett. **21**, 1134–1141 (2008)
12. Guo, S., Qi, F., Srivastava, H.M.: Necessary and sufficient conditions for two classes of functions to be logarithmically completely monotonic. Integr. Transf. Spec. Funct. **18**, 819–826 (2007)
13. Hausdorff, F.: Summationsmethoden und momentfolgen I. Math. Z. **9**, 74–109 (1921)
14. Hausdorff, F.: Momentprobleme fuř ein endliches intervall. Math. Z. **16**, 220–248 (1923)
15. Hewitt, E.: Integration by parts for Stieltjes integrals. Am. Math. Monthly **67**, 419–423 (1960)
16. Kimberling, C.H.: A probabilistic interpretation of complete monotonicity. Aequ. Math. **10**, 152–164 (1974)
17. Marshall, A.W., Olkin, I., Arnold, B.C.: Inequalities: Theory of Majorization and Its Applications, 2nd edn. Springer Series in Statistics. Springer, New York (2009)
18. Merkle, M.: Convexity, Schur-convexity and bounds for the Gamma function involving the Digamma function. Rocky Mountain J. Math. **28**(3), 1053–1066 (1998)
19. Merkle, M.: Conditions for convexity of a derivative and some applications to the Gamma function. Aequ. Math. **55**, 273–280 (1998)
20. Merkle, M.: Representation of the error term in Jensen's and some related inequalities with applications. J. Math. Anal. Appl. **231**, 76–90 (1999)
21. Merkle, M.: Reciprocally convex functions. J. Math. Anal. Appl. **293**, 210–218 (2004)
22. Merkle, M.: Convexity in the theory of the Gamma function. Int. J. Appl. Math. Stat. **11**, 103–117 (2007)
23. Merkle, M., Merkle, M.M.R.: Krull's theory for the double gamma function. Appl. Math. Comput. **218**, 935–943 (2011)
24. Miller, K.S., Samko, S.G.: Completely monotonic functions. Integr. Transf. Spec. Funct. **12**, 389–402 (2001)
25. Schilling, R., Song, R., Vondraček, Z.: Bernstein Functions: Theory and Applications. De Gruyter Studies in Mathematics, vol. 37. Walter de Gruyter & Co., Berlin (2010)
26. Widder, D.V.: Necessary and sufficient conditions for the representation of a function as a Laplace integral. Trans. Am. Math. Soc. **33**, 851–892 (1931)
27. Widder, D.V.: The inversion of the Laplace integral and the related moment problem. Trans. Am. Math. Soc. **36**, 107–200 (1934)
28. Widder, D.V.: The Laplace Transform. Princeton University Press, Princeton (1941)

New Applications of Superquadracity

Shoshana Abramovich

Dedicated to Professor Hari M. Srivastava

Abstract In the recent years numerous articles were published related to superquadracity. Here we put together some of these results, in particular those related to Jensen, Jensen–Steffensen, and Hardy's inequalities. These inequalities are analogs of inequalities satisfied by convex functions and in those cases that the superquadratic functions are nonnegative we get refinements of inequalities satisfied by convex functions.

1 Introduction

In this survey we present some new applications of superquadracity obtained after the publication of "On superquadracity" by Abramovich [2]. Here we state only a part of the results with no proofs. This may serve as an introductory work to a detailed book on superquadracity that includes all the known theorems and their proofs on superquadracity which is planned to be written by S. Abramovich, J. Pečarić, S. Banić, and S. Varošanec. All over this survey we get analogs of inequalities satisfied by convex functions and in the case that our superquadratic functions are nonnegative, we get refinements of results that use only convexity.

There is another class of functions called superquadratic functions or weakly superquadratic functions; see [2, 18, 27, 28] and their references. Here we quote results from [1–23, 25–33].

Let us start with repeating the definition and basic properties of superquadratic functions (called in [27, 28] "strongly superquadratic functions"):

S. Abramovich (✉)
Department of Mathematics, University of Haifa, Haifa, Israel
e-mail: abramos@math.haifa.ac.il

G.V. Milovanović and M.Th. Rassias (eds.), *Analytic Number Theory, Approximation Theory, and Special Functions*, DOI 10.1007/978-1-4939-0258-3_13,

Definition 1.1 ([6, 7]). A function $\varphi : [0, b) \to \mathbb{R}$ is superquadratic provided that for all $0 \leq x < b$ there exists a constant $C_\varphi(x) \in \mathbb{R}$ such that

$$\varphi(y) - \varphi(x) \geq C_\varphi(x)(y - x) + \varphi(|y - x|) \tag{1}$$

for every $y, 0 \leq y < b$.

From the definition of superquadracity we easily get:

Lemma 1.1 ([7]). *The function φ is superquadratic on $[0, b)$ if and only if*

$$\frac{1}{A_n}\sum_{i=1}^{n} a_i \varphi(x_i) - \varphi(\overline{x}) \geq \frac{1}{A_n}\sum_{i=1}^{n} a_i \varphi(|x_i - \overline{x}|) \tag{2}$$

holds, where $x_i \in [0, b)$, $i = 1, \ldots, n$, and $a_i \geq 0$, $i = 1, \ldots, n$, are such that $A_n = \sum_{i=1}^{n} a_i > 0$ and $\overline{x} = \frac{1}{A_n}\sum_{i=1}^{n} a_i x_i$.

The function φ is superquadratic on $[0, b)$, if and only if

$$\int_\Omega \varphi(f(s))\, d\mu(s) - \varphi\left(\int_\Omega f(s)\, d\mu(s)\right) \geq \int_\Omega \varphi\left(\left|f(s) - \int_\Omega f(\sigma)\, d\mu(\sigma)\right|\right) d\mu(s), \tag{3}$$

where f is any nonnegative μ-integrable function on a probability measure space (Ω, μ).

Lemma 1.2 ([7]). *Let φ be a superquadratic function with $C_\varphi(x)$ as in Definition 1.1. Then*

(i) $\varphi(0) \leq 0$.
(ii) *If $\varphi(0) = \varphi'(0) = 0$, then $C_\varphi(x) = \varphi'(x)$ whenever φ is differentiable at $0 < x < b$.*
(iii) *If $\varphi \geq 0$, then φ is convex and $\varphi(0) = \varphi'(0) = 0$.*

Lemma 1.3 ([7]). *Suppose that $\varphi : [0, b) \to \mathbb{R}$ is continuously differentiable and $\varphi(0) \leq 0$. If φ' is superadditive or $\varphi'(x)/x$ is nondecreasing, then φ is superquadratic.*

The power functions $\varphi(x) = x^p$, $x \geq 0$, are superquadratic when $p \geq 2$ and subquadratic, that is, $-\varphi$ is superquadratic when $1 \leq p \leq 2$. When $\varphi(x) = x^2$, (1) reduces to equality and therefore the same holds for (2) and (3).

In Sect. 2 we compile some inequalities related to Jensen and Jensen–Steffensen inequalities. In Sect. 3 Hardy type inequalities are presented and in Sect. 4 various other inequalities are stated. All these inequalities stem from the properties of superquadracity, as stated in its definition and in Lemmas 1.1–1.3.

2 Jensen and Jensen–Steffensen Type Inequalities for Superquadratic Functions

Now we state several results that refine and generalize Jensen's type inequalities for superquadratic functions (2) and (3) presented in the first section. We also show how we get Jensen–Steffensen type inequalities for superquadratic functions that satisfy additional conditions like being nonnegative or having superadditive derivatives. Next we consider how Jensen's type inequalities are derived for functions like $K(x) = \varphi(x) x^{\gamma}, 0 \leq x < \infty, \gamma \in \mathbb{R}_+$ where $\varphi(x)$ is superquadratic.

In cases for which our function is superquadratic and positive we always get a refinement of a similar Jensen's type inequality for convex functions.

2.1 *Improvement of the Basic Jensen's Inequality for Superquadratic Functions*

To improve (2), we first introduce as in [15] some notations. $I_n = \{1, \ldots, n\}, n \in \mathbb{N}$, I is any finite set of integers, $A_n = \sum_{i=1}^{n} a_i$ for $a_i \in \mathbb{R}, i = 1, \ldots, n$.

For a function φ defined on $\mathbb{R}_+$, $x_k \geq 0$, $a_k > 0$, $k \in I_n$ and for every $I \subseteq I_n$ we denote

$$\Delta_I = \sum_{k \in I} a_k \varphi\left(\left|\frac{x_k}{a_k} - \frac{\sum_{j \in I} x_j}{\sum_{j \in I} a_j}\right|\right), \quad \Delta_n = \sum_{k=1}^{n} a_k \varphi\left(\left|\frac{x_k}{a_k} - \frac{\sum_{j=1}^{n} x_j}{\sum_{j=1}^{n} a_j}\right|\right),$$

$$\Delta_2(i,j) = a_i \varphi\left(\left|\frac{x_i}{a_i} - \frac{x_i + x_j}{a_i + a_j}\right|\right) + a_j \varphi\left(\left|\frac{x_j}{a_j} - \frac{x_i + x_j}{a_i + a_j}\right|\right), \quad 1 \leq i \leq j \leq n,$$

$$d_I = \sum_{k \in I} a_k \varphi\left(\frac{x_k}{a_k}\right) - \left(\sum_{k \in I} a_k\right) \varphi\left(\frac{\sum_{k \in I} x_k}{\sum_{k \in I} a_k}\right),$$

$$d_n = \sum_{k=1}^{n} a_k \varphi\left(\frac{x_k}{a_k}\right) - \left(\sum_{k=1}^{n} a_k\right) \varphi\left(\frac{\sum_{k=1}^{n} x_k}{\sum_{k=1}^{n} a_k}\right).$$

Using these notations our results are obtained in the following Theorem 2.1.

Theorem 2.1 ([15]). *Let* $x_k \geq 0$, $a_k > 0$, $k = 1, \ldots, n$, $I \subseteq I_n$ *and* $\overline{I} = I_n / I$.

(a) *If* $\varphi(x)$ *is superquadratic on* $\mathbb{R}_+$, *then*

$$d_n \geq d_I + A_I \varphi \left(\left| \sum_{j=1}^{n} \frac{x_j}{A_n} - \sum_{i \in I} \frac{x_i}{A_I} \right| \right) + \sum_{i \in \overline{I}} a_i \varphi \left(\left| \frac{x_i}{a_i} - \sum_{j=1}^{n} \frac{x_j}{A_n} \right| \right).$$

(b) *If* $\varphi(x)$ *is also nonnegative on* $\mathbb{R}_+$*, then*

$$d_n \geq d_I + A_I \varphi \left(\frac{A_{\overline{I}}}{A_n} \left| \sum_{i \in I} \frac{x_i}{A_I} - \sum_{i \in \overline{I}} \frac{x_i}{A_{\overline{I}}} \right| \right) + A_{\overline{I}} \varphi \left(\frac{A_I}{A_n} \left| \sum_{i \in I} \frac{x_i}{A_I} - \sum_{i \in \overline{I}} \frac{x_i}{A_{\overline{I}}} \right| \right), \tag{4}$$

and for $n \geq 2$

$$d_n \geq \max_{I \subseteq I_n} (\Delta_I) \geq \max_{1 \leq i < j \leq n} \Delta_2 (i, j).$$

The proof follows from the identity

$$\begin{aligned} &\sum_{i=1}^{n} a_i \varphi(x_i) - A_n \varphi \left(\sum_{i=1}^{n} \frac{a_i x_i}{A_n} \right) - \left(\sum_{i \in I} a_i \varphi(x_i) - A_I \varphi \left(\frac{\sum_{i \in I} a_i x_i}{A_I} \right) \right) \\ &\quad = \sum_{i \in \overline{I}} a_i \varphi(x_i) + A_I \varphi \left(\frac{\sum_{i \in I} a_i x_i}{A_I} \right) - A_n \varphi \left(\sum_{i=1}^{n} \frac{a_i x_i}{A_n} \right), \end{aligned}$$

and from the basic Jensen's inequality for superquadratic functions (2).

The same reasoning, which leads to (4), leads also to

$$\begin{aligned} &\sum_{i=1}^{n} a_i \varphi(x_i) - A_n \varphi \left(\sum_{i=1}^{n} \frac{a_i x_i}{A_n} \right) \geq \sum_{k=1}^{n} a_k \varphi \left(\left| x_k - \sum_{i=1}^{n} \frac{a_i x_i}{A_n} \right| \right) \\ &\quad \geq A_I \varphi \left(\frac{A_{\overline{I}}}{A_n} \left| \sum_{i \in I} \frac{a_i x_i}{A_I} - \sum_{i \in \overline{I}} \frac{a_i x_i}{A_{\overline{I}}} \right| \right) + A_{\overline{I}} \varphi \left(\frac{A_I}{A_n} \left| \sum_{i \in \overline{I}} \frac{a_i x_i}{A_{\overline{I}}} - \sum_{i \in I} \frac{a_i x_i}{A_I} \right| \right) \\ &\quad \geq A_n \varphi(0) = 0 \end{aligned}$$

for nonnegative superquadratic functions.

2.2 *Jensen–Steffensen's and Slater–Pečarić Inequalities for Nonnegative Superquadratic Functions*

In [9] the authors dealt with refinements of Jensen–Steffensen's inequality and Slater–Pečarić inequality for positive superquadratic functions. These inequalities are refinements of similar inequalities for convex functions. Two of the results are as follows:

Theorem 2.2. *Let $\varphi : [0,\infty) \to \mathbb{R}$ be differentiable superquadratic and nonnegative, let $\boldsymbol{\zeta}$ be a nonnegative monotonic n-tuple in $\mathbb{R}^n$, and let* **a** *be a real n-tuple satisfying Steffensen's coefficients, that is,* $0 \leq A_j \leq A_n$, $j = 1,\ldots,n$, $A_n > 0$,

$$A_j = \sum_{i=1}^{j} a_i, \quad \overline{A}_j = \sum_{i=j}^{n} a_i, \quad j = 1,\ldots,n.$$

Let $\overline{\zeta}$ be defined by

$$\overline{\zeta} = \frac{1}{A_n} \sum_{i=1}^{n} a_i \zeta_i .$$

Then

$$\begin{aligned}
\sum_{i=1}^{n} a_i \varphi(\zeta_i) - A_n \varphi\left(\overline{\zeta}\right) &\geq \sum_{j=1}^{k-1} A_j \varphi(\zeta_{j+1} - \zeta_j) + A_k \varphi\left(\overline{\zeta} - \zeta_k\right) \\
&\quad + \overline{A}_{k+1} \varphi\left(\zeta_{k+1} - \overline{\zeta}\right) + \sum_{j=k+2}^{n} \overline{A}_j \varphi(\zeta_j - \zeta_{j-1}) \\
&\geq \left(\sum_{i=1}^{k} A_i + \sum_{i=k+1}^{n} \overline{A_i}\right) \varphi\left(\frac{\sum_{i=1}^{n} a_i \left(|\zeta_i - \overline{\zeta}|\right)}{\sum_{i=1}^{k} A_i + \sum_{i=k+1}^{n} \overline{A_i}}\right) \\
&\geq ((n-1) A_n) \varphi\left(\frac{\sum_{i=1}^{n} a_i \left(|\zeta_i - \overline{\zeta}|\right)}{(n-1) A_n}\right),
\end{aligned}$$

where $k \in \{1,\ldots,n-1\}$ satisfies $\zeta_k \leq \overline{\zeta} \leq \zeta_{k+1}$.

If also $\sum_{i=1}^{n} a_i \varphi'(\zeta_i) \neq 0$, and we define $M = \sum_{i=1}^{n} a_i \zeta_i \varphi'(\zeta_i) / \sum_{i=1}^{n} a_i \varphi'(\zeta_i)$, then, for s satisfying $\zeta_s \leq M \leq \zeta_{s+1}$, $s+1 \leq n$,

$$\begin{aligned}
\sum_{i=1}^{n} a_i \varphi(\zeta_i) \leq A_n \varphi(M) - \Bigg(&\sum_{j=1}^{s-1} A_j \varphi(\zeta_{j+1} - \zeta_j) \\
&+ A_s \varphi(M - \zeta_s) + \overline{A}_{s+1} \varphi(\zeta_{s+1} - M) + \sum_{j=s+2}^{n} \overline{A}_j \varphi(\zeta_j - \zeta_{j-1})\Bigg)
\end{aligned}$$

$$\leq A_n\varphi(M) - \left(\sum_{j=1}^{s} A_j + \sum_{j=s+1}^{n} \overline{A}_j\right)\varphi\left(\frac{\sum_{i=1}^{n} a_i |\zeta_i - M|}{\sum_{j=1}^{s} A_j + \sum_{j=s+1}^{n} \overline{A}_j}\right)$$

$$\leq A_n\varphi(M) - ((n-1)A_n)\varphi\left(\frac{\sum_{i=1}^{n} a_i |\zeta_i - M|}{(n-1)A_n}\right).$$

2.3 Generalization of the Jensen–Steffensen Inequalities for Functions with Superadditive Derivatives

In the next two theorems the main result of [16] is presented and its integral version that appears in [14].

Theorem 2.3 ([16, Theorem 1]). *Let* $\varphi : [0, b) \to \mathbb{R}$ *be continuously differentiable and* $\varphi' : [0, b) \to \mathbb{R}$ *be superadditive function. Let* $\mathbf{a} = (a_1, \ldots, a_n)$ *be a real* n*-tuple satisfying*

$$0 \leq A_j = \sum_{i=1}^{j} a_i \leq A_n, \quad j = 1, \ldots, n, \quad A_n > 0,$$

and $\mathbf{x} = (x_1, \ldots, x_n)$ *be a monotonic* n*-tuple in* $[0, b)^n$*. Then,*

(a) *The inequality*

$$\varphi(c) - \varphi(0) + \varphi'(c)(\overline{x} - c) + \frac{1}{A_n}\sum_{i=1}^{n} a_i\varphi(|x_i - c|) \leq \frac{1}{A_n}\sum_{i=1}^{n} a_i\varphi(x_i) \quad (5)$$

holds for each $c \in [0, b)$ *where* $\overline{x} = \dfrac{1}{A_n}\sum_{i=1}^{n} a_i x_i$.

Inserting $c = \overline{x}$ *in (5) we get*

$$\frac{1}{A_n}\sum_{i=1}^{n} a_i\varphi(x_i) - \varphi(\overline{x}) + \varphi(0) \geq \frac{1}{A_n}\sum_{i=1}^{n} a_i\varphi(|x_i - \overline{x}|).$$

(b) *If in addition $\varphi(0) \leq 0$, then φ is superquadratic and*

$$\frac{1}{A_n}\sum_{i=1}^{n} a_i\varphi(x_i) \geq \varphi(c) + \varphi'(c)(\overline{x} - c) + \frac{1}{A_n}\sum_{i=1}^{n} a_i\varphi(|x_i - c|). \quad (6)$$

Inserting $c = \overline{x}$ in (6) we get

$$\frac{1}{A_n}\sum_{i=1}^{n} a_i\varphi(x_i) - \varphi(\overline{x}) \geq \frac{1}{A_n}\sum_{i=1}^{n} a_i\varphi(|x_i - \overline{x}|).$$

(c) *If in addition $\varphi \geq 0$ and $\varphi(0) = \varphi'(0) = 0$, then φ is convex increasing and superquadratic and*

$$\frac{1}{A_n}\sum_{i=1}^{n} a_i\varphi(x_i) - \varphi(c) - \varphi'(c)(\overline{x} - c) \geq \frac{1}{A_n}\sum_{i=1}^{n} a_i\varphi(|x_i - c|) \geq 0.$$

The proof of (5) follows from observing that when φ is continuously differentiable *on* $[0, b)$ and φ' is superadditive on $[0, b)$ then the function $D : [0, b) \to \mathbb{R}$ defined by

$$D(y) = \varphi(y) - \varphi(z) - \varphi'(z)(y - z) - \varphi(|y - z|) + \varphi(0)$$

is nonnegative on $[0, b)$, nonincreasing on $[0, z)$, and nondecreasing on $[z, b)$, for $0 \leq z < b$, and when $x_k \leq c \leq x_{k+1}$ we get that

$$D(x_1) \geq D(x_2) \geq \cdots \geq D(x_k) \geq 0 \quad \text{and} \quad 0 \leq D(x_{k+1}) \leq D(x_{k+2}) \leq \cdots \leq D(x_n),$$

and

$$\begin{aligned}\sum_{i=1}^{n} a_i D(x_i) &= \sum_{i=1}^{k} a_i D(x_i) + \sum_{i=k+1}^{n} a_i D(x_i)\\ &= \sum_{i=1}^{k-1} A_i\left(D(x_i) - D(x_{i+1})\right) + A_k D(x_k)\\ &\quad + \overline{A}_{k+1} D(x_{k+1}) + \sum_{i=k+2}^{n} \overline{A}_i\left(D(x_i) - D(x_{i-1})\right) \geq 0.\end{aligned}$$

and hence

$$\sum_{i=1}^{n} a_i D(x_i) = \sum_{i=1}^{n} a_i\left[\varphi(x_i) - \varphi(c) - \varphi'(c)(x_i - c) - \varphi(|x_i - c|) + \varphi(0)\right] \geq 0.$$

Now we present the integral version of Theorem 2.3 which is proved by similar reasoning that leads to Theorem 2.3.

Theorem 2.4 ([14]). *Let* $f : [\alpha, \beta] \to [0, b)$ *be a continuous and monotonic function, where* $-\infty < \alpha < \beta < +\infty$. *Let* $\varphi : [0, b) \to \mathbb{R}$ *be continuously differentiable and* $\varphi' : [0, b) \to \mathbb{R}$ *be superadditive function. Let* $\lambda : [\alpha, \beta] \to \mathbb{R}$ *be either continuous or of the bounded variation satisfying*

$$\lambda(\alpha) \leq \lambda(x) \leq \lambda(\beta) \quad \textit{for all} \quad x \in [\alpha, \beta], \quad \lambda(\beta) - \lambda(\alpha) > 0$$

and

$$\overline{x} = \frac{1}{\lambda(\beta) - \lambda(\alpha)} \int_{\alpha}^{\beta} f(t)\, \mathrm{d}\lambda(t), \quad \overline{y} = \frac{1}{\lambda(\beta) - \lambda(\alpha)} \int_{\alpha}^{\beta} \varphi(f(t))\, \mathrm{d}\lambda(t). \quad (7)$$

Then,

(a) *The inequality*

$$\varphi(c) - \varphi(0) + \varphi'(c)(\overline{x} - c) + \frac{1}{\lambda(\beta) - \lambda(\alpha)} \int_{\alpha}^{\beta} \varphi(|f(t) - c|)\, \mathrm{d}\lambda(t) \leq \overline{y} \quad (8)$$

holds for each $c \in [0, b)$ *where* $\overline{x}$ *and* $\overline{y}$ *are defined in* (7).

Inserting $c = \overline{x}$ *in* (8) *we get*

$$\varphi(\overline{x}) - \varphi(0) + \frac{1}{\lambda(\beta) - \lambda(\alpha)} \int_{\alpha}^{\beta} \varphi(|f(t) - \overline{x}|)\, \mathrm{d}\lambda(t) \leq \overline{y}.$$

(b) *If in addition* $\varphi(0) \leq 0$, *the function* φ *is superquadratic and*

$$\overline{y} \geq \varphi(c) + \varphi'(c)(\overline{x} - c) + \frac{1}{\lambda(\beta) - \lambda(\alpha)} \int_{\alpha}^{\beta} \varphi(|f(t) - c|)\, \mathrm{d}\lambda(t). \quad (9)$$

Inserting $c = \overline{x}$ *in* (9) *we get*

$$\overline{y} \geq \varphi(\overline{x}) + \frac{1}{\lambda(\beta) - \lambda(\alpha)} \int_{\alpha}^{\beta} \varphi(|f(t) - c|)\, \mathrm{d}\lambda(t).$$

(c) *If in addition* $\varphi \geq 0$ *and* $\varphi(0) = \varphi'(0) = 0$, *the function* φ *is superquadratic and convex increasing and*

$$\overline{y} - \varphi(c) - \varphi'(c)(\overline{x} - c) \geq \frac{1}{\lambda(\beta) - \lambda(\alpha)} \int_{\alpha}^{\beta} \varphi(|f(t) - c|)\, \mathrm{d}\lambda(t) \geq 0.$$

2.4 Jensen's Type Inequalities for $K(x) = x^{\gamma}\varphi(x)$, Where φ Is Superquadratic

We prove now some inequalities that hold when the given function $K(x)$ satisfies $K(x) = x^{\gamma}\varphi(x)$, $\gamma \in \mathbb{R}_{+}$, where φ is a superquadratic function. These inequalities include and generalize the results related to superquadratic function $\varphi : [0, b) \to R$.

Lemma 2.1 ([4]). *Let $K(x) = x^{\gamma}\varphi(x)$, $\gamma \in \mathbb{R}_{+}$, where $\varphi(x)$ is superquadratic on $[0, b)$. Then*

$$K(y) - K(x) \geq \varphi(x)(y^{\gamma} - x^{\gamma}) + C_{\varphi}(x) y^{\gamma}(y - x) + y^{\gamma}\varphi(|y - x|), \quad (10)$$

holds for $x \in [0, b)$, $y \in [0, b)$. Moreover,

$$\begin{aligned}\sum_{i=1}^{N} a_i K(y_i) - K\left(\sum_{i=1}^{N} a_i y_i\right) \geq{}& \varphi\left(\sum_{j=1}^{N} a_j y_j\right)\left(\sum_{i=1}^{N} a_i y_i^{\gamma} - \left(\sum_{j=1}^{N} a_j y_j\right)^{\gamma}\right) \\ &+ C_{\varphi}\left(\sum_{j=1}^{N} a_j y_j\right)\sum_{i=1}^{N} a_i y_i^{\gamma}\left(y_i - \sum_{j=1}^{N} a_j y_j\right) \\ &+ \sum_{i=1}^{N} a_i y_i^{\gamma}\varphi\left(\left|y_i - \sum_{j=1}^{N} a_j y_j\right|\right) \quad (11)\end{aligned}$$

holds for $x_i \in [0, b)$, $y_i \in [0, b)$, $0 \leq a_i \leq 1$, $i = 1, \ldots, n$, and $\sum_{i=1}^{N} a_i = 1$; and

$$\begin{aligned}&\int_{\Omega} K(f(s))\, d\mu(s) - K\left(\int_{\Omega} f(s)\, d\mu(s)\right) \\ &\geq \int_{\Omega} [\varphi(x)(f^{\gamma}(s) - x^{\gamma}) + C_{\varphi}(x) f^{\gamma}(s)(f(s) - x) \\ &\quad + f^{\gamma}(s)\varphi(|f(s) - x|)]\, d\mu(s) \quad (12)\end{aligned}$$

holds, where C_{φ} is as in (1), *f is any nonnegative μ-integrable function on the probability measure space (Ω, μ) and $x = \int_{\Omega} f(s)\, d\mu(s)$.*

If φ is subquadratic, then the reverse inequality of (10), (11), *and* (12) *hold, in particular*

$$\int_{\Omega} K(f(s))\,\mathrm{d}\mu(s) - K\left(\int_{\Omega} f(s)\,\mathrm{d}\mu(s)\right)$$
$$\leq \int_{\Omega} \left[\varphi(x)\left(f^{\gamma}(s) - x^{\gamma}\right) + C_{\varphi}(x) f^{\gamma}(s)(f(s) - x)\right.$$
$$\left. + f^{\gamma}(s)\varphi(|f(s) - x|)\right] \mathrm{d}\mu(s). \tag{13}$$

Inequalities (10), (11), *and* (12) *are satisfied in particular by* $K(x) = x^p$, $p \geq \gamma + 2$. *For* $\gamma < p \leq \gamma + 2$, $\gamma \in \mathbb{R}_+$ *the reverse inequalities hold. They reduce to equalities for* $p = \gamma + 2$.

The proof is obtained by multiplying (1) by y^{γ}, and by simple manipulations we get that $K(x) = x^{\gamma}\varphi(x)$ satisfies (10) when φ is superquadratic.

By fixing in (10) a probability measure μ and a nonnegative integrable function f, setting $x = \int_{\Omega} f\,\mathrm{d}\mu$ and $C_{\varphi}(x)$ is as in the definition of superquadracity, we obtain for $K(x) = x^{\gamma}\varphi(x)$, $\gamma \in \mathbb{R}_+$, where $\varphi(x)$ is superquadratic, that (12) holds.

Inequality (11) is the discrete case of (12).

Similarly, since $-\varphi$ is superquadratic, inequality (13) and the reverse inequalities of (10) and (11) are obtained for subquadratic functions.

Lemma 2.2 ([4]). *Let* $K(x) = x^{\gamma}\varphi(x) = x^{\gamma-1}\psi(x)$, $\gamma \geq 1$, *where* φ *is a differentiable positive superquadratic function and* $\psi(x) = x\varphi(x)$. *Then the bound obtained for* $K(x) = x^{\gamma}\varphi(x)$ *is stronger than the bound obtained for* $K(x) = x^{\gamma-1}\psi(x)$, *that is:*

$$K(y) - K(x) \geq \varphi(x)\left(y^{\gamma} - x^{\gamma}\right) + \varphi'(x)\, y^{\gamma}(y - x) + y^{\gamma}\varphi(|y - x|)$$

implies that

$$K(y) - K(x) \geq \psi(x)\left(y^{\gamma-1} - x^{\gamma-1}\right) + \psi'(x)\, y^{\gamma-1}(y - x) + y^{\gamma-1}\psi(|y - x|). \tag{14}$$

Moreover, if $K(x) = x^n\varphi(x)$, $\psi_k(x) = x^k\varphi(x)$, n *is an integer,* $k = 1, 2, \ldots, n$, *and* $\varphi(x)$ *is nonnegative superquadratic, then the inequalities*

$$\int_{\Omega} K(f(s))\,\mathrm{d}\mu(s) - K\left(\int_{\Omega} f(s)\,\mathrm{d}\mu(s)\right)$$
$$\geq \int_{\Omega} \left[\varphi(x)\left(f^{n}(s) - x^{n}\right) + C_{\varphi}(x) f^{n}(s)(f(s) - x)\right.$$
$$\left. + f^{n}(s)\varphi(|f(s) - x|)\right] \mathrm{d}\mu(s)$$
$$\geq \int_{\Omega} \left[\psi_k(x)\left(f^{n-k}(s) - x^{n-k}\right) + C_{\psi_k}(x) f^{n-k}(s)(f(s) - x)\right.$$
$$\left. + f^{n-k}(s)\psi_k(|f(s) - x|)\right] \mathrm{d}\mu(s)$$
$$\geq \int_{\Omega} \psi_n(|f(s) - x|)\,\mathrm{d}\mu(s) \geq 0 \tag{15}$$

hold for all probability measure spaces (Ω, μ) *of* μ*-integrable nonnegative functions* f*, where* $x = \int_\Omega f(s)\, d\mu(s)$.

Furthermore, if $\varphi(x)$ *is positive, increasing, convex, subquadratic, and* $\varphi(0) = \varphi'(0) = 0$*, then* $x\varphi(x)$ *is superquadratic and*

$$\begin{aligned}
&\int_\Omega \left[\varphi(x)(f^n(s) - x^n) + C_\varphi(x) f^n(s)(f(s) - x) + f^n(s)\varphi(|f(s) - x|)\right] d\mu(s) \\
&\quad \geq \int_\Omega K(f(s))\, d\mu(s) - K\left(\int_\Omega f(s)\, d\mu(s)\right) \\
&\quad \geq \int_\Omega \left[\psi_k(x)\left(f^{n-k}(s) - x^{n-k}\right) + C_{\psi_k}(x) f^{n-k}(s)(f(s) - x)\right. \\
&\quad\quad \left. + f^{n-k}(s)\psi_k(|f(s) - x|)\right] d\mu(s) \\
&\quad \geq \int_\Omega \psi_n(|f(s) - x|)\, d\mu(s) \geq 0; \quad k = 1, \ldots, n.
\end{aligned} \tag{16}$$

In particular, if $\varphi(x) = x^p$, $x \geq 0$, $p \geq 1$, *then* (15) *is satisfied when* $p \geq 2$ *and* (16) *is satisfied when* $1 \leq p \leq 2$. *When* $p = 2$ *equality holds in the first inequality of* (15) *and in the first inequality of* (16).

In the proof we show that when φ is differentiable, nonnegative superquadratic or φ is differentiable, positive increasing, convex, and $\varphi(0) = \varphi'(0) = 0$, where $\psi(x) = x\varphi(x)$, then as according to Lemma 1.2, $C_\varphi(x) = \varphi'(x)$, we show that

$$\begin{aligned}
&\varphi(x)(y^\gamma - x^\gamma) + \varphi'(x) y^\gamma (y - x) + y^\gamma \varphi(|y - x|) \\
&\quad \geq \psi(x)\left(y^{\gamma-1} - x^{\gamma-1}\right) + \psi'(x) y^{\gamma-1}(y - x) + y^{\gamma-1}\psi(|y - x|)
\end{aligned}$$

leads to the results stated in the lemma.

As φ is positive superquadratic and therefore convex, this lemma gives a refinement of Jensen's inequality for convex functions.

2.5 Normalized Jensen Functional

In [3] the authors consider the normalized Jensen functional

$$J_n(f, \mathbf{x}, \mathbf{p}) = \sum_{i=1}^n p_i f(x_i) - f\left(\sum_{i=1}^n p_i x_i\right),$$

where $\sum_{i=1}^n p_i = 1$, $f : I \to \mathbb{R}$ and I is an interval in $\mathbb{R}$.

We quote here only one of the theorems there and then its generalization:

Theorem 2.5. *Let* $\mathbf{x} = (x_1, \ldots, x_n) \in I^n$, $\mathbf{p} = (p_1, \ldots, p_n)$, $\mathbf{q} = (q_1, \ldots, q_n)$ *be nonnegative n-tuples satisfying* $\sum_{i=1}^{n} p_i = 1$, $\sum_{i=1}^{n} q_i = 1$, $q_i > 0$, $i = 1, \ldots, n$. *Let*

$$m = \min_{1 \le i \le n} \left(\frac{p_i}{q_i} \right), \quad M = \max_{1 \le i \le n} \left(\frac{p_i}{q_i} \right).$$

If I *is* $[0, a)$ *or* $[0, \infty)$ *and* φ *is a superquadratic function on* I, *then*

$$\begin{aligned} & J_n(\varphi, \mathbf{x}, \mathbf{p}) - m J_n(\varphi, \mathbf{x}, \mathbf{q}) \\ & \quad \ge m\varphi\left(\left| \sum_{i=1}^{n} (q_i - p_i) x_i \right| \right) + \sum_{i=1}^{n} (p_i - m q_i) \varphi\left(\left| x_i - \sum_{j=1}^{n} p_j x_j \right| \right) \end{aligned}$$

and

$$\begin{aligned} & J_n(\varphi, \mathbf{x}, \mathbf{p}) - M J_n(\varphi, \mathbf{x}, \mathbf{q}) \\ & \quad \le -\sum_{i=1}^{n} (M q_i - p_i) \varphi\left(\left| x_i - \sum_{j=1}^{n} q_j x_j \right| \right) - \varphi\left(\left| \sum_{i=1}^{n} (p_i - q_i) x_i \right| \right). \end{aligned}$$

We state another generalization of Jensen's inequality for superquadratic functions and apply it to a generalization of a normalized Jensen's functional.

Theorem 2.6 ([3]). *Assume that* $x = (x_1, \ldots, x_n)$ *with* $x_i \ge 0$ *for* $i \in \{1, \ldots, n\}$, $p = (p_1, \ldots, p_n)$ *is a probability sequence and* $q = (q_1, \ldots, q_k)$ *is another probability sequence with* $n, k \ge 2$. *Then for any superquadratic function* $\varphi : [0, a) \to \mathbb{R}$ *we have the inequality*

$$\begin{aligned} & \sum_{i_1, \ldots, i_k = 1}^{n} p_{i_1} \cdots p_{i_k} \varphi\left(\sum_{j=1}^{k} q_j x_{i_j} \right) \\ & \quad \ge \varphi\left(\sum_{i=1}^{n} p_i x_i \right) + \sum_{i_1, \ldots, i_k = 1}^{n} p_{i_1} \cdots p_{i_k} \varphi\left(\left| \sum_{j=1}^{k} q_j x_{i_j} - \sum_{i=1}^{n} p_i x_i \right| \right). \end{aligned}$$

From this we get for superquadratic functions:

Theorem 2.7. *Let* $\mathbf{p}$, $\mathbf{q}$, $\mathbf{r}$ *be probability measures and* $r_i > 0$, $i = 1, \ldots, n$, *and let* φ *be a superquadratic function, then*

$$\sum_{i_1,\dots,i_k=1}^{n} p_{i_1}\cdots p_{i_k}\varphi\left(\sum_{j=1}^{k} q_j x_{i_j}\right)-\varphi\left(\sum_{i=1}^{n} p_i x_i\right)$$
$$-m\left(\sum r_{i_1}\cdots r_{i_k}\varphi\left(\sum_{j=1}^{k} q_j x_{i_j}\right)-\varphi\left(\sum_{i=1}^{n} r_i x_i\right)\right)$$
$$\geq m\varphi\left(\left|\sum_{i=1}^{n}(r_i-p_i)x_i\right|\right)$$
$$+\sum_{i_1,\dots,i_k=1}^{n}\left(p_{i_1}p_{i_2}\cdots p_{i_k}-mr_{i_1}\cdots r_{i_k}\right)\varphi\left(\left|\sum_{j=1}^{k} q_j x_{i_j}-\sum_{s=1}^{n} p_s x_s\right|\right)$$

and

$$\sum_{i_1,\dots,i_k=1}^{n} p_{i_1}\cdots p_{i_k}\varphi\left(\sum_{j=1}^{k} q_j x_{i_j}\right)-\varphi\left(\sum_{i=1}^{n} p_i x_i\right)$$
$$-M\left(\sum r_{i_1}\cdots r_{i_k}\varphi\left(\sum_{j=1}^{k} q_j x_{i_j}\right)-\varphi\left(\sum_{i=1}^{n} r_i x_i\right)\right)$$
$$\leq -\varphi\left(\left|\sum_{i=1}^{n}(r_i-p_i)x_i\right|\right)$$
$$-\sum_{i_1,\dots,i_k=1}^{n}\left(p_{i_1}p_{i_2}\cdots p_{i_k}-Mr_{i_1}\cdots r_{i_k}\right)\varphi\left(\left|\sum_{j=1}^{k} q_j x_{i_j}-\sum_{j=1}^{n} r_s x_s\right|\right),$$

where

$$m:=\min_{1\leq i_1,\dots,i_k\leq n}\left(\frac{p_{i_1}\cdots p_{i_k}}{r_{i_1}\cdots r_{i_k}}\right),\quad M:=\max_{1\leq i_1,\dots,i_k\leq n}\left(\frac{p_{i_1}\cdots p_{i_k}}{r_{i_1}\cdots r_{i_k}}\right).$$

In [29, 30] the author F. G. Mitroi extends these two results even more by introducing a more general functional.

2.6 Jensen's Inequality and Isotonic Linear Functionals

Jensen's inequality for superquadratic functions, like many other inequalities for that class of functions, can be generalized for isotonic linear functionals. First we define isotonic linear functionals as in [25].

Let E be a nonempty set and L be a linear class of real-valued function $f : E \to \mathbb{R}$ having the properties:

L1: $f, g \in L \Rightarrow (\alpha f + \beta g) \in L$ for all $\alpha, \beta \in \mathbb{R}$;
L2: $\mathbf{1} \in L$, i.e., if $f(t) = 1$ for $t \in E$, then $f \in L$.

Let $A : L \to \mathbb{R}$ be a functional with properties:
A1: $A(\alpha f + \beta g) = \alpha A(f) + \beta A(g)$ for $f, g \in L$, $\alpha, \beta \in \mathbb{R}$ (A is linear);
A2: $f \in L$, $f(t) \geq 0$ on $E \Rightarrow A(f) \geq 0$ (A is isotonic).

Furthermore, if the functional A has a property
A3: $A(\mathbf{1}) = 1$, where $(\mathbf{1}) = 1$ for all $t \in E$, then we say that A is normalized.

For this functionals Banić and Varošanec proved in [25] that

Theorem 2.8. *Let L satisfy conditions* L1, L2 *and A satisfy conditions* A1, A2 *on a nonempty set E. Suppose that $k \in L$ with $k \geq 0$ and $A(k) > 0$ and that $\varphi : [0, \infty) \to \mathbb{R}$ is a continuous superquadratic function. Then for all nonnegative $f \in L$ such that $kf, k\varphi(f), k\varphi\left(\left|f - \frac{A(kf)}{A(k)} \cdot \mathbf{1}\right|\right) \in L$ we have*

$$\varphi\left(\frac{A(kf)}{A(k)}\right) \leq \frac{A(k\varphi(f)) - A\left(k\varphi\left(\left|f - \frac{A(kf)}{A(k)} \cdot \mathbf{1}\right|\right)\right)}{A(k)}. \tag{17}$$

If φ is a subquadratic function, then a reversed inequality in (17) *holds.*

By suitable choice of the functions f and k in the above theorem, the authors get a refinement of the functional Hölder's inequality.

Theorem 2.9. *Let L satisfy conditions* L1, L2, *and A satisfy conditions* A1, A2 *on a nonempty set E. Let $p \geq 2$ and define q by $\frac{1}{p} + \frac{1}{q} = 1$. Then for nonnegative functions $g, h \in L$ such that gh, g^p, h^q, $\left|g - h^{q-1}\frac{A(gh)}{A(h^q)}\right|^p \in L$ and $A(h^q) > 0$ the following inequality*

$$A(gh) \leq \left[Ag^p - A\left(\left|g - h^{q-1}\frac{A(gh)}{A(h^q)}\right|^p\right)\right]^{\frac{1}{p}} A^{\frac{1}{q}}(h^q) \tag{18}$$

holds. In case $0 < p < 2$ the inequality in (18) *is reversed.*

Similarly they obtain a functional Minkowski's inequality for superquadratic functions.

Theorem 2.10. *Let L and A be as in the previous theorem. If $p \geq 2$, then for all nonnegative functions g, h, on E such that $(g+h)^p$, g^p, $h^q \in L$, we have*

$$A^{\frac{1}{p}}\left((g+h)^p\right) \le \left(A\left(g^p\right) - A\left(\left|g-(g+h)\frac{A\left(g\left(g+h\right)^{p-1}\right)}{A\left(g+h\right)^p}\right|^p\right)\right)^{\frac{1}{p}}$$

$$+\left(A\left(h^p\right) - A\left(\left|h-(g+h)\frac{A\left(h\left(g+h\right)^{p-1}\right)}{A\left(g+h\right)^p}\right|^p\right)\right)^{\frac{1}{p}}.$$

3 Superquadratic Functions and Hardy's Inequality

3.1 Refinements of Hardy's Type Inequalities

A very interesting application of the properties of superquadratic functions to get Hardy's type inequality was dealt by Oguntuase and Persson in [31]. One of Hardy's inequalities they considered is

$$\int_0^\infty x^{-k}\left(\int_0^x f(t)\,\mathrm{d}t\right)^p \mathrm{d}x \le \left(\frac{p}{k-1}\right)^p \int_0^\infty x^{p-k} f^p(x)\,\mathrm{d}x, \quad k>1, \quad p\ge 1.$$

Their refinements is as follows:

Theorem 3.1. *Let $p>1$, $k>1$, $0<b\le\infty$, and let the function f be locally integrable on $(0,b)$ such that $0<\int_0^b x^{p-k} f^p(x)\,\mathrm{d}x<\infty$.*

(i) *If $p>2$, then*

$$\int_0^b x^{-k}\left(\int_0^x f(t)\,\mathrm{d}t\right)^p \mathrm{d}x$$

$$+\frac{k-1}{p}\int_0^b\int_t^b\left|\frac{p}{k-1}\left(\frac{t}{x}\right)^{1-\frac{k-1}{p}} f(t) - \frac{1}{x}\int_0^x f(t)\,\mathrm{d}t\right|^p x^{p-k-\frac{k-1}{p}}\,\mathrm{d}x\, t^{\frac{k-1}{p}-1}\,\mathrm{d}t$$

$$\le \left(\frac{p}{k-1}\right)^p \int_0^b \left(1-\left(\frac{x}{b}\right)^{\frac{k-1}{p}}\right) x^{p-k} f^p(x)\,\mathrm{d}x. \tag{19}$$

(ii) *If $1<p\le 2$, then the inequality holds in the reversed direction.*

Now we get new scales of refined Hardy type inequalities by using the ideas and techniques of [31] and implement them on the functions $K(x)=x^\gamma\varphi(x)$, $\gamma\in\mathbb{R}_+$, where $\varphi(x)$ is superquadratic, by using Jensen type inequalities which these functions satisfy as expressed in Sect. 2.

From the definition of superquadracity we get in [4] that

Theorem 3.2. *For $\varphi(x)=x^p$, $p\ge 2$ (therefore $C_\varphi(x)=\varphi'(x)=px^{p-1}$), $\gamma\in\mathbb{R}_+$, we find that*

$$
\begin{aligned}
&\int_0^b \left(1-\frac{x}{b}\right) f^{p+\gamma}(x)\,\frac{\mathrm{d}x}{x} - \int_0^b \left(\frac{1}{x}\int_0^x f(t)\,\mathrm{d}t\right)^{p+\gamma} \frac{\mathrm{d}x}{x} \\
&\geq \int_0^b \int_t^b \left(f^{\gamma}(t) - \left(\frac{1}{x}\int_0^x f(\tau)\,\mathrm{d}\tau\right)^{\gamma}\right)\left(\frac{1}{x}\int_0^x f(\tau)\,\mathrm{d}\tau\right)^{p} \frac{\mathrm{d}x}{x^2}\,\mathrm{d}t \\
&\quad + \int_0^b \int_t^b f^{\gamma}(t)\left(f(t) - \frac{1}{x}\int_0^x f(\tau)\,\mathrm{d}\tau\right) p \left(\frac{1}{x}\int_0^x f(\tau)\,\mathrm{d}\tau\right)^{p-1} \frac{\mathrm{d}x}{x^2}\,\mathrm{d}t \\
&\quad + \int_0^b \int_t^b f^{\gamma}(t)\left(\left|f(t) - \frac{1}{x}\int_0^x f(\tau)\,\mathrm{d}\tau\right|\right)^{p} \frac{\mathrm{d}x}{x^2}\,\mathrm{d}t, \qquad (20)
\end{aligned}
$$

where f is nonnegative and locally integrable on $[0, b)$. The reverse inequality holds when $1 < p \leq 2$.

By using (20) we are now ready to derive our new scales of refined Hardy type inequalities.

Theorem 3.3. *Let $p \geq 2$, $k > 1$, $0 < b \leq \infty$, and $\gamma \in \mathbb{R}_+$, and let the function f be nonnegative and locally integrable on $(0, b)$. Then*

$$
\begin{aligned}
&\left(\frac{p+\gamma}{k-1}\right)^{p+\gamma} \int_0^b \left(1-\left(\frac{x}{b}\right)^{\frac{k-1}{p+\gamma}}\right) x^{p+\gamma-k} f^{p+\gamma}(x)\,\mathrm{d}x \\
&\quad - \int_0^b x^{-k}\left(\int_0^x f(t)\,\mathrm{d}t\right)^{p+\gamma} \mathrm{d}x \\
&\geq \left(\frac{k-1}{p+\gamma}\right) \int_0^b \int_t^b \left(\left(f(t)\,\frac{p+\gamma}{k-1}\left(\frac{t}{x}\right)^{1-\frac{k-1}{p+\gamma}}\right)^{\gamma} - \left(\frac{1}{x}\int_0^x f(\sigma)\,\mathrm{d}\sigma\right)^{\gamma}\right) \\
&\quad \times \left(\frac{1}{x}\int_0^x f(\sigma)\,\mathrm{d}\sigma\right)^{p} x^{\left(1-\frac{k-1}{p+\gamma}\right)(p+\gamma-1)-2}\, t^{\frac{k-1}{p+\gamma}-1}\,\mathrm{d}x\,\mathrm{d}t \\
&\quad + \left(\frac{k-1}{p+\gamma}\right)^{1-\gamma} \int_0^b \int_t^b \left(f(t)\,t^{1-\frac{k-1}{p+\gamma}}\right)^{\gamma} \\
&\quad \times \left(f(t)\,\frac{p+\gamma}{k-1}\left(\frac{t}{x}\right)^{1-\frac{k-1}{p+\gamma}} - \frac{1}{x}\int_0^x f(\sigma)\,\mathrm{d}\sigma\right) \\
&\quad \times p\left(\frac{1}{x}\int_0^x f(\sigma)\,\mathrm{d}\sigma\right)^{p-1} x^{\left(1-\frac{k-1}{p+\gamma}\right)(p+1)-2}\, t^{\frac{k-1}{p+\gamma}-1}\,\mathrm{d}x\,\mathrm{d}t \\
&\quad + \left(\frac{k-1}{p+\gamma}\right)^{1-\gamma} \int_0^b \int_t^b \left(f(t)\,t^{1-\frac{k-1}{p+\gamma}}\right)^{\gamma}
\end{aligned}
$$

$$\times \left| f(t) \frac{p+\gamma}{k-1} \left(\frac{t}{x}\right)^{1-\frac{k-1}{p+\gamma}} - \frac{1}{x} \int_0^x f(\sigma)\, d\sigma \right|^p$$

$$\times\, x^{\left(1-\frac{k-1}{p+\gamma}\right)(p+1)-2} t^{\frac{k-1}{p+\gamma}-1}\, dx\, dt.$$

Moreover, if γ is a nonnegative integer, then the right-hand side of (21) *is nonnegative. If $1 < p \le 2$, then inequality* (21) *is reversed. Equality holds when $p = 2$. When $\gamma = 0$, inequality* (21) *coincides with* (19).

3.2 Refined Hardy Type Inequalities with General Kernels and Measures

The following results were proved by Oguntuase et al. [33]:

Proposition 3.1. *Let $\mathbf{b} \in (0,\infty)^n$, $u : (\mathbf{0},\mathbf{b}) \to \mathbb{R}$ is a weight function which is locally integrable in $(\mathbf{0},\mathbf{b})$ and v is defined by*

$$v(t) = t_1 \cdots t_n \int_{t_1}^{b_1} \cdots \int_{t_n}^{b_n} \frac{u(\mathbf{x})}{x_1^2 \cdots x_n^2}\, d\mathbf{x}, \quad t \in (0,b).$$

Suppose $I = (a,c)$, $0 \le a < c \le \infty$, $\varphi : I \to \mathbb{R}$, and $f : (\mathbf{0},\mathbf{b}) \to \mathbb{R}$ is an integrable function, such that $f(\mathbf{x}) \in I$, for all $\mathbf{x} \in (\mathbf{0},\mathbf{b})$.

(i) *If φ is superquadratic, then the following inequality holds:*

$$\int_{t_1}^{b_1} \cdots \int_{t_n}^{b_n} u(\mathbf{x}) \varphi\left(\frac{1}{x_1 \cdots x_n}\right) \int_0^{x_1} \cdots \int_0^{x_n} f(\mathbf{t})\, d\mathbf{t} \frac{d\mathbf{x}}{x_1 \cdots x_n}$$

$$+ \int_{t_1}^{b_1} \cdots \int_{t_n}^{b_n} \int_{t_1}^{b_1} \cdots \int_{t_n}^{b_n} \varphi\left(\left| f(\mathbf{t}) - \frac{1}{x_1 \cdots x_n} \int_0^{x_1} \cdots \int_0^{x_n} f(\mathbf{t})\, d\mathbf{t} \right|\right) \frac{u(\mathbf{x})}{x_1^2 \cdots x_n^2}\, d\mathbf{x}\, d\mathbf{t}$$

$$\le \int_0^{b_1} \cdots \int_0^{b_n} v(\mathbf{x}) \varphi(f(\mathbf{x})) \frac{d\mathbf{x}}{x_1 \cdots x_n}. \tag{21}$$

(ii) *If φ is subquadratic, then* (21) *holds in the reversed direction.*

Proposition 3.2. *Let $\mathbf{b} \in [\mathbf{0},\infty)$, $u : (\mathbf{b},\infty) \to \mathbb{R}$ be a weight function which is locally integrable in $(\mathbf{0},\mathbf{b})$ and define v by*

$$v(\mathbf{t}) = \frac{1}{t_1 \cdots t_n} \int_{b_1}^{t_1} \cdots \int_{b_n}^{t_n} u(\mathbf{x})\, d\mathbf{x} < \infty, \quad \mathbf{t} \in (\mathbf{b},\infty).$$

Suppose $I = (a,c)$, $0 \le a < c \le \infty$, $\varphi : I \to \mathbb{R}$, and $f : (\mathbf{b},\infty) \to \mathbb{R}$ is an integrable function, such that $f(\mathbf{x}) \in I$, for all $\mathbf{x} \in (\mathbf{b},\infty)$.

(i) *If φ is superquadratic, then the following inequality holds:*

$$\int_{b_1}^{\infty}\cdots\int_{b_n}^{\infty} u(\mathbf{x})\varphi\left(x_1\cdots x_n\int_{x_1}^{\infty}\cdots\int_{x_n}^{\infty} f(\mathbf{t})\frac{\mathrm{d}\mathbf{t}}{t_1^2\cdots t_n^2}\right)\frac{\mathrm{d}\mathbf{x}}{x_1\cdots x_n}$$

$$+\int_{b_1}^{\infty}\cdots\int_{b_n}^{\infty}\int_{b_1}^{t_1}\cdots\int_{b_n}^{t_n}\varphi\left(\left|f(\mathbf{t})-x_1\cdots x_n\int_{x_1}^{\infty}\cdots\int_{x_n}^{\infty} f(\mathbf{t})\frac{\mathrm{d}\mathbf{t}}{t_1^2\cdots t_n^2}\right|\right)$$

$$\times\, u(\mathbf{x})\,\mathrm{d}\mathbf{x}\frac{\mathrm{d}\mathbf{t}}{t_1^2\cdots t_n^2}$$

$$\leq \int_{b_1}^{\infty}\cdots\int_{b_n}^{\infty} v(\mathbf{x})\varphi(f(\mathbf{x}))\frac{\mathrm{d}\mathbf{x}}{x_1\cdots x_n}. \tag{22}$$

(ii) *If φ is subquadratic, then the inequality sign in* (22) *is reversed.*

In [19] the authors extended the above propositions as follows:

Let $(\Omega_1,\Sigma_1,\mu_1)$, $(\Omega_2,\Sigma_2,\mu_2)$ be measure spaces and let A_k be defined as follows:

$$A_k f(x) := \frac{1}{K(x)}\int_{\Omega_2} k(x,y) f(y)\,\mathrm{d}\mu_2(y), \tag{23}$$

where $f : \Omega_2 \to \mathbb{R}$ is measurable, $k : \Omega_1\times\Omega_2 \to \mathbb{R}$ is a measurable and nonnegative kernel and

$$K(x) := \int_{\Omega_2} k(x,y)\,\mathrm{d}\mu_2(y) < \infty, \qquad x\in\Omega_1. \tag{24}$$

Their result reads:

Theorem 3.4. *Let u be a weight function, $k(x,y)\geq 0$. Assume that $\frac{k(x,y)}{K(x)}u(x)$ is locally integrable on Ω_1 for each fixed $y\in\Omega_2$. Define v by*

$$v(y) := \int_{\Omega_1}\frac{k(x,y)}{K(x)}u(x)\,\mathrm{d}\mu_1(x) < \infty.$$

Suppose $I=(0,c)$, $c\leq\infty$, $\varphi : I\to\mathbb{R}$. If φ is a superquadratic function, then the inequality

$$\int_{\Omega 1}\varphi(A_k f(x))u(x)\,\mathrm{d}\mu_1(x)$$

$$+\int_{\Omega_2}\int_{\Omega_1}\varphi\left(|f(y)-A_k f(x)|\right)\frac{u(x)k(x,y)}{K(x)}\,\mathrm{d}\mu_1(x)\,\mathrm{d}\mu_2(y)$$

$$\leq \int_{\Omega_2}\varphi(f(y))v(y)\,\mathrm{d}\mu_2(y) \tag{25}$$

holds for all measurable functions $f : \Omega_2 \to \mathbb{R}$*, such that* $Imf \subseteq I$*, where* A_k *is defined by* (23) *and* (24) *.*

If φ *is subquadratic, then the inequality sign in* (25) *is reversed.*

By using the above results with concrete kernels we can obtain a refinement of some classical inequalities. Here we only give the following complement and refinement of the following well-known inequality:

$$\int_0^\infty \left(\int_0^\infty \frac{f(x)}{x+y}\, \mathrm{d}x \right)^p \mathrm{d}y \le \left(\frac{\pi}{\sin\left(\frac{\pi}{p}\right)} \right)^p \int_0^\infty f^p(t)\, \mathrm{d}t.$$

Corollary 3.1. *Let* $p > 1$ *and* $f \in L^p(R_+)$*. If* $p \ge 2$*, then*

$$\int_0^\infty \left(\int_0^\infty \frac{f(x)}{x+y}\, dx \right)^p \mathrm{d}y$$

$$+ \left(\frac{\pi}{\sin\left(\frac{\pi}{p}\right)} \right)^{p-1} \int_0^\infty y^{-\frac{1}{p}} \int_0^\infty \left| f(y) y^{\frac{1}{p}} - \frac{\sin\left(\frac{\pi}{p}\right)}{\pi} x^{\frac{1}{p}} \int_0^\infty \frac{f(y)}{x+y}\, \mathrm{d}y \right|^p$$

$$\times \frac{x^{\frac{1}{p}-1}}{x+y}\, \mathrm{d}x\, \mathrm{d}y$$

$$\le \left(\frac{\pi}{\sin\left(\frac{\pi}{p}\right)} \right)^p \int_0^\infty f^p(y)\, \mathrm{d}y. \tag{26}$$

If $1 < p \le 2$*, then* (26) *holds in the reversed direction.*

4 More Inequalities Related to Superquadracity

4.1 Inequalities for Averages

For a function f let

$$A_n(f) = \frac{1}{n-1} \sum_{r=1}^{n-1} f\left(\frac{r}{n}\right) \quad (n \ge 2)$$

and

$$B_n(f) = \frac{1}{n+1} \sum_{r=0}^{n} f\left(\frac{r}{n}\right) \quad (n \ge 1).$$

For these $A_n(f)$ and $B_n(f)$ we get in [6] the following theorems:

Theorem 4.1. *If f is superquadratic on* $[0,1]$, *then for* $n \geq 2$

$$A_{n+1}(f) - A_n(f) \geq \sum_{r=1}^{n-1} \lambda_r f(x_r),$$

where

$$\lambda_r = \frac{2r}{n(n-1)}, \quad x_r = \frac{n-r}{n(n+1)}.$$

Further,

$$A_{n+1}(f) - A_n(f) \geq f\left(\frac{1}{3n}\right) + \sum_{r=1}^{n-1} \lambda_r f(y_r),$$

where

$$y_r = \frac{|2n-1-3r|}{3n(n+1)}.$$

If f is also nonnegative, then for $n \geq 3$,

$$A_{n+1}(f) - A_n(f) \geq f\left(\frac{1}{3n}\right) + f\left(\frac{16}{81(n+3)}\right).$$

The theorem is proved by using the identity

$$\begin{aligned}\Delta_n &= \frac{n-1}{n}\sum_{r=1}^{n} f\left(\frac{r}{n+1}\right) - \sum_{r=1}^{n-1} f\left(\frac{r}{n}\right)\\ &= \sum_{r=1}^{n-1}\frac{r}{n}\left[f\left(\frac{r+1}{n+1}\right) - f\left(\frac{r}{n}\right)\right] + \sum_{r=1}^{n-1}\frac{n-r}{n}\left[f\left(\frac{r}{n+1}\right) - f\left(\frac{r}{n}\right)\right]\end{aligned}$$

and by using the basic Jensen's inequality (2) for superquadratic functions, taking into account that

$$\frac{r+1}{n+1} - \frac{r}{n} = \frac{n-r}{n(n+1)}.$$

Theorem 4.2. *If f is superquadratic on* $[0,1]$, *then for* $n \geq 2$,

$$B_{n-1}(f) - B_n(f) \geq \sum_{r=1}^{n} \lambda_r f(x_r),$$

where

$$\lambda_r = \frac{2r}{n(n-1)}, \quad x_r = \frac{n-r}{n(n+1)}.$$

Further

$$B_{n-1}(f) - B_n(f) \geq f\left(\frac{1}{3n}\right) + \sum_{r=1}^{n} \lambda_r f(y_r),$$

where $y_r = |2n+1-3r| / (3n(n-1))$ *the opposite inequalities hold if* f *is subquadratic.*

If f *is also nonnegative, then for* $n \geq 2$,

$$B_{n-1}(f) - B_n(f) \geq f\left(\frac{1}{3n}\right) + f\left(\frac{16}{81n}\right).$$

The proof uses the identity

$$\begin{aligned}\Delta_n &= (n+1)\left[B_{n-1}(f) - B_n(f)\right] \\ &= \sum_{r=1}^{n} \frac{r}{n}\left[f\left(\frac{r-1}{n-1}\right) - f\left(\frac{r}{n}\right)\right] + \sum_{r=0}^{n-1} \frac{n-r}{n}\left[f\left(\frac{r}{n-1}\right) - f\left(\frac{r}{n}\right)\right].\end{aligned}$$

We now formulate generalized versions of the earlier results in which $f(r/n)$ is replaced by $f(a_r/a_n)$ and $1/(n \pm 1)$ is replaced by $1/c_{n\pm 1}$, under suitable conditions on the sequences (a_n) and (c_n).

Theorem 4.3. *Let* $(a_n)_{n\geq 1}$ *and* $(c_n)_{n\geq 0}$ *be sequences such that* $a_n > 0$ *and* $c_n > 0$ *for* $n \geq 1$ *and*

(A1) $c_0 = 0$ *and* c_n *is increasing,*
(A2) $c_{n+1} - c_n$ *is decreasing for* $n \geq 0$,
(A3) $c_n(a_{n+1}/a_n - 1)$ *is decreasing for* $n \geq 1$.

Given a function f, *let*

$$A_n[f,(a_n),(c_n)] = A_n(f) = \frac{1}{c_{n-1}} \sum_{r=1}^{n-1} f\left(\frac{a_r}{a_n}\right).$$

Suppose that f *is superquadratic and nonnegative. Then*

$$\begin{aligned}A_{n+1}(f) - A_n(f) \geq{}& \frac{1}{c_n c_{n-1}} \sum_{r=1}^{n-1} c_r \left(\left|\frac{a_{r+1}}{a_{n+1}} - \frac{a_r}{a_n}\right|\right) \\ &+ \frac{1}{c_n c_{n-1}} \sum_{r=1}^{n-1} (c_n - c_r)\, f\left(\left|\frac{a_r}{a_n} - \frac{a_r}{a_{n+1}}\right|\right).\end{aligned}$$

Theorem 4.4. *Let $(a_n)_{n\geq 0}$ and $(c_n)_{n\geq 0}$ be sequences such that $a_n > 0$ and $c_n > 0$ for $n \geq 1$ and*

(B1) $c_0 = 0$ *and c_n is increasing,*
(B2) $c_n - c_{n-1}$ *is increasing for $n \geq 1$,*
(B3) $c_n (1 - a_{n-1}/a_n)$ *is increasing for $n \geq 1$,*
(B4) *either $a_0 = 0$ or (a_n) is increasing.*

Given a function f, let

$$B_n [f, (a_n), (c_n)] = B_n (f) = \frac{1}{c_{n+1}} \sum_{r=0}^{n} f\left(\frac{a_r}{a_n}\right), \quad n \geq 1.$$

Suppose that f is superquadratic and nonnegative. Then

$$B_{n-1}(f) - B_n(f) \geq \frac{1}{c_n c_{n+1}} \sum_{r=1}^{n-1} c_r f\left(\left|\frac{a_r}{a_n} - \frac{a_{r-1}}{a_{n-1}}\right|\right)$$
$$+ \frac{1}{c_n c_{n+1}} \sum_{r=1}^{n-1} (c_n - c_r) f\left(\left|\frac{a_r}{a_n - 1} - \frac{a_r}{a_n}\right|\right).$$

4.2 Bohr's Inequality

An inequality of Bohr states: *for any $z, w \in \mathbb{C}$ and for any $p, q > 1$ with $\frac{1}{p} + \frac{1}{q} = 1$*

$$|z + w|^2 \leq p\,|z|^2 + q\,|w|^2$$

holds with equality iff $w = (p-1)\,z$.

This inequality has many extensions and generalizations.

In [13] Bohr's theorem is extended for $z, w \in R_+$ by replacing the power 2 with powers $r \geq 2$ and with powers $1 \leq r \leq 2$:

Theorem 4.5. *For any $A, B \in \mathbb{R}_+$ and for any $p, q \in \mathbb{R}$ with $\frac{1}{p} + \frac{1}{q} = 1$,*

(i) *For $r \geq 2$ and $1 < p \leq 2$ $(q \geq 2)$ we get the inequality:*

$$p\,|A|^r + q\,|B|^r \geq \frac{1}{2^{r-2}}\,|(p-1)\,A + B|^r + \frac{1}{2^{r-2}}\,|B - A|^r \tag{27}$$

and for q, $1 < q \leq 2$ $(p \geq 2)$ we get that

$$p\,|A|^r + q\,|B|^r \geq \frac{1}{2^{r-2}}\,|A + (q-1)\,B|^r + \frac{1}{2^{r-2}}\,|B - A|^r \tag{28}$$

equality holds in (27) *and* (28) *when* $r > 2$ *if* $p = q = 2$ *and* $A = B$. *Moreover, if* $r = 2$ *equality holds in* (27) *and* (28) *if* $p = q = 2$, *which is the parallelogram law.*

(ii) *For* $1 \leq r \leq 2$, *if* $p \geq 2$ $(1 < q \leq 2)$, *we get that*

$$p\,|A|^r + q\,|B|^r \leq \frac{1}{2^{r-2}}\,|(p-1)\,A + B|^r + \frac{1}{2^{r-2}}\,|B - A|^r$$

and analogously if $1 < p \leq 2$ $(q \geq 2)$

$$p\,|A|^r + q\,|B|^r \leq \frac{1}{2^{r-2}}\,|A + (q-1)\,B|^r + \frac{1}{2^{r-2}}\,|B - A|^r\,.$$

For $p = q = 2$, $1 \leq r \leq 2$ *we get that*

$$|A + B|^r \leq 2^{r-1}\left(|A|^r + |B|^r\right) \leq |A + B|^r + |B - A|^r$$

which we may consider as an extension of the parallelogram law.

(iii) *For* $r = 2$ *if* $1 < p \leq 2$ $(q \geq 2)$ *we get that*

$$|(p-1)\,A+B|^2 + |B-aA|^2 \leq p\,|A|^2 + q\,|B|^2 \leq |A+(q-1)\,B|^2 + |B-A|^2\,,$$

and if $2 \leq p$ $(1 < q \leq 2)$

$$|A+(q-1)\,B|^2 + |B-A|^2 \leq p\,|A|^2 + q\,|B|^2 \leq |(p-1)\,A + B|^2 + |B - A|^2\,.$$

4.3 On Exponential Convexity and Cauchy's Means

New Cauchy type means related to superquadracity are dealt with in [17,20,21,23]. These means are obtained by applying the so-called exp-convex method established in [24,34]. Here we demonstrate only Cauchy means discussed in [21].

The definition of exponential convexity is:

Definition 4.1. A function $\varphi : (a,b) \to \mathbb{R}$ is said to be exponentially convex if it is continuous and

$$\sum_{i,j=1}^{m} u_i u_j \varphi(x_i + x_j) \geq 0$$

holds for all $m \in \mathbb{N}$ and all choices $u_i \in \mathbb{R}$, $i = 1, 2, \ldots, m$ and $x_i \in (a,b)$ such that $x_i + x_j \in (a,b)$, $1 \leq i, j \leq m$.

If $\varphi : (a,b) \to \mathbb{R}_+$ is an exponentially convex function, then φ is also log-convex, that is, $\log \varphi$ is convex.

It is easy to verify that if $\varphi : (a,b) \to \mathbb{R}_+$ is a log-convex function, then for any $x_1, x_2, y_1, y_2, \in (a,b)$ such that $x_1 \le y_1$, $x_2 \le y_2$, $x_1 \ne x_2$, $y_1 \ne y_2$ the following is valid:

$$\left(\frac{\varphi(x_2)}{\varphi(x_1)}\right)^{\frac{1}{x_2-x_1}} \le \left(\frac{\varphi(y_2)}{\varphi(y_1)}\right)^{\frac{1}{y_2-y_1}}.$$

Let L be a linear class of continuous functions $\varphi : [0,b) \to \mathbb{R}$. Let $f : [\alpha,\beta] \to (0,b)$ be continuous and monotonic and $\lambda : [\alpha,\beta] \to \mathbb{R}$ be either continuous or of bounded variation satisfying $\lambda(\alpha) \le \lambda(x) \le \lambda(\beta)$. We define the functional χ on L by

$$\chi_\varphi = \frac{1}{\lambda(\beta)-\lambda(\alpha)} \int_\alpha^\beta \left[\varphi(f(t)) - \varphi\left(\left|f(t)-\overline{f}\right|\right)\right] \mathrm{d}\lambda(t) - \varphi(\overline{f}),$$

where

$$\overline{f} = \frac{1}{\lambda(\beta)-\lambda(\alpha)} \int_\alpha^\beta f(t)\mathrm{d}\lambda(t).$$

In the discrete case we define the functional Δ on L by

$$\Delta_\varphi = \frac{1}{A_n} \sum_{i=1}^n a_i \left[\varphi(x_i) - \varphi(|x_i - \overline{x}|)\right] - \varphi(\overline{x}),$$

where $\mathbf{x} \in (0,b)^n$ is a monotonic $n-$tuple, $\mathbf{a}$ is a real $n-$tuple satisfying $0 \le A_j = \sum_{i=1}^{j} a_i \le A_n$, $j = 1,\dots,n$, $A_n > 0$ and $\overline{x} = \frac{1}{A_n} \sum_{i=1}^n a_i x_i$.

Let $s \in \mathbb{R}_+$. We define the superquadratic function $\psi_s : [0,b) \to \mathbb{R}$ by

$$\psi_s(x) = \begin{cases} \dfrac{sx\mathrm{e}^{sx} - \mathrm{e}^{sx} + 1}{s^3}, & s \ne 0, \\ \dfrac{1}{3}x^3, & s = 0. \end{cases}$$

Applying the functional χ to ψ_s we have

$$\chi_{\psi_s} = \begin{cases} \dfrac{1}{s^3}\left[\dfrac{1}{\lambda(\beta)-\lambda(\alpha)} \displaystyle\int_\alpha^\beta \left(\mathcal{R}_s(f(t)) - \mathcal{R}_s(|f(t)-\overline{f}|)\right) \mathrm{d}\lambda(t) - \mathcal{R}_s(\overline{f})\right], & s \ne 0, \\ \dfrac{1}{3}\left[\dfrac{1}{\lambda(\beta)-\lambda(\alpha)} \displaystyle\int_\alpha^\beta \left(f^3(t) - |f(t)-\overline{f}|^3\right) \mathrm{d}\lambda(t) - \overline{f}^3\right], & s = 0, \end{cases} \tag{29}$$

where we denote $\mathcal{R}_s(x) = sx\mathrm{e}^{sx} - \mathrm{e}^{sx} + 1$.

Analogously, applying the functional Δ to ψ_s, we have

$$\Delta_{\psi_s} = \begin{cases} \dfrac{1}{s^3}\left[\dfrac{1}{A_n}\sum_{i=1}^{n} a_i\left(\mathcal{R}_s(x_i) - \mathcal{R}_s(|x_i - \overline{x}|)\right) - \mathcal{R}_s(\overline{x})\right], & s \neq 0, \\ \dfrac{1}{3}\left[\dfrac{1}{A_n}\sum_{i=1}^{n} a_i\left(x_i^3 - |x_i - \overline{x}|^3\right) - \overline{x}^3\right], & s = 0. \end{cases} \tag{30}$$

For χ_{ψ_s} and Δ_{ψ_s} we get

Theorem 4.6. *Let χ_{ψ_s} and Δ_{ψ_s} be defined as in* (29) *and in* (30) *. Then*

(a) *The functions $s \mapsto \chi_{\psi_s}$ and $s \mapsto \Delta_{\psi_s}$ are exponentially convex.*
(b) *If $\chi_{\psi_s} > 0$, and $\Delta_{\psi_s} > 0$, the functions $s \mapsto \chi_{\psi_s}$ and $s \mapsto \Delta_{\psi_s}$ are* log*-convex.*

To see that $s \mapsto \chi_{\psi_s}$ is exponentially convex we observe that

$$F\left(x\right) = \sum_{i,j=1}^{m} u_i u_j \psi_{p_{\frac{i+j}{2}}}\left(x\right)$$

is superadditive and $F\left(0\right) = 0$ and therefore superquadratic, hence

$$\chi_F = \sum_{i,j=1}^{m} u_i u_j \chi_{p_{\frac{i+j}{2}}}\left(x\right) \geq 0.$$

The same type of results we get when we define the superquadratic function ϕ_s : $[0,b) \to \mathbb{R}$, $s \in \mathbb{R}_+$ by

$$\phi_s(x) = \begin{cases} \dfrac{x^s}{s(s-2)}, & s \neq 2, \\ \dfrac{x^2}{2}\log x, & s = 2, \end{cases} \tag{31}$$

with the convention $0\log 0 := 0$.

Applying the functional χ to ϕ_s we have

$$\chi_{\phi_s} = \begin{cases} \dfrac{1}{s(s-2)}\left[\dfrac{1}{\lambda(\beta)-\lambda(\alpha)}\displaystyle\int_\alpha^\beta \left(f^s(t) - |\Omega|^s\right)\mathrm{d}\lambda(t) - \overline{f}^s\right], & s \neq 2, \\ \dfrac{1}{2}\left[\dfrac{1}{\lambda(\beta)-\lambda(\alpha)}\displaystyle\int_\alpha^\beta \left(f^2(t)\log f(t) - \Omega^2\log|\Omega|\right)\mathrm{d}\lambda(t) - \overline{f}^2\log\overline{f}\right], & s = 2, \end{cases} \tag{32}$$

where $\Omega = f(t) - \overline{f}$.

Similarly, if we apply the functional Δ to ϕ_s, we have

$$\Delta_{\phi_s} = \begin{cases} \dfrac{1}{s(s-2)}\left[\dfrac{1}{A_n}\displaystyle\sum_{i=1}^{n} a_i\left(x_i^s - |\mathfrak{D}|^s\right) - \overline{x}^s\right], & s \neq 2, \\ \dfrac{1}{2}\left[\dfrac{1}{A_n}\displaystyle\sum_{i=1}^{n} a_i\left(x_i^2 \log x_i - \mathfrak{D}^2 \log|\mathfrak{D}|\right) - \overline{x}^2 \log \overline{x}\right], & s = 2, \end{cases} \tag{33}$$

where $\mathfrak{D} = x_i - \overline{x}$.

From this we get the next theorem which can be proved in a similar way as Theorem 4.6.

Theorem 4.7. *Let χ_{ϕ_s} and Δ_{ϕ_s} be defined as in* (32) *and* (33). *Then*

(a) *The functions $s \mapsto \chi_{\phi_s}$ and $s \mapsto \Delta_{\phi_s}$ are exponentially convex.*
(b) *If $\chi_{\phi_s} > 0$, and $\Delta_{\phi_s} > 0$ the functions $s \mapsto \chi_{\phi_s}$ and $s \mapsto \Delta_{\phi_s}$ are* log*-convex.*

Theorem 4.7 enables us to define new means. If we choose $\varphi = \phi_s$ and $\psi = \phi_r$, where $r, s \in \mathbb{R}_+$, $r \neq s$, $r, s \neq 2$, then we have

$$\min_{\alpha \le t \le \beta} f(t) \le \left(\frac{\chi_{\phi_s}}{\chi_{\phi_r}}\right)^{\frac{1}{s-r}} \le \max_{\alpha \le t \le \beta} f(t).$$

We denote the new means by

$$M_{s,r}(f;\lambda) = \left(\frac{\chi_{\phi_s}}{\chi_{\phi_r}}\right)^{\frac{1}{s-r}}.$$

For $r, s \in \mathbb{R}_+$ we can extend this mean as follows:

$$M_{s,r}(f;\lambda) = \exp\left(\frac{r(r-2)\left(\frac{1}{\lambda(\beta)-\lambda(\alpha)}\int_\alpha^\beta (f^s(t) - |\mathfrak{Q}|^s)\mathrm{d}\lambda(t) - \overline{f}^s\right)}{s(s-2)\left(\frac{1}{\lambda(\beta)-\lambda(\alpha)}\int_\alpha^\beta (f^r(t) - |\mathfrak{Q}|^r)\mathrm{d}\lambda(t) - \overline{f}^r\right)}\right)^{\frac{1}{s-r}},$$

for $r \neq s$ $(r, s \neq 2)$;

$$M_{r,r}(f;\lambda) = \exp\left(\frac{\frac{1}{\lambda(\beta)-\lambda(\alpha)}\int_\alpha^\beta (f^r(t)\log f(t) - |\mathfrak{Q}|^r \log|\mathfrak{Q}|)\mathrm{d}\lambda(t) - \overline{f}^r \log \overline{f}}{\frac{1}{\lambda(\beta)-\lambda(\alpha)}\int_\alpha^\beta (f^r(t) - |\mathfrak{Q}|^r)\mathrm{d}\lambda(t) - \overline{f}^r} - \frac{2r-2}{r(r-2)}\right),$$

for $r \neq 2$; and

$$M_{2,2}(f;\lambda) = \exp\left(\frac{\frac{1}{\lambda(\beta)-\lambda(\alpha)}\int_\alpha^\beta (f^2(t)\log^2 f(t) - \mathfrak{Q}^2 \log^2|\mathfrak{Q}|)\mathrm{d}\lambda(t) - \overline{f}^2 \log^2 \overline{f}}{2\left(\frac{1}{\lambda(\beta)-\lambda(\alpha)}\int_\alpha^\beta (f^2(t)\log f(t) - \mathfrak{Q}^2 \log|\mathfrak{Q}|)\mathrm{d}\lambda(t) - \overline{f}^2 \log \overline{f}\right)} - \frac{1}{2}\right),$$

where $\mathfrak{Q} = f(t) - \overline{f}$.

These means satisfy:

Theorem 4.8. *Let $r, s, u, v \in \mathbb{R}_+$ such that $r \leq u$, $s \leq v$, $r \neq s$, $u \neq v$. Then*

$$M_{s,r}(f;\lambda) \leq M_{v,u}(f;\lambda).$$

This follows from the log-convexity of $s \mapsto \chi_{\phi_s}$. Then, for any $r, s, u, v \in \mathbb{R}_+$, such that $r \leq u$, $s \leq v$, $r \neq s$, $u \neq v$, we have

$$M_{s,r}(f;\lambda) = \left(\frac{\chi_{\phi_s}}{\chi_{\phi_r}}\right)^{\frac{1}{s-r}} \leq \left(\frac{\chi_{\phi_v}}{\chi_{\phi_u}}\right)^{\frac{1}{v-u}} = M_{v,u}(f;\lambda).$$

In the discrete case we use for the new means the notation

$$M_{s,r}(\mathbf{x};\mathbf{a}) = \left(\frac{\Delta_{\phi_s}}{\Delta_{\phi_r}}\right)^{\frac{1}{s-r}},$$

and for $r, s \in \mathbb{R}_+$ we define the Cauchy type means

$$M_{s,r}(\mathbf{x};\mathbf{a}) = \left(\frac{r(r-2)\left(\frac{1}{A_n}\sum_{i=1}^{n} a_i\left(x_i^s - |\mathcal{D}|^s\right) - \overline{x}^s\right)}{s(s-2)\left(\frac{1}{A_n}\sum_{i=1}^{n} a_i\left(x_i^r - |\mathcal{D}|^r\right) - \overline{x}^r\right)}\right)^{\frac{1}{s-r}}, \quad r \neq s,\ r,s \neq 2,$$

$$M_{r,r}(\mathbf{x};\mathbf{a}) = \exp\left(\frac{\frac{1}{A_n}\sum_{i=1}^{n} a_i\left(x_i^r \log x_i - |\mathcal{D}|^r \log|\mathcal{D}|\right) - \overline{x}^r \log\overline{x}}{\frac{1}{A_n}\sum_{i=1}^{n} a_i\left(x_i^r - |\mathcal{D}|^r\right) - \overline{x}^r} - \frac{2r-2}{r(r-2)}\right), \quad r \neq 2,$$

$$M_{2,2}(\mathbf{x};\mathbf{a}) = \exp\left(\frac{\frac{1}{A_n}\sum_{i=1}^{n} a_i\left(x_i^2 \log^2 x_i - \mathcal{D}^2\log^2|\mathcal{D}|\right) - \overline{x}^2\log^2\overline{x}}{2\left(\frac{1}{A_n}\sum_{i=1}^{n} a_i\left(x_i^2 \log x_i - \mathcal{D}^2\log|\mathcal{D}|\right) - \overline{x}^2\log\overline{x}\right)} - \frac{1}{2}\right),$$

where $\mathcal{D} = \mathbf{x}_i - \overline{x}$.

We can easily check that these means are also symmetric and the special cases are limits of the general case.

4.4 Fejer and Hermite–Hadamard Type Inequalities

In [12] Fejer and Hermite–Hadamard type inequalities for superquadratic functions were discussed. Here are two results presented in [12]:

Theorem 4.9. *Let f be a superquadratic integrable function on $[a,b]$ and let p be nonnegative integrable and symmetric about $x = (a+b)/2$, $0 \le a < b$. Let $P(t)$ be*

$$P(t) = \int_a^b f\left(tx + (1-t)\frac{a+b}{2}\right) p(x)\,\mathrm{d}x, \quad t \in [0,1],$$

and let $Q(t)$ be

$$Q(t) = \int_a^b \frac{1}{2}\left[f\left(\frac{1+t}{2}a + \frac{1-t}{2}x\right) p\left(\frac{x+a}{2}\right) + f\left(\frac{1+t}{2}b + \frac{1-t}{2}x\right) p\left(\frac{x+b}{2}\right)\right] \mathrm{d}x, \quad t \in [0,1],$$

then for $0 \le s \le t \le 1$, $t > 0$,

$$P(s) \le P(t) - \int_a^b \frac{t+s}{2t} f\left((t-s)\left(\left|\frac{a+b}{2} - x\right|\right)\right) p(x)\,\mathrm{d}x - \int_a^b \frac{t-s}{2t} f\left((t+s)\left(\left|\frac{a+b}{2} - x\right|\right)\right) p(x)\,\mathrm{d}x.$$

And, if $0 \le s \le t \le 1$, we get that

$$Q(s) - Q(t) \le -\int_a^b \frac{\left(1 - \frac{t+s}{2}\right)|2x-a-b| + \frac{t+s}{2}(b-a)}{(1-t)|2x-a-b| + t(b-a)} \times f\left(\frac{t-s}{2}(b-a-|a+b-2x|)\right) p(x)\,\mathrm{d}x$$
$$- \int_a^b \frac{\frac{t-s}{2}(b-a-|a+b-2x|)}{(1-t)|2x-a-b| + t(b-a)} \times f\left(\left(1 - \frac{t+s}{2}\right)|2x-a-b| + \frac{t+s}{2}(b-a)\right) p(x)\,\mathrm{d}x.$$

4.5 Refinements of Some Classical Inequalities

In [26] the authors obtained a sequence of inequalities for superquadratic functions. Especially, when the superquadratic function is also increasing and therefore convex, then refinements of classical known results are obtained.

Here we demonstrate two of their results. In Theorem 2.1 there, a converse of Jensen's inequality for superquadratic functions is proved:

Theorem 4.10. *Let (Ω, A, μ) be a measure space with $0 < \mu(\Omega) < \infty$ and let φ be a superquadratic function. If $f : \Omega \subseteq [m, \mu] \to [0, \infty)$ is such that $f, \varphi \circ f \in L_1(\mu)$, then we have*

$$\frac{1}{\mu(\Omega)} \int_r (\varphi \circ f)\, d\mu + \Delta_c \leq \frac{M - \overline{f}}{M - m}\varphi(m) + \frac{\overline{f} - m}{M - m}\varphi(\mu),$$

where $\overline{f} = \dfrac{1}{\mu(\Omega)} \displaystyle\int_\Omega f\, d\mu$ *and*

$$\Delta_c = \frac{1}{\mu(\Omega)} \frac{1}{M - m} \int_\Omega [(M - f)\varphi(f - m) + (f - m)\varphi(M - f)]\, d\mu.$$

In [26, Theorem 4], the integral version of the reversal of Jensen's inequality is proved:

Theorem 4.11. *Let (Ω, A, μ) be a measure space with $0 < \mu(\Omega) < \infty$ and let φ be a superquadratic function. If $p, g : \Omega \to [0, \infty)$ are functions and $a, u \in [0, \infty)$ are real numbers such that*

$$p, pg, p\varphi(g), p\varphi\left(\left|\frac{\int_\Omega pg\, d\mu}{\int_\Omega p\, d\mu} - g\right|\right) \in L_1(\mu), \quad 0 < \int_\Omega p\, d\mu < u,$$

and $ua - \int_\Omega pg\, d\mu \geq 0$, *then*

$$\varphi\left(\frac{ua - \int_\Omega pg\, d\mu}{u - \int_\Omega p\, d\mu}\right) \geq \frac{u\varphi(u) - \int_\Omega p\varphi(g)\, d\mu}{u - \int_\Omega p\, d\mu} + \Delta_{RJ},$$

where

$$\begin{aligned}\Delta_{RJ} = \frac{1}{u - \int_\Omega p\, d\mu} \Bigg[&\int_\Omega p\varphi\left(\left|\frac{\int_\Omega pg\, d\mu}{\int_\Omega p\, d\mu} - g\right|\right) d\mu \\ &+ \left(\int_\Omega p\, d\mu\right)\varphi\left(\left|\frac{\int_\Omega pg\, d\mu}{\int_\Omega p\, d\mu} - a\right|\right) \\ &+ \left(u - \int_\Omega p\, d\mu\right)\varphi\left(\frac{\int_\Omega p\, d\mu}{u - \int_\Omega p\, d\mu}\left|\frac{\int_\Omega pg\, d\mu}{\int_\Omega p\, d\mu} - a\right|\right)\Bigg].\end{aligned}$$

References

1. Abramovich, S.: Superquadracity of functions and rearrangements of sets. J. Inequal. Pure Appl. Math. **8**(2), Art 46, 5 (2007)
2. Abramovich, S.: On superquadracity. J. Math. Inequal. **3**, 329–339 (2009)
3. Abramovich, S., Dragomir, S.S.: Normalized Jensen functional, superquadracity and related inequalities. Inequalities and Applications. International Series of Numerical Mathematics, vol. 157, pp. 217–228. Birkhauser, Basel (2009)
4. Abramovich, S., Persson, L.-E.: Some new scales of refined hardy type inequalities via functions related to superquadracity. Math. Inequal. Appl. **16**(3), 679–695 (2013)
5. Abramovich, S., Persson, L.-E.: Some new refined Hardy type inequalities with breaking points $p = 2$ or $p = 3$ In: Boiso, M.C., Hedenmalm, H., Kaashoek, M.A., Rodriguez, A.M., Treil, S. (eds.) Concrete Operators, Spectral Theory, Operators in Harmonic Analysis and Approximation (22nd International Workshop in Operator Theory and its Applications, Sevilla, July 2011), Series: Operator Theory: Advances and Applications, vol. 236, pp. 1–10, Birkhäuser, Basel (2014)
6. Abramovich, S., Jameson, G., Sinnamon, G.: Inequalities for averages of convex and superquadratic functions. J. Inequal. Pure Appl. Math. **5**(4), Art 91, 1–14 (2004)
7. Abramovich, S., Jameson, G., Sinnamon, G.: Refining Jensen's inequality. Bull. Math. Soc. Sci. Math. Roumanie (Novel Series) **47**(95), 3–14 (2004)
8. Abramovich, S., Banić, S., Klaricić Bacula, M.: A variant of Jensen–Steffensen's inequality for convex and for superquadratic functions. J. Inequal. Pure Appl. Math. **7**(2), Art 70, 12 (2006)
9. Abramovich, S., Banić, S., Matić, M., Pečarić, J.: Jensen Steffensen's and related inequalities for superquadratic functions. Math. Inequal. Appl. **11**, 23–41 (2007)
10. Abramovich, S., Banić, S., Matić, M.: Superquadratic functions in several variables. J. Math. Anal. Appl. **327**, 1444–1460 (2007)
11. Abramovich, S., Barić, J., Pečarić, J.: A variant of Jessen's inequality of Mercer's type for superquadratic functions. J. Inequal. Pure Appl. Math. **9**(3), Art 62, 13 (2008)
12. Abramovich, S., Barić, J., Pečarić, J.: Fejer and Hermite-Hadamard type inequalities for superquadratic functions. J. Math. Anal. Appl. **344**(2), 1048–1056 (2008)
13. Abramovich, S., Barić, J., Pečarić, J.: Superquadracity, Bohr's inequality and deviation from a mean value. Aust. J. Math. Anal. Appl. **7**(1), Art 1, 9 (2009)
14. Abramovich, S., Ivelić, S., Pečarić, J.: Generalizations of Jensen-Steffensen and related integral inequalities for superquadratic functions. Central Eur. J. Math. **8**, 937–949 (2010)
15. Abramovich, S., Ivanković, B., Pečarić, J.: Improvement of Jensen's inequality for superquadratic functions. Aust. J. Math. Anal. Appl. **7**(1), Art 15, 18 (2010)
16. Abramovich, S., Ivelić, S., Pečarić, J.: Improvement of Jensen-Steffensen's inequality for superquadratic functions. Banach J. Math. Anal. **4**(1), 159–169 (2010)
17. Abramovich, S., Farid, G., Pečarić, J.: More about Hermite-Hadamard inequalities, Cauchy's means and superquadracity. J. Inequal. Appl. **2010**, Art ID 102467, 14 (2010)
18. Abramovich, S., Ivelić, S., Pečarić, J.: Refinements of inequalities related to convexity via superquadracity, weaksuperquadracity and superterzacity. Inequalities and Applications, pp. 191–207. Birkhauser, Basel (2010)
19. Abramovich, S., Krulić, K., Person, L.-E., Pečarić, J.: Some new refined Hardy type inequalities with general kernels via superquadratic and subquadratic functions. Aequ. Math. **79**(1), 157–172 (2010)
20. Abramovich, S., Farid, G., Ivelić, S., Pečarić, J.: More on Cauchy's means and generalization of Hadamard inequality via converses of Jensen's inequality and superquadracity. Int. J. Pure Appl. Math. **69**(1), 97–116 (2011)
21. Abramovich, S., Farid, G., Ivelić, S., Pečarić, J.: On exponential convexity, Jensen-Steffensen-Boas inequality, and Cauchy's means for superquadratic functions. J. Math. Inequal. **5**(2), 169–180 (2011)

22. Abramovich, S., Ivelić, S., Pečarić, J.: Extension of Euler Lagrange identity by superquadratic power functions. Int. J. Pure Appl. Math. **74**(2), 209–220 (2012). IJPAM-2011-17-454
23. Abramovich, S., Farid, G., Pečarić, J.: More about Jensen's inequality for superquadratic functions. J. Math. Inequal. **7**(1), 11–14 (2013)
24. Anwar, M., Jakšetić, J., Pečarić, J., Ur Rehman, A.: Exponential convexity, positive semi-definite matrices and fundamental inequalities. J. Math. Inequal. **4**(2), 171–189 (2010)
25. Banić, S., Varošanec, S.: Functional inequalities for superquadratic functions. Int. J. Pure Appl. Math. **43**(4), 5037–549 (2008)
26. Banić, S., Pečarić, J., Varošanec, S.: Superquadratic functions and refinements of some classical inequalities. J. Korean Math. Soc. **45**, 513–525 (2008)
27. Gilányi, A., Troczka-Pawelec, K.: Regularity of weakly subquadratic functions. J. Math. Anal. Appl. **382**, 814–821 (2011)
28. Gilányi, A., Kézi, C.G., Troczka-Pawelec, K.: On two different concept of superquadracity. Inequalities and Applications, pp. 209–215. Birkhauser, Basel (2010)
29. Mitroi, F.G.: Connection between the Jensen and the Chebychev functional. Inequalities and Applications, pp. 217–227. Birkhauser, Basel (2010)
30. Mitroi, F.G.: On the Jensen-Steffensen inequality and superquadracity. Anale Universitatii Oradea, Fasc. Mathematica, Tom XVIII, 269–275 (2011)
31. Oguntuase, J.A., Persson, L.E.: Refinement of Hardy's inequalities via superquadratic and subquadratic functions. J. Math. Anal. Appl. **339**(2), 1305–1312 (2008)
32. Oguntuase, J.A., Popoola, B.A.: Refinement of Hardy's inequalities involving many functions via superquadratic functions. An. Ştiinţ. Univ. Al. I. Cuza Iaşi. Mat. (N.S.) **57**(2), 271–283 (2011)
33. Oguntuase, J.A., Persson, L.E., Essel, E.K., Popoola, B.A.: Refined multidimensional Hardy-type inequalities via superquadracity. Banach J. Math. Anal. **2**(2), 129–139 (2008)
34. Pečarić, J.E., Proschan, F., Tong, Y.L.: Convex Functions, Partial Orderings, and Statistical Applications. Academic, New York (1992)

Green's Operator and Differential Forms

Shusen Ding and Yuming Xing

Dedicated to Professor Hari M. Srivastava

Abstract The purpose of this paper is to present an up-to-date account of the advances made in the study of L^p theory of Green's operator applied to differential forms.

1 Introduction

We all know that Green's operator G is one of key operators which are widely studied and used in several areas, including analysis and PDEs. For example, Green's operator is often applied to study the solutions of various differential equations and to define Poisson's equation for differential forms. Much progress has been made during recent years in the study of Green's operator G and some other related operators, such as the Laplacian operator Δ and the harmonic projection operator H; see [1,2,4,7,20,23,26,27]. Differential forms have become invaluable tools for many fields of sciences and engineering, including theoretical physics, general relativity, potential theory, and electromagnetism. They can be used to describe various systems of PDEs and to express different geometric structures on manifolds. In many situations, the process to study solutions of PDEs involves estimating the various norms of the operators. The purpose of this paper is to present an up-to-date account of the progress made in investigation of L^p theory of Green's operator and the related compositions that are applied to differential forms.

S. Ding (✉)
Department of Mathematics, Seattle University, Seattle, WA 98122, USA
e-mail: sding@seattleu.edu

Y. Xing
Department of Mathematics, Harbin Institute of Technology, Harbin, China
e-mail: xyuming@hit.edu.cn

G.V. Milovanović and M.Th. Rassias (eds.), *Analytic Number Theory, Approximation Theory, and Special Functions*, DOI 10.1007/978-1-4939-0258-3_14,

We keep using the traditional notation. Let Ω be a bounded domain in $\mathbb{R}^n$, $n \geq 2$, B and σB be the balls with the same center and $\operatorname{diam}(\sigma B) = \sigma \operatorname{diam}(B)$. We do not distinguish the balls from cubes, throughout this paper. We use $|E|$ to denote the n-dimensional Lebesgue measure of a set $E \subseteq \mathbb{R}^n$. For a function u, the average of u over B is expressed by $u_B = \frac{1}{|B|}\int_B u dx$. All integrals involved in this paper are the Lebesgue integrals. We say w is a weight if $w \in L^1_{loc}(\mathbb{R}^n)$ and $w > 0$ a.e. Differential forms are extensions of differentiable functions in $\mathbb{R}^n$. For example, the function $u(x_1, x_2, \ldots, x_n)$ is called a 0-form. A differential 1-form $u(x)$ in $\mathbb{R}^n$ can be expressed as $u(x) = \sum_{i=1}^n u_i(x_1, x_2, \ldots, x_n)dx_i$, where the coefficient functions $u_i(x_1, x_2, \ldots, x_n)$, $i = 1, 2, \ldots, n$, are differentiable. Similarly, a differential k-form $u(x)$ can be expressed as

$$u(x) = \sum_I u_I(x)dx_I = \sum u_{i_1 i_2 \ldots i_k}(x)dx_{i_1} \wedge dx_{i_2} \wedge \cdots \wedge dx_{i_k},$$

where $I = (i_1, i_2, \ldots, i_k)$, $1 \leq i_1 < i_2 < \cdots < i_k \leq n$; see [2] for more properties and applications of differential forms. Let $\wedge^l = \wedge^l(\mathbb{R}^n)$ be the set of all l-forms in $\mathbb{R}^n$, $D'(\Omega, \wedge^l)$ be the space of all differential l-forms in Ω, and $L^p(\Omega, \wedge^l)$ be the l-forms $u(x) = \sum_I u_I(x)dx_I$ in Ω satisfying $\int_\Omega |u_I|^p < \infty$ for all ordered l-tuples I, $l = 1, 2, \ldots, n$. We express the exterior derivative by d and the Hodge star operator by $\star$. The Hodge codifferential operator $d^\star$ is given by $d^\star = (-1)^{nl+1} \star d \star$, $l = 1, 2, \ldots, n$. If $u = \alpha_{i_1 i_2 \ldots i_k}(x_1, x_2, \ldots, x_n)dx_{i_1} \wedge dx_{i_2} \wedge \cdots \wedge dx_{i_k} = \alpha_I dx_I$, $i_1 < i_2 < \cdots < i_k$, is a differential k-form, then

$$\star u = \star \alpha_{i_1 i_2 \ldots i_k} dx_{i_1} \wedge dx_{i_2} \wedge \cdots \wedge dx_{i_k} = (-1)^{\sum(I)} \alpha_I dx_J,$$

where $I = (i_1, i_2, \ldots, i_k)$, $J = \{1, 2, \ldots, n\} - I$, and $\sum(I) = \frac{k(k+1)}{2} + \sum_{j=1}^k i_j$. For example, in $\wedge^1(\mathbb{R}^3)$, we have $\star dx_1 = (-1)^2 dx_2 \wedge dx_3 = dx_2 \wedge dx_3$. We write $\|u\|_{s,\Omega} = \left(\int_\Omega |u|^s\right)^{1/s}$ and $||u||_{s,\Omega,w} = \left(\int_\Omega |u|^s w(x)dx\right)^{1/s}$, where $w(x)$ is a weight. Let $\wedge^l \Omega$ be the lth exterior power of the cotangent bundle and $C^\infty(\wedge^l \Omega)$ be the space of smooth l-forms on Ω. We set

$$\mathcal{W}(\wedge^l \Omega) = \left\{u \in L^1_{loc}(\wedge^l \Omega) : \ u \text{ has generalized gradient}\right\}.$$

The harmonic l-fields are defined by $\mathcal{H}(\wedge^l \Omega) = \{u \in \mathcal{W}(\wedge^l \Omega) : du = d^\star u = 0, u \in L^p \text{ for some } 1 < p < \infty\}$. The orthogonal complement of $\mathcal{H}$ in L^1 is defined by

$$\mathcal{H}^\perp = \{u \in L^1 : < u, h >= 0 \text{ for all } h \in \mathcal{H}\}.$$

Then, the Green's operator G is defined as $G : C^\infty(\wedge^l \Omega) \to \mathcal{H}^\perp \cap C^\infty(\wedge^l \Omega)$ by assigning $G(u)$ as the unique element of $\mathcal{H}^\perp \cap C^\infty(\wedge^l \Omega)$ satisfying Poisson's equation $\Delta G(u) = u - H(u)$, where H is the harmonic projection operator that maps $C^\infty(\wedge^l \Omega)$ onto $\mathcal{H}$, so that $H(u)$ is the harmonic part of u. See [2,13,20,23,28] for more properties of Green's operator. For a measurable set $E \subset \mathbb{R}^n$ and $\omega \in D'(E, \wedge^l)$, the vector-valued differential form

$$\nabla\omega = \left(\frac{\partial \omega}{\partial x_1}, \ldots, \frac{\partial \omega}{\partial x_n}\right)$$

consists of differential forms $\frac{\partial \omega}{\partial x_i} \in D'(E, \wedge^l)$, where the partial differentiation is applied to the coefficients of ω. We use $W^{1,p}(E, \wedge^l)$ to denote the Sobolev space of l-forms which equals $L^p(\wedge^l E) \cap L_1^p(\wedge^l E)$ with norm

$$\|\omega\|_{W^{1,p}(E)} = \operatorname{diam}(E)^{-1}\|\omega\|_{p,E} + \|\nabla\omega\|_{p,E}.$$

For $0 < p < \infty$ and a weight $w(x)$, the weighted norm of $\omega \in W^{1,p}(E, \wedge^l)$ over E is denoted by

$$\|\omega\|_{W^{1,p}(E),\, w} = \operatorname{diam}(E)^{-1}\|\omega\|_{p,E,w} + \|\nabla\omega\|_{p,E,w}.$$

The differential equation $d^\star A(x, d\omega) = 0$ is called the homogeneous A-harmonic equation or the A-harmonic equation, and the nonlinear elliptic partial differential equation

$$d^\star A(x, d\omega) = B(x, d\omega) \tag{1}$$

is called the nonhomogeneous A-harmonic equation for differential forms, where $A : \Omega \times \wedge^l(\mathbb{R}^n) \to \wedge^l(\mathbb{R}^n)$ and $B : \Omega \times \wedge^l(\mathbb{R}^n) \to \wedge^{l-1}(\mathbb{R}^n)$ satisfy the conditions:

$$|A(x,\xi)| \le a|\xi|^{p-1}, \quad A(x,\xi)\cdot\xi \ge |\xi|^p \quad \text{and} \quad |B(x,\xi)| \le b|\xi|^{p-1} \tag{2}$$

for almost every $x \in \Omega$ and all $\xi \in \wedge^l(\mathbb{R}^n)$. Here $a, b > 0$ are constants and $1 < p < \infty$ is a fixed exponent associated with (1). A solution to (1) is an element of the Sobolev space $W^{1,p}_{\text{loc}}(\Omega, \wedge^{l-1})$ such that

$$\int_\Omega A(x, d\omega)\cdot d\varphi + B(x, d\omega)\cdot\varphi = 0$$

for all $\varphi \in W^{1,p}_{\text{loc}}(\Omega, \wedge^{l-1})$ with compact support. Let $A : \Omega \times \wedge^l(\mathbb{R}^n) \to \wedge^l(\mathbb{R}^n)$ be defined by $A(x,\xi) = \xi|\xi|^{p-2}$ with $p > 1$. Then A satisfies required conditions and $d^\star A(x, d\omega) = 0$ becomes the p-harmonic equation

$$d^\star(du|du|^{p-2}) = 0 \tag{3}$$

for differential forms. If u is a function (a 0-form), the equation (3) reduces to the usual p-harmonic equation $\operatorname{div}(\nabla \mathrm{u}|\nabla|^{\mathrm{p}-2}) = 0$ for functions. We should notice that if the operator $B = 0$ in (1), then (1) reduces to the homogeneous A-harmonic equation. Some results have been obtained in recent years about different versions of the A-harmonic equation. In real life applications, we will use differential forms that not only depend on the coordinate variable $x \in \mathbb{R}^n$ but also the time variable t. For example, when we study a force vector field $\mathbf{F}$, this vector often varies with $x \in \mathbb{R}^3$ and the time t. Hence, we need to study differential forms with the time variable t. We use

$$u(x,t) = \sum_I u_I(x,t))dx_I = \sum u_{i_1 i_2 \dots i_k}(x,t) dx_{i_1} \wedge dx_{i_2} \wedge \cdots \wedge dx_{i_k} \tag{4}$$

to denote the differential k-form with a parameter t, where the coefficients

$$u_{i_1 i_2 \dots i_k}(x,t) = u_{i_1 i_2 \dots i_k}(x_1, x_2, \dots, x_n, t), \; 1 \le i_1 < i_2 < \cdots < i_k \le n,$$

are differentiable functions of $(x_1, x_2, \dots, x_n, t)$. We always assume that $t \in [0, \infty)$ is a parameter. For differential forms with a parameter t, the nonhomogeneous A-harmonic equation (1) becomes the parametric A-harmonic equation

$$d^\star A(x, du(x,t)) = B(x, u_t(x,t)), \tag{5}$$

which can be considered as the generalized heat or diffusion equation.

2 The L^p-Estimates for Green's Operator

The purpose of this section is to present L^p norm inequalities for Green's operator and the Laplace-Beltrami operator $\Delta = dd^\star + d^\star d$ applied to differential forms. The Laplace-Beltrami operator and Green's operator play an important role in many fields, including partial differential equations, harmonic analysis, and quasiconformal mappings. The following series basic L^p norm inequalities were established in [7].

Theorem 2.1. *Let $u \in C^\infty(\wedge^l \Omega)$, $l = 0, 1, \dots, n$, and $1 < s < \infty$. Then, there exists a constant C, independent of u, such that*

$$\|\Delta(G(u))\|_{s,M} \le C\|u\|_{s,M}. \tag{6}$$

Theorem 2.2. *Let $u \in C^\infty(\wedge^l \Omega)$, $l = 0, 1, \dots, n$. Assume that $1 < s < \infty$. Then, there exists a constant C, independent of u, such that*

$$\|G(\Delta u)\|_{s,M} \le C\|u\|_{s,M}. \tag{7}$$

Theorem 2.3. *Let $u \in C^{\infty}(\wedge^l \Omega)$, $l = 0, 1, \ldots, n$. If $1 < s < \infty$, then there exists a constant C, independent of u, such that*

$$\|(G(u))_D\|_{s,D} \leq C\|u\|_{s,D} \tag{8}$$

for any convex and bounded D with $D \subset \Omega$.

Corollary 2.4. *Let $u \in C^{\infty}(\wedge^l \Omega)$, $l = 0, 1, \ldots, n$. Assume that $1 < s < \infty$. Then, for any convex and bounded D with $D \subset \Omega$, there exists a constant C, independent of u, such that*

$$\|G(u) - (G(u))_D\|_{s,D} \leq C\|G(u) - c\|_{s,D} \tag{9}$$

for any closed form c and

$$\|G(u) - (G(u))_D\|_{s,D} \leq C\|u\|_{s,D}. \tag{10}$$

The following inequality (11) is considered as an analogue of the Poincaré inequality for Green's operator.

Theorem 2.5. *Let $u \in C^{\infty}(\wedge^l \Omega)$, $l = 0, 1, \ldots, n$. Assume that $1 < s < \infty$. Then, there exists a constant C, independent of u, such that*

$$\|G(u) - (G(u))_B\|_{s,B} \leq C \operatorname{diam}(B)\|du\|_{s,B} \tag{11}$$

for all balls B with $B \subset \Omega$.

Using Theorem 2.5, the following Sobolev-Poincaré imbedding theorem about Green's operator G applied to a differential form u was also obtained in [7].

Theorem 2.6. *Let $u \in C^{\infty}(\wedge^l \Omega)$, $l = 0, 1, \ldots, n$. Assume that $1 < s < \infty$. Then, there exists a constant C, independent of u, such that*

$$\|G(u) - (G(u))_B\|_{W^{1,s}(B)} \leq C\|du\|_{s,B} \tag{12}$$

for all balls B with $B \subset \Omega$.

The study of different versions of the A-harmonic equation for differential forms has developed rapidly in recent years. Many interesting results related to the A-harmonic equations have been established recently; see [3, 8, 10–12, 14, 16, 18, 22, 25]. Next, we present the weighted norm inequalities for the solutions to the nonhomogeneous A-harmonic equation

$$A(x, g + du) = h + d^{\star}v \tag{13}$$

for differential forms, where $g, h \in D'(\Omega, \wedge^l)$ and $A : \Omega \times \wedge^l \mathbb{R}^n \to \wedge^l \mathbb{R}^n)$ satisfies the following conditions:

$$|A(x, \xi)| \leq a|\xi|^{p-1} \quad \text{and} \quad < A(x, \xi), \xi > \; \geq \; |\xi|^p \tag{14}$$

for almost every $x \in \Omega$ and all $\xi \in \wedge^l(\mathbb{R}^n)$. Here $a > 0$ is a constant and $1 < p < \infty$ is a fixed exponent associated with (13).

Definition 2.7. We call u and v a pair of conjugate A-harmonic tensor in Ω if u and v satisfy the conjugate A-harmonic equation

$$A(x, du) = d^{\star} v \tag{15}$$

in Ω. Similarly, we call u an A-harmonic tensor in Ω if u satisfies the A-harmonic equation

$$d^{\star} A(x, du) = 0. \tag{16}$$

Note that $du = d^* v$ is an analogue of a Cauchy-Riemann system in $\mathbb{R}^n$. A differential l-form $u \in D'(\Omega, \wedge^l)$ is called a closed form if $du = 0$ in Ω. Similarly, a differential $l + 1$-form $v \in D'(\Omega, \wedge^{l+1})$ is called a coclosed form if $d^{\star} v = 0$. For example, $du = d^* v$ is an analogue of a Cauchy-Riemann system in $\mathbb{R}^n$. Clearly, the A-harmonic equation is not affected by adding a closed form to u and coclosed form to v. Therefore, any type of estimates between u and v must be modulo such forms. Throughout this paper, we always assume that $\frac{1}{p} + \frac{1}{q} = 1$.

Definition 2.8. A weight $w(x)$ is called an $A_r(E)$-weight for some $r > 1$ on a subset $E \subset \mathbb{R}^n$ and write $w \in A_r(E)$, if $w(x) > 0$ a.e., and

$$\sup_B \left(\frac{1}{|B|} \int_B w dx \right) \left(\frac{1}{|B|} \int_B \left(\frac{1}{w} \right)^{1/(r-1)} dx \right)^{(r-1)} < \infty$$

for any ball $B \subset E$.

See [2] for properties of $A_r(E)$-weights. We will need the following generalized Hölder's inequality.

Theorem 2.9. *Let $u \in C^{\infty}(\wedge^l \Omega)$, $l = 1, 2, \ldots, n$, be an A-harmonic tensor in Ω. Assume that $\rho > 1$, $1 < s < \infty$, and $w \in A_r(\Omega)$ for some $r > 1$. Then, there exists a constant C, independent of u, such that*

$$\|G(\Delta u)\|_{s,B,w^{\alpha}} \leq C \|u\|_{s,\rho B,w^{\alpha}} \tag{17}$$

for any ball $B \subset \Omega$ and any real number α with $0 < \alpha \leq 1$.

Theorem 2.10. *Let $u \in C^{\infty}(\wedge^l \Omega)$, $l = 1, 2, \ldots, n$, be an A-harmonic tensor in Ω. Assume that $\rho > 1$, $1 < s < \infty$, and $w \in A_r(\Omega)$ for some $r > 1$. Then, there exists a constant C, independent of u, such that*

$$\|\Delta(G(u))\|_{s,B,w^{\alpha}} \leq C \|u\|_{s,\rho B,w^{\alpha}} \tag{18}$$

for any ball $B \subset \Omega$ and any real number α with $0 < \alpha \leq 1$.

Combining Theorems 2.9 and 2.10, we have the following $A_r(\Omega)$-weighted inequality.

Corollary 2.11. *Let $u \in C^\infty(\wedge^l \Omega)$, $l = 1, 2, \ldots, n$, be an A-harmonic tensor in Ω. Assume that $\rho > 1$, $1 < s < \infty$, and $w \in A_r(\Omega)$ for some $r > 1$. Then, there exists a constant C, independent of u, such that*

$$\|\Delta(G(u)) + G(\Delta u)\|_{s,B,w^\alpha} \leq C\,\|u\|_{s,\rho B,w^\alpha} \tag{19}$$

for any ball $B \subset \Omega$ and any real number α with $0 < \alpha \leq 1$.

The following theorem is called $A_r(\Omega)$-weighted Sobolev-Poincaré imbedding theorem for Green's operator G.

Theorem 2.12. *Let $G(u) \in C^\infty(\wedge^l \Omega)$, $l = 1, 2, \ldots, n$, be an A-harmonic tensor in Ω. Assume that $\rho > 1$, $1 < s < \infty$, and $w \in A_r(\Omega)$ for some $r > 1$. Then, there exists a constant C, independent of u, such that*

$$\|G(u) - (G(u))_B\|_{W^{1,s}(B),w} \leq C\,\|du\|_{s,\rho B,w} \tag{20}$$

for all balls B with $\rho B \subset \Omega$.

Corollary 2.13. *Let $u \in C^\infty(\wedge^l \Omega)$, $l = 1, 2, \ldots, n$, be an A-harmonic tensor in Ω. Assume that $\rho > 1$, $1 < s < \infty$, and $w \in A_r(\Omega)$ for some $r > 1$. Then, there exists a constant C, independent of u, such that*

$$\|(G(u))_B\|_{s,B,w^\alpha} \leq C_1\,\|u\|_{s,\rho B,w^\alpha}, \tag{21}$$

$$\|G(u) - (G(u))_B\|_{s,B,w^\alpha} \leq C_2\|u\|_{s,\rho B,w^\alpha} \tag{22}$$

for all balls B with $\rho B \subset \Omega$ and any real number α with $0 < \alpha \leq 1$.

Theorem 2.14. *Let $u \in C^\infty(\wedge^{l-1}\Omega)$, $l = 1, 2, \ldots, n$, and $v \in C^\infty(\wedge^{l+1}\Omega)$, $l = 0, 1, 2, \ldots, n-1$, be a pair of solutions to the conjugate A-harmonic equation (15) in Ω. Then, there exists a constant C, independent of u and v, such that*

$$\|G(du) - (G(du))_D\|^p_{p,D} \leq C\,\|d^\star v\|^q_{q,D} \tag{23}$$

for any convex and bounded domain D with $D \subset \Omega$.

We have the following global Sobolev-Poincaré-type imbedding theorem for Green's operator.

Theorem 2.15. *Let $u \in C^\infty(\wedge^{l-1}\Omega)$, $l = 1, 2, \ldots, n$, and $v \in C^\infty(\wedge^{l+1}\Omega)$, $l = 0, 1, 2, \ldots, n-1$, be a pair of solutions to the conjugate A-harmonic equation (15) in Ω. Then, there exists a constant C, independent of u and v, such that*

$$\|G(u) - (G(u))_D\|^p_{W^{1,p}(D)} \leq C\,\|d^\star v\|^q_{q,D}$$

for any bounded domain D with $D \subset \Omega$.

Theorem 2.16. *Let $u \in C^\infty(\wedge^{l-1}\Omega)$, $l = 1,2,\ldots,n$, and $v \in C^\infty(\wedge^{l+1}\Omega)$, $l = 0,1,2,\ldots,n-1$, be a pair of solutions to the conjugate A-harmonic equation (15) in Ω. Then, there exists a constant C, independent of u and v, such that*

$$\|G(d^\star v) - (G(d^\star v))_D\|_{q,D}^q \le C\|du\|_{p,D}^p$$

for any bounded domain D with $D \subset \Omega$.

3 Lipschitz and *BMO* Norm Inequalities

In this section, we will study the Lipschitz norm and BMO norm inequalities for Green's operator applied to differential forms. Throughout this section, we always assume that $M \subset \mathbb{R}^n$ is a bounded domain. Let $\omega \in L^1_{loc}(M, \wedge^l)$, $l = 0,1,\ldots,n$. We write $\omega \in \mathrm{locLip}_k(M, \wedge^l)$, $0 \le k \le 1$, if

$$\|\omega\|_{\mathrm{locLip}_k, M} = \sup_{\sigma Q \subset M} |Q|^{-(n+k)/n}\|\omega - \omega_Q\|_{1,Q} < \infty \tag{24}$$

for some $\sigma \ge 1$. Further, we write $\mathrm{Lip}_k(M, \wedge^l)$ for those forms whose coefficients are in the usual Lipschitz space with exponent k and write $\|\omega\|_{Lip_k, M}$ for this norm. Similarly, for $\omega \in L^1_{loc}(M, \wedge^l)$, $l = 0,1,\ldots,n$, we write $\omega \in \mathrm{BMO}(M, \wedge^l)$ if

$$\|\omega\|_{\star, M} = \sup_{\sigma Q \subset M} |Q|^{-1}\|\omega - \omega_Q\|_{1,Q} < \infty \tag{25}$$

for some $\sigma \ge 1$. When ω is a 0-form, (25) reduces to the classical definition of $\mathrm{BMO}(M)$.

The following four lemmas are used in [27] to prove the Lipschitz and BMO norm inequalities.

Lemma 3.1. *Let $u \in D'(M, \wedge^l)$ be a solution to the nonhomogeneous A-harmonic equation (1) in M and $\sigma > 1$ be a constant. Then there exists a constant C, independent of u, such that*

$$\|du\|_{p,B} \le C\,\mathrm{diam}\,(B)^{-1}\|u - c\|_{p,\sigma B}$$

for all balls or cubes B with $\sigma B \subset M$ and all closed forms c. Here $1 < p < \infty$.

Lemma 3.2. *Let u be a smooth differential form satisfying the nonhomogeneous A-harmonic equation in M, $\sigma > 1$ and $0 < s,t < \infty$. Then there exists a constant C, independent of u, such that*

$$\|u\|_{s,B} \le C|B|^{(t-s)/st}\|u\|_{t,\sigma B}$$

for all balls or cubes B with $\sigma B \subset M$.

Lemma 3.3. *Let $0 < \alpha < \infty$, $0 < \beta < \infty$, and $s^{-1} = \alpha^{-1} + \beta^{-1}$. If f and g are measurable functions on $\mathbb{R}^n$, then $\| fg \|_{s,E} \leq \| f \|_{\alpha,E} \cdot \| g \|_{\beta,E}$ for any $E \subset \mathbb{R}^n$.*

Using the same method developed in the proof of Propositions 5.15 and 5.17 in [20], we can prove that for any closed ball $\overline{B} = B \cup \partial B$, we have

$$\|dd^*G(u)\|_{s,\overline{B}} + \|d^*dG(u)\|_{s,\overline{B}} + \|dG(u)\|_{s,\overline{B}} + \|d^*G(u)\|_{s,\overline{B}} + \|G(u)\|_{s,\overline{B}}$$
$$\leq C(s)\|u\|_{s,\overline{B}}.$$

Note that for any Lebesgue measurable function f defined on a Lebesgue measurable set E with $|E| = 0$, we have $\int_E f dx = 0$. Thus, $\|u\|_{s,\partial B} = 0$ and

$$\|dd^*G(u)\|_{s,\partial B} + \|d^*dG(u)\|_{s,\partial B} + \|dG(u)\|_{s,\partial B} + \|d^*G(u)\|_{s,\partial B} + \|G(u)\|_{s,\partial B} = 0$$

since $|\partial B| = 0$. Therefore, we obtain

$$\|dd^*G(u)\|_{s,B} + \|d^*dG(u)\|_{s,B} + \|dG(u)\|_{s,B} + \|d^*G(u)\|_{s,MB} + \|G(u)\|_{s,B}$$
$$= \|dd^*G(u)\|_{s,\overline{B}} + \|d^*dG(u)\|_{s,\overline{B}} + \|dG(u)\|_{s,\overline{B}} + \|d^*G(u)\|_{s,\overline{B}} + \|G(u)\|_{s,\overline{B}}$$
$$\leq C(s)\|u\|_{s,\overline{B}} = C(s)\|u\|_{s,B}.$$

Hence, we have the following lemma.

Lemma 3.4 ([27]). *Let u be a smooth differential form defined in M and $1 < s < \infty$. Then, there exists a positive constant $C = C(s)$, independent of u, such that*

$$\|dd^*G(u)\|_{s,B} + \|d^*dG(u)\|_{s,B} + \|dG(u)\|_{s,B} + \|d^*G(u)\|_{s,MB} + \|G(u)\|_{s,B}$$
$$\leq C(s)\|u\|_{s,B}$$

for any ball $B \subset M$.

Lemma 3.5 ([27]). *Let $du \in L^s(M, \wedge^l)$ be a smooth form and G be Green's operator, $l = 1, \ldots, n$, and $1 < s < \infty$. Then, there exists a constant C, independent of u, such that*

$$\|G(u) - (G(u))_B\|_{s,B} \leq C|B| \operatorname{diam}(B)\|du\|_{s,B} \tag{26}$$

for all balls $B \subset M$.

We first obtained Lemma 3.5 in [27]. Then, using Lemma 3.5, we proved the following inequality for Green's operator with Lipschitz norm.

Theorem 3.6. *Let $du \in L^s(M, \wedge^l)$, $l = 1, 2, \ldots, n$, $1 < s < \infty$, be a smooth form in a domain M. Then, there exists a constant C, independent of u, such that*

$$\|G(u)\|_{\mathrm{locLip}_k, M} \leq C\|du\|_{s,M}, \tag{27}$$

where k is a constant with $0 \leq k \leq 1$.

We also proved the following norm comparison theorem between the Lipschitz norm and the BMO norm in [27].

Theorem 3.7. *Let $u \in L^s_{loc}(M, \wedge^l)$, $l = 1, 2, \ldots, n$, $1 < s < \infty$, be a solution of the A-harmonic equation (1) in a bounded domain M and G be Green's operator. Then, there exists a constant C, independent of u, such that*

$$\|G(u)\|_{\mathrm{locLip}_k, M} \le C \|u\|_{\star, M}, \tag{28}$$

where k is a constant with $0 \le k \le 1$.

We have presented some estimates for the Lipschitz norm $\|\cdot\|_{\mathrm{locLip}_k, M}$. Next, present the following inequality between the BMO norm and the Lipschitz norm for Green's operator.

Theorem 3.8. *Let $u \in L^s(M, \wedge^l)$, $l = 1, 2, \ldots, n$, $1 < s < \infty$, be a smooth form in a bounded, convex domain M and G be Green's operator. Then, there exists a constant C, independent of u, such that*

$$\|G(u)\|_{\star, M} \le C \|G(u)\|_{\mathrm{locLip}_k, M}. \tag{29}$$

Next theorem says that we estimate BMO norm $\|\cdot\|_{\star, M}$ of Green's operator in terms of L^s norm.

Theorem 3.9. *Let $du \in L^s(M, \wedge^l)$, $l = 1, 2, \ldots, n$, $1 < s < \infty$, be a smooth form in a bounded, convex domain M and G be Green's operator. Then, there exists a constant C, independent of u, such that*

$$\|G(u)\|_{\star, M} \le C \|du\|_{s, M}. \tag{30}$$

Based on the above results, we discuss the weighted Lipschitz and BMO norms. For $\omega \in L^1_{loc}(M, \wedge^l, w)$, $l = 0, 1, \ldots, n$, we write $\omega \in \mathrm{locLip}_k(M, \wedge^l, w)$, $0 \le k \le 1$, if

$$\|\omega\|_{\mathrm{locLip}_k, M, w} = \sup_{\sigma Q \subset M} (\mu(Q))^{-(n+k)/n} \|\omega - \omega_Q\|_{1, Q, w} < \infty \tag{31}$$

for some $\sigma > 1$, where M is a bounded domain, the measure μ is defined by $d\mu = w(x)dx$, and w is a weight. For convenience, we write the following simple notation $\mathrm{locLip}_k(M, \wedge^l)$ for $\mathrm{locLip}_k(M, \wedge^l, w)$. Similarly, for $\omega \in L^1_{loc}(M, \wedge^l, w)$, $l = 0, 1, \ldots, n$, we will write $\omega \in \mathrm{BMO}(M, \wedge^l, w)$ if

$$\|\omega\|_{\star, M, w} = \sup_{\sigma Q \subset M} (\mu(Q))^{-1} \|\omega - \omega_Q\|_{1, Q, w} < \infty \tag{32}$$

for some $\sigma > 1$, where the measure μ is defined by $d\mu = w(x)dx$, w is a weight, and α is a real number. Again, we shall write $\mathrm{BMO}(M, \wedge^l)$ to replace $\mathrm{BMO}(M, \wedge^l, w)$ when it is clear that the integral is weighted.

We say a pair of weights $(w_1(x), w_2(x))$ satisfies the $A_{r,\lambda}(E)$-condition in a set $E \subset \mathbb{R}^n$ and write $(w_1(x), w_2(x)) \in A_{r,\lambda}(E)$, for some $\lambda \geq 1$ and $1 < r < \infty$ with $1/r + 1/r' = 1$ if

$$\sup_{B \subset E} \left(\frac{1}{|B|} \int_B (w_1)^{\lambda} dx \right)^{\frac{1}{\lambda r}} \left(\frac{1}{|B|} \int_B w_2^{\frac{-\lambda r'}{r}} dx \right)^{\frac{1}{\lambda r'}} < \infty. \tag{33}$$

The following version of weak reverse Hölder inequality appeared in [12].

Lemma 3.10. *Suppose that u is a solution to the nonhomogeneous A-harmonic equation (1) in M, $\sigma > 1$ and $p, q > 0$. There exists a constant C, depending only on σ, n, p, a, b, and q, such that $\|du\|_{p,Q} \leq C|Q|^{(q-p)/pq}\|du\|_{q,\sigma Q}$ for all Q with $\sigma Q \subset M$.*

Using the Hölder inequality and Lemma 3.10, we extend inequality (26) into the following weighted version.

$$\|G(u) - (G(u))_B\|_{s,B,w_1} \leq C|B| \operatorname{diam}(B)\|du\|_{s,\sigma B,w_2} \tag{34}$$

for all balls B with $\sigma B \subset M$, where $(w_1(x), w_2(x)) \in A_{r,\lambda}(M)$ for some $\lambda \geq 1$, $1 < r < \infty$ and $\sigma > 1$.

Theorem 3.11. *Let $du \in L^s(M, \wedge^l, \nu)$, $l = 1, 2, \ldots, n$, $1 < s < \infty$, be a solution of the nonhomogeneous A-harmonic equation in a bounded, convex domain M and G be Green's operator, where the measure μ and ν are defined by $d\mu = w_1 dx$, $d\nu = w_2 dx$ and $(w_1(x), w_2(x)) \in A_{r,\lambda}(M)$ for some $\lambda \geq 1$ and $1 < r < \infty$ with $w_1(x) \geq \varepsilon > 0$ for any $x \in M$. Then, there exists a constant C, independent of u, such that*

$$\|G(u)\|_{\mathrm{locLip}_k, M, w_1} \leq C\|du\|_{s,M,w_2}, \tag{35}$$

where k is a constant with $0 \leq k \leq 1$.

We now estimate the $\|\cdot\|_{\star,M,w_1^{\alpha}}$ norm in terms of the L^s norm.

Theorem 3.12. *Let $du \in L^s(M, \wedge^l, \nu)$, $l = 1, 2, \ldots, n$, $1 < s < \infty$, be a solution of the nonhomogeneous A-harmonic equation in a bounded domain M and G be Green's operator, where the measures μ and ν are defined by $d\mu = w_1 dx$, $d\nu = w_2 dx$ and $(w_1(x), w_2(x)) \in A_{r,\lambda}(M)$ for some $\lambda \geq 1$ and $1 < r < \infty$ with $w_1(x) \geq \varepsilon > 0$ for any $x \in M$. Then, there exists a constant C, independent of u, such that*

$$\|G(u)\|_{\star,M,w_1} \leq C\|du\|_{s,M,w_2}. \tag{36}$$

As applications, we display some estimates for the Jacobian $J(x, f)$ of a mapping $f : \Omega \to \mathbb{R}^n$, $f = (f^1, \dots, f^n)$. We know that the Jacobian $J(x, f)$ of a mapping f is an n-form, specifically, $J(x, f)dx = df^1 \wedge \cdots \wedge df^n$, where $dx = dx_1 \wedge dx_2 \wedge \cdots \wedge dx_n$. Let $f : \Omega \to \mathbb{R}^n$, $f = (f^1, \dots, f^n)$ be a mapping, whose distributional differential $Df = [\partial f^i / \partial x_j] : \Omega \to GL(n)$ is a locally integrable function in M with values in the space $GL(n)$ of all $n \times n$-matrices. We use

$$J(x, f) = \det Df(x) = \begin{vmatrix} f^1_{x_1} & f^1_{x_2} & f^1_{x_3} & \cdots & f^1_{x_n} \\ f^2_{x_1} & f^2_{x_2} & f^2_{x_3} & \cdots & f^2_{x_n} \\ \vdots & \vdots & \vdots & \ddots & \vdots \\ f^n_{x_1} & f^n_{x_2} & f^n_{x_3} & \cdots & f^n_{x_n} \end{vmatrix}$$

to denote the Jacobian determinant of f. Let u be the subdeterminant of Jacobian $J(x, f)$, which is obtained by deleting the l rows and l columns, $l = 0, 1, \dots, n-1$, that is,

$$u = J(x_{j_1}, x_{j_2}, \dots, x_{j_{n-l}}; f^{i_1}, f^{i_2}, \dots, f^{i_{n-l}}) = \begin{vmatrix} f^{i_1}_{x_{j_1}} & f^{i_1}_{x_{j_2}} & f^{i_1}_{x_{j_3}} & \cdots & f^{i_1}_{x_{j_{n-l}}} \\ f^{i_2}_{x_{j_1}} & f^{i_2}_{x_{j_2}} & f^{i_2}_{x_{j_3}} & \cdots & f^{i_2}_{x_{j_{n-l}}} \\ \vdots & \vdots & \vdots & \ddots & \vdots \\ f^{i_{n-l}}_{x_{j_1}} & f^{i_{n-l}}_{x_{j_2}} & f^{i_{n-l}}_{x_{j_3}} & \cdots & f^{i_{n-l}}_{x_{j_{n-l}}} \end{vmatrix} \tag{37}$$

which is a subdeterminant of Jacobian $J(x, f)$, here $\{i_1, i_2, \dots, i_{n-l}\} \subset \{1, 2, \dots, n\}$ and $\{j_1, j_2, \dots, j_{n-l}\} \subset \{1, 2, \dots, n\}$. Also, it is easy to see that

$$J(x_{j_1}, x_{j_2}, \dots, x_{j_{n-l}}; f^{i_1}, f^{i_2}, \dots, f^{i_{n-l}}) dx_{j_1} \wedge dx_{j_2} \wedge \cdots \wedge dx_{j_{n-l}}$$

is an $(n-l)$-form. Thus, all estimates for differential forms are applicable to the $(n-l)$-form $J(x_{j_1}, x_{j_2}, \dots, x_{j_{n-l}}; f^{i_1}, f^{i_2}, \dots, f^{i_{n-l}}) dx_{j_1} \wedge dx_{j_2} \wedge \cdots \wedge dx_{j_{n-l}}$. For example, choosing $u = J(x, f)dx$ and applying Theorems 3.11 and 3.12, we have the following Theorems 3.13 and 3.14, respectively.

Theorem 3.13. *Let $G(J(x, f)dx) \in \mathrm{locLip}_k(\Omega, \wedge^n)$, $0 \le k \le 1$, where $J(x, f)$ is the Jacobian of the mapping $f = (f^1, \dots, f^n) : \Omega \to \mathbb{R}^n$ and Ω is a bounded domain in $\mathbb{R}^n$, and G is Green's operator. Then, $G(J(x, f)dx) \in BMO(\Omega, \wedge^n)$ and*

$$\|G(J(x, f))\|_{\star, \Omega} \le C \|G(J(x, f))\|_{\mathrm{locLip}_k, \Omega}, \tag{38}$$

where C is a constant.

Theorem 3.14. *Let $u \in L^s(M, \wedge^{n-l}, \nu)$, $1 < s < \infty$, be the $(n-l)$-form defined by (37), $l = 1, 2, \ldots, n-1$, and G be Green's operator, where the measures μ and ν are defined by $d\mu = w_1 dx$, $d\nu = w_2 dx$ and $(w_1(x), w_2(x)) \in A_{r,\lambda}(M)$ for some $\lambda \geq 1$ and $1 < r < \infty$ with $w_1(x) \geq \varepsilon > 0$ for any $x \in M$. Then, there exists a constant C, independent of u, such that*

$$\|G(u)\|_{\mathrm{locLip}_k, M, w_1} \leq C \|du\|_{s,M,w_2},$$

where k is a constant with $0 \leq k \leq 1$.

Applying Theorem 3.12 to the $(n-l)$-form defined in (37), we have the following result.

Theorem 3.15. *Let G be Green's operator and $du \in L^s(M, \wedge^{n-l+1})$, $l = 1, 2, \ldots, n-1$, $1 < s < \infty$, where u is the $(n-l)$-form defined by (37) in a bounded, convex domain M. Then, there exists a constant C, independent of u, such that*

$$\|G(u)\|_{\star,M} \leq C \|du\|_{s,M}.$$

4 Inequalities with $L^p(\log L)^\alpha$ Norms

A continuously increasing function $\varphi : [0, \infty) \to [0, \infty)$ with $\varphi(0) = 0$ and $\varphi(\infty) = \infty$ is called an Orlicz function. The Orlicz space $L^\varphi(\Omega)$ consists of all measurable functions f on Ω such that

$$\int_\Omega \varphi\left(\frac{|f|}{k}\right) dx < \infty \tag{39}$$

for some $k = k(f) > 0$. $L^\varphi(\Omega)$ is equipped with the nonlinear Luxemburg functional

$$\|f\|_\varphi = \inf\{k > 0 : \frac{1}{|\Omega|} \int_\Omega \varphi\left(\frac{|f|}{k}\right) dx \leq 1\}. \tag{40}$$

A convex Orlicz function φ is often called a Young function. A special useful Young function $\varphi : [0, \infty) \to [0, \infty)$, termed an N-function, is a continuous Young function such that $\varphi(x) = 0$ if and only if $x = 0$ and $\lim_{x\to 0} \varphi(x)/x = 0$, $\lim_{x\to\infty} \varphi(x)/x = +\infty$. If φ is a Young function, then $\|\cdot\|_\varphi$ defines a norm in $L^\varphi(\Omega)$, which is called the Luxemburg norm. If φ is a Young function, then $\|\cdot\|_\varphi$ defines a norm in $L^\varphi(\Omega)$, which is called the Luxemburg norm. The Orlicz space $L^\psi(\Omega)$ with $\psi(t) = t^p \log^\alpha(e + t/c)$ will be denoted by $L^p(\log L)^\alpha(\Omega)$ and the corresponding norm will be denoted by $\|f\|_{L^p(\log L)^\alpha(\Omega)}$, where $1 \leq p < \infty$,

$\alpha \geq 0$, and $c > 0$ are constants. The spaces $L^p(\log L)^0(\Omega)$ and $L^1(\log L)^1(\Omega)$ are usually referred to as $L^p(\Omega)$ and $L \log L(\Omega)$, respectively. From [13], we have the equivalence

$$\|f\|_{L^p(\log L)^\alpha(\Omega)} \approx \left(\int_\Omega |f|^p log^\alpha\left(e + \frac{|f|}{||f||_{p,\Omega}}\right)dx\right)^{1/p}. \qquad (41)$$

Similarly, we have

$$\|f\|_{L^p(\log L)^\alpha(\Omega,\mu)} \approx \left(\int_\Omega |f|^p log^\alpha\left(e + \frac{|f|}{||f||_{p,\Omega}}\right)d\mu\right)^{1/p}, \qquad (42)$$

where μ is a measure defined by $d\mu = w(x)dx$ and $w(x)$ is a weight. In this chapter, we simply write

$$\|f\|_{L^p(\log L)^\alpha(E,w^\alpha)} = \left(\int_E |f|^p log^\alpha\left(e + \frac{|f|}{||f||_{p,E}}\right)w^\alpha dx\right)^{1/p} \qquad (43)$$

and $\|f\|_{L^p(\log L)^\alpha(E)} = \|f\|_{L^p(\log L)^\alpha(E,1)}$, where w is a weight.

In [4], the authors prove the following three inequalities for Green's operator with $L^p(logL)^\alpha$ norms.

Theorem 4.1. *Let $\omega \in D'(E, \wedge^k)$ be a solution of the nonhomogeneous A-harmonic equation in a domain $E \subset \mathbb{R}^n$ and $d\omega \in L^p(E, \wedge^{k+1})$, $k = 0, 1, \ldots, n$, $1 < p < \infty$. G is Green's operator. Then, there is a constant C, independent of ω, such that*

$$\|G(\omega) - (G(\omega))_B\|_{L^p(\log L)^\alpha(B)} \leq C|B| \operatorname{diam}(B)\|d\omega\|_{L^p(\log L)^\alpha(\sigma B)}. \qquad (44)$$

for all balls B with $\sigma B \subset E$ and $\operatorname{diam}(B) \geq d_0$. *Here $\alpha > 0$ is any constant and $\sigma > 1$ and $d_0 > 0$ are some constants.*

Theorem 4.2. *Let $\omega \in D'(E, \wedge^k)$ be a solution of the nonhomogeneous A-harmonic equation in a domain $E \subset \mathbb{R}^n$ and $d\omega \in L^p(E, \wedge^{k+1})$, $k = 0, 1, \ldots, n$. G is Green's operator. Assume that $1 < p < \infty$ and $(w_1(x), w_2(x)) \in A_r(E)$ for some $1 < r < \infty$. Then, there is a constant C, independent of ω, such that*

$$\|G(\omega) - (G(\omega))_B\|_{L^p(logL)^\alpha(B,w_1^\beta)} \leq C|B| \operatorname{diam}(B)\|d\omega\|_{L^p(logL)^\alpha(\sigma B,w_2^\beta)}. \qquad (45)$$

for all balls B with $\sigma B \subset E$ and $\operatorname{diam}(B) \geq d_0$. *Here $\alpha > 0$ is any constant and $\sigma > 1$, $0 < \beta \leq 1$ and $d_0 > 0$ are some constants.*

Theorem 4.3. *Assume G is Green's operator and $\Omega \subset \mathbb{R}^n$ is a bounded $L^\varphi(\mu)$-domain with $\varphi(t) = t^p log^\alpha\left(e + \frac{t}{c}\right)$, where $c = \|G(\omega) - (G(\omega))_{B_0}\|_{p,\Omega}$, $1 < p < \infty$, and $B_0 \subset \Omega$ is a fixed ball. Let $\omega \in D'(\Omega, \wedge^0)$ be a solution of the* *nonhomogeneous A-harmonic equation in* Ω *and* $d\omega \in L^p(\Omega, \wedge^1)$ *as well as* $(w_1(x), w_2(x)) \in A_r(\Omega)$ *for some* $1 < r < \infty$*. Then, there exists a constant* C*, independent of* ω*, such that*

$$\|G(\omega) - (G(\omega))_\Omega\|_{L^p(\log L)^\alpha(\Omega, w_1^\beta)} \le C|\Omega| \operatorname{diam}(\Omega)\|d\omega\|_{L^p(\log L)^\alpha(\Omega, w_2^\beta)}, \tag{46}$$

where β *is a constant with* $0 < \beta \le 1$.

5 Inequalities with L^φ Norms

In the last section, we present the inequalities for Green's operator with $L^p(logL)^\alpha$ norms. In this section, we will study the Poincaré-type inequalities with unbounded factors for Green's operator on the solutions of the nonhomogeneous A-harmonic equation for differential forms

$$d^\star A(x, du) = B(x, du)$$

in $\mathbb{R}^n$. Furthermore, we discuss both local and global Poincaré inequalities with L^φ norms (Orlicz norms) for Green's operator applied to differential forms in $L^\varphi(m)$-averaging domains. These results are extensions of L^p norm inequalities for Green's operator and can be used to estimate the norms of differential forms or the norms of other operators, such as the projection operator. The Poincaré-type inequalities have been widely studied and used in PDEs, analysis, and the related areas, and different versions of the Poincaré-type inequalities have been established during the recent years; see [1, 2, 4, 12, 22]. We all know that Green's operator is one of key operators which are widely used in many areas, such as analysis and PDEs. In many situations, we often need to evaluate the integrals with unbounded factors. For instance, if the object P_1 with mass m_1 is located at the origin and the object P_2 with mass m_2 is located at (x, y, z) in $\mathbb{R}^3$, then Newton's Law of Gravitation states that the magnitude of the gravitational force between two objects P_1 and P_2 is $|\mathbf{F}| = m_1 m_2 G/d^2(P_1, P_2)$, where $d(P_1, P_2) = \sqrt{x^2 + y^2 + z^2}$ is the distance between P_1 and P_2, and G is the gravitational constant. Hence, we need to deal with an integral whenever the integrand contains $|\mathbf{F}|$ as a factor and the integral domain includes the origin. Moreover, in calculating an electric field, we will evaluate the integral

$$E(y) = \frac{1}{4\pi\epsilon_0} \int_D \rho(x) \frac{y - x}{\|y - x\|^3} dx,$$

where $\rho(x)$ is a charge density and x is the integral variable. The integrand is unbounded if $y \in D$. This is the motivation to study the Poincaré-type inequalities for Green's operator G with unbounded factors. All results presented in this section can be found in [26].

Theorem 5.1. *Let $du \in L^s_{loc}(\Omega, \wedge^l)$, $l = 1, 2, \ldots, n$, $1 < s < \infty$, be a solution of the nonhomogeneous A-harmonic equation in a bounded domain Ω and G be Green's operator. Then, there exists a constant C, independent of u, such that*

$$\left(\int_B |G(u) - (G(u))_B|^s \frac{1}{|x - x_B|^\alpha} dx\right)^{1/s}$$
$$\leq C(n, s, \alpha, \lambda, \Omega)|B|^\gamma \left(\int_{\sigma B} |du|^s \frac{1}{|x - x_B|^\lambda} dx\right)^{1/s} \quad (47)$$

for all balls B with $\sigma B \subset \Omega$ and any real numbers α and λ with $\alpha > \lambda \geq 0$, where $\gamma = \frac{1}{n} - \frac{\alpha - \lambda}{ns}$ and x_B is the center of ball B and $\sigma > 1$ is a constant.

Since

$$\frac{1}{d(x, \partial\Omega)} \leq \frac{1}{r_B - |x|}$$

for any $x \in B$, where r_B is the radius of ball $B \subset \Omega$, using the same method developed in the proof of Theorem 5.1, the author proved the following Poincaré-type inequality for Green's operator with unbounded factors in [26].

Theorem 5.2. *Let $du \in L^s_{loc}(\Omega, \wedge^l)$, $l = 1, 2, \ldots, n$, $1 < s < \infty$, be a solution of the nonhomogeneous A-harmonic equation in a bounded domain Ω, G be Green's operator. Then, there exists a constant C, independent of u, such that*

$$\left(\int_B |G(u) - (G(u))_B|^s \frac{1}{d^\alpha(x, \partial\Omega)} dx\right)^{1/s}$$
$$\leq C(n, s, \alpha, \lambda, \Omega)|B|^\gamma \left(\int_{\sigma B} |du|^s \frac{1}{|x - x_B|^\lambda} dx\right)^{1/s} \quad (48)$$

for all balls B with $\sigma B \subset \Omega$ and any real numbers α and λ with $\alpha > \lambda \geq 0$, where $\gamma = \frac{1}{n} - \frac{\alpha - \lambda}{ns}$ and x_B is the center of ball B and $\sigma > 1$ is a constant.

Definition 5.3. A proper subdomain $\Omega \subset \mathbb{R}^n$ is called a δ-John domain, $\delta > 0$, if there exists a point $x_0 \in \Omega$ which can be joined with any other point $x \in \Omega$ by a continuous curve $\gamma \subset \Omega$ so that

$$d(\xi, \partial\Omega) \geq \delta|x - \xi|$$

for each $\xi \in \gamma$. Here $d(\xi, \partial\Omega)$ is the Euclidean distance between ξ and $\partial\Omega$.

Lemma 5.4 ([18] Covering Lemma). *Each Ω has a modified Whitney cover of cubes $\mathcal{V} = \{Q_i\}$ such that*

$$\cup_i Q_i = \Omega, \quad \sum_{Q_i \in \mathcal{V}} \chi_{\sqrt{\frac{5}{4}}Q_i} \leq N\chi_\Omega$$

and some $N > 1$, and if $Q_i \cap Q_j \neq \emptyset$, then there exists a cube R (this cube need not be a member of $\mathcal{V}$) in $Q_i \cap Q_j$ such that $Q_i \cup Q_j \subset NR$. Moreover, if Ω is δ -John, then there is a distinguished cube $Q_0 \in \mathcal{V}$ which can be connected with every cube $Q \in \mathcal{V}$ by a chain of cubes $Q_0, Q_1, \ldots, Q_k = Q$ from $\mathcal{V}$ and such that $Q \subset \rho Q_i$, $i = 0, 1, 2, \ldots, k$, for some $\rho = \rho(n, \delta)$.

The following global Poincaré-type inequalities for Green's operator with unbounded factors in John domains were also proved in [26].

Theorem 5.5. *Let $u \in D'(\Omega, \wedge^0)$ be a solution of the nonhomogeneous A-harmonic equation (1) and s be a fixed exponent associated with (1). Then, there exists a constant $C(n, N, s, \alpha, \lambda, Q_0, \Omega)$, independent of u, such that*

$$\left(\int_\Omega |G(u) - (G(u))_{Q_0}|^s \frac{1}{d^\alpha(x, \partial\Omega)} dx \right)^{1/s} \leq C(n, N, s, \alpha, \lambda, Q_0, \Omega) \left(\int_\Omega |du|^s g(x) dx \right)^{1/s} \tag{49}$$

for any bounded δ-John domain $\Omega \subset \mathbb{R}^n$, where $g(x) = \sum_i \chi_{Q_i} \frac{1}{|x - x_{Q_i}|^\lambda}$. Here α and λ are constants with $0 \leq \lambda < \alpha < min\{n, s + \lambda - n\}$, $s + \lambda > n$, and the fixed cube $Q_0 \subset \Omega$, the cubes $Q_i \subset \Omega$, and the constant $N > 1$ appeared in Lemma 5.4; x_{Q_i} is the center of Q_i.

Next, we present the global norm inequality in the $L^\varphi(m)$-averaging domains, which are extension of John domains and L^s-averaging domain; see [2].

Definition 5.6 ([5]). We say a Young function φ lies in the class $G(p, q, C)$, $1 \leq p < q < \infty$, $C \geq 1$, if (i) $1/C \leq \varphi(t^{1/p})/\Phi(t) \leq C$ and (ii) $1/C \leq \varphi(t^{1/q})/\Psi(t) \leq C$ for all $t > 0$, where Φ is a convex increasing function and Ψ is a concave increasing function on $[0, \infty)$.

From [5], each of φ, Φ, and Ψ in above definition is doubling in the sense that its values at t and $2t$ are uniformly comparable for all $t > 0$ and the consequent fact that

$$C_1 t^q \leq \Psi^{-1}(\varphi(t)) \leq C_2 t^q, \quad C_1 t^p \leq \Phi^{-1}(\varphi(t)) \leq C_2 t^p, \tag{50}$$

where C_1 and C_2 are constants. Also, for all $1 \leq p_1 < p < p_2$ and $\alpha \in \mathbb{R}$, the function $\varphi(t) = t^p \log_+^\alpha t$ belongs to $G(p_1, p_2, C)$ for some constant $C = C(p, \alpha, p_1, p_2)$. Here $\log_+(t)$ is defined by $\log_+(t) = 1$ for $t \leq e$ and $\log_+(t) = \log(t)$ for $t > e$. Particularly, if $\alpha = 0$, we see that $\varphi(t) = t^p$ lies in $G(p_1, p_2, C)$, $1 \leq p_1 < p < p_2$.

We first prove the following generalized Poincaré inequality that will be used to establish the global inequality.

Theorem 5.7. *Let φ be a Young function in the class $G(p, q, C)$, $1 \leq p < q < \infty$, $C \geq 1$, Ω be a bounded domain and $q(n - p) < np$. Assume that $u \in D'(\Omega, \wedge^l)$ is* *any differential l-form, $l = 0, 1, \ldots, n-1$, and $\varphi(|du|) \in L^1_{\text{loc}}(\Omega, m)$. Then, there exists a constant C, independent of u, such that*

$$\int_B \varphi\left(|G(u) - (G(u))_B|\right) dm \leq C \int_B \varphi\left(|du|\right) dm \tag{51}$$

for all balls B with $B \subset \Omega$.

Lemma 5.8. *Let $u \in D'(\Omega, \wedge^l)$, $l = 0, 1, \ldots, n-1$, be an A-harmonic tensor on Ω. Assume that $\rho > 1$ and $1 < s < \infty$. Then, there exists a constant C, independent of u, such that*

$$\|\Delta G(u) - (\Delta G(u))_B\|_{s,B} \leq C \operatorname{diam}(B) \|du\|_{s,\rho B} \tag{52}$$

for any ball B with $\rho B \subset \Omega$.

Using Lemma 5.8 and the method developed in the proof of Theorem 5.7, the authors also prove the following version of Poincaré-type inequality for the composition of Δ and G.

Theorem 5.9. *Let φ be a Young function in the class $G(p, q, C)$, $1 \leq p < q < \infty$, $C \geq 1$, Ω be a bounded domain, and $q(n-p) < np$. Assume that $u \in D'(\Omega, \wedge^l)$ is any differential l-form, $l = 0, 1, \ldots, n-1$, and $\varphi(|du|) \in L^1_{\text{loc}}(\Omega, m)$. Then, there exists a constant C, independent of u, such that*

$$\int_B \varphi\left(|\Delta G(u) - (\Delta G(u))_B|\right) dm \leq C \int_B \varphi\left(|du|\right) dm \tag{53}$$

for all balls B with $B \subset \Omega$.

The following $L^\varphi(m)$-averaging domains can be found in [9].

Definition 5.10 ([9]). Let φ be an increasing convex function on $[0, \infty)$ with $\varphi(0) = 0$. We call a proper subdomain $\Omega \subset \mathbb{R}^n$ an $L^\varphi(m)$-averaging domain, if $m(\Omega) < \infty$ and there exists a constant C such that

$$\int_\Omega \varphi(\tau|u - u_{B_0}|)dm \leq C \sup_{B \subset \Omega} \int_B \varphi(\sigma|u - u_B|)dm \tag{54}$$

for some ball $B_0 \subset \Omega$ and all u such that $\varphi(|u|) \in L^1_{\text{loc}}(\Omega, m)$, where τ, σ are constants with $0 < \tau < \infty$, $0 < \sigma < \infty$ and the supremum is over all balls $B \subset \Omega$.

From the above definition we see that L^s-averaging domains and $L^s(m)$-averaging domains are special $L^\varphi(m)$-averaging domains when $\varphi(t) = t^s$ in Definition 5.10. Also, uniform domains and John domains are very special $L^\varphi(m)$-averaging domains; see [2, 9] for more results about domains.

Theorem 5.11. *Let φ be a Young function in the class $G(p,q,C)$, $1 \le p < q < \infty$, $C \ge 1$, Ω be a bounded $L^{\varphi}(m)$-averaging domain and $q(n-p) < np$. Assume that $u \in D'(\Omega, \wedge^0)$ and $\varphi(|du|) \in L^1(\Omega, m)$. Then, there exists a constant C, independent of u, such that*

$$\int_{\Omega} \varphi\left(|G(u) - (G(u))_{B_0}|\right) dm \le C \int_{\Omega} \varphi\left(|du|\right) dm, \tag{55}$$

where $B_0 \subset \Omega$ is some fixed ball.

Choosing $\varphi(t) = t^p \log_+^{\alpha} t$ in Theorems 5.11, we obtain the following Poincaré inequalities with the $L^p(\log_+^{\alpha} L)$-norms.

Corollary 5.12. *Let $\varphi(t) = t^p \log_+^{\alpha} t$, $1 \le p_1 < p < p_2$ and $\alpha \in \mathbb{R}$ and Ω be a bounded $L^{\varphi}(m)$-averaging domain and $p_2(n - p_1) < np_1$. Assume that $u \in D'(\Omega, \wedge^0)$, $\varphi(|du|) \in L^1(\Omega, m)$, Then, there exists a constant C, independent of u, such that*

$$\int_{\Omega} |G(u) - (G(u))_{B_0}|^p \log_+^{\alpha}\left(|G(u) - (G(u))_{B_0}|\right) dm \le C \int_{\Omega} |du|^p \log_+^{\alpha}\left(|du|\right) dm, \tag{56}$$

where $B_0 \subset \Omega$ is some fixed ball.

Note that (56) can be written as the following version with the Luxemburg norm

$$\|G(u) - (G(u))_{B_0}\|_{L^p(\log_+^{\alpha} L)(\Omega)} \le C \|du\|_{L^p(\log_+^{\alpha} L)(\Omega)}$$

provided the conditions in Corollary 5.12 are satisfied.

6 Actions on Minimizers

In this section, we discuss both local and global L^{φ}-norm inequalities for Green's operator acting on minimizers for functionals defined on differential forms in L^{φ}-averaging domains. These results are extensions of L^p-norm inequalities for Green's operator and can be used to estimate the norms of other operators applied to differential forms.

We say that a differential form $u \in W^{1,1}_{\mathrm{loc}}(\Omega, \Lambda^{\ell})$ is a k-quasi-minimizer for the functional

$$I(\Omega; v) = \int_{\Omega} \varphi(|dv|) dx \tag{57}$$

if and only if, for every $\varphi \in W^{1,1}_{\mathrm{loc}}(\Omega, \Lambda^{\ell})$ with compact support,

$$I(\operatorname{supp}\varphi; u) \le k \cdot I(\operatorname{supp}\varphi; u + \varphi),$$

where $k > 1$ is a constant. We say that φ satisfies the so-called Δ_2-condition if there exists a constant $p > 1$ such that

$$\varphi(2t) \le p\varphi(t) \tag{58}$$

for all $t > 0$, from which it follows that $\varphi(\lambda t) \le \lambda^p \varphi(t)$ for any $t > 0$ and $\lambda \ge 1$; see [15].

The results presented in this section were recently obtained in [1]. We will need the following lemma which can be found in [15] or [19].

Lemma 6.1. *Let $f(t)$ be a nonnegative function defined on the interval $[a,b]$ with $a \ge 0$. Suppose that for $s,t \in [a,b]$ with $t < s$,*

$$f(t) \le \frac{M}{(s-t)^\alpha} + N + \theta f(s)$$

holds, where M, N, α and θ are nonnegative constants with $\theta < 1$. Then, there exists a constant $C = C(\alpha, \theta)$ such that

$$f(\rho) \le C\left(\frac{M}{(R-\rho)^\alpha} + N\right)$$

for any $\rho, R \in [a,b]$ *with $\rho < R$.*

Theorem 6.2. *Let $u \in W^{1,1}_{\mathrm{loc}}(\Omega, \Lambda^\ell)$ be a k-quasi-minimizer for the functional (57), φ be a Young function in the class $G(p,q,C)$, $1 \le p < q < \infty$, $C \ge 1$ and $q(n-p) < np$, Ω be a bounded domain, and G be Green's operator. Then, there exists a constant C, independent of u, such that*

$$\int_B \varphi(|G(u) - (G(u))_B|)dx \le C \int_{2B} \varphi(|u-c|)\, dx \tag{59}$$

for all balls $B = B_r$ with radius r and $2B \subset \Omega$, where c is any closed form.

Since each of φ, Φ and Ψ is doubling, from the proof of Theorem 6.2 or directly from (50), we have

$$\int_B \varphi\left(\frac{|G(u) - (G(u))_B|}{\lambda}\right) dx \le C \int_{2B} \varphi\left(\frac{|u-c|}{\lambda}\right) dx \tag{60}$$

for all balls B with $2B \subset \Omega$ and any constant $\lambda > 0$. From definition of the Luxemburg norm and (60), the following inequality with the Luxemburg norm

$$\|G(u) - (G(u))_B\|_{\varphi(B)} \le C\,\|u-c\|_{\varphi(2B)} \tag{61}$$

holds under the conditions described in Theorem 6.2.

Note that in Theorem 6.2, c is any closed form. Hence, we may choose $c = 0$ in Theorem 6.2 and obtain the following version of φ-norm inequality which may be convenient to be used.

Corollary 6.3. *Let $u \in W^{1,1}_{\rm loc}(\Omega, \Lambda^\ell)$ be a k-quasi-minimizer for the functional (57), φ be a Young function in the class $G(p,q,C)$, $1 \le p < q < \infty$, $C \ge 1$ and $q(n-p) < np$, Ω be a bounded domain, and G be Green's operator. Then, there exists a constant C, independent of u, such that*

$$\int_B \varphi(|G(u) - (G(u))_B|)dx \le C \int_{2B} \varphi\left(|u|\right) dx \tag{62}$$

for all balls $B = B_r$ with radius r and $2B \subset \Omega$.

Theorem 6.4. *Let $u \in W^{1,1}_{\rm loc}(\Omega, \Lambda^0)$ be a k-quasi-minimizer for the functional (57), φ be a Young function in the class $G(p,q,C)$, $1 \le p < q < \infty$, $C \ge 1$ and $q(n-p) < np$, Ω be any bounded L^φ-averaging Domain, and G be Green's operator. Then, there exists a constant C, independent of u, such that*

$$\int_\Omega \varphi(|G(u) - (G(u))_{B_0}|)dx \le C \int_\Omega \varphi\left(|u - c|\right) dx, \tag{63}$$

where $B_0 \subset \Omega$ is some fixed ball and c is any closed form.

We know that any John domain is a special L^φ-averaging domain. Hence, we have the following inequality in John domain.

Theorem 6.5. *Let $u \in W^{1,1}_{\rm loc}(\Omega, \Lambda^0)$ be a k-quasi-minimizer for the functional (57), φ be a Young function in the class $G(p,q,C)$, $1 \le p < q < \infty$, $C \ge 1$ and $q(n-p) < np$, Ω be any bounded John domain, and G be Green's operator. Then, there exists a constant C, independent of u, such that*

$$\int_\Omega \varphi(|G(u) - (G(u))_{B_0}|)dx \le C \int_\Omega \varphi\left(|u - c|\right) dx, \tag{64}$$

where $B_0 \subset \Omega$ is some fixed ball and c is any closed form.

Choosing $\varphi(t) = t^p \log_+^\alpha t$ in Theorem 6.5, we obtain the following inequalities with the $L^p(\log_+^\alpha L)$-norms.

Corollary 6.6. *Let $u \in W^{1,1}_{\rm loc}(\Omega, \Lambda^0)$ be a k-quasi-minimizer for the functional (57), $\varphi(t) = t^p \log_+^\alpha t$, $\alpha \in \mathbb{R}$, $q(n-p) < np$ for $1 \le p < q < \infty$ and G be Green's operator. Then, there exists a constant C, independent of u, such that*

$$\begin{aligned} &\int_\Omega |G(u) - (G(u))_{B_0}|^p \log_+^\alpha \left(|G(u) - (G(u))_{B_0}|\right) dx \\ &\le C \int_\Omega |u - c|^p \log_+^\alpha \left(|u - c|\right) dx \end{aligned} \tag{65}$$

for any bounded L^φ-averaging domain Ω, where $B_0 \subset \Omega$ is some fixed ball and c is any closed form.

We can also write (65) as the following inequality with the Luxemburg norm

$$\|G(u) - (G(u))_{B_0}\|_{L^p(\log_+^\alpha L)(\Omega)} \le C\,\|u - c\|_{L^p(\log_+^\alpha L)(\Omega)} \tag{66}$$

provided the conditions in Corollary 6.6 are satisfied.

Similar to the local case, we may choose $c = 0$ in Theorem 6.5 and obtain the following version of L^φ-norm inequality.

Corollary 6.7. *Let $u \in W^{1,1}_{\mathrm{loc}}(\Omega, \Lambda^0)$ be a k-quasi-minimizer for the functional (57), φ be a Young function in the class $G(p,q,C)$, $1 \le p < q < \infty$, $C \ge 1$ and $q(n-p) < np$, Ω be any bounded L^φ-averaging domain, and G be Green's operator. Then, there exists a constant C, independent of u, such that*

$$\int_\Omega \varphi(|G(u) - (G(u))_{B_0}|)dx \le C \int_\Omega \varphi\,(|u|)\,dx, \tag{67}$$

where $B_0 \subset \Omega$ is some fixed ball.

It should be noticed that both of the local and global norm inequalities for Green's operator presented in this section can be used to estimate other operators applied to a k-quasi-minimizer. Here, we give an example using Theorem 6.5 to estimate the projection operator H. Similar to the case of harmonic tensors, we can prove the similar result for k-quasi-minimizers. Using the basic Poincaré inequality to $\Delta G(u)$ and noticing that d commute with Δ and G, we can prove the following Lemma 6.8.

Lemma 6.8. *Let $u \in W^{1,1}_{\mathrm{loc}}(\Omega, \Lambda^\ell)$ be a k-quasi-minimizer for the functional (57). Assume that $\rho > 1$ and $1 < s < \infty$. Then, there exists a constant C, independent of u, such that*

$$\|\Delta G(u) - (\Delta G(u))_B\|_{s,B} \le C\,\mathrm{diam}\,(B)\|du\|_{s,\rho B} \tag{68}$$

for any ball B with $\rho B \subset \Omega$.

Using Lemma 6.8 and the method developed in the proof of Theorem 6.5, we can prove the following inequality for the composition of Δ and G.

Theorem 6.9. *Let $u \in W^{1,1}_{\mathrm{loc}}(\Omega, \Lambda^\ell)$ be a k-quasi-minimizer for the functional (57), φ be a Young function in the class $G(p,q,C)$, $1 \le p < q < \infty$, $C \ge 1$ and $q(n-p) < np$, Ω be a bounded domain, and G be Green's operator. Then, there exists a constant C, independent of u, such that*

$$\int_B \varphi\,(|\Delta G(u) - (\Delta G(u))_B|)\,dx \le C \int_{2B} \varphi\,(|u - c|)\,dx \tag{69}$$

for all balls $B = B_r$ with radius r and $2B \subset \Omega$, where c is any closed form.

Now, we are ready to estimate the projection operator applied to a k-quasi-minimizer for the functional defined by (57).

Theorem 6.10. *Let $u \in W^{1,1}_{loc}(\Omega, \Lambda^\ell)$ be a k-quasi-minimizer for the functional (57), φ be a Young function in the class $G(p, q, C)$, $1 \le p < q < \infty$, $C \ge 1$ and $q(n-p) < np$, Ω be a bounded domain, and H be projection operator. Then, there exists a constant C, independent of u, such that*

$$\int_B \varphi(|H(u) - (H(u))_B|)dx \le C \int_{2B} \varphi(|u - c|)\, dx \tag{70}$$

for all balls $B = B_r$ with radius r and $2B \subset \Omega$, where c is any closed form.

Remark. (i) We know that the L^s-averaging domains and uniform domains are the special L^φ-averaging domains. Thus, Theorems 6.4 also holds if Ω is tan L^s-averaging domain or uniform domain. (ii) Theorem 6.10 can also be extended into the global case in $L^\varphi(m)$-averaging domain.

7 Compositions with the Maximal Operators

The main purpose of this section is to present some L^s-norm estimates for the compositive operators $\mathbb{M}_s \circ G$ and $\mathbb{M}_s^\sharp \circ G$. Here $\mathbb{M}_s$ is the Hardy-Littlewood maximal operator, $\mathbb{M}_s^\sharp$ is the sharp maximal operator, and G is Green's operator, applied to differential forms.

For a locally L^s-integrable form $u(y)$, the Hardy-Littlewood maximal operator $\mathbb{M}_s$ is defined by

$$\mathbb{M}_s(u) = \mathbb{M}_s u = \mathbb{M}_s u(x) = \sup_{r>0} \left(\frac{1}{|B(x,r)|} \int_{B(x,r)} |u(y)|^s dy \right)^{1/s}, \tag{71}$$

where $B(x, r)$ is the ball of radius r, centered at x, $1 \le s < \infty$. We write $\mathbb{M}(u) = \mathbb{M}_1(u)$ if $s = 1$. Similarly, for a locally L^s-integrable form u, we define the sharp maximal operator $\mathbb{M}_s^\sharp$ by

$$\mathbb{M}_s^\sharp(u) = \mathbb{M}_s^\sharp u = \mathbb{M}_s^\sharp u(x) = \sup_{r>0} \left(\frac{1}{|B(x,r)|} \int_{B(x,r)} |u(y) - u_{B(x,r)}|^s dy \right)^{1/s}. \tag{72}$$

These operators and Green's operator play an important role in many diverse fields, including partial differential equations and analysis. All results presented in this section were obtained in [10]. From [21], we know that if $u \in L^s(M, \wedge^l)$, $1 < s < \infty$, then $\mathbb{M}(u) \in L^s(M)$; specifically, we have the following lemma.

Lemma 7.1. *Let $u \in L^s(M, \wedge^l)$, $l = 0, 1, 2, \ldots, n$, $1 < s < \infty$, be a differential form in a domain M and $\mathbb{M}$ be the Hardy-Littlewood maximal operator defined in (1) with $s = 1$.*

$$\|\mathbb{M}(u)\|_{s,M} \le C \|u\|_{s,M} \tag{73}$$

for some constant C, independent of u.

We first introduce the following estimate for the Hardy-Littlewood maximal operator.

Theorem 7.2. *Let* $\mathbb{M}_s$ *be the Hardy-Littlewood maximal operator defined in (71) and* $u \in L^t(M, \wedge^l)$, $l = 1, 2, \ldots, n$, $1 \le s < t < \infty$, *be a differential form in a domain* M. *Then,* $\mathbb{M}_s(u) \in L^t(M)$ *and*

$$\|\mathbb{M}_s(u)\|_{t,M} \le C\|u\|_{t,M} \tag{74}$$

for some constant C, *independent of* u.

If we replace u by $G(u)$ in Theorem 7.2 and use Lemma 3.4, we have the following estimate for the composition of the Hardy-Littlewood maximal operator and Green's operator.

Theorem 7.3. *Let* $\mathbb{M}_s$ *be the Hardy-Littlewood maximal operator defined in (71),* $1 \le s < t < \infty$, G *be Green's operator, and* $u \in C^\infty(\wedge^l M)$, $l = 1, 2, \ldots, n$, *be a differential form in a domain* M. *Then,*

$$\|\mathbb{M}_s(G(u))\|_{t,M} \le C\|u\|_{t,M} \tag{75}$$

for some constant C, *independent of* u.

We now develop some estimates related to the sharp maximal operator $\mathbb{M}_s^\sharp$ and Green's operator and then study the relationship between $\|\mathbb{M}_s^\sharp\|_{s,M}$ and $\|\mathbb{M}_s\|_{s,M}$.

Theorem 7.4. *Let* $u \in C^\infty(\wedge^l M)$, $l = 1, 2, \ldots, n$, $1 < s < \infty$, *be a differential form in a bounded domain* M, $\mathbb{M}_s^\sharp$ *be the sharp maximal operator defined in (72), and* G *be Green's operator. Then,*

$$\|\mathbb{M}_s^\sharp(G(u))\|_{s,M} \le C|M|^{1/s}\|u\|_{s,M} \tag{76}$$

for some constant C, *independent of* u.

Theorem 7.5. *Let* $u \in L^s_{\text{loc}}(M, \wedge^l)$, $l = 0, 1, 2, \ldots, n-1$, $1 < s < \infty$, *be a smooth differential form in a bounded domain* M, $\mathbb{M}$ *be the Hardy-Littlewood maximal operator defined in 73), and* $\mathbb{M}_s^\sharp$ *be the sharp maximal operator defined in (74). Then,*

$$\|\mathbb{M}_s^\sharp u\|_{s,M} \le C\|\mathbb{M}_s du\|_{s,M} \tag{77}$$

for some constant C, *independent of* u.

Next, we introduce the fractional maximal operator of order α. Let $u(y)$ be a locally L^s-integrable form, $1 \le s < \infty$, and α be a real number. We define the *fractional maximal operator* $\mathbb{M}_{s,\alpha}$ of order α by

$$\mathbb{M}_{s,\alpha}u(x) = \sup_{r>0}\left(\frac{1}{|B(x,r)|^{1+\alpha/n}}\int_{B(x,r)}|u(y)|^s dy\right)^{1/s}. \tag{78}$$

Clearly, (78) for $\alpha = 0$ reduces to the Hardy-Littlewood maximal operator, and hence, we write $\mathbb{M}_s(u) = \mathbb{M}_{s,0}(u)$.

Theorem 7.6. *Let $u \in L^s(M, \wedge^l)$, $l = 0, 1, 2, \ldots, n$, $1 < s < \infty$, be a smooth differential form satisfying equation (1) in a bounded domain M, $\mathbb{M}_s$ be the Hardy-Littlewood maximal operator defined in (71), and $\mathbb{M}_{s,\alpha}$ be the fractional maximal operator of order α. Then,*

$$\|\mathbb{M}_s du(x)\|_{s,M} \le C\,\|\mathbb{M}_{s,\alpha}(u(x) - c)\|_{s,M} \tag{79}$$

for some constant C, independent of u, where $\alpha = s$ and c is any closed form.

Note that in Theorem 7.6, c is any closed form. Thus, we can choose $c = 0$ in Theorem 7.6, to obtain the following corollary.

Corollary 7.7. *Let $u \in L^s(M, \wedge^l)$, $l = 1, 2, \ldots, n$, $1 < s < \infty$, be a smooth differential form satisfying equation (1) in a bounded domain M, $\mathbb{M}_s$ be the Hardy-Littlewood maximal operator defined in (71), and $\mathbb{M}_{s,\alpha}$ be the fractional maximal operator of order α. Then,*

$$\|\mathbb{M}_s du(x)\|_{s,M} \le C\,\|\mathbb{M}_{s,\alpha}u(x)\|_{s,M} \tag{80}$$

for some constant C, independent of u.

Lemma 7.7 ([2]). *If $w \in A_r(M)$, then there exist constants $\beta > 1$ and C, independent of w, such that*

$$\| w \|_{\beta,B} \le C|B|^{(1-\beta)/\beta} \| w \|_{1,B} \tag{81}$$

for all balls $B \subset M$.

Theorem 7.8. *Let $\mathbb{M}_s$ be the Hardy-Littlewood maximal operator defined in (71), G be Green's operator, and $u \in C^{\infty}(\wedge^l M)$ be a solution to the A-harmonic equation (1) in a domain M, where $1 \le s < t < \infty$ and, $l = 1, 2, \ldots, n$. Assume that $w \in A_r$ for some $r > 1$. Then, there exists a constant C, independent of u, such that*

$$\|\mathbb{M}_s(G(u))\|_{t,B,w} \le C\,\|u\|_{t,\sigma B,w} \tag{82}$$

for all balls B with $\sigma B \subset M$, where $\sigma > 1$ is a constant.

Using the similar method to the proof of Theorem 7.8, we can prove the following theorem.

Theorem 7.9. *Let $u \in C^\infty(\wedge^l M)$, $l = 1, 2, \ldots, n$, $1 < s < \infty$, be a solution to the A-harmonic equation (1) in a domain M, $\mathbb{M}_s^\sharp$ be the sharp maximal operator defined in (72), and G be Green's operator. Assume that $w \in A_r$ for some $r > 1$. Then, there exists a constant C, independent of u, such that*

$$\|\mathbb{M}_s^\sharp(G(u))\|_{s,B,w} \le C\,\|u\|_{s,\sigma B,w} \tag{83}$$

for all balls B with $\sigma B \subset M$, where $\sigma > 1$ is a constant.

We say a pair of weights $(w_1(x), w_2(x))$ satisfies the $A_{r,\lambda}(E)$-condition in a set $E \subset \mathbf{R}^n$ and write $(w_1(x), w_2(x)) \in A_{r,\lambda}(E)$, for some $\lambda \ge 1$ and $1 < r < \infty$ with $1/r + 1/r' = 1$ if

$$\sup_{B \subset E} \left(\frac{1}{|B|}\int_B (w_1)^\lambda dx\right)^{1/\lambda r} \left(\frac{1}{|B|}\int_B \left(\frac{1}{w_2}\right)^{\lambda r'/r} dx\right)^{1/\lambda r'} < \infty. \tag{84}$$

The class of $A_{r,\lambda}(E)$-weights (or the two-weight) appears in [17]. It is easy to see that the $A_{r,\lambda}(E)$-weight is an extension of the usual $A_r(E)$-weight.

Theorem 7.10. *Let $\mathbb{M}_s$ be the Hardy-Littlewood maximal operator defined in (71) and $u \in L^t(M, \wedge^l)$, $l = 1, 2, \ldots, n$, be a solution to the A-harmonic equation (1) in a domain M, where $1 \le s < t < \infty$. Assume that $(w_1(x), w_2(x)) \in A_{r,\lambda}(M)$ for some $\lambda \ge 1$ and $1 < r < \infty$. Then, there exists a constant C, independent of u, such that*

$$\|\mathbb{M}_s(u)\|_{t,B,w_1^\alpha} \le C\,\|u\|_{t,\sigma B,w_2^\alpha} \tag{85}$$

or

$$\left(\int_B |\mathbb{M}_s(u)|^t w_1^\alpha dx\right)^{1/t} \le C\left(\int_{\sigma B} |u|^t w_2^\alpha dx\right)^{1/t} \tag{86}$$

for all balls B with $\sigma B \subset M$. Here α and $\sigma > 1$ are constants with $0 < \alpha < \lambda$.

Note that Theorem 7.10 contains two weights, $w_1(x)$ and $w_2(x)$, and two parameters, λ and α. These features make the inequality be more flexible and more useful. For example, if we choose $\alpha = 1$ in Theorem 7.10, we have the following corollary.

Corollary 7.11. *Let $\mathbb{M}_s$ be the Hardy-Littlewood maximal operator defined in (71) and $u \in L^t(M, \wedge^l)$, $l = 1, 2, \ldots, n$, be a solution to the A-harmonic equation (1) in a domain M, where $1 \le s < t < \infty$. Assume that $(w_1(x), w_2(x)) \in A_{r,\lambda}(M)$ for some $\lambda \ge 1$ and $1 < r < \infty$. Then, there exists a constant C, independent of u, such that*

$$\|\mathbb{M}_s(u)\|_{t,B,w_1} \leq C\|u\|_{t,\sigma B,w_2} \tag{87}$$

for all balls B with $\sigma B \subset M$. Here $\sigma > 1$ is a constant.

If we put $w_1(x) = w_2(x) = w(x)$ in (87), we have

$$\|\mathbb{M}_s(u)\|_{t,B,w} \leq C\|u\|_{t,\sigma B,w} \tag{88}$$

where $w(x) \in A_r(M)$, which is the A_r-weighted inequality.

It is easy to see that if we choose $w_1(x) = w_2(x)$ and $\lambda = 1$ in the last definition of the weights, we have

$$\sup_{B\subset E}\left(\frac{1}{|B|}\int_B wdx\right)^{1/r}\left(\frac{1}{|B|}\int_B\left(\frac{1}{w}\right)^{r'/r}dx\right)^{1/r'} < \infty,$$

that is,

$$\sup_{B\subset E}\left(\left(\frac{1}{|B|}\int_B wdx\right)\left(\frac{1}{|B|}\int_B\left(\frac{1}{w}\right)^{1/(r-1)}dx\right)^{r-1}\right)^{1/r} < \infty$$

since $r'/r = 1/(r-1)$. Thus, we see that the $A_{r,\lambda}(M)$-weight reduces to the usual $A_r(M)$-weight if $w_1(x) = w_2(x)$ and $\lambda = 1$.

Lemma 7.12 ([2]). *Let φ be a strictly increasing convex function on $[0,\infty)$ with $\varphi(0) = 0$ and D be a domain in $\mathbb{R}^n$. Assume that u is a function in D such that $\varphi(|u|) \in L^1(D;\mu)$ and $\mu(\{x \in D : |u - c| > 0\}) > 0$ for any constant c. Then, for any positive constant a, we have*

$$\int_D \varphi(\frac{1}{2}a|u-u_{D,\mu}|)d\mu \leq \int_D \varphi(a|u|)d\mu \leq C\int_D \varphi(2a|u-u_{D,\mu}|)d\mu,$$

where C is a positive constant and $u_{D,\mu} = \frac{1}{\mu(D)}\int_D ud\mu$.

Choosing $\varphi(x) = x^t$, $t > s \geq 1$, and replacing u by $\mathbb{M}_s(G(u))$ and $\mathbb{M}_s^\sharp(G(u))$, respectively, in Lemma 7.12, we can prove the following Theorem 7.13.

Theorem 7.13. *Let $\mathbb{M}_s$ be the Hardy-Littlewood maximal operator defined in (71), $\mathbb{M}_s^\sharp$ be the sharp maximal operator defined in (72), and G be Green's operator. Assume that $u \in L^t(M, \wedge^l, \mu)$, $l = 1, 2, \ldots, n$, is a differential form in a domain M, $1 \leq s < t < \infty$, and the measure $\mu(x)$ is defined by $d\mu = w(x)dx$, where w is a weight. Then,*

$$C_1\|\mathbb{M}_s(G(u)) - (\mathbb{M}_s(G(u)))_{M,\mu}\|_{t,M,w} \le \|\mathbb{M}_s(G(u))\|_{t,M,w}$$

$$\le C_2\|\mathbb{M}_s(G(u)) - (\mathbb{M}_s(G(u)))_{M,\mu}\|_{t,M,w},$$

$$C_3\|\mathbb{M}_s^\sharp(G(u)) - (\mathbb{M}_s^\sharp(G(u)))_{M,\mu}\|_{t,M,w} \le \|\mathbb{M}_s^\sharp(G(u))\|_{t,M,w}$$

$$\le C_4\|\mathbb{M}_s^\sharp(G(u)) - (\mathbb{M}_s^\sharp(G(u)))_{M,\mu}\|_{t,M,w},$$

where C_1, C_2, C_3 and C_4 are constants, independent of u.

We all know that differential forms have many applications in geometric analysis and physics; see [6, 24]. For example, we can apply our results to the Jacobian $J(x, f)$ of a mapping $f : M \to \mathbb{R}^n$, $f = (f^1, \ldots, f^n)$. It is well know that Jacobian $J(x, f)$ is an n-form, specifically,

$$J(x, f)dx = df^1 \wedge \cdots \wedge df^n,$$

where $dx = dx_1 \wedge dx_2 \wedge \cdots \wedge dx_n$. For example, let $f = (f^1, f^2)$ be a differential mapping in $\mathbb{R}^2$. Then,

$$J(x, f)dx \wedge dy = \begin{vmatrix} f_x^1 & f_y^1 \\ f_x^2 & f_y^2 \end{vmatrix} dx \wedge dy = (f_x^1 f_y^2 - f_y^1 f_x^2)dx \wedge dy,$$

$$\begin{aligned} df^1 \wedge df^2 &= \left(f_x^1 dx + f_y^1 dy\right) \wedge \left(f_x^2 dx + f_y^2 dy\right) \\ &= f_y^1 f_x^2 dy \wedge dx + f_x^1 f_y^2 dx \wedge dy \\ &= \left(f_x^1 f_y^2 - f_y^1 f_x^2\right) dx \wedge dy, \end{aligned}$$

where we have used the property

$$dx_i \wedge dx_j = \begin{cases} 0, & i = j \\ -dx_j \wedge dx_i, & i \ne j. \end{cases}$$

Clearly,

$$J(x, f)dx \wedge dy = df^1 \wedge df^2.$$

Let $f : M \to \mathbb{R}^n$, $f = (f^1, \ldots, f^n)$ be a mapping, whose distributional differential $Df = [\partial f^i/\partial x_j] : \Omega \to GL(n)$ is a locally integrable function on M with values in the space $GL(n)$ of all $n \times n$-matrices. A homeomorphism $f : M \to \mathbb{R}^n$ is said to be K-quasiconformal, $1 \le K < \infty$, if its differential matrix $Df(x)$ and the Jacobian determinant

$$J(x,f) = \det Df(x) = \begin{vmatrix} f^1_{x_1} & f^1_{x_2} & f^1_{x_3} & \cdots & f^1_{x_n} \\ f^2_{x_1} & f^2_{x_2} & f^2_{x_3} & \cdots & f^2_{x_n} \\ \vdots & \vdots & \vdots & \ddots & \vdots \\ f^n_{x_1} & f^n_{x_2} & f^n_{x_3} & \cdots & f^n_{x_n} \end{vmatrix}$$

satisfy

$$|Df(x)|^n \le KJ(x,f),$$

where $|Df(x)| = \max\{|Df(x)h| : |h| = 1\}$ denotes the norm of the Jacobi matrix $Df(x)$. Let u be the subdeterminant of Jacobian $J(x,f)$, which is obtained by deleting the k rows and k columns, $k = 0, 1, \ldots, n-1$, say,

$$J(x_{j_1}, x_{j_2}, \ldots, x_{j_{n-k}}; f^{i_1}, f^{i_2}, \ldots, f^{i_{n-k}}) = \begin{vmatrix} f^{i_1}_{x_{j_1}} & f^{i_1}_{x_{j_2}} & f^{i_1}_{x_{j_3}} & \cdots & f^{i_1}_{x_{j_{n-k}}} \\ f^{i_2}_{x_{j_1}} & f^{i_2}_{x_{j_2}} & f^{i_2}_{x_{j_3}} & \cdots & f^{i_2}_{x_{j_{n-k}}} \\ \vdots & \vdots & \vdots & \ddots & \vdots \\ f^{i_{n-k}}_{x_{j_1}} & f^{i_{n-k}}_{x_{j_2}} & f^{i_{n-k}}_{x_{j_3}} & \cdots & f^{i_{n-k}}_{x_{j_{n-k}}} \end{vmatrix},$$

which is an $(n-k) \times (n-k)$ subdeterminant of $J(x,f)$, $\{i_1, i_2, \ldots, i_{n-k}\} \subset \{1, 2, \ldots, n\}$ and $\{j_1, j_2, \ldots, j_{n-k}\} \subset \{1, 2, \ldots, n\}$. Also, it is easy to see that

$$J(x_{j_1}, x_{j_2}, \ldots, x_{j_{n-k}}; f^{i_1}, f^{i_2}, \ldots, f^{i_{n-k}}) dx_{j_1} \wedge dx_{j_2} \wedge \cdots \wedge dx_{j_{n-k}}$$

is an $(n-k)$-form. Thus, all estimates for differential forms are applicable to the $(n-k)$-form $J(x_{j_1}, x_{j_2}, \ldots, x_{j_{n-k}}; f^{i_1}, f^{i_2}, \ldots, f^{i_{n-k}}) dx_{j_1} \wedge dx_{j_2} \wedge \cdots \wedge dx_{j_{n-k}}$. For example, choosing $u = J(x,f)$ and applying Theorems 7.8 and 7.9 to u, respectively, we have the following theorems.

Theorem 7.15. *Let $\mathbb{M}_s$ be the Hardy-Littlewood maximal operator defined in (71), G be Green's operator, and $J(x,f) \in L^t(M, \wedge^n)$, $1 \le s < t < \infty$, be the Jacobian of the mapping $f = (f^1, \ldots, f^n) : M \to \mathbb{R}^n$. Then, $\mathbb{M}_s(G(J(x,f))) \in L^t(M)$ and*

$$\|\mathbb{M}_s(G(J(x,f)))\|_{t,M} \le C\|J(x,f)\|_{t,M}$$

for some constant C, independent of $J(x,f)$.

Theorem 7.16. *Let $J(x, f) \in L^s(M, \wedge^n)$ be the Jacobian of the mapping $f = (f^1, \dots, f^n) : M \to \mathbb{R}^n$ in a bounded domain M, $\mathbb{M}_s^\sharp$, $1 < s < \infty$, be the sharp maximal operator defined in (72), and G be Green's operator. Then,*

$$\|\mathbb{M}_s^\sharp(G(J(x, f)))\|_{s,M} \le C|M|^{1/s}\|J(x, f)\|_{s,M}$$

for some constant C, independent of $J(x, f)$.

Note.

(1) If we apply Theorems 7.8 and 7.9 to an $(n-k) \times (n-k)$ subdeterminant of $J(x, f)$, we will have the similar estimates for subdeterminant of $J(x, f)$.
(2) We can also apply the other results obtained in previous sections to the Jacobian $J(x, f)$ or its subdeterminants to obtain different estimates. Considering the length of the chapter, we do not include them here.

References

1. Agarwal, R.P., Ding, S.: Inequalities for Green's operator applied to the minimizers. J. Inequal. Appl. **2011**, 66 (2011)
2. Agarwal, R.P., Ding, S., Nolder, C.A.: Inequalities for Differential Forms. Springer, New York (2009)
3. Bao, G.: $A_r(\lambda)$-weighted integral inequalities for A-harmonic tensors. J. Math. Anal. Appl. **247**, 466–477 (2000)
4. Bi, H., Xing, Y.: Poincaré-type inequalities with $L^p(\log L)^\alpha$-norms for Green's operator. Comput. Math. Appl. **60**, 2764–2770 (2010)
5. Buckley, S.M., Koskela, P.: Orlicz-Hardy inequalities. Illinois J. Math. **48**, 787–802 (2004)
6. Cartan, H.: Differential Forms. Houghton Mifflin Co, Boston (1970)
7. Ding, S.: Integral estimates for the Laplace-Beltrami and Green's operators applied to differential forms. Z. Anal. Anwendungen **22**, 939–957 (2003)
8. Ding, S.: Two-weight Caccioppoli inequalities for solutions of nonhomogeneous A-harmonic equations on Riemannian manifolds. Proc. Am. Math. Soc. **132**, 2367–2375 (2004)
9. Ding, S.: $L^\varphi(\mu)$-averaging domains and the quasihyperbolic metric. Comput. Math. Appl. **47**, 1611–1618 (2004)
10. Ding, S.: Norm estimates for the maximal operator and Green's operator. Dyn. Contin. Discrete Impuls. Syst. Ser. A Math. Anal. **16**, 72–78 (2009)
11. Ding, S., Liu, B.: Global estimates for singular integrals of the composite operator. Illinois J. Math. **53**, 1173–1185 (2009)
12. Ding, S., Nolder C.A.: Weighted Poincaré-type inequalities for solutions to the A-harmonic equation. Illinois J. Math. **2**, 199–205 (2002)
13. Fang, R., Ding, S.: Singular integrals of the compositions of Laplace-Beltrami and Green's operators. J. Inequal. Appl. **2011**(1), 74 (2011). doi:10.1186/1029-242X-2011-74
14. Gariepy, R.F.: A Caccioppoli inequality and partial regularity in the calculus of variations. Proc. Roy. Soc. Edinburgh Sect. A **112**, 249–255 (1989)
15. Giaquinta, M., Giusti, E.: On the regularity of the minima of variational integrals. Acta Math. **148**, 31–46 (1982)
16. Liu, B.: $A_r^\lambda(\Omega)$-weighted imbedding inequalities for A-harmonic tensors. J. Math. Anal. Appl. **273**(2), 667–676 (2002)
17. Neugebauer, C.J.: Inserting A_p-weights. Proc. Am. Math. Soc. **87**, 644–648 (1983)

18. Nolder, C.A.: Hardy-Littlewood theorems for A-harmonic tensors. Illinois J. Math. **43**, 613–631 (1999)
19. Sbordone, C.: On some integral inequalities and their applications to the calculus of variations. Boll. Un. Mat. Ital. Analisi Funzionali e Applicazioni **5**, 73–94 (1986)
20. Scott, C.: L^p-theory of differntial forms on manifolds. Trans. Am. Soc. **347**, 2075–2096 (1995)
21. Stein, E.M.: Harmonic Analysis. Princeton University Press, Princeton (1993)
22. Wang, Y., Wu, C.: Sobolev imbedding theorems and Poincaré inequalities for Green's operator on solutions of the nonhomogeneous A-harmonic equation. Comput. Math. Appl. **47**, 1545–1554 (2004)
23. Warner, F.: Foundations of Differentiable Manifolds and Lie Groups. Springer, New York (1983)
24. Westenholz, C.: Differential Forms in Mathematical Physics. North Holland Publishing, Amesterdam (1978)
25. Xing, Y.: Weighted integral inequalities for solutions of the A-harmonic equation. J. Math. Anal. Appl. **279**, 350–363 (2003)
26. Xing, Y.: Poincaré type inequalities for Green's operator with Orlicz norms. Preprint
27. Xing, Y., Ding, S.: Inequalities for Green's operator with Lipschitz and BMO norms. Comput. Math. Appl. **58**, 273–280 (2009)
28. Xing, Y., Wu, C.: Global weighted inequalities for operators and harmonic forms on manifolds. J. Math. Anal. Appl. **294**, 294–309 (2004)

Multidimensional Discrete Hilbert-Type Inequalities, Operators and Compositions

Bicheng Yang

Dedicated to Professor Hari M. Srivastava

Abstract Hilbert-type inequalities with their operators are important in analysis and its applications. In this paper by using the methods of weight coefficients and technique of real analysis, a multidimensional discrete Hilbert-type inequality with a best possible constant factor is given. The equivalent forms, two types of reverses, a more accurate inequality with parameters, as well as a strengthened version of Hardy-Hilbert's inequality with Euler constant are obtained. We also consider the relating operators with the norms, some particular examples and the compositions of two discrete Hilbert-type operators in certain conditions.

1 Introduction

Assuming that $p > 1$, $\frac{1}{p} + \frac{1}{q} = 1$, $f(x), g(y) \geq 0$, $f \in L^p(\mathbf{R}_+)$, $g \in L^q(\mathbf{R}_+)$,

$$||f||_p = \left\{\int_0^\infty f^p(x)dx\right\}^{\frac{1}{p}} > 0,$$

$||g||_q > 0$, we have the following Hardy-Hilbert's integral inequality (cf. [10]):

$$\int_0^\infty \int_0^\infty \frac{f(x)g(y)}{x+y}dxdy < \frac{\pi}{\sin(\pi/p)}||f||_p||g||_q, \tag{1}$$

B. Yang (✉)
Department of Mathematics, Guangdong University of Education,
Guangzhou, Guangdong 510303, P. R. China
e-mail: bcyang@gdei.edu.cn; bcyang818@163.com

G.V. Milovanović and M.Th. Rassias (eds.), *Analytic Number Theory, Approximation Theory, and Special Functions*, DOI 10.1007/978-1-4939-0258-3_15,

where the constant factor $\pi/\sin(\pi/p)$ is the best possible. If $a_m, b_n \geq 0$, $a = \{a_m\}_{m=1}^{\infty} \in \ell^p$, $b = \{b_n\}_{n=1}^{\infty} \in \ell^q$,

$$||a||_p = \left\{\sum_{m=1}^{\infty} a_m^p\right\}^{\frac{1}{p}} > 0,$$

$||b||_q > 0$, then we still have the following discrete variant of the above inequality

$$\sum_{m=1}^{\infty}\sum_{n=1}^{\infty} \frac{a_m b_n}{m+n} < \frac{\pi}{\sin(\pi/p)}||a||_p||b||_q, \tag{2}$$

with the same best constant $\pi/\sin(\pi/p)$ (cf. [10]). Inequalities (1) and (2) are important in analysis and its applications (cf. [10, 20, 24, 25, 27, 29]).

In 1998, by introducing an independent parameter $\lambda \in (0, 1]$, Yang [22] gave an extension of (1) at $p = q = 2$. Recently, Yang [24, 27] gave some extensions of (1) and (2) as follows:

If $\lambda_1, \lambda_2, \lambda \in \mathbf{R}, \lambda_1 + \lambda_2 = \lambda, k_\lambda(x, y)$ is a non-negative homogeneous function of degree $-\lambda$, with

$$k(\lambda_1) = \int_0^{\infty} k_\lambda(t, 1)t^{\lambda_1 - 1}dt \in \mathbf{R}_+,$$

$\phi(x) = x^{p(1-\lambda_1)-1}$, $\psi(y) = y^{q(1-\lambda_2)-1}$, $f(x)$, $g(y) \geq 0$,

$$f \in L_{p,\phi}(\mathbf{R}_+) = \left\{f;\ ||f||_{p,\phi} := \left\{\int_0^{\infty} \phi(x)|f(x)|^p dx\right\}^{1/p} < \infty\right\},$$

$g \in L_{q,\psi}(\mathbf{R}_+)$, $||f||_{p,\phi}$, $||g||_{q,\psi} > 0$, then

$$\int_0^{\infty}\int_0^{\infty} k_\lambda(x, y)f(x)g(y)dxdy < k(\lambda_1)||f||_{p,\phi}||g||_{q,\psi}, \tag{3}$$

where the constant factor $k(\lambda_1)$ is the best possible. Moreover, if $k_\lambda(x, y)$ is finite and $k_\lambda(x, y)x^{\lambda_1 - 1}(k_\lambda(x, y)y^{\lambda_2 - 1})$ is decreasing with respect to $x > 0$ $(y > 0)$, then for $a_m, b_n \geq 0$,

$$a \in \ell_{p,\phi} = \left\{a;\ ||a||_{p,\phi} := \left\{\sum_{n=1}^{\infty} \phi(n)|a_n|^p\right\}^{1/p} < \infty\right\},$$

$b = \{b_n\}_{n=1}^{\infty} \in \ell_{q,\psi}$, $||a||_{p,\phi}$, $||b||_{q,\psi} > 0$, we have

$$\sum_{m=1}^{\infty}\sum_{n=1}^{\infty} k_\lambda(m, n)a_m b_n < k(\lambda_1)||a||_{p,\phi}||b||_{q,\psi}, \tag{4}$$

where the constant factor $k(\lambda_1)$ is still the best possible.

Clearly, for $\lambda = 1, k_1(x, y) = 1/(x + y), \lambda_1 = 1/q, \lambda_2 = 1/p$, (3) reduces to (1), while (4) reduces to (2). Some other results including multidimensional Hilbert-type integral inequalities are provided by [3, 4, 11, 13, 16, 18, 34–37, 39, 44].

About half-discrete Hilbert-type inequalities with the non-homogeneous kernels, Hardy et al. provided a few results in Theorem 351 of [10]. But they did not prove that the constant factors are the best possible. However, Yang [23] gave a result with the kernel $1/(1 + nx)^\lambda$ by introducing a variable and proved that the constant factor is the best possible. In 2011, Yang [26] gave the following half-discrete Hardy-Hilbert's inequality with the best possible constant factor $B(\lambda_1, \lambda_2)$:

$$\int_0^\infty f(x) \sum_{n=1}^\infty \frac{a_n}{(x+n)^\lambda} dx < B(\lambda_1, \lambda_2) ||f||_{p,\phi} ||a||_{q,\psi}, \tag{5}$$

where $\lambda_1 \lambda_2 > 0, 0 \leq \lambda_2 \leq 1, \lambda_1 + \lambda_2 = \lambda$,

$$B(u, v) = \int_0^\infty \frac{1}{(1+t)^{u+v}} t^{u-1} dt \quad (u, v > 0)$$

is the beta function. Zhong et al. [40–46] investigated several half-discrete Hilbert-type inequalities with particular kernels.

Using the way of weight functions and the techniques of discrete and integral Hilbert-type inequalities with some additional conditions on the kernel, a half-discrete Hilbert-type inequality with a general homogeneous kernel of degree $-\lambda \in \mathbf{R}$ and a best constant factor $k(\lambda_1)$ is obtained as follows:

$$\int_0^\infty f(x) \sum_{n=1}^\infty k_\lambda(x, n) a_n \, dx < k(\lambda_1) ||f||_{p,\phi} ||a||_{q,\psi},$$

which is an extension of (5) (see Yang and Chen [31]). At the same time, a half-discrete Hilbert-type inequality with a general non-homogeneous kernel and a best constant factor is given by Yang [28].

Remark 1.1. (1) Many different kinds of Hilbert-type discrete, half-discrete and integral inequalities with applications are presented in recent twenty years. Special attention is given to new results proved during 2009–2012. Included are many generalizations, extensions and refinements of Hilbert-type discrete, half-discrete and integral inequalities involving many special functions such as beta, gamma, hypergeometric, trigonometric, hyperbolic, zeta, Bernoulli functions, Bernoulli numbers and Euler constant.

(2) In his five books, Yang [24,25,27,29,30] presented many new results on Hilbert-type operators with general homogeneous kernels of degree of real numbers and two pairs of conjugate exponents as well as the related inequalities. These

research monographs contained recent developments of discrete, half-discrete and integral types of operators and inequalities with proofs, examples and applications.

In this paper, by using the methods of weight coefficients and techniques of real analysis, a multidimensional discrete Hilbert-type inequality with a best possible constant factor is given, which is an extension of (4). The equivalent forms, two types of reverses, a more accurate inequality with parameters and its equivalent form, as well as a strengthened version of Hardy-Hilbert's inequality with Euler constant are obtained. We also consider the operator expression with the norm, some particular examples as applications and the compositions of two discrete Hilbert-type operators in certain conditions. The lemmas and theorems in this chapter have provided an extensive account of this type of inequalities and operators.

2 Main Results, the Equivalent Forms and Reverses

If $i_0, j_0 \in \mathbf{N}$ ($\mathbf{N}$ is the set of positive integers), $\alpha, \beta > 0$, we put

$$||x||_\alpha := \left(\sum_{k=1}^{i_0} |x_k|^\alpha\right)^{\frac{1}{\alpha}} \quad (x = (x_1, \ldots, x_{i_0}) \in \mathbf{R}^{i_0}),$$

$$||y||_\beta := \left(\sum_{k=1}^{j_0} |y_k|^\beta\right)^{\frac{1}{\beta}} \quad (y = (y_1, \ldots, y_{j_0}) \in \mathbf{R}^{j_0}).$$

In the latter part of this chapter, we agree that $p \in \mathbf{R}\backslash\{0,1\}$, $\frac{1}{p} + \frac{1}{q} = 1, \lambda, \lambda_1, \lambda_2 \in \mathbf{R}$, $\lambda_1 + \lambda_2 = \lambda$, $k_\lambda(x, y)$ (≥ 0) is a finite homogeneous function of degree $-\lambda$ in $\mathbf{R}_+^2$, satisfying for any $u, x, y \in \mathbf{R}_+$,

$$k_\lambda(ux, uy) = u^{-\lambda} k_\lambda(x, y).$$

Definition 2.1. For $m = (m_1, \ldots, m_{i_0}) \in \mathbf{N}^{i_0}$, $n = (n_1, \ldots, n_{j_0}) \in \mathbf{N}^{j_0}$, define two weight coefficients $\omega_\lambda(\lambda_2, n)$ and $\varpi_\lambda(\lambda_1, m)$ as follows:

$$\omega_\lambda(\lambda_2, n) := ||n||_\beta^{\lambda_2} \sum_m k_\lambda(||m||_\alpha, ||n||_\beta) \frac{1}{||m||_\alpha^{i_0-\lambda_1}},$$

$$\varpi_\lambda(\lambda_1, m) := ||m||_\alpha^{\lambda_1} \sum_n k_\lambda(||m||_\alpha, ||n||_\beta) \frac{1}{||n||_\beta^{j_0-\lambda_2}},$$

where $\sum_m = \sum_{m_{i_0}=1}^\infty \cdots \sum_{m_1=1}^\infty$ and $\sum_n = \sum_{n_{j_0}=1}^\infty \cdots \sum_{n_1=1}^\infty$.

Lemma 2.1. *If* $a_m = a_{(m_1,\ldots,m_{i_0})} \geq 0$, *then*

(1) For $p > 1$, *we have the following inequality:*

$$J_1 := \left\{ \sum_n \frac{||n||_\beta^{p\lambda_2 - j_0}}{[\omega_\lambda(\lambda_2, n)]^{p-1}} \left(\sum_m k_\lambda(||m||_\alpha, ||n||_\beta) a_m \right)^p \right\}^{\frac{1}{p}}$$

$$\leq \left\{ \sum_m \varpi_\lambda(\lambda_1, m) ||m||_\alpha^{p(i_0 - \lambda_1) - i_0} a_m^p \right\}^{\frac{1}{p}}. \tag{6}$$

(2) For $p < 0$, *or* $0 < p < 1$, *we have the reverse of* (6).

Proof. (1) For $p > 1$, by Hölder's inequality with weight (cf. [15]), it follows

$$\sum_m k_\lambda(||m||_\alpha, ||n||_\beta) a_m$$

$$= \sum_m k_\lambda(||m||_\alpha, ||n||_\beta) \left[\frac{||m||_\alpha^{(i_0-\lambda_1)/q}}{||n||_\beta^{(j_0-\lambda_2)/p}} a_m \right] \left[\frac{||n||_\beta^{(j_0-\lambda_2)/p}}{||m||_\alpha^{(i_0-\lambda_1)/q}} \right]$$

$$\leq \left\{ \sum_m k_\lambda(||m||_\alpha, ||n||_\beta) \frac{||m||_\alpha^{(i_0-\lambda_1)(p-1)}}{||n||_\beta^{j_0-\lambda_2}} a_m^p \right\}^{\frac{1}{p}}$$

$$\times \left\{ \sum_m k_\lambda(||m||_\alpha, ||n||_\beta) \frac{||n||_\beta^{(j_0-\lambda_2)(q-1)}}{||m||_\alpha^{i_0-\lambda_1}} \right\}^{\frac{1}{q}}$$

$$= [\omega_\lambda(\lambda_2, n)]^{\frac{1}{q}} ||n||_\beta^{\frac{j_0}{p} - \lambda_2} \left\{ \sum_m k_\lambda(||m||_\alpha, ||n||_\beta) \frac{||m||_\alpha^{(i_0-\lambda_1)(p-1)}}{||n||_\beta^{j_0-\lambda_2}} a_m^p \right\}^{\frac{1}{p}}. \tag{7}$$

Then we have

$$J_1 \leq \left\{ \sum_n \sum_m k_\lambda(||m||_\alpha, ||n||_\beta) \frac{||m||_\alpha^{(i_0-\lambda_1)(p-1)}}{||n||_\beta^{j_0-\lambda_2}} a_m^p \right\}^{\frac{1}{p}}$$

$$= \left\{ \sum_m \sum_n k_\lambda(||m||_\alpha, ||n||_\beta) \frac{||m||_\alpha^{(i_0-\lambda_1)(p-1)}}{||n||_\beta^{j_0-\lambda_2}} a_m^p \right\}^{\frac{1}{p}}$$

$$= \left\{ \sum_m \varpi_\lambda(\lambda_1, m) ||m||_\alpha^{p(i_0-\lambda_1)-i_0} a_m^p \right\}^{\frac{1}{p}}.$$

Hence, (6) follows.

(2) For $p < 0$, or $0 < p < 1$, by the reverse Hölder's inequality with weight (cf. [15]), we obtain the reverse of (7). Then we still obtain the reverse of (6). □

Lemma 2.2. *If* $a_m = a_{(m_1,\ldots,m_{i_0})} \geq 0$, $b_n = b_{(n_1,\ldots,n_{j_0})} \geq 0$, *then*

(1) *For* $p > 1$, *we have the following inequality equivalent to* (6)*:*

$$I = \sum_n \sum_m k_\lambda(||m||_\alpha, ||n||_\beta) a_m b_n$$
$$\leq \left\{ \sum_m \varpi_\lambda(\lambda_1, m) ||m||_\alpha^{p(i_0-\lambda_1)-i_0} a_m^p \right\}^{\frac{1}{p}} \left\{ \sum_n \omega_\lambda(\lambda_2, n) ||n||_\beta^{q(j_0-\lambda_2)-j_0} b_n^q \right\}^{\frac{1}{q}}. \tag{8}$$

(2) *For* $p < 0$, *or* $0 < p < 1$, *we have the reverse of* (8) *equivalent to the reverse of* (6).

Proof. (1) For $p > 1$, by Hölder's inequality (cf. [15]), it follows

$$I = \sum_n \frac{||n||_\beta^{\frac{j_0}{q}-(j_0-\lambda_2)}}{[\omega_\lambda(\lambda_2, n)]^{\frac{1}{q}}} \left[\sum_m k_\lambda(||m||_\alpha, ||n||_\beta) a_m \right] \left[[\omega_\lambda(\lambda_2, n)]^{\frac{1}{q}} ||n||_\beta^{(j_0-\lambda_2)-\frac{j_0}{q}} b_n \right]$$
$$\leq J_1 \left\{ \sum_n \omega_\lambda(\lambda_2, n) ||n||_\beta^{q(j_0-\lambda_2)-j_0} b_n^q \right\}^{\frac{1}{q}}. \tag{9}$$

Then by (6), we have (8).

On the other hand, assuming that (8) is valid, we set

$$b_n := \frac{||n||_\beta^{p\lambda_2 - j_0}}{[\omega_\lambda(\lambda_2, n)]^{p-1}} \left(\sum_m k_\lambda(||m||_\alpha, ||n||_\beta) a_m \right)^{p-1}, \quad n \in \mathbf{N}^{j_0}.$$

Then it follows

$$J_1^p = \sum_n \omega_\lambda(\lambda_2, n) ||n||_\beta^{q(j_0-\lambda_2)-j_0} b_n^q.$$

If $J_1 = 0$, then (6) is trivially valid; if $J_1 = \infty$, then by (7), (6) keeps the form of equality ($= \infty$). Suppose that $0 < J_1 < \infty$. By (8), we have

$$0 < \sum_n \omega_\lambda(\lambda_2, n) ||n||_\beta^{q(j_0-\lambda_2)-j_0} b_n^q = J_1^p = I$$
$$\leq \left\{ \sum_m \varpi_\lambda(\lambda_1, m) ||m||_\alpha^{p(i_0-\lambda_1)-i_0} a_m^p \right\}^{\frac{1}{p}} \left\{ \sum_n \omega_\lambda(\lambda_2, n) ||n||_\beta^{q(j_0-\lambda_2)-j_0} b_n^q \right\}^{\frac{1}{q}} < \infty.$$

It follows

$$J_1 = \left\{ \sum_n \omega_\lambda(\lambda_2, n) ||n||_\beta^{q(j_0-\lambda_2)-j_0} b_n^q \right\}^{\frac{1}{p}} \leqslant \left\{ \sum_m \varpi_\lambda(\lambda_1, m) ||m||_\alpha^{p(i_0-\lambda_1)-i_0} a_m^p \right\}^{\frac{1}{p}},$$

and then (6) follows. Hence, inequalities (8) and (6) are equivalent.

(2) for $p < 0$, or $0 < p < 1$, by the same way, we have the reverse of (8) equivalent to the reverse of (6). The lemma is proved.

□

Setting

$$\varphi(m) := ||m||_\alpha^{p(i_0-\lambda_1)-i_0}, \ \psi(n) := ||n||_\beta^{q(j_0-\lambda_2)-j_0},$$

$$\tilde{\varphi}(m) := (1 - \theta_\lambda(m))||m||_\alpha^{p(i_0-\lambda_1)-i_0} \qquad (\theta_\lambda(m) \in (0,1); \ m \in \mathbf{N}^{i_0})$$

and

$$\tilde{\psi}(n) := (1 - \vartheta_\lambda(n))||n||_\beta^{q(j_0-\lambda_2)-j_0} \qquad (\vartheta_\lambda(n) \in (0,1); \ n \in \mathbf{N}^{j_0}),$$

by Lemmas 2.1 and 2.2, we have the following theorem:

Theorem 2.1. *Suppose that $p > 1$, there exist constants $K_i > 0$ $(i = 1, 2)$, such that*

$$\varpi_\lambda(\lambda_1, m) < K_1, \quad \omega_\lambda(\lambda_2, n) < K_2 \quad (m \in \mathbf{N}^{i_0}, \ n \in \mathbf{N}^{j_0}).$$

If $a_m = a_{(m_1,\ldots,m_{i_0})} \geq 0$, $b_n = b_{(n_1,\ldots,n_{j_0})} \geq 0$, satisfying

$$0 < ||a||_{p,\varphi} := \left\{ \sum_m ||m||_\alpha^{p(i_0-\lambda_1)-i_0} a_m^p \right\}^{\frac{1}{p}} < \infty,$$

$$0 < ||b||_{q,\psi} := \left\{ \sum_n ||n||_\beta^{q(j_0-\lambda_2)-j_0} b_n^q \right\}^{\frac{1}{q}} < \infty,$$

then we have the following equivalent inequalities:

$$\sum_n \sum_m k_\lambda(||m||_\alpha, ||n||_\beta) a_m b_n < K_1^{\frac{1}{p}} K_2^{\frac{1}{q}} ||a||_{p,\varphi} ||b||_{q,\psi}, \tag{10}$$

$$J := \left\{ \sum_n ||n||_\beta^{p\lambda_2 - j_0} \left(\sum_m k_\lambda(||m||_\alpha, ||n||_\beta) a_m \right)^p \right\}^{\frac{1}{p}} < K_1^{\frac{1}{p}} K_2^{\frac{1}{q}} ||a||_{p,\varphi}. \tag{11}$$

Theorem 2.2. *Suppose that $p < 0$, there exist constants $K_i > 0$ $(i = 1,2)$, such that*

$$\varpi_\lambda(\lambda_1, m) < K_1 \quad (m \in \mathbf{N}^{i_0}),$$

$$0 < K_2(1 - \vartheta_\lambda(n)) < \omega_\lambda(\lambda_2, n) < K_2 \quad (n \in \mathbf{N}^{j_0}).$$

If $a_m = a_{(m_1,\ldots,m_{i_0})} \geq 0,\ b_n = b_{(n_1,\ldots,n_{j_0})} \geq 0$, satisfying $0 < ||a||_{p,\varphi} < \infty$, and

$$0 < ||b||_{q,\tilde{\psi}} := \left\{ \sum_n (1 - \vartheta_\lambda(n))||n||_\beta^{q(j_0-\lambda_2)-j_0} b_n^q \right\}^{\frac{1}{q}} < \infty,$$

then we have the following equivalent inequalities:

$$\sum_n \sum_m k_\lambda(||m||_\alpha, ||n||_\beta) a_m b_n > K_1^{\frac{1}{p}} K_2^{\frac{1}{q}} ||a||_{p,\varphi} ||b||_{q,\tilde{\psi}}, \tag{12}$$

$$\left\{ \sum_n \frac{||n||_\beta^{p\lambda_2-j_0}}{(1-\vartheta_\lambda(n))^{p-1}} \left(\sum_m k_\lambda(||m||_\alpha, ||n||_\beta) a_m \right)^p \right\}^{\frac{1}{p}} > K_1^{\frac{1}{p}} K_2^{\frac{1}{q}} ||a||_{p,\varphi}. \tag{13}$$

Theorem 2.3. *Suppose that $0 < p < 1$, there exist constants $K_i > 0$ $(i = 1,2)$, such that*

$$0 < K_1(1 - \theta_\lambda(m)) < \varpi_\lambda(\lambda_1, m) < K_1 \quad (m \in \mathbf{N}^{i_0}),$$

$$\omega_\lambda(\lambda_2, n) < K_2 \quad (n \in \mathbf{N}^{j_0}).$$

If $a_m = a_{(m_1,\ldots,m_{i_0})} \geq 0,\ b_n = b_{(n_1,\ldots,n_{j_0})} \geq 0$, satisfying

$$0 < ||a||_{p,\tilde{\varphi}} := \left\{ \sum_m (1 - \theta_\lambda(m))||m||_\alpha^{p(i_0-\lambda_1)-i_0} a_m^p \right\}^{\frac{1}{p}} < \infty,$$

and $0 < ||b||_{q,\psi} < \infty$, then we have the following equivalent inequalities:

$$\sum_n \sum_m k_\lambda(||m||_\alpha, ||n||_\beta) a_m b_n > K_1^{\frac{1}{p}} K_2^{\frac{1}{q}} ||a||_{p,\tilde{\varphi}} ||b||_{q,\psi}, \tag{14}$$

$$\left\{ \sum_n ||n||_\beta^{p\lambda_2-j_0} \left(\sum_m k_\lambda(||m||_\alpha, ||n||_\beta) a_m \right)^p \right\}^{\frac{1}{p}} > K_1^{\frac{1}{p}} K_2^{\frac{1}{q}} ||a||_{p,\tilde{\varphi}}. \tag{15}$$

3 The Best Constant Factor

Lemma 3.1. *Suppose that $h(t)$ is a non-negative measurable function in $\mathbf{R}_+$, $a \in \mathbf{R}$, and there exists a constant $\delta_0 > 0$, such that for any $\delta \in [0, \delta_0)$,*

$$k(a \pm \delta) := \int_0^\infty h(t) t^{(a \pm \delta)-1} dt \in \mathbf{R}.$$

Then we have

$$k(a \pm \delta) = k(a) + o(1) \quad (\delta \to 0^+). \tag{16}$$

Proof. For any $\delta \in [0, \delta_0/2)$, it follows

$$h(t) t^{(a\pm\delta)-1} \le g(t) := \begin{cases} h(t) t^{(a-\delta_0/2)-1}, \ t \in (0,1], \\ h(t) t^{(a+\delta_0/2)-1}, \ t \in (1,\infty). \end{cases}$$

Since we find

$$\begin{aligned} 0 \le \int_0^\infty g(t) dt &= \int_0^1 h(t) t^{(a-\delta_0/2)-1} dt + \int_1^\infty h(t) t^{(a+\delta_0/2)-1} dt \\ &\le \int_0^\infty h(t) t^{(a-\delta_0/2)-1} dt + \int_0^\infty h(t) t^{(a+\delta_0/2)-1} dt \\ &= k(a - \delta_0/2) + k(\alpha + \delta_0/2) \in \mathbf{R}, \end{aligned}$$

then for any $\delta \in (0, \delta_0/2)$, by Lebesgue control convergence theorem (cf. [14]), it follows

$$k(a \pm \delta) = \int_0^\infty h(t) t^{(a\pm\delta)-1} dt = \int_0^\infty h(t) t^{a-1} dt + o(1) \quad (\delta \to 0^+),$$

and then (16) follows. The lemma is proved. □

Lemma 3.2 ([29]). *If $s \in \mathbf{N}$, γ, $M > 0$, $\Psi(u)$ is a Non-negative measurable function in $(0, 1]$, and*

$$D_M := \left\{ x \in \mathbf{R}_+^s;\ \sum_{i=1}^s x_i^\gamma \le M^\gamma \right\},$$

then we have

$$\int \cdots \int_{D_M} \Psi\left(\sum_{i=1}^s \left(\frac{x_i}{M} \right)^\gamma \right) dx_1 \cdots dx_s = \frac{M^s \Gamma^s(\frac{1}{\gamma})}{\gamma^s \Gamma(\frac{s}{\gamma})} \int_0^1 \Psi(u) u^{\frac{s}{\gamma}-1} du. \tag{17}$$

Lemma 3.3. *For $s \in \mathbf{N}, \gamma > 0$, $\varepsilon > 0$, $c = (c_1, \ldots, c_s) \in [0,1)^s$, we have*

$$\int_{\{u \in \mathbf{R}_+^s;\, u_i \geq 1\}} ||u||_\gamma^{-s-\varepsilon} du = \frac{\Gamma^s(\frac{1}{\gamma})}{\varepsilon s^{\varepsilon/\gamma} \gamma^{s-1} \Gamma(\frac{s}{\gamma})}, \tag{18}$$

$$\sum_m ||m-c||_\gamma^{-s-\varepsilon} = \frac{\Gamma^s(\frac{1}{\gamma})}{\varepsilon s^{\varepsilon/\gamma} \gamma^{s-1} \Gamma(\frac{s}{\gamma})} + O(1) \quad (\varepsilon \to 0^+). \tag{19}$$

Proof. For $M > s^{1/\gamma}$, we set

$$\Psi(u) = \begin{cases} 0, & 0 < u < \frac{s}{M^\gamma}, \\ (Mu^{1/\gamma})^{-s-\varepsilon}, & \frac{s}{M^\gamma} \leq u \leq 1. \end{cases}$$

Then by (17), it follows

$$\begin{aligned} \int_{\{u \in \mathbf{R}_+^s; u_i \geq 1\}} ||u||_\gamma^{-s-\varepsilon} du &= \lim_{M \to \infty} \int \cdots \int_{D_M} \Psi\left(\sum_{i=1}^s \left(\frac{x_i}{M}\right)^\gamma\right) dx_1 \cdots dx_s \\ &= \lim_{M \to \infty} \frac{M^s \Gamma^s(\frac{1}{\gamma})}{\gamma^s \Gamma(\frac{s}{\gamma})} \int_{s/M^\gamma}^1 (Mu^{1/\gamma})^{-s-\varepsilon} u^{\frac{s}{\gamma}-1} du \\ &= \frac{\Gamma^s(\frac{1}{\gamma})}{\varepsilon s^{\varepsilon/\gamma} \gamma^{s-1} \Gamma(\frac{s}{\gamma})}, \end{aligned}$$

namely, (18) follows. By (18), we find

$$\begin{aligned} \sum_m ||m-c||_\gamma^{-s-\varepsilon} &\geq \int_{\{x \in \mathbf{R}_+^s; x_i \geq 1+c_i\}} ||x-c||_\gamma^{-s-\varepsilon} dx \\ &= \int_{\{u \in \mathbf{R}_+^s; u_i \geq 1\}} ||u||_\gamma^{-s-\varepsilon} du = \frac{\Gamma^s(\frac{1}{\gamma})}{\varepsilon s^{\varepsilon/\gamma} \gamma^{s-1} \Gamma(\frac{s}{\gamma})}. \end{aligned}$$

For $s = 1, 0 < \sum_{m=1}^2 ||m-c||_\gamma^{-1-\varepsilon} < \infty$; for $s \geq 2$, by (18), we have

$$0 < \sum_{\{m \in \mathbf{N}^s; \exists i_0, m_{i_0} = 1,2\}} ||m-c||_\gamma^{-s-\varepsilon} \leq a + \sum_{\{m \in \mathbf{N}^{s-1}; m_i \geq 3\}} ||m-c||_\gamma^{-(s-1)-(1+\varepsilon)}$$

$$\leq a + \int_{\{x \in \mathbf{R}_+^{s-1}; x_i \geq 1+c_i\}} ||x-c||_\gamma^{-(s-1)-(1+\varepsilon)} dx$$

$$= a + \int_{\{u \in \mathbf{R}_+^s; u_i \geq 1\}} ||u||_\gamma^{-(s-1)-(1+\varepsilon)} du$$

$$= a + \frac{\Gamma^{s-1}(\frac{1}{\gamma})}{(1+\varepsilon)(s-1)^{(1+\varepsilon)/\gamma}\gamma^{s-2}\Gamma(\frac{s-1}{\gamma})} < \infty \ (a \in \mathbf{R}_+),$$

and then

$$\sum_m ||m-c||_\gamma^{-s-\varepsilon} = \sum_{\{m \in \mathbf{N}^s; \exists i_0, m_{i_0}=1,2\}} ||m-c||_\gamma^{-s-\varepsilon} + \sum_{\{m \in \mathbf{N}^s; m_i \geq 3\}} ||m-c||_\gamma^{-s-\varepsilon}$$

$$\leq O_1(1) + \int_{\{x \in \mathbf{R}_+^s; x_i \geq 1+c_i\}} ||x-c||_\gamma^{-s-\varepsilon} dx$$

$$= O_1(1) + \int_{\{u \in \mathbf{R}_+^s; u_i \geq 1\}} ||u||_\gamma^{-s-\varepsilon} du = O_1(1) + \frac{\Gamma^s(\frac{1}{\gamma})}{\varepsilon s^{\varepsilon/\gamma}\gamma^{s-1}\Gamma(\frac{s}{\gamma})}.$$

Then we have (19). The lemma is proved. □

In Theorems 3.1–3.3, we suppose that $k(\lambda_1) \in \mathbf{R}_+$, and

$$K_1 = \frac{\Gamma^{j_0}(\frac{1}{\beta})}{\beta^{j_0-1}\Gamma(\frac{j_0}{\beta})} k(\lambda_1), \quad K_2 = \frac{\Gamma^{i_0}(\frac{1}{\alpha})}{\alpha^{i_0-1}\Gamma(\frac{i_0}{\alpha})} k(\lambda_1).$$

Theorem 3.1. *With the same assumptions of Theorem* 2.1 ($p > 1$), *if there exists a constant* $\delta_0 > 0$, *such that for any* $\tilde{\lambda}_i \in (\lambda_i - \delta_0, \lambda_i + \delta_0)$ $(i = 1, 2)$, $\tilde{\lambda}_1 + \tilde{\lambda}_2 = \lambda$,

$$\tilde{K}_2(1 - \tilde{\vartheta}_\lambda(n)) < \omega_\lambda(\tilde{\lambda}_2, n) \quad (n \in \mathbf{N}^{j_0}), \tag{20}$$

where

$$\tilde{K}_2 = \frac{\Gamma^{i_0}(\frac{1}{\alpha})}{\alpha^{i_0-1}\Gamma(\frac{i_0}{\alpha})} k(\tilde{\lambda}_1) \in \mathbf{R}_+ \quad \text{and} \quad \tilde{\vartheta}_\lambda(n) = O\Big(\frac{1}{||n||_\beta^\eta}\Big) \in (0,1) \quad (\eta > 0),$$

then the constant factor $K_1^{\frac{1}{p}} K_2^{\frac{1}{q}}$ *in* (10) *and* (11) *is the best possible.*

Proof. For $0 < \varepsilon < q\delta_0$, $\tilde{\lambda}_1 = \lambda_1 + \frac{\varepsilon}{q}$, $\tilde{\lambda}_2 = \lambda_2 - \frac{\varepsilon}{q}$, we set

$$\tilde{a}_m := ||m||_\alpha^{-i_0+\lambda_1-\frac{\varepsilon}{p}} \ (m \in \mathbf{N}^{i_0}), \quad \tilde{b}_n := ||n||_\beta^{-j_0+\lambda_2-\frac{\varepsilon}{q}} \ (n \in \mathbf{N}^{j_0}).$$

Then by (19), (16) and (20), we obtain

$$||\tilde{a}||_{p,\varphi}||\tilde{b}||_{q,\psi} = \left\{\sum_m ||m||_\alpha^{p(i_0-\lambda_1)-i_0}\tilde{a}_m^p\right\}^{\frac{1}{p}} \left\{\sum_n ||n||_\beta^{q(j_0-\lambda_2)-j_0}\tilde{b}_n^q\right\}^{\frac{1}{q}}$$

$$= \left\{\sum_m ||m||_\alpha^{-i_0-\varepsilon}\right\}^{\frac{1}{p}} \left\{\sum_n ||n||_\beta^{-j_0-\varepsilon}\right\}^{\frac{1}{q}}$$

$$= \frac{1}{\varepsilon}\left[\frac{\Gamma^{i_0}(\frac{1}{\alpha})}{i_0^{\varepsilon/\alpha}\alpha^{i_0-1}\Gamma(\frac{i_0}{\alpha})} + \varepsilon O(1)\right]^{\frac{1}{p}} \left[\frac{\Gamma^{j_0}(\frac{1}{\beta})}{j_0^{\varepsilon/\beta}\beta^{j_0-1}\Gamma(\frac{j_0}{\beta})} + \varepsilon\tilde{O}(1)\right]^{\frac{1}{q}},$$

$$\tilde{I} := \sum_n \left[\sum_m k_\lambda(||m||_\alpha, ||n||_\beta)\tilde{a}_m\right]\tilde{b}_n$$

$$= \sum_n \left[||n||_\beta^{\tilde{\lambda}_2} \sum_m k_\lambda(||m||_\alpha, ||n||_\beta)||m||_\alpha^{-i_0+\tilde{\lambda}_1}\right] ||n||_\beta^{-j_0-\varepsilon}$$

$$= \sum_n \omega_\lambda(\tilde{\lambda}_2, n)||n||_\beta^{-j_0-\varepsilon} > \tilde{K}_2 \sum_n \left(1 - O(\frac{1}{||n||_\beta^\eta})\right) ||n||_\beta^{-j_0-\varepsilon}$$

$$= (K_2 + o(1))\left[\frac{\Gamma^{j_0}(\frac{1}{\beta})}{\varepsilon j_0^{\varepsilon/\beta}\beta^{j_0-1}\Gamma(\frac{j_0}{\beta})} + \tilde{O}(1) - O(1)\right].$$

If there exists a constant $K \le K_1^{\frac{1}{p}} K_2^{\frac{1}{q}}$, such that (10) is valid when replacing $K_1^{\frac{1}{p}} K_2^{\frac{1}{q}}$ by K, then in particular, we have

$$(K_2 + o(1))\left[\frac{\Gamma^{j_0}(\frac{1}{\beta})}{j_0^{\varepsilon/\beta}\beta^{j_0-1}\Gamma(\frac{j_0}{\beta})} + \varepsilon\tilde{O}(1) - \varepsilon O(1)\right] < \varepsilon\tilde{I} < \varepsilon K||\tilde{a}||_{p,\varphi}||\tilde{b}||_{q,\psi}$$

$$= K\left[\frac{\Gamma^{i_0}(\frac{1}{\alpha})}{i_0^{\varepsilon/\alpha}\alpha^{i_0-1}\Gamma(\frac{i_0}{\alpha})} + \varepsilon O(1)\right]^{\frac{1}{p}} \left[\frac{\Gamma^{j_0}(\frac{1}{\beta})}{j_0^{\varepsilon/\beta}\beta^{j_0-1}\Gamma(\frac{j_0}{\beta})} + \varepsilon\tilde{O}(1)\right]^{\frac{1}{q}}.$$

For $\varepsilon \to 0^+$, we find

$$\frac{\Gamma^{j_0}(\frac{1}{\beta})}{\beta^{j_0-1}\Gamma(\frac{j_0}{\beta})} \frac{\Gamma^{i_0}(\frac{1}{\alpha})}{\alpha^{i_0-1}\Gamma(\frac{i_0}{\alpha})} k(\lambda_1) \le K\left[\frac{\Gamma^{i_0}(\frac{1}{\alpha})}{\alpha^{i_0-1}\Gamma(\frac{i_0}{\alpha})}\right]^{\frac{1}{p}} \left[\frac{\Gamma^{j_0}(\frac{1}{\beta})}{\beta^{j_0-1}\Gamma(\frac{j_0}{\beta})}\right]^{\frac{1}{q}},$$

and then $K_1^{\frac{1}{p}} K_2^{\frac{1}{q}} \le K$. Hence, $K = K_1^{\frac{1}{p}} K_2^{\frac{1}{q}}$ is the best possible of (10).

The constant factor $K_1^{\frac{1}{p}} K_2^{\frac{1}{q}}$ in (11) is still the best possible. Otherwise, we would reach a contradiction by (9) (for $\omega_\lambda(\lambda_2, n) < K_2$) that the constant factor in (8) is not the best possible. The theorem is proved. □

Theorem 3.2. *With the same assumptions of Theorem* 2.2 $(p < 0)$, *if*

$$\vartheta_\lambda(n) = O\Big(\frac{1}{||n||_\beta^\eta}\Big) \in (0,1) \quad (\eta > 0),$$

there exists a constant $\delta_0 > 0$, *such that for any* $\tilde{\lambda}_i \in (\lambda_i - \delta_0, \lambda_i + \delta_0)$ $(i = 1,2)$, $\tilde{\lambda}_1 + \tilde{\lambda}_2 = \lambda$,

$$\omega_\lambda(\tilde{\lambda}_2, n) < \tilde{K}_2 \quad (n \in \mathbf{N}^{j_0}), \tag{21}$$

where

$$\tilde{K}_2 = \frac{\Gamma^{i_0}(\frac{1}{\alpha})}{\alpha^{i_0-1}\Gamma(\frac{i_0}{\alpha})} k(\tilde{\lambda}_1) \in \mathbf{R}_+,$$

then the constant factor $K_1^{\frac{1}{p}} K_2^{\frac{1}{q}}$ *in* (12) *and* (13) *is the best possible.*

Proof. For $0 < \varepsilon < q\delta_0$, $\tilde{\lambda}_1 = \lambda_1 + \frac{\varepsilon}{q}$, $\tilde{\lambda}_2 = \lambda_2 - \frac{\varepsilon}{q}$, we set $\tilde{a}_m$, $\tilde{b}_n$ as in Theorem 3.1. Then by (19), (21) and (16), we obtain

$$\begin{aligned}
||\tilde{a}||_{p,\varphi}||\tilde{b}||_{q,\tilde{\psi}} &= \Big\{\sum_m ||m||_\alpha^{p(i_0-\lambda_1)-i_0}\tilde{a}_m^p\Big\}^{\frac{1}{p}} \Big\{\sum_n (1-\vartheta_\lambda(n))||n||_\beta^{q(j_0-\lambda_2)-j_0}\tilde{b}_n^q\Big\}^{\frac{1}{q}} \\
&= \Big\{\sum_m ||m||_\alpha^{-i_0-\varepsilon}\Big\}^{\frac{1}{p}} \Big\{\sum_n \Big(1 - O\Big(\frac{1}{||n||_\beta^\eta}\Big)\Big)||n||_\beta^{-j_0-\varepsilon}\Big\}^{\frac{1}{q}} \\
&= \frac{1}{\varepsilon}\left(\frac{\Gamma^{i_0}(\frac{1}{\alpha})}{i_0^{\varepsilon/\alpha}\alpha^{i_0-1}\Gamma(\frac{i_0}{\alpha})} + \varepsilon O(1)\right)^{\frac{1}{p}} \left(\frac{\Gamma^{j_0}(\frac{1}{\beta})}{j_0^{\varepsilon/\beta}\beta^{j_0-1}\Gamma(\frac{j_0}{\beta})} + \varepsilon\tilde{O}(1) - \varepsilon O(1)\right)^{\frac{1}{q}},
\end{aligned}$$

$$\begin{aligned}
\tilde{I} &= \sum_n \Big[\sum_m k_\lambda(||m||_\alpha, ||n||_\beta)\tilde{a}_m\Big]\tilde{b}_n = \sum_n \omega_\lambda(\tilde{\lambda}_2, n)||n||_\beta^{-j_0-\varepsilon} < \tilde{K}_2 \sum_n ||n||_\beta^{-j_0-\varepsilon} \\
&= (K_2 + o(1))\left[\frac{\Gamma^{j_0}(\frac{1}{\beta})}{\varepsilon j_0^{\varepsilon/\beta}\beta^{j_0-1}\Gamma(\frac{j_0}{\beta})} + \tilde{O}(1)\right].
\end{aligned}$$

If there exists a constant $K \geq K_1^{\frac{1}{p}} K_2^{\frac{1}{q}}$, such that (12) is valid when replacing $K_1^{\frac{1}{p}} K_2^{\frac{1}{q}}$ by K, then we have

$$(K_2+o(1))\left[\frac{\Gamma^{j_0}(\frac{1}{\beta})}{j_0^{\varepsilon/\beta}\beta^{j_0-1}\Gamma(\frac{j_0}{\beta})}+\varepsilon\tilde{O}(1)\right] > \varepsilon\tilde{I} > \varepsilon K||\tilde{a}||_{p,\varphi}||\tilde{b}||_{q,\tilde{\psi}}$$

$$= K\left[\frac{\Gamma^{i_0}(\frac{1}{\alpha})}{i_0^{\varepsilon/\alpha}\alpha^{i_0-1}\Gamma(\frac{i_0}{\alpha})}+\varepsilon O(1)\right]^{\frac{1}{p}}\left[\frac{\Gamma^{j_0}(\frac{1}{\beta})}{j_0^{\varepsilon/\beta}\beta^{j_0-1}\Gamma(\frac{j_0}{\beta})}+\varepsilon\tilde{O}(1)-\varepsilon O(1)\right]^{\frac{1}{q}}.$$

For $\varepsilon \to 0^+$, it follows

$$\frac{\Gamma^{j_0}(\frac{1}{\beta})}{\beta^{j_0-1}\Gamma(\frac{j_0}{\beta})}\frac{\Gamma^{i_0}(\frac{1}{\alpha})}{\alpha^{i_0-1}\Gamma(\frac{i_0}{\alpha})}k(\lambda_1) \geqslant K\left[\frac{\Gamma^{i_0}(\frac{1}{\alpha})}{\alpha^{i_0-1}\Gamma(\frac{i_0}{\alpha})}\right]^{\frac{1}{p}}\left[\frac{\Gamma^{j_0}(\frac{1}{\beta})}{\beta^{j_0-1}\Gamma(\frac{j_0}{\beta})}\right]^{\frac{1}{q}},$$

and then $K_1^{\frac{1}{p}}K_2^{\frac{1}{q}} \geq K$. Hence, $K = K_1^{\frac{1}{p}}K_2^{\frac{1}{q}}$ is the best possible constant factor of (12).

The constant factor $K_1^{\frac{1}{p}}K_2^{\frac{1}{q}}$ in (13) is still the best possible. Otherwise, we would reach a contradiction by the reverse of (9) that the constant factor in (12) is not the best possible. The theorem is proved. □

Theorem 3.3. *With the same assumptions of Theorem* 2.3 $(0 < p < 1)$, *if*

$$\theta_\lambda(m) = O\left(\frac{1}{||m||_\alpha^\rho}\right) \in (0,1) \quad (\rho > 0),$$

and there exists a constant $\delta_0 > 0$, *such that for any* $\tilde{\lambda}_i \in (\lambda_i-\delta_0, \lambda_i+\delta_0)$ $(i = 1,2)$, $\tilde{\lambda}_1+\tilde{\lambda}_2 = \lambda$,

$$\omega_\lambda(\tilde{\lambda}_2, n) < \tilde{K}_2 \quad (n \in \mathbf{N}^{j_0}), \tag{22}$$

where

$$\tilde{K}_2 = \frac{\Gamma^{i_0}(\frac{1}{\alpha})}{\alpha^{i_0-1}\Gamma(\frac{i_0}{\alpha})}k(\tilde{\lambda}_1) \in \mathbf{R}_+,$$

then the constant factor $K_1^{\frac{1}{p}}K_2^{\frac{1}{q}}$ *in* (14) *and* (15) *is the best possible.*

Proof. For $0 < \varepsilon < |q|\delta_0$, $\tilde{\lambda}_1 = \lambda_1 + \frac{\varepsilon}{q}$, $\tilde{\lambda}_2 = \lambda_2 - \frac{\varepsilon}{q}$, we set $\tilde{a}_m$, $\tilde{b}_n$ as in Theorem 3.1. Then by (19), (22) and (16), we obtain

$$||\tilde{a}||_{p,\tilde{\varphi}}||\tilde{b}||_{q,\psi}$$

$$= \left\{ \sum_m (1-\theta_\lambda(m)) ||m||_\alpha^{p(i_0-\lambda_1)-i_0} \tilde{a}_m^p \right\}^{\frac{1}{p}} \left\{ \sum_n ||n||_\beta^{q(j_0-\lambda_2)-j_0} \tilde{b}_n^q \right\}^{\frac{1}{q}}$$

$$= \left\{ \sum_m \left(1 - O\left(\frac{1}{||m||_\alpha^\rho}\right)\right) ||m||_\alpha^{-i_0-\varepsilon} \right\}^{\frac{1}{p}} \left\{ \sum_n ||n||_\beta^{-j_0-\varepsilon} \right\}^{\frac{1}{q}}$$

$$= \frac{1}{\varepsilon} \left[\frac{\Gamma^{i_0}(\frac{1}{\alpha})}{i_0^{\varepsilon/\alpha} \alpha^{i_0-1} \Gamma(\frac{i_0}{\alpha})} + \varepsilon O(1) - \varepsilon \hat{O}(1) \right]^{\frac{1}{p}} \left[\frac{\Gamma^{j_0}(\frac{1}{\beta})}{j_0^{\varepsilon/\beta} \beta^{j_0-1} \Gamma(\frac{j_0}{\beta})} + \varepsilon \tilde{O}(1) \right]^{\frac{1}{q}},$$

$$\tilde{I} = \sum_n \left[\sum_m k_\lambda(||m||_\alpha, ||n||_\beta) \tilde{a}_m \right] \tilde{b}_n = \sum_n \omega_\lambda(\tilde{\lambda}_2, n) ||n||_\beta^{-j_0-\varepsilon} < \tilde{K}_2 \sum_n ||n||_\beta^{-j_0-\varepsilon}$$

$$= (K_2 + o(1)) \left[\frac{\Gamma^{j_0}(\frac{1}{\beta})}{j_0^{\varepsilon/\beta} \beta^{j_0-1} \Gamma(\frac{j_0}{\beta})} + \varepsilon \tilde{O}(1) \right].$$

If there exists a constant $K \geq K_1^{\frac{1}{p}} K_2^{\frac{1}{q}}$, such that (14) is valid when replacing $K_1^{\frac{1}{p}} K_2^{\frac{1}{q}}$ by K, then we have

$$(K_2 + o(1)) \left[\frac{\Gamma^{j_0}(\frac{1}{\beta})}{j_0^{\varepsilon/\beta} \beta^{j_0-1} \Gamma(\frac{j_0}{\beta})} + \varepsilon \tilde{O}(1) \right] > \varepsilon \tilde{I} > \varepsilon K ||\tilde{a}||_{p,\tilde{\varphi}} ||\tilde{b}||_{q,\psi}$$

$$= K \left[\frac{\Gamma^{i_0}(\frac{1}{\alpha})}{i_0^{\varepsilon/\alpha} \alpha^{i_0-1} \Gamma(\frac{i_0}{\alpha})} + \varepsilon O(1) - \varepsilon \hat{O}(1) \right]^{\frac{1}{p}} \left[\frac{\Gamma^{j_0}(\frac{1}{\beta})}{j_0^{\varepsilon/\beta} \beta^{j_0-1} \Gamma(\frac{j_0}{\beta})} + \varepsilon \tilde{O}(1) \right]^{\frac{1}{q}}.$$

For $\varepsilon \to 0^+$, we find

$$\frac{\Gamma^{j_0}(\frac{1}{\beta})}{\beta^{j_0-1} \Gamma(\frac{j_0}{\beta})} \frac{\Gamma^{i_0}(\frac{1}{\alpha})}{\alpha^{i_0-1} \Gamma(\frac{i_0}{\alpha})} k(\lambda_1) \geq K \left[\frac{\Gamma^{i_0}(\frac{1}{\alpha})}{\alpha^{i_0-1} \Gamma(\frac{i_0}{\alpha})} \right]^{\frac{1}{p}} \left[\frac{\Gamma^{j_0}(\frac{1}{\beta})}{\beta^{j_0-1} \Gamma(\frac{j_0}{\beta})} \right]^{\frac{1}{q}},$$

and then $K_1^{\frac{1}{p}} K_2^{\frac{1}{q}} \geq K$. Hence, $K = K_1^{\frac{1}{p}} K_2^{\frac{1}{q}}$ is the best possible of (14).

The constant factor $K_1^{\frac{1}{p}} K_2^{\frac{1}{q}}$ in (15) is still the best possible. Otherwise, we would reach a contradiction by the reverse of (9) that the constant factor in (14) is not the best possible. The theorem is proved. □

Corollary 3.1. *Suppose that* $k(\lambda_1) \in \mathbf{R}_+$,

$$K_1 = \frac{\Gamma^{j_0}(\frac{1}{\beta})}{\beta^{j_0-1}\Gamma(\frac{j_0}{\beta})}k(\lambda_1), \quad K_2 = \frac{\Gamma^{i_0}(\frac{1}{\alpha})}{\alpha^{i_0-1}\Gamma(\frac{i_0}{\alpha})}k(\lambda_1),$$

and

$$K_1(1-\theta_\lambda(m)) < \varpi_\lambda(\lambda_1, m) < K_1 \quad (m \in \mathbf{N}^{i_0}), \tag{23}$$

where

$$\theta_\lambda(m) = O\left(\frac{1}{||m||_\alpha^\rho}\right) \in (0,1) \quad (\rho > 0).$$

If there exists a constant $\delta_0 > 0$, *such that for any* $\tilde{\lambda}_i \in (\lambda_i - \delta_0, \lambda_i + \delta_0)$ $(i = 1,2)$, $\tilde{\lambda}_1 + \tilde{\lambda}_2 = \lambda$,

$$\tilde{K}_2(1-\tilde{\vartheta}_\lambda(n)) < \omega_\lambda(\tilde{\lambda}_2, n) < \tilde{K}_2 \quad (n \in \mathbf{N}^{j_0}), \tag{24}$$

where

$$\tilde{\vartheta}_\lambda(n) = O\left(\frac{1}{||n||_\beta^\eta}\right) \in (0,1) \quad (\eta > 0), \quad \tilde{K}_2 = \frac{\Gamma^{i_0}(\frac{1}{\alpha})}{\alpha^{i_0-1}\Gamma(\frac{i_0}{\alpha})}k(\tilde{\lambda}_1) \in \mathbf{R}_+,$$

then the constant factor $K_1^{\frac{1}{p}} K_2^{\frac{1}{q}}$ *in Theorems* 3.1–3.3 *is the best possible.*

Theorem 3.4. *If* $k_\lambda(x,y)y^{\lambda_2 - j_0}$ *is a strict decreasing function with respect to* $y \in \mathbf{R}_+$, *there exists a constant* $\delta_0 > 0$, *such that for any* $\tilde{\lambda}_1 \in (\lambda_1 - \delta_0, \lambda_1 + \delta_0)$, $k_\lambda(x,y)x^{\tilde{\lambda}_1 - i_0}$ *is strict decreasing with respect to* $x \in \mathbf{R}_+$,

$$k(\tilde{\lambda}_1) := \int_0^\infty k_\lambda(t,1)t^{\tilde{\lambda}_1 - 1}dt \in \mathbf{R},$$

and there exists a constant $\delta_1 < \lambda_1 - \delta_0$, *satisfying*

$$k_\lambda(t,1) \le \frac{L}{t^{\delta_1}} \quad (t \in (0,\infty)),$$

then the constant factor $K_1^{\frac{1}{p}} K_2^{\frac{1}{q}}$ *in Theorems* 3.1–3.3 *is the best possible with*

$$k(\lambda_1) = \int_0^\infty k_\lambda(t,1)t^{\lambda_1 - 1}dt \in \mathbf{R}_+.$$

Proof. We need to prove that the conditions of Corollary 3.1 are fulfilled. For any $\tilde{\lambda}_i \in (\lambda_i - \delta_0, \lambda_i + \delta_0)$ $(i = 1, 2)$, $\tilde{\lambda}_1 + \tilde{\lambda}_2 = \lambda$,

$$k_\lambda(||x||_\alpha, ||n||_\beta)\frac{1}{||x||_\alpha^{i_0-\tilde{\lambda}_1}} \qquad (x = (x_1, \ldots, x_{i_0}))$$

is strict decreasing with respect to $x_i \in \mathbf{R}_+$ $(i = 1, \ldots, i_0)$. Setting

$$D_M := \left\{x \in \mathbf{R}_+^{i_0};\ \sum_{i=1}^{i_0} x_i^\alpha \le M^\alpha\right\},$$

by the decreasing property and (17), we find

$$\omega_\lambda(\tilde{\lambda}_2, n) = ||n||_\beta^{\tilde{\lambda}_2} \sum_m k_\lambda(||m||_\alpha, ||n||_\beta)\frac{1}{||m||_\alpha^{i_0-\tilde{\lambda}_1}} < ||n||_\beta^{\tilde{\lambda}_2} \int_{\mathbf{R}_+^{i_0}} \frac{k_\lambda(||x||_\alpha, ||n||_\beta)}{||x||_\alpha^{i_0-\tilde{\lambda}_1}} dx$$

$$= ||n||_\beta^{\tilde{\lambda}_2} \lim_{M\to\infty} \int_{D_M} k_\lambda\left(M||\frac{x}{M}||_\alpha, ||n||_\beta\right) \frac{M^{\tilde{\lambda}_1 - i_0}}{||\frac{x}{M}||_\alpha^{i_0-\tilde{\lambda}_1}} dx$$

$$= ||n||_\beta^{\tilde{\lambda}_2} \lim_{M\to\infty} \left[\frac{M^{\tilde{\lambda}_1}\Gamma^{i_0}(\frac{1}{\alpha})}{\alpha^{i_0}\Gamma(\frac{i_0}{\alpha})} \int_0^1 k_\lambda(Mu^{\frac{1}{\alpha}}, ||n||_\beta)\frac{u^{\frac{i_0}{\alpha}-1}}{u^{(i_0-\tilde{\lambda}_1)/\alpha}} du\right]$$

$$= \frac{\Gamma^{i_0}(\frac{1}{\alpha})}{\alpha^{i_0-1}\Gamma(\frac{i_0}{\alpha})} k(\tilde{\lambda}_1) = \tilde{K}_2.$$

Hence, it follows

$$K_2 = \frac{\Gamma^{i_0}(\frac{1}{\alpha})}{\alpha^{i_0-1}\Gamma(\frac{i_0}{\alpha})} k(\lambda_1) > 0$$

and then $k(\lambda_1) \in \mathbf{R}_+$. For $M > i_0^{1/\alpha}$, we set

$$\Psi(u) = \begin{cases} 0, & 0 < u < \frac{i_0}{M^\alpha}, \\ k_\lambda(Mu^{1/\alpha}, ||n||_\beta)\frac{1}{u^{(i_0-\tilde{\lambda}_1)/\alpha}}, & \frac{i_0}{M^\alpha} \le u \le 1. \end{cases}$$

Then by the decreasing property and (17), we have

$$\omega_\lambda(\tilde{\lambda}_2, n) > ||n||_\beta^{\tilde{\lambda}_2} \int_{\{x\in\mathbf{R}_+^{i_0};x_i\ge 1\}} k_\lambda(||x||_\alpha, ||n||_\beta)\frac{1}{||x||_\alpha^{i_0-\tilde{\lambda}_1}} dx$$

$$\omega_\lambda(\tilde{\lambda}_2,n) > ||n||_\beta^{\tilde{\lambda}_2}\int_{\{x\in\mathbf{R}_+^{i_0};x_i\geq 1\}} k_\lambda(||x||_\alpha,||n||_\beta)\frac{1}{||x||_\alpha^{i_0-\tilde{\lambda}_1}}dx$$

$$= ||n||_\beta^{\tilde{\lambda}_2}\lim_{M\to\infty}\int_{D_M}\Psi\left(\sum_{i=1}^{i_0}\left(\frac{x_i}{M}\right)^\gamma\right)M^{\tilde{\lambda}_1-i_0}dx_1\cdots dx_{i_0}$$

$$= ||n||_\beta^{\tilde{\lambda}_2}\lim_{M\to\infty}\left[\frac{M^{\tilde{\lambda}_1}\Gamma^{i_0}(\frac{1}{\alpha})}{\alpha^{i_0}\Gamma(\frac{i_0}{\alpha})}\int_{\frac{i_0}{M^\alpha}}^1 k_\lambda(Mu^{\frac{1}{\alpha}},||n||_\beta)\frac{u^{\frac{i_0}{\alpha}-1}}{u^{(i_0-\tilde{\lambda}_1)/\alpha}}du\right]$$

$$= \frac{\Gamma^{i_0}(\frac{1}{\alpha})}{\alpha^{i_0-1}\Gamma(\frac{i_0}{\alpha})}\int_{i_0^{1/\alpha}/||n||_\beta}^\infty k_\lambda(v,1)v^{\tilde{\lambda}_1-1}dv \qquad (v = M||n||_\beta^{-1}u^{1/\alpha})$$

$$= \tilde{K}_2(1-\tilde{\vartheta}_\lambda(n)) > 0,$$

where

$$0 < \tilde{\vartheta}_\lambda(n) := \frac{1}{k(\tilde{\lambda}_1)}\int_0^{i_0^{1/\alpha}/||n||_\beta} k_\lambda(v,1)v^{\tilde{\lambda}_1-1}dv$$

$$\leq \frac{L}{k(\tilde{\lambda}_1)}\int_0^{i_0^{1/\alpha}/||n||_\beta} v^{\tilde{\lambda}_1-\delta_1-1}dv$$

$$\leq \frac{L}{k(\tilde{\lambda}_1)}\frac{i_0^{(\tilde{\lambda}_1-\delta_1)/\alpha}}{\tilde{\lambda}_1-\delta_1}\frac{1}{||n||_\beta^{\lambda_1-\delta_0-\delta_1}}.$$

Setting $\eta = \lambda_1-\delta_0-\delta_1 > 0$, it follows

$$\tilde{\vartheta}_\lambda(n) = O\left(\frac{1}{||n||_\beta^\eta}\right) \in (0,1).$$

By the same way, we can prove that

$$K_1 = \frac{\Gamma^{j_0}(\frac{1}{\beta})}{\beta^{j_0-1}\Gamma(\frac{j_0}{\beta})}k(\lambda_1),$$

and

$$K_1(1-\theta_\lambda(m)) < \varpi_\lambda(\lambda_1,m) < K_1 \quad (m \in \mathbf{N}^{i_0}),$$

with

$$\theta_\lambda(m) = O\left(\frac{1}{||m||_\alpha^\rho}\right) \in (0,1) \quad (\rho > 0).$$

By Corollary 3.1, the constant factor $K_1^{\frac{1}{p}} K_2^{\frac{1}{q}}$ in Theorems 3.1, 3.2 and 3.3 is the best possible. The theorem is proved. □

Corollary 3.2. *If $k_\lambda(x, y)$ is a positive decreasing function with respect to x $(y) \in \mathbf{R}_+$, $\lambda_1 < i_0, \lambda_2 < j_0$, there exists a constant $0 < \delta_0 < j_0 - \lambda_2$, such that for any $\tilde{\lambda}_1 \in (\lambda_1 - \delta_0, \lambda_1 + \delta_0)$,*

$$k(\tilde{\lambda}_1) = \int_0^\infty k_\lambda(t, 1) t^{\tilde{\lambda}_1 - 1} dt \in \mathbf{R},$$

and there exists a constant $\delta_1 < \lambda_1 - \delta_0$, satisfying

$$k_\lambda(t, 1) \leq \frac{L}{t^{\delta_1}} \quad (t \in (0, \infty)),$$

then the constant factor $K_1^{\frac{1}{p}} K_2^{\frac{1}{q}}$ in Theorems 3.1–3.3 is the best possible with

$$k(\lambda_1) = \int_0^\infty k_\lambda(t, 1) t^{\lambda_1 - 1} dt \in \mathbf{R}_+.$$

Remark 3.1. By (23) and (24), we find

$$\lim_{n \to \infty} \omega_\lambda(\lambda_2, n) = K_2 = \frac{\Gamma^{i_0}(\frac{1}{\alpha})}{\alpha^{i_0 - 1} \Gamma(\frac{i_0}{\alpha})} k(\lambda_1),$$

$$\lim_{m \to \infty} \varpi_\lambda(\lambda_1, m) = K_1 - \frac{\Gamma^{j_0}(\frac{1}{\beta})}{\beta^{j_0 - 1} \Gamma(\frac{j_0}{\beta})} k(\lambda_1).$$

4 The More Accurate Inequality and Its Equivalent Form

Definition 4.1. For $0 < \alpha, \beta \leq 1$, $\tau = (\tau_1, \ldots, \tau_{i_0}) \in (0, 1/2]^{i_0}$, $\sigma = (\sigma_1, \ldots, \sigma_{j_0}) \in (0, 1/2]^{j_0}$, $m - \tau = (m_1 - \tau_1, \ldots, m_{i_0} - \tau_{i_0}) \in \mathbf{R}_+^{i_0}$, $n - \sigma = (n_1 - \sigma_1, \ldots, n_{j_0} - \sigma_{j_0}) \in \mathbf{R}_+^{j_0}$, define two weight coefficients $w_\lambda(\lambda_2, n)$ and $W_\lambda(\lambda_1, m)$ as follows:

$$w_\lambda(\lambda_2, n) := \sum_m k_\lambda(||m - \tau||_\alpha, ||n - \sigma||_\beta) \frac{||n - \sigma||_\beta^{\lambda_2}}{||m - \tau||_\alpha^{i_0 - \lambda_1}},$$

$$W_\lambda(\lambda_1, m) := \sum_n k_\lambda(||m - \tau||_\alpha, ||n - \sigma||_\beta) \frac{||m - \tau||_\alpha^{\lambda_1}}{||n - \sigma||_\beta^{j_0 - \lambda_2}}.$$

Lemma 4.1. *If for $t > 0$, $(-1)^i h^{(i)}(t) > 0$ $(i = 1,2)$, then for $b > 0$, $0 < \alpha \le 1$, we have*

$$(-1)^i \frac{d^i}{dx^i} h((b + x^\alpha)^{1/\alpha}) > 0 \quad (x > 0;\ i = 1,2). \tag{25}$$

Proof. We find

$$\frac{d}{dx} h((b + x^\alpha)^{1/\alpha}) = h'((b + x^\alpha)^{1/\alpha})(b + x^\alpha)^{\frac{1}{\alpha}-1} x^{\alpha-1} < 0$$

and

$$\begin{aligned}
\frac{d^2}{dx^2} h((b + x^\alpha)^{1/\alpha}) &= \frac{d}{dx}[h'((b + x^\alpha)^{\frac{1}{\alpha}})(b + x^\alpha)^{\frac{1}{\alpha}-1} x^{\alpha-1}] \\
&= h''((b + x^\alpha)^{\frac{1}{\alpha}})(b + x^\alpha)^{\frac{2}{\alpha}-2} x^{2\alpha-2} \\
&\quad + \alpha\Big(\frac{1}{\alpha} - 1\Big) h'((b + x^\alpha)^{\frac{1}{\alpha}})(b + x^\alpha)^{\frac{1}{\alpha}-2} x^{2\alpha-2} \\
&\quad + (\alpha - 1) h'((b + x^\alpha)^{\frac{1}{\alpha}})(b + x^\alpha)^{\frac{1}{\alpha}-1} x^{\alpha-2},
\end{aligned}$$

i.e.,

$$\begin{aligned}
\frac{d^2}{dx^2} h((b + x^\alpha)^{1/\alpha}) &= h''((b + x^\alpha)^{\frac{1}{\alpha}})(b + x^\alpha)^{\frac{2}{\alpha}-2} x^{2\alpha-2} \\
&\quad + b(\alpha - 1) h'((b + x^\alpha)^{\frac{1}{\alpha}})(b + x^\alpha)^{\frac{1}{\alpha}-2} x^{\alpha-2} > 0.
\end{aligned}$$

Hence, (25) follows. The lemma is proved. □

Lemma 4.2. *With the same assumptions of Definition* 4.1, *if*

$$(-1)^i \frac{\partial^i}{\partial x^i}\left(k_\lambda(x, y) x^{\lambda_1 - i_0}\right) > 0, \quad (-1)^i \frac{\partial^i}{\partial y^i}\left(k_\lambda(x, y) y^{\lambda_2 - j_0}\right) > 0 \quad (i = 1,2),$$

then

(i) *We have*

$$w_\lambda(\lambda_2, n) < K_2 \quad (n \in \mathbf{N}^{j_0}), \tag{26}$$

$$W_\lambda(\lambda_1, m) < K_1 \quad (m \in \mathbf{N}^{i_0}); \tag{27}$$

(ii) *For $p > 1$, setting $\tilde{\lambda}_1 = \lambda_1 + \frac{\varepsilon}{q}$, $\tilde{\lambda}_2 = \lambda_2 - \frac{\varepsilon}{q}$, we have*

$$\tilde{K}_2(1 - \tilde{\theta}_\lambda(n)) < w_\lambda(\tilde{\lambda}_2, n),$$

where

$$\tilde{\theta}_\lambda(n) = \frac{1}{k(\tilde{\lambda}_1)} \int_0^{i_0^{1/\alpha}/||n-\sigma||_\beta} k_\lambda(v,1) v^{\tilde{\lambda}_1 - 1} dv \in (0,1),$$

$$\tilde{K}_2 = \frac{\Gamma^{i_0}(\frac{1}{\alpha})}{\alpha^{i_0-1}\Gamma(\frac{i_0}{\alpha})} k(\tilde{\lambda}_1), \quad k(\tilde{\lambda}_1) = \int_0^\infty k_\lambda(v,1) v^{\tilde{\lambda}_1 - 1} dv.$$

Proof. (i) By Lemma 4.1 and Hermite-Hadamard's inequality (cf. [15]), similarly to Theorem 3.4, it follows

$$w_\lambda(\lambda_2, n) < \int_{(1/2,\infty)^{j_0}} k_\lambda(||x-\tau||_\alpha, ||n-\sigma||_\beta) \frac{||n-\sigma||_\beta^{\lambda_2}}{||x-\tau||_\alpha^{i_0-\lambda_1}} dx$$

$$= \int_{\{u \in \mathbf{R}_+^{j_0}; u_i > \frac{1}{2} - \tau_i\}} k_\lambda(||u||_\alpha, ||n-\sigma||_\beta) \frac{||n-\sigma||_\beta^{\lambda_2}}{||u||_\alpha^{i_0-\lambda_1}} du$$

$$\leq \int_{\mathbf{R}_+^{j_0}} k_\lambda(||u||_\alpha, ||n-\sigma||_\beta) \frac{||n-\sigma||_\beta^{\lambda_2}}{||u||_\alpha^{i_0-\lambda_1}} du = K_2.$$

Hence, we have (26). By the same way, we have (27).

(ii) By the decreasing property and the same way as in Theorem 3.4, we have

$$w_\lambda(\tilde{\lambda}_2, n) > ||n-\sigma||_\beta^{\tilde{\lambda}_2} \int_{\{x \in \mathbf{R}_+^{i_0}; x_i \geq 1 + \iota_i\}} \frac{k_\lambda(||x-\tau||_\alpha, ||n-\sigma||_\beta) dx}{||x-\tau||_\alpha^{i_0 - \tilde{\lambda}_1}}$$

$$= ||n-\sigma||_\beta^{\tilde{\lambda}_2} \int_{\{u \in \mathbf{R}_+^{i_0}; u_i \geq 1\}} \frac{k_\lambda(||u||_\alpha, ||n-\sigma||_\beta) du}{||u||_\alpha^{i_0 - \tilde{\lambda}_1}}$$

$$= \frac{\Gamma^{i_0}(\frac{1}{\alpha})}{\alpha^{i_0-1}\Gamma(\frac{i_0}{\alpha})} \int_{i_0^{1/\alpha}/||n-\sigma||_\beta}^\infty k_\lambda(v,1) v^{\tilde{\lambda}_1 - 1} dv$$

$$= \tilde{K}_2(1 - \tilde{\theta}_\lambda(n)) > 0.$$

The lemma is proved. □

Setting $\Phi(m) := ||m-\tau||_\alpha^{p(i_0-\lambda_1)-i_0}$ $(m \in \mathbf{N}^{i_0})$ and $\Psi(n) := ||n-\sigma||_\beta^{q(j_0-\lambda_2)-j_0}$ $(n \in \mathbf{N}^{j_0})$, we have:

Theorem 4.1. *If* $0 < \alpha, \beta \leq 1$, $\tau \in (0, 1/2]^{i_0}$, $\sigma \in (0, 1/2]^{j_0}$,

$$(-1)^i \frac{\partial^i}{\partial x^i}(k_\lambda(x,y) x^{\lambda_1 - i_0}) > 0, \quad (-1)^i \frac{\partial^i}{\partial y^i}(k_\lambda(x,y) y^{\lambda_2 - j_0}) > 0 \quad (i = 1, 2),$$

there exists a constant $\delta_0 > 0$*, such that for any* $\tilde{\lambda}_1 \in (\lambda_1 - \delta_0, \lambda_1 + \delta_0)$*,*

$$k(\tilde{\lambda}_1) := \int_0^\infty k_\lambda(t,1)t^{\tilde{\lambda}_1 - 1}dt \in \mathbf{R},$$

and there exists a constant $\delta_1 < \lambda_1 - \delta_0$*, satisfying*

$$k_\lambda(t,1) \leq \frac{L}{t^{\delta_1}} \quad (t \in (0,\infty)),$$

then for $p > 1$*,* $a_m, b_n \geq 0$*,* $0 < ||a||_{p,\Phi}, ||b||_{q,\Psi} < \infty$*, we have the following inequality:*

$$I(\tau,\sigma) := \sum_n \sum_m k_\lambda(||m-\tau||_\alpha, ||n-\sigma||_\beta)a_m b_n < K_1^{\frac{1}{p}} K_2^{\frac{1}{q}} ||a||_{p,\Phi}||b||_{q,\Psi}, \quad (28)$$

where the constant factor

$$K_1^{\frac{1}{p}} K_2^{\frac{1}{q}} = \left[\frac{\Gamma^{j_0}(\frac{1}{\beta})}{\beta^{j_0-1}\Gamma(\frac{j_0}{\beta})}\right]^{\frac{1}{p}} \left[\frac{\Gamma^{i_0}(\frac{1}{\alpha})}{\beta^{i_0-1}\Gamma(\frac{i_0}{\alpha})}\right]^{\frac{1}{q}} k(\lambda_1)$$

is the best possible with

$$k(\lambda_1) = \int_0^\infty k_\lambda(t,1)t^{\lambda_1-1}dt \in \mathbf{R}_+.$$

Proof. As in Lemma 2.2, by Hölder's inequality, we still have

$$\begin{aligned} I(\tau,\sigma) &= \sum_n \sum_m k_\lambda(||m-\tau||_\alpha, ||n-\sigma||_\beta)a_m b_n \\ &\leq \left\{\sum_m W_\lambda(\lambda_1,m)||m-\tau||_\alpha^{p(i_0-\lambda_1)-i_0} a_m^p\right\}^{\frac{1}{p}} \\ &\quad \times \left\{\sum_n w_\lambda(\lambda_2,n)||n-\sigma||_\beta^{q(j_0-\lambda_2)-j_0} b_n^q\right\}^{\frac{1}{q}}. \end{aligned}$$

Then by (26) and (27), we have (28).

For $0 < \varepsilon < q\delta_0$, $\tilde{\lambda}_1 = \lambda_1 + \frac{\varepsilon}{q}$, $\tilde{\lambda}_2 = \lambda_2 - \frac{\varepsilon}{q}$, by the assumptions, we find

$$k(\tilde{\lambda}_1) = k(\lambda_1) + o(1)(\varepsilon \to 0^+),$$

and for $\delta_1 < \lambda_1 - \delta_0$, $\lambda_1 - \delta_1 > 0$,

$$0 \leq \tilde{\theta}_\lambda(n) = \frac{1}{k(\tilde{\lambda}_1)} \int_0^{i_0^{1/\alpha}/||n-\sigma||_\beta} k_\lambda(v,1) v^{\tilde{\lambda}_1 - 1} dv$$
$$\leq \frac{L}{k(\tilde{\lambda}_1)} \int_0^{i_0^{1/\alpha}/||n-\sigma||_\beta} v^{\tilde{\lambda}_1 - \delta_1 - 1} dv$$
$$\leq \frac{L i_0^{(\tilde{\lambda}_1 - \delta_1)/\alpha}}{k(\tilde{\lambda}_1)(\tilde{\lambda}_1 - \delta_1)} \frac{1}{||n-\sigma||_\beta^{\lambda_1 - \delta_0 - \delta_1}},$$

and then

$$\tilde{\theta}_\lambda(n) = O\left(\frac{1}{||n-\sigma||_\beta^\eta}\right) \quad (\eta = \lambda_1 - \delta_0 - \delta_1 > 0).$$

We set

$$\tilde{a}_m := ||m-\tau||_\alpha^{-i_0 + \lambda_1 - \frac{\varepsilon}{p}}, \quad \tilde{b}_n := ||n-\sigma||_\beta^{-j_0 + \lambda_2 - \frac{\varepsilon}{q}} \quad (m \in \mathbf{N}^{i_0}, n \in \mathbf{N}^{j_0}).$$

Then by (18), we obtain

$$||\tilde{a}||_{p,\Phi} ||\tilde{b}||_{q,\Psi} = \left\{\sum_m ||m-\tau||_\alpha^{p(i_0-\lambda_1)-i_0} \tilde{a}_m^p\right\}^{\frac{1}{p}} \left\{\sum_n ||n-\sigma||_\beta^{q(j_0-\lambda_2)-j_0} \tilde{b}_n^q\right\}^{\frac{1}{q}}$$
$$= \left\{\sum_m ||m-\tau||_\alpha^{-i_0-\varepsilon}\right\}^{\frac{1}{p}} \left\{\sum_n ||n-\sigma||_\beta^{-j_0-\varepsilon}\right\}^{\frac{1}{q}}$$
$$= \frac{1}{\varepsilon}\left(\frac{\Gamma^{i_0}(\frac{1}{\alpha})}{i_0^{\varepsilon/\alpha} \alpha^{i_0-1} \Gamma(\frac{i_0}{\alpha})} + \varepsilon O(1)\right)^{\frac{1}{p}} \left(\frac{\Gamma^{j_0}(\frac{1}{\beta})}{j_0^{\varepsilon/\beta} \beta^{j_0-1} \Gamma(\frac{j_0}{\beta})} + \varepsilon \tilde{O}(1)\right)^{\frac{1}{q}},$$

$$\tilde{I}(\tau,\sigma) := \sum_n \left[\sum_m k_\lambda(||m-\tau||_\alpha, ||n-\sigma||_\beta) \tilde{a}_m\right] \tilde{b}_n = \sum_n w_\lambda(\tilde{\lambda}_2, n) ||n-\sigma||_\beta^{-j_0-\varepsilon}$$
$$> \tilde{K}_2 \sum_n \left(1 - O(\frac{1}{||n||_\beta^{\lambda_1-\delta_1}})\right) ||n||_\beta^{-j_0-\varepsilon}$$
$$= \tilde{K}_2 \left[\frac{\Gamma^{j_0}(\frac{1}{\beta})}{\varepsilon j_0^{\varepsilon/\beta} \beta^{j_0-1} \Gamma(\frac{j_0}{\beta})} + \tilde{O}(1) - O(1)\right].$$

If there exists a constant $K \leq K_1^{\frac{1}{p}} K_2^{\frac{1}{q}}$, such that (28) is valid when replacing $K_1^{\frac{1}{p}} K_2^{\frac{1}{q}}$ by K, then we have

$$(K_2 + o(1)) \left[\frac{\Gamma^{j_0}(\frac{1}{\beta})}{\varepsilon j_0^{\varepsilon/\beta} \beta^{j_0-1} \Gamma(\frac{j_0}{\beta})} + \tilde{O}(1) - O(1) \right] < \varepsilon \tilde{I}(\tau,\sigma) < \varepsilon K ||\tilde{a}||_{p,\varphi} ||\tilde{b}||_{q,\psi}$$

$$= K \left[\frac{\Gamma^{i_0}(\frac{1}{\alpha})}{i_0^{\varepsilon/\alpha} \alpha^{i_0-1} \Gamma(\frac{i_0}{\alpha})} + \varepsilon O(1) \right]^{\frac{1}{p}} \left[\frac{\Gamma^{j_0}(\frac{1}{\beta})}{j_0^{\varepsilon/\beta} \beta^{j_0-1} \Gamma(\frac{j_0}{\beta})} + \varepsilon \tilde{O}(1) \right]^{\frac{1}{q}}.$$

For $\varepsilon \to 0^+$, we find

$$\frac{\Gamma^{j_0}(\frac{1}{\beta})}{\beta^{j_0-1} \Gamma(\frac{j_0}{\beta})} \frac{\Gamma^{i_0}(\frac{1}{\alpha})}{\alpha^{i_0-1} \Gamma(\frac{i_0}{\alpha})} k(\lambda_1) \leq K \left[\frac{\Gamma^{i_0}(\frac{1}{\alpha})}{\alpha^{i_0-1} \Gamma(\frac{i_0}{\alpha})} \right]^{\frac{1}{p}} \left[\frac{\Gamma^{j_0}(\frac{1}{\beta})}{\beta^{j_0-1} \Gamma(\frac{j_0}{\beta})} \right]^{\frac{1}{q}},$$

and then $K_1^{\frac{1}{p}} K_2^{\frac{1}{q}} \leq K$. Hence, $K = K_1^{\frac{1}{p}} K_2^{\frac{1}{q}}$ is the best possible constant factor of (28). The theorem is proved. □

Theorem 4.2. *With the same assumptions of Theorem* 4.1, *for* $0 < ||a||_{p,\Phi} < \infty$, *we have the following inequality equivalent to* (28) *with the best constant factor* $K_1^{\frac{1}{p}} K_2^{\frac{1}{q}}$:

$$J(\tau,\sigma) := \left\{ \sum_n ||n-\sigma||_\beta^{p\lambda_2 - j_0} \left(\sum_m k_\lambda(||m-\tau||_\alpha, ||n-\sigma||_\beta) a_m \right)^p \right\}^{\frac{1}{p}}$$
$$< K_1^{\frac{1}{p}} K_2^{\frac{1}{q}} ||a||_{p,\Phi}. \tag{29}$$

Proof. We set b_n as follows:

$$b_n := ||n-\sigma||_\beta^{p\lambda_2 - j_0} \left(\sum_m k_\lambda(||m-\tau||_\alpha, ||n-\sigma||_\beta) a_m \right)^{p-1}, \quad n \in \mathbf{N}^{j_0}.$$

Then it follows $[J(\tau,\sigma)]^p = ||b||_{q,\Psi}^q$. If $J(\tau,\sigma) = 0$, then (29) is trivially valid since $0 < ||a||_{p,\Phi} < \infty$; if $J(\tau,\sigma) = \infty$, then it is impossible since the right-hand side of (29) is finite. Suppose that $0 < J(\tau,\sigma) < \infty$. Then by (28), we find

$$||b||_{q,\Psi}^q = [J(\tau,\sigma)]^p = I(\tau,\sigma) < K_1^{\frac{1}{p}} K_2^{\frac{1}{q}} ||a||_{p,\Phi} ||b||_{q,\Psi},$$

namely,

$$||b||_{q,\Psi}^{q-1} = J(\tau,\sigma) < K_1^{\frac{1}{p}} K_2^{\frac{1}{q}} ||a||_{p,\Phi},$$

and then (29) follows.

On the other hand, assuming that (29) is valid, by Hölder's inequality, we have

$$I(\tau,\sigma) = \sum_n (\Psi(n))^{-1/q} \left[\sum_m k_\lambda(||m-\tau||_\alpha, ||n-\sigma||_\beta) a_m\right] [(\Psi(n))^{\frac{1}{q}} b_n]$$
$$\leq J(\tau,\sigma) ||b||_{q,\Psi}. \tag{30}$$

Then by (29), we have (28). Hence (29) and (28) are equivalent.

By the equivalency, the constant factor $K_1^{\frac{1}{p}} K_2^{\frac{1}{q}}$ in (29) is the best possible. Otherwise, we would reach a contradiction by (30) that the constant factor $K_1^{\frac{1}{p}} K_2^{\frac{1}{q}}$ in (28) is not the best possible. The theorem is proved. □

Remark 4.1. (1) For $\tau = \sigma = 0$, (28) reduces to (10). Hence, (28) is a more accurate of inequality of (10). We still can consider the reverses of (28) and (29) as in the front section.

(2) If $0 < \lambda_1 \leq i_0$, $0 < \lambda_2 \leq j_0$, then

$$k_\lambda(x,y) = \frac{1}{(x+y)^\lambda} \quad (\lambda > 0), \quad k_\lambda(x,y) = \frac{\ln(x/y)}{x^\lambda - y^\lambda} \quad (0 < \lambda \leq 1)$$

and

$$k_\lambda(x,y) = \prod_{k=1}^{s} \frac{1}{x^{\lambda/s} + a_k y^{\lambda/s}} \quad (0 < a_1 < \cdots < a_s,\ 0 < \lambda \leq s)$$

all satisfy the conditions of

$$(-1)^i \frac{\partial^i}{\partial x^i}(k_\lambda(x,y) x^{\lambda_1 - i_0}) > 0, \quad (-1)^i \frac{\partial^i}{\partial y^i}(k_\lambda(x,y) y^{\lambda_2 - j_0}) > 0 \quad (i = 1, 2),$$

for using Theorems 4.1 and 4.2.

5 Euler Constant in a Strengthened Version of Hardy-Hilbert's Inequality

For $i_0 = j_0 = 1, \lambda = 1, \lambda_1 = \frac{1}{r}, \lambda_2 = \frac{1}{s}$ $\left(r > 1, \frac{1}{r} + \frac{1}{s} = 1\right), k_\lambda(x, y) = \frac{1}{x+y}$ in Definition 2.1, we have the following weight coefficients:

$$\omega(s, n) := \omega_1\left(\frac{1}{s}, n\right) = n^{1/s} \sum_{m=1}^{\infty} \frac{1}{(m+n)m^{1/s}},$$

$$\varpi(r, m) := \varpi_1\left(\frac{1}{r}, m\right) = m^{1/r} \sum_{n=1}^{\infty} \frac{1}{(m+n)n^{1/r}},$$

and then in Theorem 2.1, we have the following Hardy-Hilbert's inequality with a best constant factor $\pi/\sin(\pi/r)$:

$$I_1 := \sum_{n=1}^{\infty} \sum_{m=1}^{\infty} \frac{1}{m+n} a_m b_n < \frac{\pi}{\sin(\frac{\pi}{r})} \left(\sum_{m=1}^{\infty} m^{\frac{p}{s}-1} a_m^p\right)^{\frac{1}{p}} \left(\sum_{n=1}^{\infty} n^{\frac{q}{r}-1} b_n^q\right)^{\frac{1}{q}}. \quad (31)$$

In this section, we build a strengthened version of (31) with Euler constant $\gamma = 0.57721566^+$ as follows (cf. [33]):

$$I_1 < \left\{\sum_{m=1}^{\infty} \left[\frac{\pi}{\sin(\frac{\pi}{r})} - \frac{1-\gamma}{m^{1/s}}\right] m^{\frac{p}{s}-1} a_m^p\right\}^{\frac{1}{p}} \left\{\sum_{n=1}^{\infty} \left[\frac{\pi}{\sin(\frac{\pi}{r})} - \frac{1-\gamma}{n^{1/r}}\right] n^{\frac{q}{r}-1} b_n^q\right\}^{\frac{1}{q}}. \quad (32)$$

Note. The other name of Euler constant $\gamma = \gamma_0$ is called Stieltjes constant of 0-order, which is the first term constant of the Laurent series of Riemann $\varsigma-$ function $\varsigma(s)$ in the isolated singular point $s = 1$ as follows (cf. [27]):

$$\varsigma(s) = \frac{1}{s-1} + \sum_{n=0}^{\infty} (-1)^n \frac{\gamma_n}{n!} (s-1)^n \quad (0 < |s-1| < \infty).$$

The Riemann $\varsigma-$ function, gamma function, beta function and Bernoulli function are very important in Number Theory and its applications. About the modern development of Analysis Number Theory, please refer to [1, 2, 5, 6, 9, 12, 17, 19].

Lemma 5.1 ([25, 38]). *If* $(-1)^i F^{(i)}(t) > 0 \ \ (t \in (0, \infty);\ i = 0, 1, 2, 3)$, *then we have*

$$-\frac{1}{12} F(1) < \int_1^{\infty} P_1(t) F(t) dt < -\frac{1}{12} F\left(\frac{3}{2}\right),$$

where $P_1(t) = t - [t] - \frac{1}{2}$ *is Bernoulli function of* 1*-order.*

Setting $f(t) := \frac{1}{(x+t)t^{1/r}}$ $(x \geq 1,\ t > 0)$, we find

$$f'(t) = \frac{-1}{(x+t)^2 t^{1/r}} - \frac{1}{r(x+t)t^{1+(1/r)}} = -\frac{(r+1)t + x}{r(x+t)^2 t^{1+(1/r)}}.$$

By Euler-Maclaurin summation formula (cf. [27]), it follows

$$\varpi(r,x) = x^{\frac{1}{r}} \sum_{n=1}^{\infty} \frac{1}{(x+n)n^{1/r}} = x^{\frac{1}{r}} \left[\int_1^{\infty} f(t)dt + \frac{1}{2} f(1) + \int_1^{\infty} P_1(t) f'(t) dt \right]$$

$$= x^{\frac{1}{r}} \int_0^{\infty} f(t)dt - x^{\frac{1}{r}} \int_0^{1} f(t)dt + \frac{x^{\frac{1}{r}}}{2} f(1) + x^{\frac{1}{r}} \int_1^{\infty} P_1(t) f'(t) dt$$

$$= \frac{\pi}{\sin(\frac{\pi}{r})} - \int_0^{\frac{1}{x}} \frac{u^{-\frac{1}{r}} du}{1+u} + \frac{x^{\frac{1}{r}}}{2(x+1)} - x^{\frac{1}{r}} \int_1^{\infty} P_1(t) \frac{(r+1)t + x}{r(x+t)^2 t^{1+(1/r)}} dt.$$

Setting

$$G(t,x) := \frac{(r+1)tx + x^2}{r(x+t)^2 t^{1+(1/r)}},\ A(x) := x^{1-\frac{1}{r}} \int_0^{\frac{1}{x}} \frac{u^{-\frac{1}{r}}}{1+u} du,\ B(x) := \int_1^{\infty} P_1(t) G(t,x) dt$$

and

$$\theta(r,x) := A(x) + B(x) - \frac{x}{2(x+1)} \quad (x \in [1,\infty)),$$

then we have the following decomposition:

$$\varpi(r,m) = \frac{\pi}{\sin(\frac{\pi}{r})} - \frac{\theta(r,m)}{m^{\frac{1}{r}}} \quad (m \in \mathbf{N}).$$

Lemma 5.2. *We have*

$$\min_{x \geq 1} \theta(r,x) = \theta(r,1) = \frac{\pi}{\sin(\frac{\pi}{r})} - \varpi(r,1). \tag{33}$$

Proof. By ([7]), Lemma 2.1, we have

$$\int_0^{\frac{1}{x}} \frac{u^{-\frac{1}{r}}}{1+u} du \geq \frac{r(2r-1)x^{\frac{1}{r}}}{(r-1)[(2r-1)x + r - 1]} \quad (x \geq 1).$$

Then we find

$$A'(x) = \left(1 - \frac{1}{r}\right) x^{-\frac{1}{r}} \int_0^{\frac{1}{x}} \frac{u^{-\frac{1}{r}} du}{1+u} - \frac{1}{x+1} \geq \frac{(1-\frac{1}{r})r(2r-1)}{(r-1)[(2r-1)x + r - 1]} - \frac{1}{x+1}$$

$$= \frac{(2r-1)}{(2r-1)x + r - 1} - \frac{1}{x+1} = \frac{r}{(x+1)[(2r-1)x + r - 1]}.$$

Setting $F_1(t) = \frac{1}{(x+t)^2 t^{1/r}}$ and $F_2(t) = \frac{1}{(x+t)^3 t^{1/r}}$, then by Lemma 5.1, it follows

$$B'(x) = \int_1^\infty P_1(t) G'_x(t,x)dt = \frac{r+1}{r}\int_1^\infty P_1(t)F_1(t)dt - 2x\int_1^\infty P_1(t)F_2(t)dt$$

$$> \frac{r+1}{r}\Big(-\frac{1}{12}F_1(1)\Big) + \frac{2x}{12}F_2\Big(\frac{3}{2}\Big) = -\frac{r+1}{12r(x+1)^2} + \frac{4x}{3(2x+3)^3}\Big(\frac{2}{3}\Big)^{\frac{1}{r}}.$$

Then we have

$$\begin{aligned}\theta'_x(r,x) &= A'(x) + B'(x) - \frac{1}{2(x+1)^2}\\ &= \frac{r}{(x+1)[(2r-1)x+r-1]} - \frac{r+1}{12r(x+1)^2} + \frac{4x}{3(2x+3)^3}\Big(\frac{2}{3}\Big)^{\frac{1}{r}} - \frac{1}{2(x+1)^2}\\ &= \frac{(-2r^2+5r+1)x + (5r^2+6r+1)}{12r(x+1)^2[(2r-1)x+r-1]} + \frac{4x}{3(2x+3)^3}\Big(\frac{2}{3}\Big)^{\frac{1}{r}}.\end{aligned}$$

(1) If $1 < r < 5/2, -2r^2 + 5r + 1 > 0$ and then $\theta'_x(r,x) > 0$.
(2) If $r \geq 5/2, (2/3)^{1/r} > 4/5$, then we obtain

$$\begin{aligned}\theta'_x(r,x) &> \frac{(-2r^2+5r+1)x + (5r^2+6r+1)}{12r(x+1)^2[(2r-1)x+r-1]} + \frac{16x}{15(2x+3)^3}\\ &= \frac{5[(-2r^2+5r+1)x + (5r^2+6r+1)](2x+3)^3}{60r(x+1)^2(2x+3)^3[(2r-1)x+r-1]}\\ &\quad + \frac{64rx(x+1)^2[(2r-1)x+r-1]}{60r(x+1)^2(2x+3)^3[(2r-1)x+r-1]}\\ &> \frac{(48r^2-44r+40)x^4 + (160r^2+1076r+92)x^3}{60r(x+1)^2(2x+3)^3[(2r-1)x+r-1]} > 0 \quad (x \in [1,\infty)).\end{aligned}$$

Hence, $\theta(r,x)$ is increasing with respect to $x \in [1,\infty)$, and then we have (33). The lemma is proved. □

Lemma 5.3. *For $k \in \mathbf{N},\ k \geq 5$, the function*

$$I(r,k) := \int_0^k \frac{u^{-\frac{1}{r}}du}{1+u} - \frac{k^{-\frac{1}{r}}}{2(1+k)} - \sum_{m=1}^{k-1}\frac{m^{-\frac{1}{r}}}{1+m}$$

is strict decreasing with respect to $r \in (1,\infty)$.

Proof. For $k \geq 5$, we find

$$I_r'(r,k) = \frac{1}{r^2}\left\{-\frac{k^{-\frac{1}{r}}\ln k}{2(1+k)} + \left[\int_0^4 \frac{u^{-\frac{1}{r}}\ln u}{1+u}du - \frac{\ln 2}{3\cdot 2^{\frac{1}{r}}} - \frac{\ln 3}{4\cdot 3^{\frac{1}{r}}}\right]\right.$$
$$\left. - \left[\sum_{m=4}^{k-1}\frac{m^{-\frac{1}{r}}\ln m}{1+m} - \int_4^k \frac{u^{-\frac{1}{r}}\ln u}{1+u}du\right]\right\}.$$

It is evident that for $u \geq 4$,

$$\frac{d}{du}\left(\frac{u^{-\frac{1}{r}}\ln u}{1+u}\right) = \frac{u^{-\frac{1}{r}}}{1+u}\left(-\frac{\ln u}{ru} - \frac{\ln u}{1+u} + \frac{1}{u}\right) < \frac{u^{-\frac{1}{r}}}{1+u}\left(\frac{1}{u} - \frac{\ln u}{1+u}\right) < 0,$$

and then $u^{-\frac{1}{r}}\ln u/(1+u)$ is decreasing with respect to $u \geq 4$. It follows that

$$\sum_{m=4}^{k-1}\frac{m^{-1/r}\ln m}{1+m} - \int_4^k \frac{u^{-1/r}\ln u}{1+u}du \geq 0.$$

Setting $u = e^{-y}$, we obtain

$$J(r) := \int_0^4 \frac{u^{-1/r}\ln u}{1+u}du = -\int_{-\ln 4}^{\infty}\frac{ye^{(-1+\frac{1}{r})y}}{1+e^{-y}}dy < -\frac{1}{5}\int_{-\ln 4}^{\infty} ye^{(-1+\frac{1}{r})y}dy$$
$$= \frac{r4^{1-\frac{1}{r}}}{5(r-1)}\left(\ln 4 - \frac{r}{r-1}\right) = \frac{s4^{1/s}}{5}(\ln 4 - s).$$

If $1 < s = r/(r-1) < \ln 4$, namely, $r > \ln 4/(\ln 4 - 1) = 3.5887^{+}$, then we find

$$\frac{d}{ds}\left\{\frac{s4^{\frac{1}{s}}}{5}(\ln 4 - s)\right\} = \frac{4^{\frac{1}{s}}}{5}(\ln 4 - s)\left(1 - \frac{\ln 4}{s}\right) - \frac{s4^{\frac{1}{s}}}{5} < 0,$$

and

$$\frac{s4^{\frac{1}{s}}}{5}(\ln 4 - s) < \frac{4}{5}(\ln 4 - 1).$$

In this case,

$$\int_0^4 \frac{u^{-\frac{1}{r}}\ln u}{1+u}du - \frac{\ln 2}{3\cdot 2^{\frac{1}{r}}} - \frac{\ln 3}{4\cdot 3^{\frac{1}{r}}} < \frac{4}{5}(\ln 4 - 1) - \frac{\ln 2}{3\cdot 2^{1/3.5887}} - \frac{\ln 3}{4\cdot 3^{1/3.5887}}$$
$$< -0.083996 < 0.$$

If $\ln 4 \leq r/(r-1)$, then $J(r) < 0$.

Therefore, $I_r'(r,k) < 0$ and then $I(r,k)$ is strict decreasing with respect to $r \in (1,\infty)$. The lemma is proved. □

Lemma 5.4 ([8]). *If $r > 1$, $\frac{1}{r} + \frac{1}{s} = 1$, then we have the following inequalities:*

$$\varpi(r,m) < \frac{\pi}{\sin(\frac{\pi}{r})} - \frac{1-\gamma}{m^{\frac{1}{r}}} \quad (m \in \mathbf{N}), \tag{34}$$

$$\omega(s,n) < \frac{\pi}{\sin(\frac{\pi}{r})} - \frac{1-\gamma}{n^{\frac{1}{s}}} \quad (n \in \mathbf{N}), \tag{35}$$

where $1-\gamma = 0.42278433^+$ is the best value (γ is Euler constant).

Proof. For $k \in \mathbf{N}$, $k \geq 5$, we have

$$\theta(r,1) = \frac{\pi}{\sin(\frac{\pi}{r})} - \varpi(r,1) = \int_0^\infty \frac{u^{-\frac{1}{r}}}{1+u}du - \sum_{m=1}^\infty \frac{m^{-\frac{1}{r}}}{1+m}$$

$$= \int_0^k \frac{u^{-\frac{1}{r}}}{1+u}du + \int_k^\infty \frac{u^{-\frac{1}{r}}}{1+u}du - \sum_{m=1}^{k-1} \frac{m^{-\frac{1}{r}}}{1+m} - \sum_{m=k}^\infty \frac{m^{-\frac{1}{r}}}{1+m}.$$

Setting $g(t) := 1/[(1+t)t^{1/r}]$, then by Euler-Maclaurin summation formula, we have

$$\int_k^\infty \frac{u^{-\frac{1}{r}}du}{1+u} + \frac{u^{-\frac{1}{r}}}{2(1+k)} < \sum_{m=k}^\infty \frac{m^{-\frac{1}{r}}}{1+m} < \int_k^\infty \frac{u^{-\frac{1}{r}}du}{1+u} + \frac{u^{-\frac{1}{r}}}{2(1+k)} - \frac{g'(k)}{12}.$$

It follows

$$I(r,k) + \frac{g'(k)}{12} < \theta(r,1) < I(r,k),$$

$$\inf_{r>1} I(r,k) + \frac{1}{12}\inf_{r>1} g'(k) \leqslant \inf_{r>1} \theta(r,1) \leq \inf_{r>1} I(r,k) \quad (k \geq 5).$$

Since for any $k \geq 5$,

$$0 \geq \inf_{r>1} g'(k) = -\sup_{r>1}\left[\frac{1}{(1+k)^2 k^{1/r}} + \frac{1}{r(1+k)k^{1+(1/r)}}\right]$$

$$\geq -\left[\frac{1}{(1+k)^2} + \frac{1}{(1+k)k}\right] \to 0 \quad (k \to \infty),$$

then it follows $\lim_{k\to\infty} \inf_{r>1} g'(k) = 0$. Hence by Lemma 4.1, we obtain

$$\inf_{r>1} \theta(r,1) = \lim_{k\to\infty} \inf_{r>1} I(r,k) = \lim_{k\to\infty} \lim_{r\to\infty} I(r,k)$$

$$= \lim_{k\to\infty}\left[\int_0^k \frac{du}{1+u} - \frac{1}{2(1+k)} - \sum_{m=1}^{k-1}\frac{1}{1+m}\right]$$
$$= 1 - \lim_{k\to\infty}\left[\sum_{m=1}^{k+1}\frac{1}{m} - \ln(1+k) - \frac{1}{2(k+1)}\right]$$
$$= 1-\gamma.$$

Therefore, since $\inf_{m\geq 1}\theta(r,m) = \theta(r,1)$, we have

$$\varpi(r,m) \leq \frac{\pi}{\sin(\frac{\pi}{r})} - \frac{\theta(r,1)}{m^{\frac{1}{r}}} < \frac{\pi}{\sin(\frac{\pi}{r})} - \frac{\inf_{r>1}\theta(r,1)}{m^{\frac{1}{r}}}$$
$$= \frac{\pi}{\sin(\frac{\pi}{r})} - \frac{1-\gamma}{m^{\frac{1}{r}}} \quad (m \in \mathbf{N}).$$

It is evident that the constant $1-\gamma$ in (34) is the best possible. By the same way, we still have (35). The lemma is proved. □

For $i_0 = j_0 = 1,\ \lambda = 1,\ \lambda_1 = \frac{1}{r},\ \lambda_2 = \frac{1}{s}\ \left(r > 1,\ \frac{1}{r}+\frac{1}{s} = 1\right), k_\lambda(x,y) = \frac{1}{x+y}$ in (8), we have

$$I_1 \leq \left(\sum_{m=1}^{\infty}\varpi(r,m)m^{\frac{p}{s}-1}a_m^p\right)^{\frac{1}{p}}\left(\sum_{n=1}^{\infty}\omega(s,n)n^{\frac{q}{r}-1}b_n^q\right)^{\frac{1}{q}}.$$

Then by (34) and (35), it follows:

Theorem 5.1. *If $r > 1,\ \frac{1}{r}+\frac{1}{s} = 1,\ a_m,\ b_n \geq 0$,*

$$0 < \sum_{m=1}^{\infty} m^{\frac{p}{s}-1}a_m^p < \infty, \quad 0 < \sum_{n=1}^{\infty} n^{\frac{q}{r}-1}b_n^q < \infty,$$

then we have (32) *with the best possible constant factor $\pi/\sin(\pi/r)$.*

In particular, for $r = q,\ s = p$, we have

$$I_1 < \left\{\sum_{m=1}^{\infty}\left[\frac{\pi}{\sin(\frac{\pi}{p})} - \frac{1-\gamma}{m^{1/p}}\right]a_m^p\right\}^{\frac{1}{p}}\left\{\sum_{n=1}^{\infty}\left[\frac{\pi}{\sin(\frac{\pi}{p})} - \frac{1-\gamma}{n^{1/q}}\right]b_n^q\right\}^{\frac{1}{q}}. \quad (36)$$

For $r = p, s = q$, we have the dual form of (36) *as follows:*

$$I_1 < \left\{\sum_{m=1}^{\infty}\left[\frac{\pi}{\sin(\frac{\pi}{p})} - \frac{1-\gamma}{m^{1/q}}\right]m^{p-2}a_m^p\right\}^{\frac{1}{p}}\left\{\sum_{n=1}^{\infty}\left[\frac{\pi}{\sin(\frac{\pi}{p})} - \frac{1-\gamma}{n^{1/p}}\right]n^{q-2}b_n^q\right\}^{\frac{1}{q}}.$$

Corollary 5.1 ([32]). *If* $r > 1,\ \frac{1}{r} + \frac{1}{s} = 1,\ a_m,\ b_n \geq 0,$

$$0 < \sum_{m=1}^{\infty} m^{\frac{p}{s}-1} a_m^p < \infty, \quad 0 < \sum_{n=1}^{\infty} n^{\frac{q}{r}-1} b_n^q < \infty,$$

then we have the following inequality with the best possible constant $\pi/\sin(\pi/r)$:

$$I_1 < \left\{ \sum_{m=1}^{\infty} \left[\frac{\pi}{\sin(\frac{\pi}{r})} - \frac{1}{2m^{\frac{1}{s}} + m^{-\frac{1}{r}}} \right] m^{\frac{p}{s}-1} a_m^p \right\}^{\frac{1}{p}}$$

$$\times \left\{ \sum_{n=1}^{\infty} \left[\frac{\pi}{\sin(\frac{\pi}{r})} - \frac{1-\gamma}{2n^{\frac{1}{r}} + n^{-\frac{1}{s}}} \right] n^{\frac{q}{r}-1} b_n^q \right\}^{\frac{1}{q}}.$$

Proof. By the same way of Theorem 5.1, we need only to prove the following inequality:

$$\varpi(r,m) < \frac{\pi}{\sin(\frac{\pi}{r})} - \frac{1}{2m^{1/s} + m^{-1/r}} \quad (m \in \mathbf{N}). \tag{37}$$

We find

$$A(m) = m^{1-\frac{1}{r}} \int_0^{\frac{1}{m}} \frac{u^{-\frac{1}{r}}}{1+u} du = m^{1-\frac{1}{r}} \int_0^{\frac{1}{m}} \sum_{k=0}^{\infty} (-1)^k u^{k-\frac{1}{r}} du$$

$$= m^{1-\frac{1}{r}} \sum_{k=0}^{\infty} (-1)^k \int_0^{\frac{1}{m}} u^{k-\frac{1}{r}} du$$

$$= \sum_{k=0}^{\infty} \frac{(-1)^k}{(k+\frac{1}{s})m^k} > \sum_{k=0}^{3} \frac{(-1)^k}{(k+\frac{1}{s})m^k}$$

and

$$B(m) = \int_1^{\infty} P_1(t) G(t,m) dt = \int_1^{\infty} P_1(t) \left[\frac{m}{(m+t)^2 t^{1/r}} + \frac{m}{r(m+t)t^{1+1/r}} \right] dt$$

$$> -\frac{1}{12} \left[\frac{m}{(m+1)^2} + \frac{m}{r(m+1)} \right].$$

For $m \geq 2$, we obtain

$$\frac{m}{m+1} = \left(1 + \frac{1}{m}\right)^{-1} < 1 - \frac{1}{m} + \frac{1}{m^2},$$

$$\frac{m}{(m+1)^2} = \frac{1}{m}\left(1 + \frac{1}{m}\right)^{-2} < \frac{1}{m}\left(1 - \frac{2}{m} + \frac{3}{m^2}\right);$$

for $m = 1$, the above inequalities are still valid. Then we have

$$\theta(r,m) = A(m) + B(m) - \frac{m}{2(m+1)} > f_m(s) + g_m(s) \quad (m \in \mathbf{N}),$$

where

$$f_m(s) := s + \frac{1}{12s} + \frac{1}{(1+s)m} + \frac{1}{12sm^2} + \frac{1}{3(1+3s)m^3},$$

$$g_m(s) := -\frac{1}{12sm} - \frac{1}{2(1+2s)m^2} - \frac{7}{12} - \frac{1}{2m} + \frac{1}{12m^2} - \frac{7}{12m^3}.$$

For $s > 1,\ m \in \mathbf{N}$, we find

$$\begin{aligned} f_m'(s) &= 1 - \frac{1}{12s^2} - \frac{1}{(1+s)^2 m} - \frac{1}{12s^2m^2} - \frac{1}{(1+3s)^2m^3} \\ &> 1 - \frac{1}{12} - \frac{1}{4} - \frac{1}{12} - \frac{1}{16} > 0, \\ g_m'(s) &= \frac{1}{12s^2m} + \frac{1}{(1+2s)^2m^2} > 0. \end{aligned}$$

Then we obtain

$$\theta(r,m) > f_m(s) + g_m(s) > \lim_{s \to 1+} (f_m(s) + g_m(s)) = \frac{1}{2} - \frac{1}{12m} - \frac{1}{2m^3}.$$

For $m \geq 3$, since

$$\left(\frac{1}{2} - \frac{1}{12m} - \frac{1}{2m^3}\right)\left(1 + \frac{1}{2m}\right) = \frac{1}{2} + \frac{1}{m}\left(\frac{1}{6} - \frac{1}{24m} - \frac{1}{2m^2} - \frac{1}{4m^3}\right) > \frac{1}{2},$$

we have

$$\frac{1}{2} - \frac{1}{12m} - \frac{1}{2m^3} > \frac{1}{2(1+\frac{1}{2m})} = \frac{1}{2+m^{-1}},$$

and then we find

$$\begin{aligned} \varpi(r,m) &= \frac{\pi}{\sin(\frac{\pi}{r})} - \frac{\theta(r,m)}{m^{\frac{1}{r}}} < \frac{\pi}{\sin(\frac{\pi}{r})} - \frac{1}{m^{\frac{1}{r}}}\left(\frac{1}{2} - \frac{1}{12m} - \frac{1}{2m^3}\right) \\ &< \frac{\pi}{\sin(\frac{\pi}{r})} - \frac{1}{m^{\frac{1}{r}}(2+m^{-1})} = \frac{\pi}{\sin(\frac{\pi}{r})} - \frac{1}{2m^{\frac{1}{r}} + m^{-\frac{1}{s}}} \quad (m \geq 3). \end{aligned}$$

Since $\gamma < 0.6$, $1 - \gamma > 1/3$, and $(1-\gamma)(2+2^{-1}) > 1$, then it follows

$$\varpi(r,1) < \frac{\pi}{\sin(\frac{\pi}{r})} - \frac{1-\gamma}{1^{\frac{1}{r}}} < \frac{\pi}{\sin(\frac{\pi}{r})} - \frac{1}{2\cdot 1^{\frac{1}{r}} + 1^{-\frac{1}{s}}},$$

$$\varpi(r,2) < \frac{\pi}{\sin(\frac{\pi}{r})} - \frac{1-\gamma}{2^{\frac{1}{r}}} < \frac{\pi}{\sin(\frac{\pi}{r})} - \frac{1}{2\cdot 2^{\frac{1}{r}} + 2^{-\frac{1}{s}}}.$$

Hence, (37) is valid for $m \in \mathbf{N}$. The corollary is proved. □

6 The Operator Expressions and Some Examples

For $p > 1$, $\varphi(m) = ||m||_\alpha^{p(i_0-\lambda_1)-i_0}$ $(m \in \mathbf{N}^{i_0})$, and $\psi(n) = ||n||_\beta^{q(j_0-\lambda_2)-j_0}$, wherefrom

$$[\psi(n)]^{1-p} = ||n||_\beta^{p\lambda_2-j_0} \quad (n \in \mathbf{N}^{j_0}),$$

we define two real weight normal discrete spaces $\mathbf{l}_{p,\varphi}$ and $\mathbf{l}_{q,\psi}$ as follows:

$$\mathbf{l}_{p,\varphi} := \left\{ a = \{a_m\};\ ||a||_{p,\varphi} = \left\{ \sum_m \varphi(m)|a_m|^p \right\}^{\frac{1}{p}} < \infty \right\},$$

$$\mathbf{l}_{q,\psi} := \left\{ b = \{b_n\};\ ||b||_{q,\psi} = \left\{ \sum_n \psi(n)|b_n|^q \right\}^{\frac{1}{q}} < \infty \right\}.$$

As the assumptions of Theorem 2.1, in view of the fact that

$$J < K_1^{\frac{1}{p}} K_2^{\frac{1}{q}} ||a||_{p,\varphi},$$

we have the following definition:

Definition 6.1. Define a multidimensional Hilbert-type operator $T : \mathbf{l}_{p,\varphi} \to \mathbf{l}_{p,\psi^{1-p}}$ as follows: For $a \in \mathbf{l}_{p,\varphi}$, there exists a unique representation $Ta \in \mathbf{l}_{p,\psi^{1-p}}$, satisfying

$$(Ta)(n) := \sum_m k_\lambda(||m||_\alpha, ||n||_\beta) a_m \quad (n \in \mathbf{N}^{j_0}).$$

For $b \in \mathbf{l}_{q,\psi}$, we define the following formal inner product of Ta and b as follows:

$$(Ta, b) := \sum_n \sum_m k_\lambda(||m||_\alpha, ||n||_\beta) a_m b_n.$$

Then by Theorem 3.4 (or Corollary 3.2), for $0 < ||a||_{p,\varphi}, \ ||b||_{q,\psi} < \infty$, we have the following equivalent inequalities:

$$(Ta, b) < K_1^{\frac{1}{p}} K_2^{\frac{1}{q}} ||a||_{p,\varphi} ||b||_{q,\psi}$$

$$||Ta||_{p,\psi^{1-p}} < K_1^{\frac{1}{p}} K_2^{\frac{1}{q}} ||a||_{p,\varphi}. \tag{38}$$

It follows that T is bounded with

$$||T|| := \sup_{a(\neq\theta)\in \mathbf{l}_{p,\varphi}} \frac{||Ta||_{p,\psi^{1-p}}}{||a||_{p,\varphi}} \leq K_1^{\frac{1}{p}} K_2^{\frac{1}{q}}.$$

Since the constant factor $K_1^{\frac{1}{p}} K_2^{\frac{1}{q}}$ in (38) is the best possible, we have

Corollary 6.1. *With the same assumptions of Theorem* 3.4 (*or Corollary* 3.2), T *is defined by Definition* 6.1, *it follows*

$$||T|| = K_1^{\frac{1}{p}} K_2^{\frac{1}{q}} = \left[\frac{\Gamma^{j_0}(\frac{1}{\beta})}{\beta^{j_0-1}\Gamma(\frac{j_0}{\beta})}\right]^{\frac{1}{p}} \left[\frac{\Gamma^{i_0}(\frac{1}{\alpha})}{\alpha^{i_0-1}\Gamma(\frac{i_0}{\alpha})}\right]^{\frac{1}{q}} k(\lambda_1).$$

Remark 6.1. In Corollary 6.1,

(1) If for $x > y$, $k_\lambda(x, y) = 0$, then we define the first kind Hardy-type operator as follows:

$$(T_1 a)(m) := \sum_{m \leq n} k_\lambda(||m||_\alpha, ||n||_\beta) a_m \quad (m \in \mathbf{N}^{i_0}).$$

We find

$$k_1(\lambda_1) := \int_0^1 k_\lambda(t, 1) t^{\lambda_1 - 1} dt \in \mathbf{R}_+,$$

and then

$$||T_1|| = \left[\frac{\Gamma^{j_0}(\frac{1}{\beta})}{\beta^{j_0-1}\Gamma(\frac{j_0}{\beta})}\right]^{\frac{1}{p}} \left[\frac{\Gamma^{i_0}(\frac{1}{\alpha})}{\alpha^{i_0-1}\Gamma(\frac{i_0}{\alpha})}\right]^{\frac{1}{q}} k_1(\lambda_1).$$

(2) If for $0 < x < y$, $k_\lambda(x, y) = 0$, then we define the second kind Hardy-type operator as follows:

$$(T_2 a)(m) := \sum_{m \geq n} k_\lambda(||m||_\alpha, ||n||_\beta) a_m \quad (m \in \mathbf{N}^{i_0}).$$

We find

$$k_2(\lambda_1) := \int_1^\infty k_\lambda(t,1)t^{\lambda_1-1}dt \in \mathbf{R}_+,$$

and then

$$||T_2|| = \left[\frac{\Gamma^{j_0}(\frac{1}{\beta})}{\beta^{j_0-1}\Gamma(\frac{j_0}{\beta})}\right]^{\frac{1}{p}} \left[\frac{\Gamma^{i_0}(\frac{1}{\alpha})}{\alpha^{i_0-1}\Gamma(\frac{i_0}{\alpha})}\right]^{\frac{1}{q}} k_2(\lambda_1).$$

Example 6.1. (1) We set

$$k_\lambda(x,y) = \frac{1}{x^\lambda + y^\lambda} \quad (\lambda > 0,\ 0 < \lambda_1 < i_0,\ 0 < \lambda_2 < j_0),$$

$$k_\lambda(x,y) = \frac{1}{x^\lambda + y^\lambda} \quad (\lambda > 0,\ 0 < \lambda_1 < i_0,\ 0 < \lambda_2 < j_0),$$

which is a positive decreasing function with respect to x $(y) \in \mathbf{R}_+$. For $\delta_0 = \frac{1}{2}\min\{\lambda_1, \lambda_2, i_0 - \lambda_1, j_0 - \lambda_2\} > 0$, and $\tilde{\lambda}_1 \in (\lambda_1 - \delta_0, \lambda_1 + \delta_0)$, it follows

$$k(\tilde{\lambda}_1) = \int_0^\infty \frac{1}{t^\lambda + 1} t^{\tilde{\lambda}_1 - 1} dt \quad \left(v = t^\lambda = \frac{1}{\lambda}\int_0^\infty \frac{1}{v+1} v^{\frac{\tilde{\lambda}_1}{\lambda}-1} dv\right)$$
$$= \frac{\pi}{\lambda \sin \pi(\tilde{\lambda}_1/\lambda)} \in \mathbf{R}_+.$$

Setting $\delta_1 = 0\ (< \lambda_1 - \delta_0)$, we have

$$k_\lambda(t,1) = \frac{1}{t^\lambda + 1} \le 1 = \frac{1}{t^{\delta_1}} \quad (t \in (0,\infty)).$$

Then by Corollary 6.1, we have

$$||T|| = \left[\frac{\Gamma^{j_0}(\frac{1}{\beta})}{\beta^{j_0-1}\Gamma(\frac{j_0}{\beta})}\right]^{\frac{1}{p}} \left[\frac{\Gamma^{i_0}(\frac{1}{\alpha})}{\alpha^{i_0-1}\Gamma(\frac{i_0}{\alpha})}\right]^{\frac{1}{q}} \frac{\pi}{\lambda \sin(\frac{\pi\lambda_1}{\lambda})}.$$

(2) We set $k_\lambda(x,y) = (x+y)^{-\lambda}$ $(\lambda > 0,\ 0 < \lambda_1 < i_0,\ 0 < \lambda_2 < j_0)$, which is a positive decreasing function with respect to x $(y) \in \mathbf{R}_+$. For $\delta_0 = \frac{1}{2}\min\{\lambda_1, \lambda_2, i_0 - \lambda_1, j_0 - \lambda_2\} > 0$, and $\tilde{\lambda}_1 \in (\lambda_1 - \delta_0, \lambda_1 + \delta_0)$, it follows

$$k(\tilde{\lambda}_1) = \int_0^\infty \frac{1}{(t+1)^\lambda} t^{\tilde{\lambda}_1 - 1} dt = B(\tilde{\lambda}_1, \lambda - \tilde{\lambda}_1) \in \mathbf{R}_+.$$

Setting $\delta_1 = 0\ (< \lambda_1 - \delta_0)$, we have

$$k_\lambda(t,1) = \frac{1}{(t+1)^\lambda} \le 1 = \frac{1}{t^{\delta_1}} \quad (t \in (0,\infty)).$$

Then by Corollary 6.1, we have

$$||T|| = \left[\frac{\Gamma^{j_0}(\frac{1}{\beta})}{\beta^{j_0-1}\Gamma(\frac{j_0}{\beta})}\right]^{\frac{1}{p}} \left[\frac{\Gamma^{i_0}(\frac{1}{\alpha})}{\alpha^{i_0-1}\Gamma(\frac{i_0}{\alpha})}\right]^{\frac{1}{q}} B(\lambda_1,\lambda_2).$$

(3) We set $k_\lambda(x,y) = \ln(x/y)/(x^\lambda - y^\lambda)$ $(\lambda > 0,\ 0 < \lambda_1 < i_0,\ 0 < \lambda_2 < j_0)$, which is a positive decreasing function with respect to x $(y) \in \mathbf{R}_+$ (cf. [25]). For $\delta_0 = \frac{1}{2}\min\{\lambda_1, \lambda_2, i_0 - \lambda_1, j_0 - \lambda_2\} > 0$, and $\tilde{\lambda}_1 \in (\lambda_1 - \delta_0, \lambda_1 + \delta_0)$, it follows

$$k(\tilde{\lambda}_1) = \int_0^\infty \frac{\ln t}{t^\lambda - 1} t^{\tilde{\lambda}_1 - 1} dt \quad \left(v = t^\lambda = \frac{1}{\lambda^2}\int_0^\infty \frac{\ln v}{v-1} v^{\frac{\tilde{\lambda}_1}{\lambda}-1} dt\right)$$

$$= \left[\frac{\pi}{\lambda \sin \pi(\tilde{\lambda}_1/\lambda)}\right]^2 \in \mathbf{R}_+.$$

Setting $0 < \delta_1 = \frac{\lambda_1 - \delta_0}{2} < \lambda_1 - \delta_0$, since

$$\lim_{t\to 0+} \frac{t^{\delta_1}\ln t}{t^\lambda - 1} = \lim_{t\to\infty} \frac{t^{\delta_1}\ln t}{t^\lambda - 1} = 0,$$

there exists a constant $L > 0$, such that $t^{\delta_1}\ln t/(t^\lambda - 1) \le L$, and then

$$k_\lambda(t,1) = \frac{\ln t}{t^\lambda - 1} \le \frac{L}{t^{\delta_1}} \quad (t \in (0,\infty)).$$

Then by Corollary 6.1, we have

$$||T|| = \left[\frac{\Gamma^{j_0}(\frac{1}{\beta})}{\beta^{j_0-1}\Gamma(\frac{j_0}{\beta})}\right]^{\frac{1}{p}} \left[\frac{\Gamma^{i_0}(\frac{1}{\alpha})}{\alpha^{i_0-1}\Gamma(\frac{i_0}{\alpha})}\right]^{\frac{1}{q}} \left[\frac{\pi}{\lambda \sin(\frac{\pi\lambda_1}{\lambda})}\right]^2.$$

Lemma 6.1. *If* $\mathbf{C}$ *is the set of complex numbers and* $\mathbf{C}_\infty = \mathbf{C} \cup \{\infty\}$, $z_k \in \mathbf{C} \setminus \{z \mid Re\, z \ge 0,\ Im\, z = 0\}$, $k = 1,2,\ldots,n$, *are different points, the function* $f(z)$ *is analytic in* $\mathbf{C}_\infty$ *except for* z_i $(i = 1,2,\ldots,n)$, *and* $z = \infty$ *is a zero point of* $f(z)$ *whose order is not less than* 1, *then for* $\alpha \in \mathbf{R}$, *we have*

$$\int_0^\infty f(x)x^{\alpha-1}dx = \frac{2\pi i}{1 - e^{2\pi\alpha i}} \sum_{k=1}^n Res[f(z)z^{\alpha-1}, z_k], \quad (39)$$

where $0 < Im(\ln z) = \arg z < 2\pi$. *In particular, if* z_k $(k = 1, \ldots, n)$ *are all poles of order* 1, *setting* $\varphi_k(z) = (z - z_k)f(z)$ $(\varphi_k(z_k) \neq 0)$, *then*

$$\int_0^\infty f(x)x^{\alpha-1}dx = \frac{\pi}{\sin\pi\alpha}\sum_{k=1}^n(-z_k)^{\alpha-1}\varphi_k(z_k). \tag{40}$$

Proof. By [21, p. 118], we have (39). We find

$$\begin{aligned} 1 - e^{2\pi\alpha i} &= 1 - \cos 2\pi\alpha - i\sin 2\pi\alpha \\ &= -2i\sin\pi\alpha(\cos\pi\alpha + i\sin\pi\alpha) \\ &= -2ie^{i\pi\alpha}\sin\pi\alpha. \end{aligned}$$

In particular, since $f(z)z^{\alpha-1} = \frac{1}{z-z_k}(\varphi_k(z)z^{\alpha-1})$, it is obvious that

$$\mathrm{Res}[f(z)z^{\alpha-1}, -a_k] = {z_k}^{\alpha-1}\varphi_k(z_k) = -e^{i\pi\alpha}(-z_k)^{\alpha-1}\varphi_k(z_k).$$

Then by (39), we obtain (40). The lemma is proved. □

Example 6.2. For $s \in \mathbf{N}$, we set

$$k_\lambda(x,y) = \prod_{k=1}^s \frac{1}{x^{\lambda/s} + a_k y^{\lambda/s}}\ (0 < a_1 < \cdots < a_s,\ \lambda > 0,\ 0 < \lambda_1 < i_0,\ 0 < \lambda_2 < j_0),$$

which is a positive decreasing function with respect to $x\,(y) \in \mathbf{R}_+$. For $\delta_0 = \frac{1}{2}\min\{\lambda_1, \lambda_2, i_0 - \lambda_1, j_0 - \lambda_2\} > 0$, and $\tilde{\lambda}_1 \in (\lambda_1 - \delta_0, \lambda_1 + \delta_0)$, by (40), we find

$$\begin{aligned} k(\tilde{\lambda}_1) &= \int_0^\infty \prod_{k=1}^s \frac{1}{t^{\lambda/s} + a_k} t^{\tilde{\lambda}_1 - 1}dt = \frac{s}{\lambda}\int_0^\infty \prod_{k=1}^s \frac{1}{u + a_k} u^{\frac{s\tilde{\lambda}_1}{\lambda} - 1}du \\ &= \frac{\pi s}{\lambda\sin(\frac{\pi s\tilde{\lambda}_1}{\lambda})}\sum_{k=1}^s a_k^{\frac{s\tilde{\lambda}_1}{\lambda} - 1}\prod_{\substack{j=1\\ j\neq k}}^s \frac{1}{a_j - a_k} \in \mathbf{R}_+. \end{aligned}$$

Setting $\delta_1 = 0 (< \lambda_1 - \delta_0)$, $L = \prod_{k=1}^s \frac{1}{a_k}$, we obtain

$$k_\lambda(t,1) = \prod_{k=1}^s \frac{1}{t^{\lambda/s} + a_k} \leqslant \prod_{k=1}^s \frac{1}{a_k} = L \cdot \frac{1}{t^{\delta_1}} \quad (t \in (0,\infty)).$$

Then by Corollary 6.1, we have

$$||T|| = \left[\frac{\Gamma^{j_0}(\frac{1}{\beta})}{\beta^{j_0-1}\Gamma(\frac{j_0}{\beta})}\right]^{\frac{1}{p}} \left[\frac{\Gamma^{i_0}(\frac{1}{\alpha})}{\alpha^{i_0-1}\Gamma(\frac{i_0}{\alpha})}\right]^{\frac{1}{q}} \frac{\pi s}{\lambda \sin(\frac{\pi s \lambda_1}{\lambda})} \sum_{k=1}^{s} a_k^{\frac{s\lambda_1}{\lambda}-1} \prod_{\substack{j=1 \\ j\neq k}}^{s} \frac{1}{a_j - a_k}.$$

In particular,

(1) For $s = 1$, $k_\lambda(x, y) = \frac{1}{x^\lambda + a_1 y^\lambda}$ $(a_1 > 0)$, we have

$$||T|| = \left[\frac{\Gamma^{j_0}(\frac{1}{\beta})}{\beta^{j_0-1}\Gamma(\frac{j_0}{\beta})}\right]^{\frac{1}{p}} \left[\frac{\Gamma^{i_0}(\frac{1}{\alpha})}{\alpha^{i_0-1}\Gamma(\frac{i_0}{\alpha})}\right]^{\frac{1}{q}} \frac{\pi a_1^{\frac{\lambda_1}{\lambda}-1}}{\lambda \sin(\frac{\pi\lambda_1}{\lambda})}.$$

(2) For $s = 2$,

$$k_\lambda(x, y) = \frac{1}{(x^{\lambda/2} + a_1 y^{\lambda/2})(x^{\lambda/2} + a_2 y^{\lambda/2})} \quad (0 < a_1 < a_2),$$

we have

$$||T|| = \left[\frac{\Gamma^{j_0}(\frac{1}{\beta})}{\beta^{j_0-1}\Gamma(\frac{j_0}{\beta})}\right]^{\frac{1}{p}} \left[\frac{\Gamma^{i_0}(\frac{1}{\alpha})}{\alpha^{i_0-1}\Gamma(\frac{i_0}{\alpha})}\right]^{\frac{1}{q}} \frac{2\pi}{\lambda \sin(\frac{2\pi\lambda_1}{\lambda})} \frac{a_1^{\frac{2\lambda_1}{\lambda}-1} - a_2^{\frac{2\lambda_1}{\lambda}-1}}{a_2 - a_1}.$$

Since we find

$$\lim_{\lambda_1 \to \frac{\lambda}{2}} \frac{a_1^{\frac{2\lambda_1}{\lambda}-1} - a_2^{\frac{2\lambda_1}{\lambda}-1}}{\sin(\frac{2\pi\lambda_1}{\lambda})} = \lim_{\lambda_1 \to \frac{\lambda}{2}} \frac{\frac{2}{\lambda}\left(a_1^{\frac{2\lambda_1}{\lambda}-1} \ln a_1 - a_2^{\frac{2\lambda_1}{\lambda}-1} \ln a_2\right)}{\frac{2\pi}{\lambda}\cos(\frac{2\pi\lambda_1}{\lambda})} = \frac{\ln(a_2/a_1)}{\pi},$$

then for $\lambda_1 = \lambda_2 = \lambda/2$, we obtain

$$||T|| = \left[\frac{\Gamma^{j_0}(\frac{1}{\beta})}{\beta^{j_0-1}\Gamma(\frac{j_0}{\beta})}\right]^{\frac{1}{p}} \left[\frac{\Gamma^{i_0}(\frac{1}{\alpha})}{\alpha^{i_0-1}\Gamma(\frac{i_0}{\alpha})}\right]^{\frac{1}{q}} \frac{2\ln(a_2/a_1)}{\lambda(a_2 - a_1)}.$$

(3) For $s = 3$,

$$k_\lambda(x, y) = \frac{1}{(x^{\lambda/3} + a_1 y^{\lambda/3})(x^{\lambda/3} + a_2 y^{\lambda/3})(x^{\lambda/3} + a_3 y^{\lambda/3})} \quad (0 < a_1 < a_2 < a_3),$$

we have

$$||T|| = \left[\frac{\Gamma^{j_0}(\frac{1}{\beta})}{\beta^{j_0-1}\Gamma(\frac{j_0}{\beta})}\right]^{\frac{1}{p}} \left[\frac{\Gamma^{i_0}(\frac{1}{\alpha})}{\alpha^{i_0-1}\Gamma(\frac{i_0}{\alpha})}\right]^{\frac{1}{q}} \frac{3\pi}{\lambda \sin(\frac{3\pi\lambda_1}{\lambda})} \sum_{k=1}^{3} a_k^{\frac{3\lambda_1}{\lambda}-1} \prod_{\substack{j=1 \\ j\neq k}}^{3} \frac{1}{a_j - a_k}.$$

Then for $\lambda_1 = \lambda_2 = \lambda/2$, we may find that

$$||T||=\left[\frac{\Gamma^{j_0}(\frac{1}{\beta})}{\beta^{j_0-1}\Gamma(\frac{j_0}{\beta})}\right]^{\frac{1}{p}}\left[\frac{\Gamma^{i_0}(\frac{1}{\alpha})}{\alpha^{i_0-1}\Gamma(\frac{i_0}{\alpha})}\right]^{\frac{1}{q}}\frac{3\pi}{\lambda(\sqrt{a_1}+\sqrt{a_2})(\sqrt{a_1}+\sqrt{a_3})(\sqrt{a_2}+\sqrt{a_3})}.$$

Example 6.3. (1) We set

$$k_\lambda(x,y)=\frac{1}{x^\lambda+2(xy)^{\lambda/2}\cos\gamma+y^\lambda}\ \left(\lambda>0,\ 0<\gamma\le\frac{\pi}{2},\ 0<\lambda_1<i_0,\ 0<\lambda_2<j_0\right),$$

which is a positive decreasing function with respect to x $(y) \in \mathbf{R}_+$. For $\delta_0 = \frac{1}{2}\min\{\lambda_1, \lambda_2, i_0 - \lambda_1, j_0 - \lambda_2\} > 0$, and $\tilde{\lambda}_1 \in (\lambda_1 - \delta_0, \lambda_1 + \delta_0)$, setting $z_1 = -e^{i\gamma}$, $z_2 = -e^{-i\gamma}$, by (40), it follows

$$\begin{aligned}k(\tilde{\lambda}_1) &= \int_0^\infty \frac{1}{t^\lambda+2t^{\lambda/2}\cos\gamma+1}t^{\tilde{\lambda}_1-1}dt = \frac{2}{\lambda}\int_0^\infty \frac{1}{u^2+2u\cos\gamma+1}u^{\frac{2\tilde{\lambda}_1}{\lambda}-1}du\\
&= \frac{2\pi}{\lambda\sin(\frac{2\pi\tilde{\lambda}_1}{\lambda})}\left[(e^{i\gamma})^{\frac{2\tilde{\lambda}_1}{\lambda}-1}\frac{1}{e^{-i\gamma}-e^{i\gamma}}+(e^{-i\gamma})^{\frac{2\tilde{\lambda}_1}{\lambda}-1}\frac{1}{e^{i\gamma}-e^{-i\gamma}}\right]\\
&= \frac{2\pi\sin\gamma(1-\frac{2\tilde{\lambda}_1}{\lambda})}{\lambda\sin\gamma\sin(\frac{2\pi\tilde{\lambda}_1}{\lambda})}\in\mathbf{R}_+.\end{aligned}$$

Setting $\delta_1 = 0\ (< \lambda_1 - \delta_0)$, we have

$$k_\lambda(t,1) = \frac{1}{t^\lambda+2t^{\lambda/2}\cos\gamma+1} \leqslant 1 = \frac{1}{t^{\delta_1}}\quad (t\in(0,\infty)).$$

Then by Corollary 6.1, we have

$$||T|| = \left[\frac{\Gamma^{j_0}(\frac{1}{\beta})}{\beta^{j_0-1}\Gamma(\frac{j_0}{\beta})}\right]^{\frac{1}{p}}\left[\frac{\Gamma^{i_0}(\frac{1}{\alpha})}{\alpha^{i_0-1}\Gamma(\frac{i_0}{\alpha})}\right]^{\frac{1}{q}}\frac{2\pi\sin\gamma(1-\frac{2\tilde{\lambda}_1}{\lambda})}{\lambda\sin\gamma\sin(\frac{2\pi\tilde{\lambda}_1}{\lambda})}.$$

(2) We set

$$k_\lambda(x,y) = \frac{1}{x^\lambda+b(xy)^{\lambda/2}+cy^\lambda}\ \left(\begin{array}{l}c>0,\ 0\le b\le 2\sqrt{c},\\ \lambda>0,\ 0<\lambda_1=\lambda_2=\frac{\lambda}{2}<\min\{i_0,j_0\}\end{array}\right),$$

which is a positive decreasing function with respect to x $(y) \in \mathbf{R}_+$. For $\delta_0 = \frac{1}{2}\min\{\frac{\lambda}{2}, i_0 - \frac{\lambda}{2}, j_0 - \frac{\lambda}{2}\} > 0$, and $\tilde{\lambda}_1 \in (\frac{\lambda}{2} - \delta_0, \frac{\lambda}{2} + \delta_0)$, it follows

$$k\left(\frac{\lambda}{2}\right) = \int_0^\infty \frac{1}{t^\lambda+bt^{\lambda/2}+c}t^{\frac{\lambda}{2}-1}dt = \frac{2}{\lambda}\int_0^\infty \frac{1}{u^2+bu+c}du$$

$$= \begin{cases} \frac{2\pi}{\lambda\sqrt{c}}, & b = 0, \\ \frac{8}{\lambda\sqrt{4c-b^2}} \arctan \frac{\sqrt{4c-b^2}}{b}, & 0 < b < 2\sqrt{c}, \\ \frac{4}{\lambda\sqrt{c}}, & b = 2\sqrt{c}, \end{cases}$$

and

$$0 < k(\tilde{\lambda}_1) = \int_0^\infty \frac{1}{t^\lambda + bt^{\lambda/2} + c} t^{\tilde{\lambda}_1 - 1} dt \leqslant \int_0^1 \frac{1}{c} t^{\frac{\lambda}{2} - \delta' - 1} dt + \int_1^\infty \frac{1}{t^\lambda} t^{\frac{\lambda}{2} + \delta' - 1} dt$$

$$= \Big(\frac{1}{c} + 1\Big) \frac{1}{\frac{\lambda}{2} - \delta'} < \infty.$$

Setting $\delta_1 = 0 \ (< \lambda_1 - \delta_0)$, we have

$$k_\lambda(t, 1) = \frac{1}{t^\lambda + bt^{\lambda/2} + c} \leq \frac{1}{c} = \frac{1}{ct^{\delta_1}} \quad (t \in (0, \infty)).$$

Then by Corollary 6.1, we have

$$||T|| = \left[\frac{\Gamma^{j_0}(\frac{1}{\beta})}{\beta^{j_0-1}\Gamma(\frac{j_0}{\beta})}\right]^{\frac{1}{p}} \left[\frac{\Gamma^{i_0}(\frac{1}{\alpha})}{\alpha^{i_0-1}\Gamma(\frac{i_0}{\alpha})}\right]^{\frac{1}{q}} k\Big(\frac{\lambda}{2}\Big).$$

Example 6.4. We set

$$k_\lambda(x, y) = \frac{(\min\{x, y\})^\gamma}{(\max\{x, y\})^{\lambda+\gamma}} \quad (-\gamma < \lambda_1 < i_0 - \gamma, \ -\gamma < \lambda_2 < j_0 - \gamma).$$

Then we find

$$k_\lambda(x, y) y^{\lambda_2 - j_0} = \frac{(\min\{x, y\})^\gamma}{(\max\{x, y\})^{\lambda+\gamma}} y^{\lambda_2 - j_0} = \begin{cases} \frac{y^{\gamma+\lambda_2-j_0}}{x^{\lambda+\gamma}}, & 0 < y < x, \\ \frac{x^\gamma}{y^{\lambda_1+\gamma+j_0}}, & y \geq x, \end{cases}$$

is a strict decreasing function with respect to $y \in \mathbf{R}_+$. There exists a constant

$$\delta_0 = \frac{1}{2} \min\{i_0 - \gamma - \lambda_1, \gamma + \lambda_2, \gamma + \lambda_1\} > 0,$$

such that for any $\tilde{\lambda}_1 \in (\lambda_1 - \delta_0, \lambda_1 + \delta_0)$,

$$k_\lambda(x,y)x^{\tilde{\lambda}_1-i_0} = \frac{(\min\{x,y\})^\gamma}{(\max\{x,y\})^{\lambda+\gamma}}x^{\tilde{\lambda}_1-i_0} = \begin{cases} \dfrac{x^{\gamma+\tilde{\lambda}_1-i_0}}{y^{\lambda+\gamma}}, & 0 < x < y, \\ \dfrac{y^\gamma}{x^{\lambda-\tilde{\lambda}_1+i_0+\gamma}}, & x \geq y, \end{cases}$$

is strict decreasing with respect to $x \in \mathbf{R}_+$,

$$k(\tilde{\lambda}_1) = \int_0^\infty \frac{(\min\{t,1\})^\gamma}{(\max\{t,1\})^{\lambda+\gamma}} t^{\tilde{\lambda}_1-1} dt = \int_0^1 t^{\tilde{\lambda}_1+\gamma-1} dt + \int_1^\infty \frac{1}{t^{\lambda+\gamma}} t^{\tilde{\lambda}_1-1} dt$$

$$= \frac{\lambda+2\gamma}{(\tilde{\lambda}_1+\gamma)(\lambda-\tilde{\lambda}_1+\gamma)} \in \mathbf{R}.$$

Since $\delta_0 \leq \frac{1}{2}(\gamma+\lambda_1) < \gamma+\lambda_1$ and

$$k_\lambda(t,1) = \frac{(\min\{t,1\})^\gamma}{(\max\{t,1\})^{\lambda+\gamma}} = \begin{cases} t^\gamma, & 0 < t < 1, \\ \dfrac{1}{t^{\lambda+\gamma}}, & t \geq 1, \end{cases}$$

then there exists a constant $\delta_1 = -\gamma < \lambda_1 - \delta_0$, satisfying

$$k_\lambda(t,1) \leq \frac{1}{t^{\delta_1}} \quad (t \in (0,\infty)).$$

Therefore, the assumptions of Theorem 3.4 are satisfied and by Corollary 6.1, it follows

$$||T|| = \left[\frac{\Gamma^{j_0}(\frac{1}{\beta})}{\beta^{j_0-1}\Gamma(\frac{j_0}{\beta})}\right]^{\frac{1}{p}} \left[\frac{\Gamma^{i_0}(\frac{1}{\alpha})}{\alpha^{i_0-1}\Gamma(\frac{i_0}{\alpha})}\right]^{\frac{1}{q}} \frac{\lambda+2\gamma}{(\lambda_1+\gamma)(\lambda_2+\gamma)}.$$

In particular,

(i) For $\gamma = 0$, we have

$$k_\lambda(x,y) = \frac{1}{(\max\{x,y\})^\lambda} \quad (0 < \lambda_1 < i_0,\ 0 < \lambda_2 < j_0)$$

and

$$||T|| = \left[\frac{\Gamma^{j_0}(\frac{1}{\beta})}{\beta^{j_0-1}\Gamma(\frac{j_0}{\beta})}\right]^{\frac{1}{p}} \left[\frac{\Gamma^{i_0}(\frac{1}{\alpha})}{\alpha^{i_0-1}\Gamma(\frac{i_0}{\alpha})}\right]^{\frac{1}{q}} \frac{\lambda}{\lambda_1\lambda_2}.$$

(ii) For $\gamma = -\lambda$, we have

$$k_\lambda(x,y) = \frac{1}{(\min\{x,y\})^\lambda} \quad (-j_0 < \lambda_1 < 0,\ -i_0 < \lambda_2 < 0)$$

and

$$||T|| = \left[\frac{\Gamma^{j_0}(\frac{1}{\beta})}{\beta^{j_0-1}\Gamma(\frac{j_0}{\beta})}\right]^{\frac{1}{p}} \left[\frac{\Gamma^{i_0}(\frac{1}{\alpha})}{\alpha^{i_0-1}\Gamma(\frac{i_0}{\alpha})}\right]^{\frac{1}{q}} \frac{(-\lambda)}{\lambda_1\lambda_2}.$$

(iii) For $\lambda = 0$, we have

$$k_0(x,y) = \left(\frac{\min\{x,y\}}{\max\{x,y\}}\right)^\gamma \quad (\max\{-\gamma, \gamma - j_0\} < \lambda_1 = -\lambda_2 < \min\{\gamma, i_0 - \gamma\})$$

and

$$||T|| = \left[\frac{\Gamma^{j_0}(\frac{1}{\beta})}{\beta^{j_0-1}\Gamma(\frac{j_0}{\beta})}\right]^{\frac{1}{p}} \left[\frac{\Gamma^{i_0}(\frac{1}{\alpha})}{\alpha^{i_0-1}\Gamma(\frac{i_0}{\alpha})}\right]^{\frac{1}{q}} \frac{2\gamma}{\gamma^2 - \lambda_1^2}.$$

7 Compositions of Two Discrete Hilbert-Type Operators

For $p > 1$, still setting

$$\varphi(x) = x^{p(1-\lambda_1)-1}, \quad \psi(y) = y^{q(1-\lambda_2)-1} \quad (x, y \in \mathbf{R}_+),$$

as in the front section, for $i_0 = j_0 = 1$, we define two normal spaces as follows:

$$\ell_{p,\varphi} := \left\{a = \{a_m\}_{m=1}^\infty;\ ||a||_{p,\varphi} = \left\{\sum_{m=1}^\infty \varphi(m)|a_m|^p\right\}^{\frac{1}{p}} < \infty\right\},$$

$$\ell_{q,\psi} := \left\{b = \{b_n\}_{n=1}^\infty;\ ||b||_{q,\psi} = \left\{\sum_{n=1}^\infty \psi(n)|b_n|^q\right\}^{\frac{1}{q}} < \infty\right\}.$$

Corollary 7.1. *With the same assumptions of Theorem* 3.4 (*or Corollary* 3.2) *for* $i_0 = j_0 = 1$, T *is defined by Definition* 4.1 (*for* $i_0 = j_0 = 1$), *such as*

$$(Ta)(m) = \sum_{m=1}^\infty k_\lambda(m,n)a_m \quad (m \in \mathbf{N}),$$

we have

$$||T|| = k(\lambda_1) = \int_0^\infty k_\lambda(t,1)t^{\lambda_1-1}dt \in \mathbf{R}_+.$$

In the following, we agree that $p > 1$, $\frac{1}{p} + \frac{1}{q} = 1$, $\lambda, \lambda_1, \lambda_2 \in \mathbf{R}$, $\lambda_1 + \lambda_2 = \lambda$, $k_\lambda^{(i)}(x,y)$ $(i = 1,2,3)$ are non-negative finite homogeneous functions of degree $-\lambda$ in $\mathbf{R}_+^2$, with

$$k^{(i)}(\lambda_1) := \int_0^\infty k_\lambda^{(i)}(u,1)u^{\lambda_1-1}du \in \mathbf{R}_+,$$

and $k_\lambda^{(1)}(x,y)$ is symmetric.

Definition 7.1. If $k \in \mathbf{N}$, we define two functions $\tilde{F}_k(y)$ and $\tilde{G}_k(x)$ as follows:

$$\tilde{F}_k(y) : = y^{\lambda-1}\int_1^\infty k_\lambda^{(2)}(x,y)x^{\lambda_1-\frac{1}{pk}-1}dx, \quad y \in \mathbf{R}_+,$$

$$\tilde{G}_k(x) : = x^{\lambda-1}\int_1^\infty k_\lambda^{(3)}(x,y)y^{\lambda_2-\frac{1}{qk}-1}dy, \quad x \in \mathbf{R}_+.$$

Lemma 7.1. *If there exists a constant* $\delta_0 > 0$, *such that* $k^{(i)}(\lambda_1 \pm \delta_0) \in \mathbf{R}_+$ $(i = 1,2,3)$, *and there exist constants* $\delta_1 \in (0,\delta_0)$ *and* $L > 0$, *satisfying for any* $u \in [1,\infty)$,

$$k_\lambda^{(2)}(1,u)u^{\lambda_2+\delta_1} \le L, \quad k_\lambda^{(3)}(u,1)u^{\lambda_1+\delta_1} \le L, \tag{41}$$

then for $k \in \mathbf{N}$, $k > \frac{1}{\delta_1}\max\{\frac{1}{p},\frac{1}{q}\}$, *setting functions* $F_k(y)$ *and* $G_k(x)$ *as follows:*

$$F_k(y) : = y^{\lambda_1-\frac{1}{pk}-1}k^{(2)}\Big(\lambda_1 - \frac{1}{pk}\Big) - \tilde{F}_k(y), \quad y \in \mathbf{R}_+,$$

$$G_k(x) : = x^{\lambda_2-\frac{1}{qk}-1}k^{(3)}\Big(\lambda_1 + \frac{1}{qk}\Big) - \tilde{G}_k(x), \quad x \in \mathbf{R}_+,$$

we have

$$0 \le F_k(y) = O(y^{\lambda_1-\delta_1-1}) \ (y \in [1,\infty)), \ 0 \le G_k(x) = O(x^{\lambda_2-\delta_1-1})(x \in [1,\infty)).$$

Proof. Setting $u = x/y$, we obtain

$$\tilde{F}_k(y) = y^{\lambda_1-\frac{1}{pk}-1}\int_{1/y}^\infty k_\lambda^{(2)}(u,1)u^{\lambda_1-\frac{1}{pk}-1}du$$

$$= y^{\lambda_1-\frac{1}{pk}-1}\int_0^\infty k_\lambda^{(2)}(u,1)u^{\lambda_1-\frac{1}{pk}-1}du - y^{\lambda_1-\frac{1}{pk}-1}\int_0^{1/y} k_\lambda^{(2)}(u,1)u^{\lambda_1-\frac{1}{pk}-1}du$$
$$= y^{\lambda_1-\frac{1}{pk}-1}k^{(2)}\Big(\lambda_1-\frac{1}{pk}\Big) - y^{\lambda_1-\frac{1}{pk}-1}\int_0^{1/y} k_\lambda^{(2)}(u,1)u^{\lambda_1-\frac{1}{pk}-1}du.$$

Hence, it follows

$$F_k(y) = y^{\lambda_1-\frac{1}{pk}-1}k^{(2)}\Big(\lambda_1-\frac{1}{pk}\Big) - \tilde{F}_k(y) = y^{\lambda_1-\frac{1}{pk}-1}\int_0^{1/y} k_\lambda^{(2)}(u,1)u^{\lambda_1-\frac{1}{pk}-1}du$$
$$= y^{\lambda_1-\frac{1}{pk}-1}\int_y^\infty k_\lambda^{(2)}(1,v)v^{\lambda_2+\frac{1}{pk}-1}dv \ge 0 \quad (y\in[1,\infty)).$$

In view of (41), we have

$$0 \le F_k(y) \le y^{\lambda_1-\frac{1}{pk}-1}L\int_y^\infty v^{-\lambda_2-\delta_1}v^{\lambda_2+\frac{1}{pk}-1}dv$$
$$= y^{\lambda_1-\frac{1}{pk}-1}L\int_y^\infty v^{-\delta_1+\frac{1}{pk}-1}dv = \frac{Ly^{\lambda_1-\delta_1-1}}{\delta_1-\frac{1}{pk}},$$

and then

$$F_k(y) = O(y^{\lambda_1-\delta_1-1}) \quad (y\in[1,\infty)).$$

Still setting $u = x/y$, we have

$$\tilde{G}_k(x) = x^{\lambda_2-\frac{1}{qk}-1}\int_0^x k_\lambda^{(3)}(u,1)u^{\lambda_1+\frac{1}{qk}-1}du$$
$$= x^{\lambda_2-\frac{1}{qk}-1}k^{(3)}\Big(\lambda_1+\frac{1}{qk}\Big) - x^{\lambda_2-\frac{1}{qk}-1}\int_x^\infty k_\lambda^{(3)}(u,1)u^{\lambda_1+\frac{1}{qk}-1}du.$$

Hence it follows

$$G_k(x) = x^{\lambda_2-\frac{1}{qk}-1}k^{(3)}\Big(\lambda_1+\frac{1}{qk}\Big) - \tilde{G}_k(x)$$
$$= x^{\lambda_2-\frac{1}{qk}-1}\int_x^\infty k_\lambda^{(3)}(u,1)u^{\lambda_1+\frac{1}{qk}-1}du \geqslant 0.$$

By (41), we have

$$0 \le G_k(x) \le x^{\lambda_2-\frac{1}{qk}-1}L\int_x^\infty u^{-\delta_1+\frac{1}{qk}-1}du = \frac{Lx^{\lambda_2-\delta_1-1}}{\delta_1-\frac{1}{qk}},$$

and then $G_k(x) = O(x^{\lambda_2-\delta_1-1})\quad (x \in [1,\infty))$. The lemma is proved. □

Lemma 7.2. *With the same assumptions of Lemma* 7.1, *we have*

$$L_k := \frac{1}{k}\int_1^\infty \left(\int_1^\infty k_\lambda^{(1)}(x,y)x^{\lambda_2-\frac{1}{qk}-1}y^{\lambda_1-\frac{1}{pk}-1}dx\right)dy = k^{(1)}(\lambda_1)+o(1), \quad (42)$$

when $k \to \infty$.

Proof. Setting $u = y/x$, since $k_\lambda^{(1)}(x,y)$ is symmetric, by Lemma 3.1, it follows

$$\begin{aligned}
L_k &= \frac{1}{k}\int_1^\infty y^{-\frac{1}{k}-1}\left(\int_0^y k_\lambda^{(1)}(1,u)u^{\lambda_1+\frac{1}{qk}-1}du\right)dy \\
&= \frac{1}{k}\left[\int_1^\infty y^{-\frac{1}{k}-1}\left(\int_0^1 k_\lambda^{(1)}(u,1)u^{\lambda_1+\frac{1}{qk}-1}du\right)dy\right. \\
&\quad \left. +\int_1^\infty y^{-\frac{1}{k}-1}\left(\int_1^y k_\lambda^{(1)}(u,1)u^{\lambda_1+\frac{1}{qk}-1}du\right)dy\right] \\
&= \int_0^1 k_\lambda^{(1)}(u,1)u^{\lambda_1+\frac{1}{qk}-1}du + \frac{1}{k}\int_1^\infty\left(\int_u^\infty y^{-\frac{1}{k}-1}dy\right)k_\lambda^{(1)}(u,1)u^{\lambda_1+\frac{1}{qk}-1}du \\
&= \int_0^1 k_\lambda^{(1)}(u,1)u^{\lambda_1+\frac{1}{qk}-1}du + \int_1^\infty k_\lambda^{(1)}(u,1)u^{\lambda_1-\frac{1}{pk}-1}du \\
&= \int_0^\infty k_\lambda^{(1)}(u,1)u^{\lambda_1-1}du + o(1).
\end{aligned}$$

Hence, (42) is valid. The lemma is proved. □

Lemma 7.3. *With the same assumptions of Lemma* 7.1, *if* $\lambda, \lambda_1, \lambda_2 \le 1$, $k_\lambda^{(i)}(x,y)$ $(i = 1,2,3)$ *are decreasing with respect to* x $(y) \in \mathbf{R}_+$, *setting*

$$\tilde{A}_k(n) : = n^{\lambda-1}\sum_{m_1=1}^\infty k_\lambda^{(2)}(m_1,n)m_1^{\lambda_1-\frac{1}{pk}-1},$$

$$\tilde{B}_k(m) : = m^{\lambda-1}\sum_{n_1=1}^\infty k_\lambda^{(3)}(m,n_1)n_1^{\lambda_2-\frac{1}{qk}-1},$$

then we have

$$\tilde{I}_k := \frac{1}{k}\sum_{m=1}^{\infty}\sum_{n=1}^{\infty} k_{\lambda}^{(1)}(m,n)\tilde{A}_k(n)\tilde{B}_k(m) \geq \prod_{i=1}^{3} k^{(i)}(\lambda_1) + o(1) \quad (k \to \infty). \quad (43)$$

Proof. By the decreasing property, Definition 7.1 and Lemma 7.1, it follows

$$\begin{aligned}\tilde{I}_k &\geq \frac{1}{k}\int_1^{\infty}\int_1^{\infty} k_{\lambda}^{(1)}(x,y)\tilde{F}_k(y)\tilde{G}_k(x)dxdy \\ &= \frac{1}{k}\int_1^{\infty}\int_1^{\infty} k_{\lambda}^{(1)}(x,y)\left[y^{\lambda_1-\frac{1}{pk}-1}k^{(2)}\left(\lambda_1-\frac{1}{pk}\right)-F_k(y)\right] \\ &\quad\times\left[x^{\lambda_2-\frac{1}{qk}-1}k^{(3)}\left(\lambda_1+\frac{1}{qk}\right)-G_k(x)\right]dxdy \\ &\geq I_1 - I_2 - I_3,\end{aligned}$$

where I_1, I_2, I_3 are defined by

$$I_1 := k^{(2)}\left(\lambda_1-\frac{1}{pk}\right)k^{(3)}\left(\lambda_1+\frac{1}{qk}\right)\frac{1}{k}\int_1^{\infty}\int_1^{\infty} k_{\lambda}^{(1)}(x,y)x^{\lambda_2-\frac{1}{qk}-1}y^{\lambda_1-\frac{1}{pk}-1}dxdy,$$

$$I_2 := k^{(3)}\left(\lambda_1+\frac{1}{qk}\right)\frac{1}{k}\int_1^{\infty}\left(\int_1^{\infty} k_{\lambda}^{(1)}(x,y)x^{\lambda_2-\frac{1}{qk}-1}dx\right)F_k(y)dy,$$

$$I_3 := k^{(2)}\left(\lambda_1-\frac{1}{pk}\right)\frac{1}{k}\int_1^{\infty}\left(\int_1^{\infty} k_{\lambda}^{(1)}(x,y)y^{\lambda_1-\frac{1}{pk}-1}dy\right)G_k(x)dx,$$

respectively. By Lemma 7.2, we have

$$I_1 = (k^{(1)}(\lambda_1)+o(1))k^{(2)}\left(\lambda_1-\frac{1}{pk}\right)k^{(3)}\left(\lambda_1+\frac{1}{qk}\right).$$

Since $0 \leq F_k(y) = O(y^{\lambda_1-\delta_1-1})$, there exists a constant $L_2 > 0$, such that

$$F_k(y) \leq L_2 y^{\lambda_1-\delta_1-1} \quad (y \in [1,\infty)),$$

$$\begin{aligned}0 \leq I_2 &\leq k^{(3)}\left(\lambda_1+\frac{1}{qk}\right)\frac{L_2}{k}\int_1^{\infty}\left(\int_0^{\infty} k_{\lambda}^{(1)}(x,y)x^{\lambda_2-\frac{1}{qk}-1}dx\right)y^{\lambda_1-\delta_1-1}dy \\ &= k^{(3)}\left(\lambda_1+\frac{1}{qk}\right)\frac{L_2}{k}\int_1^{\infty}\left(\int_0^{\infty} k_{\lambda}^{(1)}(u,1)u^{\lambda_1+\frac{1}{qk}-1}du\right)y^{-\delta_1-\frac{1}{qk}-1}dy \\ &= \frac{1}{k}k^{(3)}\left(\lambda_1+\frac{1}{qk}\right)k^{(1)}\left(\lambda_1+\frac{1}{qk}\right)\frac{L_2}{\delta_1+\frac{1}{qk}}.\end{aligned}$$

Hence, $I_2 \to 0$ $(k \to \infty)$.

Since $0 \leq G_k(x) = O(x^{\lambda_2-\delta_1-1})$, there exists a constant $L_3 > 0$, such that

$$G_k(x) \leq L_3 x^{\lambda_2-\delta_1-1} \quad (x \in [1,\infty)),$$

and then

$$0 \leq I_3 \leq k^{(2)}\Big(\lambda_1 - \frac{1}{pk}\Big)\frac{L_3}{k}\int_1^\infty \left(\int_0^\infty k_\lambda^{(1)}(x,y)y^{\lambda_1-\frac{1}{pk}-1}dy\right)x^{\lambda_2-\delta_1-1}dx$$

$$= k^{(2)}\Big(\lambda_1 - \frac{1}{pk}\Big)\frac{L_3}{k}\int_1^\infty \left(\int_0^\infty k_\lambda^{(1)}(u,1)u^{\lambda_1-\frac{1}{pk}-1}du\right)x^{-\delta_1-\frac{1}{pk}-1}dx$$

$$= \frac{1}{k}k^{(2)}\Big(\lambda_1 - \frac{1}{pk}\Big)k^{(1)}\Big(\lambda_1 - \frac{1}{pk}\Big)\frac{L_3}{\delta_1 + \frac{1}{pk}}.$$

Hence, $I_3 \to 0$ $(k \to \infty)$. Therefore,

$$\tilde{I}_k \geq I_1 - I_2 - I_3 \to \prod_{i=1}^{3} k^{(i)}(\lambda_1) \quad (k \to \infty),$$

and then (43) follows. The lemma is proved. □

Theorem 7.1. *Suppose that for* $\lambda_1, \lambda_2 < 1$, $\lambda \leq 1$, $k_\lambda^{(i)}(x,y)$ $(i = 1,2,3)$ *are decreasing with respect to* x $(y) \in \mathbf{R}_+$, *there exists a constant* $\delta_0 > 0$, *such that*

$$k^{(i)}(\lambda_1 \pm \delta_0) \in \mathbf{R}_+ \quad (i = 1,2,3),$$

and there exist constants $\delta_1 \in (0,\delta_0)$ *and* $L > 0$, *satisfying for any* $u \in [1,\infty)$,

$$k_\lambda^{(2)}(1,u)u^{\lambda_2+\delta_1} \leq L, \quad k_\lambda^{(3)}(u,1)u^{\lambda_1+\delta_1} \leq L.$$

If $a_{m_1}, B_n \geq 0$, $a = \{a_{m_1}\}_{m_1=1}^\infty \in \ell_{p,\varphi}$, $B = \{B_m\}_{m=1}^\infty \in \ell_{q,\psi}$, $||a||_{p,\varphi}, ||B||_{q,\psi} > 0$, *setting*

$$A_\lambda(n) := n^{\lambda-1}\sum_{m_1=1}^\infty k_\lambda^{(2)}(m_1,n)a_{m_1} \quad (n \in \mathbf{N}),$$

then we have the following equivalent inequalities:

$$I := \sum_{m=1}^\infty\sum_{n=1}^\infty k_\lambda^{(1)}(m,n)A_\lambda(n)B_m < k^{(1)}(\lambda_1)k^{(2)}(\lambda_1)||a||_{p,\varphi}||B||_{q,\psi}, \tag{44}$$

$$J := \left[\sum_{m=1}^\infty m^{p\lambda_2-1}\left(\sum_{n=1}^\infty k_\lambda^{(1)}(m,n)A_\lambda(n)\right)^p\right]^{\frac{1}{p}} < k^{(1)}(\lambda_1)k^{(2)}(\lambda_1)||a||_{p,\varphi}, \tag{45}$$

where the constant factor $k^{(1)}(\lambda_1)k^{(2)}(\lambda_1)$ *is the best possible.*

In particular, if $b_{n_1} \geq 0$, $b = \{b_{n_1}\}_{n_1=1}^{\infty} \in \ell_{q,\psi}$, $||b||_{q,\psi} > 0$, *setting*

$$B_m = B_\lambda(m) := m^{\lambda-1} \sum_{n_1=1}^{\infty} k_\lambda^{(3)}(m, n_1) b_{n_1} \quad (m \in \mathbf{N}),$$

then we still have

$$\sum_{m=1}^{\infty} \sum_{n=1}^{\infty} k_\lambda^{(1)}(m,n) A_\lambda(n) B_\lambda(m) < \prod_{i=1}^{3} k^{(i)}(\lambda_1) ||a||_{p,\varphi} ||b||_{q,\psi}, \tag{46}$$

where the constant factor $\prod_{i=1}^{3} k^{(i)}(\lambda_1)$ *is still the best possible.*

Proof. By (11) (for $i_0 = j_0 = 1$), we have

$$J \leq k^{(1)}(\lambda_1) ||A_\lambda||_{p,\varphi}, \tag{47}$$

and the following inequality:

$$\begin{aligned} ||A_\lambda||_{p,\varphi} &= \left\{ \sum_{n=1}^{\infty} n^{p(1-\lambda_1)-1} A_\lambda^p(n) \right\}^{\frac{1}{p}} \\ &= \left\{ \sum_{n=1}^{\infty} n^{p\lambda_2 - 1} \left(\sum_{m_1=1}^{\infty} k_\lambda^{(2)}(m_1, n) a_{m_1} \right)^p \right\}^{\frac{1}{p}} < k^{(2)}(\lambda_1) ||a||_{p,\varphi}, \end{aligned} \tag{48}$$

then we have (45).

By Hölder's inequality, we find

$$I = \sum_{m=1}^{\infty} \left(m^{\lambda_2 - \frac{1}{p}} \sum_{n=1}^{\infty} k_\lambda^{(1)}(m,n) A_\lambda(n) \right) \left(m^{\frac{1}{p} - \lambda_2} B_m \right) \leq J ||B||_{q,\psi}. \tag{49}$$

Then by (45), we have (44). On the other hand, assuming that (44) is valid, we set

$$B_m := m^{p\lambda_2 - 1} \left(\sum_{n=1}^{\infty} k_\lambda^{(1)}(m,n) A_\lambda(n) \right)^{p-1} \quad (m \in \mathbf{N}).$$

Then we find $||B||_{q,\psi}^q = J^p$. If $J = 0$, then (45) is trivially valid; if $J = \infty$, then by (47), it follows $||A_\lambda||_{p,\varphi} = \infty$, which contradicts (48). For $0 < J < \infty$, by (44), it follows

$$||B||_{q,\psi}^q = J^p = I < k^{(1)}(\lambda_1) k^{(2)}(\lambda_1) ||a||_{p,\varphi} ||B||_{q,\psi}.$$

Dividing out $||B||_{q,\psi}$ in the above inequality, we find

$$||B||_{q,\psi}^{q-1} = J < k^{(1)}(\lambda_1)k^{(2)}(\lambda_1)||a||_{p,\varphi},$$

and then we have (45).

Hence, inequalities (45) and (44) are equivalent.

In particular, setting $B_m = B_\lambda(m)$ in (44), since by (11) (for $i_0 = j_0 = 1$), we find

$$||B||_{q,\psi} < k^{(3)}(\lambda_1)||b||_{q,\psi},$$

then we have (46).

In the following, we prove that the constant factor in (46) is the best possible. For $k \in \mathbf{N}$, $k > \frac{1}{\delta_1}\max\{\frac{1}{p}, \frac{1}{q}\}$, we set

$$\tilde{a}_{m_1} := m_1^{\lambda_1 - \frac{1}{pk} - 1}, \tilde{b}_{n_1} := n_1^{\lambda_2 - \frac{1}{qk} - 1} \quad (m_1, n_1 \in \mathbf{N}).$$

Then it follows

$$\tilde{A}_k(n) = n^{\lambda-1}\sum_{m_1=1}^{\infty} k_\lambda^{(2)}(m_1, n)\tilde{a}_{m_1}, \quad \tilde{B}_k(m) = m^{\lambda-1}\sum_{n_1=1}^{\infty} k_\lambda^{(3)}(m, n_1)\tilde{b}_{n_1}.$$

If there exists a positive constant $K \leq \prod_{i=1}^{3} k^{(i)}(\lambda_1)$ such that (46) is valid when replacing $\prod_{i=1}^{3} k^{(i)}(\lambda_1)$ by K, then in particular, it follows that

$$\tilde{I}_k = \frac{1}{k}\sum_{m=1}^{\infty}\sum_{n=1}^{\infty} k_\lambda^{(1)}(m,n)\tilde{A}_\lambda(n)\tilde{B}_\lambda(m) < \frac{1}{k}K||\tilde{a}||_{p,\varphi}||\tilde{b}||_{q,\psi} = \frac{1}{k}K\left(1 + \sum_{n=2}^{\infty} n^{-\frac{1}{k}-1}\right),$$

i.e.,

$$\tilde{I}_k < \frac{1}{k}K\left(1 + \int_1^{\infty} y^{-\frac{1}{k}-1}dy\right) = K\left(1 + \frac{1}{k}\right).$$

By (43), we find

$$\prod_{i=1}^{3} k^{(i)}(\lambda_1) + o(1) \leq \tilde{I}_k = K\left(1 + \frac{1}{k}\right),$$

and then

$$\prod_{i=1}^{3} k^{(i)}(\lambda_1) \leq K \quad (k \to \infty).$$

Hence $K = \prod_{i=1}^{3} k^{(i)}(\lambda_1)$ is the best value of (46).

By the equivalency, the constant factor in (44) is the best possible. Otherwise, for $B_m = B_\lambda(m)$, we would reach a contradiction that the constant factor in (46) is not the best possible. In the same way, the constant factor in (45) is the best possible. Otherwise, we would reach a contradiction by (49) that the constant factor in (44) is not the best possible. The theorem is proved. □

Definition 7.2. With the same assumptions of Theorem 5.1, we define a Hilbert-type operator $T^{(1)} : \ell_{p,\varphi} \to \ell_{p,\varphi}$ as follows: For $A_\lambda = \{A_\lambda(n)\}_{n=1}^{\infty} \in \ell_{p,\varphi}$, there exists a unique representation $T^{(1)} A_\lambda \in \ell_{p,\varphi}$, satisfying

$$(T^{(1)} A_\lambda)(m) = m^{\lambda-1} \sum_{n=1}^{\infty} k_\lambda^{(1)}(m,n) A_\lambda(n) \quad (m \in \mathbf{N}).$$

We can find

$$||T^{(1)} A_\lambda||_{p,\varphi} \leq k^{(1)}(\lambda_1) ||A_\lambda||_{p,\varphi},$$

where the constant factor $k^{(1)}(\lambda_1)$ is the best possible. Hence, it follows

$$||T^{(1)}|| = k^{(1)}(\lambda_1) = \int_0^\infty k_\lambda^{(1)}(t,1) t^{\lambda_1 - 1} dt \in \mathbf{R}_+.$$

Definition 7.3. With the same assumptions of Theorem 5.1, we define a Hilbert-type operator $T^{(2)} : \ell_{p,\varphi} \to \ell_{p,\varphi}$ as follows: for $a = \{a_m\}_{m=1}^{\infty} \in \ell_{p,\varphi}$, there exists a unique representation $T^{(2)} a \in \ell_{p,\varphi}$, satisfying

$$(T^{(2)} a)(n) = A_\lambda(n) = n^{\lambda-1} \sum_{m=1}^{\infty} k_\lambda^{(2)}(m,n) a_m \quad (n \in \mathbf{N}).$$

We can find

$$||T^{(2)} a||_{p,\varphi} \leq k^{(2)}(\lambda_1) ||a||_{p,\varphi},$$

where the constant factor $k^{(2)}(\lambda_1)$ is the best possible. Hence, it follows

$$||T^{(2)}|| = k^{(2)}(\lambda_1) = \int_0^\infty k_\lambda^{(2)}(t,1) t^{\lambda_1 - 1} dt \in \mathbf{R}_+.$$

Remark 7.1. In Definition 7.2,

(i) If for $x > y$, $k_{\lambda}^{(1)}(x, y) = 0$, we define the first kind Hardy-type operator as follows:

$$(T_1^{(1)} A_{\lambda})(m) := m^{\lambda-1} \sum_{n=1}^{m} k_{\lambda}^{(1)}(m,n) A_{\lambda}(n) \quad (m \in \mathbf{N}),$$

then we have

$$||T_1^{(1)}|| = k_1^{(1)}(\lambda_1) = \int_0^1 k_{\lambda}^{(1)}(t, 1) t^{\lambda_1 - 1} dt \in \mathbf{R}_+.$$

(ii) If for $0 < x < y$, $k_{\lambda}^{(1)}(x, y) = 0$, we define the second kind Hardy-type operator as follows:

$$(T_2^{(1)} A_{\lambda})(m) := m^{\lambda-1} \sum_{n=m}^{\infty} k_{\lambda}^{(1)}(m,n) A_{\lambda}(n) \quad (m \in \mathbf{N}),$$

then we have

$$||T_2^{(1)}|| = k_2^{(1)}(\lambda_1) = \int_1^{\infty} k_{\lambda}^{(1)}(t, 1) t^{\lambda_1 - 1} dt \in \mathbf{R}_+.$$

In Definition 7.3,

(i) If for $x > y$, $k_{\lambda}^{(2)}(x, y) = 0$, we define the first kind Hardy-type operator as follows:

$$(T_1^{(2)} a)(n) = n^{\lambda-1} \sum_{m=1}^{n} k_{\lambda}^{(2)}(m,n) a_m \quad (n \in \mathbf{N}),$$

then we have

$$||T_1^{(2)}|| = k_1^{(2)}(\lambda_1) = \int_0^1 k_{\lambda}^{(2)}(t, 1) t^{\lambda_1 - 1} dt \in \mathbf{R}_+.$$

(ii) If for $0 < x < y$, $k_{\lambda}^{(2)}(x, y) = 0$, we define the second kind Hardy-type operator as follows:

$$(T_2^{(2)} a)(n) = n^{\lambda-1} \sum_{m=n}^{\infty} k_{\lambda}^{(2)}(m,n) a_m \quad (n \in \mathbf{N}),$$

then we have

$$||T_2^{(2)}|| = k_2^{(2)}(\lambda_1) = \int_1^{\infty} k_{\lambda}^{(2)}(t, 1) t^{\lambda_1 - 1} dt \in \mathbf{R}_+.$$

Definition 7.4. With the same assumptions of Theorem 5.1, we define a Hilbert-type operator $T : \ell_{p,\varphi} \to \ell_{p,\varphi}$ as follows: For $a = \{a_m\}_{m=1}^{\infty} \in \ell_{p,\varphi}$, there exists a unique representation $Ta \in \ell_{p,\varphi}$, satisfying

$$(Ta)(m) = T^{(1)}A_\lambda(m) = m^{\lambda-1}\sum_{n=1}^{\infty} k_\lambda^{(1)}(m,n)A_\lambda(n)$$

$$= m^{\lambda-1}\sum_{n=1}^{\infty} k_\lambda^{(1)}(m,n)n^{\lambda-1}\left(\sum_{m_1=1}^{\infty} k_\lambda^{(2)}(m_1,n)a_{m_1}\right) \quad (m \in \mathbf{N}).$$

Since for any $a = \{a_m\}_{m=1}^{\infty} \in \ell_{p,\varphi}$, we have

$$Ta = T^{(1)}A_\lambda = T^{(1)}(T^{(2)}a) = (T^{(1)}T^{(2)})a,$$

then it follows that $T = T^{(1)}T^{(2)}$, i.e., T is a composition of $T^{(1)}$ and $T^{(2)}$. It is obvious that

$$||T|| = ||T^{(1)}T^{(2)}|| \le ||T^{(1)}|| \cdot ||T^{(2)}|| = k^{(1)}(\lambda_1)k^{(2)}(\lambda_1).$$

By (45), we have

$$||Ta||_{p,\varphi} = ||T^{(1)}A_\lambda||_{p,\varphi} = J < k^{(1)}(\lambda_1)k^{(2)}(\lambda_1)||a||_{p,\varphi},$$

where the constant factor $k^{(1)}(\lambda_1)k^{(2)}(\lambda_1)$ is the best possible. It follows that $||T|| = k^{(1)}(\lambda_1)k^{(2)}(\lambda_1)$, and then we have the following theorem:

Theorem 7.2. *With the same assumptions of Theorem* 7.1, *the operators* $T^{(1)}$ *and* $T^{(2)}$ *are respectively defined by Definitions* 7.2 *and* 7.3, *then we have*

$$||T^{(1)}T^{(2)}|| = ||T^{(1)}|| \cdot ||T^{(2)}|| = k^{(1)}(\lambda_1)k^{(2)}(\lambda_1).$$

In particular,

(i) If $k_\lambda^{(2)}(x,y) = k_\lambda^{(1)}(x,y)$, then $T^{(2)} = T^{(1)}$ and

$$||(T^{(1)})^2|| = ||T^{(1)}||^2 = (k^{(1)}(\lambda_1))^2;$$

(ii) If $T^{(2)} = T_j^{(2)}$ $(j = 1,2)$ is a Hardy-type operator defined by Remark 7.1, then we have

$$||T^{(1)}T_j^{(2)}|| = ||T^{(1)}|| \cdot ||T_j^{(2)}|| = k^{(1)}(\lambda_1)k_j^{(2)}(\lambda_1) \quad (j = 1,2).$$

(iii) If $T^{(1)} = T_i^{(1)}$ $(i = 1, 2)$ is a Hardy-type operator defined by Remark 7.1, then we have

$$||T_i^{(1)} T^{(2)}|| = ||T_i^{(1)}|| \cdot ||T^{(2)}|| = k_i^{(1)}(\lambda_1) k^{(2)}(\lambda_1) \quad (i = 1, 2).$$

(iv) If $T^{(1)} = T_i^{(1)}$ $(i = 1, 2)$, $T^{(2)} = T_j^{(2)}$ $(j = 1, 2)$, then we have

$$||T_i^{(1)} T_j^{(2)}|| = ||T_i^{(1)}|| \cdot ||T_j^{(2)}|| = k_i^{(1)}(\lambda_1) k_j^{(2)}(\lambda_1) \quad (j, i = 1, 2).$$

Example 7.1. (i) For $0 < \lambda \leq 1$, $0 < \lambda_1, \lambda_2 < 1$,

$$k_\lambda^{(i)}(x, y) = \frac{1}{x^\lambda + y^\lambda}, \quad \frac{1}{(x + y)^\lambda}, \quad \frac{\ln(x/y)}{x^\lambda - y^\lambda}, \quad \frac{1}{(\max\{x, y\})^\lambda}$$

and

$$\prod_{k=1}^{s} \frac{1}{x^{\lambda/s} + a_k y^{\lambda/s}} \qquad (i = 1, 2, 3)$$

are satisfied using Theorem 7.1.

(ii) For

$$k_\lambda^{(1)}(x, y) = \frac{1}{x^\lambda + y^\lambda}, \quad k_\lambda^{(2)}(x, y) = \frac{1}{(\max\{x, y\})^\lambda}$$

in Definitions 7.2 and 7.3 and Remark 7.1, it follows

$$(T^{(1)} A_\lambda)(m) = m^{\lambda - 1} \sum_{n=1}^{\infty} \frac{1}{m^\lambda + n^\lambda} A_\lambda(n) \qquad (m \in \mathbf{N}),$$

$$(T^{(2)} a)(n) = n^{\lambda - 1} \sum_{m=1}^{\infty} \frac{1}{(\max\{m, n\})^\lambda} a_m \qquad (n \in \mathbf{N}),$$

$$(T_1^{(2)} a)(n) = \frac{1}{n} \sum_{m=1}^{n} a_m \qquad (n \in \mathbf{N}),$$

$$(T_2^{(2)} a)(n) = n^{\lambda - 1} \sum_{m=n}^{\infty} \frac{1}{m^\lambda} a_m \qquad (n \in \mathbf{N}),$$

then by Theorem 7.1, we have

$$||T^{(1)}T^{(2)}|| = ||T^{(1)}|| \cdot ||T^{(2)}|| = \frac{\pi}{\lambda \sin \pi(\frac{\lambda_1}{\lambda})} \frac{\lambda}{\lambda_1 \lambda_2} = \frac{\pi}{\lambda_1 \lambda_2 \sin \pi(\frac{\lambda_1}{\lambda})},$$

$$||T^{(1)}T_1^{(2)}|| = ||T^{(1)}|| \cdot ||T_1^{(2)}|| = \frac{\pi}{\lambda \sin \pi(\frac{\lambda_1}{\lambda})} \frac{1}{\lambda_1} = \frac{\pi}{\lambda \lambda_1 \sin \pi(\frac{\lambda_1}{\lambda})},$$

$$||T^{(1)}T_2^{(2)}|| = ||T^{(1)}|| \cdot ||T_2^{(2)}|| = \frac{\pi}{\lambda \sin \pi(\frac{\lambda_1}{\lambda})} \frac{1}{\lambda_2} = \frac{\pi}{\lambda \lambda_2 \sin \pi(\frac{\lambda_1}{\lambda})}.$$

Acknowledgements This work is supported by 2012 Knowledge Construction Special Foundation Item of Guangdong Institution of Higher Learning College and University (No. 2012KJCX0079).

References

1. Alladi, K., Milovanovic, G.V., Rassias, M.Th. (eds.): Analytic Number Theory, Approximation Theory and Special Functions. Springer, New York (2013, to appear)
2. Apostol, T.M.: Introduction to Analytic Number Theory. Springer, New York (1984)
3. Arpad, B., Choonghong, O.: Best constant for certain multilinear integral operator. J. Inequal. Appl. **2006**, Article ID 28582, 12 (2006)
4. Azar, L.: On some extensions of Hardy-Hilbert's inequality and Applications. J. Inequal. Appl. **2009**, Article ID 546829, 14 (2009)
5. Edwards, H.M.: Riemann's Zeta Function. Dover Publications, New York (1974)
6. Erdos, P., Suranyi, J.: Topics in the Theory of Numbers. Springer, New York (2003)
7. Gao, M.Z.: On an improvement of Hilbert's inequality extended by Hardy-Riesz. J. Math. Res. Exposition **14**(2), 255–259 (1994)
8. Gao, M.Z., Yang, B.C.: On the extended Hilbert's inequality. Proc. Am. Math. Soc. **126**(3), 751–759 (1998)
9. Hardy, G.H., Wright, E.W.: An Introduction to the Theory of Numbers, 5th edn. Clarendon Press, Oxford (1979)
10. Hardy, G.H., Littlewood, J.E., Pólya, G.: Inequalities. Cambridge University Press, Cambridge (1934)
11. Hong, Y.: On Hardy-Hilbert integral inequalities with some parameters. J. Inequal. Pure Appl. Math.**6**(4), Art 92, 1–10 (2005)
12. Iwaniec, H., Kowalski, E.: Analytic Number Theory. American Mathematical Society Colloquium Publications, vol. 53. American Mathematical Society, Rhode Island (2004)
13. Krnić, M., Pečarić, J.E.: Hilbert's inequalities and their reverses. Publ. Math. Debrecen **67**(3–4), 315–331 (2005)
14. Kuang, J.C.: Introduction to Real Analysis. Hunan Education Press, Chansha, China (1996)
15. Kuang, J.C.: Applied Inequalities. Shangdong Science Technic Press, Jinan, China (2004)
16. Kuang, J.C., Debnath, L.: On Hilbert's type inequalities on the weighted Orlicz spaces. Pacific J. Appl. Math. **1**(1), 95–103 (2007)
17. Landau, E.: Elementary Number Theory, 2nd edn. Chelsea, New York (1966)
18. Li, Y.J., He, B.: On inequalities of Hilbert's type. Bull. Aust. Math. Soc. **76**(1), 1–13 (2007)
19. Miller, S.J., Takloo-Bighash, R.: An Invitation to Modern Number Theory. Princeton University Press, Princeton (2006)
20. Mitrinović, D.S., Pečarić, J.E., Fink, A.M.: Inequalities Involving Functions and Their Integrals and Derivatives. Kluwer, Boston (1991)
21. Pan, Y.L., Wang, H.T., Wang, F.T.: On Complex Functions. Science Press, Beijing (2006)

22. Yang, B.C.: On Hilbert's integral inequality. J. Math. Anal. Appl. **220**, 778–785 (1998)
23. Yang, B.C.: A mixed Hilbert-type inequality with a best constant factor. Int. J. Pure Appl. Math. **20**(3), 319–328 (2005)
24. Yang, B.C.: Hilbert-Type Integral Inequalities. Bentham Science Publishers Ltd., Sharjah (2009)
25. Yang, B.C.: The Norm of Operator and Hilbert-Type Inequalities. Science Press, Beijing (2009)
26. Yang, B.C.: A half-discrete Hilbert-type inequality. J. Guangdong Univ. Educ. **31**(3), 1–7 (2011)
27. Yang, B.C.: Discrete Hilbert-Type Inequalities. Bentham Science Publishers Ltd., Sharjah (2011)
28. Yang, B.C.: A half-discrete Hilbert-type inequality with a non-homogeneous kernel and two variables. Mediterr. J. Math. (2012). doi:10.1007/s00009- 012- 0213-50 (online first)
29. Yang, B.C.: Hilbert-type integral operators: norms and inequalities. In: Paralos, P.M., et al. (eds.) Nonlinear Analysis, Stability, Approximation, and Inequalities, pp. 771–859. Springer, New York (2012)
30. Yang, B.C.: Two Types of Multiple Half-Discrete Hilbert-Type Inequalities. Lambert Academic Publishing, Saarbrüken (2012)
31. Yang, B.C., Chen, Q.: A half-discrete Hilbert-type inequality with a homogeneous kernel and an extension. J. Inequal. Appl. **124** (2011). doi:10.1186/1029-242X-2011-124
32. Yang, B.C., Debnath, L.: On new strengthened Hardy-Hilbert's inequality. Int. J. Math. Math. Soc. **21**(2), 403–408 (1998)
33. Yang, B.C., Gao, M.Z.: On a best value of Hardy-Hilbert's inequality. Adv. Math. **26**(2), 159–164 (1997)
34. Yang, B.C., Krnić, M.: On the norm of a multi-dimensional Hilbert-type operator. Sarajevo J. Math. **7**(20), 223–243 (2011)
35. Yang, B.C., Rassias, Th.M.: On the way of weight coefficient and research for Hilbert-type inequalities. Math. Inequal. Appl. **6**(4), 625–658 (2003)
36. Yang, B.C., Rassias, Th.M.: On a Hilbert-type integral inequality in the subinterval and its operator expression. Banach J. Math. Anal. **4**(2), 100–110 (2010)
37. Yang, B.C., Brnetić, I., Krnić, M., Pečarić, J.E.: Generalization of Hilbert and Hardy-Hilbert integral inequalities. Math. Inequal. Appl. **8**(2), 259–272 (2005)
38. Zhao, D.J.: On a refinement of Hilbert double series theorem. Math. Practices Theory, **23**(1), 85–90 (1993)
39. Zhong, W.Y.: The Hilbert-type integral inequality with a homogeneous kernel of Lambda-degree. J. Inequal. Appl. **2008**, Article ID 917392, 13 (2008)
40. Zhong, W.Y.: A mixed Hilbert-type inequality and its equivalent forms. J. Guangdong Univ. Educ. **31**(5), 18–22 (2011)
41. Zhong, W.Y.: A half discrete Hilbert-type inequality and its equivalent forms. J. Guangdong Univ. Educ. **32**(5), 8–12 (2012)
42. Zhong, J.H.: Two classes of half-discrete reverse Hilbert-type inequalities with a non-homogeneous kernel. J. Guangdong Univ. Educ. **32**(5), 11–20 (2012)
43. Zhong, W.Y., Yang, B.C.: A best extension of Hilbert inequality involving several parameters. J. Jinan Univ. (Nat. Sci.) **28**(1), 20–23 (2007)
44. Zhong, W.Y., Yang, B.C.: On multiple Hardy-Hilbert's integral inequality with kernel. J. Inequal. Appl. **2007**, Article ID 27962, 17 (2007). doi:10.1155/ 2007/27
45. Zhong, W.Y., Yang, B.C.: A reverse Hilbert's type integral inequality with some parameters and the equivalent forms. Pure Appl. Math. **24**(2), 401–407 (2008)
46. Zhong, J.H., Yang, B.C.: On an extension of a more accurate Hilbert-type inequality. J. Zhejiang Univ. (Sci. Ed.) **35**(2), 121–124 (2008)

The Function $(b^x - a^x)/x$: Ratio's Properties

Feng Qi, Qiu-Ming Luo, and Bai-Ni Guo

Dedicated to Professor Hari M. Srivastava

Abstract In the present paper, after reviewing the history, background, origin, and applications of the functions $\frac{b^t - a^t}{t}$ and $\frac{e^{-\alpha t} - e^{-\beta t}}{1 - e^{-t}}$, we establish sufficient and necessary conditions such that the special function $\frac{e^{\alpha t} - e^{\beta t}}{e^{\lambda t} - e^{\mu t}}$ is monotonic, logarithmic convex, logarithmic concave, 3-log-convex, and 3-log-concave on $\mathbb{R}$, where α, β, λ, and μ are real numbers satisfying $(\alpha, \beta) \neq (\lambda, \mu)$, $(\alpha, \beta) \neq (\mu, \lambda)$, $\alpha \neq \beta$, and $\lambda \neq \mu$.

1 Introduction

Recall from [43] that a k-times differentiable function $f(t) > 0$ is said to be k-log-convex on an interval I if

$$0 \leqslant [\ln f(t)]^{(k)} < \infty, \quad k \in \mathbb{N} \tag{1}$$

on I; if the inequality (1) reverses, then f is said to be k-log-concave on I.

F. Qi (✉) • B.-N. Guo
School of Mathematics and Informatics, Henan Polytechnic University,
Jiaozuo City, Henan Province, 454010, China
e-mail: qifeng618@gmail.com; qifeng618@hotmail.com; qifeng618@qq.com;
bai.ni.guo@gmail.com; bai.ni.guo@hotmail.com; URL: http://qifeng618.wordpress.com

Q.-M. Luo
Department of Mathematics, Chongqing Normal University, Chongqing City, 401331, China
e-mail: luomath@126.com; luomath2007@163.com

G.V. Milovanović and M.Th. Rassias (eds.), *Analytic Number Theory, Approximation Theory, and Special Functions*, DOI 10.1007/978-1-4939-0258-3_16,

For $b > a > 0$, let

$$G_{a,b}(t) = \begin{cases} \dfrac{b^t - a^t}{t}, & t \neq 0; \\ \ln b - \ln a, & t = 0. \end{cases}$$

In [37, 38], the complete monotonicity and inequality properties of $G_{a,b}(t)$ were first investigated. In [1, 9, 13, 28, 33, 34], the 3-log-convex and 3-log-concave properties of $G_{a,b}(t)$ were shown. The function $G_{a,b}(t)$ has close relationships with the incomplete gamma function [21,25]. It was ever used to prove the Schur-convex properties [13,25,34], the logarithmic convexities [4,13,20,28,34], and the monotonicity [36, 39] of the extended mean values (for more information, please refer to [2,21] and closely related references therein). It was applied in [7,29,30,40] to construct Steffensen pairs. It was also employed in [41] to verify Elezović-Giordano-Pečarić's theorem [6, Theorem 1] which is related to the monotonicity of a function involving the ratio of two gamma functions. Some more applications were further established in [31,32] recently.

For $b > a > 0$, let

$$F_{a,b}(t) = \begin{cases} \dfrac{t}{e^{bt} - e^{at}}, & t \neq 0; \\ \dfrac{1}{b-a}, & t = 0. \end{cases}$$

In [3, 15, 18, 22, 44–46], [5, p. 217], and [16, p. 295], the inequalities, monotonicity, and logarithmic convexities of the function $F_{a,b}(t)$ for $a = b - 1$ and its logarithmic derivatives of the first and second orders are established. In [18], the history, background, and origin of $F_{a,b}(t)$ for $a = b - 1$ and its first two logarithmic derivatives were cultivated. In [11, 42], the logarithmic derivative of $F_{a,b}(t)$ for $a = b - 1$ was applied to study the complete monotonicity of remainders of the first Binet formula and the psi function. In [8,17,19], the function $F_{\ln a,\ln b}(t)$ was utilized to generalize Bernoulli numbers and polynomials. In [9, 33], the 3-log-convex and 3-log-concave properties of $F_{a,b}(t)$ were shown, among other things.

For real numbers α and β satisfying $\alpha \neq \beta$, $(\alpha, \beta) \neq (0, 1)$, and $(\alpha, \beta) \neq (1, 0)$, let

$$Q_{\alpha,\beta}(t) = \begin{cases} \dfrac{e^{-\alpha t} - e^{-\beta t}}{1 - e^{-t}}, & t \neq 0; \\ \beta - \alpha, & t = 0. \end{cases}$$

In [12, 23, 41], the monotonicity and logarithmic convexities of $Q_{\alpha,\beta}(t)$ were discussed and the following conclusions were procured:

1. The function $Q_{\alpha,\beta}(t)$ is increasing on $(0, \infty)$ if and only if $(\beta-\alpha)(1-\alpha-\beta) \geqslant 0$ and $(\beta - \alpha)(|\alpha - \beta| - \alpha - \beta) \geqslant 0$.
2. The function $Q_{\alpha,\beta}(t)$ is decreasing on $(0, \infty)$ if and only if $(\beta-\alpha)(1-\alpha-\beta) \leqslant 0$ and $(\beta - \alpha)(|\alpha - \beta| - \alpha - \beta) \leqslant 0$.

3. The function $Q_{\alpha,\beta}(t)$ is increasing on $(-\infty,0)$ if and only if $(\beta-\alpha)(1-\alpha-\beta) \geqslant 0$ and $(\beta-\alpha)(2-|\alpha-\beta|-\alpha-\beta) \geqslant 0$.
4. The function $Q_{\alpha,\beta}(t)$ is decreasing on $(-\infty,0)$ if and only if $(\beta-\alpha)(1-\alpha-\beta) \leqslant 0$ and $(\beta-\alpha)(2-|\alpha-\beta|-\alpha-\beta) \leqslant 0$.
5. The function $Q_{\alpha,\beta}(t)$ is increasing on $(-\infty,\infty)$ if and only if $(\beta-\alpha)(|\alpha-\beta|-\alpha-\beta) \geqslant 0$ and $(\beta-\alpha)(2-|\alpha-\beta|-\alpha-\beta) \geqslant 0$.
6. The function $Q_{\alpha,\beta}(t)$ is decreasing on $(-\infty,\infty)$ if and only if $(\beta-\alpha)(|\alpha-\beta|-\alpha-\beta) \leqslant 0$ and $(\beta-\alpha)(2-|\alpha-\beta|-\alpha-\beta) \leqslant 0$.
7. The function $Q_{\alpha,\beta}(t)$ on $(-\infty,\infty)$ is logarithmically convex if $\beta-\alpha > 1$ and logarithmically concave if $0 < \beta-\alpha < 1$.
8. If $1 > \beta-\alpha > 0$, then $Q_{\alpha,\beta}(t)$ is 3-log-convex on $(0,\infty)$ and 3-log-concave on $(-\infty,0)$; if $\beta-\alpha > 1$, then $Q_{\alpha,\beta}(t)$ is 3-log-concave on $(0,\infty)$ and 3-log-convex on $(-\infty,0)$.

The monotonicity of $Q_{\alpha,\beta}(t)$ on $(0,\infty)$ was applied in [12, 24, 35] to present necessary and sufficient conditions such that some functions involving ratios of the gamma and q-gamma functions are logarithmically completely monotonic. The logarithmic convexities of $Q_{\alpha,\beta}(t)$ on $(0,\infty)$ were used in [10,41] to supply alternative proofs for Elezović-Giordano-Pečarić's theorem. For detailed information, please refer to [26,27] and related references therein.

The functions $G_{a,b}(t)$, $F_{a,b}(t)$, and $Q_{\alpha,\beta}(t)$ have the following relations:

$$G_{a,b}(t) = \frac{1}{F_{\ln a,\ln b}(t)}, \qquad F_{a,b}(t) = \frac{1}{G_{e^b,e^a}(t)},$$

$$Q_{\alpha,\beta}(t) = \frac{G_{e^{-\alpha},e^{-\beta}}(t)}{G_{1,e^{-1}}(t)}, \quad Q_{\alpha,\beta}(t) = \frac{F_{0,-1}(t)}{F_{-\alpha,-\beta}(t)}.$$

For real numbers α, β, λ, and μ satisfying $(\alpha,\beta) \neq (\lambda,\mu)$, $(\alpha,\beta) \neq (\mu,\lambda)$, $\alpha \neq \beta$, and $\lambda \neq \mu$, let

$$H_{\alpha,\beta;\lambda,\mu}(t) = \begin{cases} \dfrac{e^{\alpha t} - e^{\beta t}}{e^{\lambda t} - e^{\mu t}}, & t \neq 0, \\[2ex] \dfrac{\beta-\alpha}{\lambda-\mu}, & t = 0. \end{cases}$$

For positive numbers r, s, u, and v satisfying $(r,s) \neq (u,v)$, $(r,s) \neq (v,u)$, $r \neq s$, and $u \neq v$, let

$$P_{r,s;u,v}(t) = \begin{cases} \dfrac{r^t - s^t}{u^t - v^t}, & t \neq 0, \\[2ex] \dfrac{\ln r - \ln s}{\ln u - \ln v}, & t = 0. \end{cases}$$

It is clear that

$$H_{\alpha,\beta;\lambda,\mu}(t) = P_{e^\alpha,e^\beta;e^\lambda,e^\mu}(t) \tag{2}$$

and

$$P_{r,s;u,v}(t) = H_{\ln r,\ln s;\ln u,\ln v}(t).$$

In addition, the functions $H_{\alpha,\beta;\lambda,\mu}(t)$ and $P_{r,s;u,v}(t)$ can be represented as

$$H_{\alpha,\beta;\lambda,\mu}(t) = \frac{F_{\lambda,\mu}(t)}{F_{\alpha,\beta}(t)} = \frac{Q_{-\alpha,-\beta}}{Q_{-\lambda,-\mu}} \quad \text{and} \quad P_{r,s;u,v}(t) = \frac{G_{r,s}(t)}{G_{u,v}(t)},$$

the ratios of $G_{a,b}(t)$, $F_{a,b}(t)$, and $Q_{\alpha,\beta}(t)$.

Since the functions $G_{a,b}(t)$, $F_{a,b}(t)$, and $Q_{\alpha,\beta}(t)$ have a long history, a deep background, and many applications to several areas, we continue to study the monotonicity and logarithmic convexities of their ratios, $H_{\alpha,\beta;\lambda,\mu}(t)$ and $P_{r,s;u,v}(t)$.

Our main results may be stated as the following theorems.

Theorem 1.1. *For real numbers α, β, λ, and μ with $(\alpha,\beta) \neq (\lambda,\mu)$, $(\alpha,\beta) \neq (\mu,\lambda)$, $\alpha \neq \beta$, and $\lambda \neq \mu$, let*

$$\mathcal{A} = (\alpha-\beta)(\alpha+\beta-\lambda-\mu), \quad \mathcal{B} = (\alpha-\beta)(\alpha+\beta-|\alpha-\beta|-2\lambda),$$

$$\mathcal{C} = (\alpha-\beta)(\alpha+\beta+|\alpha-\beta|-2\lambda), \quad \mathcal{D} = (\alpha-\beta)(\alpha+\beta+|\alpha-\beta|-2\mu),$$

$$\mathcal{E} = (\alpha-\beta)(\alpha+\beta-|\alpha-\beta|-2\mu).$$

Then the function $H_{\alpha,\beta;\lambda,\mu}(t)$ has the following properties:

1. *The function $H_{\alpha,\beta;\lambda,\mu}(t)$ is increasing on $(0,\infty)$ if and only if either $\lambda > \mu$, $\mathcal{A} \geqslant 0$, and $\mathcal{C} \geqslant 0$ or $\lambda < \mu$, $\mathcal{A} \leqslant 0$, and $\mathcal{B} \leqslant 0$.*
2. *The function $H_{\alpha,\beta;\lambda,\mu}(t)$ is decreasing on $(0,\infty)$ if and only if either $\lambda < \mu$, $\mathcal{A} \geqslant 0$, and $\mathcal{B} \geqslant 0$ or $\lambda > \mu$, $\mathcal{A} \leqslant 0$, and $\mathcal{C} \leqslant 0$.*
3. *The function $H_{\alpha,\beta;\lambda,\mu}(t)$ is increasing on $(-\infty,0)$ if and only if either $\lambda > \mu$, $\mathcal{A} \geqslant 0$, and $\mathcal{E} \geqslant 0$ or $\lambda < \mu$, $\mathcal{A} \leqslant 0$, and $\mathcal{D} \leqslant 0$.*
4. *The function $H_{\alpha,\beta;\lambda,\mu}(t)$ is decreasing on $(-\infty,0)$ if and only if either $\lambda > \mu$, $\mathcal{A} \leqslant 0$, and $\mathcal{E} \leqslant 0$ or $\lambda < \mu$, $\mathcal{A} \geqslant 0$, and $\mathcal{D} \geqslant 0$.*
5. *The function $H_{\alpha,\beta;\lambda,\mu}(t)$ is increasing on $(-\infty,\infty)$ if and only if either $\lambda > \mu$, $\mathcal{C} \geqslant 0$, and $\mathcal{E} \geqslant 0$ or $\lambda < \mu$, $\mathcal{B} \leqslant 0$, and $\mathcal{D} \leqslant 0$.*
6. *The function $H_{\alpha,\beta;\lambda,\mu}(t)$ is decreasing on $(-\infty,\infty)$ if and only if either $\lambda > \mu$, $\mathcal{C} \leqslant 0$, and $\mathcal{E} \leqslant 0$ or $\lambda < \mu$, $\mathcal{B} \geqslant 0$, and $\mathcal{D} \geqslant 0$.*
7. *The function $H_{\alpha,\beta;\lambda,\mu}(t)$ on $(-\infty,\infty)$ is logarithmically convex if $\frac{\alpha-\beta}{\lambda-\mu} > 1$ or logarithmically concave if $0 < \frac{\alpha-\beta}{\lambda-\mu} < 1$.*

8. *The function* $H_{\alpha,\beta;\lambda,\mu}(t)$ *is 3-log-convex on* $(0,\infty)$ *and 3-log-concave on* $(-\infty,0)$ *if either* $\lambda-\mu>\alpha-\beta>0$ *or* $\alpha-\beta<\lambda-\mu<0$*; the function* $H_{\alpha,\beta;\lambda,\mu}(t)$ *is 3-log-concave on* $(0,\infty)$ *and 3-log-convex on* $(-\infty,0)$ *if either* $\alpha-\beta>\lambda-\mu>0$ *or* $\lambda-\mu<\alpha-\beta<0$.

Theorem 1.2. *For positive numbers* $r,s,u,$ *and* v *with* $(r,s)\neq(u,v)$, $(r,s)\neq(v,u)$, $r\neq s$, *and* $u\neq v$, *let*

$$\mathfrak{A}=\ln\frac{rs}{uv}\ln\frac{r}{s},\quad \mathfrak{B}=\left(\ln\frac{rs}{u^2}-\left|\ln\frac{r}{s}\right|\right)\ln\frac{r}{s},\quad \mathfrak{C}=\left(\ln\frac{rs}{u^2}+\left|\ln\frac{r}{s}\right|\right)\ln\frac{r}{s},$$

$$\mathfrak{D}=\left(\ln\frac{rs}{v^2}+\left|\ln\frac{r}{s}\right|\right)\ln\frac{r}{s},\quad \mathfrak{E}=\left(\ln\frac{rs}{v^2}-\left|\ln\frac{r}{s}\right|\right)\ln\frac{r}{s}.$$

Then the function $P_{r,s;u,v}(t)$ *has the following properties:*

1. *The function* $P_{r,s;u,v}(t)$ *is increasing on* $(0,\infty)$ *if and only if either* $u>v$, $\mathfrak{A}\geqslant0$, *and* $\mathfrak{C}\geqslant0$ *or* $u<v$, $\mathfrak{A}\leqslant0$, *and* $\mathfrak{B}\leqslant0$.
2. *The function* $P_{r,s;u,v}(t)$ *is decreasing on* $(0,\infty)$ *if and only if either* $u<v$, $\mathfrak{A}\geqslant0$, *and* $\mathfrak{B}\geqslant0$ *or* $u>v$, $\mathfrak{A}\leqslant0$, *and* $\mathfrak{C}\leqslant0$.
3. *The function* $P_{r,s;u,v}(t)$ *is increasing on* $(-\infty,0)$ *if and only if* $u>v$, $\mathfrak{A}\geqslant0$, *and* $\mathfrak{E}\geqslant0$, *or* $u<v$, $\mathfrak{A}\leqslant0$, *and* $\mathfrak{D}\leqslant0$.
4. *The function* $P_{r,s;u,v}(t)$ *is decreasing on* $(-\infty,0)$ *if and only if either* $u>v$, $\mathfrak{A}\leqslant0$, *and* $\mathfrak{E}\leqslant0$ *or* $u<v$, $\mathfrak{A}\geqslant0$, *and* $\mathfrak{D}\geqslant0$.
5. *The function* $P_{r,s;u,v}(t)$ *is increasing on* $(-\infty,\infty)$ *if and only if either* $u>v$, $\mathfrak{C}\geqslant0$, *and* $\mathfrak{E}\geqslant0$ *or* $u<v$, $\mathfrak{B}\leqslant0$, *and* $\mathfrak{D}\leqslant0$.
6. *The function* $P_{r,s;u,v}(t)$ *is decreasing on* $(-\infty,\infty)$ *if and only if either* $u>v$, $\mathfrak{C}\leqslant0$, *and* $\mathfrak{E}\leqslant0$ *or* $u<v$, $\mathfrak{B}\geqslant0$, *and* $\mathfrak{D}\geqslant0$.
7. *The function* $P_{r,s;u,v}(t)$ *on* $(-\infty,\infty)$ *is logarithmically convex if* $\frac{\ln(r/s)}{\ln(u/v)}>1$ *or logarithmically concave if* $0<\frac{\ln(r/s)}{\ln(u/v)}<1$.
8. *The function* $P_{r,s;u,v}(t)$ *is 3-log-convex on* $(0,\infty)$ *and 3-log-concave on* $(-\infty,0)$ *if* $\frac{u}{v}>\frac{r}{s}>1$ *or* $\frac{r}{s}<\frac{u}{v}<1$*; the function* $P_{r,s;u,v}(t)$ *is 3-log-concave on* $(0,\infty)$ *and 3-log-convex on* $(-\infty,0)$ *if* $\frac{r}{s}>\frac{u}{v}>1$ *or* $\frac{u}{v}<\frac{r}{s}<1$.

Remark 1.1. The monotonicity of the functions $H_{\alpha,\beta;\lambda,\mu}(t)$ and $P_{r,s;u,v}(t)$ can be described by Table 1 below.

Remark 1.2. In [23, Remark 2.1] it was remarked that the function $Q_{\alpha,\beta}(t)$ cannot be either 4-log-convex or 4-log-concave in either $(-\infty,0)$ or $(0,\infty)$, saying nothing of $(-\infty,\infty)$. Therefore, neither $H_{\alpha,\beta;\lambda,\mu}(t)$ nor $P_{r,s;u,v}(t)$ is either 4-log-convex or 4-log-concave on either $(-\infty,0)$ or $(0,\infty)$, saying nothing of $(-\infty,\infty)$.

Table 1 Monotonicity of the functions $H_{\alpha,\beta;\lambda,\mu}(t)$ and $P_{r,s;u,v}(t)$

Intervals	Monotonicity	$\mathcal{A}$ or $\mathfrak{A}$	$\mathcal{B}$ or $\mathfrak{B}$	$\mathcal{C}$ or $\mathfrak{C}$	$\mathcal{D}$ or $\mathfrak{D}$	$\mathcal{E}$ or $\mathfrak{E}$	λ and μ or u and v
$(0,\infty)$	Increasing	$\geqslant 0$		$\geqslant 0$			$\lambda>\mu$ or $u>v$
$(0,\infty)$	Increasing	$\leqslant 0$	$\leqslant 0$				$\lambda<\mu$ or $u<v$
$(0,\infty)$	Decreasing	$\geqslant 0$	$\geqslant 0$				$\lambda<\mu$ or $u<v$
$(0,\infty)$	Decreasing	$\leqslant 0$		$\leqslant 0$			$\lambda>\mu$ or $u>v$
$(-\infty,0)$	Increasing	$\geqslant 0$				$\geqslant 0$	$\lambda>\mu$ or $u>v$
$(-\infty,0)$	Increasing	$\leqslant 0$			$\leqslant 0$		$\lambda<\mu$ or $u<v$
$(-\infty,0)$	Decreasing	$\leqslant 0$				$\leqslant 0$	$\lambda>\mu$ or $u>v$
$(-\infty,0)$	Decreasing	$\geqslant 0$			$\geqslant 0$		$\lambda<\mu$ or $u<v$
$(-\infty,\infty)$	Increasing			$\geqslant 0$		$\geqslant 0$	$\lambda>\mu$ or $u>v$
$(-\infty,\infty)$	Increasing		$\leqslant 0$		$\leqslant 0$		$\lambda<\mu$ or $u<v$
$(-\infty,\infty)$	Decreasing			$\leqslant 0$		$\leqslant 0$	$\lambda>\mu$ or $u>v$
$(-\infty,\infty)$	Decreasing		$\geqslant 0$		$\geqslant 0$		$\lambda<\mu$ or $u<v$

2 Proofs of Theorems

Proof of Theorem 1.1. For $t\neq 0$, the function $H_{\alpha,\beta;\lambda,\mu}(t)$ can rewritten as

$$H_{\alpha,\beta;\lambda,\mu}(t)=\frac{e^{(\alpha-\lambda)t}-e^{(\beta-\lambda)t}}{1-e^{(\mu-\lambda)t}}=\frac{e^{-\frac{\alpha-\lambda}{\mu-\lambda}w}-e^{-\frac{\beta-\lambda}{\mu-\lambda}w}}{1-e^{-w}}=\frac{e^{-Aw}-e^{-Bw}}{1-e^{-w}},$$

where

$$A=\frac{\alpha-\lambda}{\mu-\lambda},\quad B=\frac{\beta-\lambda}{\mu-\lambda},\quad w=(\lambda-\mu)t.$$

Differentiating with respect to t yields

$$H'_{\alpha,\beta;\lambda,\mu}(t)=(\lambda-\mu)Q'_{A,B}(w), \tag{3}$$

$$\begin{aligned}[\ln H_{\alpha,\beta;\lambda,\mu}(t)]''&=\left[\frac{H'_{\alpha,\beta;\lambda,\mu}(t)}{H_{\alpha,\beta;\lambda,\mu}(t)}\right]'=(\lambda-\mu)\frac{\mathrm{d}}{\mathrm{d}t}\left[\frac{Q'_{A,B}(w)}{Q_{A,B}(w)}\right]\\&=(\lambda-\mu)^2\left[\frac{Q'_{A,B}(w)}{Q_{A,B}(w)}\right]'=(\lambda-\mu)^2[\ln Q_{A,B}(w)]'',\end{aligned} \tag{4}$$

and

$$[\ln H_{\alpha,\beta;\lambda,\mu}(t)]'''=(\lambda-\mu)^3\left[\frac{Q'_{A,B}(w)}{Q_{A,B}(w)}\right]''=(\lambda-\mu)^3[\ln Q_{A,B}(w)]'''. \tag{5}$$

By virtue of [12, Theorem 3.1] or [41, Lemma 1] and the second-order derivative (4), it is easy to deduce that the function $H_{\alpha,\beta;\lambda,\mu}(t)$ is logarithmically convex if $\frac{\beta-\alpha}{\mu-\lambda} > 1$ and logarithmically concave if $0 < \frac{\beta-\alpha}{\mu-\lambda} < 1$ on $(-\infty, \infty)$.

By virtue of [23, Theorem 1.1] and the third-order derivative (5), it is not difficult to obtain the following:

1. If $\lambda > \mu$ and $1 > \frac{\beta-\alpha}{\mu-\lambda} > 0$, then $H_{\alpha,\beta;\lambda,\mu}(t)$ is 3-log-convex on $(0, \infty)$ and 3-log-concave on $(-\infty, 0)$.
2. If $\lambda > \mu$ and $\frac{\beta-\alpha}{\mu-\lambda} > 1$, then $H_{\alpha,\beta;\lambda,\mu}(t)$ is 3-log-concave on $(0, \infty)$ and 3-log-convex on $(-\infty, 0)$.
3. If $\lambda < \mu$ and $1 > \frac{\beta-\alpha}{\mu-\lambda} > 0$, then $H_{\alpha,\beta;\lambda,\mu}(t)$ is 3-log-concave on $(0, \infty)$ and 3-log-convex on $(-\infty, 0)$.
4. If $\lambda < \mu$ and $\frac{\beta-\alpha}{\mu-\lambda} > 1$, then $H_{\alpha,\beta;\lambda,\mu}(t)$ is 3-log-convex on $(0, \infty)$ and 3-log-concave on $(-\infty, 0)$.

Direct computation gives

$$(B-A)(1-A-B) = \frac{\mathcal{A}}{(\lambda-\mu)^2}; \quad (B-A)(|A-B|-A-B) = \begin{cases} \frac{\mathcal{B}}{(\lambda-\mu)^2}, & \lambda < \mu, \\ \frac{\mathcal{C}}{(\lambda-\mu)^2}, & \lambda > \mu; \end{cases}$$

$$(B-A)(2-|A-B|-A-B) = \begin{cases} \frac{\mathcal{D}}{(\lambda-\mu)^2}, & \lambda < \mu, \\ \frac{\mathcal{E}}{(\lambda-\mu)^2}, & \lambda > \mu. \end{cases}$$

Consequently, utilization of [12, Theorem 2.3] and the first-order derivative (3) yields the following conclusions:

1. The function $H_{\alpha,\beta;\lambda,\mu}(t)$ is increasing on $(0, \infty)$ if and only if $\lambda > \mu$, $\mathcal{A} \geqslant 0$, and $\mathcal{C} \geqslant 0$; the function $H_{\alpha,\beta;\lambda,\mu}(t)$ is decreasing on $(0, \infty)$ if and only if $\lambda < \mu$, $\mathcal{A} \geqslant 0$, and $\mathcal{B} \geqslant 0$.
2. The function $H_{\alpha,\beta;\lambda,\mu}(t)$ is decreasing on $(0, \infty)$ if and only if $\lambda > \mu$, $\mathcal{A} \leqslant 0$, and $\mathcal{C} \leqslant 0$; the function $H_{\alpha,\beta;\lambda,\mu}(t)$ is increasing on $(0, \infty)$ if and only if $\lambda < \mu$, $\mathcal{A} \leqslant 0$, and $\mathcal{B} \leqslant 0$.
3. The function $H_{\alpha,\beta;\lambda,\mu}(t)$ is increasing on $(-\infty, 0)$ if and only if $\lambda > \mu$, $\mathcal{A} \geqslant 0$, and $\mathcal{E} \geqslant 0$; the function $H_{\alpha,\beta;\lambda,\mu}(t)$ is decreasing on $(-\infty, 0)$ if and only if $\lambda < \mu$, $\mathcal{A} \geqslant 0$, and $\mathcal{D} \geqslant 0$.
4. The function $H_{\alpha,\beta;\lambda,\mu}(t)$ is decreasing on $(-\infty, 0)$ if and only if $\lambda > \mu$, $\mathcal{A} \leqslant 0$, and $\mathcal{E} \leqslant 0$; the function $H_{\alpha,\beta;\lambda,\mu}(t)$ is increasing on $(-\infty, 0)$ if and only if $\lambda < \mu$, $\mathcal{A} \leqslant 0$, and $\mathcal{D} \leqslant 0$.

5. The function $H_{\alpha,\beta;\lambda,\mu}(t)$ is increasing on $(-\infty,\infty)$ if and only if $\lambda > \mu$, $\mathcal{C} \geqslant 0$, and $\mathcal{E} \geqslant 0$; the function $H_{\alpha,\beta;\lambda,\mu}(t)$ is decreasing on $(-\infty,\infty)$ if and only if $\lambda < \mu$, $\mathcal{B} \geqslant 0$, and $\mathcal{D} \geqslant 0$.
6. The function $H_{\alpha,\beta;\lambda,\mu}(t)$ is decreasing on $(-\infty,\infty)$ if and only if $\lambda > \mu$, $\mathcal{C} \leqslant 0$, and $\mathcal{E} \leqslant 0$; the function $H_{\alpha,\beta;\lambda,\mu}(t)$ is increasing on $(-\infty,\infty)$ if and only if $\lambda < \mu$, $\mathcal{B} \leqslant 0$, and $\mathcal{D} \leqslant 0$.

The proof of Theorem 1.1 is complete.

Proof of Theorem 1.2. This follows directly from the combination of Theorem 1.1 with equations in (2). Theorem 1.2 is proved. □

Remark 2.1. This article is a slightly revised version of the preprint [14].

Acknowledgements The first author was partially supported by the China Scholarship Council and the Science Foundation of Tianjin Polytechnic University. The second author was supported in part by the Natural Science Foundation Project of Chongqing under Grant CSTC2011JJA00024, the Research Project of Science and Technology of Chongqing Education Commission under Grant KJ120625, and the Fund of Chongqing Normal University under Grant 10XLR017 and 2011XLZ07, China.

References

1. Berenhaut, K.S., Chen, D.-H.: Inequalities for 3-log-convex functions. J. Inequal. Pure Appl. Math. **9**(4), (2008) Art. 97. http://www.emis.de/journals/JIPAM/article1033.html.
2. Bullen, P.S.: Handbook of means and their inequalities. Mathematics and its Applications (Dordrecht) vol. 560, Kluwer Academic Publishers, Dordrecht (2003)
3. Chen, C.-P., Qi, F.: Best constant in an inequality connected with exponential functions. Octogon Math. Mag. **12**(2), 736–737 (2004)
4. Cheung, W.-S., Qi, F.: Logarithmic convexity of the one-parameter mean values. Taiwanese J. Math. **11**(1), 231–237 (2007)
5. de Souza, P.N., Silva, J.-N.:Berkeley problems in mathematics. Problem Books in Mathematics, 2nd edn., Springer, New York (2001)
6. Elezović, N., Giordano, C., Pečarić, J.: The best bounds in Gautschi's inequality. Math. Inequal. Appl. **3**, 239–252 (2000) http://dx.doi.org/10.7153/mia-03-26
7. Gauchman, H.: Steffensen pairs and associated inequalities. J. Inequal. Appl. **5**(1), 53–61 (2000) http://dx.doi.org/10.1155/S1025583400000047
8. Guo, B.-N., Qi, F.: Generalization of bernoulli polynomials. Internat. J. Math. Ed. Sci. Tech. **33**(3), 428–431 (2002). http://dx.doi.org/10.1080/002073902760047913.
9. Guo, B.-N., Qi, F.: A simple proof of logarithmic convexity of extended mean values. Numer. Algorithms **52**(1), 89–92 (2009). http://dx.doi.org/10.1007/s11075-008-9259-7.
10. Guo, B.-N., Qi, F.: An alternative proof of Elezović-Giordano-Pečarić's theorem. Math. Inequal. Appl. **14**(1), 73–78 (2011) http://dx.doi.org/10.7153/mia-14-06
11. Guo, S., Qi, F.: A class of completely monotonic functions related to the remainder of Binet's formula with applications. Tamsui Oxf. J. Math. Sci. **25**(1), 9–14 (2009)
12. Guo, B.-N., Qi, F.: Properties and applications of a function involving exponential functions. Commun. Pure Appl. Anal. **8**(4), 1231–1249 (2009). http://dx.doi.org/10.3934/cpaa.2009.8.1231

13. Guo, B.-N., Qi, F.: The function $(b^x - a^x)/x$: Logarithmic convexity and applications to extended mean values. Filomat **25**(4), 63–73 (2011). http://dx.doi.org/10.2298/FIL1104063G
14. Guo, B.-N., Qi, F.: The function $(b^x - a^x)/x$: Ratio's properties. http://arxiv.org/abs/0904.1115
15. Guo, B.-N., Liu, A.-Q., Qi, F.: Monotonicity and logarithmic convexity of three functions involving exponential function. J. Korea Soc. Math. Educ. Ser. B Pure Appl. Math. **15**(4), 387–392 (2008)
16. Kuang, J.-C.: Chángyòng Bùděngshì (Applied Inequalities), 3rd ed., Shandong Science and Technology Press, Ji'nan City, Shandong Province, China (2004) (Chinese)
17. Luo, Q.-M., Qi, F.: Relationships between generalized Bernoulli numbers and polynomials and generalized Euler numbers and polynomials. Adv. Stud. Contemp. Math. (Kyungshang) **7**(1), 11–18 (2003)
18. Liu, A.-Q., Li, G.-F., Guo, B.-N., Qi, F.: Monotonicity and logarithmic concavity of two functions involving exponential function. Internat. J. Math. Ed. Sci. Tech. **39**(5), 686–691 (2008). http://dx.doi.org/10.1080/00207390801986841
19. Luo, Q.-M., Guo, B.-N., Qi, F., Debnath, L.: Generalizations of Bernoulli numbers and polynomials. Int. J. Math. Math. Sci. **2003**(59), 3769–3776 (2003). http://dx.doi.org/10.1155/S0161171203112070
20. Qi, F.: Logarithmic convexity of extended mean values. Proc. Amer. Math. Soc. **130**(6), 1787–1796 (2002). http://dx.doi.org/10.1090/S0002-9939-01-06275-X
21. Qi, F.: The extended mean values: Definition, properties, monotonicities, comparison, convexities, generalizations, and applications. Cubo Mat. Educ. **5**(3), 63–90 (2003)
22. Qi, F.: A monotonicity result of a function involving the exponential function and an application, RGMIA Res. Rep. Coll. **7**(3), Art. 16, 507–509 (2004). http://rgmia.org/v7n3.php
23. Qi, F.: Three-log-convexity for a class of elementary functions involving exponential function. J. Math. Anal. Approx. Theory **1**(2), 100–103 (2006)
24. Qi, F.: A class of logarithmically completely monotonic functions and the best bounds in the first Kershaw's double inequality. J. Comput. Appl. Math. **206**(2), 1007–1014 (2007). http://dx.doi.org/10.1016/j.cam.2006.09.005
25. Qi, F.: A note on Schur-convexity of extended mean values. Rocky Mountain J. Math. **35** (5), 1787–1793 (2005). http://dx.doi.org/10.1216/rmjm/1181069663
26. Qi, F., Luo, Q.-M.: Bounds for the ratio of two gamma functions From Wendel's and related inequalities to logarithmically completely monotonic functions. Banach J. Math. Anal. **6**(2), 132–158 (2012)
27. Qi, F., Luo, Q.-M.: Bounds for the ratio of two gamma functions: from Wendel's asymptotic relation to Elezović-Giordano-Pečarić's theorem. J. Inequalities Appl. **2013**(542), 20 (2013). http://dx.doi.org/10.1186/1029-242X-2013-542
28. Qi, F., Cerone, P., Dragomir, S.S., Srivastava, H.M.: Alternative proofs for monotonic and logarithmically convex properties of one-parameter mean values. Appl. Math. Comput. **208**(1), 129–133 (2009). http://dx.doi.org/10.1016/j.amc.2008.11.023
29. Qi, F., Cheng, J.-X.: Some new Steffensen pairs. Anal. Math. **29**(3), 219–226 (2003). http://dx.doi.org/10.1023/A:1025467221664
30. Qi, F., Guo, B.-N.: On Steffensen pairs. J. Math. Anal. Appl. **271**(2), 534–541 (2002). http://dx.doi.org/10.1016/S0022-247X(02)00120-8
31. Qi, F., Guo, B.-N.: Some properties of extended remainder of Binet's first formula for logarithm of gamma function. Math. Slovaca **60**(4), 461–470 (2010). http://dx.doi.org/10.2478/s12175-010-0025-7
32. Qi, F., Guo, B.-N.: Some properties of extended remainder of Binet's first formula for logarithm of gamma function, http://arxiv.org/abs/0904.1118.
33. Qi, F., Guo, B.-N.: The function $(b^x - a^x)/x$: Logarithmic convexity. RGMIA Res. Rep. Coll. **11**(1) (2008), Art. 5. http://rgmia.org/v11n1.php
34. Qi, F., Guo, B.-N.: The function $(b^x - a^x)/x$: Logarithmic convexity and applications to extended mean values, http://arxiv.org/abs/0903.1203.

35. Qi, F., Guo, B.-N.: Wendel's and Gautschi's inequalities: Refinements, extensions, and a class of logarithmically completely monotonic functions. Appl. Math. Comput. **205**(1), 281–290 (2008). http://dx.doi.org/10.1016/j.amc.2008.07.005
36. Qi, F., Luo, Q.-M.: A simple proof of monotonicity for extended mean values. J. Math. Anal. Appl. **224**(2), 356–359 (1998). http://dx.doi.org/10.1006/jmaa.1998.6003
37. Qi, F., Xu, S.-L.: Refinements and extensions of an inequality, II. J. Math. Anal. Appl. **211**(2), 616–620 (1997). http://dx.doi.org/10.1006/jmaa.1997.5318
38. Qi, F., Xu, S.-L.: The function $(b^x - a^x)/x$: Inequalities and properties. Proc. Amer. Math. Soc. **126**(11), 3355–3359 (1998). http://dx.doi.org/10.1090/S0002-9939-98-04442-6
39. Qi, F., Xu, S.-L., Debnath, L.: A new proof of monotonicity for extended mean values. Int. J. Math. Math. Sci. **22**(2), 417–421 (1999). http://dx.doi.org/10.1155/S0161171299224179
40. Qi, F., Cheng, J.-X., Wang, G.: New Steffensen pairs. Inequality Theory and Applications. vol. 1, 273–279, Nova Science Publishers, Huntington (2001)
41. Qi, F., Guo, B.-N., Chen, C.-P.: The best bounds in Gautschi-Kershaw inequalities. Math. Inequal. Appl. **9**(3), 427–436 (2006) http://dx.doi.org/10.7153/mia-09-41
42. Qi, F., Niu, D.-W., Guo, B.-N.: Monotonic properties of differences for remainders of psi function. Int. J. Pure Appl. Math. Sci. **4**(1), 59–66 (2007)
43. Qi, F., Guo, S., Guo, B.-N., Chen, S.-X.: A class of k-log-convex functions and their applications to some special functions. Integral Transforms Spec. Funct. **19**(3), 195–200 (2008). http://dx.doi.org/10.1080/10652460701722627
44. Wu, Y.-D., Zhang, Z.-H.: The best constant for an inequality, Octogon Math. Mag. **12**(1), 139–141 (2004)
45. Wu, Y.-D., Zhang, Z.-H.: The best constant for an inequality, RGMIA Res. Rep. Coll. **7**(1), (2004) Art. 19. http://rgmia.org/v7n1.php.
46. Zhang, S.-Q., Guo, B.-N., Qi, F.: A concise proof for properties of three functions involving the exponential function. Appl. Math. E-Notes **9** 177–183 (2009)

On the Approximation and Bounds of the Gini Mean Difference

Pietro Cerone

Dedicated to Professor Hari M. Srivastava

Abstract A variety of mathematical inequalities are utilised to obtain approximation and bounds of the Gini mean difference. The Gini mean difference or the related index is a widely used measure of inequality in numerous areas such as health, finance and population attributes arenas. The paper provides a review of recent developments in the area with an emphasis on work with which the author has been involved.

1 Introduction

Let $f : \mathbb{R} \to [0, \infty)$ be a *probability density function* (pdf), meaning that f is integrable on $\mathbb{R}$ and $\int_{-\infty}^{\infty} f(t)\,dt = 1$, and define

$$F(x) := \int_{-\infty}^{x} f(t)\,dt,\ x \in \mathbb{R} \quad \text{and} \quad E(f) := \int_{-\infty}^{\infty} x f(x)\,dx, \tag{1}$$

to be its *cumulative function* or distribution and the *expectation* provided that the integrals exist and are finite.

The *mean difference*

$$R_G(f) := \frac{1}{2}\int_{-\infty}^{\infty}\int_{-\infty}^{\infty} |x - y|\,dF(x)\,dF(y) \tag{2}$$

was proposed by Gini in 1912 [14], after whom it is usually named, but it was discussed by Helmert and other German writers in the 1870s (cf. David [12], see

P. Cerone (✉)
La Trobe University, Melbourne, Victoria, Australia
e-mail: p.cerone@latrobe.edu.au

G.V. Milovanović and M.Th. Rassias (eds.), *Analytic Number Theory, Approximation Theory, and Special Functions*, DOI 10.1007/978-1-4939-0258-3_17,

also [21, p. 48]). The mean difference has a certain theoretical attraction, being dependent on the spread of the variate values among themselves rather than on the deviations from some central value ([21, p. 48]). Further, its defining integral (2) may converge when the *variance* $\sigma^2(f)$,

$$\sigma^2(f) := \int_{-\infty}^{\infty} (x - E(f))^2 \, dF(x), \tag{3}$$

does not. It can, however, be more difficult to compute than (3).

Another useful concept is the *mean deviation* $M_D(f)$, defined by [21, p. 48]:

$$M_D(f) := \int_{-\infty}^{\infty} |x - E(f)| \, dF(x) = 2 \int_{\mu}^{\infty} (x - \mu) \, dF(x). \tag{4}$$

As G.M. Giorgi noted in [16], some of the many reasons for the success and the relevance of the Gini mean difference or *Gini index* $I_G(f)$,

$$I_G(f) = \frac{R_G(f)}{E(f)}, \tag{5}$$

are their simplicity, certain interesting properties and useful decomposition possibilities, and these attributes have been analysed in an earlier work by Giorgi [15]. For a bibliographic portrait of the Gini index, see [16] where numerous references are given.

The Gini index given by (5) is a measure of relative inequality since it is a ratio of the Gini mean difference, a measure of dispersion, to the average value $\mu = E(f)$. Other measures are the coefficient of variation $V = \sigma/\mu$ and half the relative mean deviation $M_D(f)/(2\mu)$, where $M_D(f)$ is as defined in (4).

From (1), $F(x)$ is assumed to increase on its support and its mean $\mu = E(f)$ exist. These assumptions imply that $F^{-1}(p)$ is well defined and is the population's p^{th} quantile. The theoretical Lorenz curve (Gastwirth [13]) corresponding to a given $F(x)$ is defined by

$$L(p) = \frac{1}{\mu} \int_0^p F^{-1}(x) \, dx, \qquad 0 \le p \le 1. \tag{6}$$

Now $F^{-1}(x)$ is nondecreasing and so from (6) $L(p)$ is convex and $L'(p) = 1$ at $p = F(\mu)$.

The area between the Lorenz curve and the line p is known as the area of concentration.

The most common measure of inequality is the Gini index defined by (5) which may be shown to be equivalent to twice the area of concentration ([13])

$$C = \int_0^1 c(p) \, dp, \quad c(p) = p - L(p). \tag{7}$$

$c(p)$ vanishes at $p = 0$ or 1 and is concave since $L(p)$ is convex. Further, there is a *point of maximum discrepancy* p^* between the Lorenz curve and the line of equality which satisfies

$$c(p^*) \geq c(p) \qquad \text{for all } p \in [0, 1]. \tag{8}$$

The point $p^* = F(\mu)$ and $c(p^*) = \frac{M_D(f)}{/}(2\mu)$, where $M_D(f)$ is given by (4).

The study of income inequality has gained considerable importance and the Lorenz curve and the associated Gini mean or Gini index are certainly the most popular measures of income inequality. These have also however found application in many other problems within the health, finance and population arenas.

In a sequence of four papers, Cerone and Dragomir ([6–10]) developed approximation and bounds from identities involving the Gini mean difference $R_G(f)$. Some of these results involved using the well-known Sonin and Korkine identities. Cerone [4] procured some approximations and bounds utilising the Steffensen and Karamata inequalities and some of the results are presented in Sects. 4 and 5.

The characteristics of the Lorenz curve, $L(p)$, and its connection to the Gini index via (7) to obtain upper and lower bounds for both $L(p)$ and $I_G(f)$ were analysed by the author in [5]. This is accomplished by utilising the well-known Young's integral inequality and some less well-known reverse inequalities. These are discussed in Sect. 6 and applied in Sect. 7.

In the final section, generalisations and extensions of the Iyengar inequality to allow the approximation and bounds of Riemann-Stieltjes integrals and weighted integrals in a less restrictive framework developed in [3] are presented. These developments enabled the procurement of novel results for the approximation and bounds of the Gini mean difference which are summarised here with a brief sketch of proofs.

2 Some Identities and Inequalities for the Gini Mean Difference

Some identities for the Gini mean difference $R_G(f)$ through which results for the Gini index $I_G(f)$ may be procured via the relationship (5) will be stated here. These have been used in [6–10] to obtain approximations and bounds. The reader is referred to the book [21], Exercise 2.9, p. 94, or [10].

The following result holds (see for instance [21, p. 54] or [10]).

Theorem 2.1. *With the above notation, the identities*

$$R_G(f) = \int_{-\infty}^{\infty} (1 - F(y)) F(y) \, dy = 2 \int_{-\infty}^{\infty} x f(x) F(x) \, dx - E(f) \tag{9}$$

hold.

The following result was obtained in [6] using the well-known Sonin identity (see [23, p. 246]) for the case of univariate real functions.

Theorem 2.2. *With the above assumptions for f and F, we have the identity*

$$R_G(f) = 2\int_{-\infty}^{\infty} (x - E(f))(F(x) - \gamma) f(x)\,dx$$
$$= 2\int_{-\infty}^{\infty} (x - \delta)\left(F(x) - \frac{1}{2}\right) f(x)\,dx \tag{10}$$

for any $\gamma, \delta \in \mathbb{R}$.

The following result was studied in [7] using the Korkine identity (see [23, p. 242]) for the case of univariate real functions.

Theorem 2.3. *With the above assumptions for f and F, we have the following representation for the Gini mean difference:*

$$R_G(f) = \int_{-\infty}^{\infty}\int_{-\infty}^{\infty} (x - y)(F(x) - F(y)) f(x) f(y)\,dxdy. \tag{11}$$

The following lemma will be proven here since it is crucial for the current work in bounding the Gini index via the Lorenz curve and the area of concentration C. The identity is also proven in [21, p. 49] in a different way.

Lemma 2.1. *The following identity holds*

$$R_G(f) = \mu I_G(f) = 2\mu C, \tag{12}$$

where the quantities are defined by (2), (5), (6)–(7).

Proof. From (6) and (7) we have

$$2\mu C = 2\int_0^1 \left[p\mu - \int_0^p F^{-1}(x)\,dx\right] dp = 2\int_0^1\int_0^p \left[E(f) - F^{-1}(x)\right] dxdp.$$

An interchange of the order of integration and a substitution $x = F(t)$ produces

$$2\mu C = 2\int_{-\infty}^{\infty} (t - E(f))\, F(t)\, dF. \tag{13}$$

Now (13) is equivalent to identity (10) with $\gamma = 0$ and so $2\mu C = R_G(f)$ and hence the identity (12) is proved.

3 Bounds for the Lorenz Curve and Gini index

3.1 Inequalities for $R_G(f)$

The following result compares the Gini mean difference with the mean deviation defined by (4) which was obtained in [6] using (10).

Theorem 3.1. *With the above assumptions, we have the bounds*

$$\frac{1}{2}M_D(f) \leq R_G(f) \leq 2\sup_{x\in\mathbb{R}}|F(x)-\gamma|\, M_D(f) \leq M_D(f),$$

for any $\gamma \in [0,1]$, where $F(\cdot)$ is the cumulative distribution of f and $M_D(f)$ is the mean deviation defined by (4).

It was pointed out by J.L. Gastwirth in [13], using inequality 105 from the book [17] by Hardy, Littlewood, and Polya and the fact that F is increasing, that one can state the following results.

Theorem 3.2. *Assume that F is supported on a finite interval (a,b). Then*

$$0 \leq R_G(f) \leq \frac{1}{b-a}(b-E(f))(E(f)-a). \tag{14}$$

3.2 Inequalities via Grüss and Sonin Type Results

The following representation for the Gini mean difference

$$R_G(f) = \int_a^b F(x)(1-F(x))\,dx, \tag{15}$$

holds provided that F is supported on $[a,b]$, a finite interval.

Bounds for the quantity $R_G^*(f)$, involving $R_G(f)$ and defined here for simplicity

$$R_G^*(f) := \frac{1}{b-a}[b-E(f)][E(f)-a] - R_G(f), \tag{16}$$

will be obtained below.

Utilising the well-known Grüss inequality the following simple bound for the Gini mean difference was obtained in [9].

Theorem 3.3. *If f is defined on the finite interval $[a,b]$ and $R_G^*(f)$ is given by* (16), *then*

$$0 \le R_G^* (f) \le \frac{1}{4} (b-a) . \tag{17}$$

The following improvement of Theorem 3.3 was obtained in [9] using an improvement (see [7] and [11]) of the Grüss inequality.

Theorem 3.4. *If f is defined on the finite interval $[a,b]$ and $R_G^* (f)$ is given by* (16)*, then*

$$\begin{aligned} 0 \le R_G^* (f) &\le \frac{1}{2} \cdot \int_a^b \left| F(x) - \frac{b - E(f)}{b-a} \right| dx \\ &\le \frac{1}{2} \left[\int_a^b \left(F(x) - \frac{b - E(f)}{b-a} \right)^2 dx \right]^{\frac{1}{2}} \le \frac{1}{4} (b-a) . \end{aligned} \tag{18}$$

The Sonin identity [23, p. 246] on (15) producing the result

$$R_G^* (f) = \int_a^b \left(F(t) - \frac{b - E(f)}{b-a} \right) (F(t) - \lambda) \, dt \tag{19}$$

was used in [8] to obtain the following theorem.

Theorem 3.5. *Assume that f is defined on the finite interval $[a,b]$ and $R_G^* (f)$ is given by* (16)*, then*

$$\begin{aligned} (0 \le) R_G^* (f) &\le \inf_{\lambda \in \mathbb{R}} \| F - \lambda \|_\infty \int_a^b \left| F(t) - \frac{b - E(f)}{b-a} \right| dt \\ &\le \frac{1}{2} \int_a^b \left| F(t) - \frac{b - E(f)}{b-a} \right| dt. \end{aligned} \tag{20}$$

Taking $\lambda = \frac{b-E(f)}{b-a}$ in (19), the following simple bound was also obtained in [8]:

$$R_G^* (f) \le \frac{1}{b-a} \left[\frac{b-a}{2} + \left| E(f) + \frac{a+b}{2} \right| \right]^2 . \tag{21}$$

The following identity was obtained in [8] from using the Korkine identity [23, p. 242] on (15):

$$R_G^* (f) = \frac{1}{2(b-a)} \int_a^b \int_a^b (F(y) - F(x))^2 \, dx dy, \tag{22}$$

where $R_G^* (f)$ is as given by (16).

If upper and lower bounds for the density function f are known, then we have the following result which was obtained from (2.17) in [8].

Theorem 3.6. *If f is supported on $[a,b]$ and there exist the constants $0 < m, M < \infty$ such that*

$$m \le f(x) \le M \quad \text{for a.e. } x \in [a,b], \tag{23}$$

then

$$\frac{1}{12}m^2(b-a)^3 \le \frac{1}{b-a}[b-E(f)][E(f)-a] - R_G(f) \le \frac{1}{12}M^2(b-a)^3. \tag{24}$$

4 Results from Steffensen's Inequality

The following theorem is due to Steffensen [24] (see also [2] and [1]).

Theorem 4.1. *Let $h : [a,b] \to \mathbb{R}$ be a nondecreasing mapping on $[a,b]$ and $g : [a,b] \to \mathbb{R}$ be an integrable mapping on $[a,b]$ with*

$$-\infty < \phi \le g(x) \le \Phi < \infty \quad \text{for all } x \in [a,b],$$

then

$$\frac{1}{\lambda}\int_a^{a+\lambda} h(x)\,dx \le \frac{\int_a^b G(x)h(x)\,dx}{\int_a^b G(x)\,dx} \le \frac{1}{\lambda}\int_{b-\lambda}^b h(x)\,dx. \tag{25}$$

where

$$\lambda = \int_a^b G(x)\,dx, \quad 0 \le G(x) = \frac{g(x)-\phi}{\Phi-\phi} \le 1, \quad \Phi \ne \phi. \tag{26}$$

Remark 4.1. Equation (25) has a very pleasant interpretation, as observed by Steffensen, that the weighted integral mean of $h(x)$ is bounded by the integral means over the end intervals of length λ, the total weight.

The following results have been obtained in [4].

Theorem 4.2. *Let f be supported on the interval $[a,b]$ and $E(f)$ exist. Then the Gini mean difference $R_G(f)$ satisfies*

$$\int_a^{a+\lambda} (a+\lambda-x) f(x)\,dx \le R_G(f) \le \lambda - \int_{b-\lambda}^b [x-(b-\lambda)] f(x)\,dx, \tag{27}$$

where $\lambda = E(f) - a$.

Remark 4.2. We note that the result (27) may be compared with that of Gastwirth, namely , $\frac{\lambda}{2} \leq R_G(f) \leq \lambda$ with $\lambda = E(f) - a$, which was derived under the assumption that F is defined on (a, ∞) and satisfies a DHR property.

We notice that the upper bound in (27) is always less than λ. It is uncertain however as to whether the lower bound is greater or less than $\frac{\lambda}{2}$.

The following theorems assume that the pdfs are defined over the finite interval $[a, b]$ and so from (9) we have the identity:

$$\frac{R_G(f) + E(f)}{2} = \int_a^b x f(x) F(x) \, dx. \tag{28}$$

Theorem 4.3. *Let $f(x)$ be a pdf on $[a, b]$, $0 < \alpha \leq x f(x) \leq \beta$ and $\lambda = \frac{E(f) - \alpha(b-a)}{\beta - \alpha}$, then the Gini mean difference $R_G(f)$ satisfies*

$$(\beta - \alpha) \int_a^{a+\lambda} (a + \lambda - x) f(x) \, dx \leq \frac{R_G(f) + E(f)}{2} - \alpha(b - E(f)) \tag{29}$$

$$\leq (\beta - \alpha) \left[\lambda - \int_{b-\lambda}^b (x - (b - \lambda)) f(x) \, dx \right].$$

Theorem 4.4. *Let f be supported on the positive interval $[a, b]$ with $0 \leq a < b$ and $\phi \leq f(x) \leq \Phi$, $x \in [a, b]$ and $E(f)$ exist. With $\lambda = \frac{1 - (b-a)\phi}{\Phi - \phi}$, then the Gini mean difference $R_G(f)$ satisfies*

$$(\Phi - \phi) \int_a^{a+\lambda} \left[(a + \lambda)^2 - x^2 \right] f(x) \, dx$$

$$\leq R_G(f) + E(f) - \phi \int_a^b (b^2 - x^2) f(x) \, dx$$

$$\leq (\Phi - \phi) \left\{ \lambda (2b - \lambda) - \int_{b-\lambda}^b \left[x^2 - (b - \lambda)^2 \right] f(x) \, dx \right\}. \tag{30}$$

5 Results with Karamata's Inequality

In an interesting but not well-known paper [22], Alexandru Lupaş generalised some results due to Karamata. These are presented and applications to bounding the Gini mean difference are demonstrated in the current section.

First some notation.

Let $-\infty < a < b < +\infty$ and $e_0(x) = 1$, $x \in [a, b]$. Further, let X be a real linear space with elements being real functions defined on $[a, b]$. By $\mathcal{F} : X \to \mathbb{R}$ we denote a positive linear functional normalised by $\mathcal{F}(e_0) = 1$. The following three results were obtained by Lupaş in [22].

Theorem 5.1. *Let $h, g \in X$ with*

$$m_1 \leq h(x) \leq M_1 \quad (M_1 \neq m_1), \qquad 0 < m_2 \leq g(x) \leq M_2 \quad x \in [a,b]. \tag{31}$$

If $D(h) = M_1 - \mathcal{F}(h)$, $d(h) = \mathcal{F}(h) - m_1$, then

$$\frac{m_1 M_2 D(h) + M_1 m_2 d(h)}{M_2 D(h) + m_2 d(h)} \leq \frac{\mathcal{F}(hg)}{\mathcal{F}(g)} \leq \frac{M_1 M_2 d(h) + m_1 m_2 D(h)}{M_2 d(h) + m_2 D(h)}. \tag{32}$$

The bounds in (32) are best possible.

Theorem 5.2. *Let h, g be elements from X which satisfy (31). If $\Delta(x) = M_1 - h(x)$, $\delta(x) = h(x) - m_1$, then*

$$\begin{aligned} &v\left|\mathcal{F}(h)\mathcal{F}(g) - \mathcal{F}(hg)\right| \\ &\quad \leq \frac{M_2 - m_2}{(M_1 - m_1)(M_2 + m_2)}\left[\mathcal{F}(\Delta)\mathcal{F}(\delta g) + \mathcal{F}(\delta)\mathcal{F}(\Delta g)\right]. \end{aligned} \tag{33}$$

Theorem 5.3. *Let $h, g \in X$ with*

$$0 < m_1 \leq h(x) \leq M_1, \qquad 0 < m_2 \leq g(x) \leq M_2 \quad x \in [a,b]. \tag{34}$$

If

$$K = \frac{\sqrt{m_1 m_2} + \sqrt{M_1 M_2}}{\sqrt{m_1 M_2} + \sqrt{M_1 m_2}},$$

then

$$\frac{1}{K^2} \leq \frac{\mathcal{F}(hg)}{\mathcal{F}(h)\mathcal{F}(g)} \leq K^2. \tag{35}$$

We note that Karamata established (32) and (35) in [19] and [20] for $\mathcal{F}(h) = \int_0^1 h(t)\,dt$.

Further, h and g in Theorem 5.3 are assumed to be strictly positive and bounded, whereas in Theorems 5.1 and 5.2, h is not allowed to be constant and the requirement for positivity is removed.

The following three theorems assume that the normalised positive linear functional $\mathcal{F}(\cdot)$ is given by

$$\mathcal{F}(h) = \frac{1}{b-a}\int_a^b h(x)\,dx \tag{36}$$

and the identity (28) is used.

Theorem 5.4. *Let $f(x)$ be a pdf on $[a,b]$ and $0<\alpha\le xf(x)\le\beta$, then the Gini mean difference $R_G(f)$ satisfies*

$$\left(\frac{1-pz}{1+pz}\right)E(f)\le R_G(f)\le\left(\frac{p-z}{p+z}\right)E(f), \tag{37}$$

where $p=\frac{\beta}{\alpha}$ and $z=\frac{E(f)-a}{b-E(f)}$.

Theorem 5.5. *Let $f(x)$ be a pdf on $[a,b]$ and $0<m\le f(x)\le M$, then the Gini mean difference $R_G(f)$ satisfies*

$$\frac{2b\xi}{b\rho-(\rho-1)\xi}-E(f)\le R_G(f)\le\frac{2b\rho\xi}{b+(\rho-1)\xi}-E(f), \tag{38}$$

where

$$\rho=\frac{M}{m},\quad \xi=\frac{b^2-\mathcal{M}_2}{2(b-a)}\quad\text{and}\quad \mathcal{M}_2=\int_a^b x^2f(x)\,dx. \tag{39}$$

Theorem 5.6. *Let $f(x)$ be a pdf on $[a,b]$ and $0<m\le f(x)\le M$, then the Gini mean difference $R_G(f)$ satisfies*

$$\begin{aligned}\frac{E(f)+2M\left[a\left(\frac{a+b}{2}\right)-bE(f)\right]}{2bM-1}&\le R_G(f) \\ &\le\frac{E(f)+2M\left[b\left(\frac{a+b}{2}\right)-aE(f)\right]}{2aM-1}.\end{aligned} \tag{40}$$

Theorem 5.7. *Let $f(x)$ be a pdf on $[a,b]$ with $a>0$ and $0<m\le f(x)\le M$, $x\in[a,b]$, then the Gini mean difference $R_G(f)$ satisfies*

$$\left(\frac{1-\rho\zeta}{1+\rho\zeta}\right)E(f)\le R_G(f)\le\left(\frac{\rho-\zeta}{\rho+\zeta}\right)E(f), \tag{41}$$

where $\rho=\frac{M}{m}$, $\zeta=\frac{\mathcal{M}_2-a^2}{b^2-\mathcal{M}_2}$ and $\mathcal{M}_2=\int_a^b x^2f(x)\,dx$, the second moment about zero.

Remark 5.1. The lower bounds in (37) and (41) are only useful when they are greater than 0 since $R_G(f)$ is known to be non-negative. This occurs for $E(f)<\frac{\alpha\beta+b\alpha}{\alpha+\beta}$ and $\mathcal{M}_2<\frac{Ma^2+mb^2}{M+m}$.

Further Theorem 5.7 uses the normalised linear functional $\mathcal{F}(h)=\frac{\int_a^b\omega(x)h(x)dx}{\int_a^b\omega(x)dx}$ in Theorem 5.1 with $\omega(x)=x,\ h(x)=F(x)$ and $g(x)=f(x)$.

6 Young's Integral Inequality and Reverses

The famous Young's integral inequality states that

Theorem 6.1. *If $h : [0, A] \to \mathbb{R}$ is continuous and a strictly increasing function satisfying $h(0) = 0$, then for every positive $0 < a \leq A$ and $0 < b \leq h(A)$*

$$Y(h; a, b) := \int_0^a h(t)\, dt + \int_0^b h^{-1}(t)\, dt \geq ab \tag{42}$$

holds with equality if and only if $b = h(a)$.

In the 1912 paper in fact Young [26] proved (42), assuming differentiability of the functions. The inequality (42) has a geometric interpretation involving the areas of the two functions and the rectangular area. There has been much work on different proofs and generalisations of (42).

We notice that in (42), ab is a lower bound for the Young functional $Y(h; a, b)$. In 2007, Witkowski [25] gave two simple proofs for Theorem 6.1. The first utilises the fact that since h is strictly increasing, then its anti-derivative is strictly convex. The second uses the mean value theorem. The second proof will be replicated here to highlight the fact that this approach does not just provide a proof for Young's inequality (42) but it also gives its reverse.

Theorem 6.2. *Let the conditions of Theorem* 6.1 *hold. Then*

$$ab \leq Y(h; a, b) \leq ah(a) + h^{-1}(b)(b - h(a)) \tag{43}$$

with equality if and only if $b = h(a)$.

Remark 6.1. We note that the upper bound in (43) provides a reverse of Young's integral inequality (42). Equation (43) can be written in the appealing form

$$ab \leq Y(h; a, b) \leq ab + (b - h(a))\left(h^{-1}(b) - a\right) \tag{44}$$

or

$$0 \leq Y(h; a, b) - ab \leq (b - h(a))\left(h^{-1}(b) - a\right) \tag{45}$$

We notice that $(b - h(a))\left(h^{-1}(b) - a\right) \geq 0$ with equality holding only for $b = h(a)$ (equivalently, $a = h^{-1}(b)$).

Theorem 6.3. *Let the conditions of Theorem* 6.1 *persist. Then the inequality*

$$\alpha(a, b) \int_0^a h(t)\, dt + \beta(a, b) \int_0^b h^{-1}(t)\, dt \leq ab \tag{46}$$

holds, where

$$\alpha(a,b)=\min\left\{1,\frac{b}{h(a)}\right\}\quad\text{and}\quad\beta(a,b)=\min\left\{1,\frac{a}{h^{-1}(b)}\right\},\tag{47}$$

with equality holding if and only if $b=h(a)$.

It is instructive to compare the upper bounds for $\int_0^b h^{-1}(t)\,dt$ provided from the results of Witkowski from (43) and (46) and (47). The following Lemma was obtained in [5].

Lemma 6.1. *From Theorems 6.2 and 6.3, the following upper bounds are tighter. Namely,*

$$\int_0^b h^{-1}(t)\,dt$$
$$<\begin{cases}\frac{b}{h(a)}\left[ah(a)-\int_0^a h(t)\,dt\right] & \text{for } \Delta>0,\ b<h(a);\\ ab+\left(h^{-1}(b)-a\right)(b-h(a))-\int_0^a h(t)\,dt & \text{for } \Delta<0 \text{ or } b>h(a),\end{cases}\tag{48}$$

where $\Delta:=ah(a)-\int_0^a h(t)\,dt-h(a)h^{-1}(b)$.

7 Bounds for the Lorenz Curve and Gini via Young Type Inequalities

We are now in a position to investigate bounds for both the Lorenz curve and through the relationship (12) for the Gini index using the results of Sect. 6 based on Young type inequalities. Firstly, however, we state a result of Gastwirth [13] for bounding the Lorenz curve.

Theorem 7.1. *Let* $F(x)$ *be a distribution function with mean* μ *and support* (a,b). *Then its Lorenz curve,* $L(p)$, *satisfies*

$$B(p)\le L(p)\le p\tag{49}$$

where

$$B(p)=\begin{cases}\frac{ap}{\mu}, & p<r;\\ \frac{ar}{\mu}+\frac{b}{\mu}(p-r), & p>r\end{cases}\tag{50}$$

and r *is determined by the relation* $ra+(1-r)b=\mu$. *Here the random variable* X *generating the Lorenz curve* $B(p)$ *takes on the value* a *with probability* r *and* b *with probability* $(1-r)$.

The following technical lemma will prove useful subsequently.

Lemma 7.1. *Let $F(\cdot)$ be a distribution function defined on $(0, A]$ and its inverse $F^{-1}(\cdot)$ exists, then for $a \in (0, A]$*

$$\int_0^1 (p - F(a))\left(F^{-1}(p) - a\right) dp = \frac{1}{2}\left[A - a - \int_0^A F^2(t)\,dt\right] + (a - \mu) F(a). \tag{51}$$

Proof. Firstly, we note that for $h(0) = 0$,

$$\int_0^A h(t)\,dt = Ah(A) - \int_0^A th'(t)\,dt = Ah(A) - \int_0^{h(A)} h^{-1}(t)\,dt. \tag{52}$$

If we then associate $h(\cdot)$ with $F(\cdot)$, noting that $F(A) = 1$, then from (52)

$$\mu L(1) = \int_0^1 F^{-1}(p)\,dp = A - \int_0^A F(t)\,dt = \mu \tag{53}$$

since $L(1) = 1$.

Further, a substitution of $p = F(t)$ and integration by parts gives

$$\int_0^1 pF^{-1}(p)\,dp = \int_0^A tF(t)F'(t)\,dt = \frac{A}{2} - \frac{1}{2}\int_0^A F^2(t)\,dt. \tag{54}$$

Now,

$$\begin{aligned}&\int_0^1 (p - F(a))\left(F^{-1}(p) - a\right) dp\\ &\quad = \int_0^1 pF^{-1}(p)\,dp + aF(a) - F(a)\int_0^1 F^{-1}(p)\,dp - \frac{a}{2}.\end{aligned} \tag{55}$$

Substitution of (53) and (54) into (55) gives the stated result (51).

The following theorem uses the results of Witkowski [25] as given by (43) to procure bounds for the Lorenz curve.

Theorem 7.2. *Let $L(p)$ be the Lorenz curve defined by (6) corresponding to a given distribution (cumulative) function $F(a)$ with $F(0) = 0$, $0 < a \le A$ and $0 < p \le F(A) = 1$. Then*

$$\begin{aligned}\frac{1}{\mu}\left[ap - \int_0^a F(t)\,dt\right] &\le L(p)\\ &\le \frac{1}{\mu}\left[ap - \int_0^a F(t)\,dt\right] + \frac{1}{\mu}(p - F(a))\left(F^{-1}(p) - a\right)\end{aligned} \tag{56}$$

with equality if and only if $p = F(a)$.

Remark 7.1. The lower bound is only useful for $p > \frac{1}{a}\int_0^a F(t)\,dt$ since zero is a lower bound for $L(p)$. The upper bound is useful if it is less than p. The following Corollary results based on these observations:

Corollary 7.1. *Let the condition of Theorem 7.2 hold. Then*

$$l(p) \le L(p) \le u(p), \tag{57}$$

where

$$l(p) = \begin{cases} 0, & p < 1 - \frac{\mu}{A}; \\ \frac{A}{\mu}\left[p - \left(1 - \frac{\mu}{A}\right)\right] & , p > 1 - \frac{\mu}{A} \end{cases} \tag{58}$$

and

$$u(p) = \begin{cases} p, & 0 < p < p^*; \\ 1 + \frac{F^{-1}(p)}{\mu}(p-1), & p^* < p < 1, \end{cases} \tag{59}$$

where $p^* = F(\mu)$ is the point of maximum discrepancy satisfying (8).

Remark 7.2. It may be noticed that by taking $a = 0$ and $b = A$ in Theorem 7.1 we have $r = 1 - \mu/A$ and so Corollary 7.1 recaptures the lower bound obtained by Gastwirth [13]. The upper bound given by (59) provides a refinement of that obtained by Gastwirth [13] and shown here as (49).

Theorem 7.3. *Let the conditions of Theorem 7.2 hold. Then the Gini index defined by (5) or, equivalently, (7) satisfies*

$$\left(1 - \frac{a}{\mu}\right) + \frac{2}{\mu}\int_0^a F(t)\,dt + 2\left(1 - \frac{a}{\mu}\right)F(a) - \frac{1}{\mu}\left[A - a - \int_0^A F^2(t)\,dt\right]$$
$$\le I_G(f) \le \left(1 - \frac{a}{\mu}\right) + \frac{2}{\mu}\int_0^a F(t)\,dt. \tag{60}$$

Proof. From (56) we have, since, as shown in Lemma 2.1, the Gini index $I_G(f)$ of (5) is equivalent to twice the area of concentration, namely, $2C$. Now, (56) gives

$$\left(1 - \frac{a}{\mu}\right)p + \frac{1}{\mu}\int_0^a F(t)\,dt - \frac{1}{\mu}(p - F(a))\left(F^{-1}(p) - a\right)$$
$$\le p - L(p) \le \left(1 - \frac{a}{\mu}\right)p + \frac{1}{\mu}\int_0^a F(t)\,dt$$

so that from (12) and (7)

$$\left(1-\frac{a}{\mu}\right)+\frac{2}{\mu}\int_0^a F(t)\,dt-\frac{2}{\mu}\int_0^1 (p-F(a))\left(F^{-1}(p)-a\right)dp$$
$$\le I_G(f)\le\left(1-\frac{a}{\mu}\right)+\frac{2}{\mu}\int_0^a F(t)\,dt.$$

Using (51) from Lemma 7.1 produces the inequality as stated in (60).

Corollary 7.2. *Let the conditions of Theorems 7.2 and 7.3 hold. Then the Gini index bounds from (60) are the tightest bounds on* $(0, A]$ *at* $a = \mu$ *and* $a = m$ *for the lower and upper bounds, respectively. These are given by*

$$\frac{1}{\mu}\int_0^{\mu} F(t)\,dt < I_G(f) < \left(1-\frac{m}{\mu}\right)+\frac{2}{\mu}\int_0^m F(t)\,dt, \tag{61}$$

where $m = F^{-1}\left(\frac{1}{2}\right)$ *is the median and* μ *is the mean.*

Proof. Since $F(t)$ is defined for $t \in [0, A]$ and $F(0) = 0$, we have from (9) that

$$I_G(f)=\frac{1}{\mu}\int_0^A F(t)\,(1-F(t))\,dt. \tag{62}$$

We notice that the lower bound in (60) approaches $I_G(f)$ as $a \to 0^+$ and the upper bound tends to 1. Further, if we denote the lower bound in (60) by $\kappa(a)$, then

$$\kappa'(a)=2\left(1-\frac{a}{\mu}\right)f(a)\begin{cases} >0,\ 0<a<\mu \\ <0,\ \mu<a<A.\end{cases}$$

The maximum occurs at $a = \mu$ so that

$$\sup_{a\in(0,A)}\kappa(a)=\kappa(\mu)=\frac{2}{\mu}\int_0^{\mu} F(t)\,dt-\frac{1}{\mu}\left[A-\mu-\int_0^a F^2(t)\,dt\right] \tag{63}$$
$$=\frac{2}{\mu}\int_0^{\mu} F(t)\,dt-\frac{1}{\mu}\int_0^A F(t)\,(1-F(t))\,dt$$

since from (53), $A-\mu=\int_0^A F(t)\,dt$. We now have from (63) and using (62) that

$$\sup_{a\in(0,A)}\kappa(a)=\kappa(\mu)=\frac{2}{\mu}\int_0^{\mu} F(t)\,dt-I_G(f), \tag{64}$$

as the best choice for the lower bounds in (60) from which the lower bound in (61) results.

Further, the minimum upper bound in (60) occurs when $2F(a) - 1 = 0$, namely, at $a = m = F^{-1}\left(\frac{1}{2}\right)$, producing the upper bound in (61) from (60).

Remark 7.3. In Cerone [4, Theorem 13] the Steffensen inequality was utilised together with the property that $F(x)$ is nondecreasing to obtain

$$\frac{1}{\mu}\int_a^{a+\lambda} F(x)\,dx \leq I_G(f) \leq \frac{1}{\mu}\int_{b-\lambda}^{b} F(x)\,dx,$$

where $\lambda = \mu - a$ and f is supported on $[a,b]$. That is, taking $a = 0$ and $b = A$, we have

$$\frac{1}{\mu}\int_0^{\mu} F(x)\,dx \leq I_G(f) \leq \frac{1}{\mu}\int_{A-\mu}^{A} F(x)\,dx. \tag{65}$$

We notice that the lower bound here is recaptured by (61); however, the upper bounds differ.

Corollary 7.3. *Let the conditions of Theorem 7.2 hold. The Gini index, $I_G(f)$ satisfies*

$$\frac{1}{2\mu}\int_0^{\mu} F(x)(2 - F(x))\,dx \leq I_G(f) \leq 1 - \frac{\mu}{A}. \tag{66}$$

Sketch. Use Corollary 7.2 and Lemma 2.1.

Remark 7.4. The upper bound given in (66) was also obtained in Gastwirth [13] using a result from Hardy et al. [17]. The lower bound obtained in [13] was zero which is smaller than that given in (66).

8 Iyengar Inequality for Riemann-Stieltjes Integrals and Application to Gini

In 1938 Iyengar using geometric arguments developed the following result in the paper [18].

Theorem 8.1. *Let $h : [a,b] \to \mathbb{R}$ be a differentiable function such that for all $x \in [a,b]$ and for $M > 0$ we have $|h'(x)| \leq M$, then*

$$\left|\int_a^b h(x)\,dx - \frac{h(a)+h(b)}{2}(b-a)\right| \leq \frac{M}{4}(b-a)^2 - \frac{(h(b)-h(a))^2}{4M}. \tag{67}$$

Remark 8.1. It should be noted that for $m \leq h^{'}(x) \leq M$ then $\left|h^{'}(x) - \frac{m+M}{2}\right| \leq \frac{M-m}{2}$ and then Iyengar's result may be extended by applying it to $k(x) = h(x) - \frac{m+M}{2}x$ with bound $M_k = \frac{M-m}{2}$.

The following three results developed in [3] extend the Iyengar inequality to involve Riemann-Stieltjes integrals while also relaxing the differentiability condition.

Theorem 8.2. *Let $h, g : [a,b] \to \mathbb{R}$ be such that g is a nondecreasing function, and for all $x \in [a,b]$ and $M > 0$, the following conditions hold:*

$$|h(x) - h(a)| \leq M(x-a) \text{ and } |h(x) - h(b)| \leq M(b-x). \tag{68}$$

Then for any $t \in [a,b]$

$$\left|\int_a^b h(x)\, dg(x) - \{[g(t) - g(a)]\, h(a) + [g(b) - g(t)]\, h(b)\}\right|$$
$$\leq M\left[\int_a^t (x-a) dg(x) + \int_t^b (b-x) dg(x)\right]. \tag{69}$$

Using integration by parts of the Riemann-Stieltjes integrals from an intermediate result within the proof of Theorem 8.2, the following theorem was obtained in [3].

Theorem 8.3. *Let $h, g : [a,b] \to \mathbb{R}$ be such that g is nondecreasing and differentiable for all $x \in [a,b]$ and for $M > 0$ the following conditions hold:*

$$|h(x) - h(a)| \leq M(x-a) \text{ and } |h(x) - h(b)| \leq M(b-x). \tag{70}$$

Then for $t \in [a,b]$ the tightest bound is given by

$$-MD(t^*) \leq \int_a^b h(x)\, dg(x) - [h(b)g(b) - h(a)g(a)] \leq MD(t_*), \tag{71}$$

or

$$-2M\delta(t_m)g(t_m) \leq \int_a^b h(x)\, dg(x) - [h(b)g(b) - h(a)g(a)] \leq 2M\Delta(t_m)g(t_m), \tag{72}$$

where for

$$\alpha = \frac{a+b}{2}, \quad \beta = \frac{f(b) - f(a)}{2}; \quad t^* = \alpha - \frac{\beta}{M}, \quad t_* = \alpha + \frac{\beta}{M}$$

or $D(t_m) = 0$, *with*

$$D(t) = \int_t^b g(x)dx - \int_a^t g(x)dx$$

and

$$\delta(t) = \frac{h(b)-h(a)}{2M} - \left(t - \frac{a+b}{2}\right), \quad \Delta(t) = -\left[\frac{h(b)-h(a)}{2M} + \left(t - \frac{a+b}{2}\right)\right].$$

Here, $t^* \in [a, \frac{a+b}{2}]$ *and* $t_* \in [\frac{a+b}{2}, b]$.

The following extension producing a weighted Iyengar inequality was obtained in [3].

Theorem 8.4. *Let* $h, w : [a,b] \to \mathbb{R}$ *be such that* $w(x) > 0$ *for* $x \in (a,b)$ *and for* $M > 0$ *the following conditions hold:*

$$|h(x) - h(a)| \le M(x-a) \text{ and } |h(x) - h(b)| \le M(b-x). \tag{73}$$

Then for $t \in (a,b)$ *the tightest bound is given by*

$$\left| \int_a^b w(x)h(x)\,dx - \{h(b)W(b) + M\left[I(t_*) - I(t^*)\right]\} \right| \le M\left\{ \int_a^b (b-x)w(x)dx - \left[I(t^*) + I(t_*)\right] \right\}, \tag{74}$$

where for

$$\alpha = \frac{a+b}{2}, \quad \beta = \frac{f(b)-f(a)}{2}; \quad t^* = \alpha - \frac{\beta}{M}, \quad t_* = \alpha + \frac{\beta}{M},$$

with

$$I(t) = \int_a^t (t-x)w(x)dx\,. \tag{75}$$

If $w(a) = 0$, *then the bounds at* $t = a$ *need to be compared with* $L(t^*)$ *and* $R(t_*)$ *and similarly for* $w(b) = 0$.

Remark 8.2. It should be noted that taking $w(x) = 1$ in Theorem 8.4 recaptures the Iyengar result of Theorem 8.1 under less restrictive conditions (73) rather than $|h'(x)| < M$. It should be further emphasised that for $m \le \frac{h(x)-h(a)}{x-a} \le M$ and $m \le \frac{h(b)-h(x)}{b-x} \le M$ the above results may be extended by taking $k(x) = h(x) - \frac{M+m}{2}$

to produce the conditions of the above results for $|k(x) - k(a)| \leq \frac{M-m}{2}(x-a)$ and $|k(x) - k(b)| \leq \frac{M-m}{2}(b-x)$.

With the above results, we are now in a position to obtain bounds utilising the Iyengar type inequalities developed above to obtain approximation and bounds for the Gini mean difference. The details may be seen in [3] but we will only provide a sketch of the proofs. We shall make use of the following identities, where f is the pdf and F its corresponding distribution:

$$R_G(f) = \int_a^b (1 - F(x)) F(x)\, dx = 2\int_a^b x f(x) F(x)\, dx - E(f). \quad (76)$$

Theorem 8.5. *Let $f(x)$ be a pdf on $[a,b]$, $f(x) \leq M$ and $F(x) = \int_a^x f(u)du$, then the Gini mean difference $R_G(f)$ satisfies*

$$\left| R_G(f) + E(f) - 2\left\{bf(b)E(f) + M\left[I(t^*) - I(t_*)\right]\right\}\right| \leq 2M\left\{\int_a^b (b-x)xf(x)dx - \left[I(t_*) + I(t^*)\right]\right\}, \quad (77)$$

where

$$t^* = \frac{a+b}{2} - \frac{bf(b) - af(a)}{2M}, \quad t_* = \frac{a+b}{2} + \frac{bf(b) - af(a)}{2M},$$

and

$$I(t) = \int_a^t (t-x)xf(x)dx.$$

For $f(a) = 0$ we have

$$|R_G(f) - [2bf(b) - 1]\, E(f)| \leq 2M\int_a^b (b-x)xf(x)dx. \quad (78)$$

For $f(b) = 0$ we have

$$|R_G(f) - [2af(a) - 1]\, E(f)| \leq 2M\int_a^b (x-a)xf(x)dx. \quad (79)$$

Finally, for $f(a) = f(b) = 0$ we have

$$|R_G(f) + E(f)| \leq 2M\left\{\frac{b-a}{2}E(f) - \left|\mathcal{M}_2 - \frac{b+a}{2}E(f)\right|\right\}, \quad (80)$$

where $\mathcal{M}_2 = \int_a^b x^2 f(x)dx$, the second moment of $f(x)$.

Sketch. In Theorem 8.4 let $w(x) = xf(x)$ and $h(x) = F(x)$ so that $|h'(x)| = f(x) \leq M$.

Now from (76) we have

$$\frac{R_G(f) + E(f)}{2} = \int_a^b xf(x) F(x) \, dx, \tag{81}$$

and so there are two cases to consider. Namely, $w(x) = xf(x) > 0$ or $w(x) = xf(x) = 0$ for $x \in [a, b]$ with the first producing (77) and the other results covering the three possibilities shown in the results.

Theorem 8.6. *Let $f(x)$ be a pdf on $[a, b]$, $f(x) \leq M$ and $F(x) = \int_a^x f(u) du$, then the Gini mean difference $R_G(f)$ satisfies*

$$\left| R_G(f) - \left\{ E(f) + M \left[J(t_*) - J(t^*) \right] \right\} \right| \leq M \left\{ \frac{1}{2} \left[(b-a)^2 - (t_* - a)^2 - (t^* - a)^2 \right] \right. \\ \left. - J(b) + \left[J(t_*) + J(t^*) \right] \right\}, \tag{82}$$

where

$$t^* = \frac{a+b}{2} - \frac{1}{2M}, \quad t_* = \frac{a+b}{2} + \frac{1}{2M}, \quad J(t) = \int_a^t (t-x) F(x) dx.$$

Further, for $F(b) = 1$ we have

$$|R_G(f)| \leq \frac{M}{2} \int_a^b (x-a)^2 f(x) dx = \frac{M}{2} \left\{ \mathcal{M}_2 - a \left[2E(f) - a \right] \right\}. \tag{83}$$

where $\mathcal{M}_2 = \int_a^b x^2 f(x) dx$.

Sketch. In Theorem 8.4 let $w(x) = 1 - F(x)$ and $h(x) = F(x)$ so that $|h'(x)| = f(x) \leq M$. Now from (76) we have

$$R_G(f) = \int_a^b (1 - F(x)) F(x) \, dx, \tag{84}$$

and so considering the two possibilities, namely, $w(x) = 1 - F(x) > 0$ for $x \in [a, b)$ and $w(b) = 0$, developed in [3].

An investigation of bounds for the Gini mean difference from the Iyengar inequality (67) and the identity depicted in Lemma 2.1 reproduces the results (14) obtained in Theorem 3.2, by Gastwirth [13, p. 48] by a different approach and will thus not be elaborated further.

Acknowledgements Most of the work for this article was undertaken while at Victoria University, Melbourne Australia.

References

1. Cerone, P.: On some generalisations of Steffensen's inequality and related results. J. Ineq. Pure and Appl. Math. **2**(3), Art. 28 (2001). [ONLINE http://jipam.vu.edu.au/v2n3/]
2. Cerone, P.: On an identity for the Chebychev functional and some ramifications. J. Ineq. Pure and Appl. Math. **3**(1), Art. 4 (2002). [ONLINE http://jipam.vu.edu.au/v3n1/]
3. Cerone, P.: On Gini Mean Difference Bounds Via Generalised Iyengar Results, Submitted.
4. Cerone, P.: Bounding the Gini mean difference, Inequalities and Applications. In: Bandle, C., Losonczi, A., Pales, Zs., Plum, M. (eds.) International Series of Numerical Mathematics, vol. 157, pp. 77–89, Birkhüser Verlag, Basel (2008)
5. Cerone, P.: On Young's Inequality and its Reverse for Bounding the Lorenz Curve and Gini mean. J. Math. Inequalities **3**(3), 369–381 (2009)
6. Cerone, P., Dragomir, S.S.: Bounds for the Gini mean difference via the Sonin identity. Comp. Math. Modelling **50**, 599–609 (2005)
7. Cerone, P., Dragomir, S.S.: Bounds for the Gini mean difference via the Korkine identity. J. Appl. Math. & Computing (Korea) **22**(3), 305–315 (2006)
8. Cerone, P., Dragomir, S.S.: Bounds for the Gini mean difference of continuous distributions defined on finite intervals (II). Comput. Math. Appl. **52**(10–11) 1555–1562 (2006)
9. Cerone, P., Dragomir, S.S.: Bounds for the Gini mean difference of continuous distributions defined on finite intervals (I). Appl. Math. Lett. **20**, 782–789 (2007)
10. Cerone, P., Dragomir, S.S.: A survey on bounds for the Gini Mean Difference. In: Barnett, N.S., Dragomir, S.S. (eds.) Advances in Inequalities from Probability Theory and Statistics. pp. 81–111. Nova Science Publishers, New York (2008)
11. Cheng, X.-L., Sun, J.: A note on the perturbed trapezoid inequality. J. Inequal. Pure & Appl. Math. **3**(2) (2002) Article. 29 [ONLINE http://jipam.vu.edu.au/article.php?sid=181]
12. David, H.A., Gini's mean difference rediscovered. Biometrika, **55**, 573 (1968)
13. Gastwirth, J.L.: The estimation of the Lorentz curve and Gini index. Rev. Econom. Statist. **54** 305–316 (1972)
14. Gini, C.: Variabilità e Metabilità, contributo allo studia della distribuzioni e relationi statistiche. Studi Economica-Gicenitrici dell' Univ. di Coglani **3**, 1–158 (1912) art 2
15. Giorgi, G.M.: Alcune considerazioni teoriche su di un vecchio ma per sempre attuale indice: il rapporto di concentrazione del Gini, Metron, **XLII**(3–4), 25–40 (1984)
16. Giorgi, G.M.: Bibliographic portrait of the Gini concentration ratio. Metron **XLVIII**(1–4), 103–221 (1990)
17. Hardy, G.H., Littlewood, J.E., Polya, G.: Inequalities. Cambridge University Press, Cambridge
18. Iyengar, K.S.K.: Note on an inequality. Math. Student **6**, 75–76 (1938)
19. Karamata, J.: O prvom stavu srednjih vrednosti odredenih integrala. Glas Srpske Kraljevske Akademje **CLIV**, 119–144 (1933)
20. Karamata, J.: Sur certaines inégalités relatives aux quotients et à la différence de $\int fg$ et $\int f \cdot \int g$, Acad. Serbe Sci. Publ. Inst. Math. **2**, 131–145 (1948)
21. Kendall, M., Stuart, A.: The Advanced Theory of Statistics, Distribution Theory, vol. 1, 4th edn. Charles Griffin & Comp. Ltd., London (1977)
22. Lupaş, A.: On two inequalities of Karamata. Univ. Beograd. Publ. Elektrotechn. Fak. Ser. Mat. Fiz. 602–633, 119–123 (1978)
23. Mitrinović, D.S., Pečarić, J.E., Fink, A.M.: Classical and New Inequalities in Analysis. Kluwer Academic Publishers, London (1993)

24. Steffensen, J.F.: On certain inequalities between mean values and their application to actuarial problems. Skandinavisk Aktuarietidskrift **1**, 82–97 (1918)
25. Witkowski, A.: On Young's inequality. J. Ineq. Pure and Appl. Math. **7**(5), Art. 164 (2007). [ONLINE http://jipam.vu.edu.au/article.php?sid=782]
26. Young, W.H.: On classes of summable functions and their Fourier series. Proc. Roy. Soc. London (A), **87**, 225–229 (1912)

On Parametric Nonconvex Variational Inequalities

Muhammad Aslam Noor

Dedicated to Professor Hari M. Srivastava

Abstract In this paper, we consider the parametric nonconvex variational inequalities and parametric nonconvex Wiener–Hopf equations. Using the projection technique, we establish the equivalence between the parametric nonconvex variational inequalities and parametric nonconvex Wiener–Hopf equations. We use this alternative equivalence formulation to study the sensitivity analysis for the parametric nonconvex variational inequalities. Our results can be considered as a significant extension of previously known results. The ideas and techniques may be used to stimulate further research for multivalued nonconvex variational inequalities and their variant forms.

1 Introduction

Variational inequalities theory, which was introduced by Stampacchia [37], provides us with a simple, natural, general, and unified framework to study a wide class of problems arising in pure and applied sciences; see [1–42]. Variational inequalities have been generalized and extended in several directions using novel and innovative techniques. It is worth mentioning that almost all the results regarding the existence and iterative schemes for variational inequalities have been investigated and considered in the setting of convexity. This is because all the techniques are based on the properties of the projection operator over convex sets. These results may not hold for nonconvex sets. These facts and observations have motivated to consider the variational inequalities and related optimization problems on the nonconvex sets. Noor [29] has introduced and considered a new class of variational inequalities

M.A. Noor (✉)
Mathematics Department, COMSATS Institute of Information Technology, Islamabad, Pakistan
e-mail: noormaslam@hotmail.com

G.V. Milovanović and M.Th. Rassias (eds.), *Analytic Number Theory, Approximation Theory, and Special Functions*, DOI 10.1007/978-1-4939-0258-3_18,

on the uniformly prox-regular sets, which is called the nonconvex variational inequalities. We remark that the uniformly prox-regular sets are nonconvex and include the convex sets as a special case; see [4, 36]. It is well known that the behavior of such problem solutions as a result of changes in the problem data is always of concern. In recent years, much attention has been given to study the sensitivity analysis of variational inequalities. We remark that sensitivity analysis is important for several reasons. First, since estimating problem data often introduces measurement errors, sensitivity analysis helps in identifying sensitive parameters that should be obtained with relatively high accuracy. Second, sensitivity analysis may help to predict the future changes of the equilibrium as a result of changes in the governing systems. Third, sensitivity analysis provides useful information for designing or planning various equilibrium systems. Furthermore, from mathematical and engineering points of view, sensitivity analysis can provide new insight regarding problems being studied and can stimulate new ideas for problem solving. Over the last decade, there has been increasing interest in studying the sensitivity analysis of variational inequalities and variational inclusions. Sensitivity analysis for variational inclusions and inequalities has been studied extensively; see [1,5,8–11,17,21,31–33,37,40–42]. The techniques suggested so far vary with the problem being studied. Dafermos [5] used the fixed-point formulation to consider the sensitivity analysis of the classical variational inequalities. This technique has been modified and extended by many authors for studying the sensitivity analysis of other classes of variational inequalities and variational inclusions. In this paper, we develop the general framework of sensitivity analysis for the nonconvex variational inequalities. For this purpose, we first establish the equivalence between nonconvex variational inequalities and the Wiener–Hopf equations by using the projection technique. This fixed-point formulation is obtained by a suitable and appropriate rearrangement of the Wiener–Hopf equations. We would like to point out that the Wiener–Hopf equations technique is quite general, unified, and flexible and provides us with a new approach to study the sensitivity analysis of nonconvex variational inequalities and related optimization problems. We use this equivalence to develop sensitivity analysis for the nonconvex variational inequalities without assuming the differentiability of the given data. Our results can be considered as significant extensions of the results of Dafermos [5], Moudafi and Noor [11], Noor and Noor [31,33], and others in this area.

2 Preliminaries

Let H be a real Hilbert space whose inner product and norm are denoted by $\langle \cdot, \cdot \rangle$ and $\| \cdot \|$, respectively. Let K be a nonempty and convex set in H.

We, first of all, recall the following well-known concepts from nonlinear convex analysis and nonsmooth analysis [4, 36].

Definition 2.1. The proximal normal cone of K at $u \in H$ is given by

$$N_K^P(u) := \{\xi \in H : u \in P_K[u + \alpha\xi]\},$$

where $\alpha > 0$ is a constant and

$$P_K[u] = \{u^* \in K : d_K(u) = \|u - u^*\|\}.$$

Here $d_K(\cdot)$ is the usual distance function to the subset K, that is,

$$d_K(u) = \inf_{v \in K} \|v - u\|.$$

The proximal normal cone $N_K^P(u)$ has the following characterization.

Lemma 2.1. *Let K be a nonempty, closed, and convex subset in H. Then $\zeta \in N_K^P(u)$ if and only if there exists a constant $\alpha > 0$ such that*

$$\langle \zeta, v - u \rangle \leq \alpha \|v - u\|^2, \quad \forall v \in K.$$

Definition 2.2. The Clarke normal cone, denoted by $N_K^C(u)$, is defined as

$$N_K^C(u) = \overline{co}[N_K^P(u)],$$

where $\overline{co}$ means the closure of the convex hull.

Clearly $N_K^P(u) \subset N_K^C(u)$, but the converse is not true. Note that $N_K^P(u)$ is always closed and convex, whereas $N_K^C(u)$ is convex, but may not be closed [4, 36].

Poliquin et al. [36] and Clarke et al. [4] have introduced and studied a new class of nonconvex sets, which are called uniformly prox-regular sets. This class of uniformly prox-regular sets has played an important part in many nonconvex applications such as optimization, dynamic systems, and differential inclusions.

Definition 2.3. For a given $r \in (0, \infty]$, a subset K_r is said to be normalized uniformly r-prox-regular if and only if every nonzero proximal normal to K_r can be realized by an r-ball, that is, $\forall u \in K_r$ and $0 \neq \xi \in N_{K_r}^P(u)$ and $\|\xi\| = 1$, one has

$$\langle \xi, v - u \rangle \leq (1/2r)\|v - u\|^2, \quad \forall v \in K_r.$$

It is clear that the class of normalized uniformly prox-regular sets is sufficiently large to include the class of convex sets, p-convex sets, $C^{1,1}$submanifolds (possibly with boundary) of H, the images under a $C^{1,1}$ diffeomorphism of convex sets, and many other nonconvex sets; see [4, 35]. It is clear that if $r = \infty$, then uniformly prox-regularity of K_r is equivalent to the convexity of K. It is known that if K_r is a uniformly prox-regular set, then the proximal normal cone $N_{K_r}^P(u)$ is closed as a set-valued mapping.

For a given nonlinear operator T, we consider the problem of finding $u \in K_r$ such that

$$\langle Tu, v - u\rangle + \gamma\|v - u\|^2 \geq 0, \quad \forall v \in K_r, \tag{1}$$

which is called the *nonconvex variational inequality*, introduced and studied by Noor [29]. See also [3, 22–30, 32] for the variant forms of nonconvex variational inequalities.

We note that, if $K_r \equiv K$, the convex set in H, then problem (1) is equivalent to finding $u \in K$ such that

$$\langle Tu, v - u\rangle \geq 0, \qquad \forall v \in K. \tag{2}$$

Inequality of type (2) is called the *variational inequality*, which was introduced and studied by Stampacchia [39] in 1964. It turned out that a number of unrelated obstacle, free, moving, unilateral, and equilibrium problems arising in various branches of pure and applied sciences can be studied via variational inequalities; see [1–42] and the references therein.

It is well known that problem (2) is equivalent to finding $u \in K$ such that

$$0 \in Tu + N_K(u), \tag{3}$$

where $N_K(u)$ denotes the normal cone of K at u in the sense of convex analysis. Problem (3) is called the variational inclusion associated with variational inequality (2).

If K_r is a nonconvex (uniformly prox-regular) set, then problem (2.1) is equivalent to finding $u \in K_r$ such that

$$0 \in Tu + N^P_{K_r}(u), \tag{4}$$

where $N^P_{K_r}(u)$ denotes the normal cone of K_r at u in the sense of nonconvex analysis. Problem (4) is called the nonconvex variational inclusions problem associated with nonconvex variational inequality (1). This implies that the variational inequality (1) is equivalent to finding a zero of the sum of two monotone operators (4). This equivalent formulation plays a crucial and basic part in this paper. We would like to point out this equivalent formulation allows us to use the projection operator technique for solving the nonconvex variational inequality (1).

We now recall the well-known proposition which summarizes some important properties of the uniform prox-regular sets.

Lemma 2.2. *Let K be a nonempty closed subset of H, $r \in (0, \infty]$, and set $K_r = \{u \in H : d(u, K) < r\}$. If K_r is uniformly prox-regular, then*

(i) $\forall u \in K_r$, $P_{K_r}(u) \neq \emptyset$.
(ii) $\forall r' \in (0, r)$, *P_{K_r} is Lipschitz continuous with constant $\delta = r/(r - r')$ on $K_{r'}$.*
(iii) *The proximal normal cone is closed as a set-valued mapping.*

Noor [29] has established the equivalence between nonconvex variational inequality (1) and the fixed-point problem using the projection operator technique. This alternative formulation is used to discuss the existence of a solution of the problem (1) and to suggest and analyze an iterative method for solving the nonconvex variational inequality (1).

Lemma 2.3 ([29]). *$u \in K_r$ is a solution of the nonconvex variational inequality (1), if and only if $u \in K_r$ satisfies the relation*

$$u = P_{K_r}[u - \rho T u], \tag{5}$$

where P_{K_r} is the projection of H onto the uniformly prox-regular set K_r.

Lemma 2.3 implies that the nonconvex variational inequality (1) is equivalent to the fixed-point problem (5). This alternative equivalent formulation is very useful from the numerical and theoretical point of view.

Related to the nonconvex variational inequality (1), we consider the problem of finding $z \in H$ such that

$$T P_{K_r} z + \rho^{-1} Q_{K_r} z = 0, \tag{6}$$

where $\rho > 0$ is a constant and $Q_{K_r} = I - P_{K_r}$. Here I is the identity operator and P_{K_r} is the projection operator. The equations of the type (6) are called the nonconvex Wiener–Hopf equations. Note that for $K_r \equiv K$, the convex set, we obtain the original Wiener–Hopf equations, considered and studied by Shi [38]. For the formulation and applications of the Wiener–Hopf equations , see [11, 16–33].

We now consider the parametric versions of the problem (1) and (6). To formulate the problem, let M be an open subset of H in which the parameter λ takes values. Let $T(u,\lambda)$ be a given operator defined on $H \times H \times M$ and take value in $H \times H$.

From now onward, we denote $T_\lambda(\cdot) \equiv T(\cdot,\lambda)$ unless otherwise specified.

The parametric convex variational inequality problem is to find $(u,\lambda) \in H \times M$ such that

$$\langle \rho T_\lambda u, v - u\rangle + \gamma\|v - u\|^2 \geq 0, \forall v \in K_r. \tag{7}$$

We also assume that for some $\overline{\lambda} \in M$, problem (4) has a unique solution $\overline{u}$.

Related to the parametric nonconvex variational inequality (7), we consider the parametric nonconvex Wiener–Hopf equations. We consider the problem of finding $(z,\lambda) \in H \times M$, such that

$$T_\lambda P_{K_r} z + \rho^{-1} Q_{K_r} z = 0, \tag{8}$$

where $\rho > 0$ is a constant and $Q_{K_r} z$ is defined on the set of (z,λ) with $\lambda \in M$ and takes values in H. The equations of the type (8) are called the parametric Wiener–Hopf equations.

One can establish the equivalence between the problems (7) and (8) by using the projection operator technique; see Noor [16, 17, 21].

Lemma 2.4. *The parametric nonconvex variational inequality* (7) *has a solution* $(u, \lambda) \in H \times M$ *if and only if the parametric Wiener–Hopf equations* (8) *have a solution* $(z, \lambda) \in H \times M$, *where*

$$u = P_{K_r} z \tag{9}$$

$$z = u - \rho T_\lambda(u). \tag{10}$$

From Lemma 2.4, we see that the parametric nonconvex variational inequalities (7) and the parametric Wiener–Hopf equations (8) are equivalent. We use this equivalence to study the sensitivity analysis of the nonconvex variational inequalities. We assume that for some $\overline{\lambda} \in M$, problem (8) has a solution $\overline{z}$ and X is a closure of a ball in H centered at $\overline{z}$. We want to investigate those conditions under which, for each λ in a neighborhood of $\overline{\lambda}$, problem (8) has a unique solution $z(\lambda)$ near $\overline{z}$ and the function $z(\lambda)$ is (Lipschitz) continuous and differentiable.

Definition 2.4. Let $T_\lambda(\cdot)$ be an operator on $X \times M$. Then, the operator $T_\lambda(\cdot)$ is said to be:

(a) *Locally strongly monotone* if there exists a constant $\alpha > 0$ such that

$$\langle T_\lambda(u) - T_\lambda(v), u - v\rangle \geq \alpha\|u - v\|^2, \quad \forall \lambda \in M, u, v \in X$$

(b) *Locally Lipschitz continuous* if there exists a constant $\beta > 0$ such that

$$\|T_\lambda(u) - T_\lambda(v)\| \leq \beta\|u - v\|, \quad \forall \lambda \in M, u, v \in X$$

3 Main Results

We consider the case, when the solutions of the parametric Wiener–Hopf equations (5) lie in the interior of X. Following the ideas of Dafermos [3] and Noor [13, 14], we consider the map

$$\begin{aligned} F_\lambda(z) &= P_{K_r} z - \rho T_\lambda(P_{K_r} z), \quad \forall (z, \lambda) \in X \times M \\ &= u - \rho T_\lambda(u), \end{aligned} \tag{11}$$

where

$$u = P_{K_r} z. \tag{12}$$

We have to show that the map $F_\lambda(z)$ has a fixed point, which is a solution of the parametric Wiener–Hopf equations (8). First of all, we prove that the map $F_\lambda(z)$, defined by (11), is a contraction map with respect to z uniformly in $\lambda \in M$.

Lemma 3.1. *Let $T_\lambda(\cdot)$ be a locally strongly monotone with constant $\alpha > 0$ and locally Lipschitz continuous with constant $\beta > 0$. Then, for all $z_1, z_2 \in X$ and $\lambda \in M$, we have*

$$\|F_\lambda(z_1) - F_\lambda(z_2)\| \le \theta \|z_1 - z_2\|,$$

where

$$\theta = \delta\sqrt{1 - 2\alpha\rho + \beta^2\rho^2} \tag{13}$$

for

$$\left|\rho - \frac{\alpha}{\beta^2}\right| < \frac{\sqrt{\delta^2\alpha^2 - \beta^2(\delta^2 - 1)}}{\delta\beta^2}, \quad \delta\alpha > \beta\sqrt{\delta^2 - 1}. \tag{14}$$

Proof. For all $z_1, z_2 \in X$, $\lambda \in M$, we have, from (11),

$$\|F_\lambda(z_1) - F_\lambda(z_2)\| = \|u_1 - u_2 - \rho(T_\lambda(u_1) - T_\lambda(u_2))\|. \tag{15}$$

Using the strongly monotonicity and Lipschitz continuity of the operator T_λ, we have

$$\begin{aligned} ||u_1 - u_2 - \rho(T_\lambda(u_1) - T_\lambda(u_2))||^2 &\le ||u_1 - u_2||^2 - 2\rho\langle T_\lambda(u_1) - T_\lambda(u_2), u_1 - u_2\rangle \\ &\quad + \rho^2||T_\lambda(u_1) - T_\lambda(u_2)||^2 \\ &\le (1 - 2\rho\alpha + \rho^2\beta^2)||u_1 - u_2||^2, \end{aligned} \tag{16}$$

where $\alpha > 0$ is the strongly monotonicity constant and $\beta > 0$ is the Lipschitz continuity constant of the operator T_λ, respectively.

From (15) and (16), we have

$$\|F_\lambda(z_1) - F_\lambda(z_2)\| \le \sqrt{1 - 2\alpha\rho + \beta^2\rho^2}\|u_1 - u_2\|. \tag{17}$$

From (12) and using the Lipschitz continuity of the operator P_{K_r}, we have

$$\|u_1 - u_2\| \le \|P_{K_r}z_1 - P_{K_r}z_2\| \le \delta\|z_1 - z_2\|. \tag{18}$$

Combining (17), (18), and using (13), we have

$$\begin{aligned} \|F_\lambda(z_1) - F_\lambda(z_2)\| &\le \delta\sqrt{1 - 2\rho\alpha + \rho^2\beta^2}\|z_1 - z_2\| \\ &= \theta\|z_1 - z_2\|. \end{aligned}$$

From (14), it follows that $\theta < 1$ and consequently the map $F_\lambda(z)$ defined by (11) is a contraction map and has a fixed point $z(\lambda)$, which is the solution of the Wiener–Hopf equation (5). □

Remark 3.1. From Lemma 3.1, we see that the map $F_\lambda(z)$ defined by (11) has a unique fixed point $z(\lambda)$, that is, $z(\lambda) = F_\lambda(z)$. Also, by assumption, the function $\overline{z}$, for $\lambda = \overline{\lambda}$ is a solution of the parametric Wiener–Hopf equations (8). Again using Lemma 3.1, we see that $\overline{z}$, for $\lambda = \overline{\lambda}$, is a fixed point of $F_\lambda(z)$ and it is also a fixed point of $F_{\overline{\lambda}}(z)$. Consequently, we conclude that

$$z(\overline{\lambda}) = \overline{z} = F_{\overline{\lambda}}(z(\overline{\lambda})).$$

Using Lemma 3.1 and technique of Noor [17, 21], we can prove the continuity of the solution $z(\lambda)$ of the parametric Wiener–Hopf equations (8). However, for the sake of completeness and to convey an idea of the techniques involved, we give its proof.

Lemma 3.2. *Assume that the operator $T_\lambda(\cdot)$ is locally Lipschitz continuous with respect to the parameter λ. If the operator $T_\lambda(\cdot)$ is locally Lipschitz continuous and the map $\lambda \to P_{K_{r\lambda}}z$ is continuous (or Lipschitz continuous), then the function $z(\lambda)$ satisfying* (8) *is (Lipschitz) continuous at $\lambda = \overline{\lambda}$.*

Proof. For all $\lambda \in M$, invoking Lemma 3.1 and the triangle inequality, we have

$$\begin{aligned}\|z(\lambda) - z(\bar{\lambda})\| &\leq \|F_\lambda(z(\lambda)) - F_{\bar{\lambda}}(z(\bar{\lambda})\| + \|F_\lambda(z(\bar{\lambda})) - F_{\bar{\lambda}}(z(\bar{\lambda}))\| \\ &\leq \theta\|z(\lambda) - z(\bar{\lambda})\| + \|F_\lambda(z(\bar{\lambda})) - F_{\bar{\lambda}}(z(\bar{\lambda}))\|. \end{aligned} \tag{19}$$

From (11) and the fact that the operator T_λ is a Lipschitz continuous with respect to the parameter λ, we have

$$\begin{aligned}\|F_\lambda(z(\bar{\lambda})) - F_{\bar{\lambda}}(z(\bar{\lambda}))\| &= \|u(\bar{\lambda}) - u(\bar{\lambda}) + \rho(T_\lambda(u(\bar{\lambda}), u(\bar{\lambda})) - T_{\bar{\lambda}}(u(\bar{\lambda}), u(\bar{\lambda})))\| \\ &\leq \rho\mu\|\lambda - \bar{\lambda}\|. \end{aligned} \tag{20}$$

Combining (19) and (20), we obtain

$$\|z(\lambda) - z(\bar{\lambda})\| \leq \frac{\rho\mu}{1-\theta}\|\lambda - \bar{\lambda}\|, \quad \text{for all } \lambda, \bar{\lambda} \in M,$$

from which the required result follows. □

We now prove the main result of this paper and is the motivation of our next result.

Theorem 3.1. *Let $\overline{u}$ be the solution of the parametric general variational inequality* (7) *and $\overline{z}$ be the solution of the parametric Wiener–Hopf equations* (8) *for $\lambda = \overline{\lambda}$. Let $T_\lambda(u)$ be the locally strongly monotone Lipschitz continuous operator for all*

$u, v \in X$. If the map $\lambda \to P_{K_r}$ is (Lipschitz) continuous at $\lambda = \overline{\lambda}$, then there exists a neighborhood $N \subset M$ of $\overline{\lambda}$ such that for $\lambda \in N$, the parametric Wiener–Hopf equations (8) *have a unique solution $z(\lambda)$ in the interior of X, $z(\overline{\lambda}) = \overline{z}$ and $z(\lambda)$ is (Lipschitz) continuous at $\lambda = \overline{\lambda}$.*

Proof. Its proof follows from Lemmas 3.1, 3.2 and Remark 3.1. □

4 Conclusion

In this paper, we have introduced and studied a new class of variational inequalities, which is called the nonconvex variational inequality. We have shown that the parametric nonconvex variational inequalities are equivalent to parametric nonconvex Wiener–Hopf equations. This alternative equivalence formulation has been used to develop the general framework of the sensitivity analysis of the parametric nonconvex variational inequalities. Results proved in this paper can be extended for the nonconvex multivalued variational inequalities. This is another direction for future direction. We expect that the ideas and techniques of this paper will motivate and inspire the interested readers to explore its novel and other applications in various fields.

Acknowledgements The author would like to express his sincere gratitude to Dr. M. Junaid Zaidi, Rector, CIIT, for providing excellent research facilities. The author would like to thank Prof. Dr. Th. M. Rassias for his kind invitation and his encouragement.

References

1. Agarwal, R.P., Cho, J.J., Huang, N.J.: Sensitivity analysis for strongly nonlinear quasi variational inclusions. Appl. Math. Lett. **13**, 19–24 (2000)
2. Baiocchi, C., Capelo, A.: Variational and Quasi Variational Inequalities. Wiley, London (1984)
3. Bounkhel, M., Tadji, L., Hamdi, A.: Iterative schemes to solve nonconvex variational problems. J. Inequal. Pure Appl. Math. **4**, 1–14 (2003)
4. Clarke, F.H., Ledyaev, Y.S., Wolenski, P.R.: Nonsmooth Analysis and Control Theory. Springer, Berlin (1998)
5. Dafermos, S.: Sensitivity analysis in variational inequalities. Math. Oper. Res. **13**, 421–434 (1988)
6. Giannessi, F., Maugeri, A.: Variational Inequalities and Network Equilibrium Problems. Plenum Press, New York (1995)
7. Kinderlehrer, D., Stampacchia, G.: An Introduction to Variational Inequalities and Their Applications. SIAM, Philadelphia (2000)
8. Kyparisis, J.: Sensitivity analysis framework for variational inequalities. Math. Program. **38**, 203–213 (1987)
9. Kyparisis, J.: Sensitivity analysis for variational inequalities and nonlinear complementarity problems. Ann. Oper. Res. **27**, 143–174 (1990)

10. Lions, J.L., Stampacchia, G.: Variational inequalities. Commun. Pure Appl. Math. **20**, 493–512 (1967)
11. Liu, J.: Sensitivity analysis in nonlinear programs and variational inequalities via continuous selections. SIAM J. Control Optim. **33**, 1040–1068 (1995)
12. Moudafi, A., Noor, M.A.: Sensitivity analysis for variational inclusions by Wiener–Hopf equations technique. J. Appl. Math. Stoch. Anal. **12**, 223–232 (1999)
13. Noor, M.A.: On variational inequalities. Ph.D. Thesis, Brunel University, London, UK (1975)
14. Noor, M.A.: General variational inequalities. Appl. Math. Lett. **1**, 119–121 (1988)
15. Noor, M.A.: Quasi variational inequalities. Appl. Math. Lett. **1**, 367–370 (1988)
16. Noor, M.A.: Wiener-Hopf equations and variational inequalities. J. Optim. Theory Appl. **79**, 197–206 (1993)
17. Noor, M.A.: Sensitivity analysis for quasi variational inequalities. J. Optim. Theory Appl. **95**, 399–407 (1997)
18. Noor, M.A.: Some recent advances in variational inequalities, part I, basic concepts. New Zealand J. Math. **26**, 53–80 (1997)
19. Noor, M.A.: Some recent advances in variational inequalities, part II, other concepts. New Zealand J. Math. **26**, 229–255 (1997)
20. Noor, M.A.: New approximation schemes for general variational inequalities. J. Math. Anal. Appl. **251**, 217–229 (2000)
21. Noor, M.A.: Some developments in general variational inequalities. Appl. Math. Comput. **152**, 199–277 (2004)
22. Noor, M.A.: Fundamentals of mixed quasi variational inequalities. Int. J. Pure Appl. Math. **15**, 137–258 (2004)
23. Noor, M.A.: Iterative schemes for nonconvex variational inequalities. J. Optim. Theory Appl. **121**, 385–395 (2004)
24. Noor, M.A.: Fundamentals of equilibrium problems. Math. Inequal. Appl. **9**, 529–566 (2006)
25. Noor, M.A.: Merit functions for general variational inequalities. J. Math. Anal. Appl. **316**, 736–752 (2006)
26. Noor, M.A.: Differentiable nonconvex functions and general variational inequalities. Appl. Math. Comput. **199**, 623–630 (2008)
27. Noor, M.A.: On a class of general variational inequalities. J. Adv. Math. Stud. **1**, 75–86 (2008)
28. Noor, M.A.: Extended general variational inequalities. Appl. Math. Lett. **22**, 182–186 (2009)
29. Noor, M.A.: Projection methods for nonconvex variational inequalities. Optim. Lett. **3**, 411–418 (2009)
30. Noor, M.A.: Some iterative methods for general nonconvex variational inequalities. Comput. Math. Model. **21**, 97–109 (2010)
31. Noor, M.A., Noor, K.I.: Sensitivity analysis for quasi variational inclusions. J. Math. Anal. Appl. **236**, 290–299 (1999)
32. Noor, M.A., Noor, K.I.: Auxiliary principle technique for solving split feasibility problems. Appl. Math. Inform. Sci. **7**(1), 181–187 (2013)
33. Noor, M.A., Noor, K.I.: Sensitivity analysis of some quasi variational inequalities. J. Adv. Math. Stud. **6**(1), 43–52 (2013)
34. Noor, M.A., Noor, K.I., Rassias, Th.M.: Some aspects of variational inequalities. J. Comput. Appl. Math. **47**, 285–312 (1993)
35. Noor, M.A., Noor, K.I., Al-Said, E.: Iterative methods for solving nonconvex equilibrium variational inequalities. Appl. Math. Inform. Sci. **6**(1), 65–69 (2012)
36. Poliquin, R.A., Rockafellar, R.T., Thibault, L.: Local differentiability of distance functions. Trans. Amer. Math. Soc. **352**, 5231–5249 (2000)
37. Qiu, Y., Magnanti, T.L.: Sensitivity analysis for variational inequalities defined on polyhedral sets. Math. Oper. Res. **14**, 410–432 (1989)
38. Shi, P.: Equivalence of Wiener-Hopf equations with variational inequalities. Proc. Amer. Math. Soc. **111**, 339–346 (1991)
39. Stampacchia, G.: Formes bilineaires coercitives sur les ensembles convexes. C. R. Acad. Sci. Paris **258**, 4413–4416 (1964)

40. Tobin, R.L.: Sensitivity analysis for variational inequalities. J. Optim. Theory Appl. **48**, 191–204 (1986)
41. Yen, N.D.: Holder continuity of solutions to a parametric variational inequality. Appl. Math. Optim. **31**, 245–255 (1995)
42. Yen, N.D., Lee, G.M.: Solution sensitivity of a class of variational inequalities. J. Math. Anal. Appl. **215**, 46–55 (1997)

Part III
Approximation of Functions and Quadratures

Simultaneous Approximation for Stancu-Type Generalization of Certain Summation–Integral-Type Operators

N.K. Govil and Vijay Gupta

Dedicated to Professor Hari M. Srivastava

Abstract Srivastava and Gupta (Math. and Comput. Model. 37:1307–1315, 2003) introduced a general sequence of summation–integral-type operators, which in the literature have sometimes been termed as Srivastava–Gupta operators. In this paper we consider Stancu-type generalization of these operators and obtain moments of these operators by method of hypergeometric series. Also, for these operators we derive the asymptotic formula and error estimation in simultaneous approximation.

1 Introduction

In Approximation Theory after the well-known Bernstein polynomials, from time to time, several new operators have been introduced and their approximation properties studied, for example, there are exponential-type operators, which include some operators of discrete type such as Bernstein, Baskakov, and Szász operators. In 1976, May [10] studied exponential-type operators and established direct, inverse, and saturation results for the linear combinations of these operators. It is known that the discrete-type operators are not able to approximate integrable functions, and in this context, Kantorovich [9] first proposed the integral modification of well-known Bernstein polynomials. Later in 1967, Durrmeyer [3] considered a more general integral modification of the Bernstein polynomials. In 1985, Sahai

N.K. Govil (✉)
Department of Mathematics and Statistics, Auburn University, Auburn, AL 36849-5108, USA
e-mail: govilnk@auburn.edu

V. Gupta
Department of Mathematics, Netaji Subhas Institute of Technology, Sector 3,
Dwarka, New Delhi-110078, India
e-mail: vijaygupta2001@hotmail.com

G.V. Milovanović and M.Th. Rassias (eds.), *Analytic Number Theory, Approximation Theory, and Special Functions*, DOI 10.1007/978-1-4939-0258-3_19,

and Prasad [13] introduced Baskakov–Durrmeyer operators and estimated direct results in simultaneous approximation (approximation of derivatives of functions by corresponding order derivatives of operators), which were later improved by Sinha et al. [14]. Also in the same year Mazhar and Totik [11] introduced Szász–Mirakyan–Durrmeyer operators and estimated some approximation properties, including some direct results. In 1989, Heilmann [6] proposed a general sequence of operators, from which as special cases one can obtain Bernstein–Durrmeyer operators, Baskakov–Durrmeyer operators, and Szász–Mirakyan–Durrmeyer operators.

In Approximation Theory the genuine operators are also very important, as they are defined implicitly with values of functions at end points of the interval in which the operators are defined. Phillips [12] in 1954 introduced such operators, and later Mazhar and Totik [11] discussed these operators in different forms. In the year 2003, Srivastava and Gupta [15] introduced a general sequence of linear positive operators $G_{n,c}(f,x)$ which when applied to f are defined as

$$G_{n,c}(f,x) = n\sum_{k=1}^{\infty} p_{n,k}(x;c)\int_0^{\infty} p_{n+c,k-1}(t;c)\, f(t)\, \mathrm{d}t + p_{n,0}(x;c)\, f(0)\,, \quad (1)$$

where

$$p_{n,k}(x;c) = \frac{(-x)^k}{k!}\phi_{n,c}^{(k)}(x) \quad (2)$$

and

$$\phi_{n,c}(x) = \begin{cases} \mathrm{e}^{-nx}, & c = 0, \\ (1+cx)^{-n/c}, & c \in \mathbb{N} := \{1,2,3,\ldots\}, \\ (1-x)^n, & c = -1. \end{cases}$$

Here $\{\phi_{n,c}(x)\}_{n=1}^{\infty}$ is a sequence of functions, defined on the closed interval $[0,b]$ $(b>0)$ and satisfying the following properties for every $n \in \mathbb{N}$ and $k \in \mathbb{N}_0 := \mathbb{N} \cup \{0\}$:

1. $\phi_{n,c} \in C^{\infty}([a,b])$ $(b > a \geq 0)$.
2. $\phi_{n,c}(0) = 1$.
3. $\phi_{n,c}(x)$ is completely monotone, that is, $(-1)^k\,\phi_{n,c}^{(k)}(x) \geq 0$ $(0 \leq x \leq b)$.
4. There exists an integer c such that

$$\phi_{n,c}^{(k+1)}(x) = -n\phi_{n+c,c}^{(k)}(x) \quad (n > \max\{0,-c\};\ x \in [0,b])\,.$$

In the literature, these operators have sometimes been termed as Srivastava–Gupta operators (see [2, 8, 17]). The authors in [8] considered the Bézier variant of these operators and estimated the rate of convergence for functions of bounded variation.

The following are some of the special cases of the operators $G_{n,c}(f,x)$ defined in (1), which have the following forms:

1. If $c = 0$, then by simple computation one has $p_{n,k}(x;0) = e^{-nx}\frac{(nx)^k}{k!}$ and operators become the Phillips operators $G_{n,0}(f,x)$, introduced by Phillips in [12], which for $x \in [0,\infty)$ are defined by

$$G_{n,0}(f,x) = n\sum_{k=1}^{\infty} p_{n,k}(x;0)\int_0^{\infty} p_{n,k-1}(t;0)\,f(t)\,\mathrm{d}t + p_{n,0}(x;0)\,f(0).$$

2. If $c = 1$, then by simple computation one has

$$p_{n,k}(x;1) = \binom{n+k-1}{k}\frac{x^k}{(1+x)^{n+k}},$$

and operators become the Durrmeyer-type Baskakov operators $G_{n,1}(f,x)$, which were introduced by Gupta et al. in [5] and which for $x \in [0,\infty)$ are defined as

$$G_{n,1}(f,x) = n\sum_{k=1}^{\infty} p_{n,k}(x;1)\int_0^{\infty} p_{n+1,k-1}(t;1)\,f(t)\,\mathrm{d}t + p_{n,0}(x;1)\,f(0).$$

3. If $c = -1$, then by simple computation one gets

$$p_{n,k}(x;-1) = \binom{n}{k}x^k(1-x)^{n-k},$$

and operators become the Bernstein–Durrmeyer type $G_{n,-1}(f,x)$. In this case summation runs from 1 to n, integration from 0 to 1, and $x \in [0,1]$, and $G_{n,-1}(f,x)$ are defined as

$$G_{n,-1}(f,x) = n\sum_{k=1}^{n} p_{n,k}(x;-1)\int_0^{1} p_{n-1,k-1}(t;-1)\,f(t)\,\mathrm{d}t + p_{n,0}(x;-1)\,f(0).$$

The q-analogue of this case was studied in [4].

Based on two parameters α, β satisfying the conditions $0 \le \alpha \le \beta$, in 2003 the Stancu-type generalization of Bernstein operators was given in [16]. Recently, Bykyazici and Atakut [1] studied the Stancu-type generalization of the q-analogue of the classical Baskakov operators. Motivated by this recent work of Bykyazici and Atakut [1] on Stancu-type operators, here in this paper firstly we consider the Stancu-type generalization of the Srivastava–Gupta operators for $0 \le \alpha \le \beta$ as

$$G_{n,c}^{\alpha,\beta}(f,x) = n\sum_{k=1}^{\infty} p_{n,k}(x;c)\int_0^{\infty} p_{n+c,k-1}(t;c)\,f\Big(\frac{nt+\alpha}{n+\beta}\Big)\,\mathrm{d}t + p_{n,0}(x;c)\,f\Big(\frac{\alpha}{n+\beta}\Big),$$

where $p_{n,k}(x;c)$ is given in (2).

Then, in this paper we study the simultaneous approximation for the case $c = 1$ of the above defined operators $G_{n,c}^{\alpha,\beta}(f,x)$ and establish Voronovskaja type asymptotic formula and an estimation of error. We obtain the moments by using the concept of hypergeometric series, a technique developed recently by Ismail and Simeonov [7], who obtained direct and inverse results for Jacobi weights of Beta operators. The other cases, i.e., $c = 0$ and $c = -1$, will be discussed later in forthcoming papers. It may be mentioned that recently, Deo [2] has studied a different form of Stancu operators and obtained moments and direct results in ordinary approximation.

2 Alternate Form and Auxiliary Results

The operators $G_{n,c}^{\alpha,\beta}(f,x)$ for the case $c = 1$ can be written in the following alternate form. For simplicity, we will denote $G_{n,1}^{\alpha,\beta}(f,x)$ by $G_n^{\alpha,\beta}(f,x)$ and for $\alpha = \beta = 0$, $G_{n,1}^{0,0}(f,x)$ by $G_n(f,x)$. Also, we will denote $p_{n,k}(x;1)$ simply by $p_{n,k}(x)$. Thus, for $x \in [0,\infty)$, we have the following form:

$$G_n^{\alpha,\beta}(f,x) = \int_0^\infty K_n(x,t) f\left(\frac{nt+\alpha}{n+\beta}\right) \mathrm{d}t \tag{3}$$

$$= \sum_{k=1}^{\infty} p_{n,k}(x) \int_0^\infty b_{n,k-1}(t) f\left(\frac{nt+\alpha}{n+\beta}\right) dt + (1+x)^{-n} f\left(\frac{\alpha}{n+\beta}\right),\ x \in [0,\infty), \tag{4}$$

where the kernel $K_n(x,t) = \sum_{k=1}^{\infty} p_{n,k}(x) b_{n,k-1}(t) + (1+x)^{-n}\delta(t)$, with $\delta(t)$ the Dirac delta function. As is easy to see, Baskakov and Beta basis functions for these are given by

$$p_{n,k}(x) = \binom{n+k-1}{k} \frac{x^k}{(1+x)^{n+k}} = \frac{(n)_k}{k!} \frac{x^k}{(1+x)^{n+k}}$$

and

$$b_{n,k-1}(t) = \frac{1}{B(n,k)} \frac{t^{k-1}}{(1+t)^{n+k}} = \frac{(n)_k}{(k-1)!} \frac{t^{k-1}}{(1+t)^{n+k}},$$

where the Pochhammer symbol $(n)_k$ is defined as

$$(n)_k = n(n+1)(n+2)(n+3)\cdots(n+k-1),$$

and $B(n,k)$ is the usual Beta function.

Note that the operator given in (4) can be written as

$$G_n^{\alpha,\beta}(f,x) = \sum_{k=1}^{\infty} \frac{(n)_k}{k!}\frac{x^k}{(1+x)^{n+k}} \int_0^{\infty} \frac{(n)_k}{(k-1)!}\frac{t^{k-1}}{(1+t)^{n+k}} f\left(\frac{nt+\alpha}{n+\beta}\right) \mathrm{d}t$$
$$+ p_{n,0}(x) f\left(\frac{\alpha}{n+\beta}\right)$$
$$= \int_0^{\infty} \frac{f\left(\frac{nt+\alpha}{n+\beta}\right)x}{[(1+x)(1+t)]^{n+1}} \sum_{k=1}^{\infty} \frac{(n)_k (n)_k}{(k-1)!k!} \frac{(xt)^{k-1}}{[(1+x)(1+t)]^{k-1}} \mathrm{d}t$$
$$+ p_{n,0}(x) f\left(\frac{\alpha}{n+\beta}\right)$$
$$= n^2 \int_0^{\infty} \frac{f\left(\frac{nt+\alpha}{n+\beta}\right)x}{[(1+x)(1+t)]^{n+1}} \sum_{k=0}^{\infty} \frac{(n+1)_k (n+1)_k}{(2)_k k!} \frac{(xt)^{k}}{[(1+x)(1+t)]^{k}} \mathrm{d}t$$
$$+ p_{n,0}(x) f\left(\frac{\alpha}{n+\beta}\right).$$

Using the hypergeometric series ${}_2F_1(a,b;c;x) = \sum_{k=0}^{\infty} \frac{(a)_k (b)_k}{(c)_k k!} x^k$, we get

$$G_n^{\alpha,\beta}(f,x) = n^2 \int_0^{\infty} \frac{f\left(\frac{nt+\alpha}{n+\beta}\right)x}{[(1+x)(1+t)]^{n+1}} \, {}_2F_1\left(n+1, n+1; 2; \frac{xt}{(1+x)(1+t)}\right) \mathrm{d}t$$
$$+ p_{n,0}(x) f\left(\frac{\alpha}{n+\beta}\right),$$

which on applying Pfaff–Kummer transformation

$${}_2F_1(a,b;c;x) = (1-x)^{-a} \, {}_2F_1\left(a, c-b; c; \frac{x}{x-1}\right)$$

gives

$$G_n^{\alpha,\beta}(f,x) = n^2 \int_0^{\infty} \frac{f\left(\frac{nt+\alpha}{n+\beta}\right)x}{(1+x+t)^{n+1}} \, {}_2F_1\left(n+1, 1-n; 2; \frac{-xt}{1+x+t}\right) \mathrm{d}t$$
$$+ p_{n,0}(x) f\left(\frac{\alpha}{n+\beta}\right), \tag{5}$$

and the above gives an alternate form of the operators (4) in terms of hypergeometric functions.

As a special case, if $\alpha = \beta = 0$ the operators (5) reduce to the Baskakov–Beta operators, that is, the case $c = 1$ of (1).

Now we present the following lemmas, which would be needed for the proofs of the theorems concerning direct estimates given in Sect. 3.

Lemma 2.1. *For $n > 0$ and $r \geq 1$, we have*

$$G_n(t^r, x) = \frac{\Gamma(n-r)\Gamma(r+1)}{\Gamma(n)} nx(1+x)^{r-1}\, {}_2F_1\left(1-n, 1-r; 2; \frac{x}{1+x}\right). \tag{6}$$

Moreover,

$$G_n(t^r, x) = \frac{(n+r-1)!(n-r-1)!}{((n-1)!)^2} x^r + r(r-1)\frac{(n+r-2)!(n-r-1)!}{((n-1)!)^2} x^{r-1} + O(n^{-2}). \tag{7}$$

Proof. Taking $f(t) = t^r$, then making the transformation $t = (1+x)u$, and using Pfaff–Kummer transformation, we get

$$\begin{aligned} G_n(t^r, x) &= n^2 \int_0^\infty \frac{x(1+x)^r u^r}{((1+x)(1+u))^{n+1}} (1+x) \sum_{k=0}^\infty \frac{(n+1)_k (1-n)_k}{(2)_k k!} \frac{(-x(1+x)u)^k}{((1+x)(1+u))^k} du \\ &= n^2 \sum_{k=0}^\infty \frac{(n+1)_k (1-n)_k}{(2)_k k!} (-x)^k x (1+x)^{r-n} \int_0^\infty \frac{u^{r+k}}{(1+u)^{n+k+1}} du \\ &= n^2 \sum_{k=0}^\infty \frac{(n+1)_k (1-n)_k}{(2)_k k!} (-x)^k x (1+x)^{r-n} B(r+k+1, n-r) \\ &= n^2 \sum_{k=0}^\infty \frac{(n+1)_k (1-n)_k}{(2)_k k!} (-x)^k x (1+x)^{r-n} \frac{\Gamma(r+k+1)\Gamma(n-r)}{\Gamma(n+k+1)}, \end{aligned}$$

which, on using $\Gamma(n+k+1) = \Gamma(n+1)(n+1)_k$, gives

$$\begin{aligned} G_n(t^r, x) &= n^2 \sum_{k=0}^\infty \frac{(n+1)_k (1-n)_k}{(2)_k k!} (-x)^k x (1+x)^{r-n} \frac{\Gamma(r+1)(r+1)_k \Gamma(n-r)}{\Gamma(n+1)(n+1)_k} \\ &= n^2 x (1+x)^{r-n} \frac{\Gamma(r+1)\Gamma(n-r)}{\Gamma(n+1)} \sum_{k=0}^\infty \frac{(r+1)_k (1-n)_k}{(k!)^2} (-x)^k \\ &= n^2 x (1+x)^{r-n} \frac{\Gamma(r+1)\Gamma(n-r)}{\Gamma(n+1)}\, {}_2F_1(1-n, r+1; 2; -x). \end{aligned}$$

Finally, on using ${}_2F_1(a,b;c;x) = (1-x)^{-a}\,{}_2F_1\left(a, c-b; c; \frac{x}{x-1}\right)$, we get

$$G_n(t^r, x) = \frac{\Gamma(n-r)\Gamma(r+1)}{\Gamma(n)} nx(1+x)^{r-1}\,{}_2F_1\left(1-n, 1-r; 2; \frac{x}{1+x}\right).$$

The other consequence (7) follows from the above equation by writing the expansion of hypergeometric series. □

Lemma 2.2. *For $0 \le \alpha \le \beta$, we have*

$$\begin{aligned}
G_n^{\alpha,\beta}(t^r, x) &= x^r \frac{n^r}{(n+\beta)^r} \frac{(n+r-1)!(n-r-1)!}{((n-1)!)^2} \\
&+ x^{r-1}\left\{ r(r-1)\frac{n^r}{(n+\beta)^r} \frac{(n+r-2)!(n-r-1)!}{((n-1)!)^2} \right. \\
&\left. + r\alpha \frac{n^{r-1}}{(n+\beta)^r} \frac{(n+r-2)!(n-r)!}{((n-1)!)^2} \right\} \\
&+ x^{r-2}\left\{ r(r-1)(r-2)\alpha \frac{n^{r-1}}{(n+\beta)^r} \frac{(n+r-3)!(n-r)!}{((n-1)!)^2} \right. \\
&\left. + \frac{r(r-1)\alpha^2}{2} \frac{n^{r-2}}{(n+\beta)^r} \frac{(n+r-3)!(n-r+1)!}{((n-1)!)^2} \right\} + O(n^{-2}).
\end{aligned}$$

Proof. The relation between operators $G_n(f, x)$ and (5) can be defined as

$$\begin{aligned}
G_n^{\alpha,\beta}(t^r, x) &= \sum_{j=0}^{r} \binom{r}{j} \frac{n^j \alpha^{r-j}}{(n+\beta)^r} G_n(t^j, x) \\
&= \frac{n^r}{(n+\beta)^r} G_n(t^r, x) + r\alpha \frac{n^{r-1}}{(n+\beta)^r} G_n(t^{r-1}, x) \\
&+ \frac{r(r-1)\alpha^2}{2} \frac{n^{r-2}}{(n+\beta)^r} G_n(t^{r-2}, x) + \cdots + \frac{\alpha^r}{(n+\beta)^r} G_n(1, x),
\end{aligned}$$

which on using (7) gives the required result. □

Lemma 2.3. *Let $m \in \mathbb{N}\bigcup\{0\}$, and*

$$U_{n,m}(x) = \sum_{k=0}^{\infty} p_{n,k}(x) \left(\frac{k}{n} - x\right)^m.$$

Then $U_{n,0}(x) = 1, U_{n,1}(x) = 0$, and we have the recurrence relation:

$$nU_{n,m+1}(x) = x(1+x)\left[U'_{n,m}(x) + mU_{n,m-1}(x)\right].$$

Consequently, $U_{n,m}(x) = O\left(n^{-[(m+1)/2]}\right)$, *where* $[m]$ *is integral part of* m.

The above result is due to Sinha et al. [14], and we omit the proof as one can find it in [14, p. 219].

Lemma 2.4. *For* $m \in \mathbb{N}^0$, *if we define the central moments as*

$$\begin{aligned}\mu_{n,m}(x) &= G_n^{\alpha,\beta}((t-x)^m, x)\\ &= \sum_{k=1}^{\infty} p_{n,k}(x) \int_0^{\infty} b_{n,k-1}(t) \left(\frac{nt+\alpha}{n+\beta} - x\right)^m dt + \left(\frac{\alpha}{n+\beta} - x\right)^m p_{n,0}(x),\end{aligned}$$

then for $n > m+1$, *we have the following recurrence relation:*

$$\begin{aligned}&(n-m-1)\left(\frac{n+\beta}{n}\right)\mu_{n,m+1}(x)\\ &= x(1+x)\left[\mu'_{n,m}(x) + m\mu_{n,m-1}(x)\right]\\ &\quad + m\left(\frac{\alpha}{n+\beta} - x\right)\left[\left(\frac{\alpha}{n+\beta} - x\right)\left(\frac{n+\beta}{n}\right) - 1\right]\mu_{n,m-1}(x)\\ &\quad + \left[(nx-1) + (n-2m-1)\frac{n+\beta}{n}\left(\frac{\alpha}{n+\beta} - x\right) + (m+1)\right]\mu_{n,m}(x).\end{aligned}$$

Further, one can obtain first two moments as

$$\mu_{n,0}(x) = 1, \quad \mu_{n,1}(x) = \frac{x\,(n+\beta(1-n)) + \alpha(n-1)}{(n+\beta)(n-1)},$$

and from the recurrence relation, it can be easily verified that for all $x \in [0, \infty)$, *we have*

$$\mu_{n,m}(x) = O(n^{-[(m+1)/2]}).$$

Proof. Note that from the definition of the operators in (3), we have $\mu_{n,0}(x) = 1$. The other moments follow from the recurrence relation. Now we prove the recurrence relation as follows:

Using the identities

$$x(1+x)p'_{n,k}(x) = (k - nx)p_{n,k}(x)$$

and

$$t(1+t)b'_{n,k}(t) = (k - (n+1)t)\, b_{n,k}(t),$$

we have

$$x(1+x)\mu'_{n,m}(x) = \sum_{k=1}^{\infty}(k-nx)p_{n,k}(x)\int_0^{\infty} b_{n,k}(t)\left(\frac{nt+\alpha}{n+\beta}-x\right)^m \mathrm{d}t$$
$$- mx(1+x)\mu_{n,m-1}(x) - nx(1+x)^{-n}\left(\frac{\alpha}{n+\beta}-x\right)^m.$$

Thus,

$$\begin{aligned}
& x(1+x)\left[\mu'_{n,m}(x)+m\mu_{n,m-1}(x)\right] \\
&= \sum_{k=1}^{\infty} p_{n,k}(x)\int_0^{\infty}(k-nx)b_{n,k-1}(t)\left(\frac{nt+\alpha}{n+\beta}-x\right)^m \mathrm{d}t - nx(1+x)^{-n}\left(\frac{\alpha}{n+\beta}-x\right)^m \\
&= \sum_{k=1}^{\infty} p_{n,k}(x)\int_0^{\infty}[\{(k-1)-(n+1)t\}+(n+1)t+(1-nx)]b_{n,k-1}(t)\left(\frac{nt+\alpha}{n+\beta}-x\right)^m \mathrm{d}t \\
&\quad -nx(1+x)^{-n}\left(\frac{\alpha}{n+\beta}-x\right)^m \\
&= \sum_{k=1}^{\infty} p_{n,k}(x)\int_0^{\infty} t(1+t)b'_{n,k-1}(t)\left(\frac{nt+\alpha}{n+\beta}-x\right)^m \mathrm{d}t \\
&\quad + (n+1)\sum_{k=1}^{\infty} p_{n,k}(x)\int_0^{\infty} b_{n,k-1}(t)t\left(\frac{nt+\alpha}{n+\beta}-x\right)^m \mathrm{d}t \\
&\quad + (1-nx)\sum_{k=1}^{\infty} p_{n,k}(x)\int_0^{\infty} b_{n,k-1}(t)\left(\frac{nt+\alpha}{n+\beta}-x\right)^m \mathrm{d}t \\
&\quad - nx(1+x)^{-n}\left(\frac{\alpha}{n+\beta}-x\right)^m,
\end{aligned} \tag{8}$$

which, on using the identity $t = \frac{n+\beta}{n}\left[\frac{nt+\alpha}{n+\beta}-x-\left(\frac{\alpha}{n+\beta}-x\right)\right]$, gives

$$\begin{aligned}
& x(1+x)\left[\mu'_{n,m}(x)+m\mu_{n,m-1}(x)\right] \\
&= \frac{n+\beta}{n}\sum_{k=1}^{\infty} p_{n,k}(x)\int_0^{\infty} b'_{n,k-1}(t)\left(\frac{nt+\alpha}{n+\beta}-x\right)^{m+1} \mathrm{d}t \\
&\quad - \frac{n+\beta}{n}\left(\frac{\alpha}{n+\beta}-x\right)\sum_{k=1}^{\infty} p_{n,k}(x)\int_0^{\infty} b'_{n,k-1}(t)\left(\frac{nt+\alpha}{n+\beta}-x\right)^m \mathrm{d}t \\
&\quad + \left(\frac{n+\beta}{n}\right)^2\Bigg[\sum_{k=1}^{\infty} p_{n,k}(x)\int_0^{\infty} b'_{n,k-1}(t)\left(\frac{nt+\alpha}{n+\beta}-x\right)^{m+2} \mathrm{d}t \\
&\quad + \left(\frac{\alpha}{n+\beta}-x\right)^2\sum_{k=1}^{\infty} p_{n,k}(x)\int_0^{\infty} b'_{n,k-1}(t)\left(\frac{nt+\alpha}{n+\beta}-x\right)^m \mathrm{d}t
\end{aligned}$$

$$
\begin{aligned}
&- 2\left(\frac{\alpha}{n+\beta}-x\right)\sum_{k=1}^{\infty} p_{n,k}(x)\int_0^{\infty} b'_{n,k-1}(t)\left(\frac{nt+\alpha}{n+\beta}-x\right)^{m+1} \mathrm{d}t\Bigg]\\
&+ (n+1)\left(\frac{n+\beta}{n}\right)\sum_{k=1}^{\infty} p_{n,k}(x)\int_0^{\infty} b_{n,k-1}(t)\left(\frac{nt+\alpha}{n+\beta}-x\right)^{m+1} \mathrm{d}t\\
&- (n+1)\left(\frac{n+\beta}{n}\right)\left(\frac{\alpha}{n+\beta}-x\right)\sum_{k=1}^{\infty} p_{n,k}(x)\int_0^{\infty} b_{n,k-1}(t)\left(\frac{nt+\alpha}{n+\beta}-x\right)^{m} \mathrm{d}t\\
&+ (1-nx)\left[\mu_{n,m}(x)-\left(\frac{\alpha}{n+\beta}-x\right)^m(1+x)^{-n}\right]-nx(1+x)^{-n}\left(\frac{\alpha}{n+\beta}-x\right)^m .
\end{aligned}
$$

If we now integrate by parts the first five terms in the right-hand side of the above expression and do some simple computations, we get

$$
\begin{aligned}
&(n-m-1)\left(\frac{n+\beta}{n}\right)\mu_{n,m+1}(x)\\
&= x(1+x)\left[\mu'_{n,m}(x)+m\mu_{n,m-1}(x)\right]\\
&\quad + m\left(\frac{\alpha}{n+\beta}-x\right)\left[\left(\frac{\alpha}{n+\beta}-x\right)\left(\frac{n+\beta}{n}\right)-1\right]\mu_{n,m-1}(x)\\
&\quad +\left[(nx-1)+(n-2m-1)\frac{n+\beta}{n}\left(\frac{\alpha}{n+\beta}-x\right)+(m+1)\right]\mu_{n,m}(x),
\end{aligned}
$$

and the Lemma 2.4 is thus proved. □

Lemma 2.5 ([14, p. 220]). *There exist polynomials $q_{i,j,r}(x)$ on $[0,\infty)$, independent of n and k such that*

$$
x^r(1+x)^r\frac{\mathrm{d}^r}{\mathrm{d}x^r}p_{n,k}(x)=\sum_{\substack{2i+j\leqslant r\\ i,j\geqslant 0}} n^i(k-nx)^j q_{i,j,r}(x)p_{n,k}(x).
$$

3 Direct Estimates

In this section, we present some direct results, which include asymptotic formula and an error estimation in simultaneous approximation. Let $C_\gamma[0,\infty)$ be defined as

$$
C_\gamma[0,\infty)=\{f\in C[0,\infty): f(t)=O(t^\gamma),\ \gamma>0\}.
$$

Then the operators $G_n^{\alpha,\beta}(f,x)$ are well defined for $f\in C_\gamma[0,\infty)$, and we have

Theorem 3.1. *Let $f\in C_\gamma[0,\infty)$ be bounded on every finite subinterval of $[0,\infty)$ admitting the derivative of order $(r+2)$ at a fixed $x\in(0,\infty)$. If $f(t)=O(t^\gamma)$ as $t\to\infty$ for some $\gamma>0$, then*

$$\lim_{n\to\infty} n\left(\left(G_n^{\alpha,\beta}\right)^{(r)}(f,x) - f^{(r)}(x)\right)$$
$$= r(r-\beta) f^{(r)}(x) + [(2r+\alpha) + x(1+r-\beta)] f^{(r+1)}(x) + x(1+x) f^{(r+2)}(x).$$

Proof. By Taylor's expansion of f, we have

$$f(t) = \sum_{i=0}^{r+2} \frac{f^{(i)}(x)}{i!}(t-x)^i + \varepsilon(t,x)(t-x)^{r+2},$$

where $\varepsilon(t,x) \to 0$ as $t \to x$ and $\varepsilon(t,x) = o((t-x)^\delta)$ as $t \to \infty$ for some $\delta > 0$. Using Taylor's expansion, we can write

$$n\left[\left(G_n^{\alpha,\beta}\right)^{(r)}(f,x) - f^{(r)}(x)\right] = n\left[\sum_{i=0}^{r+2} \frac{f^{(i)}(x)}{i!}\left(G_n^{\alpha,\beta}\right)^{(r)}((t-x)^i, x) - f^{(r)}(x)\right]$$
$$+ n\left(G_n^{\alpha,\beta}\right)^{(r)}(\varepsilon(t,x)(t-x)^{r+2}, x)$$
$$=: J_1 + J_2.$$

By Lemma 2.2, we have

$$J_1 = n\sum_{i=0}^{r+2} \frac{f^{(i)}(x)}{i!} \sum_{j=r}^{i} \binom{i}{j} (-x)^{i-j} \left(G_n^{\alpha,\beta}\right)^{(r)}(t^j, x) - nf^{(r)}(x)$$
$$= \frac{f^{(r)}(x)}{r!} n\left(\left(G_n^{\alpha,\beta}\right)^{(r)}(t^r, x) - r!\right)$$
$$+ \frac{f^{(r+1)}(x)}{(r+1)!} n\left\{(r+1)(-x)\left(G_n^{\alpha,\beta}\right)^{(r)}(t^r, x) + \left(G_n^{\alpha,\beta}\right)^{(r)}(t^{r+1}, x)\right\}$$
$$+ \frac{f^{(r+2)}(x)}{(r+2)!} n\left\{\frac{(r+2)(r+1)}{2} x^2 \left(G_n^{\alpha,\beta}\right)^{(r)}(t^r, x)\right.$$
$$\left. + (r+2)(-x)\left(G_n^{\alpha,\beta}\right)^{(r)}(t^{r+1}, x) + \left(G_n^{\alpha,\beta}\right)^{(r)}(t^{r+2}, x)\right\}$$
$$= n\left[\frac{n^r(n+r-1)!(n-r-1)!}{(n+\beta)^r\ ((n-1)!)^2} - 1\right] f^{(r)}(x)$$
$$+ n\frac{f^{(r+1)}(x)}{(r+1)!}\left\{(r+1)(-x)\frac{n^r(n+r-1)!(n-r-1)!}{(n+\beta)^r\ ((n-1)!)^2} r!\right.$$
$$+ \frac{n^{r+1}(n+r)!(n-r-2)!}{(n+\beta)^{r+1}\ ((n-1)!)^2}(r+1)!x + \frac{r(r+1)n^{r+1}(n+r-1)!(n-r-2)!}{(n+\beta)^{r+1}\ ((n-1)!)^2} r!$$
$$\left. + (r+1)\alpha \frac{n^r(n+r-1)!(n-r-1)!}{(n+\beta)^{r+1}\ ((n-1)!)^2} r!\right\}$$

$$
\begin{aligned}
&+ n\frac{f^{(r+2)}(x)}{(r+2)!}\left\{\frac{(r+2)(r+1)}{2}x^2\frac{n^r(n+r-1)!(n-r-1)!}{(n+\beta)^r\ ((n-1)!)^2}r!\right.\\
&- (r+2)x\left(\frac{n^{r+1}(n+r)!(n-r-2)!}{(n+\beta)^{r+1}\ ((n-1)!)^2}(r+1)!x\right.\\
&+ \frac{r(r+1)n^{r+1}(n+r-1)!(n-r-2)!}{(n+\beta)^{r+1}\ ((n-1)!)^2}r!\\
&+ \left.(r+1)\alpha\frac{n^r(n+r-1)!(n-r-1)!}{(n+\beta)^{r+1}\ ((n-1)!)^2}r!\right)\\
&+ \frac{n^{r+2}(n+r+1)!(n-r-3)!}{(n+\beta)^{r+2}\ ((n-1)!)^2}\frac{(r+2)!}{2}x^2\\
&+ \frac{(r+1)(r+2)n^{r+2}(n+r)!(n-r-3)!}{(n+\beta)^{r+2}\ ((n-1)!)^2}(r+1)!x\\
&+ (r+2)\alpha\frac{n^{r+1}(n+r)!(n-r-2)!}{(n+\beta)^{r+2}\ ((n-1)!)^2}(r+1)!x\\
&+ \frac{r(r+1)(r+2)n^{r+1}\alpha(n+r-1)!(n-r-2)!}{(n+\beta)^{r+2}\ ((n-1)!)^2}r!\\
&+ \left.\frac{(r+2)(r+1)\alpha^2 n^r(n+r-1)!(n-r-1)!}{2\,(n+\beta)^{r+2}\ ((n-1)!)^2}r!\right\} + O(n^{-2}).
\end{aligned}
$$

The coefficients of $f^{(r)}(x)$, $f^{(r+1)}(x)$, and $f^{(r+2)}(x)$ in the above expression are respectively $r(r-\beta)$, $(2r+\alpha)+x(1+r-\beta)$, and $x(1+x)$, which follow easily by using induction hypothesis on r and then taking the limits as $n\to\infty$. Hence in order to complete the proof of this theorem it is sufficient to show that $J_2\to 0$ as $n\to\infty$, and for this note that by using Lemma 2.5, we have

$$
\begin{aligned}
|J_2| &\leqslant n\sum_{\substack{2i+j\leqslant r\\ i,j\geqslant 0}} n^i\frac{|q_{i,j,r}(x)|}{x^r(1+x)^r}\sum_{k=1}^{\infty}p_{n,k}(x)|k-nx|^j\int_0^{\infty}b_{n,k-1}(t)|\varepsilon(t,x)|\left|\frac{nt+\alpha}{n+\beta}-x\right|^{r+2}\mathrm{d}t\\
&\quad + (-1)^r\frac{(n+r-1)!}{n!}|\varepsilon(0,x)|\left|\frac{\alpha}{n+\beta}-x\right|^{r+2}\\
&=: J_3 + J_4.
\end{aligned}
$$

Since $\varepsilon(t,x)\to 0$ as $t\to x$, hence for a given $\varepsilon>0$ there exists $\delta>0$ such that $|\varepsilon(t,x)|<\varepsilon$ whenever $|t-x|<\delta$. Further if λ is any integer $\geqslant\max\{\gamma, r+2\}$, then we find a constant $K>0$ independent of t, such that

$$|\varepsilon(t,x)|\left|\frac{nt+\alpha}{n+\beta}-x\right|^{r+2}\leqslant K\left|\frac{nt+\alpha}{n+\beta}-x\right|^{\gamma},$$

for $|t-x|\geqslant\delta$. Hence

$$\begin{aligned}|J_3| &= C_1\sum_{\substack{2i+j\leqslant r\\ i,j\geqslant 0}} n^{i+1}\sum_{k=1}^{\infty}p_{n,k}(x)|k-nx|^j\left\{\int_{|t-x|<\delta}\varepsilon b_{n,k-1}(t)\left|\frac{nt+\alpha}{n+\beta}-x\right|^{r+2}\mathrm{d}t\right.\\ &\quad\left.+\int_{|t-x|\geqslant\delta}Kb_{n,k-1}(t)\left|\frac{nt+\alpha}{n+\beta}-x\right|^{\lambda}dt\right\}\\ &=: J_5+J_6,\end{aligned}$$

where

$$C_1=\sum_{\substack{2i+j\leqslant r\\ i,j\geqslant 0}}\frac{|q_{i,j,r}(x)|}{x^r(1+x)^r}.$$

Now, on applying Schwarz's inequality for the integration and summation to the above, we get

$$\begin{aligned}|J_5| &\leqslant \varepsilon C_1\sum_{\substack{2i+j\leqslant r\\ i,j\geqslant 0}} n^{i+1}\sum_{k=1}^{\infty}p_{n,k}(x)|k-nx|^j\\ &\quad\times\left(\int_0^{\infty}b_{n,k-1}(t)\,\mathrm{d}t\right)^{\frac{1}{2}}\left(\int_0^{\infty}b_{n,k-1}(t)\left(\frac{nt+\alpha}{n+\beta}-x\right)^{2r+4}\mathrm{d}t\right)^{\frac{1}{2}}\\ &\leqslant \varepsilon C_1\sum_{\substack{2i+j\leqslant r\\ i,j\geqslant 0}} n^{i+1}\left(\sum_{k=1}^{\infty}p_{n,k}(x)(k-nx)^{2j}\right)^{\frac{1}{2}}\\ &\quad\times\left(\sum_{k=1}^{\infty}p_{n,k}(x)\int_0^{\infty}b_{n,k-1}(t)\left(\frac{nt+\alpha}{n+\beta}-x\right)^{2r+4}\mathrm{d}t\right)^{\frac{1}{2}},\end{aligned}$$

which on using Lemmas 2.3 and 2.4 gives

$$|J_5|\leqslant\varepsilon C_1\sum_{\substack{2i+j\leqslant r\\ i,j\geqslant 0}} n^{i+1}\cdot O(n^{j/2})\cdot O(n^{-(r+2)/2})\leqslant\varepsilon O(1),$$

and because ε is arbitrary, this obviously implies $J_5=o(1)$.

Now again, if we apply Schwarz's inequality for the integration and summation and use Lemmas 2.3 and 2.4 to the expression for J_6, we get

$$\begin{aligned}
|J_6| &\leqslant C_2 \sum_{\substack{2i+j\leqslant r\\ i,j\geqslant 0}} n^{i+1} \sum_{k=1}^{\infty} p_{n,k}(x)|k-nx|^j \int_{|t-x|\geqslant\delta} b_{n,k-1}(t)\left|\frac{nt+\alpha}{n+\beta}-x\right|^{\lambda} dt\\
&\leqslant C_2 \sum_{\substack{2i+j\leqslant r\\ i,j\geqslant 0}} n^{i+1} \left(\sum_{k=1}^{\infty} p_{n,k}(x)(k-nx)^{2j}\right)^{\frac{1}{2}}\\
&\quad\times\left(\sum_{k=1}^{\infty} p_{n,k}(x)\int_0^{\infty} b_{n,k-1}(t)\left(\frac{nt+\alpha}{n+\beta}-x\right)^{2\lambda} dt\right)^{\frac{1}{2}}\\
&= \sum_{\substack{2i+j\leqslant r\\ i,j\geqslant 0}} n^{i+1}\cdot O(n^{j/2})\cdot O(n^{-\lambda/2})\\
&= O(n^{(r+2-\lambda)/2}) = o(1).
\end{aligned}$$

Thus, $J_3 \to 0$ as $n\to\infty$, and because it is obvious that $J_4\to 0$ as $n\to\infty$, we get $J_2 = o(1)$. Now, finally on combining the estimates of J_1 and J_2, we get the desired result, and the proof of Theorem 3.1 is thus complete. □

Theorem 3.2. *Let $f\in C_\gamma[0,\infty)$ for some $\gamma>0$ and $r\leqslant m\leqslant r+2$. If $f^{(m)}$ exists and is continuous on $(a-\eta,b+\eta)\subset(0,\infty)$, $\eta>0$, then for n sufficiently large*

$$\begin{aligned}
&\left\|\left(G_n^{\alpha,\beta}\right)^{(r)}(f,x) - f^{(r)}(x)\right\|_{C[a,b]}\\
&\leqslant C_1 n^{-1}\sum_{i=r}^{m}\|f^{(i)}\|_{C[a,b]} + C_2 n^{-1/2}\omega(f^{(m)},n^{-1/2}) + O(n^{-2}),
\end{aligned}$$

where C_1, C_2 are constants independent of f and n, $\omega(f,\delta)$ is the modulus of continuity of f on $(a-\eta,b+\eta)$, and $\|\cdot\|_{C[a,b]}$ denotes the sup *−norm on $[a,b]$.*

Proof. By Taylor's expansion of f, we have

$$f(t) = \sum_{i=0}^{m}\frac{f^{(i)}(x)}{i!}(t-x)^i + \frac{f^{(m)}(\xi)-f^{(m)}(x)}{m!}(t-x)^m\chi(t) + h(t,x)(1-\chi(t)),$$

where ξ lies between t and x and $\chi(t)$ is the characteristic function on the interval $(a-\eta,\ b+\eta)$. Now,

$$
\begin{aligned}
\left(G_n^{\alpha,\beta}\right)^{(r)}(f,x) - f^{(r)}(x) &= \left\{\sum_{i=0}^{m} \frac{f^{(i)}(x)}{i!} \left(G_n^{\alpha,\beta}\right)^{(r)}((t-x)^i, x) - f^{(r)}(x)\right\} \\
&\quad + \left(G_n^{\alpha,\beta}\right)^{(r)}\left(\frac{f^{(m)}(\xi) - f^{(m)}(x)}{m!}(t-x)^m \chi(t), x\right) \\
&\quad + \left(G_n^{\alpha,\beta}\right)^{(r)}(h(t,x)(1-\chi(t)), x) \\
&=: S_1 + S_2 + S_3.
\end{aligned}
$$

By using Lemma 2.2, we get

$$
\begin{aligned}
S_1 &= \sum_{i=0}^{m} \frac{f^{(i)}(x)}{i!} \sum_{j=0}^{i} \binom{i}{j} (-x)^{i-j} \frac{d^r}{dx^r} \left[\frac{n^j (n+j-1)!(n-j-1)!}{(n+\beta)^j\ ((n-1)!)^2} x^j \right. \\
&\quad + \left(\frac{j(j-1)n^j (n+j-2)!(n-j-1)!}{(n+\beta)^j\ ((n-1)!)^2} + \frac{j\alpha n^{j-1}(n+j-2)!(n-j)!}{(n+\beta)^j\ ((n-1)!)^2} \right) x^{j-1} \\
&\quad + \left(\frac{j(j-1)(j-2)\alpha n^{j-1}\alpha(n+j-3)!(n-j)!}{(n+\beta)^j\ ((n-1)!)^2} \right. \\
&\quad \left. \left. + \frac{j(j-1)\alpha^2 n^{j-2}(n+j-3)!(n-j+1)!}{2\,(n+\beta)^j\ ((n-1)!)^2} \right) x^{j-2} + O(n^{-2}) \right] - f^{(r)}(x).
\end{aligned}
$$

Consequently,

$$
\|S_1\|_{C[a,b]} \leqslant C_1 n^{-1} \sum_{i=r}^{m} \|f^{(i)}\|_{C[a,b]} + O(n^{-2}), \qquad \text{uniformly on } [a,b].
$$

Next, we estimate S_2 as follows:

$$
\begin{aligned}
|S_2| &\leqslant \int_0^\infty |K_n^{(r)}(x,t)| \left\{ \left| \frac{f^{(m)}(\xi) - f^{(m)}(x)}{m!} \right| \left| \frac{nt+\alpha}{n+\beta} - x \right|^m \chi(t) \right\} \mathrm{d}t \\
&\leqslant \frac{\omega(f^{(m)},\delta)}{m!} \int_0^\infty |K_n^{(r)}(x,t)| \left(1 + \frac{\left|\frac{nt+\alpha}{n+\beta} - x\right|}{\delta} \right) \left| \frac{nt+\alpha}{n+\beta} - x \right|^m \mathrm{d}t \\
&\leqslant \frac{\omega(f^{(m)},\delta)}{m!} \left[\sum_{k=1}^{\infty} |p_{n,k}^{(r)}(x)| \int_0^\infty b_{n,k-1}(t) \right. \\
&\quad \times \left(\left| \frac{nt+\alpha}{n+\beta} - x \right|^m + \delta^{-1} \left| \frac{nt+\alpha}{n+\beta} - x \right|^{m+1} \right) \mathrm{d}t \\
&\quad \left. + \frac{(n+r-1)!}{(n-1)!}(1+x)^{-n-r} \left(\left| \frac{\alpha}{n+\beta} - x \right|^m + \delta^{-1} \left| \frac{\alpha}{n+\beta} - x \right|^{m+1} \right) \right].
\end{aligned}
$$

Using Schwarz's inequality for integration and summation, we get

$$
\begin{aligned}
&\sum_{k=1}^{\infty} p_{n,k}(x)|k-nx|^j \int_0^{\infty} b_{n,k-1}(t)\left|\frac{nt+\alpha}{n+\beta}-x\right|^m \mathrm{d}t \\
&\quad\leqslant \sum_{k=1}^{\infty} p_{n,k}(x)|k-nx|^j \left(\int_0^{\infty} b_{n,k-1}(t)\,\mathrm{d}t\right)^{\frac{1}{2}} \left(\int_0^{\infty} b_{n,k-1}(t)\left(\frac{nt+\alpha}{n+\beta}-x\right)^{2m} \mathrm{d}t\right)^{\frac{1}{2}} \\
&\quad\leqslant \left(\sum_{k=1}^{\infty} p_{n,k}(x)(k-nx)^{2j}\right)^{\frac{1}{2}} \times \left(\sum_{k=1}^{\infty} p_{n,k}(x)\int_0^{\infty} b_{n,k-1}(t)\left(\frac{nt+\alpha}{n+\beta}-x\right)^{2m} \mathrm{d}t\right)^{\frac{1}{2}} \\
&\quad= O(n^{j/2})\cdot O(n^{-m/2}) \\
&\quad= O(n^{(j-m)/2}), \quad \text{uniformly on } [a,b]. \qquad (9)
\end{aligned}
$$

Therefore, by Lemma 2.5 and (9), we get

$$
\begin{aligned}
&\sum_{k=1}^{\infty} |p_{n,k}^{(r)}(x)| \int_0^{\infty} b_{n,k-1}(t)\left|\frac{nt+\alpha}{n+\beta}-x\right|^m \mathrm{d}t \\
&\quad\leqslant \sum_{k=1}^{\infty} \sum_{\substack{2i+j\leqslant r\\ i,j\geqslant 0}} n^i |k-nx|^j \frac{|q_{i,j,r}(x)|}{x^r(1+x)^r} p_{n,k}(x) \int_0^{\infty} b_{n,k-1}(t)\left|\frac{nt+\alpha}{n+\beta}-x\right|^m \mathrm{d}t \\
&\quad\leqslant \left(\sup_{\substack{2i+j\leqslant r\\ i,j\geqslant 0}} \frac{|q_{i,j,r}(x)|}{x^r(1+x)^r}\right) \sum_{\substack{2i+j\leqslant r\\ i,j\geqslant 0}} n^i \left(\sum_{k=1}^{\infty} p_{n,k}(x)|k-nx|^j \int_0^{\infty} b_{n,k-1}(t)\left|\frac{nt+\alpha}{n+\beta}-x\right|^m \mathrm{d}t\right) \\
&\quad= C\sum_{\substack{2i+j\leqslant r\\ i,j\geqslant 0}} n^i O(n^{(j-m)/2}) = O(n^{(r-m)/2}), \quad \text{uniformly on } [a,b], \qquad (10)
\end{aligned}
$$

where

$$C = \sup_{\substack{2i+j\leqslant r\\ i,j\geqslant 0}} \sup_{x\in[a,b]} \frac{|q_{i,j,r}(x)|}{x^r(1+x)^r}.$$

Now, if we choose $\delta = n^{-1/2}$ and apply (10), we obtain

$$
\begin{aligned}
||S_2||_{C[a,b]} &\leqslant \frac{\omega(f^{(m)}, n^{-1/2})}{m!}\left[O(n^{(r-m)/2}) + n^{1/2}O(n^{(r-m-1)/2}) + O(n^{-m}\right) \\
&\leqslant C_2 n^{-(r-m)/2}\omega(f^{(m)}, n^{-1/2}).
\end{aligned}
$$

Since $t \in [0,\infty) \setminus (a-\eta, b+\eta)$, we can choose δ such that $|t-x| \geqslant \delta$ for all $x \in [a,b]$. Thus, by Lemma 2.5, we get

$$|S_3| \leqslant C \sum_{\substack{2i+j\leqslant r\\ i,j\geqslant 0}} n^i \sum_{k=1}^{\infty} p_{n,k}(x)|k-nx|^j \int_{|t-x|\geq\delta} b_{n,k-1}(t)|h(t,x)|\,\mathrm{d}t$$

$$+ \frac{(n+r-1)!}{(n-1)!}(1+x)^{-n-r}\,|h\,(0,x)|\,.$$

For $|t-x| \geqslant \delta$, we can find the a constant M such that

$$|h(t,x)| \leqslant M\left|\frac{nt+\alpha}{n+\beta}-x\right|^{\beta},$$

where β is an integer $\geqslant \{\gamma, m\}$. Hence, using Schwarz's inequality for both integration and summation, Lemmas 2.3 and 2.4, it easily follows that $S_3 = O(n^{-s})$ for any $s > 0$, uniformly on $[a,b]$.

Combining the estimates of S_1, S_2, S_3, the required result is immediate. □

References

1. Büyükyazici, I., Atakut, C.: On Stancu type generalization of q-Baskakov operators. Math. Comput. Model. **52**(5–6), 752–759 (2010)
2. Deo, N.: Faster rate of convergence on Srivastava–Gupta operators. Appl. Math. Comput. **218**(21), 10486–10491 (2012)
3. Durrmeyer, J.L.: Une formule d' inversion de la Transformee de Laplace, Applications a la Theorie des Moments. Thése de 3e Cycle, Faculté des Sciences de l' Université de Paris, Paris (1967)
4. Govil, N.K., Gupta, V.: Some approximation properties of integrated Bernstein operators. In: Baswell A.R. (eds.) Advances in Mathematics Research, vol. 11, Chapter 8. Nova Science Publishers Inc., New York (2009)
5. Gupta, V., Gupta, M.K., Vasishtha, V.: Simultenaous approximations by summation–integral type operators. Nonlinear Funct. Anal. Appl. **8**(3), 399–412 (2003)
6. Heilmann, M.: Direct and converse results for operators of Baskakov-Durrmeyer type. Approx. Theory Appl. **5**(1), 105–127 (1989)
7. Ismail, M., Simeonov, P.: On a family of positive linear integral operators. In: Notions of Positivity and the Geometry of Polynomials. Trends in Mathematics, pp. 259–274. Springer, Basel (2011)
8. Ispir, N., Yuksel, I.: On the Bézier variant of Srivastava–Gupta operators. Appl. Math. E - Notes **5**, 129–137 (2005)
9. Kantorovich, L.V.: Sur certaines developments suivant les polynômes de la forme de S. Bernstein I–II. C.R. Acad. Sci. USSR **20**(A), 563–568; 595–600 (1930)
10. May, C.P.: Saturation and inverse theorems for combinations of a class of exponential type operators. Can. J. Math. **28**, 1224–1250 (1976)
11. Mazhar, S.M., Totik, V.: Approximation by modified Szász operators. Acta Sci. Math. **49**, 257–269 (1985)
12. Phillips, R.S.: An inversion formula for semi–groups of linear operators. Ann. Math. **59**(Ser. 2), 352–356 (1954)

13. Sahai, A., Prasad, G.: On the rate of convergence for modified Szász–Mirakyan operators on functions of bounded variation. Publ. Inst. Math. (Beograd) (N.S.) **53**(67), 73–80 (1993)
14. Sinha, R.P., Agrawal, P.N., Gupta, V.: On simultaneous approximation by modified Baskakov operators. Bull. Soc. Math. Belg. Ser. B **43**(2), 217–231 (1991)
15. Srivastava, H.M., Gupta, V.: A certain family of summation integral type operators. Math. Comput. Model. **37**(12–13), 1307–1315 (2003)
16. Stancu, D.D.: Approximation of functions by means of a new generalized Bernstein operator. Calcolo **20**, 211–229 (1983)
17. Verma, D.K., Agarwal, P.N.: Convergence in simultaneous approximation for Srivastava–Gupta operators. Math. Sci. **6**, 22 (2012). doi:10.1186/2251-7456-6-22

Korovkin-Type Approximation Theorem for Functions of Two Variables Via Statistical Summability $(C, 1, 1)$

M. Mursaleen and S.A. Mohiuddine

Dedicated to Professor Hari M. Srivastava

Abstract The concept of statistical summability $(C, 1, 1)$ has recently been introduced by Moricz (J. Math. Anal. Appl. 286:340–350, 2003). In this paper, we use this notion of summability to prove the Korovkin-type approximation theorem for functions of two variables.

1 Introduction and Preliminaries

The concept of statistical convergence for sequences of real numbers was introduced by Fast [8] and further studied by many others.

Let $K \subseteq \mathbb{N}$ and $K_n = \{k \leq n : k \in K\}$. Then the *natural density* of K is defined by $\delta(K) = \lim_n n^{-1}|K_n|$ if the limit exists, where $|K_n|$ denotes the cardinality of K_n.

A sequence $x = (x_k)$ of real numbers is said to be *statistically convergent* to L provided that for every $\epsilon > 0$ the set $K_\epsilon := \{k \in \mathbb{N} : |x_k - L| \geq \epsilon\}$ has natural density zero, i.e., for each $\epsilon > 0$,

$$\lim_n \frac{1}{n}|\{j \leq n : |x_j - L| \geq \epsilon\}| = 0.$$

M. Mursaleen (✉)
Department of Mathematics, Aligarh Muslim University, Aligarh 202002, India
e-mail: mursaleenm@gmail.com

S.A. Mohiuddine
Department of Mathematics, King Abdulaziz University, P.O. Box 80203, Jeddah 21589, Saudi Arabia
e-mail: mohiuddine@gmail.com

G.V. Milovanović and M.Th. Rassias (eds.), *Analytic Number Theory, Approximation Theory, and Special Functions*, DOI 10.1007/978-1-4939-0258-3_20,

By the convergence of a double sequence we mean the convergence in Pringsheim's sense [20]. A double sequence $x = (x_{jk})$ is said to be *Pringsheim's convergent* (or *P-convergent*) if for given $\epsilon > 0$ there exists an integer N such that $|x_{jk} - \ell| < \epsilon$ whenever $j, k > N$. In this case, ℓ is called the Pringsheim limit of $x = (x_{jk})$ and it is written as $P\text{-}\lim x = \ell$.

A double sequence $x = (x_{jk})$ is said to be *bounded* if there exists a positive number M such that $|x_{jk}| < M$ for all j, k.

Note that, in contrast to the case for single sequences, a convergent double sequence need not be bounded.

The idea of statistical convergence for double sequences was introduced by Mursaleen and Edely [16] and further studied in [11, 17, 18].

Let $K \subseteq \mathbb{N} \times \mathbb{N}$ be a two-dimensional set of positive integers and let $K_{m,n} = \{(j,k) : j \leq m, k \leq n\}$. Then the two-dimensional analogue of natural density can be defined as follows:

In case the sequence $(K(m,n)/mn)$ has a limit in Pringsheim's sense, then we say that K has a *double natural density* and is defined as

$$P\text{-}\lim_{m,n} \frac{K(m,n)}{mn} = \delta^{(2)}\{K\}.$$

For example, let $K = \{(i^2, j^2) : i, j \in \mathbb{N}\}$. Then

$$\delta^{(2)}\{K\} = P\text{-}\lim_{m,n} \frac{K(m,n)}{mn} \leq P\text{-}\lim_{m,n} \frac{\sqrt{m}\sqrt{n}}{mn} = 0,$$

i.e., the set K has double natural density zero, while the set $\{(i, 2j) : i, j \in \mathbb{N}\}$ has double natural density $1/2$.

A real double sequence $x = (x_{jk})$ is said to be *statistically convergent* to the number L if for each $\epsilon > 0$, the set

$$\{(j,k), j \leq m \text{ and } k \leq n : | x_{jk} - L | \geq \epsilon\}$$

has double natural density zero. In this case we write $\text{st}^{(2)}\text{-}\lim_{j,k} x_{jk} = L$.

Remark 1.1. Note that if $x = (x_{jk})$ is P-convergent, then it is statistically convergent but not conversely. See the following example:

Example 1.1. The double sequence $x = (x_{jk})$ defined by

$$x_{jk} = \begin{cases} 1\,, & \text{if } j \text{ and } k \text{ are squares;} \\ 0\,, & \text{otherwise.} \end{cases} \tag{1}$$

Then x is statistically convergent to zero but not P-convergent.

In [6], some Tauberian theorems have been obtained to get convergence from statistical convergence for double sequences.

Moricz [14] introduced the idea of statistical summability $(C,1,1)$.

We say that a double sequence $x = (x_{jk})$ is *statistically summability* $(C,1,1)$ to some number L if $\text{st}^{(2)}\text{-}\lim_{m,n} \sigma_{mn} = L$, where

$$\sigma_{mn} = \frac{1}{mn} \sum_{j=1}^{m} \sum_{k=1}^{n} x_{jk}. \tag{2}$$

In this case, we write $\text{st}_{(C,1,1)}\text{-}\lim x = L$. It is trivial that $\text{st}^{(2)}\text{-}\lim_{j,k} x_{jk} = L$ implies $\text{st}^{(2)}\text{-}\lim_{m,n} \sigma_{mn} = L$. Moricz [14] obtained the Tauberian conditions for the reverse implication.

Let $C[a,b]$ be the space of all functions f continuous on $[a,b]$. We know that $C[a,b]$ is a Banach space with norm

$$\|f\|_{C[a,b]} := \sup_{x\in[a,b]} |f(x)|,\ f \in C[a,b].$$

The classical Korovkin approximation theorem states as follows [10]:

Let (T_n) be a sequence of positive linear operators from $C[a,b]$ into $C[a,b]$. Then $\lim_n \|T_n(f,x) - f(x)\|_{C[a,b]} = 0$ for all $f \in C[a,b]$ if and only if

$$\lim_n \|T_n(e_i,x) - e_i(x)\|_{C[a,b]} = 0 \text{ for } i = 0,1,2,$$

where $e_0(x) = 1, e_1(x) = x$, and $e_2(x) = x^2$.

Quite recently, such type of approximation theorems have been established for functions of one and/or two variables, by using statistical convergence [5,9], generalized statistical convergence [7, 15, 21], A-statistical convergence [4], statistical A-summability [2, 3], weighted statistical convergence [19], almost convergence [1, 12], and statistical summability $(C,1)$ [13]. In this paper, we extend the result of [22] by using the notion of statistical summability $(C,1,1)$ and show that our result is stronger than those proved by Taşdelen and Erençin [22] and Dirik and Demirci [4].

2 Main Result

Let $I = [0,A]$, $J = [0,B]$, $A, B \in (0,1)$, and $K = I \times J$. We denote by $C(K)$ the space of all continuous real-valued functions on K. This space is a equipped with norm

$$\|f\|_{C(K)} := \sup_{(x,y)\in K} |f(x,y)|,\ f \in C(K).$$

Let $H_\omega(K)$ denote the space of all real-valued functions f on K such that

$$| f(s,t) - f(x,y) | \leq \omega\left(f; \sqrt{\left(\frac{s}{1-s} - \frac{x}{1-x}\right)^2 + \left(\frac{t}{1-t} - \frac{y}{1-y}\right)^2} \right),$$

where ω is the modulus of continuity, i.e.,

$$\omega(f;\delta) = \sup_{(s,t),(x,y)\in K} \left\{ | f(s,t) - f(x,y) | : \sqrt{(s-x)^2 + (t-y)^2} \leq \delta \right\}.$$

It is to be noted that any function $f \in H_\omega(K)$ is continuous and bounded on K.

The following result was given by Taşdelen and Erençin [22].

Theorem A. *Let $(T_{j,k})$ be a double sequence of positive linear operators from $H_\omega(K)$ into $C(K)$. Then for all $f \in H_\omega(K)$,*

$$P\text{-}\lim_{j,k\to\infty} \left\| T_{j,k}(f;x,y) - f(x,y) \right\|_{C(K)} = 0. \tag{3}$$

if and only if

$$P\text{-}\lim_{j,k\to\infty} \left\| T_{j,k}(f_i;x,y) - f_i \right\|_{C(K)} = 0 \quad (i = 0,1,2,3), \tag{4}$$

where

$$f_0(x,y) = 1,\ f_1(x,y) = \frac{x}{1-x},\ f_2(x,y) = \frac{y}{1-y},\ f_3(x,y) = \left(\frac{x}{1-x}\right)^2 + \left(\frac{y}{1-y}\right)^2.$$

Recently, Dirik and Demirci [4] proved the following theorem.

Theorem B. *Let $(T_{j,k})$ be a double sequence of positive linear operators from $H_\omega(K)$ into $C(K)$. Then for all $f \in H_\omega(K)$,*

$$\mathrm{st}^{(2)}\text{-}\lim_{j,k\to\infty} \left\| T_{j,k}(f;x,y) - f(x,y) \right\|_{C(K)} = 0. \tag{5}$$

if and only if

$$\mathrm{st}^{(2)}\text{-}\lim_{j,k\to\infty} \left\| T_{j,k}(f_i;x,y) - f_i \right\|_{C(K)} = 0 \quad (i = 0,1,2,3). \tag{6}$$

We prove the following result:

Theorem 2.1. *Let $(T_{j,k})$ be a double sequence of positive linear operators from $H_\omega(K)$ into $C(K)$. Then for all $f \in H_\omega(K)$,*

$$\mathrm{st}_{(C,1,1)}\text{-}\lim \left\| T_{j,k}(f;x,y) - f(x,y) \right\|_{C(K)} = 0. \tag{7}$$

if and only if

$$\text{st}_{(C,1,1)}\text{-}\lim \left\| T_{j,k}(1;x,y) - 1 \right\|_{C(K)} = 0, \tag{8}$$

$$\text{st}_{(C,1,1)}\text{-}\lim \left\| T_{j,k}\left(\frac{s}{1-s};x,y\right) - \frac{x}{1-x} \right\|_{C(K)} = 0, \tag{9}$$

$$\text{st}_{(C,1,1)}\text{-}\lim \left\| T_{j,k}\left(\frac{t}{1-t};x,y\right) - \frac{y}{1-y} \right\|_{C(K)} = 0, \tag{10}$$

$$\text{st}_{(C,1,1)}\text{-}\lim \left\| T_{j,k}\left(\left(\frac{s}{1-s}\right)^2 + \left(\frac{t}{1-t}\right)^2;x,y\right) - \left(\left(\frac{x}{1-x}\right)^2 + \left(\frac{y}{1-y}\right)^2\right) \right\|_{C(K)} = 0. \tag{11}$$

Proof. Since each

$$1,\ \frac{x}{1-x},\ \frac{y}{1-y},\ \left(\frac{x}{1-x}\right)^2 + \left(\frac{y}{1-y}\right)^2$$

belongs to $H_\omega(K)$, conditions (8)–(11) follow immediately from (7). Let $f \in H_\omega(K)$ and $(x,y) \in K$ be fixed. Then after using the properties of f, a simple calculation gives that

$$\begin{aligned}
&|T_{j,k}(f;x,y) - f(x,y)| \\
&\quad \le T_{j,k}(|f(s,t) - f(x,y)|;x,y) + |f(x,y)|\,|T_{j,k}(f_0;x,y) - f_0(x,y)| \\
&\quad \le \varepsilon + \left(\varepsilon + N + \frac{2N}{\delta^2}\right)|T_{j,k}(f_0;x,y) - f_0(x,y)| + \frac{4N}{\delta^2}|T_{j,k}(f_1;x,y) - f_1(x,y)| \\
&\quad\quad + \frac{4N}{\delta^2}|T_{j,k}(f_2;x,y) - f_2(x,y)| + \frac{2N}{\delta^2}|T_{j,k}(f_3;x,y) - f_3(x,y)| \\
&\quad \le \varepsilon + M\Big\{|T_{j,k}(f_0;x,y) - f_0(x,y)| + |T_{j,k}(f_1;x,y) - f_1(x,y)| \\
&\quad\quad + |T_{j,k}(f_2;x,y) - f_2(x,y)| + |T_{j,k}(f_3;x,y) - f_3(x,y)|\Big\},
\end{aligned}$$

where $N = \|f\|_{C(K)}$ and

$$M = \max\left\{\varepsilon + N + \frac{2N}{\delta^2}\left(\left(\frac{A}{1-A}\right)^2 + \left(\frac{B}{1-B}\right)^2\right), \frac{4N}{\delta^2}\left(\frac{A}{1-A}\right), \frac{4N}{\delta^2}\left(\frac{B}{1-B}\right), \frac{2N}{\delta^2}\right\}.$$

Now, replacing $T_{j,k}(f;x,y)$ by $\frac{1}{mn}\sum_{j=1}^{m}\sum_{k=1}^{n}T_{j,k}(f;x,y)$ and taking $\sup_{(x,y)\in K}$, we get

$$\begin{aligned}
&\left\|\frac{1}{mn}\sum_{j=1}^{m}\sum_{k=1}^{n}T_{j,k}(f;x,y)-f(x,y)\right\|_{C(K)}\\
&\quad\leq M\left(\left\|\frac{1}{mn}\sum_{j=1}^{m}\sum_{k=1}^{n}T_{j,k}(f_0;x,y)-f_0(x,y)\right\|_{C(K)}\right.\\
&\qquad+\left\|\frac{1}{mn}\sum_{j=1}^{m}\sum_{k=1}^{n}T_{j,k}(f_1;x,y)-f_1(x,y)\right\|_{C(K)}\\
&\qquad+\left\|\frac{1}{mn}\sum_{j=1}^{m}\sum_{k=1}^{n}T_{j,k}(f_2;x,y)-f_2(x,y)\right\|_{C(K)}\\
&\qquad\left.+\left\|\frac{1}{mn}\sum_{j=1}^{m}\sum_{k=1}^{n}T_{j,k}(f_3;x,y)-f_3(x,y)\right\|_{C(K)}\right)+\varepsilon. \qquad (12)
\end{aligned}$$

For a given $r>0$ choose $\varepsilon>0$ such that $\varepsilon<r$. Define the following sets:

$$D:=\left\{(j,k),\, j\leq m \text{ and } k\leq n:\left\|\frac{1}{mn}\sum_{j=1}^{m}\sum_{k=1}^{n}T_{j,k}(f;x,y)-f(x,y)\right\|_{C(K)}\geq r\right\},$$

$$D_1:=\left\{(j,k),\, j\leq m \text{ and } k\leq n:\left\|\frac{1}{mn}\sum_{j=1}^{m}\sum_{k=1}^{n}T_{j,k}(f_0;x,y)-f_0(x,y)\right\|_{C(K)}\geq \frac{r-\varepsilon}{4K}\right\},$$

$$D_2:=\left\{(j,k),\, j\leq m \text{ and } k\leq n:\left\|\frac{1}{mn}\sum_{j=1}^{m}\sum_{k=1}^{n}T_{j,k}(f_1;x,y)-f_1(x,y)\right\|_{C(K)}\geq \frac{r-\varepsilon}{4K}\right\},$$

$$D_3:=\left\{(j,k),\, j\leq m \text{ and } k\leq n:\left\|\frac{1}{mn}\sum_{j=1}^{m}\sum_{k=1}^{n}T_{j,k}(f_2;x,y)-f_2(x,y)\right\|_{C(K)}\geq \frac{r-\varepsilon}{4K}\right\},$$

$$D_4:=\left\{(j,k),\, j\leq m \text{ and } k\leq n:\left\|\frac{1}{mn}\sum_{j=1}^{m}\sum_{k=1}^{n}T_{j,k}(f_3;x,y)-f_3(x,y)\right\|_{C(K)}\geq \frac{r-\varepsilon}{4K}\right\}.$$

Then from (12), we see that $D \subset D_1 \cup D_2 \cup D_3 \cup D_4$ and therefore

$$\delta^{(2)}\{D\} \leq \delta^{(2)}\{D_1\} + \delta^{(2)}\{D_2\} + \delta^{(2)}\{D_3\} + \delta^{(2)}\{D_4\}.$$

Hence conditions (8)–(11) imply the condition (7).

This completes the proof of the theorem. □

3 Rate of Statistical Summability $(C, 1, 1)$

In this section we study the rate of statistical summability $(C, 1, 1)$ of a sequence of positive linear operators defined on $C(K)$.

We now present the following definition.

Definition 3.1. Let (α_{mn}) be a positive nonincreasing double sequence. A double sequence $x = (x_{mn})$ is statistically summability $(C, 1, 1)$ to a function L with the rate $o(a_{mn})$ if for every $\epsilon > 0$,

$$P\text{-}\lim_{m,n\to\infty} \frac{G(\epsilon)}{\alpha_{mn}} = 0$$

where

$$G(\epsilon) = \frac{1}{mn}\left|\left\{p \leq m, q \leq n : \mid \sigma_{pq} - L \mid \geq \varepsilon\right\}\right|,$$

and σ_{pq} is defined by (2). In this case, it is denoted by $x_{mn} - L = \mathrm{st}_{(C,1,1)} \quad o(\alpha_{mn})$ as $m, n \to \infty$.

It is easy to prove the following basic lemma.

Lemma 3.1. *Let $x = (x_{mn})$ and $x = (y_{mn})$ be double sequences. Assume that*

$$x_{mn} - L_1 = \mathrm{st}_{(C,1,1)} - o(\alpha_{mn}) \quad \text{and} \quad y_{mn} - L_2 = \mathrm{st}_{(C,1,1)} - o(\beta_{mn})$$

on X. Let $\gamma_{mn} = \max\{\alpha_{mn}, \beta_{mn}\}$. Then the following statement holds:

(i) $(x_{mn} \pm y_{mn}) - (L_1 + L_2) = \mathrm{st}_{(C,1,1)} - o(\gamma_{mn})$,
(ii) $(x_{mn} - L_1)(y_{mn} - L_2) = \mathrm{st}_{(C,1,1)} - o(\alpha_{mn}\beta_{mn})$,
(iii) $\mu(x_{mn} - L_1) = \mathrm{st}_{(C,1,1)} - o(\alpha_{mn})$ *for any real number* μ.

Theorem 3.1. *Let $(T_{m,n})$ be a double sequence of positive linear operators from $H_\omega(K)$ into $C(K)$. Assume that the following conditions hold:*

(a) $\|T_{m,n}(f_0) - (f_0)\|_{C(K)} = \mathrm{st}_{(C,1,1)} - o(\alpha_{mn})$ *as* $m, n \to \infty$;

(b) $w(f; \delta_{mn}) = \mathrm{st}_{(C,1,1)} - o(\beta_{mn})$ *as* $m, n \to \infty$, *where* $\delta_{mn} = \sqrt{\|T_{mn}(\phi)\|_{C(K)}}$ *with*

$$\phi(u,v)=\Big(\frac{u}{1-u}-\frac{x}{1-x}\Big)^2+\Big(\frac{v}{1-v}-\frac{y}{1-y}\Big)^2.$$

Then, for all $f\in H_\omega(K)$, we have

$$\|T_{m,n}(f)-(f)\|_{C(K)}=\mathrm{st}_{(C,1,1)}-o(\gamma_{mn})\quad as\quad m,n\to\infty. \tag{13}$$

Proof. Let $f\in H_\omega(K)$ and $(x,y)\in K$. Then

$$\begin{aligned}&|T_{m,n}(f;x,y)-f(x,y)|\\&=|T_{m,n}((f(u,v)-f(x,y);x,y)-f(x,y)(T_{m,n}(f_0;x,y)-f_0(x,y))|\\&\le T_{m,n}(|f(u,v)-f(x,y)|;x,y)+M\mid T_{m,n}(f_0;x,y)-f_0(x,y)|,\end{aligned}$$

where $M=\|f\|_{C(X)}$. This yields that

$$\begin{aligned}&|T_{m,n}(f;x,y)-f(x,y)|\\&\le w(f;\delta)T_{mn}\left(1+\frac{\sqrt{(\frac{u}{1-u}-\frac{x}{1-x})^2+(\frac{v}{1-v}-\frac{y}{1-y})^2}}{\delta};x,y\right)\\&\quad+M\mid T_{m,n}(f_0;x,y)-f_0(x,y)|\\&\le w(f;\delta)\mid T_{m,n}(f_0;x,y)-f_0(x,y)|+\frac{w(f;\delta)}{\delta^2}T_{m,n}(\phi;x,y)\\&\quad+w(f;\delta)+M|T_{m,n}(f_0;x,y)-f_0(x,y)|\end{aligned}$$

Now, taking $\sup_{(x,y)\in K}$, we obtain $\|T_{m,n}(f)-f\|_{C(K)}$

$$\begin{aligned}\|T_{m,n}(f)-f\|_{C(K)}&\le w(f;\delta)\|T_{m,n}(f_0)-f_0\|_{C(K)}+\frac{w(f;\delta)}{\delta^2}\|T_{m,n}(\phi)\|_{C(K)}\\&\quad+w(f;\delta)+M\|T_{m,n}(f_0)-f_0\|_{C(K)}.\end{aligned}$$

Now, let $\delta:=\delta_{mn}:=\sqrt{\|T_{mn}(\phi)\|_{C(K)}}$. Then

$$\|T_{m,n}(f)-f\|_{C(K)}\leqslant w(f;\delta)\|T_{m,n}(f_0)-f_0\|_{C(K)}$$

$$+2w(f;\delta)+M\|T_{m,n}(f_0)-f_0\|_{C(K)} \tag{14}$$

$$\le N\Big\{w(f;\delta)\|T_{m,n}(f_0)-f_0\|_{C(K)}+w(f;\delta)+\|T_{m,n}(f_0)-f_0\|_{C(K)}\Big\}, \tag{15}$$

where $N = \max\{2, M\}$. For a given $r > 0$, define the following sets:

$$E := \left\{(m,n) \in \mathbb{N} \times \mathbb{N} : \|T_{m,n}(f) - f\|_{C(K)} \geq r\right\},$$

$$E_1 := \left\{(m,n) \in \mathbb{N} \times \mathbb{N} : w(f;\delta)\|T_{m,n}(f_0) - f_0\|_{C(K)} \geq \frac{r}{3N}\right\},$$

$$E_2 := \left\{(m,n) \in \mathbb{N} \times \mathbb{N} : w(f;\delta) \geq \frac{r}{3N}\right\},$$

$$E_3 := \left\{(m,n) \in \mathbb{N} \times \mathbb{N} : \|T_{m,n}(f_0) - f_0\|_{C(K)} \geq \frac{r}{3N}\right\}.$$

Then by (14), we get $E \subset E_1 \cup E_2 \cup E_3$. Further, we define the sets

$$E_4 := \left\{(m,n) \in \mathbb{N} \times \mathbb{N} : w(f;\delta) \geq \sqrt{\frac{r}{3N}}\right\}$$

and

$$E_5 := \left\{(m,n) \in \mathbb{N} \times \mathbb{N} : \|T_{m,n}(f_0) - f_0\|_{C(K)} \geq \sqrt{\frac{r}{3N}}\right\}.$$

Then $E_1 \subset E_4 \cup E_5$ and hence $E \subset E_2 \cup E_3 \cup E_4 \cup E_5$. Therefore, by using the conditions (a) and (b), we get (13).

This completes the proof of the theorem. □

4 Example and the Concluding Remark

We show that the following double sequence of positive linear operators satisfies the conditions of Theorem 3.1 but does not satisfy the conditions of Theorems A and B.

Example 4.1. Consider the following Meyer-König and Zeller operators:

$$B_{m,n}(f;x,y) := (1-x)^{m+1}(1-y)^{n+1} \times \sum_{j=0}^{\infty}\sum_{k=0}^{\infty} f\Big(\frac{j}{j+m+1}, \frac{k}{k+n+1}\Big)\binom{m+j}{j}\binom{n+k}{k}x^j y^k,$$

where $f \in H_\omega(K)$ and $K = [0,A] \times [0,B]$, $A, B \in (0,1)$.

Since, for $x \in [0, A]$, $A \in (0, 1)$,

$$\frac{1}{(1-x)^{m+1}} = \sum_{j=0}^{\infty} \binom{m+j}{j} x^j,$$

it is easy to see that

$$B_{m,n}(f_0; x, y) = f_0(x, y).$$

Also, we obtain

$$\begin{aligned} B_{m,n}(f_1; x, y) &= (1-x)^{m+1}(1-y)^{n+1} \sum_{j=0}^{\infty}\sum_{k=0}^{\infty} \frac{j}{m+1} \binom{m+j}{j} \binom{n+k}{k} x^j y^k \\ &= (1-x)^{m+1}(1-y)^{n+1} x \sum_{j=0}^{\infty}\sum_{k=0}^{\infty} \frac{1}{m+1} \frac{(m+j)!}{m!(j-1)!} \binom{n+k}{k} x^{j-1} y^k \\ &= (1-x)^{m+1}(1-y)^{n+1} x \frac{1}{(1-x)^{m+2}} \frac{1}{(1-y)^{n+1}} = \frac{x}{1-x}, \end{aligned}$$

and similarly

$$B_{m,n}(f_2; x, y) = \frac{y}{1-y}.$$

Finally, we get

$$\begin{aligned} &B_{m,n}(f_3; x, y) \\ &= (1-x)^{m+1}(1-y)^{n+1} \sum_{j=0}^{\infty}\sum_{k=0}^{\infty} \left\{ \left(\frac{j}{m+1}\right)^2 + \left(\frac{k}{n+1}\right)^2 \right\} \binom{m+j}{j} \binom{n+k}{k} x^j y^k \\ &= (1-x)^{m+1}(1-y)^{n+1} \frac{x}{m+1} \sum_{j=0}^{\infty}\sum_{k=0}^{\infty} \frac{j}{m+1} \frac{(m+j)!}{m!(j-1)!} \binom{n+k}{k} x^{j-1} y^k \\ &\quad + (1-x)^{m+1}(1-y)^{n+1} \frac{y}{n+1} \sum_{j=0}^{\infty}\sum_{k=0}^{\infty} \frac{k}{n+1} \binom{m+j}{j} \frac{(n+k)!}{n!(k-1)!} x^j y^{k-1} \\ &= (1-x)^{m+1}(1-y)^{n+1} \frac{x}{m+1} \left\{ x \sum_{j=0}^{\infty}\sum_{k=0}^{\infty} \frac{(m+j+1)!}{(m+1)!(j-1)!} \binom{n+k}{k} x^{j-1} y^k \right. \end{aligned}$$

$$+\sum_{j=0}^{\infty}\sum_{k=0}^{\infty}\binom{m+j+1}{j}\binom{n+k}{k}x^j y^k\Bigg\}$$

$$+(1-x)^{m+1}(1-y)^{n+1}\frac{y}{n+1}\Bigg\{y\sum_{j=0}^{\infty}\sum_{k=0}^{\infty}\frac{(n+k+1)!}{(n+1)!(k-1)!}\binom{m+j}{j}x^j y^{k-1}$$

$$+\sum_{j=0}^{\infty}\sum_{k=0}^{\infty}\binom{n+k+1}{k}\binom{m+j}{j}x^j y^k\Bigg\}$$

$$=\frac{m+2}{m+1}\left(\frac{x}{1-x}\right)^2+\frac{1}{m+1}\frac{x}{1-x}+\frac{n+2}{n+1}\left(\frac{y}{1-y}\right)^2+\frac{1}{n+1}\frac{y}{1-y},$$

i.e.,

$$B_{m,n}(f_3;x,y)\to\left(\frac{x}{1-x}\right)^2+\left(\frac{y}{1-y}\right)^2.$$

Therefore, the conditions of Theorem A are satisfied and we get for all $f\in H_\omega(K)$ that

$$P\text{-}\lim_{j,k\to\infty}\|T_{j,k}(f;x,y)-f(x,y)\|_{C(K)}=0.$$

Now, define $w=(w_{mn})$ by $w_{mn}=(-1)^m$ for all n. Then, this sequence is neither P-convergent nor statistically convergent, but $\mathrm{st}_{(C,1,1)}$-$\lim w=0$ (since $(C,1,1)$-$\lim w=0$).

Let $L_{m,n}:H_\omega(K)\to C(K)$ be defined by

$$L_{m,n}(f;x,y)=(1+w_{mn})B_{m,n}(f;x,y).$$

It is easy to see that the sequence $(L_{m,n})$ satisfies the conditions (8)–(11). Hence by Theorem 2.1, we have

$$\mathrm{st}^{(2)}_{C(1,1)}\text{-}\lim_{m,n\to\infty}\|L_{m,n}(f;x,y)-f(x,y)\|=0.$$

On the other hand, the sequence $(L_{m,n})$ does not satisfy the conditions of Theorems A and B, since $(L_{m,n})$ is neither P-convergent nor A-statistically convergent. That is, Theorems A and B do not work for our operators $L_{m,n}$. Hence our Theorem 3.1 is stronger than Theorems A and B.

References

1. Anastassiou, G.A., Mursaleen, M., Mohiuddine, S.A.: Some approximation theorems for functions of two variables through almost convergence of double sequences. J. Comput. Anal. Appl. **13**, 37–40 (2011)
2. Belen, C., Mursaleen, M., Yildirim, M.: Statistical A-summability of double sequences and a Korovkin type approximation theorem. Bull. Korean Math. Soc. **49**, 851–861 (2012)
3. Demirci, K., Karakuş, S.: Korovkin-type approximation theorem for double sequences of positive linear operators via statistical A-summability. Results Math. doi:10.1007/s00025-011-0140-y
4. Dirik, F., Demirci, K.: A Korovkin type approximation theorem for double sequences of positive linesr operators of two variables in A-statistical sense. Bull. Korean Math. Soc. **47**, 825–837 (2010)
5. Dirik, F., Demirci, K.: Korovkin type approximation theorem for functions of two variables in statistical sense. Turk. J. Math. **34**, 73–83 (2010)
6. Edely, O.H.H., Mursaleen, M.: Tauberian theorems for statistically convergent double sequences. Inform. Sci. **176**, 875–886 (2006)
7. Edely, O.H.H., Mohiuddine, S.A., Noman, A.K.: Korovkin type approximation theorems obtained through generalized statistical convergence. Appl. Math. Lett. **23**, 1382–1387 (2010)
8. Fast, H.: Sur la convergence statistique. Colloq. Math. **2**, 241–244 (1951)
9. Gadjiev, A.D., Orhan, C.: Some approximation theorems via statistical convergence. Rocky Mountain J. Math. **32**, 129–138 (2002)
10. Korovkin, P.P.: Linear operators and approximation theory. Hindustan Publishing Co., Delhi (1960)
11. Kumar, V., Mursaleen, M.: On (λ, μ)-statistical convergence of double sequences on intuitionistic fuzzy normed spaces. Filomat **25**, 109–120 (2011)
12. Mohiuddine, S.A.: An application of almost convergence in approximation theorems. Appl. Math. Lett. **24**,1856–1860 (2011)
13. Mohiuddine, S.A., Alotaibi, A., Mursaleen, M.: Statistical summability $(C, 1)$ and a Korovkin type approximation theorem. J. Inequal. Appl. **2012**, 172 (2012)
14. Móricz, F.: Tauberian theorems for double sequences that are statistically summability $(C, 1, 1)$. J. Math. Anal. Appl. **286**, 340–350 (2003)
15. Mursaleen, M., Alotaibi, A.: Statistical summability and approximation by de la Vallée-Pousin mean. Appl. Math. Lett. **24**, 320–324 (2011) [Erratum: Appl. Math. Lett. **25**, 665 (2012)]
16. Mursaleen, M., Edely, O.H.H.: Statistical convergence of double sequences. J. Math. Anal. Appl. **288**, 223–231 (2003)
17. Mursaleen, M., Mohiuddine, S.A.: Statistical convergence of double sequences in intuitionistic fuzzy normed spaces. Chaos Solitons Fract. **41**, 2414–2421 (2009)
18. Mursaleen, M., Çakan, C., Mohiuddine, S.A., Savaş, E.: Generalized statistical convergence and statitical core of double sequences. Acta Math. Sinica **26**, 2131–2144 (2010)
19. Mursaleen, M., Karakaya, V., Ertürk, M., Gürsoy, F.: Weighted statistical convergence and its application to Korovkin type approximation theorem. Appl. Math. Comput. **218**, 9132–9137 (2012)
20. Pringsheim, A.: Zur theorie der zweifach unendlichen Zahlenfolgen. Math. Z. **53**, 289–321 (1900)
21. Srivastava, H.M., Mursaleen, M., Khan, A.: Generalized equi-statistical convergence of positive linear operators and associated approximation theorems. Math. Comput. Model. **55**, 2040–2051 (2012)
22. Taşdelen, F., Erençin, A.: The generalization of bivariate MKZ operators by multiple generating functions. J. Math. Anal. Appl. **331**, 727–735 (2007)

Reflections on the Baker–Gammel–Wills (Padé) Conjecture

Doron S. Lubinsky

Dedicated to Professor Hari M. Srivastava

Abstract In 1961, Baker, Gammel, and Wills formulated their famous conjecture that if a function f is meromorphic in the unit ball and analytic at 0, then a subsequence of its diagonal Padé approximants converges uniformly in compact subsets to f. This conjecture was disproved in 2001, but it generated a number of related unresolved conjectures. We review their status.

1 Introduction

Let

$$f(z) = \sum_{j=0}^{\infty} a_j z^j$$

be a formal power series, with complex coefficients. Given integers $m, n \geq 0$, the (m, n) Padé approximant to f is a rational function

$$[m/n] = P/Q$$

where P, Q are polynomials of degree at most m, n, respectively, such that Q is not identically 0, and such that

$$(fQ - P)(z) = O\left(z^{m+n+1}\right). \tag{1}$$

D.S. Lubinsky (✉)
School of Mathematics, Georgia Institute of Technology, Atlanta, GA 30332-0160, USA
e-mail: lubinsky@math.gatech.edu

G.V. Milovanović and M.Th. Rassias (eds.), *Analytic Number Theory, Approximation Theory, and Special Functions*, DOI 10.1007/978-1-4939-0258-3_21,

By this, we mean that the coefficients of $1, z, z^2, \ldots, z^{m+n}$ in the formal power series on the left-hand side vanish. In the special case $n = 0$, $[m/0]$ is just the mth partial sum of the power series.

It is easily seen that $[m/n]$ exists: we can reformulate (1) as a system of $m + n + 1$ homogeneous linear equations in the $(m + 1) + (n + 1)$ coefficients of the polynomials P and Q. As there are more unknowns than equations, there is a nontrivial solution, and it follows from (1) that Q cannot be identically 0 in any nontrivial solution. While P and Q are not separately unique, the ratio $[m/n]$ is.

It was C. Hermite who gave his student Henri Eugene Padé the approximant to study in the 1890s. Although the approximant was known earlier, by, amongst others, Jacobi and Frobenius, it was perhaps Padé's thorough investigation of the structure of the Padé table, namely, the array

$$\begin{array}{ccccc} [0/0] & [0/1] & [0/2] & [0/3] & \ldots \\ [1/0] & [1/1] & [1/2] & [1/3] & \ldots \\ [2/0] & [2/1] & [2/2] & [2/3] & \ldots \\ [3/0] & [3/1] & [3/2] & [3/3] & \ldots \\ \vdots & \vdots & \vdots & \vdots & \ddots \end{array}$$

that has ensured the approximant bearing his name.

Padé approximants have been applied in proofs of irrationality and transcendence in number theory, in practical computation of special functions, and in analysis of difference schemes for numerical solution of partial differential equations. However, the application which really brought them to prominence in the 1960s and 1970s was in location of singularities of functions: in various physical problems, for example, inverse scattering theory, one would have a means for computing the coefficients of a power series f. One could use just $2n + 1$ of these coefficients to compute the $[n/n]$ Padé approximants to f and use the poles of the approximants as predictors of the location of poles or other singularities of f. Moreover, under certain conditions on f, which were often satisfied in physical examples, this process could be theoretically justified.

In addition to their wide variety of applications, they are also closely associated with continued fraction expansions, orthogonal polynomials, moment problems, and the theory of quadrature, amongst others. See [6] and [5] for a detailed development of the theory and [10] for their history.

One of the fascinating features of Padé approximants is the complexity of their convergence theory. The convergence properties vary greatly, depending on how one traverses the table. When the denominator degree is kept fixed as n and the underlying function f is analytic in a ball center 0, except for poles of total multiplicity n, the "column" sequence $\{[m/n]\}_{m=1}^{\infty}$ converges uniformly in compact subsets omitting these poles. This is de Montessus de Ballore's theorem [6], which has been extended and explored in multiple settings.

In this paper, we focus more on the "diagonal" sequence $\{[n/n]\}_{n=1}^{\infty}$. Uniform convergence of diagonal sequences of Padé approximants has been established, for example, for Polya frequency series [3] and series of Stieltjes/Markov/Hamburger [5]. The former have the form

$$f(z) = a_o e^{\gamma z} \prod_{j=1}^{\infty} \frac{1+\alpha_j z}{1-\beta_j z},$$

where $a_0 > 0, \gamma \geq 0$, all $\alpha_j, \beta_j \geq 0$, and

$$\sum_j (\alpha_j + \beta_j) < \infty.$$

The latter have the form

$$f(z) = \int_{-\infty}^{\infty} \frac{d\mu(t)}{1-tz} = \sum_{j=0}^{\infty} z^j \int t^j d\mu(t),$$

and μ is a positive measure supported on the real line, with all finite power moments $\int t^j d\mu(t)$. When μ has non-compact support, the corresponding power series has zero radius of convergence. Nevertheless, the diagonal Padé approximants $\{[n/n]\}_{n=1}^{\infty}$ still converge off the real line to f, at least when μ is a *determinate* measure. The latter means that μ is the only positive measure having moments $\int t^j d\mu(t)$. If μ is supported on $[0,\infty)$ (the so-called Stieltjes case) and is determinate, the diagonal sequence converges uniformly in compact subsets of $\mathbb{C}\backslash(-\infty, 0]$. It is Stieltjes series that often arise in physical applications.

Various modifications of Stieltjes series have also been successfully investigated—for example, when μ' has a sign change or when a rational function is added to the Stieltjes function or multiplies it. See, for example, [1, 15, 16, 28, 41, 43, 50, 53].

Convergence has also been established for classes of special functions such as hypergeometric functions [5, 6] and q-series, even in the singular case when $|q| = 1$ [17]. For functions with "smooth" coefficients, one expects that their Padé approximants should behave well. For rapidly decaying smooth Taylor series coefficients, this has been established in [31]: if $a_j \neq 0$ for j is large enough, and

$$\lim_{j\to\infty} \frac{a_{j-1}a_{j+1}}{a_j^2} = q,$$

where $|q| < 1$, then the full diagonal sequence $\{[n/n]\}_{n=1}^{\infty}$ converges locally uniformly in compact subsets of the plane.

In stark contrast to the positive results above, there are entire functions f for which

$$\limsup_{n\to\infty} |[n/n](z)| = \infty$$

for all $z \in \mathbb{C}\backslash\{0\}$, as established by Hans Wallin [55]. Wallin's function is a somewhat extreme example of the phenomenon of *spurious poles*: approximants can have poles which in no way are related to those of the underlying function. This phenomenon was observed in the early days of Padé approximation, in a simpler form, by Dumas [21].

Physicists such as George Baker in the 1960s endeavored to surmount the problem of spurious poles. They noted that these typically affect convergence only in a small neighborhood and there were usually very few of these "bad" approximants. Thus, one might compute $[n/n]$, $n = 1, 2, 3, \ldots, 50$, and find a definite convergence trend in 45 of the approximants, with 5 of the 50 approximants displaying pathological behavior. Moreover, the 5 bad approximants could be distributed anywhere in the 50 and need not be the first few. Nevertheless, after omitting the "bad" approximants, one obtained a clear convergence trend. This seemed to be a characteristic of the Padé method and led to a famous conjecture [4].

Baker–Gammel–Wills Conjecture (1961). *Let f be meromorphic in the unit ball and analytic at 0. There is an infinite subsequence $\{[n/n]\}_{n \in \mathcal{S}}$ of the diagonal sequence $\{[n/n]\}_{n=1}^{\infty}$ that converges uniformly in all compact subsets of the unit ball omitting poles of f.*

Thus, there is an infinite sequence of "good" approximants. In the first form of the conjecture, f was required to have a nonpolar singularity on the unit circle, but this was subsequently relaxed (cf. [6, p. 188 ff.]). In other forms of the conjecture, f is assumed to be analytic in the unit ball. There is also apparently a cruder form of the conjecture due to Padé himself, dating back to the 1900s; the author must thank J. Gilewicz for this historical information.

2 Reflections

A decade after the Baker–Gammel–Wills conjecture, John Nuttall realized that convergence in measure is a perhaps more appropriate mode of convergence than uniform convergence. In a short 1970 paper [40], he established the celebrated

Nuttall's Theorem. *Let f be meromorphic in $\mathbb{C}$ and analytic at 0. Then the diagonal sequence $\{[n/n]\}_{n=1}^{\infty}$ converges in meas (planar Lebesgue measure) in compact subsets of the plane. That is, given $r, \varepsilon > 0$,*

$$\text{meas}\Big\{z : |z| \leq r \text{ and } |f - [n/n]|(z) \geq \varepsilon\Big\} \to 0 \text{ as } n \to \infty.$$

One consequence is that a subsequence converges a.e. In his 1974 paper [55] containing his counterexample, Wallin also gave conditions on the size of the power series coefficients for convergence a.e. of the full diagonal sequence. Nuttall's theorem was soon extended by Pommerenke, using the concept of cap (logarithmic capacity). For a compact set K, we define

$$\operatorname{cap}(K) = \lim_{n\to\infty} \left(\inf\left\{\|P\|_{L_\infty(K)} : P \text{ a monic polynomial of degree } n\right\}\right)^{1/n},$$

and we extend this to arbitrary sets E as inner capacity

$$\operatorname{cap}(E) = \sup\{\operatorname{cap}(K) : K \subset E,\, K \text{ compact}\}.$$

The capacity of a ball is its radius, and the capacity of a line segment is a quarter of its length. It is a "thinner" set function than planar measure. In fact any set of capacity 0 has Hausdorff dimension 0, and the usual Cantor set has positive logarithmic capacity. The exact value of this for the Cantor set is a well-known, and difficult, problem. Those requiring more background can consult [25,45,46].

Pommerenke [42] proved:

Pommerenke's Theorem. *Let f be analytic in $\mathbb{C}\backslash E$ and analytic at* 0, *where* $\operatorname{cap}(E) = 0$. *Then, given $r, \varepsilon > 0$*

$$\operatorname{cap}\{z : |z| \le r \text{ and } |f - [n/n]|(z) \ge \varepsilon^n\} \to 0 \text{ as } n \to \infty.$$

Since any countable set has capacity 0, Pommerenke's theorem implies Nuttall's. The two are often combined and called the Nuttall–Pommerenke theorem.

While E above may be uncountable, it cannot include branch points. The latter require far deeper techniques, developed primarily by Herbert Stahl in a rigorous form, building on earlier ideas from Nuttall. Stahl showed that one can cut the plane joining the branch points in a certain way, yielding a set of minimal capacity, outside which the Padé approximants converge in capacity. This celebrated and deep theory is expounded in [47–50,52]. Stahl's work gave some hope that BGW might be true for algebraic functions, and indeed, he formulated several conjectures [51], one of which is [51, p. 291]

Stahl's Conjecture for Algebraic Functions. *Let f be an algebraic function, so that for some $m \ge 1$, and polynomials $P_0, P_1, \ldots, P_m$, not all* 0,

$$P_0 + P_1 f + P_2 f^2 + \cdots + P_m f^m \equiv 0.$$

Assume also that f is meromorphic in the unit ball. Then there is a subsequence of $\{[n/n]\}_{n=1}^{\infty}$ that converges uniformly to f in compact subsets of the unit ball, omitting poles of f.

Stahl formulated a more general conjecture, where the unit ball is replaced by the "convergence domain" or "extremal domain" for f. This is the largest domain inside which $\{[n/n]\}_{n=1}^{\infty}$ converges in capacity. Stahl's conjecture was established for a large class of hyperelliptic functions by Suetin [53]. Some very impressive recent related work due to Aptekarev, Baratchart, and Yattselev appears in [2, 9] and due to Martinez–Finkelshtein, Rakhmanov, and Suetin appears in [39]. Deep Riemann–Hilbert techniques play a key role in these papers.

While the positive and negative results of the 1970s cast doubt on the truth of the Baker–Gammel–Wills conjecture, a counterexample remained elusive. It is very difficult to show pathological behavior of a *full sequence* of Padé approximants. The author looked for a long time for a counterexample among the explicitly known Padé approximants to q-series, in the exceptional case where $|q| = 1$. Of course, q-series are usually considered for $|q| < 1$ or $|q| > 1$.

In [38], E. B. Saff and the author investigated the Padé table and continued fraction for the partial theta function $\sum_{j=0}^{\infty} q^{j(j-1)/2} z^j$ when $|q| = 1$. Subsequently K.A. Driver and the author [17–20] undertook a detailed study of the Padé table and continued fraction for the more general Wynn's series [57]

$$\sum_{j=0}^{\infty}\left[\prod_{\ell=0}^{j-1}(A-q^{\ell+\alpha})\right]z^j;\quad \sum_{j=0}^{\infty}\frac{z^j}{\prod_{\ell=0}^{j-1}(C-q^{\ell+\alpha})};\quad \sum_{j=0}^{\infty}\left[\prod_{\ell=0}^{j-1}\frac{A-q^{\ell+\alpha}}{C-q^{\ell+\gamma}}\right]z^j.$$

Here A, C, α, and γ are suitably restricted parameters. All of these satisfy the Baker–Gammel–Wills conjecture.

Finally in 2001 [36], the author found a counterexample in the continued fraction of Rogers–Ramanujan. For q not a root of unity, let

$$G_q(z) := \sum_{j=0}^{\infty}\frac{q^{j^2}}{(1-q)(1-q^2)\cdots(1-q^j)}z^j$$

denote the Rogers–Ramanujan function, and

$$H_q(z) = G_q(z)/G_q(qz).$$

Meromorphic Counterexample. *Let* $q := \exp(2\pi i\tau)$ *where* $\tau := \frac{2}{99+\sqrt{5}}$. *Then* H_q *is meromorphic in the unit ball and analytic at* 0. *There does not exist any subsequence of* $\{[n/n]\}_{n=1}^{\infty}$ *that converges uniformly in all compact subsets of* $\mathcal{A} := \{z : |z| < 0.46\}$ *omitting poles of* H_q.

It did not take long for A. P. Buslaev to improve on this, by finding a function analytic in the unit ball, for which the Baker–Gammel–Wills conjecture and Stahl's conjecture for algebraic functions both fail [11,12]. Buslaev was part of the Russian school of Padé approximation, led by A.A. Goncar. One of their important foci was inverse theory: given certain properties of a sequence of Padé approximants formed from a formal power series, what can we deduce about the analytic properties of the underlying function? For example, if a ball contains none of the poles of the approximants, does it follow that the underlying function is analytic there? Some references to their work are [14,22–24,43,54].

Buslaev's Analytic Counterexample. *Let*

$$f(z) = \frac{-27 + 6z^2 + 3(9+j)z^3 + \sqrt{81(3-(3+j)z^3)^2 + 4z^6}}{2z(9+9z+(9+j)z^2)},$$

where $j = -\frac{1}{2} + \frac{\sqrt{3}}{2}i$. *The branch of the* $\sqrt{\ }$ *is chosen so that* $f(0) = 0$. *Then for some* $R > 1 > r > 0$, f *is analytic in* $\{z : |z| < R\}$, *but for large enough* n, $[n/n]$ *has a pole in* $|z| < r$, *and consequently no subsequence of* $\{[n/n]\}_{n=1}^{\infty}$ *converges uniformly in all compact subsets of* $\{z : |z| < 1\}$.

Buslaev later showed [13] that for q a suitable root of unity, the Rogers–Ramanujan function above, is also a counterexample to both BGW and Stahl's conjecture. Although this resolves the conjecture, it raises further questions. In both the above counterexamples, uniform convergence fails due to the persistence of spurious poles in a specific compact subset of the unit ball. Moreover, in both the above examples, given any point of analyticity of f in the unit ball, some subsequence converges in some neighborhood of the unit ball. In fact, just two subsequences are enough to provide uniform convergence throughout the unit ball, as pointed out by Baker in [7]. It is perhaps with this in mind that in 2005, George Baker modified his 1961 conjecture [8]:

George Baker's "Patchwork" Conjecture. *Let* f *be analytic in the unit ball, except for at most finitely many poles, none at* 0. *Then there exist a finite number of subsequences of* $\{[n/n]\}_{n=1}^{\infty}$ *such that for any given point of analyticity* z *in the ball, at least one of these subsequences converges pointwise to* $f(z)$.

It seems that if true in this form, the convergence would be uniform in some neighborhood of z. Baker also includes poles amongst the permissible z, with the understanding that the corresponding subsequence diverges to ∞.

An obvious question is why we restrict ourselves to uniform convergence of subsequences—perhaps convergence in some other mode is more appropriate. However, there is no possible analogue of the Nuttall–Pommerenke theorem for functions with finite radius of meromorphy. Indeed, the author and E.A. Rakhmanov [29, 44] independently showed that there are functions analytic in the unit ball for which the diagonal sequence $\{[n/n]\}_{n=1}^{\infty}$ does not converge in measure, let alone in capacity. But this does not exclude:

Conjecture on Convergence in Capacity of a Subsequence. *Let* f *be analytic or meromorphic in the unit ball and analytic at* 0. *There exists a subsequence of* $\{[n/n]\}_{n=1}^{\infty}$ *and* $r > 0$ *such that the subsequence converges in measure or capacity to* f *in* $\{z : |z| < r\}$.

Notice that we are not even asking for convergence in capacity throughout the unit ball nor for the r to be independent of f. Weak results in this direction appear in [33,35,37]. In [51, p. 289], this was formulated in the stronger form where $r = 1$. Another obvious point is that all the counterexamples involve a function with finite radius of meromorphy. What about entire functions or functions meromorphic in the whole plane?

Baker–Gammel–Wills Conjecture for Functions Defined in the Plane. *Let f be entire or meromorphic in $\mathbb{C}$ and analytic at 0. Then there exists $r > 0$ and a subsequence of $\{[n/n]\}_{n=1}^{\infty}$ that converges uniformly to f in compact subsets of $\{z : |z| < r\}$.*

This seems like an especially relevant addendum to the 1961 conjecture. A stronger form would be that some subsequence converges uniformly in compact subsets of the plane omitting poles of f. Of course if the r above is independent of f, the stronger form would follow.

Another relevant direction is to restrict the growth of the entire function and try establish convergence. The best growth condition is due to the author [34] but is very weak:

Theorem 2.1. *Assume that the series coefficients $\{a_n\}$ of f satisfy*

$$\limsup_{n\to\infty} |a_n|^{1/n^2} < 1. \tag{2}$$

Then there exists a subsequence of $\{[n/n]\}_{n=1}^{\infty}$ that converges uniformly in compact subsets of the plane to f.

In fact, in that paper, the Maclaurin series coefficients were replaced by errors of rational approximation on a disk, center 0, of radius $\sigma > 0$,

$$E_{nn}(\sigma) = \inf\left\{\|f - R\|_{L_\infty(|z|\leq\sigma)} : R \text{ of type } (n,n)\right\}$$

and the hypothesis was

$$\limsup_{n\to\infty} E_{nn}(\sigma)^{1/n^2} < 1,$$

while f was allowed to be meromorphic rather than entire. It seems appropriate to suggest:

Growth Conditions for the Truth of BGW. *Find the slowest rate of decay of the coefficients of an entire function that guarantees truth of BGW, or at least find a more general condition than* (1).

In an earlier related paper [30], an even weaker result was used to show that the Baker–Gammel–Wills conjecture is usually true in the sense of category. That is, if we place the topology of locally uniform convergence on the space of all entire functions, the set of entire functions for which the conjecture is false is a countable union of nowhere dense sets (i.e., is of "first category").

One can of course go beyond classical Padé approximants in looking for uniform convergence. For example, one can fix the poles of the approximants, leading to what are called Padé-type approximants. We shall not attempt to survey or reference this very extensive topic. While this avoids spurious poles, one sacrifices the degree of interpolation, and the optimal location of poles becomes an issue.

Another path is to interpolate at multiple points rather than 0, while still leaving the poles free. Here there can still be spurious poles, but one hopes that the freedom in choice of interpolation points ameliorates this. It can also help to ensure better approximation on noncircular regions [56]. It is a classical result of E. Levin [26, 27] that the best L_2 rational approximant of type (n, n) (i.e., with numerator, denominator degree $\leq n$) interpolates the approximated function f in at least $2n + 1$ points. As a consequence, for functions analytic in the closed unit ball, there is always a full sequence of diagonal multipoint approximants that converges uniformly in the closed ball to f.

In the special case where one keeps previous interpolation points as one increases the numerator and denominator degree, multipoint Padé approximation is called Newton–Padé approximation. If one allows these interpolation points to depend on the approximated function, then for functions meromorphic in the plane, one can find a full diagonal sequence of Newton–Padé approximants that converge uniformly in compact subsets omitting poles [32].

While Padé approximation may not be such a hot topic as in the period 1970–2000, it is clear that there are significant and challenging problems that are still unresolved and worthy of the efforts of young researchers.

Acknowledgements Research supported by NSF grant DMS1001182 and US-Israel BSF grant 2008399.

References

1. Amiran, A., Wallin, H.: Padé-Type Approximants of Markov and Meromorphic Functions. J. Approx. Theory **88**, 354–369 (1997)
2. Aptekarev, A., Yattselev, M.: Padé Approximants for functions with branch points - strong asymptotics of Nuttall-Stahl polynomials, manuscript.
3. Arms, R.J., Edrei, A.: The Padé tables and continued fractions generated by totally positive sequences, (in) mathematical essays. In: Shankar, H. (ed.) pp. 1–21. Ohio University Press, Athens, Ohio (1970)
4. Baker, G.A., Gammel, J.L., Wills, J.G.: An investigation of the applicability of the Padé approximant method. J. Math. Anal. Appl. **2**, 405–418 (1961)
5. Baker, G.A., Graves-Morris, P.R.: Padé Approximants. Encyclopaedia of Mathematics and its Applications, vol. 59, 2nd edn., Cambridge University Press, Cambridge (1996)
6. Baker, G.A.: Essentials of Padé Approximants, Academic Press, New York (1975)
7. Baker, G.A. Jr.: Some structural properties of two counter-examples to the Baker-Gammel-Wills conjecture. J. Comput. Appl. Math. **161**, 371–391 (2003)
8. Baker, G.A. Jr.: Counter-examples to the Baker-Gammel-Wills conjecture and patchwork convergence. J. Comput. Appl. Math. **179**(1–2), 1–14 (2005)
9. Baratchart, L., Yattselev, M.: Asymptotics of Padé approximants to a certain class of elliptic–type functions (to appear in J. Anal. Math.)
10. Brezinski, C.: A History of Continued Fractions and Padé Approximants, Springer, Berlin (1991)
11. Buslaev, V.I.: Simple counterexample to the Baker-Gammel-Wills conjecture. East J. Approximations **4**, 515–517 (2001)

12. Buslaev, V.I.: The Baker-Gammel-Wills conjecture in the theory of Padé approximants. Math. USSR. Sbornik **193**, 811–823 (2002)
13. Buslaev, V.I.: Convergence of the Rogers-Ramanujan continued fraction. Math. USSR. Sbornik **194** 833–856 (2003)
14. Buslaev, V.I., Goncar, A.A., Suetin, S.P.: On Convergence of subsequences of the mth row of a Padé table. Math. USSR Sbornik **48**, 535–540 (1984)
15. Derevyagin, M., Derkach, V.: On the convergence of Padé approximants of generalized Nevanlinna functions. Trans. Moscow. Math. Soc. **68**, 119–162 (2007)
16. Derevyagin, M., Derkach, V.: Convergence of diagonal Padé approximants for a class of definitizable functions, (in) Recent Advances in Operator Theory in Hilbert and Krein spaces. Operator Theory: Adv. Appl. **198** 97–124 (2010)
17. Driver, K.A.: Convergence of Padé Approximants for some q-Hypergeometric Series (Wynn's Power Series I, II, III), Thesis, University of the Witwatersrand (1991)
18. Driver, K.A., Lubinsky, D.S.: Convergence of Padé Approximants for Wynn's Power Series II, Colloquia Mathematica Societatis Janos Bolyai, vol. 58, pp. 221–239. Janos Bolyai Math Society (1990)
19. Driver, K.A., Lubinsky, D.S.: Convergence of Padé Approximants for a q-Hypergeometric Series (Wynn's Power Series I). Aequationes Mathematicae **42**, 85–106 (1991)
20. Driver, K.A., Lubinsky, D.S.: Convergence of Padé Approximants for Wynn's Power Series III. Aequationes Mathematicae **45**, 1–23 (1993)
21. Dumas, S.: Sur le développement des fonctions elliptiques en fractions continues, Ph.D. Thesis, Zurich (1908)
22. Goncar, A.A.: On Uniform Convergence of Diagonal Padé Approximants. Math. USSR Sbornik **46**, 539–559 (1983)
23. Goncar, A.A., Lungu, K.N.: Poles of Diagonal Padé Approximants and the Analytic Continuation of Functions. **39**, 255–266 (1981)
24. Goncar, A.A., Rakhmanov, E.A., Suetin, S.P.: On the Convergence of Pade Approximations of Orthogonal Expansions. Proc. Steklov Inst. Math. **2**, 149–159 (1993)
25. Landkof, N.S.: Foundations of Modern Potential Theory, Springer, Berlin (1972)
26. Levin, E.: The distribution of poles of rational functions of best approximation and related questions. Math. USSR Sbornik **9**, 267–274 (1969)
27. Levin, E.: The distribution of the poles of the best approximating rational functions and the analytical properties of the approximated function. Israel J. Math. **24**, 139–144 (1976)
28. Lopez Lagomasino, G.: Convergence of Padé approximants of Stieltjes type meromorphic functions and comparative asymptotics for orthogonal polynomials. Math. USSR Sbornik **64**, 207–227 (1989)
29. Lubinsky, D.S.: Diagonal Padé Approximants and Capacity. J. Math. Anal. Applns. **78**, 58–67 (1980)
30. Lubinsky, D.S.: Padé Tables of a Class of Entire Functions. Proc. Amer. Math. Soc. **94**, 399–405 (1985)
31. Lubinsky, D.S.: Padé tables of Entire Functions of very Slow and Smooth Growth II. Constr. Approx. **4**, 321–339 (1988)
32. Lubinsky, D.S.: On uniform convergence of rational, Newton-Padé interpolants of type (n, n) with free poles as $n \to \infty$. Numer. Math. **55**, 247–264 (1989)
33. Lubinsky, D.S.: Convergence of Diagonal Padé Approximants for Functions Analytic near 0. Trans. Amer. Math. Soc. **723**, 3149–3157 (1995)
34. Lubinsky, D.S.: On the diagonal Padé approximants of meromorphic functions. Indagationes Mathematicae **7**, 97–110 (1996)
35. Lubinsky, D.S.: Diagonal Padé Sequences for functions meromorphic in the unit ball approximate well near 0, in trends in approximation theory. In: Kopotun, K., Lyche, T., Neamtu, M. (eds.) pp. 297–305. Vanderbilt University Press, Nashville (2001)
36. Lubinsky, D.S.: Rogers -Ramanujan and the Baker-Gammel-Wills (Padé) Conjecture. Ann. Math. **157**, 847–889 (2003)

37. Lubinsky, D.S.: Weighted maximum over minimum modulus of polynomials, applied to Ray Sequences of Padé Approximants. Constr. Approx. **18**, 285–308 (2002)
38. Lubinsky, D.S., Saff, E.B.: Convergence of Padé approximants of partial theta functions and the Rogers-Szegö polynomials. Constr. Approx. **3**, 331–361 (1987)
39. Martinez-Finkelshtein, A., Rakhmanov, E.A., Suetin, S.P.: Heine, Hilbert, Padé, Riemann, and Stieltjes: John Nuttall's work 25 years later. Contemporary Math. **578**, 165–193 (2012)
40. Nuttall, J.: The convergence of Padé Approximants of Meromorphic Functions. J. Math. Anal. Applns. **31**, 147–153 (1970)
41. Nuttall, J., Singh, S.R.: Orthogonal Polynomials and Padé Approximants associated with a system of arcs. Constr. Approx. **2**, 59–77 (1986)
42. Pommerenke, Ch.: Padé Approximants and Convergence in Capacity. J. Math. Anal. Applns. **41**, 775–780 (1973)
43. Rakhmanov, E.A.: Convergence of diagonal Padé approximants. Math. USSR Sbornik **33**, 243–260 (1977)
44. Rakhmanov, E.A.: On the Convergence of Padé Approximants in Classes of Holomorphic Functions. Math. USSR Sbornik **40**, 149–155 (1981)
45. Ransford, T.: Potential Theory in the Complex Plane, Cambridge University Press, Cambridge (1995)
46. Saff, E.B., Totik, V.: Logarithmic Potential with External Fields, Springer, Berlin (1997)
47. Stahl, H.: Extremal domains associated with an analytic function. I, II. Complex Variables Theory Appl. **4**, 311–324, 325–338 (1985)
48. Stahl, H.: The structure of extremal domains associated with an analytic function. Complex Variables Theory Appl. **4**(4), 339–354 (1985)
49. Stahl, H.: General convergence results for Padé approximants, (in) Approximation theory VI. In: Chui, C.K., Schumaker, L.L., Ward, J.D. (eds.) pp. 605–634. Academic Press, San Diego (1989)
50. Stahl, H.: Diagonal Padé Approximants to hyperelliptic functions. Ann. Fac. Sci. Toulouse, Special Issue, **6**, 121–193 (1996)
51. Stahl, H.: Conjectures around the Baker-Gammel-Wills conjecture: research problems 97–2. Constr. Approx. **13**, 287–292 (1997)
52. Stahl, H.: The convergence of Padé approximants to functions with branch points. J. Approx. Theory **91**, 139–204 (1997)
53. Suetin, S.P.: On the uniform convergence of diagonal Padé approximants for hyperelliptic functions. Math. Sbornik **191**, 81–114 (2000)
54. Suetin, S.P.: Padé Approximants and the effective analytic continuation of a power series. Russian Math. Surveys **57**, 43–141 (2002)
55. Wallin, H.: The convergence of Padé approximants and the size of the power series coefficients. Applicable Anal. **4**, 235–251 (1974)
56. Wallin, H.: Potential theory and approximation of analytic functions by rational interpolation. In: Proceedings of Colloquium on Complex Analysis, Joensuu, Finland, 1978, Springer Lecture Notes in Mathematics, Vol. 747, pp. 434–450. Springer, Berlin (1979)
57. Wynn, P.: A General system of orthogonal polynomials. Quart. J. Math. Oxford Ser. **18**, 81–96 (1967)

Optimal Quadrature Formulas and Interpolation Splines Minimizing the Semi-Norm in the Hilbert Space $K_2(P_2)$

Abdullo R. Hayotov, Gradimir V. Milovanović, and Kholmat M. Shadimetov

Dedicated to Professor Hari M. Srivastava

Abstract In this paper we construct the optimal quadrature formulas in the sense of Sard, as well as interpolation splines minimizing the semi-norm in the space $K_2(P_2)$, where $K_2(P_2)$ is a space of functions φ which φ' is absolutely continuous and φ'' belongs to $L_2(0,1)$ and $\int_0^1(\varphi''(x)+\omega^2\varphi(x))^2dx<\infty$. Optimal quadrature formulas and corresponding interpolation splines of such type are obtained by using S.L. Sobolev method. Furthermore, order of convergence of such optimal quadrature formulas is investigated, and their asymptotic optimality in the Sobolev space $L_2^{(2)}(0,1)$ is proved. These quadrature formulas and interpolation splines are exact for the trigonometric functions $\sin\omega x$ and $\cos\omega x$. Finally, a few numerical examples are included.

A.R. Hayotov (✉)
Institute of Mathematics, National University of Uzbekistan, Do'rmon yo'li str., 29, 100125 Tashkent, Uzbekistan

Facultade de Matematicas, University of Santiago de Compostela, Campus Vida, 15782 Santiago de Compostela, Spain
e-mail: hayotov@mail.ru

G.V. Milovanović
Institute of Mathematics & Serbian Academy of Sciences and Arts, Belgrade, Serbia
e-mail: gvm@mi.sanu.ac.rs

K.M. Shadimetov
Institute of Mathematics, National University of Uzbekistan, Do'rmon yo'li str., 29, 100125 Tashkent, Uzbekistan
e-mail: kholmatshadimetov@mail.ru

G.V. Milovanović and M.Th. Rassias (eds.), *Analytic Number Theory, Approximation Theory, and Special Functions*, DOI 10.1007/978-1-4939-0258-3_22,

1 Introduction

Quadrature formulas and interpolation splines provide basic and important tools for the numerical solution of integral and differential equations, as well as for approximation of functions in some spaces.

This survey paper is devoted to construction of optimal quadrature formulas and interpolation splines in the space $K_2(P_2)$, which is the Hilbert space

$$K_2(P_2) := \Big\{\varphi : [0,1] \to \mathbb{R} \;\Big|\; \varphi' \text{ is absolutely continuous and } \varphi'' \in L_2(0,1)\Big\},$$

and equipped with the norm

$$\|\varphi \,|K_2(P_2)\| = \left\{\int_0^1 \left(P_2\left(\frac{\mathrm{d}}{\mathrm{d}x}\right)\varphi(x)\right)^2 \mathrm{d}x\right\}^{1/2}, \tag{1}$$

where

$$P_2\left(\frac{\mathrm{d}}{\mathrm{d}x}\right) = \frac{\mathrm{d}^2}{\mathrm{d}x^2} + \omega^2,\ \ \omega > 0, \qquad \text{and} \qquad \int_0^1 \left(P_2\left(\frac{\mathrm{d}}{\mathrm{d}x}\right)\varphi(x)\right)^2 \mathrm{d}x < \infty.$$

The equality (1) is semi-norm and $\|\varphi\| = 0$ if and only if $\varphi(x) = c_1 \sin\omega x + c_2 \cos\omega x$.

It should be noted that for a linear differential operator of order m, $L := P_m(\mathrm{d}/\mathrm{d}x)$, Ahlberg, Nilson, and Walsh in the book [1, Chap. 6] investigated the Hilbert spaces in the context of generalized splines. Namely, with the inner product

$$\langle \varphi, \psi \rangle = \int_0^1 L\varphi(x) \cdot L\psi(x)\mathrm{d}x,$$

$K_2(P_m)$ is a Hilbert space if we identify functions that differ by a solution of $L\varphi = 0$. Also, such type of spaces of periodic functions and optimal quadrature formulae were discussed in [10].

The paper is organized as follows. In Sect. 2 we investigate optimal quadrature formulas in the sense of Sard in $K_2(P_2)$ space. In Sect. 2.1 we give the problem of construction of optimal quadrature formulas. In Sect. 2.2 we determine the extremal function which corresponds to the error functional $\ell(x)$ and give a representation of the norm of the error functional. Section 2.3 is devoted to a minimization of $\|\ell\|^2$ with respect to the coefficients C_ν. We obtain a system of linear equations for the coefficients of the optimal quadrature formula in the sense of Sard in the space $K_2(P_2)$. Moreover, the existence and uniqueness of the corresponding solution is proved. Explicit formulas for coefficients of the optimal quadrature formula of the form (2) are presented in Sect. 2.4. In Sect. 2.5 we give the norm of the error functional (3) of the optimal quadrature formula (2). Furthermore, we give an asymptotic analysis of this norm. Section 3 is devoted to interpolation splines minimizing the semi-norm (1) in the space $K_2(P_2)$, including an algorithm for constructing such kind of splines, as well as some numerical examples.

2 Optimal Quadrature Formulas in the Sense of Sard

2.1 *The Problem of Construction of Optimal Quadrature Formulas*

We consider the following quadrature formula:

$$\int_0^1 \varphi(x)\mathrm{d}x \cong \sum_{\nu=0}^{N} C_\nu \varphi(x_\nu), \tag{2}$$

with an error functional given by

$$\ell(x) = \chi_{[0,1]}(x) - \sum_{\nu=0}^{N} C_\nu \delta(x - x_\nu), \tag{3}$$

where C_ν and x_ν ($\in$ $[0,1]$) are coefficients and nodes of the formula (2), respectively; $\chi_{[0,1]}(x)$ is the characteristic function of the interval $[0,1]$; and $\delta(x)$ is Dirac's delta-function. We suppose that the functions $\varphi(x)$ belong to the Hilbert space $K_2(P_2)$.

The corresponding error of the quadrature formula (2) can be expressed in the form

$$R_N(\varphi) = \int_0^1 \varphi(x)\mathrm{d}x - \sum_{\nu=0}^{N} C_\nu \varphi(x_\nu) = (\ell, \varphi) = \int_{\mathbb{R}} \ell(x)\varphi(x)\mathrm{d}x \tag{4}$$

and it is a linear functional in the conjugate space $K_2^*(P_2)$ to the space $K_2(P_2)$.

By the Cauchy-Schwarz inequality

$$|(\ell, \varphi)| \le \|\varphi\,|K_2(P_2)\| \cdot \|\ell\,|K_2^*(P_2)\|$$

the error (4) can be estimated by the norm of the error functional (3), i.e.,

$$\|\ell\,|K_2^*(P_2)\| = \sup_{\|\varphi\,|K_2(P_2)\|=1} |(\ell, \varphi)|\,.$$

In this way, the error estimate of the quadrature formula (2) on the space $K_2(P_2)$ can be reduced to finding a norm of the error functional $\ell(x)$ in the conjugate space $K_2^*(P_2)$.

Obviously, this norm of the error functional $\ell(x)$ depends on the coefficients C_ν and the nodes x_ν, $\nu = 0, 1, \ldots, N$. The problem of finding the minimal norm of the error functional $\ell(x)$ with respect to coefficients C_ν and nodes x_ν is called as *Nikol'skii's problem*, and the obtained formula is called *optimal quadrature formula in the sense of Nikol'skii*. This problem is first considered by Nikol'skii [36] and continued by many authors (see, e.g., [6, 7, 9, 10, 37, 61] and references therein). A minimization of the norm of the error functional $\ell(x)$ with respect only to coefficients C_ν, when nodes are fixed, is called as *Sard's problem*. The obtained formula is called the *optimal quadrature formula in the sense of Sard*. This problem was first investigated by Sard [39].

There are several methods of construction of optimal quadrature formulas in the sense of Sard (see, e.g., [6, 53]). In the space $L_2^{(m)}(a,b)$, based on these methods, Sard's problem was investigated by many authors (see, e.g., [4,6,9,11,12,20,30,31, 33,35,41–43,45,46,48,50,52–54,59,60] and references therein). Here, $L_2^{(m)}(a,b)$ is the Sobolev space of functions, with a square integrable m-th generalized derivative.

It should be noted that a construction of optimal quadrature formulas in the sense of Sard, which are exact for solutions of linear differential equations, was given in [20, 31], using the Peano kernel method, including several examples for some number of nodes.

Optimal quadrature formulas in the sense of Sard were constructed in [47] for $m = 1, 2$ and in [51] for any $m \in \mathbb{N}$, using Sobolev method in the space $W_2^{(m,m-1)}(0,1)$, with the norm defined by

$$\left\|\varphi \,|\, W_2^{(m,m-1)}(0,1)\right\| = \left\{\int_0^1 \left(\varphi^{(m)}(x) + \varphi^{(m-1)}(x)\right)^2 \mathrm{d}x\right\}^{1/2}.$$

In this section we give the solution of Sard's problem in the space $K_2(P_2)$, using Sobolev method for an arbitrary number of nodes $N+1$. Namely, we find the coefficients C_ν (and the error functional $\overset{\circ}{\ell}$) such that

$$\left\|\overset{\circ}{\ell} \,|\, K_2^*(P_2)\right\| = \inf_{C_\nu}\left\|\ell \,|\, K_2^*(P_2)\right\|. \tag{5}$$

Thus, in order to construct an optimal quadrature formula in the sense of Sard in $K_2(P_2)$, we need to solve the following two problems:

Problem 1. Calculate the norm of the error functional $\ell(x)$ for the given quadrature formula *(2)*.

Problem 2. Find such values of the coefficients C_ν such that the equality *(5)* be satisfied with fixed nodes x_ν.

2.2 *The Extremal Function and Representation of the Norm of the Error Functional*

To solve Problem 1, i.e., to calculate the norm of the error functional (3) in the space $K_2^*(P_2)$, we use a concept of the extremal function for a given functional. The function $\psi_\ell(x)$ is called the *extremal* for the functional $\ell(x)$ (cf. [52]) if the following equality is fulfilled:

$$(\ell, \psi_\ell) = \|\ell\,|K_2^*(P_2)\| \cdot \|\psi_\ell\,|K_2(P_2)\|\,.$$

Since $K_2(P_2)$ is a Hilbert space, the extremal function $\psi_\ell(x)$ in this space can be found using the Riesz theorem about general form of a linear continuous functional on Hilbert spaces. Then, for the functional $\ell(x)$ and for any $\varphi \in K_2(P_2)$, there exists such a function $\psi_\ell \in K_2(P_2)$, for which the following equality

$$(\ell, \varphi) = \langle \psi_\ell, \varphi \rangle \tag{6}$$

holds, where

$$\langle \psi_\ell, \varphi \rangle = \int_0^1 \left(\psi_\ell''(x) + \omega^2 \psi_\ell(x)\right)\left(\varphi''(x) + \omega^2 \varphi(x)\right) \mathrm{d}x \tag{7}$$

is an inner product defined on the space $K_2(P_2)$.

Following [25], we investigate the solution of the equation (6).

Let first $\varphi \in \mathring{C}^{(\infty)}(0,1)$, where $\mathring{C}^{(\infty)}(0,1)$ is a space of infinity-differentiable and finite functions in the interval $(0,1)$. Then from (7), an integration by parts gives

$$\langle \psi_\ell, \varphi \rangle = \int_0^1 \left(\psi_\ell^{(4)}(x) + 2\omega^2 \psi_\ell''(x) + \omega^4 \psi_\ell(x)\right)\varphi(x)\mathrm{d}x. \tag{8}$$

According to (6) and (8) we conclude that

$$\psi_\ell^{(4)}(x) + 2\omega^2 \psi_\ell''(x) + \omega^4 \psi_\ell(x) = \ell(x). \tag{9}$$

Thus, when $\varphi \in \mathring{C}^{(\infty)}(0,1)$ the extremal function $\psi_\ell(x)$ is a solution of the equation (9). But, we have to find the solution of (6) when $\varphi \in K_2(P_2)$.

Since the space $\mathring{C}^{(\infty)}(0,1)$ is dense in $K_2(P_2)$, then functions from $K_2(P_2)$ can be uniformly approximated as closely as desired by functions from the space $\mathring{C}^{(\infty)}(0,1)$. For $\varphi \in K_2(P_2)$ we consider the inner product $\langle \psi_\ell, \varphi \rangle$. Now, an integration by parts gives

$$\langle \psi_\ell, \varphi \rangle = \left(\psi_\ell''(x) + \omega^2 \psi_\ell(x)\right)\varphi'(x)\,\Big|_0^1 - \left(\psi_\ell'''(x) + \omega^2 \psi_\ell'(x)\right)\varphi(x)\,\Big|_0^1$$
$$+ \int_0^1 \left(\psi_\ell^{(4)}(x) + 2\omega^2 \psi_\ell''(x) + \omega^4 \psi_\ell(x)\right)\varphi(x)\mathrm{d}x.$$

Hence, taking into account arbitrariness $\varphi(x)$ and uniqueness of the function $\psi_\ell(x)$ (up to functions $\sin\omega x$ and $\cos\omega x$), keeping in mind (9), it must fulfill the following equation:

$$\psi_\ell^{(4)}(x) + 2\omega^2\psi_\ell''(x) + \omega^4\psi_\ell(x) = \ell(x), \tag{10}$$

with boundary conditions

$$\psi_\ell''(0) + \omega^2\psi_\ell(0) = 0, \quad \psi_\ell''(1) + \omega^2\psi_\ell(1) = 0, \tag{11}$$

$$\psi_\ell'''(0) + \omega^2\psi_\ell'(0) = 0, \quad \psi_\ell'''(1) + \omega^2\psi_\ell'(1) = 0. \tag{12}$$

Thus, we conclude that the extremal function $\psi_\ell(x)$ is a solution of the boundary value problem (10)–(12).

Taking the convolution of two functions f and g, i.e.,

$$(f * g)(x) = \int_{\mathbb{R}} f(x-y)g(y)\mathrm{d}y = \int_{\mathbb{R}} f(y)g(x-y)\mathrm{d}y, \tag{13}$$

we can prove the following result:

Theorem 2.1. *The solution of the boundary value problem* (10)–(12) *is the extremal function* $\psi_\ell(x)$ *of the error functional* $\ell(x)$ *and it has the following form:*

$$\psi_\ell(x) = (G * \ell)(x) + d_1 \sin\omega x + d_2 \cos\omega x,$$

where d_1 *and* d_2 *are arbitrary real numbers and*

$$G(x) = \frac{\operatorname{sgn}(x)}{4\omega^3}\,(\sin\omega x - \omega x\ \cos\omega x) \tag{14}$$

is the solution of the equation

$$\psi_\ell^{(4)}(x) + 2\omega^2\psi_\ell''(x) + \omega^4\psi_\ell(x) = \delta(x).$$

Proof. The general solution of a nonhomogeneous differential equation can be represented as a sum of its particular solution and the general solution of the corresponding homogeneous equation. In our case, the general solution of the homogeneous equation for (10) is given by

$$\psi_\ell^h(x) = d_1 \sin\omega x + d_2\cos\omega x + d_3 x\ \sin\omega x + d_4 x\ \cos\omega x,$$

where d_k, $k = 1,2,3,4$ are arbitrary constants. It is not difficult to verify that a particular solution of the equation (10) can be expressed as a convolution of the functions $\ell(x)$ and $G(x)$ defined by (13). The function $G(x)$ is the fundamental solution of the equation (10), and it is determined by (14).

It should be noted that the following rule for finding a fundamental solution of a linear differential operator

$$P_m\left(\frac{\mathrm{d}}{\mathrm{d}x}\right) := \frac{\mathrm{d}^m}{\mathrm{d}x^m} + a_1\frac{\mathrm{d}^{m-1}}{\mathrm{d}x^{m-1}} + \cdots + a_m,$$

where a_j are real numbers, is given in [57, pp. 88–89]. This rule needs to replace $\frac{\mathrm{d}}{\mathrm{d}x}$ by p, and then instead of the operator $P_m(\frac{\mathrm{d}}{\mathrm{d}x})$, we get the polynomial $P_m(p)$. Then we expand the expression $1/P_m(p)$ into the partial fractions, i.e.,

$$\frac{1}{P_m(p)} = \prod_j (p-\lambda)^{-k_j} = \sum_j \left[c_{j,k_j}(p-\lambda_j)^{-k_j} + \cdots + c_{j,1}(p-\lambda_j)^{-1}\right]$$

and to every partial fraction $(p-\lambda)^{-k}$, we correspond the expression $\frac{x^{k-1}\operatorname{sgn} x}{2(k-1)!}\cdot e^{\lambda x}$.

Using this rule, we find the function $G(x)$ which is the fundamental solution of the operator $\frac{\mathrm{d}^4}{\mathrm{d}x^4} + 2\omega^2\frac{\mathrm{d}^2}{\mathrm{d}x^2} + \omega^4$ and it has the form (14).

Thus, we have the following general solution of the equation (10):

$$\psi_\ell(x) = (\ell * G)(x) + d_1 \sin\omega x + d_2\cos\omega x + d_3 x\ \sin\omega x + d_4 x\ \cos\omega x. \quad (15)$$

In order that the function $\psi_\ell(x)$ be unique in the space $K_2(P_2)$ (up to the functions $\sin\omega x$ and $\cos\omega x$), it has to satisfy the conditions (11) and (12), where derivatives are taken in a generalized sense. In computations we need first three derivatives of the function $G(x)$:

$$G'(x) = \frac{\operatorname{sgn} x}{4\omega}x\ \sin\omega x,$$
$$G''(x) = \frac{\operatorname{sgn} x}{4\omega}(\sin\omega x + \omega x\ \cos\omega x),$$
$$G'''(x) = \frac{\operatorname{sgn} x}{4}(2\cos\omega x - \omega x\ \sin\omega x),$$

where we used the following formulas from the theory of generalized functions [57]:

$$(\operatorname{sgn} x)' = 2\delta(x), \quad \delta(x)f(x) = \delta(x)f(0).$$

Further, using the well-known formula

$$\frac{\mathrm{d}}{\mathrm{d}x}(f*g)(x) = (f'*g)(x) = (f*g')(x),$$

we get

$$\psi_\ell'(x) = (\ell * G')(x) + (d_3 - d_2\omega)\sin\omega x + (d_4 + d_1\omega)\cos\omega x - d_4\omega x\sin\omega x + d_3\omega x\cos\omega x,$$

$$\psi_\ell''(x) = (\ell * G'')(x) - (2d_4\omega + d_1\omega^2)\sin\omega x + (2d_3\omega - d_2\omega^2)\cos\omega x - d_3\omega^2 x\sin\omega x - d_4\omega^2 x\cos\omega x,$$

$$\psi_\ell'''(x) = (\ell * G''')(x) - (3d_3\omega^2 - d_2\omega^3)\sin\omega x - (3d_4\omega^2 + d_1\omega^3)\cos\omega x + d_4\omega^3 x\sin\omega x - d_3\omega^3 x\cos\omega x.$$

Now, using these expressions and (15), as well as expressions for $G^{(k)}(x)$, $k = 0, 1, 2, 3$, the boundary conditions (11) and (12) reduce to

$$\begin{cases} (\ell(y), \sin\omega y) + 4d_3\omega^2 = 0, \\ \sin\omega \cdot (\ell(y), \cos\omega y) - \cos\omega \cdot (\ell(y), \sin\omega y) + 4d_3\omega^2\cos\omega - 4d_4\omega^2\sin\omega = 0, \\ (\ell(y), \cos\omega y) + 4d_4\omega^2 = 0, \\ \cos\omega \cdot (\ell(y), \cos\omega y) + \sin\omega \cdot (\ell(y), \sin\omega y) - 4d_3\omega^2\sin\omega - 4d_4\omega^2\cos\omega = 0. \end{cases}$$

Hence, we have $d_3 = 0$, $d_4 = 0$, and therefore

$$(\ell(y), \sin\omega y) = 0, \quad (\ell(y), \cos\omega y) = 0. \tag{16}$$

Substituting these values into (15) we get the assertion of this statement. □

The equalities (16) provide that our quadrature formula is exact for functions $\sin\omega x$ and $\cos\omega x$. The case $\omega = 1$ has been recently considered in [25].

Now, using Theorem 2.1, we immediately obtain a representation of the norm of the error functional

$$\|\ell|K_2^*(P_2)\|^2 = (\ell, \psi_\ell) = \sum_{\nu=0}^{N}\sum_{\gamma=0}^{N} C_\nu C_\gamma\, G(x_\nu - x_\gamma) - 2\sum_{\nu=0}^{N} C_\nu \int_0^1 G(x - x_\nu)\mathrm{d}x + \int_0^1\int_0^1 G(x - y)\mathrm{d}x\mathrm{d}y. \tag{17}$$

In the sequel we deal with Problem 2.

2.3 Existence and Uniqueness of Optimal Coefficients

Let the nodes x_ν of the quadrature formula (2) be fixed. The error functional (3) satisfies the conditions (16). Norm of the error functional $\ell(x)$ is a multidimensional function of the coefficients C_ν ($\nu = 0, 1, \ldots, N$). For finding its minimum under the conditions (16), we apply the Lagrange method. Namely, we consider the function

$$\Psi(C_0, C_1, \ldots, C_N, d_1, d_2) = \|\ell\|^2 - 2d_1(\ell(x), \sin\omega x) - 2d_2(\ell(x), \cos\omega x)$$

and its partial derivatives equating to zero, so that we obtain the following system of linear equations:

$$\sum_{\gamma=0}^{N} C_\gamma G(x_\nu - x_\gamma) + d_1 \sin\omega x_\nu + d_2 \cos\omega x_\nu = f(x_\nu), \quad \nu = 0, 1, \ldots, N, \tag{18}$$

$$\sum_{\gamma=0}^{N} C_\gamma \sin\omega x_\gamma = \frac{1-\cos\omega}{\omega}, \quad \sum_{\gamma=0}^{N} C_\gamma \cos\omega x_\gamma = \frac{\sin\omega}{\omega}, \tag{19}$$

where $G(x)$ is determined by (14) and

$$f(x_\nu) = \int_0^1 G(x - x_\nu)\mathrm{d}x.$$

The system (18) and (19) has the unique solution and it gives the minimum to $\|\ell\|^2$ under the conditions (19).

The uniqueness of the solution of the system (18) and (19) is proved following [54, Chap. I]. For completeness we give it here.

First, we put $\mathbf{C} = (C_0, C_1, \ldots, C_N)$ and $\mathbf{d} = (d_1, d_2)$ for the solution of the system of equations (18) and (19), which represents a stationary point of the function $\Psi(\mathbf{C}, \mathbf{d})$.

Setting $C_\nu = \overline{C}_\nu + C_{1\nu}$, $\nu = 0, 1, \ldots, N$, (17) and the system (18) and (19) become

$$\|\ell\|^2 = \sum_{\nu=0}^{N}\sum_{\gamma=0}^{N} \overline{C}_\nu \overline{C}_\gamma G(x_\nu - x_\gamma) - 2\sum_{\nu=0}^{N} (\overline{C}_\nu + C_{1\nu}) \int_0^1 G(x - x_\nu)\mathrm{d}x$$

$$+ \sum_{\nu=0}^{N}\sum_{\gamma=0}^{N} \left(2\overline{C}_\nu C_{1\gamma} + C_{1\nu}C_{1\gamma}\right) G(x_\nu - x_\gamma) + \int_0^1\int_0^1 G(x-y)\mathrm{d}x\mathrm{d}y \tag{20}$$

and

$$\sum_{\gamma=0}^{N} \overline{C}_\gamma G(x_\nu - x_\gamma) + d_1 \sin \omega x_\nu + d_2 \cos \omega x_\nu = F(x_\nu), \quad \nu = 0, 1, \ldots, N, \tag{21}$$

$$\sum_{\gamma=0}^{N} \overline{C}_\gamma \sin \omega x_\gamma = 0, \quad \sum_{\gamma=0}^{N} \overline{C}_\gamma \cos \omega x_\gamma = 0, \tag{22}$$

respectively, where

$$F(x_\nu) = f(x_\nu) - \sum_{\gamma=0}^{N} C_{1\gamma} G(x_\nu - x_\gamma)$$

and $C_{1\gamma}$, $\gamma = 0, 1, \ldots, N$ are particular solutions of the system (19).

Hence, we directly get that the minimization of (17) under the conditions (16) by C_ν is equivalent to the minimization of the expression (20) by $\overline{C}_\nu$ under the conditions (22). Therefore, it is sufficient to prove that the system (21) and (22) has the unique solution with respect to $\overline{\mathbf{C}} = (\overline{C}_0, \overline{C}_1, \ldots, \overline{C}_N)$ and $\mathbf{d} = (d_1, d_2)$, and this solution gives the conditional minimum for $\|\ell\|^2$. From the theory of the conditional extremum, we need the positivity of the quadratic form

$$\Phi(\overline{\mathbf{C}}) = \sum_{\nu=0}^{N} \sum_{\gamma=0}^{N} \frac{\partial^2 \Psi}{\partial \overline{C}_\nu \partial \overline{C}_\gamma} \overline{C}_\nu \overline{C}_\gamma \tag{23}$$

on the set of vectors $\overline{\mathbf{C}} = (\overline{C}_0, \overline{C}_1, \ldots, \overline{C}_N)$, under the condition

$$S\overline{\mathbf{C}} = 0, \tag{24}$$

where S is the matrix of the system of equations (22),

$$S = \begin{pmatrix} \sin \omega x_0 & \sin \omega x_1 & \cdots & \sin \omega x_N \\ \cos \omega x_0 & \cos \omega x_1 & \cdots & \cos \omega x_N \end{pmatrix}.$$

Now, we show that in this case the condition is satisfied.

Theorem 2.2. *For any nonzero vector* $\overline{\mathbf{C}} \in \mathbb{R}^{N+1}$, *lying in the subspace* $S\overline{\mathbf{C}} = 0$, *the function* $\Phi(\overline{\mathbf{C}})$ *is strictly positive.*

Proof. Using the definition of the function $\Psi(\mathbf{C}, \mathbf{d})$ and the previous equations, (23) reduces to

$$\Phi\left(\overline{\mathbf{C}}\right) = 2 \sum_{\nu=0}^{N} \sum_{\gamma=0}^{N} G(x_\nu - x_\gamma) \overline{C}_\nu \overline{C}_\gamma. \tag{25}$$

Consider now a linear combination of shifted delta functions

$$\delta_{\overline{\mathbf{C}}}(x) = \sqrt{2}\sum_{\nu=0}^{N} \overline{C}_\nu \delta(x - x_\nu). \tag{26}$$

By virtue of the condition (24), this functional belongs to the space $K_2^*(P_2)$. So, it has an extremal function $u_{\overline{\mathbf{C}}}(x) \in K_2(P_2)$, which is a solution of the equation

$$\left(\frac{\mathrm{d}^4}{\mathrm{d}x^4} + 2\omega^2\frac{\mathrm{d}^2}{\mathrm{d}x^2} + \omega^4\right) u_{\overline{\mathbf{C}}}(x) = \delta_{\overline{\mathbf{C}}}(x). \tag{27}$$

As $u_{\overline{\mathbf{C}}}(x)$ we can take a linear combination of shifts of the fundamental solution $G(x)$:

$$u_{\overline{\mathbf{C}}}(x) = \sqrt{2}\sum_{\nu=0}^{N} \overline{C}_\nu G(x - x_\nu),$$

and we can see that

$$\left\|u_{\overline{\mathbf{C}}}|K_2(P_2)\right\|^2 = \left(\delta_{\overline{\mathbf{C}}}, u_{\overline{\mathbf{C}}}\right) = 2\sum_{\nu=0}^{N}\sum_{\gamma=0}^{N} \overline{C}_\nu \overline{C}_\gamma G(x_\nu - x_\gamma) = \Phi\left(\overline{\mathbf{C}}\right).$$

Thus, it is clear that for a nonzero $\overline{\mathbf{C}}$ the function $\Phi\left(\overline{\mathbf{C}}\right)$ is strictly positive and Theorem 2.2 is proved. □

If the nodes $x_0, x_1, \ldots, x_N$ are selected such that the matrix S has the right inverse, then the system of equations (21) and (22) has the unique solution, as well as the system of equations (18) and (19).

Theorem 2.3. *If the matrix S has the right inverse matrix, then the main matrix Q of the system of equations* (21) *and* (22) *is nonsingular.*

Proof. We denote by M the matrix of the quadratic form $\frac{1}{2}\Phi\left(\overline{\mathbf{C}}\right)$, given in (25). It is enough to consider the homogeneous system of linear equations

$$Q\begin{pmatrix}\overline{\mathbf{C}}\\ \mathbf{d}\end{pmatrix} = \begin{pmatrix} M & S^* \\ S & 0\end{pmatrix}\begin{pmatrix}\overline{\mathbf{C}}\\ \mathbf{d}\end{pmatrix} = 0 \tag{28}$$

and prove that it has only the trivial solution.

Let $\overline{\mathbf{C}}$, $\mathbf{d}$ be a solution of (28). Consider the function $\delta_{\overline{\mathbf{C}}}(x)$, defined before by (26). As an extremal function for $\delta_{\overline{\mathbf{C}}}(x)$, we take the following function:

$$u_{\overline{\mathbf{C}}}(x) = \sqrt{2}\sum_{\nu=0}^{N} \overline{C}_\nu G(x - x_\nu) + d_1 \sin\omega x + d_2 \cos\omega x.$$

This is possible because $u_{\overline{\mathbf{C}}}$ belongs to the space $K_2(P_2)$ and it is a solution of the equation (27). The first $N+1$ equations of the system (28) mean that $u_{\overline{\mathbf{C}}}(x)$ takes the value zero at all nodes x_ν, $\nu = 0, 1, \ldots, N$. Then, for the norm of the functional $\delta_{\overline{\mathbf{C}}}(x)$ in $K_2^*(P_2)$, we have

$$\left\|\delta_{\overline{\mathbf{C}}}|K_2^*(P_2)\right\|^2 = \left(\delta_{\overline{\mathbf{C}}}, u_{\overline{\mathbf{C}}}\right) = \sqrt{2}\sum_{\nu=0}^{N} \overline{C}_\nu\, u_{\overline{\mathbf{C}}}(x_\nu) = 0,$$

which is possible only when $\overline{\mathbf{C}} = 0$. According to this fact, from the first $N+1$ equations of the system (28), we obtain $S^*\mathbf{d} = 0$. Since the matrix S is a right inverse (by the hypotheses of this theorem), we conclude that S^* has the left inverse matrix, and therefore $\mathbf{d} = 0$, i.e., $d_1 = d_2 = 0$, which completes the proof. □

According to (17) and Theorems 2.2 and 2.3, it follows that at fixed values of the nodes x_ν, $\nu = 0, 1, \ldots, N$, the norm of the error functional $\ell(x)$ has the unique minimum for some concrete values of $C_\nu = \overset{\circ}{C}_\nu$, $\nu = 0, 1, \ldots, N$. As we mentioned in the first section, the quadrature formula with such coefficients $\overset{\circ}{C}_\nu$ is called *the optimal quadrature formula in the sense of Sard*, and $\overset{\circ}{C}_\nu$, $\nu = 0, 1, \ldots, N$, are the *optimal coefficients*. In the sequel, for convenience, the optimal coefficients $\overset{\circ}{C}_\nu$ will be denoted only as C_ν.

2.4 Coefficients of Optimal Quadrature Formula

In this subsection we solve the system (18) and (19) and find an explicit formula for the coefficients C_ν. We use a similar method offered by Sobolev [53, 54] for finding optimal coefficients in the space $L_2^{(m)}(0,1)$. Here, we mainly use a concept of functions of a discrete argument and the corresponding operations (see [52] and [54]). For completeness we give some of the definitions.

Let nodes x_ν be equally spaced, i.e., $x_\nu = \nu h$, $h = 1/N$. Assume that $\varphi(x)$ and $\psi(x)$ are real-valued functions defined on the real line $\mathbb{R}$.

Definition 2.1. The function $\varphi(h\nu)$ is a *function of discrete argument* if it is given on some set of integer values of ν.

Definition 2.2. The *inner product* of two discrete functions $\varphi(h\nu)$ and $\psi(h\nu)$ is given by

$$[\varphi, \psi] = \sum_{\nu=-\infty}^{\infty} \varphi(h\nu) \cdot \psi(h\nu),$$

if the series on the right- hand side converges absolutely.

Definition 2.3. The *convolution* of two functions $\varphi(h\nu)$ and $\psi(h\nu)$ is the inner product

$$\varphi(h\nu) * \psi(h\nu) = [\varphi(h\gamma), \psi(h\nu - h\gamma)] = \sum_{\gamma=-\infty}^{\infty} \varphi(h\gamma) \cdot \psi(h\nu - h\gamma).$$

Suppose that $C_\nu = 0$, when $\nu < 0$ and $\nu > N$. Using these definitions, the system (18) and (19) can be rewritten in the convolution form

$$G(h\nu) * C_\nu + d_1 \sin(h\omega\nu) + d_2 \cos(h\omega\nu) = f(h\nu), \quad \nu = 0, 1, \ldots, N, \tag{29}$$

$$\sum_{\nu=0}^{N} C_\nu \sin(h\omega\nu) = \frac{1 - \cos\omega}{\omega}, \quad \sum_{\nu=0}^{N} C_\nu \cos(h\omega\nu) = \frac{\sin\omega}{\omega}, \tag{30}$$

where

$$\begin{aligned} f(h\nu) = \frac{1}{4\omega^4} \Big[& 4 - (2 + 2\cos\omega + \omega\sin\omega)\cos(h\omega\nu) \\ & - (2\sin\omega - \omega\cos\omega)\sin(h\omega\nu) + \sin\omega(h\omega\nu)\cos(h\omega\nu) \\ & - (1 + \cos\omega)(h\omega\nu)\sin(h\omega\nu) \Big]. \end{aligned} \tag{31}$$

Now, we consider the following problem:

Problem 3. For a given $f(h\nu)$ find a discrete function C_ν and unknown coefficients d_1, d_2, which satisfy the system *(29)* and *(30)*.

Further, instead of C_ν we introduce the functions $v(h\nu)$ and $u(h\nu)$ by

$$v(h\nu) = G(h\nu) * C_\nu \quad \text{and} \quad u(h\nu) = v(h\nu) + d_1 \sin(h\omega\nu) + d_2 \cos(h\omega\nu).$$

In this statement it is necessary to express C_ν by the function $u(h\nu)$. For this we have to construct such an operator $D(h\nu)$, which satisfies the equation

$$D(h\nu) * G(h\nu) = \delta(h\nu), \tag{32}$$

where $\delta(h\nu)$ is equal to 0 when $\nu \neq 0$ and is equal to 1 when $\nu = 0$, i.e., $\delta(h\nu)$ is a discrete delta function.

In connection with this, a discrete analogue $D(h\nu)$ of the differential operator

$$\frac{d^4}{dx^4} + 2\omega^2 \frac{d^2}{dx^2} + \omega^4, \tag{33}$$

which satisfies (32), was constructed in [24], and some properties were investigated.

Following [24] we have:

Theorem 2.4. *The discrete analogue of the differential operator* (33) *satisfying the equation* (32) *has the form*

$$D(h\nu) = p \begin{cases} A_1 \lambda_1^{|\nu|-1}, \ |\nu| \geq 2, \\ 1 + A_1, \quad |\nu| = 1, \\ C + \dfrac{A_1}{\lambda_1}, \ \nu = 0, \end{cases} \tag{34}$$

where $p = 2\omega^3/(\sin h\omega - h\omega \cos h\omega)$,

$$A_1 = \frac{(2h\omega)^2 \sin^4(h\omega)\lambda_1^2}{(\lambda_1^2 - 1)(\sin h\omega - h\omega \cos h\omega)^2}, \quad C = \frac{2h\omega \cos(2h\omega) - \sin(2h\omega)}{\sin h\omega - h\omega \cos h\omega}, \tag{35}$$

and

$$\lambda_1 = \frac{2h\omega - \sin(2h\omega) - 2\sin(h\omega)\sqrt{h^2\omega^2 - \sin^2(h\omega)}}{2(h\omega \cos(h\omega) - \sin(h\omega))} \tag{36}$$

is a zero of the polynomial

$$Q_2(\lambda) = \lambda^2 + \frac{2h\omega - \sin(2h\omega)}{\sin h\omega - h\omega \cos h\omega}\lambda + 1, \tag{37}$$

and $|\lambda_1| < 1$, *h is a small parameter,* $\omega > 0$, $|h\omega| < 1$.

Theorem 2.5. *The discrete analogue* $D(h\nu)$ *of the differential operator* (33) *satisfies the following equalities:*

1) $D(h\nu) * \sin(h\omega\nu) = 0$,
2) $D(h\nu) * \cos(h\omega\nu) = 0$,
3) $D(h\nu) * (h\omega\nu)\sin(h\omega\nu) = 0$,
4) $D(h\nu) * (h\omega\nu)\cos(h\omega\nu) = 0$,
5) $D(h\nu) * G(h\nu) = \delta(h\nu)$.

Here $G(h\nu)$ *is the function of discrete argument, corresponding to the function* $G(x)$ *defined by* (14), *and* $\delta(h\nu)$ *is the discrete delta function.*

Then, taking into account (32) and Theorems 2.4 and 2.5, for optimal coefficients, we have

$$C_\nu = D(h\nu) * u(h\nu). \tag{38}$$

Thus, if we find the function $u(h\nu)$, then the optimal coefficients can be obtained from (38). In order to calculate the convolution (38) we need a representation of the function $u(h\nu)$ for all integer values of ν. According to (29) we get that $u(h\nu) = f(h\nu)$ when $h\nu \in [0,1]$. Now, we need a representation of the function $u(h\nu)$ when $\nu < 0$ and $\nu > N$.

Since $C_\nu = 0$ for $h\nu \notin [0,1]$, then $C_\nu = D(h\nu) * u(h\nu) = 0$, $h\nu \notin [0,1]$. Now, we calculate the convolution $v(h\nu) = G(h\nu) * C_\nu$ when $h\nu \notin [0,1]$.

Let $\nu < 0$, then, taking into account equalities (14) and (30), we have

$$v(h\nu) = G(h\nu) * C_\nu = \sum_{\gamma=-\infty}^{\infty} C_\gamma \, G(h\nu - h\gamma)$$

$$= \sum_{\gamma=0}^{N} C_\gamma \frac{\operatorname{sgn}(h\nu - h\gamma)}{4\omega^3} \big(\sin(h\omega\nu - h\omega\gamma) - (h\omega\nu - h\omega\gamma)\cos(h\omega\nu - h\omega\gamma)\big)$$

$$= -\frac{1}{4\omega^3}\Bigg[(\sin(h\omega\nu) - h\omega\nu\cos(h\omega\nu))\frac{\sin\omega}{\omega}$$

$$- (\cos(h\omega\nu) + h\omega\nu\sin(h\omega\nu))\frac{(1-\cos\omega)}{\omega}$$

$$+ \cos(h\omega\nu)\sum_{\gamma=0}^{N} C_\gamma \, h\omega\gamma\cos(h\omega\gamma) + \sin(h\omega\nu)\sum_{\gamma=0}^{N} C_\gamma \, h\omega\gamma\sin(h\omega\gamma)\Bigg].$$

Denoting

$$b_1 = \frac{1}{4\omega^3}\sum_{\gamma=0}^{N} C_\gamma \, h\omega\gamma\sin(h\omega\gamma) \quad \text{and} \quad b_2 = \frac{1}{4\omega^3}\sum_{\gamma=0}^{N} C_\gamma \, h\omega\gamma\cos(h\omega\gamma),$$

we get for $\nu < 0$

$$v(h\nu) = -\frac{1}{4\omega^3}\Bigg[(\sin(h\omega\nu) - h\omega\nu\cos(h\omega\nu))\frac{\sin\omega}{\omega} - (\cos(h\omega\nu)$$

$$+ h\omega\nu\sin(h\omega\nu))\frac{(1-\cos\omega)}{\omega} + 4\omega^3 b_1\sin(h\omega\nu) + 4\omega^3 b_2\cos(h\omega\nu)\Bigg],$$

and for $\nu > N$

$$v(h\nu) = \frac{1}{4\omega^3}\Bigg[(\sin(h\omega\nu) - h\omega\nu\cos(h\omega\nu))\frac{\sin\omega}{\omega} - (\cos(h\omega\nu)$$

$$+ h\omega\nu\sin(h\omega\nu))\frac{(1-\cos\omega)}{\omega} + 4\omega^3 b_1\sin(h\omega\nu) + 4\omega^3 b_2\cos(h\omega\nu)\Bigg].$$

Now, setting

$$d_1^- = d_1 - b_1, \quad d_2^- = d_2 - b_2, \quad d_1^+ = d_1 + b_1, \quad d_2^+ = d_2 + b_2$$

we formulate the following problem:

Problem 4. Find the solution of the equation

$$D(h\nu) * u(h\nu) = 0, \quad h\nu \notin [0,1] \tag{39}$$

in the form

$$u(h\nu) = \begin{cases} -\frac{\sin\omega}{4\omega^4}\big[\sin(h\omega\nu) - h\omega\nu\cos(h\omega\nu)\big] + \frac{1-\cos\omega}{4\omega^4}\big[\cos(h\omega\nu) \\ +h\omega\nu\sin(h\omega\nu)\big] + d_1^-\sin(h\omega\nu) + d_2^-\cos(h\omega\nu), & \nu < 0, \\ f(h\nu), & 0 \leqslant \nu \leqslant N, \\ \frac{\sin\omega}{4\omega^4}\big[\sin(h\omega\nu) - h\omega\nu\cos(h\omega\nu)\big] - \frac{1-\cos\omega}{4\omega^4}\big[\cos(h\omega\nu) \\ \quad +h\omega\nu\sin(h\omega\nu)\big] + d_1^+\sin(h\omega\nu) + d_2^+\cos(h\omega\nu), & \nu > N, \end{cases} \tag{40}$$

where d_1^-, d_2^-, d_1^+, d_2^+ are unknown coefficients.

It is clear that

$$d_1 = \frac{1}{2}\left(d_1^+ + d_1^-\right), \; b_1 = \frac{1}{2}\left(d_1^+ - d_1^-\right), \; d_2 = \frac{1}{2}\left(d_2^+ + d_2^-\right), \; b_2 = \frac{1}{2}\left(d_2^+ - d_2^-\right).$$

These unknowns d_1^-, d_2^-, d_1^+, d_2^+ can be found from the equation (39), using the function $D(h\nu)$. Then, the explicit form of the function $u(h\nu)$ and optimal coefficients C_ν can be obtained. Thus, in this way Problem 4, as well as Problem 3, can be solved.

However, instead of this, using $D(h\nu)$ and $u(h\nu)$ and taking into account (38), we find here expressions for the optimal coefficients C_ν, $\nu = 1, \ldots, N-1$. For this purpose we introduce the following notations:

$$m = \sum_{\gamma=1}^{\infty} \frac{A_1 p}{\lambda_1}\lambda_1^\gamma \Big[-\frac{\sin\omega}{4\omega^4}(\sin(-h\omega\gamma) + h\omega\gamma\cos(h\omega\gamma)) - f(-h\gamma)$$

$$+ \frac{1-\cos\omega}{4\omega^4}(\cos(h\omega\gamma) + h\omega\gamma\,\sin(h\omega\gamma)) - d_1^-\sin(h\omega\gamma) + d_2^-\cos(h\omega\gamma)\Big],$$

$$n = \sum_{\gamma=1}^{\infty} \frac{A_1 p}{\lambda_1}\lambda_1^\gamma \Big[\frac{\sin\omega}{4\omega^4}(\sin((N+\gamma)h\omega) - (N+\gamma)h\omega\cos((N+\gamma)h\omega)) - f((N+\gamma)h)$$

$$- \frac{1-\cos\omega}{4\omega^4}(\cos((N+\gamma)h\omega) + (N+\gamma)h\omega\sin((N+\gamma)h\omega))$$

$$+ d_1^+ \sin((N+\gamma)h\omega)[1mm] + d_2^+ \cos((N+\gamma)h\omega)\Big].$$

The series in the previous expressions are convergent because $|\lambda_1| < 1$.

Theorem 2.6. *The coefficients of optimal quadrature formulas in the sense of Sard of the form* (2) *in the space* $K_2(P_2)$ *have the following representation:*

$$C_\nu = \frac{4(1-\cos h\omega)}{\omega\cdot(h\omega+\sin h\omega)} + m\lambda_1^\nu + n\lambda_1^{N-\nu}, \quad \nu = 1,\ldots,N-1, \tag{41}$$

where m *and* n *are defined above and* λ_1 *is given in Theorem* 2.4.

Proof. Let $\nu \in \{1,\ldots,N-1\}$. Then from (38), using (34) and (40), we have

$$C_\nu = D(h\nu) * u(h\nu) = \sum_{\gamma=-\infty}^{\infty} D(h\nu - h\gamma)u(h\gamma)$$

$$= \sum_{\gamma=-\infty}^{-1} D(h\nu - h\gamma)u(h\gamma) + \sum_{\gamma=0}^{N} D(h\nu - h\gamma)u(h\gamma) + \sum_{\gamma=N+1}^{\infty} D(h\nu - h\gamma)u(h\gamma)$$

$$= D(h\nu) * f(h\nu) + \sum_{\gamma=1}^{\infty} \frac{A_1 p}{\lambda_1}\lambda_1^{\nu+\gamma}\Big[-\frac{\sin\omega}{4\omega^4}(\sin(-h\omega\gamma) + h\omega\gamma\cos(h\omega\gamma))$$

$$+ \frac{1-\cos\omega}{4\omega^4}(\cos(h\omega\gamma) + h\omega\gamma\sin(h\omega\gamma)) - d_1^- \sin(h\omega\gamma)$$

$$+ d_2^- \cos(h\omega\gamma) - f(-h\gamma)\Big]$$

$$+ \sum_{\gamma=1}^{\infty} \frac{A_1 p}{\lambda_1}\lambda_1^{N+\gamma-\nu}\Big[\frac{\sin\omega}{4\omega^4}(\sin((N+\gamma)h\omega) - (N+\gamma)h\omega\cos((N+\gamma)h\omega))$$

$$- \frac{1-\cos\omega}{4\omega^4}(\cos((N+\gamma)h\omega) + (N+\gamma)h\omega\sin((N+\gamma)h\omega))$$

$$+ d_1^+ \sin((N+\gamma)h\omega) + d_2^+ \cos((N+\gamma)h\omega) - f((N+\gamma)h)\Big].$$

Hence, taking into account the previous notations, we get

$$C_\nu = D(h\nu) * f(h\nu) + m\lambda_1^\nu + n\lambda_1^{N-\nu}. \tag{42}$$

Now, using Theorems 2.4 and 2.5 and equality (31), we calculate the convolution $D(h\nu) * f(h\nu)$,

$$D(h\nu) * f(h\nu) = D(h\nu) * \frac{1}{\omega^4} = \frac{1}{\omega^4} \sum_{\gamma=-\infty}^{\infty} D(h\gamma)$$

$$= \frac{1}{\omega^4}\left(D(0) + 2D(h) + 2\sum_{\gamma=2}^{\infty} D(h\gamma) \right)$$

$$= \frac{4(1-\cos h\omega)}{\omega \cdot (h\omega + \sin h\omega)}.$$

Substituting this convolution into (42) we obtain (41). □

According to Theorem 2.6, it is clear that in order to obtain the exact expressions of the optimal coefficients C_ν, we need only m and n. They can be found from an identity with respect to $(h\nu)$, which can be obtained by substituting the equality (41) into (29). Namely, equating the corresponding coefficients on the left- and the right-hand sides of the equation (29), we find m and n. The coefficients C_0 and C_N follow directly from (30). Now we can formulate and prove the following result:

Theorem 2.7. *The coefficients of the optimal quadrature formulas in the sense of Sard of the form* (2) *in the space* $K_2(P_2)$ *are*

$$C_\nu = \begin{cases} \dfrac{2\sin h\omega - (h\omega + \sin h\omega)\cos h\omega}{(h\omega + \sin h\omega)\omega \sin h\omega} + \dfrac{(h\omega - \sin h\omega)(\lambda_1 + \lambda_1^{N-1})}{(h\omega + \sin h\omega)\omega \sin h\omega (1+\lambda_1^N)}, \\ \nu = 0, N, \\ \dfrac{4(1-\cos h\omega)}{\omega(h\omega + \sin h\omega)} + \dfrac{2h(h\omega - \sin h\omega)\sin h\omega (\lambda_1^\nu + \lambda_1^{N-\nu})}{(h\omega + \sin h\omega)(h\omega\cos h\omega - \sin h\omega)(1+\lambda_1^N)}, \\ \nu = 1, \ldots, N-1, \end{cases}$$

where λ_1 *is given in Theorem* 2.4 *and* $|\lambda_1| < 1$.

Proof. First from equations (30) we have

$$C_0 = \frac{\sin\omega}{\omega} - \frac{\cos\omega(1-\cos\omega)}{\omega\sin\omega} - \sum_{\gamma=1}^{N-1} C_\gamma \cos(h\omega\gamma) + \frac{\cos\omega}{\sin\omega}\sum_{\gamma=1}^{N-1} C_\gamma \sin(h\omega\gamma),$$

$$C_N = \frac{1-\cos\omega}{\omega\sin\omega} - \frac{1}{\sin\omega}\sum_{\gamma=1}^{N-1} C_\gamma \sin(h\omega\gamma).$$

Hence, using (41), after some simplifications we get

$$C_0 = \frac{(h\omega - \sin h\omega)(1-\cos\omega) + 2\sin\omega\cdot(1-\cos h\omega)}{\omega\sin\omega\cdot(h\omega+\sin h\omega)}$$

$$-m\,\frac{\lambda_1(\sin\omega\ \cos h\omega - \cos\omega\ \sin h\omega) + \lambda_1^{N+1}\sin h\omega - \lambda_1^2\sin\omega}{(\lambda_1^2+1-2\lambda_1\cos h\omega)\ \sin\omega}$$

$$-n\,\frac{\lambda_1^{N+1}(\sin\omega\ \cos h\omega - \sin h\omega\ \cos\omega) + \lambda_1\sin h\omega - \lambda_1^N\sin\omega}{(\lambda_1^2+1-2\lambda_1\cos h\omega)\ \sin\omega},$$

$$C_N = \frac{(h\omega - \sin h\omega)(1-\cos\omega) + 2\sin\omega\cdot(1-\cos h\omega)}{\omega\sin\omega\cdot(h\omega+\sin h\omega)}$$

$$-m\,\frac{\lambda_1^{N+1}(\sin\omega\ \cos h\omega - \sin h\omega\ \cos\omega) + \lambda_1\sin h\omega - \lambda_1^N\sin\omega}{(\lambda_1^2+1-2\lambda_1\cos h\omega)\ \sin\omega}$$

$$-n\,\frac{\lambda_1(\sin\omega\ \cos h\omega - \cos\omega\ \sin h\omega) + \lambda_1^{N+1}\sin h\omega - \lambda_1^2\sin\omega}{(\lambda_1^2+1-2\lambda_1\cos h\omega)\ \sin\omega}.$$

Further, we consider the convolution $G(h\nu) * C_\nu$ in equation (29), i.e.,

$$\begin{aligned} G(h\nu) * C_\nu &= \sum_{\gamma=0}^{N} C_\gamma G(h\nu - h\gamma) \\ &= \sum_{\gamma=0}^{N} C_\gamma \frac{\text{sign}(h\nu - h\gamma)}{4\omega^3}\,[\sin(h\omega\nu - h\omega\gamma) \\ &\qquad -(h\omega\nu - h\omega\gamma)\cos(h\omega\nu - h\omega\gamma)] \\ &= S_1 - S_2, \end{aligned} \tag{43}$$

where

$$S_1 = \frac{1}{2\omega^3}\sum_{\gamma=0}^{\nu} C_\gamma\,[\sin(h\omega\nu - h\omega\gamma) - (h\omega\nu - h\omega\gamma)\cos(h\omega\nu - h\omega\gamma)]$$

and

$$S_2 = \frac{1}{4\omega^3}\sum_{\gamma=0}^{N} C_\gamma\,[\sin(h\omega\nu - h\omega\gamma) - (h\omega\nu - h\omega\gamma)\cos(h\omega\nu - h\omega\gamma)]\,.$$

Using (41) we obtain

$$S_1 = \frac{1}{2\omega^3} C_0 [\sin(h\omega\nu) - h\omega\nu\cos(h\omega\nu)]$$
$$+ \frac{1}{2\omega^3} \sum_{\gamma=0}^{\nu-1} (k + m\lambda_1^{\nu-\gamma} + n\lambda_1^{N+\gamma-\nu})[\sin(h\omega\gamma) - h\omega\gamma\cos(h\omega\gamma)],$$

where $k = 4(1-\cos h\omega)/(\omega(h\omega + \sin h\omega))$. After some calculations and simplifications, S_1 can be reduced to the following form:

$$S_1 = \frac{1}{\omega^4}[1-\cos(h\nu)] + \left[\frac{(h\omega - \sin h\omega)(1-\cos\omega)}{2\omega^4 \sin\omega\,(h\omega + \sin h\omega)} \right.$$
$$+ \frac{m}{2\omega^3}\left(\frac{(\lambda_1 \cos\omega - \lambda_1^{N+1})\sin h\omega}{\sin\omega\,(\lambda_1^2 + 1 - 2\lambda_1 \cos h\omega)} + \frac{(\lambda_1 - \lambda_1^3)\,h\omega\sin h\omega}{(\lambda_1^2 + 1 - 2\lambda_1\cos h\omega)^2} \right)$$
$$\left. + \frac{n}{2\omega^3}\left(\frac{(\lambda_1^{N+1}\cos\omega - \lambda_1)\sin h\omega}{\sin\omega\,(\lambda_1^2 + 1 - 2\lambda_1 \cos h\omega)} + \frac{(\lambda_1^{N+3} - \lambda_1^{N+1})h\omega\sin h\omega}{(\lambda_1^2 + 1 - 2\lambda_1\cos h\omega)^2} \right) \right] \sin(h\omega\nu)$$
$$- \left[\frac{\sin h\omega}{\omega^4\,(h\omega + \sin h\omega)} + \frac{\lambda_1(m + n\lambda_1^N)\sin h\omega}{2\omega^3(\lambda_1^2 + 1 - 2\lambda_1\cos h\omega)} \right](h\omega\nu)\sin(h\omega\nu)$$
$$+ \left[\frac{(h\omega - \sin h\omega)(\cos\omega - 1)}{\omega\,(h\omega + \sin h\omega)} - \frac{\lambda_1[(m + n\lambda_1^N)\cos\omega - (n + m\lambda_1^N)]\sin h\omega}{\lambda_1^2 + 1 - 2\lambda_1\cos h\omega} \right]$$
$$\times \frac{h\omega\nu\cos(h\omega\nu)}{2\omega^3\sin\omega},$$

where we used the fact that λ_1 is a zero of the polynomial $Q_2(\lambda)$ defined by (37).

Now, keeping in mind (30), for S_2 we get the following expression:

$$S_2 = \frac{1}{4\omega^3}\left[\frac{\sin\omega}{\omega}\sin(h\omega\nu) - \frac{1-\cos\omega}{\omega}\cos(h\omega\nu) - \frac{\sin\omega}{\omega}(h\omega\nu)\cos(h\omega\nu) \right.$$
$$- \frac{1-\cos\omega}{\omega}(h\omega\nu)\sin(h\omega\nu) + \cos(h\omega\nu)\sum_{\gamma=1}^{N} C_\gamma (h\omega\gamma)\cos(h\omega\gamma)$$
$$\left. + \sin(h\omega\nu)\sum_{\gamma=1}^{N} C_\gamma (h\omega\gamma)\sin(h\omega\gamma) \right].$$

Now, substituting (43) into equation (29), we get the following identity with respect to $(h\nu)$:

$$S_1 - S_2 + d_1 \sin(h\omega\nu) + d_2 \cos(h\omega\nu) = f(h\nu), \tag{44}$$

where $f(h\nu)$ is defined by (31).

Unknowns in (44) are m, n, d_1, and d_2. Equating the corresponding coefficients of $(h\omega\nu)\sin(h\omega\nu)$ and $(h\omega\nu)\cos(h\omega\nu)$ of both sides of the identity (44), for unknowns m and n, we get the following system of linear equations:

$$\begin{cases} m + \lambda_1^N n = \dfrac{2h \sin h\omega\ (h\omega - \sin h\omega)}{(h\omega + \sin h\omega)(h\omega \cos h\omega - \sin h\omega)}, \\ \lambda_1^N m + n = \dfrac{2h \sin h\omega\ (h\omega - \sin h\omega)}{(h\omega + \sin h\omega)(h\omega \cos h\omega - \sin h\omega)}, \end{cases}$$

from which

$$m = n = \frac{2h \sin h\omega\ (h\omega - \sin h\omega)}{(h\omega + \sin h\omega)(h\omega \cos h\omega - \sin h\omega)(1 + \lambda_1^N)}. \tag{45}$$

The coefficients d_1 and d_2 can be found also from (44) by equating the corresponding coefficients of $\sin(h\omega\nu)$ and $\cos(h\omega\nu)$. In this way the assertion of Theorem 2.7 is proved. □

Proving Theorem 2.7 we have just solved Problem 3, which is equivalent to Problem 2. Thus, Problem 2 is solved, i.e., the coefficients of the optimal quadrature formula (2) in the sense of Sard in the space $K_2(P_2)$ for equally spaced nodes are found.

Remark 2.1. Theorem 2.7 for $N = 2$, $\omega = 1$ gives the result of the example (h) in [31] when $[a, b] = [0, 1]$.

2.5 The Norm of the Error Functional of the Optimal Quadrature Formula

In this subsection we calculate the square of the norm of the error functional (3) for the optimal quadrature formula (2). Furthermore, we give an asymptotic analysis of this norm.

Theorem 2.8. *For the error functional* (3) *of the optimal quadrature formula* (2) *on the space* $K_2(P_2)$, *the following equality*

$$\| \overset{\circ}{\ell} \|^2 = \frac{3\omega - \sin\omega}{2\omega^5} + \frac{h\sin\omega - \sin h\omega}{\omega^4(h\omega + \sin h\omega)} + \frac{4(1-\cos h\omega)(h-1)}{\omega^5(h\omega+\sin h\omega)h}$$
$$- \frac{4\sin h\omega - 2(h\omega + \sin h\omega)\cos h\omega}{\omega^5(h\omega + \sin h\omega)\sin h\omega}$$
$$+ \frac{m}{2\omega^4}\left[\frac{(1-\lambda_1^2)(1-\lambda_1^N)(h\omega\cos h\omega - \sin h\omega)}{(\lambda_1^2 + 1 - 2\lambda_1\cos h\omega)\sin h\omega}\right.$$
$$\left. - \frac{(\lambda_1 + \lambda_1^{N+1})(\sin\omega + \omega)\sin h\omega + 4(\lambda_1^2 + \lambda_1^N)}{\lambda_1^2 + 1 - 2\lambda_1\cos h\omega} - \frac{4(\lambda_1 - \lambda_1^N)}{1-\lambda_1}\right],$$

holds, where λ_1 is given in Theorem 2.4, $|\lambda_1| < 1$, and m is defined by (45).

Proof. In the equal spaced case of the nodes, using (14), we can rewrite the expression (17) in the following form:

$$\|\ell\|^2 = \sum_{\nu=0}^{N} C_\nu\left[\sum_{\gamma=0}^{N} C_\gamma G(h\nu - h\gamma) - f(h\nu)\right] - \sum_{\nu=0}^{N} C_\nu f(h\nu)$$
$$+\frac{1}{2\omega^4}\left[2 + \cos\omega - \frac{3}{\omega}\sin\omega\right],$$

where $f(h\nu)$ is defined by (31).

Hence, taking into account equality (29), we get

$$\|\ell\|^2 = \sum_{\nu=0}^{N} C_\nu\left(-d_1\sin(h\omega\nu) - d_2\cos(h\omega\nu)\right)$$
$$-\sum_{\nu=0}^{N} C_\nu f(h\nu) + \frac{1}{2\omega^4}\left[2 + \cos\omega - \frac{3}{\omega}\sin\omega\right].$$

Using equalities (30) and (31), after some simplifications, we obtain

$$\|\ell\|^2 = \frac{d_1(\cos\omega - 1) - d_2\sin\omega}{\omega} - \frac{1}{4\omega^4}\left[4\sum_{\nu=0}^{N} C_\nu + \sin\omega\sum_{\nu=0}^{N} C_\nu(h\omega\nu)\cos(h\omega\nu)\right.$$
$$\left. - (1+\cos\omega)\sum_{\nu=0}^{N} C_\nu(h\omega\nu)\sin(h\omega\nu)\right] + \frac{1}{4\omega^4}\left[5 + \cos\omega - \frac{2}{\omega}\sin\omega\right]. \tag{46}$$

Now, from (44), equating the corresponding coefficients of $\sin(h\omega\nu)$ and $\cos(h\omega\nu)$, for d_1 and d_2, we find the following expressions:

$$d_1 = \frac{\omega\cos\omega - \sin\omega}{4\omega^4} + \frac{1}{4\omega^3}\sum_{\gamma=0}^{N} C_\gamma(h\omega\gamma)\sin(h\omega\gamma)$$
$$- \frac{h(h\omega - \sin h\omega)(\lambda_1^2 - 1)(\lambda_1^N - 1)}{2\omega^3(h\omega + \sin h\omega)(1+\lambda_1^N)(\lambda_1^2 + 1 - 2\lambda_1\cos h\omega)},$$

$$d_2 = \frac{1 - \cos\omega - \omega\sin\omega}{4\omega^4} + \frac{1}{4\omega^3}\sum_{\gamma=0}^{N} C_\gamma(h\omega\gamma)\cos(h\omega\gamma).$$

Substituting these expressions in (46) we get

$$\|\ell\|^2 = \frac{3\omega - \sin\omega}{2\omega^5} + \frac{h(1-\cos\omega)(h\omega - \sin h\omega)(\lambda_1^2 - 1)(\lambda_1^N - 1)}{2\omega^4(h\omega + \sin h\omega)(1+\lambda_1^N)(\lambda_1^2 + 1 - 2\lambda_1\cos h\omega)}$$
$$+\frac{\cos\omega}{2\omega^4}\sum_{\gamma=1}^{N} C_\gamma(h\omega\gamma)\sin(h\omega\gamma) - \frac{\sin\omega}{2\omega^4}\sum_{\gamma=1}^{N} C_\gamma(h\omega\gamma)\cos(h\omega\gamma) - \frac{1}{\omega^4}\sum_{\gamma=1}^{N} C_\gamma.$$

Finally, using the expression for optimal coefficients C_γ from Theorem 2.7, after some calculations and simplifications, we get the assertion of Theorem 2.8. □

Theorem 2.9. *The norm of the error functional* (3) *for the optimal quadrature formula* (2) *has the form*

$$\| \mathring{\ell} \,|K_2^*(P_2)\|^2 = \frac{1}{720}h^4 + O(h^5) \quad as \quad N \to \infty. \tag{47}$$

Proof. Since

$$\lambda_1 = \frac{2h\omega - \sin(2h\omega) - 2\sin h\omega\sqrt{h^2\omega^2 - \sin^2 h\omega}}{2(h\omega\cos h\omega - \sin h\omega)} = (\sqrt{3} - 2) + O(h^2)$$

and $|h\omega| < 1, \omega > 0$, then $|\lambda_1| < 1$ and $\lambda_1^N \to 0$ as $N \to \infty$. Thus, when $N \to \infty$ the expansion of the expression for $\| \mathring{\ell} \|^2$ (from Theorem 2.8) in a power series in h gives the assertion of Theorem 2.9. □

The next theorem gives an asymptotic optimality for our optimal quadrature formula.

Theorem 2.10. *Optimal quadrature formula of the form* (2) *with the error functional* (3) *in the space* $K_2(P_2)$ *is asymptotic optimal in the Sobolev space* $L_2^{(2)}(0,1)$, *i.e.,*

$$\lim_{N\to\infty} \frac{\| \mathring{\ell} \,|K_2^*(P_2)\|^2}{\| \mathring{\ell} \,|L_2^{(2)*}(0,1)\|^2} = 1. \tag{48}$$

Proof. Using Corollary 5.2 from [48] (for $m = 2$ and $\eta_0 = 0$), for square of the norm of the error functional (3) for the optimal quadrature formula (2) on the Sobolev space $L_2^{(2)}(0,1)$, we get the following expression:

$$\| \overset{\circ}{\ell} | L_2^{(2)*}(0,1)\|^2 = \frac{1}{720}h^4 - \frac{h^5}{12} d \sum_{i=1}^{4} \frac{q^{N+i} + (-1)^i q}{(1-q)^{i+1}} \Delta^i 0^4 = \frac{1}{720}h^4 + O(h^5), \tag{49}$$

where d is known, $q = \sqrt{3} - 2$, $\Delta^i \gamma^4$ is the finite difference of order i of γ^4, and $\Delta^i 0^4 = \Delta^i \gamma^4|_{\gamma=0}$.

Using (47) and (49) we obtain (48) and proof is finished. □

As we said in Sect. 2.1, the error (4) of the optimal quadrature formula of the form (2) in the space $K_2(P_2)$ can be estimated by the Cauchy-Schwarz inequality

$$|R_N(\varphi)| \leq \|\varphi|K_2(P_2)\| \cdot \| \overset{\circ}{\ell} | K_2^*(P_2)\|.$$

Hence, taking into account Theorem 2.9, we get

$$|R_N(\varphi)| \leq \|\varphi|K_2(P_2)\| \left(\frac{\sqrt{5}}{60}h^2 + O(h^{5/2}) \right),$$

from which we conclude that the order of the convergence of our optimal quadrature formula is $O(h^2)$.

3 Interpolation Splines Minimizing the Semi-Norm

3.1 Statement of the Problem

In order to find an approximate representation of a function φ by elements of a certain finite dimensional space, it is possible to use values of this function at some points x_β, $\beta = 0, 1, \ldots, N$. The corresponding problem is called *the interpolation problem*, and the points x_β are *interpolation nodes*.

Polynomial and spline interpolations are very wide subjects in approximation theory (cf. DeVore and Lorentz [15], Mastroianni and Milovanović [34]). The theory of splines as a relatively new area has undergone a rapid progress. Many books are devoted to the theory of splines, for example, Ahlberg *et al* [1], Arcangeli *et al* [2], Attea [3], Berlinet and Thomas-Agnan [5], Bojanov *et al* [8], de Boor [14], Eubank [17], Green and Silverman [22], Ignatov and Pevniy [28], Korneichuk *et al* [29], Laurent [32], Nürnberger [38], Schumaker [44], Stechkin and Subbotin [55], Vasilenko [56], and Wahba [58].

If the exact values $\varphi(x_\beta)$ of an unknown function $\varphi(x)$ are known, it is usual to approximate φ by minimizing

$$\int_a^b (g^{(m)}(x))^2 dx \tag{50}$$

on the set of interpolating functions (i.e., $g(x_\beta) = \varphi(x_\beta)$, $\beta = 0, 1, \ldots, N$) of the Sobolev space $L_2^{(m)}(a,b)$. It turns out that the solution is the natural polynomial spline of degree $2m-1$ with knots $x_0, x_1, \ldots, x_N$. It is called the interpolating D^m-spline for the points $(x_\beta, \varphi(x_\beta))$. In the nonperiodic case this problem was first investigated by Holladay [27] for $m = 2$, and the result of Holladay was generalized by de Boor [13] for any m. In the Sobolev space $\widetilde{L_2^{(m)}}$ of periodic functions, the minimization problem of integrals of type (50) was investigated by Schoenberg [40], Golomb [21], Freeden [18, 19], and others.

In the Hilbert space $K_2(P_2)$, defined in Sect. 1 with the semi-norm (1), we consider the following interpolation problem:

Problem 5. Find the function $S(x) \in K_2(P_2)$ which gives minimum to the semi-norm *(1)* and satisfies the interpolation condition

$$S(x_\beta) = \varphi(x_\beta), \quad \beta = 0, 1, \ldots, N,$$

for any $\varphi \in K_2(P_2)$, where $x_\beta \in [0, 1]$ are the nodes of interpolation.

From [56, p. 45–47] it follows that the solution $S(x)$ of Problem 5 exists uniquely for $N \geq \omega$.

We give a definition of the interpolation spline function in the space $K_2(P_2)$ following [32, Chap. 4, pp. 217–219].

Let $\Delta : 0 = x_0 < x_1 < \cdots < x_N = 1$ be a mesh on the interval [0, 1]. Then the interpolation spline function with respect to Δ is a function $S(x) \in K_2(P_2)$ and satisfies the following conditions:

(i) $S(x)$ is a linear combination of functions $\sin \omega x$, $\cos \omega x$, $x \sin \omega x$, and $x \cos \omega x$ on each open mesh interval $(x_\beta, x_{\beta+1})$, $\beta = 0, 1, \ldots, N-1$;
(ii) $S(x)$ is a linear combination of functions $\sin \omega x$ and $\cos \omega x$ on intervals $(-\infty, 0)$ and $(1, \infty)$;
(iii) $S^{(\alpha)}(x_\beta^-) = S^{(\alpha)}(x_\beta^+)$, $\alpha = 0, 1, 2$, $\beta = 0, 1, \ldots, N$;
(iv) $S(x_\beta) = \varphi(x_\beta)$, $\beta = 0, 1, \ldots, N$, for any $\varphi \in K_2(P_2)$.

We consider the fundamental solution $G(x)$ defined by (14) of the differential operator

$$\frac{\mathrm{d}^4}{\mathrm{d}x^4} + 2\omega^2 \frac{\mathrm{d}^2}{\mathrm{d}x^2} + \omega^4.$$

It is clear that the third derivative of the function

$$G(x - x_\gamma) = \frac{\operatorname{sgn}(x - x_\gamma)}{4\omega^3}\big[\sin(\omega x - \omega x_\gamma) - \omega(x - x_\gamma)\cos(\omega x - \omega x_\gamma)\big]$$

has a discontinuity equal to 1 at the point x_γ, and the first and the second derivatives of $G(x - x_\gamma)$ are continuous. Suppose a function $p_\gamma(x)$ coincides with the spline $S(x)$ on the interval $(x_\gamma, x_{\gamma+1})$, i.e.,

$$p_\gamma(x) := p_{\gamma-1}(x) + C_\gamma G(x - x_\gamma), \quad x \in (x_\gamma, x_{\gamma+1}),$$

where C_γ is the jump of the function $S'''(x)$ at x_γ:

$$C_\gamma = S'''(x_\gamma^+) - S'''(x_\gamma^-),$$

then the spline $S(x)$ can be written in the following form:

$$S(x) = \sum_{\gamma=0}^{N} C_\gamma G(x - x_\gamma) + p_{-1}(x), \tag{51}$$

where $p_{-1}(x) = d_1 \sin \omega x + d_2 \cos \omega x$, with real constants d_1 and d_2.

Furthermore, the function $S(x)$ satisfies the condition (ii) if the function

$$\frac{1}{4\omega^3}\sum_{\gamma=0}^{N} C_\gamma\big[\sin(\omega x - \omega x_\gamma) - \omega(x - x_\gamma)\cos(\omega x - \omega x_\gamma)\big]$$

is a linear combination of the functions $\sin \omega x$ and $\cos \omega x$. Hence, we get the following conditions for C_γ:

$$\sum_{\gamma=0}^{N} C_\gamma \sin(\omega x_\gamma) = 0, \quad \sum_{\gamma=0}^{N} C_\gamma \cos(\omega x_\gamma) = 0.$$

Taking into account the last two equations and the interpolation condition (iv) for the coefficients C_γ, $\gamma = 0, 1, 2, \ldots, N$, d_1, and d_2 in (51), we obtain the following system of $N + 3$ linear equations:

$$\sum_{\gamma=0}^{N} C_\gamma G(x_\beta - x_\gamma) + d_1 \sin(\omega x_\beta) + d_2 \cos(\omega x_\beta) = \varphi(x_\beta),$$

$$\beta = 0, 1, \ldots, N, \tag{52}$$

$$\sum_{\gamma=0}^{N} C_\gamma \sin(\omega x_\gamma) = 0, \tag{53}$$

$$\sum_{\gamma=0}^{N} C_\gamma \cos(\omega x_\gamma) = 0, \tag{54}$$

where $\varphi \in K_2(P_2)$.

Note that the analytic representation (51) of the interpolation spline $S(x)$ and the system of equations (52)–(54) for the coefficients can be also obtained from [56, pp. 45–47, Theorem 2.2].

It should be noted that systems for the coefficients of D^m-splines similar to the system (52)–(54) were investigated, for example, in [2, 16, 28, 32, 56].

In [49], using S.L. Sobolev method, it was constructed the interpolation splines minimizing the semi-norm in the space $W_2^{(m,m-1)}(0,1)$, where $W_2^{(m,m-1)}(0,1)$ is the space of functions φ, which $\varphi^{(m-1)}$ is absolutely continuous and $\varphi^{(m)}$ belongs to $L_2(0,1)$ and

$$\int_0^1 (\varphi^{(m)}(x) + \varphi^{(m-1)}(x))^2 dx < \infty.$$

Our main aim here is to solve Problem 5, i.e., to solve the system of equations (52)–(54) for equally spaced nodes $x_\beta = h\beta$, $\beta = 0, 1, \ldots, N$, $h = 1/N$, $N \geq \omega > 0$ and to find analytic formulas for the spline coefficients C_γ, $\gamma = 0, 1, \ldots, N$, d_1, d_2.

3.2 *Algorithm for Computing Coefficients of Interpolation Splines*

In this subsection we give an algorithm for solving the system of equations (52)–(54), when the nodes x_β are equally spaced. Here we use similar method proposed by Sobolev [53, 54] for finding the coefficients of optimal quadrature formulas in the space $L_2^{(m)}$. Also we use the concept of discrete argument functions and operations on them (see Sect. 2.4).

Suppose that $C_\beta = 0$, when $\beta < 0$ and $\beta > N$. Using Definition 2.3, we write (52)–(54) as follows:

$$G(h\beta) * C_\beta + d_1 \sin(h\omega\beta) + d_2 \cos(h\omega\beta) = \varphi(h\beta), \quad \beta = 0, 1, \ldots, N, \tag{55}$$

$$\sum_{\beta=0}^{N} C_\beta \sin(h\omega\beta) = 0, \tag{56}$$

$$\sum_{\beta=0}^{N} C_\beta \cos(h\omega\beta) = 0, \tag{57}$$

where $G(h\beta)$ is the function of discrete argument corresponding to the function G given in (14).

Thus, we have the following problem:

Problem 6. Find the coefficients C_β, $\beta = 0, 1, \ldots, N$ and the constants d_1 and d_2, which satisfy the system of equations *(55)–(57)*.

Further we investigate Problem 6 which is equivalent to Problem 5. Namely, instead of C_β, we introduce the following functions:

$$v(h\beta) = G(h\beta) * C_\beta, \tag{58}$$

$$u(h\beta) = v(h\beta) + d_1 \sin(h\omega\beta) + d_2 \cos(h\omega\beta). \tag{59}$$

In such a statement it is necessary to express the coefficients C_β by the function $u(h\beta)$. For this we use the operator $D(h\beta)$ which is given in Theorem 2.4.

Then, taking into account (59) and Theorems 2.4 and 2.5, for the coefficients C_β of the spline $S(x)$, we have

$$C_\beta = D(h\beta) * u(h\beta). \tag{60}$$

Thus, if we find the function $u(h\beta)$, then the coefficients C_β can be obtained from equality (60). In order to calculate the convolution (60), we need a representation of the function $u(h\beta)$ for all integer values of β. From equality (55) we get that $u(h\beta) = \varphi(h\beta)$ when $h\beta \in [0,1]$. Now, we need a representation of the function $u(h\beta)$ when $\beta < 0$ and $\beta > N$.

Since $C_\beta = 0$ when $h\beta \notin [0,1]$, then $C_\beta = D(h\beta) * u(h\beta) = 0$, $h\beta \notin [0,1]$. Now, we calculate the convolution $v(h\beta) = G(h\beta) * C_\beta$ when $\beta \le 0$ and $\beta \ge N$.

Supposing $\beta \le 0$ and taking into account equalities (58), (56), (57), we have

$$\begin{aligned}
v(h\beta) &= \sum_{\gamma=-\infty}^{\infty} C_\gamma G(h\beta - h\gamma) \\
&= \sum_{\gamma=0}^{N} C_\gamma \frac{\operatorname{sign}(h\beta-h\gamma)}{4\omega^3}\big[\sin(h\omega\beta-h\omega\gamma)-(h\omega\beta-h\omega\gamma)\cos(h\omega\beta-h\omega\gamma)\big] \\
&= -\frac{1}{4\omega^3}\sum_{\gamma=0}^{N} C_\gamma \Big\{\sin(h\omega\beta)\cos(h\omega\gamma)-\cos(h\omega\beta)\sin(h\omega\gamma) \\
&\quad -(h\omega\beta)\big[\cos(h\omega\beta)\cos(h\omega\gamma)+\sin(h\omega\beta)\sin(h\omega\gamma)\big] \\
&\quad +(h\omega\gamma)\big[\cos(h\omega\beta)\cos(h\omega\gamma)+\sin(h\omega\beta)\sin(h\omega\gamma)\big]\Big\} \\
&= -\frac{\cos(h\omega\beta)}{4\omega^3}\sum_{\gamma=0}^{N} C_\gamma (h\omega\gamma)\cos(h\omega\gamma) - \frac{\sin(h\omega\beta)}{4\omega^3}\sum_{\gamma=0}^{N} C_\gamma (h\omega\gamma)\sin(h\omega\gamma).
\end{aligned}$$

Denoting

$$b_1 = \frac{1}{4\omega^3}\sum_{\gamma=0}^{N} C_\gamma (h\omega\gamma)\sin(h\omega\gamma) \quad \text{and} \quad b_2 = \frac{1}{4\omega^3}\sum_{\gamma=0}^{N} C_\gamma (h\omega\gamma)\cos(h\omega\gamma),$$

we get for $\beta \le 0$

$$v(h\beta) = -b_1 \sin(h\omega\beta) - b_2 \cos(h\omega\beta)$$

and for $\beta \ge N$

$$v(h\beta) = b_1 \sin(h\omega\beta) + b_2 \cos(h\omega\beta).$$

Now, setting

$$d_1^- = d_1 - b_1,\ d_2^- = d_2 - b_2,\ d_1^+ = d_1 + b_1,\ d_2^+ = d_2 + b_2$$

we can formulate the following problem:

Problem 7. Find the solution of the equation

$$D(h\beta) * u(h\beta) = 0, \quad h\beta \notin [0,1] \tag{61}$$

in the form

$$u(h\beta) = \begin{cases} d_1^- \sin(h\omega\beta) + d_2^- \cos(h\omega\beta), & \beta \le 0, \\ \varphi(h\beta), & 0 \le \beta \le N, \\ d_1^+ \sin(h\omega\beta) + d_2^+ \cos(h\omega\beta), & \beta \ge N, \end{cases} \tag{62}$$

with coefficients $d_1^-, d_2^-, d_1^+, d_2^+$.

It is clear that

$$d_1 = \frac{1}{2}\left(d_1^+ + d_1^-\right),\ d_2 = \frac{1}{2}\left(d_2^+ + d_2^-\right), \tag{63}$$

$$b_1 = \frac{1}{2}\left(d_1^+ - d_1^-\right),\ b_2 = \frac{1}{2}\left(d_2^+ - d_2^-\right).$$

These unknowns d_1^-, d_2^-, d_1^+, d_2^+ can be found from (61), using the function $D(h\beta)$. Explicit forms of the function $u(h\beta)$ and coefficients C_β, d_1, d_2 can be found. Thus, Problem 7 and respectively Problems 6 and 5 can be solved.

In the next subsection we realize this algorithm for computing the coefficients C_β, $\beta = 0, 1, \ldots, N$, d_1, and d_2 of the interpolation spline (51).

3.3 Computation of Coefficients of the Interpolation Spline

In this subsection using the procedure from the previous subsection, we obtain explicit formulae for coefficients of the interpolation spline (51) which is the solution of Problem 5.

It should be noted that the interpolation spline (51), which is the solution of Problem 5, is exact for the functions $\sin \omega x$ and $\cos \omega x$.

Theorem 3.1. *Coefficients of the interpolation spline* (51) *which minimizes the semi-norm* (1) *with equally spaced nodes in the space* $K_2(P_2)$ *have the following forms:*

$$C_0 = Cp\varphi(0) + p[\varphi(h) - d_1^- \sin(h\omega) + d_2^- \cos(h\omega)]$$

$$+ \frac{A_1 p}{\lambda_1}\left[\sum_{\gamma=0}^{N} \lambda_1^{\gamma}\varphi(h\gamma) + M_1 + \lambda_1^N N_1\right]$$

$$C_\beta = Cp\varphi(h\beta) + p[\varphi(h(\beta-1)) + \varphi(h(\beta+1))]$$

$$+ \frac{A_1 p}{\lambda_1}\left[\sum_{\gamma=0}^{N} \lambda_1^{|\beta-\gamma|}\varphi(h\gamma) + \lambda_1^{\beta} M_1 + \lambda_1^{N-\beta} N_1\right], \quad \beta = 1, 2, \ldots, N-1,$$

$$C_N = Cp\varphi(1) + p[\varphi(h(N-1)) + d_1^+ \sin(\omega + h\omega) + d_2^+ \cos(\omega + h\omega)]$$

$$+ \frac{A_1 p}{\lambda_1}\left[\sum_{\gamma=0}^{N} \lambda_1^{N-\gamma}\varphi(h\gamma) + \lambda_1^N M_1 + N_1\right],$$

$$d_1 = \frac{1}{2}(d_1^+ + d_1^-), \qquad d_2 = \frac{1}{2}(d_2^+ + d_2^-),$$

where p, A_1, C, *and* λ_1 *are defined by* (35), (36),

$$M_1 = \frac{\lambda_1[d_2^-(\cos(h\omega) - \lambda_1) - d_1^- \sin(h\omega)]}{\lambda_1^2 + 1 - 2\lambda_1\cos(h\omega)}, \tag{64}$$

$$N_1 = \frac{\lambda_1[d_2^+(\cos(\omega + h\omega) - \lambda_1\cos\omega) + d_1^+(\sin(\omega + h\omega) - \lambda_1 \sin\omega)]}{\lambda_1^2 + 1 - 2\lambda_1\cos(h\omega)}, \tag{65}$$

and d_1^+, d_1^-, d_2^+, d_2^- *are defined by* (66) *and* (72).

Proof. First, we find the expressions for d_2^- and d_2^+. From (62), when $\beta = 0$ and $\beta = N$, we get

$$d_2^- = \varphi(0), \quad d_2^+ = \frac{\varphi(1)}{\cos\omega} - d_1^+ \tan\omega. \tag{66}$$

We have now two unknowns d_1^- and d_1^+ and they can be found from (61) when $\beta = -1$ and $\beta = N+1$.

Taking into account (62) and Definition 2.3, from (61), we obtain

$$\sum_{\gamma=-\infty}^{-1} D(h\beta - h\gamma)[d_1^- \sin(h\omega\gamma) + d_2^- \cos(h\omega\gamma)] + \sum_{\gamma=0}^{N} D(h\beta - h\gamma)\varphi(h\gamma)$$

$$+ \sum_{\gamma=N+1}^{\infty} D(h\beta - h\gamma)[d_1^+ \sin(h\omega\gamma) + d_2^+ \cos(h\omega\gamma)] = 0,$$

where $\beta < 0$ and $\beta > N$.

Hence, for $\beta = -1$ and $\beta = N+1$, we get the following system of equations for d_1^-, d_1^+, d_2^-, d_2^+:

$$-d_1^- \sum_{\gamma=1}^{\infty} D(h\gamma - h)\sin(h\omega\gamma) + d_2^- \sum_{\gamma=1}^{\infty} D(h\gamma - h)\cos(h\omega\gamma)$$

$$+d_1^+ \sum_{\gamma=1}^{\infty} D(h(N+\gamma)+h)\sin(\omega+h\omega\gamma) + d_2^+ \sum_{\gamma=1}^{\infty} D(h(N+\gamma)+h)\cos(\omega+h\omega\gamma)$$

$$= - \sum_{\gamma=0}^{N} D(h\gamma+h)\varphi(h\gamma), \tag{67}$$

$$-d_1^- \sum_{\gamma=1}^{\infty} D(h(N+\gamma)+h)\sin(h\omega\gamma) + d_2^- \sum_{\gamma=1}^{\infty} D(h(N+\gamma)+h)\cos(h\omega\gamma)$$

$$+d_1^+ \sum_{\gamma=1}^{\infty} D(h\gamma - h)\sin(\omega + h\omega\gamma) + d_2^+ \sum_{\gamma=1}^{\infty} D(h\gamma - h)\cos(\omega + h\omega\gamma)$$

$$= - \sum_{\gamma=0}^{N} D(h(N+1) - h\gamma)\varphi(h\gamma). \tag{68}$$

Since $|\lambda_1| < 1$, the series in the system of equations (67)–(68) are convergent.

Using (66) and taking into account (34), after some calculations and simplifications, from this system we obtain

$$B_{11}d_1^- + B_{12}d_1^+ = T_1, \qquad B_{21}d_1^- + B_{22}d_1^+ = T_2,$$

where

$$B_{11} = \lambda_1 \sin(h\omega), \qquad B_{12} = -\frac{\lambda_1^{N+1}\sin(h\omega)}{\cos\omega},$$
$$B_{21} = \lambda_1^{N+1}\sin(h\omega), \ B_{22} = -\frac{\lambda_1 \sin(h\omega)}{\cos\omega}, \tag{69}$$

$$
\begin{aligned}
T_1 &= \frac{2h\omega\lambda_1 \sin^2(h\omega)}{h\omega\cos(h\omega)-\sin(h\omega)} \sum_{\gamma=0}^{N} \lambda_1^{\gamma}\varphi(h\gamma) \\
&+[\lambda_1\cos(h\omega)-1]\varphi(0)+\lambda_1^{N+1}[\cos(h\omega)-\lambda_1-\tan\omega\sin(h\omega)]\varphi(1),
\end{aligned} \tag{70}
$$

$$
\begin{aligned}
T_2 &= \frac{2h\omega\lambda_1 \sin^2(h\omega)}{h\omega\cos(h\omega)-\sin(h\omega)} \sum_{\gamma=0}^{N} \lambda_1^{N-\gamma}\varphi(h\gamma) \\
&+\lambda_1^{N+1}[\cos(h\omega)-\lambda_1]\varphi(0)+[\lambda_1\cos(h\omega)-1-\lambda_1\tan\omega\sin(h\omega)]\varphi(1).
\end{aligned} \tag{71}
$$

Hence, we get

$$
d_1^- = \frac{T_1B_{22}-T_2B_{12}}{B_{11}B_{22}-B_{12}B_{21}}, \quad d_1^+ = \frac{T_2B_{11}-T_1B_{21}}{B_{11}B_{22}-B_{12}B_{21}}, \tag{72}
$$

where $B_{11}, B_{12}, B_{21}, B_{22}, T_1$, and T_2 are defined by (69)–(71).

Combining (63), (66), and (72), we obtain d_1 and d_2 which are given in the statement of Theorem 3.1.

Now, we calculate the coefficients C_β, $\beta = 0, 1, \dots, N$. Taking into account (62) from (60) for C_β, we have

$$
\begin{aligned}
C_\beta &= D(h\beta) * u(h\beta) \\
&= \sum_{\gamma=-\infty}^{\infty} D(h\beta - h\gamma)u(h\gamma) \\
&= \sum_{\gamma=1}^{\infty} D(h\beta+h\gamma)[-d_1^-\sin(h\omega\gamma)+d_2^-\cos(h\omega\gamma)] + \sum_{\gamma=0}^{N} D(h\beta-h\gamma)\varphi(h\gamma) \\
&+\sum_{\gamma=1}^{\infty} D(h(N+\gamma)-h\beta)[d_1^+\sin(\omega+h\omega\gamma)+d_2^+\cos(\omega+h\omega\gamma)],
\end{aligned}
$$

from which, using (34) and taking into account notations (64), (65), when $\beta = 0, 1, \dots, N$, for C_β we get expressions from the statement of Theorem 3.1.

Remark 3.1. The case $\omega = 1$ is considered in [26].

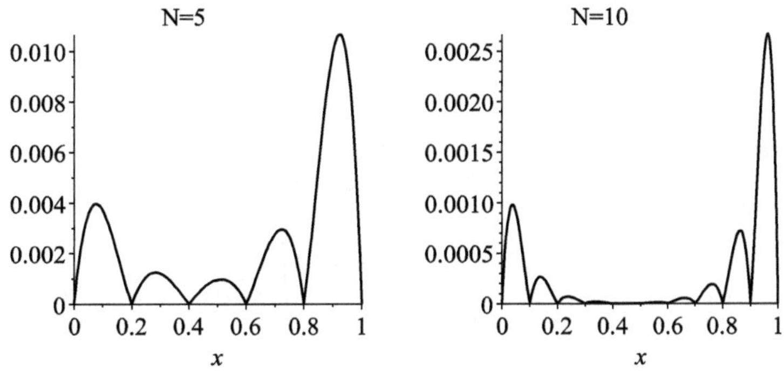

Fig. 1 Graphs of absolute errors $|f_1(x) - S_N(f_1; x)|$ for $N = 5$ and $N = 10$

3.4 Numerical Results

In this subsection, in numerical examples, we compare the interpolation spline (51) with the natural cubic spline (D^2-spline).

It is known that (cf. [2, 13, 16, 27, 28, 32, 56]) the natural cubic spline minimizes the integral $\int_0^1 (\varphi''(x))^2 dx$ in the Sobolev space $L_2^{(2)}(0,1)$ of functions with a square integrable 2nd generalized derivative. For convenience we denote the natural cubic spline as $Scubic(x)$. In numerical examples we use the standard function "`spline(X,Y,x,cubic)`" of the MAPLE package for the natural cubic spline.

Here first we consider the case $\omega = 1$ and give some numerical results which have been presented also in [26].

We apply the interpolation spline (51) and the natural cubic spline to approximation of the functions

$$f_1(x) = \mathrm{e}^x, \quad f_2(x) = \tan x, \quad f_3(x) = \frac{313x^4 - 6900x^2 + 15120}{13x^4 + 660x^2 + 15120}.$$

Using Theorem 3.1 and the standard Maple function "`spline(X,Y,x,cubic)`," with $N = 5$ and $N = 10$, we get the corresponding interpolation splines denoted by $S_N(f_k; x)$, $k = 1, 2, 3$ for the interpolation spline (51) and $Scubic_N(f_k; x)$, $k = 1, 2, 3$, for the natural cubic spline.

The corresponding absolute errors $|f_k(x) - S_N(f_k; x)|$ and $|f_k(x) - Scubic_N(f_k; x)|$ on [0, 1], for $k = 1, 2$, and 3, are displayed in Figs. 1 and 2, 3 and 4, and 5 and 6, respectively.

As we can see the smallest errors in these cases appear in Fig. 5 because $f_3(x)$ is a rational approximation for the function $\cos x$ (cf. [23, p. 66]) and the interpolation spline (51) is exact for the trigonometric functions $\sin x$ and $\cos x$.

In order to test the optimal quadrature formula in the sense of Sard in the space $K_2(P_2)$ (see [25])

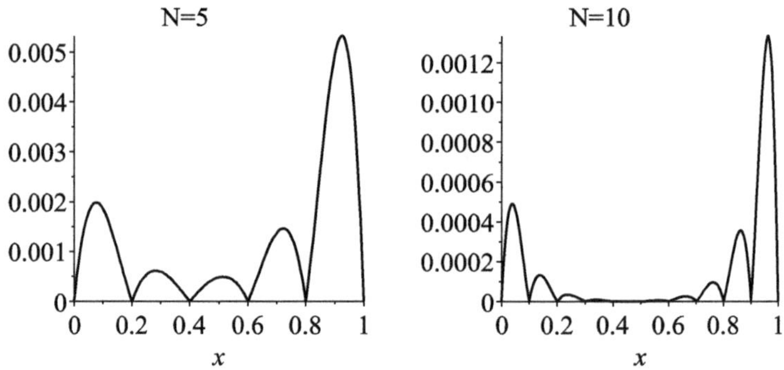

Fig. 2 Graphs of absolute errors $|f_1(x) - Scubic_N(f_1; x)|$ for $N = 5$ and $N = 10$

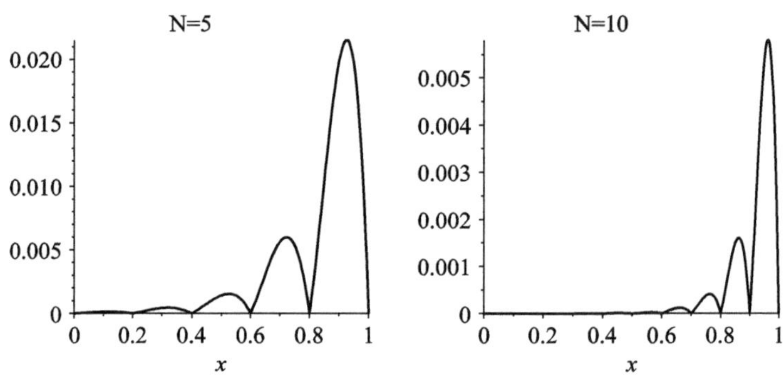

Fig. 3 Graphs of absolute errors $|f_2(x) - S_N(f_2; x)|$ for $N = 5$ and $N = 10$

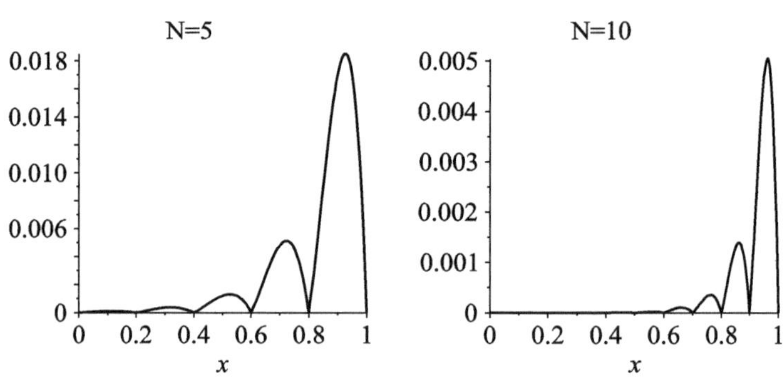

Fig. 4 Graphs of absolute errors $|f_2(x) - Scubic_N(f_2; x)|$ for $N = 5$ and $N = 10$

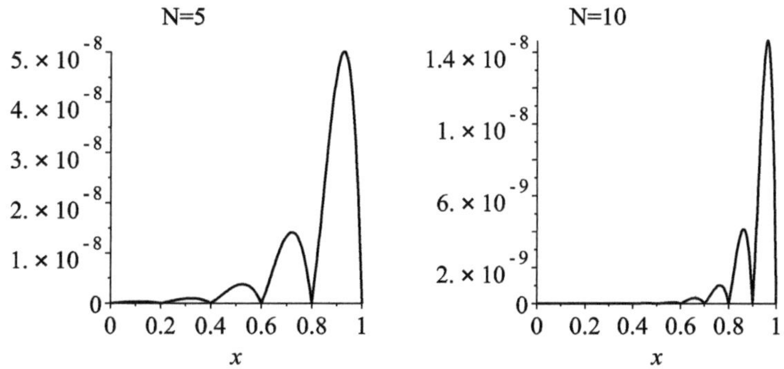

Fig. 5 Graphs of absolute errors $|f_3(x) - S_N(f_3; x)|$ for $N = 5$ and $N = 10$

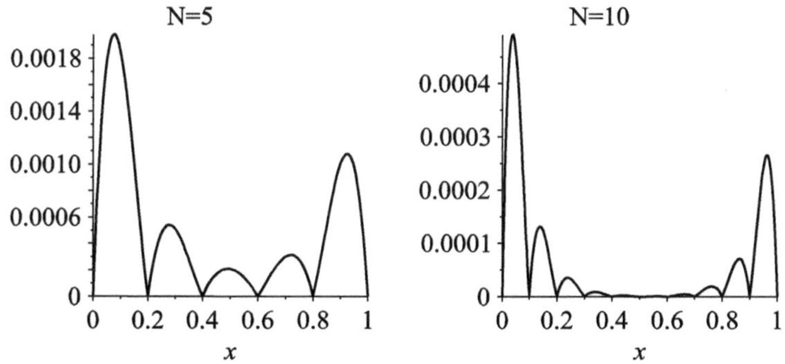

Fig. 6 Graphs of absolute errors $|f_3(x) - Scubic_N(f_3; x)|$ for $N = 5$ and $N = 10$

$$I(\varphi) := \int_0^1 \varphi(x)\mathrm{d}x \cong \sum_{\nu=0}^{N} C_\nu \varphi(x_\nu) =: Q_N(\varphi), \tag{73}$$

we use the same functions $f_k(x)$, $k = 1, 2, 3$.

The weight coefficients in (73) are

$$C_0 = C_N = \frac{2\sin h - (h + \sin h)\cos h}{(h + \sin h)\sin h} + \frac{h - \sin h}{(h + \sin h)\sin h(1 + \lambda_1^N)}(\lambda_1 + \lambda_1^{N-1})$$

and

$$C_\nu = \frac{4(1 - \cos h)}{h + \sin h} + \frac{2h(h - \sin h)\sin h}{(h + \sin h)(h\cos h - \sin h)(1 + \lambda_1^N)}(\lambda_1^\nu + \lambda_1^{N-\nu}),$$

for $\nu = 1, \ldots, N - 1$, where λ_1 is given as in (36), with $\omega = 1$ and $|\lambda_1| < 1$.

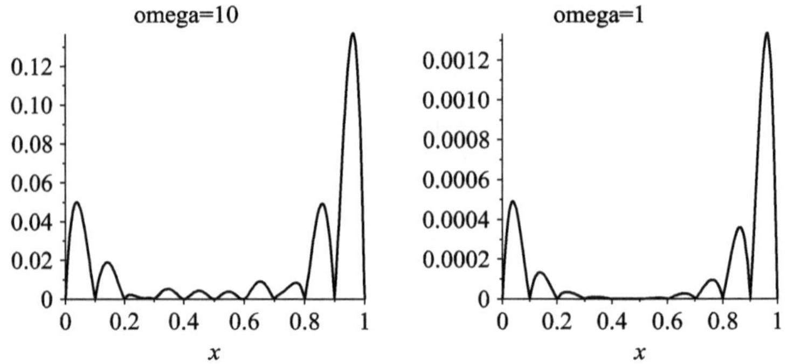

Fig. 7 Graphs of absolute errors $|S_{10}(f_1;x) - Scubic_{10}(f_1;x)|$ for $\omega = 10$ and $\omega = 1$

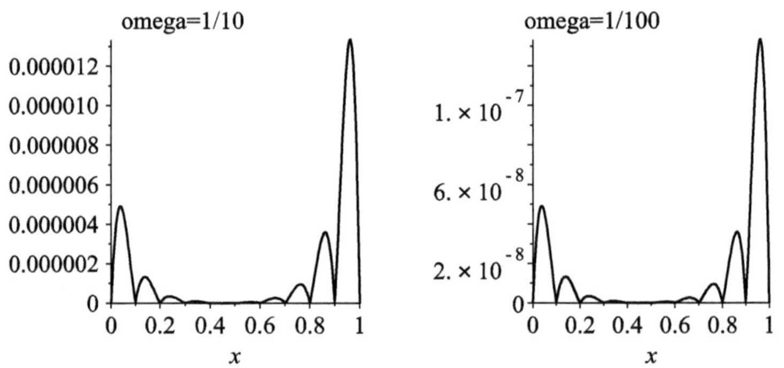

Fig. 8 Graphs of absolute errors $|S_{10}(f_1;x) - Scubic_{10}(f_1;x)|$ for $\omega = 0.1$ and $\omega = 0.01$

In [25] we have obtained the approximate numerical values

$$Q_N(f_k) = \sum_{\nu=0}^{N} C_\nu f_k(x_\nu)$$

for the corresponding integrals $I(f_k)$, $k = 1, 2, 3$, taking $N = 10, 100$, and 1000. These approximate values we can also obtain if we integrate the corresponding interpolation splines $S_N(f_k;x)$ over [0, 1], i.e., $Q_N(f_k) = I(S_N(f_k;x))$.

We consider now the values of the difference $|S_N(f_1;x) - Scubic_N(f_1;x)|$ (with $N + 1 = 11$ nodes) in cases when $\omega = 10, 1, 0.1$, and 0.01.

Graphs in Figs. 7 and 8 show that $S_N(f_1;x)$ tends to $Scubic_N(f_1;x)$ as $\omega \to 0$.

Acknowledgements The work of the first author (A.R. Hayotov) was supported by the program Erasmus Mundus Action 2, Stand 1, Lot 10, Marco XXI (project number: 204513-EM-1-2011-1-DE-ERA MUNDUS-EMA21). He thanks Professor A. Cabada for his hospitality and useful discussion during his visit at the University of Santiago de Compostela, Spain.

The second author (G.V. Milovanović) was supported in part by the Serbian Ministry of Education, Science and Technological Development (grant number #174015).

References

1. Ahlberg, J.H., Nilson, E.N., Walsh, J.L.: The Theory of Splines and Their Applications, Academic Press, New York (1967)
2. Arcangeli, R., Lopez de Silanes, M.C., Torrens, J.J.: Multidimensional Minimizing Splines, Kluwer Academic publishers, Boston (2004)
3. Attea, M.: Hilbertian kernels and spline functions. In: Brezinski, C., Wuytack, L. (eds.) Studies in Computational Matematics 4, North-Holland, Amsterdam (1992)
4. Babuška, I.: Optimal quadrature formulas (Russian). Dokladi Akad. Nauk SSSR **149**, 227–229 (1963)
5. Berlinet, A., Thomas-Agnan, C.: Reproducing Kernel Hilbert Spaces in Probability and Statistics, Kluwer Academic Publisher (2004)
6. Blaga, P., Coman, Gh.: Some problems on optimal quadrature. Stud. Univ. Babeş-Bolyai Math. **52**(4), 21–44 (2007)
7. Bojanov, B.: Optimal quadrature formulas (Russian). Uspekhi Mat. Nauk **606**(366), 33–52 (2005)
8. Bojanov, B.D., Hakopian, H.A., Sahakian, A.A.: Spline Functions and Multivariate Interpolations, Kluwer, Dordrecht, (1993)
9. Catinaş, T., Coman, Gh.: Optimal quadrature formulas based on the ϕ-function method. Stud. Univ. Babeş-Bolyai Math. **51**(1), 49–64 (2006)
10. Chakhkiev, M.A.: Linear differential operators with real spectrum, and optimal quadrature formulas (Russian). Izv. Akad. Nauk SSSR Ser. Mat. **48**(5), 1078–1108 (1984)
11. Coman, Gh.: Monosplines and optimal quadrature formulae in L_p. Rend. Mat. **5**(6), 567–577 (1972)
12. Coman, Gh.: Quadrature formulas of Sard type (Romanian). Studia Univ. Babeş-Bolyai Ser. Math.-Mech. **17**(2), 73–77 (1972)
13. de Boor, C.: Best approximation properties of spline functions of odd degree. J. Math. Mech. **12**, 747–749 (1963)
14. de Boor, C.: A Practical Guide to Splines, Springer, New York (1978)
15. DeVore, R.A., Lorentz, G.G.: Constructive Approximation, Grundlehren der mathematischen Wissenschaften, vol. 303, Springer, Berlin (1993)
16. Duchon, J.: Splines Minimizing Rotation-Invariant Semi-Norms in Sobolev Spaces, pp. 85–100. Springer, Berlin (1977)
17. Eubank, R.L.: Spline Smoothing and Nonparametric Regression. Marcel-Dekker, New–York (1988)
18. Freeden, W.: Spherical spline interpolation-basic theory and computational aspects. J. Comput. Appl. Math., **11**, 367–375, (1984)
19. Freeden, W.: Interpolation by multidimensional periodic splines. J. Approx. Theory **55**, 104–117 (1988)
20. Ghizzetti, A., Ossicini, A.: Quadrature Formulae, Akademie Verlag, Berlin (1970)
21. Golomb, M.: Approximation by periodic spline interpolants on uniform meshes. J. Approx. Theory **1**, 26–65 (1968)
22. Green, P.J., Silverman: Nonparametric regression and generalized linear models. A roughness penalty approach. Chapman and Hall, New York (1994)
23. Hart, J.F. et al.: Computer Approximations, John Wiley & Sons, Inc., New York (1968)
24. Hayotov, A.R.: Discete analogues of some differential operators (Russian). Uzbek. Math. Zh. **2012**(1), 151–155 (2012)
25. Hayotov, A.R., Milovanović, G.V., Shadimetov, Kh.M.: On an optimal quadrature formula in the sense of Sard. Numer. Algorithms **57**, 487–510 (2011)
26. Hayotov, A.R., Milovanović, G.V., Shadimetov, Kh.M.: Interpolation splines minimizing a semi-norm, Calcolo (2013) DOI: 10.1007/s10092-013-0080-x
27. Holladay, J.C.: Smoothest curve approximation. Math. Tables Aids Comput. **11**, 223–243 (1957)

28. Ignatev, M.I., Pevniy, A.B.: Natural Splines of Many Variables, Nauka, Leningrad (in Russian) (1991)
29. Korneichuk, N.P., Babenko, V.F., Ligun, A.A.: Extremal Properties of Polynomials and Splines, Naukovo Dumka, Kiev, (in Russian) (1992)
30. Köhler, P.: On the weights of Sard's quadrature formulas. Calcolo **25**, 169–186 (1988)
31. Lanzara, F.: On optimal quadrature formulae. J. Ineq. Appl. **5** 201–225 (2000),
32. Laurent, P.-J.: Approximation and Optimization, Mir, Moscow (in Russian) (1975). Translation of Laurent, P.-J.: Approximation et optimisation, Hermann, Paris (1972)
33. Maljukov, A.A., Orlov, I.I.: Construction of coefficients of the best quadrature formula for the class $W_{L_2}^{(2)}(M; ON)$ with equally spaced nodes. Optimization methods and operations research, applied mathematics (Russian), pp. 174–177, 191. Akad. Nauk SSSR Sibirsk. Otdel. Sibirsk. Ènerget. Inst., Irkutsk (1976)
34. Mastroianni, G., Milovanović, G.V.: Interpolation Processes–Basic Theory and Applications, Springer Monographs in Mathematics, Springer, Berlin (2008)
35. Meyers, L.F., Sard, A.: Best approximate integration formulas. J. Math. Physics **29**, 118–123 (1950)
36. Nikol'skii, S.M.: To question about estimation of approximation by quadrature formulas (Russian). Uspekhi Matem. Nauk **5**2(36), 165–177 (1950)
37. Nikol'skii, S.M.: Quadrature Formulas (Russian). Nauka, Moscow (1988)
38. Nürnberger, G.: Approximation by Spline Functions, Springer, Berlin (1989)
39. Sard, A.: Best approximate integration formulas; best approximation formulas. Amer. J. Math. **71**, 80–91 (1949).
40. Schoenberg, I.J.: On trigonometric spline interpolation. J. Math. Mech. **13**, 795–825 (1964)
41. Schoenberg, I.J.: On monosplines of least deviation and best quadrature formulae. J. Soc. Indust. Appl. Math. Ser. B Numer. Anal. **2**, 144–170 (1965)
42. Schoenberg, I.J.: On monosplines of least square deviation and best quadrature formulae II. SIAM J. Numer. Anal. **3**, 321–328 (1966)
43. Schoenberg, I.J., Silliman, S.D.: On semicardinal quadrature formulae. Math. Comp. **28**, 483–497 (1974).
44. Schumaker, L.: Spline Functions: Basic Theory, John Wiley, New-York (1981)
45. Shadimetov, Kh.M.: Optimal quadrature formulas in $L_2^m(\Omega)$ and $L_2^m(R^1)$ (Russian). Dokl. Akad. Nauk UzSSR **1983**(3), 5–8 (1983)
46. Shadimetov, Kh.M.: Construction of weight optimal quadrature formulas in the space $L_2^{(m)}(0, N)$ (Russian). Siberian J. Comput. Math. **5**(3), 275–293 (2002)
47. Shadimetov, Kh.M., Hayotov, A.R.: Computation of coefficients of optimal quadrature formulas in the space $W_2^{(m,m-1)}(0, 1)$ (Russian). Uzbek. Math. Zh. **2004**(3), 67–82 (2004)
48. Shadimetov, Kh.M., Hayotov, A.R.: Optimal quadrature formulas with positive coefficients in $L_2^{(m)}(0, 1)$ space. J. Comput. Appl. Math. **235**, 1114–1128 (2011)
49. Shadimetov, Kh.M., Hayotov, A.R.: Construction of interpolation splines minimizing seminorm in $W_2^{(m,m-1)}(0, 1)$ space. BIT Numer. Math. (2012) DOI: 10.1007/s10543-012-0407-z
50. Shadimetov, Kh.M., Hayotov, A.R., Nuraliev, F.A.: On an optimal quadrature formula in Sobolev space $L_2^{(m)}(0, 1)$. J. Comput. Appl. Math. **243**, 91–112 (2013)
51. Shadimetov, Kh.M., Hayotov, A.R.: Optimal quadrature formulas in the sense of Sard in $W_2^{(m,m-1)}$ space. Calcolo (2013) DOI: 10.1007/s10092-013-0076-6
52. Sobolev, S.L.: Introduction to the Theory of Cubature Formulas (Russian). Nauka, Moscow (1974)
53. Sobolev, S.L.: The coefficients of optimal quadrature formulas. Selected Works of S.L. Sobolev, pp. 561–566, Springer (2006)
54. Sobolev, S.L., Vaskevich, V.L.: The Theory of Cubature Formulas. Kluwer Academic Publishers Group, Dordrecht (1997)
55. Stechkin, S.B., Subbotin, Yu.N.: Splines in Computational Mathematics, Nauka, Moscow (in Russian) (1976)

56. Vasilenko, V.A.: Spline functions: Theory, Algorithms, Programs, Nauka, Novosibirsk (in Russian) (1983)
57. Vladimirov, V.S.: Generalized Functions in Mathematical Physics (Russian). Nauka, Moscow (1979)
58. Wahba, G.: Spline Models for Observational Data. CBMS 59, SIAM, Philadelphia (1990)
59. Zagirova, F.Ya.: On construction of optimal quadrature formulas with equal spaced nodes (Russian). 28 p. Novosibirsk (1982), (Preprint No. 25, Institute of Mathematics SD of AS of USSR)
60. Zhamalov, Z.Zh., Shadimetov, Kh.M.: About optimal quadrature formulas (Russian). Dokl. Akademii Nauk UzSSR **7**, 3–5 (1980)
61. Zhensikbaev, A.A.: Monosplines of minimal norm and the best quadrature formulas (Russian). Uspekhi Matem. Nauk. **36**, 107–159 (1981)

Numerical Integration of Highly Oscillating Functions

Gradimir V. Milovanović and Marija P. Stanić

Dedicated to Professor Hari M. Srivastava

Abstract Some specific nonstandard methods for numerical integration of highly oscillating functions, mainly based on some contour integration methods and applications of some kinds of Gaussian quadratures, including complex oscillatory weights, are presented in this survey. In particular, Filon-type quadratures for weighted Fourier integrals, exponential-fitting quadrature rules, Gaussian-type quadratures with respect to some complex oscillatory weights, methods for irregular oscillators, as well as two methods for integrals involving highly oscillating Bessel functions are considered. Some numerical examples are included.

1 Introduction

By a highly oscillating function, we mean one with a large number of local maxima and minima over some interval. The computation of integrals of highly oscillating functions is one of the most important issues in numerical analysis since such integrals abound in applications in many branches of mathematics as well as in other sciences, e.g., quantum physics, fluid mechanics, and electromagnetics. The principal examples of highly oscillating integrands occur in various transforms, e.g., Fourier transform and Fourier–Bessel transform. The standard methods of

G.V. Milovanović
Mathematical Institute of the Serbian Academy of Sciences and Arts,
Belgrade, Serbia
e-mail: gvm@mi.sanu.ac.rs

M.P. Stanić (✉)
Department of Mathematics and Informatics, Faculty of Science, University of Kragujevac,
Radoja Domanovića 12, 34000 Kragujevac, Serbia
e-mail: stanicm@kg.ac.rs

G.V. Milovanović and M.Th. Rassias (eds.), *Analytic Number Theory, Approximation Theory, and Special Functions*, DOI 10.1007/978-1-4939-0258-3_23,

numerical integration often require too much computation work and cannot be successfully applied. Because of that, for integrals of highly oscillating functions, there are a large number of special approaches, which are effective. In this paper we give a survey of some special quadrature methods for different types of highly oscillating integrands.

The earliest formulas for numerical integration of highly oscillating functions were given by Filon [12] in 1928. Filon's approach for the Fourier integral on the finite interval,

$$I[f;\omega] = \int_a^b f(x)\mathrm{e}^{\mathrm{i}\omega x}\,\mathrm{d}x,$$

is based on the piecewise approximation of $f(x)$ by parabolic arcs on the integration interval. The resulting integrals over subintervals are then integrated exactly. One can divide interval $[a,b]$ into $2N$ subintervals of equal length $h = (b-a)/(2N)$. Let $x_k = a + kh$, $k = 0,1,\ldots,2N$. Filon's formula is based on a quadratic fit of function $f(x)$ on every subinterval $[x_{2k-2}, x_{2k}]$, $k = 1,\ldots,N$, by interpolation at the mesh points. The error estimate was given by Håvie [19] and Ehrenmark [7].

It can be said that Filon's idea is one of the most fruitful in topic of integration of highly oscillating functions, because of a wide range of improvements of the previous technique. Luke [36] in 1954 approximated the function $f(x)$ in a certain interval by a polynomial of at most 10th degree. Flinn [13] used 5th degree polynomials in order to approximate $f(x)$ taking values of function and values of its derivative at the points x_{2k-2}, x_{2k-1}, and x_{2k}. Stetter [60] used the idea of approximating the transformed function by polynomials in $1/t$. Tuck [61] suggested the so-called Filon–trapezoidal rule, where polygonal arches were used instead of parabolic arches. The Filon modification in such a rule is nothing more than a simple multiplicative factor applied to the results of the crude trapezoidal rule. Einarsson [8] derived the so-called Filon–spline rule by passing cubic splines through functional values. Shampine [56] proposed method based on a smooth cubic spline, implemented in a MATLAB program. An adaptive implementation of his method deals with functions f that have peaks. His basic method can be adjusted to deal effectively with functions f that have a moderate singularity at one or both ends of $[a,b]$. Miklosko [38] used an interpolatory quadrature formula with the Chebyshev nodes. Van de Vooren and van Linde [63] obtained the Fourier integral quadrature rules which for the real part are exact if f is of at most 7th degree, and for the imaginary part if f is of at most 8th degree.

Ixary and Paternoster [25,26] derived exponential-fitting approach for the Fourier integral on $[-1,1]$, designed to be exact when the integrand is some suitably chosen combination of exponential functions, e.g., with polynomial terms, or products of polynomials and exponentials.

Recently, Ledoux and Van Daele [29] have made connection between Filon-type and exponential-fitting methods. By introducing some $S-$shaped functions, they constructed Gauss-type rules for the Fourier integral $I[f;\omega]$ on $[-1,1]$ interpolating f in frequency-dependent nodes along with Chebyshev nodes. In such a way they derived rules with an optimal asymptotic rate of decay of the error with increasing frequency, which are effective for small or moderate frequencies, too.

Very simple methods can be obtained by integration between the zeros. If the zeros of the oscillatory part of the integrand are located in the points $a \leqslant x_1 < x_2 < \cdots < x_m \leqslant b$, then the integral on each subinterval $[x_k, x_{k+1}]$ can be calculated by an appropriate rule. A rule of Gauss–Lobatto type (cf. [37, pp. 330–332]) is very good for this purpose because of using the end points of the integration subintervals, where the integrand is zero, so that, more accuracy can be obtained without additional computation (see [6]).

Several authors (see, e.g., Zamfirescu [68], Gautschi [14], Piesens [49, 50], Piesens and Haegemans [53], Davis and Rabinowitz [6]) considered usage of Gaussian formulae for oscillatory weights. Considering the following nonnegative weight functions on $[-1, 1]$,

$$c_k(t) = \frac{1}{2}(1 + \cos k\pi t), \quad s_k(t) = \frac{1}{2}(1 + \sin k\pi t),$$

it is easy to see that

$$\frac{1}{\pi}\int_{-\pi}^{\pi} f(x)\cos kx \,\mathrm{d}x = 2\int_{-1}^{1} f(\pi t)c_k(t)\,\mathrm{d}t - \int_{-1}^{1} f(\pi t)\,\mathrm{d}t,$$

$$\frac{1}{\pi}\int_{-\pi}^{\pi} f(x)\sin kx \,\mathrm{d}x = 2\int_{-1}^{1} f(\pi t)s_k(t)\,\mathrm{d}t - \int_{-1}^{1} f(\pi t)\,\mathrm{d}t.$$

Gauss-type rules can now be constructed for the first integrals on the right-hand sides of the previous equalities.

Goldberg and Varga [16] (cf. [34, 35]) proposed a method for the computation of Fourier coefficients based on Möbius inversion of Poisson summation formula.

Milovanović [39] proposed complex integration method. Let us for $\delta > 0$ denote

$$G_\delta = \{z \in \mathbb{C} \mid -1 \leqslant \operatorname{Re} z \leqslant 1,\ 0 \leqslant \operatorname{Im} z \leqslant \delta\}, \quad \Gamma_\delta = \partial G_\delta.$$

Consider the Fourier integral on the finite interval

$$I[f;\omega] = \int_{-1}^{1} f(x)\mathrm{e}^{\mathrm{i}\omega x}\,\mathrm{d}x,$$

where f is an analytic real-valued function in the half-strip of the complex plane, $-1 \leqslant \operatorname{Re} z \leqslant 1$, $\operatorname{Im} z \geqslant 0$, with singularities z_ν, $\nu = 1, \ldots, m$, in the region $\operatorname{int}\Gamma_\delta$, and

$$2\pi\mathrm{i}\sum_{\nu=1}^{m} \operatorname*{Res}_{z=z_\nu}\Big(f(z)\mathrm{e}^{\mathrm{i}\omega z}\Big) = P + \mathrm{i}Q.$$

If there exist the constants $M > 0$ and $\xi < \omega$ such that

$$\int_{-1}^{1} |f(x + \mathrm{i}\delta)| \,\mathrm{d}x \leqslant M \mathrm{e}^{\xi\delta},$$

then (see [39, Theorem 2.1])

$$\int_{-1}^{1} f(x) \cos \omega x \,\mathrm{d}x = P + \frac{2}{\omega} \int_{0}^{+\infty} \mathrm{Im}\Big[\mathrm{e}^{\mathrm{i}\omega} f_e\Big(1 + \mathrm{i}\frac{t}{\omega}\Big)\Big] \mathrm{e}^{-t} \,\mathrm{d}t,$$

$$\int_{-1}^{1} f(x) \sin \omega x \,\mathrm{d}x = Q - \frac{2}{\omega} \int_{0}^{+\infty} \mathrm{Re}\Big[\mathrm{e}^{\mathrm{i}\omega} f_o\Big(1 + \mathrm{i}\frac{t}{\omega}\Big)\Big] \mathrm{e}^{-t} \,\mathrm{d}t,$$

where $f_o(z)$ and $f_e(z)$ are the odd and even part in $f(z)$, respectively. The obtained integrals can be calculated efficiently by using Gauss–Laguerre rule.

The Fourier integral on $(0, +\infty)$,

$$F[f;\omega] = \int_{0}^{+\infty} f(x) \mathrm{e}^{\mathrm{i}\omega x} \,\mathrm{d}x,$$

can be transformed to

$$F[f;\omega] = \frac{1}{\omega} \int_{0}^{+\infty} f(x/\omega) \mathrm{e}^{\mathrm{i}x} \,\mathrm{d}x = F[f(\cdot/\omega); 1].$$

Thus, it is enough to consider only the case $\omega = 1$.

In order to calculate $F[f;1]$, for a chosen positive number a, one can write

$$F[f;1] = \int_{0}^{a} f(x) \mathrm{e}^{\mathrm{i}x} \,\mathrm{d}x + \int_{a}^{+\infty} f(x) \mathrm{e}^{\mathrm{i}x} \,\mathrm{d}x = L_1[f] + L_2[f],$$

where

$$L_1[f] = a \int_{0}^{1} f(at) \mathrm{e}^{\mathrm{i}at} \,\mathrm{d}t \quad \text{and} \quad L_2[f] = \int_{a}^{+\infty} f(x) \mathrm{e}^{\mathrm{i}x} \,\mathrm{d}x.$$

If the function $f(z)$ is defined and holomorphic in the region $D = \{z \in \mathbb{C} \mid \mathrm{Re}\, z \geqslant a > 0,\ \mathrm{Im}\, z \geqslant 0\}$, such that $|f(z)| \leqslant A/|z|$ when $|z| \to +\infty$, for some positive constant A, then (see [39, Theorem 2.2])

$$L_2[f] = \mathrm{i}\mathrm{e}^{\mathrm{i}a} \int_{0}^{+\infty} f(a + \mathrm{i}y) \mathrm{e}^{-y} \,\mathrm{d}y \quad (a > 0).$$

In the numerical implementation Gauss–Legendre rule on $(0, 1)$ and Gauss–Laguerre rule can be used for calculating $L_1[f]$ and $L_2[f]$, respectively.

In this paper we consider some specific nonstandard methods for numerical integration of highly oscillating functions, mainly based on some contour integration methods and applications of some kinds of Gaussian quadratures, including complex oscillatory weights. The paper is organized as follows. Filon-type quadratures for weighted Fourier integrals and exponential-fitting quadrature rules are studied in Sects. 2 and 3, respectively. Gaussian-type quadratures with respect to some complex oscillatory weights are given in Sect. 4. Section 5 is devoted to more general highly oscillating integrands known as irregular oscillators. Asymptotic methods, as well as Filon-type and Levin-type methods, are included. Finally, two class of methods (Levin-type and Chen's method) for integrals involving highly oscillating Bessel functions are considered in Sect. 6.

2 Filon-Type Quadrature Rules for Weighted Fourier Integral

In this section we describe and analyze Filon-type method for generalized Fourier integral in the sense that a weight function is allowed (see Iserles [21]). Let $\mathcal{P}$ be the linear space of all algebraic polynomials and $\mathcal{P}_n$ be the linear space of all algebraic polynomials of degree at most n.

For a nonnegative sufficiently smooth nonzero weight function $w \in L[0,1]$ and $h > 0$, we consider the following integral

$$I_h[f] = \int_0^h f(x)\mathrm{e}^{\mathrm{i}\omega x} w(x/h)\,\mathrm{d}x = h\int_0^1 f(hx)\mathrm{e}^{\mathrm{i}h\omega x} w(x)\,\mathrm{d}x, \tag{1}$$

where $f \in L[0,h]$ is sufficiently smooth function. Let us choose n distinct points $0 \leqslant \tau_1 < \tau_2 < \cdots < \tau_n \leqslant 1$, and interpolate function f by a polynomial of degree $n-1$,

$$f(x) \approx P_n(x;f) = \sum_{k=1}^{n} \ell_k(x/h) f(h\tau_k),$$

where $\ell_k \in \mathcal{P}_{n-1}$, $k = 1,2\ldots,n$, are fundamental polynomials of Lagrange interpolation,

$$\ell_k(x) = \frac{\prod\limits_{\substack{\nu=1\\ \nu\neq k}}^{n} (x-\tau_\nu)}{\prod\limits_{\substack{\nu=1\\ \nu\neq k}}^{n} (\tau_k-\tau_\nu)}.$$

Replacing f by $P_n(x; f)$ in (1) the Filon-type quadrature rule is obtained,

$$Q_n^h[f] = I_h[P_n(x; f)] = h \sum_{k=1}^{n} \sigma_k(\mathrm{i}h\omega) f(h\tau_k),$$

where the weights are given by

$$\sigma_k(\mathrm{i}h\omega) = \int_0^1 \ell_k(x) \mathrm{e}^{\mathrm{i}h\omega x} w(x)\,\mathrm{d}x, \quad k = 1, 2, \ldots, n.$$

Obviously, for the remainder term $R_n^h[f] = Q_n^h[f] - I_h[f]$, we have $R_n^h[f] = 0$ for all $f \in \mathcal{P}_{n-1}$. Hence, for sufficiently smooth function f, we have $R_n^h[f] = \mathcal{O}(h^{n+1})$.

Remark 2.1. Let us notice that the same weights can be obtained by solving the following Vandermonde system

$$\sum_{k=1}^{n} \sigma_k(\mathrm{i}h\omega) \tau_k^m = \mu_m(h\omega), \quad m = 0, 1, \ldots, n-1,$$

where μ_m are the corresponding moments,

$$\mu_m(\theta) = \int_0^1 x^m \mathrm{e}^{\mathrm{i}\theta x} w(x)\,\mathrm{d}x, \quad m \in \mathbb{N}_0 = \mathbb{N} \cup \{0\}.$$

If we set $\theta = h\omega$ and

$$\delta_m(\theta) = \sum_{k=1}^{n} \sigma_k(\mathrm{i}\theta) \tau_k^m - \mu_m(\theta), \quad m \in \mathbb{N}_0,$$

then $\delta_m = 0$ for $m = 0, 1, \ldots, n-1$.

Let p be the order of the corresponding Gauss–Christoffel quadrature rule with nodes $0 \leqslant \tau_1 < \tau_2 < \cdots < \tau_n \leqslant 1$. Thus, $p \in \{n, n+1, \ldots, 2n\}$ (see [15]), which means that quadrature rule is exact for all polynomials of degree less than or equal to $p-1$. The maximal algebraic degree od exactness is $2n-1$, i.e., the maximal order is $2n$, if nodes are zeros of the corresponding orthogonal polynomial of n–th degree. It is important to point out that here we talk about Gauss–Christoffel quadrature rule with respect to the weight function w on [0, 1]. In Sect. 4 we consider quadrature rules with maximal algebraic degree of exactness with respect to the complex oscillatory weight function of the form $\mathrm{e}^{\mathrm{i}\xi x} w(x)$.

We present estimates of error term $R_n^h[f]$ for sufficiently smooth function f (see [21]) in three situations: $0 < h\omega \ll 1$ (non-oscillatory); $h\omega = \mathcal{O}(1)$ (mildly oscillatory); $h\omega \gg 1$ (highly oscillatory).

Let f be an analytic function in the disc $|z| < r$ for some $r > 0$, and let its Taylor series be

$$f(z) = \sum_{m=0}^{\infty} \frac{f_m}{m!} z^m.$$

Since $R_n^h[x^m] = 0$, for $m = 0, 1, \ldots, n-1$, one can assume without loss of generality that $f_m = 0$, $m = 0, 1, \ldots, n-1$. The function

$$\tilde{f}(z) = \sum_{m=n}^{\infty} \frac{f_m}{m!} z^m$$

is the essential part of the function f, and $R_n^h[f] = R_n^h[\tilde{f}]$ regardless of the size of h and ω. Due to analyticity of f one can easily obtain that

$$R_n^h[f] = \sum_{m=n}^{\infty} \frac{f_m}{m!} h^{m+1} \delta_m(\theta). \tag{2}$$

For analytic function f, for fixed $\omega > 0$ and $0 < h \ll 1$, we have (see [21]) $R_n^h[f] = \mathcal{O}(h^{p+1})$, where p is order of the corresponding Gauss–Christoffel quadrature rule, while in the case when $h\omega = \mathcal{O}(1)$ the error term behaves like $\mathcal{O}(h^{n+1})$.

Now, we pay our attention to the highly oscillatory situation, when the standard Gauss–Christoffel quadrature rules became useless. Let

$$p(t) = \prod_{k=1}^{n} (t - \tau_k) = \sum_{k=0}^{n} a_k t^k$$

be the nodal polynomial. Let $h > 0$ be small and characteristic frequency $\theta = h\omega$ large. The main idea presented in [21] is to keep $h > 0$ fixed and consider the asymptotic expansion of the error term in negative powers of θ. It is easy to get the following asymptotic expansions for the moments:

$$\mu_0(\theta) \sim \frac{w(1)e^{i\theta} - w(0)}{i\theta} + \frac{w'(1)e^{i\theta} - w'(0)}{\theta^2} + \mathcal{O}(\theta^{-3}),$$

$$\mu_1(\theta) \sim \frac{w(1)e^{i\theta}}{i\theta} + \frac{(w(1) + w'(1))e^{i\theta} - w(0)}{\theta^2} + \mathcal{O}(\theta^{-3}),$$

$$\mu_m(\theta) \sim \frac{w(1)e^{i\theta}}{i\theta} + \frac{(mw(1) + w'(1))e^{i\theta}}{\theta^2} + \mathcal{O}(m^2\theta^{-3}), \quad m \geqslant 2.$$

By using the fact that $p(\tau_k) = 0$, $k = 1, 2, \ldots, n$, and the obtained asymptotic expansions for the moments, it was shown (see [21, Proposition 3]) that there exist two sequences of numbers $\{\alpha_m\}_{m\in\mathbb{N}_0}$ and $\{\beta_m\}_{m\in\mathbb{N}_0}$ such that

$$\delta_m(\theta) \sim \frac{\alpha_m w(0)p(0) - \beta_m w(1)p(1)\mathrm{e}^{\mathrm{i}\theta}}{\mathrm{i}\theta} + \mathcal{O}(m\theta^{-2}), \quad m \in \mathbb{N}_0. \tag{3}$$

Here, $\alpha_m = \beta_m = 0$ for $m = 0, 1, \ldots, n-1$, $\alpha_n = \beta_n = 1$. If $w(0)p(0) \neq 0$ and $w(1)p(1) \neq 0$, then α_m and β_m satisfy recurrence relations

$$\sum_{k=0}^{n} a_k \alpha_{k+m} = 0, \quad \sum_{k=0}^{n} a_k \beta_{k+m} = -1, \quad m \geqslant 1.$$

The general solutions of these equations are

$$\alpha_m = \sum_{k=1}^{n} c_k \tau_k^m, \quad \beta_m = \sum_{k=1}^{n} d_k \tau_k^m - \frac{1}{p(1)}, \quad m \geqslant 0,$$

where the constants c_k and d_k, $k = 1, 2, \ldots, n$, can be determined from the initial values by solving a Vanredmonde linear algebraic system. If $w(0)p(0) = 0$, then $\alpha_m = 0$, while if $w(1)p(1) = 0$, then $\beta_m = 0$.

Finally, from (2), (3) and the fact that $R_n^h[f] = R_n^h[\tilde{f}]$, the following result can be proved (see [21, Theorem 2]).

Theorem 2.1. *Let function f be analytic and $\theta = h\omega \gg 1$. If both $\tau_1 w(0) = 0$ and $(1-\tau_n)w(1) = 0$, then $R_n^h[f] \sim \mathcal{O}(h^{n+1}\theta^{-2})$; otherwise, $R_n^h[f] \sim \mathcal{O}(h^{n+1}\theta^{-1})$.*

According to the previous theorem, we can conclude that for general weight function the best choice of nodes for the three considered situations is that of Lobatto points (see [6]).

Disadvantages of Filon-type method will be pointed out in Sect. 5 where Filon-type method for more general integrals will be presented.

3 Exponential-Fitting Quadrature Rules

The first results on exponentially fitting quadrature rules for oscillating integrands were given in [25]. Those ideas led to Gauss-type quadrature rules for oscillatory integrands considered in [26]. Namely, we considered the following quadrature formula

$$\int_{-1}^{1} f(x)\,\mathrm{d}x = \sum_{k=1}^{n} \sigma_k f(x_k) + R_n[f], \tag{4}$$

where the nodes x_k and the weights σ_k, $k = 1,\ldots,n$, are chosen such that this quadrature formula is exact for all functions from $\mathcal{F}_{2n}(\zeta)$, which is the linear span of the set $\{x^k \cos\zeta x, x^k \sin\zeta x \mid k = 0,1,\ldots,n-1, \zeta \in \mathbb{R}\}$. Let us notice that for $\zeta \neq 0$ we have $\dim \mathcal{F}_{2n}(\zeta) = 2n$. Obviously, it is enough to consider only the case $\zeta > 0$, because $\mathcal{F}_{2n}(-\zeta) = \mathcal{F}_{2n}(\zeta)$. The case $\zeta = 0$ is trivial, since $\mathcal{F}_{2n}(0)$ reduces to a pure polynomial set, i.e., $\mathcal{F}_{2n}(0) = \mathcal{P}_{n-1}$ (the set of algebraic polynomials of degree at most $n-1$).

Ixary and Paternoster [26] presented numerical method for constructing such quadrature rules with antisymmetric nodes in $(-1,1)$ and symmetric weights, but they did not prove the existence of such quadrature rules. The existence were proved partially in [43] in the case when all nodes are positive (or all negative). In the sequel we briefly explain that proof of existence.

For a given $n \in \mathbb{N}$ and the set of nodes $\{x_1,\ldots,x_n\}$, we denote $\mathbf{x} = (x_1,\ldots,x_n)$ and introduce the nodal polynomial $\omega(x) = \prod_{k=1}^{n}(x - x_k)$. For $\nu,\mu = 1,\ldots,n$, we use the following notation:

$$\omega_\nu(x) = \frac{\omega(x)}{x - x_\nu} = \prod_{\substack{k=1\\k\neq\nu}}^{n}(x - x_k), \quad \omega_{\nu,\mu}(x) = \frac{\omega(x)}{(x - x_\nu)(x - x_\mu)} = \prod_{\substack{k=1\\k\neq\nu,\mu}}^{n}(x - x_k),$$

and $\ell_\nu(x) = \omega_\nu(x)/\omega_\nu(x_\nu)$, as well as

$$\Phi_\nu(\mathbf{x}) = \int_{-1}^{1} \omega_\nu(x) \sin\zeta(x - x_\nu)\,\mathrm{d}x, \quad \nu = 1,\ldots,n.$$

Suppose we are given mutually different nodes x_ν, $\nu = 1,\ldots,n$, of the quadrature rule (4). Then the weights can be expressed as follows (see [43, Theorem 2.1]):

$$\sigma_\nu = \int_{-1}^{1} \ell_\nu(x) \cos\zeta(x - x_\nu)\,\mathrm{d}x, \quad \nu = 1,\ldots,n. \tag{5}$$

Therefore, the weights are unique for the given set of nodes, and the weights can be considered as continuous functions of nodes on any closed subset of $\mathbb{R}^n$ which does not contain points with some pair of the same coordinates. The following result is very important for the proof of existence of quadrature rules.

Theorem 3.1. *The nodes x_ν, $\nu = 1,\ldots,n$, of the quadrature rule (4) satisfy the following system of equations*

$$\int_{-1}^{1} \omega_\nu(x) \sin\zeta(x - x_\nu)\,\mathrm{d}x = 0, \quad \nu = 1,\ldots,n. \tag{6}$$

Suppose that $\mathbf{x} = (x_1, \ldots, x_n)$ *is a solution of the system of equations* (6), *under the assumption* $x_k \neq x_j$, $k \neq j$, $k, j = 1, \ldots, n$, *we have that* x_ν, $\nu = 1, \ldots, n$, *are the nodes of the quadrature rule* (4).

Of course, we are interested only on solutions of (6) which are nodes of quadrature rule (4). Let x_ν, $\nu = 1, \ldots, n$, be the nodes of the quadrature rule (4). It was proved in [43] that

$$|\partial_{x_k} \Phi_\nu(\mathbf{x})|_{\nu,k=1}^n = \left(\prod_{k=1}^n \sigma_k \omega_k(x_k)\right) \left| \frac{\sin\zeta(x_k - x_\nu)}{x_k - x_\nu} \right|_{\nu,k=1}^n,$$

as well as that the function $\sin\zeta x/x$ in x is strictly positive definite. Supposing that $\sigma_k \neq 0, k = 1, \ldots, n$, it follows that the determinant of the Jacobian at the solution is not equal to zero.

The case when $\sigma_\mu = 0$ for some $\mu = 1, \ldots, n$, is not important since it produces a quadrature rule which does not depend on x_μ at all.

For a fixed ζ, let us consider the following two equations:

$$\int_{-1}^{1} \left(\prod_{\nu=1}^{n} (x - x_\nu)\right) \cos\zeta x \,\mathrm{d}x = 0, \qquad \int_{-1}^{1} \left(\prod_{\nu=1}^{n} (x - x_\nu)\right) \sin\zeta x \,\mathrm{d}x = 0,$$

with unknowns x_ν, $\nu = 1, \ldots, n$, and let us denote the sets of their solutions by C_n and S_n, respectively. For the proof of existence theorem, we need the following properties of the sets C_n and S_n (see Theorem 2.8 and Theorem 2.9 from [43]).

Lemma 3.1. *The set* C_n, $n \geqslant 2$, *is closed, symmetric with respect to the origin and if* $\sin 2\zeta \geqslant 0$, *we have* $C_n \cap \{\mathbf{x} \in \mathbb{R}^n \mid x_\nu > 0,\ \nu = 1, \ldots, n\} = \varnothing$.

The set S_n, $n \geqslant 3$, *is closed, symmetric with respect to the origin and if* $\sin 2\zeta \leqslant 0$ *we have* $S_n \cap \{\mathbf{x} \in \mathbb{R}^n \mid x_\nu > 0,\ \nu = 1, \ldots, n\} = \varnothing$.

We are now ready to present the main result from [43] and give the sketch of the proof.

Theorem 3.2. *In the case* $\sin 2\zeta \geqslant 0$ *for* $2 \leqslant n < \zeta/\pi - 1/2$, *system of equations* (6) *has at least* $2\binom{[\zeta/\pi - 1/2]}{n}$ *solutions which nodes are all positive or all negative.*

In the case $\sin 2\zeta \leqslant 0$ *for* $3 \leqslant n < \zeta/\pi - 1$, *system of equations* (6) *has at least* $2\binom{[\zeta/\pi - 1]}{n}$ *solutions which nodes are all positive or all negative.*

Proof. First, we consider the case $\sin 2\zeta \geqslant 0$. For the solutions which satisfy the condition $\int_{-1}^{1} \omega_\nu(x) \cos\zeta x \,\mathrm{d}x \neq 0$, $\nu = 1, \ldots, n$, the system of equations (6) can be rewritten in the form

$$x_\nu = \psi_\nu^C(\mathbf{x}) = \frac{1}{\zeta}\left(\arctan\frac{\int_{-1}^{1} \omega_\nu(x)\sin\zeta x \,\mathrm{d}x}{\int_{-1}^{1} \omega_\nu(x)\cos\zeta x \,\mathrm{d}x} + k_\nu \pi\right), \quad \nu = 1, \ldots, n, \quad (7)$$

where $k_\nu \in \mathbb{Z}$. Defining the functions $p_\nu : \mathbb{R}^n \to \mathbb{R}^n$, $\nu = 1, \dots, n$, by

$$p_\nu(x_1, \dots, x_\nu, \dots, x_n) = (x_1, \dots, x_n, \dots, x_\nu),$$

the set of solutions of $\int_{-1}^{1} \omega_\nu(x) \cos \zeta x \, dx = 0$, $\nu = 1, \dots, n$, can be described as $p_\nu(C_{n-1} \times \mathbb{R})$, $\nu = 1, \dots, n$. Thus, the transformation holds true for all the solutions which belong to the set $\mathbb{R}^n \backslash (\cup_{\nu=1}^{n} p_\nu(C_{n-1} \times \mathbb{R}))$. Since the set C_n has empty intersection with the set $\{\mathbf{x} \mid x_\nu > 0, \ \nu = 1, \dots, n\}$, it follows that $\{\mathbf{x} \mid x_\nu > 0, \ \nu = 1, \dots, n\} \subset \mathbb{R}^n \backslash \left(\cup_{\nu=1}^{n} p_\nu(C_{n-1} \times \mathbb{R})\right)$. This means that any solution of the system (6) with all positive nodes will be also the solution of the system (7). Since the set C_{n-1} is symmetric with respect to the origin, the same holds for $\cup_{\nu=1}^{n} p_\nu(C_{n-1} \times \mathbb{R})$, and in the same way one can consider quadrature rule with all negative nodes.

Let us choose some fixed vector, with strictly increasing coordinates, of positive integers $\mathbf{k} = (k_1, \dots, k_n)$, with the property $k_n < \zeta/\pi - 1/2$. The functions $\psi_\nu^C(\mathbf{x})$, $\nu = 1, \dots, n$, are continuous in $\mathbf{x}$ for $x_\nu > 0$, $\nu = 1, \dots, n$. The mapping $\Psi_{\mathbf{k}}^C : \mathbb{R}^n \to \mathbb{R}^n$ defined by $\Psi_{\mathbf{k}}^C(\mathbf{x}) = (\psi_1^C(\mathbf{x}), \dots, \psi_n^C(\mathbf{x}))$ is continuous in $\mathbf{x}$, for $x_\nu > 0$, $\nu = 1, \dots, n$. The mapping $\Psi_{\mathbf{k}}^C$ maps continuously the closed convex set $A_{\mathbf{k}} = \mathop{\times}_{\nu=1}^{n} \left[\left(k_\nu - \frac{1}{2}\right)\frac{\pi}{\zeta}, \left(k_\nu + \frac{1}{2}\right)\frac{\pi}{\zeta}\right]$ into itself. According to the Brouwer fixed point theorem (see, e.g., [47]), the map $\Psi_{\mathbf{k}}^C$ has a fixed point $\mathbf{x}_{\mathbf{k}} \in A_{\mathbf{k}}$. According to the fact that $\int_{-1}^{1} \omega_\nu(x) \cos \zeta x \, dx \neq 0$, it follows that we cannot have the solution with ν-th coordinate equal to $(k_\nu \pm 1/2)\pi/\zeta$, which means that all coordinates of the solution $\mathbf{x}_{\mathbf{k}}$ are different, according to the fact that the coordinates of the vector $\mathbf{k}$ are different. Thus, $\mathbf{x}_{\mathbf{k}}$ are the nodes of the quadrature rule (4). At this solution, all the weights are different from zero.

For the case $\sin 2\zeta \leqslant 0$, one can rewrite the system of equations (6), in the form

$$x_\nu = \psi_\nu^S(\mathbf{x}) = \frac{1}{\zeta}\left(\operatorname{arccot} \frac{\int_{-1}^{1} \omega_\nu(x) \cos \zeta x \, dx}{\int_{-1}^{1} \omega_\nu(x) \sin \zeta x \, dx} + k_\nu \pi\right), \quad \nu = 1, \dots, n, \tag{8}$$

where $k_\nu \in \mathbb{Z}$, and using the similar arguments as in the previous case prove that the mapping $\Psi_{\mathbf{k}}^S : \mathbb{R}^n \to \mathbb{R}^n$, defined by $\Psi_{\mathbf{k}}^S(\mathbf{x}) = (\psi_1^S(\mathbf{x}), \dots, \psi_n^S(\mathbf{x}))$, has a fixed point in the set $B_{\mathbf{k}} = \mathop{\times}_{\nu=1}^{n} [k_\nu \pi/\zeta, (k_\nu + 1)\pi/\zeta]$.

The number of the solutions can be easily obtained. □

The nodes x_k, $k = 1, \dots, n$, of the quadrature formula (4) can be obtained by using Newton–Kantorovich method for the system (6) with appropriately chosen starting values. Once nodes are constructed, weights σ_k, $k = 1, \dots, n$, can be computed by using formula (5). In Table 1 we give two different quadrature rules with all positive nodes for the case $n = 10$, $\zeta = 10000$ (numbers in parenthesis indicate decimal exponents). All computations are performed by using the MATHEMATICA package `OrthogonalPolynomials` [5].

Table 1 Nodes x_k and weights σ_k, $k = 1, \ldots, 10$, $\zeta = 10000$

k	x_k	σ_k
1	0.7363923439571446(−1)	−2.804303173754735
2	0.2225507250414224	2.970983118291112(1)
3	0.3673781455056259	1.583146376664096(2)
4	0.4999533548408196	−5.186895526861798(2)
5	0.6234179456200185	−1.207492403800276(3)
6	0.7330595288408521	−2.103809780509887(3)
7	0.8269931488869863	−2.821892746331952(3)
8	0.8998780982412568	2.717303933673183(3)
9	0.9554842880629808	1.718285983811184(3)
10	0.9887851701205169	−5.392769580168855(2)
1	0.1000286064833346	−1.463766698478331(2)
2	0.1499799293036977	−5.661615769092509(2)
3	0.2492542565212690	3.197802441947143(3)
4	0.3010905350216144	6.614998760647455(3)
5	0.3500993801768735	−4.923021376270480(3)
6	0.4500020261498017	1.443577768367478(3)
7	0.5499046722324926	−3.931520115104385(3)
8	0.7500241241587095	−5.332095232080515
9	0.8499267711774567	2.037155323640304(1)
10	0.9507718999014823	2.664975853402439

By using transformed systems (7) and (8) of nonlinear equations, the existence of quadrature rule (4) which has both the positive and negative nodes was proved in [44] under two conjectures, one for the case $\sin 2\zeta < 0$ and the second one for $\sin 2\zeta > 0$. We present those conjectures here, while for the proof of existence of mentioned quadrature rule we refer readers to [44].

First, we consider the case $\sin 2\zeta < 0$. Let us denote

$$b_\nu = (N - \nu + 1)\frac{\pi}{\zeta}, \quad \nu = 1, \ldots, N, \quad N = [\zeta/\pi],$$

and

$$I_n^S = \operatorname{sgn}(\sin\zeta) \int_{-1}^{1} t \prod_{\nu=1}^{n} (t^2 - b_\nu^2) \sin\zeta t \, \mathrm{d}t, \quad n = 0, 1, \ldots, N.$$

For $\sin 2\zeta > 0$, we denote

$$a_\nu = \left(N - \nu + \frac{1}{2}\right)\frac{\pi}{\zeta}, \quad \nu = 1, \ldots, N, \quad N = [\zeta/\pi],$$

and

$$I_n^C = \operatorname{sgn}(\sin\zeta)\int_{-1}^{1}\prod_{\nu=1}^{n}(t^2-a_\nu^2)\cos\zeta t\,dt, \quad n=0,1,\ldots,N.$$

It is easy to see that for $\zeta > 0$ and $\sin 2\zeta < 0$ inequality

$$I_0^S = \operatorname{sgn}(\sin\zeta)\int_{-1}^{1} t\sin\zeta t\,dt = \frac{1}{|\sin\zeta|}\frac{-\zeta\sin 2\zeta + 2\sin^2\zeta}{\zeta^2} > 0$$

holds, while for $\zeta > 0$ and $\sin 2\zeta < 0$ holds

$$I_0^C = \operatorname{sgn}(\sin\zeta)\int_{-1}^{1}\cos\zeta t\,dt = \frac{2|\sin\zeta|}{\zeta} > 0.$$

The mentioned conjectures are the following.

Conjecture 1. If $\zeta > 0$ and $\sin 2\zeta < 0$, then $I_n^S > 0$ for each $n = 1,\ldots,N$.

Conjecture 2. If $\zeta > 0$ and $\sin 2\zeta > 0$, then $I_n^C > 0$ for each $n = 1,\ldots,N$.

Under condition that Conjecture 1 is true in [44], it was proved that in the case when $\zeta > 0$ and $\sin 2\zeta < 0$ for all

$$\mathbf{x} = (x_1,\ldots,x_{2n+1}) \in \mathop{\times}_{\nu=1}^{n}([-b_\nu,0]\times[0,b_\nu])\times[-b_{n+1},b_{n+1}], \quad n < N,$$

the following inequality

$$\operatorname{sgn}(\sin\zeta)\int_{-1}^{1}\prod_{\nu=1}^{2n+1}(t-x_\nu)\sin\zeta t\,dt > 0$$

holds. This inequality implies the existence of the quadrature rule (4) in general, for the case $\sin 2\zeta < 0$.

Analogously as in the case $\sin 2\zeta < 0$, under condition that Conjecture 2 is true, it can be proved that in the case when $\zeta > 0$ and $\sin 2\zeta > 0$ for all

$$\mathbf{x} = (x_1,\ldots,x_{2n}) \in \mathop{\times}_{\nu=1}^{n}([-a_\nu,0]\times[0,a_\nu]), \quad n = 1,\ldots,N,$$

the following inequality

$$\operatorname{sgn}(\sin\zeta)\int_{-1}^{1}\prod_{\nu=1}^{2n}(t-x_\nu)\cos\zeta t\,dt > 0$$

holds. As the consequence of this inequality we have the existence of the quadrature rule (4) in general, for the case $\sin 2\zeta > 0$.

Kim, Cools, and Ixaru in [27] and [28] considered the quadrature rules which include derivatives. Van Daele, Vanden Berghe, and Vande Vyver [62] took into account the both polynomial and exponential aspects. Assuming symmetric weights and antisymmetric nodes they considered quadrature rules suited to integrate functions that can be expressed in the form $f(x) = f_1(x) + f_2(x)\cos\zeta x + f_3(x)\sin\zeta x$, where f_1, f_2, and f_3 are assumed smooth enough to be well approximated by polynomials on the wanted interval. The readers can find more details on this topic in the recent survey paper [48].

4 Gaussian Rules with Respect to Some Complex Oscillatory Weights

In this section we consider quadrature rules of Gaussian type

$$\int_{-1}^{1} f(x)w(x)\mathrm{e}^{\mathrm{i}\zeta x}\,\mathrm{d}x = \sum_{k=1}^{n} w_k^{(n)} f(x_k^{(n)}) + R_n(f), \quad \zeta \in \mathbb{R},$$

where $R_n(f) = 0$ for each $f \in \mathcal{P}_{2n-1}$. Thus, we have to consider the following complex measure

$$\mathrm{d}\mu(x) = w(x)\mathrm{e}^{\mathrm{i}\zeta x}\chi_{[-1,1]}(x)\,\mathrm{d}x, \quad \zeta \in \mathbb{R}, \tag{9}$$

supported on the interval $[-1, 1]$ (χ_A is the characteristic function of the set A). The existence of the corresponding orthogonal polynomials is not guaranteed. In order to check existence of orthogonal polynomials with respect to complex oscillatory measure $\mathrm{d}\mu(x)$, we need the general concept of orthogonal polynomials with respect to a moment functional (see [3, 37]).

Let a linear functional $\mathcal{L}$ be given on the linear space $\mathcal{P}$ of all algebraic polynomials, i.e., let the functional $\mathcal{L}$ satisfy following equality

$$\mathcal{L}[\alpha P + \beta Q] = \alpha\mathcal{L}[P] + \beta\mathcal{L}[Q], \quad \alpha, \beta \in \mathbb{C}, \quad P, Q \in \mathcal{P}.$$

Because of linearity, the value of the linear functional $\mathcal{L}$ at every polynomial is known if the values of $\mathcal{L}$ at the set of all monomials are known. The corresponding values of the linear functional $\mathcal{L}$ at the set of monomials are called the moments, and we denote them by μ_k, $k \in \mathbb{N}_0$. Thus, $\mathcal{L}[x^k] = \mu_k$, $k \in \mathbb{N}_0$.

A sequence of polynomials $\{P_n(x)\}_{n=0}^{+\infty}$ is called the polynomial sequence orthogonal with respect to a moment functional $\mathcal{L}$, provided for all nonnegative integers m and n,

- $P_n(x)$ is polynomial of degree n,
- $\mathcal{L}[P_n(x)P_m(x)] = 0$ for $m \neq n$,
- $\mathcal{L}[P_n^2(x)] \neq 0$.

If the sequence of orthogonal polynomials exists for a given linear functional $\mathcal{L}$, then $\mathcal{L}$ is called quasi-definite or regular linear functional. Under the condition $\mathcal{L}[P_n^2(x)] > 0$, the functional $\mathcal{L}$ is called positive definite.

By using only linear algebraic tools the following theorem can be proved (see [3, p. 11]).

Theorem 4.1. *The necessary and sufficient conditions for the existence of a sequence of orthogonal polynomials with respect to the linear functional $\mathcal{L}$ are that for each $n \in \mathbb{N}$ the Hankel determinants*

$$\Delta_n = \begin{vmatrix} \mu_0 & \mu_1 & \mu_2 & \cdots & \mu_{n-1} \\ \mu_1 & \mu_2 & \mu_3 & \cdots & \mu_n \\ \mu_2 & \mu_3 & \mu_4 & \cdots & \mu_{n+1} \\ \vdots & \vdots & \vdots & & \vdots \\ \mu_{n-1} & \mu_n & \mu_{n+1} & \cdots & \mu_{2n-2} \end{vmatrix} \neq 0.$$

We use previous theorem to prove existence of orthogonal polynomials with respect to some linear functionals defined by complex oscillatory measures (9) as follows:

$$\mathcal{L}[f] = \int_{\mathbb{R}} f(x)\,\mathrm{d}\mu(x) = \int_{-1}^{1} f(x)w(x)\mathrm{e}^{\mathrm{i}\zeta x}\,\mathrm{d}x, \quad f \in \mathcal{P}, \quad \zeta \in \mathbb{R}. \tag{10}$$

1° The case $w(x) = x$, $\zeta = m\pi \neq 0$, for an integer m, was considered by Milovanović and Cvetković in [40]. Here, the measure is

$$\mathrm{d}\mu_m(x) = x\mathrm{e}^{\mathrm{i}m\pi x}\chi_{[-1,1]}\,\mathrm{d}x, \quad m \in \mathbb{Z} \setminus \{0\},$$

thus, orthogonal polynomials with respect to the moment functional

$$\mathcal{L}[f] = \int_{-1}^{1} f(x)x\mathrm{e}^{\mathrm{i}m\pi x}\,\mathrm{d}x, \quad f \in \mathcal{P}, \tag{11}$$

i.e., with respect to the following (quasi) inner product

$$(f,g) = \int f(x)g(x)x\mathrm{e}^{\mathrm{i}m\pi x}\,\mathrm{d}x, \quad f,g \in \mathcal{P}, \tag{12}$$

must be considered.

By using an integration by parts it is easy to obtain the following recurrence relation for the moments

$$\mu_{k+1} = \frac{(-1)^m}{\mathrm{i}\zeta}(1-(-1)^{k+2}) - \frac{k+2}{\mathrm{i}\zeta}\mu_k, \quad \mu_0 = 2\frac{(-1)^m}{\mathrm{i}\zeta}.$$

The moments can be expressed explicitly as follows:

$$\mu_k = \frac{(-1)^{m+k}(k+1)!}{(\mathrm{i}\zeta)^{k+1}} \sum_{\nu=0}^{k} \frac{(1+(-1)^\nu)(-\mathrm{i}\zeta)^\nu}{(\nu+1)!}.$$

In [40] the following theorem was proved.

Theorem 4.2. *For every nonzero integer m, the sequence of orthogonal polynomials with respect to the linear functional* (11), *i.e., the sequence of orthogonal polynomials with respect to the weight function $x\mathrm{e}^{\mathrm{i}m\pi x}$, supported on the interval* $[-1, 1]$, *exists uniquely.*

Let us notice that in general, if $m \notin \mathbb{Z}$, the existence of orthogonal polynomials is not assured (e.g., the smallest positive solution of equation $\Delta_3 = 0$ is $\zeta \approx 7.134143996368961\ldots$).

As a consequence of the following property $(xf, g) = (f, xg)$ of the inner product (12), we have that the monic orthogonal polynomials with respect to the weight function $x\mathrm{e}^{\mathrm{i}m\pi x}$ on $[-1, 1]$ satisfy the following three-term recurrence relation

$$p_{n+1}(x) = (x - \mathrm{i}\alpha_n)p_n(x) - \beta_n p_{n-1}(x), \quad n = 0, 1, \ldots,$$

with $p_0(x) = 1$ and $p_{-1}(x) = 0$. The recursion coefficients α_n and β_n can be expressed in terms of Hankel determinants,

$$\mathrm{i}\alpha_n = \frac{\Delta'_{n+1}}{\Delta_{n+1}} - \frac{\Delta'_n}{\Delta_n}, \quad n \in \mathbb{N}_0; \quad \beta_n = \frac{\Delta_{n+1}\Delta_{n-1}}{\Delta_n^2}, \quad n \in \mathbb{N},$$

where Δ'_n is the Hankel determinant Δ_{n+1} with the penultimate column and the last row removed.

In [40] the first four recursion coefficients were given explicitly. Also, the numerical calculation of recursion coefficients was analyzed. Based on extensive numerical computations, which were done by using a combination of the Chebyshev algorithm and the Stieltjes–Gautschi procedure, applying package of routines written in MATHEMATICA (see [5]), the following conjecture was stated.

Conjecture 3. For recursion coefficients the following asymptotic relations are true

$$\alpha_k \to 0, \quad \beta_k \to \frac{1}{4}, \quad k \to +\infty.$$

Numerical calculations indicate that all of the nodes of orthogonal polynomials with respect to the weight function $x\mathrm{e}^{\mathrm{i}m\pi x}$ on $[-1, 1]$ are simple, but it was not proved. In the case of multiple zeros of orthogonal polynomials, the Gaussian quadrature rule has the following form:

$$G_n[f] = \sum_{\nu=1}^{n} \sum_{k=0}^{m_\nu - 1} w_{\nu,k}^{(n)} f^{(k)}(x_\nu^{(n)}).$$

As a matter of fact, it was proved that at most two nodes in the previous rule may have multiplicity greater than one. As it was said, in all numerical experiments, the nodes were simple, so, the Gaussian quadrature rule has standard form

$$G_n[f] = \sum_{\nu=1}^{n} w_\nu^{(n)} f(x_\nu^{(n)}). \tag{13}$$

Methods for numerical calculation of nodes and weights of Gaussian rule were also described in [40].

We present here an application of these quadrature rules for the calculation of Fourier coefficients. Namely,

$$F_m[f] = C_m[f] + \mathrm{i}S_m[f] = \int_{-1}^{1} f(x)\mathrm{e}^{\mathrm{i}m\pi x}\,\mathrm{d}x = \int_{-1}^{1} \frac{f(x) - f(0)}{x} x\mathrm{e}^{\mathrm{i}m\pi x}\,\mathrm{d}x,$$

so, we can compute it by using Gaussian quadrature rules (13) for the function g given by

$$g(x) = \frac{f(x) - f(0)}{x}, \quad g(0) = f'(0).$$

If function f is analytic in some domain $D \supset [-1, 1]$, then g is also analytic in D. Therefore, for some analytic function f, the Fourier coefficients can be calculated as follows:

$$F_m[f] = \int_{-1}^{1} f(x)\mathrm{e}^{\mathrm{i}m\pi x}\,\mathrm{d}x \approx \sum_{\nu=1}^{n} \frac{w_\nu^{(n)}}{x_\nu^{(n)}} (f(x_\nu^{(n)}) - f(0)).$$

2° The case $w(x) = x(1 - x^2)^{-1/2}$, $\zeta \in \mathbb{R} \setminus \{0\}$, was considered in [41]. In this case the linear functional $\mathcal{L}$ is given by

$$\mathcal{L}[f] = \int_{-1}^{1} f(x)x(1 - x^2)^{-1/2}\mathrm{e}^{\mathrm{i}\zeta x}\,\mathrm{d}x, \quad \zeta \in \mathbb{R} \setminus \{0\}, \quad f \in \mathcal{P}. \tag{14}$$

Let $\mu_k(\zeta)$, $k \in \mathbb{N}_0$, be the corresponding sequence of moments. It is easy to see that for each $k \in \mathbb{N}_0$ the following equality

$$\mu_k(\zeta) = \int_{-1}^{1} x^{k+1}(1 - x^2)^{-1/2}\mathrm{e}^{\mathrm{i}\zeta x}\,\mathrm{d}x = \overline{\int_{-1}^{1} x^{k+1}(1 - x^2)^{-1/2}\mathrm{e}^{-\mathrm{i}\zeta x}\,\mathrm{d}x} = \overline{\mu_k(-\zeta)}$$

holds, which means that it is enough to consider only the case $\zeta > 0$, since the corresponding results for $\zeta < 0$ can be obtained by a simple conjugation. The case $\zeta = 0$ is excluded, because for that value the linear functional $\mathcal{L}$, given by (14), is not regular ($\mu_0 = \Delta_0 = 0$).

Let J_ν be the Bessel function of the order ν, defined by (cf. [64, p. 40])

$$J_\nu(z) = \sum_{m=0}^{+\infty} \frac{(-1)^m (z/2)^{\nu+2m}}{m!\Gamma(\nu+m+1)}. \tag{15}$$

The sequence of moments $\mu_k(\zeta)$, $k \in \mathbb{N}_0$, satisfy the following recurrence relation (see [41, Theorem 2]):

$$\mu_{k+2}(\zeta) = -\frac{k+2}{\mathrm{i}\zeta}\mu_{k+1}(\zeta) + \mu_k(\zeta) + \frac{k+1}{\mathrm{i}\zeta}\mu_{k-1}(\zeta), \quad k \in \mathbb{N},$$

with the initial conditions

$$\begin{aligned}
\mu_0(\zeta) &= \mathrm{i}\pi J_1(\zeta),\\
\mu_1(\zeta) &= \frac{\pi}{\zeta}\big(\zeta J_0(\zeta) - J_1(\zeta)\big),\\
\mu_2(\zeta) &= \frac{\mathrm{i}\pi}{\zeta^2}\big(\zeta J_0(\zeta) + (\zeta^2-2)J_1(\zeta)\big).
\end{aligned}$$

Unfortunately, the sequence of orthogonal polynomials does not exist for all positive ζ. It is not hard to check that Hankel determinant Δ_3 in this case is given by

$$\Delta_3 = \frac{\mathrm{i}\pi^3 J_1^3}{\zeta^6}\left(7\zeta^3\frac{J_0^3}{J_1^3} + (2\zeta^2-21)\zeta^2\frac{J_0^2}{J_1^2} + \zeta(5\zeta^2+12)\frac{J_0}{J_1} + 2\zeta^4 - 15\zeta^2 + 4\right).$$

The smallest positive solution of the equation $\Delta_3 = 0$ is given by

$$\zeta = 6.4590081519947834555317213970325025438057106691208 82\ldots,$$

and for this ζ the sequence of orthogonal polynomials does not exist. So, the task is to find ζ for which the existence of orthogonal polynomials with respect to the linear functional (14) is ensured. For that purpose we notice that for the moment sequence we have the following representation:

$$\mu_k(\zeta) = \frac{\mathrm{i}\pi}{(\mathrm{i}\zeta)^k}\big(P_k J_1(\zeta) + \zeta Q_k J_0(\zeta)\big), \quad k \in \mathbb{N}_0,$$

where P_k and Q_k are polynomials in ζ^2 with integer coefficients of degrees $2[k/2]$ and $2[(k-1)/2]$, respectively (see [41, Theorem 3]). This expression can be easily obtained from the recurrence relation for the moments. Let ζ be any positive zero of the Bessel function $J_0(\zeta)$. Then $J_1(\zeta) \neq 0$, due to the interlacing property of the positive zeros of the Bessel functions (see [64, p. 479]), and sequence of moments becomes

$$\mu_k(\zeta) = \frac{\mathrm{i}\pi}{(\mathrm{i}\zeta)^k} P_k J_1(\zeta).$$

The following theorem was proved in [41].

Theorem 4.3. *If ζ is a positive zero of the Bessel function J_0, then the sequence of polynomials orthogonal with respect to the functional $\mathcal{L}$, given by* (14)*, exists.*

Remark 4.1. With a matrix Riemann–Hilbert problem formulation of the orthogonality relations, Aptekarev and Van Assche [1] considered the linear functional of the form $\mathcal{L}[f] = \int_{-1}^{1} f(x)\rho(x)(1-x^2)^{-1/2}\,\mathrm{d}x$, where ρ is a complex valued, nonvanishing on $[-1,1]$, which is holomorphic in some domain containing the interval $[-1,1]$. In the special case $\rho(x) = \mathrm{e}^{\mathrm{i}\zeta x}$, the linear functional (10) with $w(x) = (1-x^2)^{-1/2}$ is obtained.

3° The case $w(x) = (1-x^2)^{\lambda-1/2}$, for $\lambda > -1/2$, and $\zeta \in \mathbb{R}\backslash\{0\}$ was considered in [45]. In this case the linear functional (10) becomes

$$\mathcal{L}[f] = \int_{-1}^{1} f(x)(1-x^2)^{\lambda-1/2}\mathrm{e}^{\mathrm{i}\zeta x}\,\mathrm{d}x, \quad f \in \mathcal{P}. \tag{16}$$

As in the case of the previous weight, it is enough to consider only the case $\zeta > 0$, since the case $\zeta < 0$ can be obtain under substitution $x := -x$.

The corresponding moments $\mu_k^\lambda(\zeta)$ can be expressed in the form

$$\mu_k^\lambda(\zeta) = \frac{A}{(\mathrm{i}\zeta)^k}\left(P_k^\lambda(\zeta)J_\lambda(\zeta) + Q_k^\lambda(\zeta)J_{\lambda-1}(\zeta)\right), \quad k \in \mathbb{N}_0,$$

where $A = (2/\zeta)^\lambda \sqrt{\pi}\,\Gamma(\lambda+1/2)$, J_ν is the Bessel function of the order ν, given by (15), and P_k^λ and Q_k^λ are polynomials in ζ, which satisfy the following four-term recurrence relation:

$$y_{k+2} = -(k+2\lambda+1)y_{k+1} - \zeta^2 y_k - k\zeta^2 y_{k-1},$$

with the initial conditions $P_0^\lambda(\zeta) = 1$, $P_1^\lambda(\zeta) = -2\lambda$, $P_2^\lambda(\zeta) = 2\lambda(2\lambda+1)-\zeta^2$ and $Q_0^\lambda(\zeta) = 0$, $Q_1^\lambda(\zeta) = \zeta$, $Q_2^\lambda(\zeta) = -(2\lambda+1)\zeta$, respectively (see [45, Theorem 2.1]).

It is obvious that for each $\lambda > -1/2$, if $\zeta > 0$ is an arbitrary zero of the Bessel function J_λ, the polynomials π_n orthogonal with respect to (16) do not exist, because $\Delta_0 = \mu_0 = AJ_\lambda(\zeta) = 0$. In [45] the following result was proved.

Theorem 4.4. *If λ is a positive rational number and ζ is a positive zero of the Bessel function $J_{\lambda-1}$, then the polynomials π_n orthogonal with respect to* (16) *exist.*

Suppose that parameters λ and ζ are such that provide the existence of orthogonal polynomials with respect to linear functional (16). Due to the property $(zp,q) = (p,zq)$ of the (quasi) inner product $(p,q) := \mathcal{L}(pq)$, for $\mathcal{L}$ given by (16), the corresponding (monic) orthogonal polynomials $\{\pi_n\}_{n\in\mathbb{N}_0}$ satisfy the fundamental three-term recurrence relation

$$\pi_{n+1}(x) = (x - \mathrm{i}\alpha_n)\pi_n(x) - \beta_n\pi_{n-1}(x), \quad n \in \mathbb{N},$$

with $\pi_0(x) = 1$, $\pi_{-1}(x) = 0$. The recursion coefficients α_n and β_n can be expressed in terms of Hankel determinants as

$$\mathrm{i}\alpha_n = \frac{\Delta'_{n+1}}{\Delta_{n+1}} - \frac{\Delta'_n}{\Delta_n} = \frac{1}{\mathrm{i}\zeta}\left(\frac{H'_{n+1}}{H_{n+1}} - \frac{H'_n}{H_n}\right), \quad \beta_n = \frac{\Delta_{n+1}\Delta_{n-1}}{\Delta_n^2} = \frac{1}{(\mathrm{i}\zeta)^2}\frac{H_{n+1}H_{n-1}}{H_n^2},$$

where

$$H_n = \begin{vmatrix} P_0^\lambda(\zeta) & \dots & P_{n-1}^\lambda(\zeta) \\ P_1^\lambda(\zeta) & \dots & P_n^\lambda(\zeta) \\ \vdots & \vdots & \vdots \\ P_{n-1}^\lambda(\zeta) & \dots & P_{2n-2}^\lambda(\zeta) \end{vmatrix}, \quad H'_n = \begin{vmatrix} P_0^\lambda(\zeta) & P_1^\lambda(\zeta) & \dots & P_{n-2}^\lambda(\zeta) & P_n^\lambda(\zeta) \\ P_1^\lambda(\zeta) & P_2^\lambda(\zeta) & \dots & P_{n-1}^\lambda(\zeta) & P_{n+1}^\lambda(\zeta) \\ \vdots & \vdots & & \vdots & \vdots \\ P_{n-1}^\lambda(\zeta) & P_n^\lambda(\zeta) & \dots & P_{2n-3}^\lambda(\zeta) & P_{2n-1}^\lambda(\zeta) \end{vmatrix}.$$

The coefficient β_0 can be chosen arbitrary, but it is convenient to take $\beta_0 = \mu_0^\lambda(\zeta) = AJ_\lambda(\zeta)$.

Recursion coefficients can be calculated by using the Chebyshev algorithm, implemented in the software package `OrthogonalPolynomials` [5], similarly as in the case $w(x) = x$. According to very extensive numerical calculations, the conjecture that the recursion coefficients satisfy the following asymptotic relations

$$\alpha_n \to 0, \quad \beta_n \to \frac{1}{4}, \quad n \to +\infty,$$

was stated in [45]. Let us notice that for $\lambda = 0$, from the result given in [1], it follows that $\alpha_n \to 0$ and $\beta_n \to 1/4$, $n \to +\infty$.

It is easy to see that $\overline{\mu_k^\lambda}(\zeta) = (-1)^k\mu_k^\lambda(\zeta)$, $k \in \mathbb{N}_0$. Using that fact, it can be proved that if the sequence of monic orthogonal polynomials $\{\pi_n\}_{n\in\mathbb{N}_0}$ exists, then $\pi_n(z) = (-1)^n\overline{\pi_n(-\overline{z})}$ and the coefficients α_n and β_n are real. This implies that the zeros $x_k^{(n)}$, $k = 1,\dots,n$, of π_n are distributed symmetrically with respect to the imaginary axis. Some properties of the corresponding orthogonal polynomials were given in [42].

By using functions implemented in package `OrthogonalPolynomials` [5] in extended arithmetics we are able to construct Gaussian rules

$$\int_{-1}^{1} f(x)(1-x^2)^{\lambda-1/2}\mathrm{e}^{\mathrm{i}\zeta x}\,\mathrm{d}x = \sum_{k=1}^{n} w_k^{(n)} f(x_k^{(n)}) + R_n[f], \qquad (17)$$

where $R_n[f] = 0$ for each $f \in \mathcal{P}_{2n-1}$, which can be successfully applied for numerical calculation of certain type of highly oscillating integrals. We illustrate this applying Gaussian rule to the integral

$$I(\zeta) = \mathrm{Im}\left\{\int_{-1}^{1} \frac{1}{x-\mathrm{i}}(1-x^2)^{1/4}\mathrm{e}^{\mathrm{i}\zeta x}\,\mathrm{d}x\right\} \approx G_n(\zeta) = \mathrm{Im}\left\{\sum_{k=1}^{n} \frac{w_k^{(n)}}{x_k^{(n)}-\mathrm{i}}\right\},$$

for $\zeta \in \{\zeta_1, \zeta_2\}$, where $\zeta_1 = 99.35381121792450$ and $\zeta = 1000.990052907274$ (here, $\lambda = 3/4$ and ζ_1, ζ_2 are zeros of $J_{-1/4}(z)$). The imaginary parts of the corresponding integrands are displayed in Fig. 1.

The exact values of $I(\zeta)$ are

$$I(\zeta_1) = 0.0034446765944009118078224282065986452635896 79\ldots,$$
$$I(\zeta_2) = 0.00019149147544459801260257921097705042525703 7\ldots.$$

In Table 2, for some selected number of nodes n, the relative errors in Gaussian approximations, $r_n = |(G_n(\zeta_\nu) - I(\zeta_\nu))/I(\zeta_\nu)|$, $\nu = 1, 2$, are given, as well as the relative errors r_n^G in Gauss–Gegenbauer approximations with respect to the weight function $x \mapsto (1-x^2)^{1/4}$ (numbers in parenthesis indicate decimal exponents).

Numerical experiments indicate that our Gaussian quadrature rule (17) becomes more efficient when ζ increases, while Gauss–Gegenbauer rule becomes unusable.

4° The case $w(x) = (1-x)^{\alpha-1/2}(1+x)^{\beta-1/2}$, where $\alpha, \beta > -1/2$ are real numbers such that $\ell = |\beta - \alpha|$ is a positive integer, and $\zeta \in \mathbb{R} \setminus \{0\}$ was considered in [58]. Thus, we are concerned with the following measure:

$$\mathrm{d}\mu(x) = (1-x)^{\alpha-1/2}(1+x)^{\beta-1/2}\mathrm{e}^{\mathrm{i}\zeta x}\chi_{[-1,1]}(x)\,\mathrm{d}x$$

supported on the interval $[-1, 1]$. This measure can be written in the following form:

$$\mathrm{d}\mu(x) = \begin{cases} (1+x)^\ell(1-x^2)^{\alpha-1/2}\mathrm{e}^{\mathrm{i}\zeta x}\chi_{[-1,1]}(x)\,\mathrm{d}x, & \beta > \alpha, \\ (1-x)^\ell(1-x^2)^{\beta-1/2}\mathrm{e}^{\mathrm{i}\zeta x}\chi_{[-1,1]}(x)\,\mathrm{d}x, & \alpha > \beta. \end{cases}$$

Therefore, we consider the measures

$$\mathrm{d}\mu^{\pm}(x) = (1 \pm x)^\ell(1-x^2)^{\alpha-1/2}\mathrm{e}^{\mathrm{i}\zeta x}\chi_{[-1,1]}(x)\,\mathrm{d}x,$$

where $\alpha > -1/2$ and ℓ is a positive integer, i.e., we consider orthogonality with respect to the linear functional

$$\mathcal{L}^{\pm}[f] = \int_{-1}^{1} f(x)(1 \pm x)^\ell(1-x^2)^{\alpha-1/2}\mathrm{e}^{\mathrm{i}\zeta x}\,\mathrm{d}x, \quad f \in \mathcal{P}. \tag{18}$$

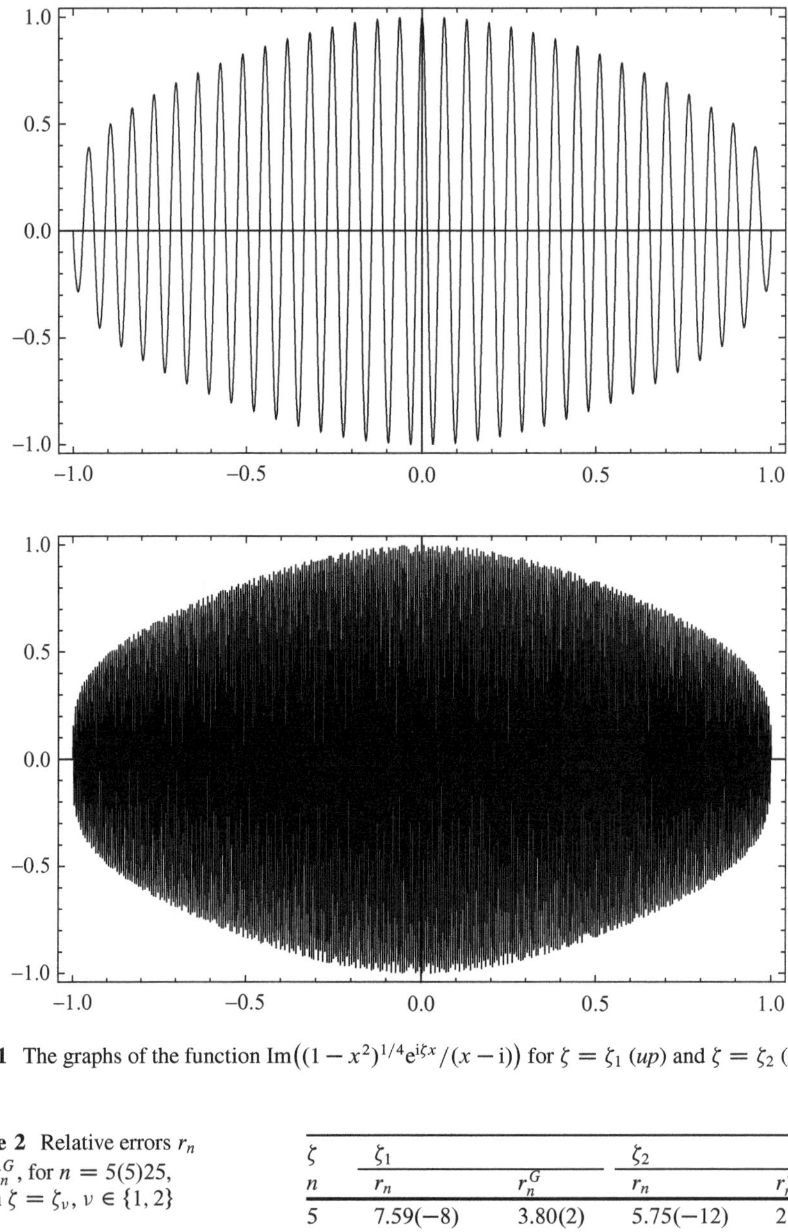

Fig. 1 The graphs of the function $\mathrm{Im}\big((1-x^2)^{1/4}\mathrm{e}^{\mathrm{i}\zeta x}/(x-\mathrm{i})\big)$ for $\zeta=\zeta_1$ (*up*) and $\zeta=\zeta_2$ (*down*)

Table 2 Relative errors r_n and r_n^G, for $n=5(5)25$, when $\zeta=\zeta_\nu$, $\nu\in\{1,2\}$

ζ	ζ_1		ζ_2	
n	r_n	r_n^G	r_n	r_n^G
5	7.59(−8)	3.80(2)	5.75(−12)	2.13(2)
10	5.82(−16)	2.24(2)	2.63(−26)	6.65(2)
15	1.16(−19)	2.72(2)	5.58(−35)	6.23(2)
20	4.11(−26)	6.32(1)	3.50(−47)	1.08(3)
25	1.79(−29)	8.46(1)	7.99(−55)	5.14(2)

Again, we restrict our attention to the case $\zeta > 0$, since the corresponding results for $\zeta < 0$ can be obtained by a simple conjugation.

The moments $\mu_k^{\pm} = \mathcal{L}^{\pm}[x^k]$, $k \in \mathbb{N}_0$, can be expressed in the form

$$\mu_k^{\pm} = \frac{A}{(\mathrm{i}\zeta)^{k+\ell}} \sum_{j=0}^{\ell} \binom{\ell}{j} (\pm 1)^j (\mathrm{i}\zeta)^{\ell-j} \left(P_{k+j}^{\alpha}(\zeta) J_{\alpha}(\zeta) + Q_{k+j}^{\alpha}(\zeta) J_{\alpha-1}(\zeta) \right),$$

where $A = (2/\zeta)^{\alpha} \sqrt{\pi}\, \Gamma(\alpha+1/2)$, J_{ν} is Bessel function of the order ν, and P_k^{α} and Q_k^{α} are polynomials in ζ, which satisfy the following four-term recurrence relation

$$y_{k+2} = -(k + 2\alpha + 1) y_{k+1} - \zeta^2 y_k - k\zeta^2 y_{k-1},$$

with the initial conditions $P_0^{\alpha}(\zeta) = 1$, $P_1^{\alpha}(\zeta) = -2\alpha$, $P_2^{\alpha}(\zeta) = 2\alpha(2\alpha+1) - \zeta^2$ and $Q_0^{\alpha}(\zeta) = 0$, $Q_1^{\alpha}(\zeta) = \zeta$, $Q_2^{\alpha}(\zeta) = -(2\alpha+1)\zeta$, respectively (see [58, Theorem 2.1]). When the existence of orthogonal polynomials with respect to the linear functional (18) is in question, the following result was proved in [58].

Theorem 4.5. *If $\alpha > -1/2$ is a rational number, ℓ is a positive integer, and ζ is a positive zero of the Bessel function $J_{\alpha-1}$, then the polynomials $\pi_n^{\pm}$ orthogonal with respect to the linear functionals $\mathcal{L}^{\pm}$, given by* (18), *exist.*

The (quasi) inner product $(p, q) = \mathcal{L}^{\pm}[pq]$ has the property $(zp, q) = (p, zq)$, which implies that the corresponding (monic) orthogonal polynomials $\{\pi_n\}_{n \in \mathbb{N}_0}$ satisfy the fundamental three-term recurrence relation

$$\pi_{n+1}(x) = (x - \alpha_n)\pi_n(x) - \beta_n \pi_{n-1}(x), \quad n \in \mathbb{N},$$

with $\pi_0(x) = 1$, $\pi_{-1}(x) = 0$. Knowing three-term recurrence coefficients, by using functions implemented in the software package `OrthogonalPolynomials` [5] in extended arithmetics we are able to construct the corresponding quadrature rules of Gaussian type

$$\int_{-1}^{1} f(x)(1-x)^{\alpha-1/2}(1+x)^{\beta-1/2} \mathrm{e}^{\mathrm{i}\zeta x}\, \mathrm{d}x = \sum_{k=1}^{n} w_k^{(n)} f(x_k^{(n)}) + R_n[f], \tag{19}$$

where $R_n[f] = 0$ for each polynomial of degree at most $2n-1$. Such rules can be efficiently applied for numerical integration of highly oscillating functions. Analogously as in the case of oscillatory modification of Gegenbauer measure, numerical experiments indicate that Gaussian quadrature rule (19) becomes more efficient when ζ increases, while Gauss–Jacobi rule with respect to weight $x \mapsto (1-x)^{\alpha-1/2}(1+x)^{\beta-1/2}$ becomes unusable (see [58] for some examples).

5 Irregular Oscillators

In this section we consider the more general highly oscillating integrand,

$$I[f;g] = \int_a^b f(x) \mathrm{e}^{\mathrm{i}\omega g(x)}\,\mathrm{d}x, \tag{20}$$

where $-\infty < a < b < +\infty$, $|\omega|$ is large, and both f and g are sufficiently smooth functions. The integrand of (20) is often called an irregular oscillator. Such integrals occur in a wide range of practical problems. There are a large number of articles where problems of numerical calculation of such integrals are treated (see [9–11, 18,20,22–24,33,46,55,57], etc.). In the case $g(x) = x$, we get the so-called regular oscillators, which have been already considered through the paper. In this section we briefly describe asymptotic method, Filon-type methods, and Levin-type methods for numerical integration of (20). We consider only the case when $g'(x) \neq 0$ for $a \leqslant x \leqslant b$, i.e., the case when g has no stationary points. Notice that from the van der Corput lemma it follows that $I[f;g] = \mathcal{O}(\omega^{-1})$, $|\omega| \to \infty$ (see [59]).

5.1 Asymptotic Methods

Asymptotic method was presented by Iserles and Nørsett [24]. Starting by the following simple transformation

$$I[f;g] = \int_a^b f(x)\mathrm{e}^{\mathrm{i}\omega g(x)}\,\mathrm{d}x = \frac{1}{\mathrm{i}\omega}\int_a^b \frac{f(x)}{g'(x)}\frac{\mathrm{d}}{\mathrm{d}x}\mathrm{e}^{\mathrm{i}\omega g(x)}\,\mathrm{d}x,$$

and applying an integration by parts, we obtain

$$I[f;g] = \frac{1}{\mathrm{i}\omega}\left(\frac{f(x)}{g'(x)}\mathrm{e}^{\mathrm{i}\omega g(x)}\right)\Bigg|_a^b - \frac{1}{\mathrm{i}\omega}\int_a^b \frac{\mathrm{d}}{\mathrm{d}x}\left(\frac{f(x)}{g'(x)}\right)\mathrm{e}^{\mathrm{i}\omega g(x)}\,\mathrm{d}x.$$

Denoting

$$Q^A[f;g] = \frac{1}{\mathrm{i}\omega}\left(\frac{f(x)}{g'(x)}\mathrm{e}^{\mathrm{i}\omega g(x)}\right)\Bigg|_a^b,$$

we have

$$I[f;g] = Q^A[f;g] - \frac{1}{\mathrm{i}\omega} I\left[\frac{\mathrm{d}}{\mathrm{d}x}\left(\frac{f(x)}{g'(x)}\right);g\right].$$

According to the van der Corput lemma, $I[f;g] - Q^A[f;g] = \mathcal{O}(\omega^{-2})$. Now, we can approximate the error term by the same rule, so, we approximate $I[f;g]$ by

$$Q^A[f;g] - \frac{1}{\mathrm{i}\omega} Q^A\left[\frac{\mathrm{d}}{\mathrm{d}x}\left(\frac{f}{g'}\right); g\right].$$

In this approximation of $I[f;g]$ the error is $\mathcal{O}(\omega^{-3})$. Continuing in this manner, after s steps, we obtain approximation of $I[f;g]$ with error $\mathcal{O}(\omega^{-s-1})$. Thus, we have derived the following asymptotic expansion.

Theorem 5.1. *Let $f \in C^\infty$ and $g'(x) \neq 0$ for $a \leqslant x \leqslant b$. Let*

$$\sigma_1[f](x) = \frac{f(x)}{g'(x)}, \quad \sigma_{k+1}[f](x) = \frac{1}{g'(x)} \frac{\mathrm{d}\sigma_k[f](x)}{\mathrm{d}x}, \; k = 0, 1, \ldots.$$

Then, for $\omega \to \infty$,

$$I[f;g] \sim -\sum_{k=1}^{\infty} \frac{1}{(-\mathrm{i}\omega)^k} \left(\sigma_k[f](b)\mathrm{e}^{\mathrm{i}\omega g(b)} - \sigma_k[f](a)\mathrm{e}^{\mathrm{i}\omega g(a)}\right).$$

Taking the s–th partial sum of the asymptotic expansion we obtain the *asymptotic method*

$$Q_s^A[f;g] = -\sum_{k=1}^{s} \frac{1}{(-\mathrm{i}\omega)^k} \left(\sigma_k[f](b)\mathrm{e}^{\mathrm{i}\omega g(b)} - \sigma_k[f](a)\mathrm{e}^{\mathrm{i}\omega g(a)}\right).$$

It is easy to see that

$$I[f;g] - Q_s^A[f;g] = \frac{1}{(-\mathrm{i}\omega)^s} \int_a^b g'(x)\sigma_{s+1}[f](x)\mathrm{e}^{\mathrm{i}\omega g(x)}\,\mathrm{d}x \sim \mathcal{O}(\omega^{-s-1}).$$

The following result follows from Theorem 5.1 (see [46]).

Lemma 5.1. *Suppose $0 = f^{(k)}(a) = f^{(k)}(b)$, for all $k = 0, 1, \ldots, s-1$ for some positive integer s, and that f depend on ω as well as that the every function in the set $\{f, f', \ldots, f^{(s+1)}\}$ is of asymptotic order $\mathcal{O}(\omega^{-n})$, $\omega \to \infty$, for some fixed n. Then, $I[f;g] \sim \mathcal{O}(\omega^{-n-s-1})$, $\omega \to \infty$.*

The drawback of the asymptotic method is that for fixed ω, in general $Q_s^A[f;g]$ diverges as $s \to \infty$. Also, numerical examples show that asymptotic method may produce very inaccurate approximation for small values of ω in general.

5.2 Filon-Type Methods

Here we describe Filon-type method for numerical computation of (20), presented in [24]. The main idea is to interpolate function f for fixed set of prescribed nodes by using Hermite interpolation and then integrate interpolating polynomial.

Let $\{x_k\}_{k=0}^{\nu}$ be a set of prescribed nodes, such that

$$a = x_0 < x_1 < \cdots < x_\nu = b.$$

Having chosen multiplicities $n_0, n_1, \ldots, n_\nu \in \mathbb{N}$, by $H_n(x) = \sum_{k=0}^{n} a_k x^k$, we denote polynomial of degree n, where $n = \sum_{k=0}^{\nu} n_k - 1$, such that

$$H_n^{(j)}(x_k) = f^{(j)}(x_k), \quad j = 0, 1, \ldots, n_k - 1, \quad k = 0, 1, \ldots, \nu. \tag{21}$$

For $s = \min\{n_0, n_\nu\}$ we define

$$Q_s^F[f;g] = I[H_n;g] = \sum_{k=0}^{n} a_k I[x^k;g].$$

The function $f - H_n$ satisfies the conditions given in Lemma 5.1 due to (21). Thus, we have

$$I[f;g]-Q_s^F[f;g] = I[f;g]-I[H_n;g] = I[f-H_n;g] \sim \mathcal{O}(\omega^{-s-1}), \quad \omega \to \infty.$$

Therefore, the asymptotic and the Filon-type methods have the same asymptotic order. In many situations (but it is not always) the accuracy of Filon-type method is significantly higher than that of the asymptotic method (see some examples in [24,46]).

Unfortunately, there are two problems with Filon-type methods. The first one is obvious from the definition of method. By definition, the Filon-type methods given above require the computation of the moments $I[x^k;g]$ analytically, which is not possible in general. The second problem is connected with the fact that the Filon-type method is based on interpolation and the accuracy of $Q_s^F[f;g]$ is directly related to accuracy of interpolation. The good example for that is Runge's example from 1901 (see [37, p. 60]). For non-oscillatory functions $f_a(x) = 1/(1+(x/a)^2)$, $x \in [-1,1]$, for sufficiently small a, interpolation polynomials with equally spaced nodes are oscillating (see Fig. 2). For such functions the Filon-type methods (especially, when only function values are used) produce less accurate results.

The magnitude of Runge's phenomenon could be reduced by using Chebyshev interpolating points. Another idea is to use cubic spline, but in that case the order is at most $\mathcal{O}(\omega^{-3})$.

Hascelik [18] modified the Filon-type methods such that they can be applied in the cases when f and g have algebraic singularity.

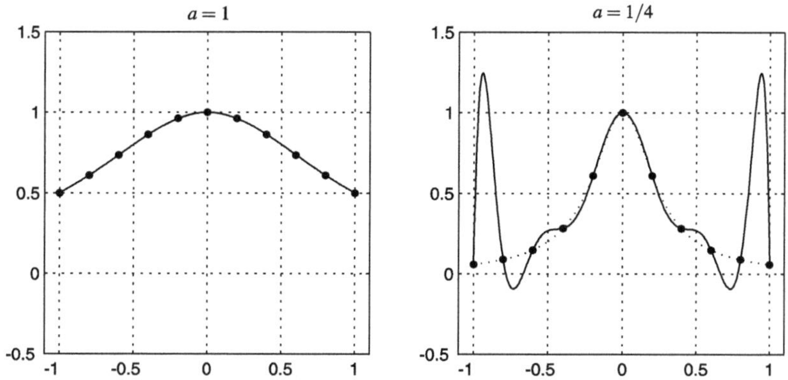

Fig. 2 The Runge's example for $n = 11$ equally spaced nodes, for $a = 1$ (*left*) and $a = 1/4$ (*right*)

5.3 Levin-Type Methods

Now, we explain method which does not require the computation of moments, introduced by Levin (see [30–32]). Levin method can be applied to more general problems, which will be presented in Sect. 6.

Suppose that $F(x)$ is a function such that

$$\frac{\mathrm{d}}{\mathrm{d}x}\left(F(x)\mathrm{e}^{\mathrm{i}\omega g(x)}\right) = f(x)\mathrm{e}^{\mathrm{i}\omega g(x)}. \tag{22}$$

It is obvious that $I[f;g] = \left(F(x)\mathrm{e}^{\mathrm{i}\omega g(x)}\right)\Big|_a^b$. The idea is to approximate F by some function V, which gives method

$$Q^L[f;g] = \left(V(x)\mathrm{e}^{\mathrm{i}\omega g(x)}\right)\Big|_a^b = V(b)\mathrm{e}^{\mathrm{i}\omega g(b)} - V(a)\mathrm{e}^{\mathrm{i}\omega g(a)}.$$

From (22), we obtain equation $L[F](x) = f(x)$, where L is the operator defined by $L[F] = F' + \mathrm{i}\omega g' F$. If $V(x) = \sum_{k=0}^{n} a_k x^k$ is the collocation polynomial, satisfying system of equations $L[V](x_k) = f(x_k)$ at points $a = x_0 < x_1 < \cdots < x_\nu = b$, then $I[f;g] - Q^L[f;g] \sim \mathcal{O}(\omega^{-2})$.

There are two natural generalizations of Levin method (see [18, 46]). The first one is to use a polynomial V such that not only the values of f and $L[V]$ are the same at nodes but also the values of their derivatives up to the given multiplicity. The second generalization is obtained allowing V to be a linear combination of a set of suitable basis functions, not only polynomial.

The following result was proved in [46].

Theorem 5.2. *Suppose that $g'(x) \neq 0$ for $x \in [a,b]$. Let $\psi = \{\psi_k\}_{k=0}^{n}$ be a basis of functions independent of ω, let $\{x_k\}_{k=0}^{\nu}$ be a set of nodes such that $a = x_0 < x_1 < \cdots < x_\nu = b$, let $\{n_k\}_{k=0}^{\nu}$ be a set of multiplicities associated with nodes, and $s = \min\{n_0, n_\nu\}$. Further, suppose that $V = \sum_{k=0}^{n} a_k \psi_k$, where $n = \sum_{k=0}^{\nu} n_k - 1$, is the solution of the system of collocation equations*

$$\frac{\mathrm{d}^j L[V]}{\mathrm{d}x^j}(x_k) = f^{(j)}(x_k), \quad j = 0,1,\ldots,n_k - 1; \quad k = 0,1,\ldots,\nu,$$

where $L[V] = V' + \mathrm{i}\omega g' V$. Define

$$\mathbf{g}_k = [(g'\psi_k)(x_0) \ \cdots (g'\psi_k)^{(n_0-1)}(x_0) \ \cdots \ (g'\psi_k)(x_\nu) \ \cdots (g'\psi_k)^{(n_\nu-1)}(x_\nu)]^T. \tag{23}$$

If the vectors $\{\mathbf{g}_0, \ldots, \mathbf{g}_n\}$ are linearly independent, then the system has a unique solution, and for

$$Q_\psi^L[f;g] = \left(V(x)\mathrm{e}^{\mathrm{i}\omega g(x)}\right)\Big|_a^b = V(b)\mathrm{e}^{\mathrm{i}\omega g(b)} - V(a)\mathrm{e}^{\mathrm{i}\omega g(a)},$$

we have $I[f;g] - Q_\psi^L[f;g] \sim \mathcal{O}(\omega^{-s-1})$, $\omega \to \infty$.

Olver [46] proved that if $\{\psi_k\}_{k=0}^{n}$ is a Chebyshev set, then the conditions on $\{\mathbf{g}_k\}_{k=0}^{n}$ of the previous theorem are satisfied for all choices of $\{x_k\}_{k=0}^{\nu}$ and $\{n_k\}_{k=0}^{\nu}$. He showed that it is possible to obtain higher asymptotic order of Levin-type method by choosing the basis in the following way:

$$\psi_0 = 1, \quad \psi_1 = \frac{f}{g'}, \quad \psi_{k+1} = \frac{\psi_k'}{g'}, \quad k = 1,2,\ldots. \tag{24}$$

Suppose that $\{x_k\}_{k=0}^{\nu}$, $\{n_k\}_{k=0}^{\nu}$, and $\{\psi_k\}_{k=0}^{n}$, where $n = \sum_{k=0}^{\nu} n_k - 1$, satisfy the conditions of Theorem 5.2. Then, for $s = \min\{n_0, n_\nu\}$, we have (see [46, Theorem 5.1])

$$I[f;g] - Q_\psi^L[f;g] \sim \mathcal{O}(\omega^{-n-s-1}).$$

Levin-type method $Q_\psi^L[f;g]$ with basis $\{\psi_k\}$ given by (24), is significant improvement over $Q^F[f;g]$ and $Q^L[f;g]$ (with standard polynomials basis), when the same nodes and multiplicities are used, and ω is sufficiently large. Also, since $Q_\psi^L[f;g]$ does not require polynomial interpolation, the Runge's phenomenon does not occur.

In general, accuracy of asymptotic, Filon-type, and Levin-type methods depends on f and g. Olver [46] presented several examples for comparisons of these three types of methods, including Levin-type method with polynomial basis, and Levin-type method with basis (24).

6 Integrals Involving Highly Oscillating Bessel Function

In this section we consider integrals of the form

$$I[f] = \int_a^b f(x) J_\nu(rx)\,\mathrm{d}x, \tag{25}$$

where $J_\nu(rx)$ is Bessel function of the first kind of order ν for some positive real number ν, $r \in \mathbb{R}$ is large, and $0 < a < b \leqslant +\infty$. Such integrals appear in many areas of science and technology and several efficient methods for their numerical calculations are derived (see, e.g., [4,39,51,52,54,65–67]). Here we present Levin-type methods [31, 32] for finite b, and Chen's method [2] for both the finite and infinite b.

6.1 Levin-Type Methods

In Sect. 5 it was explained how Levin-type method [31, 32] can be applied to irregular oscillators, as well as Olver's generalization [46]. Levin's collocation method is applicable to a wide class of oscillating integrals with weight functions satisfying certain differential conditions. It can be efficiently used for computing integral (25) with finite b.

Let $F(x) = [f_1(x) \;\; f_2(x) \;\; \cdots \;\; f_m(x)]^T$ be an $m-$vector of non-oscillating functions, $W(r,x) = [w_1(r,x) \;\; w_2(r,x) \;\; \cdots \;\; w_m(r,x)]^T$ be an $m-$vector of linearly independent highly oscillating function, depending on r, and let "$\cdot$" denotes the inner product. Let us consider general class of highly oscillatory integrals of the form

$$I[F] = \int_a^b \sum_{k=1}^{m} f_k(x) w_k(r,x)\,\mathrm{d}x \equiv \int_a^b F(x)\text{A} \cdot W(r,x)\,\mathrm{d}x. \tag{26}$$

Assume that $W'(r,x) = A(r,x)W(r,x)$, where derivative is with respect to x, and $A(r,x)$ is $m \times m$ matrix of non-oscillating functions. If F were of the form

$$F(x) = Q'(x) + A^T(r,x)Q(x),$$

then the integral (26) could be evaluated as

$$\begin{aligned} I[F] &= \int_a^b \left(Q'(x) + A^T(r,x)Q(x)\right) \cdot W(r,x)\,\mathrm{d}x \\ &= \int_a^b (Q(x) \cdot W(r,x))'\,\mathrm{d}x = Q(b) \cdot W(r,b) - Q(a) \cdot W(r,a). \end{aligned}$$

The main idea of Levin method is to select linearly independent basis function $\{\psi_k\}_{k=1}^n$ and determine

$$P(x) = \left[\sum_{k=1}^{n} a_k^{(1)} \psi_k(x) \quad \cdots \quad \sum_{k=1}^{n} a_k^{(m)} \psi_k(x) \right]^T$$

such that the following system of equations is satisfied

$$P'(x_j) + A^T(r, x_j) P(x_j) = F(x_j), \quad j = 1, 2, \ldots, n,$$

at nodes $x_1, x_2, \ldots, x_n$. The Levin's approximation of $I[F]$ is

$$\begin{aligned} Q_n^L[F] &= \int_a^b \left(P'(x) + A^T(r, x) P(x) \right) \cdot W(r, x) \, \mathrm{d}x \\ &= P(b) \cdot W(r, b) - P(a) \cdot W(r, a). \end{aligned}$$

Levin [32] presented an error analysis for the composite collocation method $Q_{n,h}^L$ with n selected nodes in each subinterval of length h including the endpoints of each subinterval. His error estimate is given in the following theorem (see also [65, Theorem 1.1]).

Theorem 6.1. *Let $F \in C^{2n+1}[a,b]$, $B(r,x) = (A(x,r)/C(r))^{-1}$ exists, $B \in C^{2n+1}[a,b]$, and its $2n+1$ derivatives are bounded uniformly in r for $C(r) \geqslant \alpha_0$. Then*

$$\left| I[F] - Q_{n,h}^L[F] \right| < \frac{M(b-a)h^{n-2}}{C(r)^2},$$

for $C(r) \geqslant \beta > 0$, where h is the length of each subinterval and M is a constant independent of r and h.

For integral (25),

$$m = 2, \quad W(r, x) = [J_{\nu-1}(rx) \;\; J_\nu(rx)]^T, \quad \text{and} \quad F(x) = [0 \;\; f(x)]^T. \tag{27}$$

Then,

$$A(r, x) = \begin{bmatrix} \dfrac{\nu - 1}{x} & -r \\ r & -\dfrac{\nu}{x} \end{bmatrix}. \tag{28}$$

Define $C(r) = r$, then by Theorem 6.1 for $h \to 0$ and $r \to \infty$ we obtain

$$I[f] - Q_{n,h}^L[f] = \mathcal{O}\left(\frac{h^{n-2}}{r^2} \right).$$

Xiang, Gui, and Moa [67] extended Levin's method by using multiple nodes. Let $\{m_k\}_{k=1}^n$ be multiplicities associated with the nodes $a = x_1 < x_2 < \cdots < x_n = b$, $s = \min\{m_1, m_n\}$ and $\{\psi_k\}_{k=0}^N$, where $N = \sum_{k=1}^n m_k - 1$, be a set of linearly independent basis functions such that the matrix $[\mathbf{a}_0 \cdots \mathbf{a}_N]$ is nonsingular, with $\mathbf{a}_k = [\psi_k(x_1)\ \psi_k'(x_1) \cdots \psi_k^{(m_1-1)}(x_1) \cdots \psi_k(x_n)\ \psi_k'(x_n) \cdots \psi_k^{(m_n-1)}(x_n)]^T$, $k = 0, 1, \ldots, N$. Let $P(x) = \left[\sum_{k=0}^N a_k^{(1)}\psi_k(x) \ \cdots \ \sum_{k=0}^N a_k^{(m)}\psi_k(x)\right]^T$ satisfies the following equations:

$$P'(x_j) + A^T(r, x_j)P(x_j) = F(x_j), \quad j = 1, 2, \ldots, n,$$

$$\left[P'(x) + A^T(r, x)P(x)\right]_{x=x_j}^{(k)} = F^{(k)}(x_j), \ k = 1, 2, \ldots, m_j - 1; \ j = 1, 2, \ldots, n.$$

Levin's approximation of integral $I[F]$, given by (26), is the following

$$Q_s^L = \int_a^b \left(P'(x) + A^T(r, x)P(x)\right) \cdot W(r, x)\,\mathrm{d}x = P(b) \cdot W(r, b) - P(a) \cdot W(r, a).$$

Let $W'(r, x) = A(r, x)W(r, x)$, where $A(r, x)$ is a nonsingular $m \times m$ matrix, and $B(r, x) = (A(r, x)/C(r))^{-1}$ for $r \gg 1$. If

- $W(r, x), B(r, x) \in C^\infty[a, b]$,
- $A(r, x)/C(r)$ and $A^{(k)}(r, x)$, $k = 1, 2, \ldots, \max_{1 \leqslant j \leqslant n} m_j - 1$, are uniformly bounded for $r \gg 1$ and all $x \in [a, b]$,
- $B(r, x)$ and its $s + 1$ derivatives are uniformly bounded for $r \gg 1$ and all $x \in [a, b]$,

then (see [67, Theorem 4.1])

$$I[F] - Q_s^L[F] = \mathcal{O}\left(\frac{\|W(r, x)\|_\infty}{C(r)^{s+1}}\right).$$

Let us now go back to our integral $I[f]$, given by (25), and denote the corresponding Levin's approximation by $Q_s^L[f]$. For the basis $\{\psi_k(x)\}$, we choose the standard polynomial basis. Here $C(r) = r$, and it is easy to see from (27) and (28) that $\frac{1}{r}A(x, r)$, $B(r, x)$, $A^{(k)}(r, x)$, and $B^{(k)}(r, x)$, $k = 1, 2, \ldots, s + 1$, are uniformly bounded for $r \gg 1$ and all $x \in [a, b]$. Therefore, according to previous general estimate and the fact that $\|W(r, x)\|_\infty = \mathcal{O}(r^{-1/2})$ for $f \in C^1[a, b]$ and $r \gg 1$ (see [67, Theorem 2.1]), the following error estimate

$$I[f] - Q_s^L[f] = \mathcal{O}(r^{-s-3/2})$$

holds. Notice that $I[f] = \mathcal{O}(r^{-3/2})$ for $f \in C^1[a, b]$ and $r \gg 1$.

6.2 Chen's Method

Chen's [2] presented method for numerical computing of integral $I[f]$ given by (25) following ideas of Milovanović [39] and Huybrechs and Vandewalle [20]. By using integral form of Bessel function and its analytic continuation, Chen transformed highly oscillating integral (25) into non-oscillating integral on $[0, +\infty)$, which could be computed efficiently applying Gauss–Laguerre quadrature rule.

Substituting integral representation of Bessel function (see [64])

$$J_\nu(x) = \frac{(x/2)^\nu}{\sqrt{\pi}\,\Gamma(\nu+1/2)} \int_{-1}^{1} (1-t^2)^{\nu-1/2} \mathrm{e}^{\mathrm{i}xt}\,\mathrm{d}t$$

in (25), we obtain

$$I[f] = \int_a^b f(x) \frac{(rx/2)^\nu}{\sqrt{\pi}\,\Gamma(\nu+1/2)} \int_{-1}^{1} (1-t^2)^{\nu-1/2} \mathrm{e}^{\mathrm{i}rxt}\,\mathrm{d}t\,\mathrm{d}x. \tag{29}$$

The function $(1-t^2)^{\nu-1/2}\mathrm{e}^{\mathrm{i}rxt}$ is analytic in the half-strip of the complex plane, $-1 \leqslant \mathrm{Re}\,z \leqslant 1$, $\mathrm{Im}\,z \geqslant 0$. By using complex integration method (see [2]), it can be proved that

$$\begin{aligned}\int_{-1}^{1} (1-t^2)^{\nu-1/2} \mathrm{e}^{\mathrm{i}rxt}\,\mathrm{d}t = {} & \frac{\mathrm{i}\mathrm{e}^{-\mathrm{i}rx}}{(rx)^{2\nu}} \int_0^{+\infty} \left(u^2 + 2\mathrm{i}rxu\right)^{\nu-1/2} \mathrm{e}^{-u}\,\mathrm{d}u \\ & - \frac{\mathrm{i}\mathrm{e}^{\mathrm{i}rx}}{(rx)^{2\nu}} \int_0^{+\infty} \left(u^2 - 2\mathrm{i}rxu\right)^{\nu-1/2} \mathrm{e}^{-u}\,\mathrm{d}u,\end{aligned}$$

which together with (29) gives

$$\begin{aligned}I[f] = {} & \frac{\mathrm{i}}{2^\nu r^\nu \sqrt{\pi}\,\Gamma(\nu+1/2)} \sum_{j=0}^{\nu} (2\mathrm{i}r)^j \binom{\nu}{j} \left(\int_a^b \frac{f(x)\mathrm{e}^{-\mathrm{i}rx}}{x^{\nu-j}} \int_0^{+\infty} \frac{u^{2\nu-j}\mathrm{e}^{-u}}{\sqrt{u^2+2\mathrm{i}rxu}}\,\mathrm{d}u\,\mathrm{d}x \right. \\ & \left. - (-1)^j \int_a^b \frac{f(x)\mathrm{e}^{\mathrm{i}rx}}{x^{\nu-j}} \int_0^{+\infty} \frac{u^{2\nu-j}\mathrm{e}^{-u}}{\sqrt{u^2-2\mathrm{i}rxu}}\,\mathrm{d}u\,\mathrm{d}x \right).\end{aligned}$$

Integrals

$$I_1(\nu, j, rx) = \frac{1}{\sqrt{\pi}} \int_0^{+\infty} \frac{u^{2\nu-j}\mathrm{e}^{-u}}{\sqrt{u^2+2\mathrm{i}rxu}}\,\mathrm{d}u, \quad I_2(\nu, j, rx) = \frac{1}{\sqrt{\pi}} \int_0^{+\infty} \frac{u^{2\nu-j}\mathrm{e}^{-u}}{\sqrt{u^2-2\mathrm{i}rxu}}\,\mathrm{d}u$$

can be represented by Whittaker W function (see [17]) as follows:

$$I_1(\nu, j, rx) = \prod_{\ell=0}^{2\nu-j-1} (2\ell+1) \frac{(rx\mathrm{i})^{(2\nu-j-1)/2} \mathrm{e}^{rx\mathrm{i}}}{2^{(\nu-j+1)/2}} W_{-(\nu-j/2),\nu-j/2}(2rx\mathrm{i}),$$

$$I_2(\nu, j, rx) = \prod_{\ell=0}^{2\nu-j-1} (2\ell+1) \frac{(-rx\mathrm{i})^{(2\nu-j-1)/2} \mathrm{e}^{-rx\mathrm{i}}}{2^{(\nu-j+1)/2}} W_{-(\nu-j/2),\nu-j/2}(-2rx\mathrm{i}).$$

It is known that Wittaker W function $W_{\alpha,\beta}(z)$ for $|\arg z| < \pi$ has the following asymptotic expansion (see [17, p. 1016]):

$$W_{\alpha,\beta}(z) \sim z^{\alpha} \mathrm{e}^{-z/2} \left(1 + \sum_{k=1}^{+\infty} \frac{\prod_{\ell=1}^{k} (\beta^2 - (\alpha - \ell + 1/2)^2)}{k! z^k} \right), \quad |z| \to \infty.$$

Taking a few terms in the corresponding expansions, integrals $I_1(\nu, j, rx)$ and $I_2(\nu, j, rx)$ can be approximated efficiently for large r. Therefore, our integral $I[f]$ is now reduced to the following:

$$I[f] = \frac{\mathrm{i}}{2^{\nu} r^{\nu} \Gamma(\nu + 1/2)} \sum_{j=0}^{\nu} (2\mathrm{i}r)^j \binom{\nu}{j} \left(\int_a^b \frac{f(x) \mathrm{e}^{-\mathrm{i}rx}}{x^{\nu-j}} I_1(\nu, j, rx)\, \mathrm{d}x \right.$$
$$\left. - (-1)^j \int_a^b \frac{f(x) \mathrm{e}^{\mathrm{i}rx}}{x^{\nu-j}} I_2(\nu, j, rx)\, \mathrm{d}x \right). \tag{30}$$

In the case when $b < +\infty$, by using complex integration method (see [2]), the integrals in (30) can be transformed as follows:

$$\int_a^b \frac{f(x)\mathrm{e}^{\ \mathrm{i}rx}}{x^{\nu-j}} I_1(\nu, j, rx)\, \mathrm{d}x$$
$$= \left(\frac{\mathrm{i}\mathrm{e}^{-\mathrm{i}rq}}{r} \int_0^{+\infty} \frac{f(q - \mathrm{i}y/r) I_1(\nu, j, r(q - \mathrm{i}y/r))}{(q - \mathrm{i}y/r)^{\nu-j}} \mathrm{e}^{-y}\, \mathrm{d}y \right) \Bigg|_{q=a}^{q=b},$$

$$\int_a^b \frac{f(x)\mathrm{e}^{\mathrm{i}rx}}{x^{\nu-j}} I_2(\nu, j, rx)\, \mathrm{d}x$$
$$= \left(\frac{\mathrm{i}\mathrm{e}^{\mathrm{i}rq}}{r} \int_0^{+\infty} \frac{f(q + \mathrm{i}y/r) I_2(\nu, j, r(q + \mathrm{i}y/r))}{(q + \mathrm{i}y/r)^{\nu-j}} \mathrm{e}^{-y}\, \mathrm{d}y \right) \Bigg|_{q=b}^{q=a}.$$

Finally, applying a n-point Gauss–Leguerre quadrature rule to the previous integrals, we get the approximation of $I[f]$, which we denote by Q_n^G. If f is an analytic function in the strip of the complex plane $a \leqslant \operatorname{Re} z \leqslant b$, then the following error estimate

$$I[f] - Q_n^G[f] = \mathcal{O}\left(\frac{(n!)^2}{(2n)!r^{2n+3/2}}\right), \quad r \gg 1,$$

holds (see [2, Theorem 2.1]).

Suppose now that $b = +\infty$ and that exists constant C such that f satisfies the condition $|f(x)| \leqslant C$ for $x \in [a, +\infty)$. Transforming the integrals on the right-hand side on (30) (see [2]), $I[f]$ can be written in the form

$$I[f] = \frac{1}{2^\nu r^\nu \Gamma(\nu + 1/2)} \sum_{j=0}^{\nu} (2\mathrm{i}r)^j \binom{\nu}{j}$$
$$\times \left(\frac{\mathrm{e}^{-\mathrm{i}ra}}{r} \int_0^{+\infty} \frac{f(a - \mathrm{i}y/r) I_1(\nu, j, r(a - \mathrm{i}y/r))}{(a - \mathrm{i}y/r)^{\nu - j}} \mathrm{e}^{-y}\, \mathrm{d}y \right.$$
$$\left. + (-1)^j \frac{\mathrm{e}^{\mathrm{i}ra}}{r} \int_0^{+\infty} \frac{f(a + \mathrm{i}y/r) I_2(\nu, j, r(a + \mathrm{i}y/r))}{(a + \mathrm{i}y/r)^{\nu - j}} \mathrm{e}^{-y}\, \mathrm{d}y \right).$$

Applying again a n-point Gauss–Leguerre quadrature rule to the integrals on the right-hand side of the previous equation, we get approximation Q_n^G of $I[f]$. For and analytic function f in $\{0 \leqslant |\arg z| \leqslant \pi/2\}$, the following error estimate

$$I[f] - Q_n^G[f] = \mathcal{O}\left(\frac{(n!)^2}{(2n)!r^{2n+3/2}}\right), \quad r \gg 1,$$

holds in this case, too (see [2, Theorem 2.2]).

Numerical examples given in [2] show that for $a < b < +\infty$ Chen's method gives better approximation for integral $I[f]$ in comparison with Levin-type method.

Acknowledgements The authors were supported in part by the Serbian Ministry of Education, Science and Technological Development (grant number #174015).

References

1. Aptekarev, A.I., Assche, W.V.: Scalar and matrix Riemann–Hilbert approach to the strong asymptotics of Pade approximants and complex orthogonal polynomials with varying weight. J. Approx. Theory. **129**, 129–166 (2004)
2. Chen, R: Numerical approximations to integrals with a highly oscillatory Bessel kernel. Appl. Numer. Math. **62**, 636–648 (2012)
3. Chihara, T.S.: An Introduction to Orthogonal Polynomials. Gordon and Breach, New York (1978)
4. Chung, K.C., Evans, G.A., Webster, J.R.: A method to generate generalized quadrature rules for oscillatory integrals. Appl. Numer. Math. **34**, 85–93 (2000)
5. Cvetković, A.S., Milovanović, G.V.: The Mathematica Package "OrthogonalPolynomials". Facta Univ. Ser. Math. Inform. **19**, 17–36 (2004)

6. Davis, P.J., Rabinowitz, P.: Methods of Numerical Integration, 2nd edn. Academic Press, Inc. San Diego (1984)
7. Ehrenmark, U.T.: On the error term of the Filon quadrature formulae. BIT **27**, 85–97 (1987)
8. Einarsson, B.: Numerical calculation of Fourier integrals with cubic splines. BIT **8**, 279–286 (1968)
9. Evans, G.A.: Two robust methods for irregular oscillatory integrals over a finite range. Appl. Numer. Math. **14**, 383–395 (1994)
10. Evans, G.A.: An expansion method for irregular oscillatory integrals. Internat. J. Comput. Math. **63**, 137–148 (1997)
11. Evans, G.A., Webster, J.R.: A high order, progressive method for the evaluation of irregular oscillatory integrals. Appl. Numer. Math. **23**, 205–218 (1997)
12. Filon, L.N.G.: On a quadrature formula for trigonometric integrals. Proc. Roy. Soc. Edinburgh **49**, 38–47 (1928)
13. Flinn, E.A.: A modification of Filon's method of numerical integration. J. Assoc. Comput. Mach. **7**, 181–184 (1960)
14. Gautschi, W.: Construction of Gauss–Christoffel quadrature formulas. Math. Comput. **22**, 251–270 (1968)
15. Gautschi, W.: A survey of Gauss–Christoffel quadrature formulae. In: Butzer, P.L., Fehér, F. (eds.) E.B. Christoffel–The Influence of His Work on Mathematics and the Physical Sciences pp. 72–147. Birkhäuser, Basel (1981)
16. Goldberg, R.R., Varga, R.S.: Moebius inversion of Fourier transforms. Duke Math. J. **23**, 553–559 (1956)
17. Gradshteyn, I.S., Ryzhik, I.M.: Table of Integrals, Series, and Products. Sixth edition, Academic Press, London (2000)
18. Hascelik, A.I.: Suitable Gauss and Filon–type methods for oscillatory integrals with an algebraic singularity. Appl. Numer. Math. **59**, 101–118 (2009)
19. Håvie, T.: Remarks on an expansion for integrals of rapidly oscillating functions. BIT **13**, 16–29 (1973)
20. Huybrechs, D., Vandewalle, S.: On the evaluation of highly oscillatory integrals by analytic continuation. SIAM J. Numer. Anal. **44**, 1026–1048 (2006)
21. Iserles, A.: On the numerical quadrature of highly–oscillating integrals I: Fourier transforms. IMA J. Numer. Anal. **24**, 365–391 (2004)
22. Iserles, A., Nørsett, S.P.: On quadrature methods for highly oscillatory integrals and their implementation. BIT **44**, 755–772 (2004)
23. Iserles, A.: On the numerical quadrature of highly–oscillating integrals II: Irregular oscillators. IMA J. Numer. Anal. **25**, 25–44 (2005)
24. Iserles, A., Nørsett, S.P.: Efficient quadrature of highly–oscillatory integrals using derivatives. Proc. R. Soc. Lond. Ser. A Math. Phys. Eng. Sci. **461**, 1383–1399 (2005)
25. Ixaru, L.Gr.: Operations on oscillatory functions. Comput. Phys. Comm. **105**, 1–19 (1997)
26. Ixaru, L.Gr., Paternoster, B.: A Gauss quadrature rule for oscillatory integrands. Comput. Phys. Comm. **133**, 177–188 (2001)
27. Kim, K.J., Cools, R., Ixaru, L.Gr.: Quadrature rules using first derivatives for oscillatory integrands. J. Comput. Appl. Math. **140**, 479–497 (2002)
28. Kim, K.J., Cools, R., Ixaru, L.Gr.: Extended quadrature rules for oscillatory integrands. Appl. Numer. Math. **46**, 59–73 (2003)
29. Ledoux, V., Van Daele, M.: Interpolatory quadrature rules for oscillatory integrals. J. Sci. Comput. **53**, 586–607 (2012)
30. Levin, D.: Procedures for computing one and two dimensional integrals of functions with rapid irregular oscillations. Math. Comp. **38** (158), 531–538 (1982)
31. Levin, D.: Fast integration of rapidly oscillatory functions. J. Comput. Appl. Math. **67**, 95–101 (1996)
32. Levin, D.: Analysis of a collocation method for integrating rapidly oscillatory functions. J. Comput. Appl. Math. **78**, 131–138 (1997)

33. Li, J., Wang, X., Wang, T., Xiao, S.: An improved Levin quadrature method for highly oscillatory integrals. Appl. Numer. Math. **60**, 833–842 (2010)
34. Lyness, J.N.: The calculation of Fourier coefficients. SIAM J. Numer. Anal. **4**, 301–315 (1967)
35. Lyness, J.N.: The calculation of Fourier coefficients by the Mobius inversion of the Poisson summation formula. Pt. I: Functions whose early derivatives are continuous. Math. Comp. **24**, 101–135 (1970)
36. Luke, Y.L.: On the computation of oscillatory integrals. Proc. Camb. Phil. Soc. **50**, 269–277 (1954)
37. Mastroianni, G., Milovanović, G.V.: Interpolation Processes – Basic Theory and Applications. Springer Monographs in Mathematics, Springer, Berlin (2008)
38. Miklosko, J.: Numerical integration with weight functions $\cos kx$, $\sin kx$ on $[0, 2\pi/t]$, $t = 1, 2, \ldots$. Apl. Mat. **14**, 179–194 (1969)
39. Milovanović, G.V.: Numerical calculation of integrals involving oscillatory and singular kernels and some applications of quadratures. Comput. Math. Appl. **36**, 19–39 (1998)
40. Milovanović, G.V., Cvetković, A.S.: Orthogonal polynomials and Gaussian quadrature rules related to oscillatory weight functions. J. Comput. Appl. Math. **179**, 263–287 (2005)
41. Milovanović, G.V., Cvetković, A.S.: Orthogonal polynomials related to the oscillatory–Chebyshev weight function. Bull. Cl. Sci. Math. Nat. Sci. Math. **30**, 47–60 (2005)
42. Milovanović, G.V., Cvetković, A.S., Marjanović, Z.M.: Orthogonal polynomials for the oscillatory–Gegenbauer weight. Publ. Inst. Math. (Beograd) (N.S.) **84**(98), 49–60 (2008)
43. Milovanović, G.V., Cvetković, A.S., Stanić, M.P.: Gaussian quadratures for oscillatory integrands. Appl. Math. Letters **20**, 853–860 (2007)
44. Milovanović, G.V., Cvetković, A.S., Stanić, M.P.: Two conjectures for integrals with oscillatory integrands. FACTA UNIVERSITATIS Ser. Math. Inform. **22**(1), 77–90 (2007)
45. Milovanović, G.V., Cvetković, A.S., Stanić, M.P.: Orthogonal polynomials for modified Gegenbauer weight and corresponding quadratures. Appl. Math. Letters **22**, 1189–1194 (2009)
46. Olver, S.: Moment–free numerical integration of highly oscillatory functions. IMA J. Numer. Anal. **26**, 213–227 (2006)
47. Ortega, J.M., Rheinboldt, W.C.: Iterative solution of nonlinear equations in several variables. In: Classics in Applied Mathematics. vol. 30, SIAM, Philadelphia, PA (2000)
48. Paternoster, B.: Present state–of–the–art in exponential fitting. A contribution dedicated to Liviu Ixaru on his 70th birthday. Comput. Phys. Comm. **183**, 2499–2512 (2012)
49. Piessens, R.: Gaussian quadrature formulas for the integration of oscillating functions. Z. Angew. Math. Mech. **50**, 698–700 (1970)
50. Piessens, R.: Gaussian quadrature formulas for the evaluation of Fourier–cosine coefficients. Z. Angew. Math. Mech. **52**, 56–58 (1972)
51. Piessens, R.: Automatic computation of Bessel function integrals. Comput. Phys. Commun. **25**, 289–295 (1982)
52. Piessens, R., Branders, M.: Modified Clewshaw–Curtis method for the computation of Bessel function integrals. BIT. **23**, 370–381 (1983)
53. Piessens, R., Haegemans, A.: Numerical calculation of Fourier–transform integrals, Electron. Lett. **9**, 108–109 (1973)
54. Puoskari, M.: A method for computing Bessel function integrals. J. Comput. Phys. **75**, 334–344 (1988)
55. Sauter, T.: Computation of irregularly oscillating integrals. Appl. Numer. Math. **35**, 245–264 (2000)
56. Shampine, L.F.: Integrating oscillatory functions in Matlab. Int. J. Comput. Math. **88** (11), 2348–2358 (2011)
57. Shampine, L.F.: Integrating oscillatory functions in Matlab, II. Electron. Trans. Numer. Anal. **39**, 403–413 (2012)
58. Stanić, M.P., Cvetković, A.S.: Orthogonal polynomials with respect to modified Jacobi weight and corresponding quadrature rules of Gaussian type. Numer. Math. Theor. Meth. Appl. **4** (4), 478–488 (2011)

59. Stein, E.: Harmonic Analysis: Real–Variable Methods, Orthogonality, and Oscillatory Integrals. Princeton University Press, Princeton, NJ (1993)
60. Stetter, H.J.: Numerical approximation of Fourier transforms. Numer. Math. **8**, 235–249 (1966)
61. Tuck, E.O.: A simple Filon–trapezoidal rule. Math. Comp. **21**, 239–241 (1967)
62. Van Daele, M., Vanden Berghe, G., Vande Vyver, H.: Exponentially fitted quadrature rules of Gauss type for oscillatory integrands. Appl. Numer. Math. **53**, 509–526 (2005)
63. van de Vooren, A.I., van Linde, H.J.: Numerical calculation of integrals with strongly oscillating integrand. Math. Comp. **20**, 232–245 (1966)
64. Watson, G.N.: A Treatise of the Theory of Bessel Functions. Cambridge University Press, cambridge (1922)
65. Xiang, S.: Numerical analysis on fast integration of highly oscillatory functions. BIT **47**, 469–482 (2007)
66. Xiang, S., Gui, W.: On generalized quadrature rules for fast oscillatory integrals. Appl. Math. Comput. **197**, 60–75 (2008)
67. Xiang, S., Gui, W., Moa P.: Numerical quadrature for Bessel transformations. Appl. Numer. Math. **58**, 1247–1261 (2008)
68. Zamfirescu, I.: An extension of Gauss's method for the calculation of improper integrals. Acad. R.P. Romîne Stud. Cerc. Mat. **14**, 615–631 (1963) (in Romanian).

Part IV
Orthogonality, Transformations, and Applications

Asymptotic Reductions Between the Wilson Polynomials and the Lower Level Polynomials of the Askey Scheme

Chelo Ferreira, José L. López, and Ester Pérez Sinusía

Dedicated to Professor Hari M. Srivastava

Abstract In López and Temme (Meth. Appl. Anal. **6**, 131–146 (1999); J. Math. Anal. Appl. **239**, 457–477 (1999); J. Comp. Appl. Math. **133**, 623–633 (2001)), the authors introduced a new technique to analyse asymptotic relations in the Askey scheme. They obtained asymptotic and, at the same time, finite exact representations of orthogonal polynomials of the Askey tableau in terms of Hermite and Laguerre polynomials. That analysis is continued in Ferreira et al. (Adv. Appl. Math. **31**(1), 61–85 (2003); Acta Appl. Math. **103**(3), 235–252 (2008); J. Comput. Appl. Math. **217**(1), 88–109 (2008)), where the authors derived new finite and asymptotic relations between polynomials located in the four lower levels of the Askey tableau. In this paper we complete that analysis obtaining finite exact representations of the Wilson polynomials in terms of the hypergeometric polynomials of the other four levels of the Askey scheme. Using an asymptotic principle based on the "matching" of their generating functions, we prove that these representations have an asymptotic character for large values of certain parameters and provide information on the zero distribution of the Wilson polynomials. A new limit of the Wilson polynomials in terms of Hermite polynomials is obtained as a consequence. Some numerical experiments illustrating the accuracy of the approximations are given.

C. Ferreira • E.P. Sinusía
Departamento de Matemática Aplicada, IUMA, Universidad de Zaragoza, Zaragoza, Spain
e-mail: cferrei@unizar.es; ester.perez@unizar.es

J.L. López (✉)
Departamento de Ingeniería Matemática e Informática, Universidad Pública de Navarra, Pamplona, Spain
e-mail: jl.lopez@unavarra.es

G.V. Milovanović and M.Th. Rassias (eds.), *Analytic Number Theory, Approximation Theory, and Special Functions*, DOI 10.1007/978-1-4939-0258-3_24,

1 Introduction

The Askey tableau contains a classification and provides a graphical hierarchy of the hypergeometric orthogonal polynomials. This scheme places these polynomials in different levels depending on the number of parameters of each polynomial, in such a way that polynomials in a certain level contain the same number of parameters and, at the same time, one more parameter than the polynomials located in an immediately lower level (see Fig. 1).

In 1998, Roelof Koekoek and René F. Swarttouw published the review *The Askey-scheme of hypergeometric orthogonal polynomials and its q-analogue* [7] that contains a description of all families of hypergeometric orthogonal polynomials appearing in the Askey scheme. Among others, it includes the definition of these polynomials in terms of hypergeometric functions, the orthogonality relation, some generating functions, the three-term recurrence relation, and Rodrigues-type formula. It also includes some limit relations between the families of orthogonal polynomials contained in different levels.

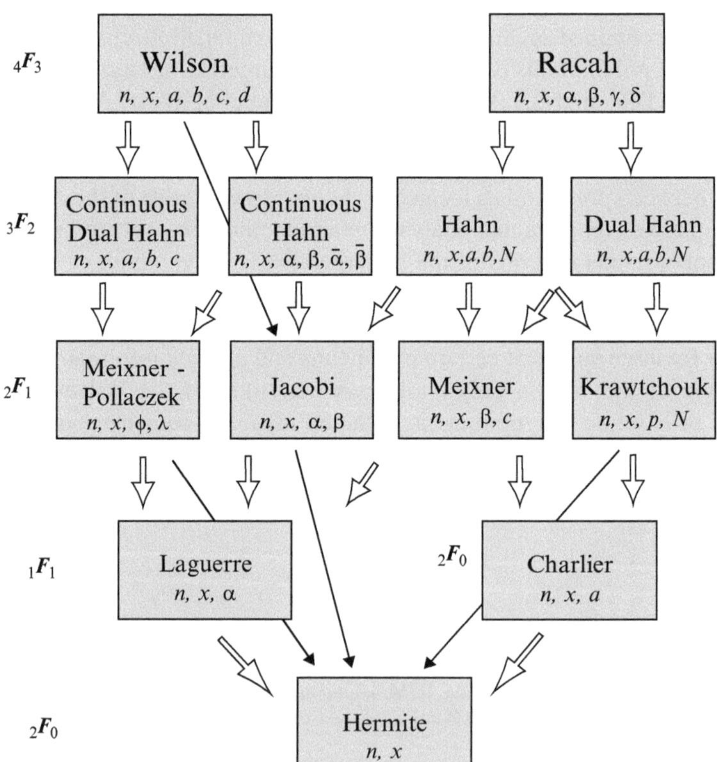

Fig. 1 The Askey scheme for hypergeometric orthogonal polynomials included in [7] with indicated limit relations between the polynomials

In 1999, José L. López and Nico M. Temme proposed a study of the Askey scheme from an asymptotic point of view. In several papers [9–11], they proposed a method to obtain asymptotic and, at the same time, finite exact representations of orthogonal polynomials of the Askey tableau in terms of Hermite and Laguerre polynomials. More precisely, the method to approximate orthogonal polynomials in terms of Hermite polynomials is described in [9], whereas [11] introduces the approximation in terms of Laguerre polynomials. All these representations have an asymptotic character for large values of certain parameters and provide information on the zero distribution of the polynomials. From these expansions, some known and unknown limits were derived.

The main idea of the asymptotic method developed by José L. López and Nico M. Temme is based on the availability of a generating function for the polynomials and is different from the techniques described in [5,6]. The techniques used in [5,6] are based on a connection problem and give deeper information on the limit relations between classical discrete and classical continuous orthogonal polynomials. On the other hand, the method developed in [9–11] gives asymptotic expansions of polynomials situated at any level of the tableau in terms of polynomials located at lower levels. This method is also different from the sophisticated uniform methods considered, for example, in [4] or [12], where asymptotic expansions of the Meixner $M_n(nx;b,c)$ or Charlier $C_n(nx;a)$ polynomials, respectively, are given for large values of n and fixed a, b, c, x. In the method presented in [9–11], the degree n keeps fixed and some parameter(s) of the polynomial are allowed to go to infinity. The asymptotic method introduced in [9–11] is also different from the technique introduced in [8]. In this paper Tom H. Koornwinder presents, for Wilson and Racah polynomials, a complete study of the limit relations existing between these polynomials and the ones placed in lower levels using the three-term recurrence relations satisfied by the polynomials.

The asymptotic study of the Askey scheme initiated in [9–11] has been continued in the papers [1–3]. In [1], the method to approximate orthogonal polynomials in terms of Charlier polynomials is described. In this reference, asymptotic expansions of Meixner–Pollaczek, Jacobi, Meixner and Krawtchouk polynomials in terms of Laguerre, Charlier and Hermite polynomials are given and four new limit relations obtained (see Fig. 2). In [2], twelve asymptotic expansions of the Hahn-type polynomials in terms of Hermite, Laguerre and Charlier polynomials are obtained and five new limits found (see Fig. 3). Finally, the study of the Hahn-type polynomials was completed in [3]. In this paper, the method to approximate orthogonal polynomials in terms of Meixner–Pollaczek, Jacobi, Meixner and Krawtchouk polynomials is described and sixteen asymptotic relations and three new limits are obtained (see Fig. 4).

In this paper, we give one more step towards the completion of the asymptotic study of the Askey scheme obtaining asymptotic relations between the Wilson polynomials (first level) and the polynomials located in lower levels. In Sect. 2, we summarize the asymptotic expansions and limit relations obtained in this study. In Sect. 3, we briefly resume the principles of the asymptotic approximations in

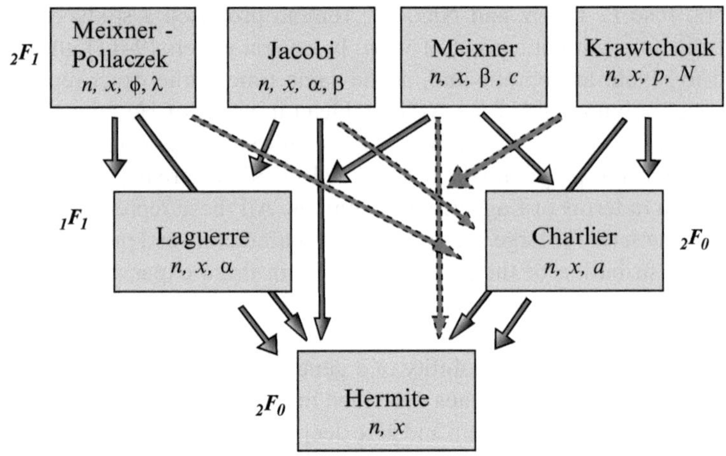

Fig. 2 Thick arrows indicate known limits and thick dashed arrows new limits obtained in [1]

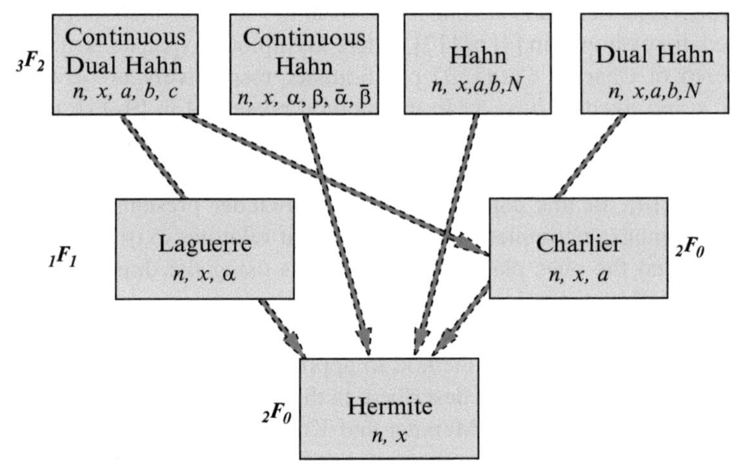

Fig. 3 Thick dashed arrows indicate new limits obtained in [2]

terms of the three lower levels introduced in [1, 9, 10]. The expansions in terms of Hahn-type polynomials are new. We give details for the case in which the basic approximants are the Continuous Dual Hahn polynomials and resume the main formulas for the remaining cases. Sect. 4 contains the proof of the formulas given in Sect. 2. Some numerical experiments illustrating the accuracy of the approximations are given in Sect. 5.

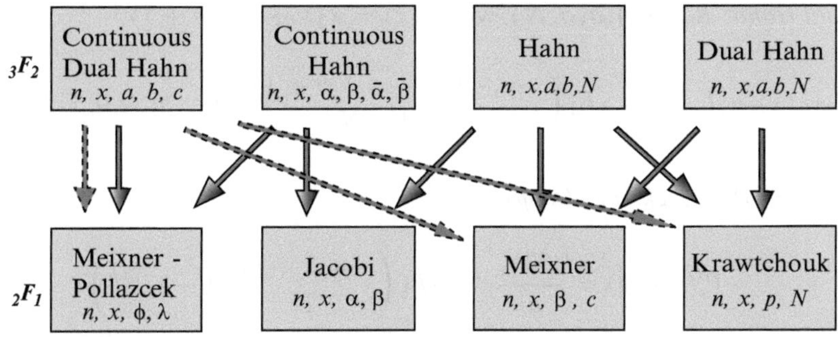

Fig. 4 Thick arrows indicate known limits and thick dashed arrows new limits obtained in [3]

2 Descending Asymptotic Expansions and Limits

Throughout this paper, we will use the notation and the definitions of the hypergeometric orthogonal polynomials of the Askey scheme introduced in [7]:

Wilson: $W_n(x^2; a, b, c, d)$

$$\frac{W_n(x^2; a, b, c, d)}{(a+b)_n(a+c)_n(a+d)_n} = {}_4F_3\left(\left.\begin{matrix} -n, n+a+b+c+d-1, a+ix, a-ix \\ a+b, a+c, a+d \end{matrix}\right| 1\right).$$

Continuous Dual Hahn: $S_n(x^2; a, b, c)$

$$\frac{S_n(x^2; a, b, c)}{(a+b)_n(a+c)_n} = {}_3F_2\left(\left.\begin{matrix} -n, a+ix, a-ix \\ a+b, a+c \end{matrix}\right| 1\right).$$

Continuous Hahn: $P_n(x; \alpha, \beta, \overline{\alpha}, \overline{\beta})$, α, $\beta \in \mathbf{C}$ ($\alpha = a + ic$, $\beta = b + id$, $a, b, c, d \in \mathbf{R}$)

$$P_n(x; \alpha, \beta, \overline{\alpha}, \overline{\beta}) = i^n \frac{(\alpha+\overline{\alpha})_n(\alpha+\overline{\beta})_n}{n!} \times {}_3F_2\left(\left.\begin{matrix} -n, n+\alpha+\beta+\overline{\alpha}+\overline{\beta}-1, \alpha+ix \\ \alpha+\overline{\alpha}, \alpha+\overline{\beta} \end{matrix}\right| 1\right).$$

Hahn: $Q_n(x; a, b, N)$

$$Q_n(x; a, b, N) = {}_3F_2\left(\left.\begin{matrix} -n, n+a+b+1, -x \\ a+1, -N \end{matrix}\right| 1\right), \quad n = 0, 1, 2, \ldots, N.$$

Dual Hahn: $R_n(\lambda(x); a, b, N)$, with $\lambda(x) = x(x + a + b + 1)$

$$R_n(\lambda(x); a, b, N) = {}_3F_2\left(\begin{matrix} -n, -x, x + a + b + 1 \\ a + 1, -N \end{matrix} \middle| 1 \right), \quad n = 0, 1, 2, \ldots, N.$$

Meixner–Pollaczek: $P_n^{(\lambda)}(x; \phi)$

$$P_n^{(\lambda)}(x; \phi) = \frac{(2\lambda)_n}{n!} e^{in\phi} {}_2F_1\left(\begin{matrix} -n, \lambda + ix \\ 2\lambda \end{matrix} \middle| 1 - e^{-2i\phi} \right).$$

Jacobi: $P_n^{(\alpha,\beta)}(x)$

$$P_n^{(\alpha,\beta)}(x) = \frac{(\alpha + 1)_n}{n!} {}_2F_1\left(\begin{matrix} -n, n + \alpha + \beta + 1 \\ \alpha + 1 \end{matrix} \middle| \frac{1 - x}{2} \right).$$

Meixner: $M_n(x; \beta, c)$

$$M_n(x; \beta, c) = {}_2F_1\left(\begin{matrix} -n, -x \\ \beta \end{matrix} \middle| 1 - \frac{1}{c} \right).$$

Krawtchouk: $K_n(x; p, N)$

$$K_n(x; p, N) = {}_2F_1\left(\begin{matrix} -n, -x \\ -N \end{matrix} \middle| \frac{1}{p} \right), \quad n = 0, 1, 2, \ldots, N.$$

Laguerre: $L_n^{(\alpha)}(x)$

$$L_n^{(\alpha)}(x) = \frac{(\alpha + 1)_n}{n!} {}_1F_1\left(\begin{matrix} -n \\ \alpha + 1 \end{matrix} \middle| x \right).$$

Charlier: $C_n(x; a)$

$$C_n(x; a) = {}_2F_0\left(\begin{matrix} -n, -x \\ - \end{matrix} \middle| -\frac{1}{a} \right).$$

Hermite: $H_n(x)$

$$H_n(x) = (2x)^n {}_2F_0\left(\begin{matrix} -n/2, -(n - 1)/2 \\ - \end{matrix} \middle| -\frac{1}{x^2} \right).$$

The orthogonality property of the polynomials of the Askey tableau only holds when the variable x and other parameters which appear in the polynomials are restricted to certain real intervals [7]. The expansions that we resume below are valid for larger domains of the variable and the parameters and for any $n \in \mathbf{N}$. Nevertheless, for the sake of clearness, we will consider that the variable and the parameters are restricted to the orthogonality intervals given in [7]. All the square roots that appear in what follows assume real positive values for real positive argument. The coefficients c_k given below are the coefficients of the Taylor expansion at $w = 0$ of the given functions $f(x, w)$:

$$c_k = \frac{1}{k!} \left. \frac{\partial^k f(x,w)}{\partial w^k} \right|_{w=0}. \tag{1}$$

The first three coefficients c_k are $c_0 = 1$, $c_1 = c_2 = 0$. Higher coefficients c_k, $k \geq 3$ can be obtained recurrently from a differential equation satisfied by $f(x, w)$ or directly from their definition (1) (using computer algebra programs like Mathematica or Maple). In our previous works about polynomials located in the first levels of the Askey tableau [1, 2], we have given recurrent formulas for c_k using a differential equation satisfied by $f(x, w)$. However, the functions $f(x, w)$ involved in this paper are more complicated, and analytic formulas for c_k are too cumbersome to be written down here.

2.1 *Wilson to Hermite*

2.1.1 Asymptotic Expansion for Large a, b, c and d

$$\frac{W_n(x^2; a, b, c, d)}{(a+b)_n (c+d)_n n!} = \sum_{k=0}^{n} \frac{c_k}{B^{k-n}(n-k)!} H_{n-k}(X), \tag{2}$$

$$B = \sqrt{\frac{1}{2} p_1^2(x) - p_2(x)}, \qquad X = \frac{p_1(x)}{2B}, \tag{3}$$

where

$$\begin{aligned} p_1(x) &= \frac{(a+ix)(b+ix)}{a+b} + \frac{(c-ix)(d-ix)}{c+d}, \\ p_2(x) &= \frac{(a+ix)(b+ix)(1+a+ix)(1+b+ix)}{2(a+b)(1+a+b)} \\ &\quad + \frac{(a+ix)(b+ix)(c-ix)(d-ix)}{(a+b)(c+d)} \\ &\quad + \frac{(c-ix)(d-ix)(1+c-ix)(1+d-ix)}{2(c+d)(1+c+d)} \end{aligned} \tag{4}$$

and c_k are the coefficients of the Maclaurin expansion of

$$f(x,\omega) = e^{-2BX\omega+B^2\omega^2}{}_2F_1\left(\left.\begin{matrix} a+ix, b+ix \\ a+b \end{matrix}\right| \omega\right) {}_2F_1\left(\left.\begin{matrix} c-ix, d-ix \\ c+d \end{matrix}\right| \omega\right).$$

2.1.2 Asymptotic Property

$$\frac{c_k}{B^{k-n}(n-k)!} H_{n-k}(X) = \mathcal{O}(a^{n+\lfloor k/3 \rfloor - k}), \tag{5}$$

when $a \to \infty$ and $a \sim b \sim c \sim d$.

2.1.3 New Limit

$$\lim_{a,b,c,d \to \infty} \frac{W_n(\tilde{x}^2; a,b,c,d)}{(a+b)_n (c+d)_n B(\tilde{x})^n} = H_n(-x) = (-1)^n H_n(x), \tag{6}$$

where

$$\tilde{x} = -\sqrt{\frac{abc+abd+acd+bcd}{a+b+c+d} + x \frac{A}{(a+b+c+d)^2}}$$

and

$$A = \sqrt{2(a+c)(b+c)(a+d)(b+d)(1+a+b)(1+c+d)(2+a+b+c+d)}.$$

2.2 *Wilson to Laguerre*

2.2.1 Asymptotic Expansion for Large *a*, *b*, *c* and *d*

$$\frac{W_n(x^2; a,b,c,d)}{(a+b)_n (c+d)_n n!} = \sum_{k=0}^{n} c_k L_{n-k}^{(X)}(A), \tag{7}$$

$$A = p_1(x)^2 + p_1(x) - 2p_2(x), \qquad X = A + p_1(x) - 1,$$

where $p_1(x)$ and $p_2(x)$ are given in (4) and c_k are the coefficients of the Maclaurin expansion of

$$f(x,\omega) = e^{\frac{A\omega}{1-\omega}}(1-\omega)^{X+1}{}_2F_1\left(\left.\begin{matrix} a+ix, b+ix \\ a+b \end{matrix}\right| \omega\right){}_2F_1\left(\left.\begin{matrix} c-ix, d-ix \\ c+d \end{matrix}\right| \omega\right).$$

2.2.2 Asymptotic Property

$$c_k L_{n-k}^{(X)}(A) = \mathcal{O}(a^{n+\lfloor k/3\rfloor-k}) \quad \text{as} \quad a\to\infty,\ a\sim b\sim c\sim d. \tag{8}$$

2.3 *Wilson to Charlier*

2.3.1 Asymptotic Expansion for Large a, b, c and d

$$\frac{W_n(x^2; a,b,c,d)}{(a+b)_n(c+d)_n n!} = \sum_{k=0}^{n} \frac{c_k}{A^{k-n}} \frac{C_{n-k}(X; A)}{(n-k)!}, \tag{9}$$

$$A = p_1(x)^2 + p_1(x) - 2p_2(x), \qquad X = A - p_1(x),$$

where $p_1(x)$ and $p_2(x)$ are given in (4) and c_k are the coefficients of the Maclaurin expansion of

$$f(x,\omega) = e^{-A\omega}(1-\omega)^{-X}{}_2F_1\left(\left.\begin{matrix} a+ix, b+ix \\ a+b \end{matrix}\right| \omega\right){}_2F_1\left(\left.\begin{matrix} c-ix, d-ix \\ c+d \end{matrix}\right| \omega\right).$$

2.3.2 Asymptotic Property

$$\frac{c_k}{A^{k-n}} \frac{C_{n-k}(X; A)}{(n-k)!} = \mathcal{O}(a^{n+\lfloor k/3\rfloor-k}) \quad \text{as} \quad a\to\infty,\ a\sim b\sim c\sim d. \tag{10}$$

2.4 *Wilson to Meixner–Pollaczek*

2.4.1 Asymptotic Expansion for Large a, b, c and d

$$\frac{W_n(x^2; a,b,c,d)}{(a+b)_n(c+d)_n n!} = \sum_{k=0}^{n} c_k P_{n-k}^{(C)}(X; A), \tag{11}$$

where $A \neq m\pi$, $m \in \mathbf{Z}$ is an arbitrary constant,

$$C = p_1(x)\cos A + \frac{1}{2}p_1(x)^2 - p_2(x),$$
$$X = \frac{-1}{2\sin A}\left[p_1(x)\cos 2A + (p_1(x)^2 - 2p_2(x))\cos A\right],$$

$p_1(x)$ and $p_2(x)$ are given in (4) and c_k are the coefficients of the Maclaurin expansion of

$$f(x,\omega) = \left(1-e^{iA}\omega\right)^{C-iX}\left(1-e^{-iA}\omega\right)^{C+iX} \times {}_2F_1\left(\left.\begin{matrix} a+ix, b+ix \\ a+b \end{matrix}\right|\omega\right) {}_2F_1\left(\left.\begin{matrix} c-ix, d-ix \\ c+d \end{matrix}\right|\omega\right).$$

2.4.2 Asymptotic Property

$$c_k P_{n-k}^{(C)}(X;A) = \mathcal{O}(a^{n+\lfloor k/3\rfloor-k}) \quad \text{as} \quad a\to\infty,\ a\sim b\sim c\sim d. \tag{12}$$

2.5 *Wilson to Jacobi*

2.5.1 Asymptotic Expansion for Large *a*, *b*, *c* and *d*

$$\frac{W_n(x^2;a,b,c,d)}{(a+b)_n(c+d)_n n!} = \sum_{k=0}^{n} c_k P_{n-k}^{(A,C)}(X), \tag{13}$$

where $X \neq \pm 1$ is an arbitrary constant,

$$A = \frac{1}{X+1}\left[2p_1(x)^2 + p_1(x) + 3p_1(x)X - 4p_2(x) + X^2 - X - 2\right],$$

$$C = \frac{1}{X-1}\left[2p_1(x)^2 - p_1(x) + 3p_1(x)X - 4p_2(x) + X^2 + X - 2\right],$$

$p_1(x)$ and $p_2(x)$ are given in (4) and c_k are the coefficients of the Maclaurin expansion of

$$f(x,\omega) = \frac{R(1+R-\omega)^A(1+R+\omega)^C}{2^{A+C}} \times {}_2F_1\left(\left.\begin{matrix} a+ix, b+ix \\ a+b \end{matrix}\right| \omega\right) {}_2F_1\left(\left.\begin{matrix} c-ix, d-ix \\ c+d \end{matrix}\right| \omega\right),$$

with $R = \sqrt{1 - 2X\omega + \omega^2}$.

2.5.2 Asymptotic Property

$$c_k P_{n-k}^{(A,C)}(X) = \mathcal{O}(a^{n+\lfloor k/3 \rfloor - k}) \quad \text{as} \quad a \to \infty,\ a \sim b \sim c \sim d. \tag{14}$$

2.6 Wilson to Meixner

2.6.1 Asymptotic Expansion for Large *a*, *b*, *c* and *d*

$$\frac{W_n(x^2; a, b, c, d)}{(a+b)_n (c+d)_n n!} = \sum_{k=0}^{n} \frac{c_k (A)_{n-k}}{(n-k)!} M_{n-k}(X; A, C), \tag{15}$$

where $C \neq 0, 1$ is an arbitrary constant,

$$A = (1+C)p_1(x) + Cp_1(x)^2 - 2Cp_2(x), \quad X = \frac{C^2}{1-C}\left[p_1(x)^2 + p_1(x) - 2p_2(x)\right],$$

$p_1(x)$ and $p_2(x)$ are given in (4), and c_k are the coefficients of the Maclaurin expansion of

$$f(x,\omega) = \left(1 - \frac{\omega}{C}\right)^{-X} (1-\omega)^{X+A} {}_2F_1\left(\left.\begin{matrix} a+ix, b+ix \\ a+b \end{matrix}\right| \omega\right) {}_2F_1\left(\left.\begin{matrix} c-ix, d-ix \\ c+d \end{matrix}\right| \omega\right).$$

2.6.2 Asymptotic Property

$$\frac{c_k (A)_{n-k}}{(n-k)!} M_{n-k}(X; A, C) = \mathcal{O}(a^{n+\lfloor k/3 \rfloor - k}) \quad \text{as} \quad a \to \infty,\ a \sim b \sim c \sim d. \tag{16}$$

2.7 *Wilson to Krawtchouk*

2.7.1 Asymptotic Expansion for Large *a*, *b*, *c* and *d*

$$\frac{W_n(x^2;a,b,c,d)}{(a+b)_n(c+d)_n n!} = \sum_{k=0}^{n} c_k \binom{C}{n-k} K_{n-k}(X;A,C), \tag{17}$$

where $A \neq 0, 1$ is an arbitrary constant,

$$X = \frac{A^2}{1-A}\left[p_1(x)^2 - p_1(x) - 2p_2(x)\right], \qquad C = p_1(x) + \frac{X}{A},$$

$p_1(x)$ and $p_2(x)$ are given in (4), and c_k are the coefficients of the Maclaurin expansion of

$$f(x,\omega) = (1+\omega)^{X-C}\left(1 - \frac{1-A}{A}\omega\right)^{-X}$$
$$\times {}_2F_1\left(\left.\begin{matrix} a+ix, b+ix \\ a+b \end{matrix}\right|\omega\right) {}_2F_1\left(\left.\begin{matrix} c-ix, d-ix \\ c+d \end{matrix}\right|\omega\right).$$

2.7.2 Asymptotic Property

$$c_k \binom{C}{n-k} K_{n-k}(X;A,C) = \mathcal{O}(a^{n+\lfloor k/3\rfloor-k}) \quad \text{as} \quad a \to \infty,\ a \sim b \sim c \sim d. \tag{18}$$

2.8 *Wilson to Continuous Dual Hahn*

2.8.1 Asymptotic Expansion for Large *a*, *b*, *c* and *d*

$$\frac{W_n(x^2;a,b,c,d)}{(a+b)_n(c+d)_n n!} = \sum_{k=0}^{n} c_k \frac{S_{n-k}(X^2;A,B,C)}{(A+B)_{n-k}(n-k)!}, \tag{19}$$

where $A = \tilde{A}a$, $B = \tilde{B}a$ with $\tilde{A}$ and $\tilde{B}$ arbitrary constants,

$$C = p_1(x) - \frac{A+B}{2} + \frac{1}{2}\sqrt{(A-B)^2 + 4(1+A+B)(p_1^2(x) + p_1(x) - 2p_2(x))},$$

$$X=\frac{1}{\sqrt{2}}\sqrt{-A^2-B^2+(A+B)\sqrt{(A-B)^2+4(1+A+B)(p_1^2(x)+p_1(x)-2p_2(x))}},$$

$p_1(x)$ and $p_2(x)$ are given in (4), and c_k are the coefficients of the Maclaurin expansion of

$$f(x,\omega) = (1-\omega)^{C-iX}\,{}_2F_1\left(\left.\begin{matrix}a+ix, b+ix\\ a+b\end{matrix}\right|\omega\right)\,{}_2F_1\left(\left.\begin{matrix}c-ix, d-ix\\ c+d\end{matrix}\right|\omega\right)$$
$$\times\left[{}_2F_1\left(\left.\begin{matrix}A+iX, B+iX\\ A+B\end{matrix}\right|\omega\right)\right]^{-1}.$$

2.8.2 Asymptotic Property

$$c_k\frac{S_{n-k}(X^2;A,B,C)}{(A+B)_{n-k}(n-k)!} = \mathcal{O}(a^{n+\lfloor k/3\rfloor-k}) \quad \text{as} \quad a\to\infty,\ a\sim b\sim c\sim d. \tag{20}$$

2.9 *Wilson to Continuous Hahn*

2.9.1 Asymptotic Expansion for Large *a*, *b*, *c* and *d*

$$\frac{W_n(x^2;a,b,c,d)}{(a+b)_n(c+d)_n n!} = \sum_{k=0}^{n} c_k\frac{P_{n-k}(X;\alpha,\beta,\overline{\alpha},\overline{\beta})}{(\alpha+\overline{\alpha})_{n-k}(\beta+\overline{\beta})_{n-k}}, \tag{21}$$

where $\alpha = A+iC$, $\beta = \tilde{B}a+iD$, $X = \tilde{X}a^2$ with D, $\tilde{B}$, $\tilde{X}$ arbitrary constants,

$$\begin{aligned}A = {} & \big[B^3(1+8p_2(x)) + (D+X)^2 + B(D+X)(D-2p_1(x)+X)\\ & + B^2(1-4(Dp_1(x)+p_2(x)+p_1(x)X)\big]\\ & /\big[B^2(-1+4p_1^2(x)-8p_2(x)) + 8B^3(p_1^2(x)-2p_2(x)) - (D+X)^2)\big],\end{aligned}$$

$$\begin{aligned}C = {} & -\big[B(D-4p_1(x))(D+X)^2 + (D+X)^3 - B^2(D+X)\\ & \times(-1+6Dp_1(x)-4p_1^2(x)-4p_2(x)+6p_1(x)X)\\ & + 2B^4(-p_1(x)(1+8p_2(x)) + 4p_1^2(x)X - 8p_2(x)X)\\ & + B^3(D(1+8p_1^2(x)+8p_2(x)) + 2p_1(x)(-1-4p_2(x)+6p_1(x)X))\big]\\ & /\big[B(B^2(-1+4p_1^2(x)-8p_2(x)) + 8B^3(p_1^2(x)-2p_2(x)) - (D+X)^2))\big],\end{aligned}$$

$p_1(x)$ and $p_2(x)$ are given in (4), and c_k are the coefficients of the Maclaurin expansion of

$$f(x,\omega) = {}_2F_1\left(\left.\begin{matrix} a+ix, b+ix \\ a+b \end{matrix}\right|\omega\right) {}_2F_1\left(\left.\begin{matrix} c-ix, d-ix \\ c+d \end{matrix}\right|\omega\right)$$
$$\times\left[{}_1F_1\left(\left.\begin{matrix} \alpha+iX \\ \alpha+\overline{\alpha} \end{matrix}\right|-i\omega\right) {}_1F_1\left(\left.\begin{matrix} \overline{\beta}-iX \\ \beta+\overline{\beta} \end{matrix}\right|i\omega\right)\right]^{-1}.$$

2.9.2 Asymptotic Property

$$c_k \frac{P_{n-k}(X;\alpha,\beta,\overline{\alpha},\overline{\beta})}{(\alpha+\overline{\alpha})_{n-k}(\beta+\overline{\beta})_{n-k}} = \mathcal{O}(a^{n+\lfloor k/3 \rfloor - k}) \quad \text{as} \quad a\to\infty,\ a\sim b\sim c\sim d. \tag{22}$$

2.10 Wilson to Hahn

2.10.1 Asymptotic Expansion for Large *a*, *b*, *c* and *d*

$$\frac{W_n(x^2;a,b,c,d)}{(a+b)_n(c+d)_n n!} = \sum_{k=0}^{n} c_k \frac{(-N)_{n-k}}{(B+1)_{n-k}(n-k)!} Q_{n-k}(X;A,B,N), \tag{23}$$

where $A = \tilde{A}a$, $B = \tilde{B}a$ with $\tilde{A}$ and $\tilde{B}$ arbitrary constants,

$$N = -\frac{(A+B+3)^2 - 1 + 2p_1(x)[B(B+3) - A(A+3)] + (A+B+2)\sqrt{D}}{2(A+B+4)},$$

$$X = -(A+1)\frac{A+B+4-2p_1(x)(A+2)+\sqrt{D}}{2(A+B+4)},$$

with

$$\begin{aligned} D = {} & 2A[4+20p_1^2(x) + B(1+14p_1^2(x) - 32p_2(x)) \\ & +2B^2(p_1^2(x) - 2p_2(x)) - 48p_2(x)] + B^2[1+8p_1^2(x) - 16p_2(x)] \\ & +8B[1+5p_1^2(x) - 12p_2(x)] + 16[1+3p_1^2(x) - 8p_2(x)] \\ & +A^2[1+4(2+B)p_1^2(x) - 8(2+B)p_2(x)], \end{aligned}$$

$p_1(x)$ and $p_2(x)$ are given in (4), and c_k are the coefficients of the Maclaurin expansion of

$$f(x,\omega) = {}_2F_1\left(\begin{matrix} a+ix, b+ix \\ a+b \end{matrix}\middle|\omega\right) {}_2F_1\left(\begin{matrix} c-ix, d-ix \\ c+d \end{matrix}\middle|\omega\right) \times\left[{}_1F_1\left(\begin{matrix} -X \\ A+1 \end{matrix}\middle|-\omega\right) {}_1F_1\left(\begin{matrix} X-N \\ B+1 \end{matrix}\middle|\omega\right)\right]^{-1}.$$

2.10.2 Asymptotic Property

$$c_k \frac{(-N)_{n-k}}{(B+1)_{n-k}(n-k)!} Q_{n-k}(X;A,B,N) = \mathcal{O}(a^{n+\lfloor k/3\rfloor-k}) \text{ as } a\to\infty,\ a\sim b\sim c\sim d. \tag{24}$$

2.11 *Wilson to Dual Hahn*

2.11.1 Asymptotic Expansion for Large *a*, *b*, *c* and *d*

$$\frac{W_n(x^2;a,b,c,d)}{(a+b)_n(c+d)_n n!} = \sum_{k=0}^{n} c_k \frac{(-N)_{n-k}}{(n-k)!} R_{n-k}(\lambda(X);A,B,N), \tag{25}$$

where $\lambda(X) = X(X+A+B+1)$, $A=\tilde{A}a$, $B=\tilde{B}a$ with $\tilde{A}$ and $\tilde{B}$ arbitrary constants,

$$N = -\frac{B+2p_1(x)+\sqrt{B^2+4(2+A)(p_1^2(x)+p_1(x)-2p_2(x))}}{2},$$

$$X = -\frac{A+B+1}{2} - \frac{\sqrt{(A+1)^2+B^2-2(A+1)\sqrt{B^2+4(2+A)(p_1^2(x)+p_1(x)-2p_2(x))}}}{2},$$

$p_1(x)$ and $p_2(x)$ are given in (4), and c_k are the coefficients of the Maclaurin expansion of

$$f(x,\omega) = (1-\omega)^{X-N}\,{}_2F_1\left(\left.\begin{matrix} a+ix, b+ix \\ a+b \end{matrix}\right|\omega\right)\,{}_2F_1\left(\left.\begin{matrix} c-ix, d-ix \\ c+d \end{matrix}\right|\omega\right) \times\left[{}_2F_1\left(\left.\begin{matrix} -X, -X-B \\ A+1 \end{matrix}\right|\omega\right)\right]^{-1}.$$

2.11.2 Asymptotic Property

$$c_k\frac{(-N)_{n-k}}{(n-k)!}R_{n-k}(\lambda(X);A,B,N) \;=\mathcal{O}(a^{n+\lfloor k/3\rfloor-k}) \quad \text{as} \quad a\to\infty,\ a\sim b\sim c\sim d. \tag{26}$$

3 Principles of the Asymptotic Approximations

The asymptotic expansions of polynomials in terms of polynomials listed above follow from an asymptotic principle based on the "matching" of their generating functions [9]. Some of these formulas have already been proved in previous works. The expansions in terms of the Hahn-type polynomials are new. We give below details for the case in which the basic approximants are the Continuous Dual Hahn polynomials and resume the main formulas for the remaining cases.

3.1 Expansions in Terms of Hermite Polynomials

To prove the results of Sect. 2.1, we need the following formulas derived in [9]. If $F(x,w)$ is the generating function of the polynomials $p_n(x)$, then

$$p_n(x) = B^n\sum_{k=0}^{n}\frac{c_k}{B^k}\frac{H_{n-k}(X)}{(n-k)!}, \tag{27}$$

where the coefficients c_k follow from

$$f(x,\omega) = \sum_{k=0}^{\infty}c_k\omega^k, \qquad f(x,\omega) = e^{-2BX\omega+B^2\omega^2}F(x,\omega).$$

The choice of X and B is based on our requirement that $c_1 = c_2 = 0$. This happens if we take

$$B = \sqrt{\frac{1}{2} p_1^2(x) - p_2(x)}, \qquad X = \frac{p_1(x)}{2B}$$

and we assume that $F(x,0) = p_0(x) = 1$ (which implies $c_0 = 1$).

The quantities X and B may depend on x, and if B turns out to be zero for a special $x-$value x_0, we write $p_n(x_0) = \sum_{k=0}^{n} \frac{c_k}{(n-k)!} p_1^{n-k}(x_0)$.

3.2 Expansions in Terms of Laguerre Polynomials

The results of Sect. 2.2 can be obtained from the formulas derived in [11]. If $F(x,w)$ is the generating function of the polynomials $p_n(x)$, then

$$p_n(x) = \sum_{k=0}^{n} c_k L_{n-k}^{(X)}(A), \tag{28}$$

where the coefficients c_k follow from

$$f(x,\omega) = \sum_{k=0}^{\infty} c_k \omega^k, \qquad f(x,\omega) = (1-\omega)^{X+1} e^{A\omega/(1-\omega)} F(x,\omega).$$

The choice of A and X is based on our requirement that $c_1 = c_2 = 0$. This happens if we take

$$A = p_1(x) + p_1^2(x) - 2p_2(x), \qquad X = A + p_1(x) - 1 \tag{29}$$

and we assume that $F(x,0) = p_0(x) = 1$ (which implies $c_0 = 1$).

3.3 Expansions in Terms of Charlier Polynomials

The following formulas derived in [11] provide a proof of the results of Sect. 2.3. If $F(x,w)$ is the generating function of the polynomials $p_n(x)$, then

$$p_n(x) = A^n \sum_{k=0}^{n} \frac{c_k}{A^k} \frac{C_{n-k}(X;A)}{(n-k)!}, \tag{30}$$

where the coefficients c_k follow from

$$f(x,\omega) = \sum_{k=0}^{\infty} c_k \omega^k, \qquad f(x,\omega) = (1-\omega)^{-X}\, e^{-A\omega} F(x,\omega).$$

The choice of A and X is based on our requirement that $c_1 = c_2 = 0$. This happens if we take

$$A = p_1(x) + p_1^2(x) - 2p_2(x), \qquad X = p_1^2(x) - 2p_2(x) \tag{31}$$

and we assume that $F(x,0) = p_0(x) = 1$ (which implies $c_0 = 1$). If A turns out to be zero for a special $x-$value x_0, then we write

$$p_n(x_0) = \sum_{k=0}^{n} c_k (-1)^{n-k} \binom{X}{n-k}.$$

3.4 Expansions in Terms of Meixner–Pollaczek Polynomials

To prove the results of Sect. 2.4, we need the following formulas derived in [3]. If $F(x,w)$ is the generating function of the polynomials $p_n(x)$, then

$$p_n(x) = \sum_{k=0}^{n} c_k\, P_{n-k}^{(C)}(X;A), \tag{32}$$

where $P_n^C(X;A)$ are the Meixner–Pollaczek polynomials, $A \neq m\pi$, $m \in \mathbf{Z}$ is an arbitrary constant, and the coefficients c_k follow from

$$f(x,\omega) = \sum_{k=0}^{\infty} c_k \omega^k$$

with

$$f(x,\omega) = (1-e^{iA}\omega)^{C-iX}(1-e^{-iA}\omega)^{C+iX} F(x,\omega),$$

$$c_1 = c_2 = 0, \qquad C = p_1(x)\cos A + \frac{1}{2} p_1(x)^2 - p_2(x),$$

$$X = \frac{-1}{2\sin A}\left[p_1(x)\cos 2A + (p_1(x)^2 - 2p_2(x))\cos A\right],$$

and we assume that $F(x,0) = p_0(x) = 1$ (which implies $c_0 = 1$).

3.5 Expansions in Terms of Jacobi Polynomials

To prove the results of Sect. 2.5, we need the following formulas derived in [3]. If $F(x,w)$ is the generating function of the polynomials $p_n(x)$, then

$$p_n(x) = \sum_{k=0}^{n} c_k P_{n-k}^{(A,C)}(X), \tag{33}$$

where $P_n^{A,C}(X)$ are the Jacobi polynomials, $X \neq \pm 1$ is an arbitrary constant, and the coefficients c_k follow from

$$f(x,\omega) = \sum_{k=0}^{\infty} c_k \omega^k, \quad \text{with} \quad f(x,\omega) = \frac{R(1+R-\omega)^A(1+R+\omega)^C}{2^{A+C}} F(x,\omega),$$

$$R = \sqrt{1-2X\omega+\omega^2}, \qquad c_1 = c_2 = 0,$$

$$A = \frac{1}{X+1}\left[2p_1(x)^2 + p_1(x) + 3p_1(x)X - 4p_2(x) + X^2 - X - 2\right],$$

$$C = \frac{1}{X-1}\left[2p_1(x)^2 - p_1(x) + 3p_1(x)X - 4p_2(x) + X^2 + X - 2\right],$$

and we assume that $F(x,0) = p_0(x) = 1$ (which implies $c_0 = 1$).

3.6 Expansions in Terms of Meixner Polynomials

To prove the results of Sect. 2.6, we need the following formulas derived in [3]. If $F(x,w)$ is the generating function of the polynomials $p_n(x)$, then

$$p_n(x) = \sum_{k=0}^{n} \frac{c_k (A)_{n-k}}{(n-k)!} M_{n-k}(X; A, C), \tag{34}$$

where $M_n(x; A, C)$ are the Meixner polynomials, $n \in \mathbf{N}$, $C \neq 0, 1$, and the coefficients c_k follow from

$$f(x,\omega) = \sum_{k=0}^{\infty} c_k \omega^k \quad \text{with} \quad f(x,\omega) = \left(1 - \frac{\omega}{C}\right)^{-X} (1-\omega)^{X+A} F(x,\omega),$$

where $C \neq 0, 1$ is an arbitrary constant. The choice of A and X is based on our requirement that $c_1 = c_2 = 0$. This happens if we take

$$A = (1+C)p_1(x)+Cp_1(x)^2-2Cp_2(x), \quad X = \frac{C^2}{1-C}\left[p_1(x)^2 + p_1(x) - 2p_2(x)\right]$$

and we assume that $F(x, 0) = p_0(x) = 1$ (which implies $c_0 = 1$).

3.7 Expansions in Terms of Krawtchouk Polynomials

To prove the results of Sect. 2.7, we need the following formulas derived in [3]. If $F(x, w)$ is the generating function of the polynomials $p_n(x)$, then

$$p_n(x) = \sum_{k=0}^{n} \binom{C}{n-k} c_k K_{n-k}(X; A, C), \tag{35}$$

where $K_n(X; A, C)$ are the Krawtchouk polynomials, $A \neq 0, 1$ is an arbitrary constant, and the coefficients c_k follow from

$$f(x, \omega) = \sum_{k=0}^{\infty} c_k \omega^k, \quad \text{with} \quad f(x, \omega) = \left(1 - \frac{1-A}{A}\omega\right)^{-X} (1+\omega)^{X-C} F(x, \omega),$$

$$X = \frac{A^2}{1-A}\left[p_1(x)^2 - p_1(x) - 2p_2(x)\right],$$

$$C = p_1(x) + \frac{A}{1-A}\left[p_1(x)^2 - p_1(x) - 2p_2(x)\right],$$

$c_1 = c_2 = 0$, and we assume that $F(x, 0) = p_0(x) = 1$ (which implies $c_0 = 1$).

3.8 Expansions in Terms of Continuous Dual Hahn Polynomials

The Continuous Dual Hahn polynomials $S_n(x^2; a, b, c)$ for $n \in \mathbf{N}$ follow from the generating function [7, (1.3.12)]

$$(1-\omega)^{-c+ix}\, {}_2F_1\left(\left. \begin{matrix} a+ix, b+ix \\ a+b \end{matrix} \right| \omega \right) = \sum_{n=0}^{\infty} \frac{S_n(x^2; a, b, c)}{(a+b)_n n!} \omega^n. \tag{36}$$

This formula gives the following Cauchy-type integral for the Continuous Dual Hahn polynomials

$$\frac{S_n(x^2;a,b,c)}{(a+b)_n n!} = \frac{1}{2\pi i}\int_{\mathcal{C}} (1-\omega)^{-c+ix}{}_2F_1\left(\begin{matrix} a+ix, b+ix \\ a+b \end{matrix}\,\middle|\, \omega\right)\frac{d\omega}{\omega^{n+1}}, \quad (37)$$

where $\mathcal{C}$ is a circle around the origin and the integration is in the positive direction.

All of the polynomials $p_n(x)$ of the Askey tableau have a generating function of the form

$$F(x,\omega) = \sum_{n=0}^{\infty} p_n(x)\omega^n, \quad (38)$$

where $F(x,\omega)$ is analytic with respect to ω in a domain that contains the origin $\omega = 0$ and $p_n(x)$ is independent of ω.

The relation (38) gives for $p_n(x)$ the Cauchy-type integral

$$p_n(x) = \frac{1}{2\pi i}\int_{\mathcal{C}} F(x,\omega)\frac{d\omega}{\omega^{n+1}},$$

where $\mathcal{C}$ is a circle around the origin inside the domain where $F(x,\omega)$ is analytic as a function of ω. We define $f(x,\omega)$ by means of the formula

$$F(x,\omega) = (1-\omega)^{-C+iX}{}_2F_1\left(\begin{matrix} A+iX, B+iX \\ A+B \end{matrix}\,\middle|\, \omega\right) f(x,\omega), \quad (39)$$

where C and X are arbitrary constants and A and B are free parameters that do not depend on ω. This gives

$$p_n(x) = \frac{1}{2\pi i}\int_{\mathcal{C}} (1-\omega)^{-C+iX}{}_2F_1\left(\begin{matrix} A+iX, B+iX \\ A+B \end{matrix}\,\middle|\, \omega\right)\frac{d\omega}{\omega^{n+1}}. \quad (40)$$

Since $f(x,\omega)$ is also analytic as a function of ω at $\omega = 0$, we can expand $f(x,\omega) = \sum_{k=0}^{\infty} c_k\omega^k$. Substituting this expansion in (40) and taking into account (37), we obtain

$$p_n(x) = \sum_{k=0}^{n} c_k \frac{S_n(X^2;A,B,C)}{(A+B)^{n-k}(n-k)!}. \quad (41)$$

The choice of C and X is based on our requirement that $c_1 = c_2 = 0$. This happens if we take

$$C = p_1(x) - \frac{A+B}{2} + \frac{1}{2}\sqrt{(A-B)^2 + 4(1+A+B)(p_1^2(x) + p_1(x) - 2p_2(x))},$$

$$X = \frac{1}{\sqrt{2}}\sqrt{-A^2 - B^2 + (A+B)\sqrt{(A-B)^2 + 4(1+A+B)(p_1^2(x) + p_1(x) - 2p_2(x))}},$$

and we assume that $F(x,0) = p_0(x) = 1$ (which implies $c_0 = 1$). The special choice of C and X is crucial for obtaining asymptotic properties.

The expansions in terms of the other three polynomials of the Hahn level can be deduced in a similar way. We only give the main formulas in the following three sections.

3.9 Expansions in Terms of Continuous Hahn Polynomials

To prove the results of Sect. 2.9, we need the following formulas. If $F(x,\omega)$ is the generating function of the polynomials $p_n(x)$, then

$$p_n(x) = \sum_{k=0}^{n} c_k \frac{P_{n-k}(X;\alpha,\beta,\overline{\alpha},\overline{\beta})}{(\alpha+\overline{\alpha})_{n-k}(\beta+\overline{\beta})_{n-k}}, \tag{42}$$

where $P_n(X;\alpha,\beta,\overline{\alpha},\overline{\beta})$ are the Continuous Hahn polynomials; $\alpha = A + iC$, $\beta = B + iD$ with $A, B, C, D \in \mathbf{R}$, B, D and X are arbitrary constants; and the coefficients c_k follow from

$$f(x,\omega) = \sum_{k=0}^{\infty} c_k \omega^k,$$

with

$$F(x,\omega) = {}_1F_1\left(\begin{matrix}\alpha + iX \\ \alpha + \overline{\alpha}\end{matrix}\middle| -i\omega\right) {}_1F_1\left(\begin{matrix}\overline{\beta} - iX \\ \beta + \overline{\beta}\end{matrix}\middle| i\omega\right) f(x,\omega),$$

$$\begin{aligned} A = {} & \left[B^3(1 + 8p_2(x)) + (D+X)^2 + B(D+X)(D - 2p_1(x) + X)\right. \\ & \left. + B^2(1 - 4(Dp_1(x) + p_2(x) + p_1(x)X)\right] \\ & / \left[B^2(-1 + 4p_1^2(x) - 8p_2(x)) + 8B^3(p_1^2(x) - 2p_2(x)) - (D+X)^2)\right], \end{aligned}$$

$$
\begin{aligned}
C = &-\big[B(D-4p_1(x))(D+X)^2+(D+X)^3-B^2(D+X)\\
&\times(-1+6Dp_1(x)-4p_1^2(x)-4p_2(x)+6p_1(x)X)\\
&+2B^4(-p_1(x)(1+8p_2(x))+4p_1^2(x)X-8p_2(x)X)\\
&+B^3(D(1+8p_1^2(x)+8p_2(x))+2p_1(x)(-1-4p_2(x)+6p_1(x)X))\big]\\
&/\big[B(B^2(-1+4p_1^2(x)-8p_2(x))+8B^3(p_1^2(x)-2p_2(x))-(D+X)^2))\big],
\end{aligned}
$$

$c_1 = c_2 = 0$ and we assume that $F(x,0) = p_0(x) = 1$ (which implies $c_0 = 1$).

3.10 Expansions in Terms of Hahn Polynomials

To prove the results of Sect. 2.10, we need the following formulas. If $F(x,\omega)$ is the generating function of the polynomials $p_n(x)$, then

$$
p_n(x) = \sum_{k=0}^{n} c_k \frac{(-N)_{n-k}}{(B+1)_{n-k}(n-k)!} Q_{n-k}(X;A,B,N), \tag{43}
$$

where $Q_n(X;A,B,N)$ are the Hahn polynomials, A and B are arbitrary constants, and the coefficients c_k follow from

$$
f(x,\omega) = \sum_{k=0}^{\infty} c_k \omega^k,
$$

with

$$
F(x,\omega) = {}_1F_1\left(\begin{matrix}-X\\A+1\end{matrix}\,\middle|\,-\omega\right){}_1F_1\left(\begin{matrix}X-N\\B+1\end{matrix}\,\middle|\,\omega\right)f(x,\omega),
$$

$$
N = -\frac{(A+B+3)^2-1+2p_1(x)(B(B+3)-A(A+3))+(A+B+2)\sqrt{D}}{2(A+B+4)}
$$

$$
X = -(A+1)\frac{A+B+4-2p_1(x)(A+2)+\sqrt{D}}{2(A+B+4)},
$$

with

$$
\begin{aligned}
D = &\,2A[4+20p_1^2(x)+B(1+14p_1^2(x)-32p_2(x))\\
&+2B^2(p_1^2(x)-2p_2(x))-48p_2(x)]+B^2[1+8p_1^2(x)-16p_2(x)]\\
&+8B[1+5p_1^2(x)-12p_2(x)]+16[1+3p_1^2(x)-8p_2(x)]\\
&+A^2[1+4(2+B)p_1^2(x)-8(2+B)p_2(x)],
\end{aligned}
$$

$c_1 = c_2 = 0$ and we assume that $F(x,0) = p_0(x) = 1$ (which implies $c_0 = 1$).

3.11 Expansions in Terms of Dual Hahn Polynomials

To prove the results of Sect. 2.11, we need the following formulas. If $F(x,\omega)$ is the generating function of the polynomials $p_n(x)$, then

$$p_n(x) = \sum_{k=0}^{n} c_k \frac{(-N)_{n-k}}{(n-k)!} R_{n-k}(\lambda(X); A, B, N) \tag{44}$$

where $\lambda(X) = X(X + A + B + 1)$ and $R_n(\lambda(X); A, B, N)$ are the Dual Hahn polynomials, A and B are arbitrary constants, and the coefficients c_k follow from

$$f(x,\omega)=\sum_{k=0}^{\infty} c_k\omega^k, \text{ with } F(x,\omega)=(1-\omega)^{N-X}{}_2F_1\left(\begin{matrix}-X,-X-B\\ A+1\end{matrix}\middle|\,\omega\right) f(x,\omega),$$

$$N = -\frac{B + 2p_1(x) + \sqrt{B^2 + 4(2 + A)(p_1^2(x) + p_1(x) - 2p_2(x))}}{2},$$

$$X = -\frac{A+B+1}{2}$$

$$-\frac{\sqrt{(A+1)^2+B^2-2(A+1)\sqrt{B^2+4(2+A)(p_1^2(x)+p_1(x)-2p_2(x))}}}{2},$$

$c_1 = c_2 = 0$ and we assume that $F(x,0) = p_0(x) = 1$ (which implies $c_0 = 1$).

3.12 Asymptotic Properties of the Coefficients c_k

The asymptotic nature of the expansions (2), (7), (9), (11), (13), (15), (17), (19), (21), (23), and (25) for large values of some of the parameters of the polynomial $p_n(x)$ depends on the asymptotic behaviour of the coefficients c_k. To prove the asymptotic character of the expansions given in Sect. 2, we will need the following lemma proved in [11]:

Lemma 3.1. *Let $\phi(\omega)$ be an analytic function at $\omega = 0$, with Maclaurin expansion of the form*

$$\phi(\omega) = \mu^s\omega^m(a_0 + a_1\omega + a_2\omega^2 + a_3\omega^3 + \ldots),$$

where m is a positive integer, s is an integer number, and a_k are complex numbers that satisfy $a_k = \mathcal{O}(1)$ when $\mu \to \infty$, $a_0 \neq 0$. Let c_k denote the coefficients of the power series of $f(\omega) = e^{\phi(\omega)}$, that is,

$$f(\omega) = e^{\phi(\omega)} = \sum_{k=0}^{\infty} c_k\omega^k.$$

Then $c_0 = 1$, $c_k = 0$, $k = 1, 2, \ldots, m-1$ *and* $c_k = \mathcal{O}\left(\mu^{\lfloor sk/m \rfloor}\right)$ *if* $s > 0$, $c_k = \mathcal{O}(\mu^s)$ *if* $s \leq 0$ *when* $\mu \to \infty$.

To obtain the asymptotic character of the function $\phi(\omega)$ in the different cases, it will be useful to consider the following results.

Lemma 3.2. *The function* $y(\omega) = \log {}_2F_1\left(\begin{matrix} a, b \\ c \end{matrix} \middle| \omega\right)$ *satisfies the following differential equation:*

$$\omega(1-\omega)[y''(\omega) + y'^2(\omega)] + [c - (a+b+1)\omega]\, y'(\omega) - ab = 0. \tag{45}$$

The coefficients of the Maclaurin series of the function $y(\omega) = \sum_{k=1}^{\infty} y_k \omega^k$ *verify*

$$y_1 = \frac{ab}{c}, \qquad y_2 = ab\frac{c(a+b+1) - ab}{2c^2(c+1)}$$

and for $k \geq 2$

$$y_{k+1} = \frac{1}{(k+1)(k+c)}\Bigg\{k(a+b+k)y_k - k y_k y_1$$
$$+ \sum_{j=0}^{k-2} (j+1) y_{j+1}[(k-j-1)y_{k-j-1} - (k-j)y_{k-j}]\Bigg\}.$$

Proof. Equation (45) follows from the differential equation satisfied by the hypergeometric function $z(\omega) = {}_2F_1\left(\begin{matrix} a, b \\ c \end{matrix} \middle| \omega\right)$:

$$\omega(1-\omega)z''(\omega) + [c - (a+b+1)\omega]\, z'(\omega) - ab\, z(\omega) = 0.$$

The coefficients y_k of the Maclaurin expansion are obtained by substituting the expansion $y(\omega) = \sum_{k=1}^{\infty} y_k \omega^k$ into (45) and identifying the coefficients of ω^k for $k \geq 1$. □

Lemma 3.3. *The function* $y(\omega) = \log {}_1F_1\left(\begin{matrix} a \\ b \end{matrix} \middle| \omega\right)$ *satisfies the following differential equation:*

$$\omega[y''(\omega) + y'^2(\omega)] + (b - \omega)y'(\omega) - a = 0. \tag{46}$$

The coefficients of the Maclaurin series of the function $y(\omega) = \sum_{k=1}^{\infty} y_k \omega^k$ *verify*

$$y_1 = \frac{a}{b}, \qquad y_2 = \frac{ab - a^2}{2b^2(1+b)}$$

and for $k \geq 2$

$$y_{k+1} = \frac{1}{(k+1)(k+b)} \left\{ k y_k - k y_k y_1 - \sum_{j=0}^{k-2} (j+1)(k-j) y_{j+1} y_{k-j} \right\}. \quad (47)$$

Proof. Equation (46) follows from the differential equation satisfied by the hypergeometric function $z(\omega) = {}_1F_1\left(\begin{matrix} a \\ b \end{matrix} \middle| \omega\right)$:

$$\omega z''(\omega) + (b - \omega) z'(\omega) - a\, z(\omega) = 0.$$

The coefficients y_k of the Maclaurin expansion are obtained by substituting the expansion $y(\omega) = \sum_{k=1}^{\infty} y_k \omega^k$ into (46) and identifying the coefficients of ω^k for $k \geq 1$. □

4 Proofs of Formulas Given in Sect. 2

4.1 Proofs of Formulas in Sect. 2.1

Substitute

$$F(x, \omega) = {}_2F_1\left(\begin{matrix} a + ix, b + ix \\ a + b \end{matrix} \middle| \omega\right) {}_2F_1\left(\begin{matrix} c - ix, d - ix \\ c + d \end{matrix} \middle| \omega\right)$$

and

$$p_n(x) = \frac{W_n(x^2; a, b, c, d)}{(a+b)_n (c+d)_n n!}$$

in the formulas of Sect. 3.1 to obtain (2).

Applying Lemma 3.2 to the function $y(\omega) = \log {}_2F_1\left(\begin{matrix} a + ix, b + ix \\ a + b \end{matrix} \middle| \omega\right)$, we deduce that $y(\omega)$ has the following Maclaurin series $y(\omega) = \sum_{k=1}^{\infty} y_k \omega^k$ where

$$y_1 = \frac{(a+ix)(b+ix)}{a+b},$$

$$y_2 = (a+ix)(b+ix) \frac{(a+b)(a+b+2ix+1) - (a+ix)(b+ix)}{2(a+b)^2(a+b+1)},$$

and

$$y_{k+1} = \frac{1}{(k+1)(k+a+b)} \left\{ k(k+a+b+2ix)y_k - k y_k y_1 \right.$$
$$\left. + \sum_{j=0}^{k-2} (j+1) y_{j+1} \left[(k-j-1) y_{k-j-1} - (k-j) y_{k-j} \right] \right\}.$$

Then, $y_1 = \mathcal{O}(a)$, $y_2 = \mathcal{O}(a)$ and using the above Recurrence we can show by induction over k that $y_k = \mathcal{O}(a)$ for $k > 2$. The same lemma can be applied to the function $z(\omega) = \log {}_2F_1\left(\begin{matrix} c-ix, d-ix \\ c+d \end{matrix} \middle| \omega \right)$. For this function, we deduce that $z(\omega) = \sum_{k=1}^{\infty} z_k \omega^k$ with $z_1 = \mathcal{O}(c)$, $z_2 = \mathcal{O}(c)$, and $z_k = \mathcal{O}(c)$ for $k > 2$. On the other hand, we have $B = \mathcal{O}(\sqrt{a})$ and $X = \mathcal{O}(\sqrt{a})$. Therefore, the function $\phi(\omega) = \log f(x, \omega)$ verifies Lemma 3.1 with $\mu = a$, $s = 1$ and $m = 3$, and we have $c_k = \mathcal{O}(a^{\lfloor k/3 \rfloor})$ for $a \to \infty$, $a \sim b \sim c \sim d$. On the other hand, $H_{n-k}(X) = \mathcal{O}(a^{\frac{n-k}{2}})$, and we obtain (5) in Sect. 2.1.2. The limit (6) follows from the first term of the expansion (2) after solving (3) for $x(X)$.

4.2 Proofs of Formulas in Sect. 2.2

Substitute

$$F(x,\omega) = {}_2F_1\left(\begin{matrix} a+ix, b+ix \\ a+b \end{matrix} \middle| \omega \right) {}_2F_1\left(\begin{matrix} c-ix, d-ix \\ c+d \end{matrix} \middle| \omega \right)$$

and

$$p_n(x) = \frac{W_n(x^2; a, b, c, d)}{(a+b)_n (c+d)_n n!}$$

in the formulas of Sect. 3.2 to obtain (7).

Using the result obtained in Sect. 4.1 for $\log F(x, \omega)$ and taking into account that $X = \mathcal{O}(a)$ and $A = \mathcal{O}(a)$, it is easy to check that the function $\phi(\omega) = \log f(x, \omega)$ verifies Lemma 3.1 with $\mu = a$, $s = 1$ and $m = 3$. Thus, we have $c_k = \mathcal{O}(a^{\lfloor k/3 \rfloor})$ for $a \to \infty$, $a \sim b \sim c \sim d$. On the other hand, taking into account that $\lim_{a\to\infty} A/X \neq 1$, we have $L_{n-k}^{(X)}(A) = \mathcal{O}(a^{n-k})$ and we obtain the asymptotic behaviour (8) in Sect. 2.2.2.

4.3 Proofs of Formulas in Sect. 2.3

Substitute

$$F(x,\omega) = {}_2F_1\left(\left.\begin{matrix} a+ix, b+ix \\ a+b \end{matrix}\right| \omega\right) {}_2F_1\left(\left.\begin{matrix} c-ix, d-ix \\ c+d \end{matrix}\right| \omega\right)$$

and

$$p_n(x) = \frac{W_n(x^2; a,b,c,d)}{(a+b)_n(c+d)_n n!}$$

in the formulas of Sect. 3.3 to obtain (9).

Using the result obtained in Sect. 4.1 for $\log F(x,\omega)$ and taking into account that $X = \mathcal{O}(a)$ and $A = \mathcal{O}(a)$, it is easy to check that the function $\phi(\omega) = \log f(x,\omega)$ verifies Lemma 3.1 with $\mu = a$, $s = 1$ and $m = 3$. Thus, we have $c_k = \mathcal{O}(a^{\lfloor k/3 \rfloor})$ for $a \to \infty$, $a \sim b \sim c \sim d$. On the other hand,

$$C_0(X;A) = \mathcal{O}(a^0), \quad C_1(X;A) = \frac{A-X}{A} = \frac{p_1(x)}{A} = \mathcal{O}(a^0),$$

and by induction over the recurrence relation [7]

$$AC_{n+1}(X;A) + (X-A-n)C_n(X;A) + nC_{n-1}(X;A) = 0,$$

we have $C_{n-k}(X;A) = \mathcal{O}(a^0)$ and we obtain the asymptotic behaviour (10) in Sect. 2.3.2.

4.4 Proofs of Formulas in Sect. 2.4

Substitute

$$F(x,\omega) = {}_2F_1\left(\left.\begin{matrix} a+ix, b+ix \\ a+b \end{matrix}\right| \omega\right) {}_2F_1\left(\left.\begin{matrix} c-ix, d-ix \\ c+d \end{matrix}\right| \omega\right)$$

and

$$p_n(x) = \frac{W_n(x^2; a,b,c,d)}{(a+b)_n(c+d)_n n!}$$

in the formulas of Sect. 3.4 to obtain (11).

Using the result obtained in Sect. 4.1 for $\log F(x,\omega)$ and taking into account that $C = \mathcal{O}(a)$ and $X = \mathcal{O}(a)$, it is easy to check that the function $\phi(\omega) = \log f(x,\omega)$

verifies Lemma 3.1 with $\mu = a$, $s = 1$ and $m = 3$. Thus, we have $c_k = \mathcal{O}(a^{\lfloor k/3 \rfloor})$ for $a \to \infty$, $a \sim b \sim c \sim d$. On the other hand,

$$P_0^{(C)}(X; A) = \mathcal{O}(a^0), \quad P_1^{(C)}(X; A) = 2C \cos A + 2X \sin A = \mathcal{O}(a),$$

and by induction over the recurrence relation [7]

$$(n+1)P_{n+1}^{(C)}(X; A) - 2[X \sin A + (n+C) \cos A] P_n^{(C)}(X; A) + (n+2C-1) P_{n-1}^{(C)}(X; A) = 0,$$

we have $P_{n-k}^{(C)}(X; A) = \mathcal{O}(a^{n-k})$ and we obtain the asymptotic behaviour (12) in Sect. 2.4.2.

4.5 Proofs of Formulas in Sect. 2.5

Substitute

$$F(x, \omega) = {}_2F_1\left(\left. \begin{matrix} a + ix, b + ix \\ a + b \end{matrix} \right| \omega \right) {}_2F_1\left(\left. \begin{matrix} c - ix, d - ix \\ c + d \end{matrix} \right| \omega \right)$$

and

$$p_n(x) = \frac{W_n(x^2; a, b, c, d)}{(a+b)_n (c+d)_n n!}$$

in the formulas of Sect. 3.5 to obtain (13).

Using the result obtained in Sect. 4.1 for $\log F(x, \omega)$ and taking into account that $A = \mathcal{O}(a)$ and $C = \mathcal{O}(a)$, it is easy to check that the function $\phi(\omega) = \log f(x, \omega)$ verifies Lemma 3.1 with $\mu = a$, $s = 1$ and $m = 3$. Thus, we have $c_k = \mathcal{O}(a^{\lfloor k/3 \rfloor})$ for $a \to \infty$, $a \sim b \sim c \sim d$. On the other hand,

$$P_0^{(A,C)}(X) = \mathcal{O}(a^0), \quad P_1^{(A,C)}(X) = \frac{1}{2}((2 + A + C)X + (C - A)) = \mathcal{O}(a),$$

and by induction over the recurrence relation [7]

$$\begin{aligned} X P_n^{(A,C)}(X) &= \frac{2(n+1)(n+A+C+1)}{(2n+A+C+1)(2n+A+C+2)} P_{n+1}^{(A,C)}(X) \\ &+ \frac{C^2 - A^2}{(2n+A+C)(2n+A+C+2)} P_n^{(A,C)}(X) \\ &+ \frac{2(n+A)(n+C)}{(2n+A+C)(2n+A+C+1)} P_{n-1}^{(A,C)}(X), \end{aligned}$$

we have $P_{n-k}^{(A,C)}(X) = \mathcal{O}(a^{n-k})$ and we obtain the asymptotic behaviour (14) in Sect. 2.5.2.

4.6 Proofs of Formulas in Sect. 2.6

Substitute

$$F(x,\omega) = {}_2F_1\left(\left.\begin{matrix} a+ix, b+ix \\ a+b \end{matrix}\right| \omega\right) {}_2F_1\left(\left.\begin{matrix} c-ix, d-ix \\ c+d \end{matrix}\right| \omega\right)$$

and

$$p_n(x) = \frac{W_n(x^2; a, b, c, d)}{(a+b)_n (c+d)_n n!}$$

in the formulas of Sect. 3.6 to obtain (15).

Using the result obtained in Sect. 4.1 for $\log F(x,\omega)$ and taking into account that $A = \mathcal{O}(a)$ and $X = \mathcal{O}(a)$, it is easy to check that the function $\phi(\omega) = \log f(x,\omega)$ verifies Lemma 3.1 with $\mu = a$, $s = 1$ and $m = 3$. Thus, we have $c_k = \mathcal{O}(a^{\lfloor k/3 \rfloor})$ for $a \to \infty$, $a \sim b \sim c \sim d$. On the other hand,

$$M_0(X; A, C) = \mathcal{O}(a^0), \quad M_1(X; A, C) = \frac{AC + (C-1)X}{AC} = \mathcal{O}(a^0),$$

and by induction over the recurrence relation [7]

$$\begin{aligned} &C(n+A)M_{n+1}(X; A, C) \\ &\quad = [(C-1)X + (n + (n+A)C)]M_n(X; A, C) - nM_{n-1}(X; A, C), \end{aligned}$$

we have $M_{n-k}(X; A, C) = \mathcal{O}(a^0)$ and we obtain the asymptotic behaviour (16) in Sect. 2.6.2.

4.7 Proofs of Formulas in Sect. 2.7

Substitute

$$F(x,\omega) = {}_2F_1\left(\left.\begin{matrix} a+ix, b+ix \\ a+b \end{matrix}\right| \omega\right) {}_2F_1\left(\left.\begin{matrix} c-ix, d-ix \\ c+d \end{matrix}\right| \omega\right)$$

and

$$p_n(x) = \frac{W_n(x^2; a, b, c, d)}{(a+b)_n(c+d)_n n!}$$

in the formulas of Sect. 3.7 to obtain (17).

Using the result obtained in Sect. 4.1 for $\log F(x, \omega)$ and taking into account that $C = \mathcal{O}(a)$ and $X = \mathcal{O}(a)$, it is easy to check that the function $\phi(\omega) = \log f(x, \omega)$ verifies Lemma 3.1 with $\mu = a$, $s = 1$ and $m = 3$. Thus, we have $c_k = \mathcal{O}(a^{\lfloor k/3 \rfloor})$ for $a \to \infty$, $a \sim b \sim c \sim d$. On the other hand,

$$K_0(X; A, C) = \mathcal{O}(a^0), \quad K_1(X; A, C) = \frac{AC - X}{AC} = \mathcal{O}(a^0),$$

and by induction over the recurrence relation [7]

$$A(C-n)K_{n+1}(X; A, C) = [A(C-n) + n(1-A) - X]K_n(X; A, C) \\ -n(1-A)K_{n-1}(X; A, C),$$

we have $K_{n-k}(X; A, C) = \mathcal{O}(a^0)$ and we obtain the asymptotic behaviour (18) in Sect. 2.7.2.

4.8 *Proofs of Formulas in Sect. 2.8*

Substitute

$$F(x, \omega) = {}_2F_1\left(\left.\begin{matrix} a+ix, b+ix \\ a+b \end{matrix}\right| \omega\right) {}_2F_1\left(\left.\begin{matrix} c-ix, d-ix \\ c+d \end{matrix}\right| \omega\right)$$

and

$$p_n(x) = \frac{W_n(x^2; a, b, c, d)}{(a+b)_n(c+d)_n n!}$$

in the formulas of Sect. 3.8 to obtain (19).

In this case, $A = \mathcal{O}(a)$, $B = \mathcal{O}(a)$, $C = \mathcal{O}(a)$ and $X = \mathcal{O}(a)$. Lemma 3.2 can be applied to the function

$$y(\omega) = \log {}_2F_1\left(\left.\begin{matrix} A+iX, B+iX \\ A+B \end{matrix}\right| \omega\right)$$

to deduce that $y(\omega) = \sum_{k=1}^{\infty} y_k \omega^k$ with $y_k = \mathcal{O}(a)$ for $k \geq 1$. Using this result and the behaviour obtained in Sect. 4.1 for $\log F(x, \omega)$, it is easy to check that the function $\phi(\omega) = \log f(x, \omega)$ verifies Lemma 3.1 with $\mu = a$, $s = 1$ and $m = 3$. Thus, we have $c_k = \mathcal{O}(a^{\lfloor k/3 \rfloor})$ for $a \to \infty$, $a \sim b \sim c \sim d$. On the other hand,

$$\tilde{S}_0(X^2; A, B, C) = \mathcal{O}(a^0), \quad \tilde{S}_1(X^2; A, B, C) = \frac{BC + A(B + C) - X^2}{(A + B)(A + C)} = \mathcal{O}(a^0),$$

and by induction over the recurrence relation [7]

$$\begin{aligned} A_n \tilde{S}_{n+1}(X^2; A, B, C) = &\left[(A_n + C_n) - (A^2 + X^2)\right] \tilde{S}_n(X^2; A, B, C) \\ &- C_n \tilde{S}_{n-1}(X^2; A, B, C), \end{aligned}$$

with

$$\tilde{S}_n(X^2; A, B, C) = \frac{S_n(X^2; A, B, C)}{(A + B)_n (A + C)_n},$$
$$A_n = (n + A + B)(n + A + C), \quad C_n = n(n + B + C - 1),$$

we have $\tilde{S}_n(X^2; A, B, C) = \mathcal{O}(a^0)$ and $S_{n-k}(X^2; A, B, C) = \mathcal{O}(a^{2n-2k})$, and we finally obtain the asymptotic behaviour (20) in Sect. 2.8.2.

4.9 Proofs of Formulas in Sect. 2.9

Substitute

$$F(x, \omega) = {}_2F_1\left(\left.\begin{matrix} a + ix, b + ix \\ a + b \end{matrix}\right| \omega\right) {}_2F_1\left(\left.\begin{matrix} c - ix, d - ix \\ c + d \end{matrix}\right| \omega\right)$$

and

$$p_n(x) = \frac{W_n(x^2; a, b, c, d)}{(a + b)_n (c + d)_n n!}$$

in the formulas of Sect. 3.9 to obtain (21).

In this case, $A = \mathcal{O}(a)$, $B = \mathcal{O}(a)$, $C = \mathcal{O}(a^2)$, and $X = \mathcal{O}(a^2)$. Lemma 3.3 can be applied to the functions $y(\omega) = \log {}_1F_1\left(\left.\begin{matrix} \alpha + iX \\ \alpha + \overline{\alpha} \end{matrix}\right| -i\omega\right)$ and $z(\omega) = \log {}_1F_1\left(\left.\begin{matrix} \overline{\beta} - iX \\ \beta + \overline{\beta} \end{matrix}\right| i\omega\right)$ to deduce that $y(\omega) = \sum_{k=1}^{\infty} y_k \omega^k$ and $z(\omega) = \sum_{k=1}^{\infty} z_k \omega^k$ with $y_k = \mathcal{O}(a)$ for $k \geq 1$ and $z_k = \mathcal{O}(a)$ for $k \geq 1$. Using this result and the behaviour obtained in Sect. 4.1 for $\log F(x,\omega)$, it is easy to check that the function $\phi(\omega) = \log f(x,\omega)$ verifies Lemma 3.1 with $\mu = a$, $s = 1$ and $m = 3$. Thus, we have $c_k = \mathcal{O}(a^{\lfloor k/3 \rfloor})$ for $a \to \infty$, $a \sim b \sim c \sim d$. On the other hand,

$$\tilde{P}_0(X;\alpha,\beta,\overline{\alpha},\overline{\beta}) = \mathcal{O}(a^0),$$

$$\tilde{P}_1(X;\alpha,\beta,\overline{\alpha},\overline{\beta}) = 1 - \frac{(\alpha+\overline{\alpha}+\beta+\overline{\beta})(\alpha+iX)}{(\alpha+\overline{\alpha})(\alpha+\overline{\beta})} = \mathcal{O}(a^0),$$

and by induction over the recurrence relation [7]

$$A_n\tilde{P}_{n+1}(X;\alpha,\beta,\overline{\alpha},\overline{\beta}) = (A_n + C_n - \alpha - iX)\tilde{P}_n(X;\alpha,\beta,\overline{\alpha},\overline{\beta})$$
$$-C_n\tilde{P}_{n-1}(X;\alpha,\beta,\overline{\alpha},\overline{\beta}),$$

with

$$\tilde{P}_n(X;\alpha,\beta,\overline{\alpha},\overline{\beta}) = \frac{n!}{i^n(\alpha+\overline{\alpha})_n(\alpha+\overline{\beta})_n} P_n(X;\alpha,\beta,\overline{\alpha},\overline{\beta}),$$

$$A_n = -\frac{(n+\alpha+\overline{\alpha}+\beta+\overline{\beta}-1)(n+\alpha+\overline{\alpha})(n+\alpha+\overline{\beta})}{(2n+\alpha+\overline{\alpha}+\beta+\overline{\beta}-1)(2n+\alpha+\overline{\alpha}+\beta+\overline{\beta})},$$

$$C_n = \frac{n(n+\overline{\alpha}+\beta-1)(n+\beta+\overline{\beta}-1)}{(2n+\alpha+\overline{\alpha}+\beta+\overline{\beta}-2)(2n+\alpha+\overline{\alpha}+\beta+\overline{\beta}-1)},$$

we have $\tilde{P}_n(X;\alpha,\beta,\overline{\alpha},\overline{\beta}) = \mathcal{O}(a^0)$ and $P_{n-k}(X;\alpha,\beta,\overline{\alpha},\overline{\beta}) = \mathcal{O}(a^{3n-3k})$, and we obtain the asymptotic behaviour (22) in Sect. 2.9.2.

4.10 Proofs of Formulas in Sect. 2.10

Substitute

$$F(x,\omega) = {}_2F_1\left(\left.\begin{matrix} a+ix, b+ix \\ a+b \end{matrix}\right| \omega\right) {}_2F_1\left(\left.\begin{matrix} c-ix, d-ix \\ c+d \end{matrix}\right| \omega\right)$$

and

$$p_n(x) = \frac{W_n(x^2;a,b,c,d)}{(a+b)_n(c+d)_n n!}$$

in the formulas of Sect. 3.10 to obtain (23).

In this case, $A = \mathcal{O}(a)$, $B = \mathcal{O}(a)$, $X = \mathcal{O}(a^2)$, and $N = \mathcal{O}(a^2)$. Lemma 3.3 can be applied to the functions $y(\omega) = \log {}_1F_1\left(\begin{matrix}-X\\A+1\end{matrix}\middle| -\omega\right)$ and $z(\omega) = \log {}_1F_1\left(\begin{matrix}X-N\\B+1\end{matrix}\middle| \omega\right)$ to deduce that $y(\omega) = \sum_{k=1}^{\infty} y_k\omega^k$ and $z(\omega) = \sum_{k=1}^{\infty} z_k\omega^k$ with $y_k = \mathcal{O}(a)$ for $k \geq 1$ and $z_k = \mathcal{O}(a)$ for $k \geq 1$. Using this result and the behaviour obtained in 4.1 for $\log F(x,\omega)$, it is easy to check that the function $\phi(\omega) = \log f(x,\omega)$ verifies Lemma 3.1 with $\mu = a$, $s = 1$ and $m = 3$. Thus, we have $c_k = \mathcal{O}(a^{\lfloor k/3 \rfloor})$ for $a \to \infty$, $a \sim b \sim c \sim d$. On the other hand,

$$Q_0(X; A, B, N) = \mathcal{O}(a^0), \quad Q_1(X; A, B, N) = 1 - \frac{(2+A+B)X}{(1+A)N} = \mathcal{O}(a^0),$$

and by induction over the recurrence relation [7]

$$A_n Q_{n+1}(X; A, B, N) = (A_n + C_n - X)Q_n(X; A, B, N) - C_n Q_{n-1}(X; A, B, N),$$

with

$$A_n = \frac{(n+A+B+1)(n+A+1)(N-n)}{(2n+A+B+1)(2n+A+B+2)},$$
$$C_n = \frac{n(n+A+B+N+1)(n+B)}{(2n+A+B)(2n+A+B+1)},$$

we have $Q_{n-k}(X; A, B, N) = \mathcal{O}(a^0)$ and we obtain the asymptotic behaviour (24) in Sect. 2.10.2.

4.11 Proofs of Formulas in Sect. 2.11

Substitute

$$F(x,\omega) = {}_2F_1\left(\begin{matrix}a+ix, b+ix\\a+b\end{matrix}\middle|\omega\right) {}_2F_1\left(\begin{matrix}c-ix, d-ix\\c+d\end{matrix}\middle|\omega\right)$$

and

$$p_n(x) = \frac{W_n(x^2; a, b, c, d)}{(a+b)_n(c+d)_n n!}$$

in the formulas of Sect. 3.11 to obtain (25).

In this case, $A = \mathcal{O}(a)$, $B = \mathcal{O}(a)$, $X = \mathcal{O}(a)$ and $N = \mathcal{O}(a)$. Lemma 3.2 can be applied to the function $y(\omega) = \log {}_2F_1\left(\begin{matrix}-X, -X-B\\ A+1\end{matrix}\middle|\,\omega\right)$ to deduce that $y(\omega) = \sum_{k=1}^{\infty} y_k \omega^k$ with $y_k = \mathcal{O}(a)$ for $k \geq 1$. Using this result and the behaviour obtained in Sect. 4.1 for $\log F(x, \omega)$, it is easy to check that the function $\phi(\omega) = \log f(x, \omega)$ verifies Lemma 3.1 with $\mu = a$, $s = 1$ and $m = 3$. Thus, we have $c_k = \mathcal{O}(a^{\lfloor k/3 \rfloor})$ for $a \to \infty$, $a \sim b \sim c \sim d$. On the other hand,

$$R_0(\lambda(X); A, B, N) = \mathcal{O}(a^0),$$

$$R_1(\lambda(X); A, B, N) = \frac{N + AN - X - AX - BX - X^2}{(1 + A)N} = \mathcal{O}(a^0),$$

and by induction over the recurrence relation [7]

$$\begin{aligned} A_n R_{n+1}(\lambda(X); A, B, N) &= [\lambda(X) - A_n - C_n]\, R_n(\lambda(X); A, B, N) \\ &\quad - C_n R_{n-1}(\lambda(X); A, B, N), \end{aligned}$$

with

$$A_n = (n + A + 1)(n - N), \quad C_n = n(n - B - N - 1),$$

we have $R_{n-k}(\lambda(X); A, B, N) = \mathcal{O}(a^0)$ and we obtain the asymptotic behaviour (26) in Sect. 2.11.2.

5 Numerical Experiments

The following graphics illustrate the approximation supplied by some of the expansions given in Sect. 2. It is worthwhile to note the accuracy obtained in the approximation of the zeros of the polynomials. In all the graphics, the degree of the polynomials is $n = 5$, continuous lines represent the exact polynomial and dashed lines represent the first order approximation given by the corresponding expansion (Figs. 5, 6, 7, and 8).

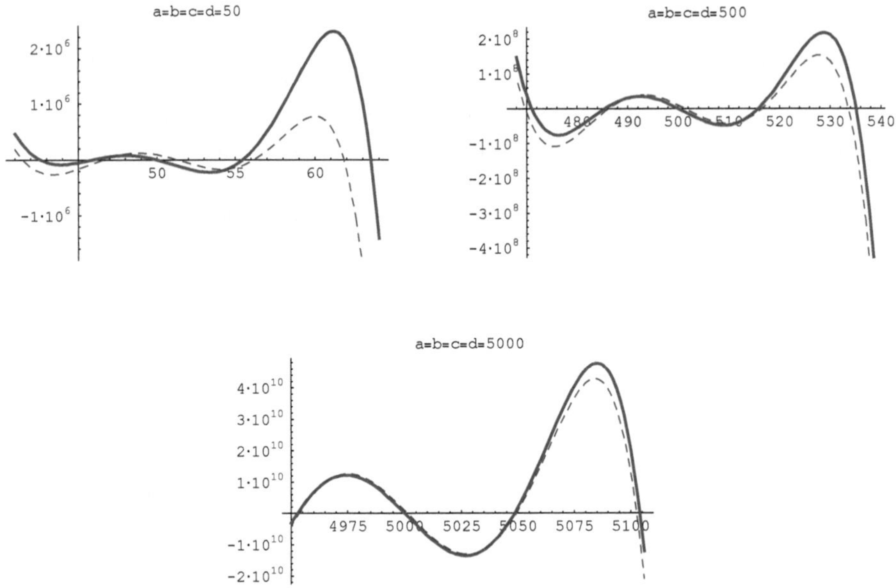

Fig. 5 Expansion (2): $W_5(x^2; a, b, c, d)/[(a+b)_5(c+d)_5]$ versus $B^5 H_5(X)$

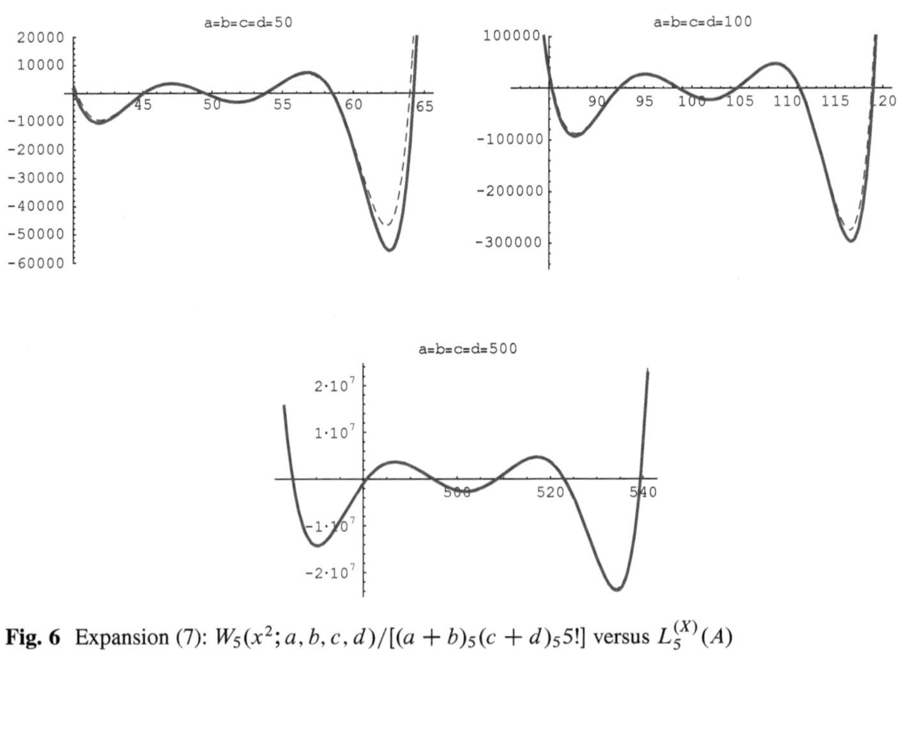

Fig. 6 Expansion (7): $W_5(x^2; a, b, c, d)/[(a+b)_5(c+d)_5 5!]$ versus $L_5^{(X)}(A)$

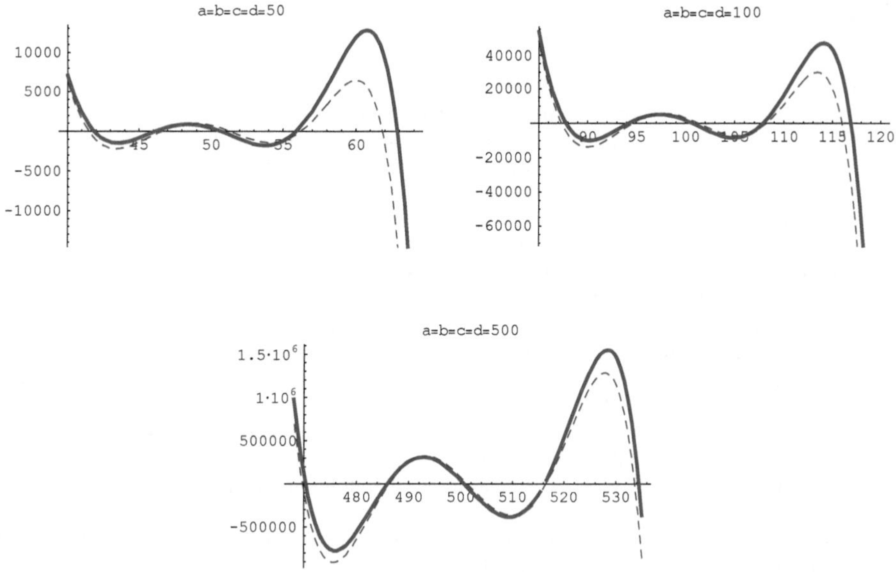

Fig. 7 Expansion (11): $W_5(x^2; a, b, c, d)/[(a+b)_5(c+d)_5 5!]$ versus $P_5^{(C)}(X; \pi/3)$

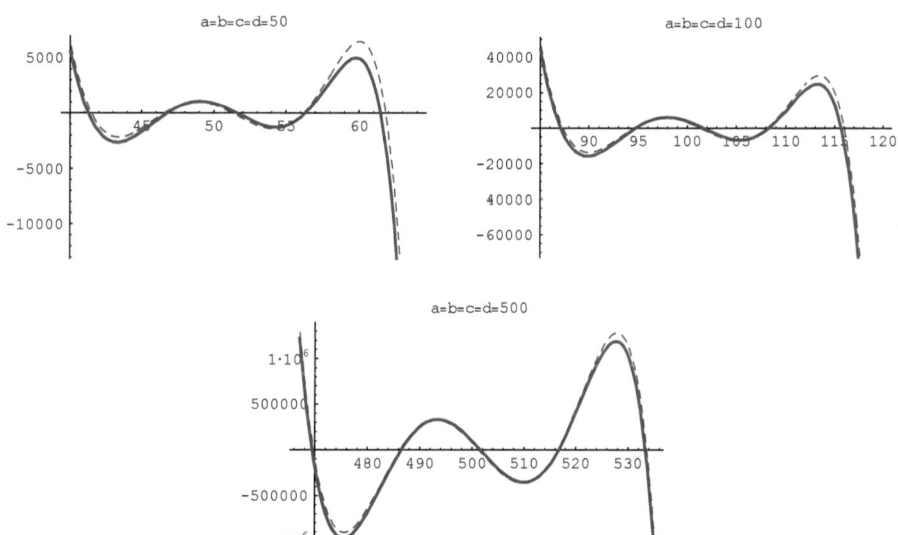

Fig. 8 Expansion (19): $W_5(x^2; a, b, c, d)/[(a+b)_5(c+d)_5 5!]$ versus $S_5(X^2; A, B, C)/[(A+B)_5 5!]$

References

1. Ferreira, C., López, J.L., Mainar, E.: Asymptotic relations in the Askey scheme for hypergeometric orthogonal polynomials. Adv. Appl. Math. **31**(1), 61–85 (2003)
2. Ferreira, C., López, J.L., Pagola, P.: Limit relations between the Hahn polynomials and the Hermite, Laguerre and Charlier polynomials. Acta Appl. Math. **103**(3), 235–252 (2008)
3. Ferreira, C., López, J.L., Pérez Sinusía, E.: Asymptotic relations between the Hahn-type polynomials and Meixner–Pollazcek, Jacobi, Meixner and Krawtchouk polynomials. J. Comput. Appl. Math. **217**(1), 88–109 (2008)
4. Frenzen, C.L., Wong, R.: Uniform asymptotic expansions of Laguerre polynomials. SIAM J. Math. Anal. **19**, 1232–1248 (1998)
5. Godoy, E., Ronveaux, A., Zarzo, A., Área, I.: On the limit relations between classical continuous and discrete orthogonal polynomials. J. Comput. Appl. Math. **91**, 97–105 (1998)
6. Godoy, E., Ronveaux, A., Zarzo, A., Área, I.: Transverse limits in the Askey tableau. J. Comput. Appl. Math. **99**, 327–335 (1998)
7. Koekoek, R., Swarttouw, R.F.: Askey scheme or hypergeometric orthogonal polynomials (1999). http://aw.twi.tudelft.nl/koekoek/askey
8. Koornwinder, T.H.: The Askey scheme as a four-manifold with corners. Ramanujan J. **20**(3), 409–439 (2009)
9. López, J.L., Temme, N.M.: Approximations of orthogonal polynomials in terms of Hermite polynomials. Method. Appl. Anal. **6**, 131–146 (1999)
10. López, J.L., Temme, N.M.: Hermite polynomials in asymptotic representations of generalized Bernoulli, Euler, Bessel and Buchholz polynomials. J. Math. Anal. Appl. **239**, 457–477 (1999)
11. López, J.L., Temme, N.M.: The Askey scheme for hypergeometric orthogonal polynomials viewed from asymptotic analysis. J. Comput. Appl. Math. **133**, 623–633 (2001)
12. Rui, B., Wong, R.: Uniform asymptotic expansions of Charlier polynomials. Method. Appl. Anal. **1**, 294–313 (1994)

On a Direct Uvarov-Chihara Problem and Some Extensions

K. Castillo, L. Garza, and F. Marcellán

Dedicated to Professor Hari M. Srivastava

Abstract In this paper, we analyze a perturbation of a nontrivial probability measure $d\mu$ supported on an infinite subset on the real line, which consists on the addition of a time-dependent mass point. For the associated sequence of monic orthogonal polynomials, we study its dynamics with respect to the time parameter. In particular, we determine the time evolution of their zeros in the special case when the measure is semiclassical. We also study the dynamics of the Verblunsky coefficients, i.e., the recurrence relation coefficients of a polynomial sequence, orthogonal with respect to a nontrivial probability measure supported on the unit circle, induced from $d\mu$ through the Szegő transformation.

1 Introduction

Let us consider the classical mechanical problem of a 1-dimensional chain of particles with neighbor interactions. Assume that the system is homogeneous (contains no impurities) and that the mass of each particle is m. We denote by y_n, the displacement of the n-th particle, and by $\varphi(y_{n+1} - y_n)$, the interaction potential

K. Castillo
Universidade Estadual Paulista – UNESP, São José do Rio Preto - SP, Brazil
e-mail: kenier@ibilce.unesp.br

L. Garza
Universidad de Colima, Las Vivoras, 28040 Colima, México
e-mail: garzaleg@gmail.com

F. Marcellán (✉)
Universidad Carlos III de Madrid, Madrid, Spain
e-mail: pacomarc@ing.uc3m.es

G.V. Milovanović and M.Th. Rassias (eds.), *Analytic Number Theory, Approximation Theory, and Special Functions*, DOI 10.1007/978-1-4939-0258-3_25,

between neighboring particles. We can consider this system as a chain of infinitely many particles joined together with nonlinear springs. Therefore, if

$$F(r) = -\varphi'(r)$$

is the force of the spring when it is stretched by the amount r and $r_n = y_{n+1} - y_n$ is the mutual displacement, then, according to the Newton's law, the equation that governs the evolution is

$$m\ddot{y}_n = \varphi'(y_{n+1} - y_n) - \varphi'(y_n - y_{n-1}),$$

where, as usual, $\dot{y}$ denotes the derivative with respect to the time. If $F(r)$ is proportional to r, that is, when $F(r)$ obeys the Hooke's law, the spring is linear and the potential can be written as $\varphi(r) = (\kappa/2)r^2$. Thus, the equation of motion is

$$m\ddot{y}_n = \kappa(y_{n-1} - 2y_n + y_{n+1}),$$

and the solutions $y_n^{(\ell)}$, $\ell \in \mathbb{N}$, are given by a linear superposition of the normal modes. In particular, when the particles located at y_0 and y_{N+1} are fixed,

$$y_n^{(\ell)} = C_n \sin\left(\frac{\pi\ell}{N+1}\right)\cos\left(\omega_\ell t + \delta_\ell\right), \quad \ell = 1, 2, \ldots, N,$$

where $\omega_\ell = 2\sqrt{\kappa/m}\sin(\pi\ell/(2N+2))$, the amplitude C_n of each mode is a constant determined by the initial conditions. In this case there is no transfer of energy between the modes. Therefore, the linear lattice is non-ergodic and cannot be an object of statistical mechanics unless some modification is made. In the early 1950s, the general belief was that if a nonlinearity is introduced in the model, then the energy flows between the different modes, eventually leading to a stable state of statistical equilibrium [5]. This phenomenon was explained by the connection to solitons[1].

There are nonlinear lattices which admit periodic behavior at least when the energy is not too high. Lattices with exponential interaction have the desired properties. The Toda lattice [18] is given by setting

$$\varphi(r) = e^{-r} + r - 1.$$

Flaschka [6] (see also [14, 15]) proved the complete integrability for the Toda lattice by recasting it as a Lax equation for Jacobi matrices. Later, Van Moerbeke [19], following a similar work [13] on Hill's equation [10], used the Jacobi matrices to define the Toda hierarchy for the periodic Toda lattices and to find the corresponding Lax pairs.

[1]In mathematics and physics, a soliton is a self-reinforcing solitary wave (a wave packet or pulse) that maintains its shape while it travels at constant speed. Solitons are caused by a cancellation of nonlinear and dispersive effects in the medium.

Flaschka's change of variable is given by

$$a_n = \frac{1}{2}e^{-(y_{n+1}-y_n)/2}, \quad b_n = \frac{1}{2}\dot{y}_n.$$

Hence the new variables obey the evolution equations

$$\dot{a}_n = a_n(b_{n+1} - b_n), \tag{1}$$

$$\dot{b}_n = 2(a_n^2 - a_{n-1}^2), \quad a_{-1} = 0, \quad n \geqslant 0, \tag{2}$$

with initial data $b_n^0 = b_n(0) = \overline{b_n(0)}$, $a_n^0 = a_n(0) > 0$, which we suppose uniformly bounded.

Let $\mathbf{J}_t$ be the semi-infinite Jacobi matrix associated with the system (1) and (2), that is,

$$\mathbf{J}_t = \begin{bmatrix} b_0(t) & a_0(t) & 0 & 0 & \cdots \\ a_0(t) & b_1(t) & a_1(t) & 0 & \\ 0 & a_1(t) & b_2(t) & a_2(t) & \\ 0 & 0 & a_2(t) & b_3(t) & \ddots \\ \vdots & & & \ddots & \ddots \end{bmatrix}.$$

If μ is a nontrivial probability measure supported on some interval $E \subset \mathbb{R}$, then it is very well known that there exists a unique sequence of polynomials $\{p_n\}_{n\geqslant 0}$, assuming the leading coefficient of p_n is positive, satisfying

$$\int_E p_n(x)p_m(x)d\mu(x) = \delta_{n,m}, \quad n, m \geqslant 0.$$

$\{p_n\}_{n\geqslant 0}$ is then said to be the sequence of orthonormal polynomials with respect to μ. $\{p_n\}_{n\geqslant 0}$ satisfies the three-term recurrence relation

$$xp_n(x) = a_n p_{n+1}(x) + b_n p_n(x) + a_{n-1}p_{n-1}(x), \quad n \geqslant 0,$$

with the initial condition $p_{-1}(x) = 0$, $p_0(x) = 1$. Notice that the matrix representation of the recurrence relation is the Jacobi matrix defined above. We use the notation $\mathbf{J}_\mu = \mathbf{J}_0$, i.e., with entries $a_n(0) = a_n^0$ and $b_n(0) = b_n^0$. Favard's theorem says that, given any Jacobi matrix $\tilde{\mathbf{J}}$, there exists a measure μ on the real line for which $\tilde{\mathbf{J}} = \mathbf{J}_\mu$. In general, μ is not unique.

Flaschka's main observation is that the equations (1)–(2) can be reformulated in terms of the Jacobi matrix $\mathbf{J}_t$ as the Lax pair

$$\dot{\mathbf{J}}_t = [\mathbf{A}, \mathbf{J}_t] = \mathbf{A}\mathbf{J}_t - \mathbf{J}_t\mathbf{A},$$

with

$$\mathbf{A} = \begin{bmatrix} 0 & a_0(t) & 0 & 0 & \cdots \\ -a_0(t) & 0 & a_1(t) & 0 & \\ 0 & -a_1(t) & 0 & a_2(t) & \\ 0 & 0 & -a_2(t) & 0 & \ddots \\ \vdots & & & \ddots & \ddots \end{bmatrix} = (\mathbf{J}_t)_+ - (\mathbf{J}_t)_-,$$

where we use the standard notation $(\mathbf{J}_t)_+$ (resp. $(\mathbf{J}_t)_-$) for the upper-triangular (resp. lower triangular) projection of the matrix $\mathbf{J}_t$, and $[\cdot,\cdot]$ denotes the commutator. At the same time, the corresponding orthogonality measure $d\mu(\cdot,t)$ goes through a simple spectral transformation,

$$d\mu(x,t) = e^{-tx} d\mu(x,0), \quad t > 0. \tag{3}$$

Notice that spectral transformations of orthogonal polynomials on the real line play a central role in the solution of the problem. Indeed, the solution of Toda lattice is a combination of the inverse spectral problem from $\{a_n^0\}_{n\geqslant 0}$, $\{b_n^0\}_{n\geqslant 0}$ associated with the measure $d\mu = d\mu(\cdot,0)$, the spectral transformation (3), and the direct spectral problem from $\{a_n(t)\}_{n\geqslant 0}$, $\{b_n(t)\}_{n\geqslant 0}$ associated with the measure $d\mu(\cdot,t)$. A generalization of the perturbation (3) has been analyzed in [9], where the authors also describe the time evolution of the zeros of such polynomials.

In this contribution, we are interested in the analysis of the dynamical properties of the family of orthogonal polynomials $P_n(x,t)$ with respect to the measure

$$d\tilde{\mu}(x) = (1 - J(t))d\mu(x) + J(t)\delta(x), \tag{4}$$

where μ is a symmetric (i.e., $d\mu(x) = \omega(x)dx$ with $\omega(x) = \omega(-x)$ and $\mathrm{supp}(d\mu(x))$ symmetric) nontrivial probability measure supported on the real line and $J : \mathbb{R}_+ \to [0,1]$ is a positive C^1 function. In other words, a time-dependent mass $J(t)$ is added to μ, in such a way that the new measure $\tilde{\mu}$ is also normalized. This problem has been analyzed in [21], where the authors describe the dynamics of the corresponding orthogonal polynomials and the recurrence relation coefficients and the connection of this problem with the Darboux transformation. These kinds of perturbations are particular examples of the so-called Uvarov perturbations. They have been extensively studied in [4], where end mass points are considered and in [1, 11] in a more general framework. In [8], the author deals with an electrostatic interpretation of the zeros of the orthogonal polynomials associated to the perturbed measure, when it is assumed that μ is a measure satisfying some extra conditions.

The manuscript is organized as follows. In Sect. 2, we extend the results in [21] for nonsymmetric measures, using a symmetrization process. In Sect. 3, we analyze the dynamical behavior of the zeros of $P_n(x,t)$, when the orthogonality measure is semiclassical. Some representative examples of such dynamics are shown, when μ is a symmetric classical measure. Finally, in Sect. 4 we deal with a similar transformation for orthogonal polynomials with respect to measures supported on the unit circle.

2 Time Dependence of Orthogonal Polynomials and Symmetrization Problems

Let $\{P_n\}_{n\geqslant 0}$ be the sequence of monic orthogonal polynomials with respect to a symmetric measure μ supported on a symmetric infinite subset of the real line. If we denote by $\{P_n(x,t)\}_{n\geqslant 0}$, the sequence of monic orthogonal polynomials associated with $\tilde{\mu}$ defined in (4), then (see [1, 11]):

$$P_n(x,t) = P_n(x) - \frac{J(t)P_n(0)}{1 - J(t) + J(t)K_{n-1}(0,0)} K_{n-1}(x,0), \tag{5}$$

where $K_n(x,y)$ is the n-th reproducing kernel defined by

$$K_n(x,y) = \sum_{k=0}^{n} \frac{P_k(x)P_k(y)}{\|P_k\|^2} = \frac{P_{n+1}(x)P_n(y) - P_n(x)P_{n+1)}(y)}{\|P_n\|^2(x-y)},$$

where the expression in the right-hand side is known as the Christoffel-Darboux formula and it is valid if $x \neq y$. Notice that $P_n(x) = P_n(x,0)$, i.e., the perturbed polynomials at zero time. Since μ is symmetric, we have $P_{2n+1}(0) = 0$, so that

$$P_{2n+1}(x,t) = P_{2n+1}(x), \quad n \geqslant 0,$$

$$P_{2n}(x,t) = P_{2n}(x) - \frac{J(t)P_{2n}(0)}{1 - J(t) + J(t)K_{2n-2}(0,0)} K_{2n-2}(x,0), \quad n \geqslant 0,\ t > 0.$$

In other words, the odd degree polynomials are invariant under time. Our interest is to find the differential equation satisfied by $P_n(x,t)$ with respect to the time parameter. Obviously, $\dot{P}_{2n+1}(x,t) = 0$. Differentiating $P_{2n}(x,t)$ with respect to the time we have

$$\dot{P}_{2n}(x,t) = -\frac{\dot{J}(t)P_{2n}(0)}{[1 - J(t) + J(t)K_{2n-2}(0,0)]^2} K_{2n-2}(x,0),$$

and using the Christoffel-Darboux formula, we get

$$\dot{P}_{2n}(x,t) = r_n \frac{P_{2n-1}(x)}{x}, \tag{6}$$

with

$$r_n = -\frac{\dot{J}(t) P_{2n}(0) P_{2n-2}(0)}{\| P_{2n-2} \|^2 [1 - J(t) + J(t) K_{2n-2}(0,0)]^2}. \tag{7}$$

Furthermore, since

$$K_{2n-2}(x,0) = \frac{P_{2n-1}(x) P_{2n-2}(0)}{\| P_{2n-2} \|^2 x},$$

we have

$$K_{2n-2}(0,0) = \frac{P'_{2n-1}(0) P_{2n-2}(0)}{\| P_{2n-2} \|^2},$$

so that

$$r_n = -\frac{\dot{J}(t) P_{2n}(0) P_{2n-2}(0) \| P_{2n-2} \|^2}{[(1 - J(t)) \| P_{2n-2} \|^2 + J(t) P'_{2n-1}(0) P_{2n-2}(0)]^2}. \tag{8}$$

In [21], the authors show that in this case, the dynamics of the coefficients of the recurrence relation is given by

$$\dot{d}_{2n} = r_n, \quad \dot{d}_{2n+1} = -r_{n+1},$$

where $d_n = a_n^2$. This represents a nonlocal integrable chain with continuous time and discrete space variable. It is related to the so-called Uvarov-Chihara problem in the theory of orthogonal polynomials (see [20]).

The dynamics of the sequence of polynomials P_n with respect to the time can be easily obtained for the general (nonsymmetric) case using a symmetrization process. Given a measure μ, we can define a linear functional u in the linear space of polynomials with real coefficients $\mathbb{P}$ such that

$$u[q(x)] = \int_E q(x) d\mu(x), \quad q \in \mathbb{P}.$$

If μ is a probability measure, then u is said to be positive definite. In a more general framework, it is enough for u to be quasi definite (i.e., the principal leading submatrices of its Gram matrix with respect to the canonical basis $\{x^n\}_{n\geqslant 0}$ are nonsingular) for the existence of a sequence monic polynomials with respect to u to be guaranteed. Let denote such a sequence by $\{P_n\}_{n\geqslant 0}$, and define the linear functional u_s as

$$u_s[x^{2n}] := u[x^n], \quad u_s[x^{2n+1}] := 0, \quad n \geqslant 0.$$

That is, the linear functional u_s is symmetric. Thus, it is well known ([3]) that, if we denote by $\{Q_n\}_{n\geqslant 0}$ the sequence of monic polynomials orthogonal with respect to u_s, then

$$Q_{2n}(x) = P_n(x^2), \quad Q_{2n+1}(x) = x\tilde{P}_n(x^2), \quad n \geqslant 0,$$

where $\{\tilde{P}_n\}_{n\geqslant 0}$ is the sequence of monic polynomials orthogonal with respect to the linear functional $\tilde{u} = xu$ (i.e., $\tilde{u}[q] = u[xq]$) for any $q \in \mathbb{P}$. $\{\tilde{P}_n\}_{n\geqslant 0}$ is the sequence of *kernel* polynomials of parameter 0 (see [3]), and they can be expressed in terms of $\{P_n(x)\}_{n\geqslant 0}$ by

$$\tilde{P}_n(x) = \frac{1}{x}\left(P_{n+1}(x) - \frac{P_{n+1}(0)}{P_n(0)}P_n(x)\right), \quad n \geqslant 0.$$

A necessary and sufficient condition for their existence is that $P_n(0) \neq 0$, $n \geqslant 0$ (in the positive definite case, that $0 \notin \operatorname{supp}(\mu)$).

Therefore, if u is a (not necessarily symmetric) positive definite linear functional, then let $\{P_n(x,t)\}_{n\geqslant 0}$ be the sequence of monic polynomials orthogonal with respect to the linear functional $u_t := (1 - J(t))u + J(t)\delta(x)$. Thus,

$$P_n(x^2,t) = Q_{2n}(x,t), \quad n \geqslant 0,$$

where $\{Q_n(x,t)\}_{n\geqslant 0}$ are symmetric polynomials orthogonal with respect to the linear functional u_s obtained from the symmetrization of u_t and, therefore,

$$\dot{P}_n(x^2,t) = \dot{Q}_{2n}(x,t) = r_n\frac{Q_{2n-1}(x,t)}{x} = r_n\frac{x\tilde{P}_{n-1}(x^2,t)}{x},$$

where r_n is computed using the polynomials Q_n and the polynomials $\tilde{P}_n(x,t)$ are orthogonal with respect to the linear functional xu_t, provided $P_n(0,t) \neq 0$, $n \geqslant 1$. Then,

$$\dot{P}_n(x,t) = r_n\tilde{P}_{n-1}(x,t), \quad n \geqslant 1.$$

Furthermore, from $Q_{2n+1}(x,t) = x\tilde{P}_n(x^2,t)$, we get

$$\dot{Q}_{2n+1}(x,t) = x\dot{\tilde{P}}_n(x^2,t) = 0,$$

and we get $\dot{\tilde{P}}_n(x,t) = 0$, $n \geqslant 0$. As a consequence

Proposition 2.1. *Let $\{P_n(x)\}_{n\geqslant 0}$ be the sequence of monic polynomials with respect to a nontrivial probability measure $d\mu$. Let $d\tilde{\mu}$ be defined as in* (4) *and denote by $\{P_n(x,t)\}_{n\geqslant 0}$ its corresponding sequence of monic orthogonal polynomials. Then,*

$$\dot{P}_n(x,t) = \frac{r_n}{x}\left(P_n(x,t) - \frac{P_n(0,t)}{P_{n-1}(0,t)}P_{n-1}(x,t)\right), \quad n \geqslant 1.$$

3 Time Evolution of Zeros of Semiclassical Orthogonal Polynomials

Let us consider a symmetric positive definite linear functional u which is semiclassical, i.e.,

$$\mathcal{D}(\phi(x)u) = \Psi(x)u,$$

for some polynomials ϕ and Ψ, which are even and odd functions, respectively, with $\deg \Psi \geqslant 1$, and let us define the linear functional

$$\tilde{u} = (1 - J(t))u + J(t)\delta(x). \tag{9}$$

Here, as above, $J : \mathbb{R}_+ \to [0, 1]$ is a positive C^1 function. Then, we have

$$x^2\phi(x)\tilde{u} = (1 - J(t))x^2\phi(x)u.$$

Applying the derivative operator in both sides, we get

$$\begin{aligned}\mathcal{D}[x^2\phi(x)\tilde{u}] &= (1 - J(t))\mathcal{D}[x^2\phi(x)u] \\ &= (1 - J(t))[2x\phi(x)u + x^2\mathcal{D}(\phi u)] \\ &= 2x\phi\tilde{u} + (1 - J(t))x^2\Psi u \\ &= (2x\phi + x^2\Psi)\tilde{u}.\end{aligned}$$

Thus, $\tilde{u}$ is also semiclassical, and then its corresponding sequence of monic orthogonal polynomials, $\{P_n(x,t)\}_{n\geqslant 0}$, satisfies the structure relation ([11, 12])

$$x^2\phi(x)\frac{\partial}{\partial x}P_n(x;t) = A_n(x;t)P_n(x;t) + B_n(x;t)P_{n-1}(x;t), \tag{10}$$

where the functions $A_n(x;t)$, $B_n(x;t)$ can be calculated explicitly using the measure associated with u and its corresponding sequence or orthogonal polynomials (see [2, 9, 12]). Let $x_{n,k}(t)$ be the k-th zero of $P_n(x;t)$, i.e.,

$$P_n(x_{n,k}(t), t) = 0.$$

Following [9], differentiating the last equation with respect t, we obtain

$$\left.\frac{\partial}{\partial x}P_n(x;t)\right|_{x=x_{n,k}} \dot{x}_{n,k} + \dot{P}_n(x_{n,k}, t) = 0.$$

Thus, evaluating (10) with $n = 2m$ at $x = x_{2m,k}(t)$ we get

$$x_{2m,k}^2(t)\phi(x_{2m,k}(t))\frac{\partial}{\partial x}P_{2m}(x_{2m,k}(t);t) = B_n(x_{2m,k}(t);t)P_{2m-1}(x_{2m,k}(t);t),$$

and, as a consequence, from (6) we obtain

$$\dot{x}_{2m,k}(t) = -r_m\frac{x_{2m,k}(t)\phi(x_{2m,k}(t))}{B_{2m}(x_{2m,k}(t))}. \tag{11}$$

Next, we consider two examples of classical families (semiclassical of class zero) of orthogonal polynomials that are symmetric, namely, the Gegenbauer (with parameter $\alpha = \beta = 1$) and Hermite polynomials. In both cases, since their structure relations are known, $A_n(x,t)$ and $B_n(x,t)$ can be easily obtained directly from the structure and recurrence relations.

First, notice that from (5), we have

$$P_{2n}'(x,t) = P_{2n}'(x) - \frac{J(t)P_{2n}(0)P_{2n-2}(0)}{1-J(t)+J(t)K_{2n-2}(0,0)}\frac{xP_{2n-1}'(x)-P_{2n-1}(x)}{\|P_{2n-2}\|^2x^2}, \tag{12}$$

where P' denotes the derivative with respect to x. Thus,

$$x^2\phi(x)P_{2n}'(x,t) = x^2\phi(x)P_{2n}'(x) - M(t)\phi(x)[xP_{2n-1}'(x) - P_{2n-1}(x)], \tag{13}$$

where

$$M(t) = \frac{J(t)P_{2n}(0)P_{2n-2}(0)}{[1-J(t)+J(t)K_{2n-2}(0,0)]\|P_{2n-2}\|^2}.$$

1. For the Gegenbauer polynomials with $\alpha = \beta = 1$, we have $\phi(x) = 1 - x^2$ and (see [12])

$$\phi(x)P_n'(x) = a_nP_{n+1} + c_nP_{n-1}(x), \tag{14}$$

$$xP_n(x) = P_{n+1}(x) - \gamma_nP_{n-1}(x), \tag{15}$$

where a_n, c_n, γ_n are given by

$$a_n = -n,$$

$$c_n = \frac{4n(n+1)^2(n+2)(n+3)}{(2n+1)(2n+2)^2(2n+3)},$$

$$\gamma_n = \frac{4n(n+1)^2(n+2)}{(2n+1)(2n+2)^2(2n+3)}.$$

As a consequence,

$$(1-x^2)P_n'(x) = a_n x P_n(x) - (a_n\gamma_n - c_n)P_{n-1}(x). \tag{16}$$

Thus, from (13) and (16), it is straightforward to show that

$$x^2(1-x^2)P_{2n}'(x,t) = A_n(x,t)P_{2n}(x,t) + B_n(x,t)P_{2n-1}(x,t),$$

with

$$A_n(x,t) = a_{2n}x^3 - M(t)\frac{a_{2n-1}\gamma_{2n-1} - c_{2n-1}}{\gamma_{2n-1}}x,$$

$$B_n(x,t) = -(a_{2n-1}\gamma_{2n-1} - c_{2n-1} + M(t))x^2 - M(t)\left(a_{2n-1}x^2 - \phi(x) - \frac{A_n(x,t)}{x}\right),$$

which can be reduced after some calculations to

$$A_n(x,t) = a_{2n}x^3 - M(t)\frac{(4n-1)(2n-1)^2}{n^2}x,$$

$$B_n(x,t) = \left[\frac{(4n^2-1)(4n-1)^2}{4n+1} - a_{2n} - (2+a_{2n-1})M(t)\right]x^2$$

$$+\frac{(4n-1)(2n-1)^2}{n^2}M^2(t) + M(t).$$

Notice that in this case, $A_n(x,t)$ and $B_n(x,t)$ are polynomials in x. Thus, from (11), the dynamics of the zeros of $P_n(x,t)$ can be described as

$$\dot{x}_{2m,k}(t) = -r_m\frac{x_{2m,k}(t)(1-x_{2m,k}^2(t))}{B_{2m}(x_{2m,k}(t))}.$$

2. Now, we consider the Hermite polynomials H_n. In this case, we have $\phi(x) = 1$, $H_n'(x) = nH_{n-1}(x)$ and

$$H_{n+1}(x) = xH_n(x) - \frac{1}{2}nH_{n-1}(x).$$

Thus, proceeding as above, we get

$$x^2H_{2n}'(x,t) = A_n(x,t)H_{2n}(x,t) + B_n(x,t)H_{2n-1}(x,t),$$

with

$$A_n(x,t) = \frac{2(2n-1)}{n} M(t)x,$$

$$B_n(x,t) = 2\left(n - \frac{2n-1}{n} M(t)\right)x^2 + \left(1 - \frac{2(2n-1)}{n} M(t)\right)M(t).$$

Again, since we have a classical family, $A_n(x,t)$ and $B_n(x,t)$ are polynomials in x. As a consequence, the behavior of the zeros of $H_{2m}(x,t)$ can be described as

$$\dot{x}_{2m,k}(t) = -r_m \frac{x_{2m,k}(t)}{B_{2m}(x_{2m,k}(t))}.$$

4 Time Dependence of Verblunsky Coefficients for OPUC

Given a nontrivial probability measure σ supported on the unit circle $\mathbb{T}$, there exists a sequence of monic polynomials $\{\Phi_n\}_{n\geqslant 0}$ which is orthogonal with respect to σ, i.e.,

$$\int_{\mathbb{T}} \Phi_n(z)\overline{\Phi_m(z)}d\sigma(z) = \kappa_n \delta_{n,m}, \quad \kappa_n > 0, \quad n,m \geqslant 0.$$

They are called orthogonal polynomials on the unit circle (OPUC). These polynomials satisfy the recurrence relation (see [16, 17])

$$\Phi_{n+1}(z) = z\Phi_n(z) + \Phi_{n+1}(0)\Phi_n^*(z), \quad n \geqslant 1,$$

where $\Phi_n^*(z) = z^n\overline{\Phi_n(1/\bar{z})}$ is called the reversed polynomial, and the complex numbers $\{\Phi_n(0)\}_{n\geq 1}$ satisfy $|\Phi_n(0)| < 1$. They are called Verblunsky (reflection, Schur, Szegő) coefficients.

On the other hand, if μ is a nontrivial probability measure supported on $[-1,1]$, then it is very well known ([17]) that it induces a nontrivial positive measure σ supported on the unit circle. This process is called the Szegő transformation. On the other hand, if σ is induced through the Szegő transformation, then their corresponding orthogonal polynomials Φ_n have real coefficients, and the Verblunsky coefficients are also real. In this case, consider the perturbation

$$d\tilde{\sigma}(z) = (1 - J(t))d\sigma(z) + J(t)\delta(z-1),$$

i.e., a time-dependent mass is added at the point $z = 1$, where $J : \mathbb{R}_+ \to [0,1]$ is a positive $\mathcal{C}^1$ function. Notice that this is the same perturbation defined in the previous sections for orthogonal polynomials on the real line, although the symmetry requirement has been removed. As before, if $\Phi_n(z;t)$ is the MOPS with respect to $\tilde{\sigma}$, then

$$\Phi_n(z;t) = \Phi_n(z) - \frac{J(t)\Phi_n(1)}{1 - J(t) + J(t)K_{n-1}(1,1)} K_{n-1}(z,1), \tag{17}$$

where $K_n(z,y)$, the reproducing kernel, is now defined as (see [16, 17])

$$K_n(z,y) = \sum_{k=0}^{n} \frac{\Phi_k(z)\overline{\Phi_k(y)}}{\|\Phi_k\|^2} = \frac{\Phi_{n+1}^*(z)\overline{\Phi_{n+1}^*(y)} - \Phi_{n+1}(z)\overline{\Phi_{n+1)}(y)}}{\|\Phi_{n+1}\|^2(1 - z\bar{y})},$$

provided $z\bar{y} \neq 1$. Therefore,

$$\Phi_n(0;t) = \Phi_n(0) - \frac{J(t)\Phi_n(1)}{1 - J(t) + J(t)K_{n-1}(1,1)} K_{n-1}(0,1), \tag{18}$$

and since we have real coefficients and $\Phi_n^*(0) = 1$,

$$\Phi_n(0;t) = \Phi_n(0) - \frac{J(t)\Phi_n^2(1)(1 - \Phi_n(0))}{\|\Phi_n\|^2[1 - J(t) + J(t)K_{n-1}(1,1)]}. \tag{19}$$

Thus,

$$\dot{\Phi}_n(0;t) = -\frac{\dot{J}(t)\Phi_n^2(1)(1 - \Phi_n(0))}{\|\Phi_n\|^2[1 - J(t) + J(t)K_{n-1}(1,1)]^2},$$

which describes the dynamic behavior of the Verblunsky coefficients of the perturbed measure with respect to the time. We will show that $\Phi_n(1)$ and $K_{n-1}(1,1)$ can be expressed in terms of the previous Verblunsky coefficients. Notice that, from the recurrence relation, we have

$$\Phi_n(1) = \Phi_{n-1}(1) + \Phi_n(0)\Phi_{n-1}^*(1),$$

but since Φ_{n-1} has real coefficients, we get

$$\Phi_n(1) = [1 + \Phi_n(0)]\Phi_{n-1}(1),$$

and, recursively,

$$\Phi_n(1) = \prod_{k=1}^{n}(1 + \Phi_k(0)).$$

On the other hand,

$$K_{n-1}(1,1)=\sum_{k=0}^{n-1}\frac{\Phi_k^2(1)}{\|\Phi_k\|^2}=\sum_{k=0}^{n-1}\frac{\prod_{j=1}^{k}(1+\Phi_j(0))^2}{\prod_{j=1}^{k}(1-\Phi_j^2(0))}=\sum_{k=0}^{n-1}\frac{\prod_{j=1}^{k}(1+\Phi_j(0))}{\prod_{j=1}^{k}(1-\Phi_j(0))}.$$

As a consequence, in order to describe the dynamics of $\Phi_n(0;t)$, the values of $\{\Phi_k(0)\}_{k=1}^n$ are required. The situation can be simplified if symmetric measures are considered. As an example, consider the perturbation of the Lebesgue measure on the real line defined by

$$d\tilde{\mu}(x,t)=dx+\frac{1}{J(t)}\delta(x+1)+\frac{1}{J(t)}\delta(x-1).$$

Notice that $d\tilde{\mu}(x,t)$ is symmetric. Applying the Szegő transformation to $d\tilde{\mu}(x)$ it will induce a measure $d\sigma(z,t)$ on the unit circle which is also symmetric. It was shown in [7] that in such a case, the Verblunsky coefficients associated with $d\sigma$ are

$$\Phi_{2n}(0,t)=\frac{-1}{2n+1}\frac{3n^2(n+1)^2+2n(n+1)J(t)-J^2(t)}{n^2(n+1)^2+2n(n+1)J(t)+J^2(t)},$$

$$\Phi_{2n+1}(0,t)=0.$$

In other words, the dynamics of $\Phi_n(0)$ can be obtained easily only in terms of $J(t)$.

Acknowledgements The research of K. Castillo was supported by CNPq Program/Young Talent Attraction, Ministério da Ciência, Tecnologia e Inovação of Brazil, Project 370291/2013–1. The research of K. Castillo and F. Marcellán was supported by Dirección General de Investigación, Ministerio de Economía y Competitividad of Spain, Grant MTM2012–36732–C03–01. The research of L. Garza was supported by Conacyt (México) grant 156668 and PROMEP.

References

1. Álvarez-Nodarse, R., Marcellán, F., Petronilho J.: WKB approximation and Krall-type orthogonal polynomials. Acta Appl. Math. **54**, 27–58 (1998)
2. Chen, Y., Ismail, M.E.H.: Ladder operators and differential equations for orthogonal polynomials. J. Phys. A: Math. Gen. **30**, 7817–7829 (1997)
3. Chihara, T.S.: An Introduction to Orthogonal Polynomials, Gordon and Breach, New York (1978)
4. Chihara, T.S.: Orthogonal polynomials and measures with end points masses. Rocky Mountain J. Math. **15**, 705–719 (1985)
5. Fermi, E., Pasta, J., Ulam, S.: Studies of Nonlinear Problemas, University of Chicago Press, Chicago (1965)

6. Flaschka, H.: Discrete and periodic illustrations of some aspects of the inverse method. Dynamical Systems, Theory and Applications (Rencontres, Battelle Seattle Research, Seattle, Washington, 1974). vol. 38, pp. 441–466. Lecture Notes in Physics (1975)
7. García-Lázaro, P., Marcellán, F., Tasis, C.: On a Szegő result: Generating sequences of orthogonal polynomials on the unit circle. In: Brezinski, C., et al, (ed.) Proceedings Erice International Symposium on Orthogonal Polynomials and Their Applications. IMACS Annals on Computer Application Mathematics, vol. 9, pp. 271–274. J. C. Baltzer AG, Basel (1991)
8. Ismail, M.E.H.: More on electrostatic models for zeros of orthogonal polynomials, Proceedings of the International Conference on Fourier Analysis and Applications (Kuwait, 1998). Numer. Funct. Anal. Optim. **21**(1–2), 191–204 (2000)
9. Ismail, M.E.H., Ma, W.-X.: Equations of motion for zeros of orthogonal polynomials related to the Toda lattices. Arab. J. Math. Sciences **17**, 1–10 (2011)
10. Magnus, W., Winkler, S.: Hill's Equation, Interscience Publishers John Wiley and Sons, New York (1966)
11. Marcellán, F., Maroni, P.: Sur l'adjonction d'une masse de Dirac à une forme regulière et semiclassique. Ann. Mat. Pura Appl (4) **162**, 1–22 (1992)
12. Maroni, P.: Une théorie algébrique des polynômes orthogonaux: Applications aux polynômes orthogonaux semi-classiques. In: Brezinski, C., et.al (eds.) Orthogonal Polynomials and their Applications, Annals on Computing and Applied Mathematics vol. 9, pp. 98–130. J.C. Baltzer AG, Basel (1991)
13. McKean, H., Van Moerbeke, P.: The spectrum of Hill's equation. Invent. Math. **30**, 217–274 (1975)
14. Moser, J.: Three integrable Hamiltonian systems connected with isospectral deformations. Adv. Math. **16**, 197–220 (1975a)
15. Moser, J.: Finitely many mass points on the line under the influence of an exponential potential - an integrable system, Dynamical Systems, Theory and Applications (Rencontres, Battelle Res. Inst., Seattle, Wash., 1974). vol. 38, pp. 467–497 Lecture Notes in Physics, Springer, Berlin (1975b)
16. Simon, B.: Orthogonal polynomials on the unit circle. 2 vols. Amererican Mathematical Society Colloquium Publications Series, vol. 54, American Mathematical Society, Providence Rhode Island (2005)
17. Szegő, G.: Orthogonal Polynomials, American Mathematical Society Colloquium Publications Series. vol. 23, 4th edn. American Mathematical Society, Providence, Rhode Island (1975)
18. Toda, M.: Theory of Nonlinear Lattices, Springer, Berlin (1989)
19. Van Moerbeke, P.: The spectrum of Jacobi matrices. Invent. Math. **37**, 45–81 (1976)
20. Uvarov, V.B.: The connection between systems of polynomials orthogonal with respect to different distribution functions. USSR Comput. Math. Math. Phys. **9**, 25–36 (1969)
21. Vinet, L., Zhedanov, A.: An integrable system connected with the Chihara-Uvarov problem for orthogonal polynomials. J. Phys. A: Math. Gen. **31**, 9579–9591 (1998)

On Especial Cases of Boas-Buck-Type Polynomial Sequences

Ana F. Loureiro and S. Yakubovich

Dedicated to Professor Hari M. Srivastava

Abstract After a slight modification, the Kontorovich-Lebedev transform is an automorphism in the vector space of polynomials. The action of this transformation over special cases of Boas-Buck-type polynomial sequences is under analysis.

1 Introduction and Preliminary Results

This work aims to give a humble contribution to polynomial sequences generated by Boas-Buck-type generating function or Boas-Buck structure. Under analysis will be the action of the so-called Kontorovich-Lebedev transform over certain Boas-Buck polynomial sequences [3, 4].

Throughout the text, $\mathbb{N}$ will denote the set of all positive integers, $\mathbb{N}_0 = \mathbb{N} \cup \{0\}$, whereas $\mathbb{R}$ and $\mathbb{C}$ the field of the real and complex numbers, respectively. The notation $\mathbb{R}_+$ corresponds to the set of all positive real numbers. The present investigation is primarily targeted at analysis of sequences of polynomials whose degrees equal its order, which will be shortly called as PS. Whenever the leading coefficient of each of its polynomials equals 1, the PS is said to be an MPS (*monic polynomial sequence*). A PS or an MPS forms a basis of the vector space of polynomials with coefficients in $\mathbb{C}$, here denoted as $\mathcal{P}$.

A.F. Loureiro (✉)
School of Mathematics, Statistics & Actuarial Science (SMSAS), University of Kent, Cornwallis Building, Canterbury, Kent CT2 7NF, UK
e-mail: A.Loureiro@kent.ac.uk

S. Yakubovich
Department of Mathematics, Faculty Sciences of University of Porto, Rua do Campo Alegre 687, 4169-007 Porto, Portugal
e-mail: syakubov@fc.up.pt

G.V. Milovanović and M.Th. Rassias (eds.), *Analytic Number Theory, Approximation Theory, and Special Functions*, DOI 10.1007/978-1-4939-0258-3_26,

The aforementioned Kontorovich-Lebedev (hereafter, KL) arises as the simplest case of a very general index integral transform, the Wimp transform, named after his work in 1964,

$$F(\tau) = \int_0^\infty G^{m,n+2}_{p+2,q}\left(x; \begin{matrix} 1-\mu+i\tau, 1-\mu-i\tau, (a_p) \\ (b_q) \end{matrix}\right) f(x)dx \tag{1}$$

whose inversion formula was established in 1985 [15] by the second author (see [16–18])

$$f(x) = \frac{1}{\pi^2}\int_0^\infty \tau\sinh(2\pi\tau)F(\tau)\, G^{q-m,p-n+2}_{p+2,q}\left(x; \begin{matrix} \mu+i\tau, \mu-i\tau, -(a_p^{n+1}), -(a_n) \\ -(b_q^{m+1}), -(b_m) \end{matrix}\right) d\tau. \tag{2}$$

This transformation has the Meijer G-function as a kernel

$$G^{m,n}_{p,q}\left(z; \begin{matrix} \mathbf{a} \\ \mathbf{b} \end{matrix}\right) := G^{m,n}_{p,q}\left(z; \begin{matrix} a_1,\ldots,a_p \\ b_1,\ldots,b_q \end{matrix}\right).$$

It is a very general function including most of the known special functions as particular cases (like all the generalized hypergeometric functions ${}_pF_q$ and Mathieu functions), and it can be defined via the Mellin-Barnes integral (the reciprocal formula of the Mellin transform)

$$G^{m,n}_{p,q}\left(z; \begin{matrix} \mathbf{a} \\ \mathbf{b} \end{matrix}\right) := G^{m,n}_{p,q}\left(z; \begin{matrix} a_1,\ldots,a_p \\ b_1,\ldots,b_q \end{matrix}\right)$$
$$= \frac{1}{2\pi i}\int_L \frac{\prod_{l=1}^{m}\Gamma(b_l-s)\prod_{l=1}^{n}\Gamma(1-a_l+s)}{\prod_{l=m}^{q-1}\Gamma(1-b_{l+1}+s)\prod_{l=n}^{p-1}\Gamma(a_{l+1}-s)} z^s ds, \tag{3}$$

as long as $0 \leqslant m \leqslant q$ and $0 \leqslant n \leqslant p$, where m, n, p, and q are integer numbers, $a_k - b_i \neq 1, 2, 3, \ldots$ for $k = 1, 2, \ldots, n$ and $j = 1, 2, \ldots, m$, which implies that no pole of any $\Gamma(b_j - s)$, $j = 1, 2, \ldots, m$, coincides with any pole of any $\Gamma(1 - a_k + s)$, $k = 1, 2, \ldots, n$, and $z \neq 0$.

So, the KL-transform arises upon the choice of parameters $m = n = p = q = 0$. The reciprocal pair of transformations (1)–(2) incorporates all the existent index transforms in the literature such as Mehler-Fock, Olevski-Fourier-Jacobi, Whittaker, and Lebedev's transform with a combination of modified Bessel functions, among others. We notice that all these index transforms can be obtained through the composition of the KL with the Mellin type convolution transforms [16, 18].

Recently [10], we have shown that the modified KL_α-*transform*, with $\alpha \geqslant 0$, defined by

$$\mathrm{KL}_\alpha[f(x)](\tau) = 2\left|\Gamma\left(\alpha+1+\frac{i\tau}{2}\right)\right|^{-2}\int_0^\infty f(x)x^\alpha K_{i\tau}(2\sqrt{x})dx, \tag{4}$$

is an automorphism in the vector space of polynomials, and we have as well characterized all the orthogonal polynomial sequences that are mapped into d-orthogonal polynomial sequences (a broadened concept of orthogonality). This transformation essentially bridges monomials into central factorials insofar as [10, 11]

$$\mathrm{KL}_\alpha[x^n](\tau) = \left(\alpha+1-\frac{i\tau}{2}\right)_n\left(\alpha+1+\frac{i\tau}{2}\right)_n = \left|\left(\alpha+1+\frac{i\tau}{2}\right)_n\right|^2, \quad n \geqslant 0.$$

The kernel of such transformation is the modified Bessel function (also called *Macdonald function*) $K_{2i\tau}(2\sqrt{x})$ of purely imaginary index, which is real valued and can be defined by integrals of Fourier type

$$K_{i\tau}(2\sqrt{x}) = \int_0^\infty \mathrm{e}^{-2\sqrt{x}\cosh(u)}\cos(\tau\, u)du, \quad x \in \mathbb{R}_+,\ \tau \in \mathbb{R}_+. \tag{5}$$

Moreover it is an eigenfunction of the operator

$$\mathcal{A} = x^2\frac{d^2}{dx^2} + x\frac{d}{dx} - x = x\frac{d}{dx}x\frac{d}{dx} - x \tag{6}$$

insofar as

$$\mathcal{A}K_{i\tau}(2\sqrt{x}) = -\left(\frac{\tau}{2}\right)^2 K_{i\tau}(2\sqrt{x}), \tag{7}$$

which is valid for any continuous function $f \in L_1\left(\mathbb{R}_+, K_0(2\mu\sqrt{x})dx\right)$, $0 < \mu < 1$, in a neighborhood of each $x \in \mathbb{R}_+$ where $f(x)$ has bounded variation.

Naturally, the identity

$$\mathrm{KL}_\alpha[x^n](\tau) = \left(\alpha+1-\frac{i\tau}{2}\right)_n\left(\alpha+1+\frac{i\tau}{2}\right)_n = \left|\left(\alpha+1+\frac{i\tau}{2}\right)_n\right|^2 \tag{8}$$

holds, enhancing the fact that KL_α is an isomorphism in the vector space $\mathcal{P}$, essentially performing the passage between the canonical basis $\{x^n\}_{n\geqslant 0}$ and the central factorial basis $\left\{\left|\left(\alpha+1+\frac{i\tau}{2}\right)_n\right|^2\right\}_{n\geqslant 0}$.

Besides, from the definition (4), we readily observe that

$$\mathrm{KL}_{\alpha+\beta}[f(x)](\tau) = \frac{\left|\Gamma\left(\alpha+1+\frac{i\tau}{2}\right)\right|^2}{\left|\Gamma\left(\alpha+\beta+1+\frac{i\tau}{2}\right)\right|^2}\mathrm{KL}_\alpha[x^\beta f(x)](\tau), \tag{9}$$

and, in particular, when $\beta = n \in \mathbb{N}_0$, we have

$$\mathrm{KL}_\alpha[x^n f(x)](\tau) = \left|\left(\alpha+1+\frac{i\tau}{2}\right)_n\right|^2 \mathrm{KL}_{\alpha+n}[f(x)](\tau) = \mathrm{KL}_\alpha[x^n](\tau)\,\mathrm{KL}_{\alpha+n}[f(x)](\tau). \tag{10}$$

As a matter of fact, the action of the KL_α operator acting on $\mathcal{P}$ can be viewed as the passage from differential relations into central difference relations, as it can be perceived from (8). To be more specific, let us represent the central difference operator by δ_ω, defined through

$$(\delta_\omega f)(\tau) := \frac{f(\tau+\omega) - f(\tau-\omega)}{2\omega\tau} \tag{11}$$

for some complex number $\omega \neq 0$.

Lemma 1.1 ([10]). *For any $f \in \mathcal{P}$, the following identities hold*

$$\left((\alpha+1)^2 + \frac{\tau^2}{4}\right)\delta_i^2\left(KL_\alpha[f(x)](\tau)\right) = KL_\alpha\left[x\frac{d^2}{dx^2}f(x)\right](\tau), \quad n \geqslant 0, \tag{12}$$

while

$$KL_{\alpha+1/2}\left[\frac{d}{dx}f(x)\right](\tau) = \delta_i\Big(KL_\alpha[f(x)](\tau)\Big), \tag{13}$$

$$\begin{aligned} &KL_\alpha\left[\left(\frac{1}{x}\mathcal{A}x + 2\alpha\frac{d}{dx}x\right)^m x^n f(x)\right](\tau) \\ &\quad = (-1)^m\left(\frac{\tau^2}{4}+\alpha^2\right)^m \left|\left(\alpha+1+\frac{i\tau}{2}\right)_n\right|^2 KL_{\alpha+n}[f](\tau), \end{aligned} \tag{14}$$

where $\mathcal{A}$ represents the operator (6).

This text has no pretension of completeness, neither the reference list. It is organized as follows. In Sect. 2, after defining the Boas-Buck-type sequences (see Definition 2.1), we analyze the image of those whose generating function involves the Meijer G-function as a component. For the particular case of the Brenke-type sequences, we provide differential relations for the corresponding polynomial sequences. When further structural properties like orthogonality or d-orthogonality are known, one might have additional relations that can lead us to the determination of the sequence. We draw some clues in this direction, but we do not enter in this kind of details. Making use of the properties of the KL_α-transform, we determine the generating function of the KL_α-transformed sequence to which we provide the corresponding properties derived from those of the original sequence. Finally, in Sect. 3 some examples of some hypergeometric-type sequences along with the corresponding KL_α-images will be given.

2 About Boas-Buck-Type Polynomial Sequences

As the central object of this work, we begin with the description of the so-called Boas-Buck-type sequences, named after the work [4].

Definition 2.1. An MPS $\{P_n\}_{n\geqslant 0}$ is said to be a Boas-Buck polynomial sequence or to have a generating function of Boas-Buck type if there exists a sequence of nonzero numbers $\{\rho_n\}_{n\geqslant 0}$ and $a, b \in \mathbb{R}$ such that for $x \in [a, b]$

$$G(x,t) = A(t)B(xg(t)) = \sum_{n\geqslant 0} P_n(x)\frac{t^n}{\rho_n}, \tag{15}$$

where

$$\{A(t), B(t), g(t)\} = \sum_{n\geqslant 0}\{a_n, b_n, g_n\}t^n \text{satisfying } a_0 \cdot b_n \cdot g_1 \neq 0 \tag{16}$$

for all $n \in \mathbb{N}_0$, and $g_0 = 0$.

The Brenke-type polynomials arise when $g(t) = t$, whereas the choice $B(y) = e^y$ brings the Sheffer-type polynomials with the Appell polynomials included (upon the additional condition of $g(t) = t$).

The Boas-Buck polynomials, named after the work [4] by the two authors, were at that stage called as the generalized Appell polynomials. As this nomenclature is used nowadays with respect to other types of polynomials, we avoid such names.

In order to describe the KL_α-transform of Boas-Buck-type polynomial sequences, we need to ensure that a corresponding generating function can actually be described as the KL_α-transform of the original one. Theorem 2.1 provides the necessary conditions, but first we need the following result.

Lemma 2.1. *Let $g : [a, b] \to \mathbb{R}_+$ with $a, b \in \mathbb{R}$ and $B(x) \in L_1(\mathbb{R}_+; x^{-\gamma}dx)$, with $\gamma > -\alpha$. If $G(x, t) = A(t)B(xg(t))$, with A, B, and g realizing conditions* (16), *then the KL_α-transform of $G(x, t)$ with respect to x can be calculated by the formula*

$$KL_\alpha[G(x,t)](\tau) = \frac{A(t)}{2\pi i\left|\Gamma\left(\alpha+1+\frac{i\tau}{2}\right)\right|^2} \times \int_{1-\gamma-i\infty}^{1-\gamma+i\infty} \Gamma\left(1-s+\alpha+\frac{i\tau}{2}\right)\Gamma\left(1-s+\alpha-\frac{i\tau}{2}\right)B^\star(s)\big(g(t)\big)^{-s}ds, \tag{17}$$

where $\Gamma(z)$ is the Euler Gamma function and $B^*(s)$ is the Mellin transform of $B(x)$:

$$B^\star(s) = \int_0^\infty B(x)x^{s-1}dx, \qquad \gamma = \mathrm{Re}(s) > -\alpha, \tag{18}$$

where the latter integral converges absolutely.

Proof. The condition $B(x) \in L_1(\mathbb{R}_+; x^{-\gamma}dx)$ guarantees the existence of the Mellin transform (18) (see [14]). Bearing in mind the Stirling formula for the asymptotic at infinity of the Gamma function [9, vol. I] and the following integral representation of the product of Gamma functions as the Mellin transform of the modified Bessel function $K_{i\tau}(2\sqrt{x})$ [13]

$$\Gamma\left(\alpha+s+\frac{i\tau}{2}\right)\Gamma\left(\alpha+s-\frac{i\tau}{2}\right)=2\int_0^\infty K_{i\tau}(2\sqrt{x})x^{\alpha+s-1}dx,\ \ \mathrm{Re}(s)>-\alpha,$$

then, via Parseval equality for the Mellin transform (see the analog of Theorem 35 in [14]), and according to the definition of the KL_α-transform (4), we have

$$\begin{aligned}\mathrm{KL}_\alpha[G(x,t)](\tau) &= 2A(t)\left|\Gamma\left(\alpha+1+\frac{i\tau}{2}\right)\right|^{-2}\int_0^\infty x^\alpha K_{i\tau}(2\sqrt{x})B(xg(t))dx\\ &= \frac{1}{2\pi i}A(t)\left|\Gamma\left(\alpha+1+\frac{i\tau}{2}\right)\right|^{-2}\\ &\times\int_{\gamma-i\infty}^{\gamma+i\infty}\Gamma\left(s+\alpha+\frac{i\tau}{2}\right)\Gamma\left(s+\alpha-\frac{i\tau}{2}\right)B^\star(1-s)\Big(g(t)\Big)^{s-1}ds.\end{aligned}$$

The result now follows after an elementary substitution in the latter integral. □

Theorem 2.1. *Let $\{P_n\}_{n\geqslant 0}$ be an MPS generated by $G(x,t)=A(t)B(xg(t))$, with A, B, and g realizing conditions* (16) *and such that for each $\tau \in \mathbb{R}_+$, the integral of*

$$KL_\alpha\left[\left(\frac{\partial^n}{\partial u^n}B(u)\right)\Big|_{u=xt}x^n\right](\tau)$$

converges uniformly by $t\in[0,\delta]$, $\delta>0$.

Then under assumptions $B(x)\in L_1(\mathbb{R}_+,x^{-\gamma}dx)$, $\gamma>-\alpha$, and $(s)_nB^\star(s)\in L_1(1-\gamma-i\infty,1-\gamma+i\infty)$ for any $n\in\mathbb{N}$, the MPS $\{Q_n := Q_n(\cdot;\alpha)\}_{n\geqslant 0}$ defined by

$$KL_\alpha[P_n(x)](\tau)=Q_n(\tau;\alpha)$$

is generated by $A(t)KL_\alpha[B(xg(t))](\tau)$, which can be calculated via (17).

Proof. We begin by showing that for any integer $n\geqslant 0$ and fixed nonnegative τ, the following identity holds:

$$\mathrm{KL}_\alpha\left[\frac{\partial^n}{\partial t^n}B(xt)\right](\tau)=\mathrm{KL}_\alpha\left[\left(\frac{\partial^n}{\partial u^n}B(u)\right)\Big|_{u=xt}x^n\right](\tau)=\frac{\partial^n}{\partial t^n}\mathrm{KL}_\alpha\left[B(xt)\right](\tau),\ t>0. \tag{19}$$

Indeed, since $B(x)\in L_1(\mathbb{R}_+,x^{-\gamma}dx)$ and $(s)_nB^\star(s)\in L_1(1-\gamma-i\infty,1-\gamma+i\infty)$ for any $n\in\mathbb{N}$, this means that $B(x)$ is infinite times differentiable and can be represented via the reciprocal integral of the inverse Mellin transform

$$B(x) = \frac{1}{2\pi i} \int_{1-\gamma-i\infty}^{1-\gamma+i\infty} B^\star(s) x^{-s} ds, \quad x > 0.$$

Moreover, its nth derivative is given accordingly

$$\frac{\partial^n}{\partial x^n} B(x) = \frac{(-1)^n}{2\pi i} \int_{1-\gamma-i\infty}^{1-\gamma+i\infty} (s)_n B^\star(s) x^{-s-n} ds,$$

where the integral is absolutely convergent. Hence, the boundedness of the Gamma product

$$\Gamma\left(1 - s + \alpha + \frac{i\tau}{2}\right) \Gamma\left(1 - s + \alpha - \frac{i\tau}{2}\right)$$

under the condition $\gamma > -\alpha$ allows us to differentiate n times with respect to $t \geqslant t_0 > 0$ under the integral sign of (17) with $g(t) = t$, owing to the absolute and uniform convergence. Thus, we obtain

$$\begin{aligned} &\frac{\partial^n}{\partial t^n} \mathrm{KL}_\alpha \left[B(xt)\right](\tau) \\ &= \frac{(-1)^n}{2\pi i \left|\Gamma\left(\alpha + 1 + \frac{i\tau}{2}\right)\right|^2} \int_{1-\gamma-i\infty}^{1-\gamma+i\infty} \Gamma\left(1 - s + \alpha + \frac{i\tau}{2}\right) \Gamma\left(1 - s + \alpha - \frac{i\tau}{2}\right) B^\star(s)(s)_n t^{-s-n} ds. \end{aligned}$$

On the other hand, applying again the Parseval identity for the Mellin transform in the right-hand side of the latter equality, we deduce

$$\begin{aligned} &\frac{(-1)^n}{2\pi i \left|\Gamma\left(\alpha + 1 + \frac{i\tau}{2}\right)\right|^2} \int_{1-\gamma-i\infty}^{1-\gamma+i\infty} \Gamma\left(1 - s + \alpha + \frac{i\tau}{2}\right) \Gamma\left(1 - s + \alpha - \frac{i\tau}{2}\right) B^\star(s)(s)_n t^{-s-n} ds \\ &= \frac{t^{-n}}{\left|\Gamma\left(\alpha + 1 + \frac{i\tau}{2}\right)\right|^2} \int_0^\infty x^\alpha K_{i\tau}(2\sqrt{x})(xt)^n \frac{\partial^n}{\partial (xt)^n} B(xt) dx \\ &= \frac{1}{\left|\Gamma\left(\alpha + 1 + \frac{i\tau}{2}\right)\right|^2} \int_0^\infty x^\alpha K_{i\tau}(2\sqrt{x}) \frac{\partial^n}{\partial t^n} B(xt) dx = \mathrm{KL}_\alpha \left[\frac{\partial^n}{\partial t^n} B(xt)\right](\tau), \end{aligned}$$

which completes the proof of (19). Moreover, passing to the limit under the integral sign in (19), when $t \to 0+$ due to the uniform convergence on $[0, \delta]$, we obtain

$$\begin{aligned} \lim_{t\to 0+} \mathrm{KL}_\alpha \left[\frac{\partial^n}{\partial t^n} B(xt)\right](\tau) &= \mathrm{KL}_\alpha \left[\lim_{t\to 0+} \frac{\partial^n}{\partial t^n} B(xt)\right](\tau) = \lim_{t\to 0+} \frac{\partial^n}{\partial t^n} \mathrm{KL}_\alpha \left[B(xt)\right](\tau) \\ &= \left(\frac{\partial^n}{\partial t^n} \mathrm{KL}_\alpha \left[B(xt)\right](\tau)\right)\Bigg|_{t=0}. \end{aligned} \tag{20}$$

Further, we have

$$\frac{n!}{\rho_n} P_n(x) = \frac{\partial^n}{\partial t^n} G(x,t)\Big|_{t=0} = \sum_{\nu=0}^{n} \binom{n}{\nu} A_{n-\nu} \, \frac{\partial^\nu}{\partial t^\nu} B\big(xg(t)\big)\Big|_{t=0},$$

where $A_\sigma = \frac{\partial^\sigma}{\partial t^\sigma} A(t)\big|_{t=0}$, for any positive integer σ. Now the *Faa di Bruno's formula* permits us to formally write the successive derivatives of the function $B\big(xg(t)\big)$ at the point $t = 0$—see [7, pp. 137–140]. Precisely,

$$\frac{\partial^\nu}{\partial t^\nu} B\big(xg(t)\big)\Big|_{t=0} = \sum_{\mu=0}^{\nu} \frac{\partial^\mu B(u)}{\partial u^\mu}\Big|_{u=0} \mathrm{B}_{\nu,\mu}(g_1,\ldots,g_{\nu-\mu+1})\; x^\mu$$

where $\mathrm{B}_{\nu,\mu}(g_1,\ldots,g_{\nu-\mu+1})$ corresponds to the Bell polynomials evaluated at the successive derivatives of the function g. Here, $g_\mu = \frac{\partial^\mu g(u)}{\partial u^\mu}\Big|_{u=0}$, for any positive integer μ, and, for instance, the case where $g(t) = t$ implies $\mathrm{B}_{\nu,\mu}(g_1,\ldots,g_{\nu-\mu+1}) = 1$. Thus,

$$\begin{aligned}\frac{n!}{\rho_n} P_n(x) &= \sum_{\nu=0}^{n} \binom{n}{\nu} A_{n-\nu} \sum_{\mu=0}^{\nu} \mathrm{B}_{\nu,\mu}(g_1,\ldots,g_{\nu-\mu+1}) \left[\frac{\partial^\mu B(u)}{\partial u^\mu}\Big|_{u=0}\right] x^\mu \\ &= \sum_{\nu=0}^{n} \binom{n}{\nu} A_{n-\nu} \sum_{\mu=0}^{\nu} \mathrm{B}_{\nu,\mu}(g_1,\ldots,g_{\nu-\mu+1}) \left(\frac{\partial^\mu}{\partial t^\mu} B(xt)\right)\Big|_{t=0}, \quad n \geqslant 0.\end{aligned}$$

The action of the operator KL_α on both sides of the first and last members of the precedent equalities, along with the linearity of this operator, leads to

$$\frac{n!}{\rho_n} Q_n(\tau) = \sum_{\nu=0}^{n} \binom{n}{\nu} A_{n-\nu} \sum_{\mu=0}^{\nu} \mathrm{B}_{\nu,\mu}(g_1,\ldots,g_{\nu-\mu+1})\; \mathrm{KL}_\alpha \left[\frac{\partial^\mu}{\partial t^\mu} B(xt),\Big|_{t=0}\right](\tau),$$

which can be equivalently written like

$$\begin{aligned}&\frac{n!}{\rho_n} Q_n(\tau) \\ &= \sum_{\nu=0}^{n} \binom{n}{\nu} A_{n-\nu} \sum_{\mu=0}^{\nu} \left[\frac{\partial^\mu B(u)}{\partial u^\mu}\Big|_{u=0}\right] \mathrm{B}_{\nu,\mu}(g_1,\ldots,g_{\nu-\mu+1}) \left|\left(\alpha+1+\frac{i\tau}{2}\right)_\mu\right|^2,\end{aligned}$$

where $n \geqslant 0$.

Now, the identity (19) permits us to write

$$\frac{n!}{\rho_n} Q_n(x)$$
$$= \sum_{\nu=0}^{n} \binom{n}{\nu} A_{n-\nu} \sum_{\mu=0}^{\nu} \mathrm{B}_{\nu,\mu}(g_1, \ldots, g_{\nu-\mu+1}) \left(\frac{\partial^\mu}{\partial t^\mu} \mathrm{KL}_\alpha \left[B(xt) \right] (\tau) \right) \bigg|_{t=0}, \; n \geqslant 0.$$

Arguing, once again, with the Faa Di Bruno's formula followed by the Leibniz rule, we come out with

$$\frac{n!}{\rho_n} Q_n(x) = \sum_{\nu=0}^{n} \binom{n}{\nu} A_{n-\nu} \left(\frac{\partial^\nu}{\partial t^\nu} \mathrm{KL}_\alpha \left[B(xg(t)) \right] (\tau) \right) \bigg|_{t=0}$$
$$= \frac{\partial^n}{\partial t^n} \left(A(t) \, \mathrm{KL}_\alpha \left[B(xg(t)) \right] (\tau) \right) \bigg|_{t=0}, \quad n \geqslant 0,$$

whence the result. □

Some particular cases of the aforementioned Bell polynomials are worth to be shown [7]:

- If $g(t) = \mathrm{e}^t$, then $\mathrm{B}_{n,k}(1,1,\ldots,1) = S(n,k)$, $1 \leqslant k \leqslant n$, with $S(n,k)$ representing the Stirling numbers of second kind.
- If $g(t) = t\mathrm{e}^t$, then $\mathrm{B}_{n,k}(1,2,3,4,\ldots) = \binom{n}{k} k^{n-k}$, $1 \leqslant k \leqslant n$, which correspond to the idempotent numbers.
- If $g(t) = \frac{t}{1-t}$, then $\mathrm{B}_{n,k}(1!,2!,3!,4!,\ldots) = \binom{n-1}{k-1} \frac{n!}{k!}$, $1 \leqslant k \leqslant n$, which correspond to the Lah numbers.

Another necessary condition over a generating function of a given MPS is given in the following result, which, indeed supplies a generating function of the KL_α-transformed sequence.

Proposition 2.1. *Let $G(x,t)$ be the generating function given in* (15) *of the MPS $\{P_n\}_{n\geqslant 0}$. If*

$$\sum_{n\geqslant 0} \left| P_n(x) \frac{t^n}{\rho_n} \right| \in L_1(\mathbb{R}_+; x^\alpha K_0(2\sqrt{x}) dx), \tag{21}$$

then the MPS $\{S_n(\cdot) := KL_\alpha[P_n(x)](\cdot)\}_{n\geqslant 0}$ is generated by

$$A(t) KL_\alpha[B(xg(t))](\tau) = \sum_{n\geqslant 0} S_n(\tau) \frac{t^n}{\rho_n}.$$

Proof. The condition (21) ensures the integrability of $G(x,t)$ with respect to the measure $x^\alpha K_0(2\sqrt{x})dx$. Therefore, we consider the action of the KL_α-transform on both members of (15), and we obtain

$$\begin{aligned}\mathrm{KL}_\alpha[G(x,t)](\tau) &= \mathrm{KL}_\alpha\left[\lim_{n\to\infty}\sum_{j=0}^{n} P_n(x)\frac{t^n}{\rho_n}\right](\tau)\\ &= 2\left|\Gamma\left(\alpha+1+\frac{i\tau}{2}\right)\right|^{-2}\int_0^\infty\left(\lim_{n\to\infty}\sum_{j=0}^{n} P_n(x)\frac{t^n}{\rho_n}\right)x^\alpha K_{i\tau}(2\sqrt{x})dx.\end{aligned}$$

The result now follows in the light of the dominated convergence theorem. □

Particular choices of the function $B(\cdot)$ permit to explicitly express the generating function of the corresponding KL_α-transformed MPS.

Proposition 2.2. *If an MPS* $\{P_n\}_{n\geqslant 0}$ *is generated by* (15), *subject to the conditions* (16), *with*

$$B(xg(t)) = G^{m,n}_{p,q}\left(xg(t);\begin{matrix}\mathbf{a}\\ \mathbf{b}\end{matrix}\right), \tag{22}$$

then the MPS $\{S_n\}_{n\geqslant 0}$ *is generated by*

$$\begin{aligned}&2A(t)\left|\Gamma\left(\alpha+1+\frac{i\tau}{2}\right)\right|^{-2} G^{m+2,n}_{p+2,q}\left(g(t);\begin{matrix}-\alpha-i\tau/2,-\alpha+i\tau/2,\mathbf{a}\\ \mathbf{b}\end{matrix}\right)\\ &=\sum_{n\geqslant 0} S_n(\tau^2/4)\frac{t^n}{\prod_{\sigma=1}^n \rho_\sigma}.\end{aligned}$$

Proof. Let $G(x,t)$ be the generating function given in (15). Based on the relation (11) in [9, p. 215, vol. I], we obtain

$$\mathrm{KL}_\alpha[G(\cdot,t)](\tau) = 2A(t)\left|\Gamma\left(\alpha+1+\frac{i\tau}{2}\right)\right|^{-2} G^{m+2,n}_{p+2,q}\left(g(t);\begin{matrix}-\alpha-i\tau/2,-\alpha+i\tau/2,\mathbf{a}\\ \mathbf{b}\end{matrix}\right)$$

and now the result follows due to Proposition 2.1. □

Sometimes, when certain differential and structural properties are known about a given MPS, we are able to deduce those of the so-called reversed sequence. We recall the concept.

Definition 2.2. Given an MPS $\{P_n(\cdot)\}_{n\geqslant 0}$ such that $P_n(0)=\lambda_n\neq 0$ for all $n\in\mathbb{N}_0$, it is possible to construct another MPS $\{R_n(\cdot)\}_{n\geqslant 0}$ defined by

$$R_n(x) = \frac{1}{\lambda_n}x^n P_n\left(\frac{1}{x}\right),\ n\in\mathbb{N}_0, \tag{23}$$

to which we will refer to as the *reversed polynomial sequence* or simply as the *reversed polynomials*.

For instance, an Appell sequence can be generated by $G(x,t) = A(t)e^{xt}$ (a Sheffer-type sequence), with $A(0)\neq 0$, and the corresponding reversed sequence would then be generated by $H(x,t) = e(t)A(xt)$; it is a Brenke-type sequence. Within the same spirit, we can also deal with more general generating functions.

Theorem 2.2. *Suppose* $\{P_n(\cdot) := P_n\left(\cdot; {}^{\mathbf{a}}_{\mathbf{b}}\right)\}_{n\geqslant 0}$ *is an MPS possibly depending on* $p+q$ *parameters* $a_1,\ldots,a_p,b_1,\ldots,b_q$ *with* $a_1 \neq -1$ *and such that* $P_n(0)=\lambda_n \neq 0$. *Let* $\{R_n\left(\cdot; {}^{\mathbf{a}}_{\mathbf{b}}\right)\}_{n\geqslant 0}$ *be the corresponding reversed polynomial sequence defined in* (23). *The following statements are equivalent:*

(a) *The MPS* $\{P_n(\cdot) := P_n\left(\cdot; {}^{\mathbf{a}}_{\mathbf{b}}\right)\}_{n\geqslant 0}$ *is the Brenke-type generated by* $G(x,t)$, *that is,*

$$A(t)\, G^{m,n}_{p,q}\left(xt; \begin{matrix}\mathbf{a}\\ \mathbf{b}\end{matrix}\right) = \sum_{k\geqslant 0} P_k\left(x; \begin{matrix}\mathbf{a}\\ \mathbf{b}\end{matrix}\right) \frac{t^k}{\prod_{\sigma=1}^{k}\rho_\sigma},$$

where $\rho_k = k/(a_1+1)$ *or* $\rho_k = k$, $k\in\mathbb{N}$, *when there is no dependence on* $\mathbf{a}$.

(b) *The MPS* $\{P_n(\cdot) := P_n\left(\cdot; {}^{\mathbf{a}}_{\mathbf{b}}\right)\}_{n\geqslant 0}$ *fulfills*

$$\frac{d}{dx}\, P_{k+1}\left(x; \begin{matrix}\mathbf{a}+1\\ \mathbf{b}+1\end{matrix}\right) = \frac{k+1}{a_1+1}\left\{ P_k\left(x; \begin{matrix}a_1-1,a_2,\ldots,a_p\\ b_1,\ldots,b_q\end{matrix}\right) + a_1\, P_k\left(x; \begin{matrix}\mathbf{a}\\ \mathbf{b}\end{matrix}\right)\right\} \tag{24}$$

for $k\in\mathbb{N}_0$, *with* $P_0\left(\cdot; {}^{\mathbf{a}}_{\mathbf{b}}\right)=1$.

(c) *The sequence* $\{R_n(\cdot) := R_n\left(\cdot; {}^{\mathbf{a}}_{\mathbf{b}}\right)\}_{n\geqslant 0}$ *fulfills*

$$\begin{aligned}\frac{d}{dx}x\, R_{k+1}\left(x; \begin{matrix}\mathbf{a}+1\\ \mathbf{b}+1\end{matrix}\right) &= (k+2)\, R_{k+1}\left(x; \begin{matrix}\mathbf{a}+1\\ \mathbf{b}+1\end{matrix}\right)\\ &-\frac{k+1}{a_1+1}\frac{\lambda_k}{\lambda_{k+1}}\left\{ R_k\left(x; \begin{matrix}a_1-1,a_2,\ldots,a_p\\ b_1,\ldots,b_q\end{matrix}\right) + a_1\, R_k\left(x; \begin{matrix}\mathbf{a}\\ \mathbf{b}\end{matrix}\right)\right\}\end{aligned} \tag{25}$$

with $R_0\left(\cdot; {}^{\mathbf{a}}_{\mathbf{b}}\right)=1$.

(d) *The MPS* $\{R_n\left(\cdot; {}^{\mathbf{a}}_{\mathbf{b}}\right)\}_{n\geqslant 0}$ *is generated by*

$$G^{m,n}_{p,q}\left(t; \begin{matrix}\mathbf{a}\\ \mathbf{b}\end{matrix}\right) A(xt) = \sum_{k\geqslant 0} \lambda_k R_k\left(x; \begin{matrix}\mathbf{a}\\ \mathbf{b}\end{matrix}\right) \frac{t^k}{\prod\limits_{\sigma=1}^{k}\rho_\sigma}.$$

Proof. The change of variable $x \to 1/x$ and $t \to xt$ permits to readily conclude the equivalence of statements (a) and (d), along with the equivalence between statements (b) and (c).

On the grounds of the properties of the Meijer G-function (see [9, vol. I, p. 210] or [8, Sect. 16.19]), we have

$$\frac{d}{dz}\, G^{n,m}_{p,q}\left(z; \begin{matrix}\mathbf{a}+1\\ \mathbf{b}+1\end{matrix}\right) = G^{m,n}_{p,q}\left(z\; \begin{matrix}a_1-1,a_2,\ldots,a_p\\ b_1,\ldots,b_q\end{matrix}\right) + a_1\, G^{m,n}_{p,q}\left(z; \begin{matrix}a_1,\ldots,a_p\\ b_1,\ldots,b_q\end{matrix}\right), \tag{26}$$

we are able to successively write the identities

$$\begin{aligned}
&\sum_{k\geqslant 1}\left(\frac{d}{dx}P_k\left(x;\begin{matrix}\mathbf{a}+1\\ \mathbf{b}+1\end{matrix}\right)\right)\frac{t^k}{\rho_k}\\
&=\frac{d}{dx}\left(A(t)\,G^{m,n}_{p,q}\left(xt;\begin{matrix}\mathbf{a}+1\\ \mathbf{b}+1\end{matrix}\right)\right)\\
&=t\,A(t)\left\{G^{m,n}_{p,q}\left(xt;\begin{matrix}a_1-1,a_2,\ldots,a_p\\ b_1,\ldots,b_q\end{matrix}\right)+a_1\,G^{m,n}_{p,q}\left(xt;\begin{matrix}a_1,\ldots,a_p\\ b_1,\ldots,b_q\end{matrix}\right)\right\}\\
&=\sum_{k\geqslant 0}\left\{P_k\left(x;\begin{matrix}a_1-1,a_2,\ldots,a_p\\ b_1,\ldots,b_q\end{matrix}\right)+a_1\,P_k\left(x;\begin{matrix}\mathbf{a}\\ \mathbf{b}\end{matrix}\right)\right\}\frac{t^{k+1}}{\rho_k}
\end{aligned}$$

ensuring the equivalence between (a) and (b). □

Let us respectively denote by $\{\hat{P}_n\}_{n\geqslant 0}$ and $\{\hat{R}_n\}_{n\geqslant 0}$ the images by the KL_α-transform of the two MPSs $\{P_n\}_{n\geqslant 0}$ and $\{R_n\}_{n\geqslant 0}$ respectively given in statements (a)–(b) and (c)–(d) in Theorem 2.2. Here, we mean

$$\hat{P}_n(\cdot):=\hat{P}_n\left(x;\begin{matrix}\mathbf{a}\\ \mathbf{b}\end{matrix};\alpha\right)=\mathrm{KL}_\alpha\left[P_n\left(x;\begin{matrix}\mathbf{a}\\ \mathbf{b}\end{matrix}\right)\right](\tau),\quad n\in\mathbb{N}_0,$$

and

$$\hat{R}_n(\cdot):=\hat{R}_n\left(x;\begin{matrix}\mathbf{a}\\ \mathbf{b}\end{matrix};\alpha\right)=\mathrm{KL}_\alpha\left[R_n\left(x;\begin{matrix}\mathbf{a}\\ \mathbf{b}\end{matrix}\right)\right](\tau),\quad n\in\mathbb{N}_0.$$

Thus, under these notations, from (24) and in the light of the property (13) of the KL_α-transform, it follows that $\{\hat{P}_n\}_{n\geqslant 0}$ fulfills

$$\begin{aligned}
&\delta_i\left(\hat{P}_{k+1}\left(x;\begin{matrix}\mathbf{a}+1\\ \mathbf{b}+1\end{matrix};\alpha\right)\right)\\
&=\frac{k+1}{a_1+1}\left\{\hat{P}_k\left(x;\begin{matrix}a_1-1,a_2,\ldots,a_p\\ b_1,\ldots,b_q\end{matrix};\alpha+\frac{1}{2}\right)+a_1\,\hat{P}_k\left(x;\begin{matrix}\mathbf{a}\\ \mathbf{b}\end{matrix};\alpha+\frac{1}{2}\right)\right\},\quad k\in\mathbb{N}_0,
\end{aligned}$$

where δ_ω represents the operator given by (11).

Likewise, due to (13) the relation (25) realized by the MPS $\{R_n\}_{n\geqslant 0}$ induces a corresponding one fulfilled by $\{\hat{R}_n\}_{n\geqslant 0}$:

$$\begin{aligned}
&\left((\alpha+1)^2+\frac{\tau^2}{4}\right)\delta_i\left(\hat{R}_{k+1}\left(x;\begin{matrix}\mathbf{a}+1\\ \mathbf{b}+1\end{matrix};\alpha+\frac{1}{2}\right)\right)=(k+1)\,\hat{R}_{k+1}\left(x;\begin{matrix}\mathbf{a}+1\\ \mathbf{b}+1\end{matrix};\alpha\right)\\
&-\frac{k+1}{a_1+1}\frac{\lambda_k}{\lambda_{k+1}}\left\{\hat{R}_k\left(x;\begin{matrix}a_1-1,a_2,\ldots,a_p\\ b_1,\ldots,b_q\end{matrix};\alpha\right)+a_1\,\hat{R}_k\left(x;\begin{matrix}\mathbf{a}\\ \mathbf{b}\end{matrix};\alpha\right)\right\},\quad k\in\mathbb{N}_0.
\end{aligned}$$

Other relations for the two MPSs $\{\hat{P}_n\}_{n\geqslant 0}$ and $\{\hat{R}_n\}_{n\geqslant 0}$ may be obtained after straightforward computations on the grounds of the relations (13)–(14).

3 Some Examples of Generalized Hypergeometric-Type Polynomials

The generalized hypergeometric functions are particular realizations of the Meijer G-function, and these include elementary functions and some well-known special functions. For further reading, we refer to [12, Chap. VI] or [9]. For instance, the hypergeometric functions can be expressed as

$$\left(\frac{\prod_{k=1}^{p}\Gamma(a_k)}{\prod_{k=1}^{q}\Gamma(b_k)}\right) {}_pF_q\left(\begin{matrix}a_1,\ldots,a_p\\ b_1,\ldots,b_q\end{matrix};z\right) = G^{1,p}_{p,q+1}\left(-z;\begin{matrix}1-a_1,\ldots,1-a_p\\ 0,1-b_1,\ldots,1-b_q\end{matrix}\right)$$
$$= G^{p,1}_{q+1,p}\left(-\frac{1}{z};\begin{matrix}1,b_1,\ldots,b_q\\ a_1,\ldots,a_p\end{matrix}\right)$$

for $p \leqslant q$ or $p = q+1$ and $|z| < 1$.

The KL_α-transform of an MPS $\{P_n\}_{n\geqslant 0}$ of hypergeometric type defined by

$$P_n(x) = (-1)^n \frac{\left(\prod_{\nu=1}^{q}(b_\nu)_n\right)}{\left(\prod_{\nu=1}^{p}(a_\nu)_n\right)}\ {}_{p+1}F_q\left(\left.\begin{matrix}-n,a_1,\ldots,a_p\\ b_1,\ldots,b_q\end{matrix}\right| x\right) \tag{27}$$

where the coefficients a_j, b_k with $j = 1,\ldots,p$ and $k = 1,\ldots,q$ do not depend on x but possibly depending on n is again a hypergeometric polynomial type MPS, say, $\{S_n\}_{n\geqslant 0}$, and is given by

$$S_n\left(\frac{\tau^2}{4}\right) = (-1)^n \frac{\prod_{\nu=1}^{q}(b_\nu)_n}{\prod_{\nu=1}^{p}(a_\nu)_n}\ {}_{p+3}F_q\left(\left.\begin{matrix}-n,a_1,\ldots,a_p,\alpha+1-\frac{i\tau}{2},\alpha+1+\frac{i\tau}{2}\\ b_1,\ldots,b_q\end{matrix}\right| 1\right). \tag{28}$$

We drive the attention toward some examples of polynomial sequences whose generating function involves hypergeometric functions, which are nothing but a particular case of the Meijer G-function.

Example 3.1. The polynomial sequences generated by hypergeometric-type functions next considered can be found in [9, vol. III, pp. 266–267]. Their KL_α-images can be computed straightforwardly, and they just give rise to other polynomial sequences. We will adopt the notation $\{p_n\}_{n\geqslant 0}$ instead of $\{P_n\}_{n\geqslant 0}$, if we are referring to a polynomial sequence whose elements are not monic.

1. Just as noticed by Rainville (1947), the polynomial sequence $\{p_n\}_{n\geqslant 0}$ with

$$p_n(x) = {}_{p+1}F_q\left(\begin{matrix}-n, a_1, \ldots, a_p \\ b_1, \ldots, b_q\end{matrix}\,\middle|\, x\right)$$

is generated by

$$e^t\, {}_pF_q\left(\begin{matrix}a_1, \ldots, a_p \\ b_1, \ldots, b_q\end{matrix}\,\middle|\, -xt\right) = \sum_{n\geqslant 0} p_n(x)\frac{t^n}{n!}, \tag{29}$$

which is mapped by the KL_α-transform into the sequence of polynomials $\{q_n(\cdot;\alpha)\}_{n\geqslant 0}$ whose explicit expression is

$$q_n(\tau^2/4;\alpha) = {}_{p+3}F_q\left(\begin{matrix}-n, a_1, \ldots, a_p, \alpha+1+i\frac{\tau}{2}, \alpha+1-i\frac{\tau}{2} \\ b_1, \ldots, b_q\end{matrix}\,\middle|\, 1\right), \quad n \geqslant 0$$

and generated by

$$e^t\, {}_{p+2}F_q\left(\begin{matrix}a_1, \ldots, a_p, \alpha+1+i\frac{\tau}{2}, \alpha+1-i\frac{\tau}{2} \\ b_1, \ldots, b_q\end{matrix}\,\middle|\, -t\right) = \sum_{n\geqslant 0} q_n(\tau^2/4;\alpha)\frac{t^n}{n!}. \tag{30}$$

The polynomial sequences $\{p_n\}_{n\geqslant 0}$ and $\{q_n(\cdot;\alpha)\}_{n\geqslant 0}$ are both examples of d-orthogonal sequences [1,2]. For instance, the orthogonal polynomial sequence of the Continuous Dual Hahn polynomials is the image of a 2-orthogonal sequence of Laguerre type [10].

2. In the sequel of the works of Fasenmeyr (1947) and Brafman [5], it turns out that

$$\frac{1}{1-t}\, {}_pF_q\left(\begin{matrix}a_1, \ldots, a_p \\ b_1, \ldots, b_q\end{matrix}\,\middle|\, -\frac{4xt}{(1-t)^2}\right) = \sum_{n\geqslant 0} {}_{p+2}F_{q+2}\left(\begin{matrix}-n, n+1, a_1, \ldots, a_p \\ 1/2, 1, b_1, \ldots, b_q\end{matrix}\,\middle|\, x\right) t^n.$$

The action of KL_α over both sides of the precedent equality provides

$$\frac{1}{1-t}\, {}_{p+2}F_q\left(\begin{matrix}a_1, \ldots, a_p, \alpha+1-i\tau/2, \alpha+1+i\tau/2 \\ b_1, \ldots, b_q\end{matrix}\,\middle|\, -\frac{4t}{(1-t)^2}\right)$$

$$= \sum_{n\geqslant 0} {}_{p+4}F_{q+2}\left(\begin{matrix}-n, n+1, a_1, \ldots, a_p, \alpha+1-i\tau/2, \alpha+1+i\tau/2 \\ 1/2, 1, b_1, \ldots, b_q\end{matrix}\,\middle|\, 1\right) t^n.$$

3. Analogously, the sequence of polynomials $\{p_n := p_n(\cdot;\lambda)\}_{n\geqslant 0}$ defined by

$$p_n(x;\lambda) = (\lambda)_n\, {}_{p+1}F_q\left(\begin{matrix}-n, a_1, \ldots, a_p \\ b_1, \ldots, b_q\end{matrix}\,\middle|\, x\right), \quad n \geqslant 0$$

and generated by [6]

$$(1-t)^{-\lambda}\ {}_{p+1}F_q\left(\begin{matrix}\lambda, a_1, \ldots, a_p \\ b_1, \ldots, b_q\end{matrix}\,\middle|\, -\frac{xt}{1-t}\right) = \sum_{n\geqslant 0} p_n(x;\lambda)\frac{t^n}{n!}$$

is mapped, by the KL$_\alpha$-transform, into the sequence $\{q_n := p_n(\cdot;\lambda,\alpha)\}_{n\geqslant 0}$ given by

$$q_n\Big(\frac{\tau^2}{4};\lambda,\alpha\Big) = \frac{t^n}{n!}\ {}_{p+3}F_q\left(\begin{matrix}-n, a_1, \ldots, a_p, \alpha+1-i\tau/2, \alpha+1+i\tau/2 \\ b_1, \ldots, b_q\end{matrix}\,\middle|\, 1\right)$$

and generated by

$$(1-t)^{-\lambda}\ {}_{p+3}F_q\left(\begin{matrix}\lambda, a_1, \ldots, a_p, \alpha+1-i\tau/2, \alpha+1+i\tau/2 \\ b_1, \ldots, b_q\end{matrix}\,\middle|\, -\frac{t}{1-t}\right)$$
$$= \sum_{n\geqslant 0} q_n(\tau^2/4;\lambda,\alpha)\frac{t^n}{n!}\,.$$

As pointed out in [10], the KL$_\alpha$-transform maps d-orthogonal polynomial sequence into other $\tilde{d}$-orthogonal polynomial sequences, with $d,\ \tilde{d} = 1,2,3,\ldots$ with $d + \tilde{d} \geqslant 3$. Based on the description of all the d-orthogonal polynomial sequences provided in [2], it would be worth to analyze the "KL$_\alpha$-connections" between such sequences.

Acknowledgements Work of AFL was supported by Fundação para a Ciência e a Tecnologia via the grant SFRH/BPD/63114/2009. Research was partially funded by the European Regional Development Fund through the programme COMPETE and by the Portuguese Government through the FCT (Fundação para a Ciência e a Tecnologia) under the project PEst-C/MAT/UI0144/2011.

References

1. Ben Cheikh, Y., Chaggara, H.: Connection coefficients between Boas-Buck polynomial sets. J. Math. Anal. Appl. **319**, 665–689 (2006)
2. Ben Cheikh, Y., Lamiri, I., Ouni, A.: On Askey-scheme and d-orthogonality, I: A characterization theorem. J. Comput. Appl. Math. **233**(3), 621–629 (2009)
3. Boas, Jr., R.P., Buck, R.C.: Polynomials defined by generating relations. Amer. Math. Monthly **63**, 626–632 (1956)
4. Boas, R.P., Buck, R.C.: Polynomial Expansions of Analytic Functions. Springer, Berlin (1964)
5. Brafman, F.: Generating functions of Jacobi and related polynomials. Proc. Amer. Math. Soc. **2**, 942–949 (1951)
6. Chaunday, T.X.: An extension of hypergeometric functions (I). Quart. J. Math. Oxford **14**, 55–78 (1943)
7. Comtet, L.: Advanced Combinatorics - The Art of Finite and Infinite Expansions. D. Reidel Publishing Co., Dordrecht (1974)

8. Digital Library of Mathematical Functions, 2011-07-01. National Institute of Standards and Technology from http://dlmf.nist.gov/24.
9. Erdélyi, A., Magnus, W., Oberhettinger, F., Tricomi, F.G.: Higher Transcendental Functions, vol. I, II, III. McGraw-Hill, New York (1953)
10. Loureiro, A.F., Yakubovich, S.: The Kontorovich-Lebedev transform as a map between d-orthogonal polynomials. arXiv:1206.4899
11. Loureiro, A.F., Yakubovich, S.: Central factorials under the Kontorovich-Lebedev transform of polynomials. Integr. Transf. Spec. Funct. **24**(3), 217–238 (2013)
12. Luke, Y.L.: Special Functions and Their Approximations, vol. I. Academic, New York (1969)
13. Prudnikov, A.P., Brychkov, Y.A., Marichev, O.I.: Integral Transforms: Special Functions, vol. II. Gordon and Breach, New York (1986)
14. Titchmarsh, E.C.: An Introduction to the Theory of Fourier Integrals. Clarendon, Oxford (1937)
15. Yakubovich, S.B.: A remark on the inversion formula for Wimp's integral transformation with respect to the index. (Russian) Differentsial'nye Uravneniya **21**(6), 1097–1098 (1985)
16. Yakubovich, S.B.: Index Transforms. World Scientific Publishing Company, Singapore (1996)
17. Yakubovich, S.: Encyclopedia of Mathematics. http://www.encyclopediaofmath.org/index.php?title=Kontorovich-Lebedev_transform&oldid=22663
18. Yakubovich, S.B., Luchko, Y.F.: The Hypergeometric Approach to Integral Transforms and Convolutions. Mathematics and Applications, vol. 287. Kluwer Academic, Dordrecht (1994)

Goursat's Hypergeometric Transformations, Revisited

Per W. Karlsson

Dedicated to Professor Hari M. Srivastava

Abstract In a number of cases pairs of Goursat's higher-order hypergeometric transformations lend themselves to eliminations of the right-hand members; one just has to make his or her variables equal by solving certain algebraic equations of degree up to six. This is carried out by the aid of MAPLE, in most cases supplemented by suitable one-to-one variable substitutions.

1 Introduction

Goursat obtained [4] quite a few cubic, quartic and sextic transformations of the Gaussian hypergeometric function, and some of them have an interesting property which we shall consider in this article. Indeed, it is readily noticed that certain transformations in [4] may be combined into subsets that have parametrically identical ${}_2F_1$'s on the right-hand sides. So, from a pair of such transformations we may eliminate the said ${}_2F_1$ by adjusting the variables and arrive at a new transformation, or several such ones. This 'adjustment' means solving algebraic equations of degree up to six, and such a scheme could not be carried out by hand. Nowadays, MAPLE can handle such equations, but a direct approach turns out to be in most cases inconvenient or even useless. Fortunately, introduction of suitable auxiliary variables helps MAPLE finish the job.

Each transformation obtained contains one free parameter, and the variable on the right-hand side is explicitly given by the one on the left. Some of the transformations may be discarded because they turn out to be particular cases of well-known transformations. Still, a number of interesting cases do remain. A few have been considered previously [5].

P.W. Karlsson (deceased)

G.V. Milovanović and M.Th. Rassias (eds.), *Analytic Number Theory, Approximation Theory, and Special Functions*, DOI 10.1007/978-1-4939-0258-3_27,

Linear transformations may be applied to all higher-order transformations, and Goursat does in fact consider such variations to some extent. We shall omit such listings in the sequel.

Berndt et al. [1] and Garvan [3] have, from a different point of view, considered transformations of a similar appearance as far as the hypergeometric parameters are concerned; the variables, however, are in most cases given parametrically rather than explicitly.

Formula numbers from (75) onwards refer to [4]. Twenty-three transformations are to be considered as follows: (75), (76), (77) and (118) through (137). Note also the quadratic transformations listed in [2, Sect. 2.11].

For roots we choose the branch that is positive when the radicand is real and positive. In other words, $(1+z)^{\gamma} = {}_1F_0[-\gamma;\ ;-z]$.

2 Twelve Goursat Transformations

We first consider the largest subset, namely, the twelve transformations: (118) to (121), (126) to (129) and (134) to (137). They may be written in the common form

$$ {}_2F_1\left[\begin{matrix} a,b; \\ c; \end{matrix}\ x\right] = Q(x)^{-3\alpha}\,{}_2F_1\left[\begin{matrix} \alpha,\alpha+\frac{1}{3}; \\ 2\alpha+\frac{5}{6}; \end{matrix}\ \frac{P(x)}{Q(x)^3}\right], \tag{1} $$

where P and Q are polynomials given in the table below, and *a, b and c* depend upon α. Actually, we consider only seven transformations, because the others are obtained by taking $x = -\xi/(1-\xi)$ in (1), followed by application of a Euler transformation (L) to the left-hand side; this is indicated in the rightmost column. A few errors of sign in [4] have been corrected.

No.	a,b,c	$Q(x)$	$P(x)$	(L) $\rightarrow$
(118)	$3\alpha, 3\alpha+\frac{1}{2}, 4\alpha+\frac{2}{3}$	$1-\frac{3}{4}x$	$\frac{27}{64}x^2(1-x)$	(120)
(119)	$3\alpha, 3\alpha+\frac{1}{2}, 2\alpha+\frac{5}{6}$	$1+3x$	$27x(1-x)^2$	(121)
(126)	$4\alpha, 4\alpha+\frac{1}{3}, 6\alpha+\frac{1}{2}$	$1-\frac{8}{9}x$	$\frac{64}{729}x^3(1-x)$	(128)
(127)	$4\alpha, 4\alpha+\frac{1}{3}, 2\alpha+\frac{5}{6}$	$1+8x$	$64x(1-x)^3$	(129)
(134)	$6\alpha, 2\alpha+\frac{1}{3}, 4\alpha+\frac{2}{3}$	$1-x+x^2$	$-\frac{27}{4}x^2(1-x)^2$	itself
(136)	$6\alpha, 4\alpha+\frac{1}{6}, 2\alpha+\frac{5}{6}$	$1+14x+x^2$	$108x(1-x)^4$	(135)
(137)	$6\alpha, 4\alpha+\frac{1}{6}, 8\alpha+\frac{1}{3}$	$1-x+\frac{1}{16}x^2$	$\frac{27}{1024}x^4(1-x)$	itself

For each pair $(m)\,\&\,(n)$ of these Goursat transformations, we shall consider Eq. (1) for (m) together with a similar equation with tilded quantities for (n), yet $\tilde{\alpha}$ is the same as α. Writing y in place of $\tilde{x}$ we then have a new transformation:

$$ {}_2F_1\left[\begin{matrix} a,b; \\ c; \end{matrix} x\right] = \left[\frac{Q(x)}{\tilde{Q}(y)}\right]^{-3\alpha} {}_2F_1\left[\begin{matrix} \tilde{a},\tilde{b}; \\ \tilde{c}; \end{matrix} y\right] \tag{2} $$

provided that the condition

$$ \frac{P(x)}{Q(x)^3} = \frac{\tilde{P}(y)}{\tilde{Q}(y)^3} \tag{3} $$

is satisfied. We rewrite (3) as

$$ M(x,y) = 0, \tag{4} $$

where

$$ M(x,y) = L\left[P(x)\tilde{Q}(y)^3 - \tilde{P}(y)Q(x)^3\right], \tag{5} $$

and the constant L is, for convenience, so chosen that the polynomial M has integral coefficients. We shall consider only transformations that are, like Goursat's, valid in a neighbourhood of the origin. Accordingly, the *relevant* roots of (4) are those for which we have an asymptotic expression of the form

$$ y \simeq Kx^{\beta} \quad \text{for} \quad x \to 0, \tag{6} $$

where K and β are constants, easily obtained from (4) by inspection. The actual regions of validity for the transformations established will not be considered here; a few examples are found in [5].

So, the first step will consist in solving (4) with respect to y and selecting the relevant root(s) by comparison with (6). It turns out that a direct approach by MAPLE may leave us with inconvenient or even enormous expressions, or just `RootOf`-statements. However, introduction of suitable auxiliary variables by one-to-one substitutions of the form $x{=}\varphi(z_{\text{aux}})$ leads to reasonable expressions for the roots. It is then a simple matter to find the $\tilde{Q}$-polynomials and establish the resulting transformations (2).The adjacent table gives an overview of the results. Nine pairs of Goursat transformations lead to trivialities: particular cases of quadratic or linear transformations or the identical transformation. For brevity, these cases are not considered below. The remaining nineteen pairs lead to the interesting, at least cubic, transformations C1, C2, ... and C19. Some of these are accompanied by trivialities as indicated, and some are in fact two transformations.

3 The First Case

We first consider the pair (136) and (118), for which we find the polynomial

No.	(118)	(119)	(126)	(127)	(134)	(136)	(137)
(118)	C3, I						
(119)	C4	I					
(126)	C16	C10	C7, I				
(127)	C17	C11	C6	I			
(134)	Q	Q	C14	C15	L, I		
(136)	C1	Q	C12	C13	Q	I	
(137)	C5, Q	C2	C18	C19	Q	C8	C9, L, I

$$M = 4x(1-x)^4(4-3y)^3 - y^2(1-y)\left(1+14x+x^2\right)^3$$

and the asymptotic expression $y^2 \simeq 64x$. MAPLE finds the three roots, but the relevant ones involve $\sqrt{x}$. To avoid a branch point we substitute simply $x = u^2$. (Incidentally, one of the quadratic transformations, viz. [2, 2.11(5)], is also written in this way.) MAPLE now finds the roots $\left(1-u^2\right)^2 / \left(1+u^2\right)^2$, which are not relevant, and

$$y_{\pm} = \frac{\pm 16u(1 \pm u)^2}{\left(1 \pm 6u + u^2\right)^2}.$$

It follows that

$$4 - 3y_{\pm} = \frac{4\left(1+14u^2+u^4\right)}{\left(1 \pm 6u + u^2\right)^2};$$

hence,

$$\frac{1+14x+x^2}{1-\frac{3}{4}y_{\pm}} = \left(1 \pm 6u + u^2\right)^2,$$

and we have the transformations

$${}_2F_1\left[\begin{matrix}6\alpha, 4\alpha+\frac{1}{6};\\ 2\alpha+\frac{5}{6};\end{matrix}\ u^2\right] = \left(1 \pm 6u + u^2\right)^{-6\alpha} {}_2F_1\left[\begin{matrix}3\alpha, 3\alpha+\frac{1}{2};\\ 4\alpha+\frac{2}{3};\end{matrix}\ y_{\pm}\right].$$

Similar intermediate steps are carried out in the other cases; for brevity, such details are omitted in the sequel.

4 Transformations $C2$ to $C5$

For these pairs MAPLE finds the relevant roots as rational functions of x and $\sqrt{1-x}$. Simpler expressions arise if we substitute $x = 1 - u^2$, but we shall prefer the auxiliary variable $w = 1 - u$, which is small when x is small. In other words, we introduce the substitution

$$x = 2w - w^2, \qquad w = 1 - \sqrt{1-x}, \qquad w \simeq \frac{1}{2}x \quad \text{for} \quad x \to 0. \tag{7}$$

After substitution into (4) the resulting equation is solved for y in terms of w. It is noted that MAPLE finds *all* roots. Actually, these four transformations can also be proved by applying successively *two* quadratic transformations to the left-hand member and imposing one condition upon the parameters. This idea is due to Kummer, who obtained [6] many more instances.

4.1 C2 – (137) & (119)

Polynomial and relevant root:

$$M = 4x^4(1-x)(1+3y)^3 - y(1-y)^2\left(16 - 16x + x^2\right)^3, \quad y_0 = \frac{w^4}{\left(8 - 8w + w^2\right)^2}.$$

Transformation:

$${}_2F_1\left[\begin{matrix} 6\alpha, 4\alpha + \frac{1}{6}; \\ 8\alpha + \frac{1}{3}; \end{matrix} x\right] = \left(1 - w + \frac{1}{8}w^2\right)^{-6\alpha} {}_2F_1\left[\begin{matrix} 3\alpha, 3\alpha + \frac{1}{2}; \\ 2\alpha + \frac{5}{6}; \end{matrix} y_0\right],$$

$$w = 1 - \sqrt{1-x}.$$

Alternative: (31) and (4) in [2, Sect. 2.11].

4.2 C3 – (118) & (118)

Polynomial:

$$M = x^2(1-x)(4-3y)^3 - y^2(1-y)(4-3x)^3.$$

The relevant roots are x, which is trivial, and

$$y_0 = \frac{-8w(1-w)}{(2-3w)^2},$$

which leads to the transformation:

$$ {}_2F_1\left[\begin{matrix} 3\alpha, 3\alpha+\frac{1}{2}; \\ 4\alpha+\frac{2}{3}; \end{matrix} x\right] = \left(1-\frac{3}{2}w\right)^{-6\alpha} {}_2F_1\left[\begin{matrix} 3\alpha, 3\alpha+\frac{1}{2}; \\ 4\alpha+\frac{2}{3}; \end{matrix} y_0\right], $$

$$ w = 1-\sqrt{1-x}. $$

Alternative: (6) and (20) in [2, Sect. 2.11].

4.3 C4 – (118) & (119)

Polynomial and relevant root:

$$ M = x^2\,(1-x)\,(1+3y)^3 - y\,(1-y)^2\,(4-3x)^3, \quad y_0 = \left(\frac{w}{4-3w}\right)^2. $$

Transformation:

$$ {}_2F_1\left[\begin{matrix} 3\alpha, 3\alpha+\frac{1}{2}; \\ 4\alpha+\frac{2}{3}; \end{matrix} x\right] = \left(1-\frac{3}{4}w\right)^{-6\alpha} {}_2F_1\left[\begin{matrix} 3\alpha, 3\alpha+\frac{1}{2}; \\ 2\alpha+\frac{5}{6}; \end{matrix} y_0\right], $$

$$ w = 1-\sqrt{1-x}. $$

Alternative: (6) and (28) in [2, Sect. 2.11]. Since y_0 is a square, one might try the substitution $y = U^2$, and indeed (2.31) in [3] is obtained in this way.

4.4 C5 – (137) & (118)

Polynomial and relevant roots:

$$ M = 4x^4\,(1-x)\,(4-3y)^3 - y^2\,(1-y)\,\left(16-16x+x^2\right)^3, $$

$$ y_0 = -\frac{w^2\,(1-w)}{\left(1-w-\frac{1}{4}w^2\right)^2}, \qquad y_1 = \frac{x^2}{(2-x)^2}. $$

The latter leads to a particular case of (4) in [2, Sect. 2.11]. The former yields the transformation:

$$ {}_2F_1\left[\begin{matrix} 6\alpha, 4\alpha+\frac{1}{6}; \\ 8\alpha+\frac{1}{3}; \end{matrix} x\right] = \left(1 - w - \frac{1}{4}w^2\right)^{-6\alpha} {}_2F_1\left[\begin{matrix} 3\alpha, 3\alpha+\frac{1}{2}; \\ 4\alpha+\frac{2}{3}; \end{matrix} y_0\right], $$

$$ w = 1 - \sqrt{1-x}. $$

Alternative: (31) and (20) in [2, Sect. 2.11].

5 Transformations C6 to C9

It turns out that a modified version of the above method works in four cases: we set $x = 1 - u^k$ and $u = 1 - w$, where k equals 3 or 4. In other words, the substitution reads:

$$ k = 3: \; x = 3w - 3w^2 + w^3, \qquad w = 1 - \sqrt[3]{1-x}, $$
$$ k = 4: \; x = 4w - 6w^2 + 4w^3 - w^4, \; w = 1 - \sqrt[4]{1-x}. $$

5.1 C6 – (126) & (127)

Exponent $k = 3$. Polynomial and relevant root:

$$ M = x^3(1-x)(1+8y)^3 - y(1-y)^3(9-8x)^3, \quad y_0 = \left(\frac{w}{3-2w}\right)^3. $$

Transformation:

$$ {}_2F_1\left[\begin{matrix} 4\alpha, 4\alpha+\frac{1}{3}; \\ 6\alpha+\frac{1}{2}; \end{matrix} x\right] = \left(1 - \frac{2}{3}w\right)^{-12\alpha} {}_2F_1\left[\begin{matrix} 4\alpha, 4\alpha+\frac{1}{3}; \\ 2\alpha+\frac{5}{6}; \end{matrix} y_0\right], $$

$$ w = 1 - \sqrt[3]{1-x}. $$

Since y is a cube, one might try the substitution $y = U^3$. This would lead to (2.28) in [3].

5.2 C7 – (126) & (126)

Exponent $k = 3$. Polynomial:

$$ M = x^3(1-x)(9-8y)^3 - y^3(1-y)(9-8x)^3. $$

The relevant roots are x, which is trivial, and

$$y_{\pm} = \frac{36w\,(w-1)\left(3 \pm \mathrm{i}\sqrt{3} - 2w\right)}{\left(3 \pm \mathrm{i}\sqrt{3} - 4w\right)^3},$$

which leads to the transformations:

$${}_2F_1\left[\begin{matrix} 4\alpha, 4\alpha + \frac{1}{3}; \\ 6\alpha + \frac{1}{2}; \end{matrix} x\right] = \left[1 - \left[1 \mp \frac{\mathrm{i}}{\sqrt{3}}\right] w\right]^{-12\alpha} {}_2F_1\left[\begin{matrix} 4\alpha, 4\alpha + \frac{1}{3}; \\ 6\alpha + \frac{1}{2}; \end{matrix} y_{\pm}\right],$$

$$w = 1 - \sqrt[3]{1-x}.$$

All upper [lower] signs are taken together. Compare [5].

5.3 C8 – (137) & (136)

Exponent $k = 4$. Polynomial and relevant root:

$$M = x^4\,(1-x)\left(1+14y+y^2\right)^3 - y\,(1-y)^4\left(16-16x+x^2\right)^3, \quad y_0 = \left(\frac{w}{2-w}\right)^4.$$

Transformation:

$${}_2F_1\left[\begin{matrix} 6\alpha, 4\alpha + \frac{1}{6}; \\ 8\alpha + \frac{1}{3}; \end{matrix} x\right] = \left(1 - \frac{1}{2}w\right)^{-24\alpha} {}_2F_1\left[\begin{matrix} 6\alpha, 4\alpha + \frac{1}{6}; \\ 2\alpha + \frac{5}{6}; \end{matrix} y_0\right],$$

$$w = 1 - \sqrt[4]{1-x}.$$

5.4 C9 – (137) & (137)

Exponent $k = 4$. Polynomial:

$$M = y^4\,(1-y)\left(16-16x+x^2\right)^3 - x^4\,(1-x)\left(16-16y+y^2\right)^3.$$

The relevant roots are x and $-x/(1-x)$, which are trivial, and

$$y_{\pm} = \frac{\pm 8\mathrm{i}w\,(1-w)\,(2-w)}{(1 \mp \mathrm{i} - w)^4},$$

which leads to the transformations:

$$ {}_2F_1\left[\begin{matrix}6\alpha, 4\alpha+\frac{1}{6};\\ 8\alpha+\frac{1}{3};\end{matrix}\, x\right] = \left(1-\frac{1}{2}\,(1\pm \mathrm{i})\,w\right)^{-24\alpha} {}_2F_1\left[\begin{matrix}6\alpha, 4\alpha+\frac{1}{6};\\ 8\alpha+\frac{1}{3};\end{matrix}\, y_{\pm}\right], $$

$$ w = 1-\sqrt[4]{1-x}. $$

All upper [lower] signs are taken together. Compare [5].

6 Transformations C10 to C13

In each of these four cases we arrive at the result by means of an auxiliary variable:

$$ t = \sin^2\left(\frac{1}{3}\,\mathrm{Arcsin}\sqrt{x}\right). \tag{8} $$

It satisfies

$$ x = t\,(3-4t)^2\,. \tag{9} $$

It is noted that the branch point at the origin is an apparent one since the sine is squared and, moreover, that (8) and (9) apply all along. In all cases MAPLE gives us the relevant root(s) and in most cases all roots. The transformations are more complicated now, and for clarity auxiliary functions are introduced when suitable.

6.1 C10 – (126) & (119)

Polynomial and relevant root:

$$ M = 64x^3\,(1-x)\,(1+3y)^3 - 27y\,(1-y)^2\,(9-8x)^3\,,\quad y_0 = \frac{64t^3(1-t)}{3(3-12t+8t^2)^2}. $$

Transformation:

$$ {}_2F_1\left[\begin{matrix}4\alpha, 4\alpha+\frac{1}{3};\\ 6\alpha+\frac{1}{2};\end{matrix}\, x\right] = \left(1-4t+\frac{8}{3}t^2\right)^{-6\alpha} {}_2F_1\left[\begin{matrix}3\alpha, 3\alpha+\frac{1}{2};\\ 2\alpha+\frac{5}{6};\end{matrix}\, y_0\right]. $$

6.2 *C11 – (127) & (119)*

Polynomial and relevant root:

$$M = 64x(1-x)^3(1+3y)^3 - 27y(1-y)^2(1+8x)^3, \quad y_0 = \frac{64t(1-t)^3}{3(1+4t-8t^2)^2}.$$

Transformation:

$${}_2F_1\left[\begin{matrix}4\alpha, 4\alpha+\frac{1}{3};\\ 2\alpha+\frac{5}{6};\end{matrix}\; x\right] = \left(1+4t-8t^2\right)^{-6\alpha}\, {}_2F_1\left[\begin{matrix}3\alpha, 3\alpha+\frac{1}{2};\\ 2\alpha+\frac{5}{6};\end{matrix}\; y_0\right].$$

6.3 *C12 – (126) & (136)*

Polynomial and auxiliary functions:

$$\begin{aligned}
M &= 16x^3(1-x)\left(1+14y+y^2\right)^3 - 27y(1-y)^4(9-8x)^3,\\
G &= 27 - 216t + 576t^2 - 608t^3 + 224t^4,\\
H &= \left(3-12t+8t^2\right)(3-4t)\sqrt{3(1-4t)(3-4t)}.
\end{aligned}$$

MAPLE finds only two roots explicitly, but one of these is the relevant root:

$$y_0 = \frac{G-H}{32t^3(1-t)}.$$

However, $G^2 - H^2 = 1024t^6(1-t)^2$; hence $y_0 = 32t^3(1-t)/(G+H)$, and, moreover,

$$1 + 14y_0 + y_0^2 = \frac{3(9-8x)(G-H)}{512t^6(1-t)^2} = \frac{6(9-8x)}{G+H}.$$

A similar step applies in the following cases. The transformation obtained reads:

$${}_2F_1\left[\begin{matrix}4\alpha, 4\alpha+\frac{1}{3};\\ 6\alpha+\frac{1}{2};\end{matrix}\; x\right] = \left(\frac{G+H}{54}\right)^{-3\alpha}\, {}_2F_1\left[\begin{matrix}6\alpha, 4\alpha+\frac{1}{6};\\ 2\alpha+\frac{5}{6};\end{matrix}\; y_0\right].$$

6.4 C13 – (127) & (136)

Polynomial and auxiliary functions:

$$M = 16x\,(1-x)^3\left(1+14y+y^2\right)^3 - 27y\,(1-y)^4\,(1+8x)^3,$$
$$G = 3 - 8t + 96t^2 - 288t^3 + 224t^4,$$
$$H = \left(1+4t-8t^2\right)\sqrt{3\,(3-4t)\,(1-4t)^3}.$$

MAPLE finds only two roots explicitly, but one of these is the relevant root:

$$y_0 = \frac{32t\,(1-t)^3}{G+H}.$$

The transformation obtained reads:

$${}_2F_1\left[\begin{matrix}4\alpha, 4\alpha+\frac{1}{3};\\ 2\alpha+\frac{5}{6};\end{matrix}\; x\right] = \left(\frac{G+H}{6}\right)^{-3\alpha}\; {}_2F_1\left[\begin{matrix}6\alpha, 4\alpha+\frac{1}{6};\\ 2\alpha+\frac{5}{6};\end{matrix}\; y_0\right].$$

A variant with x and y_0 expressed as functions of a parameter is given as (2.33) in [3].

7 Transformations 14 and 15

The substitution (9) would in two cases lead to results involving $\sqrt{t}$. To avoid the branch point at the origin, we set $x = u^2$ and introduce the auxiliary variable:

$$v = \sin\left(\frac{1}{3}\mathrm{Arcsin}(u)\right);$$

hence, $u = v\left(3-4v^2\right)$. MAPLE finds all roots in these two cases.

7.1 C14 – (126) & (134)

Polynomial, auxiliary functions and relevant roots:

$$M = 256x^3\,(1-x)\left(1-y+y^2\right)^3 - 27y^2\,(1-y)^2\,(9-8x)^3,$$
$$G = 27 - 216v^2 + 576v^4 - 512v^6 + 128v^8,$$

$$H=\left(3-12v^{2}+8v^{4}\right)\sqrt{3\left(1-v^{2}\right)},\quad y_{\pm}\frac{16v^{3}\left(-8v^{3}(1-v^{2})\pm H\right)}{\left(3-4v^{2}\right)^{3}\left(1-4v^{2}\right)}.$$

Transformations:

$${}_2F_1\left[\begin{matrix}4\alpha,4\alpha+\frac{1}{3};\\6\alpha+\frac{1}{2};\end{matrix}u^{2}\right]=\left(\frac{G\pm 16v^{3}H}{27}\right)^{-3\alpha}{}_2F_1\left[\begin{matrix}6\alpha,2\alpha+\frac{1}{3};\\4\alpha+\frac{2}{3};\end{matrix}y_{\pm}\right].$$

All upper [lower] signs are taken together. A variant with x and y_0 expressed as functions of a parameter is given as (2.29) in [3].

7.2 *C15 – (127) & (134)*

Polynomial, auxiliary functions and relevant roots:

$$M=256x\left(1-x\right)^{3}\left(1-y+y^{2}\right)^{3}-27y^{2}\left(1-y\right)^{2}\left(1+8x\right)^{3},$$

$$G=3+88v^{2}-192v^{4}-128v^{8},\quad H=\left(1+4v^{2}-8v^{4}\right)\sqrt{3\left(1-v^{2}\right)},$$

$$y_{\pm}=\frac{16v\left(1-v^{2}\right)\left(-8v\left(1-v^{2}\right)^{2}\pm H\right)}{\left(3-4v^{2}\right)\left(1-4v^{2}\right)^{3}}.$$

Transformations:

$${}_2F_1\left[\begin{matrix}4\alpha,4\alpha+\frac{1}{3};\\2\alpha+\frac{5}{6};\end{matrix}u^{2}\right]=\left(\frac{G\pm 16v\left(1-v^{2}\right)H}{3}\right)^{-3\alpha}{}_2F_1\left[\begin{matrix}6\alpha,2\alpha+\frac{1}{3};\\4\alpha+\frac{2}{3};\end{matrix}y_{\pm}\right].$$

All upper [lower] signs are taken together.

8 Transformations C16 to C19

This type is akin to the preceding one. In each case we introduce the auxiliary variable:

$$\tau=4\sin\left(\frac{1}{3}\pi+\frac{1}{3}\operatorname{Arcsin}\sqrt{x}\right)\sin\left(\frac{1}{3}\operatorname{Arcsin}\sqrt{x}\right);$$

it satisfies $x=\tau^{2}\left(3-\tau\right)/4$. After substitution, MAPLE can find roots. But now the origin is actually a branch point, and the plane has to be cut along the negative real axis. The branch point may be avoided by taking $x = u^2$. For brevity, we do not consider this option. MAPLE finds all roots in the cases C16 and C17 but not in the last two cases.

8.1 C16 – (126) & (118)

Polynomial, auxiliary functions and relevant roots:

$$M = 64x^3(1-x)(4-3y)^3 - 27y^2(1-y)(9-8x)^3,$$

$$G = 45 - 30\tau^2 + 12\tau^3 - \tau^4,$$

$$H = (3-\tau)\left(3-\tau^2\right)\sqrt{3(1+\tau)(3-\tau)}, \quad S = 9 - 6\tau^2 + 6\tau^3 - 2\tau^4,$$

$$y_0 = \frac{2\left[(3-\tau)^3(1+\tau)S - (9-8x)H\right]}{3\left(18 - 12\tau^2 + 6\tau^3 - 2\tau^4\right)^2}, \quad y_1 = -\frac{4\tau^3(2-\tau)}{3\left(3-\tau^2\right)^2}.$$

Transformations:

$${}_2F_1\left[\begin{matrix}4\alpha, 4\alpha+\frac{1}{3};\\ 6\alpha+\frac{1}{2};\end{matrix}\, x\right] = \left(1-\tfrac{1}{3}\tau^2\right)^{-6\alpha}\, {}_2F_1\left[\begin{matrix}3\alpha, 3\alpha+\frac{1}{2};\\ 4\alpha+\frac{2}{3};\end{matrix}\, y_1\right],$$

$${}_2F_1\left[\begin{matrix}4\alpha, 4\alpha+\frac{1}{3};\\ 6\alpha+\frac{1}{2};\end{matrix}\, x\right] = \left(\frac{G-H}{18}\right)^{-3\alpha}\, {}_2F_1\left[\begin{matrix}3\alpha, 3\alpha+\frac{1}{2};\\ 4\alpha+\frac{2}{3};\end{matrix}\, y_0\right].$$

8.2 C17 – (127) & (118)

Polynomial, auxiliary functions and relevant roots:

$$M = 64x(1-x)^3(4-3y)^3 - 27y^2(1-y)(1+8x)^3,$$

$$G = 5 + 8\tau + 18\tau^2 - 4\tau^3 - \tau^4,$$

$$H = (1+\tau)\left(1-4\tau+\tau^2\right)\sqrt{3(1+\tau)(3-\tau)}, \quad S = 1 + 16\tau - 18\tau^2 + 10\tau^3 - 2\tau^4,$$

$$y_0 = \frac{2\left[(3-\tau)(1+\tau)^3 S - (1+8x)H\right]}{3\left(2 + 8\tau + 2\tau^3 - \tau^4\right)^2}, \quad y_1 = \frac{-4\tau(2-\tau)^3}{3\left(1 - 4\tau + \tau^2\right)^2}.$$

Transformations:

$$ {}_2F_1\left[\begin{matrix}4\alpha, 4\alpha+\frac{1}{3};\\ 2\alpha+\frac{5}{6};\end{matrix}\, x\right] = \left(1-4\tau+\tau^2\right)^{-6\alpha}\; {}_2F_1\left[\begin{matrix}3\alpha, 3\alpha+\frac{1}{2};\\ 4\alpha+\frac{2}{3};\end{matrix}\, y_1\right], $$

$$ {}_2F_1\left[\begin{matrix}4\alpha, 4\alpha+\frac{1}{3};\\ 2\alpha+\frac{5}{6};\end{matrix}\, x\right] = \left(\frac{G-H}{2}\right)^{-3\alpha}\; {}_2F_1\left[\begin{matrix}3\alpha, 3\alpha+\frac{1}{2};\\ 4\alpha+\frac{2}{3};\end{matrix}\, y_0\right]. $$

8.3 C18 – (126) & (137)

Polynomial, asymptotic expression and auxiliary functions:

$$ \begin{aligned} M &= 16x^3\,(1-x)\left(16-16y+y^2\right)^3 - 27y^4\,(1-y)\,(9-8x)^3\,,\\ y &\simeq \varepsilon\frac{16x^{\frac{3}{4}}}{3^{\frac{9}{4}}} \simeq \varepsilon\frac{4\sqrt{2}\tau^{\frac{3}{2}}}{3\sqrt{3}}, \quad \text{where} \quad \varepsilon\in\{1,\mathrm{i},-1,-\mathrm{i}\}\,,\\ G &= 27-18\tau^2-8\tau^3+7\tau^4, \quad H = \left(3-\tau^2\right)\sqrt{3\tau\,(2-\tau)}. \end{aligned} $$

MAPLE finds only the two roots corresponding to $\varepsilon = \pm\mathrm{i}$,, viz.

$$ y_\pm = \frac{4\tau\left[2\tau^2\,(2-\tau)\pm\mathrm{i}H\right]}{(1+\tau)\,(3-\tau)^3}. $$

Thus, two of the relevant roots *are missed.* The transformations obtained read:

$$ {}_2F_1\left[\begin{matrix}4\alpha, 4\alpha+\frac{1}{3};\\ 6\alpha+\frac{1}{2};\end{matrix}\, x\right] = \left(\frac{G\pm 4\mathrm{i}\tau H}{27}\right)^{-3\alpha}\; {}_2F_1\left[\begin{matrix}6\alpha, 4\alpha+\frac{1}{6};\\ 8\alpha+\frac{1}{3};\end{matrix}\, y_\pm\right]. $$

All upper [lower] signs are taken together. A variant with x and y_0 expressed as functions of a parameter is given as (2.34) in [3].

8.4 C19 – (127) & (137)

Polynomial, asymptotic expression and auxiliary functions:

$$ M = 16x\,(1-x)^3\left(16-16y+y^2\right)^3 - 27y^4\,(1-y)\,(1+8x)^3\,, $$

$$y \simeq \varepsilon\frac{16x^{\frac{1}{4}}}{3^{\frac{3}{4}}} \simeq \varepsilon\frac{8\sqrt{2}\sqrt{\tau}}{\sqrt{3}}, \quad \text{where} \quad \varepsilon \in \{1, \mathrm{i}, -1, -\mathrm{i}\},$$

$$G = 3 - 56\tau + 102\tau^2 - 48\tau^3 + 7\tau^4, \quad H = \left(1 - 4\tau + \tau^2\right)\sqrt{3\tau\,(2-\tau)}.$$

Again, MAPLE finds only the two roots corresponding to $\varepsilon = \pm\mathrm{i}$, viz.

$$y_{\pm} = \frac{4\,(2-\tau)\left[2\tau\,(2-\tau)^2 \pm \mathrm{i}H\right]}{(1+\tau)^3\,(3-\tau)},$$

and so again two relevant roots are missed. The transformations obtained read:

$${}_2F_1\left[\begin{matrix}4\alpha, 4\alpha+\frac{1}{3};\\ 2\alpha+\frac{5}{6};\end{matrix}\,x\right] = \left(\frac{G \pm 4\mathrm{i}\,(2-\tau)\,H}{3}\right)^{-3\alpha} {}_2F_1\left[\begin{matrix}6\alpha, 4\alpha+\frac{1}{6};\\ 8\alpha+\frac{1}{3};\end{matrix}\,y_{\pm}\right].$$

All upper [lower] signs are taken together.

9 Concluding Remarks

Of the transformations mentioned in the introduction, eleven have not yet been investigated: four pairs and one triple.

The pairs are (122) & (123), (124) & (125), (130) & (131) and (132) & (133). It turns out that in these cases eliminations lead only to linear transformations.

For the triple (75), (76) and (77) we have analogues of (1) and its accompanying table. The result of the eliminations may be stated as follows:

No.	(75)	(76)	(77)
(75)	I		
(76)	Q	I	
(77)	L	Q	I

So, further interesting transformations did not emerge.

References

1. Berndt, B.C., Bhargava, S., Garvan, F.G.: Ramanujan's theories of elliptic functions to alternative bases. Trans. Amer. Math. Soc. **347**, 4163–4244 (1995)
2. Erdélyi, A., Magnus, W., Oberhettinger, F., Tricomi, F.G.: Higher Transcendental Functions, vol. I. McGraw-Hill, New York (1953)

3. Garvan, F.G.: Ramanujan's theories of elliptic functions to alternative bases – a symbolic excursion. J. Symbolic Comput. **20**, 517–536 (1995)
4. Goursat, É.: Sur l'équation différentielle linéaire qui admet pour intégrale la série hypergéométrique. Ann. Sci. École Norm. Sup. **10**(2), S3–S142 (1881)
5. Karlsson, P.W.: On some hypergeometric transformations. Panam. Math. J. **10**(4), 59–69 (2000)
6. Kummer, E.E.: Über die hypergeometrische Reihe.... Crelles J. **15**, 39–83, 127–172 (1836)

Convolution Product and Differential and Integro: Differential Equations

Adem Kılıçman

Dedicated to Professor Hari M. Srivastava

Abstract In this paper, we consider partial differential equations with convolution term. Further, by using the convolution we propose a new method to solve the partial differential equations and compare the several properties before and after the convolution. In this new method when the operator has some singularities, then we multiply the partial differential operator with continuously differential functions by using the convolution to remove the singularity. We also study the existence and uniqueness of the new equations. In order to show numerical examples, the following types of problem will be considered:

$$G(x, y) * P(D)u = f(x, y),$$

where P(D) is a differential operator. For computational purpose the computer algebra package can be used to solve recurrence relations with associated boundary conditions.

1 Introduction

The partial differential equations (PDEs) is a very important subject, yet there is no general method to solve all types of the PDEs. The behavior of the solutions very much depends essentially on the classification of PDEs.

It is also well known that some of the second-order linear partial differential equations can be classified as parabolic, hyperbolic, or elliptic; however, if a

A. Kılıçman (✉)
Department of Mathematics and Institute for Mathematical Research, University Putra Malaysia, 43400 UPM, Serdang, Selangor, Malaysia
e-mail: akilicman@putra.upm.edu.my

G.V. Milovanović and M.Th. Rassias (eds.), *Analytic Number Theory, Approximation Theory, and Special Functions*, DOI 10.1007/978-1-4939-0258-3_28,

PDE has coefficients which are not constant, it is rather a mixed type. In many applications of partial differential equations the coefficients need not necessarily be constant; in fact, they might be a function of two or more independent variables and possible dependent variables. Therefore the analysis to describe the solution may not be held globally for equations with variable coefficients that we have for the equations having constant coefficients.

In the literature, there are some very useful physical problems in which their type can be changed. One of the best-known examples is the transonic flow, where the equation is in the form of

$$\left(1-\frac{u^2}{c^2}\right)\phi_{xx}-\frac{2uv}{c^2}\phi_{xy}+\left(1-\frac{v^2}{c^2}\right)\phi_{yy}+f(\phi)=0,$$

where u and v are the velocity components and c is a constant; see [4].

Similarly, partial differential equations with variable coefficients are also used in finance, for example, the arbitrage-free value C of many derivatives

$$\frac{\partial C}{\partial \tau}+s^2\frac{\sigma^2(s,\tau)}{2}\frac{\partial^2 C}{\partial s^2}+b(s,\tau)\frac{\partial C}{\partial s}-r(s,\tau)C=0,$$

with three variable coefficients $\sigma(s,\tau)$, $b(s,\tau)$, and $r(s,\tau)$. In fact this partial differential equation holds whenever C is twice differentiable with respect to s and once with respect to τ; see [30].

However, in the literature there was no systematic way to generate a partial differential equation with variable coefficients by using the equations with constant coefficients. Recently, Kılıçman and Eltayeb in [20] studied the classifications of hyperbolic and elliptic equations with nonconstant coefficients and extended in [22] to the finite product of convolutions and classifications of hyperbolic and elliptic PDEs where the authors consider the coefficients of polynomials with positive coefficients. Thus, in [25], the same authors proposed a systematic way to generate PDEs with variable coefficients by using the convolution product.

2 Convolutions

The convolutions are important in the development of differential equation and difference equation. The classical definition for the convolution product of two functions f and g is as follows:

Definition 2.1. Let f and g be functions. Then the convolution product $f * g$ is defined by the equation

$$(f*g)(x)=\int_{-\infty}^{\infty} f(t)g(x-t)\,\mathrm{d}t$$

for all points x for which the integral exists, that is, an integral which expresses the amount of overlap of one function g as it is shifted over another function f. The discrete version of the convolution is given by

$$(f * g)(m) = \sum_n f(n)g(m - n).$$

When multiplying two polynomials, the coefficients of the product are given by the convolution of the original coefficient sequences. Now if the convolution $h = f * g$ exists for all x, then h is a continuous function.

The convolution operator differs from the other multiplication operator in that $1 * f \neq f$ and $f * f \neq f^2$, for example, the square of the any number is positive for the ordinary function this also true, however, the convolution of a function f with itself might be negative. Now, let f, g, and h be arbitrary functions and a a constant, then convolution has many properties of ordinary multiplication. For example,

$$\begin{aligned}
f * (g * h) &= (f * g) * h, \\
f * (g + h) &= (f * g) + (f * h), \\
a(f * g) &= (af) * g = f * (ag), \\
f * 0 &= 0 * f = 0.
\end{aligned}$$

In more general

$$\int_a^x \int_a^x f(t)\,\mathrm{d}t\,\mathrm{d}x = \int_a^x (x - t) f(t)\,\mathrm{d}t$$

also gives a convolution. However, it is not true in general that $f * 1$ is equal to f. To see this,

$$(f * 1) = \int_0^t f(t - u) \cdot 1\,\mathrm{d}u = \int_0^t f(t - u)\,\mathrm{d}u.$$

In particular, if $f(t) = \cos t$, then

$$(\cos *1) = \int_0^t \cos(t - u)\,\mathrm{d}u = \sin(t - u)\Big|_{u=0}^{u=t} = \sin 0 - \sin t = -\sin t.$$

Of course now it is obvious that $(f * 1)(t) \neq f(t)$. Similarly, it is not necessary that $f * f$ is not negative. Thus, it follows easily from the definition that if $f * g$ exists, then $g * f$ exists and

$$f * g = g * f.$$

Similarly, if $(f * g)'$ and $f * g'$ (or $f' * g$) exist, then it can be shown that

$$(f * g)' = f * g' \quad (\text{or } f' * g).$$

That is, the derivative of a convolution satisfies a very important property of convolutions that derivatives of a convolution may be placed on either factor but not both. In the one-variable case,

$$\frac{\mathrm{d}}{\mathrm{d}x}(f * g) = \frac{\mathrm{d}f}{\mathrm{d}x} * g = f * \frac{\mathrm{d}g}{\mathrm{d}x},$$

where $\mathrm{d}/\,\mathrm{d}x$ is the derivative. More generally, in the case of functions of several variables, an analogous formula holds with the partial derivative:

$$\frac{\partial}{\partial x_i}(f * g)(x) = \frac{\partial f}{\partial x_i} * g = f * \frac{\partial g}{\partial x_i}.$$

A particular consequence of this is that the convolution can be viewed as a "smoothing" operation: the convolution of f and g is differentiable as many times as f and g are together. In fact, we can say that because of this rule, the convolutions that are important in the solutions to differential equations are often given by convolutions where one factor is the given function and the other is a special kernel.

The area under a convolution is the product of areas under the factors,

$$\begin{aligned}\int_{-\infty}^{\infty}(f * g)\,\mathrm{d}x &= \int_{-\infty}^{\infty}\left[\int_{-\infty}^{\infty} f(u)g(x-u)\,\mathrm{d}u\right] \\ \mathrm{d}x &= \int_{-\infty}^{\infty} f(u)\left[\int_{-\infty}^{\infty} g(x-u)\,\mathrm{d}x\right]\mathrm{d}u \\ &= \left[\int_{-\infty}^{\infty} f(u)\,\mathrm{d}u\right]\left[\int_{-\infty}^{\infty} g(x)\,\mathrm{d}x\right].\end{aligned}$$

Thus, if T is a linear operator then

$$T(f * g) = T(f)\cdot T(g).$$

Furthermore there is no algebra of functions that possesses an identity element for the convolution. The lack of identity is typically not a major inconvenience, since most collections of functions on which the convolution is performed can be convolved with a delta distribution or, at the very least, admit approximations to the identity. The linear space of compactly supported distributions does, however, admit an identity under the convolution. Specifically,

$$f * \delta = f,$$

where δ is the delta distribution.

We also note that the convolution is more often taken over an infinite range,

$$(f * g)\,(t) \equiv \int_{-\infty}^{\infty} f(\tau)g(t-\tau)\,\mathrm{d}\tau = \int_{-\infty}^{\infty} g(\tau)f(t-\tau)\,\mathrm{d}\tau.$$

Of course we can ask the question whether this integral always exists. The answer for the question is negative. However, if we now let f, g be locally summable functions and suppose that supp $f \subseteq [a, b]$ then if G is a primitive of g and $[c, d]$ is any interval, thus

$$\int_c^d g(x-t)\,\mathrm{d}x = G(d-t) - G(c-t).$$

This implies that the function $\int_c^d g(x-t)\,\mathrm{d}x$ is bounded on the interval $[a, b]$, and so $f(t)\int_c^d g(x-t)\,\mathrm{d}x$ is a locally summable function. This proves that

$$(f * g)(x) = \int_{-\infty}^{\infty} f(t)g(x-t)\,\mathrm{d}t = \int_a^b f(t)g(x-t)\,\mathrm{d}t$$

exists and further

$$\int_c^d (f*g)(x)\,\mathrm{d}x = \int_c^d \left\{\int_a^b f(t)g(x-t)\,\mathrm{d}t\right\}\mathrm{d}x = \int_a^b f(t)\left\{\int_c^d g(x-t)\,\mathrm{d}x\right\}\mathrm{d}t,$$

proving that $f * g$ is a locally summable function if f has compact support. Similarly, $f * g$ is a locally summable function if g has compact support, and in either case $f * g = g * f$.

In the discrete form, the Cauchy product of two sequences a_n and b_n is the discrete convolution of the two sequences, thus the sequence c_n whose general term is given by

$$c_n = \sum_{k=0}^{n} a_k b_{n-k}.$$

Let f and g have the power series representations

$$f(x) = \sum_{n=0}^{\infty} a_n (x-x_0)^n \quad \text{and} \quad g(x) = \sum_{n=0}^{\infty} b_n (x-x_0)^n,$$

then

$$f(x)g(x) = \sum_{n=0}^{\infty} c_n (x-x_0)^n, \tag{1}$$

where

$$c_n = \sum_{k=0}^{n} a_k b_{n-k}.$$

Equation (1) is known as the Cauchy product of the series for $f(x)$ and $g(x)$.

Note that the convolution does not only differ from the ordinary product; further, it also has many useful applications such as on the solution of the differential and integro-differential equations.

Example 2.1. The integral equation

$$f(t) = 2\cos t - \int_0^t (t-\tau) f(\tau)\,\mathrm{d}\tau$$

and the solution can be found by letting $F(s) = L(f(t))$, and on using $L(t) = \frac{1}{s^2}$ in the convolution theorem, we can obtain

$$F(s) = \frac{2s}{s^2+1} - \frac{1}{s^2} F(s).$$

Now solving for $F(s)$ yields

$$F(s) = \frac{2s^3}{(s^2+1)^2} = \frac{2s}{s^2+1} - \frac{2s}{(s^2+1)^2},$$

and the solution is given by $f(t) = 2\cos t - t\sin t$.

Example 2.2. If $\lambda, \mu > -1$, then $x_+^\lambda * x_+^\mu = B(\lambda+1, \mu+1) x_+^{\lambda+\mu+1}$. Equivalently, $f_+^\lambda * f_+^\mu = f_+^{\lambda+\mu+1}$. In particular,

$$x_+^\lambda * H(x) = \frac{x_+^{\lambda+1}}{\lambda+1} = \int_{-\infty}^x x_+^\lambda\,\mathrm{d}x.$$

Further, if $\lambda, \mu > -1 > \lambda + \mu$, then

$$x_-^\lambda * x_+^\mu = B(\lambda+1, -\lambda-\mu-1) x_+^{\lambda+\mu+1} + B(\mu+1, -\lambda-\mu-1) x_-^{\lambda+\mu+1},$$

where B is the beta function and H is the Heaviside function, respectively; see [16].

Some of the functions have an inverse element for the convolution, $f^{(-1)}$, which is defined by

$$f^{(-1)} * f = \delta.$$

The set of invertible functions forms an Abelian group under the convolution. Further, the derivative of a convolution satisfies a very important property of convolutions that derivatives of a convolution may be placed on either factor but not both. This means that the derivatives of a function f can be expressed as convolutions, using the derivatives of the δ distribution which is strange but useful:

$$f = \delta * f, \quad f' = \delta' * f, \quad f'' = \delta'' * f.$$

Thus, if the nth-order linear differential equation has constant coefficients, we may write it as $f * x = b$ by introducing the distribution

$$f = \delta^{(n)} + a_{n-1}\delta^{(n-1)} + \cdots + a_3\delta^{(3)} + a_2\delta'' + a_1\delta' + a_0\delta.$$

Further, if we have a function such that $f * g = \delta$, we will obtain a special solution of the inhomogeneous equation as $g * b$.

By using the convolution method several initial value problems (IVPs) can also be solved. For example, the unique solution to the initial value problem

$$ay''(t) + by'(t) + cy(t) = g(t), \quad \text{with } y(0) = y_0 \text{ and } y'(0) = y_1$$

is given by

$$y(t) = u(t) + (h * g)(t),$$

where $u(t)$ is the solution to the homogeneous part of the equation

$$au''(t) + bu'(t) + cu(t) = 0 \quad \text{with } u(0) = y_0 \text{ and } u'(0) = y_1$$

and $h(t)$ has the integral transform such as Laplace transform and given by

$$H(s) = \frac{1}{as^2 + bs + c}.$$

Then the general solution is $y(t) = u(t) + v(t) = u(t) + (h * g)(t)$. In order to verify that thc initial conditions are met, we compute

$$y(0) = u(0) + v(0) = y_0 + 0 = y_0 \quad \text{and} \quad y'(0) = u'(0) + v'(0) = y_1 + 0 = y_1.$$

Example 2.3. By using the convolution method the following initial value problem

$$y''(t) + y(t) = \tan t, \quad \text{with } y(0) = 1 \text{ and } y'(0) = 2$$

can be solved $u''(t) + u(t) = 0$, with $u(0) = 1$ and $u'(0) = 2$. Taking the Laplace transform it yields $s^2U(s) - s - 2 + U(s) = 0$. Then $U(s) = (s+2)/(s^2+1)$ and it follows that $u(t) = \cos t + 2\sin t$.

Second, we observe that $H(s) = \dfrac{1}{s^2+1}$ and $h(t) = \sin\ t$ so that

$$v(t) = (h * g)(t) = \int_0^t \sin(t-s)\tan(s)\,ds = \left[\cos(t)\ln\left(\frac{\cos s}{1+\sin s}\right)\right.$$

$$\left.- \sin(t-s)\right]\Bigg|_{s=0}^{s=t} = \cos(t)\ln\left(\frac{\cos t}{1+\sin t}\right) + \sin(t).$$

Therefore, the solution is

$$y(t) = u(t) + v(t) = \cos t + 3\sin t + \cos(t)\ln\left(\frac{\cos t}{1+\sin\, t}\right).$$

In practical applications it is very common to use delta sequences rather than the delta itself. Thus, we have the following definition in the literature; see [16].

Definition 2.2. A sequence $\delta_n : \mathbb{R} \to \mathbb{R}$ is called a delta sequence of ordinary functions which converges to the singular distribution $\delta(x)$ and satisfies the following conditions:

(i) $\delta_n(x) \geq 0$ for all $x \in \mathbb{R}$.

(ii) δ_n is continuous and integrable over $\mathbb{R}$ with $\int_{-\infty}^{\infty} \delta_n(x)\,\mathrm{d}x = 1$.

(iii) given any $\epsilon > 0$,

$$\lim_{n\to\infty} \int_{-\infty}^{-\epsilon} + \int_{\epsilon}^{\infty} \delta_n(x)\,\mathrm{d}x = 0.$$

Example 2.4. Let $\delta_n(xt) = \dfrac{n}{\pi(n^2t^2+1)}$. Then

$$\int_a^b \delta_n(t)\,\mathrm{d}t = \int_a^b \frac{n}{\pi(n^2t^2+1)}\,\mathrm{d}t = \frac{1}{\pi}\left[\arctan(nb) - \arctan(an)\right].$$

Now, if we let $n \to \infty$, then it follows that δ_n is a delta sequence and converges to the Dirac delta function. Similarly show that all the following sequences are the delta sequences:

$$\delta_n(x) = \begin{cases} 0, \; x < -\frac{1}{2n}, \\ n, \; -\frac{1}{2n} < x < \frac{1}{2n}, \\ 0, \; x > \frac{1}{2n} \end{cases} \to \delta(x);$$

$$\mu_n(x) = \frac{n}{\sqrt{\pi}}\mathrm{e}^{-n^2x^2} \to \delta(x);$$

$$\epsilon_n(x) = \frac{n}{\pi}\sin\mathrm{c}(ax) \equiv \frac{\sin(nx)}{\pi x} \to \delta(x);$$

$$\lambda_n(x) = \frac{1}{\pi x}\frac{\mathrm{e}^{inx} - \mathrm{e}^{-inx}}{2\mathrm{i}} \to \delta(x);$$

$$\beta_n(x) = \frac{1}{2\pi\mathrm{i}x}\left[\mathrm{e}^{\mathrm{i}xt}\right]_{-n}^{n} \to \delta(x);$$

$$h_n(x) = \frac{1}{2\pi}\int_{-n}^{n} \mathrm{e}^{\mathrm{i}xt}\,\mathrm{d}t \to \delta(x);$$

$$m_n(x) = \frac{1}{2\pi}\frac{\sin\left[\left(n+\frac{1}{2}\right)x\right]}{\sin\left(\frac{1}{2}x\right)} \to \delta(x).$$

The above sequences are also known as approximate identities for convolution operation which are used in some situations such as numerical analysis to approximate the piecewise linear functions.

Example 2.5. In general if we let ϕ be continuous and nonnegative, $\phi(x) = 0$ for all $|x| \geq 1$ and $\int_{-1}^{1} \phi(x)\,\mathrm{d}x = 1$, if we set $\delta_n(x) = n\phi(nx)$. Then one can show that δ_n is a delta sequence.

Thus, the above examples show that there are several ways to construct a delta sequence. The convolution operation can be extended to the generalized functions. If f and g are generalized functions such that at least one of them has compact support and if ϕ is a test function, then $f * g$ is defined by

$$\langle f * g, \phi\rangle = \langle f(x) \times g(y), \phi(x+y)\rangle$$

where $\times$ is the direct product of f and g, that is, the functional on the space of test functions of two independent variables given by every infinitely differentiable function of compact support; for further details and properties, we refer to [3] and [10]. The idea of the delta sequences can also be extended to the multiple-dimensional form; for that see [34]. By using the derivatives of the δ distribution, then we obtain the following strange but very useful statements:

$$f = \delta * f, \quad f' = \delta' * f, \quad f'' = \delta'' * f.$$

Note that one can study in the type of convergence and speed of convergence by using the delta sequences.

In the case of functions of several variables, an analogous formula holds with the partial derivative

$$\frac{\partial}{\partial x_i}(f * g)(x) = \frac{\partial f}{\partial x_i} * g = f * \frac{\partial g}{\partial x_i}.$$

A particular consequence of this is that the convolution can be viewed as a "smoothing" operation: the convolution of f and g is differentiable as many times as f and g are together. In fact, we can say that because of this rule, the convolutions that are important in the solutions to differential equations are often given by convolutions where one factor is the given function and the other is a special kernel.

In the literature, there are several important partial differential equations which are of the form

$$\mathcal{P}(\mathcal{D})\,u = f(x,y).$$

In order to solve these equations, one might consider the following cases:

1. The solution $u(x, y)$ is a smooth function such that the operation can be performed as in the classical sense, and the resulting equation is an identity. Then $u(x, y)$ is a classical solution.
2. The solution $u(x, y)$ is not smooth enough so that the operation cannot be performed but satisfies as a distribution.
3. The solution $u(x, y)$ is a singular distribution; then, the solution is a distributional solution; see Kılıçman [22].

In particular, let us try to solve the linear second-order partial differential equations as follows:

$$au_{xx} + bu_{xy} + cu_{yy} + du_x + eu_y + fu = G(x, y)$$

under boundary conditions

$$u(x,0) = f_1(x) * f_2(x), \quad u(0, y) = w_1(y) * w_2(y)$$

$$u_x(x,0) = \frac{d}{\mathrm{d}x}(f_1(x)*f_2(x)), \; u_y(0, y) = \frac{d}{\mathrm{d}y}(w_1(y)*w_2(y)), \text{ and } u(0,0) = 0,$$

where the symbol $*$ is the convolution [8] and a, b, c, d, e, and f are constant coefficients.

Now consider the equation

$$au_{xx} + 2bu_{xy} + cu_{yy} + F(x, y, u, u_x, u_y) = 0 \tag{2}$$

where a, b, c, d, e, f are of class $C^2(\Omega)$ and $\Omega \subseteq \mathbb{R}^2$ is the domain and $(a, b, c) \neq (0, 0, 0)$ and the expression $au_{xx} + 2bu_{xy} + cu_{yy}$ is called the principal part of Eq. (2), and since the principal part mainly determines the properties of the solution, it is well known that

1. If $b^2 - 4ac > 0$, Eq. (2) is called a hyperbolic equation.
2. If $b^2 - 4ac < 0$, Eq. (2) is called a parabolic equation.
3. If $b^2 - 4ac = 0$, Eq. (2) is called an elliptic equation.

If this is true at all points, then the equation is said to be hyperbolic, parabolic, or elliptic in the domain. Later we generalized the classification of hyperbolic and elliptic equations by using the convolution method where we assume that the nonconstant coefficients are polynomials; see [22].

Now if we have the partial differential equations which are in the form of

$$P(D)u = f(x)$$

then if we multiply the differential operator with a function by using the convolution, that is,

$$(Q(x) * P(D))\, u = f(x).$$

In particular if we consider $Q(x) = \delta_n(x)$, that is, the delta sequence converges to Dirac delta δ, that new equation is equivalent to the original equations; see [20]. **Question:** Now if we multiply the differential operator by a function, what will happen to the classification? That is a new classification problem of the

$$(Q(x,t) * *\mathcal{P}(\mathcal{D}))\, u = f(x,t).$$

However the classification theorem guarantees that every second-order linear PDE with constant coefficients can be transformed into exactly one of the above forms.

Further, in the literature there was no systematic way to generate partial differential equations by using the equations with constant coefficients; most of the partial differential equations with variable coefficients depend on the nature of particular problems.

The classification depends upon the signature of the eigenvalues of the coefficient matrix.

1. ELLIPTIC: The eigenvalues are all positive or all negative.
2. PARABOLIC: The eigenvalues are all positive or all negative, save one that is zero.
3. HYPERBOLIC: There is only one negative eigenvalue and all the rest are positive, or there is only one positive eigenvalue and all the rest are negative.
4. ULTRAHYPERBOLIC: There is more than one positive eigenvalue and more than one negative eigenvalue, and there are no zero eigenvalues. There is only limited theory for ultrahyperbolic equations (cf. Courant and Hilbert [5]).

On the other side there are some very useful physical problems where its type can be changed. One of the best-known examples is the transonic flow, where the equation is in the form of

$$\left(1 - \frac{u^2}{c^2}\right)\phi_{xx} - \frac{2uv}{c^2}\phi_{xy} + \left(1 - \frac{v^2}{c^2}\right)\phi_{yy} + f(\phi) = 0$$

where u and v are the velocity components and c is a constant (see [4]). Similarly, partial differential equations with variable coefficients are also used in finance, for example, the arbitrage-free value C of many derivatives

$$\frac{\partial C}{\partial \tau} + s^2 \frac{\sigma^2(s,\tau)}{2}\frac{\partial^2 C}{\partial s^2} + b(s,\tau)\frac{\partial C}{\partial s} - r(s,\tau)C = 0,$$

with three variable coefficients $\sigma(s,\tau)$, $b(s,\tau)$, and $r(s,\tau)$. In fact this partial differential equation holds whenever C is twice differentiable with respect to s and once with respect to τ; see [30].

However, in the literature there was no systematic way to generate partial differential equations with variable coefficients by using the equations with constant coefficients; most of the partial differential equations with variable coefficients depend on the nature of particular problems.

Question: How do we generate a PDE with variable coefficients from the PDE with constant coefficients? In order to answer the above questions we extend the classification of partial differential equations further by using the convolution products.

For example, in the following diffusion equation,

$$u_t = u_{xx} + \delta(x-a)k(u(x,t)),$$

where $0 \leq x \leq l$, $0 \leq a \leq l$, with zero boundary and initial conditions. All we know is that this is used to model physical scenarios where the energy that is being put into the system is highly spatially localized. **Question:** Is the solution of such PDE possible? In a classical way or in another way (delta function is not real function)? Or maybe the solution of such PDE exists in a normal sense (delta function is a limit of the so-called delta sequences, which are sequences of ordinary functions).

For example, if we consider the wave equation in the following example,

$$\begin{aligned} u_{tt} - u_{xx} &= G(x,t), \quad (x,t) \in \mathbb{R}^2_+, \\ u(x,0) &= f_1(x), \quad u_t(x,0) = g_1(x), \\ u(0,t) &= f_2(t), \quad u_x(0,t) = g_2(t). \end{aligned}$$

Now, if we consider multiplying the left-hand side of the above equation by the nonconstant coefficient $Q(x,t)$ by using the double convolution with respect to x and t, respectively, then the equation becomes

$$\begin{aligned} Q(x,t) ** (u_{tt} - u_{xx}) &= G(x,t), \quad (t,x) \in \mathbb{R}^2_+, \\ u(x,0) &= f_1(x), \quad u_t(x,0) = g_1(x), \\ u(0,t) &= f_2(t), \quad u_x(0,t) = g_2(t). \end{aligned}$$

Thus, the relationship between the solutions' partial differential equations with constant coefficients and nonconstant coefficients was studied in [22]. Note that in particular case, if $\lim_{n\to\infty} Q_n(x,t) = \delta(x,t)$, then it will be an approximate identity which plays a significant role in convolution algebra as the same role as a sequence of function approximations to the Dirac delta function that is the identity element for convolution.Further, delta functions often arise as convolution semigroups. This amounts to the further constraint that the convolution of δ_n with δ_m must satisfy

$$\delta_n ** \delta_m = \delta_{n+m}$$

for all $n, m > 0$. Thus, the convolution semigroups in L^1 that form a delta function are always an approximation to the identity in the above sense; however, the semigroup condition is quite a strong restriction. In fact semigroups approximating the delta function arise as fundamental solutions or Green's functions to physically motivated elliptic or parabolic partial differential equations.

Note that there is no general method that can solve all types of the differential equations; each might require different methods and techniques.

Now let us consider the general linear second-order partial differential equation with nonconstant coefficients in the form of

$$a(x,y)u_{xx}+b(x,y)u_{xy}+c(x,y)u_{yy}+d(x,y)u_x+e(x,y)u_y+f(x,y)u=0$$

and almost linear equation in two variables

$$au_{xx}+bu_{xy}+cu_{yy}+F(x,y,u,u_x,u_y)=0, \tag{3}$$

where $a,b,c,$ are polynomials defined by

$$a(x,y)=\sum_{\beta=1}^{n}\sum_{\alpha=1}^{m}a_{\alpha\beta}x^{\alpha}y^{\beta},\ b(x,y)=\sum_{\zeta=1}^{n}\sum_{\eta=1}^{m}b_{\zeta\eta}x^{\zeta}y^{\eta},\ c(x,y)=\sum_{l=1}^{n}\sum_{k=1}^{m}c_{kl}x^{k}y^{l}$$

and $(a,b,c)\neq(0,0,0)$ and where the expression $au_{xx}+2bu_{xy}+cu_{yy}$ is called the principal part of Eq. (3), since the principal part mainly determines the properties of the solution. Throughout this paper we also use the following notations:

$$|a_{mn}|=\sum_{\beta=1}^{n}\sum_{\alpha=1}^{m}|a_{\alpha\beta}|,\quad |b_{mn}|=\sum_{\zeta=1}^{n}\sum_{\eta=1}^{m}|b_{\zeta\eta}|,\quad \text{and}\quad |c_{mn}|=\sum_{l=1}^{n}\sum_{k=1}^{m}|c_{kl}|.$$

Now in order to generate new PDEs, we convolute Eq. (3) by a polynomial with single convolution as $p(x)*^x$ where $p(x)=\sum_{i=1}^{m}p_ix^i$, then Eq. (3) becomes

$$p(x)*^x\left[a(x,y)u_{xx}+b(x,y)u_{xy}+c(x,y)u_{yy}+F(x,y,u,u_x,u_y)\right]=0, \tag{4}$$

where the symbol $*^x$ indicates single convolution with respect to x, and we shall classify Eq. (4) instead of Eq. (3) by considering and examining the function

$$D(x,y)=(p(x)*^x b(x,y))^2-(p(x)*^x a(x,y))\,(p(x)*^x c(x,y)). \tag{5}$$

From Eq. (5), one can see that if D is positive, then Eq. (4) is called hyperbolic; if D is negative then Eq. (4) is called elliptic; otherwise, it is parabolic.

First of all, we compute and examine the coefficients of the principal part of Eq. (4) as follows:

$$A_1(x,y)=p(x)*^x a(x,y)=\sum_{i=1}^{m}p_ix^i *^x\sum_{\beta=1}^{n}\sum_{\alpha=1}^{m}a_{\alpha\beta}x^{\alpha}y^{\beta}.$$

By using the single convolution definition and integration by parts, thus we obtain the first coefficient of Eq. (4) in the form of

$$A_1(x,y) = \sum_{\beta=1}^{n}\sum_{i=1}^{m}\sum_{\alpha=1}^{m} \frac{p_i a_{\alpha\beta} i!x^{\alpha+i+1}y^{\beta}}{((\alpha+1)((\alpha+2)\cdots(\alpha+i+1))}.$$

Similarly, for the coefficients of the second part in Eq. (4), we have

$$B_1(x,y) = \sum_{j=1}^{n}\sum_{\zeta=1}^{m}\sum_{i=1}^{m} \frac{p_i b_{\zeta\eta} i!x^{\zeta+i+1}y^{\eta}}{((\zeta+1)((\zeta+2)\cdots(\zeta+i+1))}.$$

Also the last coefficient of Eq. (4) is given by

$$C_1(x,y) = \sum_{l=1}^{n}\sum_{k=1}^{m}\sum_{i=1}^{m} \frac{p_i c_{kl} i!x^{k+i+1}y^{l}}{((k+1)((k+2)\cdots(k+i+1))},$$

then one can easily set up

$$D_1(x,y) = B_1^2(x,y) - A_1(x,y)C_1(x,y). \tag{6}$$

Then there are several cases, and the classification of partial differential equations with polynomial coefficients depends very much on the signs of the coefficients; see [8]. In fact, this analysis can also be carried out for the convolutional product with respect to the $*^y$ as well as the double convolution.

Now we demonstrate how to generate a PDE with variable coefficients by using the convolutions. For example, in particular we can have

$$x^3 *^x x^2y^3u_{xx} + x^3 *^x x^3y^4u_{xy} + x^3 *^x x^4y^5u_{yy} = f(x,y) *^x g(x,y). \tag{7}$$

The first coefficients of Eq. (7) is given by

$$A_1(x,y) = x^3 *^x x^2y^3 = y^5\int_0^x (x-\theta)^3\theta^2\,\mathrm{d}\theta = \frac{1}{60}y^3x^6. \tag{8}$$

Similarly, the second coefficient is given by

$$B_1(x,y) = x^3 *^x x^3y^4 = \frac{1}{140}y^4x^7. \tag{9}$$

By the same way we get the last coefficients of Eq. (7)

$$C_1(x,y) = x^3 *^x x^4y^5 = \frac{1}{280}y^5x^8. \tag{10}$$

By using Eqs. (6), (8), (9), and (10), we obtain

$$D_1(x,y) = -\frac{1}{117600}y^8x^{14}. \tag{11}$$

We can easily see from Eq. (11) that Eq. (7) is an elliptic equation for all $(x_{0,}y)$.

In the same way, if we multiply Eq. (5) by polynomial with a single convolution as $h(y)*^y$ where $h(y) = \sum_{j=1}^{n} y^j$, then Eq. (5) becomes

$$h(y) *^y \left[a(x,y)u_{xx} + b(x,y)u_{xy} + c(x,y)u_{yy} + F(x,y,u,u_x,u_y)\right] = 0, \tag{12}$$

where the symbol $*^y$ indicates single convolution with respect to y, and we shall classify Eq. (12) as. First of all, let us compute the coefficients of Eq. (12). By using the definition of single convolution with respect to y and integral by part, we obtain the first coefficient of Eq. (12) as follows:

$$A_2(x,y) = h(y) *^y a(x,y) = \sum_{\beta=1}^{n}\sum_{j=1}^{n}\sum_{\alpha=1}^{m} \frac{j!x^{\alpha}y^{\beta+j+1}}{((\beta+1)((\beta+2)\ldots(\beta+j+1))},$$

and the second coefficient of Eq. (12) is given by

$$B_2(x,y) = h(y) *^y b(x,y) = \sum_{j=1}^{n}\sum_{\eta=1}^{n}\sum_{i=1}^{m} \frac{j!x^{\xi}y^{\eta+j+1}}{((\eta+1)((\eta+2)\ldots(\eta+i+1))}.$$

Similarly, the last coefficient of Eq. (12) is given by

$$C_2(x,y) = h(y) *^y c(x,y) = \sum_{j=1}^{n}\sum_{l=1}^{n}\sum_{i=1}^{m} \frac{j!x^{k}y^{l+j+1}}{((l+1)((l+2)\cdots(l+i+1))}.$$

Similar to the previous case, in particular, let us classify the following example:

$$y^7 *^y x^2y^3u_{xx} + y^7 *^y x^3y^4u_{xy} + y^7 *^y x^4y^5u_{yy} = f(x,y) *^y g(x,y). \tag{13}$$

The symbol $*^y$ means single convolution with respect to y. If we follow the same technique used above, then the first coefficient of Eq. (13) is given by

$$A_2(x,y) = y^7 *^y x^2y^3 = \frac{1}{1320}x^2y^{11}, \tag{14}$$

the second coefficient of (13) is given by

$$B_2(x,y) = y^7 *^y x^3y^4 = \frac{1}{3860}x^3y^{12} \tag{15}$$

and the last coefficient is given by

$$C_2(x,y) = y^7 *^y x^4 y^5 = \frac{1}{10296} x^4 y^{13}. \tag{16}$$

Now on using Eqs. (6), (14), (15), and (16), we have

$$D_2(x,y) = -\frac{1}{101930400} x^6 y^{24}. \tag{17}$$

We can easily see from Eq. (17) that Eq. (13) is an elliptic equation for all (x, y_0). In this study first we consider linear partial differential equations with constant coefficients; then by applying the convolution, we can generate the partial differential equations having variable coefficients; and then we solve the new equation and compare the two solutions; see [9].

In the case of singular boundary problems, we consider the nonhomogeneous wave equation in the form

$$\begin{aligned} f_{tt} - f_{xx} &= \frac{1}{2}e^{x+t} - \frac{1}{2}\cos(x)e^t - \frac{1}{2}e^x\cos(t) + \frac{1}{2}\cos(x+t), \\ f(0,t) &= \delta(t), \quad f_t(0,t) = \delta'(t), \\ f(x,0) &= \delta(x), \quad f_x(x,0) = \delta'(x). \end{aligned} \tag{18}$$

Then we note that all the initial conditions have a singularity at $x = t = 0$, $(t,x) \in \mathbb{R}_+^2$. It is easy to see that the nonhomogeneous term of Eq. (18) can be written in the form of the double convolution as follows:

$$\sin(x+t) ** e^{x+t} = \frac{1}{2}e^{x+t} - \frac{1}{2}\cos x e^t - \frac{1}{2}e^x \cos t + \frac{1}{2}\cos(x+t).$$

Now, by applying the double Sumudu transform for the wave equation, then

$$F_{tt} - F_{xx} = -3e^{2x+t}, \quad (x,t) \in \mathbb{R}_+^2, \tag{19}$$

$$F(x,0) = e^{2x} + e^x, \quad F_t(x,0) = e^{2x} + e^x, \tag{20}$$

$$F(0,t) = 2e^t, \quad F_x(0,t) = 3e^t. \tag{21}$$

By taking the double Sumudu transform for Eq. (19) and single Sumudu transform of Eqs. (20) and (21) with u, v as transform variables for x, t, respectively, on using Eqs. (5) and (6), after some little arrangements, we obtain

$$F(u,v) = \frac{u^2\,[2-3u]\,(v+1)}{(1-u)\,(1-2u)\,[u^2-v^2]} - \frac{v^2\,(2+3u)}{(1-v)\,[u^2-v^2]} - \frac{3u^2v^2}{(1-2u)\,(1-v)\,[u^2-v^2]}.$$

Now, we consider multiplying the left-hand side equation of (19) by a nonconstant coefficient $x^3t^4 **$ where the symbol $**$ means a double convolution with respect to x and t respectively, then Eq. (19) becomes

$$xt^2 ** (F_{tt} - F_{xx}) = -3e^{2x+t}, \quad (x,t) \in \mathbb{R}^2_+, \tag{22}$$

$$F(x,0) = e^{2x} + e^x, \quad F_t(x,0) = e^{2x} + e^x, \tag{23}$$

$$F(0,t) = 2e^t, \quad F_x(0,t) = 3e^t. \tag{24}$$

Similarly, we apply the double Sumudu transform technique for Eq. (22) and single Sumudu transform for Eqs. (23) and (24); we obtain

$$F(u,v) = \frac{u^2[2-3u](v+1)}{(1-u)(1-2u)[u^2-v^2]} - \frac{v^2(2+3u)}{(1-v)[u^2-v^2]} - \frac{3}{2v(1-2u)(1-v)[u^2-v^2]}. \tag{25}$$

Now, by taking double inverse Sumudu transform for both sides of Eq. (25), we obtain the solution of Eq. (22) as follows:

$$F_1(x,t) = \frac{17}{4}e^{-2t+2x} - \frac{45}{4}e^{2t+2x} + e^{t+x} + 2e^{t+2x}.$$

3 Integro-Differential Equations

Consider the following example:

Example 3.1. Find the solution of the integro-differential equation

$$y'' + 2y' - y = h(t), \quad y(0) = A, \quad y'(0) = B,$$

for arbitrary constants A and B and arbitrary function $h(t) = (f * g)(t)$.

Solution. When we take Laplace transforms

$$[s^2Y - As - B] + 2[sY - A] - Y = F(s)G(s)$$

and solve for Y, the result is

$$Y(s) = \frac{F(s)G(s)}{s^2 + 2s - 1} + \frac{As + B + 2A}{s^2 + 2s - 1}.$$

To find the inverse transform of this function, we first note that

$$L^{-1}\left\{\frac{1}{s^2+2s-1}\right\} = L^{-1}\left\{\frac{1}{(s+1)^2-2}\right\} = e^{-t}L^{-1}\left\{\frac{1}{s^2-1}\right\} = \frac{1}{\sqrt{2}}e^{-t}\sin h\sqrt{2}t.$$

Using the convolution on the first term of $Y(s)$ now yields

$$\begin{aligned} y(t) &= \int_0^t (f*g)(u)\frac{1}{\sqrt{2}}\mathrm{e}^{-(t-u)}\sin h\sqrt{2}(t-u)\mathrm{d}u + L^{-1}\left\{\frac{A(s+1)+(B+A)}{(s+1)^2-2}\right\} \\ &= \frac{1}{\sqrt{2}}\int_0^t (f*g)(u)\mathrm{e}^{-(t-u)}\sinh\sqrt{2}(t-u)\mathrm{d}u + \mathrm{e}^{-t}L^{-1}\left\{\frac{As+(B+A)}{(s^2-2)}\right\} \\ &= \frac{1}{\sqrt{2}}\int_0^t (f*g)(u)\mathrm{e}^{-(t-u)}\sinh\sqrt{2}(t-u)\mathrm{d}u \\ &\quad +\mathrm{e}^{-t}\left(A\cosh\sqrt{2}t + \frac{A+B}{\sqrt{2}}\sinh\sqrt{2}t\right). \end{aligned}$$

The particular modified form of the above example is the so-called Hermite"s equation of order n in the following equation:

$$y'' - 2xy' - 2ny = 0.$$

Then by using the series solution method, we can obtain the solution

$$y(x) = 1 - 4x^2 + \frac{4}{3}x^4.$$

The above example also indicates that the convolution method can also be used to solve the integro-differential equations in order to generate the equation with variable coefficients. For example, we consider the problem

$$\begin{aligned} &p(x) * \left(a_2y'' + a_1y' + a_0y\right) = f(x) + \int_b^a g(t)y(t)\,\mathrm{d}t; \quad a \le x \le b \\ &y(a) = y_0, \quad y(b) = y_1, \quad y'(c) = 0, \quad y^{(3)}\left(\frac{a+b}{2}\right) = d, \end{aligned} \tag{26}$$

where $p(x)$ is a polynomial and f and g are known functions. Then we can see that the right-hand side of Eq. (26) can be shown as the convolution. A particular case of the above Eq. (26) can be given as the well-known Bratu's nonlinear boundary value problem

$$u''(t) + \lambda \mathrm{e}^{u(t)} = 0, \quad t \in (0,1),$$

with the boundary conditions

$$u(0) = 0, \quad u(1) = 0,$$

which has an analytical solution given in the following form:

$$u(t) = -2\ln\left[\frac{\cosh((t-\frac{1}{2})\frac{\theta}{2})}{\cosh\left(\frac{\theta}{4}\right)}\right],$$

where θ is the solution of $\theta = \sqrt{2\lambda}\cosh\frac{\theta}{4}$ (see [18]).

It is also well known that the Bratu's problem has zero, one, or two solutions when $\lambda > \lambda_c$, $\lambda = \lambda_c$ and $\lambda < \lambda_c$, respectively, where the critical value λ_c satisfies the equation $1 = \frac{1}{4}\sqrt{2\lambda_c}\sinh(\theta_c/4)$ and it was obtained in [1, 2] that the critical value λ_c is given by $\lambda_c = 3.513830719$.

$$F(x) * P(D)y = (f * g)(t) = \int_0^t f(x)g(x-t)\,\mathrm{d}x; \quad a \le x \le b,$$

$$y(a) = y(b) = c, \quad y(0) = y_0, \quad y'(0) = y_0'.$$

In order to model the real-world application, the fractional differential equations should be considered by using the fractional derivatives. There are many different starting points for the discussion of classical fractional calculus; see [14] and [32]. We also note that the convolution further can be used in the definition of the fractional integrals. For example, there are many different starting points for the discussion of classical fractional calculus; see [19]. One can begin with a generalization of repeated integration. If $f(t)$ is absolutely integrable on $[0, b)$, it can be found that [14, 32, 33].

$$\int_0^t \mathrm{d}t_n \int_0^{t_n} \mathrm{d}t_{n-1} \cdots \int_0^{t_3} \mathrm{d}t_2 \int_0^{t2} f(t_1)\,\mathrm{d}t_1 = \frac{1}{(n+1)!}\int_0^t (t-t_1)^{n-1} f(t_1)\,\mathrm{d}t_1$$
$$= \frac{1}{(n+1)!}t^{n-1} * f(t),$$

where $n = 1, 2, \ldots$, and $0 \le t \le b$. On writing $\Gamma(n) = (n-1)!$, an immediate generalization in the form of the operation I^α defined for $\alpha > 0$ is

$$(I^\alpha f)(t) = \frac{1}{\Gamma(\alpha)}\int_0^t (t-t_1)^{\alpha-1} f(t_1)\,\mathrm{d}t_1 = \frac{1}{\Gamma(\alpha)}t^{\alpha-1} * f(t), \quad 0 \le t < b, \tag{27}$$

where $\Gamma(\alpha)$ is the gamma function and

$$t^{\alpha-1} * f(t) = \int_0^t f(t-t_1)^{\alpha-1}(t_1)\,\mathrm{d}t_1$$

is the convolution product of $t^{\alpha-1}$ and $f(t)$. Equation (27) is called the Riemann–Liouville fractional integral of order α for the function $f(t)$. Similarly, by using the convolution we can also generate an ODE with variable coefficients from the ODE with constant coefficients; see [17].

Now if $f(t)$ is expanded in block pulse functions, the Riemann–Liouville fractional integral becomes

$$(I^\alpha f)(t) = \frac{1}{\Gamma(\alpha)} t^{\alpha-1} * f(t) \simeq \xi^T \frac{1}{\Gamma(\alpha)} \left\{ t^{\alpha-1} * \phi_m(t) \right\}$$

Thus, if $t^{\alpha-1} * \phi_m(t)$ can be integrated and then expanded in block pulse functions, the Riemann–Liouville fractional integral is solved via the block pluse functions. Thus, we note that Kronecker convolution product can be expanded in order to define the Riemann–Liouville fractional integrals for matrices by using the block pulse operational matrix as follows:

$$\frac{1}{\Gamma(\alpha)} \int_0^t (t - t_1)^{\alpha-1} \phi_m(t_1)\, \mathrm{d}t_1 \simeq F_\alpha \phi_m(t),$$

where

$$F_\alpha = \left(\frac{b}{m}\right)^\alpha \frac{1}{\Gamma(\alpha+2)} \begin{bmatrix} 1 & \xi_2 & \xi_3 & \dots & \xi_m \\ 0 & 1 & \xi_2 & \dots & \xi_{m-1} \\ 0 & 0 & 1 & \dots & \xi_{m-2} \\ 0 & 0 & 0 & \ddots & \vdots \\ 0 & 0 & 0 & 0 & 1 \end{bmatrix}$$

(cf. [19]). Further a particular consequence of this is that the convolution can be viewed as a “smoothing” operation: the convolution of Č and g is differentiable as many times as Č and g are together. Thus, if we have a differential operator having singularity, then convolution can be used for regularization.

An important feature of the convolution is known that if f and g both decay rapidly, then $f * g$ also decays rapidly. Further, the convolution is also a finite measure, whose total variation satisfies the inequality

$$||f * g|| \leq ||f|| \cdot ||g||$$

then we have the following theorem:

Theorem 3.1. *Let the differential equation $P(D)u = f(x)$ have a solution, and if g is an arbitrary function having bounded support, then*

$$||g * P(D)|| \leq ||g|| \cdot ||P(D)||, \tag{28}$$

has also a solution.

The proof of the theorem was held in [17]. Further note the above theorem suggests that if the original equation has a solution, then the new equation also has a solution;

the relations and the comparisons between the solutions were studied in [17]. However (28) does not indicate how to find the solution. In order to find the solution explicitly, we have to follow some of the other techniques such as in the following theorem:

Theorem 3.2. *The general differential transformation for nonlinear nth-order BVPs,*

$$y^{(n)}(x) = e^{-\lambda x} * (y(x))^m \tag{29}$$

is given by

$$Y(n+k) = \frac{k!}{(n+k)!}\left[\sum_{k_m=0}^{k}\sum_{k_{m-1}=0}^{k_m}\cdots\sum_{k_1=0}^{k_2}\frac{(-\lambda)^{k_1}}{k_1!}\left(\left[\prod_{i=2}^{m} Y(k_i-k_{i-1})\right]Y(k-k_m)\right)\right].$$

We note that by applying the convolutional derivative to Eq. (29), then we easily obtain

$$y^{(n+2)}(x) = -\lambda m e^{-\lambda x} * y'\,(y(x))^{m-1}.$$

References

1. Aregbesola, Y.A.S.: Numerical solution of Bratu problem using the method of weighted residual. Electron. J. South. Afr. Math. Sci. **3**(1), 1–7 (2003)
2. Boyd, J.P.: Chebyshev polynomial expansions for simultaneous approximation of two branches of a function with application to the one-dimensional Bratu equation. Appl. Math. Comput. **143**, 189–200 (2003)
3. Brychkov, Y.A., Glaeske, H.J., Prudnikov, A.P., Tuan, V.K.: Multidimensional Integral Transformations. Gordon and Breach Science Publishers, Philadelphia (1992)
4. Chen, G.Q., Feldman, M.: Multidimensional transonic shocks and free boundary problems for nonlinear equations of mixed type. J. Amer. Math. Soc. **16**, 461–494 (2003)
5. Courant, R., Hilbert, D.: Methods of Mathematical Physics. Vol. II: Partial differential equations. Interscience Publishers (a division of John Wiley & Sons), New York-London (1962)
6. Ditkin, V.A., Prudnikov, A.P.: Operational Calculus in Two Variables and its Applications. Pergamon, New York (1962)
7. Eltayeb, H., Kılıçman, A.: On some applications and new integral transform. Int. J. Math. Anal. **4**(3), 123–132 (2010)
8. Eltayeb, H., Kılıçman, A.: A note on solutions of wave, Laplace's and heat equations with convolution terms by using double Laplace transform. Appl. Math. Lett. **21**, 1324–1329 (2008)
9. Eltayeb, H., Kılıçman, A., Agarwal, R.P.: An analysis on classifications of hyperbolic and elliptic PDEs. Mathematical Sciences (to appear)
10. Eltayeb, H., Kılıçman, A., Fisher, B.: A new integral transform and associated distributions. Integral Transforms Spec. Funct. **21**(5), 367–379 (2010)
11. Estrada, R., Fulling, S.A.: How singular functions define distributions. J. Phys. A. **35**(13), 3079–3089 (2002)

12. Fisher, B., Jolevsaka-Tuneska, B., Kılıçman, A.: On defining the incomplete Gamma function. Integral Transforms Spec. Funct. **14**(4), 293–299 (2003)
13. Kanwal, R.P.: Generalized Functions Theory and Applications. Birkhauser, Boston (2007)
14. Kilbas, A.A., Srivastava, H.M., Trujillo, J.J.: Theory and Applications of Fractional Differential Equations. North-Holland Mathematics Studies, vol. 204. Elsevier, Amsterdam (2006)
15. Kılıçman, A.: A note on Mellin transform and distributions. J. Math. Comput. Appl. **9**(1), 65–72 (2004)
16. Kılıçman, A.: A remark on the products of distributions and semigroups. J. Math. Sci. Adv. Appl. **1**(2), 423–430 (2008)
17. Kılıçman, A., Altun, O.: On the solution of some boundary value problems by using differential transform method with convolution term. Filomat **26**(5), 917–928 (2012)
18. Kılıçman, A., Altun, O.: On higher-order boundary value problems by using differential transformation method with convolution terms. J. Franklin Inst. (to appear). http://dx.doi.org/10.1016/j.jfranklin.2012.08.010
19. Kılıçman, A., Al Zhour, Z.A.A.: Kronecker operational matrices for fractional calculus and some applications. Appl. Math. Comput. **187**(1), 250–265 (2007)
20. Kılıçman, A., Eltayeb, H.: A note on defining singular integral as distribution and partial differential equations with convolution terms. Math. Comput. Modelling. **49**(1–2), 327–336 (2009)
21. Kılıçman, A., Eltayeb, H.: On finite product of convolutions and classifications of hyperbolic and elliptic equations. Math. Comput. Modelling. **54**(9–10), 2211–2219 (2011)
22. Kılıçman, A., Eltayeb, H.: A note on classification of hyperbolic and elliptic equations with polynomial coefficients. Appl. Math. Lett. **21**(11), 1124–1128 (2008)
23. Kılıçman, A., Eltayeb, H.: A note on integral transforms and partial differential equations. Appl. Math. Sci. **4**(3), 109–118 (2010)
24. Kılıçman, A., Gadain, H.E.: An application of double Laplace transform and double Sumudu transform. Lobachevskii J. Math. **30**(3), 214–223 (2009)
25. Kılıçman, A., Eltayeb, H.: Note on partial differential equations with non-constant coefficients and convolution method. Appl. Math. Inf. Sci. **6**(1), 59–63 (2012)
26. Kılıçman, A., Eltayeb, H., Agarwal, R.P.: On Sumudu transform and system of differential equations. Abstr. Appl. Anal. **2010**, Article ID 598702, 11 pp (2010)
27. Li, S., Liao, S.J.: An analytic approach to solve multiple solutions of a strongly nonlinear problem. Appl. Math. Comput. **169**, 854–865 (2005)
28. Liao, S., Tan, Y.: A general approach to obtain series solutions of nonlinear differential equations. Stud. Appl. Math. **119**, 297–354 (2007)
29. McGough, J.S.: Numerical continuation and the Gelfand problem. Appl. Math. Compu. **89**, 225–239 (1998)
30. Merton, R.C.: Theory of rational option pricing. Bell J. Econ. Manag. Sci. **4**, 141–183 (1973)
31. Mounim, A.S., de Dormale, B.M.: From the fitting techniques to accurate schemes for the Liouville-Bratu-Gelfand problem. Numer. Methods Partial. Differ. Equ. **22**(4), 761–775 (2006)
32. Ross, B.: Fractional Calculus and its Applications. Springer, Berlin (1975)
33. Sumita, H.: The matrix Laguerre transform. Appl. Math. Comput. **15**, 1–28 (1984)
34. Susarla, V., Walter, G.: Estimation of a multivariate density function using delta sequences. Ann. Statist. **9**(2), 347–355 (1981)
35. Syam, M.I., Hamdan, A.: An efficient method for solving Bratu equations. Appl. Math. Comput. **176**, 704–713 (2006)
36. Wazwaz, A.M.: Adomian decomposition method for a reliable treatment of the Bratu-type equations. Appl. Math. Comput. **166**, 652–663 (2005)
37. Zayed, A.I.: Handbook of Functions and Generalized Function Transformations, CRC Press, Boca Raton (1966)
38. Zemanian, A.H.: Distribution Theory and Transform Analysis: An Introduction to Generalized Functions With Applications. Dover, New York (1987)
39. Zhou, J.K.: Differential Transformation and Its Application for Electrical Circuits. Huazhong University Press, Wuhan (1986)

Orthogonally Additive: Additive Functional Equation

Choonkil Park

Dedicated to Professor Hari M. Srivastava

Abstract Using fixed point method, we prove the Hyers–Ulam stability of the orthogonally additive–additive functional equation

$$f\left(\frac{x}{2}+y\right)+f\left(\frac{x}{2}+z\right)=f(x)+f(y)+f(z)$$

for all x, y, z with $x \perp y$, in orthogonality Banach spaces and in non-Archimedean orthogonality Banach spaces.

1 Introduction and Preliminaries

In 1897, Hensel [23] introduced a normed space which does not have the Archimedean property. It turned out that non-Archimedean spaces have many nice applications (see [12, 31, 32, 39]).

A *valuation* is a function $|\cdot|$ from a field K into $[0,\infty)$ such that 0 is the unique element having the 0 valuation, $|rs| = |r|\cdot|s|$, and the triangle inequality holds, i.e.,

$$|r+s| \leqslant |r|+|s|, \qquad \forall r, s \in K.$$

A field K is called a *valued field* if K carries a valuation. Throughout this paper, we assume that the base field is a valued field; hence, call it simply a field. The usual absolute values of $\mathbb{R}$ and $\mathbb{C}$ are examples of valuations.

C. Park (✉)
Department of Mathematics, Hanyang University, Seoul 133-791, Republic of Korea
e-mail: baak@hanyang.ac.kr

G.V. Milovanović and M.Th. Rassias (eds.), *Analytic Number Theory, Approximation Theory, and Special Functions*, DOI 10.1007/978-1-4939-0258-3_29,

Let us consider a valuation which satisfies a stronger condition than the triangle inequality. If the triangle inequality is replaced by

$$|r+s| \leqslant \max\{|r|,|s|\}, \qquad \forall r,s \in K,$$

then the function $|\cdot|$ is called a *non-Archimedean valuation*, and the field is called a *non-Archimedean field*. Clearly $|1| = |-1| = 1$ and $|n| \leqslant 1$ for all $n \in \mathbb{N}$. A trivial example of a non-Archimedean valuation is the function $|\cdot|$ taking everything except for 0 into 1 and $|0| = 0$.

Definition 1.1 ([38]). Let X be a vector space over a field K with a non-Archimedean valuation $|\cdot|$. A function $\|\cdot\| : X \to [0,\infty)$ is said to be a *non-Archimedean norm* if it satisfies the following conditions:

(i) $\|x\| = 0$ if and only if $x = 0$
(ii) $\|rx\| = |r|\|x\| \quad (r \in K, x \in X)$
(iii) The strong triangle inequality

$$\|x+y\| \leqslant \max\{\|x\|,\|y\|\}, \qquad \forall x,y \in X$$

holds. Then $(X, \|\cdot\|)$ is called a *non-Archimedean normed space.*

Definition 1.2.

(i) Let $\{x_n\}$ be a sequence in a non-Archimedean normed space X. Then the sequence $\{x_n\}$ is called *Cauchy* if for a given $\varepsilon > 0$ there is a positive integer N such that

$$\|x_n - x_m\| \leqslant \varepsilon$$

for all $n, m \geqslant N$.
(ii) Let $\{x_n\}$ be a sequence in a non-Archimedean normed space X. Then the sequence $\{x_n\}$ is called *convergent* if for a given $\varepsilon > 0$ there are a positive integer N and an $x \in X$ such that

$$\|x_n - x\| \leqslant \varepsilon$$

for all $n \geqslant N$. Then we call $x \in X$ a limit of the sequence $\{x_n\}$ and denote by $\lim_{n\to\infty} x_n = x$.
(iii) If every Cauchy sequence in X converges, then the non-Archimedean normed space X is called a *non-Archimedean Banach space.*

Assume that X is a real inner product space and $f : X \to \mathbb{R}$ is a solution of the orthogonal Cauchy functional equation $f(x+y) = f(x) + f(y), \langle x, y\rangle = 0$. By the Pythagorean theorem $f(x) = \|x\|^2$ is a solution of the conditional equation. Of course, this function does not satisfy the additivity equation everywhere. Thus orthogonal Cauchy equation is not equivalent to the classic Cauchy equation on the whole inner product space.

G. Pinsker [44] characterized orthogonally additive functionals on an inner product space when the orthogonality is the ordinary one in such spaces. K. Sundaresan [54] generalized this result to arbitrary Banach spaces equipped with the Birkhoff–James orthogonality. The orthogonal Cauchy functional equation,

$$f(x + y) = f(x) + f(y), \qquad x \perp y,$$

in which $\perp$ is an abstract orthogonality relation, was first investigated by S. Gudder and D. Strawther [22]. They defined $\perp$ by a system consisting of five axioms and described the general semicontinuous real-valued solution of conditional Cauchy functional equation. In 1985, J. Rätz [51] introduced a new definition of orthogonality by using more restrictive axioms than of S. Gudder and D. Strawther. Moreover, he investigated the structure of orthogonally additive mappings. J. Rätz and Gy. Szabó [52] investigated the problem in a rather more general framework.

Let us recall the orthogonality in the sense of J. Rätz; cf. [51].

Suppose X is a real vector space with $\dim X \geq 2$ and $\perp$ is a binary relation on X with the following properties:

(O_1) totality of $\perp$ for zero: $x \perp 0, 0 \perp x$ for all $x \in X$;
(O_2) independence: if $x, y \in X - \{0\}$, $x \perp y$, then x, y are linearly independent;
(O_3) homogeneity: if $x, y \in X, x \perp y$, then $\alpha x \perp \beta y$ for all $\alpha, \beta \in \mathbb{R}$;
(O_4) the Thalesian property: if P is a 2-dimensional subspace of $X, x \in P$ and $\lambda \in \mathbb{R}_+$, which is the set of nonnegative real numbers, then there exists $y_0 \in P$ such that $x \perp y_0$ and $x + y_0 \perp \lambda x - y_0$.

The pair $(X, \perp)$ is called an orthogonality space. By an orthogonality normed space, we mean an orthogonality space having a normed structure.

Some interesting examples are

(i) The trivial orthogonality on a vector space X defined by (O_1), and for nonzero elements $x, y \in X, x \perp y$ if and only if x, y are linearly independent.
(ii) The ordinary orthogonality on an inner product space $(X, \langle\cdot,\cdot\rangle)$ given by $x \perp y$ if and only if $\langle x, y\rangle = 0$.
(iii) The Birkhoff–James orthogonality on a normed space $(X, \|\cdot\|)$ defined by $x \perp y$ if and only if $\|x + \lambda y\| \geq \|x\|$ for all $\lambda \in \mathbb{R}$.

The relation $\perp$ is called symmetric if $x \perp y$ implies that $y \perp x$ for all $x, y \in X$. Clearly examples (i) and (ii) are symmetric but example (iii) is not. It is remarkable to note, however, that a real normed space of dimension greater than 2 is an inner product space if and only if the Birkhoff–James orthogonality is symmetric. There are several orthogonality notions on a real normed space such as Birkhoff–James, Boussouis, Singer, Carlsson, unitary-Boussouis, Roberts, Phythagorean, isosceles, and Diminnie (see [1, 3, 7, 14, 28]).

The stability problem of functional equations originated from the following question of Ulam [56]: *Under what condition does there exist an additive mapping near an approximately additive mapping?* In 1941, Hyers [24] gave a partial affirmative answer to the question of Ulam in the context of Banach spaces. In 1978, Th.M. Rassias [46] extended the theorem of Hyers by considering the unbounded Cauchy difference

$$\|f(x+y)-f(x)-f(y)\| \leq \varepsilon(\|x\|^p+\|y\|^p) \quad (\varepsilon>0,\ p\in[0,1)).$$

The first author treating the stability of the quadratic equation was F. Skof [53] by proving that if f is a mapping from a normed space X into a Banach space Y satisfying

$$\|f(x+y)+f(x-y)-2f(x)-2f(y)\| \leq \varepsilon$$

for some $\varepsilon>0$, then there is a unique quadratic mapping $g: X\to Y$ such that

$$\|f(x)-g(x)\| \leq \frac{\varepsilon}{2}.$$

P. W. Cholewa [8] extended the Skof's theorem by replacing X by an abelian group G. The Skof's result was later generalized by S. Czerwik [9] in the spirit of Ulam–Hyers–Rassias. The stability problem of functional equations has been extensively investigated by some mathematicians (see [10, 11, 16–19, 25, 29, 43, 47–50]).

R. Ger and J. Sikorska [21] investigated the orthogonal stability of the Cauchy functional equation $f(x+y)=f(x)+f(y)$, namely, they showed that if f is a mapping from an orthogonality space X into a real Banach space Y and

$$\|f(x+y)-f(x)-f(y)\| \leq \varepsilon$$

for all $x,y\in X$ with $x\perp y$ and some $\varepsilon>0$, then there exists exactly one orthogonally additive mapping $g: X\to Y$ such that

$$\|f(x)-g(x)\| \leq \frac{16}{3}\varepsilon$$

for all $x\in X$.

The orthogonally quadratic equation

$$f(x+y)+f(x-y)=2f(x)+2f(y), \quad x\perp y$$

was first investigated by F. Vajzović [57] when X is a Hilbert space, Y is the scalar field, f is continuous and $\perp$ means the Hilbert space orthogonality. Later, H. Drljević [15], M. Fochi [20], M.S. Moslehian [35, 36], and Gy. Szabó [55] generalized this result. See also [37, 40].

Let X be a set. A function $d: X\times X\to[0,\infty]$ is called a *generalized metric* on X if d satisfies

(1) $d(x,y)=0$ if and only if $x=y$;
(2) $d(x,y)=d(y,x)$ for all $x,y\in X$;
(3) $d(x,z)\leqslant d(x,y)+d(y,z)$ for all $x,y,z\in X$.

We recall a fundamental result in fixed point theory.

Theorem 1.1 ([4, 13]). *Let (X, d) be a complete generalized metric space and let $J : X \to X$ be a strictly contractive mapping with Lipschitz constant $\alpha < 1$. Then for each given element $x \in X$, either*

$$d(J^n x, J^{n+1} x) = \infty$$

for all nonnegative integers n or there exists a positive integer n_0 such that

(1) $d(J^n x, J^{n+1} x) < \infty, \quad \forall n \geqslant n_0;$
(2) *the sequence $\{J^n x\}$ converges to a fixed point y^* of J;*
(3) *y^* is the unique fixed point of J in the set*

$$Y = \{y \in X \mid d(J^{n_0} x, y) < \infty\};$$

(4) $d(y, y^*) \leqslant \frac{1}{1-\alpha} d(y, Jy)$ *for all $y \in Y$.*

In 1996, G. Isac and Th.M. Rassias [26] were the first to provide applications of stability theory of functional equations for the proof of new fixed point theorems with applications. By using fixed point methods, the stability problems of several functional equations have been extensively investigated by a number of authors (see [5, 6, 30, 34, 41, 42, 45]).

This paper is organized as follows: In Sect. 2, we prove the Hyers–Ulam stability of the orthogonally additive–additive functional equation (1) in orthogonality spaces. In Sect. 3, we prove the Hyers–Ulam stability of the orthogonally additive–additive functional equation (1) in non-Archimedean orthogonality spaces.

2 Hyers–Ulam Stability of the Orthogonally Additive–Additive Functional Equation

Throughout this section, assume that $(X, \perp)$ is an orthogonality space and that $(Y, \|.\|_Y)$ is a real Banach space.

In this section, applying some ideas from [21, 25], we deal with the stability problem for the orthogonally additive–additive functional equation

$$Df(x, y, z) := f\left(\frac{x}{2} + y\right) + f\left(\frac{x}{2} + z\right) - f(x) - f(y) - f(z) = 0$$

for all $x, y, z \in X$ with $x \perp y$ in orthogonality spaces.

If f is a mapping with $Df(x, y, z) = 0$, then

$$f\left(\frac{x}{2} + y\right) + f\left(\frac{x}{2}\right) = f(x) + f(y)$$

and so $f(x+y) = f(x) + f(y)$ for all $x, y \in X$ with $x \perp y$, and

$$f\left(\frac{x}{2}\right) + f\left(\frac{x}{2} + z\right) = f(x) + f(z)$$

and so

$$f(x+z) = f(x) + f(z)$$

for all $x, z \in X$. That is, f is orthogonally additive and additive.

Definition 2.1. A mapping $f : X \to Y$ is called an *orthogonally additive–additive mapping* if

$$f\left(\frac{x}{2} + y\right) + f\left(\frac{x}{2} + z\right) = f(x) + f(y) + f(z)$$

for all $x, y, z \in X$ with $x \perp y$.

Theorem 2.1. *Let $\varphi : X^3 \to [0, \infty)$ be a function such that there exists an $\alpha < 1$ with*

$$\varphi(x, y, z) \leqslant 2\alpha\varphi\left(\frac{x}{2}, \frac{y}{2}, \frac{z}{2}\right) \tag{1}$$

for all $x, y, z \in X$ with $x \perp y$. Let $f : X \to Y$ be a mapping satisfying

$$\|Df(x, y, z)\|_Y \leqslant \varphi(x, y, z) \tag{2}$$

for all $x, y, z \in X$ with $x \perp y$. Then there exists a unique orthogonally additive–additive mapping $L : X \to Y$ such that

$$\|f(x) - L(x)\|_Y \leqslant \frac{\alpha}{1-\alpha}\varphi(x, 0, 0) \tag{3}$$

for all $x \in X$.

Proof. Putting $y = z = 0$ in (3), we get

$$\left\|2f\left(\frac{x}{2}\right) - f(x)\right\|_Y \leqslant \varphi(x, 0, 0) \tag{4}$$

for all $x \in X$, since $x \perp 0$. So

$$\left\|f(x) - \frac{1}{2}f(2x)\right\|_Y \leqslant \frac{1}{2}\varphi(2x, 0, 0) \leqslant \alpha\varphi(x, 0, 0) \tag{5}$$

for all $x \in X$.

Consider the set

$$S := \{h : X \to Y\}$$

and introduce the generalized metric on S:

$$d(g,h) = \inf\{\mu \in \mathbb{R}_+ : \|g(x) - h(x)\|_Y \leqslant \mu\varphi(x,0,0), \quad \forall x \in X\},$$

where, as usual, $\inf\phi = +\infty$. It is easy to show that (S,d) is complete (see [33]).

Now we consider the linear mapping $J : S \to S$ such that

$$Jg(x) := \frac{1}{2}g(2x)$$

for all $x \in X$.

Let $g, h \in S$ be given such that $d(g,h) = \varepsilon$. Then

$$\|g(x) - h(x)\|_Y \leqslant \varphi(x,0,0)$$

for all $x \in X$. Hence

$$\|Jg(x) - Jh(x)\|_Y = \left\|\frac{1}{2}g(2x) - \frac{1}{2}h(2x)\right\|_Y \leqslant \alpha\varphi(x,0,0)$$

for all $x \in X$. So $d(g,h) = \varepsilon$ implies that $d(Jg, Jh) \leqslant \alpha\varepsilon$. This means that

$$d(Jg, Jh) \leqslant \alpha d(g,h)$$

for all $g, h \in S$.

It follows from (6) that $d(f, Jf) \leqslant \alpha$.

By Theorem 1.1, there exists a mapping $L : X \to Y$ satisfying the following:

(1) L is a fixed point of J, i.e.,

$$L(2x) = 2L(x) \tag{6}$$

for all $x \in X$. The mapping L is a unique fixed point of J in the set

$$M = \{g \in S : d(h,g) < \infty\}.$$

This implies that L is a unique mapping satisfying (7) such that there exists a $\mu \in (0,\infty)$ satisfying

$$\|f(x) - L(x)\|_Y \leqslant \mu\varphi(x,0,0)$$

for all $x \in X$;

(2) $d(J^n f, L) \to 0$ as $n \to \infty$. This implies the equality

$$\lim_{n\to\infty} \frac{1}{2^n} f(2^n x) = L(x)$$

for all $x \in X$;

(3) $d(f, L) \leqslant \frac{1}{1-\alpha} d(f, Jf)$, which implies the inequality

$$d(f, L) \leqslant \frac{\alpha}{1-\alpha}.$$

This implies that the inequality (4) holds.

It follows from (2) and (3) that

$$\begin{aligned}\|DL(x,y,z)\|_Y &= \lim_{n\to\infty} \frac{1}{2^n} \|Df(2^n x, 2^n y, 2^n z)\|_Y \\ &\leqslant \lim_{n\to\infty} \frac{1}{2^n} \varphi(2^n x, 2^n y, 2^n z) \leqslant \lim_{n\to\infty} \frac{2^n \alpha^n}{2^n} \varphi(x,y,z) = 0\end{aligned}$$

for all $x, y, z \in X$ with $x \perp y$. Hence

$$L\left(\frac{x}{2} + y\right) + L\left(\frac{x}{2} + z\right) = L(x) + L(y) + L(z)$$

for all $x, y, z \in X$ with $x \perp y$. So $L : X \to Y$ is an orthogonally additive–additive mapping.

Thus, $L : X \to Y$ is a unique orthogonally additive–additive mapping satisfying (4), as desired. □

Corollary 2.1. *Assume that $(X, \perp)$ is an orthogonality normed space. Let θ be a positive real number and p a real number with $0 < p < 1$. Let $f : X \to Y$ be a mapping satisfying*

$$\|Df(x,y,z)\|_Y \leq \theta(\|x\|^p + \|y\|^p + \|z\|^p) \tag{7}$$

for all $x, y, z \in X$ with $x \perp y$. Then there exists a unique orthogonally additive–additive mapping $L : X \to Y$ such that

$$\|f(x) - L(x)\|_Y \leqslant \frac{2^p \theta}{2 - 2^p} \|x\|^p$$

for all $x \in X$.

Proof. The proof follows from Theorem 2.1 by taking

$$\varphi(x,y,z) = \theta(\|x\|^p + \|y\|^p + \|z\|^p)$$

for all $x, y, z \in X$ with $x \perp y$. Then we can choose $\alpha = 2^{p-1}$ and we get the desired result. □

Theorem 2.2. *Let $f : X \to Y$ be a mapping satisfying* (3) *for which there exists a function $\varphi : X^3 \to [0, \infty)$ such that*

$$\varphi(x, y, z) \leqslant \frac{\alpha}{2}\varphi(2x, 2y, 2z)$$

for all $x, y, z \in X$ with $x \perp y$. Then there exists a unique orthogonally additive–additive mapping $L : X \to Y$ such that

$$\|f(x) - L(x)\|_Y \leqslant \frac{1}{1-\alpha}\varphi(x, 0, 0) \tag{8}$$

for all $x \in X$.

Proof. Let (S, d) be the generalized metric space defined in the proof of Theorem 2.1.

Now we consider the linear mapping $J : S \to S$ such that

$$Jg(x) := 2g\left(\frac{x}{2}\right)$$

for all $x \in X$.

It follows from (5) that $d(f, Jf) \leqslant 1$. So

$$d(f, L) \leqslant \frac{1}{1-\alpha}.$$

Thus we obtain the inequality (9).

The rest of the proof is similar to the proof of Theorem 2.1. □

Corollary 2.2. *Assume that $(X, \perp)$ is an orthogonality normed space. Let θ be a positive real number and p a real number with $p > 1$. Let $f : X \to Y$ be a mapping satisfying* (8)*. Then there exists a unique orthogonally additive–additive mapping $L : X \to Y$ such that*

$$\|f(x) - L(x)\|_Y \leqslant \frac{2^p\theta}{2^p - 2}\|x\|^p$$

for all $x \in X$.

Proof. The proof follows from Theorem 2.2 by taking

$$\varphi(x, y, z) = \theta(\|x\|^p + \|y\|^p + \|z\|^p)$$

for all $x, y, z \in X$ with $x \perp y$.

Then we can choose $\alpha = 2^{1-p}$ and we get the desired result. □

3 Hyers–Ulam Stability of the Orthogonally Additive–Additive Functional Equation in Non-Archimedean Orthogonality Spaces

Throughout this section, assume that $(X, \perp)$ is a non-Archimedean orthogonality space and that $(Y, \|\cdot\|_Y)$ is a real non-Archimedean Banach space. Assume that $|2| \neq 1$.

In this section, applying some ideas from [21,25], we deal with the stability problem for the orthogonally additive–additive functional equation $Df(x, y, z) = 0$, given in the previous section, in non-Archimedean orthogonality spaces.

Theorem 3.1. *Let $\varphi : X^3 \to [0, \infty)$ be a function such that there exists an $\alpha < 1$ with*

$$\varphi(x, y, z) \leqslant |2|\alpha\varphi\left(\frac{x}{2}, \frac{y}{2}, \frac{z}{2}\right)$$

for all $x, y, z \in X$ with $x \perp y$. Let $f : X \to Y$ be a mapping satisfying

$$\|Df(x, y, z)\|_Y \leqslant \varphi(x, y, z) \tag{9}$$

for all $x, y, z \in X$ with $x \perp y$. Then there exists a unique orthogonally additive–additive mapping $L : X \to Y$ such that

$$\|f(x) - L(x)\|_Y \leqslant \frac{\alpha}{1-\alpha}\varphi(x, 0, 0) \tag{10}$$

for all $x \in X$.

Proof. Putting $y = z = 0$ in (10), we get

$$\left\|2f\left(\frac{x}{2}\right) - f(x)\right\|_Y \leqslant \varphi(x, 0, 0) \tag{11}$$

for all $x \in X$, since $x \perp 0$. So

$$\left\|f(x) - \frac{1}{2}f(2x)\right\|_Y \leqslant \frac{1}{|2|}\varphi(2x, 0, 0) \leqslant \alpha\varphi(x, 0, 0) \tag{12}$$

for all $x \in X$.

Let (S, d) be the generalized metric space defined in the proof of Theorem 2.1. Now we consider the linear mapping $J : S \to S$ such that

$$Jg(x) := \frac{1}{2}g(2x)$$

for all $x \in X$.

It follows from (13) that $d(f, Jf) \leqslant \alpha$. Thus, we obtain the inequality (11).
The rest of the proof is similar to the proof of Theorem 2.1. □

Corollary 3.1. *Assume that $(X, \perp)$ is a non-Archimedean orthogonality normed space. Let θ be a positive real number and p a real number with $0 < p < 1$. Let $f : X \to Y$ be a mapping satisfying (8). Then there exists a unique orthogonally additive–additive mapping $L : X \to Y$ such that*

$$\|f(x) - L(x)\|_Y \leqslant \frac{|2|\theta}{|2|^p - |2|}\|x\|^p$$

for all $x \in X$.

Proof. The proof follows from Theorem 3.1 by taking

$$\varphi(x, y, z) = \theta(\|x\|^p + \|y\|^p + \|z\|^p)$$

for all
Then we can choose $\alpha = |2|^{1-p}$ and we get the desired result. □

Theorem 3.2. *Let $f : X \to Y$ be a mapping satisfying* (10) *for which there exists a function $\varphi : X^3 \to [0, \infty)$ such that*

$$\varphi(x, y, z) \leqslant \frac{\alpha}{|2|}\varphi(2x, 2y, 2z)$$

for all $x, y, z \in X$ with $x \perp y$. Then there exists a unique orthogonally additive–additive mapping $L : X \to Y$ such that

$$\|f(x) - L(x)\|_Y \leqslant \frac{1}{1-\alpha}\varphi(x, 0, 0) \tag{13}$$

for all $x \in X$.

Proof. Let (S, d) be the generalized metric space defined in the proof of Theorem 2.1.

Now we consider the linear mapping $J : S \to S$ such that

$$Jg(x) := 2g\left(\frac{x}{2}\right)$$

for all $x \in X$.
It follows from (12) that $d(f, Jf) \leqslant 1$. So

$$d(f, L) \leqslant \frac{1}{1-\alpha}.$$

Thus we obtain the inequality (13).
The rest of the proof is similar to the proof of Theorem 2.1. □

Corollary 3.2. *Assume that* $(X, \perp)$ *is a non-Archimedean orthogonality normed space. Let* θ *be a positive real number and* p *a real number with* $p > 1$. *Let* $f : X \to Y$ *be an odd mapping satisfying* (8). *Then there exists a unique orthogonally additive–additive mapping* $L : X \to Y$ *such that*

$$\|f(x) - L(x)\|_Y \leqslant \frac{|2|\theta}{|2| - |2|^p}\|x\|^p$$

for all $x \in X$.

Proof. The proof follows from Theorem 3.2 by taking

$$\varphi(x, y, z) = \theta(\|x\|^p + \|y\|^p + \|z\|^p)$$

for all $x, y, z \in X$ with $x \perp y$.

Then we can choose $\alpha = |2|^{p-1}$ and we get the desired result. □

Acknowledgements This work was supported by Basic Science Research Program through the National Research Foundation of Korea funded by the Ministry of Education, Science and Technology (NRF-2009-0070788).

References

1. Alonso, J., Benítez, C.: Orthogonality in normed linear spaces: a survey I. Main properties. Extracta Math. **3**, 1–15 (1988)
2. Alonso, J., Benítez, C.: Orthogonality in normed linear spaces: a survey II. Relations between main orthogonalities. Extracta Math. **4**, 121–131 (1989)
3. Birkhoff, G.: Orthogonality in linear metric spaces. Duke Math. J. **1**, 169–172 (1935)
4. Cădariu, L., Radu, V.: Fixed points and the stability of Jensen's functional equation. J. Inequal. Pure Appl. Math. **4**(1), Art. ID 4 (2003)
5. Cădariu, L., Radu, V.: On the stability of the Cauchy functional equation: a fixed point approach. Grazer Math. Ber. **346**, 43–52 (2004)
6. Cădariu, L., Radu, V.: Fixed point methods for the generalized stability of functional equations in a single variable. Fixed Point Theory Appl. **2008**, Art. ID 749392 (2008).
7. Carlsson, S.O.: Orthogonality in normed linear spaces. Ark. Mat. **4**, 297–318 (1962)
8. Cholewa, P.W.: Remarks on the stability of functional equations. Aequationes Math. **27**, 76–86 (1984)
9. Czerwik, S.: On the stability of the quadratic mapping in normed spaces. Abh. Math. Sem. Univ. Hamburg **62**, 59–64 (1992)
10. Czerwik, S.: Functional Equations and Inequalities in Several Variables. World Scientific Publishing Company, New Jersey (2002)
11. Czerwik, S.: Stability of Functional Equations of Ulam–Hyers–Rassias Type. Hadronic Press, Palm Harbor (2003)
12. Deses, D.: On the representation of non-Archimedean objects. Topology Appl. **153**, 774–785 (2005)
13. Diaz, J., Margolis, B.: A fixed point theorem of the alternative for contractions on a generalized complete metric space. Bull. Amer. Math. Soc. **74**, 305–309 (1968)

14. Diminnie, C.R.: A new orthogonality relation for normed linear spaces. Math. Nachr. **114**, 197–203 (1983)
15. Drljević, F.: On a functional which is quadratic on A-orthogonal vectors. Publ. Inst. Math. (Beograd) **54**, 63–71 (1986)
16. Eshaghi Gordji, M.: A characterization of (σ, τ)-derivations on von Neumann algebras. Politehn. Univ. Bucharest Sci. Bull. Ser. A Appl. Math. Phys. **73**(1), 111–116 (2011)
17. Eshaghi Gordji, M., Bodaghi, A., Park, C.: A fixed point approach to the stability of double Jordan centralizers and Jordan multipliers on Banach algebras. Politehn. Univ. Bucharest Sci. Bull. Ser. A Appl. Math. Phys. **73**(2), 65–74 (2011)
18. Eshaghi Gordji, M., Khodaei, H., Khodabakhsh, R.: General quartic-cubic-quadratic functional equation in non-Archimedean normed spaces. Politehn. Univ. Bucharest Sci. Bull. Ser. A Appl. Math. Phys. **72**(3), 69–84 (2010)
19. Eshaghi Gordji, M., Savadkouhi, M.B.: Approximation of generalized homomorphisms in quasi-Banach algebras. An. Stiint. Univ. Ovidius Constanta Ser. Mat. **17**(2), 203–213 (2009)
20. Fochi, M.: Functional equations in A-orthogonal vectors. Aequationes Math. **38**, 28–40 (1989)
21. Ger, R., Sikorska, J.: Stability of the orthogonal additivity. Bull. Polish Acad. Sci. Math. **43**, 143–151 (1995)
22. Gudder, S., Strawther, D.: Orthogonally additive and orthogonally increasing functions on vector spaces. Pacific J. Math. **58**, 427–436 (1975)
23. Hensel, K.: Ubereine news Begrundung der Theorie der algebraischen Zahlen. Jahresber. Deutsch. Math. Verein **6**, 83–88 (1897)
24. Hyers, D.H.: On the stability of the linear functional equation. Proc. Nat. Acad. Sci. U.S.A. **27**, 222–224 (1941)
25. Hyers, D.H., Isac, G., Rassias, Th.M.: Stability of Functional Equations in Several Variables. Birkhäuser, Basel (1998)
26. Isac, G., Rassias, Th.M.: Stability of ψ-additive mappings: Applications to nonlinear analysis. Internat. J. Math. Math. Sci. **19**, 219–228 (1996)
27. James, R.C.: Orthogonality in normed linear spaces. Duke Math. J. **12**, 291–302 (1945)
28. James, R.C.: Orthogonality and linear functionals in normed linear spaces. Trans. Amer. Math. Soc. **61**, 265–292 (1947)
29. Jung, S.: Hyers–Ulam–Rassias Stability of Functional Equations in Mathematical Analysis. Hadronic Press, Palm Harbor (2001)
30. Jung, Y., Chang, I.: The stability of a cubic type functional equation with the fixed point alternative. J. Math. Anal. Appl. **306**, 752–760 (2005)
31. Katsaras, A.K., Beoyiannis, A.: Tensor products of non-Archimedean weighted spaces of continuous functions. Georgian Math. J. **6**, 33–44 (1999)
32. Khrennikov, A.: Non-Archimedean analysis: quantum paradoxes, dynamical systems and biological models. In: Mathematics and its Applications, vol. 427. Kluwer Academic Publishers, Dordrecht (1997)
33. Miheţ, D., Radu, V.: On the stability of the additive Cauchy functional equation in random normed spaces. J. Math. Anal. Appl. **343**, 567–572 (2008)
34. Mirzavaziri, M., Moslehian, M.S.: A fixed point approach to stability of a quadratic equation. Bull. Braz. Math. Soc. **37**, 361–376 (2006)
35. Moslehian, M.S.: On the orthogonal stability of the Pexiderized quadratic equation. J. Difference Equat. Appl. **11**, 999–1004 (2005)
36. Moslehian, M.S.: On the stability of the orthogonal Pexiderized Cauchy equation. J. Math. Anal. Appl. **318**, 211–223 (2006)
37. Moslehian, M.S., Rassias, Th.M.: Orthogonal stability of additive type equations. Aequationes Math. **73**, 249–259 (2007)
38. Moslehian, M.S., Sadeghi, Gh.: A Mazur-Ulam theorem in non-Archimedean normed spaces. Nonlinear Anal. TMA **69**, 3405–3408 (2008)
39. Nyikos, P.J.: On some non-Archimedean spaces of Alexandrof and Urysohn. Topology Appl. **91**, 1–23 (1999)

40. Paganoni, L., Rätz, J.: Conditional function equations and orthogonal additivity. Aequationes Math. **50**, 135–142 (1995)
41. Park, C.: Fixed points and Hyers–Ulam–Rassias stability of Cauchy–Jensen functional equations in Banach algebras. Fixed Point Theory Appl. **2007**, Art. ID 50175 (2007)
42. Park, C.: Generalized Hyers–Ulam–Rassias stability of quadratic functional equations: a fixed point approach. Fixed Point Theory Appl. **2008**, Art. ID 493751 (2008)
43. Park, C., Park, J.: Generalized Hyers–Ulam stability of an Euler–Lagrange type additive mapping. J. Difference Equat. Appl. **12**, 1277–1288 (2006)
44. Pinsker, A.G.: Sur une fonctionnelle dans l'espace de Hilbert. C. R. (Dokl.) Acad. Sci. URSS, n. Ser. **20**, 411–414 (1938)
45. Radu, V.: The fixed point alternative and the stability of functional equations. Fixed Point Theory **4**, 91–96 (2003)
46. Rassias, Th.M.: On the stability of the linear mapping in Banach spaces. Proc. Amer. Math. Soc. **72**, 297–300 (1978)
47. Rassias, Th.M.: On the stability of the quadratic functional equation and its applications. Studia Univ. Babeş-Bolyai Math. **43**, 89–124 (1998)
48. Rassias, Th.M.: The problem of S.M. Ulam for approximately multiplicative mappings. J. Math. Anal. Appl. **246**, 352–378 (2000)
49. Rassias, Th.M.: On the stability of functional equations in Banach spaces. J. Math. Anal. Appl. **251**, 264–284 (2000)
50. Rassias, Th.M. (ed.): Functional Equations, Inequalities and Applications. Kluwer Academic Publishers, Dordrecht (2003)
51. Rätz, J.: On orthogonally additive mappings. Aequationes Math. **28**, 35–49 (1985)
52. Rätz, J., Szabó, Gy.: On orthogonally additive mappings IV. Aequationes Math. **38**, 73–85 (1989)
53. Skof, F.: Proprietà locali e approssimazione di operatori. Rend. Sem. Mat. Fis. Milano **53**, 113–129 (1983)
54. Sundaresan, K.: Orthogonality and nonlinear functionals on Banach spaces. Proc. Amer. Math. Soc. **34**, 187–190 (1972)
55. Szabó, Gy.: Sesquilinear-orthogonally quadratic mappings. Aequationes Math. **40**, 190–200 (1990)
56. Ulam, S.M.: Problems in Modern Mathematics. Wiley, New York (1960)
57. Vajzović, F.: Über das Funktional H mit der Eigenschaft: $(x, y) = 0 \Rightarrow H(x + y) + H(x - y) = 2H(x) + 2H(y)$. Glasnik Mat. Ser. III **2**(22), 73–81 (1967)

Part V
Special and Complex Functions and Applications

Alternating Mathieu Series, Hilbert–Eisenstein Series and Their Generalized Omega Functions

Árpád Baricz, Paul L. Butzer, and Tibor K. Pogány

Dedicated to Professor Hari M. Srivastava

Abstract In this paper our aim is to generalize the complete Butzer–Flocke–Hauss (BFH) Ω-function in a natural way by using two approaches. Firstly, we introduce the generalized Omega function via alternating generalized Mathieu series by imposing Bessel function of the first kind of arbitrary order as the kernel function instead of the original cosine function in the integral definition of the Ω. We also study the following set of questions about generalized BFH Ω_ν-function: **(i)** two different sets of bounding inequalities by certain bounds upon the kernel Bessel function; **(ii)** linear ordinary differential equation of which particular solution is the newly introduced Ω_ν-function, and by virtue of the Čaplygin comparison theorem another set of bounding inequalities are given.

In the second main part of this paper we introduce another extension of BFH Omega function as the counterpart of generalized BFH function in terms of the positive integer order Hilbert–Eisenstein (HE) series. In this study we realize by exposing basic analytical properties, recurrence identities and integral representation formulae of Hilbert–Eisenstein series. Series expansion of these generalized BFH functions is obtained in terms of Gaussian hypergeometric function and some bridges are derived between Hilbert–Eisenstein series and alternating generalized Mathieu series. Finally, we expose a Turán-type inequality for the HE series $\mathfrak{h}_r(w)$.

Á. Baricz
Department of Economics, Babeş-Bolyai University, Cluj-Napoca 400591, Romania
e-mail: bariczocsi@yahoo.com

P.L. Butzer
Lehrstuhl A für Mathematik, RWTH Aachen, D-52056 Aachen, Germany
e-mail: butzer@rwth-aachen.de

T.K. Pogány (✉)
Faculty of Maritime Studies, University of Rijeka, Rijeka 51000, Croatia
e-mail: poganj@pfri.hr

G.V. Milovanović and M.Th. Rassias (eds.), *Analytic Number Theory, Approximation Theory, and Special Functions*, DOI 10.1007/978-1-4939-0258-3_30,

1 Invitation to Ω-Function and Alternating Mathieu Series

The *complex-index* Euler function $\mathfrak{E}_\alpha(z)$ is defined by [6, Definition 2.1]

$$\mathfrak{E}_\alpha(z) := \frac{\Gamma(\alpha+1)}{\pi\mathrm{i}} \int_{\mathfrak{C}_r} \frac{\mathrm{e}^{zu}}{\mathrm{e}^u+1}\, u^{-\alpha-1}\,\mathrm{d}u \qquad \alpha \in \mathbb{C};\ z \in \mathbb{C}\setminus\mathbb{R}_0^-,$$

where $\mathfrak{C}_r$ denotes the positively oriented loop around the negative real axis $\mathbb{R}^-$, which is composed of a circle $C(0;r)$ centered at the origin and of radius $r \in (0,\pi)$ together with the lower and upper edges C_1 and C_2 of the complex plane cut along the negative real axis.

The *complex-index* Bernoulli function $\mathfrak{B}_\alpha(z)$ is given by [4, Definition 2.3(a)]

$$\mathfrak{B}_\alpha(z) := \frac{\Gamma(\alpha+1)}{2\pi\mathrm{i}} \int_{\mathfrak{C}_\rho} \frac{\mathrm{e}^{zv}}{\mathrm{e}^v-1}\, v^{-\alpha}\,\mathrm{d}v, \qquad \alpha \in \mathbb{C};\ z \in \mathbb{C}\setminus\mathbb{R}_0^-.$$

Here, $\mathfrak{C}_\rho$ denotes the same shape integration contour as above with $\rho \in (0,2\pi)$. For the connections between these two functions by way of their Hilbert transforms $\mathfrak{E}_\alpha^\sim(z)$ and $\mathfrak{B}_\alpha^\sim(z)$, the interested reader is referred to [4].

Almost twenty years ago, in their investigation of the complex-index Euler function $E_\alpha(z)$, Butzer, Flocke and Hauss [6] introduced the following special function:

$$\Omega(w) = 2\int_{0+}^{\frac{1}{2}} \sinh(wu)\cot(\pi u)\,\mathrm{d}u, \qquad w \in \mathbb{C},$$

which they called the *complete Omega function* [4, Definition 7.1], [6]. On the other hand, in view of the definition of the Hilbert transform, the complete Omega function $\Omega(w)$ is the Hilbert transform $\mathscr{H}[\mathrm{e}^{-wx}]_1(0)$ at 0 of the 1-periodic function $(\mathrm{e}^{-wx})_1$ defined by the periodic continuation of the following exponential function [4, p. 67]:

$$\mathrm{e}^{-xw}, \qquad |x| \leqslant \tfrac{1}{2};\ w \in \mathbb{C},$$

that is,

$$\mathscr{H}[\mathrm{e}^{-\cdot w}]_1(0) := \mathrm{PV}\int_{-\frac{1}{2}}^{\frac{1}{2}} \mathrm{e}^{wu}\cot(\pi u)\,\mathrm{d}u = \Omega(w), \qquad w \in \mathbb{C},$$

where the integral is to be understood in the sense of Cauchy's principal value (PV) at zero. The highly important role of the Omega function in deep considerations and generating-function description of the Euler and Bernoulli functions was pointed out rather precisely by Butzer and his collaborators in their recent investigations [4–6]. There is also a basic association of the Omega function $\Omega(x)$ with the

Eisenstein series for circular functions and Hilbert–Eisenstein series introduced by Hauss [23]. For this matter as well as for some related open conjectures, see the work by Butzer [4, Sect. 9]. Further integral representations have been derived for the complete, real argument Omega function. So we mention the result by Butzer, Pogány and Srivastava [8, Theorem 2]:

$$\Omega(x) = \frac{2}{\pi} \sinh\left(\frac{x}{2}\right) \int_0^\infty \cos\left(\frac{xt}{2\pi}\right) \frac{\mathrm{d}t}{\mathrm{e}^t + 1}, \qquad x \in \mathbb{R}. \tag{1}$$

Tomovski and Pogány [46, Theorem 3.3] proved that

$$\Omega(x) = 2\sqrt{\frac{2}{\pi}} \sinh\left(\frac{x}{2}\right) \mathrm{PV} \int_0^\infty \sinh\left(\frac{xt}{\pi}\right) \tan t \, \mathrm{d}t, \qquad x \in \mathbb{R}. \tag{2}$$

Also, for the sake of completeness, we mention the Pogány–Srivastava integral representation [39, p. 589, Theorem 1]:

$$\Omega(x) = 16\pi^3 \sinh\left(\frac{x}{2}\right) \int_1^\infty \frac{\sin^2\left(\frac{\pi}{2}[\sqrt{u}]\right) - [\sqrt{u}]\cos\left(\pi[\sqrt{u}]\right)}{(4\pi^2 u + x^2)^2} \, \mathrm{d}u, \tag{3}$$

for all $x \in \mathbb{R}$. Here $[a]$ stands for the integer part of some real a. Secondly, bounding inequalities have been established for $\Omega(x)$ Butzer, Pogány and Srivastava [8, Theorem 3]:

$$\frac{1}{\pi} \sinh\left(\frac{x}{2}\right) \log\left(\frac{\zeta(3)x^2 + 8\pi^2}{3x^2 + 2\pi^2}\right) \leqslant \Omega(x) \leqslant \frac{1}{\pi} \sinh\left(\frac{x}{2}\right) \log\left(\frac{3x^2 + 8\pi^2}{\zeta(3)x^2 + 2\pi^2}\right).$$

The above inequalities are valid for $x > 0$, and for $x < 0$, the opposite inequalities hold true. Here $\zeta(3) = 1.20205690\ldots$ stands for the celebrated Apéry's constant. Thus, the Omega function behaves asymptotically like

$$\left(\frac{1}{\pi} \log \frac{\zeta(3)}{3}\right) \cdot \mathrm{e}^{\frac{x}{2}} \leqslant \Omega(x) \leqslant \left(\frac{1}{\pi} \log \frac{3}{\zeta(3)}\right) \cdot \mathrm{e}^{\frac{x}{2}}, \qquad x \to \infty.$$

Applying the Čaplygin comparison theorem [9, 10, 34], Pogány and Srivastava [39] obtained a bilateral bounding inequality for Ω function, by means of a linear ODE given earlier in [8]. Finally, different types of bounding inequalities were established by Alzer, Brenner and Ruehr; Draščić and Pogány; Mortici; Pogány, Srivastava and Tomovski; and others for the so-called generalized Mathieu series (see [41] and the references therein). Employing also the Čaplygin comparison theorem, Pogány, Tomovski and Leškovski [41] established very recently a set of bilateral inequalities for the real parameter complete Ω function considering the so-called alternating generalized Mathieu series' bounds; see [41] and the exhaustive companion references list.

The Omega function possesses an elegant and useful partial fraction representation [4, Theorem 1.3], [5, Theorem 1.24]:

$$\frac{\pi\Omega(2\pi w)}{\sinh(\pi w)} = \sum_{n=1}^{\infty} \frac{(-1)^{n-1}\,2n}{n^2+w^2}, \qquad w \in \mathbb{C}\setminus \mathrm{i}\mathbb{Z}\,. \tag{4}$$

On the other hand, the generalized Mathieu series was introduced by Guo [21] in the form

$$S_\nu(r) = \sum_{n\geqslant 1} \frac{2n}{(n^2+r^2)^\nu}, \qquad \nu > 1,\ r > 0\,,$$

posing the problem as to whether there is an integral representation for $S_\nu(r)$. The problem was solved by Cerone and Lenard [14, Theorem 1], who gave the integral representation

$$S_\nu(r) = \frac{\sqrt{\pi}}{(2r)^{\nu-\frac{3}{2}}\,\Gamma(\nu)} \int_0^\infty \frac{t^{\nu-\frac{1}{2}}}{\mathrm{e}^t-1}\, J_{\nu-\frac{3}{2}}(rt)\,\mathrm{d}t, \qquad r>0,\ \nu>1\,.$$

In [40] Pogány, Srivastava and Tomovski proved that the alternating generalized Mathieu series

$$\tilde{S}_\nu(r) = \sum_{n\geqslant 1} \frac{(-1)^{n-1}\,2n}{(n^2+r^2)^\nu}, \qquad \nu > 0,\ r > 0\,, \tag{5}$$

posses the integral representation formula

$$\tilde{S}_\nu(r) = \frac{\sqrt{\pi}}{(2r)^{\nu-\frac{3}{2}}\,\Gamma(\nu)} \int_0^\infty \frac{t^{\nu-1/2}}{\mathrm{e}^t+1}\, J_{\nu-\frac{3}{2}}(rt)\,\mathrm{d}t, \qquad r>0,\ \nu>0, \tag{6}$$

where J_a stands for the Bessel function of the first kind of order a. Obviously, letting here $\nu \to 1$, which, in view of the following relationship [1, p. 202, Eq. (4.6.4)]:

$$J_{-\frac{1}{2}}(z) = \sqrt{\frac{2}{\pi z}}\,\cos z,$$

leads us easily to the integral representation (1). A number of other fashion integral representations for S_2, $\tilde{S}_2$ and for S_3 have been presented by Choi and Srivastava in [15, pp. 865–866, Theorem 3, Corollary 2]. The background of the relation (6) is very interesting. Namely, consider the classical Gegenbauer's formula [19] (see also [20, p. 712], [47, p. 386, Eq. 13.(6)]):

$$\int_0^\infty \mathrm{e}^{-\alpha x} x^{\mu+1} J_\nu(\beta x)\mathrm{d}x = \frac{2\alpha(2\beta)^\mu\,\Gamma\big(\mu+\frac{3}{2}\big)}{\sqrt{\pi}\,(\alpha^2+\beta^2)^{\mu+3/2}}, \tag{7}$$

which is valid for all $\mathrm{Re}\{\mu\} > -1, \mathrm{Re}\{\alpha\} > \left|\mathrm{Im}\{\beta\}\right|$. Setting $\alpha = n,\ \mu = \nu - \frac{3}{2}$ and $\beta = r$ in (7), multiplying this relation with $(-1)^{n-1}$ and then summing up with respect to $n \in \mathbb{N}$, we clearly arrive at (6). But, noting (4) and Sect. 4 below,

$$\frac{\pi\Omega(2\pi x)}{\sinh(\pi x)} = \tilde{S}_1(x) = 2\int_0^\infty \frac{\cos(xt)}{\mathrm{e}^t+1}\,\mathrm{d}t, \tag{8}$$

the integral representation only being valid for $w = x$ real therefore now follows (1) when we replace $2\pi x \mapsto x$.

Let us now introduce a new generalized Omega function, namely, $\Omega_\nu(\cdot)$, defined in terms of

$$\frac{\pi\Omega_\nu(2\pi w)}{\sinh(\pi w)} = \tilde{S}_\nu(w), \qquad w \in \mathbb{C}\setminus \mathrm{i}\mathbb{Z}, \tag{9}$$

an extensive counterpart of (8).

In order to apply the foregoing results of Pogány, Srivastava and Tomovski [40], we here need to restrict ourselves to $w = x \in \mathbb{R}$. Thus, (6) gives an analytic definition in matters of (9) as

$$\Omega_\nu(x) = \frac{\pi^{\nu-2}}{\Gamma(\nu)\,x^{\nu-\frac{3}{2}}}\sinh\left(\frac{x}{2}\right)\int_0^\infty \frac{t^{\nu-1/2}}{\mathrm{e}^t+1}\,J_{\nu-\frac{3}{2}}\left(\frac{xt}{2\pi}\right)\mathrm{d}t, \quad \Omega_1(x) \equiv \Omega(x)\,, \tag{10}$$

where $x \neq 0, \nu > 0$. In what follows, we call Ω_ν the *complete generalized* BFH *Omega function of the order* ν.

2 Bounds for Ω_ν by Using Results on J_μ

The main purpose here is to establish a bounding inequality of Ω_ν in terms of J_μ. Rearranging the integral representation (10) of the complete generalized BFH Omega function of the order ν, we deduce

$$\begin{aligned}\left|\Omega_\nu(x)\right| &\leqslant \frac{\pi^{\nu-2}}{\Gamma(\nu)\,|x|^{\nu-\frac{3}{2}}}\left|\sinh\left(\frac{x}{2}\right)\right|\int_0^\infty \frac{t^{\nu-1/2}}{\mathrm{e}^t+1}\left|J_{\nu-\frac{3}{2}}\left(\frac{xt}{2\pi}\right)\right|\mathrm{d}t \\ &= \frac{2^{\nu+\frac{1}{2}}\,\pi^{2\nu-\frac{3}{2}}}{\Gamma(\nu)\,|x|^{2\nu-1}}\left|\sinh\left(\frac{x}{2}\right)\right|\int_0^\infty \frac{t^{\nu-1/2}}{\mathrm{e}^{2\pi t/x}+1}\left|J_{\nu-\frac{3}{2}}(t)\right|\mathrm{d}t\,. \end{aligned} \tag{11}$$

Now, we are confronted with the problem of bounding $\left|J_\mu\right|$ by certain sharp bound on the positive real half-axis. Firstly, we will inspect the appropriate bounds' literature. Fortunately, there are numerous suitable bounds for the modulus of the Bessel function of the first kind, like Hansen's, valid for positive integral order Bessel function [22, pp. 107 et seq.], [47, p. 31]

$$|J_0(t)| \leqslant 1, \ \ |J_r(t)| \leqslant \frac{1}{\sqrt{2}}, \qquad r \in \mathbb{N}, t \in \mathbb{R}; \tag{12}$$

von Lommel extended [33, pp. 548–549], [47, p. 406] this results to

$$|J_\mu(t)| \leqslant 1, \ \ |J_{\mu+1}(t)| \leqslant \frac{1}{\sqrt{2}}, \qquad \mu > 0, t \in \mathbb{R}. \tag{13}$$

Simple, efficient bounding inequality was proved by Minakshisundaram and Szász [35, p. 37, Corollary]:

$$|J_\mu(x)| \leqslant \frac{1}{\Gamma(\mu+1)} \left(\frac{|x|}{2}\right)^\mu, \qquad x \in \mathbb{R}; \tag{14}$$

obviously, this bound reduces Hansen's for $\mu = 0$.

More sophisticated bounds were given by Landau [31], who gave in a sense best possible bounds for the first kind Bessel function $J_\nu(x)$ with respect to ν and x, which read as follows:

$$|J_\mu(x)| \leqslant b_L\, \mu^{-1/3}, \qquad b_L = \sqrt[3]{2} \sup_{x \in \mathbb{R}_+} \mathrm{Ai}(x), \tag{15}$$

$$|J_\mu(x)| \leqslant c_L\, |x|^{-1/3}, \qquad c_L = \sup_{x \in \mathbb{R}_+} x^{1/3} J_0(x), \tag{16}$$

where $\mathrm{Ai}(\cdot)$ stands for the familiar Airy function

$$\mathrm{Ai}(x) := \frac{\pi}{3}\sqrt{\frac{x}{3}}\Big(J_{-1/3}\{2(x/3)^{3/2}\} + J_{1/3}\{2(x/3)^{3/2}\}\Big).$$

In fact Krasikov [30] pointed out that these bounds are sharp only in the transition region, i.e. for x around $j_{\mu,1}$, the first positive zero of $J_\mu(x)$.

In his recent article Olenko [37, Theorem 1] established the following sharp upper bound:

$$\sup_{x \geqslant 0} \sqrt{x}|J_\mu(x)| \leqslant b_L \sqrt{\mu^{1/3} + \frac{\alpha_1}{\mu^{1/3}} + \frac{3\alpha_1^2}{10\mu}} := d_O, \qquad \mu > 0, \tag{17}$$

where α_1 is the smallest positive zero of the Airy function $\mathrm{Ai}(x)$ and b_L is the Landau constant in (15). For further reading and detailed discussion, consult [37, Sect. 3].

Krasikov also established a uniform bound for $|J_\mu|$. Let $\mu > -1/2$, then

$$J_\mu^2(t) \leqslant \frac{4\big(4t^2 - (2\mu+1)(2\mu+5)\big)}{\pi\big((4t^2-\lambda)^{3/2} - \lambda\big)} =: \mathfrak{K}_\mu(t), \tag{18}$$

for all

$$t > \tfrac{1}{2}\sqrt{\lambda + \lambda^{2/3}}, \; \lambda := (2\mu+1)(2\mu+3)\,.$$

The estimate is sharp in certain sense, consult [30, Theorem 2]. Moreover, Krasikov mentioned that (18) provides sharp bound in the whole oscillatory region; however, in the transition region, this estimate becomes very poor and should be replaced with another estimate. Having in mind Krasikov's discussion, we propose to combine Krasikov's bound with Olenko's one. This approach was used by Srivastava and Pogány in [45]. Let us denote $\chi_S(x)$ the characteristic (or indicator) function of a set S, that is, $\chi_S(x) = 1$ for all $x \in S$ and $\chi_S(x) = 0$ otherwise. Since the integration domain coincides with the positive real half-axis, we need an efficient bound for $|J_\mu(t)|$ on $(0, A]$, $A > \sqrt{\lambda + \lambda^{2/3}}/2$. Therefore, we introduce the bounding function

$$|J_\mu(t)| \leqslant \mathscr{V}_\mu(t) := \frac{d_O}{\sqrt{t}}\, \chi_{(0,A_\lambda]}(t) + \sqrt{\mathfrak{K}_\mu(t)}\,\big(1 - \chi_{(0,A_\lambda]}(t)\big)\,, \tag{19}$$

where, by simplicity reasons, our choice would be

$$A_\lambda = \tfrac{1}{2}\big(\lambda + (\lambda+1)^{2/3}\big),$$

because $\mathfrak{K}_\nu(t)$ is positive and monotonous decreasing for $t \in \frac{1}{2}\big((\lambda + \lambda^{2/3}), \infty\big)$, compare [45, Sect. 3]. Moreover, we point out that for A_λ, we can take any $\frac{1}{2}\big(\lambda + (\lambda+\eta)^{2/3}\big)$, $\eta > 0$.

Next, Pogány derived a different fashion bound for $|J_\mu|$, when the argument of the considered Bessel function is coming from a closed Cassinian oval from $\mathbb{C}$. To recall this result, we need the following definitions. Let us denote $\mathbb{D}_\eta = \{z\colon |z| \leqslant \eta\}$ the closed centered disc having diameter 2η, while the open unit disc $\mathbb{D} = \{z\colon |z| < 1\}$ and the closed Cassinian oval [38]

$$\mathfrak{C}_{\mu,\lambda} := \left\{ z\colon \big|z^2 - j_{\mu,1}^2\big| \leqslant j_{\mu,1}^2\, \frac{1-\lambda}{1+\lambda} \right\}, \qquad \lambda \in [0,1]\,.$$

The famous von Lommel theorem "*$J_\nu(z)$ has an infinity of real zeros, for any given real value of ν*", [47, p. 478], ensures the existence of such $j_{\nu,1}$. Thus [38, Theorem 1]

$$\big|J_\mu(z)\big| \leqslant \frac{|z|^\mu}{2^\mu \Gamma(\mu+1)} \exp\left\{ -\frac{\lambda |z|^2}{4(\mu+1)} \right\}, \qquad \lambda \in (0,1),\; \mu > 0,\; z \in \mathfrak{C}_{\mu,\lambda}. \tag{20}$$

Here we mention the inequality [25, p. 215]

$$|J_\mu(t)| \leqslant \frac{t^\mu}{2^\mu \Gamma(\mu+1)} \exp\left\{-\frac{t^2}{4(\mu+1)}\right\}, \qquad t>0,\ \mu \geqslant 0. \tag{21}$$

Note that Watson [47, p. 16] originated back to Cauchy a weaker variant of this inequality (the exponential term contains $-t^2/4$), for integer order μ, see [11, p. 687], [12, p. 854].

Ifantis–Siafarikas improved (21) for the domain $t \in (0, j_{\mu,1}), \mu > -1$ in the following form [25, Eq. (3.15)]:

$$J_\mu(t) < \frac{t^\mu}{2^\mu \Gamma(\mu+1)} \exp\left\{-\frac{t^2}{4(\mu+1)} - \frac{t^4}{32(\mu+1)^2(\mu+2)}\right\}. \tag{22}$$

It is worth to mention Sitnik's paper [43], in which he reported stronger but more complicated bounds involving Rayleigh sums of negative powers of Bessel function zeros; his results concern Bessel function bounds inside the open unit disc $\mathbb{D}$. Interesting upper bound was established also by Lee and Shah for complex variable, integer order Bessel function $J_r(t)$; see [32, p. 148]. Finally, we refer to Cerone's book chapter [13, Sect. 2] for an inequality accomplished by bounds on a Čebyšev functional.

Theorem 2.1. *The following bounding inequalities hold true:*

a. *For all $x \geqslant 0, \nu - \frac{3}{2} = r \in \mathbb{N}_0$, we have*

$$|\Omega_{r+\frac{3}{2}}(x)| \leqslant \frac{\pi^{r-\frac{1}{2}}\,[1+\delta_{0r}(\sqrt{2}-1)]\,(r+1)!\,\eta(r+2)}{\sqrt{2}\,\Gamma(r+\frac{3}{2})\,x^r} \sinh\left(\frac{x}{2}\right), \tag{23}$$

where δ_{ab} stands for the Kronecker delta, while

$$\eta(p) = \sum_{n\geqslant 1} \frac{(-1)^{n-1}}{n^p}, \qquad \mathrm{Re}\{p\} > 0,$$

denotes the Dirichlet Eta function.

b. *For all $x \geqslant 0, \nu > \frac{1}{2}$, it is*

$$|\Omega_\nu(x)| \leqslant \frac{2}{\pi}\,\eta(2\nu-1)\,\sinh\left(\frac{x}{2}\right). \tag{24}$$

c. *Let us denote b_L, c_L the Landau constants given in* (15), (16). *Then for all $x > 0$, we have*

$$|\Omega_\nu(x)| \leqslant \begin{cases} \dfrac{b_L \pi^{\nu-2}\,\Gamma\left(\nu+\frac{1}{2}\right)\,\eta\left(\nu+\frac{1}{2}\right)}{\Gamma(\nu)\,\left(\nu-\frac{3}{2}\right)^{\frac{1}{3}}\,x^{\nu-\frac{3}{2}}} \sinh\left(\dfrac{x}{2}\right), & \nu > \frac{3}{2}, \\ \dfrac{c_L 2^{\frac{1}{3}}\,\pi^{\nu-\frac{5}{3}}\,\Gamma\left(\nu+\frac{1}{6}\right)\,\eta\left(\nu+\frac{1}{6}\right)}{\Gamma(\nu)\,x^{\nu-\frac{7}{6}}} \sinh\left(\dfrac{x}{2}\right), & \nu > -\frac{1}{6}. \end{cases} \tag{25}$$

d. *Let d_O be the Olenko coefficient in* (17). *Then*

$$|\Omega_\nu(x)| \leqslant \frac{d_O\,\sqrt{2}\,\pi^{\nu-\frac{3}{2}}\,\eta(\nu)}{x^{\nu-1}}\,\sinh\left(\frac{x}{2}\right), \qquad x>0,\ \nu>\tfrac{3}{2}. \tag{26}$$

Proof. **a.** Consider the integral [20, p. 349, Eq. 3.411.3]

$$\mathscr{I}(\alpha,\beta) := \int_0^\infty \frac{t^{\alpha-1}}{e^{\beta t}+1}\,dt = \beta^{-\alpha}\,\Gamma(\alpha)\,\eta(\alpha), \qquad \min\big(\mathrm{Re}\{\alpha\},\mathrm{Re}\{\beta\}\big)>0\,.$$

Applying Hansen's bound (12) to the Bessel function $J_{\nu-\frac{3}{2}}(t)$ appearing in (11), we conclude

$$|\Omega_\nu(x)| \leqslant \frac{2^\nu\,\pi^{2\nu-\frac{3}{2}}[1+\delta_{0,\nu-\frac{3}{2}}(\sqrt{2}-1)]}{\Gamma(\nu)\,x^{2\nu-1}}\,\sinh\left(\frac{x}{2}\right)\mathscr{I}\left(\nu+\frac{1}{2},\frac{2\pi}{x}\right) \qquad x>0.$$

Substituting $r=\nu-\frac{3}{2}\in\mathbb{N}_0$ and reducing the previous bound, we arrive at (23).
b. & **c.** & **d.** Similarly to the case **a**, estimating $|J_{\nu-\frac{3}{2}}|$ with the aid of the bounds (14), (15), (16) and (17), we derive appropriate respective specifications. For the four consequent bounding inequalities (24), (25) and (26). Observe that (25) consists from TWO upper bounds. □

Remark 2.1. We point out that von Lommel's extension (13) of Hansen's bounds (12) will give substantially more general but in form equivalent bound upon $\Omega_\nu(x)$; therefore, it is not necessary to consider this case separately.

Obviously, being the integration domain for $\Omega_\nu(x)$ the positive real half-axis, Krasikov's bound itself is automatically eliminated as a candidate to be employed in estimating the Bessel function in the kernel of the integrand of $\Omega_\nu(x)$. Therefore, instead of Krasikov's, the synthetized Olenko–Krasikov bound $\mathscr{V}_\mu(t)$ (19) shall we apply. The *lower incomplete Gamma function* $\gamma(s,\omega)$ [20, **8.350** 1.] one defines truncating the integration domain of Eulerian Gamma function to $[0,\alpha]$, i.e.

$$\gamma(s,\alpha) := \int_0^\alpha t^{s-1}e^{-t}\,dt\,.$$

Also, the *upper incomplete Gamma function* or *complementary incomplete Gamma function* [20, **8.350** 2.] is given by

$$\Gamma(s,\alpha) := \Gamma(s)-\gamma(s,\alpha) = \int_\alpha^\infty t^{s-1}e^{-t}\,dt.$$

For both incomplete Gamma functions, $\mathrm{Re}\{s\}>0$, $|\arg(\alpha)|\leqslant\pi-\epsilon$, $\epsilon\in(0,\pi)$. Let us remark that for certain fixed α, $\Gamma(s,\alpha)$ is an entire function of s, while $\gamma(s,\alpha)$ is a meromorphic function of α with simple poles at $s\in\mathbb{Z}_0^-$.

Theorem 2.2. *Let $\nu > 1, \lambda = 4\nu(\nu-1)$ and let x be positive real. Then we have the following bounding inequality:*

$$\left|\Omega_\nu(x)\right| \leqslant \frac{\sqrt{2}\,\pi^{\nu-\frac{3}{2}}}{\Gamma(\nu)\,x^{\nu-1}}\,\sinh\left(\frac{x}{2}\right)\left\{\frac{d_O\cdot\gamma\left(\nu,\frac{2\pi}{x}A_\lambda\right)}{1+\exp\{-\frac{2\pi}{x}A_\lambda\}}\right.$$
$$\left.+\frac{2^{\nu-1}\sqrt{x\,\mathfrak{K}_{\nu-\frac{3}{2}}(A_\lambda)}\cdot\Gamma\left(\nu+\frac{1}{2},\frac{\pi}{x}A_\lambda\right)}{\sqrt{\pi}\,\cosh\left(\frac{\pi}{x}A_\lambda\right)}\right\}, \tag{27}$$

where $A_\lambda = \frac{1}{2}\left(\lambda+(\lambda+1)^{2/3}\right)$ *and*

$$\mathfrak{K}_{\nu-\frac{3}{2}}(A_\lambda) = \frac{4}{\pi}\left\{\left(\lambda+(\lambda+1)^{2/3}\right)^2-4(\nu^2-1)\right\}, \qquad x \geqslant A_\lambda\,.$$

Proof. By (11) and (19), it follows that

$$\left|\Omega_\nu(x)\right| \leqslant \frac{2^{\nu+\frac{1}{2}}\,\pi^{2\nu-\frac{3}{2}}}{\Gamma(\nu)\,x^{2\nu-1}}\,\sinh\left(\frac{x}{2}\right)\int_0^\infty \frac{t^{\nu-1/2}}{e^{2\pi t/x}+1}\,\mathscr{V}_{\nu-\frac{3}{2}}(t)\,dt$$
$$= \frac{2^{\nu+\frac{1}{2}}\,\pi^{2\nu-\frac{3}{2}}}{\Gamma(\nu)\,x^{2\nu-1}}\,\sinh\left(\frac{x}{2}\right)\left(d_O\int_0^{A_\lambda}\frac{t^{\nu-1}}{e^{2\pi t/x}+1}\,dt\right. \qquad (=: \mathscr{I}_1)$$
$$\left.+\int_{A_\lambda}^\infty \frac{t^{\nu-\frac{1}{2}}}{e^{2\pi t/x}+1}\sqrt{\mathfrak{K}_{\nu-\frac{3}{2}}(t)}\,dt\right) \qquad (=: \mathscr{I}_2)\,. \tag{28}$$

Now, for the integral $\mathcal{I}_1$, we calculate in the following manner:

$$\mathcal{I}_1 = \int_0^{A_\lambda}\frac{t^{\nu-1}\,e^{-\beta t}}{1+e^{-\beta t}}\,dt \leqslant \frac{1}{1+e^{-\beta A_\lambda}}\int_0^{A_\lambda} t^{\nu-1}\,e^{-\beta t}\,dt = \frac{\gamma(\nu,\beta A_\lambda)}{\beta^\nu(1+e^{-\beta A_\lambda})}\,. \tag{29}$$

The fact that $\mathfrak{K}_\mu$ decreases on $[A_\lambda,\infty)$ has been established already in [45, p. 199]; hence,

$$\mathfrak{K}_{\nu-\frac{3}{2}}(x) \leqslant \mathfrak{K}_{\nu-\frac{3}{2}}(A_\lambda) = \frac{4}{\pi}\left[\left(\lambda+(\lambda+1)^{2/3}\right)^2-4(\nu^2-1)\right], \qquad x \geqslant A_\lambda\,.$$

Accordingly,

$$\mathcal{I}_2 = \frac{1}{2}\int_{A_\lambda}^\infty \frac{t^{\nu-\frac{1}{2}}\,e^{-\frac{\beta}{2}t}}{\cosh\left(\frac{\beta}{2}t\right)}\,dt \leqslant \frac{1}{2\cosh\left(\frac{\beta}{2}A_\lambda\right)}\int_{A_\lambda}^\infty t^{\nu-\frac{1}{2}}\,e^{-\frac{\beta}{2}t}\,dt$$
$$= \frac{2^{\nu-\frac{1}{2}}\,\Gamma\left(\nu+\frac{1}{2},\frac{1}{2}\beta\,A_\lambda\right)}{\beta^{\nu+\frac{1}{2}}\,\cosh\left(\frac{\beta}{2}A_\lambda\right)} \tag{30}$$

in both integrals $\beta := 2\pi x^{-1}$. Now, obvious transformations of (28), (29) and (30) lead to the asserted bound (27). □

Considering further bounds (20), (21) and (22) upon the Bessel function of the first kind, we see that only the bound (21) possesses the property of direct applicability since the integration domain of $(0,\infty)$ in defining $\Omega_\nu(x)$. Here, and in what follows, ${}_p\Psi_q$ denotes the Fox–Wright generalization of the hypergeometric ${}_pF_q$ function with p numerator and q denominator parameters defined by (cf. e.g. [44], [45, p. 197, Eq. (7)])

$${}_p\Psi_q\left[\begin{matrix}(a_1,\alpha_p),\ldots,(a_p,\alpha_p)\\(b_1,\beta_1),\ldots,(b_q,\beta_q)\end{matrix}\middle|\,x\right]={}_p\Psi_q\left[\begin{matrix}(a_p,\alpha_p)\\(b_q,\beta_q)\end{matrix}\middle|\,x\right]:=\sum_{m=0}^{\infty}\frac{\prod_{\ell=1}^{p}\Gamma(a_\ell+\alpha_\ell m)}{\prod_{\ell=1}^{q}\Gamma(b_\ell+\beta_\ell m)}\,\frac{x^m}{m!},$$

where

$$\alpha_\ell\in\mathbb{R}_+,\ \ell=\overline{1,p};\quad \beta_j\in\mathbb{R}_+,\ j=\overline{1,q};\quad \Delta:=1+\sum_{\ell=1}^{q}\beta_\ell-\sum_{j=1}^{p}\alpha_j\geqslant 0\,,$$

and in the case of equality $\Delta = 0$, the absolute convergence holds for suitably bounded values of x given by

$$|x|<\nabla=\prod_{j=1}^{q}\beta_j^{\beta_j}\,\prod_{j=1}^{p}\alpha_j^{-\alpha_j}\,,$$

while in the case $|x| = \nabla$, the condition

$$\mathrm{Re}\left\{\sum_{\ell=1}^{q}b_\ell-\sum_{j=1}^{p}a_j\right\}+\frac{p-q-1}{2}>0$$

suffices for the absolute convergence of the series ${}_p\Psi_q[x]$.

Next, we introduce the Krätzel function, which is defined for $u > 0$, $\rho\in\mathbb{R}$ and $\nu\in\mathbb{C}$, being such that $\mathrm{Re}\{\nu\} < 0$ for $\rho\leqslant 0$, by the integral

$$Z_\rho^\nu(u)=\int_0^\infty t^{\nu-1}\mathrm{e}^{-t^\rho-\frac{u}{t}}\,\mathrm{d}t\,. \tag{31}$$

For $\rho\geqslant 1$ the function (31) was introduced by E. Krätzel [29] as a kernel of the integral transform

$$\left(K_\nu^\rho f\right)(u)=\int_0^\infty Z_\rho^\nu(ut)\,f(t)\,\mathrm{d}t,$$

which was applied to the solution of some ordinary differential equations. The study of the Krätzel function (31) and the above integral transform was continued, for example, in the paper by Kilbas, Saxena and Trujilló [27], in which the authors deduced explicit forms of Z_ρ^ν in terms of the generalized Wright function, or in the paper [2] by Baricz, Jankov and Pogány devoted among others to convexity property research and Laguerre- and Turán-type inequalities for the Krätzel function.

Theorem 2.3. *Let $\nu > \frac{1}{2}$. Then for all $x > 0$, the following inequality is valid:*

$$\begin{aligned}\big|\Omega_\nu(x)\big| &\leqslant \frac{\pi^{2\nu-\frac{3}{2}}\,(4\nu-2)^{\nu-\frac{1}{2}}\,\sinh\left(\frac{x}{2}\right)}{\Gamma(\nu)\Gamma(\nu-\frac{1}{2})\,x^{2\nu-1}}\;{}_1\Psi_0\left[\begin{matrix}(\nu-\frac{1}{2},\frac{1}{2})\\ -\end{matrix}\,\middle|\,-\frac{\pi}{x}\sqrt{4\nu-2}\right]\\ &= \frac{2\pi^{2\nu-\frac{3}{2}}\,(4\nu-2)^{\nu-\frac{1}{2}}\,\sinh\left(\frac{x}{2}\right)}{\Gamma(\nu)\Gamma(\nu-\frac{1}{2})\,x^{2\nu-1}}\;Z_{-2}^{1-2\nu}\left(\frac{\pi}{x}\sqrt{4\nu-2}\right).\end{aligned}$$

Proof. By (11) and (21) we conclude the estimate

$$\big|\Omega_\nu(x)\big| \leqslant \frac{4\pi^{2\nu-\frac{3}{2}}\,\sinh\left(\frac{x}{2}\right)}{\Gamma(\nu)\Gamma(\nu-\frac{1}{2})\,x^{2\nu-1}}\int_0^\infty \frac{t^{2\nu-2}}{e^{2\pi t/x}+1}\,\exp\left\{-\frac{t^2}{2(2\nu-1)}\right\}\,dt.$$

Denoting

$$\mathcal{I}_3(\alpha,\beta,\gamma) = \int_0^\infty \frac{t^{\alpha-1}\,e^{-\gamma t^2}}{e^{\beta t}+1}\,dt,$$

we estimate the value of this integral:

$$\begin{aligned}\mathcal{I}_3(\alpha,\beta,\gamma) &\leqslant \frac{1}{2}\int_0^\infty \frac{t^{\alpha-1}}{\cosh\left(\frac{\beta}{2}t\right)}\,e^{-\frac{\beta}{2}t-\gamma t^2}\,dt\\ &\leqslant \frac{1}{2}\int_0^\infty t^{\alpha-1}\,e^{-\frac{\beta}{2}t-\gamma t^2}\,dt && (=: \mathcal{I}_4)\\ &= \frac{1}{4}\sum_{n\geqslant 0}\frac{\left(-\frac{\beta}{2}\right)^n}{n!}\int_0^\infty t^{\frac{\alpha+n}{2}-1}\,e^{-\gamma t}\,dt\\ &= \frac{1}{4\gamma^{\frac{\alpha}{2}}}\sum_{n\geqslant 0}\Gamma\left(\frac{\alpha}{2}+\frac{1}{2}n\right)\frac{\left(-\frac{\beta}{2\sqrt{\gamma}}\right)^n}{n!}\\ &= \frac{1}{4\gamma^{\frac{\alpha}{2}}}\;{}_1\Psi_0\left[\begin{matrix}(\frac{\alpha}{2},\frac{1}{2})\\ -\end{matrix}\,\middle|\,-\frac{\beta}{2\sqrt{\gamma}}\right]; && (32)\end{aligned}$$

the Fox–Wright function converges absolutely since $\Delta = \frac{1}{2} > 0$.

On the other hand, considering the integral $\mathfrak{I}_4 = \mathfrak{I}_4(\alpha, \beta, \gamma)$, we can express it *via* the Krätzel function:

$$\mathfrak{I}_4 = \frac{1}{\gamma^{\frac{\alpha}{2}}}\, Z_{-2}^{-\frac{\alpha}{2}}\left(\frac{\beta}{2\sqrt{\gamma}}\right). \tag{33}$$

Indeed, the substitution $t^{-1} \mapsto t$ results in $Z_\rho^\nu(u) = \int_0^\infty t^{-\nu-1}\, \mathrm{e}^{-t^{-\rho}-ut}\, \mathrm{d}t$, and

$$\mathfrak{I}_4 = \frac{1}{\gamma^{\frac{\alpha}{2}}} \int_0^\infty t^{\alpha-1}\, \mathrm{e}^{-t^2-\frac{\beta}{2\sqrt{\gamma}}\, t}\, \mathrm{d}t = \frac{1}{\gamma^{\frac{\alpha}{2}}} \int_0^\infty t^{-(-\alpha)-1}\, \mathrm{e}^{-t^{-(-2)}-\frac{\beta}{2\sqrt{\gamma}}\, t}\, \mathrm{d}t\,,$$

so the relationship (33) is proved.

Now, it remains to specify in both formulae (32) and (33)

$$\alpha = 2\nu - 1, \quad \beta = \frac{2\pi}{x}, \quad \gamma = \frac{1}{4\nu - 2},$$

because of

$$\begin{aligned}
\left|\Omega_\nu(x)\right| &\leqslant \frac{4\pi^{2\nu-\frac{3}{2}} \sinh\left(\frac{x}{2}\right)}{\Gamma(\nu)\Gamma(\nu-\frac{1}{2})\, x^{2\nu-1}}\, \mathfrak{I}_3\left(2\nu-1, \frac{2\pi}{x}, \frac{1}{4\nu-2}\right) \\
&= \frac{4\pi^{2\nu-\frac{3}{2}} \sinh\left(\frac{x}{2}\right)}{\Gamma(\nu)\Gamma(\nu-\frac{1}{2})\, x^{2\nu-1}}\, \mathfrak{I}_4\left(2\nu-1, \frac{2\pi}{x}, \frac{1}{4\nu-2}\right) \\
&= \frac{\pi^{2\nu-\frac{3}{2}} (4\nu-2)^{\nu-\frac{1}{2}} \sinh\left(\frac{x}{2}\right)}{\Gamma(\nu)\Gamma(\nu-\frac{1}{2})\, x^{2\nu-1}}\, {}_1\Psi_0\left[\begin{matrix}(\nu-\frac{1}{2}, \frac{1}{2}) \\ -\end{matrix}\,\middle|\, -\frac{\pi}{x}\sqrt{4\nu-2}\right] \\
&= \frac{2\pi^{2\nu-\frac{3}{2}} (4\nu-2)^{\nu-\frac{1}{2}} \sinh\left(\frac{x}{2}\right)}{\Gamma(\nu)\Gamma(\nu-\frac{1}{2})\, x^{2\nu-1}}\, Z_{-2}^{1-2\nu}\left(\frac{\pi}{x}\sqrt{4\nu-2}\right).
\end{aligned}$$

The proof is complete. □

Remark 2.2. As a by-product of the proof of Theorem 2.3 we get a relationship between Fox–Wright generalized hypergeometric function ${}_1\Psi_0[\cdot]$ and the Krätzel function $Z_\rho^\nu(\cdot)$:

$$\rho\, Z_\rho^\nu(u) = {}_1\Psi_0\left[\begin{matrix}(\frac{\nu}{\rho}, -\frac{1}{\rho}) \\ -\end{matrix}\,\middle|\, -u\right].$$

To prove this result we can apply the same methods as in the previous proof.

3 Bilateral Bounds Deduced via the Čaplygin Comparison Theorem

Two-sided bounding inequalities for the complete Butzer–Flocke–Hauss Omega function by the Čaplygin comparison theorem have been established for the first time by Pogány and Srivastava in [39, p. 591, Theorem 3]. Following this approach Pogány, Tomovski and Leškovski devoted the whole article [41] to this subject, deriving a few sets of bilateral inequalities for the BFH Omega function via alternating Mathieu series, which is closely connected in their proper fraction representation.

In this section we shall obtain some fashion bilateral bounding inequalities for the generalized BFH Omega function Ω_ν, adapting the Čaplygin differential inequality procedure developed in [39, 41]. Firstly, we consider a linear nonhomogeneous ordinary differential equation, of which a particular solution is Ω_ν, mentioning that the case $\nu = 1$ has been extensively studied by Butzer, Pogány and Srivastava [8, pp. 1074–1075, Theorem 1].

Theorem 3.1. *For all $\nu > \frac{1}{2}$, $x \in \mathbb{R}$, the generalized complete BFL $\Omega_\nu(x)$ function is a particular solution of the following ordinary differential equation:*

$$y' = \frac{1}{2}\coth\left(\frac{x}{2}\right) y - \frac{\nu x}{2\pi^3}\sinh\left(\frac{x}{2}\right) h(x)\,,$$

where

$$h(x) = \begin{cases} \tilde{S}_{\nu+1}\left(\dfrac{x}{2\pi}\right), & x \neq 0,\\ 2\,\eta(2\nu-1), & x = 0. \end{cases}$$

Here $\tilde{S}_{\nu+1}(\cdot)$ stands for the generalized alternating Mathieu series of order $\nu + 1$, while $\eta(\cdot)$ denotes the Dirichlet Eta function.

Proof. Rewriting (9), we get

$$\Omega_\nu(x) = \frac{1}{\pi}\sinh\left(\frac{x}{2}\right)\tilde{S}_\nu\left(\frac{x}{2\pi}\right).$$

Differentiating this formula we obtain

$$\begin{aligned}\pi\Omega'_\nu(x) &= \frac{1}{2}\cosh\left(\frac{x}{2}\right)\tilde{S}_\nu\left(\frac{x}{2\pi}\right) - \frac{\nu x}{2\pi^2}\sum_{n\geq 1}\frac{(-1)^{n-1}2n}{(n^2+(\frac{x}{2\pi})^2)^{\nu+1}}\\ &= \frac{\pi}{2}\coth\left(\frac{x}{2}\right)\frac{1}{\pi}\sinh\left(\frac{x}{2}\right)\tilde{S}_\nu\left(\frac{x}{2\pi}\right) - \frac{\nu x}{2\pi^2}\tilde{S}_{\nu+1}\left(\frac{x}{2\pi}\right),\end{aligned}$$

which is in fact equivalent to the asserted ordinary differential equation, because

$$\lim_{\xi \to 0} \tilde{S}_\nu(\xi) = 2\,\eta(2\nu - 1)\,.$$

The proof is complete. □

Consider the Cauchy problem given by

$$y' = f(x, y) \quad \text{and} \quad y(x_0) = y_0\,. \tag{34}$$

For a given interval $\mathfrak{J} \subseteq \mathbb{R}$, let $x_0 \in \mathfrak{J}$ and let the functions $\varphi, \psi \in \mathrm{C}^1(\mathfrak{J})$. We say that φ and ψ are the *lower* and the *upper functions*, respectively, if

$$\varphi'(x) \leqslant f\big(x, \varphi(x)\big) \quad \text{and} \quad \psi'(x) \geqslant f\big(x, \psi(x)\big) \qquad x \in \mathfrak{J};$$
$$\varphi(x_0) = \psi(x_0) = y_0\,.$$

Suppose also that the function $f(x, y)$ is continuous on some domain $\mathscr{D}$ in the (x, y)-plane containing the interval $\mathfrak{J}$ with the lower and upper functions φ and ψ, respectively. Then the solution $y(x)$ of the Cauchy problem (34) satisfies the following two-sided inequality:

$$\varphi(x) \leqslant y(x) \leqslant \psi(x), \qquad x \in \mathfrak{J}\,.$$

This is actually the so-called *Čaplygin-type differential inequality* or the *Čaplygin comparison theorem* [9, 10, 34] (also see [3, p. 202] and [36, pp. 3–4]).

We divide into four steps the derivation of two-sided bounds: **A.** obtaining guard functions couple $\tilde{L}_\nu(x), \tilde{R}_\nu(x)$ for $\tilde{S}_{\nu+1}(x)$, **B.** fixing the domain $\mathfrak{J}$ of solution and the initial condition $\varphi_\nu(x_0) = \psi_\nu(x_0) = y_0$, **C.** solving the boundary ordinary differential equations for lower and upper guard functions and finally **D.** considering the particular solutions $\varphi_\nu(x), \psi_\nu(x)$.

A. Keeping in mind the definition (9) of $\Omega_\nu(x)$, the natural choice of domain is $\mathfrak{J} = [0, \infty)$. On the other hand, since $\Omega_\nu(x)$ behaves near to the origin like

$$\Omega_\nu(x) = \frac{2}{\pi}\,\eta(2\nu - 1)\,x\,(1 + o(1)), \qquad x \to 0,$$

we pick up the initial condition of our Cauchy problem

$$\varphi_\nu(0) = \psi_\nu(0) = 0\,.$$

B. Let $L_\mu(x), R_\mu(x)$ denote the guard functions for the generalized Mathieu series $S_\mu(x)$, $x \in \mathfrak{J}$. By the arithmetic mean–geometric mean inequality, we have

$$S_\mu(x) \leqslant \begin{cases} 2^{1-\mu}\,x^{-\mu}\,\zeta(\mu - 1) & x > 0 \\ 2\,\zeta(2\mu - 1) & x = 0 \end{cases} \qquad := R_\mu(x)\,,$$

and

$$S_\mu(x) \geqslant \int_1^\infty \frac{2t}{(t^2+x^2)^\mu}\,\mathrm{d}t = \frac{1}{(\mu-1)(1+x^2)^{\mu-1}} = L_\mu(x)\,.$$

Rewriting the fractional representation of the alternating generalized Mathieu series into

$$\begin{aligned}\tilde{S}_\mu(x) &= \sum_{n\geqslant 1} \frac{(-1)^{n-1}\,2n}{(n^2+x^2)^\mu} \\ &= \sum_{n\geqslant 1} \frac{2n}{(n^2+x^2)^\mu} - 4\cdot \sum_{n\geqslant 1} \frac{2n}{(4n^2+x^2)^\mu} = S_\mu(x) - 4^{1-\mu}\, S_\mu\left(\frac{x}{2}\right),\end{aligned}$$

we clearly deduce that

$$\tilde{L}_\mu(x) := L_\mu(x) - 4^{1-\mu} R_\mu\left(\frac{x}{2}\right) \leqslant \tilde{S}_\mu(x) \leqslant R_\mu(x) - 4^{1-\mu} L_\mu\left(\frac{x}{2}\right) =: \tilde{R}_\mu(x)\,. \tag{35}$$

Therefore, writing $\mu = \nu + 1$ throughout in (35), we get

$$\tilde{L}_{\nu+1}(x) = \frac{1}{\nu(1+x^2)^\nu} - \frac{\zeta(\nu)}{2^{2\nu-1}\,x^{\nu+1}}, \qquad \tilde{R}_{\nu+1}(x) = \frac{\zeta(\nu)}{2^\nu\, x^{\nu+1}} - \frac{1}{\nu(4+x^2)^\nu}\,.$$

C. Following the lines of Theorem 3.1, the lower function's ODE will be

$$\begin{aligned}\varphi_\nu' - \frac{1}{2}\coth\left(\frac{x}{2}\right)\varphi_\nu &= -\frac{\nu x}{2\pi^3}\sinh\left(\frac{x}{2}\right)\tilde{R}_{\nu+1}\left(\frac{x}{2\pi}\right) \\ &= \pi^{\nu-2}\sinh\left(\frac{x}{2}\right)\left\{\frac{2^{2\nu-1}\pi^{\nu-1}\,x}{(16\pi^2+x^2)^\nu} - \frac{\nu\,\zeta(\nu)}{x^\nu}\right\}\,;\end{aligned}$$

hence

$$\varphi_\nu(x) = \sinh\left(\frac{x}{2}\right)\left\{c_\varphi + \pi^{\nu-2}\left(\frac{4^{\nu-1}\pi^{\nu-1}}{(1-\nu)(x^2+16\pi^2)^{\nu-1}} + \frac{\nu\,\zeta(\nu)}{(\nu-1)\,x^{\nu-1}}\right)\right\}\,.$$

The upper function's ODE reads as follows:

$$\begin{aligned}\psi_\nu' - \frac{1}{2}\coth\left(\frac{x}{2}\right)\psi_\nu &= -\frac{\nu x}{2\pi^3}\sinh\left(\frac{x}{2}\right)\tilde{L}_{\nu+1}\left(\frac{x}{2\pi}\right) \\ &= \pi^{\nu-2}\sinh\left(\frac{x}{2}\right)\left\{\frac{\nu\,\zeta(\nu)}{2^{\nu-2}\,x^\nu} - \frac{2^{2\nu-1}\,\pi^{\nu-1}x}{(4\pi^2+x^2)^\nu}\right\}\,,\end{aligned}$$

accordingly,

$$\psi_\nu(x) = \sinh\left(\frac{x}{2}\right)\left\{c_\psi + \frac{\pi^{\nu-2}\,\nu\,\zeta(\nu)}{2^{\nu-2}\,(1-\nu)\,x^{\nu-1}} + \frac{4^{\nu-1}\pi^{2\nu-3}}{(\nu-1)(x^2+4\pi^2)^{\nu-1}}\right\}.$$

Here $\zeta(\nu), \nu \in \left(\frac{1}{2},1\right)$ signifies the analytic continuation of the Riemann ζ function by the widely known formula

$$\zeta(\nu) = \frac{\eta(\nu)}{1-2^{1-\nu}}, \qquad \mathrm{Re}\{\nu\} > 0\,.$$

D. Bearing in mind the asymptotics of lower and upper functions and that of $\Omega_\nu(x)$ near to zero, we immediately get

$$\frac{\varphi_\nu(x)}{\sinh\left(\frac{x}{2}\right)} \sim c_\varphi + \frac{4}{\pi(1-\nu)} = \frac{2}{\pi}\,\eta(2\nu-1) = c_\psi - \frac{1}{\pi(1-\nu)} \sim \frac{\psi_\nu(x)}{\sinh\left(\frac{x}{2}\right)},$$

which determines the values of integration constants

$$c_\varphi = \frac{2}{\pi}\,\eta(2\nu-1) - \frac{4}{\pi(1-\nu)},$$
$$c_\psi = \frac{2}{\pi}\,\eta(2\nu-1) + \frac{1}{\pi(1-\nu)}.$$

Thus, the proof of the following result is given.

Theorem 3.2. *Let $\nu \in \left(\frac{1}{2},1\right)$. Then for all $x \in \mathfrak{I} = \mathbb{R}_+$, we have the following two-sided inequality:*

$$\varphi_\nu(x) \leqslant \Omega_\nu(x) \leqslant \psi_\nu(x),$$

where

$$\varphi_\nu(x) = \sinh\left(\frac{x}{2}\right)\left\{\frac{2}{\pi}\,\eta(2\nu-1) - \frac{4}{\pi(1-\nu)} + \frac{4^{\nu-1}\pi^{2\nu-3}}{(1-\nu)(x^2+16\pi^2)^{\nu-1}} - \frac{\pi^{\nu-2}\,\nu\,\zeta(\nu)}{(1-\nu)\,x^{\nu-1}}\right\},$$

$$\psi_\nu(x) = \sinh\left(\frac{x}{2}\right)\left\{\frac{2}{\pi}\,\eta(2\nu-1) + \frac{1}{\pi(1-\nu)} + \frac{\pi^{\nu-2}\,\nu\,\zeta(\nu)}{2^{\nu-2}(1-\nu)x^{\nu-1}} - \frac{4^{\nu-1}\pi^{2\nu-3}}{(1-\nu)(x^2+4\pi^2)^{\nu-1}}\right\}.$$

Here $\zeta(\nu)$, stands for the analytic continuation of the Riemann ζ function to $\nu \in \left(\frac{1}{2},1\right)$.

4 Hilbert–Eisenstein Series and Their Basic Properties

4.1 In the sequel we give a brief introduction to certain important properties of the Hilbert–Eisenstein series. One of their bases is the Eisenstein theory of circular functions, founded by Gotthold Eisenstein [16] in 1847. They play an important role in number theory, especially their extension to elliptic functions; see, e.g. Weil [48, 49] and Iwaniec and Kowalski [26]. The series $\varepsilon_r(w)$ defined for all $w \in \mathbb{C} \setminus \mathbb{Z}$ and all integer $r \geqslant 2$ by

$$\varepsilon_r(w) = \sum_{k \in \mathbb{Z}} \frac{1}{(w+k)^r},$$

is called the *Eisenstein series of order* r. The $\varepsilon_r(w)$ are normally convergent and represent meromorphic functions in $\mathbb{C}$, are holomorphic in $\mathbb{C} \setminus \mathbb{Z}$ and posses poles in $k \in \mathbb{Z}$ (of order r and principal part $(w-k)^{-r}$). Recall that a series $\sum_k f_k$ of functions $f_k \colon X \mapsto \mathbb{C}$ is *normally convergent in* X if to each point $x \in X$ there exists a neighbourhood U such that $\sum_k |f_k| < \infty$. If the series is normally convergent in X, then $\sum |f_k|$ converges compactly in X. The converse is valid for domains $X = D \subseteq \mathbb{C}$, if all f_k are holomorphic in D. If such a series converges compactly in D, so does the series of its derivatives and, under weak assumptions, also the series of its primitives; see [42, pp. 92–95, 224].

For $r = 1$, the definition reads in Eisenstein's principal value notation

$$\begin{aligned}\varepsilon_1(w) &= \sum_{k \in \mathbb{Z}}{}_e \frac{1}{w+k} := \lim_{N \to \infty} \sum_{|k| \leqslant N} \frac{1}{w+k} \\ &= \frac{1}{w} + \sum_{k \geqslant 1} \left(\frac{1}{w+k} + \frac{1}{w-k} \right) \\ &= \frac{1}{w} + \sum_{k \geqslant 1} \frac{2w}{w^2 - k^2} = \pi \cot(\pi w), \qquad w \in \mathbb{C} \setminus \mathrm{i}\mathbb{Z}. \end{aligned} \tag{36}$$

This partial fraction expansion of the cotangent function (due to Euler [17]), which is essentially the "alternating" generating function of the classical Bernoulli numbers $B_{2k} := B_{2k}(0)$ ($B_n(x)$ being the Bernoulli polynomials, $n \in \mathbb{N}_0$), namely,

$$\frac{w}{2} \cot \frac{w}{2} = \sum_{k \geqslant 0} \frac{(-1)^k}{(2k)!} B_{2k}\, w^{2k}, \qquad |w| < 2\pi,$$

is regarded by Konrad Knopp [28, p. 207] as the "*most remarkable expansion in partial fractions*".

Differentiating the normally convergent series (36), observing that $(\cot w)' = -\sin^{-2} w$, there follows

$$\varepsilon_2(w) = \frac{\pi^2}{\sin^2(\pi w)}, \qquad \varepsilon_3(w) = \frac{\pi^3 \cot(\pi w)}{\sin^2(\pi w)},$$

giving the surprising result [42, p. 303]

$$\varepsilon_3(w) = \varepsilon_1(w) \cdot \varepsilon_2(w) .$$

Eisenstein series' two essential properties are [48, pp. 6–13] and [42, p. 303]

$$\varepsilon_r'(w) = -r\varepsilon_{r+1}(w), \qquad \varepsilon_r(w) = \frac{(-1)^{r-1}}{(r-1)!}\,\varepsilon_1^{(r-1)}(w), \qquad r \in \mathbb{N}_2 ,$$

as well as their 1-periodicity in the sense that $\varepsilon_r(w+k) = \varepsilon_r(w)$ for all $w \in \mathbb{C}$, $k \in \mathbb{Z}$.

Differentiating the Fourier series expansion of $\cot \pi w$ [18, p. 386 et seq.], namely of

$$\cot \pi w = \mathrm{i} \begin{cases} -1 - 2\sum\limits_{k\geqslant 1} \mathrm{e}^{2\pi \mathrm{i} wk}, & \mathrm{Im}\{w\} > 0 \\ 1 + 2\sum\limits_{k\leqslant -1} \mathrm{e}^{2\pi \mathrm{i} wk}, & \mathrm{Im}\{w\} < 0 \end{cases},$$

iteratively, there follows that the $\varepsilon_r(w)$ posses Fourier expansions in the upper and lower half-planes with period π.

The basis to the Hilbert–Eisenstein series also includes the background to the so-called "Basler" problem, an open question since 1690, namely, whether there exists a counterpart of Euler's famous result on the closed expression for the Riemann Zeta function for even arguments, thus

$$\zeta(2m) = (-1)^{m+1} 2^{2m-1} \pi^{2m} \frac{B_{2m}}{(2m)!}, \qquad m \in \mathbb{N} ,$$

to the case of odd arguments, namely, $\zeta(2m+1)$.

Theorem A (Counterpart of Euler's formula for $\zeta(2m+1)$). *For $m \in \mathbb{N}$, there holds*

$$\zeta(2m+1) = (-1)^m 4^m \pi^{2m+1} \frac{B^{\sim}_{2m+1}(0)}{(2m+1)!} .$$

Thus, the solution consists in replacing the Bernoulli numbers B_{2m} in Euler's formula by the conjugate Bernoulli numbers $B^{\sim}_{2m+1}$ which are defined in terms of the Hilbert transform and were introduced by Butzer, Hauss and Leclerc in [7].

Starting with the 1-periodic Bernoulli polynomials $\mathscr{B}_n(x)$ defined as the periodic extension of $\mathscr{B}_n(x) = B_n(x), x \in (0,1]$, one can—using Hilbert transforms—introduce 1-periodic conjugate Bernoulli "polynomials" $\mathscr{B}_n^{\sim}(x), x \in \mathbb{R}$ ($x \notin \mathbb{Z}$ if $n = 1$) by

$$\mathscr{B}_n^{\sim}(x) := \mathscr{H}\left[\mathscr{B}_n(\cdot)\right]_1(x), \qquad n \in \mathbb{N}.$$

These conjugate periodic functions $\mathfrak{B}_n^{\sim}(x)$ can be used to define the non-periodic functions $B_n^{\sim}(x)$ in a form such that their properties are similar to those of the classical Bernoulli polynomials $B_n(x)$. For details, see Butzer and Hauss [5, p. 22] and Butzer [4, pp. 37–56]. The conjugate Bernoulli numbers in question, the $B_{2m+1}^{\sim}$, are the $B_{2m+1}^{\sim}(0)(= B_{2m+1}^{\sim}(1))$ for which

$$B_{2m+1}^{\sim}\left(\frac{1}{2}\right) = \left(4^{-m} - 1\right) \cdot B_{2m+1}^{\sim}(1).$$

Some values of the conjugate Bernoulli numbers are

$$B_{2m+1}^{\sim}\left(\frac{1}{2}\right) = \begin{cases} -\dfrac{\log 2}{\pi} & m=0 \\ \dfrac{\log 2}{4\pi} - 2\displaystyle\int_{0+}^{\frac{1}{2}} u^2 \cot(\pi u) du & m=1 \\ \dfrac{11}{8}\displaystyle\int_{0+}^{\frac{1}{2}} u \cot \pi du + \dfrac{5}{3}\displaystyle\int_{0+}^{\frac{1}{2}} u^3 u \cot \pi du - 2\displaystyle\int_{0+}^{\frac{1}{2}} u^5 \cot \pi u \, du & m=2 \end{cases}.$$

Now, the indirect basis of the Hilbert–Eisenstein series is the Omega function $\Omega(\cdot)$. One arrives at it through the counterpart for the $B_n^{\sim}(x)$ of the exponential generating function of the classical polynomials $B_n(x)$, namely,

$$\sum_{n \geqslant 0} B_n(x) \frac{w^n}{n!} = \frac{w\, e^{wx}}{e^w - 1}, \qquad w \in \mathbb{C}, \ |w| < 2\pi, \ x \in \mathbb{R}. \tag{37}$$

Theorem B (Exponential generating function of $B_k^{\sim}(\frac{1}{2})$). *For $|w| < 2\pi$, there holds*

$$\sum_{n \geqslant 0} B_n^{\sim}\left(\frac{1}{2}\right) \frac{w^n}{n!} = \frac{w\, e^{wx}}{e^w - 1} \cdot \Omega(w).$$

The proofs of Theorems A and B are connected with the Hilbert transform versions of the Euler–Maclaurin and Poisson summation formulae established by Hauss [24] (see also [5, p. 21–29] and [4, pp. 37–38, 78–80]). Observe that Theorem B tells us that the Hilbert transform of (37) essentially results in multiplying $w\, e^{wx}(e^w - 1)^{-1}$ by the Omega function $\Omega(w)$, taken at $x = \frac{1}{2}$.

4.2 The direct basis to what follows is the partial fraction expansion of $\Omega(w)$, thus a Hilbert-type version of the basic partial fraction expansion of $\cot \pi w$ in (36). It reads,

Theorem C (Partial fraction expansion of $\Omega(w)$). *For $w \in \mathbb{C} \setminus \mathrm{i}\mathbb{Z}$, one has*

$$\Omega(2\pi w) = \frac{1}{\pi}\left(\mathrm{e}^{-\pi w} - \mathrm{e}^{\pi w}\right) \sum_{k\geqslant 1} \frac{(-1)^k\, k}{w^2 + k^2} = \frac{\sinh(\pi w)}{\mathrm{i}w} \sum_{k\in\mathbb{Z}} \frac{(-1)^k \operatorname{sgn}(k)}{w + \mathrm{i}k}\,.$$

Definition 4.1. The Hilbert–Eisenstein (HE) series $\mathfrak{h}_r(w)$ are defined for $w \in \mathbb{C}\setminus\mathrm{i}\mathbb{Z}$ and $r \in \mathbb{N}_2$ by

$$\mathfrak{h}_r(w) := \sum_{k\in\mathbb{Z}} \frac{(-1)^k \operatorname{sgn}(k)}{(w+\mathrm{i}k)^r} = \sum_{k\geqslant 1} (-1)^k \left(\frac{1}{(w+\mathrm{i}k)^r} - \frac{1}{(w-\mathrm{i}k)^r} \right), \tag{38}$$

and

$$\mathfrak{h}_1(w) := {\sum_{k\in\mathbb{Z}}}_e \frac{(-1)^k \operatorname{sgn}(k)}{w+\mathrm{i}k} = \frac{\mathrm{i}\pi\, \Omega(2\pi w)}{\sinh \pi w} = \mathrm{i} \sum_{k\in\mathbb{Z}} \frac{(-1)^k}{w+\mathrm{i}k} \cdot \Omega(2\pi w)\,. \tag{39}$$

For $w = 0$, $\mathfrak{h}_1(0)$ can, since $\operatorname{sgn}(0) = 0$, be taken as

$$\mathfrak{h}_1(0) = {\sum_{k\in\mathbb{Z}\setminus\{0\}}}_e \frac{(-1)^k \operatorname{sgn}(k)}{\mathrm{i}k} = 2\mathrm{i}\,\log 2\,.$$

Observe that the partial fraction expansion of $\pi(\sinh \pi w)^{-1}$ follows by replacing w by $\mathrm{i}w$ and recalling $\sinh w = -\mathrm{i} \sin \mathrm{i}w$ in the well-known expansion

$$\frac{\pi}{\sin \pi w} = \sum_{k\in\mathbb{Z}} \frac{(-1)^k}{w+k}, \qquad w \in \mathbb{C} \setminus \mathbb{Z}\,,$$

and $\pi(\sinh \pi w)^{-1}$ also possesses the Dirichlet series expansion for the right half-plane (see, e.g. [18, p. 405 et seq.])

$$\frac{\pi}{\sinh \pi w} = 2\pi \sum_{k\geqslant 0} \mathrm{e}^{-2\pi(2k+1)w}, \qquad \operatorname{Re}\{w\} > 0\,.$$

The case $r = 1$ of Definition 4.1 reveals that to achieve the Hilbert-type version of $\pi(\sinh \pi w)^{-1}$, that is, $\mathfrak{h}_1(w)$, one multiplies it (or its partial fraction expansion) by the complete Omega function $\Omega(2\pi w)$ and i. The counterparts of the corresponding properties of $\varepsilon_r(w)$ read

Proposition 4.1. *There holds for $w \in \mathbb{C} \setminus \mathrm{i}\mathbb{Z}$ and $r \in \mathbb{N}_2$, $m \in \mathbb{N}$*

$$\mathfrak{h}_r^{(m)}(w) = (-1)^m \, (r)_m \, \mathfrak{h}_{r+m}(w)\,, \tag{40}$$

where

$$(r)_m := \frac{\Gamma(r+m)}{\Gamma(r)} = r(r+1)\cdots(r+m-1)\,, \quad (r)_0 \equiv 1\,,$$

stands for the Pochhammer symbol (or shifted, rising factorial); moreover,

$$\mathfrak{h}_r(w) = \frac{(-1)^r}{\Gamma(r)}\, \mathfrak{h}_2^{(r-2)}(w), \qquad r \in \mathbb{N}_2\,, \tag{41}$$

as well as

$$\mathfrak{h}_r(w) + \mathfrak{h}_r(w+\mathrm{i}) = w^{-r} - (w+\mathrm{i})^{-r}\,. \tag{42}$$

Proof. Here the differentiability properties follow readily from Definition 4.1, and the i-periodicity-type formula we conclude from

$$\begin{aligned}\sum_{|k|\leqslant N} \frac{(-1)^k \operatorname{sgn}(k)}{(w+\mathrm{i}+\mathrm{i}k)^r} &= \sum_{k=2}^{N+1} \frac{(-1)^{k-1}\operatorname{sgn}(k-1)}{(w+\mathrm{i}k)^r} + \frac{1}{w^r} + \sum_{k=-N+1}^{-1} \frac{(-1)^{k-1}\operatorname{sgn}(k-1)}{(w+\mathrm{i}k)^r}\\ &= -\sum_{k=-N+1}^{N-1} \frac{(-1)^k \operatorname{sgn}(k)}{(w+\mathrm{i}k)^r} + \frac{1}{w^r} - \frac{1}{(w+\mathrm{i})^r} - \frac{(-1)^{N+1}}{(w+\mathrm{i}(N+1))^r} - \frac{(-1)^N}{(w+\mathrm{i}N)^r}\,.\end{aligned}$$

Letting $N \to \infty$, we immediately arrive at (42). □

Theorem 4.1. *The* HE *series $\mathfrak{h}_r(w)$ possesses for $x \in \mathbb{R}$ and $r \in \mathbb{N}$ the integral representation*

$$\mathfrak{h}_r(x) = \frac{2\mathrm{i}(-1)^{r-2}}{\Gamma(r)} \int_0^\infty \frac{u^{r-1}}{\mathrm{e}^u+1} \sin\Big(\frac{r-2}{2}\pi + xu\Big)\mathrm{d}u\,. \tag{43}$$

Specifically, we have for $r = 1$ and 2,

$$\mathfrak{h}_1(x) = 2\mathrm{i} \int_0^\infty \frac{\cos(xu)}{\mathrm{e}^u+1}\,\mathrm{d}u \tag{44}$$

and

$$\mathfrak{h}_2(x) = 2\mathrm{i} \int_0^\infty u\,\frac{\sin(xu)}{\mathrm{e}^u+1}\,\mathrm{d}u\,. \tag{45}$$

Proof. Beginning with the representation

$$\mathfrak{h}_2(w) = \sum_{k\in\mathbb{Z}}^{e} \frac{(-1)^k \mathrm{sgn}(k)}{(w+\mathrm{i}k)^2} = 2\mathrm{i} \sum_{k\geqslant 1} \frac{(-1)^{k-1} 2kw}{(w^2+k^2)^2}, \tag{46}$$

we try to express the fraction $2kw(w^2+k^2)^{-2}$ as a Laplace transform, thus *via*

$$\frac{2sw}{(w^2+s^2)^2} = \mathscr{L}_u[u \sin(wu)](s) = \int_0^\infty \mathrm{e}^{-su} u \sin(wu)\,\mathrm{d}u\,.$$

Now, this transform is correct for $\mathrm{Re}\{s\} > |\,\mathrm{Im}\{w\}|$. But for the needed $s = k \in \mathbb{N}$, this inequality requires that $|\,\mathrm{Im}\{w\}| = 0$, so that w must be real, i.e. $w = x$.

Noting

$$\sum_{k\geqslant 1} (-1)^{k-1} \mathrm{e}^{-ku} = \frac{1}{\mathrm{e}^u + 1},$$

one has

$$\mathfrak{h}_2(x) = 2\mathrm{i} \sum_{k\geqslant 1} (-1)^{k-1} \int_0^\infty \mathrm{e}^{-ku} u \sin(xu)\,\mathrm{d}u = 2\mathrm{i} \int_0^\infty \frac{u \sin(xu)}{\mathrm{e}^u+1}\,\mathrm{d}u\,, \tag{47}$$

where, being $|\sin(xu)| \leqslant 1$, the interchange of sum and the integral is legitimate. This proves (45).

Repeated $r-2$-fold differentiation of $\mathfrak{h}_2(x)$ with respect to x according to (40), that is, (41), delivers

$$\begin{aligned}\mathfrak{h}_r(x) &= \frac{2\mathrm{i}(-1)^r}{\Gamma(r)} \frac{\mathrm{d}^{r-2}}{\mathrm{d}x^{r-2}} \int_0^\infty u \frac{\sin(xu)}{\mathrm{e}^u+1}\,\mathrm{d}u \\ &= \frac{2\mathrm{i}(-1)^r}{\Gamma(r)} \int_0^\infty \frac{u^{r-1}}{\mathrm{e}^u+1} \sin\Big(\frac{r-2}{2}\pi + xu\Big)\mathrm{d}u,\end{aligned}$$

which is (43), for all $r \in \mathbb{N}_2$.

It remains the case $r = 1$, which has to be considered separately. In turn, we have to connect the Hilbert–Eisenstein series $\mathfrak{h}_1(z)$, which converges in the sense of Eisenstein summation (but does not converges normally), and the normally convergent HE series $\mathfrak{h}_r(z), r \geqslant 2$, which is termwise integrable. Thus,

$$\begin{aligned}\int_0^x \mathfrak{h}_2(t)\,\mathrm{d}t &= \sum_{k\in\mathbb{Z}}^{e} (-1)^k \mathrm{sgn}(k) \int_0^x \frac{\mathrm{d}t}{(t+\mathrm{i}k)^2} \\ &= \sum_{k\in\mathbb{Z}}^{e} \frac{(-1)^k \mathrm{sgn}(k)}{ik} - \sum_{k\in\mathbb{Z}}^{e} \frac{(-1)^k \mathrm{sgn}(k)}{x+ik} \\ &= 2\mathrm{i} \sum_{k\in\mathbb{N}} \frac{(-1)^k}{k} - \mathfrak{h}_1(x) = 2\mathrm{i} \log 2 - \mathfrak{h}_1(x)\,.\end{aligned}$$

On the other hand, the legitimate integration order exchange in (47) leads to

$$\begin{aligned}\int_0^x \mathfrak{h}_2(u)\mathrm{d}u &= 2\mathrm{i}\int_0^\infty \frac{u}{\mathrm{e}^u+1}\left\{\int_0^x \sin(xu)\,\mathrm{d}x\right\}\,\mathrm{d}u \\ &= 2\mathrm{i}\left\{\int_0^\infty \frac{1}{\mathrm{e}^u+1}\,\mathrm{d}u - \int_0^\infty \frac{\cos(xu)}{\mathrm{e}^u+1}\,\mathrm{d}u\right\} \\ &= 2\mathrm{i}\,\log 2 - 2\mathrm{i}\int_0^\infty \frac{\cos(xu)}{\mathrm{e}^u+1}\,\mathrm{d}u\,.\end{aligned}$$

The rest is clear. □

At this point let us recall integral representations which have been derived for the complete, real argument Omega function $\Omega(x)$. Among others we mentioned in Sect. 1 the results by Butzer, Pogány and Srivastava (1); consult [8, Theorem 2], by Pogány and Tomovski (2), and see [46, Theorem 3.3] and the Pogány–Srivastava integral representation (3) [39, p. 589, Theorem 1].

However, having in mind Theorem C and the differentiability property (41) in conjunction with Theorem 4.1, we now obtain a new integral representation formula for the complete Ω function and its derivatives *via* the rth order Hilbert–Eisenstein series $\mathfrak{h}_r$.

Theorem 4.2. *For $x \in \mathbb{R}$ and $r \in \mathbb{N}$, there holds true*

$$\Omega^{(r)}(x) = \frac{1}{2^r\,\pi}\int_0^\infty \frac{\mathrm{Re}\,\{\kappa_r(x;u)\}}{\mathrm{e}^u+1}\,\mathrm{d}u, \tag{48}$$

where

$$\kappa_r(x;u) = \left(\mathrm{e}^{(1+\frac{\mathrm{i}u}{\pi})\frac{x}{2}} - (-1)^r\,\mathrm{e}^{-(1+\frac{\mathrm{i}u}{\pi})\frac{x}{2}}\right)\left(1+\frac{\mathrm{i}u}{\pi}\right)^r.$$

Proof. Theorem C in conjunction with (44) implies

$$\Omega(2\pi x) = \frac{2}{\pi}\,\sinh(\pi x)\int_0^\infty \frac{\cos(x\,u)}{\mathrm{e}^u+1}\,\mathrm{d}u\,; \tag{49}$$

actually, by replacing $(2\pi x \mapsto x)$, we reobtained the integral expression (1). Differentiating this formula r times with respect to x, we get by virtue of property (41),

$$\Omega^{(r)}(x) = \frac{2}{\pi}\sum_{m=0}^r \binom{r}{m}\left\{\int_0^\infty \cos\left(\frac{xu}{2\pi}\right)\frac{\mathrm{d}u}{\mathrm{e}^u+1}\right\}^{(m)}\cdot\left\{\sinh\left(\frac{x}{2}\right)\right\}^{(r-m)}. \tag{50}$$

Differentiating the integral m times with respect to x, we get

$$\int_0^\infty \left(\frac{\mathrm{d}}{\mathrm{d}x}\right)^m \cos\left(\frac{xu}{2\pi}\right) \frac{\mathrm{d}u}{\mathrm{e}^u+1} = \frac{\mathrm{i}^m}{2(2\pi)^m} \int_0^\infty \frac{u^m}{\mathrm{e}^u+1} \left(\mathrm{e}^{\mathrm{i}\frac{xu}{2\pi}} + (-1)^m \mathrm{e}^{-\mathrm{i}\frac{xu}{2\pi}}\right) \mathrm{d}u.$$

On the other hand,

$$\left\{\sinh\left(\frac{x}{2}\right)\right\}^{(r-m)} = \frac{1}{2^{r-m+1}}\left(\mathrm{e}^{\frac{x}{2}} - (-1)^{r-m}\mathrm{e}^{-\frac{x}{2}}\right).$$

Replacing both derivatives into (50), after suitable reduction of the material and summing up all terms inside the integrand, we arrive at the expression

$$\begin{aligned}\Omega^{(r)}(x) = \frac{1}{2^{r+1}\pi} \int_0^\infty &\left\{\left(\mathrm{e}^{(1+\frac{\mathrm{i}u}{\pi})\frac{x}{2}} - (-1)^r\, \mathrm{e}^{-(1+\frac{\mathrm{i}u}{\pi})\frac{x}{2}}\right)\left(1+\frac{\mathrm{i}u}{\pi}\right)^r\right. \\ &\left. + \left(\mathrm{e}^{(1-\frac{\mathrm{i}u}{\pi})\frac{x}{2}} - (-1)^r\, \mathrm{e}^{-(1-\frac{\mathrm{i}u}{\pi})\frac{x}{2}}\right)\left(1-\frac{\mathrm{i}u}{\pi}\right)^r\right\} \frac{\mathrm{d}u}{\mathrm{e}^u+1},\end{aligned}$$

which is equivalent to the statement. □

We mention that the HE series $\mathfrak{h}_r(w)$, or better still $w^r\, \mathfrak{h}_r(w)$, posses a Taylor series expansion, the coefficients even involving the Dirichlet Eta function values $\eta(2k+1)$, where the Dirichlet Eta function

$$\eta(s) = \sum_{k\geqslant 1}(-1)^{k-1}k^{-s}, \qquad \mathrm{Re}\{s\} > 0.$$

Indeed,

Theorem 4.3 ([4, p. 83, Theorem 9.1]). *For $w \in \mathbb{C}$, $|w| < 1$ and $r \in \mathbb{N}$, one has*

$$w\,\mathfrak{h}_1(w) = 2\mathrm{i} \sum_{k\geqslant 0}(-1)^k\, \eta(2k+1)\, w^{2k+1}$$

and

$$w^r\, \mathfrak{h}_r(w) = 2\mathrm{i}\,(-1)^{r-1} \sum_{k\geqslant [r/2]} (-1)^k \binom{2k}{r-1} \eta(2k+1)\, w^{2k+1},$$

with $\mathfrak{h}_1(0) = 2\mathrm{i}\log 2$.

Thus, $\mathfrak{h}_r(w)$ is holomorphic in $\mathbb{C}\setminus \mathrm{i}\mathbb{Z}$.

4.3 Let us briefly consider some connections between the $\mathfrak{h}_r(w)$ and alternating Mathieu series $\tilde{S}_r(w)$.

Firstly, according to Proposition 4.1 (or evaluating $\mathfrak{h}_2(w)$ directly from its definition), having in mind (5) again, we have

$$\mathfrak{h}_2(w) = 2\mathrm{i}w \sum_{k\geqslant 1} \frac{(-1)^{k-1}2k}{(w^2+k^2)^2} = 2\mathrm{i}w\tilde{S}_2(w)\,.$$

As to the next step,

$$\mathfrak{h}_3(w) = -\mathrm{i}\sum_{k\geqslant 1} \frac{(-1)^{k-1}2k(k^2-3w^2)}{(w^2+k^2)^3}\,, \tag{51}$$

$$\tilde{S}_3(w) = \sum_{k\geqslant 1} \frac{(-1)^{k-1}2k}{(w^2+k^2)^3}\,.$$

Although the two look incomparable, see nevertheless Theorem 5.1 in Sect. 5.

For further systematic connections between $\tilde{S}_r(w)$ and $\mathfrak{h}_r(w)$, see Sect. 5.

4.4 We finally turn to a counterpart of the function $\Omega_\nu(w)$ introduced in (9). In regard to $\mathfrak{h}_r(w)$ we have seen that

$$\frac{\pi\,\Omega(2\pi w)}{\sinh(\pi w)} = -\mathrm{i}\mathfrak{h}_1(w)\,.$$

We define a new function, $\tilde{\Omega}_r(\cdot)$, say, in terms of the HE series $\mathfrak{h}_r(w)$ as follows. This answers a conjecture raised in [4, p. 82].

Definition 4.2. For all $w \in \mathbb{C}\setminus \mathrm{i}\mathbb{Z}$ and for all $r \in \mathbb{N}$, the extended Omega function $\tilde{\Omega}_r(\cdot)$ of positive integer order r is the function which satisfies equation

$$\frac{\pi\,\tilde{\Omega}_r(2\pi w)}{\sinh(\pi w)} = -\mathrm{i}\,\mathfrak{h}_r(w).$$

In fact, the new special function

$$\tilde{\Omega}_r(w) = -\frac{\mathrm{i}}{\pi}\,\sinh\left(\frac{w}{2}\right)\,\mathfrak{h}_r\left(\frac{w}{2\pi}\right), \qquad w \in \mathbb{C}\setminus \mathrm{i}\mathbb{Z},\ r\in\mathbb{N} \tag{52}$$

is the positive integer order counterpart of the function $\Omega_r(\cdot)$, defined already in terms of the alternating generalized Mathieu series $\tilde{S}_\nu(\cdot)$:

$$\frac{\pi\,\Omega_\nu(2\pi w)}{\sinh(\pi w)} = \sum_{k\geqslant 1} \frac{(-1)^{k-1}2k}{(k^2+w^2)^\nu} =: \tilde{S}_\nu(w)\,,$$

studied in [8] and here in Sect. 2 and Sect. 3, even for $\nu \in \mathbb{R}_+$.

Connecting Definition 4.2 and Theorem 4.1, we clearly arrive at the integral representation of the real variable alternating extended Omega function, recalling the integral expression for the Eta function

$$\eta(r) = \int_0^\infty \frac{u^{r-1}}{e^u + 1}\, du\,.$$

Theorem 4.4. *For any $r \in \mathbb{N}$ one has for $x \in \mathbb{R}$*

$$\tilde{\Omega}_r(x) = \frac{2(-1)^{r-1}}{\pi\, \Gamma(r)} \sinh\left(\frac{x}{2}\right) \int_0^\infty \frac{u^{r-1}}{e^u + 1} \sin\left(\frac{r-2}{2}\pi + \frac{xu}{2\pi}\right) du\,.$$

Moreover, in the same range of parameters we have the estimate

$$\left|\tilde{\Omega}_r(x)\right| \leqslant \frac{2}{\pi}\, \eta(r)\, \left|\sinh\left(\frac{x}{2}\right)\right|\,. \tag{53}$$

Remark 4.1. We recognize that the bound in the inequality (53) for $\tilde{\Omega}_r(x)$ is the same fashion result as the bound (24) achieved for the generalized BFH $\Omega_r(w)$.

5 Some Bridges Between $\Omega_\nu(w)$, $\tilde{\Omega}_r(w)$, $\tilde{S}_\nu(x)$ and $\mathfrak{h}_r(w)$

5.1 The alternating Mathieu series (5)

$$\tilde{S}_\nu(w) = \sum_{n\geqslant 1} \frac{(-1)^{n-1}\, 2n}{(n^2 + w^2)^\nu}, \qquad \nu > 0, w > 0\,,$$

and the Hilbert–Eisenstein series (38)

$$\mathfrak{h}_r(w) = \sum_{k\in\mathbb{Z}} \frac{(-1)^k\, \mathrm{sgn}(k)}{(w + \mathrm{i}k)^r}, \qquad w \in \mathbb{C} \setminus \mathrm{i}\mathbb{Z}, r \in \mathbb{N}$$

are intimately connected by (39), and relationship (9), e.g. we have seen in the previous section that

$$\mathfrak{h}_2(w) = 2\mathrm{i}\, w\, \tilde{S}_2(w)\,.$$

However, higher-order alternating Mathieu and HE series do not coincide, but there is a close connection between them:

Theorem 5.1. *For all $w > 0, r \in \mathbb{N}_2$ there holds true*

$$\mathfrak{h}_r(w) = \frac{\mathrm{i}\,(-1)^{r-1}}{w^{r+1}} \sum_{\substack{0\leqslant m\leqslant r\\ r-m\equiv 1 (\mathrm{mod}\ 2)}} \sum_{j=\frac{r+m+1}{2}}^{r} \binom{r}{m}\binom{\frac{r+m+1}{2}}{r-j}(-w^2)^j\, \tilde{S}_j(w)\,. \tag{54}$$

Proof. Beginning with Definition 4.1, one transforms

$$\begin{aligned}\mathfrak{h}_r(w) &= \sum_{k\geqslant 1} \frac{(-1)^k}{(w+\mathrm{i}k)^r} \sum_{m=0}^{r} \binom{r}{m}((-1)^{r-m}-1)w^m(\mathrm{i}k)^{r-m} \\ &= \mathrm{i}\sum_{k\geqslant 1} \frac{2(-1)^{k-1}k}{(w^2+k^2)^r} \sum_{\substack{0\leqslant m\leqslant r\\ r-m \text{ odd}}} \binom{r}{m} w^m(-1)^{(r-m+1)/2}(k^2)^{(r-m-1)/2} \\ &= \mathrm{i}\sum_{\substack{0\leqslant m\leqslant r\\ r-m \text{ odd}}} \binom{r}{m} \sum_{j=0}^{(r-m-1)/2} \binom{\frac{r-m-1}{2}}{j}(-1)^{j-1}w^{r-1-2j}\,\tilde{S}_{r-j}(w)\,;\end{aligned}$$

changing the summation order, we get the asserted expression. □

The next few low-order particular cases, coming after $\mathfrak{h}_2(w)$, are

$$\mathfrak{h}_3(w) = -5w^2\mathrm{i}\,\tilde{S}_3(w) + 2\mathrm{i}\,\tilde{S}_2(w),$$

$$\mathfrak{h}_4(w) = -8w^3\mathrm{i}\,\tilde{S}_4(w) + 4w\mathrm{i}\,\tilde{S}_3(w).$$

The opposite question also arises, that is, how can we express alternating generalized Mathieu series *via* a linear combination of Hilbert–Eisenstein series of up to the same order? Collecting formulae for $\mathfrak{h}_j(w)$ in terms of $\tilde{S}_j(w)$, $j = 2, 3, 4$ and solving it with respect to Mathieu series $\tilde{S}_j(w)$, we get

$$\begin{aligned}\tilde{S}_2(w) &= -\frac{\mathrm{i}}{2w}\,\mathfrak{h}_2(w), \\ \tilde{S}_3(w) &= -\frac{\mathrm{i}}{5w^3}\,\mathfrak{h}_2(w) + \frac{\mathrm{i}}{5w^3}\,\mathfrak{h}_3(w), \\ \tilde{S}_4(w) &= -\frac{\mathrm{i}}{10w^5}\,\mathfrak{h}_2(w) + \frac{\mathrm{i}}{10w^3}\,\mathfrak{h}_3(w) + \frac{\mathrm{i}}{8w^3}\,\mathfrak{h}_4(w), \qquad \text{etc.}\end{aligned}$$

As an immediate consequence of Theorem 5.1, we get by means of Definition 4.2 (of the HE series) the following connection between the generalized complete positive integer order BFH Omega function $\Omega_r(w)$ and its counterpart $\tilde{\Omega}_r(w)$.

Theorem 5.2. *Let the situation be the same as in previous* Theorem 5.1. *Then we have the following representation:*

$$\tilde{\Omega}_r(w) = \frac{(-1)^{r-1}}{w^{r+1}} \sum_{\substack{0\leqslant m\leqslant r \\ r-m\equiv 1 (\text{mod } 2)}} \sum_{j=\frac{r+m+1}{2}}^{r} \binom{r}{m} \binom{\frac{r+m+1}{2}}{r-j} (-w^2)^j \, \Omega_j(w) .$$

Proof. Transforming representation (54) by Definition 4.2, (52) from one and Theorem 5.1 from another side, we derive

$$\begin{aligned}\tilde{\Omega}_r(w) &= -\frac{\mathrm{i}}{\pi} \sinh\left(\frac{w}{2}\right) \mathfrak{h}_r\left(\frac{w}{2\pi}\right) \\ &= \sum_{\substack{0\leqslant m\leqslant r \\ r-m \text{ odd}}} \sum_{j=0}^{(r-m-1)/2} \binom{r}{m} \binom{\frac{r-m-1}{2}}{j} w^{r-1-2j} (-1)^{j-1} \\ &\quad \times \underbrace{\frac{1}{\pi} \sinh\left(\frac{w}{2}\right) \tilde{S}_{r-j}\left(\frac{w}{2\pi}\right)}_{\Omega_{r-j}(w)},\end{aligned}$$

which is equivalent to the assertion. □

Now, for $w > 0$ it is not hard to compile the formulae:

$$\tilde{\Omega}_2(w) = \frac{2w}{\pi} \sinh\left(\frac{w}{2}\right) \tilde{S}_2\left(\frac{w}{2\pi}\right) = 2w\, \Omega_2(w),$$

$$\begin{aligned}\tilde{\Omega}_3(w) &= \frac{1}{\pi} \sinh\left(\frac{w}{2}\right) \left\{2\tilde{S}_2\left(\frac{w}{2\pi}\right) - 5w^2 \tilde{S}_3\left(\frac{w}{2\pi}\right)\right\} \\ &= 2\Omega_2(w) - 5w^2\, \Omega_3(w)\end{aligned}$$

$$\begin{aligned}\tilde{\Omega}_4(w) &= \frac{4w}{\pi} \sinh\left(\frac{w}{2}\right) \left\{\tilde{S}_3\left(\frac{w}{2\pi}\right) - 2w^2 \tilde{S}_4\left(\frac{w}{2\pi}\right)\right\} \\ &= 4w\left\{\Omega_3(w) - 2w^2\, \Omega_4(w)\right\}.\end{aligned}$$

We point out that both Theorems 5.1 and 5.2 ensure a good tool for further bilateral bounding inequalities upon $\tilde{\Omega}_r(w)$. Namely, we establish in Sects. 2 and 3 numerous two-sided bounding inequalities for the generalized BFH Omega function $\Omega_j(w),\ j \in \mathbb{N}$.

5.2 In the following, we study a series representation of the Hilbert–Eisenstein series in terms of the Gaussian hypergeometric function ${}_2F_1$. For the values w of the argument coming from the open unit disc $\mathbb{D} := \{w : |w| < 1\}$, we have the following expansion:

$$\begin{aligned}\mathfrak{h}_r(w) &= \mathrm{i}^{-r} \sum_{k\geqslant 1} \frac{(-1)^k}{k^r} \left\{ \left(1+\frac{w}{\mathrm{i}k}\right)^{-r} - (-1)^r \left(1-\frac{w}{\mathrm{i}k}\right)^{-r} \right\} \\ &= \mathrm{i}^{-r} \sum_{k\geqslant 1} \frac{(-1)^k}{k^r} \sum_{j\geqslant 0} \binom{-r}{j} (1-(-1)^{r+j}) \left(\frac{w}{\mathrm{i}k}\right)^j \\ &= \mathrm{i}^{-r} \sum_{k\geqslant 1} \frac{(-1)^k}{k^r} \sum_{j\geqslant 0} \frac{(-1)^j (1-(-1)^{r+j}) \Gamma(r+j)}{\Gamma(r)\, j!} \left(\frac{w}{\mathrm{i}k}\right)^j .\end{aligned}$$

If r is either even or odd, we have more specific further results.

Theorem 5.3. *For all $|w| < 1$ and $r \in \mathbb{N}$, we have*

$$\mathfrak{h}_{2r-1}(w) = 2\mathrm{i}(-1)^r \sum_{k\geqslant 1} \frac{(-1)^{k-1}}{k^{2r-1}}\, {}_2F_1 \left[\begin{matrix} r-\frac{1}{2},\, r \\ \frac{1}{2} \end{matrix} \middle| -\frac{w^2}{k^2} \right];$$

moreover,

$$\mathfrak{h}_{2r}(w) = 4wr\mathrm{i}(-1)^r \sum_{k\geqslant 1} \frac{(-1)^{k-1}}{k^{2r+1}}\, {}_2F_1 \left[\begin{matrix} r+\frac{1}{2},\, r+1 \\ \frac{3}{2} \end{matrix} \middle| -\frac{w^2}{k^2} \right].$$

Corollary 5.1. *For all $|w| < 1$ and $r \in \mathbb{N}$, there hold*

$$\mathfrak{h}_{2r-1}(w) = 2\mathrm{i}(-1)^r \sum_{k\geqslant 1} \frac{(-1)^{k-1}}{(w^2+k^2)^{r-\frac{1}{2}}} \cos\left((2r-1)\arctan\frac{w}{k}\right), \tag{55}$$

and

$$\mathfrak{h}_{2r}(w) = 2\mathrm{i}(-1)^r \sum_{k\geqslant 1} \frac{(-1)^{k-1}}{(w^2+k^2)^{r}} \sin\left(2r\arctan\frac{w}{k}\right). \tag{56}$$

Proof. By applying formulae

$${}_2F_1 \left[\begin{matrix} r-\frac{1}{2},\, r \\ \frac{1}{2} \end{matrix} \middle| -\frac{w^2}{k^2} \right] = \frac{\cos\left((2r-1)\arctan\frac{w}{k}\right)}{\left(1+\frac{w^2}{k^2}\right)^{r-\frac{1}{2}}}$$

and

$$ {}_2F_1\left[\begin{matrix} r+\frac{1}{2},\, r+1 \\ \frac{3}{2} \end{matrix}\,\middle|\, -\frac{w^2}{k^2}\right] = \frac{k\,\sin\left(2r\,\arctan\dfrac{w}{k}\right)}{2rw\left(1+\dfrac{w^2}{k^2}\right)^r}, $$

we immediately conclude the asserted results (55) and (56), respectively. □

Setting $r=2$ in the formula (56), we achieve the companion series representations to $\mathfrak{h}_3(w)$ (51), that is,

$$ \mathfrak{h}_4(w) = 8\mathrm{i}w \sum_{k\geqslant 1} \frac{(-1)^{k-1}k(k^2-w^2)}{(w^2+k^2)^4}. $$

5.3 We introduce Dirichlet's Beta function as the series

$$ \beta(s) = \sum_{k\geqslant 0} \frac{(-1)^k}{(2k+1)^s}, \qquad \mathrm{Re}\{s\} > 0. $$

We are now interested in a specific Hilbert–Eisenstein series, precisely in $\mathfrak{h}_r(\frac{\mathrm{i}}{2})$. Since

$$ \mathfrak{h}_\nu\left(\frac{\mathrm{i}}{2}\right) = \left(\frac{2}{\mathrm{i}}\right)^\nu \left\{\sum_{k\geqslant 0} \frac{(-1)^k}{(2k+1)^\nu} - 1\right\} + \left(-\frac{2}{\mathrm{i}}\right)^\nu \sum_{k\geqslant 1} \frac{(-1)^{k-1}}{(2(k-1)+1)^\nu} $$
$$ - \left(\frac{2}{\mathrm{i}}\right)^\nu \left[(1+(-1)^\nu)\,\beta(\nu) - 1\right], $$

we have

$$ \frac{\mathfrak{h}_{\nu+\mu}\left(\frac{\mathrm{i}}{2}\right)}{\mathfrak{h}_\nu\left(\frac{\mathrm{i}}{2}\right)\mathfrak{h}_\mu\left(\frac{\mathrm{i}}{2}\right)} = \frac{(1+(-1)^{\nu+\mu})\,\beta(\nu+\mu) - 1}{\left[(1+(-1)^\nu)\,\beta(\nu) - 1\right]\left[(1+(-1)^\mu)\,\beta(\mu) - 1\right]}. $$

Now, choosing $\nu = 2r-1,\ \mu = 2s-1;\ r, s \in \mathbb{N}$, it follows

$$ \beta(2r+2s-2) = \frac{1}{2}\left\{\frac{\mathfrak{h}_{2r+2s-2}\left(\frac{\mathrm{i}}{2}\right)}{\mathfrak{h}_{2r-1}\left(\frac{\mathrm{i}}{2}\right)\mathfrak{h}_{2s-1}\left(\frac{\mathrm{i}}{2}\right)} + 1\right\}, \qquad r, s \in \mathbb{N}. $$

5.4 Finally, observe that the function

$$ \mu \mapsto \Gamma(2\mu)\mathfrak{h}_{2\mu}(\mathrm{i}w) = 2\int_0^\infty \frac{u^{2\mu-1}}{\mathrm{e}^u+1}\sinh(wu)\mathrm{d}u $$

is logarithmically convex on $(0, \infty)$ for $w \in \mathbb{R} \setminus \mathbb{Z}$. This can be verified by using the classical Hölder-Rogers inequality for integrals or by using the fact that the integrand is logarithmically convex in μ and the integral preserves the logarithmical convexity. Consequently for all $\mu_1, \mu_2 > 0$, we have

$$\Gamma^2(\mu_1 + \mu_2)\mathfrak{h}^2_{\mu_1+\mu_2}(\mathrm{i}w) \leqslant \Gamma(2\mu_1)\mathfrak{h}_{2\mu_1}(\mathrm{i}w)\Gamma(2\mu_2)\mathfrak{h}_{2\mu_2}(\mathrm{i}w)$$

and choosing $\mu_1 = m - 1$ and $\mu_2 = m + 1$, we arrive at the Turán-type inequality

$$\mathfrak{h}^2_{2m}(\mathrm{i}w) \leq \frac{2m(2m+1)}{(2m-1)(2m-2)}\mathfrak{h}_{2m-2}(\mathrm{i}w)\mathfrak{h}_{2m+2}(\mathrm{i}w),$$

which holds for all $m \in \{2, 3, \dots\}$ and $w \in \mathbb{R} \setminus \mathbb{Z}$.

References

1. Andrews, G.E., Askey, R., Roy, R.: Special Functions. Cambridge University Press, Cambridge (1999)
2. Baricz, Á., Jankov, D., Pogány, T.K.: Turán type inequalities for Krätzel functions. J. Math. Anal. Appl. **388**(2), 716–724 (2012)
3. Bertolino, M.: Numerical Analysis. Naučna knjiga, Beograd (1977, in Serbo–Croatian)
4. Butzer, P.L.: Bernoulli functions, Hilbert-type Poisson summation formulae, partial fraction expansions, and Hilbert–Eisenstein series. In: He, T.-X., Shiue, P.J.-S., Li, Z.-K. (eds.) Analysis, Combinatorics and Computing, pp. 25–91. Nova Science Publishers, New York (2002)
5. Butzer, P.L., Hauss, M.: Applications of sampling theory to combinatorial analysis, Stirling numbers, special functions and the Riemann zeta function. In: Higgins, J.R., Stens, R.L. (eds.) Sampling Theory in Fourier and Signal Analysis: Advanced Topics, pp. 1–37 and 266–268. Clarendon (Oxford University) Press, Oxford (1999)
6. Butzer, P.L., Flocke, S., Hauss, M.: Euler functions $E_\alpha(z)$ with complex α and applications. In: Anastassiou, G.A., Rachev, S.T. (eds.) Approximation, Probability and Related Fields, pp. 127–150. Plenum Press, New York (1994)
7. Butzer, P.L., Hauss, M., Leclerc, M.: Bernoulli numbers and polynomials of arbitrary complex indices. Appl. Math. Lett. **5**(6), 83–88 (1992)
8. Butzer, P.L., Pogány, T.K., Srivastava, H.M.: A linear ODE for the Omega function associated with the Euler function $E_\alpha(z)$ and the Bernoulli function $B_\alpha(z)$. Appl. Math. Lett. **19**(10), 1073–1077 (2006)
9. Čaplygin, S.A.: Osnovaniya novogo sposoba približennogo integrirovanniya differenciyal'nyh uravnenij. Moskva, 1919, pp. 348–368 Sobranie sočinenij I, Gostehizdat, Moskva (1948)
10. Čaplygin, S.A.: Približennoe integrirovanie obykvennogo differencial'nogo uravneniya pervogo poryadka, Brošura, izdannaya Komissiei osobyh artil'erijskyh opytov, Moskva, 1920, pp. 402–419 Sobranie sočinenij I, Gostehizdat, Moskva (1948)
11. Cauchy, A.L.: Note sur une transcendente que renferme de devéloppement de la fonction perturbatrice relative au système planétaire. Compt. Rendus **XIII**, 682–687 (1841). (Oeuvres **(1)** vi, 341–346, 1888)
12. Cauchy, A.L.: Note sur la substitution des anomalies excentriques aux anomalies moyennes, dans le devéloppement de la fonction perturbatrice. Compt. Rendus **XIII**, 850–854 (1841). (*Oeuvres* (1) vi, 354–359, 1888)

13. Cerone, P.: Special functions approximations and bounds via integral representations. In: Cerone, P., Dragomir, S. (eds.) Advances in Inequalities for Special Functions, pp. 1–35. Nova Science Publishers, New York (2008)
14. Cerone, P., Lenard, C.T.: On integral forms of generalized Mathieu series. JIPAM J. Inequal. Pure Appl. Math. **4**(5), Article 100 1–11 (2003, electronic)
15. Choi, Y., Srivastava, H.M.: Mathieu series and associated sums involving the Zeta function. Comp. Math. Appl. **59**, 861–867 (2010)
16. Eisenstein, G.: Genaue Untersuchung der unendlichen Doppelproducte, aus welchen die elliptischen Functionen als Quotienten zusammengesetzt sind, und der mit ihnen zusammenhängenden Doppelreihen (als eine neue Begründungsweise der Theorie der elliptischen Functionen, mit besonderer Berücksichtung ihrer Analogie zu der Kreisfunctionen). J. Reine Angew. Math. (Crelle's J.) **35**, 153–274 (1847). Also see: G. Eisenstein, *Mathematische Werke* I., Chelsea Publishing Company, New York, 1975, 357–478.
17. Euler, L.: Introductio in Analysin Infinitorum, Tomus Primus. Marc-Michael Bousquet & Socios, Lausanne (1748)
18. Freitag, E., Busam, R.: Funktionentheorie. Springer, Berlin (1993)
19. Gegenbauer, L.: Über einige bestimmte Integrale, Sitzungsberichte der Kaiserlichen Akademie der Wissenschaften. Wien **LXX**(2), 433–443 (1875)
20. Gradshteyn, I.S., Ryzhik, I.M.: Table of Integrals, Series and Products. Academic, New York (1980)
21. Guo, B.-N.: Note on Mathieu's inequality. RGMIA Res. Rep. Coll. **3**(3), Article 5 (2000). Available online at http://rgmia.vu.edu.au/v3n3.html.
22. Hansen, A.P.: Ermittelung der absoluten Störungen in Ellipsen von beliebiger Excentricität und Neigung. Part 1: Welcher als Beispiel die Berechnung, Gotha (1843)
23. Hauss, M.: Verallgemeinerte Stirling, Bernoulli und Euler Zahlen, deren Anwendungen und schnell konvergente Reihen für Zeta Funktionen, Doctoral Dissertation, RWTH-Aachen, 209 pp. Verlag Shaker, Aachen (1995)
24. Hauss, M.: An Euler–Maclaurin-type formula involving conjugate Bernoulli polynomials and an application to $\zeta(2m+1)$. Commun. Appl. Anal. **1**(1), 15–32 (1997)
25. Ifantis, E.K., Siafarikas, P.D.: Inequalities involving Bessel and modified Bessel functions. J. Math. Anal. Appl. **147**, 214–227 (1990) http://functions.wolfram.com/Hypergeometric Functions/Hypergeometric2F1/03/06/04/0006
26. Iwaniec, H., Kowalski, E.: Analytic Number Theory, American Mathematical Society Colloquium Publications, vol. 53. American Mathematical Society, Providence (2004) http://functions.wolfram.com/HypergeometricFunctions/Hypergeometric2F1/03/06/04/0008
27. Kilbas, A.A., Saxena, R.K., Trujilló, J.J.: Krätzel function as the function of hypergeometric type. Frac. Calc. Appl. Anal. **9**(2), 109–131 (2006)
28. Knopp, K.: Theory und Application of Infinite Series. Blackie & Son, London (1928). (reprinted 1948)
29. Krätzel, E.: Integral transformations of Bessel type. In: Dimovski, I. (ed.) Proceedings of International Conference on Generalized Functions and Operational Calculi, Varna, 1975, pp. 148–155. Publication House of Bulgarian Academy of Sciences, Sofia (1979)
30. Krasikov, I.: Uniform bounds for Bessel functions. J. Appl. Anal. **12**(1), 83–91 (2006)
31. Landau, L.: Monotonicity and bounds on Bessel functions. In: Warchall. H. (ed.) Proceedings of the Symposium on Mathematical Physics and Quantum Field Theory(Berkeley, California: June 11–13, 1999), 147–154 (2000); Electronic Journal Differential Equations, vol. 4. Southwest Texas State University, San Marcos (2002)
32. Lee, B.–S., Shah, S.M.: An inequality involving Bessel functions and its derivatives. J. Math. Anal. Appl. **30**, 144–156 (1970)
33. von Lommel, E.C.J.: Die Beugungserscheinungen geradlinig begrenzter Schirme. Abh. der Math. Phys. Classe der k. b. Akad. der Wiss. (München) **15**, 529–664 (1884–1886)
34. Luzin, N.N.: O metode približennogo integrirovaniya akad. S. A. Čaplygina. Uspehi Matem. Nauk **6**(46), 3–27 (1951)

35. Minakshisundaram, S., Szász, O.: On absolute convergence of multiple Fourier series. Trans. Am. Math. Soc. **61**(1), 36–53 (1947)
36. Mitrinović, D.S., Pečarić, J.: Differential and Integral Inequalities. Matematički problemi i ekspozicije, vol. 13. Naučna knjiga, Beograd (1988). (in Serbo–Croatian)
37. Olenko, A.Ya.: Upper bound on $\sqrt{x}J_\nu(x)$ and its applications. Integral Transforms Spec. Funct. **17**(6), 455–467 (2006)
38. Pogány, T.: Further results on generalized Kapteyn-type expansions. Appl. Math. Lett. **22**(2), 192–196 (2009)
39. Pogány, T.K., Srivastava, H.M.: Some two-sided bounding inequalities for the Butzer–Flocke–Hauss Omega function. Math. Inequal. Appl. **10**(3), 587–595 (2007)
40. Pogány, T.K., Srivastava, H.M., Tomovski, Ž.: Some families of Mathieu a-series and alternating Mathieu a-series. Appl. Math. Comput. **173**(1), 69–108 (2006)
41. Pogány, T.K., Tomovski, Ž., Leškovski, D.: Two-sided bounds for the complete Butzer–Flocke–Hauss Omega function. Mat. Vesnik **65**(1), 104–121 (2013)
42. Remmert, R., Schumacher, G.: Funktionentheorie I, 5. Neu Bearbeitete Auflage. Springer, Berlin (2002)
43. Sitnik, S.M.: Inequalities for Bessel functions. Dokl. Akad. Nauk **340**(1), 29–32 (1995). (in Russian)
44. Srivastava, H.M., Gupta, K.C., Goyal, S.P.: The H-Functions of One and Two Variables with Applications. South Asian Publishers, New Delhi (1982)
45. Srivastava, H.M., Pogány, T.K.: Inequalities for a unified Voigt functions in several variables. Russian J. Math. Phys. **14**(2), 194–200 (2007)
46. Tomovski, Ž., Pogány, T.K.: Integral expressions for Mathieu-type power series and for Butzer–Flocke–Hauss Ω-function. Frac. Calc. Appl. Anal. **14**(4), 623–634 (2011)
47. Watson, G.N.: A treatise on the theory of Bessel functions. 1st edn. Cambridge University Press, Cambridge (1922)
48. Weil, A.: Elliptic Functions according to Eisenstein and Kronecker. Springer, Berlin (1976)
49. Weil, A.: Number Theory: An Approach through History from Hammurapi to Legendre. Birkhäuser–Verlag, Boston (1984).

Properties of the Product of Modified Bessel Functions

Árpád Baricz and Tibor K. Pogány

Dedicated to Professor Hari M. Srivastava

Abstract Discrete Chebyshev-type inequalities are established for sequences of modified Bessel functions of the first and second kind, recognizing that the sums involved are actually Neumann series of modified Bessel functions I_ν and K_ν. Moreover, new closed integral expression formulae are established for the Neumann series of second type, which occur in the discrete Chebyshev inequalities.

1 Introduction

Modified Bessel functions of the first and second kind I_ν and K_ν are frequently used in physics, applied mathematics, and engineering sciences. Their product $I_\nu K_\nu$ is also useful in some applications. We refer, for example, to the papers [23, 24] about the hydrodynamic and hydromagnetic (in)stability of different cylindrical models, in which the monotonicity of $x \mapsto P_\nu(x) := I_\nu(x)K_\nu(x)$ for $\nu > 1$ is used. See also the paper of Hasan [12], where the electrogravitational instability of non-oscillating streaming fluid cylinder under the action of the self-gravitating, capillary, and electrodynamic forces has been discussed. In these papers the authors use (without proof) the inequality

$$P_\nu(x) < \tfrac{1}{2}$$

Á. Baricz
Department of Economics, Babeş-Bolyai University, Cluj-Napoca 400591, Romania
e-mail: bariczocsi@yahoo.com

T.K. Pogány (✉)
Faculty of Maritime Studies, University of Rijeka, Rijeka 51000, Croatia
e-mail: poganj@pfri.hr

G.V. Milovanović and M.Th. Rassias (eds.), *Analytic Number Theory, Approximation Theory, and Special Functions*, DOI 10.1007/978-1-4939-0258-3_31,

for all $\nu > 1$ and $x > 0$. We note that the above inequality readily follows from the fact that $x \mapsto P_\nu(x)$ is decreasing on $(0, \infty)$ for all $\nu > -1$. More precisely, for all $x > 0$ and $\nu > 1$ we have

$$P_\nu(x) < \lim_{x\to 0} P_\nu(x) = \tfrac{1}{2\nu} < \tfrac{1}{2}.$$

For different proofs on the monotonicity of the function $x \mapsto P_\nu(x)$, we refer to the papers [2, 19, 21]. It is worth to mention that the above monotonicity property has been used also in a problem in biophysics (see [11]). Moreover, recently Klimek and McBride [17] used this monotonicity to prove that a Dirac operator (subject to Atiyah–Patodi–Singer-like boundary conditions on the solid torus) has a bounded inverse, which is actually a compact operator. In [13, 14] van Heijster et al. investigated the existence, stability, and interaction of localized structures in a one-dimensional generalized FitzHugh–Nagumo-type model. Recently, van Heijster and Sandstede [15] started to analyze the existence and stability of radially symmetric solutions in the planar variant of this model. The product of modified Bessel functions P_ν arises naturally in their stability analysis, and the monotonicity (see [7, 15]) of $\nu \mapsto P_\nu(x)$ is important to conclude (in)stability of these radially symmetric solutions.

In this paper, motivated by the above applications, we focus on Chebyshev-type discrete inequalities for Neumann series of modified Bessel functions I_ν and K_μ of the first and the second kind, respectively. Moreover, we deduce integral representations formulae for these Neumann series appearing in newly derived discrete Chebyshev inequalities in the manner of such results given recently by Baricz, Jankov, Pogány, and Süli in a set of articles [4–6, 22] for the first-type Neumann series.

According to the established nomenclatures in the sequel, we will consider *first-type Neumann series* introduced in [4] as

$$\mathfrak{M}_\nu^\mu(z) := \sum_{n\geqslant 1} \mu_n\, I_{\nu+n}(z) \quad \text{and} \quad \mathfrak{J}_\nu^\mu(z) := \sum_{n\geqslant 1} \mu_n\, K_{\nu+n}(z)\,. \tag{1}$$

In the next section our aim is to present the Chebyshev-type discrete inequality in the terminology of Neumann series (1) and its closed form integral representation. In this goal we introduce a *second-type Neumann series*

$$\mathfrak{G}_{\nu,\eta}^\mu(z) := \sum_{n\geqslant 1} \mu_n\, I_{\nu+n}(z) K_{\eta+n}(z)\,. \tag{2}$$

Our main derivation tools include Cahen's Laplace integral form of a Dirichlet series [8, p. 97] (see the exact proof in Perron's article [20]), the condensed form of Euler–Maclaurin summation formula [22, p. 2365], and certain bounding inequalities for I_ν and K_ν; see [3].

Throughout $[a]$ and $\{a\} = a - [a]$ denote the integer and fractional part of a real number a, respectively.

2 Discrete Chebyshev Inequalities

We begin with the discrete form of the celebrated Chebyshev inequality reported (in part) by Graham [10, p. 116]. Here, and in what follows, let μ be a nonnegative discrete measure, $\mu(n) \equiv \mu_n$, $n \in \mathbb{N}$. Assuming f, g are both nonnegative and same (opposite) kind monotone, then

$$\sum_{n\geqslant 1} \mu_n f(n) \sum_{n\geqslant 1} \mu_n g(n) \leqslant (\geqslant) \|\mu\|_1 \sum_{n\geqslant 1} \mu_n f(n) g(n), \tag{3}$$

where $\|\mu\|_1$ stands for the appropriate ℓ_1-norm. Let us signify throughout

$$\|\mathbb{N}^\alpha \mu\|_1 := \sum_{n\geqslant 1} n^\alpha \mu_n, \qquad \alpha \in \mathbb{R}.$$

Now, let us recall some monotonicity properties of modified Bessel functions. Jones [16] proved that $I_{\nu_1}(x) < I_{\nu_2}(x)$ holds for all $x > 0$ and $\nu_1 > \nu_2 \geq 0$, while Cochran [9] and Reudink [25] established the inequality $\partial I_\nu(x)/\partial \nu < 0$ for all $x, \nu > 0$. With other words, the function $\nu \mapsto I_\nu(x)$ is strictly decreasing on $(0, \infty)$ for all $x > 0$ fixed.

Moreover, as it was pointed out by Laforgia [18], the function $\nu \mapsto K_\nu(x)$ is strictly increasing on $(0, \infty)$ for all $x > 0$ fixed.

Finally, recall that recently in [7, 15] it was proved the function $\nu \mapsto P_\nu(x)$ is strictly decreasing on $(0, \infty)$ for all $x > 0$ fixed.

Having in mind these properties, we can see that modified Bessel functions of the first and second kind I_ν, K_η and also their equal order product P_ν are ideal candidates to establish discrete Chebyshev inequalities of the type (3).

Our first main result is the following theorem.

Theorem 2.1. *Let $\nu, \eta > 0$ and let μ be a positive discrete measure on $\mathbb{N}$ such that $\|\mu\|_1 < \infty$, not necessarily the same in different occasions. Then the following assertions are true:*

(a) *For all fixed $x \in \mathfrak{J}_0 := \big(2e^{-1} \limsup_{n\to\infty} n\mu_n^{1/n}, \infty\big)$, we have*

$$\mathfrak{M}_\nu^\mu(x)\, \mathfrak{J}_\eta^\mu(x) \geqslant \|\mu\|_1\, \mathfrak{G}_{\nu,\eta}^{\mu}(x). \tag{4}$$

(b) *For all fixed $x \in \mathfrak{J}_1 := \big(0, 2e^{-1}/ \limsup_{n\to\infty} n^{-1}\mu_n^{1/n}\big)$, it holds*

$$\|\mu\|_1\, \mathfrak{G}_{\nu,\eta}^{\mu I_\eta}(x) \geqslant \mathfrak{M}_\nu^\mu(x)\, \mathfrak{G}_{\eta,\eta}^{\mu}(x), \tag{5}$$

whenever $\|\mathbb{N}^{(\eta-\nu-1)_+} \mu\|_1 < \infty$, where $(a)_+ = \max\{0, a\}$.

(c) *Moreover, for all fixed* $x \in \mathfrak{I}_0$ *and* $\|\mathbb{N}^{(\eta-\nu-1)_+}\,\mu\|_1 < \infty$, *we have*

$$\mathfrak{J}_\nu^\mu(x)\,\mathfrak{G}_{\nu,\eta}^{\,\mu}(x) \geqslant \|\mu\|_1\,\mathfrak{G}_{\nu,\eta}^{\mu K_\nu}(x). \tag{6}$$

Proof. We apply the Chebyshev inequality (3) by choosing **(a)** $f \equiv I_\nu, g \equiv K_\eta$, **(b)** $f \equiv I_\nu, g \equiv I_\eta K_\eta$, and **(c)** $f \equiv K_\nu, g \equiv I_\nu K_\eta$. In the cases **(a)** and **(c)**, the functions f and g are opposite kind monotone, and thus we immediately conclude (4) and (6), respectively. Moreover, in the case **(b)** both f and g decrease, which imply the derived inequality (5).

It remains only to find the x-domains of the inequalities.

Observe that $\|\mu\|_1 < \infty$ suffices for the absolute and uniform convergence of the Neumann series $\mathfrak{M}_\nu^\mu(x)$. This has been established by Baricz et al. in the proof of [4, Theorem 2.1] for all $x > 0$ and $\nu > -1$. Moreover, in the same paper [4] the authors proved that $\mathfrak{I}_\eta^\mu(x)$ converges absolutely and uniformly when $\eta > 0$ and $x \in \mathfrak{I}_0$. Now, by using the inequalities [3, p. 583]

$$I_\nu(x) < \frac{\left(\frac{x}{2}\right)^\nu}{\Gamma(\nu+1)}\,e^{\frac{x^2}{4(\nu+1)}}, \qquad \nu > -1,\ x > 0,$$

and [4]

$$K_\eta(x) \leqslant \frac{2^{\eta-1}}{x^\eta}\,\Gamma(\eta), \qquad \eta > 0,\ x > 0\,,$$

applied to the summands of $\mathfrak{G}_{\nu,\eta}^{\,\mu}(x)$, we obtain

$$\left|\mathfrak{G}_{\nu,\eta}^{\,\mu}(x)\right| \leqslant \frac{1}{2}\left(\frac{x}{2}\right)^{\eta-\nu} e^{\frac{x^2}{4(\nu+2)}} \sum_{n\geqslant 1} n^{\eta-\nu-1}\,\mu_n\,. \tag{7}$$

Observe that the convergence of the right-hand-side series, that is, $\|\mathbb{N}^{\eta-\nu-1}\,\mu\|_1 < \infty$, ensures the convergence of the second kind Neumann series $\mathfrak{G}_{\nu,\eta}^{\,\mu}(x)$ for all $\nu, \eta, x > 0$. This together with the additional requirement $\|\mu\|_1 < \infty$ yields

$$\max\left\{\|\mu\|_1, \|\mathbb{N}^{\eta-\nu-1}\,\mu\|_1\right\} = \|\mathbb{N}^{(\eta-\nu-1)_+}\,\mu\|_1 < \infty.$$

Finally, let us consider the series $\mathfrak{G}_{\nu,\eta}^{\mu I_\eta}(x)$ which ensures the convergence of both left-hand-side Neumann series in (5). By virtue of the above listed upper bounds for I_ν, I_η and K_η, we conclude that

$$\begin{aligned}\mathfrak{G}_{\nu,\eta}^{\mu I_\eta}(x) &= \sum_{n\geqslant 1} \mu_n I_{\eta+n}(x) I_{\nu+n}(x) K_{\eta+n}(x)\\ &\leqslant \frac{1}{2\sqrt{2\pi}}\left(\frac{x}{2}\right)^\nu e^{\frac{x^2}{4}\left(\frac{1}{\nu+2}+\frac{1}{\eta+2}\right)} \sum_{n\geqslant 1} \frac{\mu_n}{n^{n+\nu+3/2}}\left(\frac{xe}{2}\right)^n,\end{aligned} \tag{8}$$

where the bounding power series converges for all $x \in \mathfrak{I}_1$.

Combining all these estimates, we arrive at the asserted inequality domains. □

3 Integral Form of Related Second-Type Neumann Series

Our next goal is to prove integral representations for the second-type Neumann series

$$\mathfrak{G}_{\nu,\eta}^{\mu}(x), \quad \mathfrak{G}_{\nu,\eta}^{\mu I_\eta}(x) \quad \text{and} \quad \mathfrak{G}_{\nu,\eta}^{\mu K_\nu}(x),$$

which appeared in Theorem 3.1. This will be realized on the account of procedure introduced by Pogány and Süli in [22] and further developed and promoted by Baricz et al. [4–6].

Theorem 3.1. *Let* $\mu \in \mathrm{C}^1(\mathbb{R}_+)$, $\mu|_{\mathbb{N}} = (\mu_n)_{n\in\mathbb{N}}$ *such that* $\limsup_{n\to\infty} |\mu_n|^{1/n} \leqslant 1$. *Then, for all* $x > 0$ *and* $\nu, \eta > -3/2$, *we have the integral representation*

$$\mathfrak{G}_{\nu,\eta}^{\mu}(x) = -\frac{x^{\nu-\eta}}{4}\int_1^{\infty}\int_0^{[t]} \frac{\partial}{\partial t}\left(\frac{\Gamma\left(t+\nu+\frac{1}{2}\right)}{\Gamma\left(t+\eta+\frac{1}{2}\right)}\, I_{t+\nu}(x)\, K_{t+\eta}(x)\right)$$
$$\times\, \mathfrak{d}_s\left(\frac{\mu(s)\Gamma\left(s+\eta+\frac{1}{2}\right)}{\Gamma\left(s+\nu+\frac{1}{2}\right)}\right) \mathrm{d}t\, \mathrm{d}s\,, \tag{9}$$

where

$$\mathfrak{d}_x := 1 + \{x\}\frac{\mathrm{d}}{\mathrm{d}x}.$$

Proof. First, we recall the following integral representation [26, p. 79]:

$$I_\nu(x) = \frac{2^{1-\nu}x^\nu}{\sqrt{\pi}\,\Gamma(\nu+1/2)}\int_0^1 (1-t^2)^{\nu-\frac{1}{2}}\cosh(xt)\mathrm{d}t, \quad x>0,\ \nu>-1/2, \tag{10}$$

and the integral representation formula referred to Basset [26, p. 173]

$$K_\eta(x) = \frac{2^\eta\Gamma\left(\eta+\frac{1}{2}\right)}{\sqrt{\pi}x^\eta}\int_0^\infty \frac{\cos(xt)}{(1+t^2)^{\eta+\frac{1}{2}}}\,\mathrm{d}t, \qquad x>0, \eta>-1/2\,. \tag{11}$$

Applying (10) and (11) to $\mathfrak{G}_{\nu,\eta}^{\mu}(x)$, we conclude

$$\mathfrak{G}_{\nu,\eta}^{\mu}(x) = \frac{x^{\nu-\eta}}{2\pi}\int_0^1\int_0^\infty \frac{(1-t^2)^{\nu-\frac{1}{2}}\cosh(xt)\cos(xs)}{(1+s^2)^{\eta+\frac{1}{2}}}$$
$$\times \sum_{n\geqslant 1}\frac{\mu_n\Gamma\left(n+\eta+\frac{1}{2}\right)}{\Gamma\left(n+\nu+\frac{1}{2}\right)}\left(\frac{1-t^2}{1+s^2}\right)^n \mathrm{d}t\mathrm{d}s\,. \tag{12}$$

The inner sum we recognize as the Dirichlet series

$$\mathfrak{D}_0(t,s) = \sum_{n\geqslant 1} \frac{\mu_n \Gamma\left(n+\eta+\frac{1}{2}\right)}{\Gamma\left(n+\nu+\frac{1}{2}\right)} \exp\left(-n \ln \frac{1+s^2}{1-t^2}\right), \tag{13}$$

which parameter $\ln(1+s^2)(1-t^2)^{-1}$ is obviously positive on $(t,s) \in (0,1)\times\mathbb{R}_+$ independently of x. Also, the power series (13) has the radius of convergence

$$\rho_{\mathfrak{D}_0} = \frac{1}{\limsup\limits_{n\to\infty} |\mu_n|^{1/n}},$$

and then $\mathfrak{D}_0(t,s)$ is convergent for all $(t,s) \in (0,1)\times\mathbb{R}_+$, being $\rho_{\mathfrak{D}_0} \geqslant 1$ according to the assumption of the theorem.

Thus, by Cahen's Laplace integral formula for the Dirichlet series [8, p. 97] and by the condensed Euler–Maclaurin summation formula [22, p. 2365], we get

$$\mathfrak{D}_0(t,s) = \ln\frac{1+s^2}{1-t^2} \int_0^\infty \int_0^{[w]} \left(\frac{1-t^2}{1+s^2}\right)^w \mathfrak{d}_z \left(\frac{\mu(z)\Gamma\left(z+\eta+\frac{1}{2}\right)}{\Gamma\left(z+\nu+\frac{1}{2}\right)}\right) \mathrm{d}w\mathrm{d}z. \tag{14}$$

Substituting (14) into (12) we get

$$\mathfrak{G}_{\nu,\eta}^{\mu}(x) = -\frac{x^{\nu-\eta}}{2\pi} \int_0^\infty \int_0^{[w]} \mathfrak{d}_z \left(\frac{\mu(z)\Gamma\left(z+\eta+\frac{1}{2}\right)}{\Gamma\left(z+\nu+\frac{1}{2}\right)}\right)$$

$$\times \left(\int_0^1 \int_0^\infty \left(\frac{1-t^2}{1+s^2}\right)^{w+\nu-\frac{1}{2}} \ln\frac{1-t^2}{1+s^2} \cdot \frac{\cosh(xt)\cos(xs)}{(1+s^2)^{\eta-\nu+1}} \mathrm{d}t\mathrm{d}s\right) \mathrm{d}w\mathrm{d}z.$$

Denote

$$\mathfrak{I}(\alpha) := \int_0^1 \int_0^\infty \left(\frac{1-t^2}{1+s^2}\right)^\alpha \ln\frac{1-t^2}{1+s^2} \cdot \frac{\cosh(xt)\cos(xs)}{(1+s^2)^{\eta-\nu+1}} \mathrm{d}t\mathrm{d}s.$$

Now, having in mind (10) and (11), we deduce

$$\begin{aligned}
\int \mathfrak{I}(\alpha)\,\mathrm{d}\alpha &= \int_0^1 \int_0^\infty \left(\frac{1-t^2}{1+s^2}\right)^\alpha \frac{\cosh(xt)\cos(xs)}{(1+s^2)^{\eta-\nu+1}} \mathrm{d}t\mathrm{d}s \\
&= \int_0^1 \int_0^\infty \frac{(1-t^2)^\alpha \cosh(xt)\cos(xs)}{(1+s^2)^{\alpha+\eta-\nu+1}} \mathrm{d}t\mathrm{d}s \\
&= \frac{\pi}{2} \frac{\Gamma(\alpha+1)}{\Gamma(\alpha+\eta-\nu+1)} I_{\alpha+\frac{1}{2}}(x) K_{\alpha+\eta-\nu+\frac{1}{2}}(x),
\end{aligned}$$

that is, choosing $\alpha \mapsto w + \nu - \frac{1}{2}$ we have

$$\mathfrak{I}\left(w + \nu - \frac{1}{2}\right) = \frac{\pi}{2} \frac{\partial}{\partial w} \frac{\Gamma\left(w + \nu + \frac{1}{2}\right)}{\Gamma\left(w + \eta + \frac{1}{2}\right)} I_{w+\nu}(x) K_{w+\eta}(x)\,.$$

Hence

$$\mathfrak{G}_{\nu,\eta}^{\mu}(x) = -\frac{x^{\nu-\eta}}{4} \int_0^\infty \int_0^{[t]} \frac{\partial}{\partial t} \left(\frac{\Gamma\left(t + \nu + \frac{1}{2}\right)}{\Gamma\left(t + \eta + \frac{1}{2}\right)} I_{t+\nu}(x)\, K_{t+\eta}(x) \right)$$

$$\times\, \mathfrak{d}_s \left(\frac{\mu(s) \Gamma\left(s + \eta + \frac{1}{2}\right)}{\Gamma\left(s + \nu + \frac{1}{2}\right)} \right) \mathrm{d}t\, \mathrm{d}s,$$

which is equivalent to the asserted double integral expression (9). □

Theorem 3.2. *Let $\mu \in \mathrm{C}^1(\mathbb{R}_+)$, $\mu|_{\mathbb{N}} = (\mu_n)_{n\in\mathbb{N}}$. Then, for all $x \in \mathfrak{I}_1$, $\nu, \eta > -3/2$, there holds*

$$\mathfrak{G}_{\nu,\eta}^{\mu I_\eta}(x) = -\frac{x^{\nu-\eta}}{4} \int_1^\infty \int_0^{[t]} \frac{\partial}{\partial t} \left(\frac{\Gamma\left(t + \nu + \frac{1}{2}\right)}{\Gamma\left(t + \eta + \frac{1}{2}\right)} I_{t+\nu}(x)\, K_{t+\eta}(x) \right)$$

$$\times\, \mathfrak{d}_s \left(\frac{\mu(s) I_{s+\eta}(x) \Gamma\left(s + \eta + \frac{1}{2}\right)}{\Gamma\left(s + \nu + \frac{1}{2}\right)} \right) \mathrm{d}t\, \mathrm{d}s\,. \tag{15}$$

Moreover, for $x \in \mathfrak{I}_0$, $\nu > -1, \eta > -3/2$, we have

$$\mathfrak{G}_{\nu,\eta}^{\mu K_\nu}(x) = -\frac{x^{\nu-\eta}}{4} \int_1^\infty \int_0^{[t]} \frac{\partial}{\partial t} \left(\frac{\Gamma\left(t + \nu + \frac{1}{2}\right)}{\Gamma\left(t + \eta + \frac{1}{2}\right)} I_{t+\nu}(x)\, K_{t+\eta}(x) \right)$$

$$\times\, \mathfrak{d}_s \left(\frac{\mu(s) K_{s+\nu}(x) \Gamma\left(s + \eta + \frac{1}{2}\right)}{\Gamma\left(s + \nu + \frac{1}{2}\right)} \right) \mathrm{d}t\, \mathrm{d}s\,. \tag{16}$$

Proof. We follow the proof of (9) to get the integral representations. It remains only to remark that the Dirichlet series $\mathfrak{D}_1(t,s)$ associated with $\mathfrak{G}_{\nu,\eta}^{\mu I_\eta}(x)$ satisfies

$$|\mathfrak{D}_1(t,s)| \leqslant \sum_{n\geqslant 1} \frac{|\mu_n||I_{n+\eta}(x)|\Gamma\left(n + \eta + \frac{1}{2}\right)}{\Gamma\left(n + \nu + \frac{1}{2}\right)} \left(\frac{1-t^2}{1+s^2}\right)^n$$

$$\leqslant \frac{1}{\sqrt{2\pi}} \left(\frac{|x|}{2}\right)^\eta e^{\frac{x^2}{2(\eta+2)}} \sum_{n\geqslant 1} \frac{\mu_n}{n^{n+\nu+1/2}} \left(\frac{|x|e}{2}\, \frac{1-t^2}{1+s^2}\right)^n,$$

so x has to be from $\mathfrak{I}_1$. Similarly it can be concluded that for the Dirichlet series $\mathfrak{D}_2(t,s)$ associated with $\mathfrak{G}_{\nu,\eta}^{\mu I_\eta}(x)$ holds the estimate

$$|\mathfrak{D}_2(t,s)| \leqslant \sum_{n\geqslant 1} \frac{|\mu_n||K_{n+\nu}(x)|\Gamma(n+\nu)\Gamma\left(n+\eta+\frac{1}{2}\right)}{\Gamma\left(n+\nu+\frac{1}{2}\right)} \left(\frac{1-t^2}{1+s^2}\right)^n$$

$$\leqslant \sqrt{\frac{\pi}{2}} \left(\frac{2}{|x|}\right)^\nu \sum_{n\geqslant 1} n^{n+\eta-1/2} \mu_n \left(\frac{2}{|x|e} \frac{1-t^2}{1+s^2}\right)^n,$$

of which convergence requirement causes $x \in \mathfrak{I}_0$. □

4 Indefinite Integral Expressions for Second Kind Neumann Series $\mathfrak{G}_{\nu,\nu}^{\mu}(x)$

In this section our aim is to establish indefinite integral representation formulae for the one-parameter second kind Neumann series of the product of two modified Bessel functions of the first kind P_ν. First of all, observe that P_ν is a particular solution of the homogeneous third-order linear differential equation

$$x^2 y'''(x) + 3xy''(x) - (4\nu^2 + 4x^2 - 1)y'(x) - 4xy(x) = 0. \tag{17}$$

To see this, let us recall that I_ν and K_ν both satisfy the differential equation

$$x^2 y''(x) + xy'(x) - (x^2 + \nu^2)y(x) = 0$$

and consequently

$$x^2 I_\nu''(x) = (x^2 + \nu^2)I_\nu(x) - xI_\nu'(x) \tag{18}$$

and

$$x^2 K_\nu''(x) = (x^2 + \nu^2)K_\nu(x) - xK_\nu'(x). \tag{19}$$

Applying these relations we obtain

$$x^2 P_\nu''(x) = 2(x^2 + \nu^2)P_\nu(x) - xP_\nu'(x) + 2x^2 I_\nu'(x)K_\nu'(x).$$

Now, differentiating both sides of this equation and applying again the above relations, we arrive at

$$x^2 P_\nu'''(x) + 3xP_\nu''(x) - (4\nu^2 + 4x^2 - 1)P_\nu'(x) - 4xP_\nu(x) = 0.$$

Repeating this procedure twice in view of (18) and (19), we can show[1] that actually I_ν^2 and K_ν^2 are also particular solutions of the third-order linear differential equation (17).

Now, let us show that I_ν^2, $I_\nu K_\nu$, and K_ν^2 are independent being the Wronskian $W[I_\nu^2, I_\nu K_\nu, K_\nu^2] \neq 0$ on $\mathbb{R}$. After some computations we get

$$W[I_\nu^2, I_\nu K_\nu, K_\nu^2](x) = \begin{vmatrix} I_\nu^2(x) & I_\nu(x)K_\nu(x) & K_\nu^2(x) \\ \left(I_\nu^2(x)\right)' & (I_\nu(x)K_\nu(x))' & \left(K_\nu^2(x)\right)' \\ \left(I_\nu^2(x)\right)'' & (I_\nu(x)K_\nu(x))'' & \left(K_\nu^2(x)\right)'' \end{vmatrix}$$

$$= -\frac{1}{4}\big(I_\nu(x)K_{\nu-1}(x) + I_{\nu-1}(x)K_\nu(x) + I_{\nu+1}(x)K_\nu(x) + I_\nu(x)K_{\nu+1}(x)\big)^3$$

$$= 2\big(I_\nu(x)K_\nu'(x) - I_\nu'(x)K_\nu(x)\big)^3 = 2W^3[I_\nu, K_\nu](x) = -\frac{2}{x^3} \neq 0,$$

where we used the fact that $W[I_\nu, K_\nu](x) = -1/x$.

Thus, by the variation of constants method, we get the desired particular solution of the nonhomogeneous variant of (17), that is,

$$x^2y'''(x) + 3xy''(x) - (4\nu^2 + 4x^2 - 1)y'(x) - 4xy(x) = f(x)\,, \tag{20}$$

where f is a suitable real function. Hence, bearing in mind (18), the general solution reads as follows

$$y(x) = c_1 I_\nu^2(x) + c_2 I_\nu(x)K_\nu(x) + c_3 K_\nu^2(x)$$
$$- 4\int_1^x tf(t)\big(I_\nu(x)K_\nu(t) - I_\nu(t)K_\nu(x)\big)^2\,\mathrm{d}t\,.$$

[1] It is worth to mention here that the above procedure for modified Bessel functions is similar of the method for Bessel functions applied by Wilkins [27]. See also Andrews et al. [1] for more details. More precisely, Wilkins proved that the Hankel functions $\big(H_\nu^{(1)}\big)^2$ and $\big(H_\nu^{(2)}\big)^2$, as well as $J_\nu^2 + Y_\nu^2$, where J_ν and Y_ν stand for the Bessel functions of the first and second kind, are particular solutions of the third-order homogeneous differential equation [1, p. 225]

$$x^2y'''(x) + 3xy''(x) + (1 + 4x^2 - 4\nu^2)y'(x) + 4xy(x) = 0.$$

The above result was used to prove the celebrated Nicholson formula [1, p. 224]

$$J_\nu^2(x) + Y_\nu^2(x) = \frac{8}{\pi^2}\int_0^\infty K_0(2x\sinh t)\cosh(2\nu t)dt,$$

which generalizes the trigonometric identity $\sin^2 x + \cos^2 x = 1$.

Choosing the constants c_1, c_2, and c_3 to be zero, the particular solution y_p of the nonhomogeneous ODE (20) becomes

$$y_p(x) = -4 \int_1^x t f(t)\big(I_\nu(x)K_\nu(t) - I_\nu(t)K_\nu(x)\big)^2 \,\mathrm{d}t\,. \tag{21}$$

Now, by using (17) we have

$$x^2\, P'''_{n+\nu}(x) + 3x\, P''_{n+\nu}(x) - (4(n+\nu)^2 + 4x^2 - 1)\, P'_{n+\nu}(x) - 4x\, P_{n+\nu}(x) = 0$$

and multiplying with the weight μ_n and summing up on the set of positive integers $\mathbb{N}$, transformations lead to the nonhomogeneous third-order linear differential equation

$$\begin{aligned} x^2 \left(\mathfrak{G}_{\nu,\nu}^{\mu}(x)\right)''' + 3x \left(\mathfrak{G}_{\nu,\nu}^{\mu}(x)\right)'' - (4\nu^2 + 4x^2 - 1) \left(\mathfrak{G}_{\nu,\nu}^{\mu}(x)\right)' - 4x\, \mathfrak{G}_{\nu,\nu}^{\mu}(x) \\ = 4 \sum_{n\geqslant 1} n(n+2\nu)\mu_n I_{n+\nu}(x) K_{n+\nu}(x) := \mathfrak{H}_{\nu,\nu}^{\mu}(x)\,, \end{aligned} \tag{22}$$

where $\mathfrak{H}_{\nu,\nu}^{\mu}(x)$ stands for the *second kind equal parameter Neumann series of modified Bessel functions associated with the Neumann series* $\mathfrak{G}_{\nu,\nu}^{\mu}(x)$.

Theorem 4.1. *Let* $\mu \in \mathrm{C}^1(\mathbb{R}_+)$, $\mu|_{\mathbb{N}} = (\mu_n)_{n\in\mathbb{N}}$ *such that* $\limsup_{n\to\infty} |\mu_n|^{1/n} \leqslant 1$ *and* $\|\mathbb{N}^{-1}\,\mu\|_1 < \infty$. *Then for all* $\nu > -3/2$ *and* $x > 0$, *we have*

$$\mathfrak{G}_{\nu,\nu}^{\mu}(x) = -4 \int_1^x u \mathfrak{H}_{\nu,\nu}^{\mu}(u)\big(I_\nu(x)K_\nu(u) - I_\nu(u)K_\nu(x)\big)^2 \,\mathrm{d}u\,, \tag{23}$$

where $\mathfrak{H}_{\nu}^{\mu}(x)$ *possesses the integral representation*

$$\mathfrak{H}_{\nu,\nu}^{\mu}(x) = -\int_1^\infty \int_0^{[t]} \frac{\partial}{\partial t}\big(I_{t+\nu}(x)\, K_{t+\nu}(x)\big)\, \mathfrak{d}_s\big(s(s+2\nu)\mu(s)\big)\mathrm{d}t\, \mathrm{d}s\,. \tag{24}$$

Proof. The integral representation (24) of the associated second kind Neumann series of Bessel function $\mathfrak{H}_{\nu,\nu}^{\mu}(x)$ can be obtained by using the integral expression (9) in Theorem 3.1, just putting $\eta \equiv \nu$ for the weight function $\mu_n \mapsto 4n(n+2\nu)\mu_n$, when

$$\limsup_{n\to\infty} |4n(n+2\nu)\mu_n|^{1/n} = \limsup_{n\to\infty} |\mu_n|^{1/n} \leqslant 1\,.$$

After that by straightforward application of (21) with $f(x) \equiv \mathfrak{H}_{\nu,\nu}^{\mu}(x)$, we deduce the desired integral expression (23). □

Acknowledgements The authors are grateful to Christoph Koutschan who provided expert help in deriving the differential equation (17).

The research of Á. Baricz was supported by a grant of the Romanian National Authority for Scientific Research, CNCS-UEFISCDI, project number PN-II-RU-TE-2012-3-0190.

References

1. Andrews, G.E., Askey, R., Roy, R.: Special functions. In: Encyclopedia of Mathematics and its Applications, vol. 71. Cambridge University Press, Cambridge (1999)
2. Baricz, Á.: On a product of modified Bessel functions. Proc. Amer. Math. Soc. **137**(1), 189–193 (2009)
3. Baricz, Á.: Bounds for modified Bessel functions of the first and second kinds. Proc. Edinb. Math. Soc. **53**(3), 575–599 (2010)
4. Baricz, Á., Jankov, D., Pogány, T.K.: Integral representations for Neumann-type series of Bessel functions I_ν, Y_ν and K_ν. Proc. Amer. Math. Soc. **140**(2), 951–960 (2012)
5. Baricz, Á., Jankov, D., Pogány, T.K.: Neumann series of Bessel functions. Integral Transforms Spec. Funct. **23**(7), 529–538 (2012)
6. Baricz, Á., Pogány, T.K.: Turán determinants of Bessel functions. Forum Math. (2011 in press)
7. Baricz, Á., Ponnusamy, S.: On Turán type inequalities for modified Bessel functions. Proc. Amer. Math. Soc. **141**(2), 523–532 (2013)
8. Cahen, E.: Sur la fonction $\zeta(s)$ de Riemann et sur des fontions analogues. Ann. Sci. l'École Norm. Sup. Sér. **11**, 75–164 (1894).
9. Cochran, J.A.: The monotonicity of modified Bessel functions with respect to their order. J. Math. Phys. **46**, 220–222 (1967)
10. Graham, R.L.: Application of the FKG Inequality and its Relatives, Mathematical Programming: The State of the Art (Bonn, 1982), pp. 115–131. Springer, Berlin (1983)
11. Grandison, S., Penfold, R., Vanden-Broeck, J.M.: A rapid boundary integral equation technique for protein electrostatics. J. Comput. Phys. **224**, 663–680 (2007)
12. Hasan, A.A.: Electrogravitational stability of oscillating streaming fluid cylinder. Phys. B. **406**, 234–240 (2011)
13. van Heijster, P., Doelman, A., Kaper, T.J.: Pulse dynamics in a three-component system: stability and bifurcations. Phys. D. Nonlinear Phenomena **237**(24), 3335–3368 (2008)
14. van Heijster, P., Doelman, A., Kaper, T.J., Promislow, K.: Front interactions in a three-component system. SIAM J. Appl. Dyn. Syst. **9**, 292–332 (2010)
15. van Heijster, P., Sandstede, B.: Planar radial spots in a three-component FitzHugh-Nagumo system. J. Nonlinear Sci. **21**, 705–745 (2011)
16. Jones, A.L.: An extension of an inequality involving modified Bessel functions. J. Math. Phys. **47**, 220–221 (1968)
17. Klimek, S., McBride, M.: Global boundary conditions for a Dirac operator on the solid torus. J. Math. Phys. **52**, Article 063518, 14 pp (2011)
18. Laforgia, A.: Bounds for modified Bessel functions. J. Computat. Appl. Math. **34**(4), 263–267 (1991)
19. Penfold, R., Vanden-Broeck, J.M., Grandison, S.: Monotonicity of some modified Bessel function products. Integral Transforms Spec. Funct. **18**, 139–144 (2007)
20. Perron, O.: Zur Theorie der Dirichletschen Reihen. J. Reine Angew. Math. **134**, 95–143 (1908)
21. Phillips, R.S., Malin, H.: Bessel function approximations. Amer. J. Math. **72**, 407–418 (1950)
22. Pogány, T.K., Süli, E.: Integral representation for Neumann series of Bessel functions. Proc. Amer. Math. Soc. **137**(7), 2363–2368 (2009)
23. Radwan, A.E., Dimian, M.F., Hadhoda, M.K.: Magnetogravitational stability of a bounded gas-core fluid jet. Appl. Energy **83**, 1265–1273 (2006)

24. Radwan, A.E., Hasan, A.A.: Magneto hydrodynamic stability of self-gravitational fluid cylinder. Appl. Math. Modell. **33**, 2121–2131 (2009)
25. Reudink, D.O.: On the signs of the ν-derivatives of the modified Bessel functions $I_\nu(x)$ and $K_\nu(x)$. J. Res. Nat. Bur. Standards **B72**, 279–280 (1968)
26. Watson, G.N.: A Treatise on the Theory of Bessel Functions. Cambridge University Press, Cambridge (1922)
27. Wilkins, J.E.: Nicholson's integral for $J_n^2(z) + Y_n^2(z)$. Bull. Amer. Math. Soc. **54**, 232–234 (1948)

Mapping Properties of an Integral Operator Involving Bessel Functions

Saurabh Porwal and Daniel Breaz

Dedicated to Professor Hari M. Srivastava

Abstract The purpose of the present paper is to study the mapping properties of an integral operator involving Bessel functions of the first kind on a subclass of analytic univalent functions.

1 Introduction

Let A denote the class of functions f of the form

$$f(z) - z + \sum_{k=2}^{\infty} a_k z^k, \tag{1}$$

which are analytic in the open unit disk $U = \{z \in \mathbb{C}\colon |z| < 1\}$ and satisfy the normalization condition $f(0) = f'(0) - 1 = 0$. Further, we denote by S the subclass of A consisting of functions of the form (1) which are also univalent in U.

For $1 < \beta \leqslant 3/2$ and $z \in U$, let

$$N(\beta) = \left\{ f \in A : \operatorname{Re}\left(1 + \frac{zf''(z)}{f'(z)}\right) < \beta \right\}.$$

S. Porwal (✉)
Department of Mathematics, U.I.E.T. Campus, C.S.J.M. University, Kanpur 208024, UP, India
e-mail: saurabhjcb@rediffmail.com

D. Breaz
Department of Mathematics, "1 Decembrie 1918" University of Alba Iulia, str. N. Iorga, No. 11-13, Alba 510009, Romania
e-mail: dbreaz@uab.ro

G.V. Milovanović and M.Th. Rassias (eds.), *Analytic Number Theory, Approximation Theory, and Special Functions*, DOI 10.1007/978-1-4939-0258-3_32,

This class was extensively studied by Uralegaddi et al. [9] (see also [4, 6, 7]).

Recently, Breaz [2] studied the mapping properties of an integral operator on the class $N(\beta)$. This result was generalized by Porwal [5].

Several authors such as ([1, 3]) studied the various integral operators involving Bessel functions. Motivating with their works we have attempted to study the mapping properties of an integral operator involving Bessel functions of the first kind.

The Bessel function of the first kind of order ν is defined by the infinite series

$$J_\nu(z) = \sum_{n=0}^{\infty} \frac{(-1)^n \left(\frac{z}{2}\right)^{2n+\nu}}{n!\Gamma(n+\nu+1)},$$

where Γ stands for the Euler gamma function, $z \in \mathbb{C}$ and $\nu \in \mathbb{R}$. Recently, Szasz and Kupan [8] investigated the univalence of the normalized Bessel function of the first kind $g_\nu : U \to \mathbb{C}$, defined by

$$g_\nu(z) = 2^\nu \Gamma(\nu+1) z^{1-\nu/2} J_\nu(z^{1/2}) = z + \sum_{n=1}^{\infty} \frac{(-1)^n z^{n+1}}{4^n n!(\nu+1)(\nu+2)\ldots(\nu+n)}. \quad (2)$$

Baricz and Frasin [1] have obtained the sufficient conditions for the univalence of the various integral operators involving Bessel functions of the first kind.

In the present paper, we are mainly interested in the integral operator of the following type which involves the normalized Bessel function of the first kind:

$$F_{\nu_1,\ldots,\nu_n,\alpha_1,\ldots,\alpha_n}(z) = \int_0^z \prod_{i=1}^{n} \left(\frac{g_{\nu_i}(t)}{t}\right)^{\alpha_i} dt. \quad (3)$$

More precisely, we would like to obtain sufficient conditions for $F_{\nu_1,\ldots,\nu_n,\alpha_1,\ldots,\alpha_n}(z)$ to be in class $N(\beta)$. In particular, we obtain simple sufficient conditions for some integral operator which involve the sine and cosine functions.

To prove our main results we shall require the following lemma due to Szász and Kupan [8].

Lemma 1.1. *Let $\nu > (-5+\sqrt{5})/4$ and consider the normalized Bessel function of the first kind $g_\nu : U \to \mathbb{C}$, defined by $g_\nu(z) = 2^\nu \Gamma(\nu+1) z^{1-\nu/2} J_\nu(z^{1/2})$, where J_ν stands for the Bessel function of the first kind. Then the following inequality holds for all $z \in U$:*

$$\left| \frac{z g_\nu'(z)}{g_\nu(z)} - 1 \right| \leqslant \frac{\nu+2}{4\nu^2+10\nu+5}. \quad (4)$$

2 Main Results

We study the mapping properties for the integral operator defined by (3).

Theorem 2.1. *Let n be a positive integer number and let $\nu_1, \nu_2, \ldots, \nu_n > \left(\frac{-5+\sqrt{5}}{4}\right)$. Consider the functions $g_{\nu_i} : U \to \mathbb{C}$, defined by*

$$g_{\nu_i}(z) = 2^{\nu_i}\Gamma(\nu_i + 1)z^{1-\nu_i/2}J_{\nu_i}(z^{1/2}). \tag{5}$$

Let $\nu = \min\{\nu_1, \nu_2, \ldots, \nu_n\}$ and $\alpha_1, \alpha_2, \ldots, \alpha_n$ be positive real numbers. More over suppose that these numbers satisfy the following inequality:

$$1 < 1 + \frac{2+\nu}{4\nu^2 + 10\nu + 5}\sum_{i=1}^{n}\alpha_i \leq \frac{3}{2}.$$

Then the function $F_{\nu_1,\ldots,\nu_n,\alpha_1,\ldots,\alpha_n}(z) : U \to \mathbb{C}$ defined by (3) is in $N(\mu)$, where

$$\mu = 1 + \frac{2+\nu}{4\nu^2 + 10\nu + 5}\sum_{i=1}^{n}\alpha_i.$$

Proof. First, we observe that, since for all $i \in \{1, 2, \ldots, n\}$, we have $g_{\nu_i} \in A$, i.e., $g_{\nu_i}(0) = g'_{\nu_i}(0) - 1 = 0$, clearly $F_{\nu_1,\ldots,\nu_n,\alpha_1,\ldots,\alpha_n} \in A$, i.e.,

$$\mathrm{F}_{\nu_1,\ldots,\nu_n,\alpha_1,\ldots,\alpha_n}(0) = \mathrm{F}'_{\nu_1}, \ldots, \nu_n, \alpha_1, \ldots, \alpha_n(0) - 1 = 0.$$

On the other hand, it is easy to see that

$$F'_{\nu_1,\ldots,\nu_n,\alpha_1,\ldots,\alpha_n}(z) = \prod_{i=1}^{n}\left(\frac{g_{\nu_i}(z)}{z}\right)^{\alpha_i} \text{ and } z\frac{F''_{\nu_1,\ldots,\nu_n,\alpha_1,\ldots,\alpha_n}(z)}{F'_{\nu_1,\ldots,\nu_n,\alpha_1,\ldots,\alpha_n}(z)} = \sum_{i=1}^{n}\alpha_i\left(\frac{zg'_{\nu_i}(z)}{g_{\nu_i}(z)} - 1\right)$$

or, equivalently,

$$1 + z\frac{F''_{\nu_1,\ldots,\nu_n,\alpha_1,\ldots,\alpha_n}(z)}{F'_{\nu_1,\ldots,\nu_n,\alpha_1,\ldots,\alpha_n}(z)} = \sum_{i=1}^{n}\alpha_i\left(\frac{zg'_{\nu_i}(z)}{g_{\nu_i}(z)}\right) + 1 - \sum_{i=1}^{n}\alpha_i. \tag{6}$$

Taking the real part of both sides of (6), we have

$$\mathrm{Re}\left\{1 + z\frac{F''_{\nu_1,\ldots,\nu_n,\alpha_1,\ldots,\alpha_n}(z)}{F'_{\nu_1,\ldots,\nu_n,\alpha_1,\ldots,\alpha_n}(z)}\right\} = \sum_{i=1}^{n}\alpha_i\mathrm{Re}\left\{\frac{zg'_{\nu_i}(z)}{g_{\nu_i}(z)}\right\} + \left(1 - \sum_{i=1}^{n}\alpha_i\right). \tag{7}$$

Now by using the inequality (4) for each ν_i, where $i \in \{1,2,\ldots,n\}$, we obtain

$$\begin{aligned}
\mathrm{Re}\left\{1 + z\frac{F''_{\nu_1,\ldots,\nu_n,\alpha_1,\ldots,\alpha_n}(z)}{F'_{\nu_1,\ldots,\nu_n,\alpha_1,\ldots,\alpha_n}(z)}\right\} &= \sum_{i=1}^{n}\alpha_i \mathrm{Re}\left\{\frac{zg'_{\nu_i}(z)}{g_{\nu_i}(z)}\right\} + \left(1-\sum_{i=1}^{n}\alpha_i\right)\\
&\leqslant \sum_{i=1}^{n}\alpha_i\left(1+\frac{\nu_i+2}{4\nu_i^2+10\nu_i+5}\right) + \left(1-\sum_{i=1}^{n}\alpha_i\right)\\
&= 1+\sum_{i=1}^{n}\alpha_i\left(\frac{\nu_i+2}{4\nu_i^2+10\nu_i+5}\right)\\
&\leqslant 1+\frac{2+\nu}{4\nu^2+10\nu+5}\sum_{i=1}^{n}\alpha_i
\end{aligned}$$

for all $z \in U$ and $\nu,\nu_1,\ldots,\nu_n > (-5+\sqrt{5})/4$. Here we used that the function $\phi : ((-5+\sqrt{5})/4,\infty) \to \mathbb{R}$, defined by

$$\phi(x) = \frac{x+2}{4x^2+10x+5},$$

is decreasing, and consequently for all $i \in \{1,2,\ldots,n\}$, we have

$$\frac{\nu_i+2}{4\nu_i^2+10\nu_i+5} \leqslant \frac{\nu+2}{4\nu^2+10\nu+5}.$$

Because

$$1 < 1+\frac{2+\nu}{4\nu^2+10\nu+5}\sum_{i=1}^{n}\alpha_i \leqslant \frac{3}{2},$$

we have $F_{\nu_1,\ldots,\nu_n,\alpha_1,\ldots,\alpha_n}(z) \in N(\mu)$, where

$$\mu = 1+\frac{2+\nu}{4\nu^2+10\nu+5}\sum_{i=1}^{n}\alpha_i.$$

Thus, the proof of Theorem 2.1 is established. □

Choosing $\alpha_1 = \alpha_2 = \cdots = \alpha_n = \alpha$ in Theorem 2.1, we have the following result:

Corollary 2.1. *Let the numbers $\nu,\nu_1,\ldots,\nu_n$ be as in Theorem 2.1 and let $\alpha_1,\alpha_2,\ldots,\alpha_n$ be positive real numbers. Moreover, suppose that the functions $g_{\nu_i} \in A$ defined by* (5) *and the following inequality:*

$$\delta = 1+\frac{(2+\nu)n\alpha}{4\nu^2+10\nu+5}.$$

Observe that $g_{\frac{1}{2}}(z) = \sqrt{z}\sin\sqrt{z}$ and $g_{-1/2}(z) = z\cos\sqrt{z}$. Thus, taking $n = 1$ in Theorem 2.1 or in Corollary 2.1, we immediately obtain the following result.

Corollary 2.2. *Let $\nu > (-5+\sqrt{5})/4$ and $\alpha > 0$ be a real number. Moreover suppose that these numbers satisfy the following inequality*

$$1 < 1 + \frac{(2+\nu)\alpha}{4\nu^2 + 10\nu + 5} \leqslant \frac{3}{2}.$$

Then the function $F_{\nu,\alpha} : U \to \mathbb{C}$, defined by

$$F_{\nu,\alpha}(z) = \int_0^z \left(\frac{g_\nu(t)}{t}\right)^\alpha dt,$$

is in $N(\eta)$, where

$$\eta = 1 + \frac{(2+\nu)\alpha}{4\nu^2 + 10\nu + 5}.$$

In particular, if $0 < \alpha \leqslant 11/5$, then the function $F_{1/2,\alpha} : U \to \mathbb{C}$, defined by

$$F_{1/2,\alpha}(z) = \int_0^z \left(\frac{\sin\sqrt{t}}{\sqrt{t}}\right)^\alpha dt,$$

is in $N(\zeta)$, where $\zeta = 1 + 5\alpha/22$.

Moreover, if $0 < \alpha \leqslant 1/3$, then the function $F_{-1/2,\alpha} : U \to \mathbb{C}$, defined by $F_{-1/2,\alpha}(z) = \int_0^z \left(\cos\sqrt{t}\right)^\alpha dt$, is in $N(\lambda)$, where $\lambda = 1 + 3\alpha/2$.

References

1. Baricz, A., Frasin, B.A.: Univalence of integral operators involving Bessel functions. Appl. Math. Lett. **23**(4), 371–376 (2010). doi:10.1016/j.aml.2009.10013.
2. Breaz, D.: Certain integral operators on the classes $M(\beta_i)$ and $N(\beta_i)$. J. Inequal. Appl. Art. ID 719354, 1–4 (2008)
3. Frasin, B.A.: Sufficient conditions for integral operator defined by Bessel functions. J. Math. Inequal. **4**(3), 301–306 (2010)
4. Owa, S., Srivastava, H.M.: Some generalized convolution properties associated with certain subclasses of analytic functions. J. Inequal. Pure Appl. Math. **3**(3), 1–13 (2002)
5. Porwal, S.: Mapping properties of an integral operator. Acta Univ. Apul. **27**, 151–155 (2011)
6. Porwal, S., Dixit, K.K.: An application of certain convolution operator involving hypergeometric functions. J. Raj. Acad. Phy. Sci. **9**(2), 173–186 (2010)

7. Porwal, S., Dixit, K.K., Kumar, V., Dixit, P.: On a subclass of analytic functions defined by convolution. General Math. **19**(3), 57–65 (2011)
8. Szasz, R., Kupan, P.: About the univalence of the Bessel functions. Stud. Univ. Babes-Bolyai Math. **54**(1), 127–132 (2009)
9. Uralegaddi, B.A., Ganigi, M.D., Sarangi, S.M.: Univalent functions with positive coefficients. Tamkang J. Math. **25**(3), 225–230 (1994)

Poincaré α-Series for Classical Schottky Groups

Vladimir V. Mityushev

Dedicated to Professor Hari M. Srivastava

Abstract The Poincaré α-series ($\alpha \in \mathbb{R}^n$) for classical Schottky groups are introduced and used to solve Riemann–Hilbert problems for n-connected circular domains. The classical Poincaré θ_2-series is a partial case of the α-series when α vanishes. The real Jacobi inversion problem and its generalizations are investigated via the Poincaré α-series. In particular, it is shown that the Riemann theta function coincides with the Poincaré α-series. Relations to conformal mappings of the multiply connected circular domains onto slit domains and the Schottky–Klein prime function are established. A fast algorithm to compute Poincaré series for disks close to each other is outlined.

1 Introduction

The θ_2-series of Poincaré associated to the classical Schottky groups is used in the constructive theory of analytic functions in multiply connected domains. Such objects of multiply connected domains as the harmonic measures [26, 29, 36], the Abelian functions [1, 4, 8, 9, 18], the canonical conformal mappings [6, 11, 16, 31], the Christoffel–Schwarz formula [13–15, 17, 35], and the Bergman kernel [21] can be constructed by the Poincaré series. These objects can be also considered on the Schottky double. The Poincaré series have applications to extremal polynomials [5], to the generalized alternating method of Schwarz [25, 27], and to composites [38]. The above objects are ultimately constructed for arbitrary circular multiply connected domains [21, 31, 35, 36] via the uniformly convergent Poincaré

V.V. Mityushev (✉)
Department of Computer Sciences and Computer Methods, Pedagogical University, ul. Podchorazych 2, Krakow, 30-084, Poland
e-mail: mityu@up.krakow.pl

G.V. Milovanović and M.Th. Rassias (eds.), *Analytic Number Theory, Approximation Theory, and Special Functions*, DOI 10.1007/978-1-4939-0258-3_33,

θ_2-series [28]. The method of construction is based on Riemann–Hilbert problems and functional equations (without integral terms) [26,29,36]. It is worth noting that investigations based on the absolute convergence have geometrical restrictions on the location of the circular holes [39].

It was noted in [26, 29] that a method of functional equations [26, 29] yields more general series than the classical θ_2-series of Poincaré. In the present paper, such series, called the Poincaré α-series (shortly, the α-series), are systematically discussed. Here, α is a constant vector from $\mathbb{R}^n$. If α vanishes, we arrive at the classical Poincaré series. The Schottky–Klein prime function [10] and its α-prime counterpart are introduced similar to the Poincaré α-series. Riemann–Hilbert problems are solved in terms of the α-series. In order to simplify the presentation, the special Riemann–Hilbert is considered. Its solution is the theta function of Riemann. Hence, it can be applied to the Jacobi inversion problem and its generalizations [41,42]. A fast algorithm to compute Poincaré series is described. At the end of the paper we discuss some open problems stated by Crowdy [8].

2 Poincaré Series for Classical Schottky Groups

Consider mutually disjointed disks $D_k = \{z \in \mathbb{C} : |z - a_k| < r_k\}$ in the complex plane $\mathbb{C}$ and the multiply connected domain D, the complement of the closed disks $|z - a_k| \leq r_k$ to the extended complex plane $\hat{\mathbb{C}} = \mathbb{C} \cup \{\infty\}$ (see Fig. 1). Consider the inversion with respect to the circle $|z - a_k| = r_k$

$$z^*_{(k)} = \frac{r_k^2}{\overline{z - a_k}} + a_k.$$

Introduce the composition of successive inversions with respect to the circles

$$z^*_{(k_p k_{p-1} \ldots k_1)} := \left(z^*_{(k_{p-1} \ldots k_1)}\right)^*_{(k_p)}. \tag{1}$$

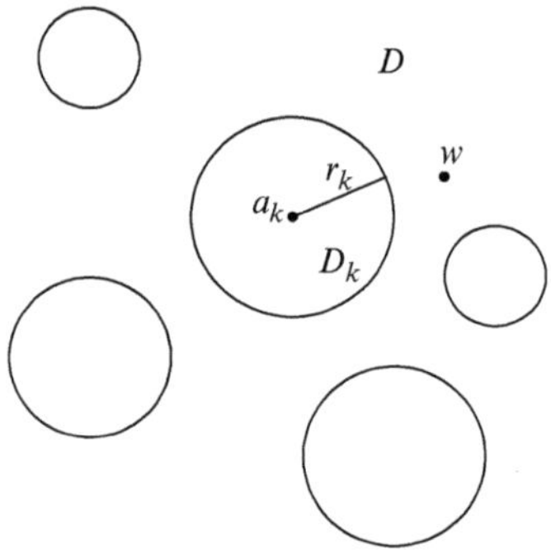

Fig. 1 Multiply connected domain D with circular inclusions D_k

In the sequence $k_1, k_2, \ldots, k_p$ no two neighboring numbers are equal. The number p is called the level of the mapping. When p is even, these are Möbius transformations. If p is odd, we have anti-Möbius transformations, i.e., Möbius transformations in $\overline{z}$. Thus, these mappings can be written in the form

$$\gamma_j(z) = \left(e_j z + b_j\right) / \left(c_j z + d_j\right), \ p \in 2\mathbb{Z}, \qquad (2)$$
$$\gamma_j(\overline{z}) = \left(e_j \overline{z} + b_j\right) / \left(c_j \overline{z} + d_j\right), \ p \in 2\mathbb{Z} + 1,$$

where the normalization $e_j d_j - b_j c_j = 1$ is taken. Here, we introduce the identical mapping with the level $p = 0$

$$\gamma_0(z) := z,$$

n simple inversions ($p = 1$)

$$\gamma_1(\overline{z}) := z^*_{(1)}, \ldots, \gamma_n(\overline{z}) := z^*_{(n)},$$

$n^2 - n$ pairs of inversions ($p = 2$)

$$\gamma_{n+1}(z) := z^*_{(12)}, \ \gamma_{n+2}(z) := z^*_{(13)}, \ldots, \ \gamma_{n^2}(z) := z^*_{(n,n-1)},$$

triples ($p = 3$)

$$\gamma_{n^2+1}(\overline{z}) := z^*_{(121)}, \quad \gamma_{n^2+2}(\overline{z}) := z^*_{(122)}, \quad \ldots$$

and so on. The set of the subscripts j of γ_j is ordered in such a way that the level p is increasing. The functions (2) generate a Schottky group $\mathcal{K}$. Thus, each element of $\mathcal{K}$ is presented in the form of the composition of inversions (1) or in the form of linearly ordered functions (2). All elements γ_j of the even levels generate a subgroup $\mathcal{E}$ of the group $\mathcal{K}$. The set of the elements γ_j of odd level $\mathcal{K} \backslash \mathcal{E}$ is denoted by $\mathcal{O}$.

Let $H(z)$ be a rational function. The following series is called the Poincaré θ_2-series:

$$\theta_2(z) := \sum_{\gamma_j \in \mathcal{E}} H[\gamma_j(z)](c_j z + d_j)^{-2} \qquad (3)$$

associated with the subgroup $\mathcal{E}$. It was proved in [28] that the series (3) converges uniformly in every compact subset not containing the limit points of $\mathcal{K}$ and poles of $H[\gamma_j(z)]$. Moreover, (3) is an automorphic function of the weight (-2):

$$\theta_2(z) = \theta_2[\gamma_j(z)](c_j z + d_j)^{-2}. \qquad (4)$$

Using the inversions (1) instead of (2), we can write (3) in an extended form. First, following [28] introduce the series

$$\Theta_2^{(1)}(z) = H(z) - \sum_{k=1}^{n} \overline{H[z_{(k)}^*]}\,\overline{(z_{(k)}^*)'} + \sum_{k=1}^{n}\sum_{k_1\neq k} H[z_{(k_1k)}^*](z_{(k_1k)}^*)' \tag{5}$$

$$-\sum_{k=1}^{n}\sum_{k_1\neq k}\sum_{k_2\neq k_1} \overline{H[z_{(k_2k_1k)}^*]}\,\overline{(z_{(k_2k_1k)}^*)'} + \cdots$$

and

$$\Theta_2^{(2)}(z) = H(z) + \sum_{k=1}^{n} \overline{H[z_{(k)}^*]}\,\overline{(z_{(k)}^*)'} + \sum_{k=1}^{n}\sum_{k_1\neq k} H[z_{(k_1k)}^*](z_{(k_1k)}^*)' \tag{6}$$

$$+\sum_{k=1}^{n}\sum_{k_1\neq k}\sum_{k_2\neq k_1} \overline{H[z_{(k_2k_1k)}^*]}\,\overline{(z_{(k_2k_1k)}^*)'} + \cdots$$

The Poincaré θ_2-series (3) can be written in the form

$$\theta_2(z) = \frac{1}{2}\left(\Theta_2^{(1)}(z) + \Theta_2^{(2)}(z)\right). \tag{7}$$

Let α_k $(k = 1, 2, \ldots, n)$ be real numbers from the segment $[0, 2\pi)$. Introduce the multi-index $\alpha = (\alpha_1, \alpha_2, \ldots, \alpha_n)$ and the series

$$\Theta_2^{(1)}(z;\alpha) = H(z) - \sum_{k=1}^{n} e^{2i\alpha_k}\overline{H[z_{(k)}^*]}\,\overline{(z_{(k)}^*)}' + \sum_{k=1}^{n}\sum_{k_1\neq k} e^{2i(\alpha_k-\alpha_{k_1})} H[z_{(k_1k)}^*](z_{(k_1k)}^*)'$$

$$-\sum_{k=1}^{n}\sum_{k_1\neq k}\sum_{k_2\neq k_1} e^{2i(\alpha_k-\alpha_{k_1}+\alpha_{k_2})}\overline{H[z_{(k_2k_1k)}^*]}\,\overline{(z_{(k_2k_1k)}^*)}' + \cdots, \tag{8}$$

$$\Theta_2^{(2)}(z;\alpha) = H(z) + \sum_{k=1}^{n} e^{2i\alpha_k}\overline{H[z_{(k)}^*]}\,\overline{(z_{(k)}^*)}' + \sum_{k=1}^{n}\sum_{k_1\neq k} e^{2i(\alpha_k-\alpha_{k_1})} H[z_{(k_1k)}^*](z_{(k_1k)}^*)'$$

$$+\sum_{k=1}^{n}\sum_{k_1\neq k}\sum_{k_2\neq k_1} e^{2i(\alpha_k-\alpha_{k_1}+\alpha_{k_2})}\overline{H[z_{(k_2k_1k)}^*]}\,\overline{(z_{(k_2k_1k)}^*)}' + \cdots \tag{9}$$

and

$$\theta_2(z;\alpha) = \frac{1}{2}\left[\Theta_2^{(1)}(z;\alpha) + \Theta_2^{(2)}(z;\alpha)\right]. \tag{10}$$

We call the series (8)–(10) by the α-series. The series (8)–(10) uniformly converge in every compact subset not containing the limit points of $\mathcal{K}$ and poles of $H[\gamma_j(z)]$ [34]. If $\alpha = (0, 0, \dots, 0)$, we arrive at the classic Poincaré series (3).

3 Riemann–Hilbert Problem

To find a function $\psi(z)$ analytic in D and continuously differentiable in $D \cup \partial D$ with the following Riemann–Hilbert boundary condition [36]

$$\mathrm{Im}\left[e^{-i\alpha_k}\frac{t-a_k}{r_k}\psi(t)\right] = 0, \quad |t-a_k| = r_k,\ k = 1, 2, \dots, n. \tag{11}$$

It is assumed that the function $\psi(z)$ is normalized at infinity:

$$\psi(\infty) = 1. \tag{12}$$

Let a function $\varphi(z)$ be a primitive of $\psi(z)$, i.e., $\varphi'(z) = \psi(z)$. Then $\varphi(z)$ satisfies the Riemann–Hilbert boundary condition:

$$\mathrm{Re}\ [e^{-i\alpha_k}\varphi(t)] = c_k, \quad |t-a_k| = r_k,\ k = 1, 2, \dots, n, \tag{13}$$

where c_k are undetermined constants. In order to prove it, we consider the parametrization of the circle $|t-a_k| = r_k$ with the natural arc parameter $s \in [0, 2\pi r_k)$

$$t(s) = a_k + r_k \exp\left(\frac{is}{r_k}\right). \tag{14}$$

One can see that the derivative can be written in the form

$$t'(s) = i\ \frac{t-a_k}{r_k}. \tag{15}$$

Differentiation (13) on s yields

$$\mathrm{Re}[e^{-i\alpha_k}\psi(t)t'(s)] = 0, \quad |t-a_k| = r_k,\ k = 1, 2, \dots, n. \tag{16}$$

Using (15) we arrive at the boundary value problem (11).

It follows from (12) that $\varphi(z)$ is analytic in D except at the infinite point where it satisfies the hydrodynamic normalization at infinity [24]:

$$\varphi(z) = z + \varphi_0 + \frac{\varphi_1}{z} + \frac{\varphi_2}{z^2} + \dots. \tag{17}$$

The function $\varphi(z)$ is multivalued in D. More precisely, it is represented in the form [36]

$$\varphi(z) = z + \varphi_0(z) + \sum_{k=1}^{n} e^{i\alpha_k} A_k \ln(z - a_k), \tag{18}$$

where $\varphi_0(z)$ is single-valued analytic in D and A_k are undetermined real constants. The logarithm $\ln(z - a_k)$ is defined in such a way that it is analytic in the complex plane except a cut connecting the points $z = a_k$ and infinity. It is assumed that the cut does not cross $|z - a_m| \leq r_m$ for $m \neq k$. The term $e^{i\alpha_k} A_k$ has such a form since the increment of the function $\mathrm{Re}[e^{-i\alpha_k}\varphi(t)]$ along $|t - a_k| = r_k$ must vanish because of (13). The problem (13) is discussed for multivalued functions as well as for single-valued functions when all $A_k = 0$.

Consider the Banach space $\mathcal{H}^\mu(L)$ consisting of functions Hölder continuous on Lyapunov's curve L endowed the norm

$$||\omega|| = \sup_{t\in L} |\omega(t)| + \sup_{t_{1,2}\in L} \frac{|\omega(t_1)| - \omega(t_2)|}{|t_1 - t_2|^\mu}, \tag{19}$$

where $0 < \mu \leq 1$. Analytic functions considered in the present paper can be continuous or continuously differentiable in the closures of the analyticity domains. The space $\mathcal{H}^{(k,\mu)}(L)$ consists of those functions which have Hölder continuous derivative of the kth order belonging to $\mathcal{H}^\mu(L)$. Let $\partial\Omega$ be the boundary of a domain Ω not necessary connected. Introduce a space $\mathcal{H}_A^\mu(\Omega)$ consisting of functions analytic in Ω and Hölder continuous in the closure of Ω endowed the norm (19). The space $\mathcal{H}_A^\mu(\Omega)$ is Banach, since the maximum principle for analytic functions implies that the norm in $\mathcal{H}_A^\mu(\Omega)$ coincides with the norm in $\mathcal{H}^\mu(\partial\Omega)$. One can consider $\mathcal{H}_A^\mu(\Omega)$ as a closed subspace of $\mathcal{H}^\mu(\partial\Omega)$. The space $\mathcal{H}_A^{(k,\mu)}(\Omega)$ is introduced in the same way as a subspace of $\mathcal{H}^{(k,\mu)}(\Omega)$. Therefore, the boundary value problems (11) and (13) are considered in the spaces $\mathcal{H}_A^{(1,\mu)}(D)$ and $\mathcal{H}_A^{(\mu)}(D)$, respectively.

Lemma 3.1 ([12]). *The problem* (13), (17) *for single-valued functions has a unique solution up to an arbitrary additive constant.*

Let $\zeta = u + iv$ denotes a complex variable on the complex plane with slits Γ_k $(k = 1, 2, \ldots, n)$ lying on the lines:

$$-\sin\alpha_k\, u + \cos\alpha_k\, v = c_k, \tag{20}$$

where c_k are the same as in (11). Let D' denote the complement of all the segments Γ_k to $\hat{\mathbb{C}}$. The conformal mapping $\tilde{\varphi}(z) = u(z) + iv(z)$ from D onto D' satisfies the boundary value problem (13), (17). It follows from Lemma 3.1 that the conformal mapping $\tilde{\varphi}(z)$ coincides with the unique solution $\varphi(z)$ of the problem (13), (17) up to an additive constant.

Lemma 3.2. *The problem* (13), (17) *for multivalued functions represented in the form* (18) *has* $(n+1)$ $\mathbb{R}$*-linear independent solutions.*

Proof. One independent solution is a constant and other n independent solutions are produced by the terms $e^{i\alpha_k} A_k \ln(z-a_k)$ in the representation (18). Another proof follows from the relation between the problems (13), (17), and (11), (12). The winding number (index) of the problem (11), (12) is equal to n [19, 31]. Hence, it has n $\mathbb{R}$-linear independent solutions. The $(n+1)$th solution is a constant obtained by integration of (11).

Remark 3.1. According to [19] the winding number $\varkappa$ of the problem (11) is equal to $(n+1)$. The number of $\mathbb{R}$-linear independent solutions is equal to $\varkappa$, and the inhomogeneous problem corresponding to (11) is always solvable. The condition (12) reduces the number of $\mathbb{R}$-linear independent solutions to n that is in agreement with the above conclusion.

The problem (13) for multivalued functions can be reduced to the $\mathbb{R}$-linear problem [31]:

$$\begin{aligned} \varphi(t) = \varphi_k(t) - e^{2i\alpha_k}\overline{\varphi_k(t)} + e^{i\alpha_k} c_k + e^{i\alpha_k} \xi_k \ln \tfrac{t-a_k}{r_k}, \\ |t-a_k| = r_k,\ k = 1, 2, \ldots, n, \end{aligned} \tag{21}$$

where $\varphi_k(z)$ is analytic in $|z-a_k| < r_k$ and continuously differentiable in $|z-a_k| \le r_k$ and real constants ξ_k are undetermined.

Lemma 3.3. (i) *Let $\varphi(z)$ and $\varphi_k(z)$ be solutions of* (21) *with arbitrarily fixed real constants ξ_k. Then $\varphi(z)$ satisfies* (13).
(ii) *Let $\varphi(z)$ be a solution of* (13) *and real constants ξ_k are arbitrarily fixed. Then there exist such functions $\varphi_k(z)$ that for each $k = 1, \ldots, n$ the $\mathbb{R}$-linear conditions* (21) *are fulfilled.*

Proof of the first assertion is evident. It is sufficient to multiply (21) by $e^{-i\alpha_k}$ and to take the real part.

Conversely, let $\varphi(z)$ satisfies (13) and a real constant ξ_k is fixed. The function $e^{-i\alpha_k}\varphi_k(z)$ can be uniquely determined up to an additive real constant from the simple Schwarz problem for the disk $|z-a_k| < r_k$ [19, 36]

$$2\,\mathrm{Im}\left[e^{-i\alpha_k}\varphi_k(t)\right] = \mathrm{Im}\left[e^{-i\alpha_k}\varphi(t) - \xi_k \ln \frac{t-a_k}{r_k}\right], \quad |t-a_k| = r_k. \tag{22}$$

It follows from the latter boundary condition that the function $\varphi_k(z)$ belongs to the spaces $\mathcal{H}_A^{(1,\alpha)}(D)$ except at a point $z = a_k$, where $\mathrm{Im}\ln(t-a_k) = \arg(t-a_k)$ has a discontinuity.

The lemma is proved. □

Differentiate (21) on s along the circles $|t-a_k|=r_k$ and divide the results by $t'(s)$ calculated with (15)

$$\psi(t)=\psi_k(t)+e^{2i\alpha_k}\left(\frac{r_k}{t-a_k}\right)^2\overline{\psi_k(t)}+\frac{e^{i\alpha_k}\xi_k}{t-a_k},\quad |t-a_k|=r_k,\ k=1,2,\dots,n, \tag{23}$$

where $\psi(z)=\varphi'(z)$ and $\psi_k(z)=\varphi_k'(z)$. Therefore, the Riemann–Hilbert problem (11) is reduced to the $\mathbb{R}$-linear problem (23).

4 Functional Equations

The $\mathbb{R}$-linear problem (23) can be reduced to functional equations. Following [31, 36] introduce the function

$$\Phi(z):=\begin{cases}\psi_k(z)-\sum\limits_{m\neq k}e^{2i\alpha_m}\left(\dfrac{r_m}{z-a_m}\right)^2\overline{\psi_m\left(z_{(m)}^*\right)}-\sum\limits_{m\neq k}\dfrac{e^{i\alpha_m}\xi_m}{z-a_m}, \\ \qquad\qquad |z-a_k|\le r_k,\quad k=1,2,\dots,n, \\ \psi(z)-\sum\limits_{m=1}^{n}e^{2i\alpha_m}\left(\dfrac{r_m}{z-a_m}\right)^2\overline{\psi_m\left(z_{(m)}^*\right)}-\sum\limits_{m=1}^{n}\dfrac{e^{i\alpha_m}\xi_m}{z-a_m}, \qquad z\in D,\end{cases}$$

analytic in $|z-a_k|<r_k$ $(k=1,2,\dots,n)$ and D. Calculate the jump across the circle $|t-a_k|=r_k$

$$\Delta_k:=\Phi^+(t)-\Phi^-(t),\ |t-a_k|=r_k,$$

where $\Phi^+(t):=\lim_{z\to t\ z\in D}\Phi(z)$, $\Phi^-(t):=\lim_{z\to t\ z\in D_k}\Phi(z)$. Application of (23) gives $\Delta_k=0$. It follows from the principle of analytic continuation that $\Phi(z)$ is analytic in the extended complex plane. Moreover, $\psi(\infty)=1$ yields $\Phi(\infty)=1$. Then Liouville's theorem implies that $\Phi(z)\equiv 1$. The definition of $\Phi(z)$ in $|z-a_k|\le r_k$ yields the following system of functional equations:

$$\psi_k(z)=\sum_{m\neq k}e^{2i\alpha_m}\left(\frac{r_m}{z-a_m}\right)^2\overline{\psi_m\left(z_{(m)}^*\right)}+1+\sum_{m\neq k}\frac{e^{i\alpha_m}\xi_m}{z-a_m}, \tag{24}$$
$$|z-a_k|\le r_k,\ k=1,2,\dots,n.$$

Let $\psi_k(z)$ $(k=1,2,\dots,n)$ be a solution of (24). Then the function $\psi(z)$ can be found from the definition of $\Phi(z)$ in D

$$\psi(z) = \sum_{m=1}^{n} e^{2i\alpha_m} \left(\frac{r_m}{z-a_m}\right)^2 \overline{\psi_m\left(z^*_{(m)}\right)} + 1 + \sum_{m=1}^{n} \frac{e^{i\alpha_m}\xi_m}{z-a_m}, \quad z \in D \cup \partial D. \tag{25}$$

Consider inhomogeneous functional equations with a given element $f \in \mathcal{H}_A(\cup_{k=1}^n D_k)$:

$$\psi_k(z) = \sum_{m\neq k} e^{2i\alpha_m} \left(\frac{r_m}{z-a_m}\right)^2 \overline{\psi_m\left(z^*_{(m)}\right)} + f(z), |z-a_k| \leq r_k, \ \ k = 1,2,\ldots,n. \tag{26}$$

Theorem 4.1 ([36]). *The system* (26) *has a unique solution for any circular multiply connected domain D. This solution can be found by the method of successive approximations convergent in the space $\mathcal{H}_A(\cup_{k=1}^n D_k)$, i.e., uniformly convergent in every disk $|z-a_k| \leq r_k$.*

The system of functional equations (24) can be decomposed onto $(n+1)$ systems:

$$\psi_k^{(1)}(z) = \sum_{m\neq k} e^{2i\alpha_m} \left(\frac{r_m}{z-a_m}\right)^2 \overline{\psi_m^{(1)}\left(z^*_{(m)}\right)} + 1, \quad |z-a_k| \leq r_k, \quad k = 1,2,\ldots,n \tag{27}$$

and

$$\Psi_k^{(\ell)}(z) = \sum_{m\neq k} e^{2i\alpha_m} \left(\frac{r_m}{z-a_m}\right)^2 \overline{\Psi_m^{(\ell)}\left(z^*_{(m)}\right)} + \frac{e^{i\alpha_\ell}}{z-a_\ell}\,\delta'_{\ell k}, \ |z-a_k| \leq r_k, \quad k = 1,2,\ldots,n, \tag{28}$$

where $\delta'_{\ell k} = 1 - \delta_{\ell k}$ and $\delta_{\ell k}$ is the Kronecker symbol. The unique solution of (24) can be represented in the form

$$\psi_k(z) = \psi_k^{(1)}(z) + \sum_{\ell=1}^{n} \xi_\ell \, \Psi_k^{(\ell)}(z). \tag{29}$$

The functions $\psi_k(z)$ can be constructed by two methods. First, they can be obtained by iterations applied to (24); second, by iterations applied separately to (27) and to (28) and further their linear combination (29). For any fixed $\psi_k(z)$, these iterations yield a series (in general conditionally convergent) with two different orders of summations. It follows from Theorem 4.1 that the result will be the same since we construct the same unique solution of (24) by two different methods. It is worth noting that (29) is a $\mathbb{C}$-linear combination of the basic functions because $\xi_\ell \in \mathbb{R}$ for $\ell = 1,2,\ldots,n$.

We now apply Theorem 4.1 to (27). Let $w \in D$ be a fixed point not equal to infinity. Introduce the functions

$$\phi_m(z) = \int_{w^*_{(m)}}^{z} \psi_m^{(1)}(t)dt + \phi_m(w^*_{(m)}), \quad |z-a_m| \le r_m, \quad m = 1,2,\ldots,n, \tag{30}$$

and

$$\omega(z) = -\sum_{m=1}^{n} e^{2i\alpha_m}\left[\overline{\phi_m\left(z^*_{(m)}\right)} - \overline{\phi_m\left(w^*_{(m)}\right)}\right]. \tag{31}$$

The functions $\omega(z)$ and $\phi_m(z)$ analytic in D and in D_m, respectively, and continuously differentiable in the closures of the domains considered. One can see from (30) that the function $\phi_m(z)$ is determined by $\psi_m(z)$ up to an additive constant which vanishes in (31). The function $\omega(z)$ vanishes at $z = w$. Investigate the function $\omega(z)$ on the boundary of D. It follows from (31) and $t = t^*_{(k)}$ ($|t-a_k| = r_k$) for each fixed k that

$$\omega(t) = -e^{2i\alpha_k}\left[\overline{\phi_k(t)} - \overline{\phi_k\left(w^*_{(k)}\right)}\right] - \Psi_k(t), \tag{32}$$

where

$$\Psi_k(z) = \sum_{m\neq k} e^{2i\alpha_m}\left[\overline{\phi_m\left(z^*_{(m)}\right)} - \overline{\phi_m\left(w^*_{(m)}\right)}\right]. \tag{33}$$

Using the relation [36]

$$\frac{d}{dz}\left[\overline{\phi_m\left(z^*_{(m)}\right)}\right] = -\left(\frac{r_m}{z-a_m}\right)^2 \overline{\phi'_m\left(z^*_{(m)}\right)}, \quad |z-a_m| > r_m, \tag{34}$$

calculate the derivative

$$\Psi'_k(z) = -\sum_{m=1}^{n} e^{2i\alpha_m}\left(\frac{r_m}{z-a_m}\right)^2 \overline{\psi_m^{(1)}\left(z^*_{(m)}\right)}. \tag{35}$$

Application of (24) yields

$$\Psi'_k(z) = 1 - \psi_k^{(1)}(z). \tag{36}$$

Then (32) and (30) implies that

$$e^{-i\alpha_k}\omega(t) = -e^{i\alpha_k}\left[\overline{\phi_k(t)} - \overline{\phi_k\left(w^*_{(k)}\right)}\right] + e^{-i\alpha_k}[\phi_k(t) - t + d_k], \quad |t-a_k| = r_k, \tag{37}$$

where d_k is a constant of integration. Calculation of the real part of (37) gives

$$\text{Re}\,[e^{-i\alpha_k}(\omega(t)+t)] = p_k, \quad |t-a_k| = r_k, \tag{38}$$

where p_k is a constant. Comparing (38) with (13) and using Lemma 3.1 we conclude that the conformal mapping D onto D' has the form

$$\tilde{\varphi}(z) = z + \omega(z) + \text{constant}, \tag{39}$$

where $\omega(z)$ is calculated by (31). Application of the method of successive approximations to (27) and integration terms by terms of the obtained uniformly convergent series yield the exact formula:

$$\varphi_k(z) = q_k + z - \sum_{k_1\neq k} e^{2i\alpha_{k_1}}\overline{(z^*_{(k_1)} - w^*_{(k_1)})} + \sum_{k_1\neq k}\sum_{k_2\neq k_1} e^{2i(\alpha_{k_1}-\alpha_{k_2})}(z^*_{(k_2k_1)} - w^*_{(k_2k_1)})$$
$$-\sum_{k_1\neq k}\sum_{k_2\neq k_1}\sum_{k_3\neq k_2} e^{2i(\alpha_{k_1}-\alpha_{k_2}+\alpha_{k_3})}\overline{(z^*_{(k_3k_2k_1)} - w^*_{(k_3k_2k_1)})} + \cdots, \quad |z-a_k|\leq r_k. \tag{40}$$

Using (31) and (40), we write the function (39) up to an arbitrary additive constant in the form

$$\tilde{\varphi}(z) = z - \sum_{k=1}^{n} e^{2i\alpha_k}\overline{(z^*_{(k)} - w^*_{(k)})} + \sum_{k=1}^{n}\sum_{k_1\neq k} e^{2i(\alpha_k-\alpha_{k_1})}(z^*_{(k_1k)} - w^*_{(k_1k)}) \tag{41}$$
$$-\sum_{k=1}^{n}\sum_{k_1\neq k}\sum_{k_2\neq k_1} e^{2i(\alpha_k-\alpha_{k_1}+\alpha_{k_2})}\overline{(z^*_{(k_2k_1k)} - w^*_{(k_2k_1k)})} + \cdots$$

Differentiation of the latter uniformly convergent series term by term yields the α-series (8) with $H(z) = 1$:

$$\psi^{(1)}(z) = 1 - \sum_{k=1}^{n} e^{2i\alpha_k}\overline{(z^*_{(k)})'} + \sum_{k=1}^{n}\sum_{k_1\neq k} e^{2i(\alpha_k-\alpha_{k_1})}(z^*_{(k_1k)})' \tag{42}$$
$$-\sum_{k=1}^{n}\sum_{k_1\neq k}\sum_{k_2\neq k_1} e^{2i(\alpha_k-\alpha_{k_1}+\alpha_{k_2})}\overline{(z^*_{(k_2k_1k)})'} + \cdots$$

A similar method can be used to construct $\Psi_k^{(\ell)}(z)$ satisfying (28) and to construct

$$\Psi^{(\ell)}(z) = \sum_{m=1}^{n} e^{2i\alpha_m}\left(\frac{r_m}{z-a_m}\right)^2 \overline{\Psi_m^{(\ell)}\left(z^*_{(m)}\right)} + \frac{e^{i\alpha_\ell}}{z-a_\ell}, \quad z\in D\cup\partial D. \tag{43}$$

We have

$$\Psi^{(\ell)}(z) = \frac{e^{i\alpha_\ell}}{z - a_\ell} - e^{-i\alpha_\ell} \sum_{k=1}^{n} \frac{e^{2i\alpha_k}}{\overline{z^*_{(k)} - a_k}} \overline{(z^*_{(k)})'} + e^{i\alpha_\ell} \sum_{k=1}^{n} \sum_{k_1 \neq k} \frac{e^{2i(\alpha_k - \alpha_{k_1})}}{z^*_{(k_1 k)} - a_{k_1}} (z^*_{(k_1 k)})'$$

$$-e^{-i\alpha_\ell} \sum_{k=1}^{n} \sum_{k_1 \neq k} \sum_{k_2 \neq k_1} \frac{e^{2i(\alpha_k - \alpha_{k_1} + \alpha_{k_2})}}{\overline{z^*_{(k_2 k_1 k)} - a_{k_2}}} \overline{(z^*_{(k_2 k_1 k)})'} + \cdots \tag{44}$$

Therefore, the general solution of the Riemann–Hilbert problem (11) has the form

$$\psi(z) = \psi^{(1)}(z) + \sum_{\ell=1}^{n} \xi_\ell \, \Psi^{(\ell)}(z), \tag{45}$$

where $\psi^{(1)}(z)$ is given by (42) and $\Psi^{(\ell)}(z)$ by (44).

Integration of (45) from w to z yields

$$\varphi(z) = \tilde{\varphi}(z) + \sum_{\ell=1}^{n} \xi_\ell \, \tilde{\varphi}^{(\ell)}(z) + \text{constant}, \tag{46}$$

where $\tilde{\varphi}(z)$ has the form (41). The function $\tilde{\varphi}^{(\ell)}(z)$ can be written explicitly as follows:

$$\tilde{\varphi}^{(\ell)}(z) = e^{i\alpha_\ell} \ln \frac{z - a_\ell}{w - a_\ell} - e^{-i\alpha_\ell} \sum_{k=1}^{n} e^{2i\alpha_k} \ln \overline{\frac{z^*_{(k)} - a_k}{w^*_{(k)} - a_k}} \tag{47}$$

$$+ e^{i\alpha_\ell} \sum_{k=1}^{n} \sum_{k_1 \neq k} e^{2i(\alpha_k - \alpha_{k_1})} \ln \frac{z^*_{(k_1 k)} - a_{k_1}}{z^*_{(k_1 k)} - a_{k_1}}$$

$$- e^{-i\alpha_\ell} \sum_{k=1}^{n} \sum_{k_1 \neq k} \sum_{k_2 \neq k_1} e^{2i(\alpha_k - \alpha_{k_1} + \alpha_{k_2})} \ln \overline{\frac{z^*_{(k_2 k_1 k)} - a_{k_2}}{w^*_{(k_2 k_1 k)} - a_{k_2}}} + \cdots$$

It is worth noting that separation of the terms with z and w in (47) can fail to converge [31].

In order to compare (46) and (18), we note that the conformal mapping $\tilde{\varphi}(z)$ coincides with $z + \varphi_0(z)$ up to an additive constant. Hence,

$$\sum_{k=1}^{n} e^{i\alpha_k} A_k \ln(z - a_k) = \sum_{\ell=1}^{n} \xi_\ell \, \tilde{\varphi}^{(\ell)}(z), \quad z \in D. \tag{48}$$

Substitution of (47) into (48) can yield relations between the constants A_k and ξ_ℓ.

5 Schottky–Klein Prime Function

The Schottky–Klein prime function was described in [1, 6, 8–11]. Following formulae (5)–(7) from Sect. 2, we represent the Schottky–Klein prime function in terms of the uniformly convergent products for arbitrary circular multiply connected domains and introduce the α-prime function.

Let ζ and w be fixed points of $(D \cup \partial D)\backslash\{\infty\}$. The following functions were introduced in [31, 36] (see formulae (40) and (41) in [31]):

$$\omega_0(z,\zeta,w) = \ln \prod_{j=1}^{\infty} \mu_j(z,\zeta,w), \tag{49}$$

where

$$\mu_j(z,\zeta,w) = \begin{cases} \dfrac{\zeta - \gamma_j(z)}{\zeta - \gamma_j(w)}, & \text{if } \gamma_j \in \mathcal{E}, \\ \\ \overline{\left(\dfrac{\zeta - \gamma_j(\overline{w})}{\zeta - \gamma_j(\overline{z})}\right)}, & \text{if } \gamma_j \in \mathcal{O}. \end{cases} \tag{50}$$

The multipliers $\mu_j(z,\zeta,w)$ in (49) are arranged in accordance with the increasing level of γ_j. The infinite product (49) converges uniformly in the variable z in every compact subset of $(D \cup \partial D)\backslash(\{\infty\},\{\zeta\},\{w\})$. The justification of these assertions is based on the application of Theorem 4.1 from Sect. 4 to the functional equations studied in [28, 31, 36]:

$$\varphi_k(z) = -\sum_{m \neq k} \left[\overline{\varphi_m\left(z^*_{(m)}\right)} - \overline{\varphi_m\left(w^*_{(m)}\right)}\right] + \ln\frac{z-\zeta}{w-\zeta}, \ |z-a_k| \le r_k, \ k = 1,\ldots,n. \tag{51}$$

Instead of (51) we can apply Theorem 4.1 to the following functional equations:

$$\varphi_k(z) = \sum_{m \neq k} \left[\overline{\varphi_m\left(z^*_{(m)}\right)} - \overline{\varphi_m\left(w^*_{(m)}\right)}\right] + \ln\frac{z-\zeta}{w-\zeta}, \ |z-a_k| \le r_k, \ k = 1,\ldots,n. \tag{52}$$

This justifies introduction of the function

$$\omega_1(z,\zeta,w) = \ln \prod_{j=1}^{\infty} \nu_j(z,\zeta,w), \tag{53}$$

where

$$
\nu_j(z,\zeta,w) = \begin{cases} \dfrac{\zeta-\gamma_j(z)}{\zeta-\gamma_j(w)}, & \text{if } \gamma_j \in \mathcal{E}, \\[2ex] \dfrac{\overline{\zeta-\gamma_j(\overline{z})}}{\overline{\zeta-\gamma_j(\overline{w})}}, & \text{if } \gamma_j \in \mathcal{O}. \end{cases} \tag{54}
$$

Similar to (10) we introduce the function

$$
\omega(z,\zeta,w) = \frac{1}{2}[\omega_0(z,\zeta,w)+\omega_1(z,\zeta,w)] = \frac{1}{2}\ln \prod_{\gamma_j\in\mathcal{E}\setminus\{\gamma_0\}} \frac{\zeta-\gamma_j(z)}{\zeta-\gamma_j(w)}. \tag{55}
$$

Hence, the following infinite product is correctly defined for z not equal to ζ, w, and infinity:

$$
\Omega(z,\zeta,w) = \prod_{\gamma_j\in\mathcal{E}\setminus\{\gamma_0\}} \frac{\zeta-\gamma_j(z)}{\zeta-\gamma_j(w)}. \tag{56}
$$

Therefore, we can introduce the function of two variables:

$$
S(z,\zeta) = (\zeta-z)\Omega(\zeta,z,z)\Omega(z,\zeta,\zeta) = (\zeta-z) \prod_{\gamma_j\in\mathcal{E}\setminus\{\gamma_0\}} \frac{z-\gamma_j(\zeta)}{z-\gamma_j(z)}\,\frac{\zeta-\gamma_j(z)}{\zeta-\gamma_j(\zeta)}. \tag{57}
$$

This is the famous Schottky–Klein function presented in the form of uniformly convergent product. More precisely, the uniform convergence is proved for $\Omega(\zeta,z,z)$ in the variable ζ in every compact subset of $(D\cup\partial D)\setminus(\{z\},\{\infty\})$ and for $\Omega(z,\zeta,\zeta)$ in the variable z in every compact subset of $(D\cup\partial D)\setminus(\{\zeta\},\{\infty\})$. The uniform convergence in the variable (z,ζ) in subsets of $\mathbb{C}^2$ could be proved by refined investigations of the corresponding functional equations [28, 31, 36].

Similar to (8)–(10) one can introduce α-prime functions

$$
S(z,\zeta,\alpha) = (\zeta-z) \prod_{\gamma_j\in\mathcal{E}\setminus\{\gamma_0\}} e^{2is_j(\alpha)} \frac{z-\gamma_j(\zeta)}{z-\gamma_j(z)}\,\frac{\zeta-\gamma_j(z)}{\zeta-\gamma_j(\zeta)}. \tag{58}
$$

where p is odd and

$$
s_j(\alpha) := \alpha_k - \alpha_{k_1} + \cdots + \alpha_{k_{p-1}} - \alpha_{k_p}. \tag{59}
$$

The correspondence between j and $(k_p, k_{p-1}, \cdots, k_1, k)$ in (59) is established via the numeration of the elements of $\mathcal{E}$, i.e., via the relation $\gamma_j(z) = z^*_{(k_p k_{p-1}\cdots k_1 k)}$.

6 Schottky Double

The Schottky double $\mathcal{S}$ is obtained from two equal multiply connected domains D and $\tilde{D} \equiv D$ glued along the circles $|t - a_k| = r_k$ $(k = 1, 2, \ldots, n)$. Analytic functions in a domain of $\mathcal{S}$ are those functions which are analytic on $D \cup \partial D$ in z and analytic on $\tilde{D} \cup \partial \tilde{D}$ in $\overline{z}$ with the condition $\Phi(t) = \tilde{\Phi}(\overline{t})$ on the joint part of ∂D. Hence, the Schottky double $\mathcal{S}$ is a compact Riemann surface of genus $(n-1)$ [42]. Let t_k be a fixed point on the circle $|t - a_k| = r_k$ and $\mathbf{a}' \subset D$ be a simple smooth curve connecting the points t_n and t_k $(k = 1, 2, \ldots, n-1)$. Introduce the symmetric curve $\tilde{\mathbf{a}'}_k \subset \tilde{D}$ connecting the points t_k and t_n and the closed curve $\mathbf{a}_k = \mathbf{a}' \cup \tilde{\mathbf{a}'}_k$ on $\mathcal{S}$. Let $\mathbf{b}_k$ denote the clockwise oriented circle $|t - a_k| = r_k$. The curves $\mathbf{a}_k$ and $\mathbf{b}_k$ $(k = 1, 2, \ldots, n-1)$ form a homology basis for $\mathcal{S}$, and any cycle on $\mathcal{S}$ is homologous to a linear combination of $\mathbf{a}_k$ and $\mathbf{b}_k$ with integer coefficients.

The harmonic measure $\omega_\ell(z)$ of the circle $|t - a_\ell| = r_\ell$ relative to the multiply connected domain D is a function harmonic in D continuous in $D \cup \partial D$ which satisfies the Dirichlet problem:

$$\omega_\ell(t) = \delta_{\ell k}, \quad |t - a_k| = r_k \quad (k = 1, 2, \ldots, n), \tag{60}$$

where $\delta_{\ell k}$ stands for the Kronecker symbol. The harmonic measures were constructed in [26, 29, 36] in terms of the Poincaré θ_2-series (3). Let $\tilde{\omega}_\ell(z)$ be a multivalued function harmonically conjugated to $\omega_\ell(z)$. The functions $w_\ell(z) = \omega_\ell(z) + i\tilde{\omega}_\ell(z)$ $(\ell = 1, 2, \ldots, n-1)$ analytic in D are called the normalized Abelian integrals of first kind in D. The differentials $dw_\ell(z)$ generate the linear space of the Abelian differentials of first kind and $dw_\ell(z)$ $(\ell = 1, 2, \ldots, n-1)$ form the basis of this space. Each differential $dw_\ell(z)$ takes pure imaginary values on ∂D. Hence, it can be analytically continued into $\tilde{D}$ in the topology of the Schottky double by the symmetry principle. Moreover, $dw_\ell(z)$ is single valued on $\mathcal{S}$.

The periods of the Abelian differentials

$$\int_{\mathbf{a}_k} dw_m(t) = 2\int_{\mathbf{a}'_k} dw_m(t), \quad B_{km} = \int_{\mathbf{b}_k} dw_m(t) \quad (k = 1, 2, \ldots, n-1)$$

form two matrices. The second one has the form iB, where $B = \{B_{km}\}$ is a real negatively determined matrix. Following [42] we consider the real Jacobi inversion problem. Let $w_k^-(t)$ denote the limit values of the Abelian integral on the curve $\mathbf{a}'_k$ when z tends to $t \in \mathbf{a}'_k$ from the right side of the curve $\mathbf{a}'_k$. The function $w_k^-(t)$ is multivalued. We fix any of its branch in the simply connected domain $D \setminus \left(\cup_{k=1}^{n-1} \mathbf{a}'_k\right)$ where it is single valued. Given constants e_k $(k = 1, 2, \ldots, n-1)$. To find the points z_m $(m = 1, 2, \ldots, n-1)$ in $D \cup \partial D$ satisfying the relation

$$\sum_{m=1}^{n-1} \operatorname{Im} w_k(z_m) \equiv e_k - \frac{1}{2} B_{kk} + \sum_{m \neq k}^{n-1} \operatorname{Im} \int_{\mathbf{a}'_k} w_k^-(t) dw_m(t) \quad (k = 1, 2, \ldots, n-1). \tag{61}$$

Here, $\equiv$ means equality modulo B-periods. The generalized real Jacobi inversion problem has the form [42]

$$\sum_{m=1}^{n-1} \operatorname{Im} w_k(z_m) \equiv \frac{1}{2\pi i}\int_{\partial D} \gamma(t)dw_m(t) \quad (k = 1,2,\ldots,n-1), \tag{62}$$

where $\gamma(t)$ is a given Hölder continuous function except at a finite number of points where finite step discontinuities are possible.

Let $\lambda(t)$ be a given Hölder continuous function on ∂D satisfying the condition $|\lambda(t)| = 1$. The Riemann–Hilbert problem

$$\operatorname{Im}[\overline{\lambda(t)}\psi(t)] = 0, \quad t \in \partial D, \tag{63}$$

was solved in terms of the α-series [26,29,36]. Let the functions $\lambda(t)$ from (63) and $\gamma(t)$ from (62) be related by formula

$$\lambda(t) = \exp[i\gamma(t)]. \tag{64}$$

We now consider the particular case (11) of the problem (63) and the corresponding generalized Jacobi inversion problem (62). We have

$$\lambda(t) = \frac{e^{i\alpha_k} r_k}{t - a_k}, \ \gamma(t) = \alpha_k - \arg\frac{t - a_k}{r_k}, \ |t - a_k| = r_k \quad (k = 1,2,\ldots,n). \tag{65}$$

The branch of the argument corresponds to the chosen branch of the logarithm $\ln(t - a_k)$ from Sect. 3. Each nontrivial solution of the problem (11) has exactly $n - 1$ zeros z_m $(m = 1,2,\ldots,n-1)$ in $D \cup \partial D$ which solve the generalized Jacobi inversion problem (62) with $\gamma(t)$ given by (65). The α-series (42) is a solution of (11).

The conditions $\xi_k = 0$ by (48) imply that all $A_k = 0$ in the representation (18). Hence, this case corresponds to the problem (13), (17) in a class of single-valued functions. The unique solution of this problem is given by (41). This function is the conformal mapping of the domain D onto the slit domain D' with the normalization (17). The function $\psi^{(1)}(z)$ given by (42) is the derivative of this conformal mapping. Hence, it cannot have zeros interior the domain D. Therefore, all the zeros z_m $(m = 1,2,\ldots,n-1)$ of $\psi(z)$ which solve the generalized Jacobi inversion problem (62) lie on the boundary ∂D. This observation can be useful to numerical solution of the Jacobi inversion problem on the Schottky double.

We now briefly explain where does disappear the Jacobi inversion problem in solution to the general Riemann–Hilbert problem by the method [26,34]

$$\operatorname{Re}[\overline{\lambda(t)}\varphi(t)] = f(t), \quad t \in \partial D. \tag{66}$$

Let the winding number $\varkappa$ of the nonvanishing Hölder continuous function $\lambda(t)$ be nonnegative. Application of the factorization method [19] to (66) reduces it to the partial problem:

$$\mathrm{Re}[e^{-i\alpha_k}\omega(t)] = h(t) + \sum_{s=1}^{2\varkappa} p_s\beta_s(t), \quad t \in \partial D. \tag{67}$$

A constructive factorization of the coefficient $\lambda(t)$ is performed by the Schwarz operator constructed in [26,34] by the classical θ_2-series. Here, the known functions $h(t)$, $\beta_s(t)$ are expressed in terms of $\lambda(t)$ and $f(t)$, the constants α_k in terms of $\lambda(t)$, and p_k are undetermined real constants. If $\varkappa < 0$, the sum $\sum_{s=1}^{2\varkappa}$ in (67) disappears and additionally the function $\omega(z)$ must have zero at infinity of order $|\varkappa|$. It is worth noting that differentiation of the homogeneous boundary conditions (67) ($h(t) = \beta_s(t) = 0$) yields (11). This is the reason why the special Riemann–Hilbert (11) is important. Further solution to the problem (67) repeats solution to the problems (13) and (11) by the use of the α-series with careful tracking of the arbitrary constants p_k and single valuedness conditions for the function $\omega(z)$. The method by Zverovich [42] contains an additional step which can be treated as the factorization of the coefficients $e^{-i\alpha_k}$ in (67) in order to reduce (67) to the Schwarz problem:

$$\mathrm{Re}\ \tilde{\omega}(t) = \tilde{h}(t), \quad t \in \partial D, \tag{68}$$

and further application of the Schwarz operator to (68). This factorization of $e^{-i\alpha_k}$ can be treated as solution to the boundary value problem (11). It is reduced to the Jacobi inversion problem solved by the theta function of Riemann. Therefore, the crucial point of the method [42] is an effective construction of the theta function of Riemann (not a multidimensional theta series, but its composition with the Abelian integrals). Such an effective construction of the theta function is described in this paper by solution to the homogeneous problem (11) in terms of the α-series. One can see that this step is redundant because the problem (67) can be directly solved in terms of the α-series [26, 34].

Remark 6.1. It is assumed in the theory of boundary value problems [19] that Riemann–Hilbert problems for a simply connected domain Ω have been solved up to a conformal mapping of Ω onto the unit disk. The construction of conformal mappings is a separate problem frequently pure numerical. The same point of view can be accepted for multiply connected domains. It is known that any multiply connected domain can be mapped onto a circular one [23]. For instance, the Schwarz–Christoffel formula for multiply connected domains bounded by polygons and its particular cases for slit domains were deduced in [12, 31, 35]. Effective numerical methods for conformal mappings were developed by many mathematicians (see [17, 24] and works cited therein) that supports the above point of view. However, the direct scheme [34] to solve Riemann–Hilbert problems for special multiply connected domains can be developed in the form of the generalized alternating method of Schwarz [25, 27, 36]. For instance, a boundary value problem for multiply connected domains bounded by nonoverlapping ellipses was solved in [30].

7 Fast Algorithm

Though the complete solution of the Riemann–Hilbert problem for an arbitrary circular multiply connected domain was written explicitly in terms of the α-series, many mathematicians apply the standard absolutely convergent scheme to the Poincaré series in partial cases and use direct methods of computation to the Poincaré series [39]. Perhaps, it is related to the fact that even absolutely convergent Poincaré series are slowly convergent for closely spaced disks. We suppose that modifications of the iterative functional equations can increase the convergence. In the present section, we discuss such a modification proposed in [37] to construct a basic solution of the problem (11). For brevity, we consider the classical Poincaré series when $\alpha_k = 0$ $(k = 1, 2, 3)$ for three equal disks $(r_k = r)$.

Consider an auxiliary problem for two disks. Let the domain G be the complement of two disjoint disks $|z - a_k| \leq r$ $(k = 1, 2)$ to the extended complex plane. The quadratic equation $z^*_{(1)} = z^*_{(2)}$ with respect to z has two roots

$$z_{12} = \frac{a_1 + a_2}{2} - \frac{a_2 - a_1}{2}\sqrt{1 - 2\frac{r_1^2 + r_2^2}{|a_2 - a_1|^2}}, \tag{69}$$

$$z_{21} = \frac{a_1 + a_2}{2} + \frac{a_2 - a_1}{2}\sqrt{1 - 2\frac{r_1^2 + r_2^2}{|a_2 - a_1|^2}},$$

The complex potential

$$\Psi_{12}(z) = \frac{1}{z - z_{12}} - \frac{1}{z - z_{21}} \tag{70}$$

describes the flux between the disks when the difference $u_1 - u_2$ of the potentials on the boundaries of the disks is equal to

$$u_1 - u_2 = \ln \frac{1 - \sqrt{1 - \frac{4r^2}{|a_2 - a_1|^2}}}{1 + \sqrt{1 - \frac{4r^2}{|a_2 - a_1|^2}}}. \tag{71}$$

The main idea of the fast method is based on the decomposition of the complex flux $\psi(z)$ onto $\psi_8(z)$ and $\psi_0(z)$ where the singular function $\psi_8(z)$ has the form

$$\psi_8(z) = \Psi_{12}(z) + \Psi_{13}(z), \tag{72}$$

where $\Psi_{13}(z)$ is introduced similar to (70) (the subscript 2 is replaced by 3). A solution of the boundary value problem (11) for $n = 3$ is looked for in the form

$$\psi(z) = \psi_0(z) + \psi_8(z), \quad z \in \mathbb{D}, \tag{73}$$

where $\psi_8(z)$ is given by (72). The boundary value problem (11) becomes

$$\operatorname{Im} \frac{t - a_k}{r} [\psi_0(t) + \psi_8(t)] = 0, \quad |t - a_k| = r, \ k = 1, 2, 3. \tag{74}$$

Introduce the functions analytic in $|z - a_k| < r$

$$f_k(z) = \begin{cases} 0 & \text{for } k = 1 \\ \Psi_{13}(z) & \text{for } k = 2, \\ \Psi_{12}(z) & \text{for } k = 3. \end{cases} \tag{75}$$

One can see that

$$\operatorname{Im} \frac{t - a_m}{r} \Psi_{12}(t) = 0, \quad |t - a_m| = r \ (m = 1, 2)$$

and

$$\operatorname{Im} \frac{t - a_m}{r} \Psi_{13}(t) = 0, \quad |t - a_m| = r \ (m = 1, 3).$$

Then (74) can be written in the form

$$\operatorname{Im} \frac{t - a_k}{r} [\psi_0(t) + f_k(t)] = 0, \quad |t - a_k| = r, \ k = 1, 2, 3. \tag{76}$$

The boundary value problem (76) is reduced to the $\mathbb{R}$-linear problem

$$\psi_0(t) = \psi_k(t) + \left(\frac{r}{t - a_k}\right)^2 \overline{\psi_k(t)} - f_k(t), \ |t - a_k| = r, \ k = 1, 2, 3. \tag{77}$$

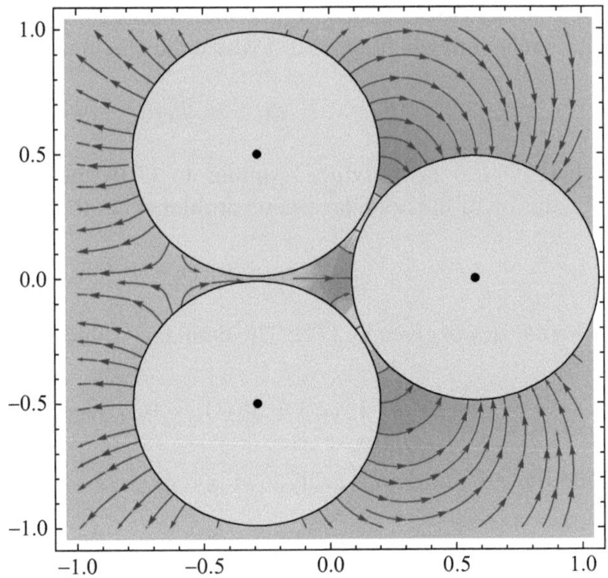

Fig. 2 Streamlines of the complex flux $\psi(z)$ computed in the sixth order approximation around three disks with the centers at $a_1 = \frac{1}{\sqrt{3}}$, $a_2 = \frac{1}{\sqrt{3}} e^{\frac{2}{3}\pi i}$, $a_3 = \frac{1}{\sqrt{3}} e^{-\frac{2}{3}\pi i}$ of the radius 0.49

The problem (77) can be reduced to the following system of functional equations (see Sect. 4):

$$\psi_k(z) = \sum_{m \neq k} \left(\frac{r_m}{z - a_m} \right)^2 \overline{\psi_m \left(z_{(m)}^* \right)} + f_k(z). \tag{78}$$

The iteration method can be applied to solve the system (24)

$$\psi_k^{(0)}(z) = f_k(z), \tag{79}$$

$$\psi_k^{(p)}(z) = \sum_{m \neq k} \left(\frac{r_m}{z - a_m} \right)^2 \overline{\psi_m^{(p-1)} \left(z_{(m)}^* \right)} + f_k(z), \; p = 1, 2, \ldots. \tag{80}$$

The p-th approximation of the complex flux is calculated by formula

$$\psi^{(p)}(z) = \sum_{m=1,2,3} \left(\frac{r_m}{z - a_m} \right)^2 \overline{\psi_m^{(p)} \left(z_{(m)}^* \right)} + \psi_8(z), \quad z \in \mathbb{D}, \tag{81}$$

where $\psi_8(z)$ is given by (72).

Figure 2 describes the flux around three closely placed disks. It follows from computations that 6 iterations are sufficient to obtain an accessible result (the respective error on the boundary is 2%).

8 Discussion

In the present paper, we introduce the Poincaré α-series. First time, the α-series were used in [26] without their deep discussion to solve Riemann–Hilbert problems for an arbitrary circular multiply connected domain. The main difference in the methods [41] and [26, 34] applied to Riemann–Hilbert problems is that [41] is based on the Jacobi inversion problem and [26, 34] is not. But [26, 34] includes the α-series that coincide with the theta function of Riemann. Hence, the solution to the Jacobi inversion problem is implicitly included in the α-series. It is worth noting that solution to Riemann–Hilbert problems in [41] and later investigations by this scheme are not completed. First of all, it was assumed in [41] that the Abelian integrals of first kind were known. Substitution of the Abelian integrals into the multidimensional theta series yielded the theta function of Riemann. The latter function was applied to investigate the Jacobi inversion problem. After this the Schwarz operator was applied to get the solution of the Riemann–Hilbert problem. Construction of the Abelian integrals (harmonic measures) and the Schwarz operator in terms of the classical θ_2-series of Poincaré [26, 34] could make this complicated scheme effective. However, the generalized Jacobi inversion problem cannot be avoided in the scheme [41].

Application of the α-series simplifies solution to Riemann–Hilbert problems by elimination of the Jacobi inversion problem and produces directly the theta function of Riemann. Moreover, the scheme [26, 34] allows to constructively solve the Jacobi inversion problem as a separately stated problem. Bojarski's linear algebraic system (see Bojarski's addition to [40]) which describes solvability of the Riemann–Hilbert problem is explicitly written in terms of the α-series. The constructive method [26, 34] is valid for the Schottky double. It is interesting to extend it to the general compact Riemann surfaces.

Crowdy [8] stated open problems of the constructive theory of functions in multiply connected domains. In particular, Crowdy wrote [8] about Schwarz–Christoffel-type conformal mappings: "The history of this particular problem also presents a paradigm for a key message of this paper: that, given modern advances in computational capability and in light of modern applications, many topics in classical geometric function theory can (and should!) be revisited and reappraised." I think that this phrase should concern the whole constructive theory of functions in multiply connected domains. In this paper, we answer some questions stated by Crowdy [8]. It is worth noting that these answers are not always complete and require further investigations.

Question 1 of Crowdy addressed to the infinite product representation (57) for the Schottky–Klein prime function which is always uniformly convergent. Some properties of the function were described by use of the absolutely convergent product (57), in particular, by changing the order of multiplication in this product that is forbidden for conditionally convergent series. In order to extend investigations to uniformly convergent products, one can use functional equations following lines of the paper [28] where the automorphic property (4) was proved without changing the order of summation.

Question 2 of Crowdy concerned effective computational methods. Such a fast method is presented in Sect. 7 and Appendix for a triply connected domain. Actually, it concerns only the Dirichlet problems with constant boundary values. Further extensions are possible by deriving computationally effective formulae for boundary value problems in doubly connected domains.

Question 3 of Crowdy concerned the complicated scheme by Zverovich [41] used by many authors for Riemann–Hilbert problems. It is explained above in this section that the method [26, 34] based on α-series is constructive and simpler than the method [41].

Crowdy in Question 4 paid attention to an alternative class of canonical multiply connected domains introduced by Bell [2, 3, 7, 22]. It can be add to this that Bell's domains have applications to neutral inclusions [20]. The latter problem is related to eigenvalue problems, Courant's nodal domain theorem, the $\mathbb{R}$-linear [33], and nonlinear Riemann–Hilbert problems discussed in [32]. It is interesting to study the $\mathbb{R}$-linear eigenvalue problems for Bell's domains in order to estimate the minimal size of the coating of invisible inclusions.

Question 5 of Crowdy addressed to the Riemann–Hilbert problem (13) and eventual use of the classical Schottky–Klein prime function. As it follows from the result of this paper, the α-prime function (58) can be applied to (13) and it is rather impossible to solve the problem (13) in terms of the classical prime function (57).

Acknowledgements The author is grateful to D. Crowdy and T. DeLillo for helpful discussions, E.A. Krushevski for discussions concerning the results [41], and A.E. Malevich for the help in preparation of the code (see Appendix).

Appendix

In this section, the code in Mathematica© is attached for symbolic computations used in Sect. 7.

```
ClearAll[Z1];
Z1[i_Integer, j_Integer, R_] :=
 (Z1[i, j, R] = (a[[i]] + a[[j]])/2 + (R[[i]]^2 - R[[j]]^2)/(2 Conjugate[a[[j]] - a[[i]]]) + (a[[i]] - a[[j]])/2
    Sqrt[1 + (R[[i]]^2 - R[[j]]^2)^2/Abs[a[[i]] - a[[j]]]^4 - 2 (R[[i]]^2 + R[[j]]^2)/Abs[a[[i]] - a[[j]]]^2]) /; 0 < i ≤ 𝒩 ∧ 0 < j ≤ 𝒩
```

```
ClearAll[F];
(**F[1,_,_]=0;**)
F[k : 1, R_, z_] := F[k, R, z] = 0;
F[k : 2, R_, z_] := F[k, R, z] = 1/(z - Z1[3, 1, R]) - 1/(z - Z1[1, 3, R]);
F[k : 3, R_, z_] := F[k, R, z] = 1/(z - Z1[2, 1, R]) - 1/(z - Z1[1, 2, R]);
ClearAll[ψ];
ψ[0, k_Integer, R_, z_] := (ψ[0, k, R, z] = F[k, R, z]) /; 0 < k ≤ 𝒩
```

```
ClearAll[Inv];
Inv[m_Integer, R_, z_] := (Inv[m, R, z] = R[[m]]^2/(z - a[[m]])* + a[[m]]) /; 0 < m ≤ 𝒩
ψ[Step_Integer?Positive, k_Integer, R_, z_] :=
 (ψ[Step, k, R, z] = F[k, R, z] + Sum[{ 0                                                     m == k
                                        R[[m]]^2 ψ[Step-1, m, R, Inv[m, R, z]]*/(z-a[[m]])^2   True }, {m, 1, 𝒩}]) /; 0 < k ≤ 𝒩
```

```
MT[z_, R_] := Min[Table[Abs[z - a[[k]]] - R[[k]], {k, 1, N}]]

ClearAll[Ψ];

Ψ1[Step_, R_, z_] := Ψ1[Step, R, z] = Sum[R[[m]]^2 ψ[Step, m, R, Inv[m, R, z]]* / (z - a[[m]])^2, {m, 1, N}] +
    (1/(z - Z1[3, 1, R]) - 1/(z - Z1[1, 3, R])) + (1/(z - Z1[2, 1, R]) - 1/(z - Z1[1, 2, R]));
Ψ[Step_, R_, z_] := Ψ[Step, R, z] = If[MT[z, R] < 0, 0, Ψ1[Step, R, z]]

ΨA[Step_, R_, z_] := ΨA[Step, R, z] = Apart[Ψ1[Step, R, z] // N] // Chop
IΨA[Step_, R_, w_] := ∫ΨA[Step, R, z] ⅆz /. z → w;
ϕ[Step_, R_, z_] := If[MT[z, R] < 0, 0, IΨA[Step, R, z]]
u[Step_, R_, z_] := Re[ϕ[Step, R, z]]
dΨ[Step_, R_, z_] := dΨ[Step, R, z] = Ψ[Step, R, z] - Ψ[Step - 1, R, z]

q[Step_, R_] := StreamDensityPlot[{Re[Ψ[Step, R, x + ⅈ y]], -Im[Ψ[Step, R, x + ⅈ y]]},
   {x, -1, 1}, {y, -1, 1}, Epilog → FigD[R], ColorFunction → "LightTerrain"]

q[3, rN]
```

References

1. Baker, H.F.: Abelian Functions: Abel's Theorem and the Allied Theory, Including the Theory of the Theta Functions. New York, Cambridge University Press (1995)
2. Bell, S.: Finitely generated function fields and complexity in potential theory in the plane. Duke Math. J. **98**, 187–207 (1999)
3. Bell, S.: A Riemann surface attached to domains in the plane and complexity in potential theory. Houston J. Math. **26**, 277–297 (2000)
4. Bobenko, A.I., Klein Ch. (Eds.): Computational Approach to Riemann Surfaces, Lecture Notes in Mathematics. Berlin, Springer (2011)
5. Bogatyrev, A.: Extremal Polynomials and Riemann Surfaces. Berlin, Springer (2012)
6. Crowdy, D.: Conformal mappings between canonical multiply connected domains. Comput. Methods Funct. Theory **6**(1), 59–76 (2006)
7. Crowdy, D.: Conformal mappings from annuli to canonical doubly connected Bell representations J. Math. Anal. Appl. **340**, 669–674 (2007)
8. Crowdy, D.: Geometric function theory: a modern view of a classical subject. Nonlinearity **21**, T205–T219 (2008)
9. Crowdy, D.: The Schwarz problem in multiply connected domains and the Schottky–Klein prime function. Complex Variables and Elliptic Equat. **53**, 221–236 (2008)
10. Crowdy, D.: The Schottky–Klein prime function on the schottky double of planar domains. Comput. Methods Funct. Theory **10**(2), 501–517 (2010)
11. Crowdy, D.G., Fokas, A.S., Green, C.C.: Conformal mappings to multiply connected polycircular arc domains. Comput. Methods Funct. Theory **11**(2), 685–706 (2011)
12. Czapla, R., Mityushev, V., Rylko, N.: Conformal mapping of circular multiply connected domains onto slit domains. Electron. Trans. Numer. Anal. **39**, 286–297 (2012)
13. DeLillo, T.K. Elcrat, A.R., Pfaltzgraff, J.A.: Schwarz–Christoffel mapping of multiply connected domains. J. d'Analyse Mathematique **94**, 17–47(2004)
14. DeLillo, T.K.: Schwarz–Christoffel mapping of bounded, multiply connected domains. Comput. Methods Funct. Theory **6**(2), 275–300 (2006)
15. DeLillo, T.K., Driscoll, T.A., Elcrat, A.R., Pfaltzgraff, J.A.: Computation of multiply connected Schwarz–Christoffel map for exterior domains. Comput. Methods Funct. Theory **6**(2), 301–315 (2006)
16. DeLillo, T.K., Driscoll, T.A., Elcrat, A.R., Pfaltzgraff, J.A.: Radial and Circular Slit Maps of Unbounded Multiply Connected Circle Domains. vol. A464, pp. 1719–1737. Proceedings of the Royal Society, London (2008)
17. Driscoll, T.A., Trefethen, L.N.: Schwarz-Christoffel Mapping. Cambridge University Press, Cambridge (2002)
18. Dubrovin, B.A.: Riemann surfaces and non-linear equations. RHD Publication, Izhevsk (2001) (in Russian)
19. Gakhov, F.D.: Boundary Value Problems. Nauka, Moscow, (1977) (3rd edn. in Russian); English translation of 1st edn.: Pergamon Press, Oxford (1966)
20. Jarczyk, P., Mityushev, V.: Neutral coated inclusions of finite conductivity. Proc. Roy. Soc. London A, **468**A, 954–970 (2012)
21. Jeong, M., Mityushev, V.: The Bergman kernel for circular multiply connected domains. Pacifc J. Math. **233**, 145–157 (2007)
22. Jeong, M., Oh, J.-W., Taniguchi, M.: Equivalence problem for annuli and Bell representations in the plane. J. Math. Anal. Appl. **325**, 1295–1305 (2007)
23. Golusin, G.M.: Geometric Theory of Functions of Complex Variable. Nauka, Moscow 1966 (2nd edn. in Russian); English translation by AMS, Providence, RI (1969)
24. Kühnau, R.: Handbook of Complex Analysis: Geometric Function Theory. Elsevier, North Holland, Amsterdam (2005)
25. Mikhlin, S.G.: Integral Equations, Pergamon Press, New York (1964)

26. Mityushev, V.V.: Solution of the Hilbert boundary value problem for a multiply connected domain. Slupskie Prace Mat.-Przyr. **9a**, 37–69 (1994)
27. Mityushev, V.V.: Generalized method of Schwarz and addition theorems in mechanics of materials containing cavities. Arch. Mech. **47**(6), 1169–1181 (1995)
28. Mityushev, V.V.: Convergence of the Poincaré series for classical Schottky groups. Proc. Amer. Math. Soc. **126**(8), 2399–2406 (1998)
29. Mityushev, V.V.: Hilbert boundary value problem for multiply connected domains. Complex Variables **35**, 283–295 (1998)
30. Mityushev, V.: Conductivity of a two–dimensional composite containing elliptical inclusions, Proc Royal Soc, London **A465**, 2991–3010 (2009)
31. Mityushev, V.: Riemann-Hilbert problems for multiply connected domains and circular slit maps. Comput. Methods Funct. Theory, **11**(2), 575–590 (2011)
32. Mityushev, V.V.: New boundary value problems and their applications to invisible materials. In: Rogosin, S.V. (ed.) Analytic Methods of Analysis and Differential Equations: AMADE-2011: materials of the 6th International Conference dedicated to the memory of prof. A.A. Kilbas, pp. 141–146. Publishing House of BSU, Minsk (2012)
33. Mityushev, V.: $\mathbb{R}$–linear and Riemann–Hilbert Problems for Multiply Connected Domains. In: Rogosin, S.V., Koroleva, A.A. (eds.) Advances in Applied Analysis. pp. 147–176. Birkhäuser, Basel (2012)
34. Mityushev, V.: Scalar Riemann-Hilbert Problem for Multiply Connected Domains, In: Rassias, Th.M., Brzdek, J. (eds.) Functional Equations in Mathematical Analysis, Springer Optimization and Its Applications, vol. 52, pp. 599–632. Springer Science+Business Media, LLC (2012)
35. Mityushev, V.: Schwarz–Christoffel formula for multiply connected domains. Comput. Methods Funct. Theory **12**(2), 449–463 (2012)
36. Mityushev, V.V., Rogosin, S.V.: Constructive methods to linear and non-linear boundary value problems of the analytic function. Theory and applications. Monographs and Surveys in Pure and Applied Mathematics, Chapman & Hall / CRC, Boca Raton etc. (2000)
37. Mityushev, V., Rylko, N.: A fast algorithm for computing the flux around non–overlapping disks on the plane. Mathematical and Computer Modelling, **57**, 1350–1359 (2013)
38. Mityushev, V., Pesetskaya, E., Rogosin, S.: Analytical Methods for Heat Conduction, in Composites and Porous Media in Cellular and Porous Materials Ochsner A., Murch G, de Lemos M. (eds.) Wiley-VCH, Weinheim (2008)
39. Schmies, M.: Computing Poincaré Theta series for Schottky groups. In Bobenko, A.I., Klein, Ch. (eds.): Computational Approach to Riemann Surfaces, Lecture Notes in Mathematics, pp. 165–182. Berlin, Springer (2011)
40. Vekua, I.N.: Generalized Analytic Functions. Nauka, Moscow (1988) (2nd edn. in Russian); English translation of 1st ed.: Pergamon Press, Oxford (1962)
41. Zverovich, E.I.: Boundary value problems in the theory of analytic functions in Hölder classes on Riemann surfaces. Russ. Math. Surv. **26**(1), 117–192 (1971)
42. Zverovich, E.I.: The inversion Jacobi problem, its generalizations and some applications. Aktual'nye problemy sovremennogo analiza, pp. 69–83. Grodno University Publication, Grodno (2009) (in Russian)

Inclusion Properties for Certain Classes of Meromorphic Multivalent Functions

Nak Eun Cho

Dedicated to Professor Hari M. Srivastava

Abstract Making use of an operator $I_{\lambda,\mu}$, which is defined by using convolution, the author introduces several new subclasses of meromorphic multivalent functions and investigates various inclusion properties of these subclasses. Some interesting applications involving these and other classes of integral operators are also considered.

1 Introduction

Let Σ_p denote the class of functions of the form

$$f(z) = \frac{1}{z^p} + \sum_{k=1}^{\infty} a_k z^{k-p} \qquad (p \in \mathbb{N} = \{1, 2, \ldots\}),$$

which are analytic in the punctured open unit disk $\mathbb{D} = \{z \in \mathbb{C} : 0 < |z| < 1\}$. If f and g are analytic in $\mathbb{U} = \mathbb{D} \cup \{0\}$, we say that f is subordinate to g, written $f \prec g$ or $f(z) \prec g(z)$, if there exists a Schwarz function w in $\mathbb{U}$ such that $f(z) = g(w(z))$. For $0 \leq \eta, \beta < 1$, we denote by $\mathcal{MS}_p(\eta)$ and $\mathcal{MK}_p(\eta)$ and $\mathcal{MC}_p(\eta, \beta)$ the subclasses of Σ_p consisting of all meromorphic functions which are, respectively, p-valent starlike of order η and p-valent convex of order η and p-valent close to convex of order β and type η in $\mathbb{U}$ (for details, see, e.g., [5]).

Let $\mathcal{N}$ be the class of univalent functions ϕ in $\mathbb{U}$ normalized by $\phi(0) = 1$ for which $\phi(\mathbb{U})$ is convex and $\mathrm{Re}\{\phi(z)\} > 0$ $(z \in \mathbb{U})$.

N.E. Cho (✉)
Department of Applied Mathematics, Pukyong National University, Busan 608-737, South Korea
e-mail: necho@pknu.ac.kr

G.V. Milovanović and M.Th. Rassias (eds.), *Analytic Number Theory, Approximation Theory, and Special Functions*, DOI 10.1007/978-1-4939-0258-3_34,

Making use of the principle of subordination between analytic functions, we introduce the subclasses $\mathcal{MS}_p(\eta, \phi)$, $\mathcal{MK}_p(\eta, \phi)$, and $\mathcal{MC}_p(\eta, \beta;\ \phi, \psi)$ of the class Σ_p for $0 \leq \eta, \beta < 1$, and $\phi, \psi \in \mathcal{N}$, which are defined by

$$\mathcal{MS}_p(\eta;\ \phi) := \left\{ f \in \Sigma_p : \ \frac{1}{p-\eta}\left(-\frac{zf'(z)}{f(z)} - \eta\right) \prec \phi(z) \text{in } \mathbb{U} \right\},$$

$$\mathcal{MK}_p(\eta;\ \phi) := \left\{ f \in \Sigma_p : \ \frac{1}{p-\eta}\left(-\left\{1 + \frac{zf''(z)}{f'(z)}\right\} - \eta\right) \prec \phi(z) \text{in } \mathbb{U} \right\},$$

and

$$\mathcal{MC}_p(\eta, \beta;\ \phi, \psi) : = \left\{ f \in \Sigma_p : \exists g \in \mathcal{MS}_p(\eta;\ \phi) \right.$$

$$\left. \text{such that } \frac{1}{1-\beta}\left(-\frac{zf'(z)}{g(z)} - \beta\right) \prec \psi(z) \text{in } \mathbb{U} \right\}.$$

We note that the classes mentioned above are motivated essentially by the familiar classes which have been used widely on the space of analytic and univalent functions in $\mathbb{U}$ (see, for details, [2, 6, 9]), and for special choices for the functions ϕ and ψ involved in these definitions, we can obtain the well-known subclasses of Σ_1 [1,4,5].

Let

$$f_\lambda(z) = \frac{1}{z^p(1-z)^{\lambda+p}} \qquad (\lambda > -p;\ z \in \mathbb{D})$$

and let $f_{\lambda,\mu}$ be defined such that

$$f_\lambda(z) * f_{\lambda,\mu}(z) = \frac{1}{z^p(1-z)^{\mu}} \qquad (\lambda > -p;\ \mu > 0;\ z \in \mathbb{D}), \tag{1}$$

where the symbol $(*)$ stands for the Hadamard product (or convolution). Then we define the operator $I_{\lambda,\mu} : \Sigma_p \to \Sigma_p$ as follows:

$$I_{\lambda,\mu} f(z) = \left(f_{\lambda,\mu} * f\right)(z) \quad (f \in \Sigma_p;\ \lambda > -p;\ \mu > 0). \tag{2}$$

In particular, we note that $I_{0,p+1} f(z) = (zf'(z) + 2f(z))/p$ and $I_{1,p+1} f(z) = f(z)$. In view of (1) and (2), we obtain the useful identities as follows:

$$z\left(I_{\lambda+1,\mu} f(z)\right)' = (\lambda + p) I_{\lambda,\mu} f(z) - (\lambda + 2p) I_{\lambda+1,\mu} f(z) \tag{3}$$

and

$$z\left(I_{\lambda,\mu} f(z)\right)' = \mu I_{\lambda,\mu+1} f(z) - (\mu + p) I_{\lambda,\mu} f(z). \tag{4}$$

For $p = 1$, the operator $I_{\lambda,\mu}$ is closely related to the Choi–Saigo–Srivastava operator for analytic and univalent functions [2], which extends the Noor integral operator studied by Liu [7] (also, see [8, 11, 12]).

Next, by using the operator $I_{\lambda,\mu}$, we introduce the following classes of meromorphic functions for $\phi, \psi \in \mathcal{N}$, $\lambda > -p$, $\mu > 0$, and $0 \leq \eta, \beta < p$:

$$\mathcal{MS}_p(\lambda, \mu;\ \eta;\ \phi) := \{f \in \Sigma_p : I_{\lambda,\mu} f \in \mathcal{MS}_p(\eta; \phi)\},$$

$$\mathcal{MK}_p(\lambda, \mu;\ \eta;\ \phi) := \{f \in \Sigma_p : I_{\lambda,\mu} f \in \mathcal{MK}_p(\eta; \phi)\},$$

and

$$\mathcal{MC}_p(\lambda, \mu;\ \eta, \beta;\ \phi, \psi) := \{f \in \Sigma_p : I_{\lambda,\mu} f \in \mathcal{MC}_p(\eta, \beta; \phi, \psi)\}.$$

We also note that

$$f(z) \in \mathcal{MK}_p(\lambda, \mu;\ \eta;\ \phi) \iff -zf'(z)/p \in \mathcal{MS}_p(\lambda, \mu;\ \eta;\ \phi). \tag{5}$$

In particular, we set

$$\mathcal{MS}_p\left(\lambda, \mu;\ \eta;\ \frac{1 + Az}{1 + Bz}\right) = \mathcal{MS}_p(\lambda, \mu;\ \eta;\ A, B)\ \ (-1 \leq B < A \leq 1)$$

and

$$\mathcal{MK}_p\left(\lambda, \mu;\ \eta;\ \frac{1 + Az}{1 + Bz}\right) = \mathcal{MK}_p(\lambda, \mu;\ \eta;\ A, B)\ \ (-1 \leq B < A \leq 1).$$

In this paper, we investigate several inclusion properties of the classes

$$\mathcal{MS}_p(\lambda, \mu;\ \eta; \phi),\quad \mathcal{MK}_p(\lambda, \mu;\ \eta; \phi),\quad \text{and}\quad \mathcal{MC}_p(\lambda, \mu;\ \eta, \beta;\ \phi, \psi)$$

associated with the operator $I_{\lambda,\mu}$. Some applications involving integral operators are also considered.

2 Inclusion Properties Involving the Operator $I_{l,\mu}$

The following results will be needed in our investigation.

Lemma 2.1 ([3]). *Let ϕ be convex univalent in $\mathbb{U}$ with $\phi(0) = 1$ and*

$$\mathrm{Re}\{\kappa\phi(z) + \nu\} > 0 \quad (\kappa, \nu \in \mathbb{C}).$$

If p is analytic in $\mathbb{U}$ with $p(0) = 1$, then

$$p(z) + \frac{zp'(z)}{\kappa p(z) + \nu} \prec \phi(z) \quad (z \in \mathbb{U})$$

implies that

$$p(z) \prec \phi(z) \quad (z \in \mathbb{U}).$$

Lemma 2.2 ([10]). *Let ϕ be convex univalent in $\mathbb{U}$ and ω be analytic in $\mathbb{U}$ with* $\mathrm{Re}\{\omega(z)\} \geq 0$. *If p is analytic in $\mathbb{U}$ and $p(0) = \phi(0)$, then*

$$p(z) + \omega(z)zp'(z) \prec \phi(z) \quad (z \in \mathbb{U})$$

implies that

$$p(z) \prec \phi(z) \quad (z \in \mathbb{U}).$$

Theorem 2.1. *Let $\phi \in \mathcal{N}$ with*

$$\max_{z\in\mathbb{U}} \mathrm{Re}\{\phi(z)\} < \min\left\{\frac{\mu + p - \eta}{p - \eta}, \frac{\lambda + 2p - \eta}{p - \eta}\right\} \quad (\lambda > -p;\ \mu > 0;\ 0 \leq \eta < p).$$

Then

$$\mathcal{MS}_p(\lambda, \mu + 1;\ \eta;\ \phi) \subset \mathcal{MS}_p(\lambda, \mu;\ \eta;\ \phi) \subset \mathcal{MS}_p(\lambda + 1, \mu;\ \eta;\ \phi).$$

Proof. First of all, we will show that

$$\mathcal{MS}_p(\lambda, \mu + 1;\ \eta;\ \phi) \subset \mathcal{MS}_p(\lambda, \mu;\ \eta;\ \phi).$$

Let $f \in \mathcal{MS}_p(\lambda, \mu + 1;\ \eta;\ \phi)$ and set

$$p(z) = \frac{1}{p - \eta}\left(-\frac{z(I_{\lambda,\mu} f(z))'}{I_{\lambda,\mu} f(z)} - \eta\right), \tag{6}$$

here $p(z)$ is analytic in $\mathbb{U}$ with $p(0) = 1$. Applying (4) and (6), we obtain

$$-\mu\frac{I_{\lambda,\mu+1} f(z)}{I_{\lambda,\mu} f(z)} = (p - \eta)p(z) - (\mu + p - \eta). \tag{7}$$

Taking the logarithmic differentiation on both sides of (7) and multiplying by z, we have

$$\frac{1}{p - \eta}\left(-\frac{z(I_{\lambda,\mu+1} f(z))'}{I_{\lambda,\mu+1} f(z)} - \eta\right) = p(z) + \frac{zp'(z)}{-(p - \eta)p(z) + \mu + p - \eta} \quad (z \in \mathbb{U}). \tag{8}$$

Since

$$\max_{z\in\mathcal{U}} \mathrm{Re}\{\phi(z)\} < \frac{\mu+p-\eta}{p-\eta},$$

we see that

$$Re\{-(p-\eta)\phi(z)+\mu+p-\eta\} > 0 \quad (z \in \mathbb{U}).$$

Applying Lemma 2.1 to (8), it follows that $p(z) \prec \phi(z)$ for $z \in \mathbb{U}$, that is, $f \in \mathcal{MS}_p(\lambda, \mu;\ \eta;\ \phi)$.

To prove the second part, let $f \in \mathcal{MS}_p(\lambda, \mu;\ \eta;\ \phi)$ and put

$$s(z) = \frac{1}{p-\eta}\left(-\frac{z(I_{\lambda+1,\mu}f(z))'}{I_{\lambda+1,\mu}f(z)} - \eta\right),$$

where s is an analytic function with $s(0) = 1$. Then, by using the arguments similar to those detailed above with (3), it follows that $s(z) \prec \phi(z)$ for $z \in \mathbb{U}$, which implies that $f \in \mathcal{MS}_p(\lambda+1, \mu;\ \eta;\ \phi)$.

Therefore, we complete the proof of this statement. □

Theorem 2.2. *Let* $\phi \in \mathcal{N}$ *with*

$$\max_{z\in\mathbb{U}} \mathrm{Re}\{\phi(z)\} < \min\left\{\frac{\mu+p-\eta}{p-\eta}, \frac{\lambda+2p-\eta}{p-\eta}\right\} \quad (\lambda > -p;\ \mu > 0;\ 0 \le \eta < p).$$

Then

$$\mathcal{MK}_p(\lambda, \mu+1;\ \eta;\ \phi) \subset \mathcal{MK}_p(\lambda, \mu;\ \eta;\ \phi) \subset \mathcal{MK}_p(\lambda+1, \mu;\ \eta;\ \phi).$$

Proof. Applying (5) and Theorem 2.1, we observe that

$$\begin{aligned} f(z) \in \mathcal{MK}_p(\lambda, \mu+1;\ \eta;\ \phi) &\Longleftrightarrow -zf'(z)/p \in \mathcal{MS}_p(\lambda, \mu+1;\ \eta;\ \phi) \\ &\Longrightarrow -zf'(z)/p \in \mathcal{MS}_p(\lambda, \mu;\ \eta;\ \phi) \\ &\Longleftrightarrow I_{\lambda,\mu}f(z) \in \mathcal{MK}_p(\eta;\ \phi) \\ &\Longleftrightarrow f(z) \in \mathcal{MK}_p(\lambda, \mu;\ \eta;\ \phi), \end{aligned}$$

and

$$\begin{aligned} f(z) \in \mathcal{MK}_p(\lambda, \mu;\ \eta;\ \phi) &\Longleftrightarrow -zf'(z)/p \in \mathcal{MS}_p(\lambda, \mu;\ \eta;\ \phi) \\ &\Longrightarrow -zf'(z)/p \in \mathcal{MS}_p(\lambda+1, \mu;\ \eta;\ \phi) \\ &\Longleftrightarrow I_{\lambda+1,\mu}f(z) \in \mathcal{MK}_p(\eta;\ \phi) \\ &\Longleftrightarrow f(z) \in \mathcal{MK}_p(\lambda+1, \mu;\ \eta;\ \phi), \end{aligned}$$

which evidently proves Theorem 2.2. □

Taking

$$\phi(z) = \frac{1 + Az}{1 + Bz} \quad (-1 \leq B < A \leq 1;\ z \in \mathbb{U})$$

in Theorems 2.1 and 2.2, we have the following result.

Corollary 2.1. *Let*

$$\frac{1 + A}{1 + B} < \min\left\{\frac{\mu + p - \eta}{p - \eta}, \frac{\lambda + 2p - \eta}{p - \eta}\right\},$$

with $\lambda > -p;\ \mu > 0;\ 0 \leq \eta < p;\ -1 < B < A \leq 1$. *Then*

$$\mathcal{MS}_p(\lambda, \mu + 1;\ \eta;\ A, B) \subset \mathcal{MS}_p(\lambda, \mu;\ \eta;\ A, B) \subset \mathcal{MS}_p(\lambda + 1, \mu;\ \eta;\ A, B)$$

and

$$\mathcal{MK}_p(\lambda, \mu + 1;\ \eta;\ A, B) \subset \mathcal{MK}_p(\lambda, \mu;\ \eta;\ A, B) \subset \mathcal{MK}_p(\lambda + 1, \mu;\ \eta;\ A, B).$$

Next, by using Lemma 2.2, we obtain the following inclusion relation for the class $\mathcal{MC}_p(\lambda, \mu;\ \eta, \beta;\ \phi, \psi)$.

Theorem 2.3. *Let* $\phi, \psi \in \mathcal{N}$ *with*

$$\max_{z \in \mathbb{U}} \mathrm{Re}\{\phi(z)\} < \min\left\{\frac{\mu + p - \eta}{p - \eta}, \frac{\lambda + 2p - \eta}{p - \eta}\right\} \quad (\lambda > -p;\ \mu > 0;\ 0 \leq \eta < p).$$

Then

$$\mathcal{MC}_p(\lambda, \mu{+}1;\ \eta, \beta;\ \phi, \psi) \subset \mathcal{MC}_p(\lambda, \mu;\ \eta, \beta;\ \phi, \psi) \subset \mathcal{MC}_p(\lambda{+}1, \mu;\ \eta, \beta;\ \phi, \psi).$$

Proof. We begin by proving that

$$\mathcal{MC}_p(\lambda, \mu + 1;\ \eta, \beta;\ \phi, \psi) \subset \mathcal{MC}_p(\lambda, \mu;\ \eta, \beta;\ \phi, \psi).$$

Let $f \in \mathcal{MC}_p(\lambda, \mu + 1;\ \eta, \beta;\ \phi, \psi)$. Then, in view of the definition of the class $\mathcal{MC}_p(\lambda, \mu + 1;\ \eta, \beta;\ \phi, \psi)$, there exists a function $r \in \mathcal{MS}_p(\eta;\ \phi)$ such that

$$\frac{1}{p - \beta}\left(-\frac{z(I_{\lambda,\mu+1} f(z))}{r(z)} - \beta\right) \prec \psi(z) \quad (z \in \mathbb{U}).$$

Choose the function g such that $I_{\lambda,\mu+1} g(z) = r(z)$. Then $g \in \mathcal{MS}_p(\lambda, \mu + 1; \eta; \phi)$ and

$$\frac{1}{p-\beta}\left(-\frac{z(I_{\lambda,\mu+1}f(z))'}{I_{\lambda,\mu+1}g(z)}-\beta\right) \prec \psi(z) \quad (z \in \mathbb{U}). \tag{9}$$

Now let

$$p(z) = \frac{1}{p-\beta}\left(\frac{z(I_{\lambda,\mu}f(z))'}{I_{\lambda,\mu}g(z)}-\beta\right), \tag{10}$$

where $p(z)$ is analytic in $\mathbb{U}$ with $p(0) = 1$. Using (4), we obtain

$$\frac{1}{p-\beta}\left(-\frac{z(I_{\lambda,\mu+1}f(z))'}{I_{\lambda,\mu+1}g(z)}-\beta\right)$$

$$= \frac{1}{p-\beta}\left(\frac{\dfrac{z(I_{\lambda,\mu}(-zf'(z)))'}{I_{\lambda,\mu}g(z)} + (\mu+p)\dfrac{I_{\lambda,\mu}(-zf'(z))}{I_{\lambda,\mu}g(z)}}{\dfrac{z(I_{\lambda,\mu}g(z))'}{I_{\lambda,\mu}g(z)} + \mu + p} - \beta\right). \tag{11}$$

Since $g \in \mathcal{MS}_p(\lambda, \mu+1;\ \eta;\ \phi) \subset \mathcal{MS}_p(\lambda, \mu;\ \eta;\ \phi)$, by Theorem 2.1, we set

$$q(z) = \frac{1}{p-\eta}\left(-\frac{z(I_{\lambda,\mu}g(z))'}{I_{\lambda,\mu}g(z)}-\eta\right),$$

where $q(z) \prec \phi(z)$ for $z \in \mathbb{U}$ with the assumption for $\phi \in \mathcal{N}$. Then, by virtue of (10) and (11), we observe that

$$I_{\lambda,\mu}(-zf'(z)) = (p-\beta)p(z)I_{\lambda,\mu}g(z) + \beta I_{\lambda,\mu}g(z) \tag{12}$$

and

$$\frac{1}{p-\beta}\left(-\frac{z(I_{\lambda,\mu+1}f(z))'}{I_{\lambda,\mu+1}g(z)}-\beta\right)$$

$$= \frac{1}{p-\beta}\left(\frac{\dfrac{z(I_{\lambda,\mu}(-zf'(z)))'}{I_{\lambda,\mu}g(z)} + (\mu+p)(p-\beta)p(z)+\beta)}{-(p-\eta)q(z)+\mu+p-\eta} - \beta\right). \tag{13}$$

Upon differentiating both sides of (12), we have

$$\frac{z(I_{\lambda,\mu}(-zf'(z)))'}{I_{\lambda,\mu}g(z)} = (p-\beta)zp'(z) - ((p-\beta)p(z)+\beta)((p-\eta)q(z)+\eta). \tag{14}$$

Making use of (9), (13), and (14), we get

$$\frac{1}{p-\beta}\left(-\frac{z(I_{\lambda,\mu+1}f(z))'}{I_{\lambda,\mu+1}g(z)}-\beta\right)$$

$$= p(z) + \frac{zp'(z)}{-(p-\eta)q(z)+\mu+p-\eta} \prec \psi(z) \ \ (z \in \mathbb{U}). \tag{15}$$

Since $\mu > 0$ and $q(z) \prec \phi(z)$ for $z \in \mathbb{U}$ with

$$\max_{z\in\mathbb{U}} \operatorname{Re}\{\phi(z)\} < \frac{\mu+p-\eta}{p-\eta},$$

$$\operatorname{Re}\{-(p-\eta)q(z)+\mu+p-\eta\} > 0 \ \ (z \in \mathbb{U}).$$

Hence, by taking

$$\omega(z) = \frac{1}{-(p-\eta)q(z)+\mu+p-\eta},$$

in (15), and applying Lemma 2.2, we can show that $p(z) \prec \psi(z)$ for $z \in \mathbb{U}$, so that $f \in \mathcal{MC}_p(\lambda,\mu;\ \eta,\beta;\ \phi,\psi)$.

For the second part, by using the arguments similar to those detailed above with (3), we obtain

$$\mathcal{MC}_p(\lambda,\mu;\ \eta,\beta;\ \phi,\psi) \subset \mathcal{MC}_p(\lambda+1,\mu;\ \eta,\beta;\ \phi,\psi).$$

Therefore, we complete the proof of Theorem 2.3. □

3 Inclusion Properties Involving the Integral Operator F_c

In this section, we consider the integral operator F_c [1,4,5] defined by

$$F_c(f) := F_c(f)(z) = \frac{c}{z^{c+1}} \int_0^z t^c f(t)dt \quad (f \in \Sigma_p;\ c > 0). \tag{16}$$

Theorem 3.1. *Let $\lambda > -p$, $\mu > 0$ and let $\phi \in \mathcal{N}$ with*

$$\max_{z\in\mathbb{U}} \operatorname{Re}\{\phi(z)\} < \frac{c+p-\eta}{p-\eta} \quad (c > 0;\ 0 \le \eta < p).$$

If $f \in \mathcal{MS}_p(\lambda,\mu;\ \eta;\ \phi)$, then $F_c(f) \in \mathcal{MS}_p(\lambda,\mu;\ \eta;\ \phi)$.

Proof. Let $f \in \mathcal{MS}_p(\lambda, \mu; \eta; \phi)$ and set

$$p(z) = \frac{1}{p-\eta}\left(-\frac{z(I_{\lambda,\mu}F_c(f)(z))'}{I_{\lambda,\mu}F_c(f)(z)} - \eta\right), \tag{17}$$

where $p(z)$ is analytic in $\mathbb{U}$ with $p(0) = 1$. From (16), we also have

$$z(I_{\lambda,\mu}F_c(f)(z))' = cI_{\lambda,\mu}f(z) - (c+p)I_{\lambda,\mu}F_c(f)(z). \tag{18}$$

Then, by using (17) and (18), we obtain

$$-c\frac{I_{\lambda,\mu}f(z)}{I_{\lambda,\mu}F_c(f)(z)} = (p-\eta)p(z) - (c+p-\eta). \tag{19}$$

The remaining part of the proof is similar to that of Theorem 2.1, and so we may omit the detailed proof involved. □

Theorem 3.2. *Let $\lambda > -p,\ \mu > 0$ and let $\phi \in \mathcal{N}$ with*

$$\max_{z\in\mathbb{U}} \mathrm{Re}\{\phi(z)\} < \frac{c+p-\eta}{p-\eta} \quad (c > 0;\ 0 \le \eta < p).$$

If $f \in \mathcal{MK}_p(\lambda, \mu; \eta; \phi)$, then $F_c(f) \in \mathcal{MK}_p(\lambda, \mu; \eta; \phi)$.

Proof. By applying Theorem 3.1, it follows that

$$\begin{aligned} f(z) \in \mathcal{MK}_p(\lambda, \mu; \eta; \phi) &\iff -zf'(z)/p \in \mathcal{MS}_p(\lambda, \mu; \eta; \phi) \\ &\implies -z(F_c(f)(z))'/p \in \mathcal{MS}_p(\lambda, \mu; \eta; \phi) \\ &\iff F_c(f)(z) \in \mathcal{MK}_p(\lambda, \mu; \eta; \phi), \end{aligned}$$

which proves Theorem 3.2. □

Corollary 3.1. *Let $\lambda > -p,\ \mu > 0$ and*

$$\frac{(p-\eta)(1+A)}{1+B} < c+p-\eta \quad (c > 0;\ -1 < B < A \le 1;\ 0 \le \eta < p).$$

Then if $f \in \mathcal{MS}_p(\lambda, \mu; \eta; A, B)$ (or $\mathcal{MK}_p(\lambda, \mu; \eta; A, B)$), we have that $F_c(f) \in \mathcal{MS}_p(\lambda, \mu; \eta; A, B)$ (or $\mathcal{MS}_p(\lambda, \mu; \eta; A, B)$).

Theorem 3.3. *Let $\lambda > -p, \mu > 0$ and let $\phi, \psi \in \mathcal{N}$ with*

$$\max_{z\in\mathbb{U}} \mathrm{Re}\{\phi(z)\} < \frac{c+p-\eta}{p-\eta} \quad (c > 0;\ 0 \le \eta < p).$$

If $f \in \mathcal{MC}_p(\lambda, \mu; \eta, \beta; \phi, \psi)$, then $F_c(f) \in \mathcal{MC}_p(\lambda, \mu; \eta, \beta; \phi, \psi)$.

Proof. Let $f \in \mathcal{MC}_p(\lambda, \mu;\ \eta, \beta;\ \phi, \psi)$.

Then, from the definition of $\mathcal{MC}_p(\lambda, \mu;\ \eta, \beta;\ \phi, \psi)$, there exists a function $g \in \mathcal{MS}_p(\lambda, \mu;\ \eta;\ \phi)$ such that

$$\frac{1}{p-\beta}\left(-\frac{z(I_{\lambda,\mu}f(z))'}{I_{\lambda,\mu}g(z)} - \beta\right) \prec \psi(z) \quad (z \in \mathbb{U}). \tag{20}$$

Thus, we set

$$p(z) = \frac{1}{p-\beta}\left(-\frac{z(I_{\lambda,\mu}F_c(f)(z))'}{I_{\lambda,\mu}F_c(g)(z)} - \beta\right),$$

where $p(z)$ is analytic in $\mathbb{U}$ with $p(0) = 1$. Applying (18), we get

$$\frac{1}{p-\beta}\left(-\frac{z(I_{\lambda,\mu}f(z))'}{I_{\lambda,\mu}g(z)} - \beta\right)$$

$$= \frac{1}{p-\beta}\left(\frac{\dfrac{z(I_{\lambda,\mu}F_c(-zf'(z))(z))'}{I_{\lambda,\mu}F_c(g)(z)} + (c+p)\dfrac{I_{\lambda,\mu}F_c(-zf'(z))(z)}{I_{\lambda,\mu}F_c(g)(z)}}{\dfrac{z(I_{\lambda,\mu}F_c(g)(z))'}{I_{\lambda,\mu}F_c(g)(z)} + c + p} - \beta\right). \tag{21}$$

Since $g \in \mathcal{MS}_p(\lambda, \mu;\ \eta;\ \phi)$, we see from Theorem 3.1 that $F_c(g) \in \mathcal{MS}_p(\lambda, \mu;\ \eta;\ \phi)$. Let us now put

$$q(z) = \frac{1}{p-\eta}\left(-\frac{z(I_{\lambda,\mu}F_c(g)(z))'}{I_{\lambda,\mu}F_c(g)(z)} - \eta\right),$$

where $q(z) \prec \phi(z)$ for $z \in \mathbb{U}$ with the assumption for $\phi \in \mathcal{N}$. Then, by using the same techniques as in the proof of Theorem 2.3, we can prove from (20) and (21) $F_c(f) \in \mathcal{MC}_p(\lambda, \mu;\ \eta, \beta;\ \phi, \psi)$. □

Remark. If we take $p = 1$, $\lambda = 1$, and $\mu = 2$ in all theorems of this section, then we extend the results by Goel and Sohi [4], which reduce the results earlier obtained by Bajpai [1].

Acknowledgements This research was supported by the Basic Science Research Program through the National Research Foundation of Korea (NRF) funded by the Ministry of Education, Science and Technology (No. 2011-0007037).

References

1. Bajpai, S.K.: A note on a class of meromorphic univalent functions. Rev. Roumaine Math. Pures Appl. **22**, 295–297 (1977)
2. Choi, J.H., Saigo, M., Srivastava, H.M.: Some inclusion properties of a certain family of integral operators. J. Math. Anal. Appl. **276**, 432–445 (2002)
3. Eenigenberg, P., Miller, S.S., Mocanu, P.T., Reade, M.O.: On a Briot–Bouquet differential subordination. In: General Inequalities, 3 (Oberwolfach, 1981). International Series of Numerical Mathematics, vol. 64, pp. 339–348. Birkhäuser Verlag, Basel (1983)
4. Goel, R.M., Sohi, N.S.: On a class of meromorphic functions. Glas. Mat. **17**(37), 19–28 (1982)
5. Kumar, V., Shukla, S.L.: Certain integrals for classes of p-valent meromorphic functions. Bull. Austral. Math. Soc. **25**, 85–97 (1982)
6. Kim, Y.C., Choi, J.H. Choi, Sugawa, T.: Coefficient bounds and convolution properties for certain classes of close-to-convex functions. Proc. Japan Acad. Ser. A Math. Sci. **76**, 95–98 (2000)
7. Liu, J.-L.: The Noor integral and strongly starlike functions. J. Math. Anal. Appl. **261**, 441–447 (2001)
8. Liu, J.-L., Noor, K.I.: Some properties of Noor integral operator. J. Natur. Geom. **21**, 81–90 (2002)
9. Ma, W.C., Minda, D.: An internal geometric characterization of strongly starlike functions. Ann. Univ. Mariae Curie-Sklodowska Sect. A **45**, 89–97 (1991)
10. Miller, S.S., Mocanu, P.T.: Differential subordinations and univalent functions. Michigan Math. J. **28**, 157–171 (1981)
11. Noor, K.I.: On new classes of integral operators. J. Natur. Geom. **16**, 71–80 (1999)
12. Noor, K.I., Noor, M.A.: On integral operators. J. Math. Anal. Appl. **238**, 341–352 (1999)

A Journey from Gross-Problem to Fujimoto-Condition

Indrajit Lahiri and Abhijit Banerjee

Dedicated to Professor Hari M. Srivastava

Abstract In the short survey we discuss the influence of Gross-problem on the set sharing of entire and meromorphic functions. We also see the impact of Fujimoto-condition on the study of uniqueness polynomials.

1 Introduction and Unique Range Sets

The value distribution theory is a prominent branch of complex analysis. In this theory one studies how an entire or a meromorphic function assumes some values and the influence of assuming certain values, in some specific manner, on a function. In other words, one studies the frequency with which a meromorphic function takes up different values in the complex plane. Perhaps the fundamental theorem of classical algebra is the most well-known value distribution theorem and the next one is Picard's theorem.

The uniqueness theory of entire and meromorphic functions has been emerged as an active subfield of the value distribution theory with distinguishable entity. This theory mainly studies conditions under which there exists essentially only one function satisfying the prescribed conditions. It is well known that in a given domain D only a single analytic function exists that assumes specified values in a sequence of points $\{z_n\}$ convergent to a point of D. This result completely characterises an analytic function in a domain D just by its behaviour in a small subset of D and can be regarded as the gateway to the uniqueness theory.

I. Lahiri (✉) • A. Banerjee
Department of Mathematics, University of Kalyani, Kalyani, West Bengal 741235, India
e-mail: ilahiri@hotmail.com; abanerjee_kal@yahoo.co.in; abanerjee_kal@rediffmail.com

G.V. Milovanović and M.Th. Rassias (eds.), *Analytic Number Theory, Approximation Theory, and Special Functions*, DOI 10.1007/978-1-4939-0258-3_35,

Let f and g be two nonconstant meromorphic functions and let a be a finite complex number. We say that f and g share the value a CM (counting multiplicities), if $f - a$ and $g - a$ have the same set of zeros with the same multiplicities. Also f and g are said to share the value a IM (ignoring multiplicities) if $f - a$ and $g - a$ have the same set of zeros, where the multiplicities are not taken into account. In addition we say that f and g share ∞ CM (IM) if $1/f$ and $1/g$ share 0 CM (IM).

Let S be a set of distinct elements of $\mathbb{C} \cup \{\infty\}$ and

$$E_f(S) = \bigcup_{a \in S} \{z : f(z) - a = 0\},$$

where each zero is counted according to its multiplicity. If we ignore the multiplicity, then this set is denoted by $\overline{E}_f(S)$. If $E_f(S) = E_g(S)$ ($\overline{E}_f(S) = \overline{E}_g(S)$), we say that f and g share the set S CM (IM). Evidently if S is a singleton, then the definition coincides with the definition of CM (IM) sharing of a value.

In 1926, R. Nevanlinna proved that a nonconstant meromorphic function is uniquely determined by the inverse image of five distinct values (ignoring multiplicities) in the extended complex plane. A few years later, he showed that when multiplicities are considered, four values are sufficient to determine a function. In this case two functions either coincide or one is a bilinear transformation of the other. The above two fundamental results, known as five and four value theorems, respectively, are the starting points of the modern uniqueness theory for entire and meromorphic functions.

In 1977 F. Gross [9] initiated the uniqueness theory under more general setup by considering preimages of sets of distinct elements (counting multiplicities). He proposed the following problem which has a significant influence on the theory and popularly known as "Gross-problem":

> Does there exist a finite set S such that for two entire functions f and g, $E_f(S) = E_g(S)$ implies $f \equiv g$?

In 1982 F. Gross and C.C. Yang [10] proved the following result:

Theorem 1.1. *Let $S = \{z \in \mathbb{C} : e^z + z = 0\}$ and f, g be two nonconstant entire functions. If $E_f(S) = E_g(S)$, then $f \equiv g$.*

In [10], a set S is termed as a unique range set for entire functions (URSE in short) if for any two non-constant entire functions f and g, the conditions $E_f(S) = E_g(S)$ imply $f \equiv g$. In a similar fashion one can define a unique range set for meromorphic functions (URSM in short).

In 1997 H.X. Yi [20] called any set $S \subset \mathbb{C} \cup \{\infty\}$ a unique range set for meromorphic (entire) functions with ignoring multiplicities (URSM-IM/URSE-IM) or a reduced unique range set for meromorphic (entire) functions (R-URSM/R-URSE) if $\overline{E}_f(S) = \overline{E}_g(S)$ implies $f \equiv g$ for any pair of nonconstant meromorphic functions f and g.

The study of relationships between two entire or meromorphic functions via the preimage sets of several distinct values in the range has a long history. We may note that the URSE as obtained by F. Gross and C.C. Yang (Theorem 1.1) is an infinite set. Since then with the inspiration of "Gross-problem", the continuous study of unique range sets (URS's) is focused mainly on two aspects : *finding a URS with minimal cardinality and characterising a URS.*

In 1994 H. X. Yi [17] settled "Gross-problem" affirmatively and exhibited a URSE with 15 elements. Just in the next year P. Li and C.C. Yang [13] exhibited a URSM with 15 elements and a URSE with 7 elements. H.X.Yi [18] also confirmed the result of H.X. Yi and P. Li–C.C. Yang for URSE. To study a URSE or a URSM, P. Li and C.C. Yang actually investigated the zero set S of a polynomial of the form $P(z) = z^n + az^{n-m} + b$, where gcd $(n, m) = 1, n > m \geq 1$ and a, b are so chosen that P has only simple zeros. When $m \geq 2$, then the set S generates a URSM, and when $m = 1$, then the set S generates a URSE. In 1996 H.X. Yi [19] further improved the result of P. Li and C.C. Yang by reducing the cardinality of URSM to 13.

Till date the URSM with 11 elements is the smallest available URSM obtained by Frank and Reinders [6], which is the zero set of the following polynomial:

$$P(z) = \frac{(n-1)(n-2)}{2} z^n - n(n-2)z^{n-1} + \frac{n(n-1)}{2} z^{n-2} - c,$$

where $n \geq 11$ and $c \neq 0, 1$.

It is observed that a URSE must contain at least 5 elements, whereas a URSM must contain at least 6 elements (see [16, p. 517 and p. 527]).

2 Anatomy of Polynomials Generating Unique Range Sets

Before 1995 the investigations to determine a finite URSM (or URSE) were solely confined to that of finding a set with certain number of distinct elements. It took up to 1995 to realise the underlying importance of the polynomial backbone of a finite URSM (or URSE). P. Li and C.C. Yang [13] first highlighted the fact that a finite URSM (or URSE) is, infact, the set of distinct zeros of a polynomial, and equal emphasis should be given to the mechanism of the polynomial.

The following two terminologies were introduced by P. Li and C. C. Yang [13]:

Definition 2.1. Let $S = \{a_1, a_2, \ldots, a_n\}$ be a subset of $\mathbb{C}$ with finite distinct elements. If S is a URSM (URSE), then any polynomial of degree n, which has S as the set of its zeros, is called a polynomial of URSM (URSE).

We call it a PURSM (PURSE) in brief.

Definition 2.2. Let P be a polynomial. If the condition $P(f) \equiv P(g)$ implies $f \equiv g$ for any two nonconstant meromorphic (entire) functions f and g, we say that P is a uniqueness polynomial for meromorphic (entire) functions. In brief we write P as a UPM (UPE).

It is easy to note that every PURSM (PURSE) is a UPM (UPE). However the converse situation needs some special attention, which we discuss later on.

Till date, to the knowledge of the authors, three types of PURSM (and so UPM) have been achieved. The principal mechanism of construction of such a polynomial is to ensure that it has only simple zeros and most importantly to choose the coefficients in such a judicious manner which ultimately yields the equality $f \equiv g$ under $P(f) \equiv P(g)$. The situation will be better understood if we discuss the anatomy of these polynomials.

First Type: Let $P_1(z) = z^n + az^{n-m} + b$, where n, m are mutually prime positive integers with $m \geq 2$ and $n \geq 2m + 9$ and a, b are two nonzero constants satisfying $(n-1)^{(n-1)} \neq b(-n/a)^n$. The hypothesis ensures that P_1 has no multiple zero. Now following the method of H.X. Yi [19], one can verify that P_1 is a PURSM. Hence for any two nonconstant meromorphic functions f and g $P_1(f) \equiv P_1(g)$ implies $f \equiv g$.

Second Type: Let

$$P_2(z) = \frac{(n-1)(n-2)}{2} z^n - n(n-2)z^{n-1} + \frac{n(n-1)}{2} z^{n-2} - c,$$

where n (≥ 11) is a positive integer and c ($\neq 0, 1$) is a constant. Now,

$$P_2'(z) = \frac{n(n-1)(n-2)}{2} (z-1)^2 z^{n-3}.$$

Clearly, $P_2(1) = 1 - c$ and $P_2(0) = -c$. So, 0 and 1 are the zeros of $P_2 + c$ and $P_2 - (1 - c)$ with respective multiplicities $n - 2$ and 3. Therefore, we can put $P_2(z) - (1 - c) = (z - 1)^3 Q_{n-3}(z)$ and $P_2(z) + c = z^{n-2} Q_2(z)$, where Q_{n-3} and Q_2 are polynomials of degree $n - 3$ and 2, respectively, with $Q_{n-3}(1) \neq 0$ and $Q_2(0) \neq 0$. Further, we note that all the zeros of Q_{n-3} and Q_2 are simple. For otherwise P_2' would have zeros other than 1 and 0. So for every $d \in \mathbb{C}\backslash\{1 - c, -c\}$, $P_2 - d$ has only simple zeros. Hence P_2 has only simple zeros.

In 1998 G. Frank and M. Reinders [6] proved that the above polynomial is a PURSM and so it is a UPM.

Third Type: Let $P_3(z) = az^n + n(n-1)z^2 + 2n(n-2)bz - (n-1)(n-2)b^2$, where n (≥ 6) is an integer and a, b are two nonzero complex numbers satisfying $ab^{n-2} \neq 1, 2$. Clearly, $P_3'(z) = (n/z)[az^n - 2(n-1)z^2 + 2(n-2)bz]$.

Now at each zero of P_3' we get

$$\begin{aligned} P_3(z) &= az^n - n(n-1)z^2 + 2n(n-2)bz - (n-1)(n-2)b^2 \\ &= 2(n-1)z^2 - 2(n-2)bz - n(n-1)z^2 + 2n(n-2)bz - (n-1)(n-2)b^2 \\ &= -(n-1)(n-2)[z^2 - 2bz + b^2] \\ &= -(n-1)(n-2)(z-b)^2. \end{aligned}$$

So, at a zero of P_3', P_3 will have a zero if $P_3'(b) = 0$. Since $P_3'(b) = nb(ab^{n-2} - 2) \neq 0$, we see that P_3 and P_3' do not have any common zero. Hence P_3 has only simple zeros. Using the technique of T.C. Alzahary [1] one can verify that P_3 is a PURSM and so a UPM.

3 Uniqueness Polynomials and Fujimoto-Condition

After introducing the idea of uniqueness polynomial in 1995, P. Li and C.C. Yang [13] studied some basic characterisations of the same. The results of Li and Yang [13] are worth mentioning as these inspired a lot of workers to investigate the uniqueness polynomials, thus opened a new avenue for research on uniqueness theory. The following four theorems (from [13]) may be considered as the initial characterisations of a uniqueness polynomial:

Theorem 3.1. *If P_1 is a* UPM (UPE) *and P_2 is a polynomial, then $P_1 \circ P_2$ is a* UPM (UPE) *if and only if P_2 is a* UPM (UPE).

Theorem 3.2. *Let P_1 be a* PURSM (PURSE) *and P_2 is a* UPM (UPE). *If $P_1 \circ P_2$ has no multiple zero, then $P_1 \circ P_2$ is a* PURSM (PURSE).

Theorem 3.3. *Any polynomial of degree* 2 *or* 3 *is not a* UPE.

Theorem 3.4. *Let $P(z) = z^4 + a_3z^3 + a_2z^2 + a_1z + a_0$. Then P is not a* UPM. *Also P is a* UPE *if and only if*

$$\left(\frac{a_3}{2}\right)^3 - \frac{a_2a_3}{2} + a_1 \neq 0.$$

A UPM (UPE) of degree less than five is completely characterised by Theorems 3.4 and 3.5. The polynomials of higher degree require further considerations, which was first studied by C.C. Yang and H.X. Hua [15]. We now mention the following two results of Yang and Hua [15]:

Theorem 3.5. *Let $P(z) = z^n + a_{n-1}z^{n-1} + \cdots + a_1z + a_0$ $(n \geq 4)$ be a monic polynomial. If there exists an integer t with $1 \leq t < n-2$ and $\gcd(n,t) = 1$ such that $a_{n-1} = \cdots = a_{t+1} = 0$ but $a_t \neq 0$, then P is a* UPE.

Theorem 3.6. *Let $P(z) = z^n + a_m z^m + a_0$ be a monic polynomial such that gcd $(n, m) = 1$ and $a_m \neq 0$. If $n \geq 5$ and $1 \leq m < n - 1$, then P is a* UPM.

Considering $m = n - 1$ with $a_m = -1$, one can see that

$$P\left(\frac{\sum_{j=1}^{n-1} h^j}{\sum_{j=0}^{n-1} h^j}\right) = P\left(\frac{\sum_{j=0}^{n-2} h^j}{\sum_{j=0}^{n-1} h^j}\right)$$

for any nonconstant meromorphic functions h (see [15]). So $P(z) = z^n - z^{n-1} + a_0$ is not a UPM. However, it is easy to see that $P(z) = z^n - z^{n-1} + a_0$ is a UPE.

While searching a sufficient condition for a polynomial to be a UPM or UPE, H. Fujimoto [7] introduced a variant of the notion of the uniqueness polynomial, which is called by T. T. H. An, J. T. Y. Wang and P. Wong [2] as a strong uniqueness polynomial.

Definition 3.1 ([7]). Let P be a nonconstant polynomial in $\mathbb{C}$. If for any two nonconstant meromorphic (entire) functions f and g, $P(f) \equiv cP(g)$ implies $f \equiv g$, where c is a suitable nonzero constant, then P is called a strong uniqueness polynomial for meromorphic (entire) functions. In short, we write SUPM and SUPE, respectively.

Clearly every UPM (UPE) is a SUPM (SUPE), but, in general, the converse is not true.

The key discovery of H. Fujimoto [7] is to highlight a special property of a polynomial, which plays a pivotal role in characterising a SUPM or a SUPE. This property was called by Fujimoto himself as "property (H)". Later on T.T.H. An, J.T.Y. Wang and P. Wang [2] and T.T.H. An [3] referred this property as "separation condition" and "injective condition", respectively.

A polynomial P is said to satisfy "Fujimoto-condition (H)" if $P(\alpha) \neq P(\beta)$ for any two distinct zeros α, β of the derivative P'. Since the zeros of P' are critical points of P, the same condition is called in [4] as the "critical injective property" and a polynomial with this property is called a "critically injective polynomial".

Let P be a nonconstant polynomial without multiple zeros. Suppose that the derivative P' has k distinct zeros $d_1, d_2, \ldots, d_k$ with respective multiplicities $q_1, q_2, \ldots, q_k$. The following results of H. Fujimoto [7] give sufficient conditions for a higher-degree polynomial to be UPM and SUPM.

Theorem 3.7. *Let P be a critically injective polynomial of degree ≥ 5. If $k \geq 4$, then P is a* UPM.

If a polynomial is not critically injective, then it may not be a UPM. In fact, for generically chosen constants $a_1, a_2, \ldots, a_n$ $(n \geq 1)$, a polynomial $P(z) = z^{2n} + a_1 z^{2n-2} + \cdots + a_{n-1} z^2 + a_n$ has no multiple zero and $P'(z)$ has $2n - 1$ distinct zeros, $0, \pm d_1, \pm d_2, \ldots, \pm d_{n-1}$. Clearly $P(d_l) = P(-d_l)$ for $1 \le l \le n-1$ and P is not a UPM because $P(f) = P(-f)$ for any nonconstant meromorphic function f (see [7]).

Theorem 3.8. *Let P be a critically injective polynomial with $k \ge 4$. If P is not a* SUPM, *then there is some permutation $(t_1, t_2, \ldots, t_k)$ of $(1, 2, \ldots, k)$, such that*

$$\frac{P(d_{t_1})}{P(d_1)} = \frac{P(d_{t_2})}{P(d_2)} = \cdots = \frac{P(d_{t_k})}{P(d_k)} \neq 1.$$

As a consequence of Theorem 3.8 we see that a critically injective polynomial with $k \ge 4$ is an SUPM if $P(d_1) + P(d_2) + \cdots + P(d_k) \neq 0$. Using this result one can improve a result of B. Shiffman [14, Theorem 3] to SUPM.

For the case of three critical values, H. Fujimoto [7] proved the following theorems:

Theorem 3.9. *Let P be a critically injective polynomial. Assume that $k = 3$ and $\min\{q_1, q_2, q_3\} \ge 2$. If any one of the three values d_1, d_2 and d_3 is not the arithmetic mean of two others, then P is a* UPM.

Theorem 3.10. *Let P be a critically injective polynomial. Assume that $k = 3$ and $\min\{q_1, q_2, q_3,\} \ge 2$. If P is not a* SUPM, *then, after suitable changes of indices of d_j's, either*

$$d_3 = \frac{d_1 + d_2}{2} \quad \text{or} \quad \frac{P(d_2)}{P(d_1)} = \frac{P(d_3)}{P(d_2)} = \frac{P(d_1)}{P(d_3)}.$$

In 2002 W. Cherry and J. T. Y. Wang [5] established a geometrical characterisation of UPE and SUPE. The characterisation is important in the sense that it gives a necessary and sufficient condition for a UPE and SUPE. In order to understand the results of Cherry and Wang [5], we need some terminologies (cf. [5]).

Let P be a nonconstant polynomial in $\mathbb{C}$ and S be its zero set, a zero being counted according to its multiplicity. A bijection $T : S \to S$ is called an affine transformation preserving S if T can be written in the form $T(z) = az + b$, where a ($\neq 0$) and b are complex constants. If $a^n = 1$, then the affine transformation T is called very special with respect to n. The set S is called affinely rigid if the only affine transformation preserving S is the identity mapping $T(z) = z$.

Let us consider two variables quadratic polynomial

$$Q(x, y) = \alpha_{2,0}x^2 + \alpha_{1,1}xy + \alpha_{0,2}y^2 + \alpha_{1,0}x + \alpha_{0,1}y + \alpha_{0,0}. \tag{1}$$

Let $S = \{s_1, s_2, \ldots, s_n\}$ be the zero set of the polynomial P. We now define a transformation T_Q on S as follows:

If $\alpha_{2,0} \neq 0$, then for each $s_j \in S$ we define $t_{j,1}$ and $t_{j,2}$ to be the two solutions in t of the equation $Q(t, s_j) = 0$; noting that in the case of repeated root, we might have $t_{j,1} = t_{j,2}$. Then we put $T_Q(S) = \{t_{1,1}, t_{1,2}, t_{2,1}, t_{2,2}, \ldots, t_{n,1}, t_{n,2}\}$.

If $\alpha_{2,0} = 0$, then $Q(t, s_j) = 0$ has a unique solution t_j, say. In this case we put $T_Q(S) = \{t_1, t_2, \ldots, t_n\}$.

We call T_Q the quadratic transformation associated to Q. One may note that the cardinality of $T_Q(S)$ is twice that of S if $\alpha_{2,0} \neq 0$ and is same as S if $\alpha_{2,0} = 0$.

We say that S is preserved by a quadratic transformation if one can find a quadratic polynomial of the form (1) such that $T_Q(S) = 2S$ if $\alpha_{2,0} \neq 0$ and $T_Q(S) = S$ if $\alpha_{2,0} = 0$, where T_Q is the quadratic transformation associated to Q and $2S$ is obtained from S by repeating its each element two times.

A quadratic polynomial Q as defined by (1) is called special with respect to a positive integer $n \geq 2$ if $\alpha_{2,0} = 1$ and $x^2 + \alpha_{1,1}xy + \alpha_{0,2}y^2$ divides $x^n - cy^n$ for some nonzero constant c. If we can take $c = 1$, then Q is called very special.

Now we state the results of W. Cherry and J.T.Y. Wang [5] on characterisation of UPE and SUPE.

Theorem 3.11. *Let P be a monic polynomial of degree at least two in $\mathbb{C}$ and S be its zero set (counted with multiplicities). Then the following are equivalent:*

(i) *P is a* UPE;
(ii) *The plane curve defined by $P(x) - P(y)$ has no linear or quadratic factors, except for the (possibly repeated) linear factor $x - y$;*
(iii) *S is not preserved by any non-trivial affine transformation very special with respect to the degree of P, and S is not preserved by any quadratic transformation associated to a non-degenerate quadratic polynomial $Q(x, y)$ very special with respect to the degree of P.*

We, in view of Theorem 3.3, note that Theorem 3.11 is meaningful only when the degree of P is at least 4.

Theorem 3.12. *Let P be a monic polynomial of degree at least two in $\mathbb{C}$ and S be its zero set (counted with multiplicities). Then the following are equivalent:*

(i) *P is a* SUPE;
(ii) *None of the plane curves defined by $P(x) - cP(y)$ for all complex numbers $c \neq 0$ have linear or quadratic factors, except for the (possibly repeated) linear factor $(x - y)$ when $c = 1$;*
(iii) *S is affinely rigid and is not preserved by any quadratic transformation associated to a non-degenerate quadratic polynomial $Q(x, y)$ special with respect to the degree of P.*

In 2003 H. Fujimoto [8] obtained the following characterisation of UPMs for critically injective polynomials:

Theorem 3.13. *Let P be a critically injective monic polynomial in $\mathbb{C}$. Further suppose that $d_1, d_2, \ldots, d_k$ are the distinct zeros of P' with respective multiplicities $q_1, q_2, \ldots, q_k$. Then P is a* UPM *if and only if*

$$\sum_{1 \leq l < m \leq k} q_l q_m > \sum_{l=1}^{k} q_l. \tag{2}$$

We note that if $k \geq 4$, then (2) is obvious. Also (2) holds when max $(q_1, q_2, q_3) \geq 2$ for the case $k = 3$ and when min $(q_1, q_2) \geq 2$ and $q_1 + q_2 \geq 5$ for the case $k = 2$.

A precise characterisation of a gap polynomial (a polynomial with some terms missing) for SUPM was established by T.T.H. An, J.T.Y. Wang and P.M. Wong [2].

Theorem 3.14. *Let* $P(z) = a_n z^n + \sum_{i=0}^{m} a_i z^i$ $(0 \leq m < n,\ a_i \in \mathbb{C}$ *and* $a_n, a_m \neq 0)$ *be a polynomial of degree* n*. Let* $I = \{i : a_i \neq 0\}$, $l = \min\{i : i \in I\}$ *and* $J = \{i - l : i \in I\}$*. Then the following statements are valid:*

(i) *If* $n - m \geq 3$*, then* P *is a strong uniqueness polynomial for rational functions if and only if the greatest common divisor of the indices in* I *is* 1 *and the greatest common divisor of the indices in* J *is also* 1.
(ii) *If* $n - m \geq 4$*, then* P *is a* SUPM *if and only if the greatest common divisor of the indices in* I *is* 1 *and the greatest common divisor of the indices in* J *is also* 1.

From the above result we see that the advantage of a gap polynomial over a usual one is that the gap polynomial enables us to avoid the critical injection hypothesis. In the future we shall note this fact again.

In the next characterisation we see that the critical injection hypothesis is required when we do not consider a gap polynomial.

Theorem 3.15. *Let* $P(z)$ *be a critically injective polynomial of degree* n *in* $\mathbb{C}$ *and* $P'(z) = \lambda(z - \alpha_1)^{m_1} \ldots (z - \alpha_l)^{m_l}$*, where* λ *is a nonzero constant and* $\alpha_i \neq \alpha_j$ *for* $1 \leq i \neq j \leq l$*. Then*

(i) P *is a* UPM *if and only if one of the following conditions is satisfied;* (a) $l \geq 3$ *except when* $n = 4$, $m_1 = m_2 = m_3 = 1$; *or* (b) $l = 2$ *and* $\min\{m_1, m_2\} \geq 2$ *except when* $n = 5$, $m_1 = m_2 = 2$.
(ii) P *is a* SUPM *if and only if the set of zeros of* P *is affinely rigid and one of the following conditions is satisfied;* (a) $l \geq 3$ *except when* $n = 4$, $m_1 = m_2 = m_3 = 1$; *or* (b) $l = 2$ *and* $\min\{m_1, m_2\} \geq 2$ *except when* $n = 5$, $m_1 = m_2 = 2$.

For polynomials of the special type $(z - \alpha)^n + a(z - \alpha)^m + b$, the following complete characterisation is obtained in [2]:

Theorem 3.16. *Let* $P(z) = (z - \alpha)^n + a(z - \alpha)^m + b$ *be a polynomial of degree* n *and* $1 \leq m \leq n - 1$*. Then*

(i) P *is a uniqueness polynomial for rational functions if and only if* $n \geq 4$, $n - m \geq 2$, $\gcd(n, m) = 1$ *and* $a \neq 0$;
(ii) P *is a strong uniqueness polynomial for rational functions if and only if* $n \geq 4$, $n - m \geq 2$, $\gcd(n, m) = 1$, $a \neq 0$ *and* $b \neq 0$;
(iii) P *is a* UPM *if and only if* $n \geq 5$, $n - m \geq 2$, $\gcd(n, m) = 1$, $a \neq 0$;
(iv) P *is a* SUPM *if and only if* $n \geq 5$, $n - m \geq 2$, $\gcd(n, m) = 1$, $a \neq 0$ *and* $b \neq 0$.

Though every PURSM (PURSE) is a UPM (UPE), the converse is not true as we see in the following examples:

Example 3.1 ([4]). Let $P(z) = az + b$ $(a \neq 0)$. Clearly P is a UPM, but for $f = -(b/a)e^z$ and $g = -(b/a)e^{-z}$, we see that $E_f(S) = E_g(S)$, where $S = \{z : az + b = 0\}$. So P is not a PURSM.

Example 3.2 ([15]). Let $P(z) = z^4 + 2z^3 - 9z^2 - 2z + 8 = (z-1)(z+1)(z-2)(z+4)$. Then the zero set of P is $S = \{1, -1, 2, -4\}$. By Theorem 3.4, P is a UPE. However, for $f = \frac{3}{2}\sqrt{5}e^z + \frac{7}{2}$ and $g = \frac{3}{2}\sqrt{5}e^{-z} + \frac{7}{2}$, we see that $E_f(S) = E_g(S)$. So P is not a PURSE.

H. Fujimoto [7] investigated the converse situation for a critically injective polynomial. Several extensions and generalisations of Fujimoto's result are available in the literature, which we do not mention. In the following result of Fujimoto [7], a polynomial is called PURSM-IM (PURSE-IM) if it generates a URSM-IM (URSE-IM).

Theorem 3.17. *Let P be a critically injective monic polynomial of degree n having only simple zeros and its derivative P' has distinct zeros $d_1, d_2, \ldots, d_k$ with multiplicities $q_1, q_2, \ldots, q_k$, respectively. Suppose further that P is* SUPM (SUPE) *and $k \geq 3$ or $k = 2$ and* $\min(q_1, q_2) \geq 2$.

If $n > 2k + 6$ $(n > 2k + 2)$, then P is a PURSM (PURSE), *and if $n > 2k + 12$ $(n > 2k + 5)$, then P is a* PURSM-IM (PURSE-IM).

In 2011 T.T.H. An [3] also made an attempt to deal the converse situation with a polynomial which is not necessarily critically injective. As we have already mentioned, a gap polynomial extended the helping hand in this case also. The following two are the results of T.T.H. An [3]:

Theorem 3.18. *Let $P(z) = a_n z^n + a_m z^m + a_{m-1} z^{m-1} + \cdots + a_0$ $(1 \leq m < n,\ a_m \neq 0)$ be a polynomial in $\mathbb{C}$ of degree n with only simple zeros and S be its zero set. Further suppose that $P'(z)$ has k distinct zeros and $I = \{i : a_i \neq 0\}$, $\lambda = \min\{i : i \in I\}$, $J = \{i - \lambda : i \in I\}$. If $n \geq \max\{m + 4, 2k + 7\}$, then the following statements are equivalent:*

(i) *S is a* URSM;
(ii) *P is a* SUPM;
(iii) *S is not preserved by any non-trivial affine transformation of $\mathbb{C}$;*
(iv) *The greatest common divisors of the indices, respectively, in I and J are both* 1.

Theorem 3.19. *Let $P(z) = a_n z^n + a_m z^m + a_{m-1} z^{m-1} + \cdots + a_p z^p + a_0$ with $n > m > p$ and $a_m a_p a_0 \neq 0$ be a polynomial of degree n in $\mathbb{C}$ having only simple zeros. Let S be its zero set and $n > 2m+8$ and $p \geq 4$. Then the following statements are equivalent:*

(i) *S is a* URSM;
(ii) *P is an* SUPM;
(iii) *S is affine rigid;*
(iv) *The greatest common divisors of the indices in $I = \{i : a_i \neq 0\}$ is* 1.

A recent development in the uniqueness theory of meromorphic functions is the introduction of the notion of weighted sharing of values and sets (see [11, 12]). This measures a gradual increment from sharing with ignoring multiplicities to sharing with counting multiplicities.

Let k be a nonnegative integer or infinity. For $a \in \mathbb{C} \cup \{\infty\}$ we denote by $E_k(a; f)$ the set of all a-points of f, where an a-point of multiplicity m is counted m times if $m \leq k$ and $k+1$ times if $m > k$.

If for two meromorphic functions f and g we have $E_k(a; f) = E_k(a; g)$, then we say that f and g share the value a with weight k.

The IM and CM sharing, respectively, correspond to weight 0 and ∞.

For $S \subset \mathbb{C} \cup \{\infty\}$ we define $E_f(S, k)$ as

$$E_f(S,k) = \bigcup_{a \in S} E_k(a; f),$$

where k is a nonnegative integer or infinity. Clearly $E_f(S) = E_f(S, \infty)$.

A set $S \subset \mathbb{C} \cup \{\infty\}$ is called a unique range set for meromorphic (entire) functions with weight k if for any two nonconstant meromorphic (entire) functions f and g, $E_f(S,k) = E_g(S,k)$ implies $f \equiv g$. We write S as URSMk (URSEk) in short.

Let k be a positive integer or infinity. We denote by $E_{k)}(a; f)$ the set of a-points of f whose multiplicities are not greater than k, and each a-point is counted according to its multiplicity. For $S \subset \mathbb{C} \cup \{\infty\}$ we put $E_{k)}(S, f) = \bigcup_{a \in S} E_{k)}(a; f)$. The set S is called a URSM$_{k)}$ (URSE$_{k)}$) if for any two nonconstant meromorphic(entire) functions f, g, $E_{k)}(S, f) = E_{k)}(S, g)$ implies $f \equiv g$.

It may be noted that the proof of Theorem 3.18 contains a gap. However, if we change the gap polynomial, then an analogous but better result can be obtained (see [4]).

Theorem 3.20. *Let $P(z) = a_n z^n + \sum_{j=2}^{m} a_j z^j + a_0$ be a polynomial of degree n having only simple zeros, where $n - m \geq 4$ and $a_p a_m \neq 0$ for some positive integer p with $2 \leq p \leq m$ and* $\gcd(p, 3) = 1$. *Suppose that S be the zero set of P. Let k be the number of distinct zeros of the derivative P'. Also suppose that $I = \{j : a_j \neq 0\}$, $\lambda = \min\{j : j \in I\}$ and $J = \{j - \lambda : j \in I\}$. If $n \geq 2k + 7$ ($n \geq 2k + 3$), then the following statements are equivalent:*

(i) *P is a* SUPM (SUPE);
(ii) *S is a* URSM2 (URSE2);
(iii) *S is a* URSM$_{3)}$ (URSE$_{3)}$);
(iv) *S is a* URSM (URSE);
(v) *P is a* UPM (UPE).
(vi) *The greatest common divisor of the indices in I is* 1 *and the greatest common divisor of the indices in J is also* 1.

The following example shows that the hypothesis of Theorem 3.20 does not ensure a polynomial to be a SUPM, rather, in addition, the condition stated in (vi) is essential.

Example 3.3. Let $P(z) = \frac{1}{20}z^{20} - \frac{1}{16}z^{16} + 1$ and $P'(z) = z^{15}(z^4 - 1)$ and $\{0, 1, -1, i, -i\}$ is the set of all distinct zeros of $P^{'}$. Since $P(0) = 1$ and $P(\pm 1) = P(\pm i) = 79/80$, we see that P has only simple zeros and P is not critically injective. Also $n = 20 > 17 = 2k + 7$, but $\gcd(20, 16) = 4$. So, by Theorem 3.14(ii), P is not a SUPM.

From the above discussion it appears that a gap polynomial has the potential to become an alternative to a critically injective polynomial. Therefore, the properties of a gap polynomial should further be explored in the context of uniqueness polynomials and unique range sets for meromorphic (entire) functions.

As it is a short survey, we cannot include a considerable number of results on the topic, which an interested reader may find in the literature. However, we expect that this article will give an idea of the stream of research done on Gross-problem and Fujimoto-condition.

References

1. Alzahary, T.C.: Meromorphic functions with weighted sharing of one set. Kyungpook Math. J. **47**, 57–68 (2007)
2. An, T.T.H., Wang, J.T.Y., Wong, P.: Strong uniqueness polynomials: the complex Case. Complex Var. Theory Appl. **49**(1), 25–54 (2004)
3. An, T.T.H.: Unique range sets for meromorphic functions constructed without an injectivity hypothesis. Taiwanese J. Math. **15**(2), 697–709 (2011)
4. Banerjee, A., Lahiri, I.: A uniqueness polynomial generating a unique range set and vise-versa. Comput. Methods Funct. Theory (CMFT) **12**(2), 527–539 (2012)
5. Cherry, W., Wang, J.T.Y.: Uniqueness polynomials for entire functions. Int. J. Math. **13**(3), 323–332 (2002)
6. Frank, G., Reinders, M.: A unique range set for meromorphic functions with 11 elements. Complex Var. Theory Appl. **37**, 185–193 (1998)
7. Fujimoto, H.: On uniqueness of meromorphic functions sharing finite sets. Am. J. Math. **122**(6), 1175–1203 (2000)
8. Fujimoto, H.: On uniqueness polynomials for meromorphic functions. Nagoya Math. J. **170**, 33–46 (2003)
9. Gross, F.: Factorization of meromorphic functions and some open problems. In: Complex Analysis. Lecture Notes in Mathematics, vol. 599, pp. 51–67. Springer, Berlin (1977)
10. Gross, F., Yang, C.C.: On preimage and range sets of meromorphic functions. Proc. Jpn. Acad. **58**, 17–20 (1982)
11. Lahiri, I.: Weighted sharing and uniqueness of meromorphic functions. Nagoya Math. J. **161**, 193–206 (2001)
12. Lahiri, I.: Weighted value sharing and uniqueness of meromorphic functions. Complex Var. Theory Appl. **46**, 241–253 (2001)
13. Li, P., Yang, C.C.: Some further results on the unique range sets of meromorphic functions. Kodai Math. J., **18**, 437–450 (1995)

14. Shiffman, B.: Uniqueness of entire and meromorphic functions sharing finite sets. Complex Var. Theory Appl. **43**, 433–449 (2001)
15. Yang, C.C., Hua, X.H.: Unique polynomials of entire and meromorphic functions. Mat. Fiz. Anal. Geom. **4**(3), 391–398 (1997)
16. Yang, C.C., Yi, H.X.: Uniqueness Theory of Meromorphic Functions. Science Press/Kluwer Academic, Dordrecht (2003)
17. Yi, H.X.: On a problem of Gross. Sci. China Ser. A **24**, 1137–1144 (1994)
18. Yi, H.X.: A question of Gross and the uniqueness of entire functions. Nagoya Math. J. **138**, 169–177 (1995)
19. Yi, H.X.: Unicity theorems for meromorphic or entire functions III. Bull. Austral. Math. Soc. **53**, 71–82 (1996)
20. Yi, H.X.: The reduced unique range sets for entire or meromorphic functions. Complex Var. Theory Appl. **32**, 191–198 (1997)

Index

G.V. Milovanović and M.Th. Rassias (eds.), *Analytic Number Theory, Approximation Theory, and Special Functions*, DOI 10.1007/978-1-4939-0258-3,

MIX
Papier aus verantwortungsvollen Quellen
Paper from responsible sources
FSC® C105338

If you have any concerns about our products, you can contact us on
ProductSafety@springernature.com

In case Publisher is established outside the EU, the EU authorized representative is:
Springer Nature Customer Service Center GmbH
Europaplatz 3, 69115 Heidelberg, Germany

Printed by Libri Plureos GmbH
in Hamburg, Germany